ROCK FOUNDATION

PROCEEDINGS OF THE INTERNATIONAL WORKSHOP ON ROCK FOUNDATION
TOKYO / JAPAN / 30 SEPTEMBER 1995

Rock Foundation

Edited by
R. YOSHINAKA
Saitama University, Japan
K. KIKUCHI
Kyoto University, Japan

A.A. BALKEMA / ROTTERDAM / BROOKFIELD / 1995

The texts of the various papers in this volume were set individually by typists under the supervision of each of the authors concerned.

Authorization to photocopy items for internal or personal use, or the internal or personal use of specific clients, is granted by A.A.Balkema, Rotterdam, provided that the base fee of US$1.50 per copy, plus US$0.10 per page is paid directly to Copyright Clearance Center, 222 Rosewood Drive, Danvers, MA 01923, USA. For those organizations that have been granted a photocopy license by CCC, a separate system of payment has been arranged. The fee code for users of the Transactional Reporting Service is: 90 5410 562 3/95 US$1.50 + US$0.10.

Published by
A.A.Balkema, P.O.Box 1675, 3000 BR Rotterdam, Netherlands (Fax: +31.10.4135947)
A.A.Balkema Publishers, Old Post Road, Brookfield, VT 05036, USA (Fax: 802.276.3837)

ISBN 90 5410 562 3
© 1995 A.A.Balkema, Rotterdam
Printed in the Netherlands

Table of contents

Preface	IX
Organization	XI

Special lectures

Stiffness of soft rocks in Tokyo metropolitan area – From laboratory tests to full-scale behaviour F.Tatsuoka, Y.Kohata, K.Ochi & T.Tsubouchi	3
30 years behavior of rock foundation of Kurobe Dam – Result of measurement at 186 m height Arch Dam T.Watanabe & T.Hayakawa	19
Foundation engineering of nuclear power station on soft rock K.Kishi, K.Terada, T.Sakai, K.Momose & E.Fukazawa	25
Design and its evaluation through displacement measurement for the Akashi Kaikyo Bridge foundation M.Yamagata, A.Nitta & S.Yamamoto	35
UHV pylon foundation design for mountainous areas – Establishment of design methods based on full scale tests Y.Yoshii	47
Compaction and subsidence in North Sea hydrocarbon fields M.Gutierrez, N.Barton & A.Makurat	57

Invited paper

Outline of Southern Hyogo Earthquake (M = 7.2) disasters and behaviors of foundations of the Akashi Kaikyo Bridge I.Yoshida & K.Tada	67

A: Mechanical properties

Deformation characteristics of weathered granite taking into account of confining pressure and strain dependence A.Nitta, S.Yamamoto, K.Miyajima & Y.Kaji	75
Investigation of dilatancy in rock massif V.A.Mansurov	81
Plastic yield criteria and their applications Xu Gancheng	85
Mechanical properties and failure mechanism of gravelly soft rocks T.Kobayashi, R.Yoshinaka & T.Mimuro	91
A study on the reasonable loading stress waveforms and loading systems of the dynamic experimental technique Liu Deshun, Li Xibing & Yang Xiangbi	99
About the in-situ determination of the rock mass deformability N.F.Grossmann	103
The comparison between dynamic and static strength of soft sedimentary rocks R.Yoshinaka & M.Osada	109
Method to calculate engineering parameters of jointed rockmass Faquan Wu	115

Large scale sampling of Akashi Gravelly Layer and its mechanical property — 119
S. Kashima, S. Yamamoto, K. Takahashi, M. Sasao & S. Yamada

Creep characteristics of the Kobe Formation of Miocene in Tertiary period — 127
M. Yamagata, K. Yamada, Y. Nishigaki & S. Matsumura

Modification to the original Hoek-Brown strength criterion — 133
V. S. Vutukuri & S. M. F. Hossaini

Failure criterion and Mohr's envelope for sedimentary soft rocks considering tensile strength — 139
K. Hattori, K. Muranaka & S. Nishie

Homogenization analysis of rock properties and rock bolts — 145
T. Kyoya, N. Tokashiki & T. Kawamoto

Influence of pre-existent joints and induced tensile cracks on pressuremeter tests conducted in rock ground — 151
K. Tani, T. Okamoto & K. Nishi

Non destructive detection of micro cracking in rock — 157
M. P. Luong

The examples of the new foundation evaluation system and permeability tests for weak rocks — 163
K. Hirata, T. Sugiyama, M. Karasawa & M. Ueda

Rock mass classification of weathered granite and evaluation of geomechanical parameters for bridge foundation of large scale structure in the Honshu-Shikoku Bridge Authority — 169
K. Ishikawa, H. Ochi & S. Takada

Geological investigation and evaluation of foundation bedrock for long-span bridges in Japan — 177
S. Yamamoto, Y. Kimura, K. Takahashi & K. Miyajima

The relationship between ground characteristics and P- and S-wave velocities measured by the PS sonic logging system — 185
T. Aizawa & K. Sasaki

A quick and accurate strength profile along rock cores with the rock strength apparatus — 191
D. M. Fourmaintraux, C. Lasserre, E. Detournay & A. Drescher

Representative elementary volume in natural discontinuous rock masses — 197
K. Suzuki, T. Kuwahara, M. Maruyama & K. Hirama

Scale effect of the rock — 203
V. A. Mansurov

Scale effect on the deformability of layered rocks — 207
Z. Y. Yang & C. S. Liu

Scale dependency of rock mass properties — 213
A. P. Cunha

The scale and creep effects on strength of welded tuff — 219
N. Yuki, S. Aoto, Y. Ogata, R. Yoshinaka & M. Terada

Mechanism of scale effect in rock joint — 223
R. Yoshinaka, S. Arisaka, K. Sasaki & J. Yoshida

B: Design and analysis

Stability analysis of the LAXIWA arch dam abutment on complex rock formations — 231
Zhou Wei Yuan, Yang Rouziong, Yan Guang Ri, Zhang Ming Yao & Wu Xi

Strength and deformability evaluation for jointed rock foundations of large dams — 235
S. A. Yufin, V. N. Burlakov & M. G. Zertsalov

Stability and deformation analysis of complex rock foundations of several large dams and hydropower stations in China — 243
Ge Xiurun, Feng Dingxiang, Gu Xianrong & Feng Shuren

Interaction between dam structure and rock foundation – A case-history and analysis — 249
Sijing Wang

Stability of foundations on jointed rock – Case studies D.C.Wyllie	253
A procedure of aseismic design for Akashi Kaikyo bridge foundations A. Nitta, S.Yamamoto, S.Masuda & R.Isoyama	259
Investigation and seismic stability evaluation of rock foundation of nuclear power plants in Japan H.Ito, Y.Momose & Y.Shibagaki	265
Stability evaluation of large-scale bridge foundations built on sedimentary soft rock K.Kanazawa, K.Kawaguchi, S.Matsumoto & T.Sanbyakuda	273
A unified design method for rock anchor foundations of super-high pylons Ö.Aydan, S.Komura, S.Ebisu & T.Kawamoto	279
Sloping rock layer foundation of bridge structure N.Ogata & S.Gose	285
Deep excavation of soft rock by vertical NATM R. Ito, K.Watanabe, A.Takagi, M.Ueno & K.Nakasita	293
State of the arts on the super-high-rise building foundation in Japan Y.Nagataki, S.Iiboshi & K.Aoshima	299
Non-linear elastic stress and coupled fluid flow analysis of anisotropic rock masses T.Yamabe, M.Oda & K.Maekawa	307
Bearing capacity of rock foundations T.Ramamurthy	311
The influence of compression and tensile loads on water permeability on rock foundations E.S.Kalustian & E.G.Gaziev	317
Allowable bearing capacity of typical rock classes depending on RMR and compression strength of intact rock A.Serrano & C.Ollala	321
Bearing capacity of soft rock foundation on in-situ bearing capacity tests under inclined load A. Nitta, S.Yamamoto, T.Sonoda & T.Hosono	327
Bearing capacity of foundation on rock mass with weak layer P.Miščević & I.Jašarević	333
Stability analysis of jointed rock foundations by discontinuous deformation analysis T.Sasaki, D.Ishii, Y.Ohnishi & R.Yoshinaka	337
Stability analysis of jointed rock foundations by FEM using the multiple yield models T.Sasaki, S.Morikawa, T.Matsukawa & R.Yoshinaka	343
In-situ seepage flow tests on jointed rock mass and its analysis K.Kikuchi, Y.Mito & M.Nakada	349
Seepage under concrete dam founded on rock formation using artificial neural networks Y.Ohnishi & M.Soliman	355

C: Monitoring and reinforcement

Strength and deformation behaviour of grout jointed sandstone – A laboratory study R.K.Srivastava, M.Singh & A.K.Tripathi	363
Behaviors of Kaore Arch Dam and its foundation rock during test filling H.Suzuki	367
On a quantitative management of dam grouting by real time analysis T.Tashiro, K.Hayashi, K.Mihashi & K.Takahashi	375
Management of the excavation of foundation of the Ohta Dams constructed on soft rock S.Yasufuku, M.Kageyama & H.Kakuhara	381
Treatment of soft rock masses used for the soil cement mixing wall at the Ohta Dams S.Yasufuku, M.Kageyama & K.Sakashita	387

Case study on the mechanical improvement of rock masses by grouting 393
K. Kikuchi, Y. Mito & T. Adachi

Behavior of soft rock foundation of the Ohta Dams during construction 399
H. Nakajima, M. Idogaki & T. Torii

Treatment methods for highly permeable and soft foundations of rockfill dam construction 405
K. Kanazawa & J. Takimoto

Settlement of a pier foundation for Akashi-Kaikyo Bridge and its numerical analysis 413
M.S.A. Siddiquee, F. Tatsuoka, A. Inoue, Y. Kohata, O. Yoshida, Y. Yamamoto & T. Tanaka

Estimation of ground properties and behavior under construction of huge suspension bridge 421
K. Izumi, M. Ogiwara, K. Nishida & H. Kameya

A consideration on the selection of excavation shape for large underground opening 427
S. Murakami

Behaviour of arch abutments anchored in rock under thermal loading 433
G. Ballivy, F.B. Slimane, M. Melouki & J. Tardif

Inspection and confirmation for undersea foundation rock of Kurushima Bridges 439
Y. Yanaka, Y. Hasegawa & N. Masui

Application of SMA splitter to the breaking of cast-in-place concrete pile 443
T. Inaba, M. Niishida, K. Kaneko & K. Yamanouchi

Assessment of structural stability through measurements 449
H. Maleki & K. Hollberg

Author index 457

Preface

Through many experiences, human beings knew rocks, compared with soils those characteristics are firmer and stronger, and durable against weathering and erosion. For this reason, they have been using rocks as supporting ground to build artificial heavy structures.

During the progress of large-heavy structures, Karl Terzaghi's theory of bearing capacity of soil was systemized in the early 1940's, and the method for stability analysis of foundations on soils was greatly developed. On the other hand, in the field of rocks, the expectation on soundness of rock masses was too predominated, and the necessity for researching in details about strength, deformation properties, permeability, etc., were not much recognized. Therefore, the investigation and testings for rock masses were behind in development. Besides there was no theory for properly analyzing the mechanical behavior of rock masses.

Under these conditions, in 1959, the disaster of Malpasset dam was caused by the failure of foundation rock. With this as the turning point, the research in mechanical properties of rock masses – especially in discontinuities of rock masses – have been greatly developed. It also brought to advancement the series on techniques of rock engineering, those which were put on the rock foundations.

In these days, the heavy and extra-large structures founded on rocks have been varied in many kinds such as high dams constructed on complex formations, highway bridges built on slope consisted of weathered and/or fissured rocks, high-rise pylons for electric transmission lines, strait crossing long-span bridges, nuclear power stations which should be considered on the possible effects that might be caused by strong earthquake.

These structures are consisted in the important infrastructure to maintain modern urban functions. Therefore they must be required in certain safeness more than ordinary structures. The foundation for structures should be in safety considered of the bearing capacity, settlement and deformation, sliding and overturning caused under the most cruel conditions that will be expected in future time, such as seismic force, water pressure, ocean wave pressure, ice pressure, wind pressure and so on.

The roles of function and social importance of these structures are differentiated, therefore, each design must meet various required conditions and aims. So every kind of structure has its own technical progress in each field and its systemized techniques have been brought in each situation. Because of the reasons considered in various conditions, that have own design standards or specifications in each category.

In foundation engineering, the investigation, analysis and design until the beginning of construction are extremely important. The reason why, foundation planning is almost impossible to be changed after the beginning of construction work. Continuously to construct after making foundation, the almost all superstructures should be built. The stream of construction foundation, is different from building structures such as tunnels, underground caverns and rock slopes, such structures mainly consisted of excavation and supporting techniques, that is indeed characteristic of foundation engineering.

In this International Workshop, through the common situations on the point of view from rock engineering, its own purpose will be discussed on the the techniques of rock foundation with the development of foundation engineering for structures such as dams, bridges, nuclear power plants etc. and we will exchange the newest research results on rock foundation.

The contents in the Proceedings are included in the whole foundation field covered in rock and rock mass; investigation, testing, physical properties of rock/rock mass, rock analysis, seepage flow, foundation design, improvement and reinforcement of rock mass, construction, monitoring for mechanical behavior of rock foundation.

The results of this workshop will greatly contribute to the development of rock foundation engineering in future time.

Professor R. Yoshinaka, Chairman
Professor K. Kikuchi, Co-Chairman

Organization

The International Workshop on Rock Foundation was organized by the Japanese Geotechnical Society, with the sponsorship of the Organizing Committee for ISRM 8th Congress and the Japan Institute of Systems Research.

ORGANIZING COMMITTEE

Yoshinaka, R., (Chairman), Prof., Saitama University
Kikuchi, K., (Co-Chairman), Prof., Kyoto University
Hattori, K., Shimizu Corporation
Ishikawa, K., Dr., Chuo Kaihatsu Corporation
Itoh, H., Dr., Central Research Institute of Electric Power Industry
Kitagawa, T., Dr., Nishimatsu Construction Co., Ltd
Kyoya, T., Associate Prof., Tohoku University
Mimuro, T., Dr., Tokyo Electric Power Services Co., Ltd
Mito, Y., Dr., Kyoto University
Ohnishi, Y., Prof., Kyoto University
Sasaki, T., Dr., Kajima Corporation
Shimizu, K., Nipponkoei Co., Ltd
Shinji, M., Dr., OYO Corporation
Suzuki, K., Dr., Obayashi Corporation
Tada, K., Honshu-Shikoku Bridge Authority
Tamura, T., Dr., Mitsui Construction Co., Ltd
Tanaka, S., OYO Corporation
Tashiro, T., Kajima Corporation
Uda, S., CTI Engineering Co., Ltd
Yamabe, T., Associate Prof., Saitama University
Yoshikawa, T., The Kansai Electric Power Co., Ltd

ADVISORY COMMITTEE

Kawamoto, T., Emer. Prof., Nagoya University
Miki, K., Dr., Kawasaki Geological Engineering Co., Ltd
Tatsuoka, F., Prof., University of Tokyo
Sakurai, S., Prof., Kobe University
Yamagata, M., Honshu-Shikoku Bridge Authority

SECRETARY

Yamabe, T., Associate Prof., Saitama University
Osada, M., Assistant, Saitama University

Ms. Saito, H.,
c/o Department of Civil and Environmental Engineering,
Saitama University,
Shimo-Ohkubo 255,
Urawa, Saitama,
Japan

Special lectures

Stiffness of soft rocks in Tokyo metropolitan area – From laboratory tests to full-scale behaviour

F. Tatsuoka
IIS, University of Tokyo, Japan

Y. Kohata
Railway Technical Research Institute, Japan (Formerly: IIS, University of Tokyo, Japan)

K. Ochi & T. Tsubouchi
Institute of Technology, Tokyu Construction, Co. Ltd, Japan

ABSTRACT: Stiffness at small strains of sedimentary and artificial soft rocks encountered in some large-scaled construction projects in the Tokyo metropolitan area is described. Elastic deformation moduli obtained from field seismic surveys and triaxial tests using high-quality samples of these soft rocks are very similar to each other. Several case histories of field full-scale behaviour are presented, which can be simulated very well by using stiffness evaluated based on elastic moduli while taking into account the dependency of stiffness on shear strain and pressure. Other related influencing factors including the accuracy of testing, sample disturbance, anisotropy and so on are discussed.

INTRODUCTION

In the Tokyo metropolitan area (**Fig. 1**), Holocene and Pleistocene uncemented soil deposits are underlain by a thick sedimentary soft rock (Kazusa Group), which is about one to two million years old of the Late Pliocene to the early Pleistocene Epoches as seen from **Fig. 2**. Kazusa Group is underlain by an older sedimentary soft rocks (Miura Group). The top of Kazusa Group becomes deeper in the north of the Tokyo Bay area, which is the center of a depressing basin. For Kazusa Group, the original soil type, clay (mud), silt or sand, changes from one site to another and from one depth to another, while soft mudstones are generally well cemented, but those of fine sand are less uncemented, or sometimes nearly uncemented. A reason (or reasons) for a high cementation of mudstone compared with its relatively young geological age is (are) not well understood. Except for those exposed in the air, the soft rock of Kazusa and Miura Groups has not been weathered, while some zones have been slightly disturbed by tectonic forces. These soft rock deposits have been considered as a very good foundation having sufficiently high strength and stiffness to support most non-huge civil engineering structures including RC buildings, bridge foundations and so on. In these cases, its detailed mechanical properties are not critically evaluated.

For the last decade, a number of large-scaled important structures were constructed on or in the sedimentary soft rock deposit of Kazusa Group, and associated large-scaled ground excavation was performed. On the other hand, soft clay deposits at a number of sites along the seashore of Tokyo Bay were improved by being mixed in-situ with cement slurry by the method called the Deep Mixing Method (DMM), while a large volume of fill was reclaimed by using cement-treated soil in the Trans-Tokyo Bay (TTB) Highway project (Fig. 1). Several other projects which will be related to these sedimentary and artificially soft rocks are under construction, at design stage, or under consideration. Corresponding to the above, many detailed geotechnical investigations for research and design were and are being performed.

This report summarizes the experiences which the authors obtained from a number of geotechnical investigations with respect to issues of the stiffness of these sedimentary and artificial soft rocks, which include;

a) small strains in the ground under working loads,
b) importance of accurate triaxial compression (TC) tests and their roles,
c) essentially the same elastic deformation moduli that are measured dynamically and statically,
d) evaluation of field stiffness based on elastic moduli from field seismic surveys,
e) dependency of stiffness on strain and pressure,
f) sample disturbance,
g) inherent and stress system-induced anisotropy, and
h) similarity between these sedimentary and artificial soft rocks.

Although there is not a universally accepted definition for "soft rocks", those referred in this report have an unconfined compressive strength q_u less than 100 kgf/cm^2 (10 MPa).

WHY ACCURATE STIFFNESS OF SOFT ROCK ?

The conventional design procedure for a given civil engineering structure includes the following two steps;
a) evaluation of the safety factor for failure, which is the ratio of the specified working load to the calculated ultimate failure load (the ultimate limit design); and
b) evaluation of the deformation of ground and the displacements of the structure under the specified design working load and a comparison of these values with specified allowable values (the serviceability limit design).

Design loads may be only static ones or both static and seismic (or dynamic) ones. In this report, mainly the issue of stiffness for static loads is discussed, but the same methodology as the one advocated in this report can be applied to the issue of stiffness for seismic loads. For relatively small-scaled non-important structures, the serviceability limit design based on approximated stiffness values, which are usually largely underestimated values, may be justified as a method which does not lose economics. In a simplified design procedure, the serviceability limit design may be omitted by assuming that a sufficiently large safety factor for failure can ensure acceptable small deformation of ground and displacements of a given structure.

For large-scaled important structures, however, the serviceability limit design becomes imperative, since it may control their structure dimensions and construction procedures. This design should be based on the stiffness of concerned geomaterials that have been evaluated as precisely as possible so as to achieve economical design and construction procedures while not losing a reasonable margin for safety. A too large margin for the serviceability limit state that may be obtained based on largely under-estimated stiffness may lead to an unduly high over-design, or even to a judgement from an economical point of view that the construction of the concerned structure is not feasible. Some examples are given below.

1) Rainbow bridge (**Fig. 3a**), a 798 m-long suspension bridge completed in 1993 (Odagiri et al., 1993, 1994): The foundations were placed directly on a sedimentary mudstone deposit of Kazusa Group without using deep pile foundations. Since the use of this foundation type was the first time for this scale of foundation, the settlement of the foundations upon their construction was one of the major engineering concerns. The measured settlements of Anchors A1 and A4 upon the construction of an anchorage block (denoted by the letter A in Fig. 3a) having a weight of 140,000 tonf on the respective caisson foundation was about 18 mm and about 15 mm. On the other hand, the settlement estimated based on compressibilities during loading obtained from oedometer tests was about thirty times as large as these measured values. The value estimated based on reloading compressibilities becomes much smaller about three times as large as these measured values, but the net stress change in the ground was loading.

Similar to the above, the settlement of Pier 3P for Akashi Kaikyo (Strait) bridge under construction measured when the pier was nearly completed with an average foundation pressure was 9.0 kgf/cm² was about 50 mm. The pier was constructed on a sedimentary soft silt-to-sandstone of Kobe Group with a geological age of about 4 million years or older. The settlement is estimated to be about five times as large as the measured value when estimated based on the average E_{50} value from unconfined compression tests (U tests) (Tatsuoka and Kohata, 1995a).

The construction of a suspension bridge at the Tokyo Bay entrance is under consideration, which will be longer than 3,910 m-long Akashi Kaikyo bridge. The foundations may be placed directly on a sedimentary soft rock of Kazusa Group (on the east side) and Miura Group (on the west side), which may range from sandstone to mudstone. Although the bridge will be longer than Akashi Strait bridge, Miura Group and Kazusa Group are, respectively, equivalent to and younger than Kobe Group. One of the main purposes of the geotechnical investigation for this bridge is, therefore, to evaluate as accurately as possible the stiffness as well as strength of these soft rocks.

2) Sagamihara test site (**Fig. 3b**) (Ochi et al., 1993, 1994): A 50 m-deep experimental shaft and a short tunnel at the bottom of the shaft were excavated from 1989 until 1992 in a sedimentary soft mudstone deposit of Kazusa Group without using a stiff support as used in other actual construction projects under similar conditions. The extension of the shaft to deeper levels is now under way. The deformation and strength characteristics of the mudstone have been investigated in great detail (Kim et al., 1994, Tatsuoka et al., 1993, 1995 a, b, c, d, Tsubouchi et al., 1994, Kohata et al., 1995). Shown also in **Fig. 3b** are the compressive strength q_{max} obtained from CU and CD triaxial compression (TC) tests on core specimens isotropically consolidated to the respective in-situ effective overburden pressure σ_v'(in-situ) and the corresponding q_u values. In the engineering practice in Japan, it is usual to determine design earth pressure for an excavation in a saturated clay deposit based on the specified earth pressure coefficients defined in total stresses. This method is often used also for an excavation in a sedimentary soft rock deposit. The distribution of the design earth pressure with depth for the shaft at Sagamihara site is shown in **Fig. 4**.

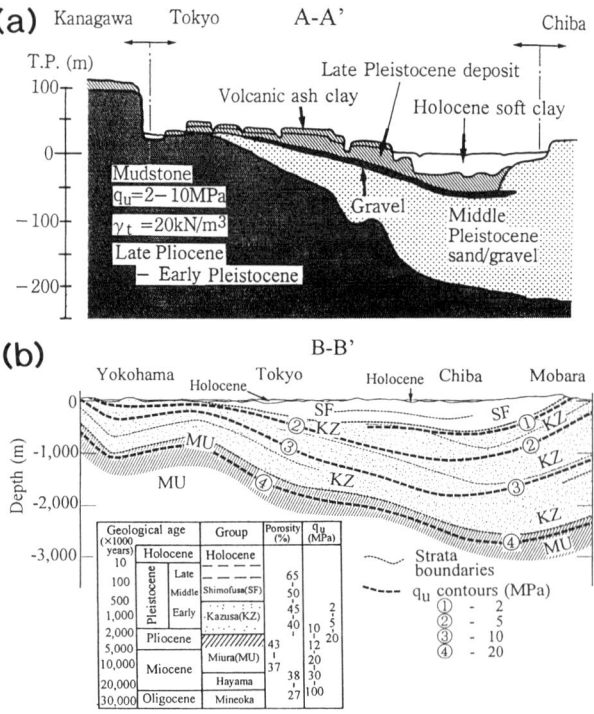

Fig. 1 Locations of some typical construction projects related to soft rock in the Tokyo metropolitan area

Fig. 2 Two typical geological cross-sections in the Tokyo metropolitan area, see Fig. 1 for the locations (Fig. b is after Hoshino, 1993)

It is to be noted that this design earth pressure is usually smaller than the earth pressure calculated using the earth pressure coefficients defined in effective stresses plus hydrostatic pressure. It has been one of the objectives of this research project to confirm that the design earth pressure based on the conventional method is too conservative for a deep excavation in this type of soft rock deposit. The full-scale behaviour showed that it is the case as seen from Fig. 4 (Ochi et al., 1993, 1994).

3) Negishi underground LNG tanks (**Fig. 3c**) (Goto and Takahashi, 1993, Ito et al., 1994): Two 85,000 kl and one 200,000 kl LNG tanks were constructed by excavating a sedimentary soft mudstone deposit of Kazusa Group from 1991. Another 200,000 kl LNG tank is now under construction. Since both a diaphragm wall and long anchors were not used while using only short nailing in these excavations, the ground movement and the wall stability upon ground excavation was one of the major concerns. Goto and Takahashi (1993) reported that the largest outward displacement at the wall face caused by ground excavation was about 15 mm for one of the two 85,000 kl tanks, the one other than that shown in Fig. 3c. This displacement was about one third as small as the value estimated based on the E_{50} value (= 3,000 kgf/cm^2) obtained from U tests. This point will be discussed again later.

It is also well known that the amount of heaving at an excavation bottom in stiff soil and soft rock ground upon excavation and the settlement of the ground upon the construction of a high-rise building is usually largely over-estimated when estimated based on the stiffness obtained from conventional types of pressuremeter test, plate loading test, U test and TC test (Akino, 1990, Akino et al., 1994, Miyazaki et al., 1994). In this paper, the data of sedimentary soft mudstone obtained from a systematic geotechnical investigation performed at the construction site of a high-rise building at Tsunashima (Fig. 1) is included in the data sets in this paper.

Perhaps, the deepest boring in which a systematic geotechnical investigation was performed in a deposit of Kazusa Group is those at Sodegaura (600 m deep) and Kajima (500 m deep) (Fig. 1) (Kawasaki et al., 1993). The results were quoted by Tatsuoka and Kohata (1995a).

4) TTB Highway under construction (**Fig. 3d**; Uchida et al., 1993): Several new ground improvement techniques by cement-treatment were used extensively for constructing man-made islands and for improving existing soft clay deposits at the sites numbered ① ～ ④ in Fig. 3d. The total volume of soil improved by DMM is 1.8 million m^3 and that of the fills constructed by reclamation using cement-mixed sandy soil is 1.93 million m^3. In the beginning stage of design, the stiffness of the cement-treated soils was evaluated based on E_{50} values from conventional U tests, the typical value of which was 3,000 kgf/cm^2 for slurry-type cement-mixed sandy soil used at the sites ① shown in Fig. 3d. It was found at the later design stage by performing accurate TC tests on this material that the deformation is nearly linear elastic for a strain range of about 0.01 % (as shown later). The Young's modulus for elastic deformation was found to be equal to around ten times as large as the E_{50} value, while the E_{tan} value at the largest in-ground strain (about 0.05 %) expected for the design seismic load

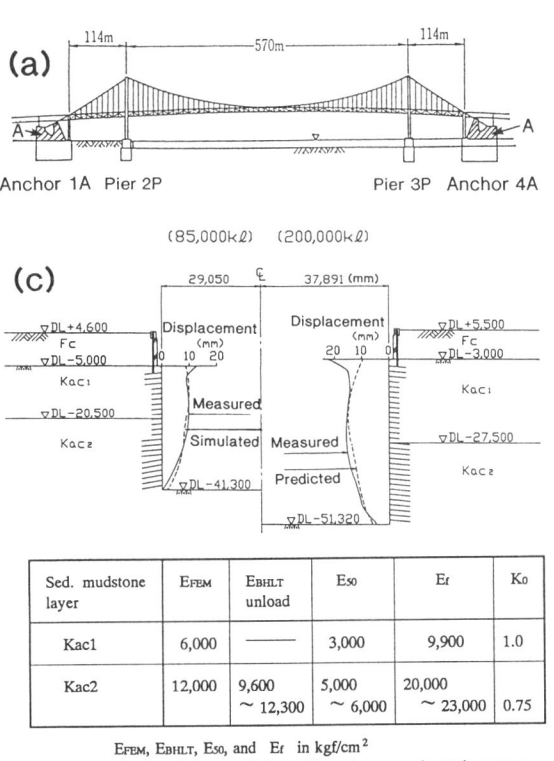
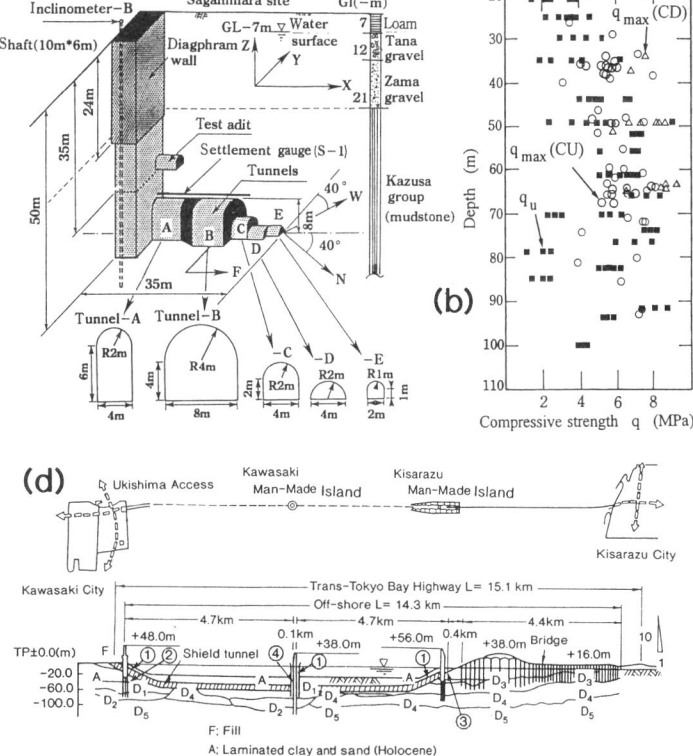

Fig. 3 Some of the structures indicated in Fig. 1;
a) Rainbow bridge (Odagiri et al., 1993, 1994)
b) Sagamihara site;
a general view and compressive strength
c) Negishi LNG tanks (Ito et al., 1994)
d) Trans-Tokyo Bay Highway project

was still about fifteen times as large as the E_{50} value (Shibuya at al., 1992, Uchida et al., 1993). This fact made the engineers in charge confident with the use of these ground improvement techniques for this project. As shown later, the strength and deformation properties of these artificial soft rocks are similar to those of the sedimentary soft rocks referred in this paper.

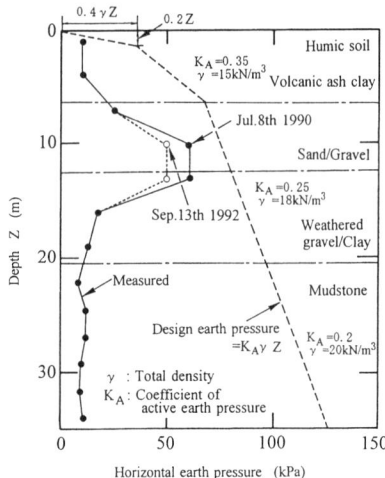

Fig. 4 Desing and measured earth pressure distributions, the shaft at Sagamihara site

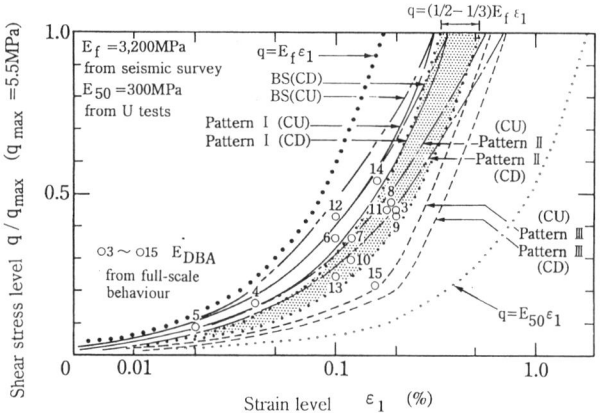

Fig. 5 $q/q_{max} \sim \varepsilon_1$ relations from the field full-scale behaviour and TC tests, Sagamihara site

WHY SMALL STRAINS ?

This paper discusses only the pre-failure deformation properties of soft rocks. As the major principal strain ε_1 at the peak stress state in TC tests on sedimentary soft rocks is around 0.5 % (Tatsuoka et al., 1995d), ε_1 values in sedimentary soft rock deposits under working loads are naturally much less than that value, as illustrated below:
1) Rainbow bridge: The estimated largest ε_1 in the sedimentary soft mudstone ground immediately beneath Anchor A4 upon the completion of the bridge is about 0.1 %.
2) Akashi Strait bridge: The largest measured ε_1 in the sedimentary soft silt-to-sandstone deposit of Kobe Group immediately beneath Pier 3P with a diameter of 68 m was about 0.2 %.
3) Sagamihara site: **Fig. 5** shows the relationship between shear stress level $q(=\sigma_1-\sigma_3)/q_{max}$ and log. of ε_1 obtained from the back-analysis of the full-scale behaviour of the shaft and tunnel upon their excavation, which are denoted by the numbers 3 through 15. In this figure, relations obtained from TC tests and others are also presented, which will be explained later. It is seen that despite the use of a very light support system in this experimental excavation, the largest ε_1 is small, about 0.2 %, at the walls of the shaft and the tunnel. The smallest local safety factor, which is the inverse of q/q_{max}, remains larger than about 2.0.
4) Negishi LNG tanks: For both the 85,000 kl and 200,000 kl tanks (Fig. 3c), the largest ε_1 value in the periphery direction at the wall face was about 0.05 %.

It is seen from the above that accurate determination of the stiffness of soft rocks at strains of less than around 0.2 % is of particular practical importance for the serviceability limit design of these structures. However, in most of the conventional geotechnical investigation methods, stiffness of soft rocks (and other geomaterials) is measured usually at strains larger than 0.2 %.

CONVENTIONAL METHODS

Figs. 6a and **b** compare the following Young's moduli obtained at Sagamihara site:
1) $E_{PLT}(D)$ from the relationships of plate load between plate settlement during primary loading obtained by plate loading tests (PLT tests), evaluated based on the conventional linear theory. In this case, stiffness values from

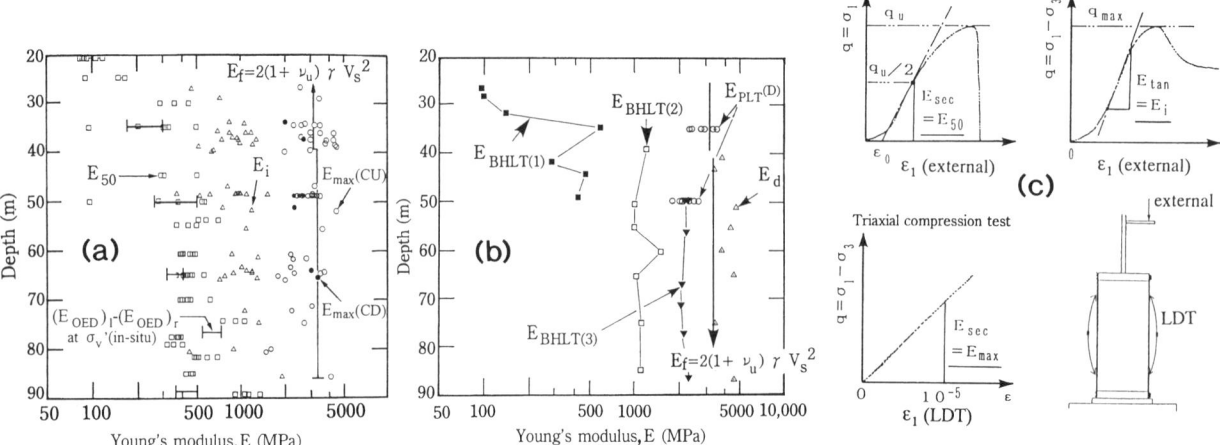

Fig. 6 Young's moduli from field and laboratory stress-strain tests, Sagamihara site (Ochi et al., 1994, Tatsuoka et al., 1995d)

unloading/reloading curves were larger only slightly (by about 20 %) than the values from primary loading curves (Ochi et al., 1993, 1994).
2) E_{BHLT} from primary loading curves obtained by three different types of bore-hole lateral loading tests (BHLT tests) or pressuremeter tests (as explained in Tatsuoka et al., 1995d), evaluated based on the conventional linear theory.
3) $E_{50} = q/\varepsilon_1$ when $q = q_{max}/2$ from U tests (**Fig. 6c**).
4) E_i (initial Young's modulus) from TC tests (Fig. 6c; see also Fig. 12a).
5) E_{OED} from compressibilities during loading and reloading obtained by oedometer tests.
In most of the conventional design procedure, the stiffness of soft rock which is used to evaluate the deformation of ground and the displacements of structures under static working loads is estimated by one or some of the methods 1) ~ 5) shown above (except by those based on SPT N values). These moduli of deformation have been conventionally called "static modulus of deformation" or "static elastic modulus".

Other Young's moduli that are shown in Fig. 6 are:
6) $E_f = 2(1+\nu_u) \cdot \rho \cdot V_s^2$ (ν_u; undrained Poisson's ratio= 0.43 in this case) from field shear wave velocities V_s obtained by the down-hole logging and the up-hole local logging called the suspension method.
7) E_d from laboratory ultra-sonic wave tests on core samples consolidated to σ'_v(in-situ). This value is used usually as an index property (n.b., the E_d values of unconfined specimens are generally much smaller than corresponding E_f values and could be very erratic, Kim et al., 1994).
These moduli of deformation, E_f and E_d, have been conventionally called "the dynamic elastic modulus". It is seen from Fig. 6 that these "static and dynamic moduli of (elastic) deformation" scatter very largely. Among them, the E_{50} values are particularly low. **Fig. 7** compares the lateral outward displacements at the shaft wall which were measured and those obtained by a 3-D FEM analysis using the average value of E_{50} (= 3,000 kgf/cm² or 300 MPa). The simulations were performed by assuming two different initial lateral pressure coefficients $K = \sigma_h/\sigma_v$ defined in total stresses equal to 0.75 and 1.5. Ochi et al. (1993, 1994) reported that $K = 1.5$ is the best estimate for this case. It is seen from Fig. 7 that the use of this E_{50} value considerably over-estimates the ground movement under static working loads. Indeed, in these simulations, the ε_1 values at the shaft wall face obtained as the ratio of "the lateral displacement at the shaft wall face" to 2.5 m (= a half of the shaft width) are close to or exceed the average ε_1 value at failure (= 0.5 %). Corresponding to the above, it is seen from Fig. 5 that the relationship between $q (= E_{50} \cdot \varepsilon_1)/q_{max}$ and ε_1 exhibits far smaller stiffness than the back-calculated relations. Furthermore, the q_u values are generally lower than the q_{max} values obtained from TC tests (Fig. 3c). Kawasaki et al. (1993) reported that the geotechnical investigation at Sodegaura and Kajima sites showed that the E_{50} and q_u values are systematically smaller than, respectively, the E_f values from the field seismic survey and the q_{max} values from CU TC tests. Many other similar case histories can be found in the literature (Tatsuoka and Kohata, 1995a). The main reasons for the above are: a) the effects of no confinement in U tests, which exaggerate the effects of sample disturbance (so, larger effects of no confinement may be seen for more pressure-sensitive materials such as sedimentary soft sandstone, Tatsuoka and Kohata, 1995a), b) the effects of bedding error (for E_{50}), and c) large strains to define the stiffness (for E_{50}). Indeed, U tests are inexpensive to perform, but design based on the results of U tests may be too conservative.

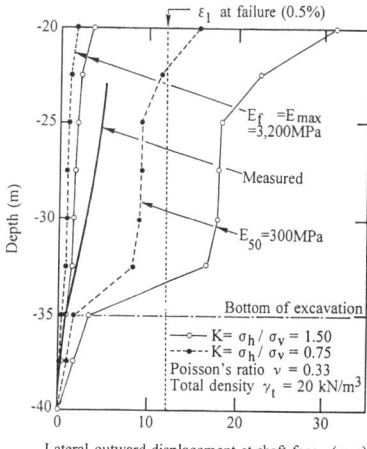

Fig. 7 Comparison between measured and simulated lateral displacements at 0.5 m behind the wall face of the 50 m-deep shaft, Sagamihara site

In some cases, compressibilities $m_v = d\varepsilon_v/d\sigma_v$ obtained from oedometer tests using a 2 cm high specimen are used to predict the settlement of a foundation constructed on sedimentary soft rock. **Fig. 8** shows results typical of those from oedometer tests using samples obtained by block sampling at Sagamihara site. For an isotropic material, the tangent Young's modulus E_{OED} can be obtained as;

$$E_{OED} = (1/m_v)(1+\nu_d)(1-2\nu_d)/(1-\nu_d) \quad (1)$$

where ν_d is the drained Poisson's ratio. For $\nu_d = 0.2$, we obtain the relationship $E_{OED} = 0.9 \cdot (1/m_v)$, which is close to the relationship $E_{OED} = 1/m_v$. The values of E_{OED} obtained from the values of m_v at σ'_v(in-situ) during loading and reloading obtained from these and other similar tests are plotted in Fig. 6a. It is seen that most of the values during loading $(E_{OED})_l$ are even smaller than the E_{50} values, while those during reloading $(E_{OED})_r$ are similar to the E_{50} values. This result shows that the

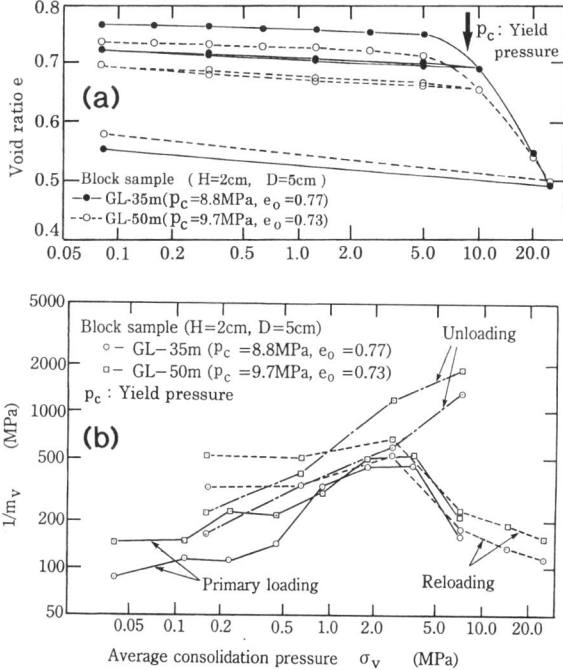

Fig. 8 Results of two typical oedometer tests on sedimentary soft mudstone from Sagamihara site

E_{OED} values are too small when used to predict field full-scale behaviour under static working loads at Sagamihara site. It is also the case in the settlement analysis of a large foundation constructed on a sedimentary soft rock such as those for Rainbow bridge (Tatsuoka and Kohata, 1995a). These too small stiffness values of soft rock from oedometer tests can be attributed to very large effects of bedding error due to a very small specimen height (2 cm). In the opinion of the authors, oedometer tests on soft rock are most suitable for the evaluation of 1-D yield pressure.

It may be seen from Fig. 6b that the values of E_{BHLT} and E_{PLT} scatter largely. This is largely due to the fact that these results are largely test-dependent, caused by large variations in the degree of disturbance of the bore hole wall, the effects of bedding errors and in-ground strains among different tests. Tatsuoka and Kohata (1995a) argued that for the last several decades, the E_{BHLT} values reported in the literature have increased largely with time, which is due seemingly to the fact that with time, the effects of system compliance have decreased, the boring technique has been improved, the use of cyclic tests has become more often, and measurable in-ground strains have decreased by using more sensitive pick-ups.

The E_i values shown in Fig. 6a were obtained from the apparent linear part of the relationship between q and externally measured ε_1 by a TC test on a specimen isotropically consolidated to σ_v(in-situ). A typical result for soft mudstone at Sagamihara site is shown in Fig. 2 of Tatsuoka et al. (1995d) (also in Fig. 12 of Tatsuoka et al., 1995b, Fig. 23 of Tatsuoka et al., 1995c, Fig. 3 of Kohata et al., 1995) (n.b., a similar result for sedimentary soft rock obtained at Kan-non-zaki is shown in Fig. 12). For TC tests on saturated sedimentary soft rock, it has been the conventional method to obtain axial strains from the axial displacements of the specimen cap or the loading piston (i.e., external ε_1). It is seen from Fig. 6a that these E_i values are noticeably larger than the E_{50} values, which is due to decreased effects of bedding error, but they are still much smaller than the E_f values.

It is seen from the discussions above that the stiffness values from a variety of the conventional static testing methods, 1) through 5), are not consistent to each other, while they are noticeably smaller than the dynamically determined values, E_f and E_d. It seems that the concept of unlinked static and dynamic moduli of deformation has stemmed from this apparent difference. It is shown below, however, that this conventional distinguishment is not correct and misleading.

A METHOD BASED ON ELASTIC MODULUS

In a TC test on a soft rock specimen, axial strains ε_1 measured locally along the lateral surface of specimen, for example by using a pair of Local Deformation Transducer (LDT; Goto et al., 1991, see Fig. 6c), are generally much smaller than externally measured ones (as seen from Figs. 12 and 13). This is mostly due to the effects of bedding error at the specimen ends. The true elastic Young's modulus E_{max} (Fig. 6c) can be obtained only from a relationship between q and local ε_1 for a range of ε_1 less than about 0.001 % ~ 0.01 %, depending on the range of linear elastic zone (see Fig. 12d) (Tatsuoka et al., 1994a, b).

It seems that the practical importance of sensitive and precise measurements of ε_1, as shown in Fig. 12, has been overlooked for a long time for soft rocks. The first reason would be the lack of data showing

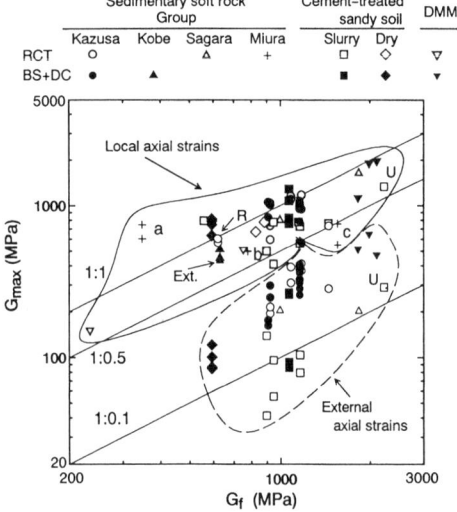

Fig. 9 Summary of the comparison between $G_f = \rho \cdot V_s^2$ and G_{max} from TC tests on samples obtained by block sampling and rotary core tube sampling for sedimentary soft rocks and cement-treated soils

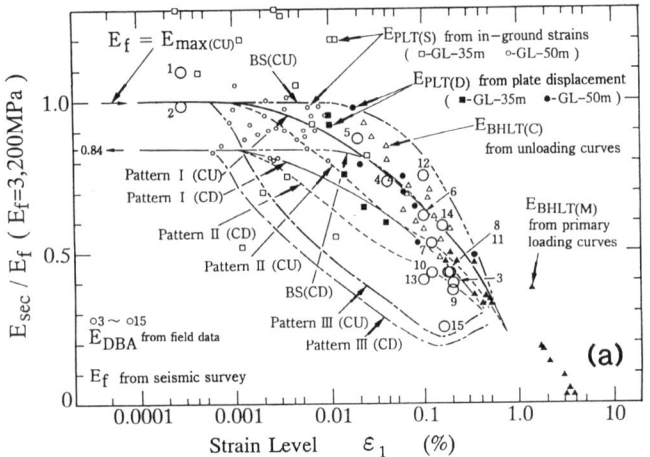
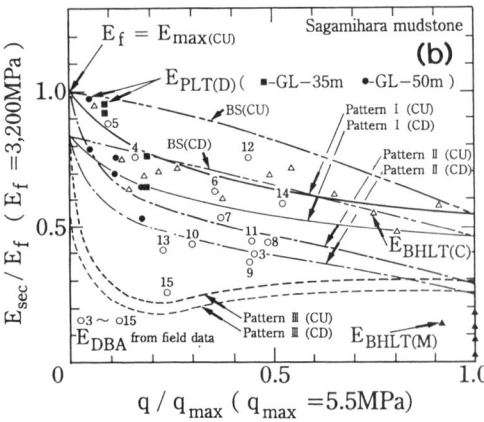

Fig. 10 a) $E_{sec}/E_f \sim \varepsilon_1$ and b) $E_{sec}/E_f \sim q/q_{max}$ relations from the back-analysis of the full-scale behaviour of the shaft and tunnel and field and laboratory stress-strain tests, Sagamihara site (Ochi et al., 1994, Tatsuoka et al., 1995d)

possibly large effects of bedding error. To measure axial strains locally, accurately and sensitively, the use of electric-resistent strain gauges has been a standard practice for tests on hard rocks. This method is, however, difficult to use with a saturated sedimentary soft rock specimen having a relatively rough and loose surface. The second reason would be the fact that sedimentary soft rock has the strength and deformation characteristics which are between those of stiff uncemented soils and hard rocks. When following the conventional soil mechanics approach, triaxial testing methods which have been developed mainly for testing relatively soft soils would be used. In that case, small strain measurements may not be a major concern. On the other hand, when following the conventional rock mechanics approach, sophisticated laboratory stress-strain tests on core samples may not be performed considering that the results of core tests do not represent the properties of a mass in the field having discontinuities.

It is seen from Fig. 6a that the values of E_{max} are on average very close to the E_f values. This result suggests that we do not need to distinguish between dynamic and static elastic moduli for this type of soft rock (and other types of sedimentary soft rock and most uncemented geomaterials) (Tatsuoka and Shibuya, 1992, Shibuya et al., 1992, Tatsuoka et al., 1994b). This point has been supported by the findings that at strains lower than about 0.001 %, the stiffness of sedimentary soft mudstone is essential strain-rate independent (see Fig. 2.60 of Tatsuoka and Kohata, 1995a, Fig. 42 of Tatsuoka et al., 1995c), and the deformation is nearly recoverable (see Figs. 12 and 13).

Corresponding values of E_{max} and E_f are very close to each other also for sedimentary soft rocks at many other sites, including Akashi Strait bridge site, Negishi LNG tank site, Tsunashima site and others for sedimentary soft rocks and several TTB Highway sites for cement-treated soils. **Fig. 9** compares the corresponding values of $G_f = \rho \cdot V_s^2$ and G_{max} from TC tests. Each data point represents the average of several number of data obtained under similar conditions. In this figure, the data of TC tests performed on core samples obtained by direct coring using a well-fixed diamond core barrel and by hand carving (block sampling), which are least disturbed, and those obtained by rotary core tube sampling, which may be more-or-less disturbed, are distinguished from each other. The G_{max} values based on local and external ε_1 values are also distinguished from each other. Fig. 9 includes

several new data sets additional to those presented in Fig. 20 of Tatsuoka et al. (1995b). They are a new data set obtained in 1993 for Sagamihara site, data sets for Rainbow bridge site (denoted by the letter R; Odagiri et al., 1993, 1994) and Kan-non-zaki site (these denoted by the letter a for the samples from depths 5~57 m deep, b; 57~94 m deep, and c; 94~120 m deep) and a data set of slurry-type cement-treated sandy soil at Ukishima Access (denoted by the letter U; see also Fig. 3d) (Sugawara et al., 1995). The following trends of behaviour may be noted from Fig. 9:
1) The values of G_{max} based external ε_1 values are much lower than the respective value of G_f, due to the effects of bedding error.
2) The values of G_{max} based on local ε_1 values of the least disturbed samples are on average very close to the respective G_f value.
3) Even based on local ε_1 values, many G_{max} values of rotary core tube samples are noticeably lower than the respective G_f value, due to the effect of sample disturbance (Tatsuoka et al., 1995b). This point will be discussed again later.

On the other hand, it can be seen from Fig. 5 that the relationship between $q(=E_f \cdot \varepsilon_1)/q_{max}$ and ε_1 exhibits stiffness larger than those back-calculated from field full-scale behavior, while most of the field data are located within the range between the curves representing $q = (1/2 - 1/3) \cdot E_f \cdot \varepsilon_1$. It can also be seen from Fig. 7 that the simulated shaft wall displacements based on the E_f value are noticeably smaller than those measured. These results indicate that the use of E_f value is relevant to evaluate the elastic stiffness in the field, but the strain dependency of stiffness should be taken into account (and the pressure level-dependency of stiffness when needed) to properly evaluate the deformation of ground and the displacements of structures under working loads.

Based on the above, the secant Young's moduli E_{sec} obtained from field and laboratory tests, excluding laboratory test data including noticeable effects of bedding error, were plotted against ε_1 and the shear stress level q/q_{max} (**Figs. 10a and b**), together with other field test data (as explained in Tatsuoka et al., 1995d). The relationships between the Young's modulus E_{DBA} and the largest in-ground ε_1 obtained from back-analysis of the field full-scale behaviour are also plotted. In so doing, these E_{sec} values were divided by the average E_f value (= 32,000 kgf/cm²). The continuous curves are obtained, by assuming that $E_{max} = E_f$, from the results of TC tests on a) the least disturbed samples obtained by direct coring and block sampling, which are denoted as BS(CU) and BS(CD), and b) those obtained by rotary core tube sampling, which may be more-or-less disturbed (Tatsuoka et al., 1995b). For the latter group of samples, depending on the degree of sample disturbance, the averaged TC stress-strain relations, Patterns I, II and III, were defined (Tsubouchi et al., 1994). Besides, different TC test relations were defined for drained and undrained conditions with the ratio of undrained to drained E_{max} values being equal to $(1+\nu_d)/(1+\nu_u) = 1.2/1.43 = 0.84$.

It may be seen from Figs. 10a and b that the Young's moduli shown in Fig. 6 have become much more consistent to each other by taking into account the dependency of stiffness on strain or shear stress level. It should be noted that the curve denoted by BS(CU) representing the relation for the least disturbed samples under undrained conditions forms the upper boundary for the data points from field full-scale behaviour. This trend is more obvious for the data at larger strains, in particular, the data point No. 15, as discussed by Tatsuoka et al. (1995d).

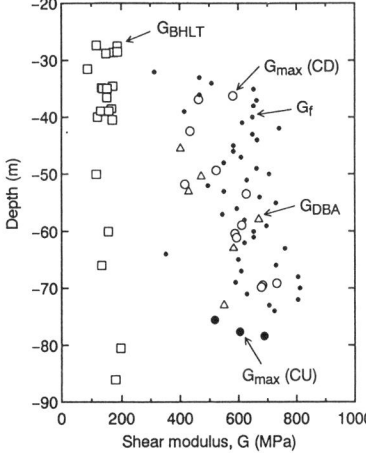

Fig. 11 Comparison of Young's moduli obtained from field and laboratory stress-strain tests, Rainbow bridge site (Odagiri et al., 1993, 1994)

This fact may be attributed to the effects on the E_{DBA} values of drainage or disturbance by excavation work or both. The latter point can be inferred by the fact that the V_S value has decreased from the original value (= 760 m/sec) measured in a bore hole before the shaft excavation to 560 m/sec measured in a 5 m zone from the shaft bottom excavated to a depth of 35 m, and from 730 m/s to 620 m/s in a 1.5 m zone from the wall surface at a depth of 35 m. These reductions in V_S are equivalent to a reduction of E_f to 54 % and 72 % of the respective original value. A similar order of reduction in E_f, a reduction to 60 % and 85 % of the respective original value, was observed in an about 2 m zone from the wall face and in an about 1.2 m zone from the bottom of excavation, respectively, for one of the 20,000 kl LNG tanks shown in Fig. 3c (Ito et al., 1994). These reductions in E_f may be due also to the pressure level-dependency of stiffness when the pressure level is nearly zero, but it may be only partly. This is because for this sedimentary soft mudstone at Sagamihara site and Negishi site, the effect of pressure level-dependency of stiffness as evaluated by TC tests is very small (as shown later; see Fig. 16) (Tatsuoka et al., 1993, Kim et al., 1994).

In engineering practice, the field seismic survey has been performed so far mainly for earthquake response analysis. Based on these findings shown above, however, the following approach can be proposed as a method to predict the deformation of ground and the displacement of structures under static working loads as well as seismic loads:
1) The in-situ elastic stiffness, E_f or $G_f = \rho \cdot V_S^2$, is obtained from field seismic surveys prior to construction.
2) The dependency of stiffness on strain (or shear stress level) and pressure level are evaluated by laboratory stress-strain tests using high-quality core samples.
3) The in-situ relationship among tangent or secant stiffness, strain (or shear stress level) and pressure is obtained from the above 1) and 2). If possible, this relation is validated by the results of relevant pressuremeter tests or plate loading tests or both.

The rationales for this methodology for sedimentary soft rocks, at least for soft rocks referred in this paper, could be summarized as:
1) elastic behaviour at very small strains,
2) relatively small strains in the ground under working loads, for which non-linearity of stiffness is still not significant,
3) relative easiness of measuring V_S in the field, and
4) relatively small or negligible effects of discontinuities and heterogeneities on the stiffness in the field (this point will be discussed later).

Some case records which support the methodology based on E_f values, other than that for Sagamihara site case, are as follows;
1) Akashi Strait bridge: The settlement of Pier 3 upon its construction on a sedimentary soft silt-to-sandstone (Kobe Group) was about 50 mm. This behaviour was successively simulated by a numerical analysis based on the E_f values, while taking into account the dependency of stiffness on shear stress and pressure levels which were obtained from CD TC tests on undisturbed samples (Fig. 7.10 of Tatsuoka and Kohata, 1995a, Fig. 15 of Siddiquee et al., 1995, this symposium).
2) Rainbow bridge (Fig. 3a): The average settlement of Anchorages 1A and 4A upon the construction of an anchor block on the crest of the respective anchorage caisson were about 18 mm and 15 mm. This behaviour was successively simulated by a numerical analysis based on E_f while taking into account the dependency of stiffness on strain which was obtained from TC tests on rotary core tube samples (Odagiri et al., 1993, 1994). In these analyses for Akashi Strait bridge and Rainbow bridge, the results of pressuremeter and plate loading tests were not utilized. **Fig. 11** compares the following shear moduli for Rainbow bridge case; a) G_{BHLT} values (based on the conventional linear theory), b) G_f (by the suspension method), c) G_{max} from CD and CU TC tests measuring axial strains locally, and d) G_{DBA} from back-analysis of the full-scale behaviour. A trend very similar to that seen in Fig. 6 may be noted. The fact that the G_{max} values and the G_{DBA} values are slightly smaller than the G_f values may be due to different reasons, which are the effects of sample disturbance and the effect of strain non-linearity, respectively.
3) Negishi LNG tanks (Fig. 3c): The lateral outward displacements of the shaft wall for one of the 200,000 kl LNG tanks was predicted, prior to ground excavation, by 3-D numerical analysis using the Young's modulus values E_{FEM} and initial K_0 values listed in the table inset in Fig. 3c. The E_{FEM} and K_0 values for the lower soft rock layer Kac2 are the same with those used for the back-analysis of the ground movement upon excavation for the 85,000 kl LNG tank depicted in Fig. 3c. It is seen that the agreement between the predicted and measured ground movement is remarkably good. These Young's modulus values E_{FEM} are about a half of the respective E_f value. This ratio (1/2) is too small when considering from the low strain level in the ground (i.e., about 0.05 % at the largest). It seems that the following reasons, among others, are responsible for this low ratio: a) The assumed original K_0 value equal to 0.75 in the total stresses means a K_0 value equal to about 0.5 in effective stresses, which may be too small for this type of sedimentary soft rock deposit. As the assumed K_0 value decreases, the back-calculated E_{DBA} value decreases for a given amount of ground movement (or the back-calculated ground movement decreases for a given E_{FEM} value, see Fig. 7). b) The effects of disturbance by construction work on the stiffness as described above may not be negligible. On the other hand, the E_{50} values from U tests are about a half of the E_{FEM} values. The E_{BHLT} value from unload curves is close to the respective E_{FEM} value. Perhaps, it may be due partly to that the strain levels involved in the unloading pressuremeter tests were small enough.

SIMILARITY BETWEEN SEDIMENTARY AND ARTIFICIAL SOFT ROCKS

Figs. 12 and **13** respectively show the results of two TC tests (CD and CU) on a rotary core tube sample of soft siltstone (Miura Group) obtained at Kan-non-zaki site and cement-treated sandy soil (slurry-type) obtained from the approach fill of Ukishima Access (Fig. 3d). It may be seen that, in a broad sense, the behaviour of the two materials is similar to each other (n.b., the effects of different drainage conditions on these results are small). It may also be seen that in both tests, the effects of bedding error are significant. Note also that in these tests, several very small unload/reload cycles were applied during triaxial compression to evaluate elastic Young's moduli during shearing (Tatsuoka et al., 1994a, b).

Figs. 14a and **b** summarize the $E_{max} \sim q_{max}$ relations obtained from TC tests on a number of sedimentary soft rocks including those in the Tokyo metropolitan area. **Fig. 15** is a similar one for cement-treated soils used for the TTB Highway project. It is seen that although the ratio E_{max}/q_{max} is slightly higher for artificial soft rocks than for sedimentary soft rocks, the order of the E_{max} and q_{max} values is similar between the two types of soft rocks. In Fig. 14a, the relationship between q_{max} and E_i (based on external ε_1) from TC tests and that between q_u and E_{50} are also plotted. It may be noted that it is difficult to discuss on the similarity or difference among different types of soft rocks based on such data.

By the DMM method, a group of cement-treated clay columns are formed in a soft clay deposit either leaving some volume of untreated soil or not (i.e., 100 % improvement) as in the TTB Highway project. Tanaka and Terashi (1986) showed that the properties are not very uniform even within each column. One may say that because of the above, it is more than sufficient to evaluate approximated strength values and deformation characteristics by the conventional U tests. However, the stiffness may be grossly underestimated when based on U test results. For the TTB Highway project, therefore, it was decided to evaluate the stiffness of cement-treated soil based on TC tests measuring axial strains locally (see Fig. 13). Considering similarities of strength and deformation characteristics between these sedimentary and artificial soft rocks, it seems that the design based on stiffness from U tests measuring axial strains only externally would restrict the possible use of a soft clay deposit improved by DMM and a fill reclaimed by using cement-treated soil as a foundation for an important permanent structure.

In some cases, the effects of discontinuities and heterogeneities in a mass of soft rock in-situ on the relationship between the elastic stiffness E_{max} or G_{max} from TC tests on high-quality undisturbed samples and E_f (or G_f) from field seismic surveys may not be negligible. It is certain that the E_{max} or G_{max} values of either core specimens which does not include discontinuities existing in the field or those retrieved from a relatively stiffer zone in a highly heterogeneous mass should be noticeably larger than the E_f value, since the E_f value represents the elastic stiffness for a relatively large mass. The data shown in Fig. 9 indicate, however, that for the sedimentary and artificial soft rocks referred in this paper, those effects on this relationship are not significant.

On the other hand, it may be seen from Fig. 6b that the E_d values of core specimens confined at σ'_v(in-situ) are slightly larger than the E_f value (thus larger than the values of E_{max}). It is likely that because of its short wave length, ultra-sonic wave measurements may over-estimate slightly even the average stiffness of a nominally homogeneous material like sedimentary soft mudstone at

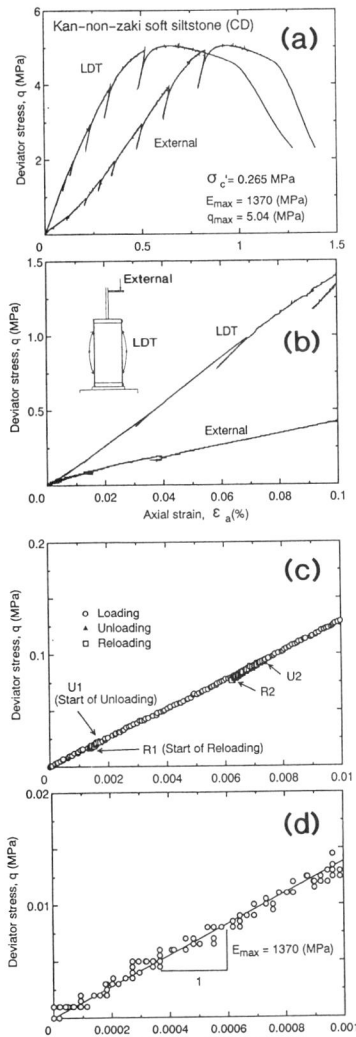

Fig. 12 Typical CD TC test result of sedimentary soft siltstone obtained by rotary core tube sampling at Kan-non-zaki

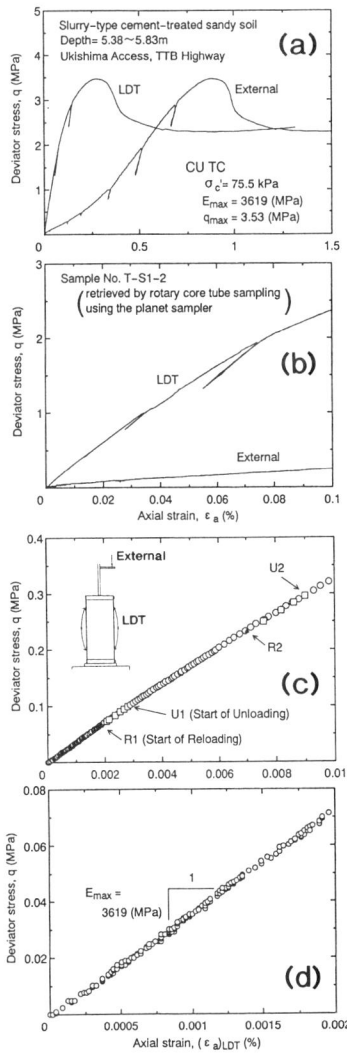

Fig. 13 Typical CU TC test result of cement-treated sandy soil (slurry-type) obtained by rotary core tube sampling from the approach fill at Ukishima Access

Sagamihara site. A more significant difference was observed between E_d values obtained from ultra-sonic wave measurements in an in-situ compacted layer of non-uniformly cement-mixed cinder at the summit of Mauna Kea of the Hawaii Island and E_{max} values obtained from the corresponding TC tests (Tatsuoka et al., 1994c, Tatsuoka and Kohata, 1995a). Therefore, ultra-sonic tests are not suggested to evaluate the average stiffness of a highly heterogeneous material like soil mixed with cement not thoroughly.

SOME IMPORTANT ISSUES OF DEFORMATION PROPERTIES

The method based on E_f values described above is more sophisticated than the conventional methods, and requires many influencing factors to be evaluated properly. These factors include; a) the effects of sample disturbance, b) the effects of discontinuities and heterogeneities, c) the effects of pressure level, or more generally the effects of stress state, d) anisotropy, e) the dependency of stiffness on strain or shear stress level, and f) the effects of drainage conditions, consolidation and creep.

Most of these factors were discussed by Tatsuoka and Shibuya (1992) and Tatsuoka and Kohata (1995a) and in the above, but they are never fully understood yet. Some discussions are made below on some of these factors based on the data obtained after these two reports above were prepared.

Effects of sample disturbance: As discussed in detail by Tsubouchi et al. (1994) and Tatsuoka et al. (1995b), the conventional rotary core tube sampling method may more-or-less disturb soft rock, which may result in a noticeable reduction in E_{max} or G_{max} (Fig. 9). In particular, it may be seen that for the data points denoted by the letters a, b, and c of rotary core tube samples retrieved from shallower to deeper levels at Kan-non-zaki site, the ratio G_{max}/G_f tends to decrease with sampling depth, seemingly due to increased degrees of sample disturbance with depth. Sample disturbance may distort the stress-strain relation, reducing to a larger extent the values of tangent modulus E_{tan} at strains exceeding elastic limit strain (about 0.001 %), particularly those at q/q_{max} equal to about 0.2 (Figs. 5, 10a and 10b). Sample disturbance is one of the urgent problems to be solved for improving our abilities of evaluating the stiffness of soft rock in the field.

Effects of stress state: **Fig. 16** summarized the relationship between E_{max} values measured at several isotropic stress states using a single specimen and the confining pressure σ'_c for several types of sedimentary soft rocks in the Tokyo

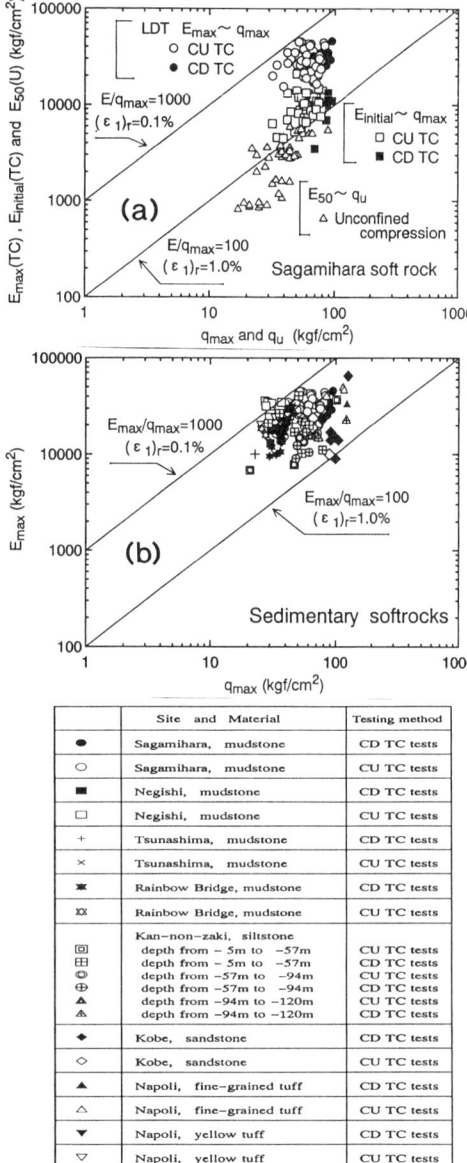

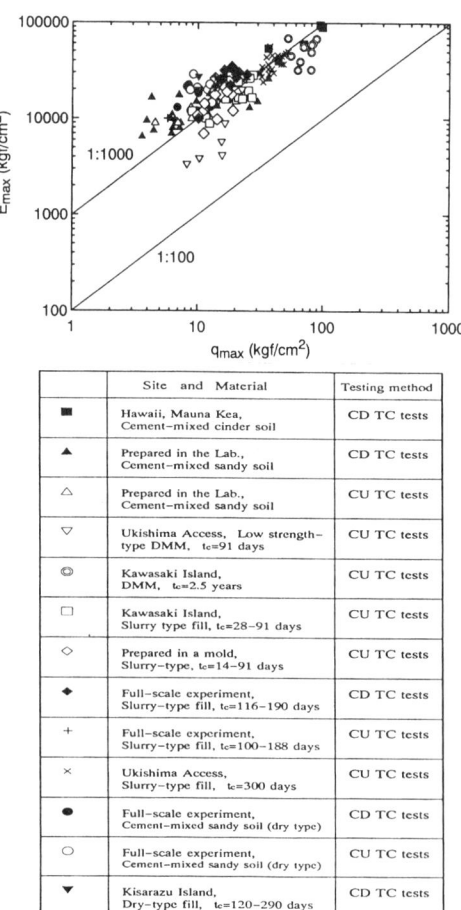

Fig. 14 $E_{max} \sim q_{max}$ relations for sedimentary soft rocks; a) Sagamihara site and b) all the sites

Fig. 15 Summary of $E_{max} \sim q_{max}$ relations for cement-treated soils

metropolitan area and others and one weathered granite (another type of soft rock). It may be seen that the stiffness of the sedimentary soft mudstones and siltstones are much insensitive to pressure changes, when compared with the sedimentary silt-to-sandstone and the weathered granite.

It is not certain, however, that a large pressure dependency of E_{max} seen for some relatively poorly cemented soft rocks, shown in Fig. 16, is totally representative of the behaviour in the field. This infer is supported by the relationship between the E_d value from laboratory ultra-sonic wave measurement and the isotropic confining pressure σ'_c of core samples (5 cm high) obtained from thin uncemented sand seams found within a soft mudstone deposit at Sagamihara site (**Fig. 17**). It is seen that the E_d values have not been recovered to the field value $E_f = 32,000$ kgf/cm^2 even at $\sigma'_c = \sigma'_v$(in-situ), perhaps due to the remaining effects of sample disturbance. The slope of the log-log relationship,

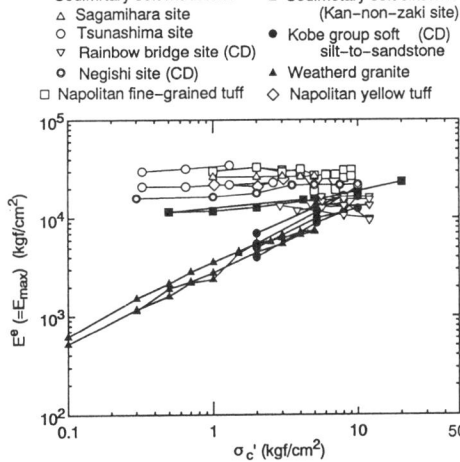

Fig. 16 Summary of $E_{max} \sim \sigma'_c$ for several types of sedimentary soft rocks and one weathered granite (CU for those other than the ones denoted by CD)

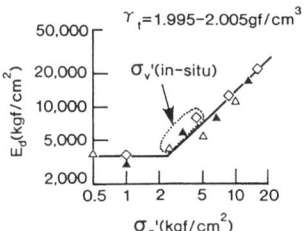

Fig. 17 $E_d \sim \sigma'_c$ relations of core samples retrieved from uncemented sand seams at depths between 50 m and 60 m at Sagamihara site

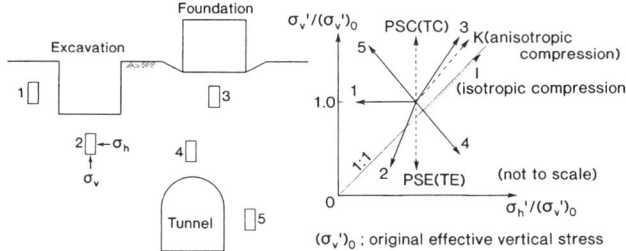

Fig. 18 Schematic figure for some typical stress paths during construction and in some laboratory tests (under plane strain conditions)

between E_d and σ'_c for a range of σ_c' exceeding about 2.5 kgf/cm^2 is as large as 0.92, which is much larger than a typical value of 0.4 for reconstituted sand specimens. It seems that this large slope is due to the closing with the increase in σ_c' of micro-cracks produced during sampling and handling.

In any case, when it is judged to be not negligible, the pressure level-dependency of stiffness should be taken into account when evaluating stiffness in the field, considering that stress state changes in different ways in different boundary value problems as illustrated in **Fig. 18**. The stress paths denoted by the numbers 1 through 5 represent some typical stress changes. Then, the manner how the stiffness depends on stress state should be known.

Tatsuoka and Kohata (1995a) reported several triaxial test data of cohesionless soils which showed that the elastic Young's modulus defined for elastic strain increments $d\varepsilon_1^e$ in a certain direction is a rather unique function of the normal stress working in that direction. Kohata et al. (1994) showed some TC test data of sedimentary soft rock which exhibited a noticeable increase in the elastic Young's modulus E_{eq} obtained from very small unload/reload cycles with the increase in σ_1 during a TC test with a constant σ_3. It is also the case with the CD TC test on sedimentary soft siltstone from Kan-non-zaki (Fig. 12). **Fig. 19** shows the relationship between the E_{eq} value and the shear stress level SL = q/q_{max}. Then, it can be naturally inferred that also for soft rocks, the elastic Young's modulus defined for $d\varepsilon_1^e$ in a certain direction is a rather unique function of the normal stress working in that direction. When it is the case, the elastic properties of soft rock become anisotropic under anisotropic stress states. Based on this assumption, Tatsuoka and Kohata (1995a) introduced a cross-anisotropic model for sedimentary soft rocks, in which elastic anisotropy is induced by different vertical and horizontal stresses. This model was used to simulate of the results of plate loading tests at Anchor 1A site and the full-scale behaviour of Pier 3P for Akashi Kaikyo bridge (Siddiquee et al., 1994, 1995).

Inherent anisotropy: Kohata et al. (1995) showed that the sedimentary soft mudstone at a depth of 50 m at Sagamihara site is slightly, by about 30 % at the largest, stiffer and stronger in certain horizontal directions than in the other directions. This degree of anisotropy is not very large, compared with other larger uncertainties as to the stiffness of soft rock in the field. Further study will be required to reach a more general conclusion.

Dependency of stiffness on strain or shear stress level: So far, $E_{sec} \sim \varepsilon_1$ plot as shown in **Fig. 20** has been the most popular method to represent the non-linearity of stiffness. This method has, however, the following limitations:
1) This relation is controlled by the definition of the origin for ε_1, which is, however, not unique. As a more general method, $E_{tan} \sim q/q_{max}$ plot as shown in Fig. 19 may be employed. This relation is independent of the origin of ε_1 as far as the E_{tan} value is a unique function of the current stress state (along a given stress path).
2) When elastic Young's moduli are noticeably pressure-dependent, E_{sec} (and E_{tan}) values also become pressure-dependent. In this case, a E_{sec} (E_{tan}) $\sim \varepsilon_1$ (or q/q_{max}) relation obtained for a specific stress path of TC is not applicable directly to more general stress paths.

The authors consider that $E_{tan}/E^e \sim$ SL= q/q_{max} plot as shown in **Fig. 21** is more general than the above two methods. In that plot, each value of E_{tan} is divided by the undamaged elastic Young's modulus E^e obtained at the isotropic stress σ'_c which is equal to σ_1 where the E_{tan} is measured. The E^e value is a function of the confining pressure under un-sheared isotropic conditions (as shown in Fig. 16). For this plot of E_{tan}/E^e shown in Fig. 21, the following equation for E^e was used:

$$E^e(\sigma_a) = E_{max} \cdot (\sigma'_a/\sigma'_c)^{0.3} \quad (2)$$

this equation is shown in Fig. 19 after having been converted into a $E^e(\sigma_a) \sim$ SL relation. When σ'_c is the confining pressure, which is σ'_a at the start of TC when E_{max} is measured. The power 0.3 was determined from the E^e and σ'_c relation, which is shown in Fig. 16, for a range of σ'_c larger than 2.7 kgf/m² obtained from another CD TC test on a sample from Kan-non-zaki site. The E_{tan}/E^e and SL relation obtained from that test with σ'_c = 4.3 kgf/m², for which both E_{tan} and E^e values were obtained by using the same simple, is presented in **Fig. 22**. In Figs. 21 and 22, the relationship between E_{eq}/E^e and q/q_{max} is also shown. It is seen that the elastic Young's modulus E_{eq} measured during TC becomes smaller than the corresponding E^e value to a larger extent with the increase in SL. This feature can be considered to be a result of the damage to the structure by shear deformation. Kohata et al. (1994) and Tatsuoka and Kohata (1995a) called the $E_{eq}/E^e \sim$ SL relation the damage function denoted by f(SL). The $E_{tan}/E_{eq} \sim$ SL relations represent how the ratio of plastic to elastic strain increments increases with shearing, which is called the plasticity function denoted by g(SL). The combined relation $E_{tan}/E^e \sim$ SL relation representing the non-linearity of stiffness is called the non-linearity function denoted by h(SL) (see Figs. 21 and 22). h(SL) is equal to $f(SL) \cdot g(SL)$.

It was found by summarizing the data of the functions of f(SL) and g(SL) for a wide variety of soft rocks that there is a correlation between them, although it is not so strong. That is, **Figs. 23a** and **b** show the relationships between f(SL) and g(SL) when SL is equal to 0.2 and 0.5, respectively. The general trend seen from these figures is that when the E_{eq} decreases to a larger extent when compared with the corresponding E^e value, the E_{tan} value also decreases to a larger extent compared with the corresponding E_{eq} value. It is a general tendency that a material which is less cemented tends to exhibit a more degree of pressure dependency of stiffness, while showing a more degree of damaging and a lager ratio of plastic to elastic strain increments with the increase in SL.

MODELING

By summarizing the discussions above, when cross-anisotropy is assumed, the following general framework for estimating the tangent Young's modulus E_{tan} at a given stress state can be proposed (Tatsuoka and Kohata, 1995a):

$$(E_{tan})_v = (E_f)_v \cdot j_v(\sigma_v) \cdot \alpha_v \cdot \beta_v \cdot f_v(SL) \cdot g_v(SL) \quad (3a)$$
$$(E_{tan})_h = (E_f)_h \cdot j_h(\sigma_h) \cdot \alpha_h \cdot \beta_h \cdot f_h(SL) \cdot g_h(SL) \quad (3b)$$

where the subscripts v and h mean the properties in the vertical and horizontal directions, $j(\sigma_v)$ is the pressure level function, and α and β are the parameters, which are 1.0 or less, representing, respectively, the effects of discontinuities and heterogeneities in the field and the effects of damage (disturbance) caused by construction works, in particular by excavation. Here, E_f is the undamaged elastic Young's modulus E^e measured prior to construction by field seismic surveys. Furthermore, it is assumed that the functions f and g

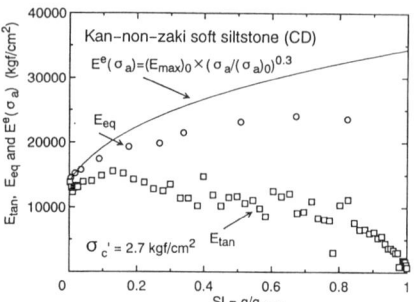

Fig. 19 E_{eq}, E_{sec} and $E_{tan} \sim q/q_{max}$ relations from the CD TC test result shown in Fig. 12

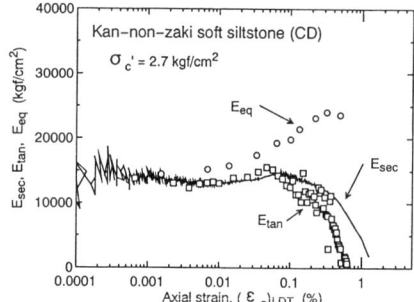

Fig. 20 E_{eq}, E_{sec} and $E_{tan} \sim \varepsilon_1$ relations from the CD TC test result shown in Fig. 12

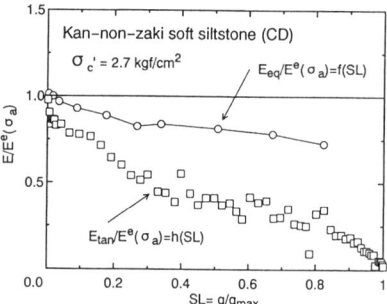

Fig. 21 E_{eq}/E^e and $E_{tan}/E^e \sim q/q_{max}$ relations from the CD TC test result shown in Fig. 12

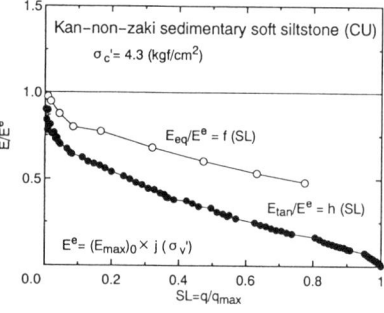

Fig. 22 E_{eq}/E^e and $E_{tan}/E^e \sim q/q_{max}$ relations from a CD TC test on a sedimentary soft siltstone from Kan-non-zaki

are independent of σ_v (or σ_h) for a given values of SL. This model was used in a simplified manner for numerically simulating the plate loading tests at Anchor A1 site and the full-scale behaviour of Pier P3 for Akashi Strait bridge (Siddiquee et al., 1994, 1995).

Depending on different dependencies of the tangent

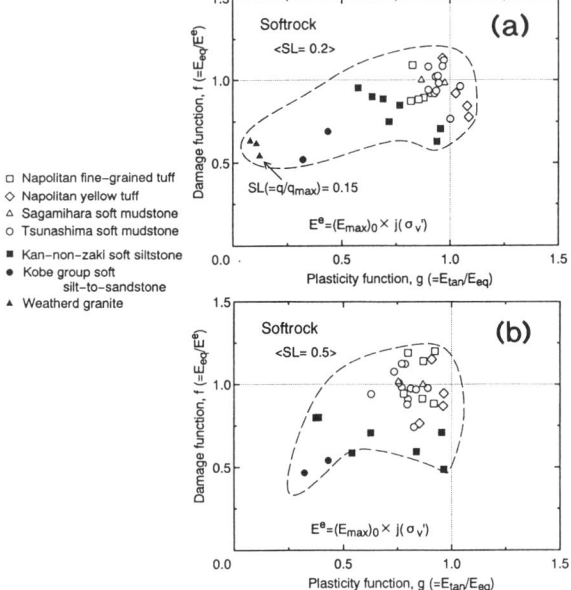

Fig. 23 f(SL)~ g(SL) relations for a variety of soft rocks; a) SL=0.2, and b) SL=0.5

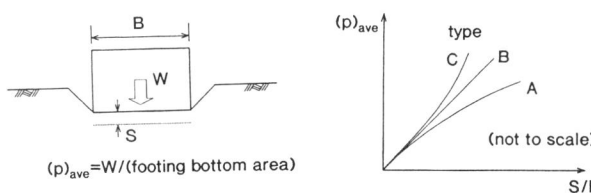

Fig. 24 Schematic figure of three pressure-settlement patterns for a footing placed on soft rock

stiffness on shear stress level and pressure level, the pattern of the relationship between the average pressure at the footing bottom $(p)_{ave}$ and the average settlement ratio, S/"the average footing width $(B)_{av}$", for a foundation constructed on soft rock becomes different. It is also the case with the plate pressure and plate settlement relations in plate loading tests. As illustrated in **Fig. 24**, the following three typical patterns are possible:

a) Type A: This type appears when both E^e and E_{tan} are pressure level-independent (i.e., $j(\sigma_v) = 1.0$), while $h(SL) = E_{tan}/E^e$ decreases noticeably with the increase in SL. The rate of increase in SL with the increase in $(p)_{ave}$ becomes more significant for a material having a smaller frictional strength component compared with the cohesion strength component. This type of behaviour was observed in plate loading tests on sedimentary soft mudstone at Sagamihara site and the full-scale behaviour of Anchorages 1A and 4A for Rainbow bridge.

b) Type B: This type appears when no yielding occurs in the ground of a linear elastic material. For actual non-linear geomaterials, this type may appear when $j(\sigma_v)$ and $h(SL)$ increases and decreases, respectively, with the increase in $(p)_{ave}$ and when the effects of pressure level dependency and non-linearity of stiffness are balanced (i.e., $j(\sigma_v) \cdot h(SL)$ is kept equal to 1.0). The full-scale behaviour of Pier 3P constructed on sedimentary soft silt-to-sandstone for Akashi Strait bridge was similar to this trend (see Fig. 7.10b of Tatsuoka and Kohata, 1995a, Fig. 15 of Siddiquee et al., 1995, this symposium).

c) Type C: This type appears when the effects of pressure dependency of stiffness overwhelms over the effects of non-linearity (i.e., $j(\sigma_v) \cdot h(SL)$ increases as $(p)_{ave}$ increases). This type may appear for originally non-linear materials having a high frictional strength component. That is, the principal stress ratio σ_1/σ_3 in the ground along the center line of the footing does not increase at a fast rate when $(p)_{ave}$ increases (stress path 3 in Fig. 18). For a highly frictional material, the SL level does not increases at a fast rate, which leads to a small effect of h(SL) on the $(p)_{ave} \sim (B)_{ave}$ relation. This type was observed noticeably in the full-scale behaviour of a foundation under construction on a weathered granite for one of Honshu-Shikoku Bridges (Tatara), core samples from which exhibited a very high pressure dependency (see Fig. 16). This type was observed marginally in the full-scale behaviour of Pier 2P constructed on a nearly uncemented gravel layer (Akashi Group) for Akashi Strait Bridge (see Fig. 7.10a of Tatsuoka and Kohata, 1995a).

There are still a number of uncertain factors as to the deformation characteristics of soft rocks. For example, our understanding of Poisson's ratios for elastic and plastic strains is still very poor. The possible dependency of the functions f and g on pressure level will have to be studied. The effects of consolidation by dissipation of excess pore water and drained creep are difficult to be distinguished from each other for soft clay deposits, which is also the case with a thick sedimentary soft rock deposit subjected to relatively fast stress changes. Furthermore, the effects of intermediate stress, cyclic loading and continuous rotation of principal stress directions on the stiffness of soft rocks are very poorly understood, but they should be known before we can use confidently the results of TC tests in cases under general 3-D stress conditions. For example, the stress paths in plane strain compression (PSC) and plane strain extension (PSE) tests are different from most of stress paths in the field (Fig. 18). Yet, these tests are difficult to perform. Therefore, simplified tests, triaxial compression (TC) tests, are usually performed. We need a more comprehensive model which can predict the behaviour of soft rock for general stress changes, hopefully based on only the results of TC tests and isotropic (or anisotropic) compression tests.

Moreover, the current ultimate limit design based on the classical limit equilibrium stability analysis, which assumes a perfectly plastic material while ignoring the effects of progressive failure, can be improved only by introducing the information as to strain localization into a shear band (or shear bands), which includes the thickness of shear band and the shear stress-shear strain-dilatancy relation in a shear band (Tatsuoka and Kim, 1994). Although scale effects in failure problems are linked to these shear band characteristics, they are only poorly understood.

CONCLUDING REMARKS

Some experiences obtained from geotechnical investigations performed in relation to the construction and design of large-scaled structures in the Tokyo metropolitan area and related research projects, the following conclusions were obtained:

1) Strains in the ground under working loads are about 0.2 % or less. The deformation properties at these small strain levels should be evaluated precisely for the rational serviceability limit design.

2) Most conventional field and laboratory static loading tests measure the stiffness at strains larger than the working strains shown above. They may largely under-estimate the stiffness in the field under working loads.

3) Elastic moduli of deformation obtained from field seismic surveys and those from triaxial compression tests on high-quality samples are nearly the same to each other.
4) Continuous stress-strain relations for a strain range from less than 0.001 % to that at the peak obtained from triaxial compression tests measuring axial strains locally is essential to construct an unified picture for non-linear stiffness for the data obtained from field seismic surveys, plate loading tests, pressuremeter tests and field full-scale behaviour.
5) Reasonably accurate stiffness values to be used for estimating the deformation of ground and the displacements of structures under working loads (static and seismic) can be evaluated based on elastic moduli of deformation obtained from field seismic surveys, while taking into account the dependency of stiffness on strain (or shear stress level) and pressure level (or stress state), among others.
6) Core samples retrieved by the conventional rotary core tube sampling may be more-or-less disturbed. The effects of sample disturbance can be particularly large on the results of conventional unconfined compression tests.
7) A model for non-linear and pressure level-dependent Young's moduli is described, which consists of functions representing pressure level-dependency, damage and plasticity.
8) Sedimentary soft rocks and cement-treated soils described in this paper have similar strength and deformation characteristics to each other.

ACKNOWLEDGEMENTS

The authors are indebted to Honshu-Shikoku Connection Bridge Authority, Metropolitan Expressway Public Corp., and the Ministry of Construction for providing us a number of field data. The authors are also grateful to Mr. Sato,T., and Mr. Wang,L. for helping us to prepare this paper. This research has been supported financially by the Ministry of Education, Science and Culture.

REFERENCES

Akino,N. 1990. Estimation of rigidity of ground and prediction of settlement of building, Journal of Struct. Constr. Engng, Architectural Institute of Japan, No.412, June, pp.109-119 (in Japanese).

Akino,N. and Sahara,M. 1994. Strain-dependency of ground stiffness based on measured ground settlement, Proc. Int. Symp. on prefailure Deformation Characteristics of Geomaterials, IS-Hokkaido '94, Balkema, Vol.I., pp.181-187.

Goto,S., Tatsuoka,F., Shibuya,S., Kim,Y.S. and Sato,T. 1991. A simple gauge for local small strain measurements in the laboratory, Soils and Foundations, 31-1, pp.169-180.

Goto,S. and Takahashi,Y. 1993. Technical development of large-scale deep excavation for LNG inground storage tank construction, Jour. of Geotechnical Engnrg, Japan Society of Civil Engineers, No.469, III-23, pp.1-13.

Hoshino,K. 1993. Geological evolution from the soil to the rock: Mechanical lithification and change of mechanical properties, Geotech. Engnrg of Hard Soils-Soft Rocks (eds. Anagnostopoulos et al.), Balkema, 1, pp.131-138.

Ito,R., Watanabe,K., Nakano,M., Ueno,M. and Nakashita, K. 1994. Analytical and observed results during cylindrical excavation of mudstone layer, Proc. 9th Japan Symp. on Rock Mech., pp.593-598 (in Japanese)

Kawasaki,S., Nishi,K. and Fujiwara,Y. 1993. Mechanical properties of deep soft rock ground in the suburbs of Tokyo, Geotech. Engnrg of Hard Soils-Soft Rocks (eds. Anagnostopoulos et al.), Balkema, 1, pp.593-600.

Kim,Y.-S., Tatsuoka,F. and Ochi,K. 1994. Deformation characteristics at small strains of sedimentary soft rocks by triaxial compression tests, Geotechnique, 44-3, pp.461-478.

Kohata,Y. Tatsuoka,F., Dong J., Teachavorasinskun,S. and Mizumoto,K. 1994. Stress-state affecting elastic deformation moduli of geomaterials, Proc. Int. Symp. on Prefailure Deformation Characteristics of Geomaterials, IS-Hokkaido '94, Balkema, Vol.I, pp.3-9.

Kohata,Y., Wang,L, Tatsuoka,F. Ochi,K., and Tsubouchi, T. 1995. Inherent and Induced Anisotropy of Sedimentary Softrock, Proc. 10th Asian Regional Conf. on SMFE, Beijing (to appear).

Miyazaki,K., Hameed,R.A., Sato,Y, Kohata,Y. and Tatsuoka,F. 1994. Deformation characteristics of undisturbed silty-sand from triaxial compression and in-situ tests and full-scale behaviour, Proc. Int. Symp. on Prefailure Deformation Characteristics of Geomaterials, IS-Hokkaido '94, Balkema, Vol.I., pp. 241-246.

Ochi,K., Tatsuoka,F. and Tsubouchi,T. 1993. Stiffness of sedimentary soft rock from in situ and laboratory tests and field behaviour, Geotech. Engnrg of Hard Soils-Soft Rocks (eds. Anagnostopoulos et al.), Balkema, Vol.I, pp.707-714.

Ochi,K., Tsubouchi,T. and Tatsuoka,F. 1994. Deformation characteristics of sedimentary soft rock evaluated by full-scale excavation, Proc. Int. Symp. on Prefailure Deformation Characteristics of Geomaterials, IS- Hokkaido '94, Balkema, Vol.I, pp. 601-607.

Odagiri,N., Ogiwara,M., Namikawa,K., Kameya,H. and Hirayama,N. 1993. Deformation characteristics of the foundation of "the Rainbow Bridge", Proc. 28th Japan National Conf. of SMFE, JSSMFE, Vol.I, pp.1361-1364, (in Japanese).

Odagiri,N., Ogiwara,M. and Kameya,H. 1994. Observed behaviour and deformation characteristics of the foundation ground of long-span suspension bridge, Jour. of Struct. Engnrg, 40a, JSCE, pp.1495-1503 (in Japanese).

Shibuya,S., Tatsuoka,F., Teachavorasinskun,S., Kong, X.-J., Abe,F., Kim,Y.-S., and Park,C.-S. 1992. Elastic deformation properties of geomaterials, Soils and Foundations, JSSMFE, 32-3, pp.26-46.

Siddiquee,M.S.A., Tatsuoka,F., Hoque,E., Tsubouchi,T, Yoshida,O., Yamamoto,S. and Tanaka,T. 1994. FEM simulation of footing settlement for stiff geomaterials, Proc. Int. Symp. on Prefailure Deformation Characteristics of Geomaterials, IS-Hokkaido '94, Balkema, Vol.I., pp.531-537.

Siddiquee,M.S.A., Tatsuoka,F., Yoshida,O., Yamamoto,Y., Tanaka,T. and Inoue,A. 1995. Settlement of a Pier Foundation for Akashi-Kaikyo Bridge and its Numerical analysis, Proc. Int. Workshop on Rock Foundation of Large-Scaled Structures, Tokyo (this symposium).

Sugawara,N., Ito,Y., Kohata,Y., Wang,L. and Suwa,K. 1995: New core boring sampler with planet gear, Proc. of Symp. on Sampling, JSSMFE, pp.9-14 (in Japanese)

Tanaka,H. and Terashi,M. 1986. Properties of treated soils formed in-situ by Deep Mixing Method, Report of the Port and Harbour Research Institute, 25-2 (in Japanese)

Tatsuoka, F. and Shibuya,S. 1992. Deformation characteristics of soils and rocks from field and laboratory tests, Proc. 9th Asian Regional Conf. on SMFE, Bangkok, 1991, Vol.2, pp.101-170.

Tatsuoka,F., Kohata,Y., Mizumoto,K., Kim,Y.-S., Ochi,K. and Shi,D. 1993. Measuring small strain stiffness of softrocks, Geotechnical Engineering of Hard Soils-Soft Rocks, Balkema, Vol.1, pp.809-816.

Tatsuoka,F., Teachavorasinskun,S., Dong,J., Kohata,Y. and Sato,T. 1994a. Importance of measuring local strains in cyclic triaxial tests on granular materials, Dynamic Geotechnical Testing: Second

Volume, ASTM STP 1213 (Edelhar, Drnevich & Kutter eds.), pp.288-302.

Tatsuoka,F., Sato,T., Park,C.S., Kim,Y.S., Mukabi,J.N. and Kohata,Y. 1994b. Measurements of elastic properties of geomaterials in laboratory compression tests, Geotechnical Testing Journal, ASTM, 17-1, pp.80-94.

Tatsuoka,F., Kohata,Y., Karoji,H. and Miyashita,A. 1994c. Stiffness of the ground improved to support the pier of JNLT atop Mauna Kea, Proc. SPIE- the Int. Society for Optical Engnrg., Kona, Hawaii, Vol.2199, pp.404-413.

Tatsuoka,F. and Kim, Y.-S. 1994. Deformation of shear zone in sedimentary soft rock observed in triaxial compression, Location and Bifurcation Theory for Soils and Rocks, Balkema, pp.181-187.

Tatsuoka,F. and Kohata,Y. 1995a. Stiffness of hard soils and soft rocks in engineering applications, Keynote Lecture, Proc. Int. Symp. on Prefailure Deformation Characteristics of Geomaterials, IS-Hokkaido '94, Balkema, Vol.II.

Tatsuoka,F., Kohata,Y.,Tsubouchi,T., Murata, K., Ochi,K. and Wang,L. 1995b, Disturbance in rotary core tube sampling, Proc. Int. Conf. on Advances in Site Investigation Practice, the Institute of Civil Engineers, March, London.

Tatsuoka, F., Lo Presti, D. & Kohata, Y. 1995c. Deformation characteristic of soils and soft rocks under monotonic and cyclic loads and their relationships. Keynote Lecture. Proc. Int. Conf. on recent Advances in Geotech. Earthquake Engnrg and Soil Dynamics. St Louis. (Prakash eds.), Vol.II, pp. 851-879.

Tatsuoka, F., Kohata, Y., Tsubouchi, T. & Ochi, K. 1995d. Stiffness of sedimentary softrocks evaluated by triaxial compression tests. Proc. 8th Int. Congress on Rock Mechanics, Tokyo (to appear).

Tsubouchi,T., Ochi,K. and Tatsuoka,F. 1994. Non-linear FEM analyses of pressuremeter tests in a sedimentary soft rock, Proc. Int. Symp. on Prefailure Deformation Characteristics of Geomaterials, IS-Hokkaido '94, Balkema, Vol.1., pp.539-544.

Uchida,K., Shioi,Y., Hirukawa,T. and Tatsuoka,F. 1993. The Trans-Tokyo Highway Project - A huge project currently under construction, Proc. of Int. Conf. on Transportation Facilities through Difficult Terrain (Wu and Barrett eds.), Balkema, pp.57-87.

30 years behavior of rock foundation of Kurobe Dam – Result of measurement at 186 m height Arch Dam

Takeshi Watanabe
The Kansai Electric Power Co., Inc., Japan

Tomoo Hayakawa
Office of Civil Engineering and Architecture, The Kansai Electric Power Co., Inc., Japan

ABSTRACT: The Kansai Electric Power Co., Ltd. has been continuing the measurement and monitoring not only on the dam main body itself but also on the foundation rock, extending over last about 30 years since a construction of the Kurobe Dam. From this observation result extending over a long time, the following two facts have been obtained as the distinctive characteristics of the foundation rock; i)the seasonal movements of the both rock mass in left and right banks affect a dam behavior, ii)some creep behaviors can be observed in a part of the data of the abutment still at present. This time these results will be introduced and concurrently an outline of some consideration results will be reported.

1. At the Outset

As for the Kurobe Dam, 30 years have elapsed since a measurement initiation, and furthermore a measurement and monitoring have been continued putting emphasis not only on the dam main body itself but also on the rock foundation.

The Kurobe Dam is a dome type arch dam with a height of 186m, a crest length of 492 m, and a dam volume of 1.6 million m^3, and it was completed in June 1963. A drawing of dam is shown in Fig-1.1,1.2,1.3 and dimensions of dam are shown in Table-1 respectively. Without waiting for the placing completion of dam concrete, a water filling has been commenced since 1960, and subsequently confirming a safety of the dam even through an increase of the water storage load, in a way to elevate the highest water level by 10 ～ 15 m every year, the first full water has been achieved in July 1969 when 9 years have passed after the water filling was commenced. From that time on the storage reservoir operation of annually adjusting type (a cycle that the highest water level is reached in around August and November, and the lowest water level is reached in the end of March every year) has been repeated, and consequently an increase and decrease of the water storage load has been experienced more than 20 times so far. (Fig-2)

The following two facts have been obtained by the measurement results performed up to now. First, it was found that a measured arch dam deflection at the crown cantilever could be explained reasonably by considering the third load fluctuating seasonally besides the water storage load and the thermal load. This load has been inferred to be caused by the movement of the both rock mass in left and right banks, and then the highly accurate distance survey of a valley width has been carried out in order to do backling it. The result was that the seasonal fluctuation of a valley width was recognized.(We call this phenomenon "mountain deformation".)

Secondly, when the deformation measurement results of the rock mass have been analyzed in detail, some creep behaviors have been recognized in a part of the data.

It is thought that these two phenomena are worthy of a special attention as the measurement results of the dam foundation rock extending over a long time, and therefore they will be introduced this time.

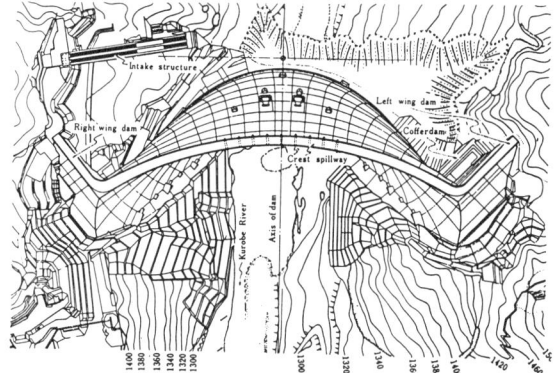

Fig. 1.1 General plan of dam

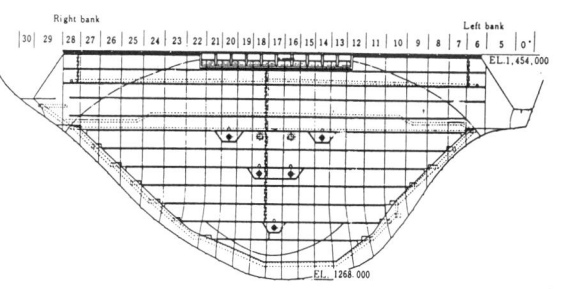

Fig. 1.2 Developed downstream elevation of the dam

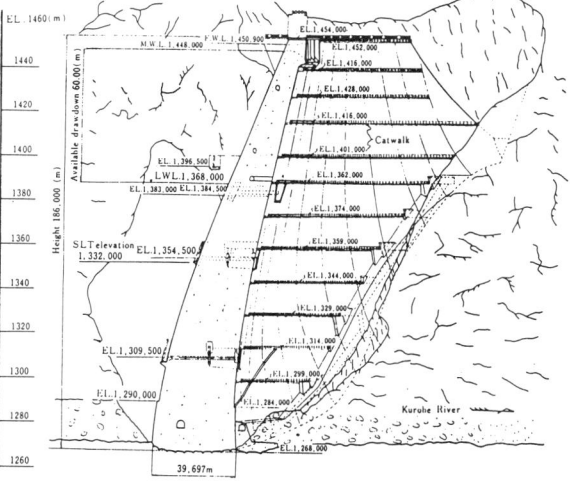

Fig. 1.3 Crown section and left side view of dam

Table 1. Dimension of Kurobe Dam

Dam	Name	Kurobe Dam
	Purpose	Hydroelectric
	Type	Dome type arch dam
	Crest length	492 m
	Height	186 m
	Enbankment volume	1,583,000 m^3
	Design flood	1,260 m^3/s
Reservoir	Catchment area	184.5 km^2
	Reservoir area	3,450 km^2
	Gross capacity	174,000,000 m^3
	Effective capacity	138,000,000 m^3
	Available depth	60 m
	Normal water level	EL. 1,448.0 m
	Surcharge water level	EL. 1,450.9 m
	Mimimum water level	EL. 1,388.0 m

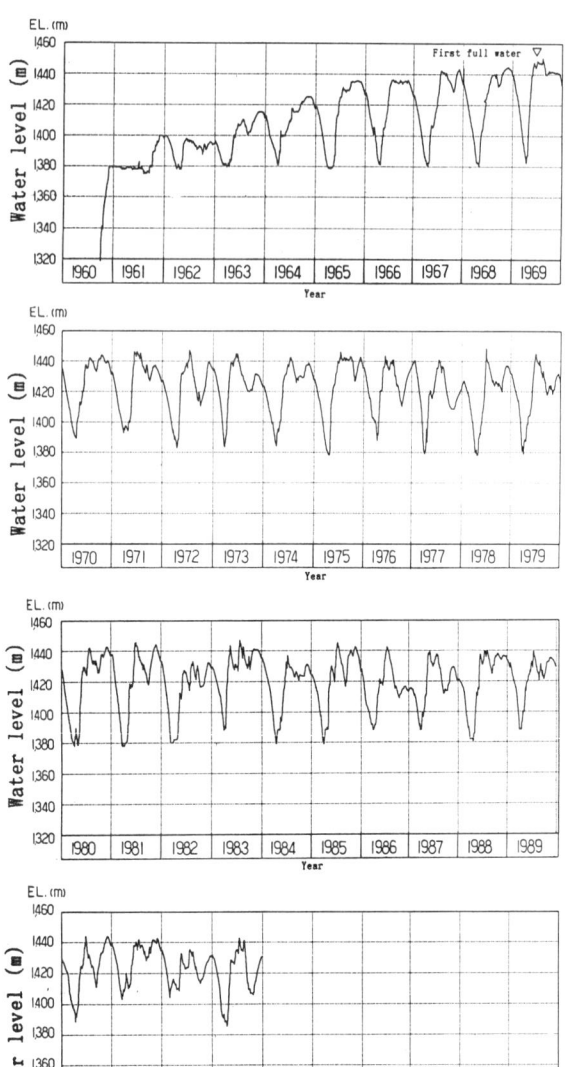

Fig. 2 Reservoir water level

grained and white predominant colored appearance. A lithological quality is in general medium hard, and a biotite becomes brown color by the weathering action, and in addition in the right bank a chloritization action is remarkable, and therefore there are many places where a biotite is discolorfed to a light green ~ light yellowish green color. In addition, the joints and the fractured zone exist in many number of them, irrespectively of a large scale or a small scale. In designing the dam, in order to grasp a strength of the rock mass adding a consideration to them, 5 kinds of in-situ rock tests with a large scale have been performed. They were such as a block shear test, rock shear test, fault shear test, rock triaxial test and fault triaxial test, and as a consequence, total number of the specimen tested reached to 25.

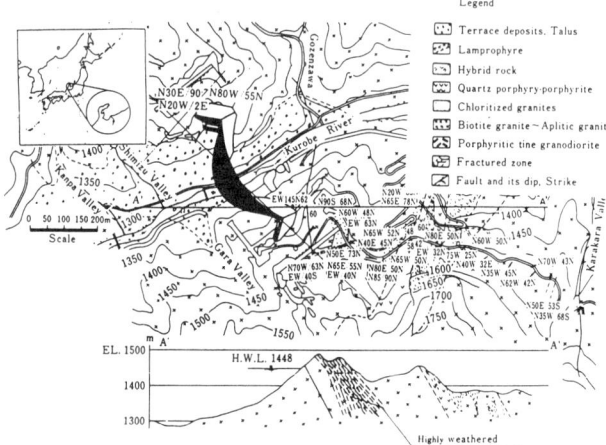

Fig. 3 Geological map of Kurobe Dam site

3. Measurement of Dam and Foundation Rock
3.1. Measurement item

In the Kurobe Dam, the measurement facilities are arranged not only on the dam main body itself, but also on the foundation rock extending over fairly wide range so as to grasp a displacement status of the rock mass accompanied with the load change.

The main measurement items are shown in Table-2, and an arrangement drawing of the measurement facilities relating to a deformation and displacement is shown in Fig-4 respectively.

2. Topography and Geology at The Dam Site

The dam site is located in the upper stream part of the Kurobe River which rises from the Northern Alps in the Central District of Japan as a headspring and finally flows into the Sea of Japan. As for a topographic condition at the dam site, there is scarcely covering materials in the right bank, and a steep slope continues up to the upper altitude. In the left bank the talus, the terrace sand and gravel formation distribute from the river bed (E.L. + 1,290 m) up to near E.L.+ 1,400 m, and consequently the slope is a little bit flatter in this part, however, it becomes steep in the upper altitude above there. In addition, an altitude of the mountain ridge line in the both banks is E.L.+ 2,600 m ~ 3,000 m. Because the gorges enter into the lower stream side of the abutment of the dam in the both left and right banks, as shown in Fig-3, the ridges at the abutment of the dam are fairly thin in a direction of the thrust from the dam . The reason why the dam abutment site was selected here was that the dam site location was limited, because a valley width becomes remarkably wider and moreover the talus becomes thicker in the left bank if the dam site is put in the more upper stream part.

The rock constituting the dam site is mainly the granite, and besides it the kinds of igneous rock such as a quartz porphyry, hybrid rock, porphyritic granodiorite, lamprophyre and so forth distribute in a small range or in a vein type. The granite distributes widely in this district all over, and it presents the coarse

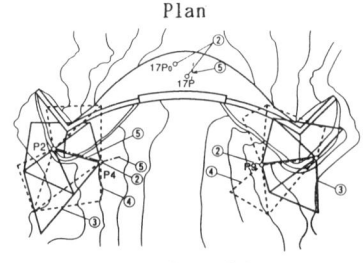

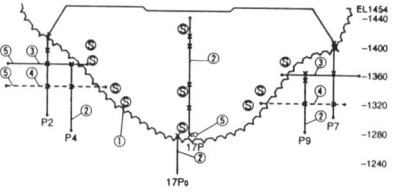

Fig. 4 Arrangement of static measuring equipment

Table 2. Main items of measurement

Items of measurement	Measuring instruments	Number of instruments	Interval of measurement	
			Initial stage	Present
Water level	Water gauge	1	Continuousement	Continuousement
Temperature of water	Thermometer	1	1 day	1 month
Atmospheric temperature	Thermometer	1	1 day	1 day
Horizontal displacement of dam and bedrock	Pendulum at dam	4(4)	2 days	2 days
	Pendulum at bedrock	18(18)	2 days	5 days
	Coordimeter	9	1 month	1 month
	Rock deformeter	91	5 days	5 days~1 month
Vertical displacement of dam	Precision level	24	1 month	4 months
	Clinometer	2(2)	5 days	1 month
Mesurement of dam concrete	Strain of dam	286(350)	1~10 days	1 month
	Compressive stress of dam	41(45)		
	Temperature of dam	38(44)		
	Opening of joint	73(100)		
Measurement of seepage flow	Seepage flow from joint	2(2)	1 day	1 day
	Seepage flow from drain holes	23(23)	10 days	1 month
Uplift	Piezometer	54(54)		1 month
Measurement of valley width	electro-optical precision distance meters (mecometer)	4 line	1 month	1 month

(Original)

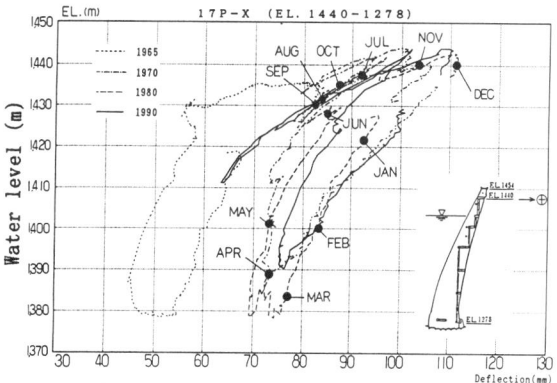

Fig. 5　Water level and radial deflection at dam crown

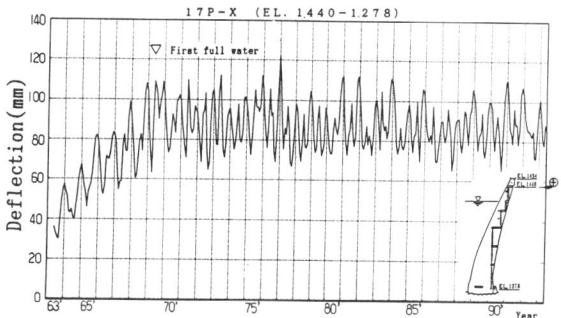

Fig. 6　Radial deflection at dam crown

3.2. Measurement results
3.2.1. Arch dam deflection at the crown cantilever

The arch dam deflection at the crown cantilever in the direction of the upper and lower stream parts of the river, is an important measurement item as a response of the arch dam to the load acting on the dam. In the Kurobe Dam this deflection is measured by the following two method; namely, the pendulums installed in the dam body, and a sight survey of the dam crest. This deflection will be stated here centering around a behavior of the pendulums by which a deflection is directly measured, of which measurement method is relatively simple, which has the characters together with a highly accurate measurement value and with a high reliability.

A variation of the measurement value of the pendulums in a direction of the upper and lower stream parts, namely, an arch dam deflection at the crown cantilever in a direction of the upper and lower stream parts is put in order with a relation to a reservoir water level, and therefore it is shown in Fig-5. A deflection toward a direction to the lower stream part is shown as a positive. As for the measurement values in this figure, the values in 1965 as a data before the first full water and the values in 1970, 1980 and 1990 as a data after the first full water are shown. As for the values in 1980, the months in which a measurement has been executed are also shown by plotting. From this figure it can be understood that the measurement values depict a loop in every year.

Although in a period when the highest water level was gradually elevated, namely, in a period when the new load acted in 1965 ~ 1970, a loop region is shifted greatly to the right side in Fig-5, and that is to say a plastic deformation behavior is seen. Since 1970 when the first full water was achieved, it is recognized that loops are depicted in almost same region in Fig-5, that is to say an elastic behavior is observed. In the same way, a change by time elapsed of the deflection after water filling has commenced is shown in Fig-6. From this figure as well, a deflection after the first full water is elastic, and furthermore an plastic behavior, an increasing trend year by year, can hardly be recognized.

Moreover as for a loop in Fig-5, in which two displacement quantities exist even at the same rservoir water level, it is considered that they are originated from a load except a water storage. It will be considered in 4. mentioned below.

3.2.2. Behavior of foundation rock

As a measurement of the behavior of the foundation rock, rock displacements of the arch abutment are measured by rock deformeters and the floating pendulums (reverse pendulums), and in addition a displacement at the crown cantilever base is measured by the floating pendulums. The structure of a rock deformeter is that a double steel tube is installed in a horizontal boring hole in the rock mass, and an outer tube is fixed at the hole wall, and an end of inner tube is fixed at the rock mass. Through the displacement quantity of this inner tube, a relative displacement between two points in the rock mass should be measured. In addition, the floating pendulums are that a lower end of the stainless steel wire is fixed at the bottom of vertical boring hole in the rock mass, and an upper end is floated in the water basin by fixing the float to the upper part. By doing so, a relative displacement of the rock mass between an upper end and a lower end of the floating pendulums can be measured.

A change of the time elapsed of the measurement values on a displacement at the crown cantilever base in a direction of an upper and lower stream parts of the river is shown in Fig-7. After the first full water in 1969, an elastic behavior has been repeated in every year, a plastic behavior, an increasing trend year by year can not be found.

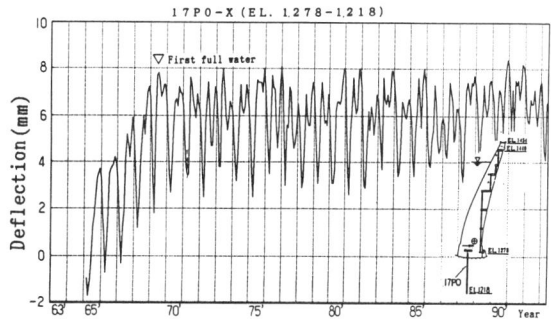

Fig. 7　Deflection of crown cantilever base

As a horizontal displacement of the arch abutment in the rock mass, out of 4 measurement values by the rock deformeter (2 in the right bank and 2 in the left bank) in a thrust direction, the data in 3 locations of 6-2, 6-3 and 7-3 are shown in Fig-8 ～ 10 as the data taken for the all period. Each data is put in order relating to the reservoir water level, and moreover + direction shows tensile direction and − direction shows compressive direction. Even in each figure, in the period until the first full water in 1965 ～ 1970, a loop region shifts to the left side, and that is to say a large plastic deformation can be seen. After its period an elastic behavior can be found depicting loops in an approximately same region, except a case of 6-3. The changes by the time elapsed of the measurement values at 3 locations since a measurement commencement are shown in Fig-11 ～ 13. From these data as well, in the measurement values at a location of 6-3, a gradual increasing trend can be seen after the first full water. By way of contrast, it is confirmed that no gradual increasing trend can be found in the data of the other 88 rock deformeters.

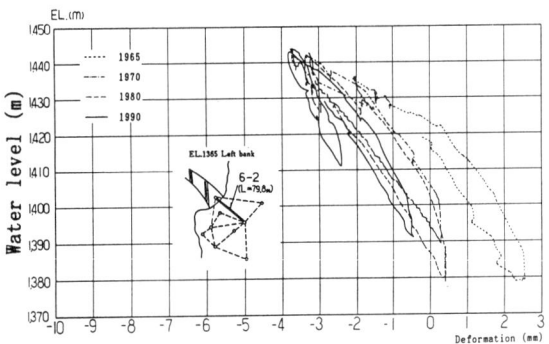

Fig. 8 Relation between water level and deformation of rock mass (6-2)

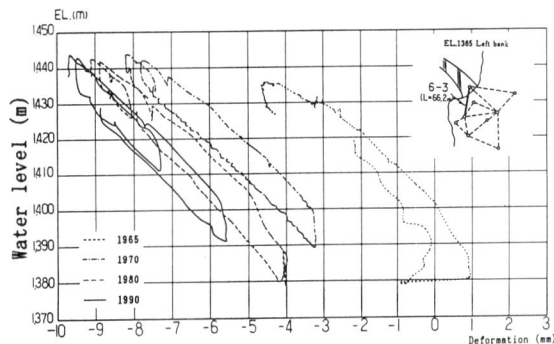

Fig. 9 Relation between water level and deformation of rock mass (6-3)

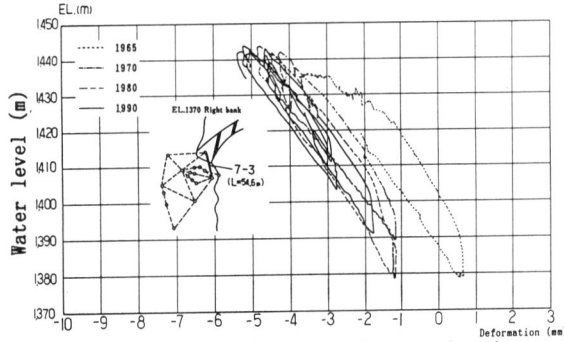

Fig. 10 Relation between water level and deformation of rock mass (7-3)

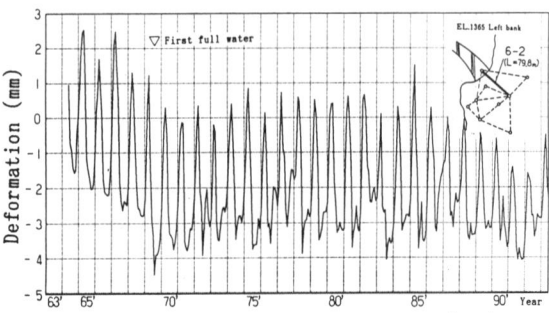

Fig. 11 Deformation of rock mass (6-2) (Left bank EL. +1365m)

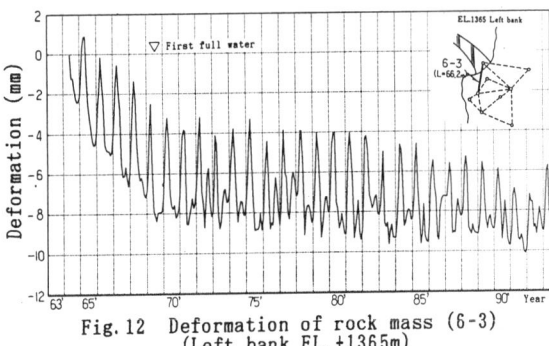

Fig. 12 Deformation of rock mass (6-3) (Left bank EL. +1365m)

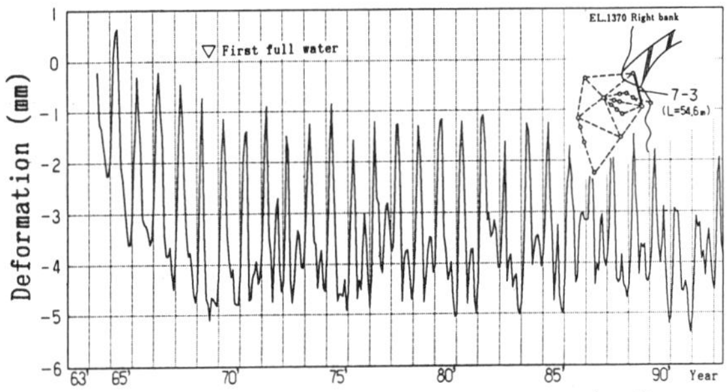

Fig. 13 Deformation of rock mass (7-3) (Right bank EL. +1370m)

4. Consideration
4.1. Mountain deformation phenomena

Measurement values of the arch dam deflection at the crown cantilever and calculation values were compared. Generally speaking, the main load acting on the dam is a hydrostatic pressure of the reservoir and thermal load. The dam and the foundation rock were modeled by 3 dimensional FEM model, and then computed an arch dam deflection quantity at the crown cantilever with the hydrostatic pressure load only. An evaluated value of the in-situ rock test results was used as a deformation modulus of the rock mass for computation. The measurement values of an arch dam deflection at the crown cantilever for last 10 years and the computed values considered the hydrostatic pressure are shown in Fig-14. In addition to them in this figure, a difference between them is also shown. In a difference between the measurement values and computed values, an amplitude in a degree of 30 mm of the annual period can be seen. It is thought that a thermal load may influence on this annual periodicity, and therefore the computed results similar to the previous computation after considering the thermal load are shown in Fig-15. Although a difference between the measurement values and the computed values becomes smaller, an annual periodicity with an amplitude of about 15 mm is still seen. Accordingly as a cause of this annual periodic fluctuation, the third force, which acted to the dam as a result of the seasonal deformation of the rock mass in the both banks, was presumed.

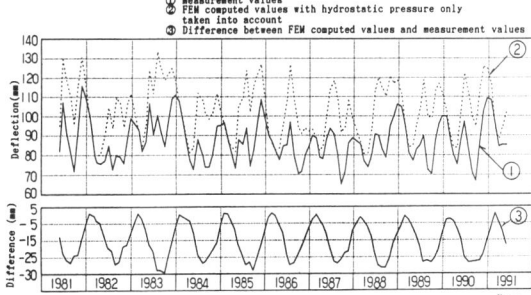

Fig. 14 Deflection of crown cantilever
(Hydrostatic pressure only taken into account)

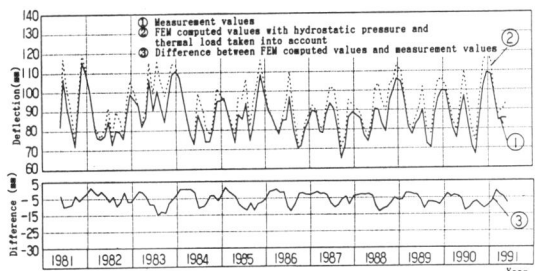

Fig. 15 Deflection of crown cantilever
(Hydrostatic pressure and thermal load taken into account)

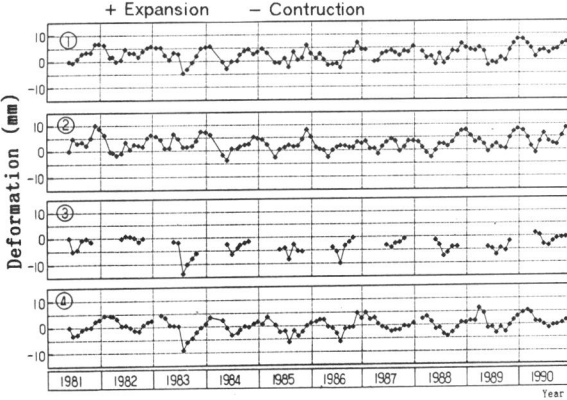

Fig. 17 Change of valley width

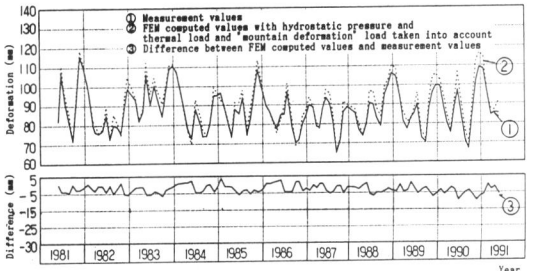

Fig. 18 Deflection of crown cantilever
(Hydrostatic pressure and thermal load and "mountain deformation" load taken into account)

In order to confirm actually this "mountain deformation" phenomena, a measurement of the arch chord length of the dam and the valley width has been executed by using an electro-optical distance meter with a high accuracy since 1981. The measurement location is shown in Fig-16 and measurement results are shown in Fig-17 respectively. According to the measurement results, even in the survey line 4, where it is thought that an influence of confining by the dam and an influence of the storage reservoir is small, a difference of a degree of about 10 mm between summer time and winter time was observed. (Valley width in summer is shorter than in winter.) Based on it it can be said that a real existence of the "mountain deformation" phenomena could be confirmed. These measurement values are given to the FEM model mentioned above as a forced displacement to the dam, and then the arch dam deflection at the crown cantilever is computed with 3 kinds of load, namely, hydrostatic pressure, thermal load and "mountain deformation", and subsequently the results are compared with the measurement values. Their comparison are shown in Fig-18. A difference between the measurement values and computed values becomes very small and furthermore annual periodicity is scarcely observed.

Based on the facts explained above, as a load giving an influence on the behavior of the Kurobe Dam, it is thought that an external force caused by the "mountain deformation" exists besides the water pressure and thermal load. In Fig-19 ~ 21, taking 1985 as an example, a comparison between the measurement values and computed values of the arch dam deflection at the crown cantilever putting in order through the reservoir water level is shown. In Fig-19 only the water pressure is considered for computation, in Fig-20 the water pressure and thermal load are considered for computation, and in Fig-21 the water pressure, thermal load and "mountain deformation" are considered respectively for computation. From this figure as well it is understood that the measurement values can be well explained by considering the load due to "mountain deformation".

In addition, because the load due to "mountain deformation" pushes the arch crown back toward the upstream and the resultant thrust on the abutment gets directed toward the mountain side in summer season when the water level is elevated and the water storage load increases toward the downstream in the present reservoir operation, it consequentry becomes to give a favorable influence for the safety of the dam and the foundation rock.

It is possible that the seasonal fluctuation of the ground water level in the mountain causes "mountain deformation" phenomena,because a correlation can be seen between the fluctuation tendency of the measurement values for the valley width and the fluctuation tendency of a gush water quantity in the traffic tunnel penetrating the mountain range in the right bank.

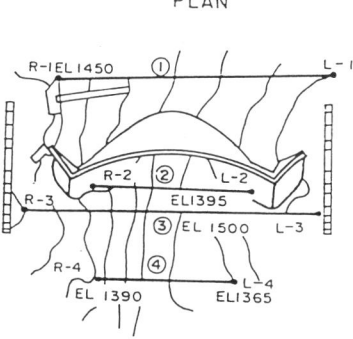

Fig. 16 Location of measuring points for width of valley and dam code length

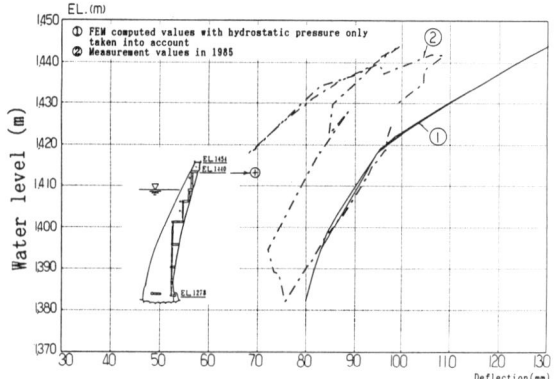

Fig. 19 Relation between water level and deflection of crown cantilever

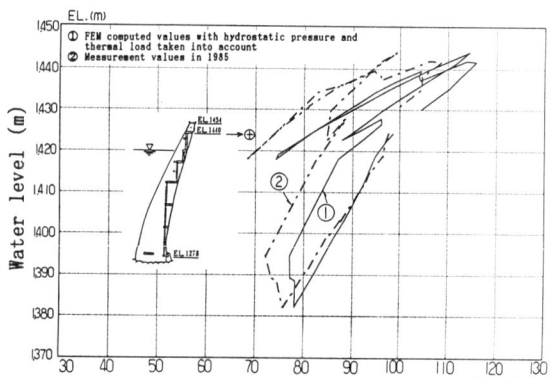

Fig. 20 Relation between water level and deflection of crown cantilever

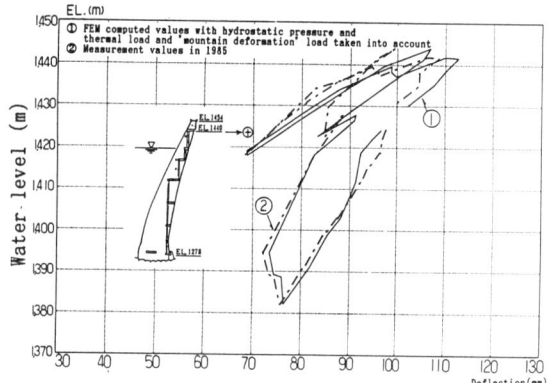

Fig. 21 Relation between water level and deflection of crown cantilever

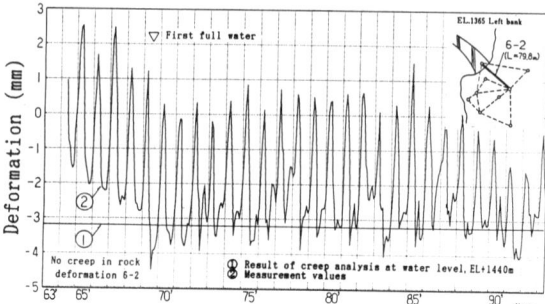

Fig. 22 Deformation of rock mass (6-2)

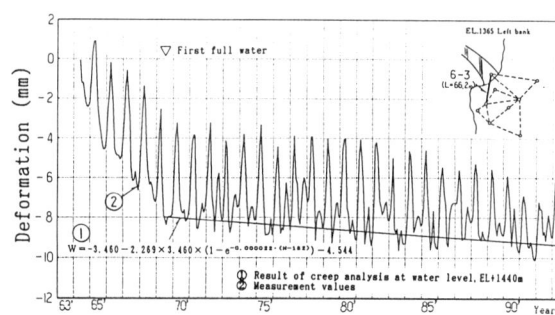

Fig. 23 Deformation of rock mass (6-3)

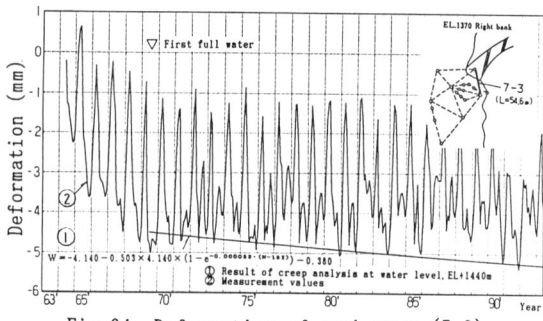

Fig. 24 Deformation of rock mass (7-3)

4.2. Creep deformation

From the measurement results extending over a long time, it was found that a major part of plastic deformation of the dam body and the foundation rock had converged in about 10 years from a water filling commencement to the first full water. Almost all of the measurement values shows the elastic behavior after the first full water, however, a part of the rock deformation data shows still tendency of the gradual increase as stated in 3.2.2.. For 3 rock deformeters mentioned above, by using those measurement values (1969 ~ 1993) at a relatively high reservoir water level (around E.L.+1,440) were adapted to the Voigt model. According to it, it is confirmed that there is a creep trend in 7-3 as well as 6-3. The results are shown in Fig-22, 23 and 24.

In addition as the confirmation of the safety, by using creep constants obtained here, 2 dimensional creep analysis has been executed for the foundation rock of the arch abutment, and subsequently a safety of the dam at the final convergence state has been evaluated. As its results, it has been confirmed that the dam will be completely safe extending over the future from a viewpoints of the maximum stress intensity, maximum strain quantity, and local safety factor of rock and concrete element.

5. Conclusion

From the measurement data of the Kurobe Dam extending over a long time, the followings have been made clear.

(1) In a period of about 9 years from the water filling commencement to the first full water, it was confirmed that a major part of the plastic deformation had converged.

(2) Even at a time elapsed more than 30 years after completion of the dam, some creep-like deformation behavior can be recognized in a part of the rock mass of the arch abutment.
However, it has been confirmed that the dam will be completely safe at the final convergence state.

(3) It has been clarified that in the Kurobe Dam the load due to "mountain deformation" caused by the movement of the rock mass in left and right banks acts, and moreover influences a behavior of the dam body besides the water pressure and thermal load.

Foundation engineering of nuclear power station on soft rock

Kiyoshi Kishi, Kenji Terada, Toshiaki Sakai & Kazuo Momose
Tokyo Electric Power Company, Japan

Eizo Fukazawa
Kajima Corporation, Japan

ABSTRACT: In this paper, geological reseaches, physical and mechanical tests of rock mass and core samples, seismistic stability of the foundation ground under nuclear power plants against earthquakes are shown, which are done at the site of Kashiwazaki Kariwa nuclear power plants.
And also artificial soft rock is introduced, which is invented for the improvement of foundation ground.

1 Introduction

The foundation ground of a nuclear power plant has to be strong enough to withstand any earthquake motion. It is stated in "Regulatory Guideline for Aseismic Design of Nuclear Power Reactor Facilities", published by Nuclear Safety Commission,
This guideline says, in principle, buildings should be rigid and important buildings should be founded on a solid stratum to ensure thet they are safe against seismic forces.
"Guideline on Licensing Examination for Geology and Ground Condition of Nuclear Power Plants", written by the Japan Committee on Examination of Reactor Safety, describes the necessary investigations and tests to evaluate seismic stability.
Based on the guideline, active fault and earthquake history are researched at and around nuclear power plant site to determine design earthquake wave.
Borehole investigations, elastic wave prospecting and many physical and mechanical tests using borehole cores are done to determine the geological structure and physical characteristics of the site.
Exploratory adits are made directly under reactor building and condition of rock mass, condition and activity of faults in the foundation rock mass are visually checked.
Considering these results (borehole investigations in the site, in-situ tests in the exploratory adit, physical and mechanical laboratory tests using bore hole core), soil structure model and physical input data are determined. Thus, static and dynamic FEM analysis is done to evaluate the seismic stability.
This paper discribes method for safety evaluation of the foundation ground under nuclear power plant against earthquake, based on many kinds of researches and tests, when the Kashiwazaki Kariwa nuclear power plant is constructed.
It also discribes artificial soft rock, invented to achieve the required quality and used as replacement material of foundation ground under the Kashiwazaki Kariwa nuclear power plant which is a very important structure founded on the ground of relatively low compression strength.

2 Researches

There are two types of researches for site selection. Both researches are necessary when building nuclear power plants. One is geological and seismological research at and around the site to determine input earthquake wave. The other is geological, physical, rock mechanical research at the construction point.
This paper mainly describes geological, physical, and rock mechanical research.

2.1 Research around the nuclear power plant site

Around the site, paper research, geological research, survey of surface geology, interpretation of aerial photographs, physical prospecting, and monitoring of earthquake wave are done.
By this result, soil and rock dispersion, regional geological structure (formation, fold, fault, etc.), history of earthquake is known. Based on the result, evaluation of active fault is done, very important to determine the standard seismic wave data. Activity and the scale of fault must be minutely researched, because they determine the value of standard seismic wave data. Evaluation of active fault is important, detailed research is necessary because it directly decide the evaluation of standard seismic wave data.

2.2 In-situ research

At the site, borehole investigations, elastic wave prospectings are done, in order to check dispersion of rock mass, rock quality, areas of faults. At the place nuclear power plant is built, exploratory adit is excavated and condition of rock mass is visually checked.
Fig-1 is the map of the site showing localities of geological survey. Kashiwazaki Kariwa nuclear power station is located at the site with an area of 4.2 million square meters across both Kashiwazaki-Kariwa cities nearby the center of Niigata prefecture, and total capacity of 8,212 MW nuclear power plants are now under construction. Unit1 to Unit5 plants (BWR types 1,100 MW×5) are on operation. Unit6 and Unit7 (ABWR types, 1,356 MW×2) are now under construction. Unit6 plant will be on operation at 1996.12 and Unit7 at 1997.7.
Tokyo Electric Power Company started the geological researches from the beginning of 1970's, to build the Kashiwazaki Kariwa Unit1 (first) plant.
Table-1 shows stratigraphic succession in the site. The oldest formation is Shiiya Formation (sand stone and sand stone-mud stone alternation, intercalated conglomerates) next is Nishiyama Formation (mud stone contains intercalated sand stone, tuffs, nodules and fossil shells, etc.) Haizume Formation (tufficious mud stone, tufficious sand stone and tuff). Shiiya, Nishiyama, Haizume Formations are made in Neogene Age and covered by Yasuda Formation, Banjin Sand bed, New Dune sand bed (these formation are made in Quaternary Age).
Fig-2 shows geologic map in the site. The geologic structure of Neogene in the site is characterlized by the fold of one anticline and one syncline. Folding does not long to Quaternary formations, so it must have occurred before the Quaternary Age. When there is folding at the site, the time folding occurred,

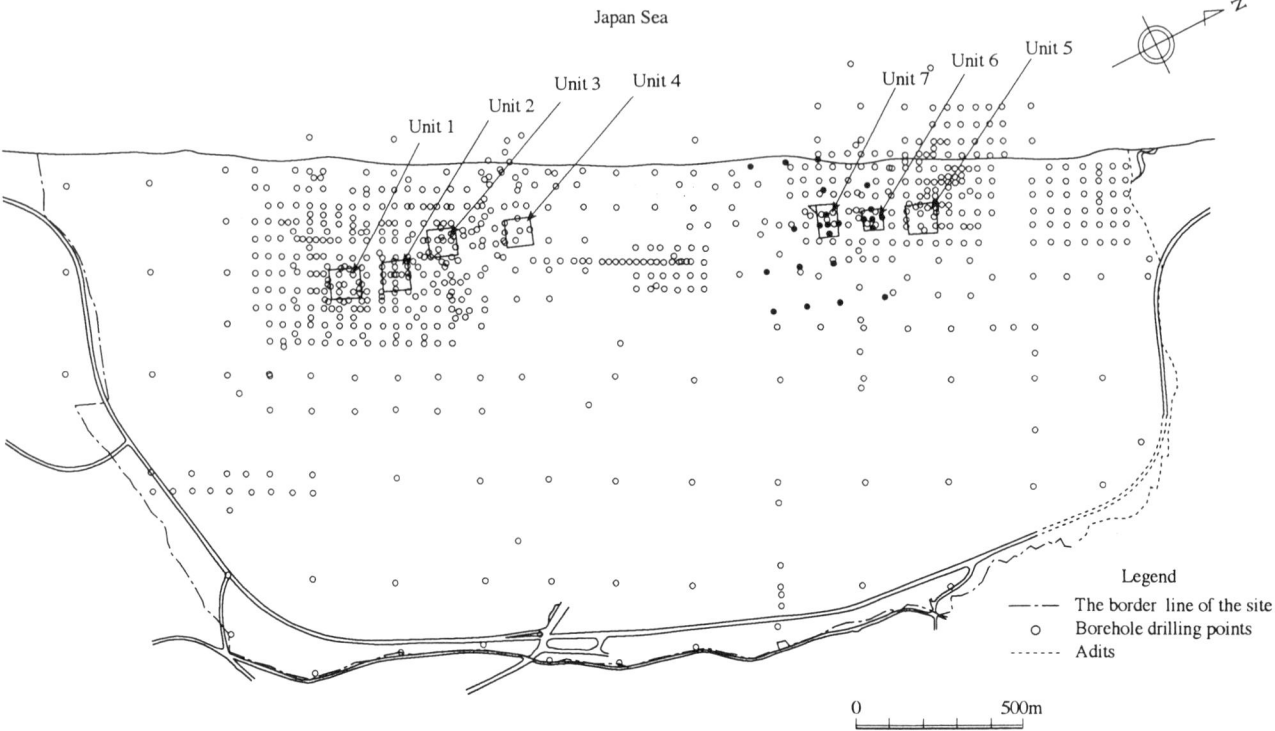

Figure 1. Map of the site, showing the localities of geological survey

activity, duration of activity must be checked. By investigating Yasuda Formations which unconformably overlies Neogene Formation, Yasuda Formation was made about 120 thousand to 130 thousand years ago.

2.3 Research at the place power plant is constructed

In Japan, based on "Regulatory Guideline For Aseismic Design Of Nuclear Power Reactor Facilities", nuclear power plants (especially reactor buildings) should be built on rock body. So reactor building is built on rock body. Before constructing reacter power plants, following to this guideline, rocks, foundations is checked (by making exploratory adit, borehole, etc.).

2.3.1 Exploratory adits

Exploratory adit is horizontally excavated and made. The length is about the width of reactor building. All adits cross at right angles. In the adit, the condition of bedrock, formations, and faults, are visually checked. To know physical, rock mechanical characteristics of foundation ground of the reactor building, the deformation test, plate loading test, strength of bearing capacity test and bedrock shear test are done. The results of these tests are used as input data when seismic stability of foundation ground is evaluated. It is very important, by using geology and technical background, to decide the place and number of exploratory adits in order to classify, judge specimen and rock mass, following to classified rock table.

Width of reactor building is about 60m to 80m and, in order to know the changes, strike directions, anisotropy of rock mass, Schmidt rock hammer test, elastic wave test are necessary. When foundation of the reactor building consists of soft rock (strength got by unconfined compression test is less than 10MPa), creep test is needed in order to check long-term settlement of reactor building.

Fig-3 is layout of exploratory adits and position of in-situ tests done at site of Kashiwazaki Kariwa Unit6 and Unit7.

Table 1. Stratigraphic succession in the site

Geological age		Formation name	Typical columnar section	Facies
Quaternary	Holocene	New Dune Sand bed		fine~medium grain light gray sand.
				fine~medium grain brown sand, contains humas and fragments of pottery.
		Banjin Sand bed (Ancient Dune)		medium~coarse grain light gray sand.
				medium~coarse grain reddish brown~brown sand. contains intercalated thin silt layers.
	Pleistocene	Yasuda Formation A4 member		clay~silt, contains a number of intercalated sand layers.
		A3 member		clay~silt, contains intercalated layered clays, organic materials and sand layer, and fossil shells.
		A2 member		clay~silt, contains intercalated sand layers, fat gravel layers and organis materials.
		A1 member		clay~silt, contains intercalated sand layers and gravel layers.
Neogene	Pliocene	Haizume Formation		tufficious mud stone, tufficious sand stone and tuff.
		Nishiyama Formation N3 member		mud stone (sandy), contains intercalated sand stone and tuffs, nodules and fossil shells.
		N2 member		mud stone (silty), contains intercalated layered mud stones and tuffs, and nodules.
		N1 member		mud stone (silty~clay), contains intercalated sand stone and tuffs, nodules and fossil siliceous sponges.
	Miocene	Shiiya Formation		sand stone and sand stone~mud stone alternation, intercalated conglomerates.

~~~ unconformity

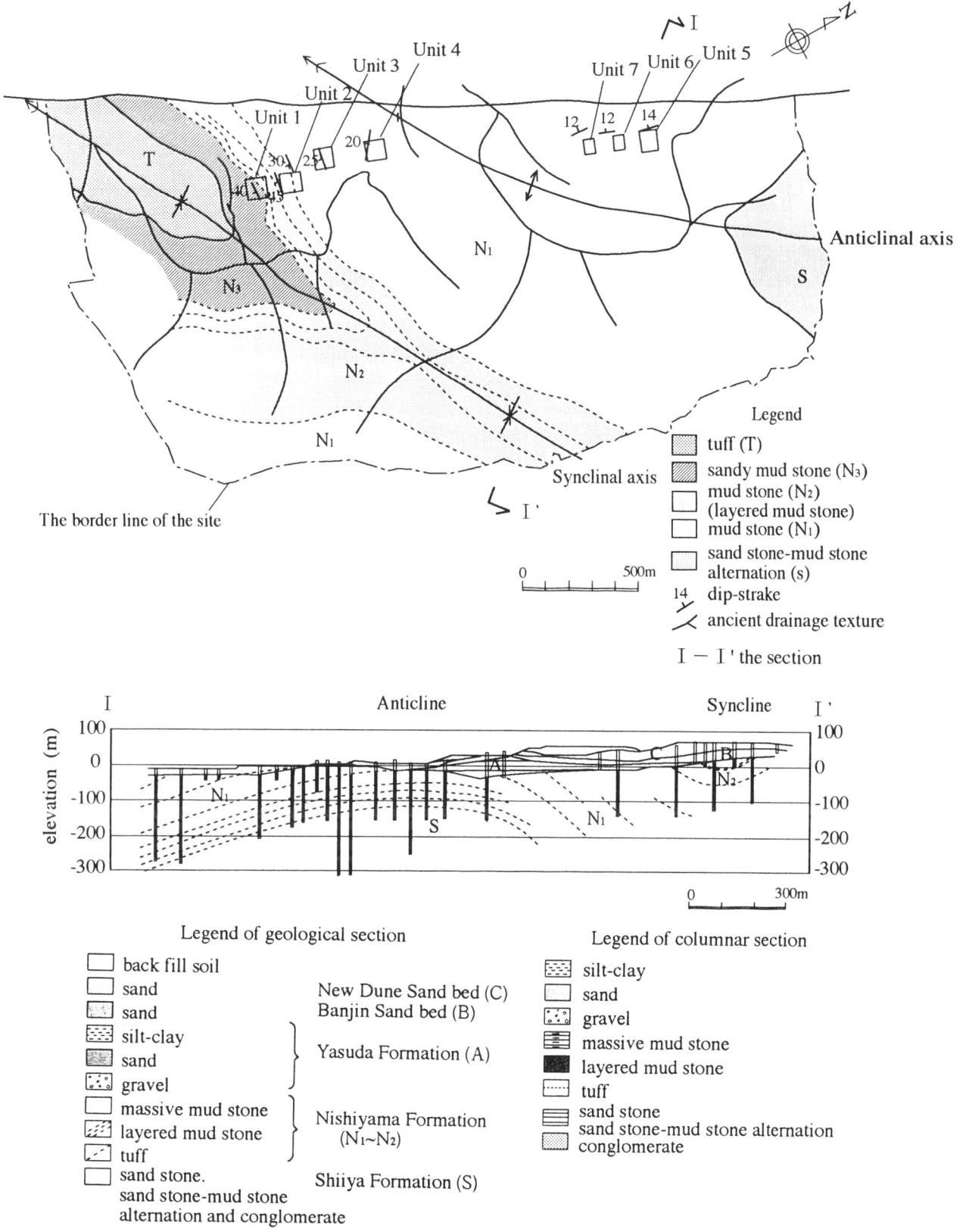

Figure 2. Geologic map in the site

2.3.2 Borehole investigations

Around the site (with 200m radius from the reactor building), minute geological researches are done in order to know geological structure (the rock mass dispersion, faults, folding, etc.).

By using the borehole sample, physical test, tri-axial compression test, ultrasonic wave velocity test

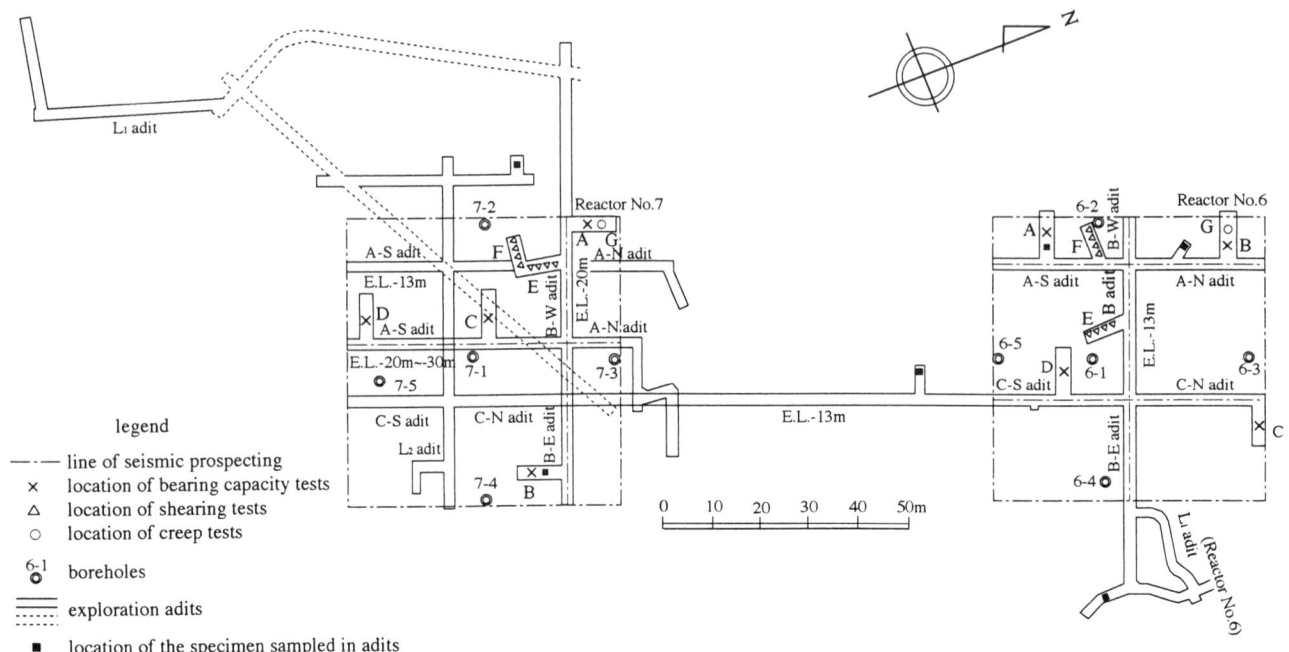

Figure 3. Layout of exploratory adits and positions of in-situ tests, according to the unit 6 and 7 project

are done in order to know mechanical characteristics of foundation rock. P-S logging, pressure meter test are done by using borehole. According to guidline, more than five boreholes are needed under one reactor building. The length of one borehole is more than the width of reactor building. The length of one borehole is needed to penetrate open free surface (definition of open free surface is, this formation ranges wide, and hard, Vs is more than 700m/s). For this reason, the length of one borehole is ususlly about 200m to 300m. Based on results of exploratory adit and borehole investigations, sectional map of the site is made (vertical and horizontal sectional map). Calculation model to evaluate stability of reactor against earthquakes is made.

Fig-4 shows the vertical sectional geologic map including Unit6 and Unit7. As for the fault near the reactor building, if fault should be activated and moved, it is hard to maintain stability of reacter buildings. In order to ensure safety of rock mass, exploratory adits are made, borehole investigations are done. If faults of rock should be found, we must check and make sure this fault will never move in the future. For the sake of this purpose, accumulation order of formation and fault is checked by using test adit, radioactive age determination in the fault is done. When fault has no probability of activating in the future, the fault will not cause slip, foundation crush, when earthquake occurs.

In Japan, most important buildings must be based on rocks. Seismic stability depends on weak layers and faults. It is necessary to minutely know dispersion, condition of rock mass and fault. TV inspections using boreholes, surface geological researches, P-S loggings, specific resistivity tests and many kinds of researches are done.

3 Physical characteristics and stability evaluation of foundation rock against earthquakes

3.1 Abstract of stability evaluation

Fig-5 is survey stage and flow. For the sake of the efficiency, stability evaluation of foundation ground is done, following to this flow chart. Flow chart consist of three types of calculation methods.
1. conventional (sliding) method (step1)
2. static FEM analysis (step2)
3. dynamic FEM analysis (step3)

The latter calculation method is more complexed, accurate and less conservative than the former one. When one calculation method shows rocks, foundations are stable to earthquakes (usually shown by safety factor:Fs), further calculation step is unnecessary, because the result of former step is considered to be safer than the latter step. The calculation results are obtained by Fs. Allowable Fs is 2.0 (when it is calculated by the conventional calculation method, or static FEM analysis) and 1.5 (calculated by dynamic FEM analysis). As the geological structure is very complexed in Japan, usually FEM analysis is done. The calculation case is selected by static FEM, dynamic FEM is done after that.

As for stability of foundation ground, the bearing capacity and settlement are also very important. But in Japan, most reactor buildings are built on rocks, so the main purpose of the stability evaluation is about rock sliding.

3.2 Seismic force

When the static seismic force is made, in order to calculate seismic safety factors, the seismic power coefficient is decided as 0.2 (Kh:horizontal power) and as 0.1 (Kv:vertical power). When dynamic seismic force is made, S2 power wave is used at the open free surface (for horizontal power). And vertical power is the same as that of the static calculation (Kv: vertical power coefficient is 0.1).

Fig-6 shows the comparison of Fs (static, dynamic calculations) of many kinds of cases (S-wave velocity and seismic force is different). By the result, when seismic acceleration is less than 500 gals, dynamic Fs is bigger than static Fs. When acceleration is more than 500 gals, dynamic Fs is smaller than static Fs. When seismic acceleration is more than 500gals, new kinds of seismic coefficient must be used as the new static seismic force (not Kh=0.1 to 0.2 but equal to the S2 earthquake wave power).

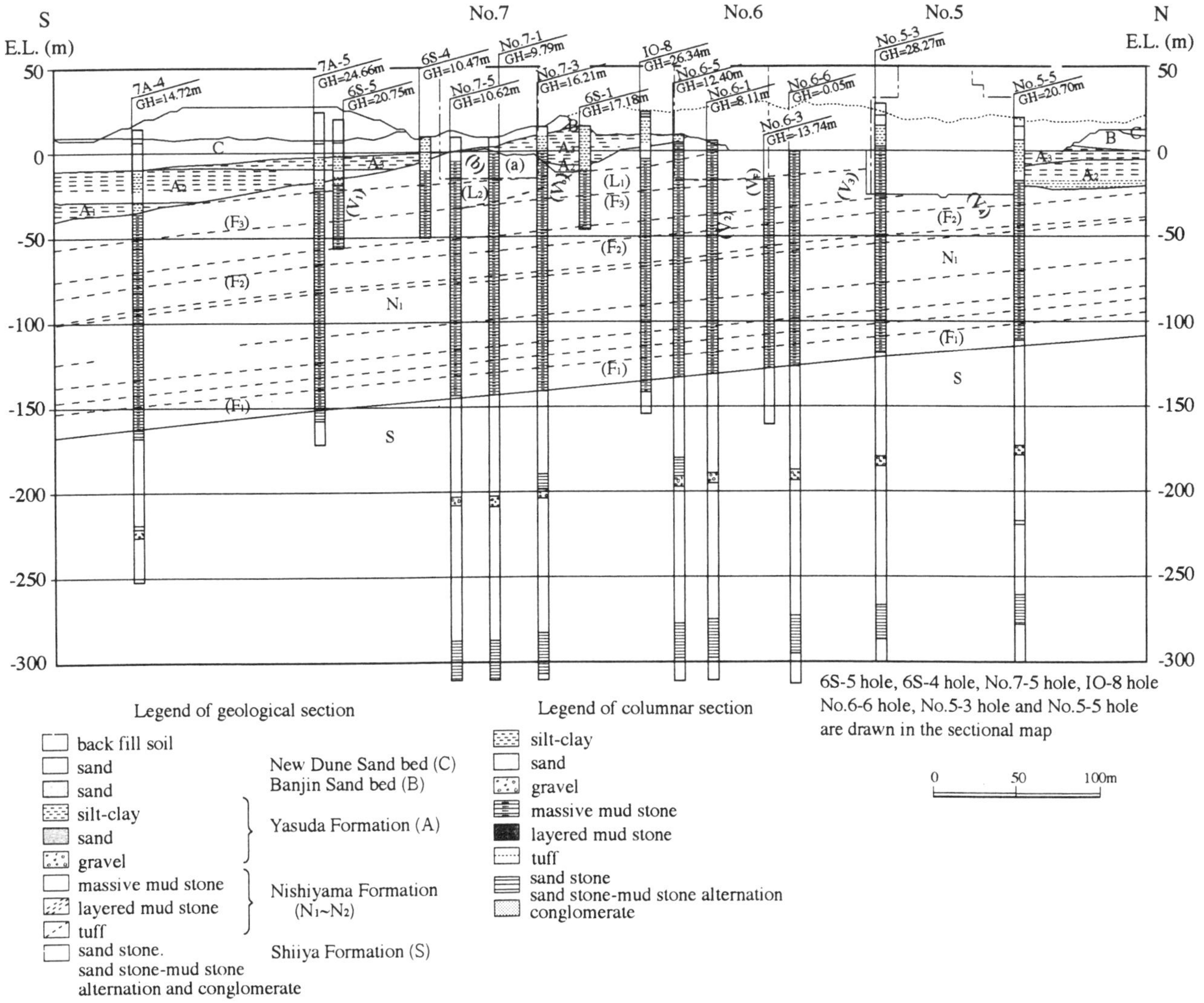

Figure 4. Vertical sectional geologic map including the unit 6 and unit 7

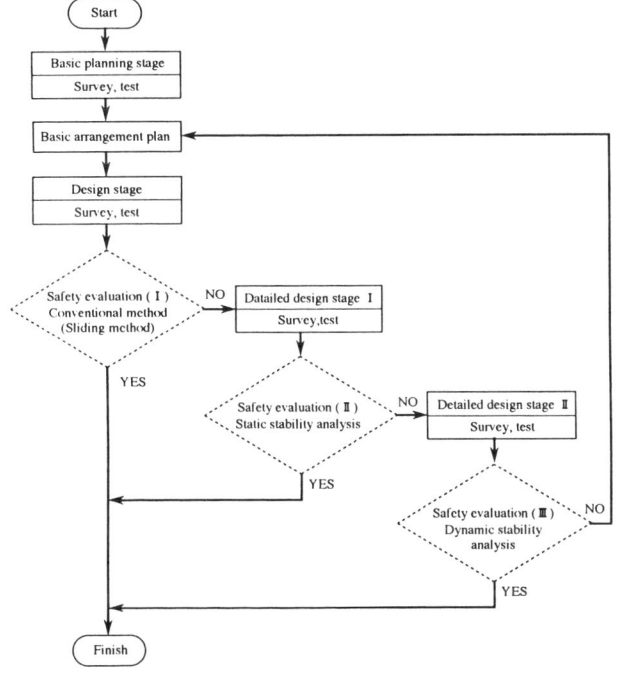

Figure 5. Survey stages and flow

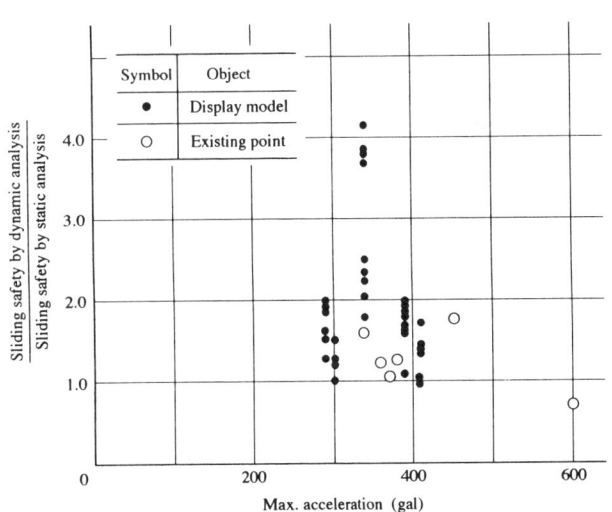

Figure 6. Relation of maximum acceleration value of design seismic motion S2 and
Sliding Safety by Dynamic Analysis / Sliding Safety by Static Analysis

3.3 Evaluation of rock material

By using borehole core samples obtained near reactor building, many kinds of tests are done (unconfined compression tests, tri-axial compression tests, split tension tests, many kinds of other strength tests).

In borehole, many kinds of tests are done (pressure meter test, P-S logging, etc.). In exploratory adit directly under the reactor building, many kinds of tests are done (deformation test, bearing capacity test, bedrock shear test, rock creep test, etc.).

Table-2 shows the combination of physical property value used to safety evaluation of foundation ground. These calculation input data are fixed, both by test results and geological backbones.

When joints of rock affect physical characteristics of rock mass, safety of this rock mass is checked by the result of exploratory adit, in-situ test. As for the deepest places where the rock materials cannot be checked by exploratory adit, characteristics must be decided by the borehole data. In this case, we must consider and compare the results given from shallower places (exploratory adit, in-situ test, etc.).

When foundation is sedimentary soft rock, physical characteristics affected by joints are negligible, so all of the characteristics can be decided only by the borehole data. But, in this case, because of the differences of loading pattern and failure process, test results between in-situ test and laboratory test is sometimes different. In such cases, the reason such kind of differences happened must be considered.

The foundation of Kashiwazaki Kariwa nuclear power plant is, as mentioned before, sedimentary soft rock, so the physical characteristics (input data for FEM analysis) are decided from the results of borehole data.

As for the strength of rock, ϕ (inner friction angle) is $0°$ (by tri-axial compression tests in the laboratory). On the other side, the strength of this rock, ϕ (inner friction angle) is $30° \sim 40°$ (from the result of in-situ rock shear test in the adit).

As for the reason of this difference, the breakage process of this rock (from the result of in-situ rock shear test in the adit) is thought to be progressive failure, so results of tri-axial compression tests in laboratory are used as input data of FEM analysis.

Fig-7 shows the relation between normal stress and shear stress (tri-axial laboratory compression test, FEM analysis, in-situ rock shear test). When using results of tri-axial compression test in laboratory (ϕ is $0°$) as the input data of FEM analysis, output data (rock strength) of FEM analysis is very likely to the result of in-situ rock mass shear test in the adit. By checking rock failure process in this simulation, we found the strength of rock, ϕ (inner friction angle) is $30° \sim 40°$ (when there is small vertical stress, like condition of in-situ rock shear test in the adit, because the tension failure starts at first). As for the deformation characteristics, static simple loading test (pressure meter test, plate loading test, tri-axial compression test, etc.), cyclic loading test (cyclic plate loading test, cyclic simple shear test, etc.) and pulse loading test (elastic wave velocity measurement in the exploratory adit, suspension P-S logging, etc.) are done.

Fig-8 shows deformation modulus by various methods. Deformation moduls is different, following to loading methods. Deformation modulus that is obtained by the static loading test, is quite different from those obtained by cyclic and pulse loading test. And the difference between laboratory test and in-situ test is very few.

Fig-9 shows relation between shear modulus and shear strain at the same depth by various methods. By this figure, the reason of this difference about elastic modulus can be explained by the strain level dependence of rock mass. Considering these results (relation of strength and deformation characteristics of rock mass), result obtained by laboratory is used as the input data for FEM analysis of Kashiwazaki Kariwa nuclear power plant.

By using these physical characteristics and input seismic forces, the seismic stability of foundation ground of nuclear power plant is checked.

Table 2. Combination of physical property value used to safety evaluation of foundation ground and slope (bedrock)

Physical property value / Analytical method	Unit Weight γ (t/m³)	Static strength c, c' (MPa)	Static strength ϕ, ϕ' (°)	Static elastic modulus E (MPa)	Static Poisson's ratio ν	Limit bearing capacity	Dynamic strength c_d, c_d' (MPa)	Dynamic strength ϕ_d, ϕ_d' (°)	Dynamic elastic modulus E_d (MPa)	Dynamic Poisson's ratio ν_d	Shear elastic modulus G, G_0 (MPa)	Damping coefficient h (%)
Conventional method (Seismic intensity method) (Analysis of sliding plane)	Boring Core sample, block sample	Bedrock shear test	Same as left									
Static analysis (Ordinary)	Same as above	Bedrock shear test Tri-axial compression	Same as left	Plate loading test (Secant) Laboratory tri-axial test (E_1, E_{50})	Uniaxial compression test Tri-axial compression test							
Static analysis (In earthquake)	Same as above	Same as above	Same as above	Same as above	Same as above				PS logging (Used locally)	PS logging	Same as the part of dynamic elastic modulus	
Dynamic analysis	Same as above	Same as above	Same as above				Static strength is used mostly.	Same as left	Elastic wave prospecting (Vp,Vs) Dynamic shear Dynamic tri-axial test	Elastic wave prospecting (Vp,Vs)	Same as above	Conventional value Dynamic tri-axial test Dynamic shear test

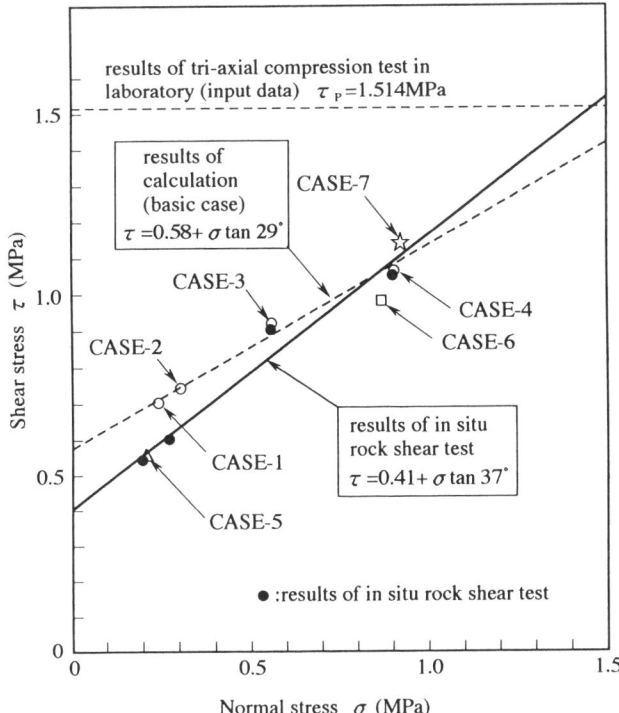

Figure 7. Relation between normal stress and shear stress

4 Improvement of foundation soft rock

4.1 Development of artificial soft rock

In Japan, in accordance with the regulatory guide for aseismic design of nuclear power reactor facilities, nuclear power plant must be constructed on bed rock.

As mentioned before, for the site selection, many kinds of tests must be done, (in order to know rock distributions, borehole investigations must be done and, at the place reactor power plant is constructed, exploratory adits are needed).

As for Kashiwazaki Kariwa nuclear power plants, as the foundation rock mass is very deep, we cut hills and made the suitable area of these nuclear power plants. We also excavate about some ten meters deep to make foundations of the nuclear power plants. The amount of waste soil is very large. Some of soil is used as backfilling soil, but this is not enough. There is about 12 million m³ remain as waste soil.

For constructing reactor building, because of many kinds of conditions (construction method, restriction of time schedule) we must dig parts of foundation of this reactor building faster and deeper. Exploratory adits excavated for research must be backfilled. So good quality artificial soft rock is needed. In this case, concrete is usually used. But, when foundation rock is soft rock, compression strength of concrete is about ten times larger than that of soft rock. So, the probability of stress concentration is very high. In this case, artificial soft rock is needed to be so soft as bed rock, and waste soil (12 million m³) must be used. Artificial soft rock is developed from the plastic soil cement using waste soil.

And quality of the artificial soft rock is checked, by using artificial soft rock as backfilling material of exploratory adits. By these results, artificial soft rock proved to be very good for the long-term stability, construction speed, cost, etc. So this artificial soft rock is used as foundation of this nuclear power plant.

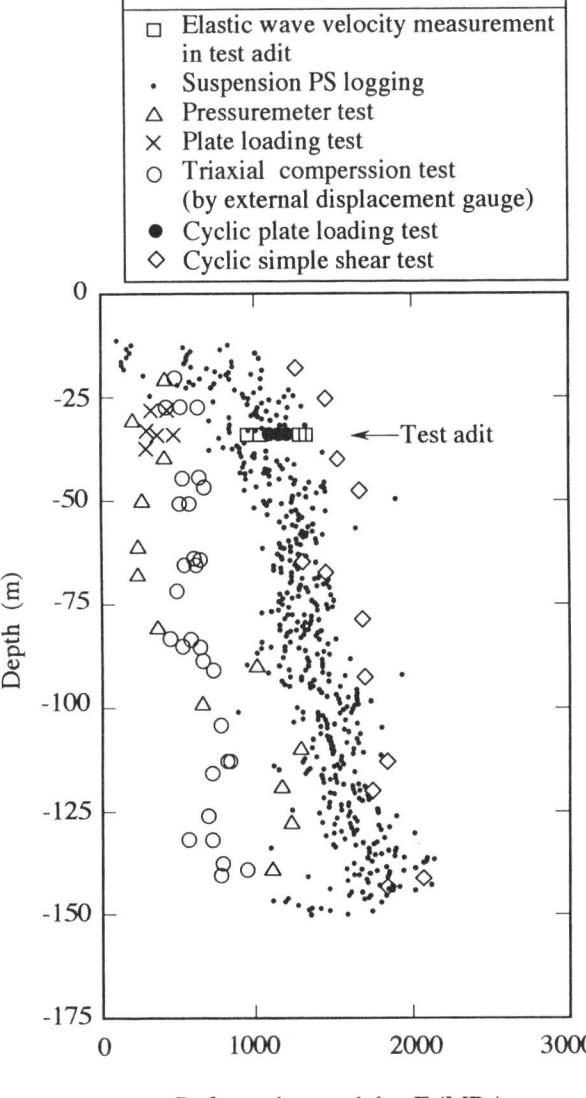

Figure 8. Deformation modulus by various methods

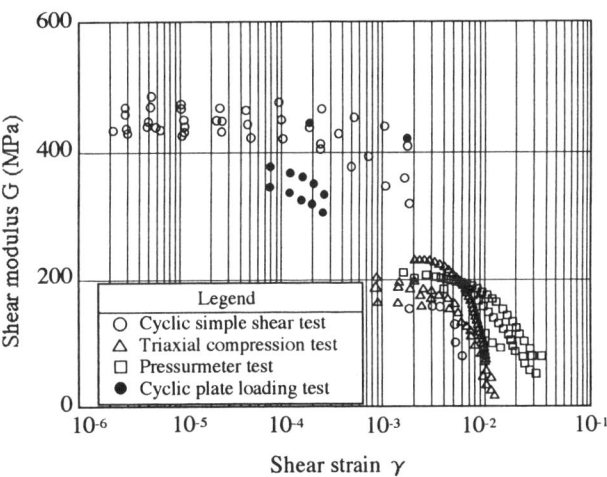

Figure 9. Relation between shear modulus and shear strain at the same depth by various methods (depth:-10m TO -35m)

4.2 Component of artificial soft rock

The artificial soft rock is usually made from clay soil, hardener, and water. By mixing and hardening them, artificial soft rock can be made.

At Kashiwazaki Kariwa nuclear plant site, there is a lot of excavated sedimentary waste soft rock. By wet crushing them, mud slurry is made as material of this artificial soft rock. In order to change the physical characteristics (rigid, weight, etc.), fine aggregate produced in this site is mixed. Mixture of clinker, water blast furnace slag, gypsum is used as hardener in order to make ettringite.

Table-3 shows mixture rate and chemical composition of hardener.

4.3 Characteristics of artificial soft rock

4.3.1 Mechanism of strength acquisition

Physical and mechanical characteristics of artificial soft rock can be obtained by the hydration activity of hardener. Clinker, water blast furnace slag, clay minerals in soft rock slurry start chemical hydration activity.

Fig-10 shows chemical hydration activities of this artificial soft rock. At first, the ettringite take much water as bonding water, and lessen water content in per cent of soil. Secondary, much needle shaped crystals are produced, and make structural backbones of this artificial soft rock. Calcium silicate hydration is made and filled, the strength of this artificial soft rock can be made.

4.3.2 Physical characteristics of artificial soft rock

Physical characteristics of this artificial soft rock appears by chemical hydration activities of hardener. Physical characteristics (unit weight), mechanical characteristics (compression strength, deformation modulus, etc.) of this artificial soft rock can be changed, within some range, by varying mix proportion of mud rock, hardener, sand, and water.

Table-4 shows mixture rate example of artificial soft rock which has the same physical characteristics used in this plant.

Table-5 is physical and mechanical characteristics of artificial soft rock (aged 91 days). The physical characteristics of artificial soft rock is almost the same as foundation ground soft rock.

4.3.3 Long term stability of artificial soft rock

Fig-11 shows flow about long-term stability of this artificial soft rock. Following to this flow chart, long-term stability of this artificial soft rock is checked.

Fig-12 shows result of unconfined compression test of artificial soft rock. The strength saturates at very early times, and seems no change of compression strength after that. X-ray, microscope inspection of hydration product is done, the stability of hydration product can be seen.

And, by many kinds of infiltration chemical tests (natrium chloride, sulface natrium, polycarbonated natrium, etc.), the conclusion that these minerals are stable to chemical environment, is obtained.

4.4 The method to make artificial soft rock

Fig-13 shows the method to make artificial soft rock. The plant is the same, when ordinary concrete is made (the stationary weight sensor, batch type compulsory concrete mixer, and mud rock crusher, etc.). Which condition most affect the quality of artificial soft

Table 3. Mixture rate and chemical composition of hardener

name of hardener	mixture rate (%)			chemical composition of hardener (%)						
	clinker	slag	gypsum	SiO_2	Al_2O_3	Fe_2O_3	CaO	MgO	SO_3	others
hardener of normal portland cement	96	0	4	22.8	5.1	3.0	63.8	1.5	1.9	1.9
hardener of portlant blast furnace cement (A type)	67	29	4	23.8	7.1	2.3	58.7	3.0	2.1	3.0
hardener of artificial soft rock	62	20	18	20.4	6.4	2.0	53.6	2.0	9.4	6.2

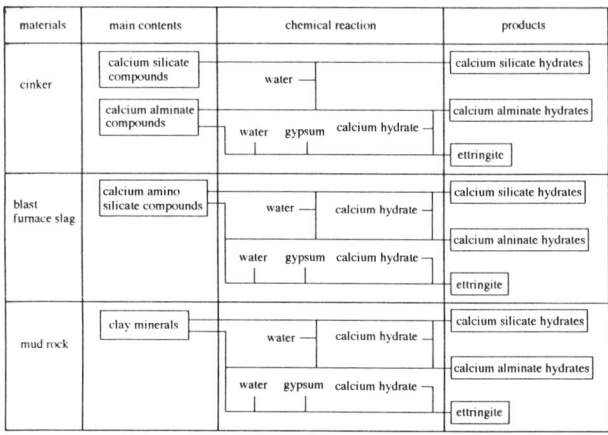

Figure 10. Chemical hydration activities of artificial soft rock

Table 4. Mixture rate example of artificial soft rock

unit weight (kg/m³)				density weight (t/m³)	table flow (mm)
mud rock	sand	hardener	water		
190~230	700~890	150~180	550~600	1.71~1.78	180~210

Table 5. Physical and mechanical characteristics of artificial soft rock

physical characteristics	unit	artificial soft rock (aged 91 days)	Nishiyama formation
density weight	t/m³	1.70~1.78	1.69
compression strength	MPa	3.3~5.0	3.0~3.4
deformation modulus	MPa	690~930	450~510
poisson ratio	—	0.43~0.44	0.46~0.48
elastic wave velocity	Vp m/s	1.800~2.200	1.600
	Vs m/s	670~810	480
creep coefficient	—	0.25~0.35	0.31

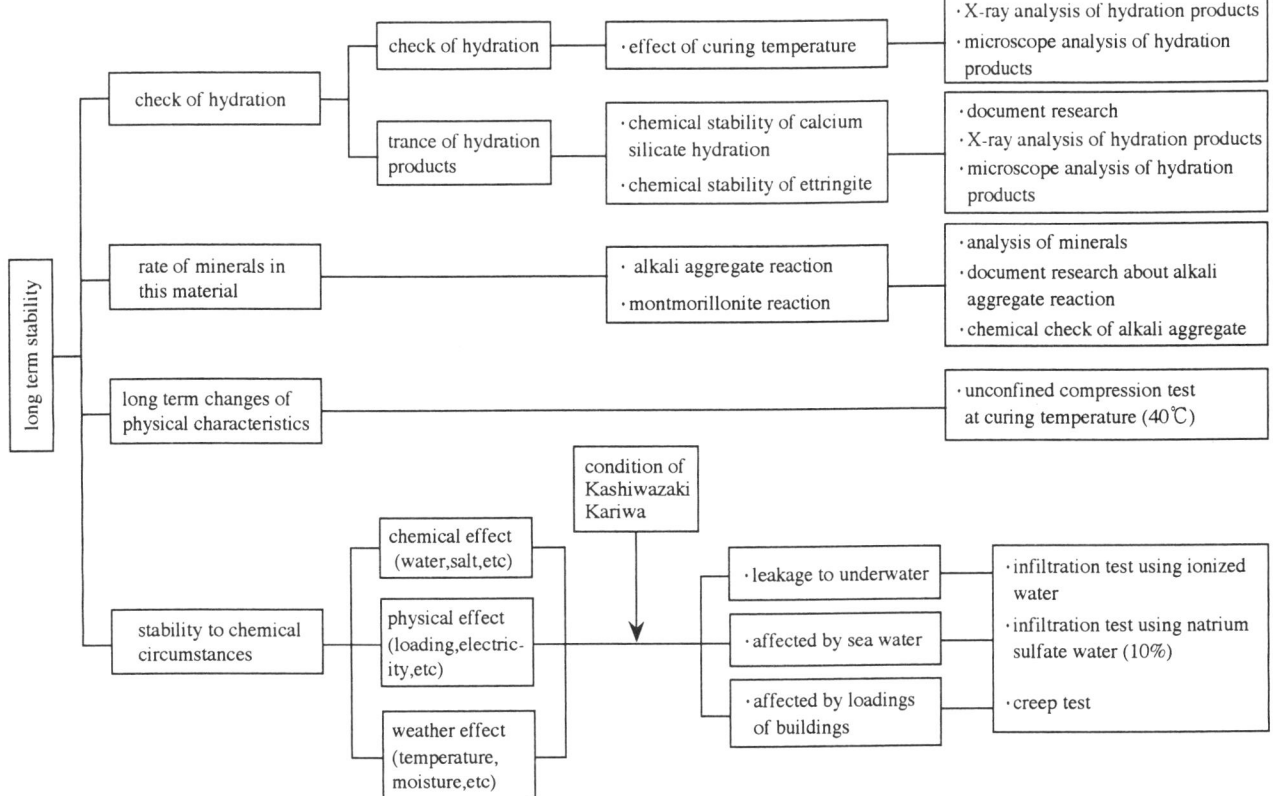

Figure 11. Flow chart about long term stability of artificial soft rock

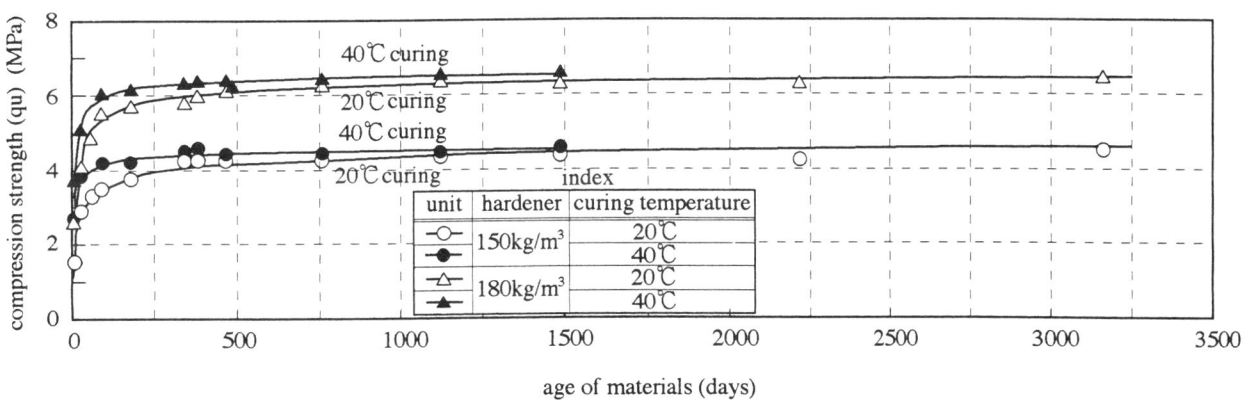

Figure 12. Result of unconfined compression test about artificial soft rock

rock is checked, and the rate mud rock is crushed, the density of slurry, most affect the quality of artificial soft rock.

When we make and maintain the quality of artificial rock, the rate mud rock is crushed and the density of slurry are important. By controlling them uniformity of artificial soft rock can be maintained. And the fluctuation of strength got by unconfined compression test is about 11% (quality of artificial soft rock is almost the same as ordinary concrete).

4.5 The method to apply artificial soft rock as the foundation of nuclear power plant

Artificial soft rock is also used as the foundation of tower cranes, backfilling of exploratory adits, and foundations of other kinds of buildings. And this artificial soft rock proved to be very good to construct the basement of buildings.

In Kashiwazaki Kariwa site, when Unit6 and Unit7 nuclear power plants are planned to be built, surface geological survey, borehole investigation, test in the adit under reactor building, are performed to get regional geological structure of this site. By the research, small gravitated rock faults are found at the surface of this foundation. In order to check more minutely, many kinds of tests in exploratory adits are performed alongside these rock faults. When the construction of Unit6 and Unit7 nuclear power plants started, this rock (soft rock with small rock faults) is also replaced into artificial soft rock.

Fig-14 shows the sectional map of foundation ground replaced by artificial soft rock. To replace this soft rock into artificial rock, many kinds of tests mentioned before are done (many kinds of physical and chemical characteristics, long-term stability, etc.), to make sure the quality of artificial soft rock. Calculation considering physical characteristics of

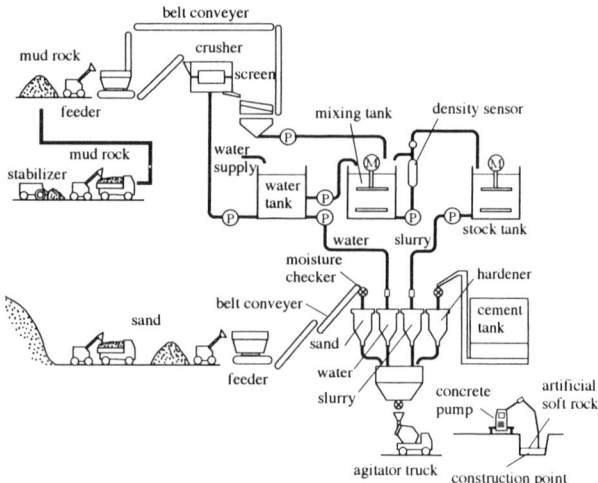

Figure 13. The method to make artificial soft rock

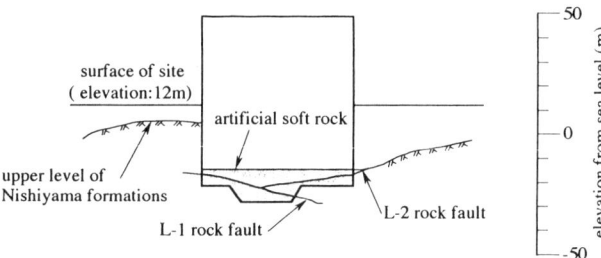

Figure 14. Sectional map of foundation ground replaced by artificial soft rock

artificial soft rock are done, and the stability of foundation ground, partly replaced by artificial soft rock, is checked. When soft rock is replaced into artificial rock, thermal stress which is produced by making a lot of artificial soft rock or concrete, is checked and there is little crack produced by this.

5 Conclusions

In this paper, we introduced many kinds of geological researches at site, mechanical, physical tests, the method to check seismic stability of the foundation ground of Kashiwazaki Kariwa nuclear power plant.

As for the construction method, artificial soft rock invented and used as the replacement material of foundation ground, is introduced.

REFERENCES

Science and Technology Agency, Nuclear Safety Bureau, Office of Nuclear Safety Policy Research 1994. Guides for licensing review nuclear safety commission

Japan Society of Civil Engineers 1985. The method to research, test soil foundations and evaluate seismistic stability of nuclear power plant

Japan Electric Association 1987. Technical guideline for aseismic design of nuclear power plants

Hideo Otsuki & Kiyoshi Kishi & Toshiaki Sakai & Hideaki Hyodo & Soichi Tanaka 1995. Evaluation of deformation characteristics of soft rock measured by cyclicic loading tests and elastic wave velocity measurement. ISRM 8th Congress

Toshiaki Sakai & Masayuki Satou & Haruhiko Uno & Sirou Fukui 1990. The strength characteristics of soft rock mass. The 25th Japan national conference on soil mechanics and foundation engineering. 1079-1082

Kiyoshi Kishi & Yoichi Nojiri 1989. Development of artificial soft rock. Cement Concrete No511. 78-86

Design and its evaluation through displacement measurement for the Akashi Kaikyo Bridge foundation

Mamoru Yamagata, Atushi Nitta & Sigeki Yamamoto
Honshu Shikoku Bridge Authority, Tokyo, Japan

ABSTRACT: This paper presents the process of determination of geotechnical constants for the Akashi Kaikyo Bridge design and the results of evaluation to the design deformation modulus through settlement measurement for the bedrock. The Akashi Kaikyo Bridge will be a suspension bridge with the total length of 3,910m upon completion. To analyse and to study the performance of bridges on soft rock, a proper understanding of the stress-strain and deformation behaviour of bedrock under high stresses anticipated was required. It was a demanding task for the bridge engineers to arrive at meaningful parameters controlling strength and deformation characteristics of bedrock.

1 Introduction

The Akashi Kaikyo Bridge, as illustrated in Fig.1, is one of the bridges in the group of Honshu-Shikoku Bridge complex connecting the Main Island (Honshu) to the Shikoku Island of Japan. It will be a three-span, two-hinge stiffening truss suspension bridge with the total length of 3,910m upon completion. The bridge consists of main tower foundations(2P,3P), anchorages(1A,4A), towers, cables and girder as shown in Fig.2.

Tower foundations were constructed by a laying-down caisson method. 2P is founded on the Akashi Formation at the foundation level of −60m while 3P on the Kobe Group at the foundation level of −57m. Each of the 2P and 3P consists of one column of 80m diameter and 70m high, 78m and 67m, respectively.

The Akashi Straits, as shown in Fig.3, has a maximum water depth of T.P.−110m at the center of the straits and shallower extension at both sides.

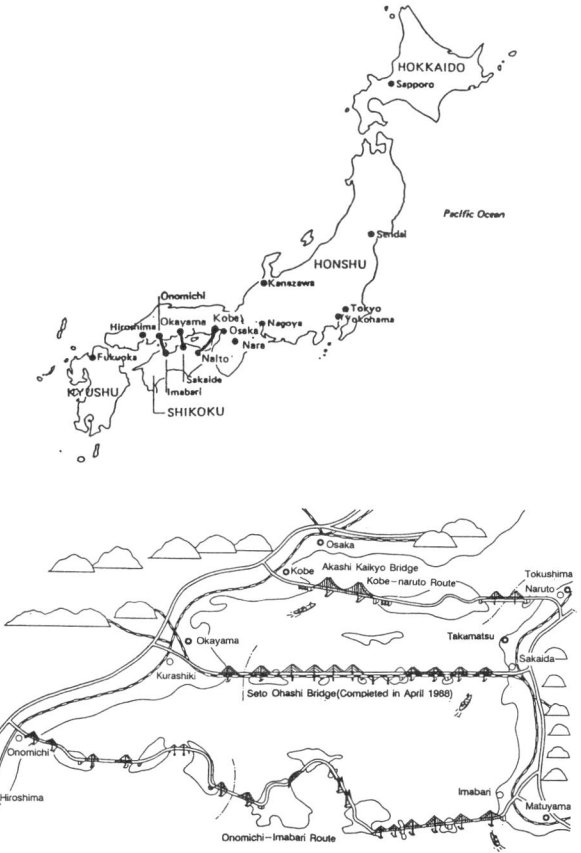

Figure 1 Schematic view of Honshu-Shikoku Bridge Routes

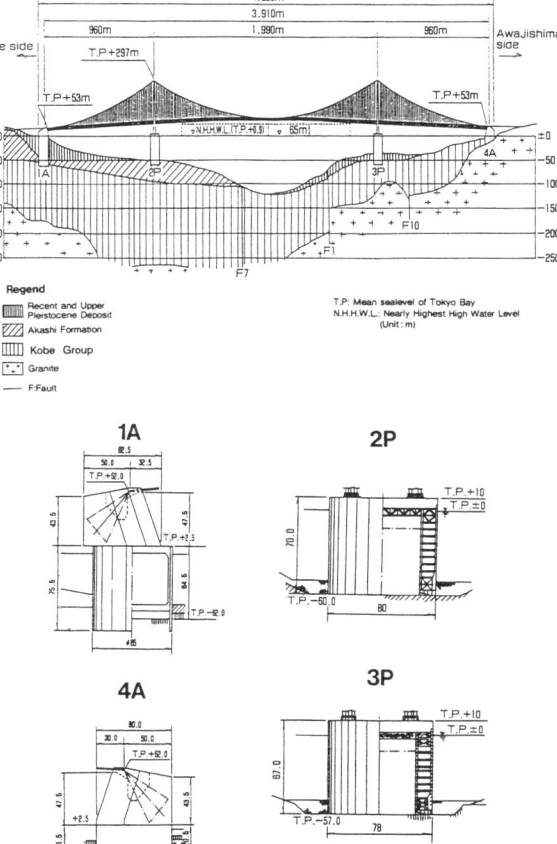

Figure 2 Geologic Profile along Akashi Kaikyo Bridge

The geological structures at the Akashi Straits are given in Fig.2. As shown in the figure, the basement consists of granite over which Kobe Group, Akashi Formation, upper diluvium and alluvium unconformably exist. Granite is assinged to the late Mesozoic and its weathering has not progressed very much. Kobe Group is made up of a soft rock of Miocene, which consists of alternative of sandstone and claystone. Akashi Formation is weakly cemented gravel and sand layer ranging from Pliocene to Pleistocene.

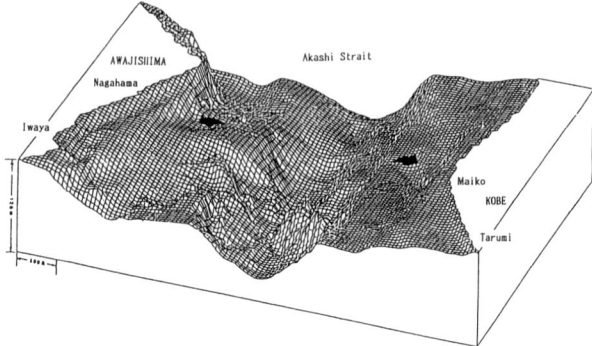

Figre 3 Bathymetric Map at the Akashi Strait

2 Design flow through the geotechnical investigation

Fig.4 illustrates a flow from the geotechnical investigation to the design of foundation. In this flow three items are considered relevant to foundation design, i.e., (1)determination of analytical methods of foundation stability and displacement, (2)modelling of bed rock, (3)determination of design geotechnical constants.
The geotechnical investigation was carried out to provide the solution for these items.

3 Geotechnical investigation

Table 1 summarizes the geotechnical investigation performed. Exploration was accomplished in a phased sequence as follows:
(1) Comprehensive exploration
(2) Preliminary investigation
(3) Detailed investigation (phase 1 and phase 2)
In the first phase of detailed investigation special emphasis was put on in-situ plate loading tests. Large plate loading tests were performed in order to reveal the mechanism of failure and deformation of bedrock.
Table 2 shows test item concerning the detailed investigation (second phase) and its quantity.
For bridge planning, it is vital to investigate the

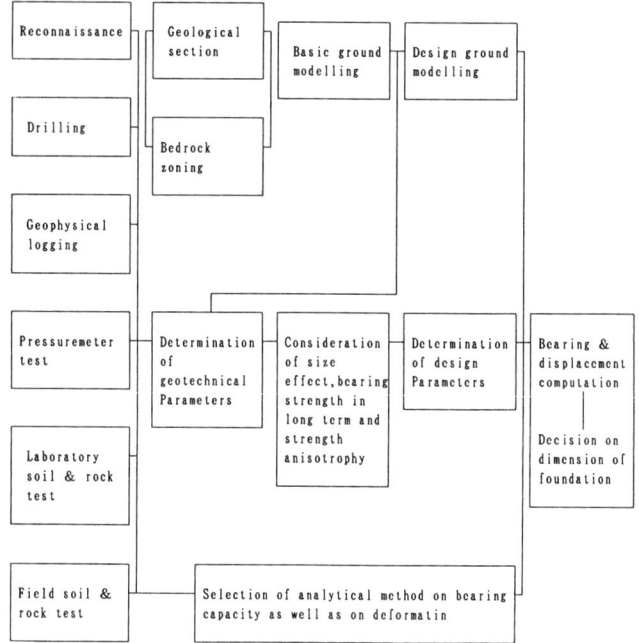

Figure 4 Foudation Design Flow Chart

Table — 1 Phased geotechnical explorations performed for the Akashi kaikyo Bridge

Investigation		Items	Resuts
comprehensive exploration	Overall Data Collection for comparison/selection of feasible routes.	#Bathymetric Survey and Topographic Survey for islands in straits. Sonic Prospecting, Geologic Reconnaissance, Echo-Sounding, Bottom Material Sampling by Dredging, Sea Floor Observation by Submarine Vessel, Aerial photographing	· Classification Map of Topography · Bathymetric Map
Preliminary investigation	To obtain basic data for route recommendation and design/construc--tion plan	#Drilling Core Drilling, Geophysical Logging, Pollen Analysis, Diatom Analysis, Foraminifera Analysis, Standard Penetration Analysis, Standard Penetration Test, Soil Test, Rock Test #In-Situ Rock Test Direct Shear Test, Static Loading Test, Rapid Loading Test, Geophysical Prospecting, Soil Test, Rock Test #Seismic Risk Survey	· Geologic Map · Geohistory · Contour Map showing surface boundary of Kobe Group · Aerophotograph
Detailed Investigation — Phase 1	Selected Drilling for revealing ground conditions at all substructure sites and for establishment of investigation standards	#Drilling Core Drilling, Geophysical Logging, Pressuremeter Test, Soil Test, Rock Test, Standard Penetration Test #In-Situ Rock Test Direct Shear Test, Creep Test, Static/Dynamic Loading Test, Triaxial Compression Test on φ 30cm Samples	· Geologic Profile · Establishment of investigation Standards
Detailed Investigation — Phase 2	Detailed Drilling for securing rock cores and rock/unit classification	#Drilling Core Drilling, Geophysical Logging, Pressuremeter Test, In-Situ Permeability Test, Soil Rock Test-Unit weight, Grain Size Analysis, Pulse Velocity, Unconfined Compression Test, Triaxial Test·CU·CD·UU, Cyclic Triaxial Test, Dynamic Triaxial Test, Creep Triaxial Test, Simplified Slaking Test	· Geologic Model at each site. · Shear Strength, Bearing Capacity Deformation Characteristic of Akashi/Kobe Formation · Geotechnical Parameter for each unit

Table 2 Summary of No. and items of Detailed Investigation(phase2)

Location	Boring		Pressuremeter test (point)	Geophysical logging (set)	Laboratory test(No.)	
	No.	Total length(m)				
1 A	6	767	113	6	Soil test	
					Unit weight test	46
					Grain size analysis	66
2 P	7	725	69	7	Triaxial compression test $\overline{CU}$	28
					CD	12
					UU	5
					Cyclic triaxial test	16
3 P	14	738	162	10	Creep triaxial test	6
					Rock test	
					Unit weight	434
					Ultra-sonic wave velocity	434
4 A	11	190	32	11	Uniaxial compression test	297
					Triaxial compression test $\overline{CU}$	214
					CD	42
					Cyclic triaxial test	73
Total	38	2420	376	34	Creep triaxial test	44
					simplified slaking test	20

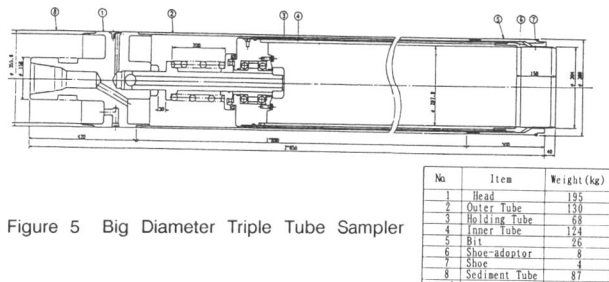

Figure 5 Big Diameter Triple Tube Sampler

proposed bridge site in order to gather information about the stratigraphic situation and about the properties of bedrock.
The boring was performed at an interval of 15-30m so as to enable preparation of a geological profile both along the bridge axis and at right angles to the bridge axis. The investigation depth was determined basically to be about the width of the assumed foundation, but 1 or 2 boreholes were extended up to the granite.
The types of in-situ tests performed in bore holes were pressuremeter test, geophysical logging and pressure test for permeability(JFT). The pressuremeter test was conducted to obtain deformation properites of bedrock at the interval of 3m in boreholes. The geophysical logging consists of three methods, seismic, electrical resistivity and density. Electrical logging furnished the useful information for bedrock profiling. PS logging, especially secondary wave velocity, was helpful not only for zoning but also for providing shearing properties of foundation rock.
Core-drilling was achieved with a triple tube sampler of 360mm in diameter for Akashi Formation which contains the gravel from 50 to 100mm or more. While in the Kobe Group, a triple tube sampler with 116mm in diameter was used. Fig.5 illustrates the structure of the sampler used.
Next item is a laboratory test. The bedrock differs from the formation on land in terms of a physical properties. Therefore in order to clarify engineering properties of each layer, laboratory tests including a triaxial compression test were performed to acquire a geotechnical constant for the analysis of bearing capacity and displacement of foundation. Table 3 shows a series of triaxial tests conducted and those test condition.
Test condition was determined as follows;
① Drainage condition
It is considered that bedrock may be under drained condition in the normal state while under undrained condition during earthquake . A shearing strength of bedrock in the normal condition can be fundamentally estimated by conducting drained triaxial compression tests. In defining the shearing strength it was decided by incorporating $\overline{CU}$ from the undrained triaxial compression tests. Shearing strength under earthquake loadings was investigated by undrained triaxial compression tests.
② Determination of a consolidation stress condition
Fig.6(a) shows the distribution of the ratio among principal stresses in bedrock by numerical analyses. A simulated stress ratio resulting from the weight of the foundation is almost in the ranged of 2 to 3 beneath the foundation, but apart from the foundation it becomes mostly less than 2. Judging from the analyses, it may be considered that the stress condition within the bedrock is in the anisotropically consolidated state. However, it is impractical to conduct tests considering such an anisotropically state existing. Besides, it has been proved that the result of isotropically consolidated tests indicates lower value compared with the result of anisotropically consolidated tests. Therefore, geotechnical constants was designated based on the result of isotropically consolidated tests.
③ Effective confining stress
The stress within the bedrock results from the overburden pressure and the stress added by application of the structure. A mean effective principal stress, $\sigma'_{mc}=(\sigma_1+2*\sigma_3)/3$, in bedrock differs from place to place. A confining pressure in the test was determined considering a field conditions by taking into account the ability of testing apparatus as well as the range of back pressure.
④ A simulated direction of principal stress σ_1 and difference of the stress condition on a slip surface.
Fig.6(b) exhibits the result of numerical analyses on the direction of principal stress. The rotation angle of major principal stress to the vertical deposition direction does not exceed 5° within the bedrock beneath the foundation, but it becomes larger within the bedrock around the foundation. For appling the test results to the actural bedrock with the rotation angle of major principal stress, the difference concerning the direction of principal stress should be considered. Because the strength is the highest when major principal stress is

Table 3 Triaxial test condition

Kind of tests	Geology		Confining pressure		Drained condition	Strain speed (%/min)	Creep load	Remarks
			effective σ_3 kgf/cm²	Back pressure B.P kgf/cm²				
Simple triaxial test	Akashi Formation		1, 2, 4, 6, 10, 14	B.P=3, $\sigma_3<14$	$\overline{CU}$	0.1		One set consists of 6 stage testing
				B.P=1, $\sigma_3<14$	CD	0.05		
			2, 4, 6, 10, 12	3	UU	0.3		
	Kobe Group	1 A	2, 5, 10, 15, 20, 25	6	$\overline{CU}$	0.05		
		2 P	2, 5, 10, 15, 20, 30		CD	0.025		
		3 P	1, 2, 5, 10, 15, 20					
		4 A	1, 2, 5, 10, 15, 20					
Creep triaxial test	Akashi Formation		4, 8	3	CD	—	20, 50, 80% of $(\sigma_1-\sigma_3)$max	7days/1stage Final stage 30 days
	Kobe Group		5, 10	6	$\overline{CU}$	—	20, 40, 60% of $(\sigma_1-\sigma_3)$max	
Cyclic triaxial test	Akashi Formation		0.5, 1, 2, 4	3	U-D	—	—	0.1Hz 10~15step
	Kobe Group		1, 5, 10	6				

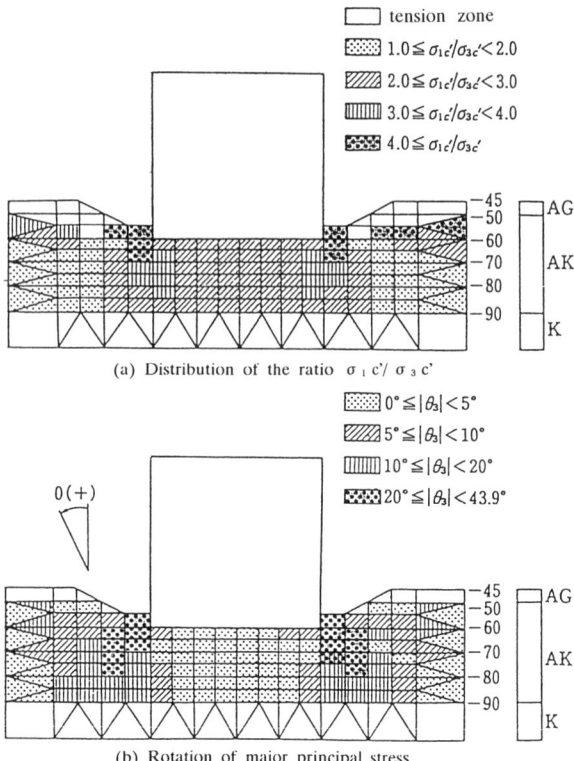

(a) Distribution of the ratio $\sigma_1 c'/\sigma_3 c'$

(b) Rotation of major principal stress

Figure 6 Stress distribution by numerical analyses

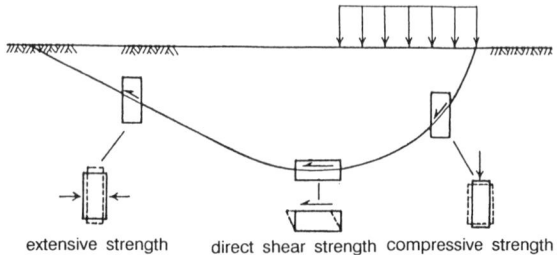

Figure 7 Strength to be developed along slip surface

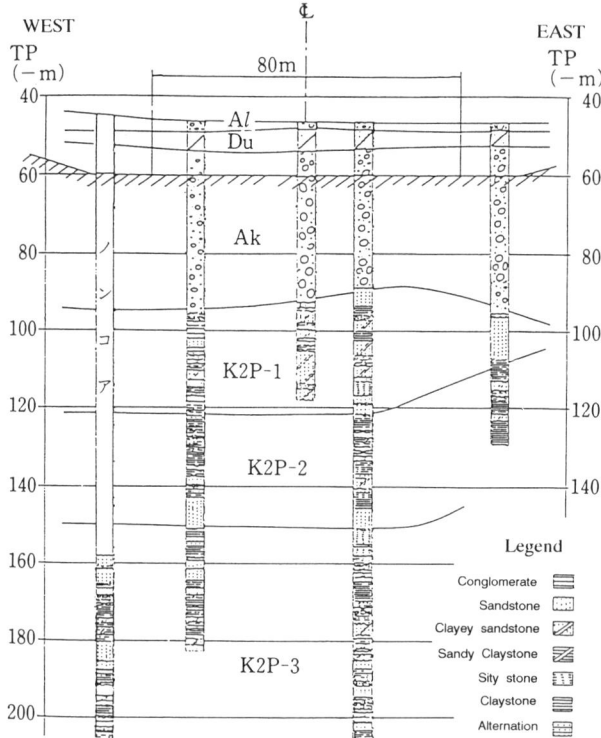

Figure 8 Geotechnical profile perpendicular to axis(2P)

perpendicular to the bedding plane. Therefore, after studing the affect of rotation through a numerical analyses, it was considered in so-called engineering judgement.

Fig.7 represents the strength to be developed along a slip surface. The strength is represented as compressive strength, direct shear strength and extensive strength due to the place. The problem to be solved was which strength can be considered as representative, on how a geotechnical constants should be determined by taking the difference among them into accout. This subject was resolved by reducing the shearing strength obtained from the triaxial compression tests which enable the confining condition.

A series of creep triaxial tests was carried out to determine creep constants for the analyses of creep deformation and to examine the creep rupture characteristics of soft rock. The creep load was applied in three stages i.e., 20,50,80% of the shear strength of the sample. At each stage, the load was maintained for seven days. For some of test specimens, the maximun loads were maintained for 30 days if creep failure did not occur.

Cyclic triaxial tests were carried out to study the relationship of shear modulus G and shear strain γ as well as damping ratio D and shear strain γ. In the cyclic loading tests the stage-shearing was imposed in a stress control using a frequency of 0.1Hz.

4 Results of geotechnical investigation

(1) Geological profile at the bridge site

2P site
Geological profile comprises a sequence of layers of granite, Kobe Group, Akashi Formation, upper diluvium and alluvium from the bottom as shown in Fig.8 . The top of granite comes at the level of T.P − 260m. Kobe Group with a thickness of about 170m consists of alternation of soft sandstone and claystone. Akashi Formation is a sort of gravel layer with about 40m which has an unconformable relationship with the underlying Kobe Group.

3P site
Kobe Group consists of an alternation of sandstone/claystone in which the sandstone is predominant as shown in Fig.9. The granite has been altered by weathering to depths of more than 30m. The Kobe Group beneath the foundation varies from 39 to 65m in thickness.

(2)Physical and mechanical properties of bedrock

a Physical characteristics

Physical characteristics are available through the laboratory tests and physical logging. Such an information is very useful for grouping the layer since it can be gotten easily and continuously.

It is very important to find out the physical properties as the mechanical properties can be assumed from the physical characteristics also.

Akashi Formation
The laboratory tests and in-situ tests(PS logging) revealed physical properties of Akashi Formation. The laboratory tests conducted consist of the test for specific gravity of the solids, water contests, density and grain size analysis. Akashi Formation at the 2P appears practically uniform with depth in terms of physical characteristics.

The grain size distribution curves of Akashi Formation

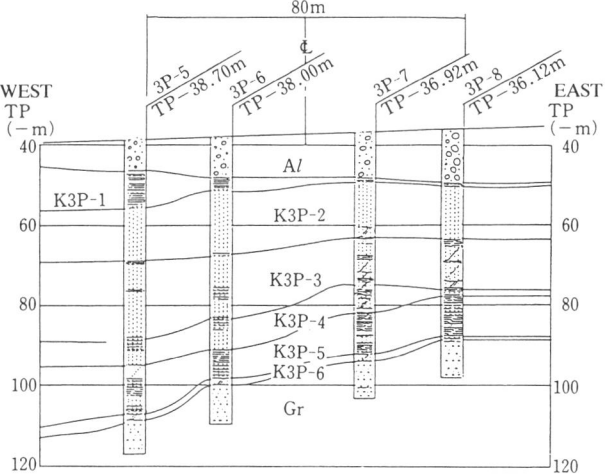

Figure 9 Geotechnical profile perpendicular to axis(3P)

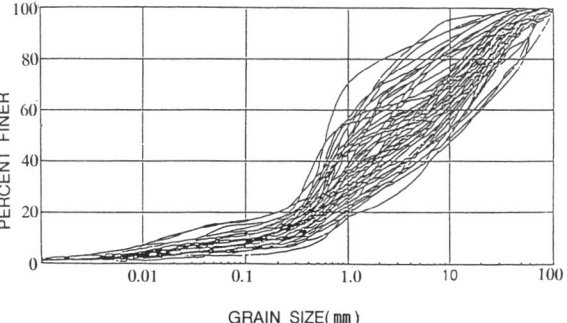

Figure 10 Grain size distribution of Akashi Formation

is represented inside a spindle as shown in Fig.10. The uniformity coefficient Cu is about 10. The coefficient curvature Uc' is in the range of 1 to 3. Akashi Formation is recognized as well-graded gravels.
The wet unit weight varies from 1.99 to 2.19gf/cm³ with the average of 2.09gf/cm³. The void ratio e lies in a range from 0.4 and 0.5(porosity is about 30%). These properties seem to be somewhat different compared with those exihibited by common geotechnical materials of the same era. The natural water content Wn scatters between 9 and 20%, mostly in the range of 10 to 15% with the average of 14%. The P and S wave velocities are in the range of Vp=1.6 to 2.1km/sec and Vs=0.3 to 0.6km/sec, respectively with little variation through the depth.

Kobe Group
The laboratory tests and in-situ tests(PS logging) revealed physical properties of Kobe Group. The laboratory tests conducted consist of the test for specific gravity of the solids, water contests, density and ultrasonic test.
The wet unit weight for the claystone is 2.25gf/cm³ and for the sandstone is 2.35gf/cm³. The void ratio e for the claystone is 0.35 and for the sandstone is 0.30 on average. The natural water content Wn for the claystone is beteween 12 and 16% and for the sandstone is in the range of 8 to 12% with the average of 11%. The P and S wave velocities measured by the PS logging are in the range of Vp=1.8 to 3.4km/sec and Vs=0.3 to 1.2km/sec, respectively.

b Shear strength properties

The shearing properties of these formations were evaluated principally with the triaxial compression test on undisturbed specimens.

Akashi Formation
Fig.11 exhibits the result of triaxial isotropically consolidated undrained tests. The graph is prepared with principal stress difference $\sigma_1 - \sigma_3$ and pore pressure U along the ordinate and axial strain ε_a along the abscissa. Most of the relationship of principal stress difference and axial strain has a hardening tendency and has no definite peak strength. The pore water pressure increases initially then decreases. The Mohr circles in terms of total stress and the failure envelope are drawn in Fig.12.
Fig.13 shows the result of triaxial isotropically consolidated drained tests. The graph is prepared with principal stress difference $\sigma_1 - \sigma_3$ and volumetric strain ε_v along the ordinate and axial strain ε_a along the abscissa. The characteristics are very similar to those exhibited by common loose sands. It is evident that the maximum strength depends upon a confining pressure.
At lower confining pressure, the volume of the specimen reduces initially but then expand, especially below

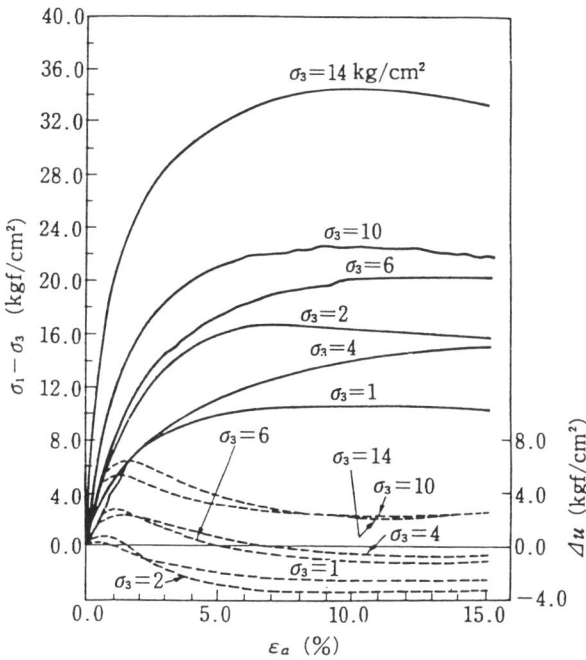

Figure 11 Results of triaxual isotropically consolidated undrained test(Akashi formation)

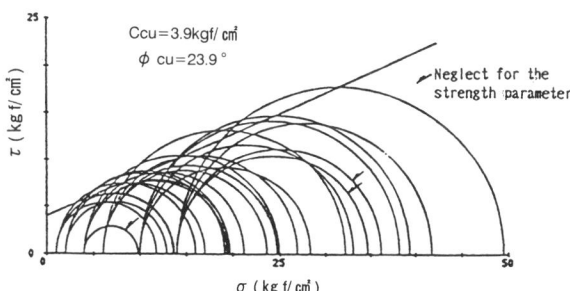

Figure 12 Mohr Circles in total Stress(Akashi Formation)

6kgf/cm². However, under higher pressures, the volume of specimen gradually reduces. The Mohr circles in terms of effective stress and the failure envelope are drawn in Fig.14.
Fig.15 shows the result of isotropically unconsolidated undrained tests. First, the chamber confining pressure of 4kgf/cm² is applied, after which the stress difference $\sigma_1 - \sigma_3$ is increased until failure occurs in unconsolidated undrained loading condition. The stress difference at failure is almost the same irrespective of the confining pressure. This refers to as the shear strength based on $\phi=0$ concept. It can not be assumed that the shear strength of rock changes with a corresponding change in the forces acting during earthquake.

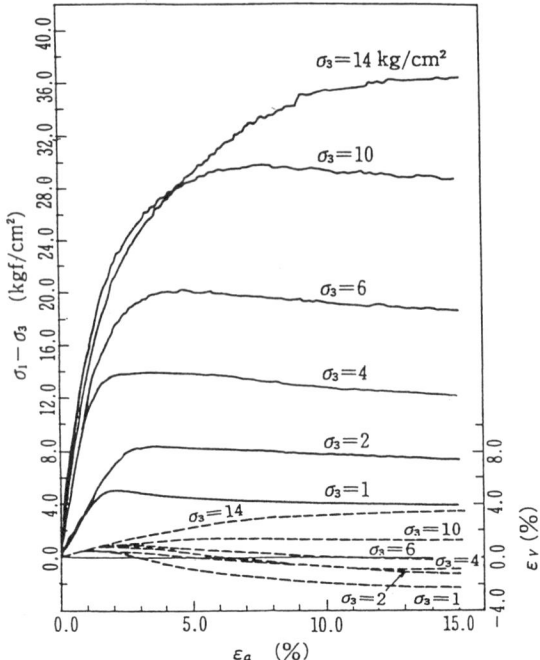

Figure 13 Results of triaxial isotropically consolidated drained test (Akashi Formation)

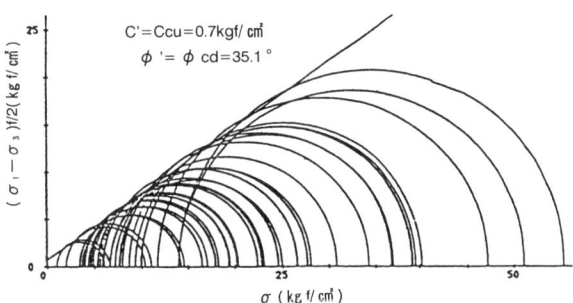

Figure 14 Mohr Circles in Effective Stress (Akashi Formation)

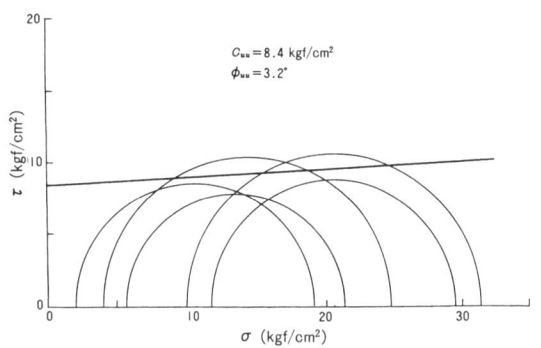

Figure 15 Mohr circles of unconsolidated undrained triaxial test (Akashi Formation)

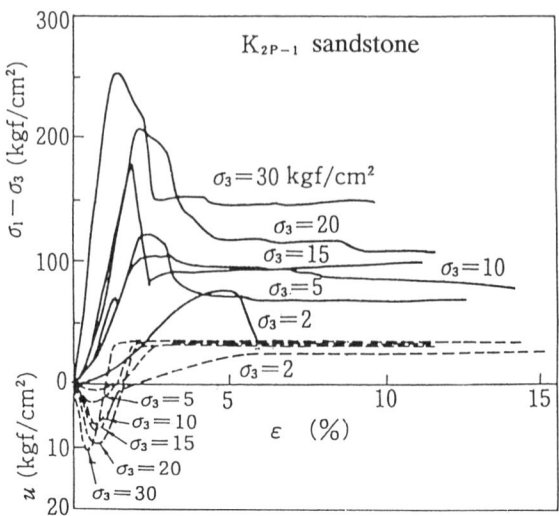

Figure 16 Results of triaxial isotropically consolidated undrained test (Sandstone in Group)

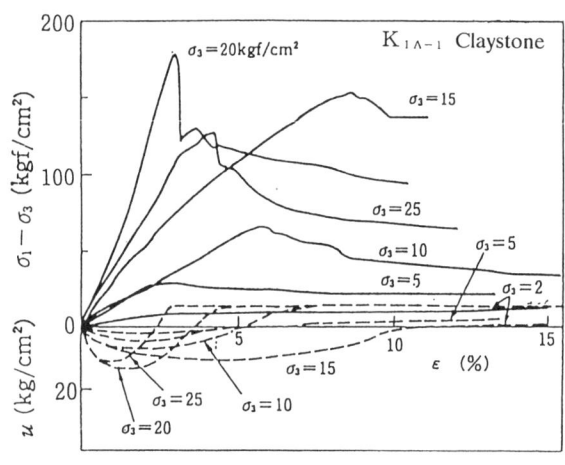

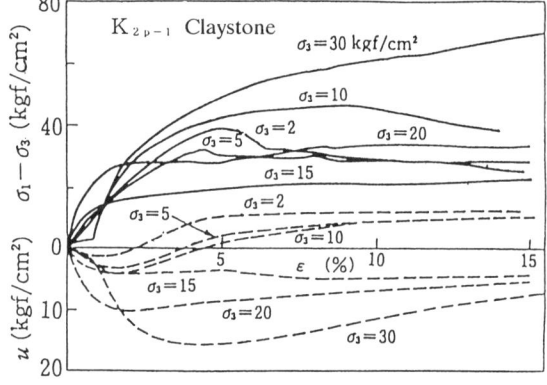

Figure 17 Results of triaxial isotropscally consolidated undrained test (Claystone in Kobe Group)

Kobe Group

Fig.16 and Fig.17 exhibit the result of triaxial isotropically consolidated undrained tests. The graph is prepared with principal stress difference $\sigma_1 - \sigma_3$ and pore pressure U along the ordinate and axial strain ε_a along the abscissa. The sandstone, as shown in Fig.16, has high strength in comparison with the claystone and this behaviour seems to be characteristic of over-consolidated soils. The stress difference increases rapidly up to a peak value at about an axial strain of 2.0%, thereafter a softening process takes place and finally some "critical state" is approached. On the other hand, the strength of the claystone, as shown in Fig.17, varies due to place and depth. The harder rocks among them are characterized by a strain-softening behaviour while the soft rocks of them exhibit a strain-hardning prosperity.

c Deformation properties

To investigate the deformation properties of Akashi Formation, the pressuremeter test in borehole, the PS logging in borehole and the laboratory test on undisturbed sample were employed. The laboratory tests include the triaxial compression test, the creep triaxial loading test and the cyclic triaxial test.

Akashi Formation
The deformation modulus Esb obtained from the pressuremeter varies from $5 * 10^2$ to $3.6 * 10^3$ kgf/cm^2. The value was assumed to increase with depth. The elastic shear modulus Gmax by the PS logging is in the range of $3 * 10^2$ to $3 * 10^3$ kgf/cm^2. There was no variation with depth.
The stress-strain curves by the triaxial compressive drained test yielded the straight line up to the strain of 1.5% as shown in Fig.13. From this figure it may be inferred that the deformation modulus Ec is dependent of the confining pressures.
Fig.18 shows the relationship of strain rate and time obtained by the creep triaxial tests. It is possible to represent it with an analogical model which is a combination in series of "Spring-Voigt Body".
The creep constants were obtained based on these figures.

Kobe Group
The deformation modulus Esb obtained from the pressuremeter varies from $5 * 10^2$ to $1 * 10^4$ kgf/cm^2 with an average of $3 * 10^3$ kgf/cm^2. The value was assumed to increase with depth.
Fig.19(a),(b) exhibits the results of the triaxial compressive drained test on sandstone and claystone. The sandstone has high peak strength with the strain of 1.0 to 1.5% in comparison with the claystone.

d Dynamic deformation characteristics

The dynamic deformation characteristics of samples were evaluated principally with the cyclic triaxial loading test. Fig.20 presents the variation of the nomalized dynamic shear modulus curves($G/Gmax - \gamma$) and dynamic shear damping curves($D \sim \gamma$). In the range of low level strain from $1 * 10^{-6}$ to $2 * 10^{-5}$ dynamic shear modulus remains almost constant, but at strain levels greater it becomes lower and lower, while the damping ratio gradually increases.

The shear modulus G is lower and the damping ratio D is higher for the Akashi Formation than for the Kobe Group in the large strain region with an axial strain more than 10^{-4}. The influence of confining pressure on both shear modulus G and damping ratio D is clearly shown in Fig.21. The larger the confining pressure, the greater the shearing modulus.

5 Application of the above results to the foundation design

(1) Moddelling of bedrock

Moddelling of bedrock is important for the foundation design of the bridge.
At first, bedrock is devided into several zones in terms of rock types and characteristics. After zoning the bedrock, geotechnical constants are assigned on the basis of available information. Lastly, the basic ground model is modified into design ground model depending on method of analysis.

a Zoning of bedrock

Bedrock is grouped into several divisions with available information such as core observation, logging data and pressuremeter value. Thereafter, grouping is adjusted the boundary and thickness focussing on the laboratory test results.

b Determination of representative geotechnical constants of each zone

Akashi Formation is relatively uniform, while Kobe Group is very complex even if in the same group. Therefore, "mean value method" was employed to decide the geotechnical value for each zone. This is the method

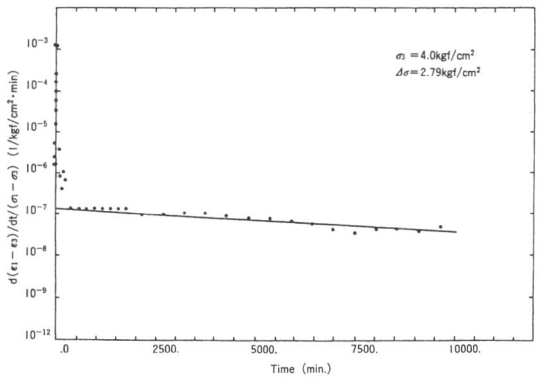

Figure 18   Creep rate — time relation ship (Akashi Formation)

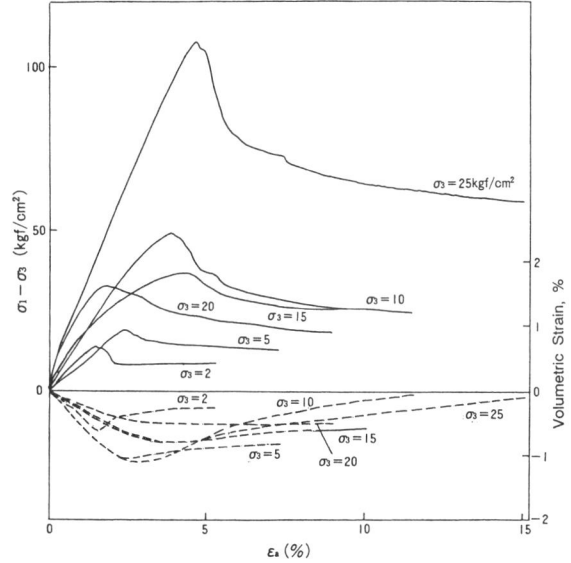

Figure 19(a)   Results of triaxial isotropically consolidated drained test (Sandstone in Kobe Group)

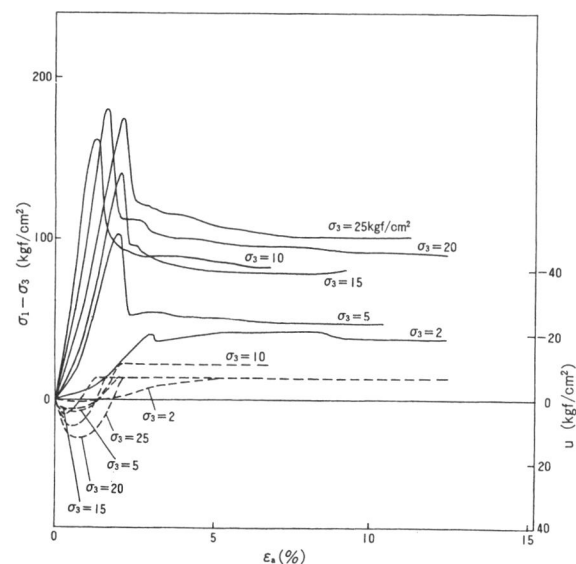

Figure 19(b)   Results of triaxial isotropically consolidated undrained test (Claystone in kobe Group)

to determine the constants from the representative values of each rock type considering a component ratio of sandstone and claystone.

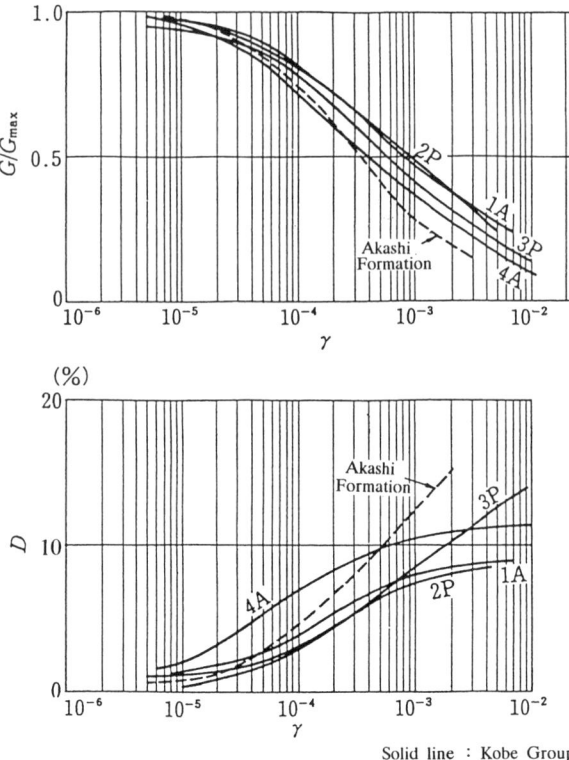

Figure 20 Shear modulus G and damping ratio D from a sequence of cyclis loading tests

c Design ground model (basic ground model, earthquake ground model)

Design ground model(basic ground model, earthquake ground model) is induced from geotechnical model. The basic ground model is accepted as a valuable tool for the stability analysis of foundation and the deformation analysis of bedock. Whereas the earthquake ground model is provided for the stability analysis of foundation during earthquake.

Basic ground model is made considering geological constitute and their properties as they are. However, some analytical method requests that the zoning is idealized by assuming that it is horizontally layered.

(2) Selection of the analytical methods for foundation stability and bedrock displacement

The foundation stability was checked on the modified Bishop Method and the Rigid Bodies Spring Model(RBSM). The deformation analysis for the bedrock was executed by a visco-elastic FEM with a series of spring-voigt model, elastic FEM and simplified creep calculation methods.

(3) Determination of design geotechnical constants

The parameters required for the foundation analysis and their testing methods used to obtain these parameters are summarized in Table 4. As shown in the table, constants are derived from the results of physical tests, triaxial tests, pressuremeter tests, PS logging and so on. In which design shearing constants was determined through a technical judgement considered a progressive failure, anisotropy, size effect, creep strength and so forth.

One of items to be considered is progressive failure.
A strain yields in the bedrock upon an application of the load. With increasing of the load up to limit, failure occurs thereafter it gradually progresses along a slip surface. Namely, each point in the bedrock does not exhibit the peak strength at the same time and exhibits the strengh correspond to the strain. Progressive failure should be particularly considered for the over-consolidated soil as shown in Fig.22.

Fig.23 shows scale effects in shear strength of joints in quartz-diorite. Shear strength, as the joint area increases, decreases to a nearly constant value known as the residual strength. Another quite interesting output from the shearing test on the quartz-diorite is described in Fig. 24. Shearing strength of joint was determined for $2 * 10^2 - 5 * 10^3$ cm^2 in area. Peak shearing strength decreases with the incresing size of the specimen. These slopes reveal almost the same. The decreasing of the strength appears to be caused by the decreasing of the apparent cohesion. It seems to be considered for the design of large scale structure.

Creep strength was also one of our concern. In creep

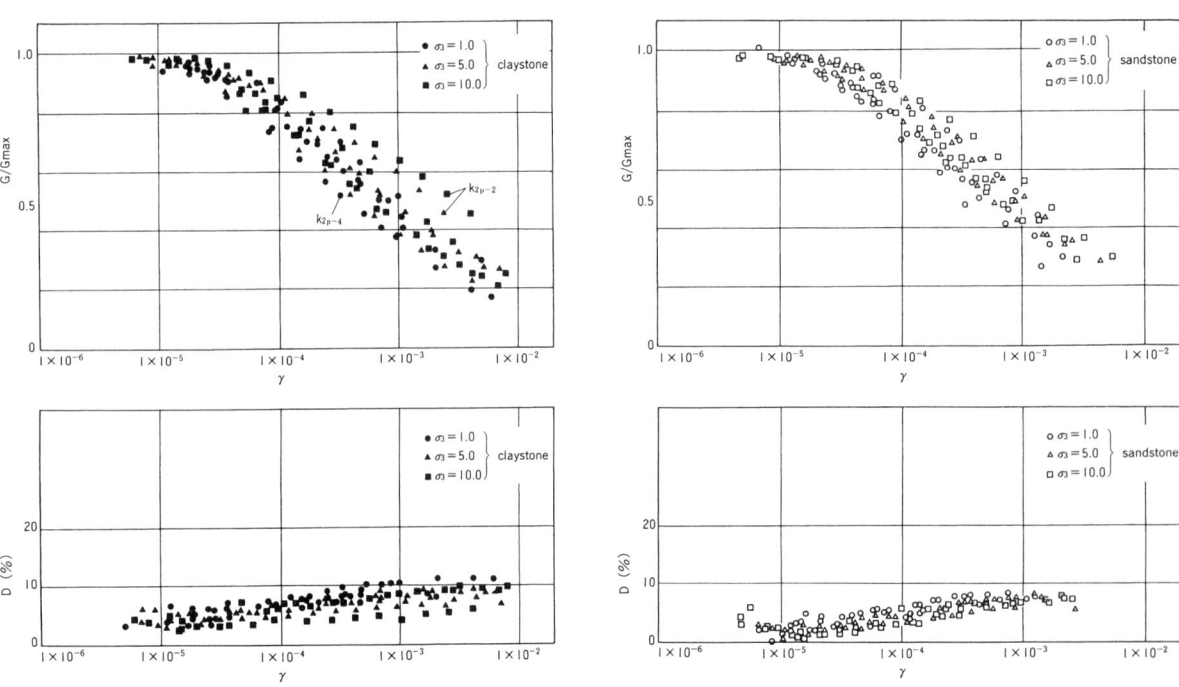

Figure 21 Strain dependency of G/G$_{max}$ and D(Kobe Group at 2P)

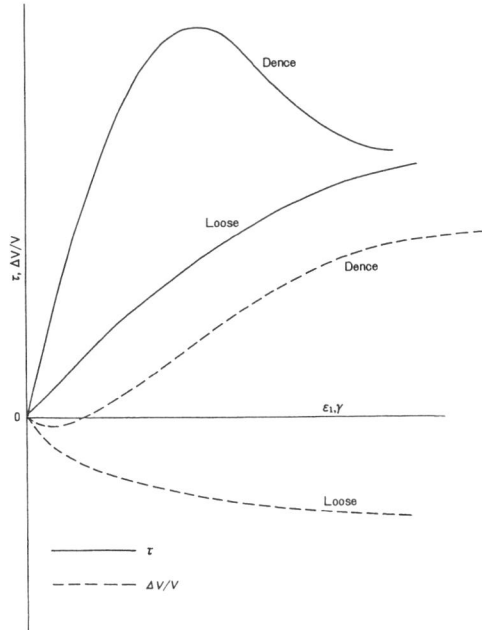

Figure 22 Drained triaxial test results for sand

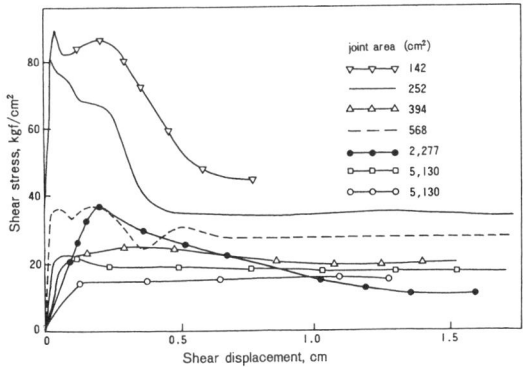

Figure 23 Scale effects in the shear strength of joints in quartz diorite (After Pratt, Biack, Brace, 1974)

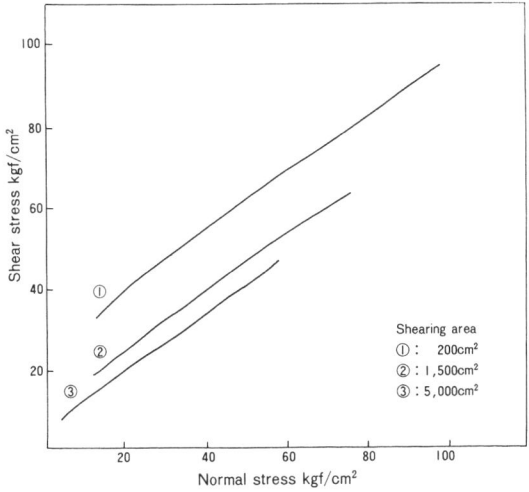

Figure 24 Scale effects in the shear strength of joints (After Pratt, Biack, 1974)

loading some sample deforms with the time and then collapses under the fairly lower stress than failure strength. Therefore after confirming the fact by the creep tests it was considered for determination of constants.
Geotechnical design parameters were determined according to the flow chart shown in Fig.25.
Table 4 shows summary of geotechnical properties and design constants at 2P bedrock.

6 Evaluation of design geotechnical constants from settlement measurement

Settlement measurement was conducted during construction of the main pier foundation. The objective of such measurement was to ensure that detrimental deformation does not occur in the ground during construction work as well as to predict subsequent displacement of the pier foundation. Another objective was to establish a prediction method for deformation through settlement measurement.
With the emphasis on the establishment of the deformation prediction method, the method is described below for the determination of the design geotechnical constants.

(1) Settlement measurement system

Sliding micrometers were set in the ground at the position corresponding to the center of the foundation and the deformation of the bedrock due to loading was measured. From these results a back analysis was performed to provide feedback for the construction work. At the same time, a measuring point was installed atop the foundation to provide a confirmation of deformation or inclination of the foundation. Sliding micrometers were installed for approximately 120m from the foundation level. These micrometers have special probes measuring changes in the mutual distance between measuring marks located at one meter intervals. (See Figure 26).

(2) Settlement measurement results and their interpretation

1) Ground deformation and foundation displacement

In Fig.27 is shown the relationship between the measurement results obtained by the sliding micrometers and the stresses acting on the foundation level.
The record of the foundation displacement obtained from the measurements is also indicated by black points on this figure.
On comparing the measurement data obtained from the sliding micrometer with the displacement data measured by survey it was found that both indicate approximately the same values, but that the displacement record by survey tends to give larger values at an initially stage. Thereafter, this trend has been reversed subsequent to completion of the placing of the inner aquaconcrete. Such trend was also observed at the behavioral measurements for the 3P foundation. This phenomenon is inferred to be the result of the fact that the lower surface of the concrete of the inner part was deflected at an initially stage and with placing of the outer part concrete the lower surface of the concrete became level by the so-called effect of a counterweight fill.

2) Vertical strain distribution

Fig.28 shows the vertical strain distribution obtained upon completion of concrete placing at 2P and 3P. The maximum strain is observed in the vicinity of the foundation with the order from 10^{-3} to 10^{-2}. The strain decreases with increasing depth and it exhibits the order of 10^{-5} at the depth of 120m(T.P.−180m) from the foundation level, that is, at the depth of about 1.5 times of the foundation width.
Somewhat different trend is observed with regard to the strain distribution at both sites. At 2P site, in the Akashi Formation the strain remains constant except 7m

43

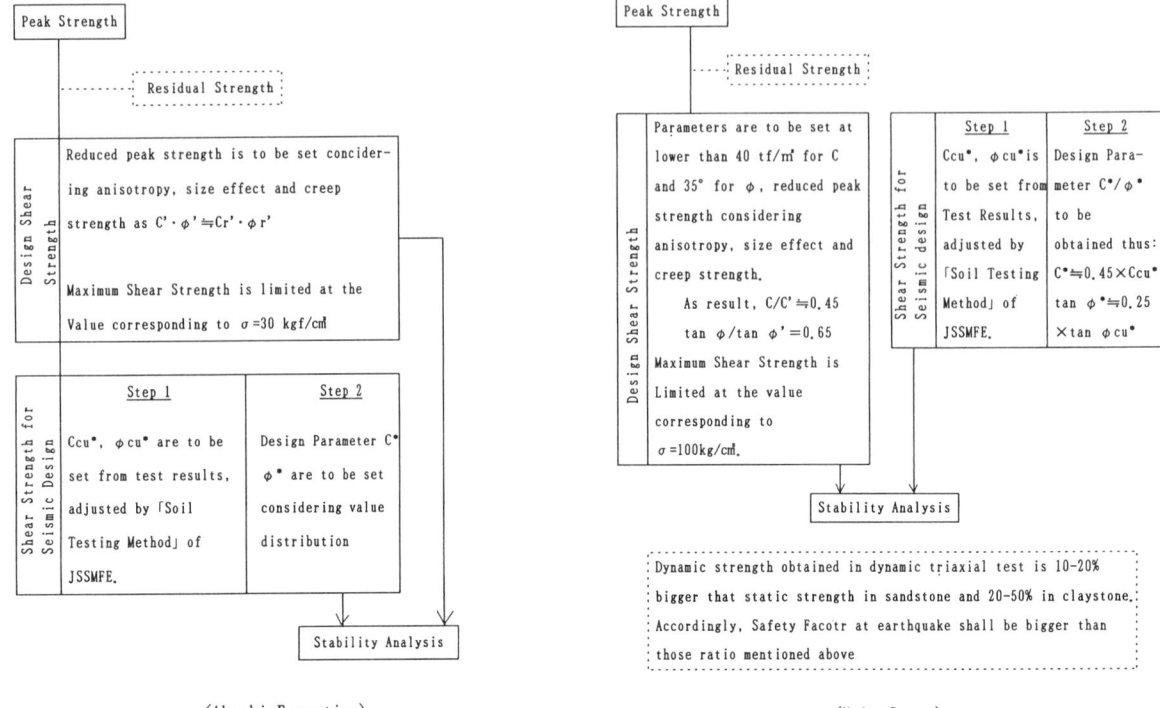

(Akashi Formation)　　　　　　　　　　　　(Kobe Group)

Figure 25　Flow of Determination on Shear Strength Parameter for Design

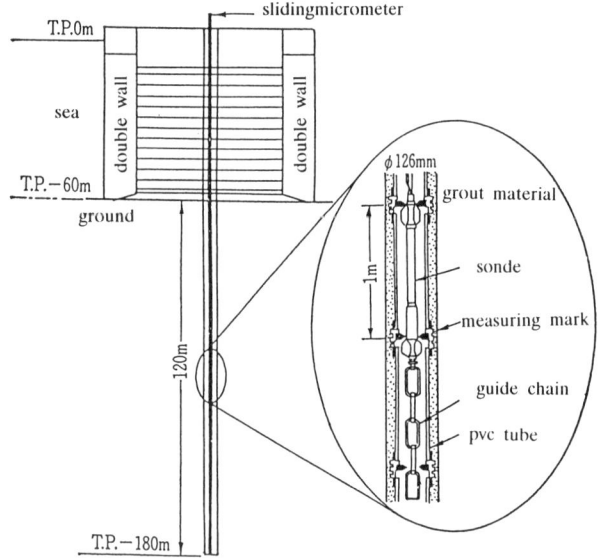

Figure 26　Setting layout of slidingmicrometer

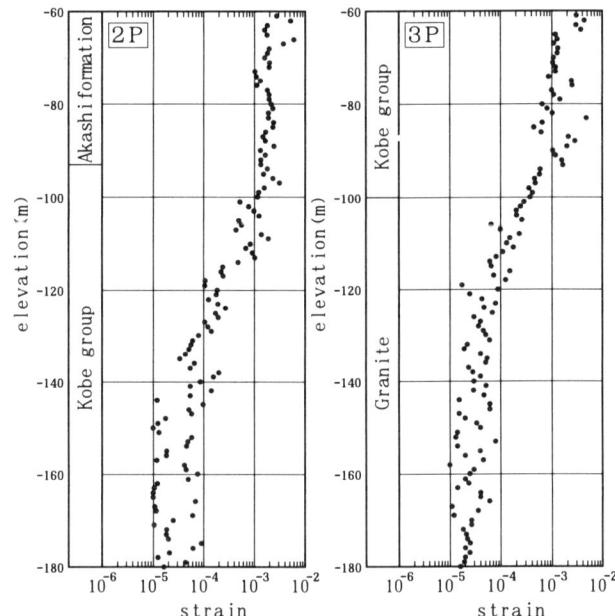

Figure 28　Strain distribution in depth of ground by slidingmicrometer at the center of foundation

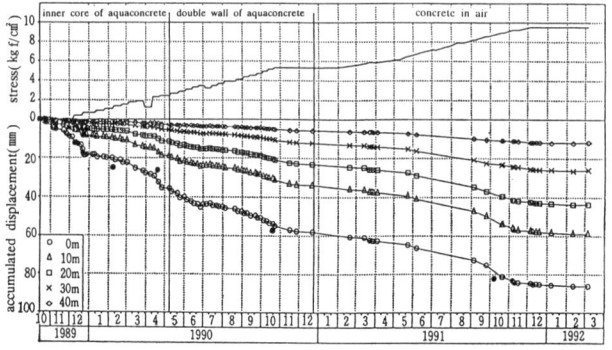

Figure 27　Time depending curve of ground displacement at 2P

section beneath the foundation while in the Kobe Group the strain decreases with the depth up to T.P.−140m. At lower level the values derived from inclinometer measurements is subjected to a scatter. On the other hand at 3P site, in Kobe Group to a depth of about 40m from the foundation level the strain is almost constant like measurements at 2P with a scatter at the depth of between 15m and 33m from the foundation level. The scatter is inferred due to alternating layers. In the granite the strain value scatters at the order of 10^{-4} to 10^{-5}. From these two case histories a general strain distribution for the displacement analysis seems to be available.

44

Table 4 Basic Ground Model for 2P Site

Geology		Elevation T.P. -m	Thickness m	Physical Parameter					Strength Parameter				Deformation Parameter						
									Usual condition		Special condition		Usual Condition				Special Condition	
Stratum	Unit			γ_{sat}	γ'	Vs	Vp	ν_0					Modulus of Deformation		Creep Parameter			Modulus of Deformation	
									Ccd	ϕcd	Ccu*	ϕcu*	Es	νs	G*	G*/G$_s$	G$_s$/η_s	Ed	ν_0
				tf/m³	tf/m³	m/s	m/s		tf/m²	·	tf/m²	·	tf/m²		tf/m²		1/min	tf/m²	
Recent~Upper Pleistocene Deposits	Al ~Du	46 53	7	1.96	0.96	360	1,700	0.48	-	-	-	-	8,400	0.38	-	-	-	16,800	0.38
Akashi F.	AK	92	39	2.15	1.15	450	1,800	0.47	5	35	50	30	46,000	0.33	4,500	0.08	1.95 x10⁻⁴	92,000	0.33
Kobe G.	K2p-1	120	28	2.27	1.27	720	2,200	0.44	25	30	130	20	79,000	0.39	21,900	0.13	0.16 x10⁻⁴	150,000	0.39
	K2p-2	150	30	2.24	1.24	710	2,200	0.44	35	25	100	20	83,000	0.33	20,700	0.21	0.08 x10⁻⁴	166,000	0.33
	K2p-3	212	62	2.28	1.28	880	2,400	0.43	40	30	220	15	86,000	0.32	27,300	0.10	0.31 x10⁻⁴	172,000	0.32
	K2p-4	258	46	2.35	1.35	1,100	3,200	0.43	40	30	305	30	90,000	0.36	33,100	0.10	0.31 x10⁻⁴	180,000	0.36
weathered Granite	Gr'	268	10	2.35	1.35	1,200	4,000	0.45	10	37	Ccd=10	ϕcd=37	90,000	0.41	31,900	0.50	5.3 x10⁻¹⁰	180,000	0.41
Granite	Gr		-	2.60	1.60	2,000	4,500	0.38	50	40	Ccd=50	ϕcd=40	100,000	0.25	64,000	0.50	4.9 x10⁻¹²	200,000	0.25

Note : Special Conditions - Earthquake, Typhoon, Vessel Collision

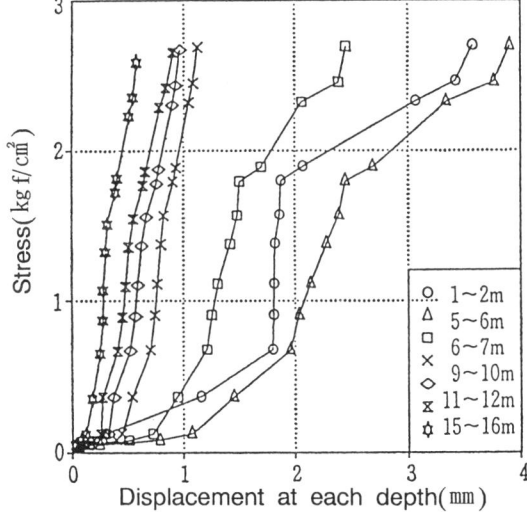

Figure 29 Relationships between stress and displacemnt

3) The extent of ground relaxation due to rebound
Fig.29 shows the relationship between the vertical stress within the ground at each depth by pseudo-elastic back analysis and the ground deformation for a 1m thick layer at that depth. The considerable ground deformation occurs in the vicinity of the foundation at the initial stage of concrete placing. As the load increases this tendency disappears. This seems to be due to the recompaction of ground. Approximately 14m was excavated from the original seabed to the foundation level. At 2P the extent of ground relaxation was assumed approximately 7m below the foundation level, while at 3P it was presumed to be about 4m. The displacement data taken at the time of initial reconsolidation was putted out of the analysis.

4) Region of influence of stress resulting from the foundation
Fig.30 shows the vertical displacement distribution at 2P.

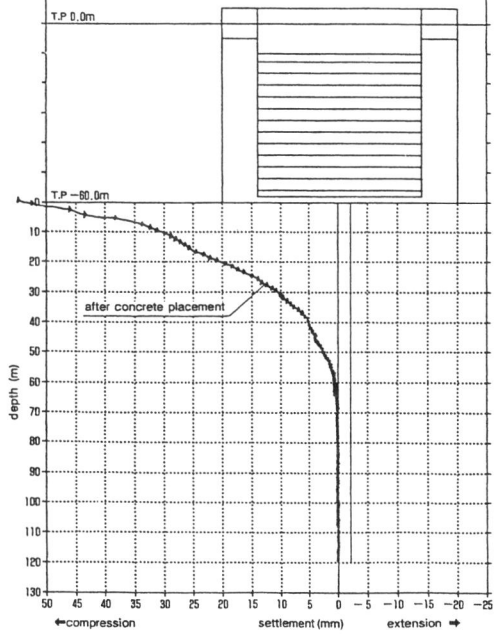

Figure 30 Settlement distribution(2P)

In the region extending a distance of 10m from the foundation level, i.e., a distance equivalent to 1/8 of the foundation width, 30% of the total settlement occurred, and 50% of the settlement occurred in the region extending 20m (1/4 of the foundation width). Approximately 70% of the total settlement occurred within the Akashi stratum.
From these facts, it is assumed that the major influence area due to the foundation will be within the depth equal to the foundation width from the foundation level.

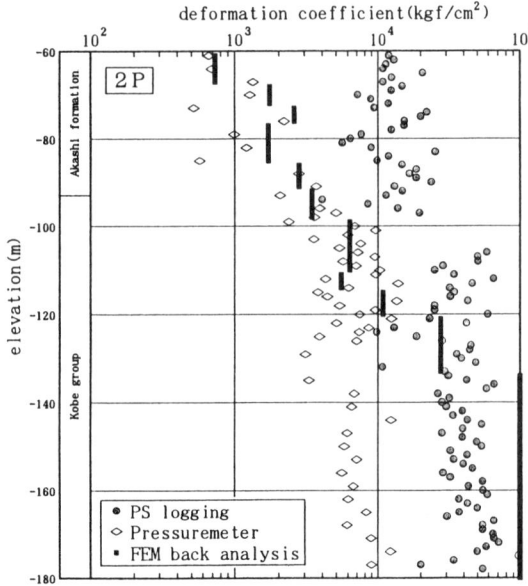

Figure 31(a) Comparison between deformation coefficient by various method at 2P

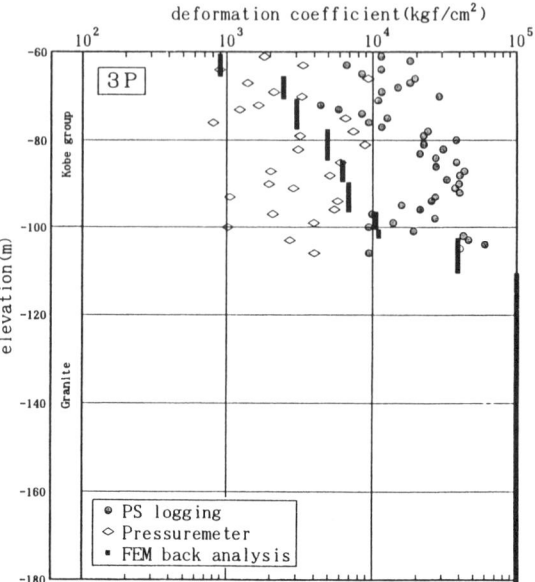

Figure 31(b) Comparison between deformation coefficient by various method at 3P

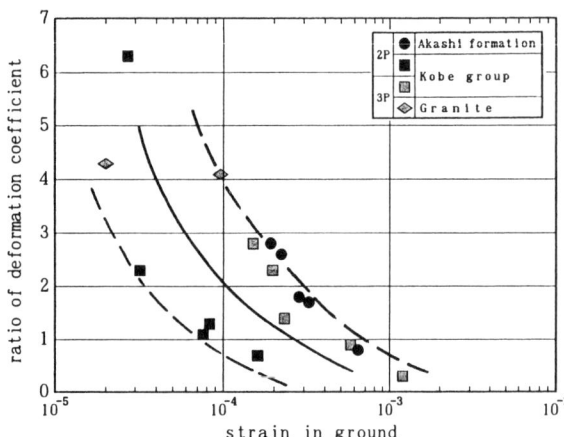

Figure 32 Relationship between deformation coefficient by FEM backanalyzed and pressuremeter

(3) Comparison of the results of back analysis and geotechnical investigation

To find a method of setting the suitable design geotechnical constants, the comparison was made of the results of back analysis and geotechnical investigation. The back analysis was performed based on the measurement data obtained on completion of concrete placing. Fig.31(a) and 31(b) shows the combined results of the deformation coefficient (elastic coefficient) derived from pressuremeter test, PS logging and the back analysis. At 2P, at depth of 20−60m, back analysis value seems to be close to the deformation coefficient obtained from the pressuremeter tests. However, their gap becomes gradually larger with increasing depth.

In Fig. 32, the ratio of the deformation coefficient obtained from back analysis and pressuremeter tests is plotted against the strain measured at the related depths. As the strain within the ground becomes smaller the ratio becomes larger.

(4) Method of determining design geotechnical constants for displacement prediction

The determination of the design geotechnical constants is of great importance in the prediction of the displacement of structure.

A method of determining the suitable design constants from the deformation coefficient based on behavioral measurements is proposed below.

Two methods for making such a determination may be available.

1) Method 1 (Using the results of pressuremeter tests)

① Provision of the strain distribution
Firstly the strain distribution within the ground is provided referring to the strain distribution obtained by field measurement as shown in Fig.28.
② Determination of the design geotechnical constants
By multiplying the ratio corresponding to the strain in Fig.32 to the value of pressuremeter test, the design geotechnical constants at its depth are obtained.

2) Method 2 (Using the results of laboratory cyclic deformation tests and PS logging)

Provision of the strain distribution
① The strain distribution beneath the foundation is derived from Fig.28 in the same way as described for Method 1 above.
② Representation of the relation between the normalized coefficient and the strain
The relation of the normalized coefficient E/E_{max} and the strain γ is obtained from cyclic loading tests for each geological zone as shown in Fig.21.
③ Determination of the design geotechnical constants
After assuming the strain distribution, utilizing such a relationship as shown in Fig.21, the deformation coefficient, i.e, the design geotechnical constant E at each zone is determined

7 Conclusions

Several conclusions can be made through the geotechnical investigation, foundation design and displacement measurement during construction.
① Displacement measurement provided useful information on a strain distribution and effective stressaffected area within a bedrock for prediction of ground displacement.
② A geotechnical investigation was established for a large−scale bridge on soft rock across the straits.
③ A method of determining the suitable design constants from the deformation coefficient based on behavioral measurements is developed.

UHV pylon foundation design for mountainous areas – Establishment of design methods based on full scale tests

Yukio Yoshii
Tokyo Electric Power Company Inc., Japan

ABSTRACT: In 1993, the Tokyo Electric Power Company Inc. completed Japan's first Ultra High Voltage (UHV) transmission lines. The design conditions for the pylon foundations stipulated that the design loads be double those of previous pylons and that they must be constructed on steep slopes of sandy soil or soft rock. In order to verify the applicability of existing design methods and to establish a foundation construction that would be cost effective, work began to formulate a new set of design criteria. Based on a series of full scale tests, support formulae were established regarding the vertical and horizontal forces. In addition, an effective method for determining the physical property values of the ground, required in design, was established.

1. INTRODUCTION

The Tokyo Electric Power Co., Inc., (TEPCO), is the largest power company in Japan, and supplies electric power to 41 million people within an area of 39,000km^2, including Metropolitan Tokyo (Refer to Figure 1). The sales of electric power in FY 1993 were 232 billion kWh, which amounts to 33% of the total demand in Japan.

The power demand has more than doubled in the last 20 years. Since the mid-1970's, TEPCO has rigorously tried to expand the 500kV power transmission line network despite the extreme difficulty in expanding power transmission routes in Japan due to the limited availability of land. Because of this, and the fact that an increased number of 500kV power transmission lines requiring short-circuit capacity countermeasures would be required, it was decided to construct UHV (1000kV) transmission lines having a capacity 3 to 4 times greater than that of conventional 500kV transmission lines. During the period from 1988 to 1993, the first UHV line was completed to transmit power from the Kashiwazaki-Kariwa Nuclear Power Station to the Tokyo Metropolitan area some 250km away. The foundations for this power transmission line were designed under the conditions that the design loads be double those of existing pylons and that they would have to be constructed on steep slopes of sandy soil or soft rock. Two design problems were involved in meeting these conditions. The first problem was technical, namely, whether or not the existing design formulae could be applied in this particular case. Secondly, there was the financial problem which involved using the existing design due to the foundation size being considerably larger. For solving these two problems, we set about establishing a new set of design formulae for the foundations that would meet the conditions described above and would solve the aforementioned problems.

This report describes the results of the full scale tests that were carried out to derive the support formulae for the vertical and horizontal loads. In addition, an effective method for determining the physical property values of the ground, required in these formulae, is also described.

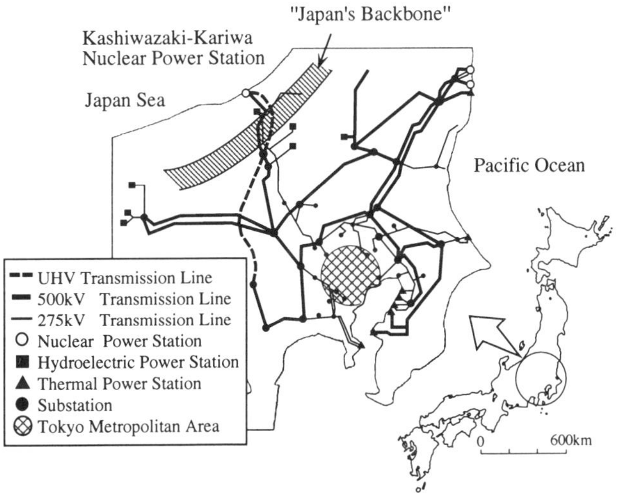

2. CONSTRUCTION OUTLINE

2.1 *Scale of the work*

For the UHV transmission line, 418 pylons are constructed over a distance of 250km. Each pylon is 110m high, 40m wide, 3.7MN in weight in average (Refer to Table 1), and supported on four foundations.

Among the 418 pylons, 359 comprise pier foundations (Refer to Figure 2) with a length of 20m and a diameter of 3.5m. The amount of concrete required for each foundation is approximately 150m^3.

Figure 1. Power transmission network

Table 1. Outline of UHV transmission lines

Length	250km
Voltage and number of circuits	1000kV design, 2 circuits
Electric wires	Aluminum conductors steel reinforced Size:610mm^2×8
Towers	Number : 418 units Height : 110m on average
Construction Period	1988 to 1993

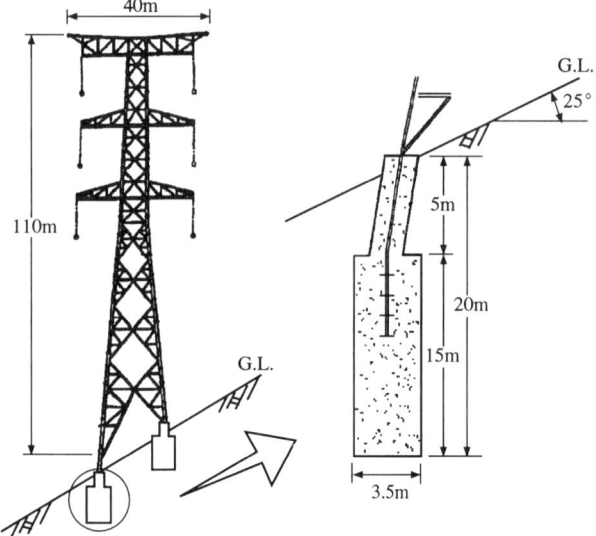

Figure 2. Outline of a pylon

2.2 Route topography

About 70% of the Japanese archipelago is mountainous. As such, the construction of this transmission line, which runs from the Japan Sea to the Tokyo Metropolitan area, involves traversing the 1500m and higher mountainous region known as "Japan's Backbone". The average slope at the construction site was 25°, the maximum being 50°. In some locations, work had to be stopped during the winter months due to snow more than 2m deep.

2.3 Route geology

The ground along the construction route comprises sandy soil and soft rock having a uniaxial compressive strength of 10MPa or less. Geologically, it can be divided into five regions from the north; a sedimentary rock region composed of sandstone and mudstone, a terrace deposit region, a tuff and andesite green tuff region, a slate and shale middle palaeozoic strata region, and a granite region.

2.4 Construction method

An area of about 5000m² is necessary for the construction of one pylon. The work was performed using methods suitable to mountain slopes. Excavation work was carried out using a pipe clamshell and manual labor. Pylons were set into the excavated hole, reinforcement was inserted, and concrete was placed to complete the foundation. The necessary construction materials were conveyed either by truck, cable suspension or by helicopter, with the method used for each particular site determined by the terrain conditions and economic considerations. As a result, trucks were used for 46% of the work, cable suspension for 44%, and helicopters for the remaining 10% of conveying.

3. LOADING

The load acting on the foundation resulted not only from the weight of the steel pylon, transmission cables and insulators, but also from the cable tension and wind loading. Wind loads were estimated at a standard wind velocity of 40m/s. For a given wind direction, three types of load were considered; compressive loads, lifting loads and lateral loads. The maximum compressive load was estimated to be 14MN, the maximum lifting load 11MN, and the maximum lateral loading 2MN.

These loads were nearly double the loads normally encountered with 500kV transmission cables and consequently, were beyond the limits of previous experience.

4. GEOLOGICAL SURVEY

The data required for design establishment comprised the geological structure and physical values such as shear strength, modulus of deformation and soil weight. As the route was 250km long and necessitated 418 pylons, the expense, in terms of both time and money, involved in mechanical testing, especially triaxial compression testing for each pylon, was considered prohibitive. Therefore, the method described hereinunder was adopted to overcome this problem (Refer to Figure 3).

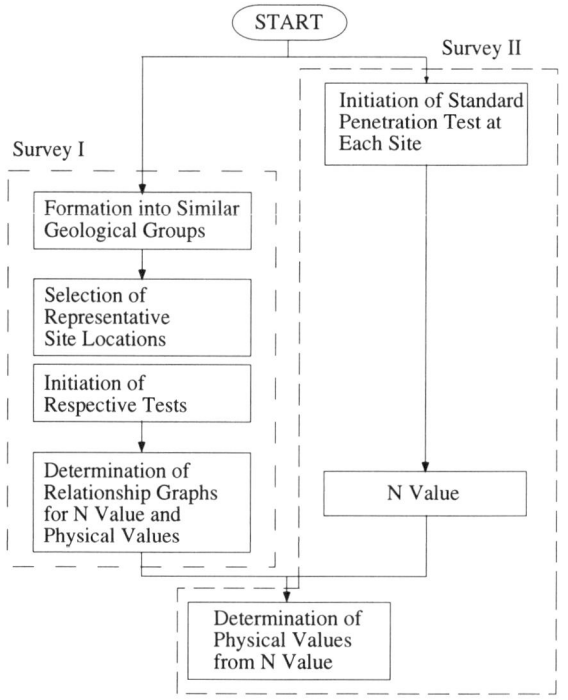

Figure 3. Geological survey flow chart

The method is based on incorporating the results of the standard penetration test, i.e. N value, to determine the physical constants for the ground indirectly. By measuring the physical properties and the N value at a representative site where a steel pylon was to be erected, a correlation diagram of the two was produced (Refer to Survey I). Then, by measuring only the N value at other similar sites, the diagram could be used to determine the relevant physical constants in a simple and cost effective manner (Refer to Survey II). In this way, a major portion of the mechanical testing was able to be omitted.

Survey I

STEP 1: Based on an analysis of related reference literature, areas having similar geological conditions were classified into 5 groups.
STEP 2: A site was chosen from each group in which the local terrain would enable conveyance of a boring machine to the location without difficulty. Then, the mechanical tests and the standard penetration test were carried out.

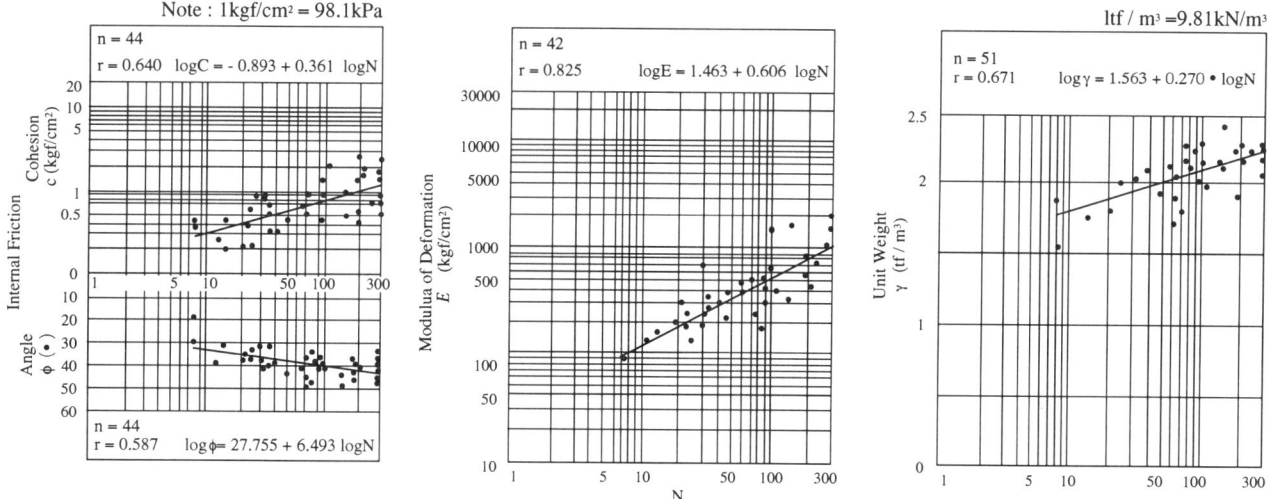

Figure 4. Ground physical values and N value correlation diagrams

STEP 3: Shear strength, modulus of deformation and unit weight of soil were found from triaxial compression tests, borehole loading tests and soil weighing, respectively.

STEP 4: A correlation diagram was devised to establish the relationships among cohesion (c), internal friction angle (ϕ), Young's modulus (E), unit weight (γ) and standard penetration test value (N). An example of this is shown in Figure 4.

Survey II

The standard penetration test was carried out at all pylon sites. Then, from the N values obtained, the physical constants were determined using the correlation diagram of STEP 4 in Survey I.

The structure of the ground at each pylon site was determined from the N value obtained in Survey II.

5. FOUNDATION DESIGN

5.1 Design to resist lateral forces

The dimensions of the foundation are determined from the magnitude of compressive, lifting and lateral forces. Of the three, lateral forces are particularly dominant on soft rock and sandy soil slopes. However, the support mechanism of foundations on severe mountainous slopes was not clear, so that in-situ tests were carried out to confirm this mechanism.

Two locations were chosen as test sites, one being soft rock and the other sand.

5.1.1 Horizontal loading tests in soft rock

(1) Test site
In most cases, the geology at the test area comprised volcanic-elastic rock, rhyolitic tuffy breccia from the Miocene epoch, and a covering layer of loamy soil of Pleistocene origin.

At depths of beyond 10m, the rock was rather hard, while at depths of less than 10m, the rock was extremely brittle due to the effects of progressive weathering. Soil properties found at the test site are shown in Table 2.

(2) Test set-up
As shown in Figure 5, 20° and 30° slopes were formed on a mountain ridge, and two RC foundations, one 10m deep and 3.0m in diameter (Test Pier I) and another 10m deep and 3.5m in diameter (Anchor Pier), were constructed.

Between the two piers, reaction beams, hydraulic jacks and PC steel bars were installed to serve as an alternating loading apparatus. Also shown in the figure are various instruments for measuring strain, pressure, deflection, and displacement of the piers and the surrounding soil.

(3) Characteristics of pier displacement
Figure 6 shows the load/pier head displacement curve up to the maximum loading level.

When the load reached 9.81MN, displacement of Test Pier I head rapidly increased. Strain data for reinforcing bars showed that, at the same loading levels, Test Pier I had not reached the highest level of resistance. Most of the ground around Test Pier I seemed to have reached a plastic condition. Displacement of the head of the Anchor Pier was a mere 33mm, indicating that the Anchor Pier had more than ample elastic strength. Therefore, it was determined that the Anchor Pier exerted no influence on the ground.

Table 2. Site geology and soil properties (I)

Depth (m)	Geology	Standard Penetration Test N-Value	Triaxial Compression Test			Plate Bearing Test
			Cohesion C_u(Mpa)	Internal Friction Angle ϕ_u (Degree)	Modulus of Deformation E_{50} (MPa)	Modulus of Deformating D (Mpa)
2	(a) Loam		0.013	24	8	
4	(b) Tuffy Clay		0.078	17	19	93
6	(c) Tuffy Breccia (Weathering)					422
8			0.26	51	324	
10	(d) Tuffy Breccia					2680

Note: 1Mpa = 10.197 kgf/cm²

3. When the load exceeded 8.83MN, the front of the pier was apparently thrust forward. Simultaneously, cracks(3) occurred in a direction perpendicular to radial cracks(2), and subsequently assumed a circular shape. Cracks(3) finally connected with cracks(1).

By visual inspection of the excavated test area, it was found that cracks(1) ran at an angle of about 45° to the horizontal surface, as shown in Figure 8(b). Cracks(3) cut vertically through the slope at a 30° angle, as shown in Figure 8(c). This angle approximately corresponds to that of the classical sliding surface, i.e., 45° - $\phi/2$, where ϕ = angle of internal friction of the ground.

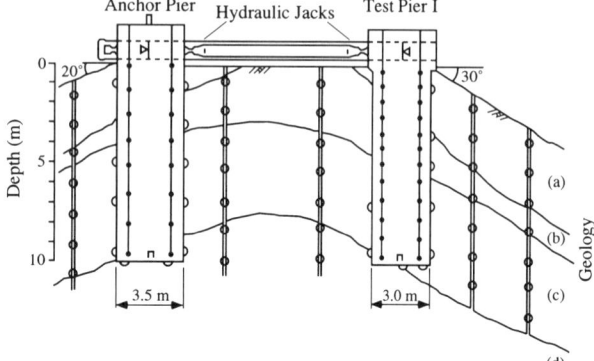

Figure 5. Testing apparatus and arrangement of instruments

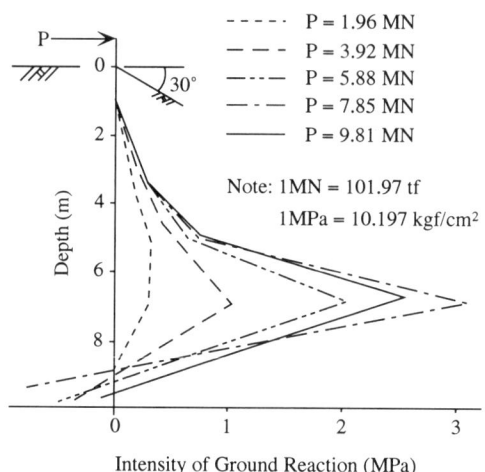

Figure 7. Distribution of ground reaction

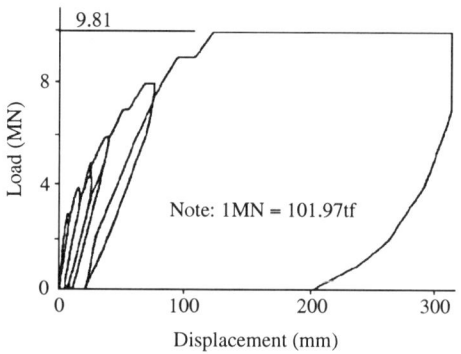

Figure 6. Load/Pier head displacement curve

(4) Characteristics of ground reaction

Figure 7 shows the distribution of ground reaction as measured by the earth pressure gauge buried in the front of Test Pier I. The ground reaction at each point initially increases linearly with the increase in load. However, the rate of increase declined for loads over 3.92MN at depths of 1.0m and 3.5m as well as for loads over 5.88MN at 5.0m. At 7.0m, ground reaction declined, amounting to 9.81MN. These data suggest that the plastic region of the soil extended gradually from the upper to the lower layers. At points between 8m ~ 9m, the reaction was in the direction opposite that of the reaction at the upper site of the pier. Thus, rotation of the pier must have occurred due to the rotational center being located within this region.

(5) Failure process of the ground

Figure 8(a) shows the final condition of the cracks occurring in the area of Test Pier I. In the figure, the different crack patterns are labelled(1), (2) and(3). The development process of these cracks is described below.
1. With a load of 3.43MN, the first tensile cracks(1) appeared, extending at an angle of 45° in relation to the load direction from the side of the pier.
2. The openings of cracks(1) enlarged as the load increased. Then, radial cracks(2) appeared towards the front of the Anchor Pier and the load increased from 4.41MN to 7.85MN.

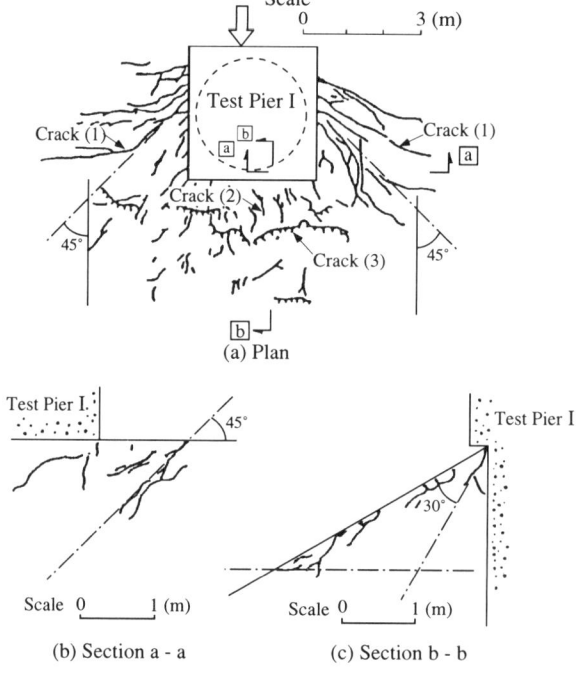

Figure 8. Crack occurrence in ground

5.1.2 Horizontal loading tests in sandy soil

(1) Test site
The geology of the test area consists of a sedimentary formation from the Pleistocene epoch. An upper layer of fine sand extends from the surface to a depth of 8.5m. Following this are alternate layers of silt and sand at depths from 8.5m through 12.0m. Below this is a lower fine sand layer at 12.0m. The upper fine sand layer has N values of 20 ~ 30, and is almost homogenous. The geology and soil properties of the site are as shown in Table 3.

Table 3. Site geology and soil properties (II)

Depth (m)	Geology	Standard Penetration Test N - Value 0 — 50	Triaxial Compression Test			Plate Bearing Test
			Cohesion C_D (Mpa)	Internal Friction Angle ϕ_D (Degree)	Modulus of Deformation E_{50} (MPa)	Modulus of Deformating D (Mpa)
2	(a) Fine Sand		0.029	35.2	40	27
4						
6			0.049	33.5	45	36
8						36
10	(b) Sand and Silt		0.049	36.9	69	
12	(c) Fine Sand		0.044	35.4	55	

Note: 1Mpa = 10.197 kgf/cm²

(2) Test set-up
30° slopes were created on a mountain slope. An RC pier foundation (Test Pier II), 10m in depth and 3.0m in diameter, was constructed, as shown in Figure 9. Instruments were placed around Test Pier II and in the ground as described in 5.1.1 (2).

(3) Characteristics of pier displacement
Figure 10 shows the load/pier head displacement curve up to the maximum loading level.
   Displacement of Pier Head II showed a marked increase when the load reached 4.66MN. Strain data for reinforcing bars showed that at the same loading levels, Test Pier II had ample elastic strength. It appears, therefore, that the ground around the pier had entered a plastic condition.

(4) Characteristics of ground reaction
Figure 11 shows the distribution of ground reaction measured by the earth pressure gauge buried at the front of Test Pier II.
   The ground reaction at each point increases with the increase in load. However, at a depth of 1.0m through 2.0m, the rate of increase drops for loads over 1.96MN. At 8.0m or more, the reaction is in the direction opposite that at the upper side of the pier. Thus, rotation, with the rotational center located at the depth of about 8.0m of the pier, must have taken place.

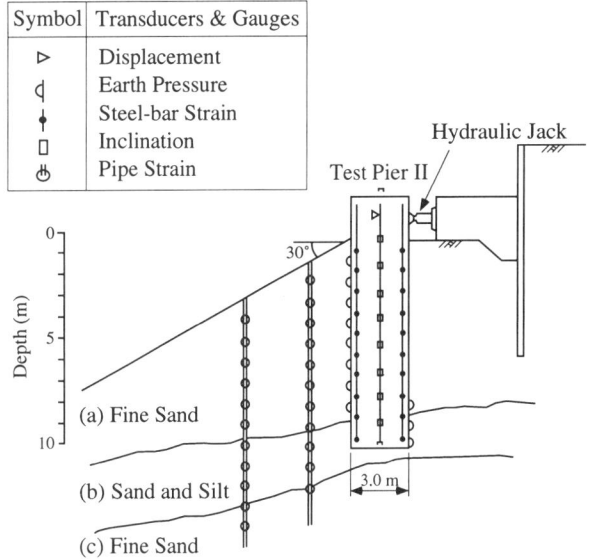

Figure 9. Testing apparatus and arrangement of instruments

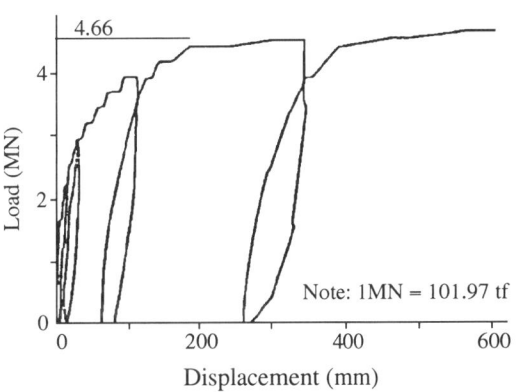

Figure 10. Load/Pier head displacement curve

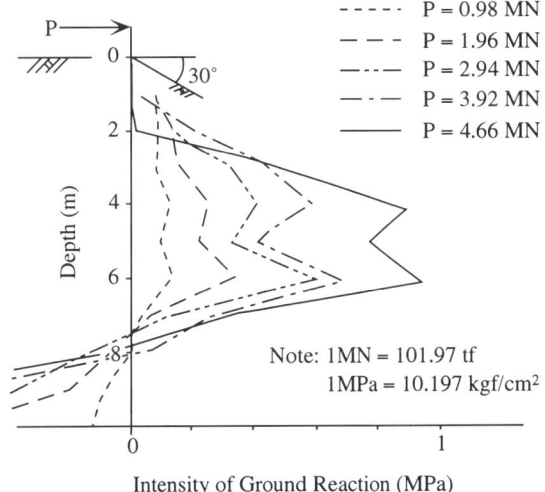

Figure 11. Distribution of ground reaction

(5) Failure process of the ground
Figure 12 shows the final condition of the cracks occurring in the area of Test Pier II. The different crack patterns are labelled (1), (2), (3) and (4) in the figure. The development process of these cracks is similar to that seen with Test Pier I, as shown in Figure 8. The difference in level between the inside and outside of (4) represents a few centimeters.

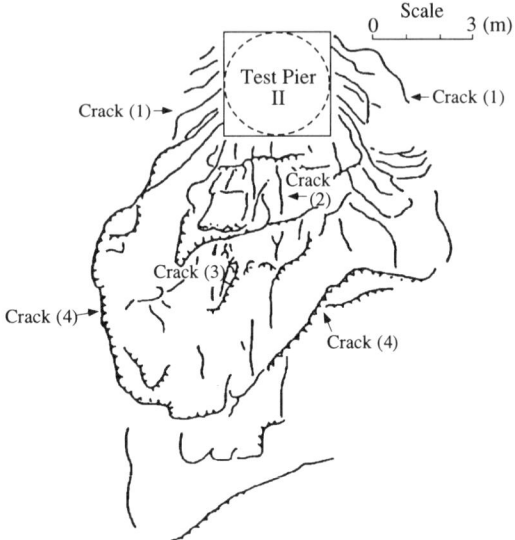

Figure 12. Crack occurrence in ground

5.1.3 *Calculation of ultimate bearing capacity*

(1) Ground sliding model
Observation of the cracking pattern shown by Test Piers I and II led to the assumption that the mass of soil, as seen in Figure 13, had thrust forwards. The outline of this mass of soil is as shown in Figure 13(b). It is described by lines extending from both sides of the pier at an angle of 45° to the load direction, and by lines whose width is three times greater than the diameter of the pier, drawn parallel to the load direction.

Figure 13(c) shows the vertical section to the load of the mass of soil. The sliding face to the surface is $45° + \phi/2 + \theta$, in relation to the vertical line.

(2) Bearing capacity and calculation results
The ultimate bearing capacity of the ground in front of the pier foundation can be expressed by applying the Mohr-Coulomb law of shear resistance to each sliding face, as shown in Figure 13.

From a practical point of view, the sliding mass of soil can be simplified as shown in Figure 14. It is assumed that a shear force developed only at the bottom sliding surface. In such a case, the ultimate bearing capacity, Fu, is simplified by the equilibrium of the shear force and the weight of the mass of soil as follows:

$$Fu = \frac{W_{(x)}(\cos\alpha + \tan\phi \cdot \sin\alpha) + C \cdot A}{\sin\alpha - \tan\phi \cdot \cos\alpha} \quad \quad (1)$$

where;
  $W_{(x)}$: Soil weight at depth X
  α : Angle between the bottom of the mass of soil and vertical line
    $(45° + \phi/2 + \theta)$
  θ : Angle of tilt of the ground surface
  C : Cohesion
  φ : Angle of internal friction
  A : Area of sliding face at the bottom of the mass of soil

Intensity of ultimate bearing capacity, pu, at each depth is obtained by taking the differential of Fu with respect to the depth, x, and dividing it by the diameter of the pier, D.

$$Pu = \frac{1}{D} \cdot \frac{dFu}{dx} \quad \quad (2)$$

Figure 15 shows the results of applying the values for C and f shown in Tables 2 and 3 to formulas (1) and (2). The intensity of ground reaction designated by the symbol "○" in this figure corresponds to the upper limit values of the ground reaction shown in Figures 7 and 11. The calculated values from the model in Figure 13 are shown in Figure 15 by the bold line, while the broken line shows the results obtained by the simplified model of Figure 14. While small differences can be seen in the values obtained from the simplified model in comparison to those obtained from actual measurement or the more precise model, these differences were considered too small to pose any problem in practical application. Therefore, the simplified model of ground failure, as shown in Figure 14, was adopted for use.

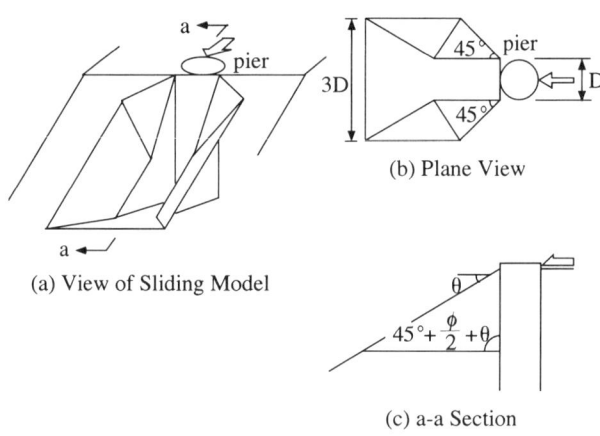

Figure 13. Precise sliding model

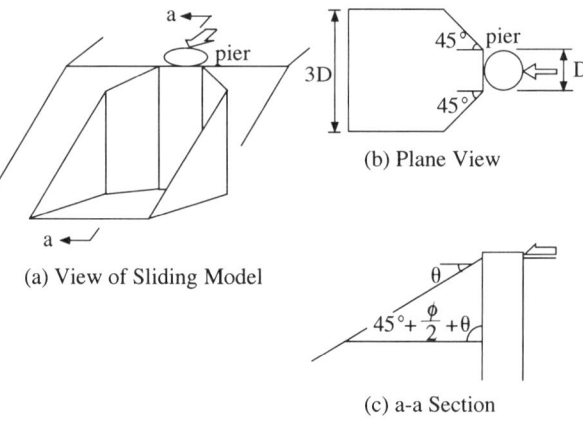

Figure 14. Simplified sliding model

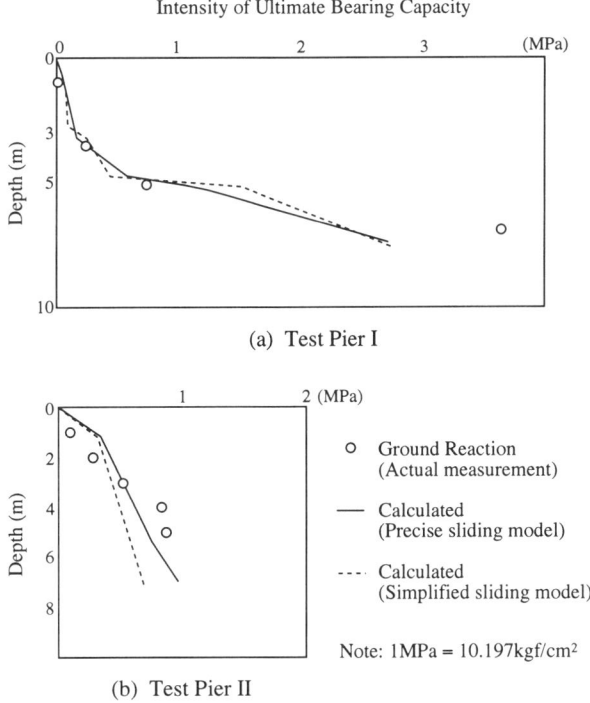

(a) Test Pier I

(b) Test Pier II

Note: 1MPa = 10.197kgf/cm²

Figure 15. Results of calculation for intensity of ultimate bearing capacity

5.1.4 Ground reaction force

In the case of hard rock, when the ground fails, the ground reaction forces decline due to the ground becoming plastic. However, no such decline was confirmed for the sandy soil or soft rock (Refer to Figure 16). When the ground reaction force was simulated using a bi-linear curve, the test load-displacement relationship showed good agreement, so that a bi-linear curve was adopted for the ground reaction force model used for the design (Refer to Figures 17 and 18).

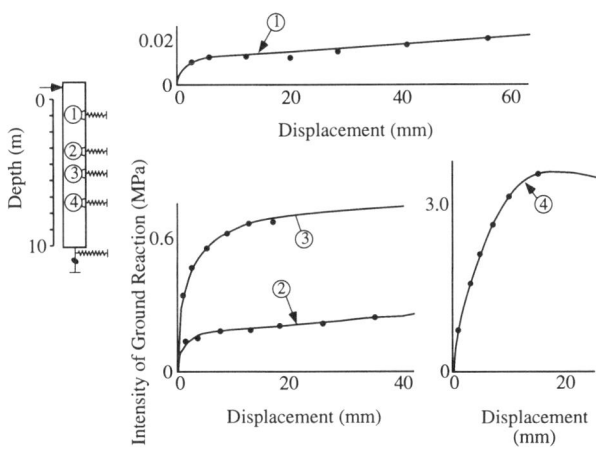

Note: 1MPa = 10.197kgf/cm²

Figure 16. Ground reaction-displacement curve (soft rock)

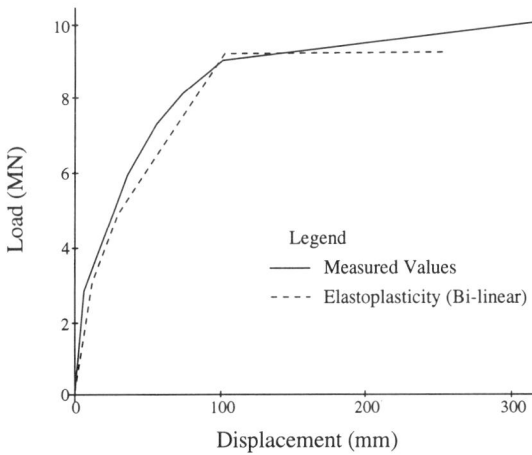

Figure 17. Load displacement curves of bi-linear model (soft rock)

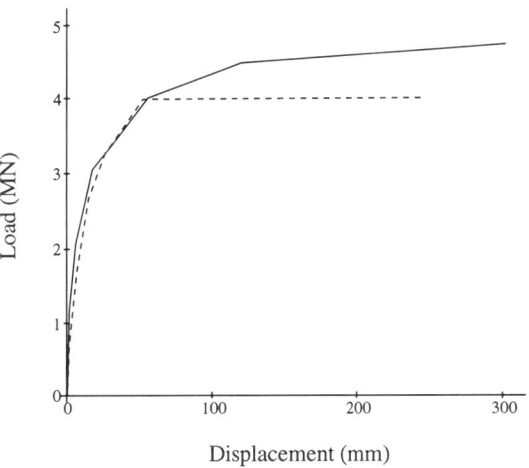

Figure 18. Load displacement curves of bi-linear model (sandy soil)

5.1.5 Security of the foundation

As the standards for determining the support force for the foundation, the conditions that (a) the elastic region of the ground exceeds the diameter of the foundation and (b) the region is more than 1/3 longer of the foundation length, must be fulfilled (Refer to Figure 19).

5.2 Design to withstand uplift forces

Regarding the design method for ensuring security of the foundation under the effect of uplift forces, it must be considered that the uplift force occurring on these transmission pylons is unique, as the design loads are greater than those of general pylons. Therefore, the pylons must be anchored deep within the ground.

Uplift loads on pylons are considerable when compared to other general structures, making it essential to take these loads into full account at the design stage. In the case of UHV pylons, this is particularly so. Therefore, design formulae for uplift were formulated based on study of the ground support mechanism and failure mechanism, and were determined by carrying out both centrifugal model tests and full scale tests. The methods used to determine the behavior of the foundation/ground are described below.

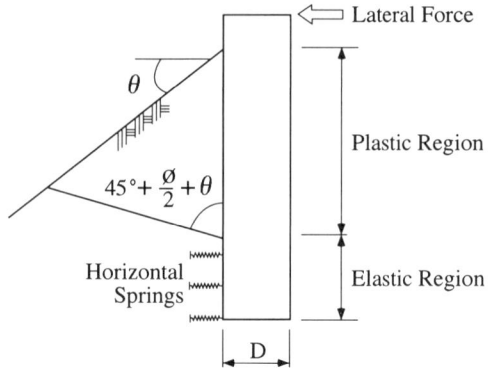

Figure 19. Foundation security

5.2.1 *Centrifugal model tests*

The general design methods for uplift forces are as shown in Figure 20.
  (1) Shear Method: ground fails along the foundation circumferential surface
  (2) Logarithmic Spiral Method: ground fails from the underside of the foundation in a logarithmic spiral mode
  (3) Earth Cone with Su Method: ground fails from the underside of the foundation in a 45° shaped mode

In order to verify the applicability of the above methods, centrifugal model tests were performed based on the scale and topographical conditions of the site, with the results shown in Figure 21. The logarithmic spiral failure mode occurs in the case of shallow foundations, while in the case of deep foundations, at deeper levels, it was found that a shear plane had formed close to the circumferential surface of the foundation, even though the same logarithmic spiral shear plane had formed near the surface.

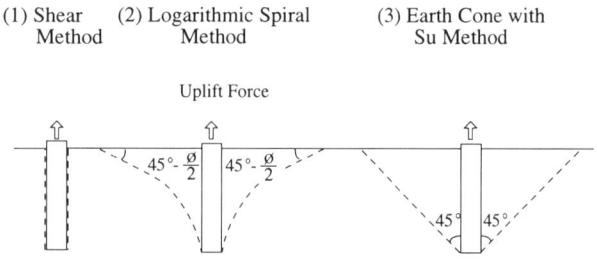

Figure 20. Design method for uplift support force

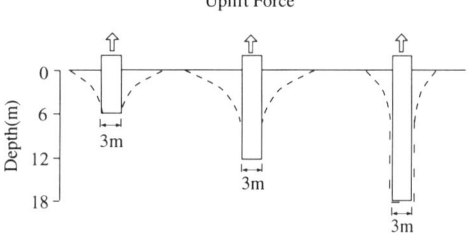

Figure 21. Failure mode (in centrifugal tests)

5.2.2 *Full scale tests*

To confirm the behavior of an actual foundation, full scale tests were performed at the four positions indicated in Table 4. Here, the latest test, involving the highest loading at position D-1, is described. The soil of the test site is weathered granite. As can be seen from Figure 22, an actual foundation (Test Pier III) having a diameter 2.5m and a length of 8m was sunk to a depth of 5m below the ground surface. Then, the structure was subjected to tensile loading via jacks attached to the upper portion of the Anchor Piers (5MN × 6 jacks).

Displacement gauges, inclinometers, strain gauges, mold gauges, earth pressure gauges, and circumferential surface friction gauges were employed to accurately determine the behavior of the foundation.

Physical values are as shown in Figure 22, with an outline of the measurement given in Figure 23.

Table 4. Table of test results

(Units: MN)

Test Case	Test Value	Calculated Value
	Lifting Force	Shear Method
A-1	4.0	3.8
-2	3.5	3.5
B-1	4.5	4.3
-2	5.0	4.4
-3	5.2	3.8
-4	11.3	7.6
C-1	2.2	2.3
-2	2.3	2.6
D-1	10.0	8.1

(1) Results of the full scale tests

As can be seen from the load/displacement curve shown in Figure 24, the vertical displacement of Test Pier III exceeded 93.5mm at an applied load of 17MN. From this point, displacement was continuous. Therefore, it was decided that this was the ultimate load. The yield load was obtained from the break point of P ~ $\Delta\delta/\Delta t$ and from $\log P$ ~ $\log \delta$. At the 1st break point load obtained from the first relation, the displacement accelerates, creep occurs, and partial failure of the ground is noticed. The 2nd break point load obtained from the latter relationship is considered to be the load that, if sustained for a long enough duration, entails the serious danger of creep failure of the ground. The 1st break point load was evaluated as Pf, and the 2nd break point load as Py.

The trends in the displacement distribution in the vicinity of the ultimate load, and the volumetric strain are represented together in Figure 25. As a result of assuming that the maximum tensile strain occurs on the failure plane, the shear plane occurs in close proximity to the foundation wall.

(2) Comparison of Test Results with Design Formulas

As a result of the failure mode tests, it was decided that the shear method is an appropriate basis for calculation. The comparison was made using not only the results of this present test, but also the results of three other tests previously carried out (Refer to Table 4). It was found that good agreement was achieved between the calculated and experimental values.

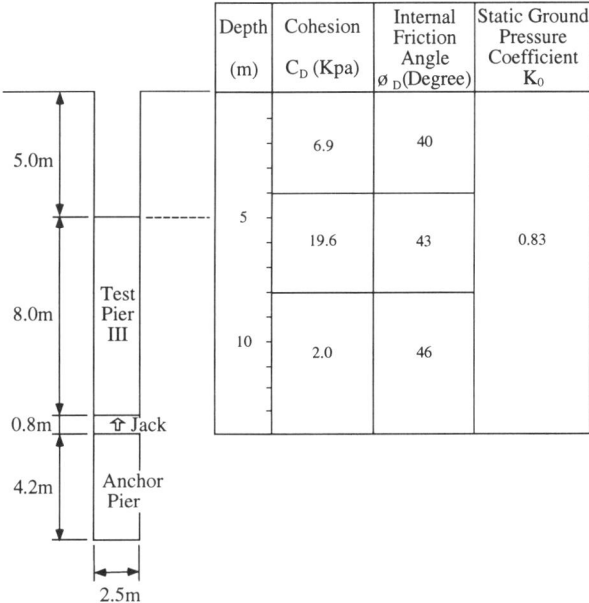

Figure 22. Testing apparatus and soil properties

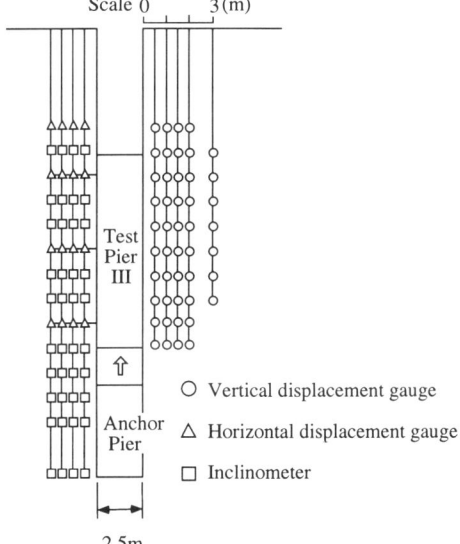

Figure 23. Outline of the measurement

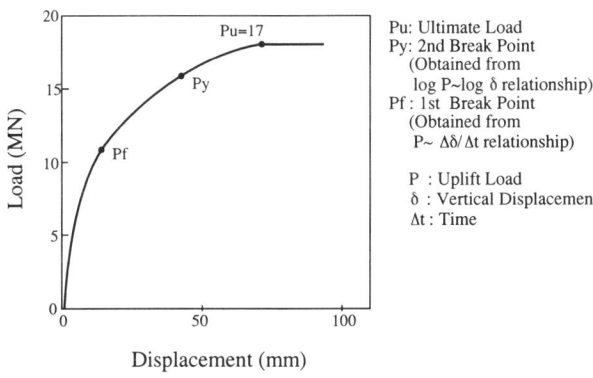

Figure 24. Load/Displacement curve

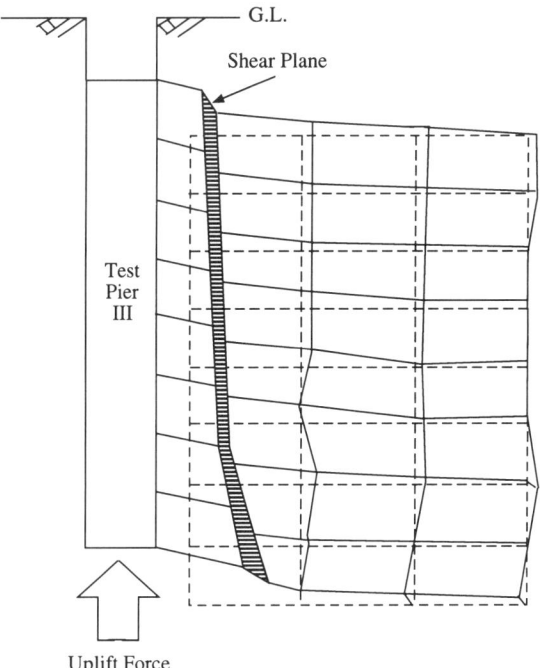

Figure 25. Displacement distribution of ultimate load

5.2.3 *Design formulas*

As explained previously, in the case of UHV pylon foundations, the shear forces around the circumference of the foundation act as uplift resisting forces, so that the following design equations for uplift were considered.

$$Q_{Ta} = \frac{1}{F} \cdot (W_c + W_s + R_f) \quad \quad \quad \quad (3)$$

where;
- Q_{Ta} : Uplift capacity
- W_c : Weight of foundation
- W_s : Weight of soil above the structure
- F : Safety factor (generally : 2.0, special cases: 1.33)
- R_f : Yield shear resistance between structure side surface and the ground

$$R_f = \frac{1}{1.5} \pi D \ell \{c + K_0 \gamma h \tan\phi\} \quad \quad (4)$$

where;
- D : Diameter
- ℓ : Uplift resistance region
- C : Cohesion
- K_0 : Static ground pressure coefficient
- γ : Unit weight of ground
- h : Depth from surface
- ϕ : Internal friction angle

The resistance region for uplift can be considered to be around the lower circumferential portion of the foundation where the ground remains elastic, as shown in Figure 19.

6. CONCLUSIONS

A design method was successfully developed for deep transmission pylon foundations located in soft rock or sandy soil ground in mountainous regions. In particular, an elastoplastic design method that is able to express the progressive failure phenomenon of the ground was used to estimate the lateral support forces. By carrying out full scale tests, design formulae of high precision were evolved. For the case of formulae for uplift resisting forces, it was possible to confirm the hitherto assumed failure mode, namely, general shear failure. As a result, it was possible to attain improvements over previously used design formulae.

At present, the design of deep foundations is being carried out in accordance with the principles described in this report. In addition, to achieve further design improvements, new construction methods and technologies, such as pier foundations incorporating ground reinforcement, are being developed.

REFERENCES

Maeda, H., "*Horizontal Behavior of Pier Foundations on a Soft Rock Slope,*" 5th ISRM: C181-84, 1983.

Uto, K., Maeda, H., Yoshii, Y., Takeuchi, M., Kinoshita, K., and Koga, A., "*Horizontal Behavior of Pier Foundations in a Shearing Type Ground Model,*" 5th ICONMIG, 1985.

Compaction and subsidence in North Sea hydrocarbon fields

Marte Gutierrez, Nick Barton & Axel Makurat
Norwegian Geotechnical Institute, Ullevål Hageby, Oslo, Norway

ABSTRACT: In the last two decades, the North Sea has been a very active area in the construction of large offshore structures for the production of oil and gas. A large number of these structures are massive gravity based platforms which are built to stand in water depths of 50 to 350 m and are designed to widthstand the extreme enviromental conditions of the North Sea. For some of these platforms, a foundation concern is the subsidence of the seabed. In fields with weak chalk and sandstone reservoirs, pressure depletion due to hydrocarbon production can cause compaction and shear failure of the reservoir. Compaction is then transferred to the seabed as subsidence which can result in significant financial and enviromental consequences. The most prominent example where subsidence has become a major problem is the Ekofisk field. This paper presents a general discussion on the problems of compaction and subsidence in North Sea hydrocarbon fields. The rock mechanical aspects of compaction and subsidence are discussed, and the results of several research work to understand and to accurately predict compaction and subsidence are presented.

1 INTRODUCTION

The discovery of oil in the North Sea in the late 1960's marked the beginning of a busy era of offshore construction in Norway. In order to produce the oil and gas below the seabed, structures were built to stand in water depths of 40 to 350 m and to withstand several storms with up to 30 m sea waves during their lifetime. Figure 1 summarizes the different platform concepts used in the North Sea. A number of these platforms are massive concrete gravity based structures which rely on their weight for stability. The CONDEEP type structure, of which the Statfjord A and the Troll platforms are two of the largest, have become the trademark of the Norwegian North Sea. To give an idea of how large these gravity based structures are, the Statfjord B platform has been superimposed on the outline of Manhattan, New York (Fig. 2). The tallest gravity based structure is the Troll platform which has just recently been installed in the Norwegian trench where the water depth is about 300 m (Fig. 3). The subsoil which consists of soft clay necessitated the use of several 36 m long skirts as the main foundation system. Total height of the platform from the tips of its seabed skirt to the top of its four towers is 370 m, and the weight without ballast is 690,000 t. The Statfjord B and the Troll platforms are in fact two of the biggest concrete structures ever built in the world.

Due to the high cost of these structures (the Troll platform cost US$1.72 billion), rigorous design and analysis of their foundations are performed to insure that they remain safe and stable during their expected lifetimes. Foundation design for typical gravity based platforms include analysis of stability, bearing capacity, settlement, earthquake response, soil-wave-structure interaction, cyclic response due to wave loading, skirt or pile resistance, scouring potential, seabed subsidence, etc. In the early days of the offshore construction activities, only preliminary analyses or no analysis at all of seabed subsidence were performed. This was probably due to lack of relevant field data or appreciation of the potential for seabed subsidence.

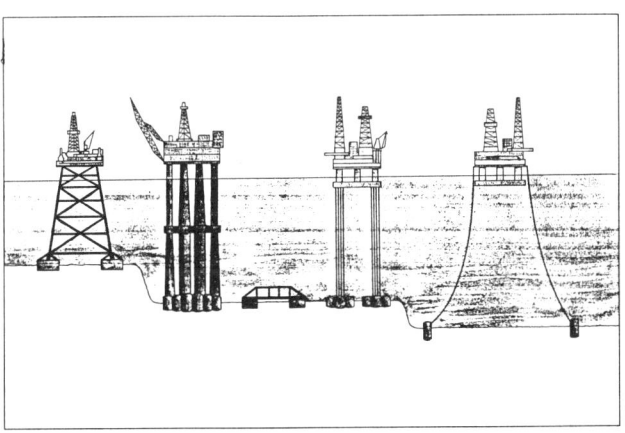

Figure 1. Types of platforms used in the North Sea.
Left to right: jacket platform, gravity based structure, subsea structure, tension leg platform (TLP), and catenary TLP.

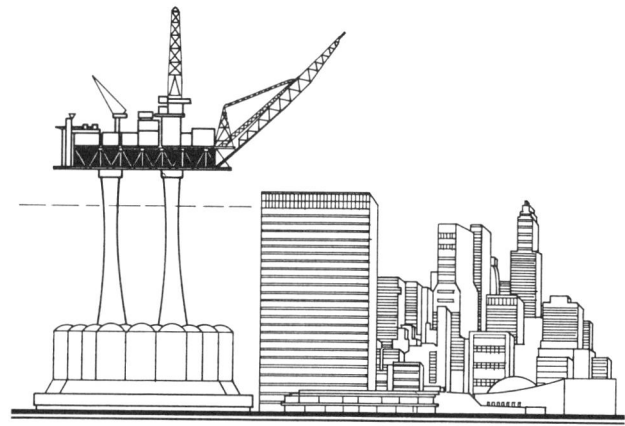

Figure 2. Statfjord B gravity structure compared to the U.N. building and the Manhattan skyline.

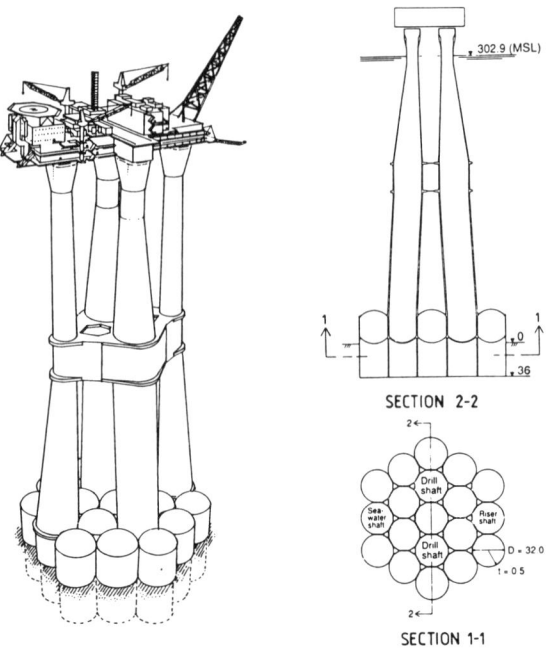

Figure 3. Troll platform.

2 PREVIOUS CASES OF SUBSIDENCE

Surface subsidence is a problem commonly associated with underground construction, mining and excessive ground water pumping. In terms of the area of influence, subsidence due to lowering of the groundwater usually covers a wider area than subsidence due to underground construction and mining. Some of the well known examples of subsidence due to groundwater utilization are: San Joaquin Valley, California, USA (with of 9m subsidence); Tokyo, Japan (subsidence = 3 m); Venice, Italy; Galveston, Texas, USA; Niigata, Japan; Ravenna, Italy; Taipei, Taiwan; Bangkok, Thailand; Las Vegas, Nevada, USA; and Mexico City, Mexico.

Cases of subsidence over oil and gas reservoirs are less common though not less important in terms of financial impact. Some prominent examples are given in Table 1.

Table 1 - Some cases of subsidence above oil/gas fields.

Location	Formation	Maximum subsidence
Goose field, Galveston, Texas, USA	Unconsolidated sands and clays reservoir depth: 350-1400 m	several m
Wilmington field, Long Beach, California, USA	Unconsolidated sand with interbedded materials reservoir depth: 500-2000 m	8 m
Inglewood field, Los Angeles, California, USA	Middle to upper pleistocene sand reservoir depth: 300-1000 m	several m
Lake Maracaibo, Venezuela	Post Eocene loose sand interbedded with clay reservoir depth: $\approx$ 1000 m	6-7 m
Groningen gas field, Netherland	Unconsolidated sandstone	<1 m

3 COMPACTION AND SUBSIDENCE IN NORTH SEA OIL AND GAS RESERVOIRS

After several years of production, it has been realized that subsidence is a concern in several North Sea chalk reservoirs. Table 2 gives the depths and sizes of these chalk reservoirs. These reservoirs are located mostly in the "Greater Ekofisk" area of the Central Graben close to the territorial boundaries of Norway, U.K. and Denmark (Fig. 4). The estimated original reserves from these 9 fields amount to 2455 M bbl ($388 \cdot 10^6$ Sm^3) of oil and 11930 B cft ($338 \cdot 10^9$ Sm^3) of gas. There are also a several smaller fields in the Danish sector (Fig. 4). The Central graben consists of mainly chalk reservoirs with porosities of 40% or higher. Two main formations are recognized: Ekofisk (Danian, early Paleocene) and Tor (Maastrichtain, late Cretaceous). The reservoir rock is comprised of reworked chalk whose main source was a cocolith ooze previously deposited along the graben margins.

The most prominent example where subsidence has become a problem is the Ekofisk Field. In the course of 25 years of

Table 2 - Chalk Fields in the Norwegian North Sea.

Field	Reservoir depth (m)	Reservoir area (km^2)	Reserves oil (M bbl)	Reserves gas (B cft)
Ekofisk	3000	50	1500	6500
Eldfisk	2700	18	350	1300
Valhall	2400	50 (?)	250	1000
West Ekofisk	3000	12	80	1000
Tor	2900	22	115	450
Albuskjell	3100	25	50	700
Tommeliten	3100	7	50	200
Hod	2600	14	35	200
Edda	3000	10	25	80

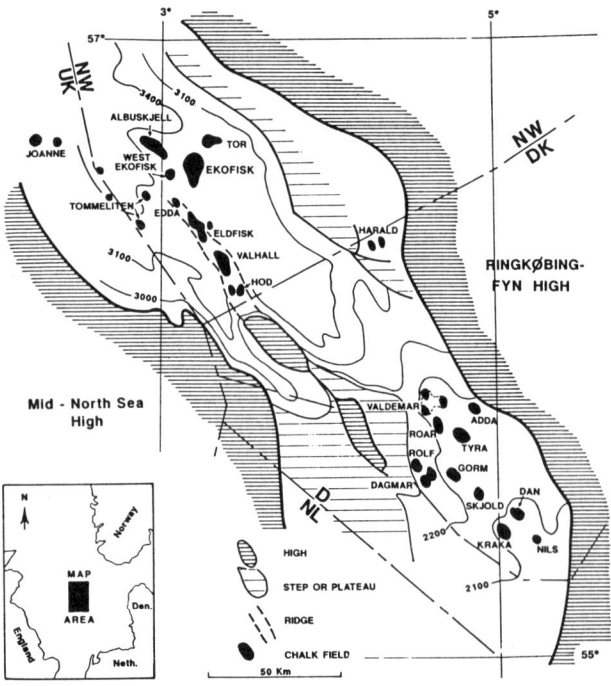

Figure 4. Location map of North Sea chalk fields.

petroleum production, the Ekofisk field has experienced subsidence involving 150 km³ of rock overburden in an area of 50 km². Subsidence was first discovered in 1984 by Phillips Petroleum Company Norway, the main Ekofisk operator. Up to that time the extent and rate of subsidence was unknown. In 1987, to ensure continuous safe operation during severe weather conditions, the platforms in the Ekofisk center were raised by 6.0 m, and in 1989, the Ekofisk tank was protected with a concrete wall (106 m high and 137 m in diameter). The total costs of these protective measures is $400 million. At the end of 1994, maximum subsidence has reached 5.5 m and is still continuing at about 40 cm/year. The protective measures performed in 1987 and 1989 have now become inadequate and most of the facilities in the Ekofisk center (including the main tank) will be abandoned and a new center will be built outside the subsidence bowl at a cost of $2.9 billion. With these figures, the Ekofisk subsidence can be classified as one of the biggest rock mechanical problems in the world both in terms of the volume of the rock and amount of money involved.

The Norwegian government has played an active role in the decision to move the Ekofisk complex out of the subsidence bowl. The reason for this is that the Ekofisk complex is the processing center for all productions from the Greater Ekofisk area, and a pivot point in the transport of oil and gas by pipelines from many fields in the Norwegian shelf to the different distribution points in Europe. Figure 5 shows the complete installations in the Ekofisk complex and the pipeline systems passing through the complex. Any subsidence induced damage to the facilities in the Ekofisk Center would therefore result in costly delay in delivery of oil and gas to Europe.

Subsidence induced damage in offshore facilities can be more serious than for comparable on-shore facilities. Damage can be due to loss of deck clearance above sea waves (and thereby reduced stability from wave action for gravity platforms), differential settlement of pile foundations for jacket platforms, mis-alignment of pipelines, and excessive surface movement of soils. Reservoir compaction and surface subsidence can also cause well casing failures. In the Ekofisk, casing failures have occurred in two-thirds of the wells (Yudovich et al., 1989). The failure of a well in an offshore field has high economic consequences (a typical North Sea well costs US$10 million to drill and complete), since wells have to be drilled in deviated directions from a few offshore installations. Aside from these damages, subsidence can cause environmental problems even for offshore sites.

However, compaction and subsidence are not entirely

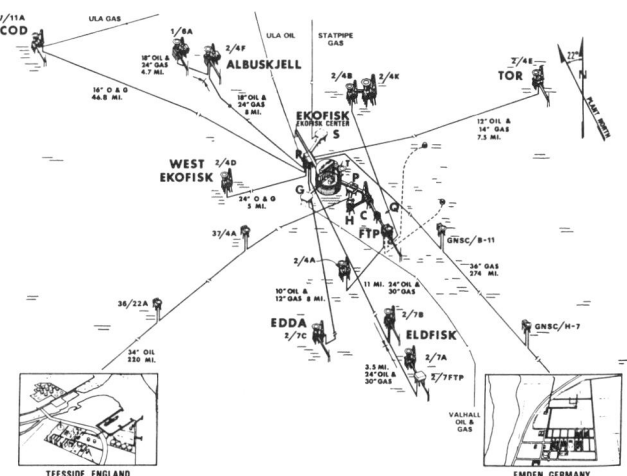

Figure 5. Structures and installations in the Ekofisk Center.

detrimental. Compaction can result in additional energy (called "compaction drive") and increase in (or at least maintenance of) pore pressure, which increases the productivity of the reservoir. Compaction drive is believed to partly account for the continued productivity of the Ekofisk field, despite anticipated reduction in permeability of the reservoir due to pore collapse during compaction (Sulak et al., 1991).

To a much lesser extent, because of their size and depth, the other fields in the Central Graben are also susceptible to subsidence. In the Valhall field, subsidence monitoring started in 1982, and for the first 3.5 years of production, subsidence was estimated to be less than 0.36 m (Ruddy, et al., 1988). Subsidence at the end of 1991 was about 2.0 m. Based on numerical simulations, maximum subsidence is expected to be about 3.3 m in the year 2011, which is safely within the platform wave height tolerance of 4.4 m (Ali and Alcock, 1992). In the recently installed Troll platform, subsidence has been considered in the design since the platform will sit on top of an unconsolidated sandstone reservoir.

4 ANALYSIS OF COMPACTION AND SUBSIDENCE

4.1 Basic Equations in Modelling of Compaction and Subsidence

The classical work of Terzaghi on one-dimensional consolidation (Terzaghi; 1943) forms a basis for all models of compaction and subsidence. This theory is expressed in terms of a differential equation describing the deformation in a fluid saturated porous media by coupling the Darcy flow equation to a linear elastic stress-strain relation via a continuity equation. The theory was later extended to three-dimensions by Biot (1956). In addition, Biot introduced a new parameter called Biot's constant which extends the theory of elasticity to porous media. Biot's constant relates the compressibility of the mineral grains to the compressibility of the skeleton of the porous media. The complete equations in Biot's formulations are:

1) The effective stress-strain relation of a porous medium, expressed as:

$$\sigma_{ij} = 2G\left[\epsilon_{ij} + \frac{\nu}{1-2\nu}\epsilon_{vol}\delta_{ij}\right] + \alpha_p p \delta_{ij} \qquad (1)$$

where

σ_{ij} = total (or bulk) stress tensor
ϵ_{ij} = strain tensor
ϵ_{vol} = volumetric strain
G = shear modulus
ν = Poisson's ratio
p = pore fluid pressure
α_p = 1 - β = Biot's constant
β = c_r/c_b = ratio of rock matrix and rock bulk (skeleton) compressibility
δ_{ij} = Kronecker delta

2) Darcy's flow equation (which may also be written for several fluid phases):

$$v_i = -\frac{k_{ij}}{\mu}\frac{\partial p}{\partial x_j} \qquad (2)$$

where v_i = fluid velocity, μ = fluid viscosity, and k_{ij} = permeability tensor.

3) The static (or dynamic) equilibrium equation:

$$\frac{\partial \sigma_{ij}}{\partial x_j} + F_i = 0 \tag{3}$$

where F_i are the applied body forces, and

4) The mass balance (or continuity) equation:

$$\frac{\partial v_i}{\partial x_i} + \epsilon_{vol} + c_b \alpha_p \phi p = 0 \tag{4}$$

where ϕ is the porosity of the material.

All the above equations are required to solve problems involving deformation and fluid flow in fluid saturated porous rock such as compaction and subsidence in hydrocarbon reservoirs. There are essentially two approaches in solving the Biot equations:

a) Uncoupled models - This involves solving the fluid flow problem (Eq. 2) to produce pressure profiles (as function of position and time) which are then used to calculate the subsidence of the formation.

b) Coupled models - Here an attempt is made to solve the above equations simultaneously. Since the formulation is complicated, the solutions can only be carried out numerically except for cases involving simple geometry with homogenous rock and single phase fluid.

A discussion of the differences between coupled and uncoupled simulations is given below.

4.2 Analytical Procedures

Analytical procedures provide easy and quick estimates of the reservoir compaction and overburden subsidence given the amount of pore pressure change. Provided the lateral dimensions of the reservoir are large compared to their thickness, the reservoir is expected to deform in an essentially vertical K_o (uniaxial strain) condition. The magnitude of the reservoir compaction is then described by a uniaxial compaction coefficient C_m (Geertsma, 1973):

$$C_m = \frac{d\epsilon_z}{dp} \quad ; \quad d\epsilon_x = d\epsilon_y = 0 \tag{5}$$

where $d\epsilon_z$ is the vertical strain and $d\epsilon_x$ and $d\epsilon_y$ are the horizontal strains. Using this parameter, the total compaction ΔH of a reservoir with thickness H to a pressure change Δp is then

$$\Delta H = \int_0^H \int_{p_i}^{p_f} C_m(p,z) dp dz \tag{6}$$

Note that the uniaxial compaction coefficient may vary within the reservoir and may be non-linear in which case it becomes dependent on the pore pressure p.

A model based on the nucleus of strain approach was introduced by Geertsma (1973) to calculate the magnitude of subsidence due to pressure reduction in a reservoir. The model gives an analytical solution for an isolated volume with reduced pore pressure in an elastically deforming half-space with a traction-free surface. The nucleus of strain approach gives the subsidence (vertical displacement) at the surface above a nucleus of very small volume V subjected to a pore pressure change of Δp as:

$$u_z(r,0) = -\frac{1}{\pi} C_m (1-\nu) \frac{D}{(r^2 + D^2)^{3/2}} \Delta p V \tag{7}$$

where r is the radial horizontal distance from the nucleus and D is the depth of the reservoir. To obtain the complete displacement field on the surface of any real reservoir, Eq. (7) needs to be integrated over the whole volume of the reservoir. For simple reservoir geometries (e.g., an embedded cylindrical disk in a half-space), a uniform pore pressure change, and homogenous reservoir and surrounding rock properties, Geertsma presented an exact analytical solution. However, for complicated geometries, pore pressure distribution and non-homogenous properties, Eq. (7) can only be integrated numerically (Evangelisti and Poggi, 1970; and Geertsma and van Opstal, 1973).

The Geertsma solution can be written as:

$$u_z(r,0) = 2 C_m (1-\nu) \Delta p H F \tag{8}$$

where $F = F(r/R, D/R)$ and R is the radius of the reservoir. Geertsma (1973) gives tabulated values of F. The major conclusions that can be obtained from Geertma's solution are:

1. The ratio of the volume of the subsidence bowl to the volume reduction in reservoir volume is:

$$\frac{V \text{ subsidence bowl}}{\Delta V \text{ reservoir}} = 2(1-\nu) \tag{9}$$

2. The maximum subsidence to maximum compaction ratio, S/C, is equal to:

$$S/C = 2(1-\nu) \left[1 - \frac{D/R}{\sqrt{1 + (D/R)^2}} \right] \tag{10}$$

Equation (10) is plotted in Fig. 6. For deep and narrow reservoirs, the subsidence to compaction ratio, S/C, is very small, while for shallow and wide reservoirs, the subsidence can be higher than the compaction for $\nu < 0.5$ (i.e., $S/C \approx 2(1-\nu)$). The work of Geertsma (1973) was modified by van Opstal (1974) to account for material inhomogeneity including the effects of a stiff underburden. With the modifications, van Opstal (1974) showed that:

1. The volume of the subsidence bowl is equal to the reduction of reservoir volume.

2. The area of subsidence is narrower than that predicted by Geertma's solution.

3. The S/C ratio is closer to 1 than that predicted by Geertma's solution.

Comparisons were made between Geertma's solution and the results of numerical calculations conducted by NGI (see discussions on numerical modelling below). For reservoirs with a rigid basement, the numerically calculated S/C ratios are about twice the calculated values from Geertsma's solution except for $D/R < 0.5$ where the Geertsma solution overpredicts the S/C ratio. The two curves in Fig. 6 give the range of expected S/C ratios for a given D/R ratio.

Figure 6 can be used to make an initial determination of the subsidence potential of a field given its reservoir area and depth. Although the Ekofisk field has a very deep reservoir, the depth to radius ratio D/R of about 0.75 (for $D = 3$ km and $R = 4$ km)

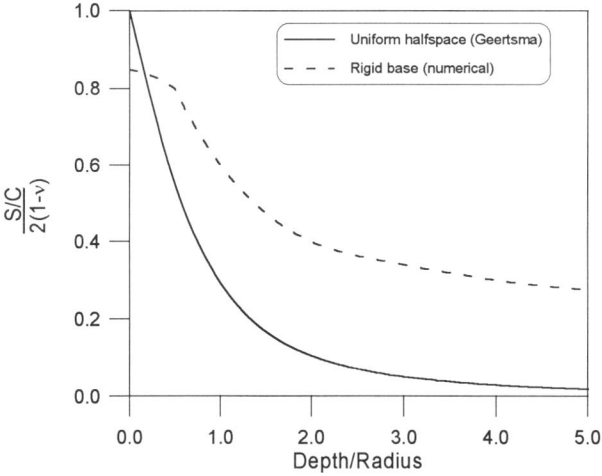

Figure 6. Subsidence to compaction, S/C, ratio from Geertma's solution and from numerical calculations.

gives $S/C=0.5$ (for uniform halfspace) and $S/C=0.8$ (with a rigid base). Since the Ekofisk has a rather high compaction to surface subsidence transfer ratio, subsidence cannot be ruled out. The other fields in the Greater Ekofisk area have reservoirs with much smaller areas and are at approximately the same depths as Ekofisk, and are therefore less susceptible to compaction. An exception is the Valhall field with a much shallower reservoir than Ekofisk. For any estimated S/C ratio, the subsidence potential will finally depend on how much the reservoir will compact during reservoir pressure reduction.

4.3 *Finite Element and Distinct Element Analyses*

For situations involving complicated geometries, heterogenous properties, non-linear rock behaviour and non-uniform pore pressure distribution, the analysis can only be carried out by numerical procedures. The most common numerical approach is the use of the finite element method. As early as 1982, a finite element formulation of Biot's equations for single phase flow was developed by Sandhu (1982). A fully-coupled finite element formulation of Biot's equation for three fluid phases was recently proposed by Lewis and Sukirman (1993).

There have been several applications of the finite element method to the analysis of compaction and subsidence. For the Ekofisk field, Chin and Boade (1990) carried out a full-field 3D subsidence analysis. Potts, Jones and Berget (1988) made axisymmetric finite element analysis of Ekofisk using a non-linear elastic material model. Another 3D analysis of the Ekofisk was done by Plischke (1994).

One of the early numerical analyses of the Ekofisk subsidence was done by NGI (Barton et al., 1986). In addition to continuum finite element analyses, NGI also performed discontinuum distinct element analyses of the Ekofisk overburden behaviour (Barton et al., 1988). The chalk compaction behaviour was analyzed using the Cam-clay model. The Cam-clay modelling emphasized the important effects of K_o-value on compaction, which was shown to increase by 0.9 m for the 300 thick Ekofisk chalk reservoir when K_o was reduced from 0.9 to 0.5. The Cam-clay calculated compaction of the reservoir was used as applied displacement boundary to the base of the overburden in the subsidence calculations.

The continuum modelling of the overburden subsidence was performed using the finite element code CONSAX (D'Orazio and Duncan, 1982). Different values of overburden stiffness were used, and the pronounced effects of Poisson's ratio on the maximum subsidence was shown. The subsidence to compaction, S/C, ratio was calculated to be about 0.52 to 0.65. A good fit between calculated and measured subsidence profile (in 1985) was shown.

The discontinuum analysis of overburden subsidence was done using the distinct element code UDEC (Cundall, 1980). Based on the interpretation of seismic surveys, two sets of fracture systems have been identified in the Ekofisk overburden: 1) a sub-vertical set of fractures (minor faults) where the individual fractures show a castellated pattern, and 2) bedding fractures. The sub-vertical fractures are continuous over distances of 100 to 300 m, and the bedding fractures show throws of 20 to 40 m at distances of a few hundred meters. The fracture geometry used in the 1988 analyses is shown in Fig. 7 (top). Values of fracture normal and shear stiffnesses were selected using the methods described by Barton (1982). Selection of friction and dilation angles were made with emphasis on large-scale, clay bearing features. The predicted overburden displacements from one of the simulations is shown in Fig. 7 (bottom). One of the main results of the discontinuum simulations is that due to potential slip along joints, faults and bedding planes, a tighter subsidence bowl and a higher S/C ratio were obtained compared to that of the continuum analyses. More recent continuum and discontinuum analyses performed at NGI continue to show good agreement between numerically calculated and measured subsidence and S/C ratios at Ekofisk.

4.4 *Fully-Coupled vs. Uncoupled Analysis*

Strictly, the analysis of reservoir compaction and surface subsidence should be fully-coupled with the analysis of multi-phase fluid flow in the reservoir. The reason for this is that for deformable reservoirs, pore pressure change and compaction occur simultaneously (compaction causes pore pressure change and vice versa). However, because full solution of the complete

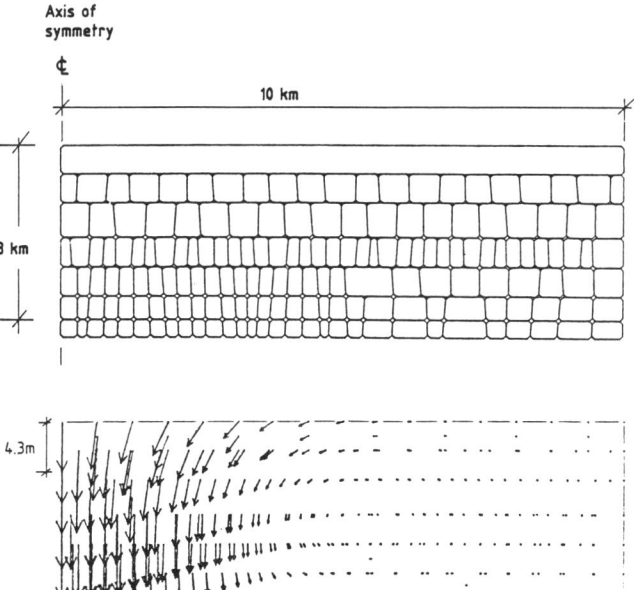

Figure 7. Discontinuum simulation of the Ekofisk overburden.

Biot's equation is extremely complicated, most solutions uncouple the fluid flow equation from the deformation equation. The usual approach is to use pressure profiles from reservoir simulations to obtain the compaction of the reservoir and the subsidence of the overburden. The calculation of compaction and subsidence is then reduced to the stress analysis of a rock mass with prescribed loads. The loads come from variations in effective stresses in the reservoir rock due to changes in pore fluid pressure. In turn, the effects of reservoir deformation on pore pressure are indirectly accounted by adjusting rock compressibility in the reservoir simulation.

Gutierrez and Hansteen (1994) have analyzed numerically and theoretically the validity of the uncoupled approach. Using a finite element formulation, the pore pressures change for a single phase reservoir simulator is calculated as:

$$\Delta p = (C + \Delta t \Phi)^{-1}(-\Delta t \Phi p + \Delta q) \quad (11)$$

where

Δp = pore pressure change
p = current pore pressure
C = compressibility matrix
Φ = permeability matrix
Δq = fluid fluxes
Δt = time increment

Note that the only rock mechanical data involved is the rock compressibility, C_r, since $C = C_f + C_r$, where C_f = fluid compressibility. For rigid reservoirs, $C_r = 0$.
For a fully coupled simulation:

$$\Delta p = (L^T K^{-1} L + \Delta t \Phi)^{-1}(-\Delta t \Phi p + \Delta q + L^T K^{-1} \Delta F) \quad (12)$$

where

K = stiffness matrix
L = coupling matrix
ΔF = load matrix

The main differences between these two equations are:

1) Pore pressure change is a function of the full rock stiffness matrix in the fully-coupled simulations, while it is a function only of the rock compressibility in the uncoupled simulation. The former accounts for the full constitutive behaviour of the reservoir-overburden system, while the latter is a scalar parameter which can be made to depend only on the pore pressure.

2) Pore pressure change is also a function of the applied load (total stress changes) in the fully-coupled analysis. Such total stress changes come from the weight of the overburden, which is transferred non-uniformly to the reservoir again according to the pore pressure distribution in the reservoir.

These results imply that adjusting the compressibility matrix as commonly done in traditional reservoir simulations is not sufficient to account for the pore pressure change due to the so-called "compaction drive". Compaction drive, Δp_{comp}, was more rigorously defined as the pore pressure change due to both rock compressibility and total stress change:

$$\Delta p_{comp} = (L^T K^{-1} L)^{-1}(-\Delta t \Phi p + \Delta q + L^T K^{-1} \Delta F) \quad (13)$$

To illustrate the differences between the uncoupled and coupled approaches, finite element simulations of a simplified model of the Ekofisk were performed. The 2D finite element model of the East-West section of the Ekofisk together with the material properties are shown in Fig. 8 (top). For the reservoir (uncoupled) simulation, a rock compressibility C_r corresponding to K_o compaction (where $K_o = \nu/(1-\nu)$) was used:

$$C_r = C_m = \frac{(1-\nu)(1-2\nu)}{E(1+\nu)} \alpha_p \quad (14)$$

The elastic parameters used were: $E = 50$ MPa, $\nu = 0.2$ and $\alpha_p = 1.0$. The same elastic parameters were used to model the reservoir in the coupled simulation. The Young's modulus was chosen to give about 6 m maximum reservoir compaction for a pressure change of 24 MPa.

The results of the two types of analyses for a pressure drawdown of 24 MPa over a period of 18.5 years from a single well (located at the reservoir center) are shown in the Fig. 8 (bottom). Different pore pressure distributions were obtained from the analyses. However, the most significant result is that pore pressure increased above the initial pore pressure of 48 MPa at the reservoir flank despite the fact that the reservoir is under pressure drawdown (production at the center well). This is a result that cannot be obtained by simply adjusting the rock compressibility in the reservoir (uncoupled) simulation. The increase in pore pressure at the reservoir flanks obtained from the coupled simulation comes from the tendency of reservoir fluids to be squeezed towards the flank by the overburden as the pressure in the center of the reservoir is reduced.

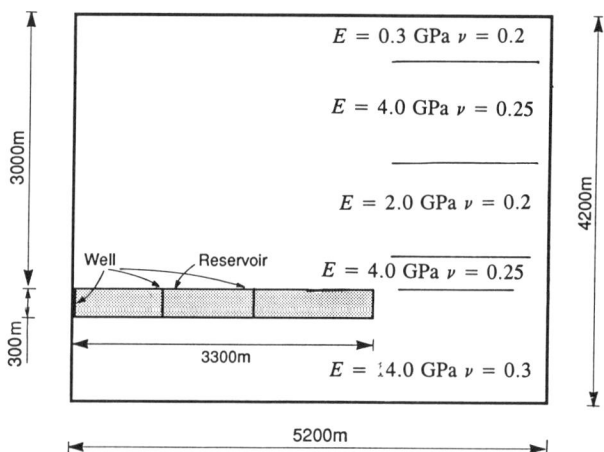

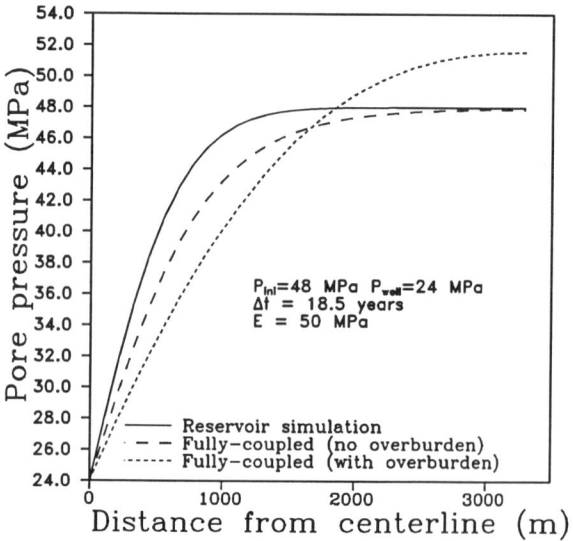

Figure 8. Top: 2D finite element model of East-West section of Ekofisk. Bottom: Pore pressures from uncoupled and fully coupled simulations.

4.5 Modelling of the Compaction Behaviour of Intact and Fractured Chalk

An important input in the analysis and modelling of reservoir compaction and surface subsidence is the model of reservoir rock constitutive behaviour. In the case of chalk, this involves modelling of the main mechanisms of deformation which are pore collapse and shear failure. For the Ekofisk chalk, fracturing complicates the modelling further. Fracturing in the Ekofisk chalk has been determined from core and FMS (formation micro scanner) logs, and two types of fracture systems have been recognized: 1) natural, and 2) compaction induced.

The behaviour of the intact Ekofisk chalk was modelled using NGI's cap plasticity model, the details of which are shown in Fig. 9. The model has a non-associated Mohr-Coulomb failure surface to model shear failure, and an elliptical cap to model pore collapse. The elliptical cap hardens during loading according to the amount of total plastic deformation $(d\epsilon_{ij}^p \cdot d\epsilon_{ij}^p)^{1/2}$. The NGI cap model is implemented in the discontinuum code UDEC.

The NGI cap model was used to simulate the results of uniaxial strain compaction tests of intact Ekofisk chalk. The model predictions agree well with the experimental stress strain curves as shown in Fig. 10 for three porosities of Ekofisk chalk. To analyze the compaction behaviour of fractured Ekofisk chalk, UDEC models of 2mx2m representative blocks of the fractured reservoirs with different porosities were analyzed (Gutierrez et al., 1994). Typical results are shown in Fig. 11. Note the non-uniform yielding (yielded zones are indicated by + signs in the top figure) and pore collapse of the fractured chalk (bottom figure). Plastic zones tend to propagate along rock bridges indicating possible secondary fracturing along the same directions. Aside from the non-uniform compaction of the fractured rock, it was also shown that fractures tend to increase the K_o-value of the rock mass, causing the fractured chalk to have an apparent higher stiffness than the intact chalk. Another result is the effect of fractures on reservoir conductivity. The

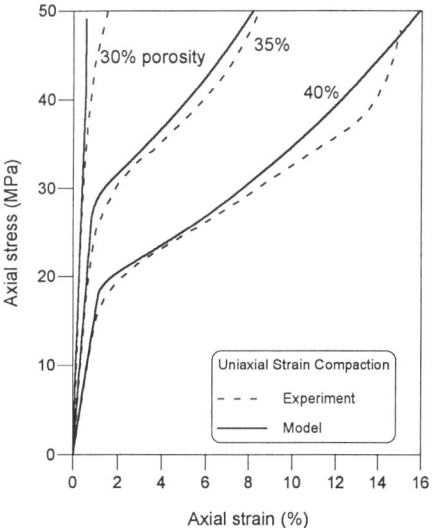

Figure 10. Calculated and measured stress-strain curves for Ekofisk chalk.

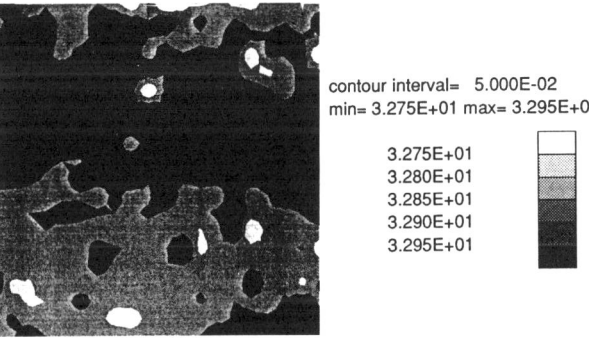

Figure 11. Top: Plastic zones (+) at σ'_v = 24 MPa. Bottom: Porosity distribution at σ'_v = 38 MPa.

discontinuum simulation showed that fracture permeability only decreased slightly during pore pressure reduction. During fluid injection into the reservoir, numerical results show considerable increase of fracture permeability over that of the initial fracture permeabilities.

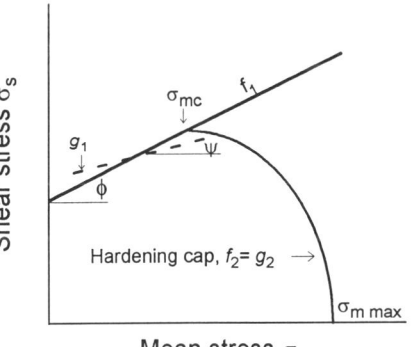

f_i = Yield function

g_i = Plastic potential

$\sigma_m = (\sigma_1 + \sigma_3)/2$

$\sigma_s = (\sigma_1 - \sigma_3)/2$

ϕ = friction angle

ψ = dilation angle

σ_{mc} = hardening parameter

c = cohesion

$\Delta\sigma_{mc} = \sigma_c(1+e)/\kappa \, \Delta\epsilon^p$

$\Delta\epsilon^p = (\Delta\epsilon^p_{ij}\Delta\epsilon^p_{ij})^{1/2}$

Figure 9. Details of the NGI cap model.

5 CONCLUSIONS

As experience from the Ekofisk has shown, compaction and subsidence can be a major problems for oil fields on weak rock formations. It is unfortunate that such experience came somewhat late for some North Sea fields. Many still ask why the Ekofisk subsidence was not foreseen. The answer is at least partly due to insufficient understanding of chalk compaction, shear failure and fracture behaviour when the Ekofisk was still being developed. Another reason was that there were just no precedent for the occurrence of subsidence from such a deeply buried reservoir.

The complexity of the problem is highlighted by the fact that the estimate of maximum subsidence in the Ekofisk of 6 m, based on which the jacking-up operation of 1987 was done, proved to be inaccurate. Also, the prediction that water injection would arrest further subsidence in the Ekofisk has, so far, not materialized. In fact, subsidence continues to this day at a rate of about 40 cm/yr despite voidage maintenance (i.e., equal volume of injected water and produced hydrocarbon).

All these point to the need for continued research to understand the compaction behaviour of chalk, and how such compaction is transferred by the overburden to the surface. Some of the research currently being done at NGI on such areas as coupled multi-phase modelling of compacting reservoirs and constitutive modelling (non-linear and time dependent behaviour) of intact and fractured reservoir and overburden rocks will hopefully contribute towards improved and reliable predictions of compaction and subsidence in hydrocarbon fields.

REFERENCES

Ali, N. and Alcock, T. 1992. Valhall Field, Norway - The First Ten Years. In *North Sea Oil and Gas Reservoirs -III*, pp. 25-40.

Barton, N. 1982. Modelling of rock joint behaviour from in situ block tests, ONWI, Columbus, Ohio, 96 p., ONWI-308, September 1982.

Barton, N., Hårvik, L. Christiansson, M., Bandis, S., Makurat, A., Chryssanthakis, P. and Vik, G. 1986. Rock mechanics modelling of the Ekofisk subsidence. Proc. 27th US Rock Mech. Symp., Univ. of Alabama

Barton, N., Makurat, A., Hårvik, L., Vik, G., Bandis, S., Christianson, M. and Addis, A., 1988. The discontinuum approach to compaction and subsidence modelling as applied to Ekofisk. Proc. of BOSS'88 the Intl. Conf. on Behaviour of Offshore Structures, Trondheim, Norway, vol. 1, pp. 129-141

Biot, M. 1956. General solution of the equations of elasticity and consolidation for a porous medium. J. Applied Mech., vol. 23, pp. 91-96.

Chin, L., and Boade, R. 1990. Full-field, 3D finite element subsidence model for Ekofisk. 3rd North Sea Chalk Symp., Copenhagen

Cundall, P.A. 1980. A generalized distinct element program for modelling of jointed rock. Report PCAR-1-80, P. Cundall Assoc., Contract DAJA37-79-C-0548, European Research Office, US Army.

D'Orazio, T.B. and Duncan, J.M. 1982. CONSAX: A computer program for axisymmetric finite element analysis of consolidation. Report No. UCB/OT/82-01, Dept. of Civil Eng., Univ. of California, Berkeley, California.

Envangelisti, G. and Poggi, B. 1979. Sopra i Fenomimi di Deformazione dei Terreni da Variazione della Pressione di Strato. Atti. Accad. Sci. Inst., Bologna, Mem. Ser., vol. II, no. 6, 124 pp.

Geertsma, J. 1973. Land subsidence above compacting oil and gas reservoirs. SPE 3730, Proc. SPE-AIME European Spring Meeting.

Geertsma, J. and van Opstal, G. 1973. A numerical technique for predicting subsidence above compacting reservoirs based on the nucleus of strain concept. Verh. Kon. Ned. Geol. Mijnbouwkundig Gennotschap, vol. 28, pp. 63-78.

Gutierrez, M. and Hansteen, H. 1994. Fully-coupled analysis of reservoir compaction and subsidence. SPE 28900, Proc. EUROPEC'94, London, UK.

Gutierrez, M., Tunbridge, L., Hansteen, H., Makurat, A. and Barton, N. 1994. Modelling of the compaction behaviour of fractured chalk. Proc. Eurock'94, pp. 803-810.

Lewis, R.W., and Sukirman, Y. 1993. Finite element modelling of three-phase flow in deforming saturated oil reservoirs. Int. J. of Num. and Anal. Meth. in Geomech., vol. 17, pp. 577-598.

van Opstal, G. .1974. The effect of base rock rigidity on subsidence due to compaction. Proc. 3rd Cong. ISRM, Denver, Co., vol. II, part B.

Plishke, B. 1994. Finite element analysis of compaction and subsidence - Experience gained from several chalk fields. Proc. Eurock'94, pp. 795-802.

Potts, D.M., Jones, M.E., and Berget, O.E. 1988. Subsidence above the Ekofisk oil reservoir. Proc. of BOSS'88 the Intl. Conf. on Behaviour of Offshore Structures, Trondheim, Norway, vol. 1, pp. 113-127.

Ruddy, I., Andersen, M.A., Patillo, P.D., Bishlawi, M. and Foged, N. 1989. Rock compressibility, compaction and subsidence in a high porosity chalk reservoir: A case of the Valhall Field. J. of Petr. Tech., July 1989, pp. 741-746.

Sandhu, R.S. 1982. Finite element analysis of subsidence due to fluid widthdrawal. Proc. of 1982 Forum on Subsidence due to Fluid Withdrawal, Oklahoma, pp. 97-197.

Sulak, R.M., Thomas, L.K. and Boade, R.R. 1991. 3D Reservoir Simulation of Ekofisk Compaction Drive. J. of Petr. Tech., October 1991, pp. 1272-1278.

Terzaghi, K. 1943. *Theoretical soil mechanics*. Wiley, N.Y., 510 pp.

Yudovich, A., Chin, L.Y. and Morgan, D.R. 1989. Casing Deformation at Ekofisk. J. of Petr. Tech., July 1989, pp. 729-740.

Invited paper

Outline of Southern Hyogo Earthquake (M = 7.2) disasters and behaviors of foundations of the Akashi Kaikyo Bridge

Iwao Yoshida
Honshu-Shikoku Bridge Engineering Corporation, Japan

Kazuo Tada
Honshu-Shikoku Bridge Authority, Tokyo, Japan

ABSTRACT : On January 17, 1995, an earthquake of M7.2 attacked areas adjacent to the Akashi kaikyo Bridge from directly underneath. This paper outlines this earthquake and reports the state of damages caused by it. The Akashi kaikyo Bridge was then under construction and thus its aseismicity was tested. This paper describes the original bridge plan and the ground conditions including the geology and faults, design earthquake, and surveys on changes in ground made after the earthquake.

1. Introduction

On the early morning of January 17, 1995, an earthquake of magnitude 7.2 in Richter scale (M7.2) broke out at an epicenter near Akashi Straits. This earthquake caused quite large damages to the area centered around Kobe City. Also, it largely affected the Akashi kaikyo Bridge, the world's longest suspension bridge then under construction, thus giving an opportunity to test the aseismicity of its foundations.

In this paper, we outline the earthquake and make a prompt report of the results of surveys made on the foundations of the Akashi kaikyo Bridge.

2. Outline of Earthquake and Damages

At 5:46 in the morning of January 17, 1995, an earthquake of M7.2 broke out at an epicenter 135 degrees three minutes of east longitude and 34 degrees 36 minutes of north latitude, at a depth of 14km.

By this earthquake trembles of 7 degree by Japan Meterological Agaucy scale, the highest rank in the seismic magnitude scale used in Japan (the "ruinous earthquake" with a building fall-down rate of more than 30 percent, causing ground crack faults) were observed in Kobe and in the Awaji Island. This earthquake was felt in wide areas ranging from the Tohoku to Kyushu regions which are some 500km away from the epicenter.

Fig. 1 shows maximum accelerations observed in different places by the Southern Hyogo Earthquake. Large earthquake motions were observed in Kobe City and its surrounding areas, with maximum horizontal accelerations amounting to 600 to over 800 gal.

Fig. 2 shows the focul region of this earthquake and major faults in its vicinity. Active faults concentrate in the zone ranging from the Awaji Island to Kobe City, and strong quakes were observed on a line extending from the Awaji Island as the epicenter north-east, in which direction the faults seems to have moved.

Fig. 3 shows records of observation at the Kobe Marine Meteorological Observatory. Here the maximum acceleration had an N-S component of 818 gal, an E-W component of 617 gal, and an U-D component of 332 gal.

It is seen that the principal shock of large amplitude lasted only 10 seconds and thus large energy attacked Kobe City from directly underneath in quite a short period of time.

This earthquake caused extremely large damages;it killed about 5,500 people, collapsed some 125,000 buildings, and tore the elevated-highways and railways at all points. It also destroyed the harbor facilities and broke the electric power, city water, and city gas supply systems (the so-called "lifelines"), thus pushing the total amount of damage up to some 10 trillion yen (11 billion dollars).

This is the largest earthguake after the Great Kanto Earthquake (September 1, 1923,M7.9) that has occurred in Japan in this century.

Photo 1 Construction work on Akashi Straits

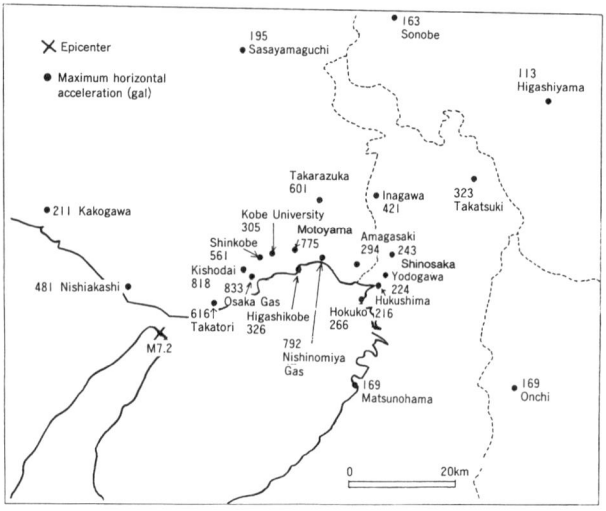

Figure 1 Distribution of Maximum Horizontal Acceleration (Note 2)

Photo 2 Collapsed elevated

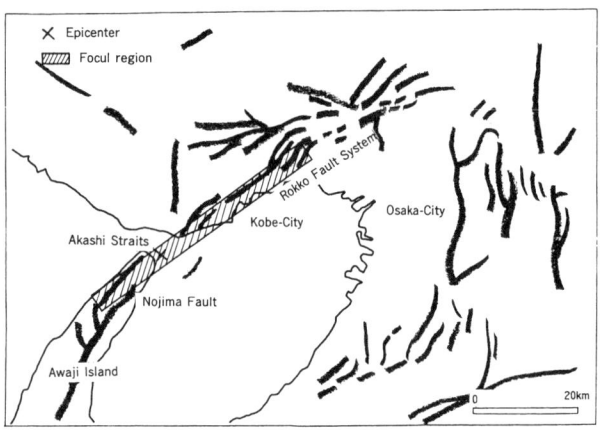

Figure 2 Focul Region and Faults

Photo 3 Fallen Steel girders

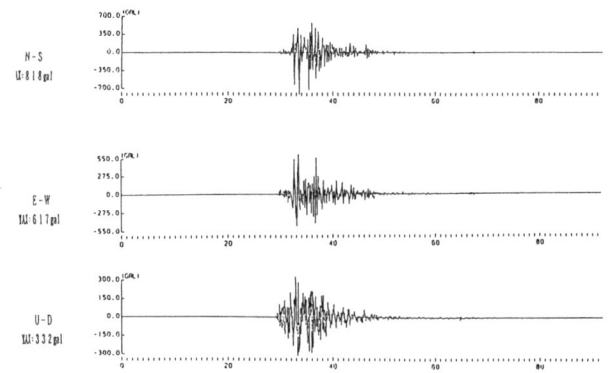

Figure 3 (1) Acceleration Waveforms (by Kobe Oceanic Meteorological Observatory) (Note 2)

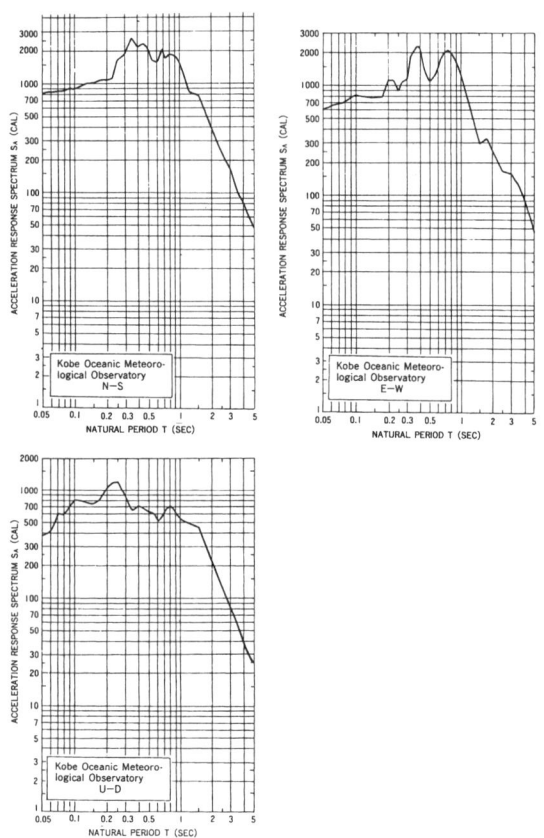

Figure 3 (2) Acceleration Response Spectra (h=0.05) (by Kobe Oceanic Meteorological Observatory) (Note 2)

Photo 4 Collapsed Shinkansen piers (Note 1)

Photo 7 Collapsed quay wall

Photo 5 Medium-rise building fallen down (Note 1)

Photo 8 Nojima Fault (Note 1)

Photo 6 Sound superhigh-rise building (Note 1)

3. Geological Conditions of Akashi Straits

The Akashi Straits has a width of about 4km and its central area is a caldron-shaped valley 110m in depth and 400m in width, with steep slopes on both sides.

Surveys of geological conditions of Akashi Straits were started in the 1950s by the Ministry of Construction and other national agencies in relation to the bridge construction project, and then taken over by the Honshu-Shikoku Bridge Authority and completed by the start-up of construction of the Akashi kaikyo Bridge.

Fig. 4 shows the results of these surveys. The soil comprises of granite dating back to the end of the mesozoic era, which is covered, in unconformity, by Kobe strata belonging to the Miocene of the Neogene period, and Akashi strata, upper diluvium, and alluvium belonging to the Diluvium epoch of the Quarternary period.

Surveys by the sonic prospecting and boring have indicated the presence of east-west faults centered on F1, F6, and F7 and south-north faults perpendicular to them (see Fig. 5). All these faults are contained in the layers deeper than the Kobe strata, not in the alluvium, upper diluvium, and Akashi strata located above it.

This means that these faults have not moved for at least the last two million years, namely that they are not active faults.

The east-west faults including F1 which does not cross the Quarternary-period layers were very clearly detectable by sonic prospecting in the center of the straits, but not so clearly at both extented sections. In order to select the location of foundations, positional relations between the extension of F5, one of the south-north faults, and tower foundation 3P, and those between the extension of F11 and anchorage foundation 4A were surveyed by sonic prospecting and other methods. It was confirmed, as a result, that neither of them extends to the area in which the foundations were to be constructed.

Thus, the foundations of the Akashi kaikyo Bridge are located, avoiding the faults existing in the Straits. The condition of seabottom faults after the Southern Hyogo Earthquake was surveyed by the Maritime Safety Agency. Seabottom faults newly found by this survey are shown in fig. 6. According to this, there is no seabottom fault newly developing in the vicinity of the Akashi kaikyo Bridge.

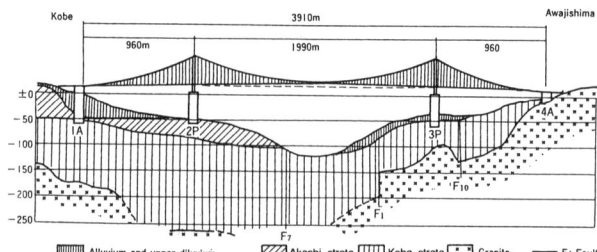

Geological time			Dates back to (X 1,000 Years)	Near Akashi Straits
Era	Period	Epoc		
Cenozoic	Quaternary	Alluvium	6 ~ 10	Alluvium
		Diluvium	2,000	Upper Diluvium Akashi strata
	Neogene	Pliocene	7,000 13,000	Kobe strata
		Miocene	20,000 26,000	
	Paloeogene			
Mesozoic	Cretaceous		65,000	Rokko Granite
	Jurassic		135,000	
	Triassic		190,000	

Figure 4 Geological Profile near Akashi Kaikyo Bridge

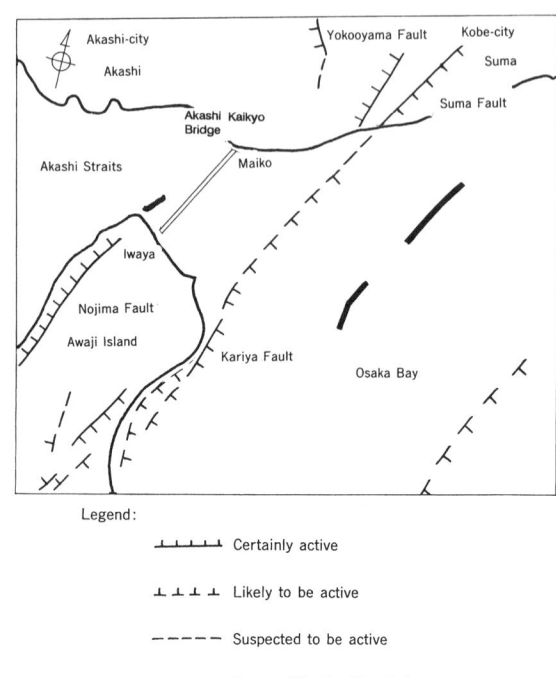

Figure 6 Faults Near Akashi Kaikyo Bridge (Notes 3& 4)

4. Foundations of Akashi kaikyo Bridge

The foundations of the Akashi kaikyo Bridge consist of anchorages 1A and 4A and tower foundations 2P and 3P.

The 2P and 3P tower foundations are constructed deep under the water (TP-60m), over which strong tidal current of 8 knots. All the foundations except for 4A are laid by a unconsolidated layer comprising soft rock and gravel, because of the large depth of the base rock.

The 2P and 3P submerged foundations are of the spread foundation type constructed by the laying-down caisson method. They have a double-wall steel caisson structure and a cylindrical shape in consideration of aseismicity and resistance to tides. They also feature a simple internal structure because of the use of special underwater concrete.

The 1A and 4A foundations are of the gravity anchorage type. Anchorage 1A consists of a caisson foundation 85m in diameter and 63.5m in depth laid by the Kobe strata and a main body 84.5m in length and 63m in width. Anchorage 4A is a spread foundation, 80m in length, 63m in width, and 23.5m in depth (deepest section) laid by as base rock granite. Fig. 7 shows overall structures of these foundations.

5. Input Design Acceleration for Akashi Kaikyo Bridge

Input design acceleration common to all the foundations of the Akashi kaikyo Bridge are defined as an envelop line of the following two acceleration response spectra (see Fig. 8).

[1] Acceleration response spectrum at the bridge construction point to an earthquake of M8.5 breaking out off the Kii Penisula(distance from epicenter:150km);and

[2] Acceleration response spectrum corresponding to a recurrence interval of 150 years based on stochastic assessment of past earthquakes (within radius of 300km and more than M6), in consideration of the possibility of earthquake forces stronger than Case [1] above.

The aseimatic design of each foundation has been studied for safety by modelling the ground conditions at the location of the foundations, considering the nonlinearlity of the soil and dynamic soil-structural interaction in case of earthquake, and obtaining response values of the foundations by dynamic analysis.

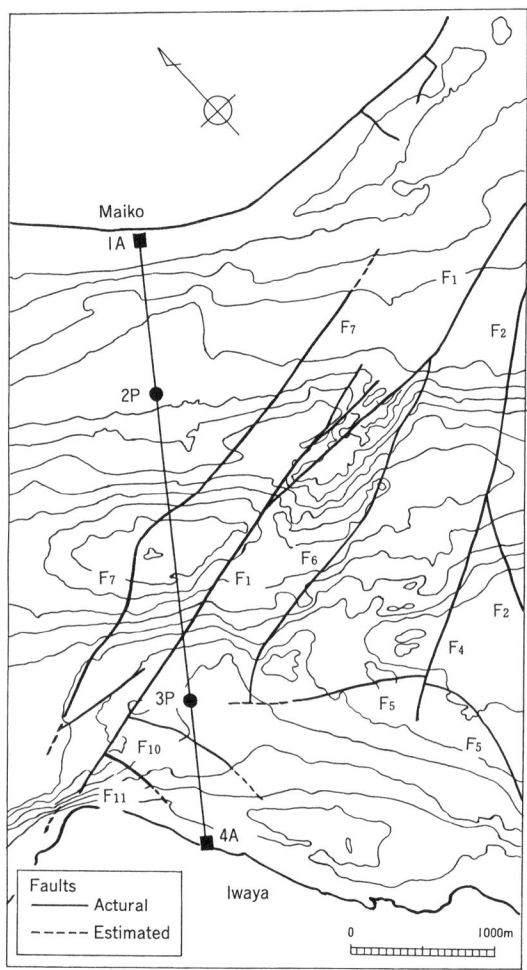

Figure 5 Submarine Geography and Estimated Fault lines Near Akashi Kaikyo Bridge

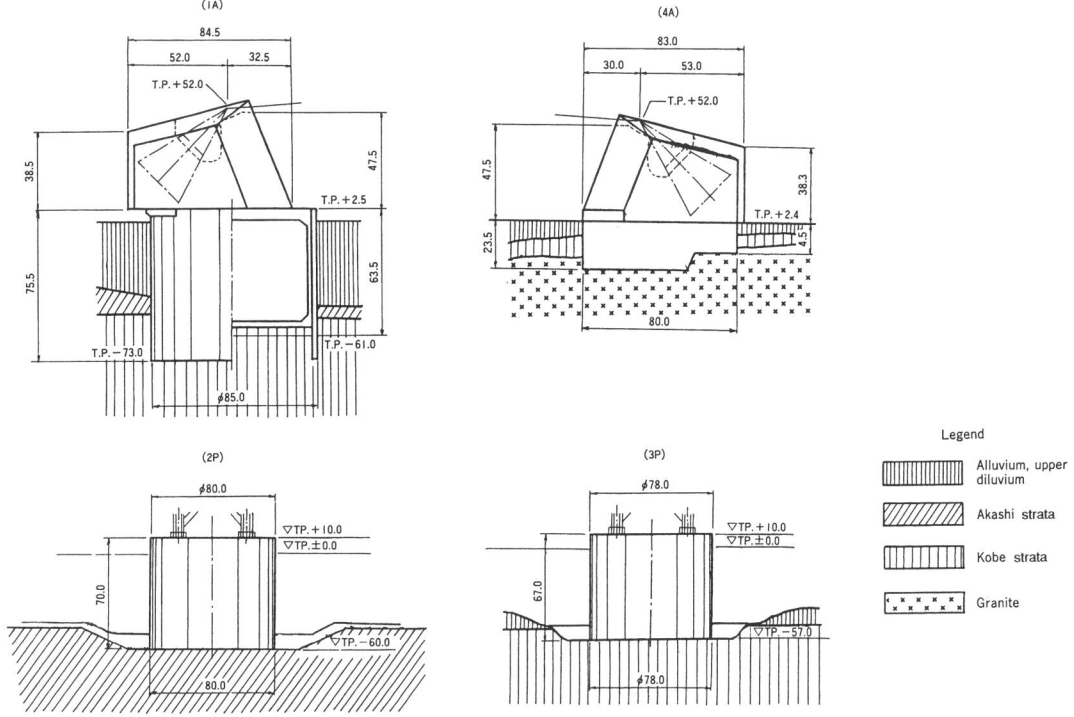

Figure 7 Overall Structure of Foundations (unit: m)

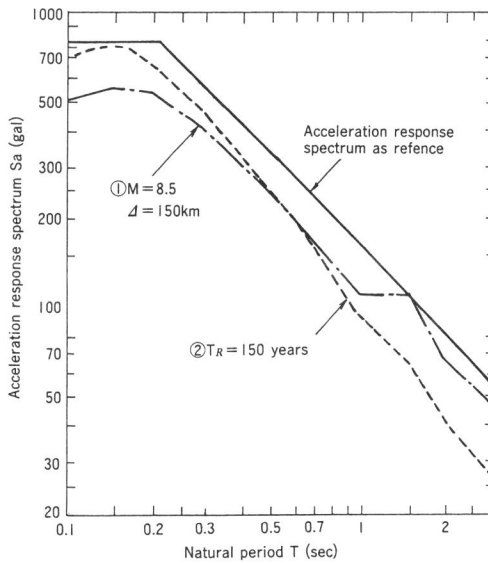

Figure 8 Design Acceleration Response Spectrum of Akashi Kaikyo Bridge (Note 5)

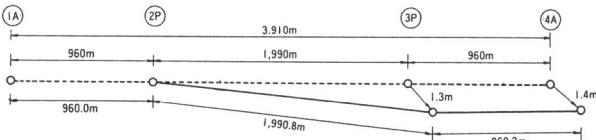

Figure 9 Location of Foundations Following Earthquake

6. Changes to Foundations Following Earthquake

On Jan. 17, when the Southern Hyogo Earthquake took place, the Akashi kaikyo Bridge had all their cable strands set in place, with squeezing work in progress.

Right after the earthquake, survey of the ground around the site was started to see the infuence of the earthquake on the bridge by checking the structures, making measurements, and taking pictures with underwater video cameras. This survey has given the following results.

[1] No damage is seen in the completed anchorages and tower foundations, and in the main towers and main cable structures that heve been erected.

[2] Location surveys by the GPS and geogemetric methods have shown that the tower foundation on the Awaji Island side (3P) relatively moved about 1.3m almost west, and the anchorage on the Awaji Island side (4A) also about 1.3m almost west.

As the result of these movements, the bridge's 1,990m center span widened to about 1990.8m, and the 960m side span on the Awaji Island side to about 960.3m. Fig. 9 shows changes in horizontal layout of the foundations caused by the earthquake. Fig. 10 shows absolute displacements of control points near the Akashi Straits prepared based on the results of surveys by the Honshu-Shikoku Bridge Aut-hority and the Geographical Survey Institute. As is shown in the figure, the survey points on the opposite sides of the Akashi Straits moved by 50 to 100cm in the directions indicated by the arrows.

It is understood from this that the relative movements of the foundations of the bridge occurred because the ground on which they were constructed themselves moved.

[3] Althogh these foundations are constructed avoiding the faults, submarine surveys within the construction work area were made to see what happened to the faults after the earthquake and whether or not new faults might have cropped up.

Surveys made by means of underwater video camera, bathymetric survey, and sonic prospecting have shown no noteworthy changes in the ground condition around the foundations.

[4] As for the tower foundations (2P and 3P), subsiding displacements of the bearing ground were measured with a sliding micrometer. The results are shown in Figs. 11 and 12.

The 2P tower foundation subsided about 20mm (strain: 4×10^{-4}) by the cause of the earthquake, probably because the bearing Akashi strata was compacted.

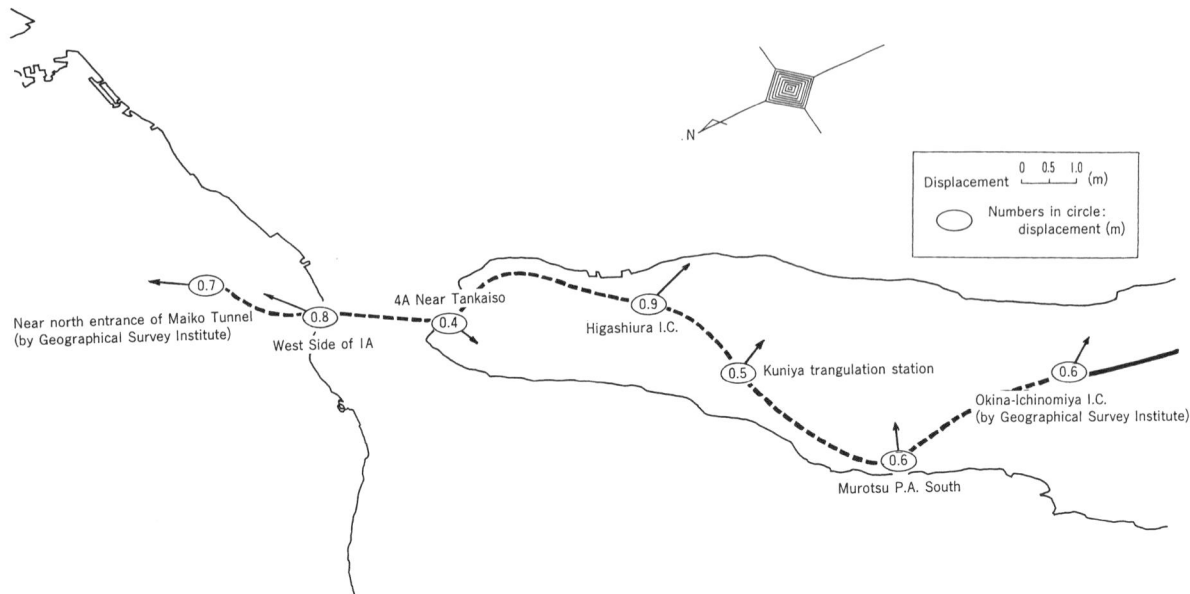

Figure 10 Absolute Displacements of Control Points Displacement

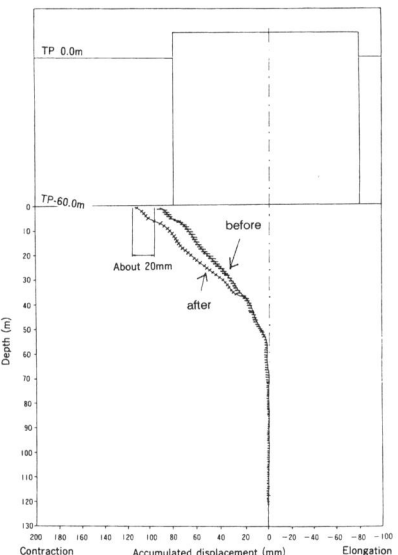

Figure 11 Results of Measurement on 2P with Sliding Micrometer before and after the Earthquake

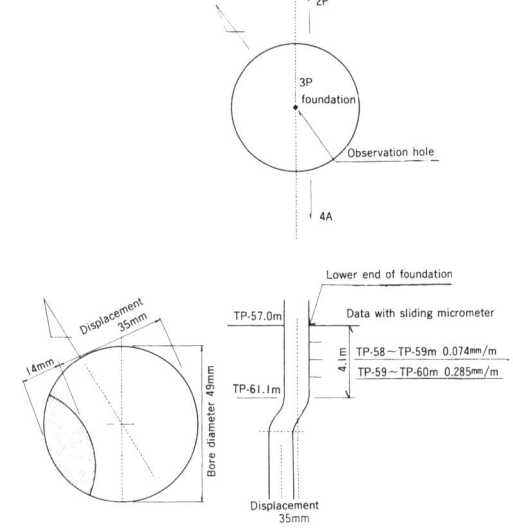

Figure 12 Position of Observation Hole

In the 3P foundation, the measurement hole located 4m below the bottom was choked, making measurement impossible. The condition of the choked section was checked with a bore hole camera, and a bend of about 35mm was observed, as is shown in Fig.12. There was no bend in sections deeper than it. This appears to be attributable to partial sliding of a mud layer which is contained in the bearing Kobe strata.

7. Conclusion

There is probably no precedent in the world's bridge history for our case where the large suspension bridge under construction was attacked by a big earthquake from directly underneath, having its foundations displaced and the spans elongated due to crustal deformations. The influence of the foundation displacements on the superstructure was studied, and it was confirmed that it was within the design tolerances in both mechanical and displacement points of view. Therefore, the elongations in girder spans were coped with by partial adjustments of the length of the stiffening girders.

As for the foundations, boring was carried out to survey changes in rock quality after the earthquake.

Also, to make a continuous survey of changes in relative positions of the foundations and to grasp behaviors of the foundations and their influence on the ground in more detail, elasto-plastic earthquake response analyses by the finite element method using obtained data are now being made.

Notes (References)
1) Report of Damages by the Southern Hyogo Earthquake (Issue 1), Kagima Corp., Feb. 1, 1995.
2) Data Compiled by the Committee on Highway Bridge Countermeasures after the Southern Hyogo Earthquake.
3) Maps of Active Faults in Japan with an Explanatory Text, Univ. of Tokyo Press, 1992.
4) Interim Prompt Report of Seabed Survey by the Maritime Safety Agency, Maritime Safety Agency, Jan. 24, 1995.

A: Mechanical properties

Deformation characteristics of weathered granite taking into account of confining pressure and strain dependence

Atsushi Nitta & Shigeki Yamamoto
Honshu Shikoku Bridge Authority, Tokyo, Japan

Keiji Miyajima & Yoshikazu Kaji
Chuo Kaihatsu Corporation, Tokyo, Japan

ABSTRACT: The general aseismic design methods used in the design of long-span bridges take into account interactions between bedrock and the structure. However, since there is little actual data on the behavior of structures and strain in rock mass during earthquakes, these methods tend to rely mainly on theoretical data. To verify this methodology, large-scale loading tests were carried out in which a large rigid-body foundation model weighing 2.9 MN was loaded onto class C to D weathered granite. The test procedure comprised (1) detailed geological investigations of the test site and (2) on-site loading tests with the large model foundation. The dynamic interactions between bedrock and foundation were determined by analysis of the measured data. This paper describes the geological investigations at the test site (determination of bedrock characteristics) and provides new technical findings on (1) techniques for sampling heavily weathered granite, (2) laboratory tests using a local deformation transducer (LDT), (3) the pressure and strain dependence of the deformation characteristics of weathered granite, and (4) the deformation characteristics of weathered granite taking into account pressure and strain dependence.

1. INTRODUCTION

The aseismic design methods used in the aseismic design of long-span bridges generally take into account bedrock-structure interactions. Actual data on the strain in rock masses and the behavior of structures during earthquakes is limited, so these methods tend to rely on theoretical data.

In particular, the kinetic elasticity of bedrock and its damping characteristics are based on various assumptions and fixed using the latest analytical techniques and technical judgment. This process, however, needs to be checked against field results using an actual structure or a large foundation model.

To verify the method, the Honshu-Shikoku Bridge Authority (HSBA) set up a test site near the proposed Takuma entrance to the Ikuchi Bridge construction site, which was under construction at the time, and carried out large-scale loading tests using a large rigid-body foundation model weighing 2.9 MN on class C to D weathered granite.

The test program consisted of (1) detailed geological investigations at the test site (including a geological survey of the weathered granite bedrock, S-wave seismic explorations, and laboratory tests on removed specimens) and (2) a loading test with the large model foundation (including free oscillation tests, vibration loading tests, reciprocal loading tests, and shearing tests).

Using the collected data, on-going simulations and waveform analysis are taking place to study the interactions between bedrock

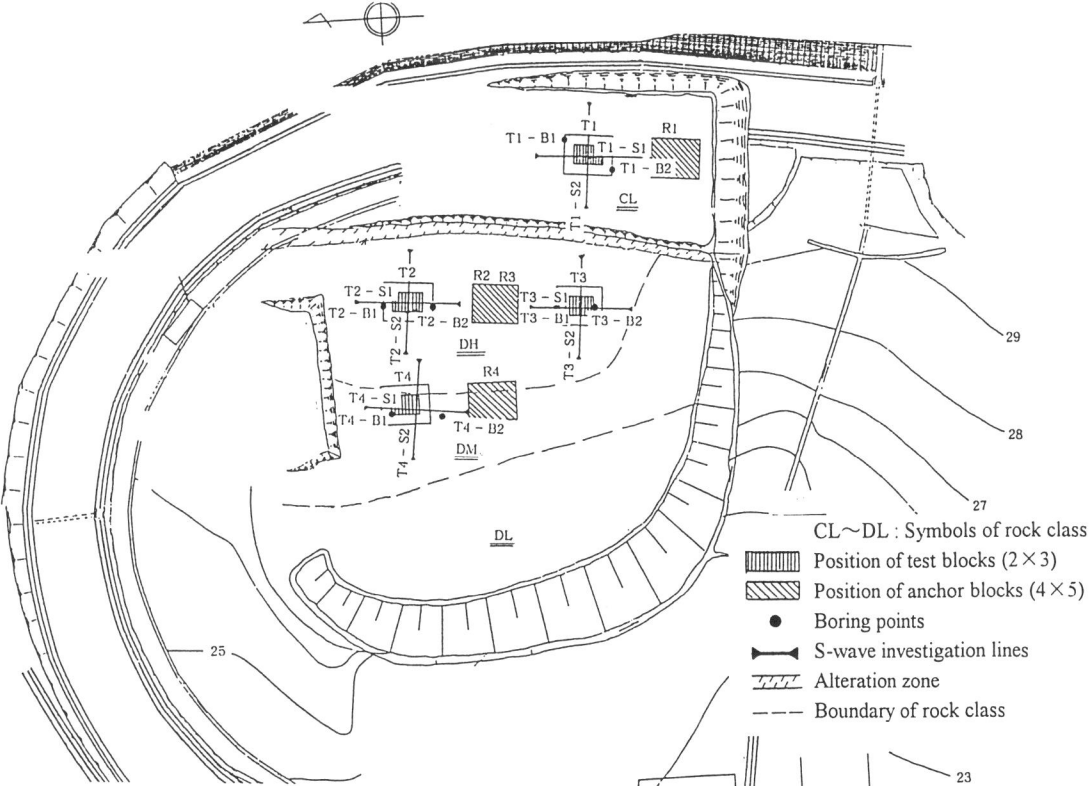

Figure 1. Test yard and test points

and the foundation.

This paper covers the detailed geological investigations, and the intention is to give an outline of the geological characteristics of the test site and the engineering characteristics of the rock. The following points receive special consideration:
(1) Sampling of heavily weathered granite

It has generally been considered difficult to obtain quality core samples from class DH to CL weathered granite using conventional methods in which mud is the circulating drilling agent.

However, in our tests we were successfully able to take samples using the Jet Foam Boring (JFB) method. This method makes use of a uniform and refined mixture consisting of compressed air and foam instead of mud.
(2) Laboratory tests with local deformation transducer (LDT)

The precision of small strain measurements was enhanced by using a very precise LDT.
(3) Deformation characteristics of weathered granite (classC to D)

New findings were obtained, including the strain and confining pressure dependence and the dependence of deformation characteristics on deviation stress.

2. OUTLINE OF DETAILED GEOLOGICAL INVESTIGATION

The granite at the test site is of the porphyritic variety classified as Hiroshima granite. It contains many plagioclase phenocrysts compared with the coarse-grained biotite granite found, for example, around the Tatara Bridge site.

The test site features a 1 m-wide sheared alteration zone stretching north to south and inclining westward at the sharp angle of 70 to 80 degrees. Both ends of this alteration zone have undergone hydrothermal alteration, and it divides the geology into class C rock to the east and class D rock to the west (see Figure 1).

Detailed geological investigations were carried out prior to the loading tests in order to (1) determine a suitable position for the loading tests using the large model foundation, (2) identify the S-wave velocity structure of the bedrock to be used in the loading tests, (3) identify the deformation characteristics of the bedrock to be used in the loading tests, and (4) prepare analytical models of the rock mass at the loading test points.

The test points and details of the geological investigations are given in Fig. 1 and Table 1, respectively. The major engineering characteristics of the bedrock at the test points (under loading surface) as determined in this investigation are given in Table 2.

Table 1. Summary of detailed geological investigations

Test items	Contents
1) Geological survey for selecting test points	
2) Seismic exploration (S-wave)	- 2 lines * 4 test points - Length of profile line = 11m - pickup interval = 0.5m
3) Boring and sampling	- Jet Foam Boring - depth = 7m - number of boreholes = 8
4) Well logging	- PS, Electrical, Density, Caliper - all boreholes
5) Borehole loading test	- Elast-meter - 17 measuring points
6) S-wave measurement under test block	
7) Laboratory test * Tests for physical properties * Monotonic loading traixial compression test * Cyclic loading triaxial test	- 9 series - 3 series (CL,DH,DM) - CUB - 6 series (CL,DH,DM) - Normal and Stepwise - CUB

3. SAMPLING METHOD

It is difficult to sample class DH to CL weathered granite using conventional boring methods, so the JFB method, as shown in Fig. 2, was adopted for sampling purposes.

This method makes use of a foam instead of mud during the boring process, and it is particularly effective for sampling unconsolidated gravel, fractured rock, and weathered granite. The following are the major characteristics of this method:
(1) Since the foam is uniform, fine-textured, and very adhesive, quality samples can be taken even from class DH to CL weathered granite with a high sampling efficiency.
(2) A borehole drilled using this method has walls of excellent stability, which helps improve the precision of data collected during in-situ tests (well logging, borehole loading tests, etc.).
(3) The foam is non toxic and so safe in use.

Photo 1 shows a core sampled using the JFB method. In the core shown, a number of specimens have been returned to their original positions in the 0 to 3 m range after laboratory mechanical tests.

Core specimens taken with the JFB method in vinyl core packing tubes were protected in half-section PVC tube to prevent disturbance. The specimens were then placed in a carrying box containing sponge mat for transport from the sampling site to the laboratory.

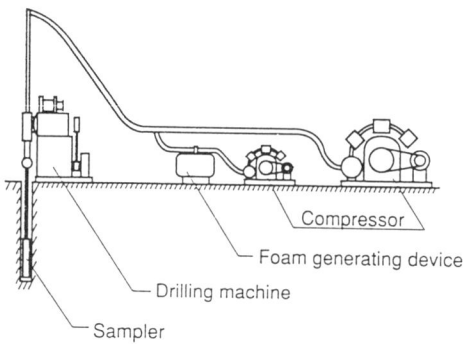

Figure 2. System structure of Jet Foam Boring (JFB) method

Photo 1. Boring core with JFB (class DH rock)

Table 2. Basic properties of bedrock at test points (Loading surface) (on-site)

Test point	Rock class	1) S-wave velocities (m/sec)	2) Rockmass densities (g/cm^3)	3) Modulus of deformation (MPa)
T1	CL	700	2.25〜2.50	188〜544
T2	DH〜DM	370	2.00〜2.35	60.8〜196
T3	DH〜DM	400	2.05〜2.30	56.9〜147
T4	DM	300	2.05〜2.30	36.3〜121

1) S-wave exploration
2) Density logging
3) Borehole loading test

4. LABORATORY MECHANICAL TESTING

(1) Test methods

1) Cyclic loading triaxial tests

In cyclic loading triaxial tests (CLT), the shear stress is varied under a predetermined confining pressure in at least 10 stages. In these tests, G-γ and h-γ relations are obtained when the shear strain (γ) is in the range 10^{-6} to 10^{-2}.
(G: shear modulus; h: damping modulus)

Eleven sine waves were loaded at each shear stress, G, h, and γ were accomplished with the tenth of these.

2) Multi-stage cyclic loading triaxial tests

Multi-stage cyclic loading triaxial tests are used to obtain the dependence of shear modulus (G_0) on confining pressure when the shear strain (γ) is 1×10^{-6}.

In these tests, the consolidated confining pressure is sequentially increased on one specimen. Cyclic loading tests are then carried out at each σ c.

In this test, 11 repetitions were implemented, and the shear stress was increased by 3 to 4 times at each σ c. Shear modulus γ was limited to 1×10^{-4} so as not to affect the results at the following confining pressure step.

Table 3. Laboratory test conditions

Test items	Monotonic loading triaxial compression test	Cyclic loading triaxial test
Water conditions of specimens	saturated (B-value > 0.9)	saturated (B-value > 0.9)
Size of specimens	Hight : 130 mm Dia. : 65 mm	Hight : 130 mm Dia. : 65 mm
Testing methods	CU-test	CU-test
Effective confining pressure (MPa)	0.20, 0.39, 0.59, 1.18, 2.35	0.20, 0.39, 0.59 0.1 - 6.0 (stepwize)
Loading speed	Strain controll 0.05 %/min	0.1 Hz
Measurement	Pressure : Inner load cell External load cell Displacement : LDT External gauge Water pressure	Pressure : Inner load cell External load cell Displacement : LDT External gauge Water pressure

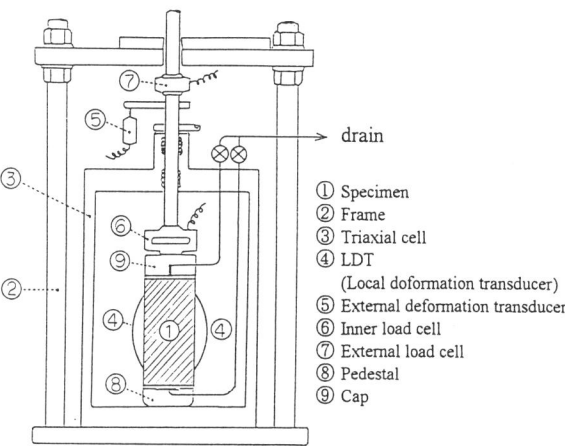

Figure 3. Schematic figure of cyclic triaxial testing system

(2) Test conditions
The test conditions are shown in Table 3.
(3) Test equipment
The test equipment is shown in Fig. 3. In the triaxial tests, a load cell was placed inside the triaxial cell. Axial strain was accurately measured with a pair of Local Deformation Transducers (LDTs) to prevent loading errors at the specimen ends.

5. DEFORMATION CHARACTERISTICS OF WEATHERED GRANITE

As is typical of soil materials, the deformation characteristics of weathered granite exhibit strain dependence and confining pressure dependence. These are discussed below. Table 4 lists the physical properties of the mechanical test specimens.

5.1 Strain dependence of shear modulus (G) and damping modulus (h)

Figure 4 shows the G/G_0-γ and h-γ relationships for class CL rock as obtained from the cyclic loading triaxial tests. On the basis of these results, analytical models for class CL rock (G/G_0-γ and h-γ relations) have been drawn up, as shown in Fig. 5.

In this figure, the solid lines are the proposed curves, while the dotted lines indicate the scatter in the measurements.

G/G_0-γ and h-γ relations for rock classes DH to DM obtained by cyclic loading triaxial tests are shown in Fig. 6. The proposed analytical models for these rock classes (G/G_0-γ and h-γ relations) are shown in Fig. 7.

(1) G/G_0-γ relationship

For class CL rock:
G/G_0 is in the range 0.7 to 0.85 when $\gamma = 1 \times 10^{-4}$
G/G_0 is in the range 0.1 to 0.3 when $\gamma = 1 \times 10^{-2}$
γ is in the range 3×10^{-4} to 1×10^{-3} when $G/G_0 = 0.5$

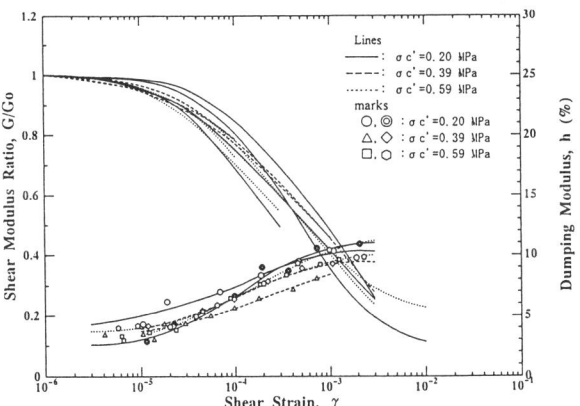

Figure 4. Relationships between G/Go, h and shear strain (measured data, class CL rock)

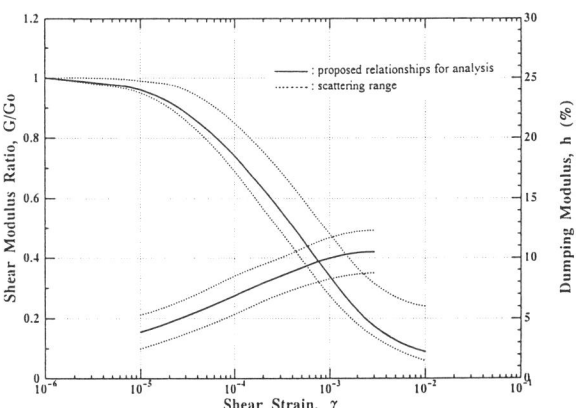

Figure 5. Relationships between G/Go, h and shear strain (proposed relationships for analysis, class CL)

For class DH to DM rock:
G/G_0 is in the range 0.8 to 0.9 when $\gamma = 1 \times 10^{-4}$
G/G_0 is in the range 0.1 to 0.2 when $\gamma = 1 \times 10^{-2}$
When $G/G_0 = 0.5$, γ is in the range 5×10^{-4} to 9×10^{-4}

(2) h-γ relationship
For class CL rock:
h is in the range 2.5 to 5.5 % when $\gamma = 1 \times 10^{-5}$

h is in the range 8 to 12 % when $\gamma = 3 \times 10^{-3}$
For class DH to DM rock:
h is in the range 2 to 4 % when $\gamma = 1 \times 10^{-5}$
h is in the range 14 to 23 % when $\gamma = 3 \times 10^{-4}$.
When γ exceeds 1×10^{-3}, the deviation of h widens.

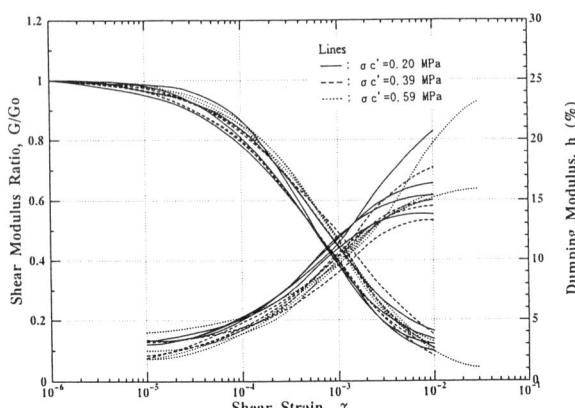

Figure 6. Relationships between G/Go, h and shear strain
(measured data ,class DH~DM rock)

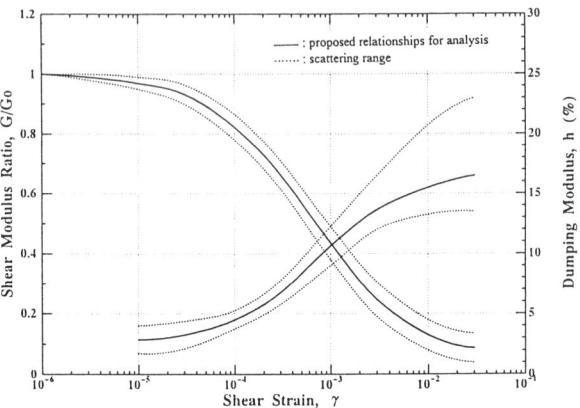

Figure 7. Relationships between G/Go, h and shear strain
(proposed relationships for analysis ,class DH~DM rock)

Table 4. Basic properties of specimens

Number of Specimen	CLT test		Sampring Bor. No.	Pysical Properties (Lab. tests)							Rock mass (in-situ)		
	$\sigma_c{'}$ (MPa)	G_0 (MPa)		ρc (g/cm³)	n_{ef} (%)	e_{ef} (%)	w (%)	S_r (%)	V_{pc} (km/s)	V_{sc} (km/s)	Rock mass class	V_{pb} (km/s)	V_{sb} (km/s)
CL-1	0.20	970	T1-B1	2.487	6.5	0.069	0.7	27.8	0.89	0.47	C_M	2.50	1.05
CL-4	0.39	1790	T1-B1	2.495	6.5	0.069	0.7	27.8	0.84	0.45	C_M	2.50	1.05
CL-5	0.59	2030	T1-B1	2.458	9.7	0.107	2.6	63.6	0.84	0.51	C_L	—	—
CL-6	0.59	1750	T1-B1	2.447	5.5	0.058	1.4	63.3	1.00	0.44	C_L	2.50	0.80
DH-1	0.20	210	T2-B1	2.097	36.4	0.572	7.1	30.8	0.51	0.28	D_H	0.35	0.22
DH-2	0.20	230	T2-B1	2.240	14.9	0.175	6.7	96.6	0.58	0.30	D_H	1.10	0.63
DH-3	0.39	380	T2-B1	2.231	22.6	0.292	6.6	57.3	0.51	0.30	D_H	1.10	0.63
DH-4	0.39	350	T2-B2	2.176	24.9	0.331	8.0	62.0	0.56	0.31	D_H	0.65	0.40
DH-5	0.59	430	T2-B2	2.143	—	—	—	—	0.48	0.26	D_M	0.52	0.26
DH-6	0.59	410	T2-B1	2.103	29.8	0.424	9.9	59.8	0.46	0.28	D_H	0.35	0.22
DM-1	0.20	220	T4-B2	2.156	22.1	0.284	6.3	56.5	0.33	0.18	D_M	1.05	0.55
DM-2	0.20	220	T4-B2	2.144	24.5	0.324	8.6	64.7	0.33	0.20	D_M	0.75	0.45
DM-3	0.39	360	T4-B2	2.046	29.9	0.426	8.0	47.5	0.34	0.21	D_M	0.75	0.45
DM-4	0.39	380	T4-B2	2.118	24.6	0.326	7.0	54.2	0.33	0.18	D_M	1.05	0.55
DM-5	0.59	410	T4-B2	2.131	26.3	0.357	8.2	58.3	0.34	0.21	D_M	0.75	0.45
DM-6	0.59	480	T4-B2	2.135	26.3	0.357	8.2	58.3	0.38	0.21	D_M	0.75	0.45
Multi-step	σ_c		Bor. No.	ρc	n_{ef}	e_{ef}	w	S_r	V_{pc}	V_{sc}	class	V_{pb}	V_{sb}
CL-M	0.10 0.3 0.6 0.15 0.4 0.20 0.5		T1-B1	2.495	6.5	0.069	0.7	27.8	0.84	0.45	C_L ~ C_M	2.50	1.05
DH-M	0.15 0.40 0.20 0.50 0.30 0.60		T2-B1	2.166	21.6	0.286	8.6	83.6	0.52	0.28	D_H	0.45	0.30
DM-M	0.01 0.10 0.40 0.03 0.15 0.50 0.05 0.20 0.60 0.07 0.30		T4-B2	2.129	25.1	0.336	8.3	64.9	0.35	0.18	D_M	0.75	0.45

5.2 Confining pressure dependence of shear modulus (G_0)

Figure 8 shows the relationship between G_0 and effective confining pressure ($\sigma'c$) when $\gamma = 1 \times 10^{-6}$. White symbols represent the results of multi-stage cyclic loading triaxial tests on a single specimen, while black symbols are the results of usual cyclic loading triaxial tests on multiple specimens. The double circles represent shear modulus (G_{vsf}) calculated from the S-wave velocity measured in the rock just under the model foundation before and after it was put in place.

The figure clearly shows a dependence of G_0 on $\sigma'c$ in the specimens. There is also a clear tendency for the in-situ shear modulus (G_{vsf}) to increase with block loading (confining pressure).

Figure 9 shows the relationship between G_0 and confining pressure as derived from the results of multi-stage cyclic loading triaxial tests based on the assumption that the relationship between G_0 and $\sigma'c$ is $G_0 = a \times \sigma'c^b$.

The regression coefficients a and b for each plot are shown in Table 5. Figure 10 also shows corresponding results for specimens sampled by grabs from the bedrock (class DH) at the 2P point of the Tatara Bridge.

Values of gradient (b) in both sets of results are almost the same, indicating that the bedrock at the test site and that at Tatara Bridge 2P have similar dependence on confining pressure.

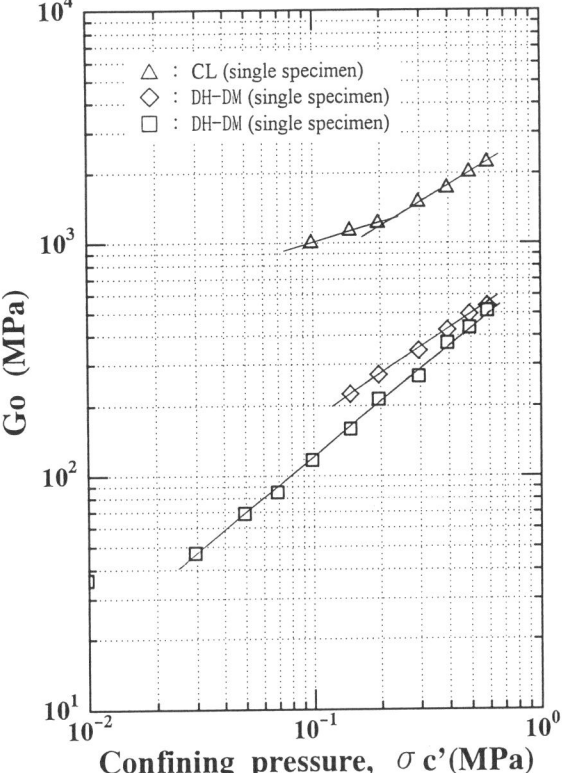

Figure 9. Relationships between Go and confinig pressure ($Go = a(\sigma c')^b$, regression)

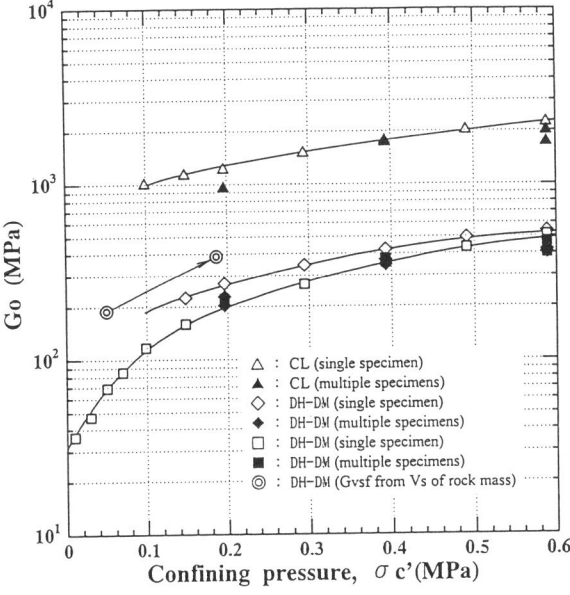

Figure 8. Relationships between Go and confinig pressure

Tabel 5. Shear modulus(Go) taking account of confined pressure (Regression coefficient a,b ; $Go = a\sigma^b$)

Rock class	Marks	Ref. figure	a	b	r^2	Remarks
CL	△	fig.9	3008.0 (1935)	0.5771 (0.2850)	0.9967 (0.9989)	test-yard
DM~DH	◇	fig.9	769.9	0.6440	0.9986	test-yard
DM~DH	□	fig.9	762.5	0.8018	0.9979	test-yard
DH	●	fig.10	432.5	0.7890	0.9992	Tatara 2P
DH	▼	fig.10	510.3	0.8644	0.9957	Tatara 2P
DH	■	fig.10	547.2	0.8149	0.9988	Tatara 2P

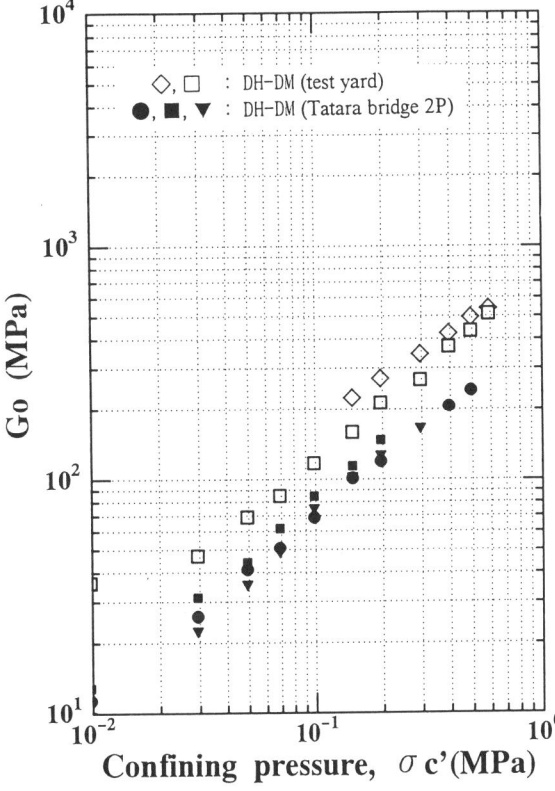

Figure 10. Relationships between Go and confinig pressure

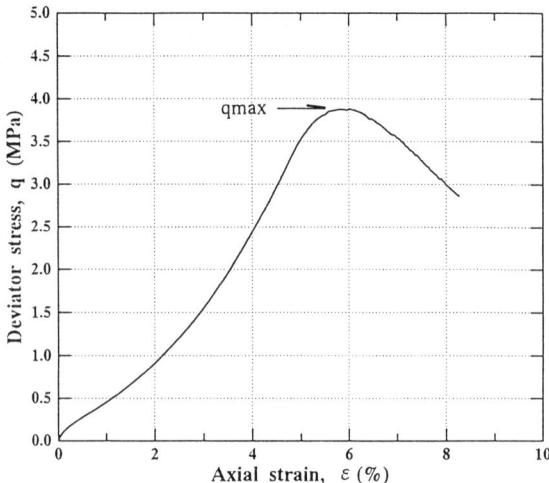

Figure 11. Stress-strain curve (class DH rock)

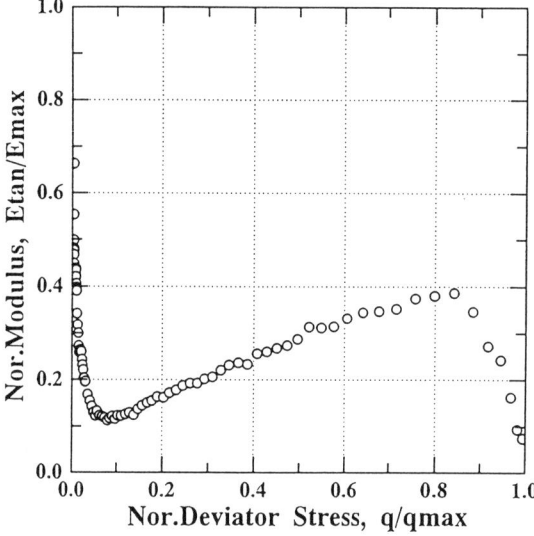

Figure 12. Relationships between E_{TAN}/E_{MAX} and q/q_{MAX} (class DH rock)

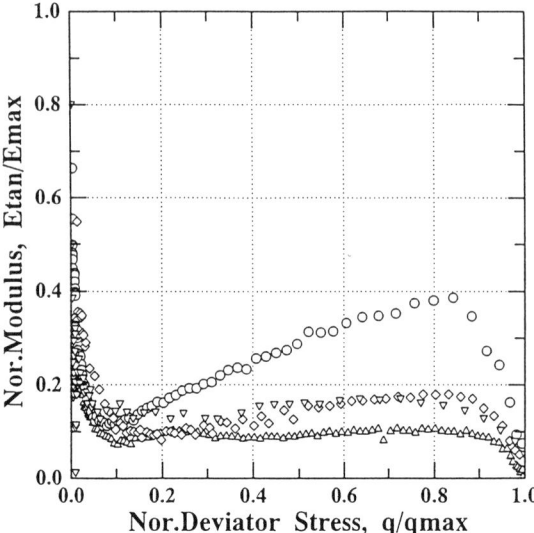

Figure 13. Relationships between E_{TAN}/E_{MAX} and q/q_{MAX} (4 specimens, class DH rocks)

5.3 Stress dependence of deformation characteristics

In the monotonic triaxial compression test on class DH weathered granite, it was observed that the deformation characteristics increased with deviation stress (q). As shown by the stress-strain curve in Fig. 11, the modulus of the tangent (E_{tan}) initially decreases with increasing q.

With further increase in q, Etan turns upwards before once again decreasing as q reaches the fracture point. To clarify this behavior, the stress-strain curve in Fig. 11 is used to calculate E_{tan}/E_{max} and q/q_{max}, and these values are plotted in Fig. 12.

The stress ratio (q/q_{max}) is the deviation stress at the time of loading (q) divided by the deviation stress at fracture (q_{max}).

The normalized deformation modulus (E_{tan}/E_{max}) is the absolute value (modulus) of the tangent of the loading curve divided by E_{max} (the initial value of E obtained from the initial linear section of the stress-strain relationship for E less than about 0.001%). Results for other rock are shown in Fig. 13.

These figures indicated that E_{tan} (E_{tan}/E_{max}) begins to increase as the stress level reaches 7% to 8% of qmax and begins to decrease again when the stress exceeds 80% of qmax just before fracture.

At the point where E_{tan} begins to rise, that is at a stress level of 7% to 8% of qmax, it is about 10% of E_{max}.

Past studies have found these characteristic S-shaped stress-strain curves in Tertiary sandstone and in Toyoura sand after cyclic loading for 120,000 times.

During loading tests with the large model on the bedrock, the loading-subsidence curve also showed a similar tendency to this S-shaped stress-strain curve in some cases.

We believe this may help in the prediction of subsidence in class DH weathered granite.

6. CONCLUSION

This report discusses the deformation characteristics of weathered granite as found in mechanical tests performed in the laboratory The relationship between in-situ deformation modulus and laboratory (specimen) deformation modulus has not been elucidated as yet. This issue will require further study.

Investigation of dilatancy in rock massif

V.A. Mansurov
Institute of Physics and Mechanics of Rocks, Bishkek, Kyrghyzstan

ABSTRACT: In the laboratory experiments was determined the energy value of elastic waves, released when a crack of determined size (unit) was formed. This characteristic for different types of rocks changes more than in ten times is physically based parameter. Determination of this characteristic was realized with the use of AE and strain measurements during loading and deformation of the rock samples. During the preparation of rock bursts and earthquakes, one of forerunners is the change of dilatancy in time. There is suggested to use the microseismic monitoring for the up - date preventive steps and forecast of current state of the controlled subject.

The received dependencies of radiation of elastic energy versus value of appearing defects, allow to make the evaluation of dilatancy in the places closed for the direct observation. Transition to the monitoring of rock massif is based on the determined consequences and laws of radiation of elastic waves on different scale levels.

Introduction

According to recent investigations of rocks failure [1,2] both laboratory samples and rock masses undergo macroscopic failure when reaching a critical concentration of cracks. Since. the last ones are three-dimensional discontinuities (flaws), the volume of the loaded body at their generation and development increases, i.e. dilatancy takes place. That is why a certain diltancy corresponds to a threshold cracks concentration. It naturally depends on structure-texture properties of the material and it stress state. In rock massif (in mines) some methods are used for evaluation of the degree of their discontinuous, like deformational measurements, rheometry, dynamic elastic tests and electrical resistivity measurements. Any of these methods has its own deficiencies and limitations. The main one is that register the degree of discontinuous on rather close to openings distances and nuclei of the preparing failure often is situated at a considerable depth.

We shall illustrate for dilatancy evaluation the possibilities of appearing and growing cracks registration by Acoustic Emission methods.

Failure Process and Experimental Methods

Failure in rock and the similar materials, come as result of: the development of deformation and failure processes, which proceed in time; the localization of these processes in space and the concentration of elastic energy in some areas, where relaxation is quite difficult. The loss of stability at such a failure nucleus is accompanied by the rapid release of accumulated elastic energy and it forms elastic impulses or waves, which can posses destructive power. For practical use it is important to estimate the energy of such phenomena, and the place and time at which they will occur. Now the definite success has been achieved in the field of indication of the most possible places of rocks failures and their energy. But the solution of the third task is still under discussion.

This paper doesn't set the task to solve these problems completely. It is supposed to discuss the physical premises which must be taken into account while the work on rock failure prediction is being put and also further work of the methods of failure processes control in the massif of rocks.

Experimental investigation of rock failure on the microscopic level showed that in the simplest cases when the stuff has only one hierarchical level of structure, i.e. it consists of structural elements equal in size and strength, the whole failure process may be divided into two main stages [3,4]. The first one is manifested by stable microcracks accumulation, the second is connected with their enlargening, leading to the failure nucleus formation and its development. It is necessary to point out that the more complex structure of rocks influences the laws of failure process. So heterogeneity of the rock structure leads to prevention of the newly generated cracks on the structural boundaries and their numerous accumulation. However, in both cases failure nucleus is formed when a certain concentration of cracks in the loaded solid volume is reached. This value may be described as the relation of average space between cracks $C^{-1/3}$ to crack sizes l_0, i.e. $K = C^{-1/3} / l_0$. K is an average space between cracks measured in units of crack sizes l_0. C is the concentration of chaotically distributed cracks in the volume. According to Zhurkov et al. [5], crack interaction and their enlargening occur when $K \approx 3$. Failure process transfer to the next scale or hierarchical level. The analysis showed its applicability for a wide range of stuffs, having different structures and sizes of initial microcracks [6,7].

Hence, for observation and control of rock failure process, we need information about sizes and concentration of cracks generated under loading. The scientists, working in the domain of failure, have a variety of direct and indirect methods. But most of them are either applicable only in a narrow range of crack dimensions or may be used by only after failure. Acoustic emission (AE) method is more universal in this respect and the most appropriate for rocks. It is based on the elastic waves emission, coming as a result of cracking and frictional sliding along a fault, being applicable for any scales of cracks from microscopic to macroscopic ones. Special methodical works [8,9] illustrated that duration of the first component of the elastic impulse T/2 generates when the crack appears or propagates in correlation with its size L. T/2 is of the elastic wave from the first time of arrival to the crossing of the time axis by it.

Figure 1 illustrates, that this correlation exists in a wide range of changes of these parameters. For comparison we involved not only laboratory data, but also that obtained in large - scale experiments on pillar loading in mines as well as for the faults in the earth crust, causing earthquakes [10].

Experimental results shown on Figure 1 (relation T/2 versus L) constitute only small part of numerous data of investigation, where reliable values of T/2 is not measured due to the complicated character and reliability of the evaluation of the first component of elastic impulse of the acoustic signal. Figure 1 illustrates also correlation between AE energy E and size of cracks L generated under loading [11]. For this case, in laboratory experiments, energy of elastic impulses (AE events) may be estimated with accuracy to the coefficient as follows:

$$E = \alpha A^2 T \qquad (1)$$

where A, T are correspondingly the amplitude and duration, measured by the envelope [12]. The energy of rock bursts and earthquakes in failure nucleus was calculated by known methods, applied in seismology and taking into account place of source of the elastic waves generation. For larger subjects of inquiry, it is necessary to know location of the AE sources and characteristics of elastic wave attenuation in the loaded volume. Besides, for complicated signals, it would be more correct to determine energy according to the expression $E = \int A^2 \, dT$. However, Figure 1 shows, that correlation between energies of elastic impulses and sizes of cracks is close enough, though in AE energy calculations for laboratory samples we used expression (1).

So, acoustic signals parameters, generated as a result of cracking enable us to get information about crack dimensions and elastic energy emitted at their nucleation. This possibility is of great significance for our further discussion, because AE method also allows the definition of number of cracks generated under loading in the tested stuff or subject of inquiry.

System of AE data acquisition and processing

For laboratory experiments the system was based on multidimensional analyser of impulses AI 4096 - 3M (produced by Ministry of Nuclear Energy of former USSR) and a computer PDP - 11. Figure 2a represent schematic diagram of loading and data acquisition systems. The elastic impulse, generated in the sample (S), is transformed into electric signal by the receiving piezotransducer (PT) and, after amplifying by amplifier (A), comes to the unit of signals processing.

Piezotransducers of P-waves were used in experiments. Their design allowed the elimination of reflections to a considerable extent. It was reached by application of damping element of copper alloy in the form of cone (Acoustic trap). It was connected with the active element (PZT ceramic) by Wood's alloy. This unit was mounted into a cylinder cell of dynamometer with oil, which minimized a back reflection of elastic wave in PZT element. The analogous digitizer unit does not permit distinguishing between mutually overlapping events, if their joint envelope does not cross the threshold of discrimination. Along with obvious faults, this approach of signal processing has its own advantages, as follow. Standardized impulses then followed to the AI, which coded every parameter and time of acoustic signal arrival. AI accumulates the necessary number of signal from 1 to 10^3, and then sends them to the computer. Time of data transmission was less than 1 s, which didn't lead to significant loss of acoustic signals. So every AE event is characterized by three initial parameters: amplitude, duration and generation time. The energy was calculated on the basis of the equation (1). Frequency range of recording made 70 kHz up to 2 MHz, which was defined by band width of the analogous digitizer and enabled to record cracks in rock, having linear dimensions of order 10^{-4} - 10^{-2} m. It is provided on one hand by structural heterogeneity of the tested rock samples, and on the other hand by commensurability of maximal crack with the sizes of the last. Transducers and units of elastic impulses amplification were specially designed for the given range. Acoustic emission research of rock failure in laboratory experiments showed that equipment sensitivity must allow to record AE signals with amplitude values from $40 \mu V$ up to 0.1V. That's why a dynamic range of the recording system must be 30 - 60 dB. In our investigations we used granite and marble samples of a cylinder shape with a ratio of 2 of the height h to the diameter d with d = 0.03m. Their loading was performed on testing machines of different stiffness. The samples were placed between two cylindrical dynamometers (load cells), in which piezotransducers were mounted, as was mentioned above. In Figure 2, only the lower one is shown. After amplification the AE signals came to the location system, which allowed to linear locate zone of elastic impulses emission due to the difference between signal arrivals. This enabled us to eliminate friction noises on butt - ends of the sample. The main analysis of acoustic emission described above was done in one of the channels. Recording, data processing and data logging were done with the help of the software, allowing to give the necessary conditions of recording and processing in the dialogue regime.

Experimental Results and Discussion

AE recording during the process of sample deformation reveals not only the qualitative, but also the quantitative relationship between the dilatancy, appearing when the rocks deforms and the

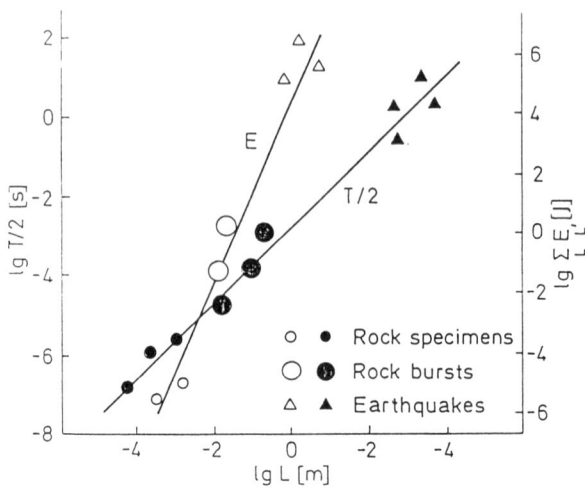

Fig.1. Connection of the duration of the first component-semi wave - (T/2) and the energy (E) of the elastic impulse generated during cracking with the size of the generated cracks (L).

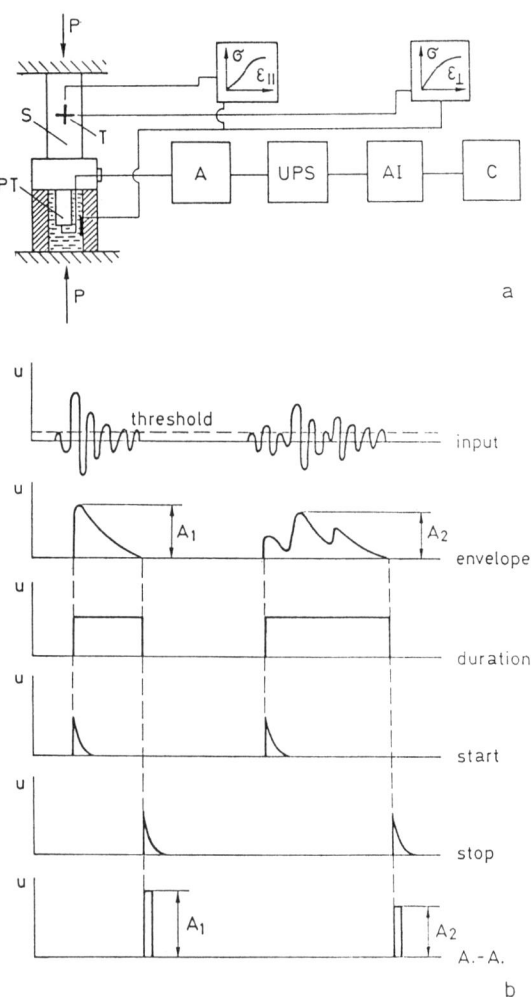

Fig. 2. Schematic diagram of loading and data acquisition systems. a: S - sample, p - the load, applied to the sample by means of loading plates, A - amplifier, UPS - unit of signals processing, AI - analyser of impulses, C - computer, "$\sigma - \varepsilon_{\parallel} - \sigma - \varepsilon_{\perp}$" plotters, T - strain gauges, PT-piezotransduser, b : time diagrams of the signal processing unit.

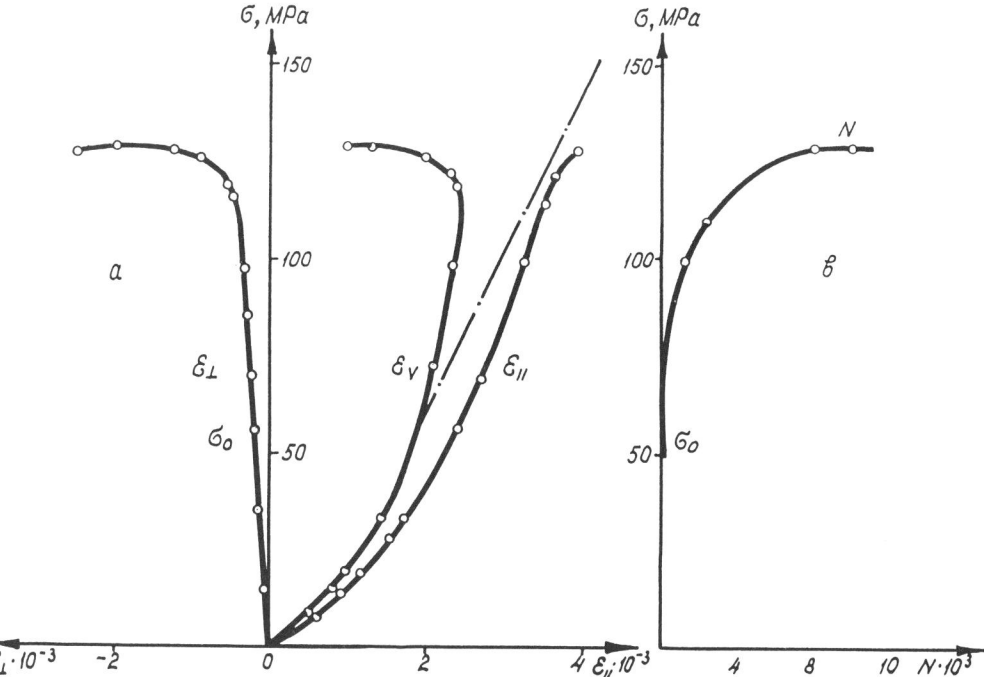

Fig. 3 Graphs of acting stress change (σ) versus (a) longitudinal ($\varepsilon_{\parallel}$), transverse ($\varepsilon_{\perp}$) and volumetric (ε_v) strains; (b) elastic impulse accumulation (N) for granite sample

AE. This is apparent from a comparison of dilatancy (Figure 3a, curve 3) and the cumulative AE curve (Figure 3b), i.e. the expansion of a sample, which qualitatively changes the run of, ε_v, begins at the moment when AE becomes noticeable. If we use the initial part of the ε_v / σ graph, when generation and growth of cracks are absent, and extrapolate it (dotted line on Figure 3a), then we can easily single out the component ε_v connected with cracks generated under loading. There is a dependence between ε_v and N, and it is nearly liner. More understandable is the dependence between ε_v and the total energy of the elastic impulses, ΣE

$$\Sigma E = \beta \varepsilon_v' \text{ Vsample.} \quad (2)$$

The premise for this is the relation between the volume of the crack Vcr and elastic energy emitted during this time Ecr [9]

$$Ecr \sim \beta \text{ Vcr} \quad (3)$$

In equations (2, 3) coefficient β characterizes the energy emitted during the formation of an individual crack volume under the loading.

Functional relations (2,3) might have been suggested apriori, since the elastic energy is emitted from the volume, in which redistribution of elastic mechanical stresses takes place at the moment of failure, commensurable with its linear sizes. Qualitatively it is substantiated by experiments [12] in a dependence of acoustic impulse amplitude versus size l of the crack, generated by loading.

In Figure 4 the dependencies of volumetric strain ε_v both on stress (curve 1) and on combined released elastic energy as acoustic signals (curve 2) is shown. Their behaviour is very complicated but it can be explained under their simultaneous analysis. Curve 1 till loads ~0.7 from strength limit σ^* is in the area of negative values ε_v, which is connected [13] with the sample sealing. Then the curves turn sharply and go into the area of positive values ε_v. The reason is the effect of cracking. The forming cracks generate acoustic signals. They explain the course of the curve 2. In its left part the sample sealing competes with its loosening while crack formation. When curve 1 crosses the ordinate axis it means that sample sealing and loosening are equal. Consequently, the volume of the crack Vo generated by the time equals to the sample volume decrease due to the starting cracks and pores sealing. If we continue the straight portion of the curve 1 (the tangent to the curve) to intersection with line, which goes through the point σ_0 of the ordinate axis we may easily evaluate ε_{v_0}. This value can be evaluated in the other way. The right part of the curve 2 is practically straight and is due to dilatancy caused by new cracks formation. If we continue this straight line portion up to the axis ε_v, we gain the value ε_{v_0} too. The values gained by the two ways are close to each other. The straight line equation may be written as:

$$E = Eo + \beta \varepsilon_{v_0} Vs \quad (4)$$

where Vs - is the starting sample volume, and parameter $\beta = Eo / \varepsilon_{v_0}$.

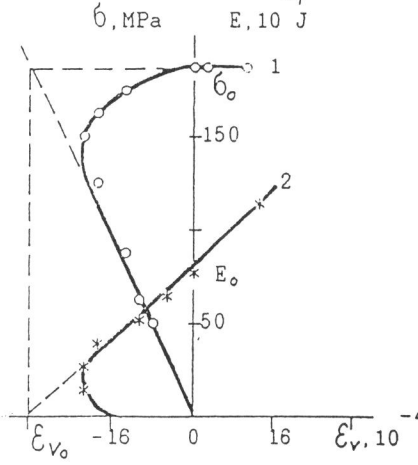

Fig. 4 Changing of volumetric strain (curve 1) and acoustic signals energy E during the increase of stress under uniaxial loading of granite sample.

It must be marced that as a rule E is measured in relative units. However the absolute value of impulses can be evaluated if the preliminary(calibration) of the loading and AE data acquisition system would be held [12]. The experiments with different types of rocks showed that this value changes from ~0.1 J/m^{-3} up to 100 J/m^{-3}.

In spite of approximate values of β, it's knowledge allows, using expression (4) on the basis of AE energy measurement to evaluate absolute values of dilatancy in the loaded body, when deformations cannot be measured, or it is difficult to do it. Proportionality ratio β has a simple physical interpretation. That is why having these values it is possible to compare them with those obtained in real conditions [14]. For many types of stuff macroscopic failure takes place at reaching dilatancy 1%. This value is well compatible with the idea about concentration enlargening of cracks. Naturally, it should be understood that elastic energy at generation of cracks with some volume Vo is emitted from some region with volume V >> Vo. From these data it is impossible to evaluate Vo.

Conclusion

The obtained parameter of the energy emitted during the formation or development of an individual crack volume under changing stress state is a very useful. It may be applicable to evaluation of dilatancy and estimation of burst prone areas in the rock massif, using similarity in the processes of deformation, failure and energy emission in a wide range of sizes of the objects being deformed [15].

References

1. Zhurkov, S.N., Kuksenko, V.S. & Petrov, V.A. 1980. Concentration criteria of solids failure. Physical processes in earthquake nuclei. Moscow. Science, 78-85 (in Russian).
2. Myachkin, V.I. 1978. Processes of earthquakes preparation. Moscow. Science, 232 (in Russian).
3. Regel, V.R., Slutsker, A.I. & Tomashevsky, A.E. 1974. Kinetic nature of solid strength. Moscow, Science, (in Russian).
4. Yanagidani, T., Ehara, S., Nishizawa, O., & Kusunose, K. 1985. Localization dilatancy in Oshima granite under constant uniaxial stresses. J. Geophys. Res., 90 (138): 6840-6855.
5. Zhurkov, S.N., Kuksenko, V.S., & Petrov, V.A. 1977 To the question of rock failure prediction. News USSR Academy of Sciences, Physics of the Earth, 6 (in Russian).
6. Sobolev, G.A., Zavialov,A.D. 1980. On concentraion criterion of seismogenic faults. Reports of USSR Academy of Sciences, Geophysics, 252(1), 69-71.(in Russian).
7. Ruzhich, V.V., Mansurov, V. A. & Babichev,A.A. 1985 On seismotectonic criterion of the Earth crust destruction in lake Baikal region at riftogenesis. Reports of USSR Academy of Sciences. Geophysics, 281 (3): 566-569 (in Russian).
8. Frolov, D.N., Kileev R.SF., Kuksenko, V.S. & Novicov, S.V., 1980. Connection between acoustic signals parameters and dimensions of faults in solid at failure in heterogeneous materials. Mechanics of Composite Materials, 5. 907-911(in Russian)
9. Stanchitz, S.A. & Tomilin N.G. 1983. Investigation of time parameters of acoustic signals while tear-off cracks generation. Earthquake prediction. Moscow- Duchanbe, Donish, 4:31-45 (in Russian).
10. Kuksenko, V.S., Liashkov, A.I. & Mirzoev, K.M. 1982 Connection between the sizes of cracks generated under loading and duration of elastic energy emission. Reports of USSR Academy of Sciences, Geophysics, 246(1), 846-848 (in Russian).
11. Kuksenko, V.S., Mansurov, V. A., Manzhikov., B. Ts. & Tomilin, N.G. et all. 1990. Similarity in the rocks failure process on different scale levels. News of USSR Academy of Sciences, Physics of the Earth, 6:66-70 (in Russian).
12. Kuksenko, V.S., Mansurov, V. A., Manzhikov., B.Ts. & Stanchits, S.A. 1985. Estimation of burst prone by their energy emission. Physico-Technical Problem of Mining, 4: 28-32 (in Russian).
13. Bieniawski, Z.T. 1967. Mechanism of Brittle Fracture of Rocks. Int. Journal Rock Mechanics and Mining Sciences, 4: 395-430.
14. Kuksenko, V.S. & Mansurov, V. A. 1986. The localization of failure in rocks in various scale levels. Physico-Technical Problems of Mining, 2: 49-55.
15. Mansurov, V.A. 1993. Laboratory experiments: Their role in the problems of rock burst prediction. Comprehensive Rock Engineering, v.3: 745 -771. Pergamon Press.

Plastic yield criteria and their applications

Xu Gancheng
Tongji University, Shanghai, People's Republic of China

ABSTRACT: In this paper, over ten yield criteria usually used in practice for rock soil material are written into a unified formula like the generalized Von Mises yield criterion. The elastoplastic finite element programme is developed the same way as the Drucker-Prager criterion, and engineering examples are given out. In addition, the equivalent-circle yield criterion whose area is equal to that of the Mohr-Coulomb irregular hexagon is advanced. The results are very close to those of the Mohr-Coulom yield criterion in terms of the equivalent-circle criterion.

1 INTRODUCTION

The friction yield criteria based on the Mohr-Coulomb friction failure law is a long-tested calculation law which is still in wide use in practice. In view of its inconvenience in computation, for perfectly plastic material elastoplastic FE programme is mostly compiled through use of the Drucker-Prager criterion on the π plane. But its curve has poor agreement with that of the Mohr-Coulomb hexagon, hence it is impossible to guarantee sufficient accuracy in calculation. In recent years, more and more yield criteria find applications in engineering, particularly the yield criteria of quadratic curve proposed by Zienkiewicz and Pande. On the π plane, however, they all use a yield curve without corners to replace the Mohr-Coulomb irregular hexagon. In this way, the calculation results cannot match those of the Mohr-Coulomb criterion. In this paper, over ten yield criteria usually used in practice for rock soil material are written into a unified formula like the generalized Von Mises yield criterion. The elastoplastic FE programme is developed the same way as the Drucker-Prager criterion, and engineering examples are given out. In addition, the equivalent-circle yield criterion whose area is equal to that of the Mohr-Coulomb irregular hexagon is advanced. The results are very close to those of the Mohr-Coulomb yield criterion in terms of the equivalent-circle criterion.

2 UNIFIED FORMULA OF YIELD CRITERIA

A general formula of yield criteria proposed by Zienkiewicz and Pande is as follows:

$$F = \beta\sigma_m^2 + \alpha_1\sigma_m \text{-} k + \bar{\sigma}_+^2 = 0 \tag{1}$$

Zhen Yinren et al (1984) develop equation (1) and write it into the following formula:

$$F = \beta\sigma_m^2 + \alpha_1\sigma_m \text{-} k + \bar{\sigma}_+^n = 0 \tag{2}$$

where $\bar{\sigma}_+ = \dfrac{\sqrt{J_2}}{g(\theta_\sigma)}$, $g(\theta_\sigma)$ is the form function of yield curve on the π plane.

But the expression is inconvenient for developing a computer code of constitutive equation. If all kinds of yield criteria are written as an expression like the generalized Von Mises yield criterion in programming, we can use the same method as the Drucker-Prager yield criterion. For non-circular yield curves on the π plane, such as the Mohr-Coulomb criterion and Zienkiewicz and Pande criterion, a series of circular curves of the generalized Von Mises criterion are used to press on toward straight lines or non-circular curves, thus we can achieve a simplicity of programming and sufficient computation accuracy. The unified expression is as follows:

$$F = \alpha' I_1 + \sqrt{J_2'} - k' = 0 \tag{3}$$

where α', J_2' and k', which are functions of stress invariant, are given in the Table 1. In the Tabe 1, $g'(\theta_\sigma)$, $g''(\theta_\sigma)$ are the form function of yield curve on the π plane defined by

$$g'(\theta_\sigma) = \dfrac{3 - \sin\varphi}{2\sqrt{3}\,(\cos\theta_\sigma - \dfrac{1}{\sqrt{3}}\sin\theta_\sigma\sin\varphi)} \tag{4}$$

for the Mohr-Coulomb yield curve, and

$$g''(\theta_\sigma) = \dfrac{2K}{(1+K) - (1-K)\sin 3\theta_\sigma}$$

$$K = \dfrac{3 - \sin\varphi}{3 + \sin\varphi} \tag{5}$$

for the Zienkiewicz and Pande rounded formula. The yield curves on the π plane are shown in Figure 1.

Yield curves on the meridian plane comprise zero degree (Von Mises and Tresca criteria), linear (Mohr-Coulomb criterion) and quadratic curves which usually include hyperbola, parabola and ellipse as shown in Figure 2(a).

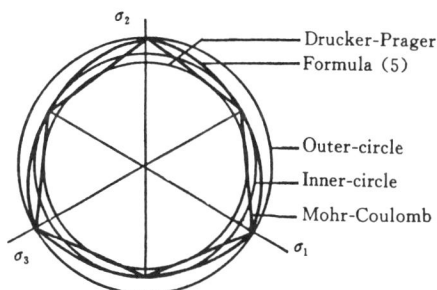

Figure 1 Two-dimensional, π plane, representation of criteria

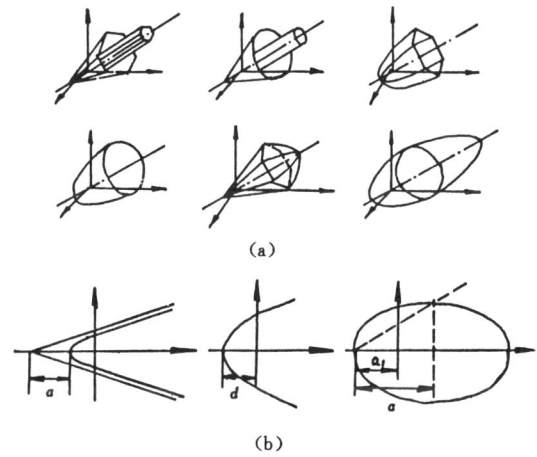

Figure 2 Geometrical represetation of yield surfaces and quadratic yield curves on the meridian plane

Table 1.

Yield Conditions		a'	k'	J_2'
Generalized Von Mises	Outer-circle	$\dfrac{2\sin\varphi}{\sqrt{3}(3-\sin\varphi)}$	$\dfrac{6C\cos\varphi}{\sqrt{3}(3-\sin\varphi)}$	J_2
	Inner-circle	$\dfrac{2\sin\varphi}{\sqrt{3}\sqrt{3+\sin^2\varphi}}$	$\dfrac{6C\cos\varphi}{\sqrt{3}(3+\sin\varphi)}$	J_2
	Inscribed-circle	$\dfrac{\sin\varphi}{\sqrt{3}(3+\sin\varphi)}$	$\dfrac{\sqrt{3}\,C\cos\varphi}{\sqrt{3+\sin^2\varphi}}$	J_2
	Equivalent-circle	$\dfrac{\sin\varphi}{\sqrt{3}(\sqrt{3}\cos\theta_\sigma^*-\sin\theta_\sigma^*\sin\varphi)}$	$\dfrac{\sqrt{3}\,C\cos\varphi}{\sqrt{3}\cos\theta_\sigma^*-\sin\theta_\sigma^*\sin\varphi}$	J_2
Von Mises		0	C	J_2
Generalized Tresca	Outer-circle	$\dfrac{\sin\varphi}{\cos\theta_\sigma(\sqrt{3}-\sin\varphi)}$	$\dfrac{3C\cos\varphi}{\cos\theta_\sigma(\sqrt{3}-\sin\varphi)}$	J_2
	Inner-circle	$\dfrac{\sin\varphi}{\cos\theta_\sigma(\sqrt{3}+\sin\varphi)}$	$\dfrac{3C\cos\varphi}{\cos\theta_\sigma(\sqrt{3}+\sin\varphi)}$	J_2
	Inscribed-circle	$\dfrac{\sin\varphi}{2\cos\theta_\sigma(\sqrt{3+\sin^2\varphi})}$	$\dfrac{3C\cos\varphi}{2\cos\theta_\sigma\sqrt{3+\sin^2\varphi}}$	J_2
Tresca		0	$\dfrac{C}{\cos\theta_\sigma}$	J_2
Mohr-Coulomb		$\dfrac{1}{3}\mathrm{tg}\bar{\varphi}g'(\theta_\sigma)$	$\bar{C}g'(\theta_\sigma)$	J_2
Zienkiewicz and Pande	Hyperbola	$\dfrac{1}{3}\mathrm{tg}\bar{\varphi}g''(\theta_\sigma)$	$\bar{C}g''(\theta_\sigma)$	$J_2+\left[\dfrac{a\,\mathrm{tg}\bar{\varphi}}{g''(\theta_\sigma)}\right]^2$
	Parabola	$\dfrac{1}{3a}\dfrac{[g''(\theta_\sigma)]^2}{\sqrt{J_2}}$	$\dfrac{d}{a}\dfrac{[g''(\theta_\sigma)]^2}{\sqrt{J_2}}$	J_2
	Ellipse	$\left[\mathrm{tg}^2\bar{\varphi}\dfrac{I_1}{9}-\dfrac{2}{3}(a-a_1)\mathrm{tg}^2\bar{\varphi}\right]\dfrac{[g''(\theta_\sigma)]^2}{\sqrt{J_2}}$	$\mathrm{tg}\bar{\varphi}(2a-\bar{C}\mathrm{tg}\bar{\varphi})\cdot\bar{C}\dfrac{[g''(\theta_\sigma)]^2}{\sqrt{J_2}}$	J_2

3 ELASTOPLASTIC STRESS AND STRAIN RELATION

After initial yielding the relationship between strain component and stress increment becomes:

$$\{d\sigma\} = [D_{ep}]\{d\varepsilon\} = ([D_e]-[D_p])\{d\varepsilon\} \tag{6}$$

where

$$[D_e] = \dfrac{E}{(1+\nu)(1-2\nu)}\begin{bmatrix} 1-\nu & \nu & 0 & \nu \\ \nu & 1-\nu & 0 & \nu \\ 0 & 0 & \dfrac{1-2\nu}{2} & 0 \\ \nu & \nu & 0 & 1-\nu \end{bmatrix}$$

$$\{d\sigma\} = \{\sigma_x \quad \sigma_y \quad \tau_{xy} \quad \sigma_z\}^T$$

$$\{d\varepsilon\} = \{\varepsilon_x \quad \varepsilon_y \quad \gamma_{xy} \quad 0\}^T$$

$$[D_p] = \dfrac{[D_e]\left\{\dfrac{\partial F}{\partial(\sigma)}\right\}\left\{\dfrac{\partial F}{\partial(\sigma)}\right\}^T[D_e]}{A+\left\{\dfrac{\partial F}{\partial(\sigma)}\right\}^T[D_e]\left\{\dfrac{\partial F}{\partial(\sigma)}\right\}^T} \tag{7}$$

For perfectly plastic case $A = 0$

$$\dfrac{\partial F}{\partial(\sigma)} = C_1\dfrac{\partial \sigma_m}{\partial(\sigma)} + C_2\dfrac{\partial\sqrt{J_2}}{\partial(\sigma)} + C_3\dfrac{\partial J_3}{\partial(\sigma)} \tag{8}$$

$$C_1 = \dfrac{\partial F}{\partial \sigma_m}, \quad C_2 = \dfrac{\partial F}{\partial\sqrt{J_2}}, \quad C_3 = \dfrac{\partial F}{\partial J_3}$$

$$\dfrac{\partial\sqrt{J_2}}{\partial(\sigma)} = \dfrac{1}{2\sqrt{J_2}}[S_x \quad S_y \quad 2\tau_{xy} \quad S_z]^T$$

$$\dfrac{\partial J_3}{\partial(\sigma)} = [S_yS_z+\dfrac{1}{3}J_2 \quad S_zS_x+\dfrac{1}{3}J_2$$

$$-2S_z\tau_{xy} \quad S_xS_y-\tau_{xy}^2+\dfrac{1}{3}J_2]^T$$

$$\dfrac{\partial\sigma_m}{\partial(\sigma)} = \dfrac{1}{3}[1\ 1\ 0\ 1]^T$$

Upon the substitution of (8) into (7) the plastic matrix becomes

$$[D_p] = \dfrac{1}{A_0}\begin{bmatrix} A_1^2 & A_1A_2 & A_1A_3 & A_1A_4 \\ A_2A_1 & A_2^2 & A_2A_3 & A_2A_4 \\ A_3A_1 & A_3A_2 & A_3^2 & A_3A_4 \\ A_4A_1 & A_4A_2 & A_4A_3 & A_4^2 \end{bmatrix} \tag{9}$$

where

$$A_0 = (\lambda+2G)(F_x^2+F_y^2+F_z^2) + GF_{xy}^2 + 2\lambda(F_yF_z+F_xF_z+F_xF_y)$$
$$A_1 = F_x(\lambda+2G)+F_y\lambda+F_z\lambda$$
$$A_2 = F_x\lambda+F_y(\lambda+2G)+F_z\lambda$$
$$A_3 = F_{xy}G$$
$$A_4 = F_x\lambda+F_y\lambda+F_z(\lambda+2G)$$

4 CONCEPT OF EQUIVALENT-CIRCLE YIELD CRITERION

From the unified formula of yield criteria yield function can be written as follows:

$$F = F(\sigma_m, \bar{\sigma}_+) \tag{10}$$

After the Lode angle θ_σ is determined the form of yield curve on the meridian plane is defined, and similarly, once σ_m is determined the form of yield curve on the π plane is certain. As shown in Figure 1, the yield curves of the generalized Von Mises criterion on the π plane include circumcircle of outer apices, circumcircle of inner apices and inscribed circle (the Drucker-Prager criterion) of the Mohr-Coulomb irregular hexagon. For the generalized Von Mises criterion $g(\theta_\sigma)$ does not vary with θ_σ, but the $g(\theta_\sigma)$ of non-circular yield curves varies as θ_σ, such as the Mohr-Coulomb, Zienkiewicz and Pande criteria. From this we can imagine that for non-circular yield curves a certain suitable θ_σ can be found, and according to this θ_σ, a $g(\theta_\sigma)$ can be determined, that is, an equivalent-circle curve can be found. In this way, the plastic zone results obtained by the equivalent-circle yield curve are very close to those achieved by the corresponding non-circular yield curve, thus such a circular yield curve described above is known as the equivalent-circle yield criterion for its corresponding yield curve.

For the Mohr-Coulomb yield criterion, according to the principle of equality of the area surrounded by the curve, we can get

$$B = \left[\frac{-\frac{2A}{3}\sin\varphi + \sqrt{\frac{4A^2}{9}\sin^2\varphi - 4\left(\frac{\sin^2\varphi}{3}+1\right)\left(\frac{A^2}{3}-1\right)}}{2\left(\frac{\sin^2\varphi}{3}+1\right)}\right]$$

$$\theta_\sigma^* = \sin^{-1}B \quad A = \sqrt{\frac{\pi(9-\sin^2\varphi)}{6\sqrt{3}}} \quad (11)$$

In terms of the formula (3), we obtain

$$\alpha' = \frac{\sin\varphi}{\sqrt{3}(\sqrt{3}\cos\theta_\sigma^* - \sin\theta_\sigma^*\sin\varphi)}$$

$$k' = \frac{\sqrt{3}\,C\cos\varphi}{\sqrt{3}\cos\theta_\sigma^* - \sin\theta_\sigma^*\sin\varphi}$$

In this way we can simplify programming because a non-circular curve is taken as a circular curve.

5 APPLICATION OF YIELD CRITERIA

We will now show four examples of plastic zone computation by use of FE program, which demonstrate the accuracy and applicability of the unified formula and the equivalent-circle criterion.

First of all, we shall present a circular excavation which is subjected to a uniform stress field of 1.55 MPa in vertical direction and 0.62 MPa in horizontal direction. The material (rock) was assumed to be perfectly plastic with

$$E = 775\text{MPa}, \quad \nu = 0.2, \quad C = 0.43\text{MPa}, \quad \varphi = 30°$$

Internal cell discretization is presented in Figure 3, where the region expected to be yielded is divided into 126 elements with 150 nodal points. Figure 3 also shows the plastic zone for each yield criterion. We can see from the results shown in Figure 3(a) that the plastic zone obtained by the Mohr-Coulomb criterion is much smaller than that by the Drucker-Prager criterion. The differences in the plastic zone are obvious because both yield curves on the π plane have great discrepancy, and their discrepancy increases with the angle of internal friction φ. The plastic zone calculated by the equivalent-circle criterion has better agreement with that by the Mohr-Coulomb criterion. This indicates that if the equivalent-circle criterion is employed in practical engineering we can not only make programming simpler, but also achieve sufficient computation accuracy.

On the part of the quadratic yield curver, they are seldom employed in practical engineering because the parameters that define the form of curves are difficult to be determined. But for rock soil material the triaxial tests indicate that a relationship between normal stress and shear stress does not exhibit linear expression defined by $\tau = C - \sigma_n \text{tg}\varphi$, but exhibits such a curve relation as hyperbola or parabola. For this reason, so long as curve parameters are suitably chosen the quadratic yield curves will be more suitable for rock soil material. After going through a series of calculations and analyses, this paper suggests the following proper parameters of quadratic yield functions in computation (Figure 2(b)):

Hyperbola: $a = \left(0 \sim \frac{1}{2}\right)\frac{\bar{C}}{\text{tg}\bar{\varphi}}$

Parabola: $a = 2.2 \sim 2.8, \quad d = \frac{\bar{C}}{\text{tg}\bar{\varphi}}$

Ellipse: $a = (8 \sim 15)\frac{\bar{C}}{\text{tg}\bar{\varphi}}, \quad a_1 = \frac{\bar{C}}{\text{tg}\bar{\varphi}}$

where $\bar{C} = \frac{6C\cos\varphi}{\sqrt{3}(3-\sin\varphi)}, \quad \text{tg}\bar{\varphi} = \frac{6\sin\varphi}{\sqrt{3}(3-\sin\varphi)}$

Figure 3(b) shows the results of plastic zone based on the quadratic yield criteria in which curves on the meridian plane are hyperbola, parabola and ellipse, and curve on the π plane is the equation (5). In addition, the results also lead to a conclusion that there is little difference between the plastic zone of the hyperbola criterion and that of the Mohr-Coulomb straight line criterion on the meridian plane; and the plastic zone of the Zienkiewicz and Pande's rounded formula on the π plane is much smaller than that of the Mohr-Coulomb criterion.

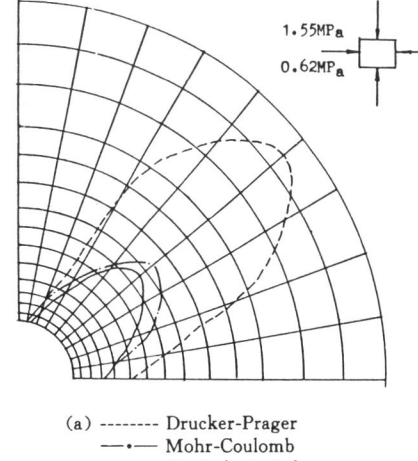

(a) -------- Drucker-Prager
—·— Mohr-Coulomb
——— Equivalent-circle

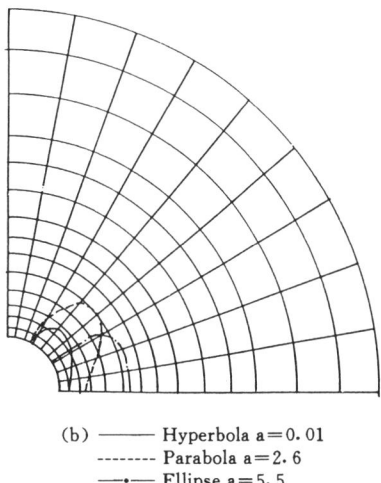

(b) ——— Hyperbola a=0.01
-------- Parabola a=2.6
—·— Ellipse a=5.5

Figure 3 Deep circle tunnel problem, discretization used for FE

In the second example, we shall calculate the plastic zone of the subterranean plant building of Laxiwa Hydropower Station located at the Yellow River, Chinghai Province. The subterranean plant building comprises the main machine room ($250 \times 29 \times 67\text{m}^3$), the main transformer room ($220 \times 23 \times 44\text{m}^3$), and the pressure regulating room of tail water ($113 \times 20 \times 56.5\text{m}^3$). Rock material at which the plant building is located is solid and intact granite with

$$E = 3100\text{MPa}, \quad \nu = 0.21, \quad C = 3.92\text{MPa}, \quad \varphi = 58°$$

The in-situ stresses are

$$\sigma_x = -14.48\text{MPa}, \quad \sigma_y = -14.92\text{MPa}, \quad \tau_{xy} = 2.68\text{MPa}$$

We employed 425 8-nodal FE, 436 nodal points in all. Figure 4 presents the calculation results of plastic zone for each yield criterion. It is clear that the changing regular pattern of plastic zone of each yield criterion is basically the same as that in example 1.

In the third example, we take the example of the irrigation engineering of Baoji Gorge in Shaanxi Province. The engineering is located at the sector of a steep slope, which consists of loess, ancient soil and calcium clay. The calculated parameters of soils (elastic modulus E, Poisson's ratio ν, angle of internal friction φ, cohesion C and unit weight γ) are listed in Table 2. We employed 135 8-nodal isoparametric FE, 472 nodal points in all. Soil material is subjected mainly to self-weight. Figure 5 shows the results of plastic zone for the Drucker-Prager and the equivalent-circle criterion. It can be seen that the spread pattern of their plastic zone is almost the same as the above mentioned two examples.

Finally we take Jianyin anchored suspension bridge foundation over Yangtze River for example in order to show the application of the proposed equivalent-circle yield criterion to foundation problem. The Nothern anchored foundation of Jianyin suspension bridge is

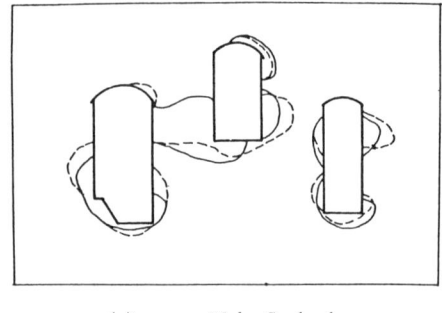
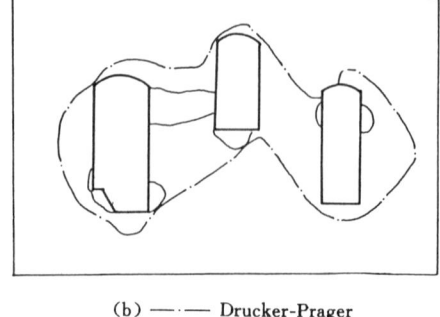

(a) ········ Mohr-Coulomb
——— Equivalent-circle

(b) —·— Drucker-Prager
——— Zienkiewicz and Pande

Figure 4 Laxiwa Hydropower Station problem, discretization used for FE, results and spread of plastic zone at the end of relaxation process

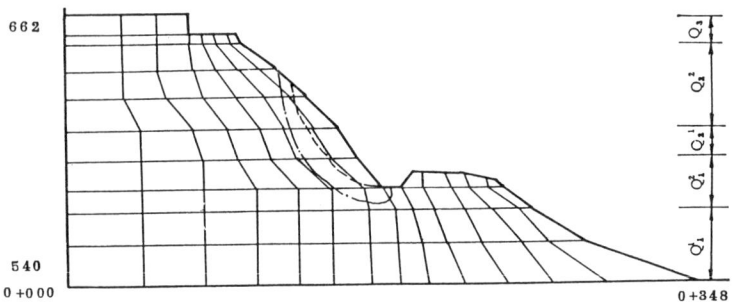

Drucker-prager
Equivalent-circle

Figure 5 Results and spread of plastic zone of the steep slope of Baoji Gorge

Table 2.

Parameters Soil layers	E (MPa)	ν	φ (degree)	C (MPa)	γ (kN/m³)
Q_1^1	454	0.3	31.65	0.11	18.91
Q_1^2	454	0.3	30.39	0.11	18.91
Q_2^1	411	0.28	27.46	0.088	17.82
Q_2^2	205	0.25	23.54	0.064	16.95
Q_3	98	0.21	26.50	0.059	17.15

built in soft soil deposit. The soft deposit consists of four layers of soil. The first layer is sandy loam with its elevation 2.4m (earth surface) to -14.1m, the second layer fine sand with its elevation -14.1m to -41.85m, the third layer loam with its elevation -41.85m to -53.4m, the fourth layer medium and coarse sand bearing gravel with its elevation -53.4m to -78.2m. The stratum below soft soil deposit is rock mass. The calculated parameters of soils (elastic modulus E, poisson's ratio ν, angle of internal friction φ, cohesion C and unit weight γ) are listed in Table 3.

Table 3.

Parameters Soil layers	E (MPa)	ν	φ (degree)	C (MPa)	γ (kN/m³)
1	4.62	0.45	23.5	7.5	18.5
2	23.98	0.368	32.0	0.0	19.4
3	23.99	0.4171	24.0	12.0	19.5
4	53.20	0.225	33.0	0.0	20.1

The anchored foundation is constructed of reinforced concrete, and belongs to Diaphragm Wall structure. The calculated parameters of reinforced concrete is: $E = 33$GPa, $\nu = 0.16$, $\varphi = 45°$, $C = 1$MPa, $\gamma = 25$kN/m³. The steel rope oblique tension acting on the the Northern anchored foundation is 640MN, the angle included between the tension and the horizontal plane is 25.39°, so the horizontal component of force is 570MN.
Finite element grid used in three-dimensioned elastoplastic analysis of anchored foundation and soil deposit is shown in Figure 6.

We employed 1630, 8-nodal isoparametric elements, 2156 nodal points in all. The equivalent-circle yield criterion is adopted, and the stress level in the foundation structure and soil deposit is assumed to be due to the steel rope oblique tension and the self-weight only. The curves of the calculated maximum principal stress σ_1 =constant and minimum principal stress σ_3 =constant are shown in Figure 7 and 8 respectively.

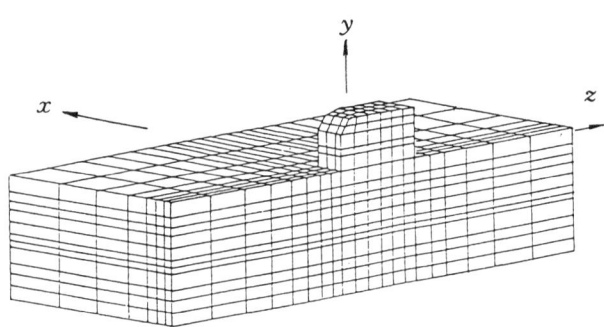

Figure 6 Finite element grids of Diaphragm Wall structure

6 CONCLUSIONS

From the above discussions, we can generalize the following useful conclusions:
At present the elastoplastic FE programme which is compiled by the Drucker-Prager yield criterion has found a wide application in practical engineering. But when compared with the plastic zone obtained by the Mohr-Coulomb yield criterion, the Drucker-Prager criterion has bigger plastic zone. As shown in the second example, for stress field that is not too large and solid intact granite material, the result that the plastic zone threads its way between tunnels is not conformable to the practical circumstances observed on scene. Therefore, the results calculated by the Drucker-Prager yield criterion is too conservative.
The Zienkiewicz and Pande's rounded formula has found more applications in practice to avoid difficulties in dealing with apex computation of the Mohr-Coulomb criterion. But from the above examples we can see that the plastic zone obtained by the Zienkiewicz

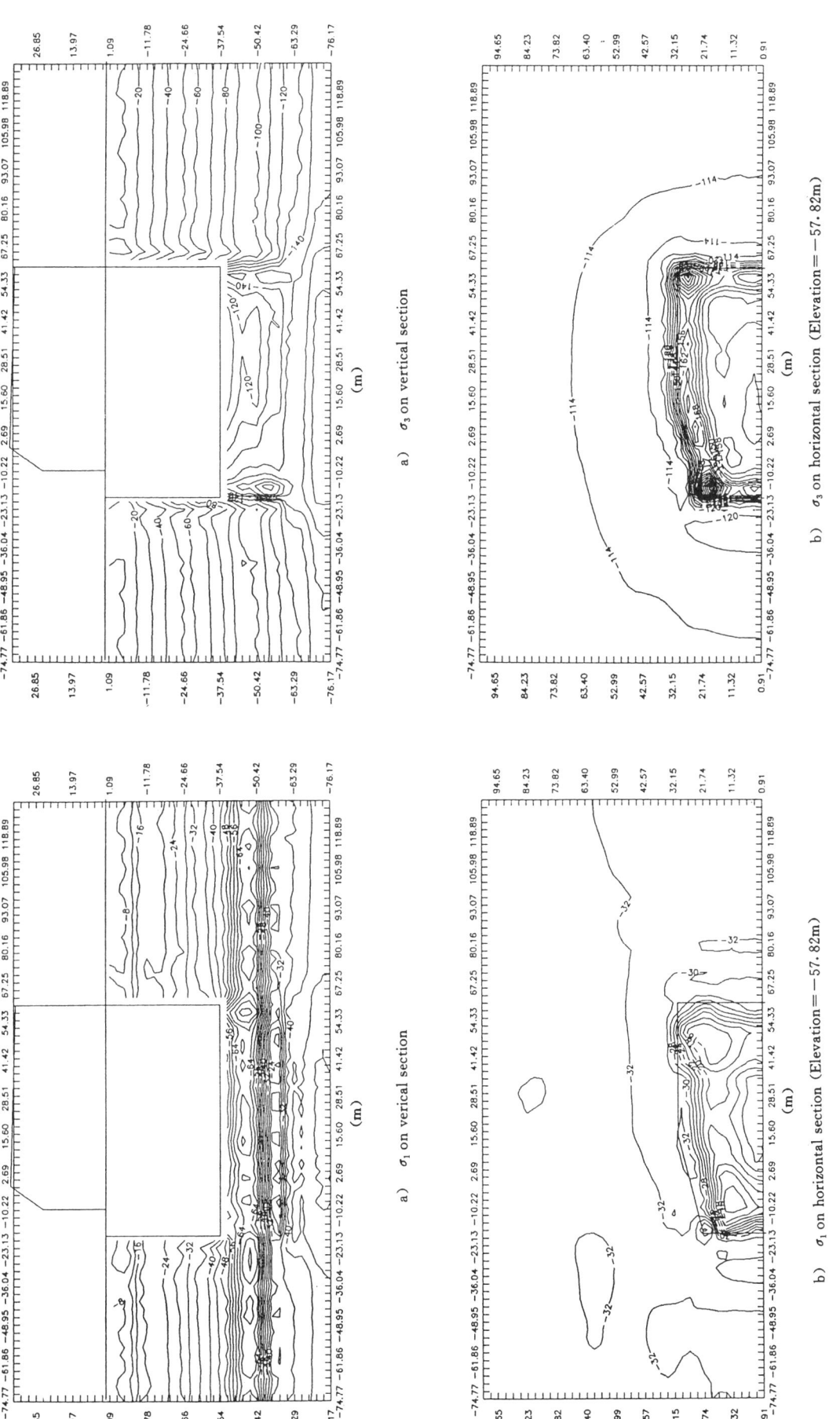

Figure 7 Contour plot of maximum principal stress σ_1 (10kPa)

Figure 8 Contour plot of minimum principal stress σ_3 (10kPa)

and Pande criterion is much smaller than that by the Mohr-Coulomb criterion, therefore, the results achieved by the Zienkiewicz and Pande yield criterion could be too risky.

Since the results of the equivalent-circle criterion approximate to those of the Mohr-Coulomb yield criterion more closely, and procedures of programming are simpler, this paper suggests that in practical engineering we recommend the equivalent-circle yield criterion or the Mohr-Coulomb yield criterion.

If the test of rock soil material indicates that the relationship between normal stress and shearing stress exhibits non-linear response, the quadratic yield criterion should be applied, and the decision to employ hyperbola, parabola or ellipse yield criteria just depends on the types of the obtained test curves.

REFERENCES

Zienkiewicz, O. C. & G. N. Pande 1977. Finite elements in geomechanics. New York: wiley.

Owen, D. R. J. & E. Hinton. 1980. Finite elements in plasticity: theory and practice. Swansee: Pineridge press.

Zhen Yinren & Chen changan. 1984. Idealized plastic geotechnical yield criterion and constitutive laws. Chinese Journal of Geotechnical Engineering. 6:13-21.

Mechanical properties and failure mechanism of gravelly soft rocks

Takashi Kobayashi & Toshiaki Mimuro
Tokyo Electric Power Services Co., Ltd, Japan

Ryunoshin Yoshinaka
Saitama University, Urawa, Japan

ABSTRACT: The present paper describes the study of the mechanical properties and its failure mechanism of gravelly soft rock obtained from FEM numerical analysis. It has demonstrated the following points: (1) Stress and strain distribution inside the gravelly soft rock is made very heterogeneous by the mixture of gravel, forming a flow of the stress threading through the gravel in the direction of the maximum principal stress. The matrix located in this flow constitutes a stress concentrated area. (2) The position and range of the stress concentrated area is determined by gravel arrangement. Its extent is determined by the elastic modulus ratio between the gravel and matrix. (3) At a low confining pressure, the failure of gravelly soft rock is mainly caused by the tensile failure of the matrix between gravels arranged orthogonally to the maximum principal stress; whereas, at a high confining pressure, it is mainly caused by the shearing failure of the matrix between gravels arranged in parallel to the maximum principal stress. (4) Such failure results from the stress level which is lower than the failure stress of the matrix alone. (5) Changes of the failure mechanism due to increase in the confining pressure are closely related to the strength of the gravelly soft rock.

1. INTRODUCTION

With regard to the heterogeneous soft rock composed of the soft continuous phase and hard dispersion phase independently of the geological components, the dispersion phase is referred to as "gravel", while the continuous phase is called "matrix"; they are called "gravelly soft rock" for the purpose of the present study.

The properties of the gravelly soft rock are characterized by a variety of factors. The major factors are considered to be the size and shape of the gravel, physical and mechanical properties of the gravel and matrix and contents of the gravel. The gravel contents is evaluated in terms of volume ratio (hereinafter referred to as "GC"). The mechanical properties of the gravelly soft rock can be handled as homogeneous from the engineering viewpoint, if the size of the gravel is much smaller than the specimen which can be sampled. When the size of the gravel is greater than the specimen, it is necessary to conduct a large test *in-site* or in the laboratory. However, when the gravel diameter is in the order of meters, the mechanical test is almost scale impossible. Accordingly, if the mechanical properties of the soft rock including large pieces of gravel can be evaluated from the practicable survey and test, its significance is considered to be very great.

The laboratory experiment using the artificial specimen has revealed that the mechanical properties of the gravelly soft rock has the following characteristics (Kobayashi 1994, Kobayashi 1994): Stress-strain relationship is basically affected by the matrix deformation characteristics (Fig. 1). Of the theoretical formulae to estimate the elastic modulus and equivalent elastic modulus of the gravel and matrix (Eshelby 1975, Hashin 1963, Hashin 1983, Mura 1987, Reuss 1929, Voigt 1989), the Ruess formula and the formula which provides the lowest limit value of Hashi-Shtrikman can be used to infer dynamic elastic modulus. The strength relationship can be effectively represented by the power function failure criterion exhibiting a

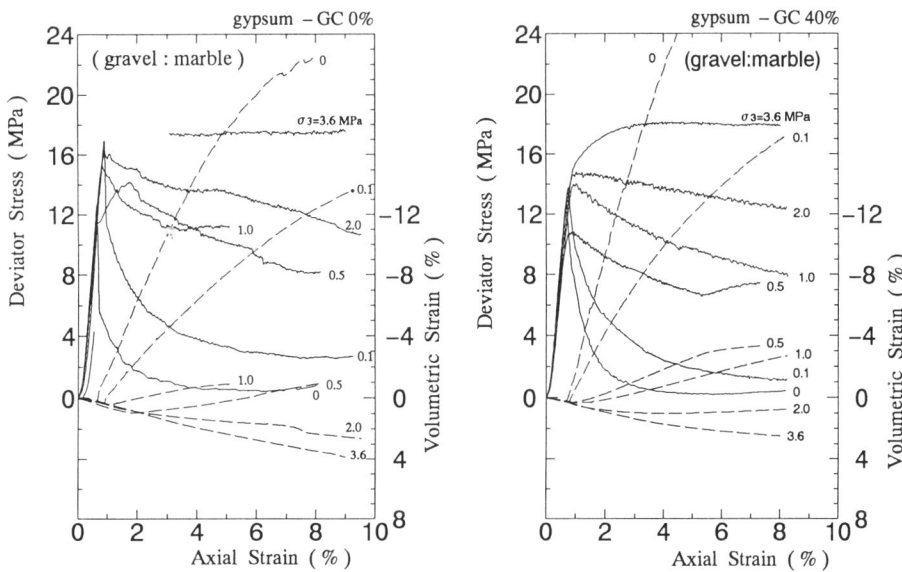

Fig. 1 Stress-Strain Curves of GYM specimen (GC: 0%, 40%)

conspicuous non-linear form. The residual strength is little affected by inclusion of gravel, and is almost the same as the matrix. The maximum strength in the low stress area exhibiting the strain softening is smaller than the matrix strength (Fig. 2) The major element dominating the maximum strength is the GC. The impact by gravel diameter is extremely small when $d/Dmax \geq 6$. The maximum strength up to 60 % in GC can be evaluated according to the maximum matrix strength, residual strength, and relationship between the GC and R_β.

The present paper studies the mechanical characteristics of the gravelly soft rock obtained from the experiment by numerical analysis based on the laboratory experiment converted into a model, as well as its failure mechanism.

2. EFFECT OF MIXED GRAVEL BY NUMERICAL ANALYSIS

Based on the experimental consideration of the mechanical characteristics of the gravelly soft rock, we used finite element method for numerical analysis of the stress-strain relationship inside the specimen which is difficult to clarify by experiment. Firstly, we used elastic analysis to study the basic effect of the GC and elastic modulus ratio of gravel and matrix on the stress-strain relationship inside the specimen. Then, we made elasto-plastic analysis for the triaxial compression test result of the specimen of gravelly rock with gypsum matrix (hereinafter referred to as "GYM specimen") as a model, and studied the failure mechanism of the gravelly soft rock.

2.1 Overview of the analysis

Analysis has been made based on the plane strain condition. The stress-strain relationship as element was assumed to be perfectly elastic in the elastic analysis while it was assumed to be perfectly elasto-plastic in the elasto-plastic analysis. Mohr-Coulomb failure condition was used as failure condition in the elastic analysis, while Drucker-Prager failure condition were used in the plastic area after failure.

The Finite element mesh used in the analysis is as shown in Fig. 3. It shows the range of analysis converted into a model on the 1/4-th cross section. The size of the gravel is assumed as $L/Dmax \geq 5$ (where L stands for the diameter of the analysis model, and Dmax represents the maximum gravel diameter). GC's of 20 % and 40 % were modeled in the elastic analysis (Fig. 3(1)), while GC's of 0 % and 20 % (Fig. 3(2)), and 40 % and 60 % were modeled in the elasto-plastic analysis. Since the analysis mesh is fixed in the elastic analysis, so GC was changed by changing the material constants of the element.

As confining conditions, the X-axis movement is allowed only in the X-axis direction, while the Y-axis movement is possible only in the Y-axis direction.

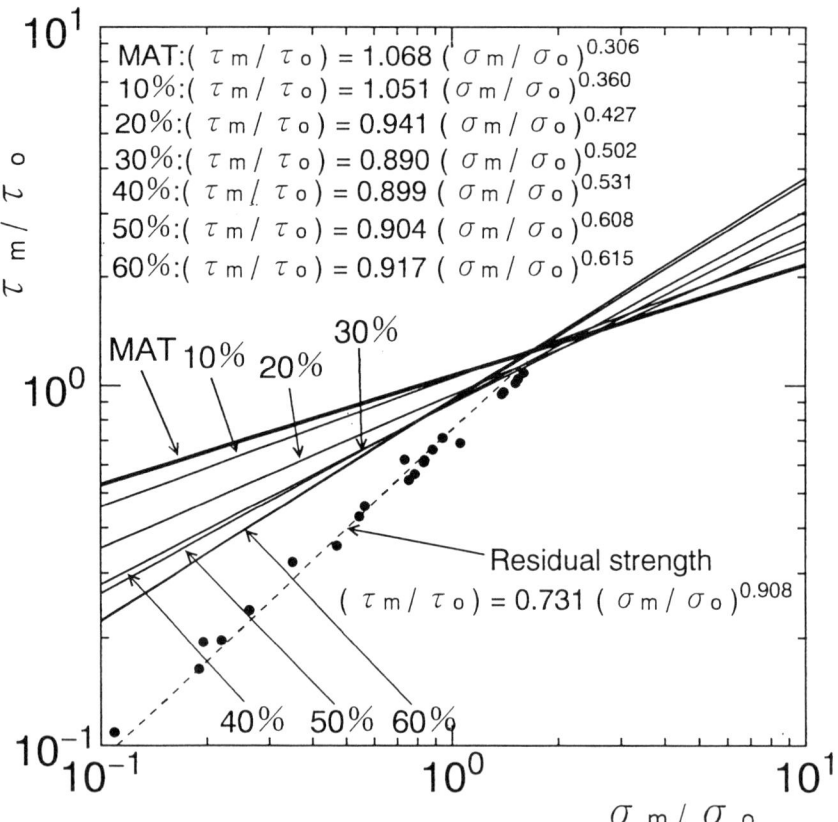

Fig. 2 Strength Relation of GYM specimen by Failure Criterion of Power Function

The loading condition is equally distributed loading. X- and Y-direction loadings in the elastic analysis are σx = 1.0 MPa and σy = 5.6 MPa, respectively. These loadings are set as critical stresses without causing failure, where the overall safety factor Fs in the specimen homogeneous only in matrix becomes 1.02. Furthermore, in the elasto-plastic analysis, loadings were added stepwise by setting up some 20 to 30 steps between the elastic range and perfect elasto-plastic range. Compression was assumed as positive for stress-strain of the element.

Table 1 illustrates the physical properties of the gravel and matrix used in the analysis. They mainly consist of the modeled physical properties of the specimen of gravelly rock with silt matrix (hereinafter referred to as "SIM specimen") and GYM specimen.

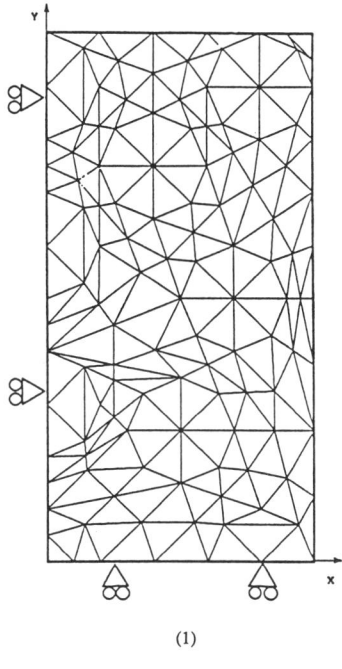
(1)

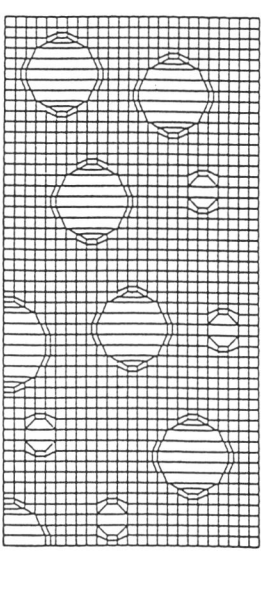
(2)

Fig. 3 Finite Elment Mesh

Table 1 Input Physical Properties

	Gravel				Matrix				Remarks
	E (MPa)	ν	C (MPa)	φ (°)	E (MPa)	ν	C (MPa)	φ (°)	
Elastic analysis	76500	0.2	10	50	770	0.35	1.5	15	effect of elastic modulus ratio
	61200	0.2	10	50	770	0.35	1.5	15	
	20400	0.2	10	50	770	0.35	1.5	15	
	12200	0.2	10	50	770	0.35	1.5	15	
	6100	0.2	10	50	770	0.35	1.5	15	
	3100	0.2	10	50	770	0.35	1.5	15	
	1500	0.2	10	50	770	0.35	1.5	15	
	90800	0.2	10	50	3200	0.35	1.5	15	dynamic modulus
Elasto-plastic analysis	20400	0.2	16 (Tensile strength) 3.1	50	2000	0.25	5.1 (Tensile strength) 1.4	20	

2.2 Result of elastic analysis

Fig. 4 illustrates the distribution of the maximum shear stress ratio inside the specimen, distribution of the maximum shear strain ratio, and local safety factor (Fs) for the GC of 40 %, an example of the result of analysis. This Figure is normalized in the maximum shear stress (2.3 MPa) and the maximum shear strain (0.81 %) of the matrix without containing gravels. The broken line in the Figure shows that the maximum shear stress ratio, the maximum shear strain ratio and the local safety factor are unity.

The domain within the bold solid lines shows that these values are unity larger, and the stress is concentrated by the mixing of gravel. The range within the small solid line shows that these values are less than unity, where stress is relieved.

These Figures reveal that the stress and strain distribution inside the specimen is very complicated under the influence of the mixture of gravel. An area subjected to great compressive stress concentration is formed in the gravel arranged in parallel to the maximum principal stress under the influence of gravel arrangement.

Furthermore, according to the local safety factor distribution, the matrix is subjected to local failure having a local safety factor of below unity. It shows that failure has occurred earlier to the specimen containing gravel than to that composed of only the matrix without it. When attention is given to the domain where the maximum shear stress ratio, the maximum shear strain ratio and local factor are unity, the effect of elastic modulus ratio between the gravel and matrix on the stress and strain distribution is found in approximately the same position and domain for any elastic modulus ratio. In other words, the position and domain of the stress concentrated area and stress relieved area depend mainly on gravel arrangement, showing that the effect of the elastic modulus ratio of the gravel and matrix is small.

On the other hand, the degree of the stress concentration and relief within the domain increases with the elastic modulus ratio of the gravel and matrix, and the domain tends to expand.

The above discussion on the stress-strain distribution within the specimen reveals that the position and domain of the stress concentrated and relieved areas are determined by gravel arrangement, and its extent depends on the elastic modulus ratio of the gravel and matrix.

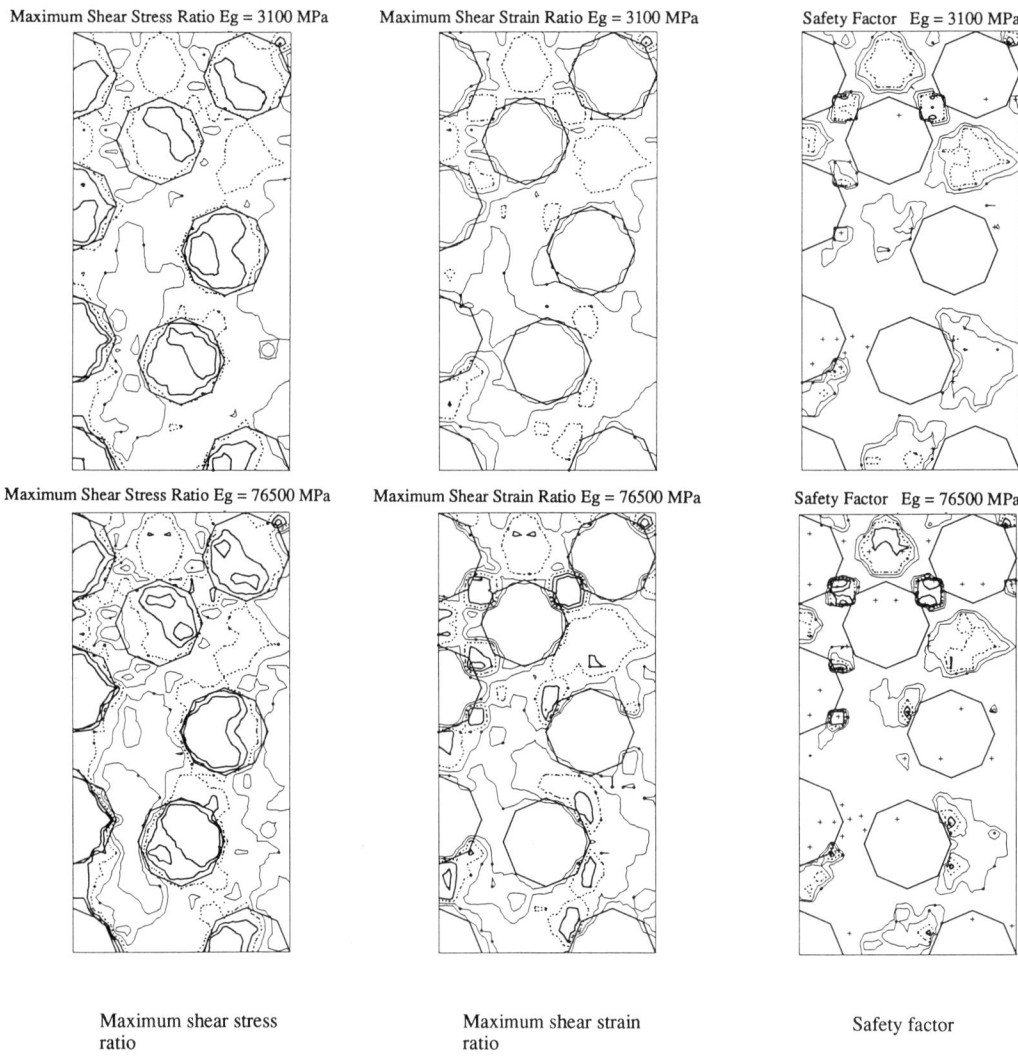

Fig. 4 Contour of Maximum Shear Stress Ratio, Maximum Shear Strain Ratio and Safety Factor

2.3 Result of elasto-plastic analysis

Fig. 5 illustrates the plastic domain for the GC of 20 % as an example of the analysis result. Its upper row (1) shows the result of σx = 0 MPa, while the lower row (2) shows the result of σx = 3.6 MPa. The Figure shows four cases of stress states as major points in studying the failure mechanism of the gravelly soft rock. ① shows the stress state (approximately 1/3 to 1/2 of the matrix failure strength) for the elastic range at GC of 0 %. ④ represents the stress state failure for at GC of 0 %. ② and ③ show the changes between ① and ④.

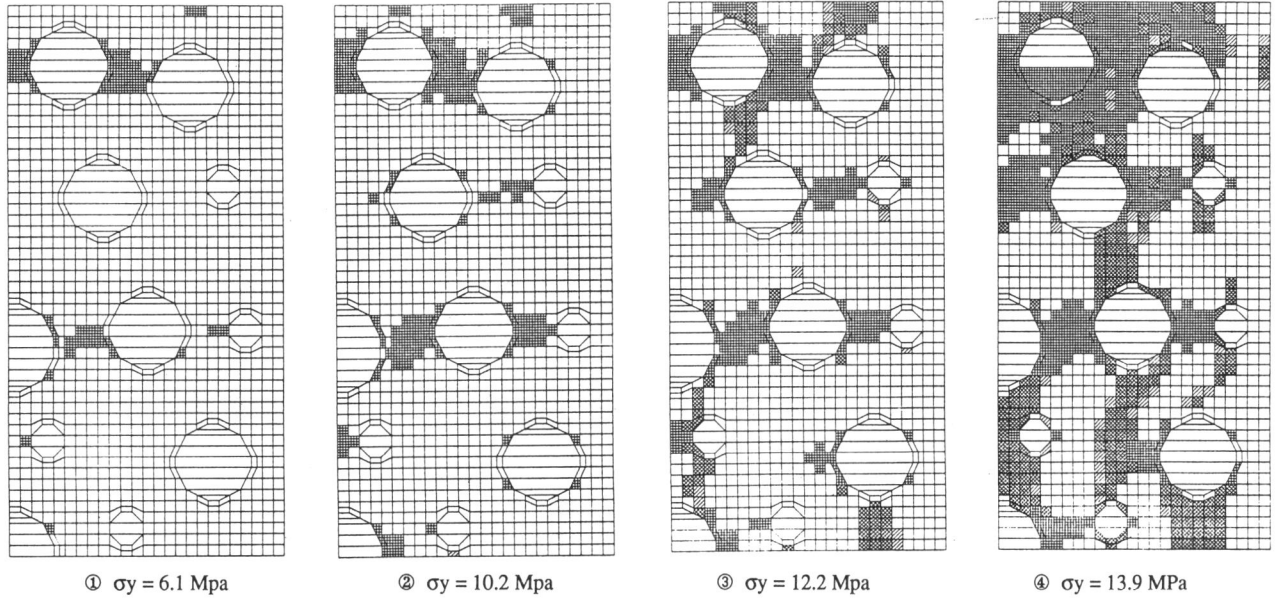

① σy = 6.1 Mpa   ② σy = 10.2 Mpa   ③ σy = 12.2 Mpa   ④ σy = 13.9 MPa

(1) Plastic range (GC 20%, σx = 0kgf/cm²)

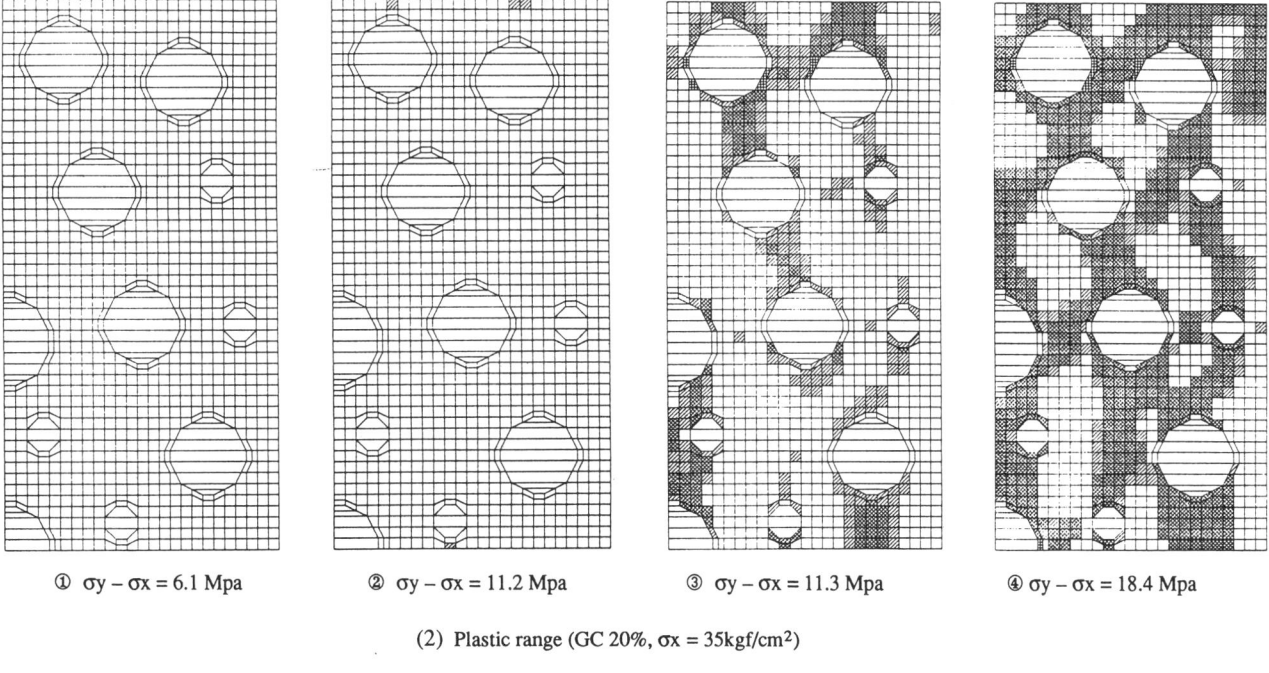

① σy − σx = 6.1 Mpa   ② σy − σx = 11.2 Mpa   ③ σy − σx = 11.3 Mpa   ④ σy − σx = 18.4 Mpa

(2) Plastic range (GC 20%, σx = 35kgf/cm²)

Immediately after shear failure   Shear failure   Tensile failure

Fig. 5 Distribution of Plastic Domain

According to the result of σx = 0 MPa, the plastic domain is formed locally from the low σy phase (approximately σ = 6.1 MPa) where no plastic domain is produced by the matrix alone without gravel. This suggests that local failure takes place from the stress level where the average stress state as specimen is small. The positions where this plastic domain occur corresponds to the matrix arranged in the direction orthogonal to the σy direction, and its failure mode is tensile failure. Furthermore, tensile failure domain tends to expand radially with the increase in σy. With the expansion of the tensile failure domain, shear failure occurs to the stress concentrated area formed by the flow of the shear stress, starting from the stage of σy = approximately 12.2 MPa. This tensile failure domain and shear failure domain are increased with increase in σy, and are connected with each other, resulting in overall failure. However, the major failure mode in the plastic domain is tensile failure even at the high σy state. From the distribution of the plastic domain, longitudinal cracks are assumed to occur to the specimen at the time of failure. This exhibits considerable agreement with the failure state of the specimen in experiment.

According to the result of σx = 3.6 MPa, the plastic range is formed locally from the low σy - σx phase where no plastic domain is produced by the matrix alone without gravel.

This suggests that local failure takes place from the stress level where the average stress state as specimen is small. The position where this plastic domain occurs is in the flow of the shear stress formed in the direction parallel to the σy direction. It is a matrix in gravel having a small ratio between gravel center distance (W) and gravel diameter (DA = (D_1 + D_2)/2). Its failure mode is shear failure. The stress state where the plastic domain by this shear occurs starts approximately from σy - σx = 10.2 to 11.2 MPa. Accordingly, the stress state where the shear failure occurs can be considered to be almost the same as σx = 0 MPa, so long as it is observed in terms of effective stress.

Furthermore, shear failure domain tends to expand radially with increase in σy - σx. From the distribution of the plastic domain, shear failure surfaces traversing the specimen in the oblique direction are assumed to occur at the time of failure. This exhibits considerable agreement with the failure state of the specimen in experiment. It should be noted that the tensile failure domain is formed locally. Almost no change in the position and domain is observed despite increase in σy - σx.

The above discussion reveals that tensile failure of the matrix between gravels arranged orthogonally to the direction of σy is the major failure mode at a low confining pressure, whereas the shear failure of the matrix between gravels arranged in parallel to the direction of σy is the major failure mode at a high confining pressure. It also demonstrates that the failure mode changes with increase in confining pressure, and that these failures occur locally from the low stress where failure does not take place in the case of the matrix alone. The failure of gravelly soft rock is considered to result in overall failure when the local failure caused by heterogeneity of the stress and strain distribution expands radially with increase in stress.

3. COMPARATIVE STUDY OF ANALYTICAL RESULT AND EXPERIMENTAL RESULT

3.1 *Comparative study of deformation characteristics*

Fig. 6 illustrates the elastic modulus calculated from the stress-strain curve obtained from FEM analysis. Fig. 7 shows the result of analysis using dynamic physical modulus values. It also shows the curve based on the experimental result and approximate formula (Eshelby 1975, Hashin 1963, Hashin 1983, Mura 1987, Reuss 1929, Voigt 1989) to estimate equivalent elastic moduli.

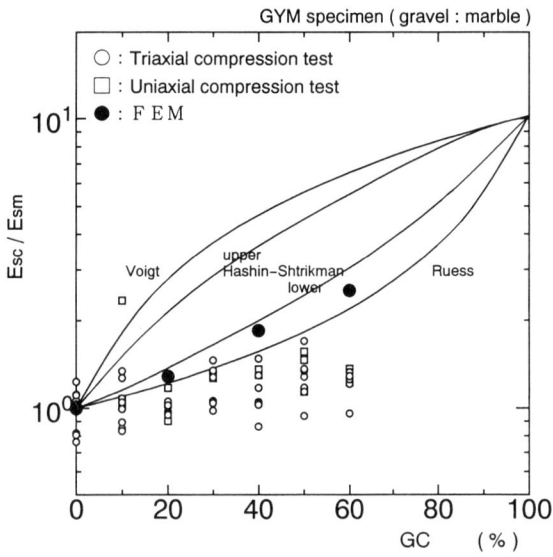

Fig. 6 Comparison Between Analytical Value and Experimental Value in Static-Elastic Moduli

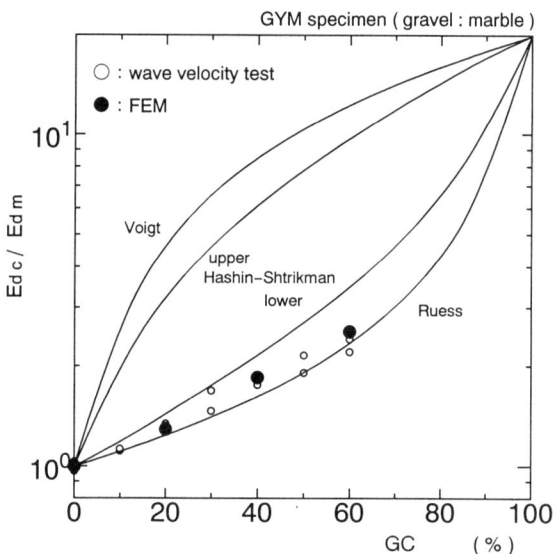

Fig. 7 Comparison Between Analytical Value and Experimental Value in Dynamic-Elastic Moduli

As will be clear from this Figure, static elastic moduli according to experiment does not tend to increase markedly despite increase in GC, and is often below the approximate curve of Ruess's equation representing the theoretical lower limit of the two-phase system. By contrast, the elastic moduli according by analysis are located between the approximate curve of Ruess's equation and the approximate curve representing the lower limit value of Hashi-Shtrikman's equation. This trend is clearly different from the experiment value.

Though the static elastic moduli according to experiment is calculated from the straight portion of the stress-strain curves, they are within the stress range corresponding to about 20 to 50 % of the failure stress, and can be said to be the value in a comparatively high stress-strain level range. Since elasto-plastic analysis has revealed that local failure has occurred or is progressing within the specimen in this stress range, it is considered highly probable that the experiment values has have been affected by the local failure within the specimen.

In the meantime, the dynamic elastic moduli for both experimental value and analytical value are located between the approximate curve of Ruess's equation and the approximate curve of Hashi-Shtrikman's equation, showing an agreement in the tendency to increase with GC.

From the above discussion, static elastic moduli have a trend different from that of the theoretical curves to estimate the equivalent elastic moduli; this is considered to be caused by the influence of the local failure having occurred or being in progress. Thus, we consider that the above discussion has succeeded in providing numerical analysis of two following points; (1) the fact that dynamic elastic modulus which features a small strain level and is not subjected to local failure can be estimated from the approximate formula of Ruess's equation or the approximate curve representing the lower limit value of Hashin-Shtrikman's equation, and GC, and (2) the deformation characteristics of the gravelly soft rock obtained from the experiment to the effect that the tendency of changes resulting from increase in GC is different between dynamic elastic moduli and static elastic moduli.

3.2 Comparative study of strength characteristics

Analytical result has revealed that local failure is clearly progressing in the gravelly specimen, even in the low stress range where failure is not caused by the matrix alone specimens. However, it can easily be imaged from the situation with the specimen at the time of uniaxial compression test that local failure does not signify overall failure.

Fig. 8 illustrates the "maximum strength" obtained from analysis. This strength can be obtained by defining it as the stress at yielding from the range showing the elastic behavior in stress-strain curve to the non-elastic behavior. It should be noted that this Figure also shows the uniaxial compressive strength obtained from experiment and the estimated values from the strength relation obtained from the triaxial compression test in experiment.

Thus, the maximum strength of $\sigma x = 0$ MPa generally corresponds to the lower limit value of the GYM specimen with marble or the upper limit value of the GYM specimen with glass beads. The maximum strength of $\sigma x = 3.6$ MPa tends to reduce with the increase in GC, but its reduction ratio is smaller than $\sigma x = 0$ MPa. This shows that the gradient of the maximum strength increases with the GC, when power function failure criteria are applied.

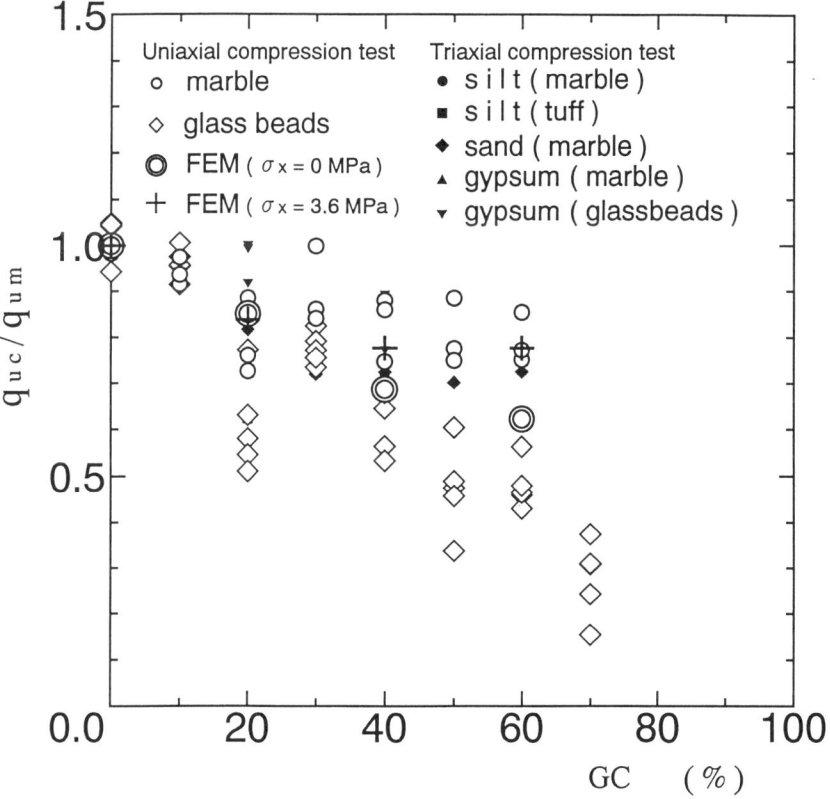

Fig. 8 Comparison Between Analytical Values and Experimental Values about Uniaxial Compressive Strength

4. CONSIDERATION OF FAILURE MECHANISM OF GRAVELLY SOFT ROCK

The changes in failure mode according to the trend of stress-strain distribution and changes in confining pressure, that described in Chapter 3 are the phenomena when the stress-strain relationship of the element is assumed as perfectly elasto-plastic. The actual gravelly soft rock exhibits the stress-strain relationship where stain softening type changes to strain hardening type with increase in confining pressure. The following discussed the failure mechanism by dividing the stress-strain relationship into three types; strain softening type, perfectly elasto-plastic type and strain hardening type:

In the low stress where stress-strain relationship shows strain softening type, tensile stress occurs in the matrix between gravels arranged orthogonally to the direction of the principal stress. The tensile stress range formed around these gravels have a greater overlapped area, as the W/D_A of the adjacent gravel is smaller, resulting in earlier occurrence of tensile failure. This shows that the strength of the specimen is smaller, as the number of the higher stress concentrated area -- the area closer to the gravel -- is greater. Furthermore, the failed area mobilizes only the stress corresponding to the residual strength, since the stress-strain relationship shows the strain softening type. The stress that is in excess is redistributed by the surrounding area. When the area adjacent to the failed area shows the stress immediately before the failure, this increasing the stress will causes new failure to take place. This phenomenon shows that the tensile failure domain will expand as a chain reaction, and signifies that overall failure as a specimen takes place all of a sudden.

In the stress state where the stress-strain relationship shows perfectly elasto-plastic type, shear stress occurs in the matrix between gravels arranged parallel to the direction of the maximum principal stress. Similar to tensile stress, this shear stress is considered to be greater, as the W/D_A is smaller. Accordingly, in the area closer to the gravel, local shear failure is estimated to occur from considerably lower stress level in the specimen. However, this failed area is capable of bearing the stress as same as the failure strength, unlike the case of strain softening type; therefore, the strength against failure is assumed to be greater than that at a low confining pressure.

In the stress state where stress-strain relationship shows to strain hardening type, the domain and failure mode of the stress concentrated area are considered to be basically those of the perfect elasto-plastic type. A large difference from the perfect elasto-plastic type is that "failure" does not occur to the matrix, since even the area having lost the initial elastic state is capable of bearing stronger stress. Therefore, the strength of the specimen is considered to be greater, as a greater amount of gravel stronger than the matrix is included, namely, the GC is greater.

From above discussion the characteristics of the strength of the gravelly specimen obtained from the experiment can be explained by the analytical consideration that is, to the fact that (1) the strength of the gravelly specimen becomes smaller than the strength of the matrix in corresponding to GC, in the state of stress where strain softening tendency is shown, and (2) the changes in the strength due to mixing of gravel are small, therefore the almost agreement is found with the matrix strength and gravelly soft rocks in the stress state where stress-strain relationship belongs to perfectly elasto-plastic type.

5. CONCLUSION

According to numerical analysis based on finite element method, we have reached the following conclusions on the stress-strain state within gravelly soft rock, the influence given by elastic moduli of gravel and matrix, and the failure mechanism:

(1) Stress and strain distribution in rock is very heterogeneous due to mixing gravels. Stress tends to concentrate on gravel, forming a flow threading through the gravel arranged in the maximum stress direction. Furthermore, the matrix located in this flow turns into a stress concentrated domain, while the matrix outside the flow turns into the stress relieved domain.

(2) The position and extent of the stress concentrated area, and stress relieved area are determined by the gravel arrangement, and its intensity depends on the elastic moduli ratio between gravel and matrix.

(3) The failure mode of the gravelly soft rock is mainly the tensile relieved of the matrix between gravels arranged orthogonally to the maximum principal stress in the low confining pressure, while it is the shear failure of the matrix between gravels arranged parallel to the maximum principal stress in the high confining pressure.

(4) The strength and deformation characteristics of the gravelly soft rock obtained from experiment can be explained by considering the changes in failure mode and stress-strain relationship from strain softening to strain hardening increase in confining pressure.

REFERENCES

Eshelby J. D.; 1975, The determination of the elastic field of an ellipsoidal inclusion, and related problems, Proc. Roy. Soc. London, Vol. A241, pp.376-396, 1975.

Hashin Z., and S.Shtrikman; 1963, A variational approach to the theory of the elastic behaviour of multiphase materials, J. Mech. Phys. Solids., Vol.11, pp.127-140, 1963.

Hashin Z.; 1983, Analysis of composit materials, J. Appl. Mech., Vol.50, pp.481-505, 1983.

Kobayashi T., R. Yoshinaka; 1994, Mechanical Properties of Gravelly Soft Rocks, JSCE, No.487/III-26, pp.31-40, 1994

Kobayashi T., R. Yoshinaka, T. Mimuro; 1994 Strength and Deformation Behabiour of Gravelly Soft Rock, IV CSMR Integral Approach to Applied Rock Mechanics, Proceedings, pp. 57-67, 1994

Mura T; 1987, Micromechanics of Defects in solids, 2nd revised ed., Martinus Nijhoff Publisher, pp.421-439, 1987.

Reuss A.; 1929, Berechnung der Fliessgrenze von Mischkristallen auf Grund der Plastizitatsbedingung für Einkristalle, Zeitschrift für Angewandte Mathematik und Mechanik, vol. 9, pp.49-58,1929.

Voigt W.; 1989, Ueber die Beziehung zwischen den beiden Elastizitätsconstanten isotroper Körper, Annalen der Physic und Chemie, vol.38, pp.573-587, 1989.

A study on the reasonable loading stress waveforms and loading systems of the dynamic experimental technique

Liu Deshun
Xiangtan Mining Institute, People's Republic of China

Li Xibing & Yang Xiangbi
Central South University of Technology, Changsha, People's Republic of China

ABSTRACT: The Split Hopkinson Pressure Bar(SHPB)has been widely used to measure the dynamic parameters of rocks. But it has been known that Pochhammer—Chree(P—C)oscillations exist in the dynamic stress—strain curves. First, this paper analyzes the reasons why P—C oscillations exist in the measurement of dynamic stressstrain curves by the SHPB subjected to rectangle loading wave. It is concluded that the reasonable loading stress waveform should be a single sine wave. Secondly, this paper introduces the Impact Discrete Inversion Technique(IDIT) proposed by authors, which can be used to design rams based on the expected stress waveforms. Third, a new ram that can produces a approaching sine wave is designed and made. Experimental results obtained by the new ram show that oscillations are effectively eliminated. Finally, it is suggested that the loading system of the SHPB be improved to use the new ram in the SHPB and the IDIT be used to design the rams that can produce the other kinds of loading stress wave.

1 INTRODUCTION

The Spilt Hopkinson Pressure Bar (SHPB) has been widely used to measure the dynamic properties of solid materials, such as metals and rocks. The loading stress wave in the SHPB is initiated by impact between a bar and a cylinder ram of the same diameter as the bar, which is a rectangle loading stress wave. Numerous inverstigators have found that oscillations exist in the incident waves, reflected waves, strain rate curves and reconstituted stress—strain curves obtained by the SHPB subjected to the rectangle loading wave(see Figure 1). Because the existence of oscillations it becomes. very difficult to obtain the exact dynamic parameters of the materials. As Pochhammer and Chree first analyzed this phenomenon, the oscillations have called as Hochhammer —Chree oscillations. The key seasons of phenomenon is because a basic assumption, which the specimen is a state of one—dimensional stress, couldn't be maintained well. Kolsky (1953), Davies and Hunters (1963) studimed this phenomenon and presented the approximate corrections for radial inertia and the criteria for the SHPB specimen design. Bertholf and Karnes (1975) examined previous study works using a comprehensive two—dimensional numerical study of the SHPB including the effects of the inertia and friction. Li and Gu discussed it based on experimental results.

As the moduli of the elasticity of rocks are smaller than that of metals, more serious oscillations exist in the measurement of the dynamic parameters of rocks by the SHPB subjected to rectangle loading stress waveform. A series of experimental results for granite specimen are shown in Figure 2—4. Thus, it is more important to eliminate P—C oscillations in the measurement of the dynamic properties of rocks. It is the objective of this paper to present a reasonable loading waveform and its corresponding loading system that can eliminate P—C oscillations.

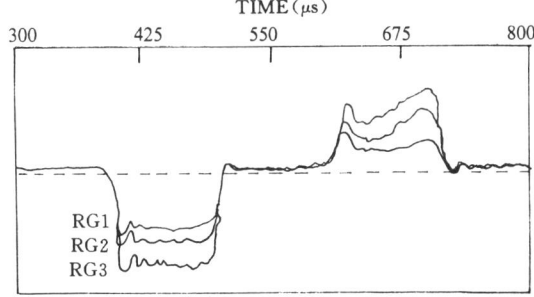

Figure 2 The incident and reflected stress waves in granite specimens subjected to rectangle loading wave

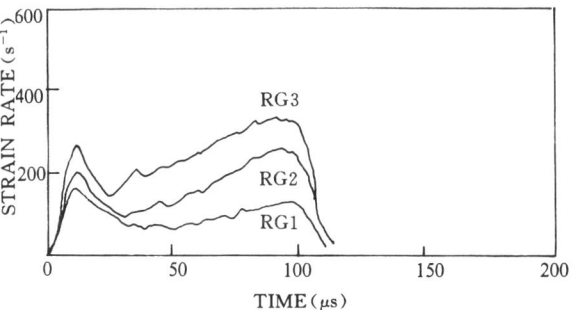

Figure 3 The time history of strain rates subjected to rectangle loading wave

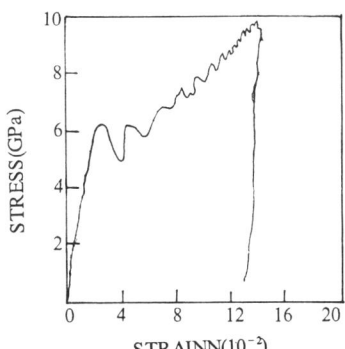

Figure 1 The dynamic stress—strain curves of metal specimen subjected to rectangle loading wave

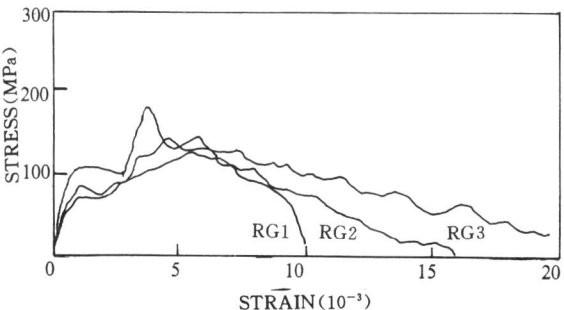

Figure 4 The dynamic stress—stain curves of granite specimens subjected to rectangle loading wave

2 THEORETICAL ANALYSIS

In order to present the method to eliminate P-C oscillations, it is necessary to make a brief explicit on the effects of inertia in the SHPB based on energy principles.

Assuming now that a very thin slice that its thickness is dx is cut out of the bar or specimen, and denoting the slice kinetic energy by E_k, the deformation energy by E_d, we have

$$\sigma \frac{d\varepsilon}{dt} = \frac{\partial E_d}{\partial t} + \frac{\partial E_k}{\partial t}$$

where

$$E_d = \frac{1}{2} E\varepsilon^2$$

$$E_k = \frac{1}{8} \rho v^2 D^2 \left(\frac{\partial \varepsilon}{\partial t}\right)^2$$

Where σ and ε are the stress and strain in the thin slice; E, ρ, υ and D are Young modulus, density, Poissons ratio and diameter, respectively, of the thin slice.
Therefore

$$\sigma = E\varepsilon + \frac{1}{4} \rho v^2 D^2 \frac{\partial^2 \varepsilon}{\partial t^2}$$

Substituting the stress (2) into motion equation

$$\rho \frac{\partial v}{\partial t} = \frac{\partial \sigma}{\partial x}$$

yields

$$\frac{\partial^2 u}{\partial t^2} - \frac{1}{4} v^2 D^2 \frac{\partial^4 u}{\partial t^2 \partial x^2} = C_0^2 \frac{\partial^2 u}{\partial x^2}$$

Where v is the velocity of the thin slice; u is the displacement of the thin slice; $C_0 = \sqrt{E/\rho}$

The above equation is the wave equation including the effects of radial inertia. Assuming a simple sine loading stress wave with the circular frequence ω, its phase velocity C can be obtained

$$C = C_0 - \frac{1}{4} v^2 D^2 \omega^2$$

The above equation shows that the stress waves with different frequences transmit with different phase velocities. Because every loading stress wave consists of some component sine waves with various frequences, the dispersion phenomenon occurs, which results in P-C oscillations. It is apparent that the wider the frequency range of principal component sine waves, the more serious the dispersion and oscillations. Therefore, the control of the oscillations is obtained by the control of the loading stress waveforms. More specifically, (half) a single sine wave does not result in the dispersion and oscillations, i.e. the reasonable loading waveform should be a single sine wave.

3 IMPACT DISCRETE INVERSION TECHNIQUE

The loading stress wave is generated by impact with the ram, Datta (1968) studied the determination of stress waveforms produced by the ram of various geometrical designs; Gupta(1978) studied impact between of a finite conical ram and a long cylindercal bar, using three-dimensional finite element model and experiments. Their study works showed that variation of the ram impact velocity produced a direct variation in amplitude of stress wave, while the ram geometry determined the profile of the stress wave. Numerous inverstigators have attempted to analyze the relationships between the stress waveforms and the geometry of the ram. To the writers' knowledge, all published analysis have been only solved the problem that the stress waveforms are determined based on the geometry of the ram given, while its inverse problem, which the geometry of the ram is designed based on the desired stress wavefom, have left to solve.

In order to design a ram that can produce a single sine loading stress wave, the Impact Discrete Inversion Technique proposed by authers is introduced here (for detailed discussions of the technique see Liu and Yang(1995)). Assuming that the profile function of the ram is $f(x)$; the waveform function is $\varphi(t)$, i.e.

$$R = f(x) \qquad 0 \leqslant x \leqslant L$$
$$P = \varphi(t) \qquad 0 \leqslant t \leqslant 2L/\alpha$$

Throughout discretion the ram profile and waveform are divided into n segments.

$$R_i = f(i\Delta x); P_i = \varphi(i\Delta t); i = 1, n$$

where $\Delta x = L/n$; $\Delta t = 2L/\alpha n$

Based on the wave theory, the following equations are obtained;

$$p_{ij} = (\mu_i - 1)q_{ij} + (2 - \mu_i)p_{i+1j-1}; i = 1, n; j = 1, n - i$$

$$q_{i+1j} = \mu_i q_{ij} + (1 - \mu_i)p_{i+1j-1}; i = 1, n; j = 1, n - i$$

$$p_{ij} = q_{ij} = 0; i = 1, n; j \leqslant 0$$

$$q_{ij} = 0; i = 0; j = 1, n$$

$$p_{0j} = P_i - P_{i-1}; i = 2, n$$

$$R_{i+1} = \sqrt{\frac{u_i}{2 - u_i}} R_i; i = 1, n - 1$$

$$R_1 = \sqrt{\frac{P_1 m_0}{(m_0 V_0 - P_1)\pi \rho \alpha}}$$

Assuming that the stress waveform function $\varphi(t)$ is given, the profile function of the ram $f(x)$ can be calculated by the above equations. Here P is force in elastic bar(N); R is radiu of the ram(m); m0 is the characteristic impedance of he bar(N.s/m); V_0 is the impact velocity(m/s); L. ρ and α are the length, the density and longtudinal Wave velocity of the ram; p_{ij} and q_{ij} are forces of forward wave and backward wave in i-th segment of the ram during j-th period (N); μ_i is transmission coefficient from (i+1)-th seqment to i-th seqment of the ram; n is the number of the segments.

Using a computer program, the profile function of the ram that can produce a half single sine wave is calculated. The results of calculation are illustrated in Figure 4-5. It must be mentioned that only half of wave functionn($0 \leqslant t \leqslant 2L/\alpha$) is exact sine function and the other of waveform is approaching sine function, because it determines the other segment of the waveform.

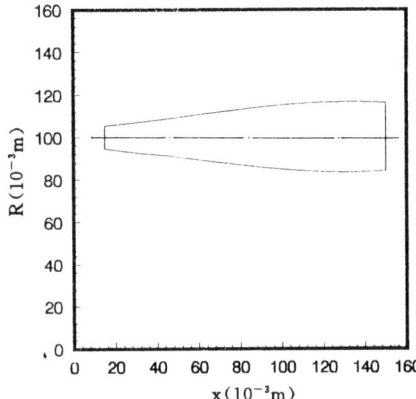

Figure 5. The theoretical profile of the ram

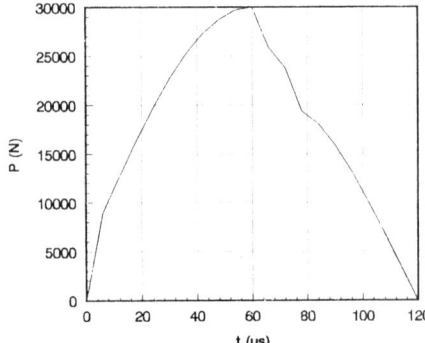

Figure 6 The theoretical loading waveform

4 EXPERIMENTAL RESULTS

Using the Impact Discrete Inversion Technique, a new ram, which produces approaching sine stress wave, is design and made. The shape of the new ram and the corresponding stress waveform are shown in Figure 7−8. Compared to the result of calculation illustrated in Figure 5−6 the practical wave is in good agreement with the desired wave.

Figure 7. The practical shape of the new ram

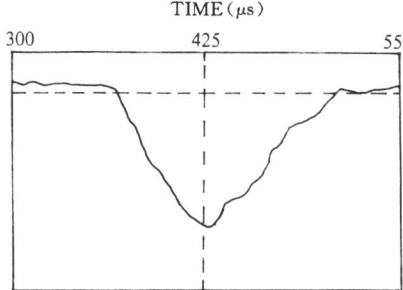

Figure 8. The practical loading waveform generated by the new ram

A serise of experiments are performed to measure the dynamic stress−strain curves of many kinds of rocks. Several experimental results for granite specimens are illustrated in Figure 9−11. Compared to the experimental results obtained subjected to rectangle loading wave shown in Figure 2−4, It is seen that the approaching sine loading wave effectively eliminates the P−C oscillations. A serise of experimental results for sandstone specimens. limestone specimens and

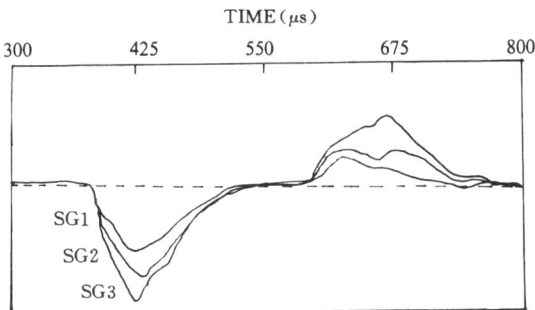

Figure 9. The incident and reflected stress waves in granits specimens subjected to approaching sine loading wave

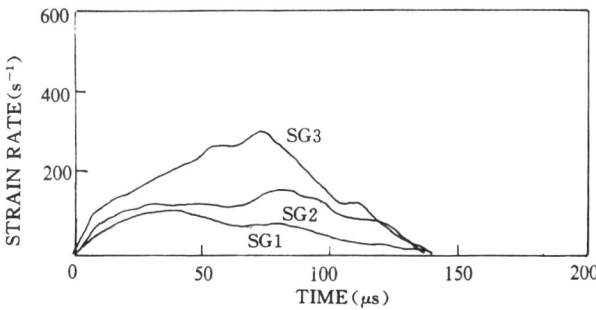

Figure 10 The time history of strain rates subjected to approaching sine loading wave

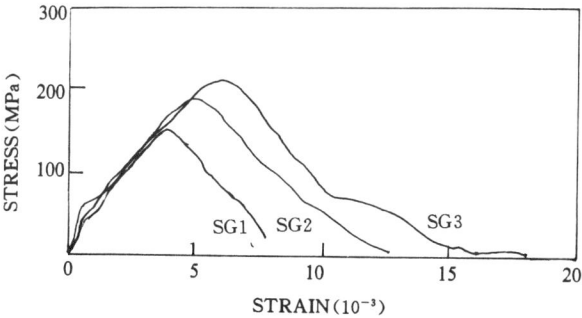

Figure 11. The dynamic stress − strain curves of granite specimens subjected to approaching sine loading wave

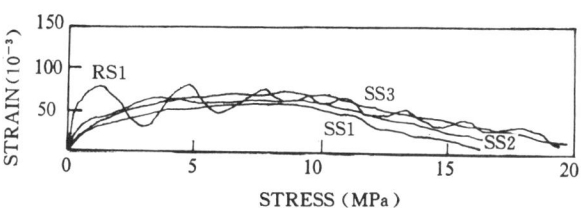

Figure 12 The dynamic stress−strain curves of sandstone specimens

marble specimens subjected to rectangle loading stress wave and approaching sine loading stress wave illustrated in Figure 12−14 (R represents rectangle loading wave and S represents approaching sine loading wave in Figures).

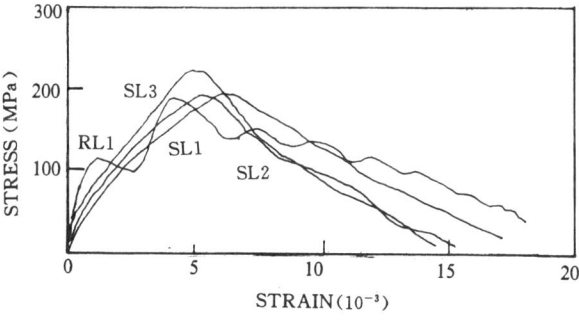

Figure 13 The dynamic stress−strain curves of limestone specimens

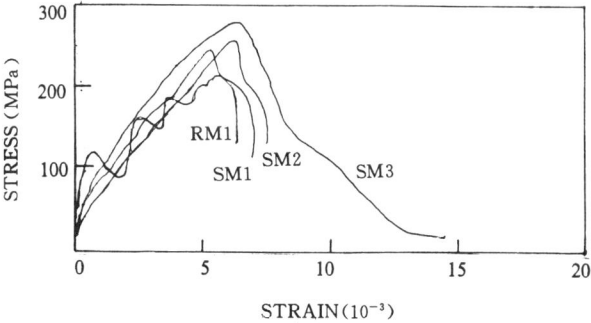

Figure 14 The dynamic stress−strain curves of marble specimens

5 CONCLUSION

From the above analysis and experimental results it can be conclued:

(1)In the measurement of dynamic parameters of rocks by the SHPB subjected to rectangle loading wave, more serious P—C oscillations exist in the dynamic stress—strain curves.

(2)The radial inertia results in P—C oscillations. A reasonable loading wave should be a single sine loading stress wave which can eliminate P—C oscillations.

(3)the new ram, that is designed based on the Impact Discrete Inversion Technique, can produce a approaching single sine loading wave. the P—C oscillations are effectively eliminated by the SHPB with the new ram.

It is suggested that the loading system of the SHPB be improved to use the new ram and the Impact Discrete Inversion Technique be used to examine the behaviors of materials subjected to different kinds of stress waveforms by the SHPB.

Acknowledgements—This study was supported financially by the National Natural Science Foundation of China

REFERENCES

Bertholf L. D. and Karnes C. H. 1975: Two—dimensional analysis of the Split Hopkinson Pressure Bar system J. Mech Phys. Solids 23 1—19

Datta P. K. 1968: The determination of stress waveforms produced by percussive drill pistons of various geometrical designs Int. J. Rock Mech. Min. Sci. 5 501—518

Davies C. D and Hunter S. C. 1963: The dynamic compression testing of solids by the method of the Split Hopkinson Pressure Bar J. Mech. Phys. Solids 11 155—179

Gupta R. B. and Nilsson L. 1978: Elastic impact between a finite conical rod and a long cylinddrecal rod J. of Sound and Vibration 60 (4) 553—563

Kolsky H. 1953: Stress waves in solids Clarendon Press Oxford

Li Xibing and Gu Desheng 1994: Rock impact dynamics Central South University of Technology China

Liu Deshun and Yang Xiangbi(1995): A study on impact inverse problem between the piston and the bar Transa. of NF Soc. 5(1)11—17 (in Chinese)

About the in-situ determination of the rock mass deformability

Nuno Feodor Grossmann
Laboratório Nacional de Engenharia Civil (LNEC), Lisboa, Portugal

ABSTRACT: The paper presents the different methods available for the in-situ determination of the rock mass deformability, and discusses their advantages and limitations. For some of these tests, the advisable test procedure is indicated, and the expressions generally used for the interpretation of their results, are introduced. A special emphasis is given to the problems of the scale effect, as well as to the correct choice of the representative value from amongst a group of different test results. The problems met with when trying to obtain correlations between the results from different types of tests, are discussed, and it is shown that, in many cases, such correlations can never be but poor. The difficulty of assessing the rock mass anisotropy, based on the performed in-situ tests, is pointed out, as well as a way to overcome it in the case of numerous borehole tests.

1 GENERAL

The deformability characteristics of a rock mass are usually determined by means of in-situ tests in which a known load is applied in a specified manner, and the corresponding rock mass deformations are measured. The results of these tests are generally interpreted in terms of the theory of elasticity of continuous, isotropic, and homogeneous media. However, due to the jointing and, sometimes, also other causes, rock masses are discontinuous, quite often anisotropic, and, if ever, only homogeneous at such a large scale that, usually, it is economically unadvisable to perform tests involving such a volume of the rock mass.

The problems caused by the insufficient test volume, are tackled with under the heading scale effect, the proceedings of the 2nd International Workshop on Scale Effects in Rock Masses (Cunha 1993), organized by the ISRM Commission on Scale Effects in Rock Mechanics, presenting the latest state of the art on that subject.

The basic scale effect problem in the deformability determination can easily be explained with the help of the theory of statistics. Whenever a rock mass deforms, the global deformation in a given direction is the sum of a large number of small, aleatory individual deformations, in the same direction, of the different constitutive elements of the rock mass (various kinds of crystals, and discontinuity surfaces of different types). Using the laws of statistics, the global deformability (the global deformation divided by the length which deforms) has a normal distribution, whose mean is independent of the number of summed individual deformations, and whose standard deviation is inversely proportional to the square root of that number. Therefore, whatever the size of the tested volume, the experimental results should always present the same mean deformability (n.b., not the same mean modulus of elasticity), and a standard deviation which is proportional to the square root of a significant length of the tested volume.

This means that, in the usually occurring rock masses, which are only homogeneous at a large scale, the results of small scale tests are much more variable than those of large scale tests, and, so, the small scale tests should always be performed in a sufficiently large number to compensate for their variability. The always existing economical constraints and, sometimes, also the imposed time schedule, unfortunately, often do not allow this last desideratum to be put in practice.

2 BOREHOLE TESTS

2.1 Introduction

The most usual small scale in-situ deformability tests are those which are performed in boreholes. This results from the fact that they are relatively cheap, because the boreholes have, nearly always, already been drilled for some other reason, and the tests, generally, do not ask for any kind of preparation of the test stretch. Furthermore, in certain cases, boreholes are the only existing access to the zone of the rock mass to be studied.

The borehole deformability tests can be grouped into 2 basic types: those in which the load is applied by means of rigid plates (e.g., the Goodman jack), and those in which the load is applied by means of flexible membranes, which adapt themselves completely to the borehole wall. The tests of this last type can be further classified into those in which only a global volumetric deformation is measured (e.g., the Ménard pressuremeter), and those in which several radial deformations are measured (e.g., the LNEC dilatometer and Amadei's directional dilatometer).

Although it is not a borehole test in the strict sense, the TIWAG radial pressure test has some similarities with the dilatometer borehole tests. Its enormous price restricts, however, its use to the specific case of important pressure tunnels, where this test corresponds to a prototype test, and, therefore, gets its justification.

The rigid plate borehole tests do not correspond to a simple theoretical load situation, and, so, the interpretation of their results is not an easy task. The greater robustness of the used equipment, in relation to the dilatometers, explains why these tests are still in use. The pressuremeters have precision limitations, as they measure volume changes and not displacements. Their field of application is, therefore, limited to the very weak rocks and soils. The most appropriate borehole tests for rock masses are, thus, the dilatometer tests, which shall be further discussed in what follows.

2.2 Dilatometer

As an example of a dilatometer, the BHD (**b**ore**h**ole **d**ilatometer) (Fig. 1), developed by the LNEC, shall be described.

The BHD has been specifically designed to carry out rock mass deformability tests in NX boreholes (≈ 76 mm diameter), by applying a uniform pressure on the borehole wall, in the zone to be tested, and measuring the induced rock mass deformation. The pressure is obtained by pumping water between the rubber jacket and the steel body of the BHD, and its value is measured, outside the borehole, with the help of a manometer. The deformation of the borehole section is measured along 4 diameters, 45° apart, with the help of 4 pairs of sensors, which are connected to differential transducers.

The advisable test procedure for a dilatometer test is the following:
1) after the installation of the dilatometer at the required depth, a low pressure, usually of 0,5 MPa, shall be applied, in order

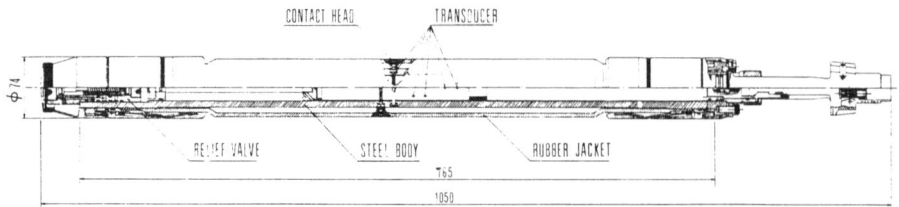

Fig. 1 — Borehole dilatometer (BHD) — Longitudinal cross-section

to ensure the adjustment of the BHD rubber jacket to the borehole wall;
2) the 1st loading-unloading cycle shall go up to its maximum pressure in, at least, 10 pressure steps of no more than 0,5 MPa, and back to the adjustment pressure in 5 pressure steps;
3) the next loading-unloading cycle(s) (usually only 1), with the same pressure limits, shall go up to the maximum pressure and back to the adjustment pressure in 5 pressure steps;
4) the following loading-unloading cycle shall go up to the maximum pressure of the previous cycles in 5 pressure steps; then, up to its own maximum pressure again in pressure steps as those used in the loading of the 1st loading-unloading cycle; and, finally, back to the adjustment pressure in 5 pressure steps; and
5) the last, usually 3, loading-unloading cycles, with the same pressure limits as the previous cycle, shall go up to the maximum pressure and back to the adjustment pressure in 5 pressure steps.

If it is necessary to test the rock mass for more than 2 maximum pressures, numbers 3) and 4) of the procedure shall be repeated with the necessary adaptations to the respective maximum pressures, and, for very simple and quick tests, numbers 3) and 4) of the procedure can even be omited.

2.3 Interpretation of the results

The interpretation of the dilatometer tests is, usually, made with the help of the formula of the theory of elasticity for the borehole with an infinite length, which is submitted to a uniform pressure on its wall

$$E = (1 + \nu) \varnothing \frac{\Delta p}{\delta} \qquad (1)$$

where E and ν are, respectively, the Young's modulus and the Poisson coefficient of the rock mass, $\varnothing$ is the borehole diameter, and δ the diametral deformation caused by the change Δp of the applied pressure.

Expression (1) is, however, only valid if the perimetric stress in the borehole wall has never ceased to be a compression, as the elastic constants of a rock mass are, generally, not the same under a compression and under a tension, and the eventually occurring tensile stresses can, if high enough, cause the rock mass to fracture, and, thus, change the geometry of the problem. This situation is, unfortunately, very frequent, especially if the tests have been performed near to the surface of the rock mass. The rather complex theoretical formulas, which have to be used to calculate the Young's modulus of the rock mass, when expression (1) is no longer valid, have been deduced by Pinto (1993).

The modulus obtained with the help of expression (1) is, therefore, in general, only the Young's modulus of an equivalent isotropic and homogeneous medium, which would present the same deformation as the one measured during the considered test, for the same load, and should, therefore, always be called dilatometric modulus.

2.4 Statistical distribution

In a homogeneous material, the individual values of the dilatometric modulus, usually, present a lognormal distribution (Graça 1974), with a standard deviation (of the logarithms of the dilatometric moduli) of about 0,4.

The most representative value of the dilatometric modulus of a rock mass is, therefore, the median (and not the mean) of the obtained results, values of the dilatometric modulus of less than about one half of the median dilatometric modulus having only a frequency of occurrence of around 5-10 %, while values of the dilatometric modulus of less than about one third of the median dilatometric modulus have a frequency of occurrence of around 1-2 %.

As examples, Fig. 2 and Fig. 3 present, respectively, the cumulative distribution function of the dilatometric moduli of the siltstones and the concrete of the Cambambe dam foundation (Grossmann et al. 1984).

2.5 Anisotropy

A dilatometer test, usually, provides independent results for several orientations normal to the borehole axis.

The differences between these results may, of course, be a con-

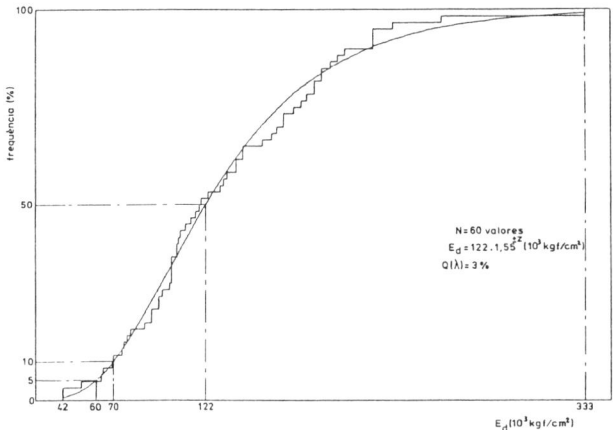

Fig. 2 — Cumulative distribution function of the dilatometric moduli of the siltstones of the Cambambe dam foundation

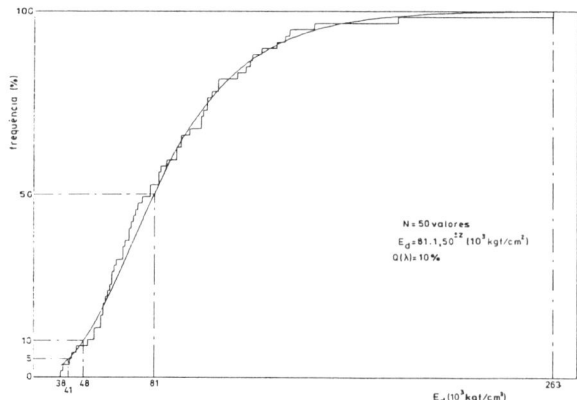

Fig. 3 — Cumulative distribution function of the dilatometric moduli of the concrete of the Cambambe dam foundation

sequence of an existing rock mass anisotropy, but it is more likely that they are caused by some aleatory discontinuity surface(s), occurring in the neighbourhood of the test stretch. The inspection of the corresponding borehole core, unfortunately, does not help very much to explain the situation, as the discontinuity surfaces with the greatest influence on the test results, are those parallel to the borehole axis, and the influence on the test results of the discontinuity surfaces normal to the borehole, is rather small.

Some information about the existing rock mass anisotropy may, even so, be obtained from the dilatometer tests, by considering not the results from the individual tests, but the means of the results, for the same direction, from all different tests performed in a homogeneous zone of the rock mass.

As an example, Fig. 4 presents the variation with the direction of the relative horizontal deformability (the ratio between the deformability determined for a given direction, and the harmonic mean of all deformabilities determined with the same test, for all directions) of the Cambambe dam foundation (Grossmann et al. 1984).

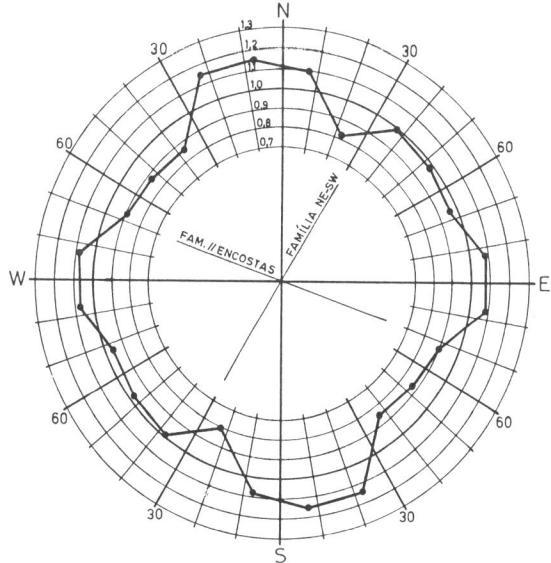

Fig. 4 — Anisotropy of the relative horizontal deformability of the Cambambe dam foundation

2.6 Correlation with borehole core characteristics

In most practical cases, a lot of different tests are performed on the borehole cores obtained during the site investigation, and, so, the temptation of trying to obtain correlations between the results of the dilatometer tests, and the results of the tests carried out on the corresponding borehole cores, always exists.

As should be expected, and the experience confirms, these correlations are, in most cases, only very poor, because the borehole core tests are only conditioned by the properties of the rock, and do not take into consideration the rock mass jointing, which may have a decisive influence on the deformability of the rock mass, and, thus, on the results obtained with the dilatometer tests.

3 LARGE SCALE DETERMINATIONS

3.1 General

The usual large scale in-situ deformability tests can be classified into 2 groups: those in which the load is applied on an already existing surface (the natural ground, or a surface in an adit), and those in which the load is applied on the walls of a slot, which is opened on purpose for the test. A minor group of rarely used, more sophisticated, and much more expensive bi- or triaxial tests should also be referred to.

The tests in which the load is applied on an already existing surface, have all the disadvantage that the response of the undisturbed rock mass to the applied load, is mixed with the response of the zone of the rock mass, which has been disturbed by the superficial stress release. The method used to avoid this mixing, is to measure only the deformations occurring at a certain distance of the loaded surface, but, due to the stress degradation with the distance to the loaded surface, it has the disadvantage that the measured deformations are rather small, and, thus, the results of the test tend to be not very precise.

The tests in which the load is applied on the walls of a slot, allow the determination of the deformability of large volumes of the rock mass in nearly undisturbed conditions. Among these tests, some have the slot opened as a series of parallel drillholes, and others have the slot opened by means of a diamond disk saw (e.g., the LNEC large flat jack test). As the smooth slot surface obtained with this last technique, even permits a direct application of the load on the rock surface, these tests can be considered the most appropriate large scale deformability tests for rock masses, and shall be further discussed in what follows.

3.2 Large flat jacks

As an example of a slot opening test, the LFJ (large flat jack) test, developed by the LNEC, shall be described.

The needed slot(s) is(are) cut with a 1 m diameter diamond-disk saw (Fig. 5), which is guided by means of a steel column with a diameter of about 15 cm. Therefore, a 168 mm diameter borehole, with a depth of about 1,5 m, must first be opened with a diamond drill, which is mounted on the same frame that will also support the LFJ slot cutting device. Both equipments are driven by an hydraulic engine, fed by an oil pump which is operated by an electric 30 kW engine.

The final dimensions of each slot (1,5 m depth, 1 m width, and 7 mm aperture) correspond exactly to those of an LFJ (Fig. 6), the borehole for the column being filled up with mortar or with wooden half-cylinders, after the insertion of the jack into the slot.

The load-transmitting zone of each flat jack is made up of a $0,75 \times 1$ m² rectangle and a semicircle with a radius of 0,5 m, the total area being, thus, 1,14 m². In order not to load the rock mass near to the surface, a 0,25 m wide strip remains unloaded. The pressure in the LFJ(s) is obtained by an oil pump, and its value controlled by a manometer.

The deformations of the rock mass are measured by means of 4 deformeters, installed inside of each LFJ. Each deformeter is basically a steel spring, which is fixed to one of the flat jack walls, and contacts the other one, due to its own force. Each spring is instrumented with 4 electrical strain-gauges, forming a full bridge, and, thus, providing an automatic temperature compensation, the changes of resistance being measured by a strain indicator.

The advisable test procedure for an LFJ test is similar to the one described for the dilatometer tests.

Fig. 5 — LFJ slot cutting equipment

Fig. 6 — Large flat jack (LFJ)

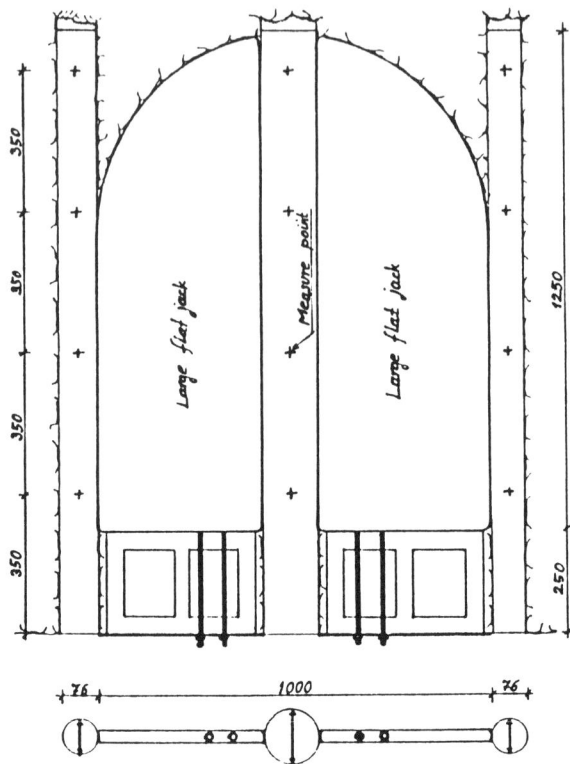

Fig. 7 — New LJF arrangement (Pinto 1993)

A new LFJ arrangement is presently under development at the LNEC (Pinto 1993). In this arrangement (Fig. 7), each slot shall contain 2 independent jacks (with about half the size of the usual LFJs), one on each side of the central, 168 mm diameter borehole. The deformeters shall no longer be inside the jacks, but in 3 independent measuring columns (with 4 deformeters each), 1 measuring column being installed in the central borehole, and the 2 others in 2 additional, NX size boreholes, which shall be opened at the 2 borders of the slot.

The advantages of the new arrangement are:
— the total testing time is reduced, because no time is lost with the hardening of the mortar in the central borehole, and, for the same amount of information, less slots have to be tested;
— the price of the tests is lower, because the measuring equipment is reusable, and, as just said, the testing time is less;
— the tests provide better information, because the readings can start before any pressure is applied (the measuring equipment is in direct contact with the rock mass), and, for each slot, the number of deformeters is more than the double of the number of deformeters in a usual LFJ; and
— long-term creep tests can be performed more easily, because the measuring equipment has no contact with the LFJ oil.

3.3 Interpretation of the results

The interpretation of the slot opening tests is, usually, made with the help of the formula of the theory of elasticity for the half-space which is submitted to a distributed normal load

$$E = (1 - v^2) k \frac{\Delta p}{\delta} \quad (2)$$

where E and v are, respectively, the Young's modulus and the Poisson coefficient of the rock mass, k is a constant, which depends on the distribution of the load, and on the boundary conditions, and δ is the deformation caused by the pressure change Δp.

For some very simple cases, the constant k can be determined by analytical methods, but in the case of the LFJ tests, although the applied load is a uniform load with a known, simple geometry, and also the slot has a known, simple geometry, the constant must be calculated with the help of numerical methods (Pinto 1981), as it is a function not only of the number of simultaneously tested LFJs, and the chosen deformeter, but also of the depth of the crack which often develops in the plane of the slot(s), during the test, due to the tensions there occurring.

Usually, the depth of this crack cannot be measured directly, and, therefore, one of the following 2 indirect methods (Pinto 1983) must be used for its determination.

The first method is based on the condition which must be fulfilled, in order that the crack does not progress any further

$$p = \varphi(h) \sigma_i + \psi(h) \sigma_t \quad (3)$$

where p is the pressure applied by the LFJ(s), σ_i the in-situ stress occurring in the rock mass, in the direction normal to the LFJ slot(s), σ_t the tensile strength of the rock mass, and $\varphi(h)$ and $\psi(h)$ are 2 functions, which depend only on the depth h of the crack. For the cases in which the in-situ stress is known or can be reasonably guessed, and also the tensile strength of the rock mass can be estimated, Pinto (1983) presents graphical solutions of equation (3).

The second method compares the actual deformation surface of the LFJ(s), obtained in the test, with the theoretical deformation surfaces, for the same applied load and different depths of the crack, and uses the method of the least squares, to find the crack depth with the best fit. The best fit corresponds to the minimum of $\sum_{i=1}^{i=N} \left[\delta_i - \frac{1-v^2}{E(h)} \Delta p \, k_i(h) \right]^2$, where the sum is extended over all N deformeters of the LFJ(s) used in the test, δ_i is the deformation measured by the deformeter i for the pressure change Δp, v is the Poisson coefficient of the rock mass, $k_i(h)$ is the constant corresponding to the deformeter i and the depth h of the crack, and $E(h)$ is the most probable modulus of elasticity for the crack depth h, given by

$$E(h) = (1 - v^2) \Delta p \frac{\sum_{i=1}^{i=N} [k_i(h)]^2}{\sum_{i=1}^{i=N} [k_i(h) \delta_i]} \qquad (4)$$

3.4 Rock mass heterogeneity

Due to the large size of the flat jacks, and the existence of several deformeters in each of them, any LFJ test discloses not only the evolution of the Young's modulus from the rock mass surface to its interior, but also the local heterogeneities which may occur in the test zone.

As an example, Fig. 8 presents the individual Young's moduli determined by each deformeter, as well as curves of equal deformability, for an LFJ test with 4 flat jacks, performed in the schists of the Ernstbachtal dam site (Grossmann 1976). The existence of a weaker zone, which crosses the 2nd jack (from the left) diagonally, approximately from deformeter 2 to deformeter 4, is clearly depicted.

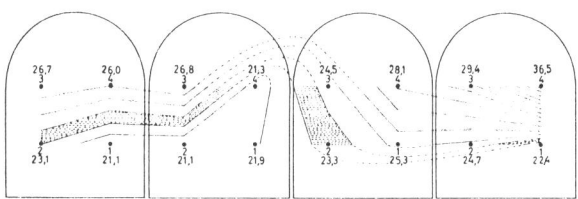

Fig. 8 — Young's moduli determined by the different deformeters, and curves of equal deformability, for an LFJ test in schists (Grossmann 1976)

REFERENCES

Cunha, A.J.V.P.D. (ed.) 1993. *Scale Effects in Rock Masses, Proceedings of the 2nd International Workshop* (Lisbon PORTUGAL, 1993 June 25). Rotterdam: A.A. Balkema.

Graça, J.G.C. 1974. *Estudo da Deformabilidade dos Maciços Rochosos com Dilatómetro* (Study of the Deformability of Rock Masses with the Dilatometer), a research officer thesis. Lisbon: Laboratório Nacional de Engenharia Civil.

Grossmann, N.F. 1976. *Druckkissenversuche fuer die Trinkwassertalsperre Ernstbachtal* (Flat Jack Tests for the Ernstbachtal Drinking Water Dam). Lisbon: Laboratório Nacional de Engenharia Civil.

Grossmann, N.F., Barroso, M.G., & Rodrigues, L.F. 1984. *Estudos Geotécnicos para o Alteamento da Barragem de Cambambe e a Construção da Central II* (Geotechnical Studies for the Heightening of the Cambambe Dam, and the Construction of the Power Plant II). Lisbon: Laboratório Nacional de Engenharia Civil.

Pinto, J.L. 1981. Determination of the Deformability Modulus of Weak Rock Masses, by Means of Large Flat Jacks (LFJ). In K. Akai, M. Hayashi, & Y. Nishimatsu (eds), *Weak Rock — Soft, Fractured, and Weathered Rock* (Proceedings of the International Symposium, Tokyo JAPAN, 1981 September 21-24), 1:447-452. Rotterdam: A.A. Balkema.

Pinto, J.L. 1983. Deformabilidade — Método LFJ (Deformability — LFJ Method). In F.M.C.P. Rodrigues (ed.), *Desenvolvimentos Recentes no Domínio da Mecânica das Rochas* (Recent Developments in the Field of Rock Mechanics), 3-27. Lisbon: Laboratório Nacional de Engenharia Civil.

Pinto, J.L. 1993. Determination of the Deformability and State of Stress in Rock Masses. In L.R.e Sousa & N.F. Grossmann (eds), *Safety and Environmental Issues in Rock Engineering//Problèmes de Sécurité et d'Environnement en Mécanique des Roches/Sicherheits- und Umweltsfragen im Felsbau* (Proceedings of the EUROCK '93 ISRM International Symposium, Lisbon PORTUGAL, 1993 June 21-24), 1:669-673. Rotterdam: A.A. Balkema.

The comparison between dynamic and static strength of soft sedimentary rocks

R. Yoshinaka & M. Osada
Saitama University, Urawa, Japan

ABSTRACT: The mechanical behavior of rocks depends on time. During earthquake, two phenomena of cyclic fatigue and rate effect should be observed in rocks. Therefore, it is necessary to consider which phenomenon dominantly appears. In this study, in order to compare dynamic strength of soft sedimentary rocks with static one, three series of triaxial compression experiments under the undrained condition and at room temperature were carried out; monotonic loading tests, fatigue tests, combined incremental and cyclic loading tests. The four rock types used in these tests were Kobe sandstone and mudstone, Yokohama siltstone, and Ohya tuff. All of them belong to different Miocene formations. The comparison between static and dynamic strength shows that dynamic strength is larger than static one. Moreover the mean ratios are almost constant regardless of initial confining pressure, and are in the range of 1.1 to 1.3 regardless of rock types. Further, it is probable that the stability of rock foundations under seismic waves can be evaluated by the failure criterion in ordinary monotonic loading tests.

1 INTRODUCTION

Soft sedimentary rocks must be often selected as the foundation ground supporting very important and large structures such as long-suspension bridges crossing straits, the nuclear power plants and others in Japan. However a lot of earthquakes will attack these structures, and therefore the stability problem of those foundations highly depends on the behavior and ultimate strength when they take place. Many experiments have been conducted to examine the deformation and strength of soft sedimentary rocks under static loading (e.g. Adachi and Ogawa, 1980, Yoshinaka and Yamabe, 1981), but few experiments have been done under dynamic loading (Nishi et al., 1983; Akai and Ohnishi, 1983; Nishi, 1984; Matsuki and Kudo, 1986; Kudo and Matsuki, 1988).

The cyclic fatigue phenomenon in various rock types, in which a material fails at a stress level lower than its static strength, has been investigated (Attewell and Farmer, 1973 for dolomite; Haimson, 1974 for sandstone and so on ; Scholz and Koczynski, 1979 for granite and diabase). For example, Haimson(1974) shows that some hard rock types are significantly fatigued by cyclic loading up to 60-80% of static strength. On the other hand, the strength of rocks is increased with a strain rate or a stress rate increasing, associated with frequency increasing (Attewell and Farmer, 1973; Nishi et al.,1983; Ishizuka et al.,1993). This phenomenon is called rate effect. It should be noted that these phenomena are the opposite effect on strength of rocks.

During earthquake, repeated load applied to rocks gradually increases in proportion to seismic acceleration. Therefore both phenomena as mentioned above should be observed in those rocks. Thus we need to consider which phenomenon dominantly appears, in particular in case of soft rocks and soils.

The purpose of this paper is to present the mechanical behavior of four soft sedimentary rocks under static and cyclic loading, and to compare between dynamic and static strength of soft sedimentary rocks.

2 ROCK TYPES AND SPECIMEN PREPARATION

Four soft sedimentary rock types were tested under static and cyclic loading; Kobe sandstone and mudstone, Yokohama siltstone, Ohya tuff. All of them belong to different Miocene formations. Kobe sandstone and mudstone, which are alternated by 1 to 2 m in thickness, were obtained from a depth of 55 m to 100 m beneath the sea floor by boring. Yokohama siltstone was sampled from the excavated deep shaft at a depth of 40 m in Yokohama City. And Ohya tuff, which is quarried for building stone, is greenish pumiceous tuff and contains volcanic rock fragments which are commonly altered to clay minerals and zeolite which are called "miso". The latter two samples were obtained as blocks.

In each rock specimens were cored from each boring core or block in the same direction because some sedimentary rocks have strong anisotropy of deformation and strength. Then they were cut to yield nearly parallel plane surface. The prepared specimens were right circular cylinders, 5 cm in diameter and about 10 cm in length. They were made saturated after being deaired in a vacuum condition for about two days. Thus B-values of more than 0.95 were observed in specimens tested. Their mean values of physical properties are listed in Table 1.

Table 1 Physical properties of samples

Rock Type	G	γt (N/m^3)	n (%)	T_0 (MPa)	qu (MPa)
Kobe Ss	2.61	2.215	24.2	0.112	2.11
Kobe Ms	2.65	2.213	26.8	0.131	1.26
Yokohama siltstone	2.68	1.991	41.5	0.217	3.41
Ohya tuff	2.31	1.754	41.6	0.565	6.82

G; specific gravity; γt; wet unit weight, n; porosity,
T_0; tensile strength, qu; uniaxial compressive strength

3 EXPERIMENTAL PROGRAM

The experimental system used in the present study mainly consists of three equipment; a compression machine with load cell, a pressure vessel, recording equipment. The compression machine is a hydraulic servo-controlled apparatus and has a loading capacity of 490 kN. The pressure vessel has a confining pressure capacity of 5 MPa, and the chamber pressure was applied by water pressure. The recording equipment can quickly collect the data at a rate of 1000 points per second. Using this system, axial load, relative displacement of piston, confining pressure and pore pressure were monitored in all specimens.

First, all specimens jacketed with the membrane of rubber were isotropically loaded up to 0.1, 0.5, 1.0, 2.0 and 3.4 MPa of confining pressure and were consolidated for about one day. Pore water was drained through the porous-metal mounted in cap and pedestal, and nylon mesh placed along the circumference of specimen to accelerate consolidation. Next, to obtain a high degree of saturation and to measure apparently negative pore pressure, a back pressure of 0.5MPa was carefully applied not to bring about the change of effective confining stress. Finally, three series of triaxial compression experiments under the undrained condition were carried out; monotonic loading tests, fatigue tests, combined incremental and cyclic loading tests. In figures and tables, these experiments are abbreviated to ML test, FT test and CICL test, respectively.

These loading patterns are shown in Fig.1. In monotonic loading tests, load was applied at a constant displacement rate, 0.01 mm/min. (Fig.1 (a)). In fatigue tests, the load was monotonically applied up to various initial deviator stress at a constant stress rate, and then repeatedly applied until specimen was ruptured by fatigue (Fig.1 (b)). The maximum deviator stress was set at between 80-120 % of the static compressive strength at the corresponding confining pressure. In combined incremental and cyclic loading tests, load was applied in a similar manner to fatigue tests, but the maximum amplitude was stepwise increased by each 5 cycles such as shown in Fig.1 (c). The increment of maximum amplitude was about one tenth of the static compressive strength by step. At this time the lower peak stress was kept near zero throughout the experiments. The cycle was stress-controlled, sine wave in shape, and had a frequency of 0.2 Hz in all cyclic loading experiments. The experimental configurations are summarized in Table 2.

4 RESULTS

4.1 Monotonic Loading test

The monotonic loading test was carried out in order to examine the mechanical properties of soft sedimentary rocks and to determine the static strength. Typical stress-strain-pore pressure relations of four rock types are shown in Fig.2. These figures show that all of rocks show the strain softening behavior in the region of applied confining pressure, but they have some appreciably different properties in deformation and strength.

In deformation characteristics, Young's moduli of Kobe sandstone and Ohya tuff are almost independent of confining pressure and rather elastic, while those of Kobe mudstone and Yokohama siltstone increase with increasing confining pressure. Peak strength of all rocks also increases with increasing confining pressure, but those of Yokohama siltstone and Ohya

Table 2. Experimental configurations

Rock Type	ML test	FT test	CICL test
KSs	0.1, 0.5, 1.0, 2.0, 3.4	-	0.1, 0.5, 1.0, 2.0, 3.4
KMs	0.1, 0.5, 1.0, 2.0, 3.4	-	0.1, 0.5, 1.0, 2.0, 3.4
YSt	0.1, 0.5, 1.0, 2.0, 3.4	5, 20	0.1, 0.5, 1.0, 2.0, 3.4
OTf	0.1, 0.5, 1.0, 2.0, 3.4	5	0.1, 0.5, 1.0, 2.0, 3.4

The figures in the table show the applied confining pressure, and are in MPa. The experiments are performed at room temperature.

tuff in higher confining pressure do not rise much owing to generating excess pore pressure. Moreover failure occurs at the almost same axial strain regardless of confining pressure, except for Kobe sandstone.

4.2 Fatigue test

As well known, a material fails at a stress level lower than its static strength under cyclic loading. This phenomenon is commonly termed "cyclic fatigue". Therefore, the fatigue test was performed in order to investigate the effect of cyclic loading on the strength of rocks. Only Yokohama siltstone and Ohya tuff were used in this test shown in Table 2. However the number of loading cycles per test was limited to 10^3.

The results are shown in Fig.3 by S-N plots which show the number of cycles (N) required to fail a specimen loaded to a certain maximum deviator stress (S) normalized by the static strength at a corresponding confining pressure determined from monotonic loading tests. In general it is difficult to determine the number of cycles to failure because failure process gradually proceeds. In our fatigue tests, the accumulated upper axial strain per cycle drastically increased just before fracture. Therefore we defined failure of specimen as a state of stress at which the accumulated upper axial strain per cycle drastically increases, and the number of cycles to failure as the corresponding number.

The data in Fig.3 indicate that the logarithm of the number of cycles to failure is inversely proportional to the maximum deviator stress. The approximate line by least square method is also shown in Fig.3. Although the scatter is considerable, these relations are described in the following equations.

$$S = 1.18 - 0.043 \, log \, N \quad (1)$$

for Ohya tuff at confining pressure of 0.5MPa,

$$S = 1.26 - 0.061 \, log \, N \quad (2)$$

for Yokohama siltstone at confining pressure of 0.5MPa,

$$S = 1.13 - 0.051 \, log \, N \quad (3)$$

for Yokohama siltstone at confining pressure of 2.0MPa. Comparing eq.(2) with eq.(3), the effect of confining pressure on fatigue is little conspicuous in the range considered.

4.3 Combined Incremental and Cyclic Loading

The combined incremental and cyclic loading test was carried out in order to examine the mechanical properties of rocks for the repeated load in which the amplitude gradually increases, and to determine the dynamic strength. An example of results is shown in Fig.4.

Fig.4(a), 4(b) and 4(c) show the behavior with time of deviator stress, axial strain and pore pressure respectively. In Fig.4(a), the deviator stress gradually decreases in the last 5 cycles. Thus we define peak stress at this time as dynamic strength against static strength. The ratios of dynamic strength to static one will be discussed in the following section. In Fig.4(b), the upper and lower axial strains were not only accumulated with time or cycle as a whole, but also in each 5 cycles. In particular, those strains were increasingly accumulated in the last 5 cycles. On the other hand, the upper and lower pore pressure is also accumulated at first, but turns to decrease at nearly half level of dynamic strength, which means the failure of internal structure.

Moreover Fig.4(d) and 4(e) show the stress-strain-pore pressure relation. Comparing these figures with Fig.2(c), it is found that the envelop of stress-strain and pore pressure-strain relation in the combined incremental and cyclic loading test has a similar

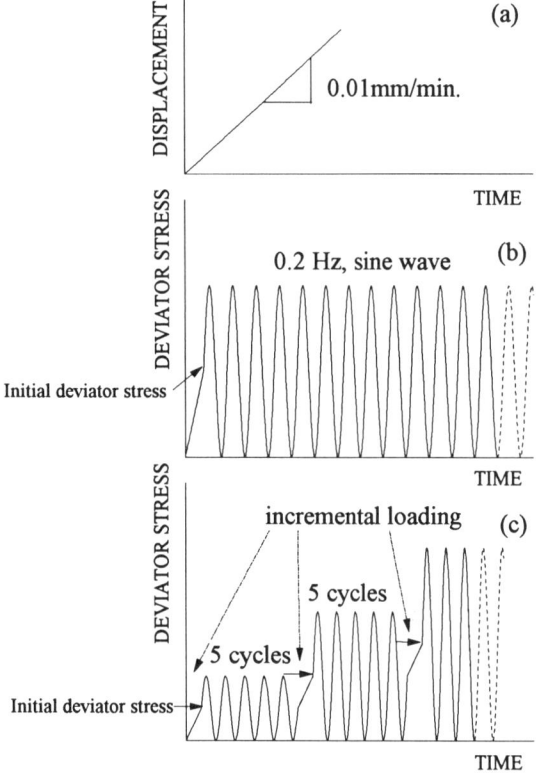

Fig.1 Loading patterns are shown. (a) Monotonic loading. (b) Cyclic loading. (c) Combined incremental and cyclic loading. The cycle was stress-controlled, sine wave in shape, and had a frequency of 0.2 Hz. (See the text in detail.)

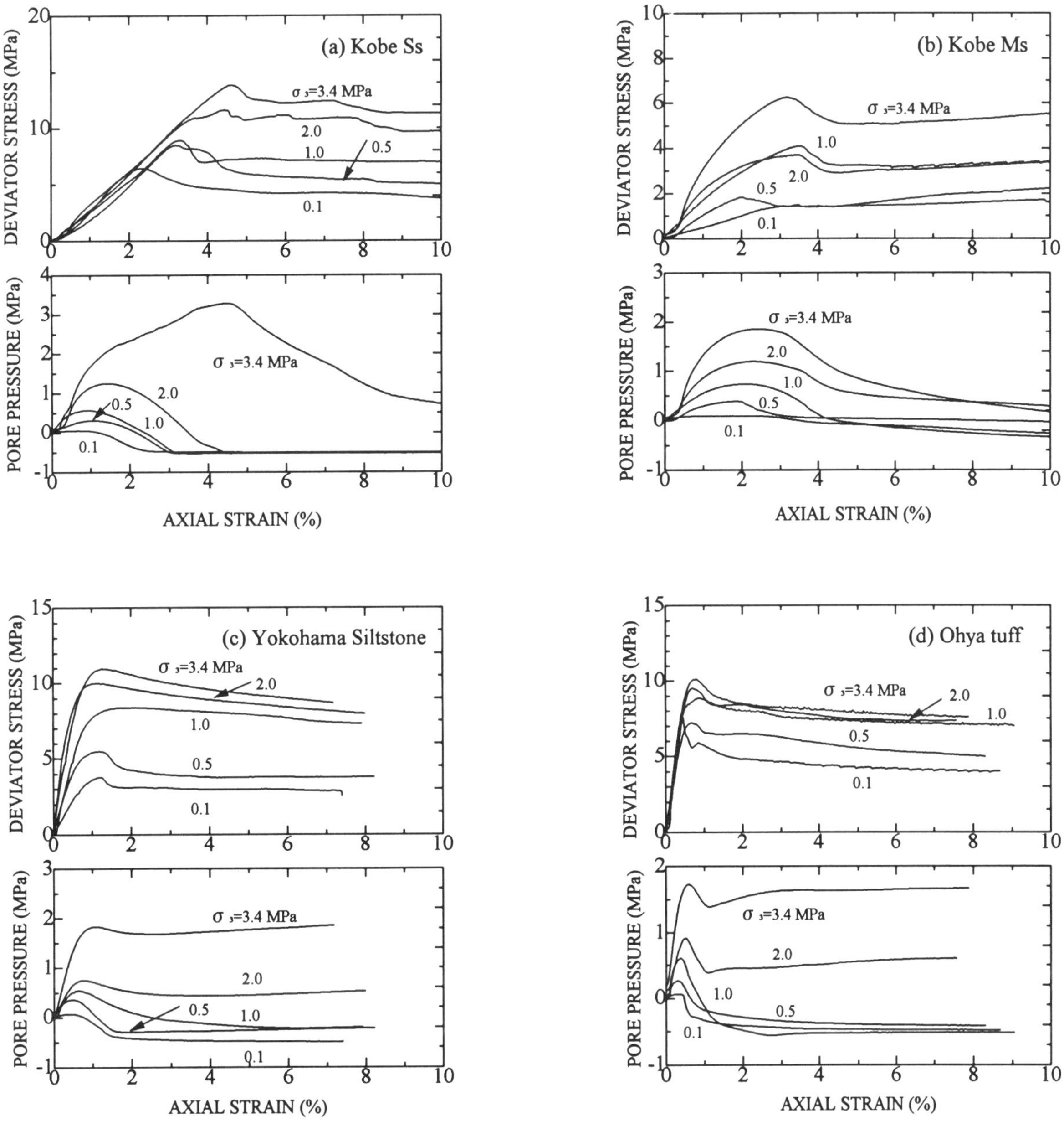

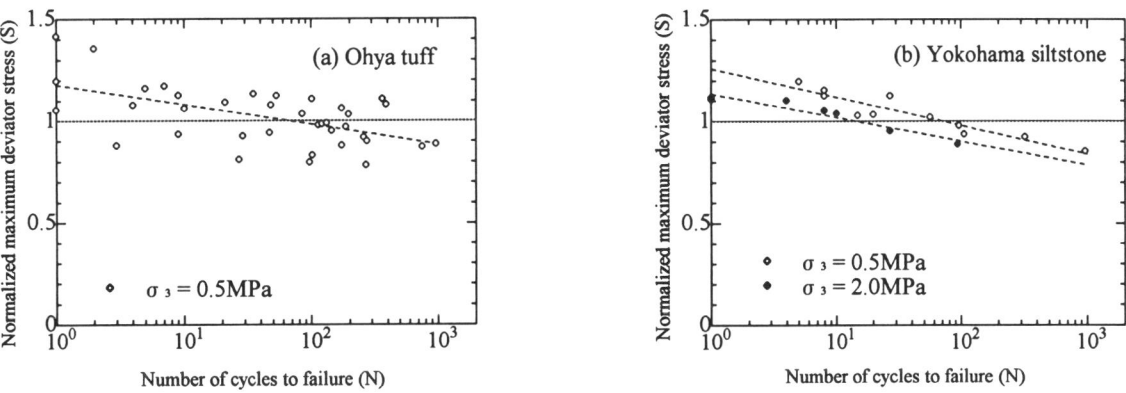

Fig.2. Typical stress-strain-pore pressure relation of specimens tested. (a) Kobe Sandstone. (b) Kobe Mudstone. (c) Yokohama Siltstone. (d) Ohya Tuff.

Fig.3. S-N plots for (a) Ohya tuff and (b) Yokohama siltstone. Both data indicate that the logarithm of the number of cycles to failure is inversely proportional to the maximum deviator stress. The dashed line show the approximate line by least square method.

trend of stress-strain and pore pressure-strain curves in monotonic loading tests.

Moreover Fig.4(d) and 4(e) show the stress-strain-pore pressure relation. Comparing these figures with Fig.2(c), it is found that the envelop of stress-strain and pore pressure-strain relation in the combined incremental and cyclic loading test has a similar trend of stress-strain and pore pressure-strain curves in monotonic loading test.

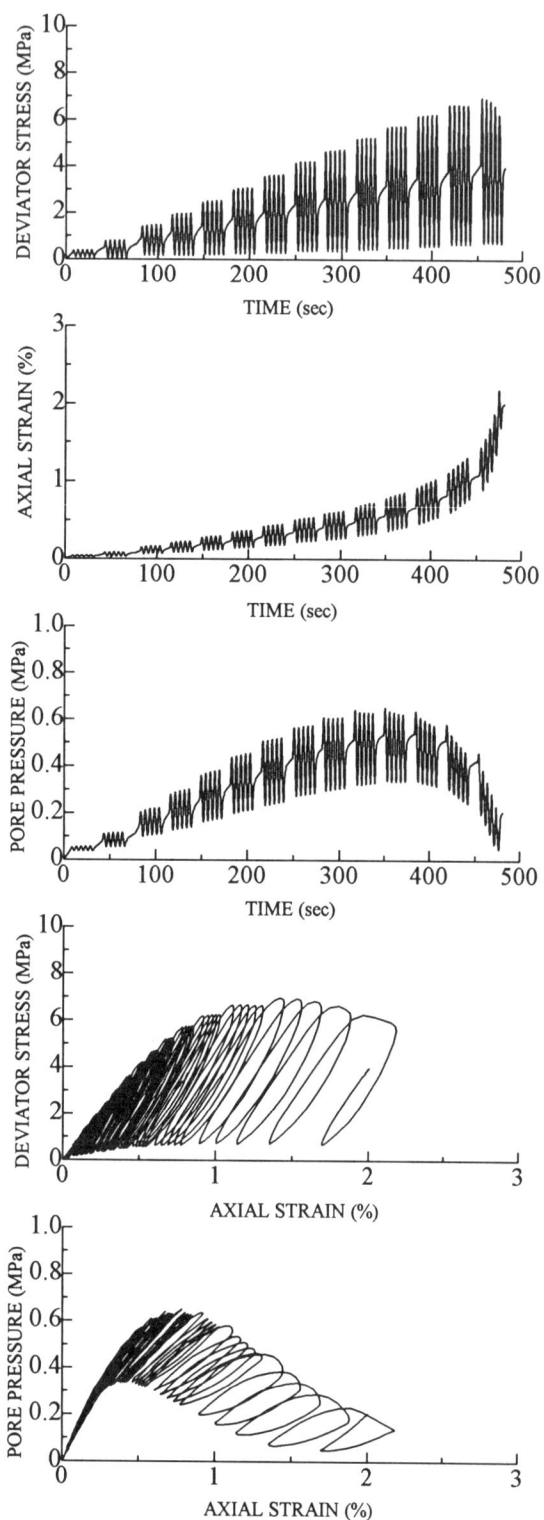

Fig.4. Example of Yokohama siltstone at confining pressure of 1.0 MPa in the combined incremental and cyclic loading test.

5 DISCUSSION

5.1 Comparison Between Static and Dynamic Strength

In order to make a comparison between the static and dynamic strength, we plotted the ratio of dynamic strength to static strength against initial confining pressure in Fig.5. Kobe mudstone and sandstone have different strength layer by layer. Therefore the ratios were calculated in each layer, but scatter. In particular only 13 specimens were tested in Kobe mudstone, so the scatter is conspicuous. Thus we will discuss the results except for Kobe mudstone.

From Fig.5 it is found that almost all the ratios are ranging from 1.0 to 1.5. The approximate line by least square method is also shown in Fig.5. Those lines indicate that although the scatter of data is considerable, the ratios are almost constant regardless of initial confining pressure and the values are in the range of 1.1-1.3 regardless of rock types. That is to say, dynamic strength is larger than static one.

Now let us consider the meaning of this ratio. In the combined incremental and cyclic loading test, the effect of loading rate on strength is not only included, but also that of fatigue due to stepwise increasing amplitude of repeated loads. Hence we need to consider which phenomenon dominantly appears. As mentioned above, the strength of rocks is increased with a stress rate increasing, associated with increasing frequency. Provided that the maximum stress rate is estimated roughly from displacement rate in monotonic loading tests, and from applied frequency in the fatigue and the combined incremental and cyclic loading test, the stress rate in both cyclic tests is about 10^3 times as large as that in the monotonic loading test. Thus it is expected that the strength in both cyclic loading tests should be larger than static one. According to the experimental results of Ishizuka et al. (1993), the relation between uniaxial compressive strength and stress rate for wet samples of Ohya tuff is described in the following equation:

$$\sigma_c = 6.83 + 0.69 \log \dot{\sigma} \qquad (4)$$

where σ_c is uniaxial compressive strength in MPa and $\dot{\sigma}$ stress rate in MPa/s. The increase of strength associated with stress rate increasing, which is calculated using the above relation, is to be 1.45 in the stress rate range of our tests. This value, however, may be somewhat small because Ishizuka et al. (1993) have pointed out that the effect of stress rate under triaxial compressive condition is smaller than that under uniaxial condition. In our tests, supposing that the stress at which the specimen reaches failure by one cycle in the fatigue test is regarded as the strength at the corresponding stress rate, the ratios of this strength to that in the monotonic loading test are consistent with the values at which the dashed line cross the ordinate in Fig.3, that is the first terms in the right-hand side of Eqs.(1), (2) and (3). The value of 1.18 for Ohya tuff is somewhat smaller than the value estimated by Eq.(4). On the other hand, these values are not very different from the ratios of dynamic strength to static strength shown in Fig.5.

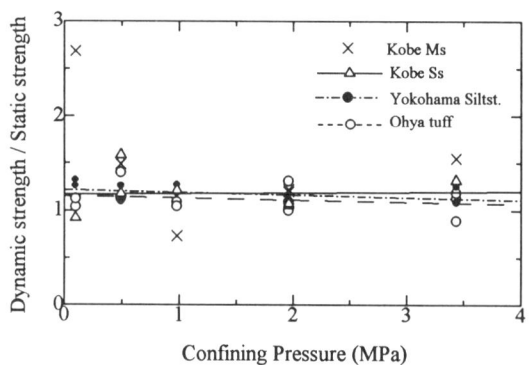

Fig.5. The ratio of dynamic strength and static strength. The data are shown with the approximate line by least square method. The approximate ratios for all rock tested except for Kobe mudstone are almost constant regardless of initial confining pressure and are in the range of 1.1-1.3 regardless of rock types.

In case of seismic wave, the distinguished frequency is in the range of about 0.1 - 1.0 Hz. The frequency used in our tests is the lower part of it. Therefore, the ratios of dynamic strength to static strength should be also evaluated as a lower value for seismic wave. Moreover in general, only several cycles, which are recorded as the peak of seismic acceleration, were observed in one earthquake. As shown in Fig.3, fatigue strength for several cycles is not smaller than static strength of Ohya tuff and Yokohama siltstone.

Consequently, it is probable that the effect of stress rate on strength dominantly appears, and that the effect of fatigue for the repeated load in which the amplitude gradually increases seems to be ignorable. This result is almost consistent with the results of Nishi (1984) in irregular cyclic loading tests using actual earthquake acceleration records. The result as mentioned above, however, can be adapted only in intact rocks. Thus further experimental work is obviously required to investigate the mechanical behavior of rock masses containing the defects such as fractures under cyclic loads.

5.2 Relation of maximum shear stress to effective mean stress using the principal stresses

We plotted the ratio of dynamic strength to static strength against initial confining pressure in Fig.5. But actually, the effective confining pressure changes as pore pressure changes. Then the relation of maximum shear stress to effective mean stress using the principal stresses is shown in Fig.6. The data points in the monotonic loading test (triangle) and in the combined incremental and cyclic loading test (solid circle) are plotted with the approximate lines by least square method. Fig.6 indicates that Mohr-Coulomb's failure criterion in the compressive region can be applicable for all rocks tested and for both tests. The coefficients of Mohr-Coulomb's failure criterion are summarized in Table 3.

The ratios of cohesion and internal friction angle for both tests, c_d/c_s and ϕ_d/ϕ_s respectively, are also shown in Table 3. The value for Ohya tuff could not be obtained because the scatter was considerable. These ratios indicate that the cohesion is subject to the influence of cyclic loads. On the other hand, the internal friction angle is not significantly affected by cyclic loads. This is partly because the effective stress path, shown by the short-dashed lines in Fig.6, is along the failure criterion near the peak of maximum shear stress. After all, the failure criterion in the combined incremental and cyclic loading test is located in the upper, or at least same region against that in the monotonic loading test. That is to say, this means that if we design for various rock structures on the basis of the failure criterion in ordinary monotonic loading tests, then we can keep them safe even if an earthquake occurs. But it is necessary to consider the effect of fatigue on strength in the cyclic loads with relatively long period such as CAES.

6 CONCLUSION

In order to investigate the mechanical behavior of four soft sedimentary rocks under static and cyclic loading, and to compare between dynamic and static strength of soft sedimentary rocks, three series of triaxial compression experiments under the undrained condition were carried out; monotonic loading tests, fatigue tests, combined incremental and cyclic loading tests.

The following results are obtained:
1) In the monotonic loading test, stress-strain-pore pressure relations of four rock types show the strain softening behavior in the region of applied confining pressure, but they have some appreciably different properties in deformation and strength.
2) In the fatigue test, the logarithm of the number of cycles to failure is inversely proportional to the maximum deviator stress.
3) The envelop of stress-strain and pore pressure-strain relation in the combined incremental and cyclic loading test has a similar trend of stress-strain and pore pressure-strain curves in the monotonic loading test.
4) The ratio of dynamic strength to static strength are almost constant regardless of initial confining pressure and are in the range of 1.1-1.3 regardless of rock types. That is to say, dynamic strength is larger than static one.
5) During earthquake, it is probable that the effect of stress rate on strength dominantly appears, and that the effect of fatigue for the repeated load in which the amplitude gradually increases

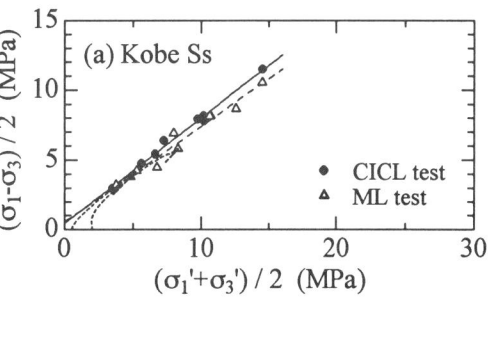

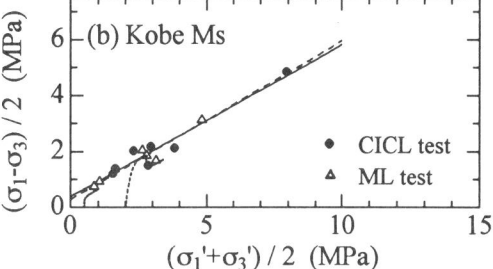

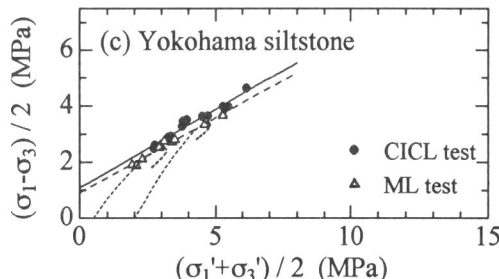

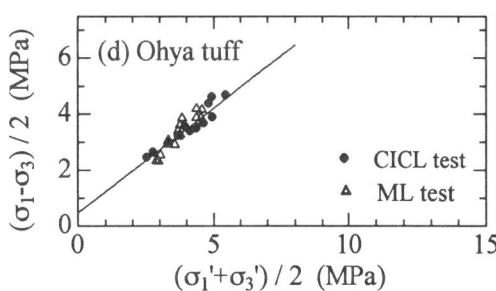

Fig.6. Relation of maximum shear stress to effective mean stress using the principal stresses. Triangle with the long-dashed line ; monotonic loading test. Solid circle with the solid line; combined incremental and cyclic loading test. Both lines are calculated by least square methods. The short-dashed lines show the typical effective stress path.

Table 3. Coefficients of Mohr-Coulomb's failure criterion

Rock Type	c_s (MPa)	ϕ_s (°)	c_d (MPa)	ϕ_d (°)	c_d/c_s	ϕ_d/ϕ_s
Kobe Ss	0.78	43	0.72	49	1.08	1.13
Kobe Ms	0.35	35	0.47	33	1.34	0.95
Yokohama Siltstone	1.09	32	1.32	34	1.21	1.05
Ohya tuff	-	-	0.72	48	-	-

The coefficients are calculated in the following equation:
$$\frac{1}{2}(\sigma_1 - \sigma_3) = c\cos\phi + \frac{1}{2}(\sigma_1' + \sigma_3')\sin\phi$$
The suffix s shows the value for the monotonic loading test and d the combined incremental and cyclic loading test.

seems to be ignorable.

6) Mohr-Coulomb's failure criterion in the compressive region can be applicable for all rocks tested, and for the monotonic loading test and the combined incremental and cyclic loading test.

7) The cohesion is more affected by cyclic loads than the internal friction angle.

8) If we design for various rock structures on the basis of the failure criterion in ordinary monotonic loading tests, then we can keep them safe even if an earthquake occurs.

Further experimental work is obviously required to investigate the mechanical behavior of rock masses containing the defects such as fractures under cyclic loads. Moreover it is necessary to discuss not only the effect of irregular cyclic loads with various frequencies but also that of cyclic loads in which the direction of principal stress change.

ACKNOWLEDGMENT

We wish our thanks to our staff and undergraduate students of Saitama University for their help during the experimental work. It is also pleasure to acknowledge Grant-in-Aid for Scientific Research (B) from the Ministry of Education, Japan. (Grant No. 03452204)

REFERENCES

Adachi, T. & Ogawa, T. 1980. Mechanical properties and failure criterion of soft sedimentary rock. Proceedings of JSCE. No.295: 51-63. (in Japanese)

Afrouz, A. & Harvey, J.M. 1974. Rheology of rocks within the soft to medium strength range. Int. J. Rock Mech. Min. Sci. & Geomech. Abstr. Vol.11: 281-290.

Akai, K. & Ohnishi, Y. 1983. Strength and deformation characteristics of soft sedimentary rock under repeated and creep loading. Proc. Int. Symp. Rock Mech. 5th: 121-124.

Attewell, P.B. & Farmer, I.W. 1973. Fatigue behavior of rock. Int. J. Rock Mech. & Min. Sci. Vol.10: 1-9.

Nishi, K., Okamoto, T. & Esashi, Y. 1983. Strength-deformation characteristics of mud-stone under some kinds of loading conditions and its unificative interpretation. Proceedings of JSCE. No.338: 149-158. (in Japanese)

Haimson, B.C. 1974. Mechanical behavior of rock under cyclic loading. Proc. of 3rd Congress of ISRM. 3-2A: 373-378.

Ishizuka, Y., Abe, T., Koyama, H. & Komura, S. 1993. Effects of strain rate and frequency on fatigue strength of rocks. Proceedings of JSCE. No.469: 15-24. (in Japanese)

Kudo, H. & Matsuki, K. 1988. Effect of pore pressure on the cyclic fatigue characteristics of water saturated rocks under confining pressure. Journal of the Mining and Metallurgical Institute of Japan. Vol.104, No.1201: 157-161. (in Japanese)

Matsuki, K. & Kudo, H. 1986. Cyclic fatigue characteristics of unsaturated rocks under confining pressure. Journal of the Mining and Metallurgical Institute of Japan. Vol.102, No.1186: 849-854. (in Japanese)

Nishi, K. 1984. Strength-deformation properties of mudstone under cyclic loading. Proceedings of JSCE. No.352: 41-50. (in Japanese)

Scholz, C.H. & Koczynski, T.A. 1979. Dilatancy anisotropy and the response of rock to large cyclic loads. Jour. Geophys. Res. Vol.84, No.B10: 5525-5534.

Yoshinaka, R. & Yamabe, T. 1981. Deformation behaviour of soft rocks. Proc. International Symposium on Weak Rock. Theme 1: 1-15.

Rock Foundation, Yoshinaka & Kikuchi (eds) © 1995 Balkema, Rotterdam. ISBN 90 5410 562 3

Method to calculate engineering parameters of jointed rockmass

Faquan Wu
China University of Geosciences, Wuhan, People's Republic of China

ABSTRACT: The paper introduces some methods to calculate the engineering parameters of jointed rockmass such as igneous rock and some thick strata based on the theory of "Statistical mechanics of rockmass". The parameters of rockmass includes' structural indexes, elastical modulus, and strength under diffirent stress states.

It is a practical but difficult task to determine engineering parameters of rockmass. In most projects in China such as hydro-power stations, minings,railway and some undergroud engineerings, it is always necessary to do large amount of in-situ testings to get data on mechanical and hydraulic properties of rockmass. Unfortunately, even it takes much money, time and material, the result is often not satisfactory and relieble. Therefore,to explore a simple and direct method to determine the parameters of rockmass has been a concerned aspect for a long period.

The most difficult problem in the work comes from the complicated mechanical effects of discontinuity network in rockmass. These effects are also the main cause leading to the size effect of sample testing. In the past years, scholrs in the world explored ways to calculate the parameters by theoretical method. Kawamoto T. and his colleagues(1988)introduced damage theory, Oda M.(1986) developed fabric tensor method, all these are remarkable.

The paper is to introduce a new theoretical method to calculate the engineering parameters of jointed rockmass based on "Statistical mechanics of rockmass" which is proposed by the author (1992,1993).The parameters includes structural indexes (such as densities, a_m, t_m, RQD etc.), elastic modulii,strength(or bearing capacity),and so on. All these data can be got in diffirent directions. To do this the only preparatory wock is to obtain enough data of uniaxial compressive strenth of rock blocks and shearing strength of cracks by laboratory work,geometry of discontinuity network, pore pressure and stress state in rockmass by field investgation and some simple calculation. This is much easier than field testing. The parameters from this way will reasonablly consider the effects of stress, fluid pressure and other environmental factors. Naturally,they should be corrected by few field testing or empirical data to confirm the realiebility.

1 STRUCTURAL PARAMETERS OF ROCKMASS

The structural parameters includes the number of joint sets, average attitude, trace length, openning, normal density of each group, area densities, volume densities and RQD in diffirent directions, and the biggest size and openning of joints with the most probability to occur in a volume.Here introduce some formulas to calculate several parameters.

1.1 Densities of cracks

We define three kinds of densities:normal density of a set of joints, area density and volume density.The normal density of a joint set is the number of cracks intersected by unit length of normal line of the set; the area desity is the number of central points of trace lengths in a unit area; and the volume density is the number of central points of cracks in a unit volume of rockmass. Here we'll not disccuss the normal density.

If the half trace lengths(l') of a joint set fit distribution h(l'), has normal desity λ ,and the central points of the cracks randomly distributes in a area,then the central point area density of the cracks λ_s is

$$\lambda_s = \frac{\lambda}{2 \int_0^\infty \int_{l'}^\infty h(l') dl' dl'} \quad (1)$$

If l' fits neg-exp (negative exponontial) distribution,then

$$\lambda_s = \mu \lambda = \frac{2\lambda}{\pi a} \quad (2)$$

where $\mu = 1/2\bar{l'}$, $\bar{l'}$ and $\bar{a}$ are the average of half trace lengths and radii respectively.

If radii(a) of a set of cracks fit distribution f(a), has normal density λ, and the central points of cracks randomly distributes in the 3 dimensional space,then the volume density is

$$\lambda_v = \frac{\lambda}{2\pi \int_0^\infty R \int_R^\infty f(a) da dR} \quad (3)$$

When a fits neg-exp distribution,then

$$\lambda_v = \frac{2}{\pi^3} \mu^2 \lambda = \frac{\lambda}{2\pi \bar{a}^2} \quad (4)$$

If there are N sets of discontinuities, the total area density and volume density are respectively

Table 1. Structural data of rockmass

group number	dip $\alpha(°)$	dip angle $\beta(°)$	density λ (1/m)	radius a (m)
1*	140	17	9.0	10.0
2	215	63	1.0	0.58
3	60	81	0.64	0.75
4	197	87	0.72	0.57

* bedding.

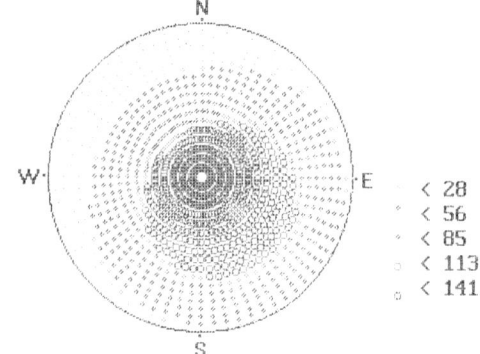

Figure 1. stereogram of λ_s of cracks
$\lambda_{smax}=141$, $\lambda_{smin}=1$, $\lambda_{save}=86$

$$\lambda_{st}=\Sigma\lambda_{si}, \qquad \lambda_{vt}=\Sigma\lambda_{vi} \qquad (5)$$

Figure 1 is the stereogram of λ_s. The basic data are shown in table 1.

1.2 Directional RQD

If the spacings(x) of intersecting points of cracks with a scanning line fit distribution $f(x)$, then the value of RQD in the direction is

$$RQD = \lambda \int_t^\infty f(x)dx \int_t^\infty x f(x) dx \, 100\% \qquad (6)$$

where $\lambda=1/x$. If a fit neg-exp distribution, then

$$RQD = (1+t\lambda)\exp(-2t\lambda)$$
$$= (1+0.1\lambda)\exp(-0.2\lambda) \, 100\% \qquad (7)$$

Figure 2 shows the directional distribution of RQD values, and the basic structural data are the same as in table 1.

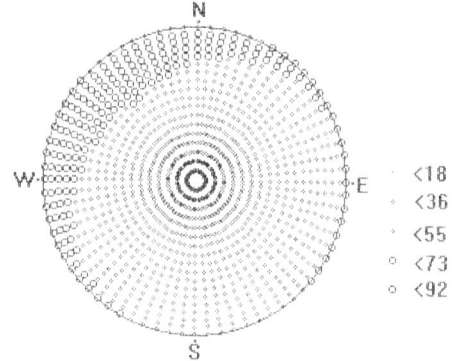

Figure 2. Stereogram of RQD of rockmass
$RQD_{ave}=47$

1.3 The biggest radius and openning of a joint set

The biggest radius(a_m) and openning(t_m) with the most probability to occur are of great importance in strength and failure probability analysis and permeability coefficient calculation.

If a factor(t) of joints, e.g. radii or opennings, fit distribution $f(t)$, then the probability density function for t to take the maximum in n samples is

$$g(t)=n f(t) [F(t)]^{n-1} \qquad (8)$$

where $F(t)=\int_0^t f(x)dx$. The modal of $g(t)$ will be the maximum of t with the most probability to occur. When t fits neg-exp distribution, then

$$t_m = t \ln(n) = t \ln(\lambda_v V) \qquad (9)$$

where $n=\lambda_v V$, V is the volume of sampling unit.

Figure 3(a)-(b) are the curves of $g(t)$ and $t_m(V)$, and the basic structural data are the same as in table 1.

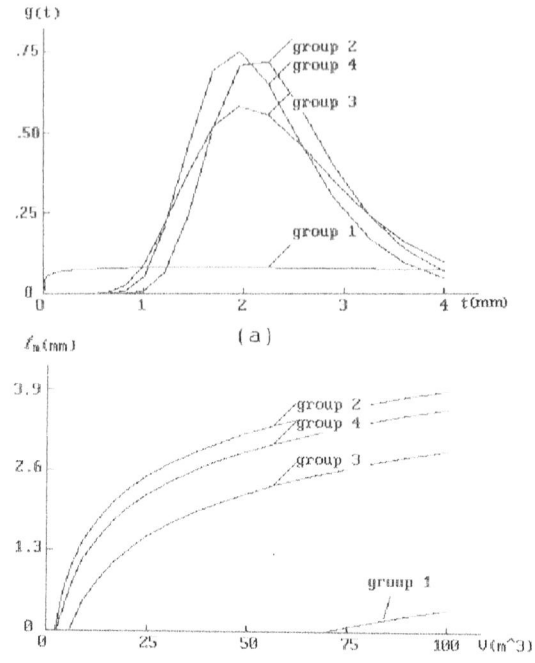

Figure 3. Chart of $g(t)$ and t_m
(a) density fuction of $g(t)$;
(b) relationship between t_m and volume.

2 ELASTIC MODULUS AND POISON RATIO

According "Statistical mechanics of rockmass" (Wu Faquan, 1993), the strain-stress relationship of jointed rockmass is

$$e_{ij}=(C_{oijst}+C_{cijst})\sigma_{st}=C_{ijst}\sigma_{st} \qquad (10)$$

where C_0 and C_c are the compliance tensor caused by rock blocks and crack networck respectively. When only one stress component is not zero, e.g. $\sigma_{11}\neq 0$, we have

$$e_{11}=C_{1111}\sigma_{11} \qquad (11)$$

therefore, the elastic modulus in this direction is

$$E_m = \frac{\sigma_{11}}{e_{11}} = \frac{1}{C_{1111}}, \qquad (12)$$

that's

$$E_m = \frac{E_0}{1+\psi\Sigma\lambda a[9(m_1^2-\nu m_1^4)+(6m_1^2+R)R]}$$
$$\psi_1 = 9\pi(2-\nu)/16(1-\nu^2), \quad \sigma\geqslant 0 \qquad (13)$$

$$E_m = \frac{E_0}{1+\psi_2\Sigma\lambda a h^2 m_1^2(1-m_1^2)}$$

$$\psi_2 = \pi(2-\nu)/32(1-\nu^2), \quad \sigma \leq 0 \quad (14)$$

On the other hand, sence

$$e_{22} = C_{2211}\sigma_{11} \quad (15)$$

so the Poison ratio is

$$\nu_{m21} = \frac{\psi_1 + 9\Sigma \lambda a\, m_1^2 m_2^2}{\psi_1 + 9\Sigma \lambda a\, m_1^2(2-\nu m_1^2)}\nu \quad \sigma \geq 0 \quad (16)$$

$$\nu_{m21} = \frac{\psi_2 \nu + \Sigma \lambda a\, h^2 m_1^2 m_2^2}{\psi_2 \nu + \nu \Sigma \lambda a\, h^2 m_1^2(1-m_1^2)}\nu \quad \sigma \leq 0 \quad (17)$$

From the above we can see:1) E_m and ν_m varies with direction(n); 2) they are affected by environmental stress, fluid pressure, and the property of rock and cracks.
Figure 4 is a stereogram of elastic moduli of rockmass, and the basic data are listed in table 1.

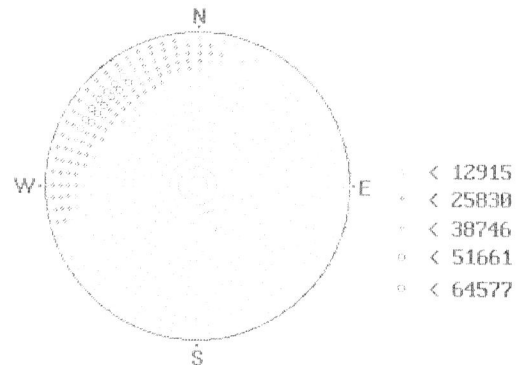

< 12915
< 25830
< 38746
< 51661
< 64577

Figure 4. stereogram of E(MPa) of rockmass
$E_{mave} = 8396.8$(MPa), $E_{m0} = 100000$(MPa)

3 STRENGTH AND FAILURE PROBABILITY OF ROCKMASS

It is pointed out by "Statistical mechanics of rockmass" that rockmass is a "weakness link system", especially for strength analysis. That's to say that the strength of a rockmass unit is controlled by the weakest link of rock blocks and cracks. Since the strengths of both rock blocks and cracks are stochastic, therefore the strength of rockmass is a subject of reliebility.

3.1 Compressive strength and failure probability of rock blocks

The compressive strength modal of rock block can be written as

$$R_m = [r_0(1-1/n)]^{1/n} \quad (18)$$

where n and r_0 are the Weibull fit parameters. The triaxial compressive strength of rock blocks can be written as the Hoek-Brown form

$$\sigma_1^* = \sigma_3^* + \sqrt{R_m^2 + m R_m \sigma_3^*} \quad (19)$$

where m is a coefficient. * stand for equivalent stress and the related strength.

$$\sigma_{st}^* = \sigma_{st} + \sigma_{ij} C_{cijkl} D_{oklst} \quad (20)$$

The failure probability of rock block is

$$P_b = 1 - \exp(-R^n/r_0) \quad (21)$$

when $R = R_m$,

$$P_b = 1 - \exp(-(1-1/n)) \quad (22)$$

and we have $P_b > 0.632$.

3.2 Strength and failure probability of discontinuities

Strength criterion of a set of cracks can be written as

$$K_{1c} = 4a \ln(\lambda_v V)[k^* \sigma_i^* \sigma_i^* + 2/(2-\nu)\tau_i^* \tau_i^*]/\pi \quad (23)$$

where K_{1c} is the failure tenacity of rock, k^* is a stress state coefficient, takes 0 while compressed and 1 while stretched, σ^* and τ^* are normal stress and residual shearing stress on cracks respectively, V is the volume of rockmass to be studied. The failure probability of the crack set is

$$P_{ci} = 1-(1-\exp(-a_c/a))^{\lambda_v V} \quad (24)$$

here a_c is critical crack radius decided by failure criterion, a is the average radius of the crack set. When $a_c = a_m = a \ln(\lambda_v V)$,

$$P_{ci} = 1-(1-1/(\lambda_v V))^{\lambda_v V} \quad (25)$$

and $0.632 \leq P_{ci} \leq 1$.

3.3 strength and failure probability of rockmass

According the "weakest link hypothesis", the strength of any rockmass unit is the minimum of that of rock blocks and cracks, and the failure probability of the unit is

$$P = 1-(1-P_b)(1-P_c) \quad (26)$$

here

$$P_c = 1-[\Pi(1-\exp(-a_c/a_i))^{\lambda_v}]^V \quad (27)$$

is the failure probability of the crack networck.
Obviously both the strength and failure probability of rockmass have volume effect or called size effect.
Through some transformation, we can get some formulas for calculating C and ϕ. Figure 5 is a

Table 2. Data for shearing strength calculation

group number	dip $\alpha(°)$	angle $\beta(°)$	density $\lambda(1/m)$	radius $a(m)$	JCS (MPa)	ϕ_j^* (°)	JRC
1	150	30	20.0	1.0	2.0	30	5
2	270	45	18.0	1.0	1.0	30	4

* frictional angle of cracks.

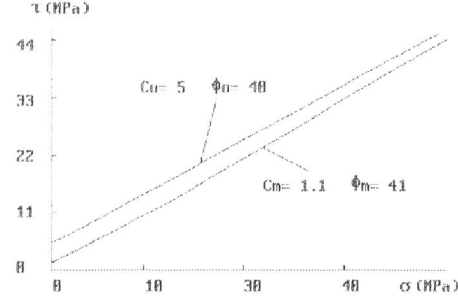

Figure 5. Shearing strength of rockmass

group of shearing strength curves by the above method, and the basic data are listed in table 2.

4 CONCLUDING REMARKS

The paper proposes a series of methods to calculate engineering parameters of rockmass and presents some examples. From the above we can see that the methods are simple and direct. Now the author has explored some other softwares to serve the practical engineering, e.g. the program to automatically trace the failure planes in slope, decide the failure range in tennel surrounding rockmass,and the program for finite element computation in rockmass engineering.

REFERENCES

Barton N. *et al.*, 1983. The shearing strength of rock joint in theory and practice. *Rock mech.* Vol.10:1-54.
Faquan W.,1992. Constitutive model and strength *Bulletin*.Vol.37.No.2:131-135.
Faquan W.,1993. *Principles of statistical mechanics of rock masses*. China University of Geosciences Press.
Faquan W.*et al.*,1994. Statistical principles in Mechanics of Rockmass.*Chinese Science Bulletin*. Vol.39.No.6:493-503.
Hoek E., 1990. Estimating Mohr-Coulomb friction and cohesion values from the Hoek-Brow failure criterion. *Int.J.Rock Mech.Min.Sci. Geomech. Abstr.*, Vol.27,No.3:227-229.

Large scale sampling of Akashi Gravelly Layer and its mechanical property

S. Kashima & S. Yamamoto
Honshu-Shikoku Bridge Authority, Tokyo, Japan

K. Takahashi, M. Sasao & S. Yamada
Kiso-Jiban Consultants Co., Ltd, Japan

ABSTRACT: Pier No. 2P of the Akashi Kaikyo Bridge is founded on a thick gravelly layer of Akashi Formation. In order to reveal mechanical properties of this gravelly layer, 300mm diameter undisturbed samples were obtained using a newly developed triple tube sampler with an outer diameter of 360mm. Consolidated-drained and consolidated-undrained triaxial compression tests were conducted on the undisturbed samples collected. It was found by the drained compression tests that peak strengths were reached before axial strain of 15% and then strengths were reduced gradually with increasing strain, i.e., strain softening. In the undrained tests, the deviator stress continued to increase till the axial strain of 15%, i.e., strain hardening and there was no clear peak deviator stress. The pore water pressure increased at the small strain and then started decreasing when the strain exceeded 0.5 to 1.5%.

1 INTRODUCTION

A gravelly layer called Akashi Formation lies under route of the Akashi Kaikyo Bridge which is now under construction. The foundation of Pier No. 2P is placed on the 45m thick Akashi Formation. Therefore, it is important to obtain accurate mechanical properties of the layer for the foundation design. In the previous investigation, disturbed samples were collected from the layer using a 100mm diameter single core tube without circulating drilling water. The layer appeared to contain 100mm or more diameter gravels and are fairly dense based on visual examination and grain size analysis. Mechanical parameters of the layer were only determined by pressuremeter tests in the previous investigation. Nishigaki et al. (1977) conducted triaxial compression tests on block samples collected in a caisson which were constructed on the Maiko Beach near the bridge. Although the block samples were collected from the Akashi Formation (Tarumi Gravel), there ma be difference in the Akashi Strait and the caisson location on the beach.

Therefore, it was crucial to collect undisturbed samples at the exact pier location and to determine the mechanical properties of the layer. Till the present investigation, there is no remarkable performance to obtain undisturbed samples from boreholes in gravelly soil. Although sampling by a frozen method is commonly used in the gravelly layer, this method is not feasible in the Akashi Kaikyo Bridge because the water depth is 46m and the current is 4.5m/sec. The samples must therefore be collected from boreholes with usual circulation fluid method. A large diameter sampler is required to collect undisturbed samples because large sizes of gravels are found in the layer. Yamagata et al (1987) developed a large diameter triple tube sampler for this project.

After several trial sampling using the large diameter sampler on shore, the sampling was conducted at the pier loctaion in the strait. As a result, it was possible to obtain good quality samples and hence reasonable test results.

This paper presents details of sampling from the gravelly layer and results on triaxial compression tests conducted in the ground investigation for the Akashi Kaikyo Bridge. The present test results are compared to the test results previously obtained from the similar gravelly layer.

2 OUTLINE OF AKASHI KAIKYO BRIDGE

The Akashi Kaikyo Bridge is a road bridge connecting Honshu Island and Awaji Island over 4km wide Akashi Strait. Fig.1

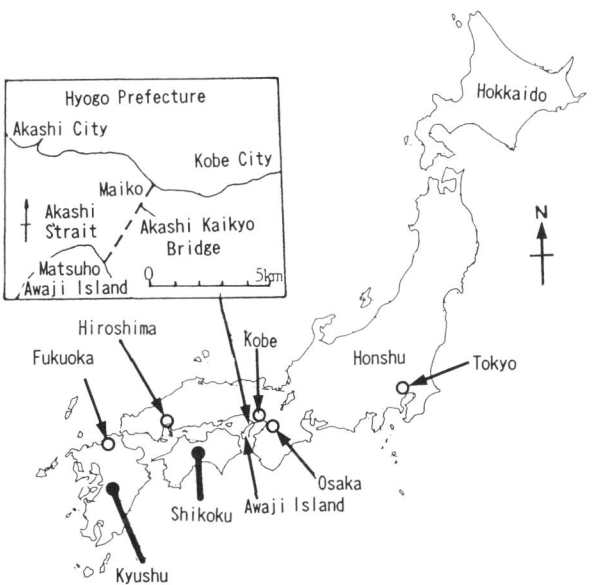

Fig.1 Location of the Akashi Kaikyo Bridge.

Table 1 Geology at the Akashi Strait

Age		Geological Formation	Thickness. (m)			
			1A	2P	3P	4A
Cenozoic	Quarternary	Holocene	Recent Deposits	5-6	2-3	8-11	2-10
		Pleistocene	Upper Pleistocene	39-42	2-3	—	—
	Tertiary	Pliocene	Akashi Formation	1-6	37-45	—	—
		Miocene	Kobe Formation	123	168	36-64	0-20
Mesozoic	Cretaceous		Granitic Rock				

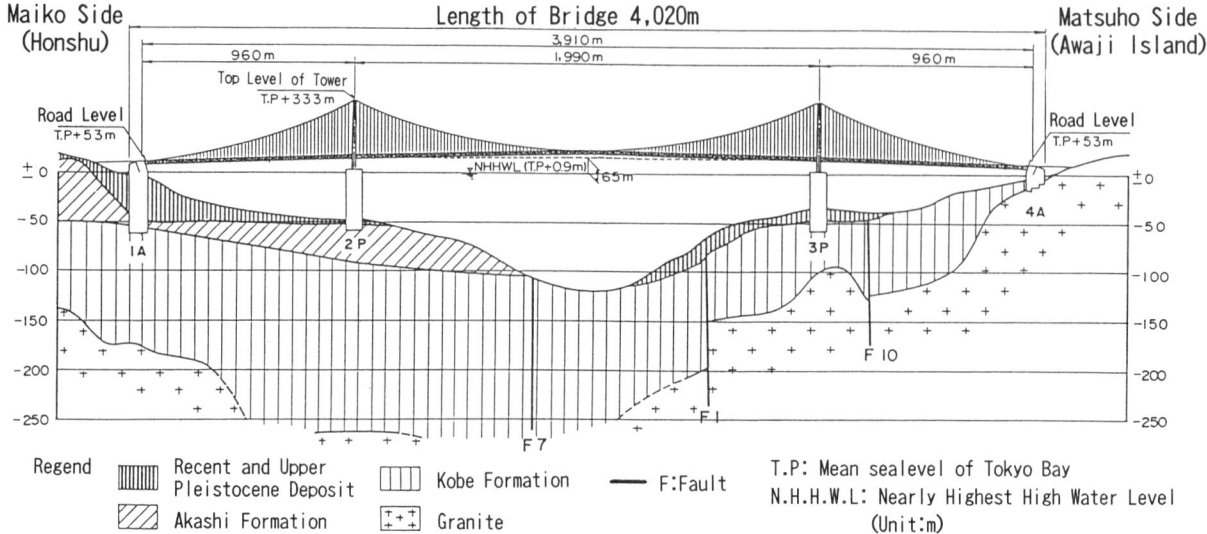

Fig.2 Geological profile along longitudinal section of the Akashi Kaikyo Bridge.

Photo.1 SEP at Site 2P.

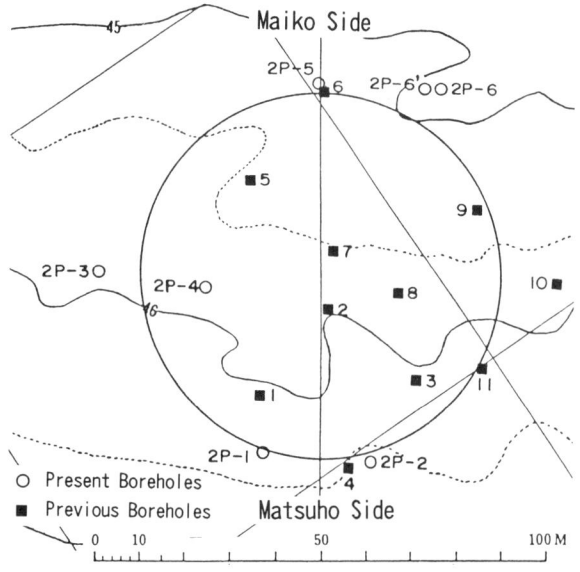

Fig.3 Location of boreholes at Site 2P.

shows location of the bridge. The bridge is a strengthened steel truss suspension structure with 3 spans and 2 hinges at 2 main piers. A total length of the bridge is 3910m and length between the 2 main piers is 1990m. The bridge will be the world longest bridge when it is completed in 1998. The Akashi Strait is located at the entrance to Osaka Bay. Number of vessels passing the strait exceed 1400 a day. The bridge is supported by 2 anchorages (1A, 4A) and 2 piers (2P, 3P) on direct foundations. The construction methods are diaphragm wall excavation at 1A, prefabricated caissons placed on dredged seabed at 2P and 3P, and open excavation using retaining system at 4A. The construction of these foundations has been completed.

Topographically, the Akashi Strait has a 110m deep submarine caldron along the strait. Both sides of the caldron form submarine terrace at several steps. Geology at the strait consists mainly of weak sedimentary rocks in Tertiary (Kobe Formation) and gravelly layer in Quarternary (Akashi Formation) as shown in Table 1 and Fig.2. Most of the long bridges in the world are founded on strong sound rocks, while the Akashi Kaikyo Bridge is on weak sedimentary rocks and semi-cemented Pleistocene gravelly layer.

Table 2 Total quantity of field and laboratory works for the Akashi Formation.

Type of Tests		Borehole No.						Total
		1	2	3	4	5	6	
Sampling	Total Length	74.48	74.38	226.96	135.98	81.32	81.33	674.45m
	φ300mm	42.46	41.05			46.28	47.69	177.48m
	φ116mm			49.46	48.88			98.34m
Pressuremeter Test		10	10		31	9	9	69回
Electrical Logging		29.40	31.30	217.97	116.25	27.88	25.81	448.61m
P & s Wave Logging		28.00	28.00	211.70	116.90	28.00	26.00	438.60m
Laboratory Test	Wet Density	10	13			11	12	46回
	Grain Size Analysis	18	16			10	9	53回
	Triaxial Compression Test — CU Test	5	7			6	6	24回
	CD Test	4	2			3	3	12回
	UU Test	2	3					5回
	Cyclic Test	4	4			4	4	16回
	Creep Test	1	2			1	12	6回

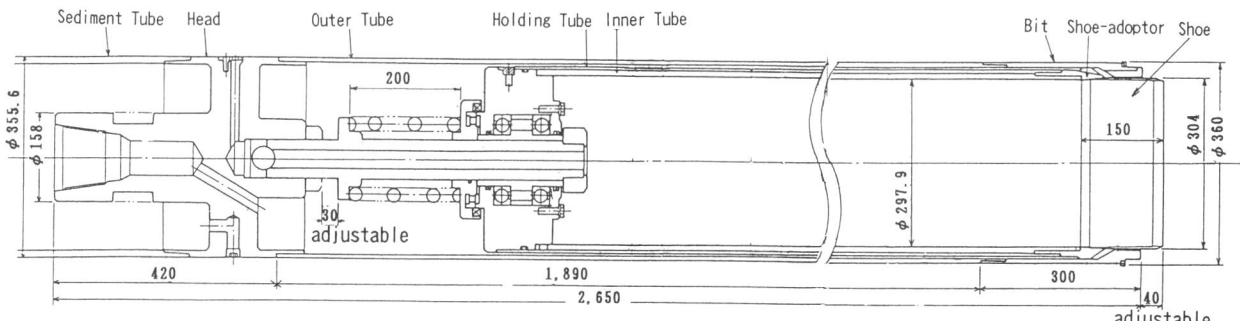

Fig.4 Large scale triple tube sampler.

Photo.2   Metal bit and cutting shoe.
Left : Metal Bit   Right : Cutting shoe

inner tube with cutting shoe is thrusted into the soil with an aid of a spring, and then the sample is led to a lining tube. The metal bit is arranged at 2 different locations as shown in Photo.2 in order to reduce the bit resistance during sampling. The cutting shoe consists of high tensile steel.

Since the sampler weight (including sample weight) exceeds 1 ton, the sampler was handled using a crane. As the cutting shoe length and inner clearance influence greatly the recovery and quality of the samples and sampling speed. Three sizes of cutting

3 UNDISTURBED SAMPLING FROM GRAVELLY LAYER IN PLEISTOCENE

The sampling was conducted on a large self elevated platform. (Photo.1) As shown in Fig.3, rotary drilling was carried at 6 locations. Undisturbed samples were collected in 4 of the 6 boreholes. Table 2 presents actual quantity of field works performed. Because of rapid current and deep water, a 1200mm diameter casing was first lowered and penetrated 3m into the seabed in order to protect the 550mm diameter drill casings.

Based on experience in the trial sampling on shore, the gravels were intended to be cut using cutting shoe in advance and followed by rotating metal bit which is located behind the shoe. However, the shoe was broken at the very beginning of the sampling at 2P because of presence of strong gravels. The gravels found at 2P location contain less decomposed gravels (20 to 30% of total gravel content) than those found on shore. Hence, the sampling method was changed to cut the gravels first using metal bit rotation.

Fig.4 shows a large scale triple tube sampler used for sampling. The metal bit attached at the tip of an outer tube cuts gravels, the

Table 3 Summary of sampling records for the Akashi Formation.

Borehole No.	Each Sample Length (m)	Total Drilling Length (m)	Nos. of Sampling (Nos.)	Time Taken for Sampling (min)	Average Drilling Speed (min/cm)	Average Bit Force (kgf)	Circulation Fluid Amount (l/min)	Circulation Fluid Pressure (kgf/cm²)	Cutting Shoe Advanced Length (mm)
1	0.10~1.65	42.5	46	65	0.74	4400	150	2	0 ~ 50
2	0.15~1.60	41.0	47	76	0.89	4100	150	3	0 ~ 50
5	0.08~1.85	46.3	51	59	0.65	4400	210	1	0 ~ 50
6	0.18~1.80	47.7	38	60	0.49	3500	180	3	5 ~ 50
Total	0.08~1.85	177.5	182	65	0.69	4100	170	0.5	0 ~ 50

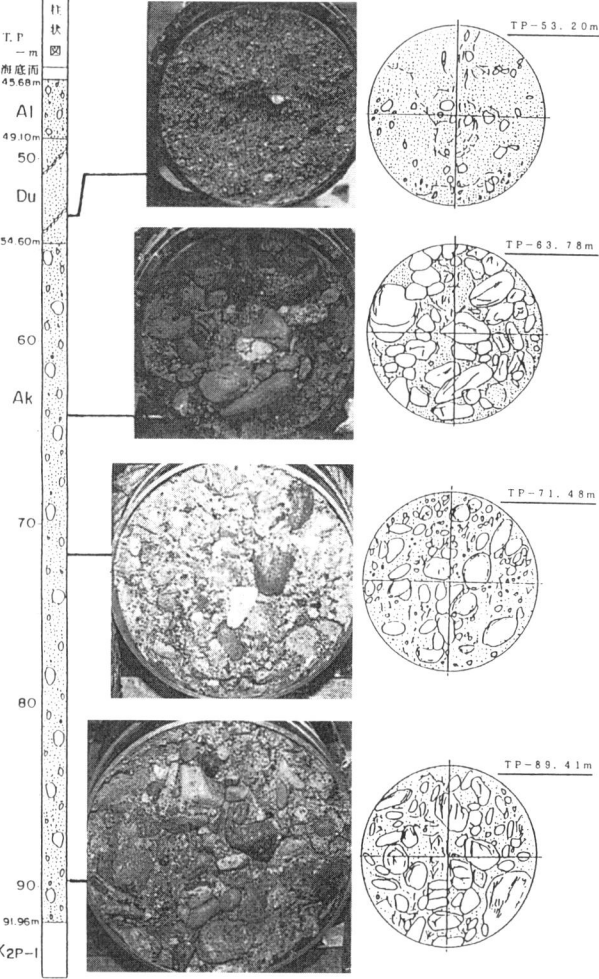

Photo.3 Photograph of sample and observation record.

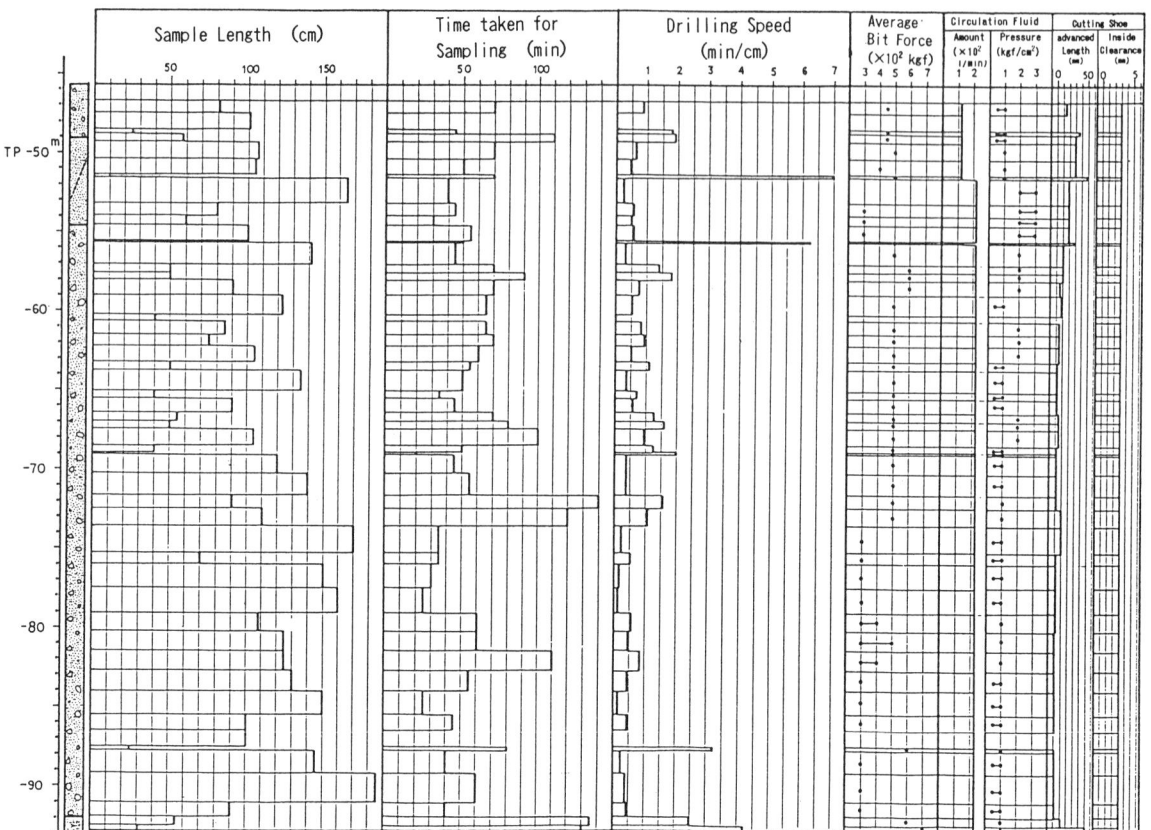

Fig.5 Sampling record in borehole No. 2P-5.

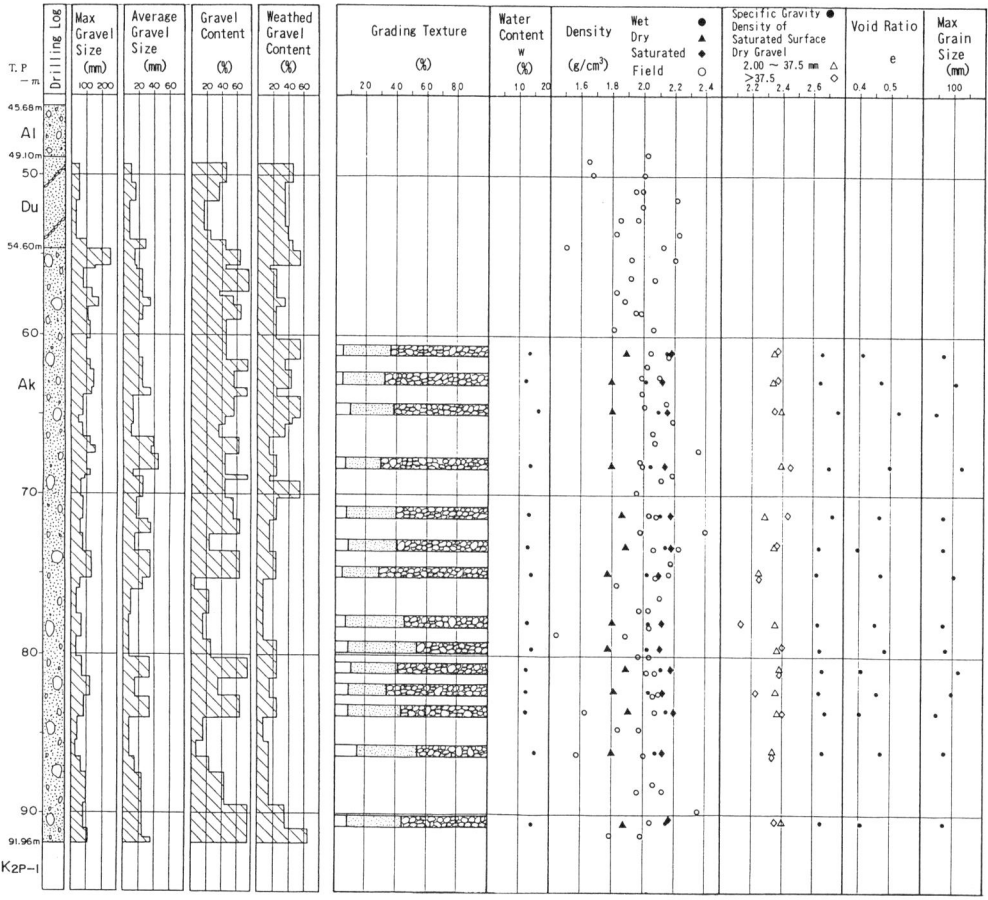

Fig.6 Visual examination and laboratory index property test results.

Table 4 Observation record of sample collected.

Borehole No.	Geological Formation	Thickness (m)	Max. Gravel Size (mm)	Average Gravel Size (mm)	Gravel Content (%)	Precentage of Weathed Gravel (%)	Total Sampling Length (m)	No. of Specimen	Recovery Ratio	Specimen Length / Sample Length (%)
1	Recent Deposite	2.68	−	−	−	−	0	0	93.9	45.5
	Upper Pleistocene Deposite	4.80	150	7~10	6~29	20~26	45.0	3		
	Akashi Formation	36.65	200	9~44	7~77	6~35	36.65	24		
2	Recent Deposite	2.58	−	−	−	−	0	0	93.7	41.7
	Upper Pleistocene Deposite	3.90	150	8~20	20~25	13~22	3.90	2		
	Akashi Formation	34.57	200	8~80	9~80	7~50	34.57	21		
5	Recent Deposite	3.42	−	−	−	−	0	0	92.0	47.7
	Upper Pleistocene Deposite	5.50	100	8~29	15~46	35~45	5.20	5		
	Akashi Formation	37.36	250	8~45	5~76	9~65	37.36	24		
6	Recent Deposite	3.33	−	−	−	−	0	0	90.7	56.6
	Upper Pleistocene Deposite	5.50	125	7~26	28~45	17~25	4.40	4		
	Akashi Formation	38.86	165	8~56	25~85	16~76	38.86	31		
Total	Recent Deposite	2.58~3.42 (3.00)	−	−	−	−	0	0	92.5	48.1
	Upper Pleistocene Deposite	3.90~5.50 (4.93)	100~150	7~29	6~46	13~45	18.30	14		
	Akashi Formation	34.57~38.86 (36.86)	165~250	8~85	5~85	6~76	147.44	100		

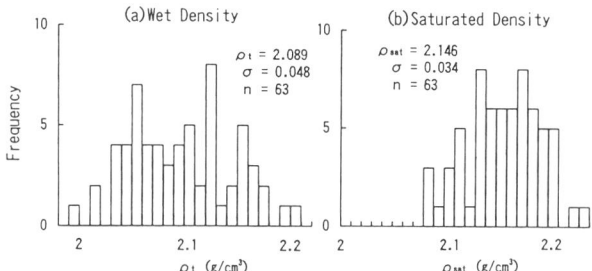

Fig.7 Histogram of density.

shoe with 3 to 4mm thickness were prepared and used depending on soil conditions. An applied load on the bit was 2000 to 3000 kg when there was little gravel. When large quantity of strong gravel had to be cut, the applied load was increased to 4000 to 7000 kg. Hence the applied load was adjusted frequently. Information obtained during sampling is summarized by Tkahashi et al (1990) in Table 3 and Fig.5.

The applied load and cutting shoe length were mainly adjusted during sampling. The sample quality will be better if more samples are taken. As compared to the sampling in boreholes No.1 and 2 which were sunk in the first stage, the sample recovery and sampling speed were improved in the boreholes No.5 and 6 in later stage due to added experience.

Undisturbed samples could not be taken from loose alluvial sand, while a total of 182 samples with a total length of samples of 177.48m were collected from upper Pleistocene gravels and the Akashi Formation. The sample recovery ratio was 100%. A total of 100 samples (48% of total sample length) could be used for laboratory mechanical tests.

As soon as the samples were retrieved from the borehole, both ends of the sample were visually examined, sketched and photographed. Photo.3 shows examples of the sample conditions recorded at the site. Table 4 summarizes sample observation

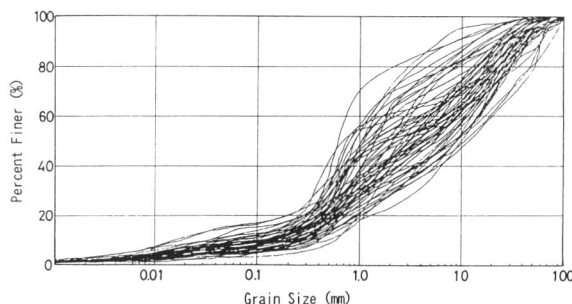

Fig.8 Grain size of distribution curves of Akashi Formation.

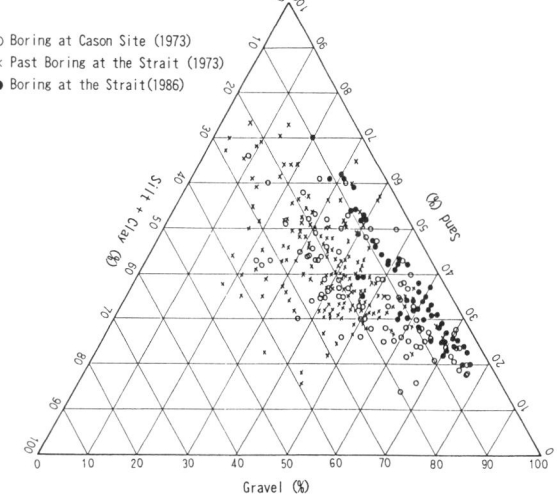

Fig.9 Grain size distribution of Akashi Formation obtained in past and present investigation.

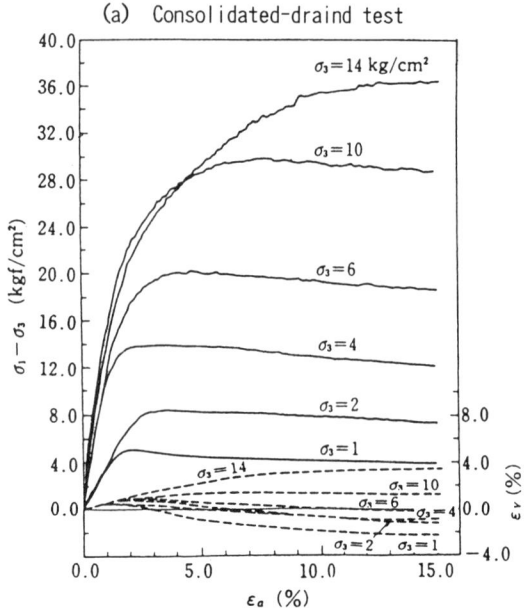
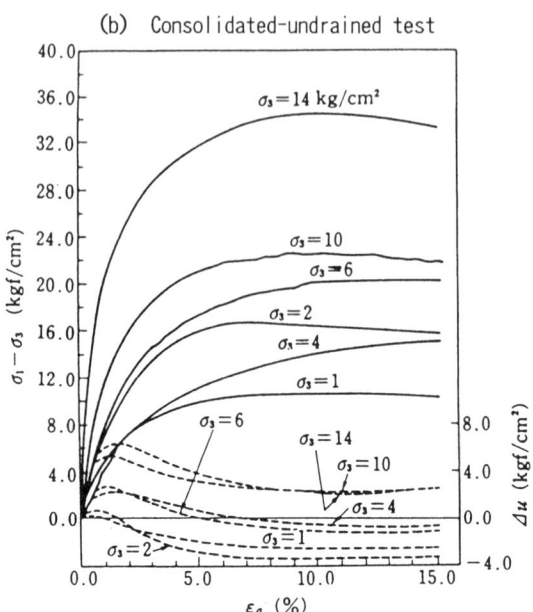

Fig.10　Results of triaxial compression tests on Akashi Formation.

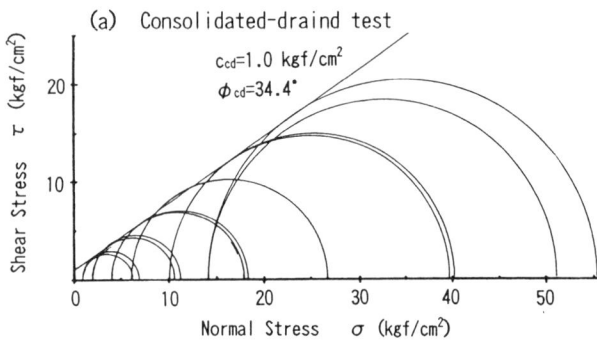

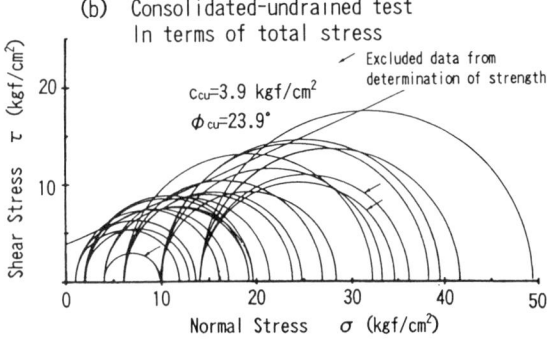

records. Fig.6 compares visual observation record and index properties of samples determined in laboratory.

The gravels are composed mainly of acidic rocks, felsite, sandstone, quartz and granite. The maximum diameter of the gravels observed was 150mm. The maximum gravel diameter was measured by a measuring tape for each sample and gravel contents were visually approximated. The information obtained by the observation provides general soil conditions such that the maximum gravel size ranges widely from 20 to 250mm in diameter and an average gravel size ranges between 10 and 40mm. Gravel contents are high above the elevation of TP-75m, while the gravel contents vary irregularily below TP-75m.

After the observation, the samples were extruded from the lining tubes with a aid of hydraulic pressure and inserted into steel containers with rubber membrane for transportation. The samples were then cut to the required length and both ends were sealed by rubber disc.

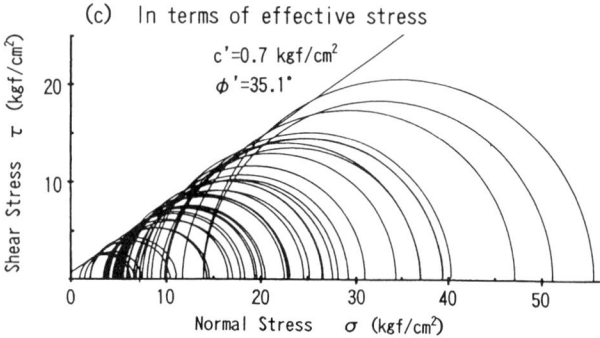

Fig.11　Mohr's circles obtained by triaxial compression.

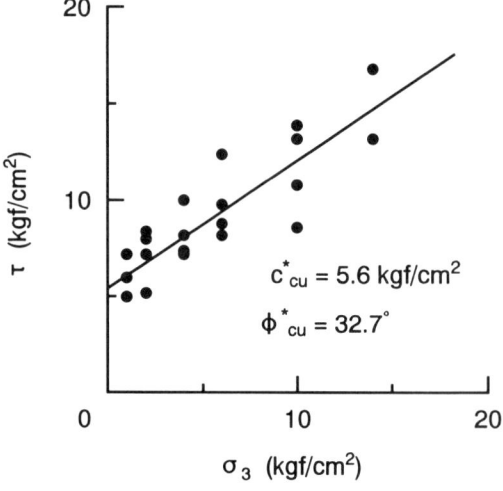

Fig.12　Shear strength on average principal stress plane.

Table 5 Results of consolidated-undrained triaxial compression.

Sample Name	Wet Density ρ_t g/cm³	Strength Parameter			
		c' kgf/cm²	φ' deg	C_{cu}^* kgf/cm²	ϕ_{cu}^* deg
Nishi-Yagi Formation	1.77	0	32.2	1.5	12.1
Tarumi Gravel B	1.91	0	33.5	3.5	16.9
Tarumi Gravel C	2.18	0	36.2	4.9	26.7
Akashi Gravel (Hill Area)	2.22	0.43	35.7	3.2	21.7
Akashi Gravel (Strait Area)	2.09	0.7	35.1	5.6	32.7

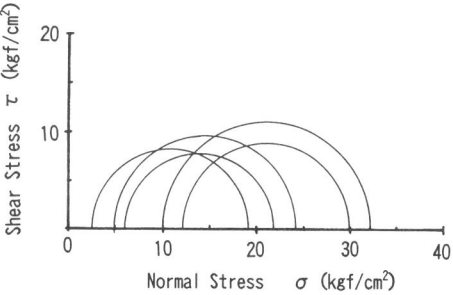

Fig.14 Mohr's circles obtained by unconsolidated-undrained triaxial compression tests.

4 LABORATORY TEST RESULTS

Only sample with a length of 680mm or longer and having maximum gravel diameter of 100mm or less was subjected to laboratory testing.

Wet density of the Akashi Formation ranges from 1.99 to 2.19 g/cm³ with an average value of 2.09 g/cm³ as shown in Fig.7(a). The wet density is smaller than the actual density because of water loss in the samples during sampling. The calculated saturated density is in a small range from 2.1 to 2.2 g/cm³ as compared to the wet density.

The grain size distribution curves of the Akashi Formation are scattered as shown in Fig.8. The maximum diameter D_{max} ranges from 36 to 136mm and a diameter at 60% passing D_{60} ranges from 0.7 to 19mm. The Akashi Formation is mostly classified into gravel with fines based on Japanese Soil Classification. The coefficient of uniformity generally exceeds 10. Gravel shape is mainly subangular and subrounded. The formation contains 20 to 30% of decomposed gravel.

Fig.9 compares grain size distribution determined by the present investigation (●), the previous investigation at the same site in 2P (×), and boring in the caisson on shore (○). The gravel contests are the highest in the present investigation and exceed 90%. The silt and clay contents are high in the samples obtained in the previous investigation at the site. The silt/clay contents may have been increased by breaking particles due to the dry sampling method.

Triaxial compression tests were conducted in consolidated-drained condition and consolidated-undrained condition with porewater pressure measurement. Back pressure was 3 kgf/cm², strain rate was 0.1 %/min., and confining pressures were 1, 2, 4, 6, 10 and 14 kgf/cm² in all the tests.

Fig.10 shows typical test results. In the drained compression tests, all the specimens reaches peak stress, the stress-strain curve gradually declines due to strain softening. Volumetric strain during shear increases till the peak stress is reached under low confining pressures of 6 kgf/cm² or smaller, while it keeps decreasing under high pressures.

There is no clear peak stress and stress strain curve increases till 15% strain in the undrained tests, i.e. strain hardening till 15% strain. The porewater pressure increases immediately upon shearing. When the strain exceeds 0.5 to 1.5%, the porewater pressure starts decreasing and negative porewater pressure is registered under the low confining pressure of 6 kgf/cm² or less.

Some samples show constant increase in porewater pressure and give low strengths (indicated by arrow mark in Fig.11(b)). These data was excluded from the determination of strength. These data and the other data have no difference when they were plotted in Mohr's circles in terms of effective stress. Therefore, it is judged that these samples were disturbed during sampling or transportation. Fig.11 shows Mohr's circles obtained by

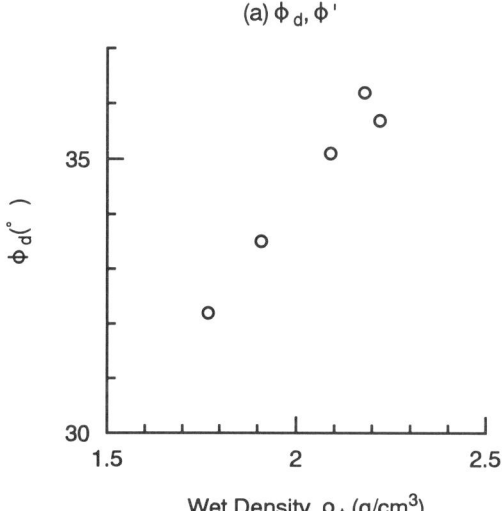

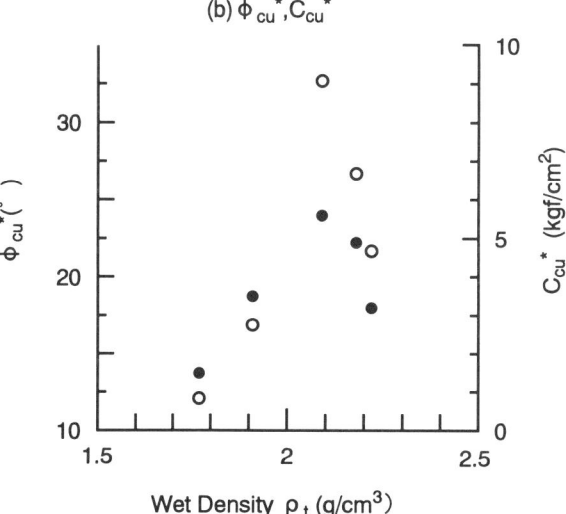

Fig.13 Relationship between wet dencity and shear strength.

consolidated-drained and consolidated-undrained triaxial compression tests.

The drained strength of $c_d = 1.0$ kgf/cm^2 and $\phi_d = 34.3°$ are determined by the drained tests. The drained strength of $c' = 0.7$ kgf/cm^2 and $\phi' = 35.1°$ is determined based on Mohr's circles in terms of effective stress by the consolidated-undrained tests together with Mohr's circles by the consolidated-drained tests.

Fig.12 shows undrained shear strengths determined on an average principle stress plane for the earthquake design. The undrained shear strength of $c_{cu}* = 5.6$ kgf/cm^2 and $\phi_{cu}* = 32.7°$ is obtained.

Table 5 summarizes shear strengths of the Akashi Formation together with those obtained in the previous studies on the similar gravelly layers in the region. The drained shear strengths have a little difference among the present and previous studies, while the undrained shear strength obtained in the present study is the highest among the data. Presented in Table 5 is an average wet density of test specimens subjected to the triaxial tests in each study. The wet density obtained in the Akashi Formation at a hilly area is the largest. It is heavily weathered and oxidized, stained in yellowish brown colour. It is expected that the void in the soil is filled with fine particles produced during the weathering and as a result, the soil possesses the high density.

The samples from the Nishiyagi Formation (upper Pleistocene gravelly layer) and Tarumi Gravel B are also weathered and contain a lot of decomposed gravels. On the contrary, Tarumi Gravel C and the Akashi Formation at the site in the present study are not oxidized as indicated by their light grey to bluish grey colour although the samples contain some decomposed gravels. The later samples are more gravelly than former samples.

The samples compared in Table 5 are not the same in their properties. Fig.13 correlates wet density and shear strength obtained in each study. As can be seen from this figure, there is a tendency that the shear strength increases with increasing wet density.

For the earthquake design, the shear strength obtained by consolidated-undrained tests was used. However, in fact, the earthquake force will induce stress in the ground in the undrained condition under the present effective overburden pressure. Therefore, a shear of the soil will occur in the unconsolidated-undrained triaxial compression tests were hence conducted with confining pressures of 2, 4, 6, 10 and 12 kgf/cm^2 after the consolidation of the specimen at the effective pressure of 4 kgf/cm^2. The results of unconsolidated-undrained tests are shown in Fig.14. The results are reasonably compatible with the undrained shear strength obtained by the consolidated-undrained tests.

5 SUMMARY

Undisturbed sampling and laboratory test conducted for the Akashi Kaikyo Bridge were introduced in this paper. Previously, it was difficult to obtain samples from the gravelly layer, even disturbed samples. In the present investigation, high quality undisturbed samples was obtained using the 360mm diameter triple tube samplers under various hard site conditions such as rapid current, deep water, etc.

The samples were successfully obtained from dense Akashi Formation but not from loose alluvial gravelly layer. For the quality and speedy sampling, it is important to adjust thrust force on the drill bit and length of cutting shoe rather than amount of circulation fluid and bit rotation speed.

Except for disturbance of some samples, reliable results were obtained from laboratory triaxial compression tests. The parameters determined by the present investigation were used for the design of pier No. 2P. The present works will certainly contribute the success completion of the world longest bridge.

ACKNOWLEDGEMENT

Authors would like to thank the Honshu-Shikoku Bridge Authority for giving us the opportunity to publish this paper, and all the staff who participated in the development of the sampler, sampling and laboratory testing.

REFERENCES

Yamagata M., Y.Nishigaki, T.Noto, M.Sasao 1987. "Undisturbed Sampling from Akashi Gravelly Layer and Laboratory Test Results" (in Japanese), The 22nd Japan National Conc. on SMFE: 111-112.

Nishigaki Y., K.Takahashi, T.Noto 1977. "Sampling and Testing of Undisturbed Diluvial Gravels", Proc. 9th ICSMF, Specialty Session 2, Soil Sampling: 103-108.

Takahashi K. & M.Sasao 1990. "Drilling in and Undisturbed Sampling of Diluvial Gravels" (in Japan), Tenth Anniversary Report of the Kansai Geotechnical Center, Kiso-Jiban Consultants Co. Ltd.: 11-16.

Creep characteristics of the Kobe Formation of Miocene in Tertiary period

M. Yamagata
Honshu-Shikoku Bridge Authority, Tokyo, Japan

K. Yamada
Choudai Co., Ltd, Japan

Y. Nishigaki & S. Matsumura
Kiso-Jiban Consultants Co., Ltd, Japan

ABSTRACT: The Akashi Kaikyo Bridge will be the world's longest upon completion. The foundation of anchorage 1A and pier 3P for this bridge is to be placed on the Kobe Formation, composed of weakly cemented soft rock with complicated alternation of thin sandstone and mudstone layers. Therefore, to evaluate the long-term stability of the foundation, laboratory and field creep tests were performed. The following conclusions were reached.
(1) The estimation of long-term strength was evaluated by using the linear relations, $t_r \sim \dot{\varepsilon}_{1min} \sim \sigma_{fc}$. Here t_r, σ_{fc} are creep rupture time, minimum creep strain rate and creep rupture stress respectively.
(2) The estimation of creep deformation was evaluated using the three-element Voight model and it was assumed that shear elastic modulus, G_2 and viscoelastic modulus η_2, can be approximately estimated by the instantaneous shear elastic modulus, G^*.

1 INTRODUCTION

The creep parameters for the long-term stability of the foundation is obtained from the result of laboratory creep tests or field creep tests. Laboratory creep tests are generally performed because field creep tests are subject to the influence of temperature and are very expensive.

Creep deformation parameters are generally obtained from laboratory creep tests by assuming a viscoelastic model. However it is very difficult to determine a suitable loading period for laboratory creep tests. Therefore, the loading period is generally determined by considering the time allowed for the creep tests or the number of specimens, without scientific support. Moreover, a creep model applied to these experimental data differs in each test case. The effects of the loading period or the applied creep model on the creep coefficient have not been discussed much.

Creep rupture stress may take place under a stress level lower than peak strength. Since the foundation of anchorage 1A is soft rock mass, it is important to examine the long term strength of soft rock samples. Although Adachi & Takase[1] propose a method to predict the long term strength of soft sedimentary rock by using drained creep test results, there are few studies of this kind because of the difficulty of testing for the long period.

The objective of this study is to resolve the above problems, and triaxial creep tests were performed on sandstone and mudstone sampled from the Kobe Formation.

2 LABORATORY CREEP TEST

In order to study the fundamental characteristics of creep behavior, triaxial creep tests were performed. Testing specimens are sandstone and mudstone which were obtained from the Kobe Formation at the foundation of anchorage 1A of the Akashi Kaikyo Bridge.

The cylindrical specimens were 5 cm in diameter and 10 cm in height. Specimens were saturated by applying vacuum to a vessel and then were submerged in water for at least 24 hours. The confining pressures were 5 and 10 kgf/cm^2 and back pressure was 8 kgf/cm^2. All of the tests were performed under undrained condition and constant temperature.

Loading patterns were as shown in Fig.1. Pattern-A is the short term test and creep stress is loaded by stages until leading to failure. Pattern-B is the long term deformation test. Here, σ_f is the triaxial strength developed under a given confining pressure.

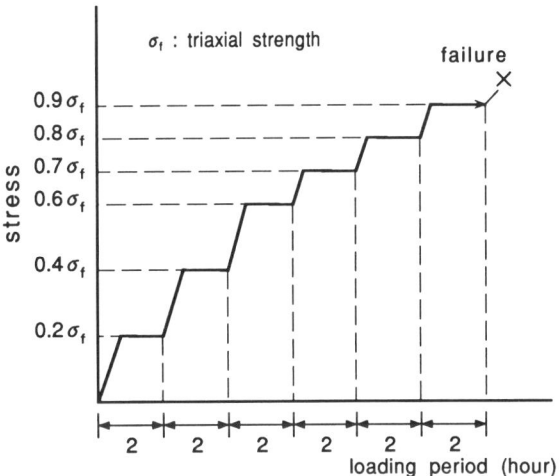

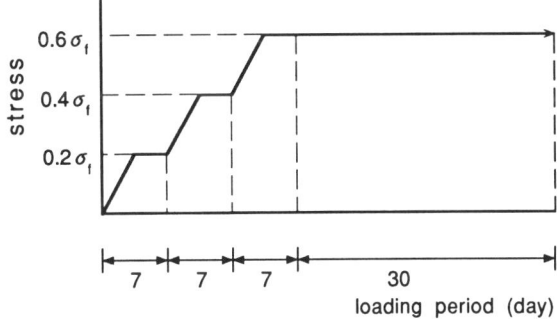

Fig.1 Loading patterns of creep test

3 STRENGTH CHARACTERISTICS

3.1 Relationship between creep strain and time

Fig.2 shows the change of axial strain, $\dot{\varepsilon}_{1min}$ ar. hydraulic pressure, Δu with the elapsed time, t for the loading pattern-A.
Here, σ_3 is a confining pressure and σ_s is a creep deviator stress. In the figure, it is clearly seen that the creep behavior of sandstone is different from that of mudstone.

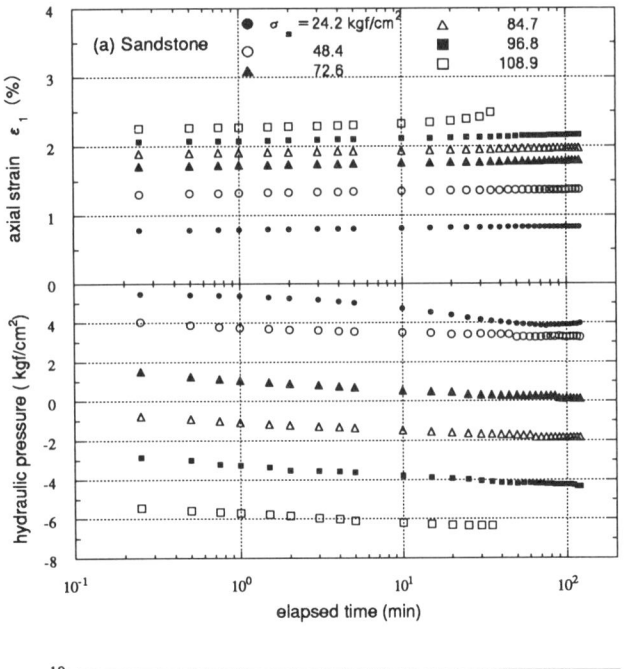

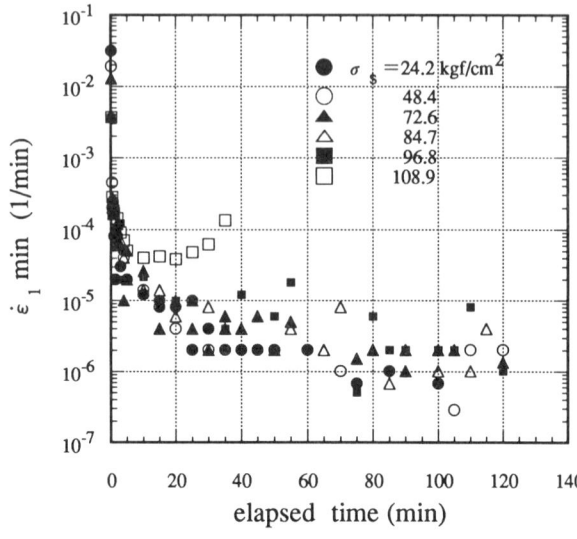

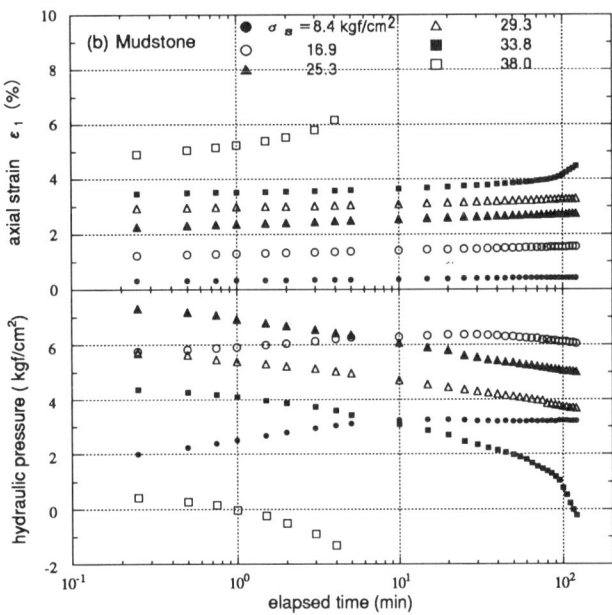

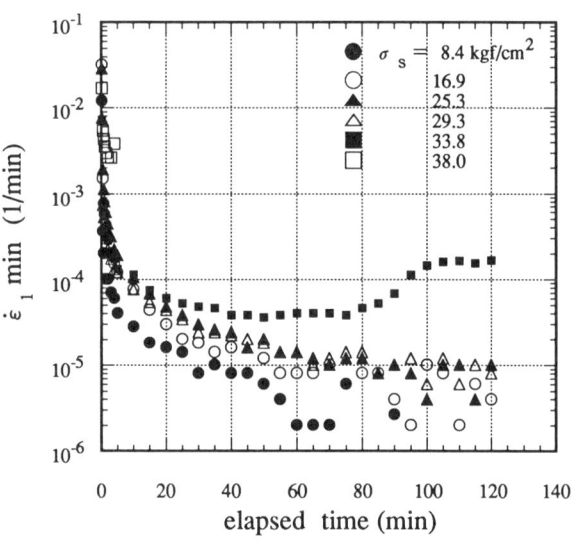

Fig.2 Typical creep curves for the Kobe Formation

Fig.3 Change of creep axial strain rate

Creep rupture stress, σ_{fc} of sandstone is 110kgf/cm^2 (failure strain, ε_f is 2.5%) and mudstone is 40 kgf/cm^2 (ε_f is 6.2%). The foundation of anchorage 1A was composed of alternate thin layers of sandstone and mudstone. So, it is important for the design to evaluate the long term stability of mudstone more than that of sandstone.

3.2 Relationship between axial strain rate and time

Fig.3 shows the change of the axial strain rate which was calculated from the data in Fig.2. The application of creep leads first to a period of transient creep in which the axial strain rate decreases continuously with time. Thereafter, the axial strain rate is constant for some period(called the steady state), and then begins to increase after hitting a minimum and finally failure occurs(called the accelerative state). The difference between the creep behavior of both rock types can be seen at steady state, and the period of steady state for mudstone(5-stage) is longer than that for sandstone(final stage).

3.3 Relationship between minimum strain rate and rupture time

Fig.4 is a well known linear relationship between the minimum axial strain rate, $\dot{\varepsilon}_{1min}$ and creep rupture time, tr (Saito & Uezawa[2]). These relations were obtained from laboratory creep tests and from plate loading tests which were performed at a site with similar geological condition to that of the foundation of anchorage 1A (Nishigaki, et al.[3]). Here, the strain distribution below the loading plate was measured with strain gauge transducers installed in the ground.

The figure lead to the following. The region of distribution of various rock types[4]~[7] lies below the region of soils as Morlier[8] indicated. The plots(○●) of sandstone and mudstone in laboratory tests are distinct from each other, and furthermore, the most important point is that the plots(△▲) of the plate loading test are included in the regions of mudstone in laboratory test. This fact indicates that the characteristics of creep deformation in the whole of the alternating layers were influenced by the behavior of mudstone.

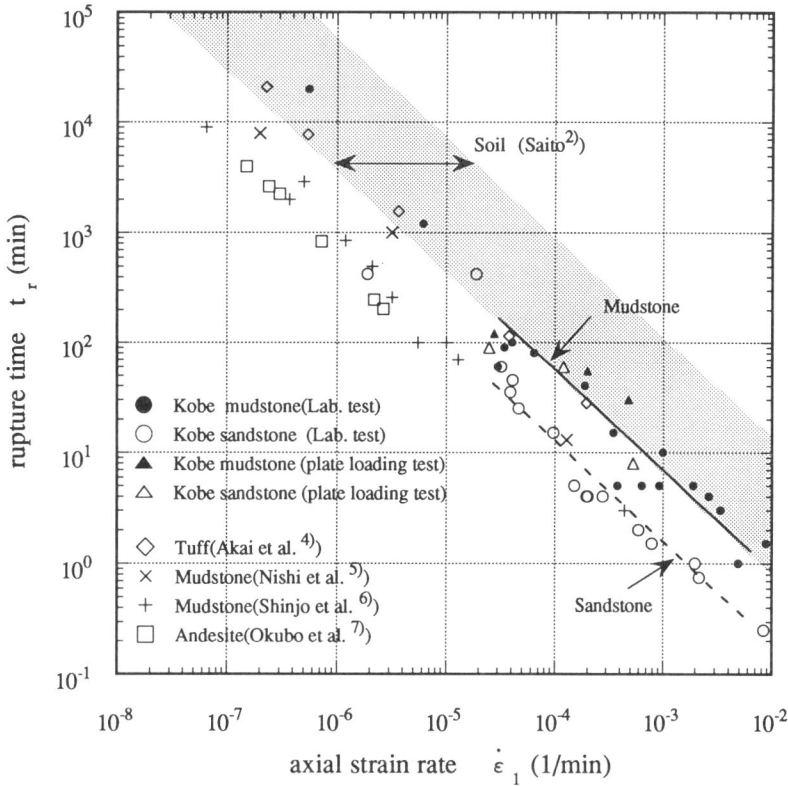

Fig.4 Relationship between minimum strain rate and rupture time

Therefore the relationship between the minimum strain rate and rupture time is written by Eq.(1) and Eq.(2) for the Kobe Formation.

sandstone:
$$\log t_r = 2.33 - 0.916 \log(\dot{\varepsilon}_{1min} \times 10^4) - 1.28 \quad (1)$$
mudstone:
$$\log t_r = 2.33 - 0.916 \log(\dot{\varepsilon}_{1min} \times 10^4) - 0.59 \quad (2)$$

3.4 Estimation of the long term strength

Fig.5 shows the relation of the minimum strain rate and creep rupture stress, σ_{fc}. In this figure, a nearly linear relationship is found between the logarithm of the minimum strain rate and creep

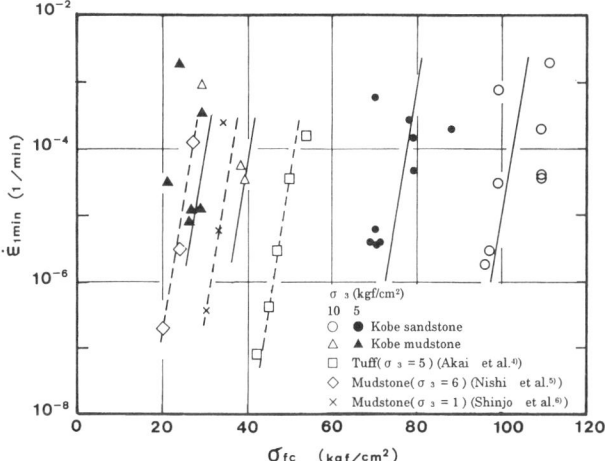

Fig.5 Relationship between minimum strain rate and creep rupture stress

rupture stress for the same confining pressure, σ_3 and the relation is written by Eq.(3)

$$\log \dot{\varepsilon}_{1min} = 0.417 \sigma_{fc} + b \quad (3)$$

sandstone:
when $\sigma_3 = 5, 10$ kgf/cm^2, b=-36.3 and -46.7
mudstone:
when $\sigma_3 = 5, 10$ kgf/cm^2, b=-16.4 and -20.9

Using the above equations, the long term strength of the Kobe Formation can be simply estimated. For example, the minimum strain rate after 100 years, $\dot{\varepsilon}_{1min\ t=100}$ can be obtained by substituting tr=5.26x10^7 into Eq.(1) or Eq.(2). Therefore, the long term strength, $\sigma_{fct=100}$ can be obtained by substituting the minimum strain rate thus obtained into Eq.(3). As the results of calculation, $\sigma_{fct=100}$ is the value of 60～90kgf/cm^2 for sandstone, and of 10～30 kgf/cm^2 for mudstone respectively, under the confining pressure of 5～10 kgf/cm^2.

4 DEFORMATION CHARACTERISTICS

4.1 Application of creep model

Generally, the Voight model has been applied as the creep model to the design. The Voight model formed by 5-elements is expressed by Eq.(4) and Fig.6.

$$\varepsilon = \varepsilon_1 - \varepsilon_3 = \varepsilon_1(1+\nu)$$
$$= \left\{ \frac{1}{2G_0} + \frac{1}{2G_1}\left(1 - e^{-\frac{G_1}{\eta_1}t}\right) + \frac{1}{2G_2}\left(1 - e^{-\frac{G_2}{\eta_2}t}\right) \right\} \sigma_c \quad (4)$$

where
- ε : shear strain
- ε_1 : axial strain
- ν : Poisson's ratio
- σ_c : stress level ($= \sigma_1 - \sigma_3$)
- G_0 : instantaneous shear elastic modulus
- G_1, G_2 : shear elastic modulus
- η_1, η_2 : viscoelastic modulus

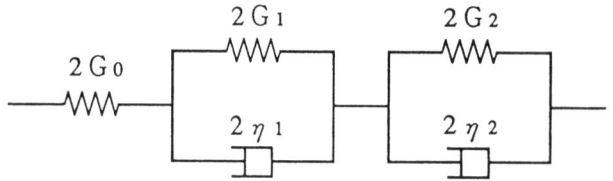

Fig.6 Voight model formed by 5-elements

Differentiating Eq.(4) with respect to time t, we get

$$\frac{\dot{\varepsilon}}{\sigma_c} = \frac{\dot{\varepsilon}_1(1+\nu)}{\sigma_c} = \frac{1}{2\eta_1}e^{-\frac{G_1}{\eta_1}t} + \frac{1}{2\eta_2}e^{-\frac{G_2}{\eta_2}t} \quad (5)$$

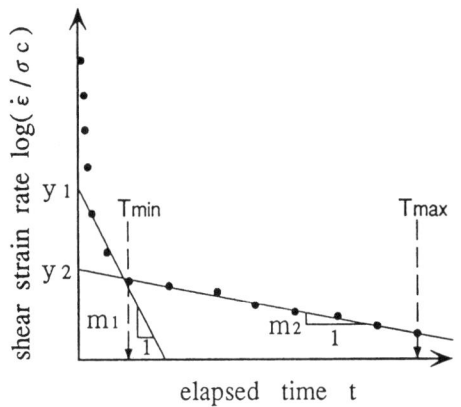

Fig.7 Method of approximation to obtain creep coefficient based on 5-elements model

and taking logarithms, we get

$$\log\left(\frac{\dot{\varepsilon}}{\sigma_c}\right) = \log\left(\frac{\dot{\varepsilon}_1(1+\nu)}{\sigma_c}\right)$$
$$= \log\frac{1}{2\eta_1} - \frac{G_1}{\eta_1}\log e \cdot t + \log\frac{1}{2\eta_2} - \frac{G_2}{\eta_2}\log e \cdot t \quad (6)$$

Eq.(6) assumes that the creep behavior follows the exponential law. Creep coefficients $(G_1, \eta_1), (G_2, \eta_2)$ are determined by the two approximate straight lines, as illustrated in Fig.7 plotted on the basis of Eq.(6), and are expressed by Eq.(7) and Eq.(8). In other words, the creep behavior of the first half and the latter half of the loading period is expressed by the two lines respectively.

$$\eta_1 = \frac{1}{2y_1} , \quad \eta_2 = \frac{1}{2y_2} \quad (7)$$

$$G_1 = -\frac{m_1\eta_1}{\log e} , \quad G_2 = -\frac{m_2\eta_2}{\log e} \quad (8)$$

G_0 can be given by Eq.(9) and an instantaneous strain ε_0 is generally defined as a strain corresponding to the elapsed time t=1 minute after loading.

$$G_0 = \frac{\sigma_c}{\varepsilon_0} \quad (9)$$

The creep behavior developed soon after loading is not so important for the long term deformation. Therefore, an instantaneous shear elastic modulus G^* may be introduced instead of G_0 and G_1 illustrated in Fig.6 and then Eq.(4) is rewritten as the following equation which is called the 3-element model.

$$\varepsilon = \left\{\frac{1}{2G^*} + \frac{1}{2G_2}\left(1 - e^{-\frac{G_2}{\eta_2}t}\right)\right\}\sigma_c \quad (10)$$

Eq.(10) is a practical equation commonly applied to the estimation of long-term deformation. In the following discussions, this model will be adopted as a standard creep model.

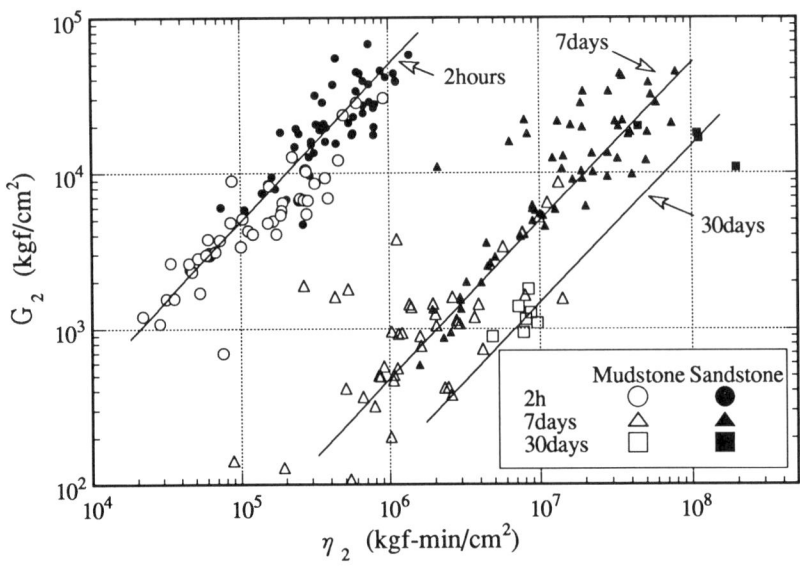

Fig.8 Relationship between shear elastic modulus G_2 and viscoelastic modulus η_2

4.2 Effect of loading period on estimation of creep coefficients

For practical application, the loading period in the creep test is generally determined by considering the time allowed for the tests, or by the number of specimens. Fig.8 is an example of the distribution of the creep coefficient, G_2 and η_2 based on the 3-element model. Specimens are Kobe mudstone and sandstone, and the loading periods are 2-hours, 7-days and 30-days respectively as shown in Fig.1. As is evident from Fig.8, the distribution of the creep coefficients has different areas because of the effect of the loading period. This fact is not caused by the characteristics of creep behavior, but is caused by the different methods used to approximate the straight lines in Fig.7 for determination of the creep coefficient.

For example, as shown in Fig.9, we usually draw the line "A" for all plotted data. However, if the creep test should end in 7-days, we draw the line "B". Then, the gradient, m_2 and the intercept, y_2 of the approximate line decrease as the loading period becomes longer. Consequently, the viscoelastic modulus η_2 increases as shown in Eq.(7). On the other hand, G_2 hardly changes, as shown in Eq.(8), because m_2 decreases and η_2 increases as the loading period becomes longer. As a result, the region of distribution of the creep coefficients, as shown in Fig.8 moves to the right with an increase in the loading period. In other words, the above fact suggests that the creep model based on the exponential law is applicable, though the creep behavior follows the logarithmic law.

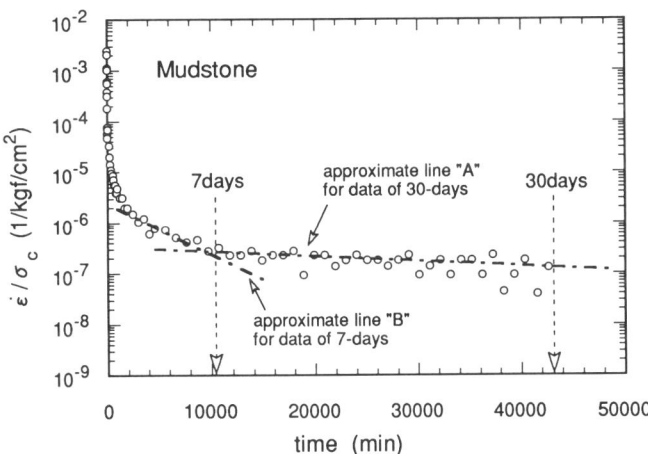

Fig.9 Method of approximation to obtain creep coefficient based on 3-element model

4.3 Estimation of retardation coefficient

In order to estimate the deformation of creep behavior more accurately, we need to make a time schedule for the creep test in which the loading period is as long as possible. However, from the above mentioned point of view, the fact that the viscoelastic modulus η_2 is affected by the loading period is a problem we should not ignore. So, we should examine how to estimate the long-term creep coefficient in the creep test performed under a restricted loading period.

The gradient of the distribution of G_2 and η_2 as shown in Fig.8 is called the retardation coefficient b, which is defined by Eq.(11).

$$\beta = G_2 / \eta_2 \qquad (11)$$

In order to examine the correlation between β and the loading period, β was calculated again for each datum by substituting G_2 and η_2 into Eq.(11). Here, G_2 and η_2 are obtained by changing the time span $T_{max}-T_{min}$ to approximate the straight line, as shown in Fig.7. Fig.10 shows β corresponding to T_{max} which was plotted for each specimen respectively. In this figure, the relationship between β and T_{max} can be represented by Eq.(12).

$$\beta \cdot T_{MAX} = \text{CONST.} \qquad (12)$$

T_{max} is equivalent to the loading period, and Eq.(12) indicates that β can be estimated for the loading period regardless of rock types.

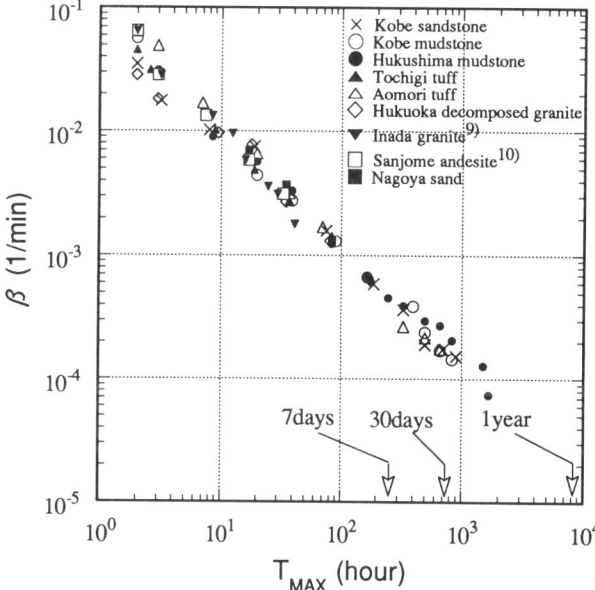

Fig.10 Relationship between retardation coefficient β and loading period T_{max} for various specimens

4.4 Estimation of shear elastic modulus

From the above mentioned point of view, the shear elastic modulus G_2 is not sensitive to changes in the loading period. Fig.11 shows the relationship between G^* and G_2 obtained for other rock types, excluding some results of the short-term creep test whose loading period is 2-hours. Although they are somewhat scattered, the distribution of these plots exist in a narrow band and can be represented by Eq.(13).

$$G_2 = A \left(G^*\right)^B , \quad A = 0.25 \sim 1.5 , \quad B = 1.3 \qquad (13)$$

In summary, we can approximately estimate the long-term creep coefficient based on the 3-element model without testing, as follows. If we can determine the instantaneous shear elastic modulus G^* in certain ways, G_2 can be determined by referring to Eq.(13). Moreover, b can be determined by using Eq.(12), where T_{max}=100 years, which is long enough in terms of engineering practice. As a result, η_2 is well-known by substituting b and G_2 for Eq.(11).

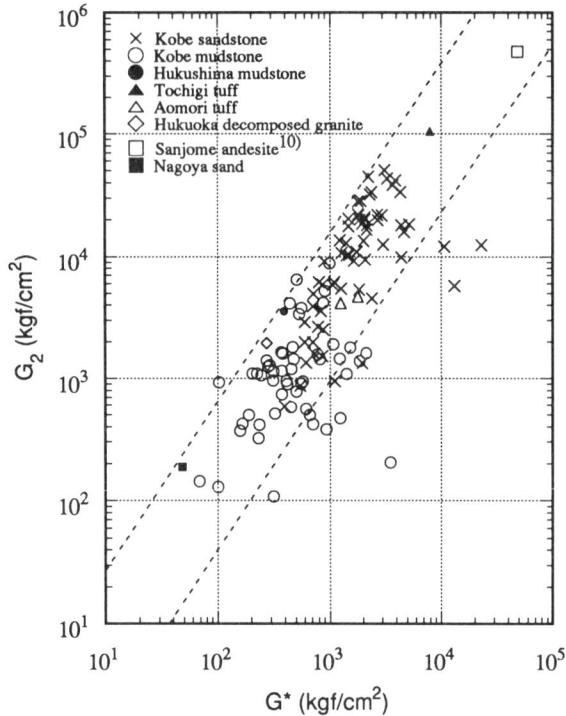

Fig.11 Relationship between shear elastic modulus G_2 and instantaneous shear elastic modulus G^* for various specimens

5 CONCLUSIONS

In this paper, in order to evaluate the long term stability of the foundation of the Akashi Kaikyo Bridge, creep characteristics of the Kobe Formation were discussed on the basis of laboratory tests and field creep tests. Firstly, the long term strength was estimated from the results of short term creep tests. Secondly, the effects of the loading period on the creep coefficient based on the Voight model were discussed, and methods to estimate the long-term creep coefficient were suggested. These results can be summarized as follows:

A well known linear relationship between creep rupture time, tr and the minimum strain rate $\dot{\varepsilon}_{1min}$ was observed not only in the laboratory tests performed on the specimens of sandstone and mudstone, but also in the plate loading tests performed on the alternating layers of the Kobe Formation.

Similarly, a linear relationship was found between the logarithm of $\dot{\varepsilon}_{1min}$ and the creep rupture stress, σ_{fc}. By using the above relations, $tr \sim \dot{\varepsilon}_{1min} \sim \sigma_{fc}$, the long-term strength of the Kobe Formation can be determined. Shear elastic modulus G_2, which is the creep coefficient based on the Voight model formed by 3-elements, is not sensitive to changes in the loading period and, on the other hand, the viscoelastic modulus η_2 increases as the loading period becomes longer.

The retardation coefficient β can be estimated from the loading period for various rock types and there is a correlation between shear elastic modulus G_2 and instantaneous shear elastic modulus G^*. In summary, we can approximately estimate without testing, the long term creep coefficients G_2, η_2 based on the 3-element model.

REFERENCES

1) Adachi,T., A.Takase 1981. Prediction of long term strength of soft sedimentary rock, Proc. of the Int. Symposium on Weak Rock, Tokyo, 99-104.

2) Saito,M. 1968. Research on forecasting the time of occurrence of slope failure, Railway Technical Research Report, No.626(267).

3) Nishigaki,Y., T.Noto & K.Takahashi 1981. Strain analysis of loading tests on multi-layered soft rocks, Proc. of the Int. Symposium on Weak Rock, Tokyo, 453-458.

4) Akai,K., T.Adachi & K.Nishi 1979. Time dependent characteristics and constitutive equation of soft sedimentary rock(porous Tuff), Proc. of J.S.C.E., No.282:75-87.

5) Nishi,K., T.Okamoto & Y.Esashi 1983. Strength-deformation characteristics of mudstone under some kinds of loading conditions and its unificative interpretation, Proc. of J.S.C.E., No.338:149-158.

6) Shinjo,T., Y.Komiya 1987. Time dependent characteristics of anisotropic mudstone, Proc. of the 7th J.S.R.M., 73-78.

7) Okubo,S., Y.Nishimatu 1986. Creep behavior and constitutive equation of Sanjome andesite and Kawazu tuff, Journal of M.M.I.J., 102(1181), 395-400.

8) Morlier,P. 1964. Etude experimentale de la deformation des roches, Revue de I'Institut Francais du Petrole, Vol.19, No.10, 1113-1147; Vol.19, No.11, 1183-1217.

9) Okubo,S., Y.Nishimatu 1985. Time dependent behavior of rock especially focused on tertiary creep, Proc. of the 17th Symposium on Rock Mechanics in Japan, 66-69.

10) Fukui,K., S.Okubo & Y.,Nishimatu 1989. Creep behavior of rock under uniaxial compression, Journal of M.M.I.J., 105, No.7, 521-526.

Modification to the original Hoek-Brown strength criterion

V.S. Vutukuri & S.M.F. Hossaini
University of New South Wales, Sydney, N.S.W., Australia

ABSTRACT: Amongst the empirical strength criteria formulated for intact rocks and rock masses, the Hoek-Brown criterion has become popular. For intact rocks, Hoek and his co-workers have suggested constant values for parameter *m* for various rock types. However, it has been found that parameter *m* is not a constant but a function of uniaxial compressive strength, σ_c for a given rock type. Relationships between parameter *m* and uniaxial compressive strength for various rock types have been given. The criterion applicable to intact rocks has been modified by replacing *m* by m_m and σ_c by σ_{cm} (subscript *m* refers to rock mass) to apply to rock mass. Relationships between $\frac{m_m}{m}$ and $\frac{\sigma_{cm}}{\sigma_c}$ for disturbed and undisturbed rock masses have been given. From laboratory model studies, relationships have also been found for plaster and sandstone. To determine σ_{cm}, a method has been suggested taking into account the effect of size and discontinuities.

1 INTRODUCTION

The theoretical triaxial strength criteria based on the actual mechanism of fracture do not fit the experimental results properly and to overcome this problem, many empirical criteria have been proposed for rocks. The strength criteria can be written in terms of either (1) principal stresses, σ_1 and σ_3 at fracture or normalised principal stresses at fracture obtained by dividing the principal stresses, σ_1 and σ_3 at fracture by the relevant uniaxial compressive strength, σ_c or (2) shear and normal stresses or normalised shear and normal stresses with respect to uniaxial compressive strength. Here, the applicability of Hoek and Brown criterion in terms of normalised principal stresses to rocks and rock masses is assessed.

2 HOEK AND BROWN CRITERION FOR INTACT ROCK

The criterion proposed by Hoek and Brown (1980a, b) is as follows:-

$$\frac{\sigma_1}{\sigma_c} = \frac{\sigma_3}{\sigma_c} + \left(1 + m\frac{\sigma_3}{\sigma_c}\right)^{0.5} \quad (1)$$

where σ_1 = major principal stress at failure;
σ_3 = minor principal stress at failure;
σ_c = uniaxial compressive strength; and
m = a constant parameter.

The revised values of constant *m* derived or estimated by Hoek et al. (1992) for various intact rocks are given in Table 1.
Substitution of $\sigma_1 = 0$ in this criterion, and solution of the resulting quadratic equation for σ_3, gives the uniaxial tensile strength, σ_t as follows:-

$$\sigma_3 = \sigma_t = \frac{1}{2}\sigma_c\left[m - (m^2 + 4)^{0.5}\right] \quad (2)$$

2.1 Methods for determination of constant parameter, m

For non linear regression analysis of triaxial test data, the selection of appropriate software is crucial in the results obtained. Earlier, Hoek and Brown (1980a) recommended a particular transformation to make their criterion a simple linear model to determine the appropriate values for the parameters by the use of a calculator. It

Table 1. Values of constant *m* in Hoek and Brown criterion for various rocks (after Hoek et al., 1992).

Rock	Value of *m*
Amphibolite	31.2
Andesite	18.9
Anhydrite	13.2
Basalt	17
Chalk	7.2
Chert	19.3
Claystone	3.4
Conglomerate	20
Dolerite	15.2
Dolomite	10.1
Gabbro	25.8
Gneiss	29.2
Granite	32.7
Gypstone	15.5
Limestone	8.4
Marble	9.3
Norite	21.7
Quartzite	23.7
Rhyolite	20
Sandstone	9.6
Slate	11.4

must be mentioned here that linearisation of a non linear model does not produce an equivalent model. Hoek and his co-workers (Shah and Hoek, 1992) now use the simplex reflection technique which has definite advantage over linear regression analysis for fitting a non linear strength criterion to laboratory data. They concluded that although linear regression analysis may, at best, give similar estimates of the parameters of any non linear criterion, the use of the simplex technique will always give better estimates of the parameters of any non linear criterion. Vutukuri and Hossaini (1992) also reported similar conclusions. As a matter of fact, non linear regression analysis through the minimisation of the sum of the squares of the relative errors fits the results better particularly in the lower range of the independent variable i.e. confining pressure, σ_3.

2.2 *Values of parameter m*

For each rock type, a unique value has been recommended by Hoek and his co-workers (Table 1). However, it has been found that the values of parameter not only vary with rock type but also depend upon the uniaxial compressive strength of the rock. Table 2 gives the relationships between m and σ_c for various rock types. Figures 1 to 7 depict the relationships. As σ_c increases, m decreases in all cases except for shale. In the case of shale, as σ_c increases, m also increases. Johnston (1985) also came to the conclusion that the values of the parameters in his criterion are not only dependent upon the rock type but also upon the uniaxial compressive strength of the rock.

Table 2. Relationships between m and σ_c for various rocks for Hoek and Brown criterion.

Rock	m
Claystone and liparite	$-460.74 + 478.88\,(\log\sigma_c)^{-0.024}$ Figure 1
Coal	$62.903 - 34.213\,(\log\sigma_c)^{0.9772}$ Figure 2
Granite	$-971 + 1055.3\,(\log\sigma_c)^{-0.06}$ Figure 3
Granodiorite	$7.86 + 8.4934e^7\,(\log\sigma_c)^{-18.36}$ Figure 4
Limestone	$-1.6115 + 62.05\,(\log\sigma_c)^{-2.7421}$ Figure 5
Sandstone	$-575.96 + 600.95\,(\log\sigma_c)^{-0.0221}$ Figure 6
Shale	$16.111 - 41.543\,(\log\sigma_c)^{-2.4388}$ Figure 7

3 HOEK AND BROWN CRITERION FOR ROCK MASS

The original Hoek and Brown criterion for rock mass proposed in 1980 is as follows:-

$$\frac{\sigma_1}{\sigma_c} = \frac{\sigma_3}{\sigma_c} + \left(s + m_m \frac{\sigma_3}{\sigma_c}\right)^{0.5} \quad (3)$$

where s and m_m are rock mass constants.

The uniaxial compressive strength of the rock mass, σ_{cm} can be obtained by substituting $\sigma_3 = 0$ into equation (3), namely,

$$\sigma_{cm} = \sqrt{s}\,\sigma_c \quad (4)$$

In 1988, they suggested the following set of approximate relationships between the constants m_m and s and the rock mass rating (RMR) developed by Bieniawski (1974):-

Disturbed rock masses:-

$$\frac{m_m}{m} = \exp\left(\frac{RMR - 100}{14}\right) \quad (5)$$

$$s = \exp\left(\frac{RMR - 100}{6}\right) \quad (6)$$

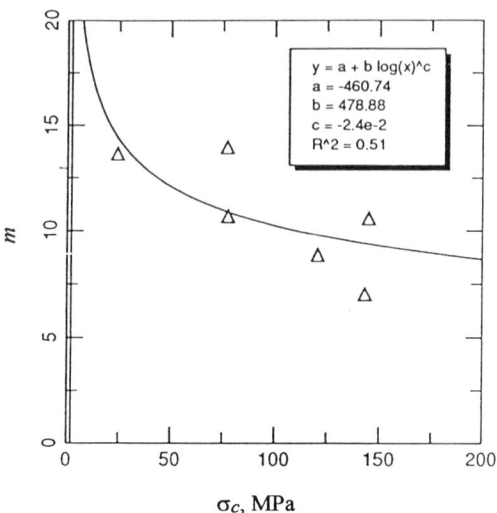

Figure 1. m in the Hoek and Brown criterion versus σ_c for claystone and liparite.

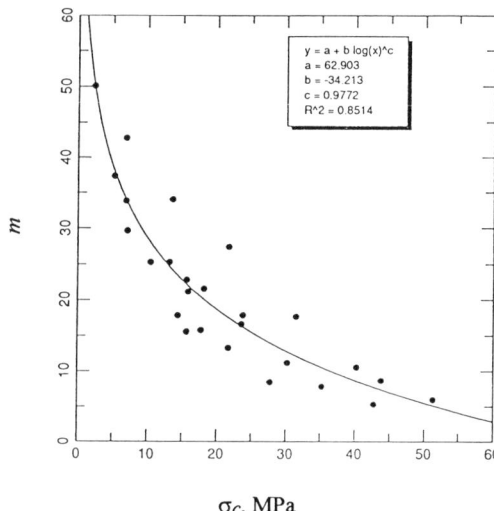

Figure 2. m in the Hoek and Brown criterion versus σ_c for coal.

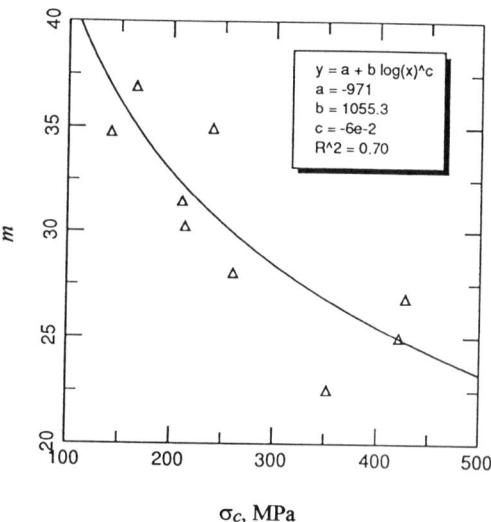

Figure 3. m in the Hoek and Brown criterion versus σ_c for granite.

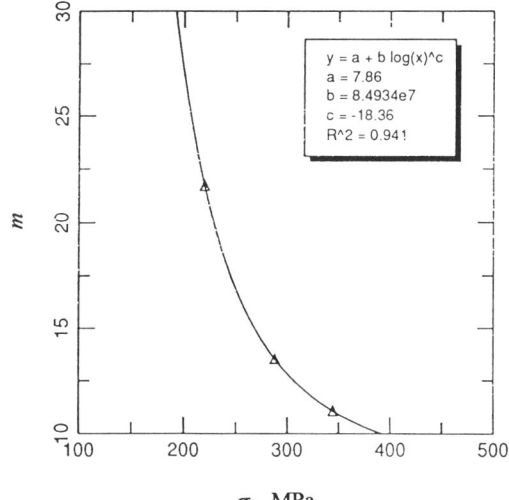

Figure 4. *m* in the Hoek and Brown criterion versus σ_c for granodiorite.

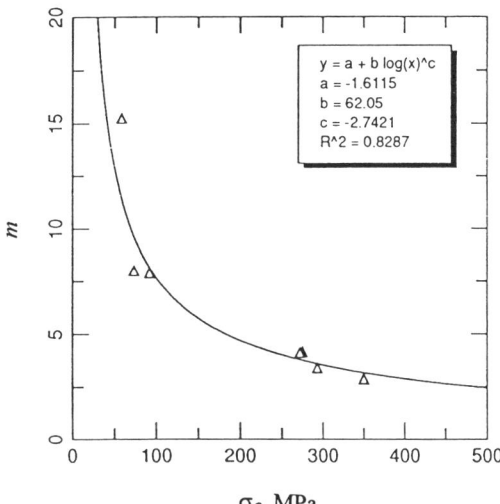

Figure 5. *m* in the Hoek and Brown criterion versus σ_c for limestone.

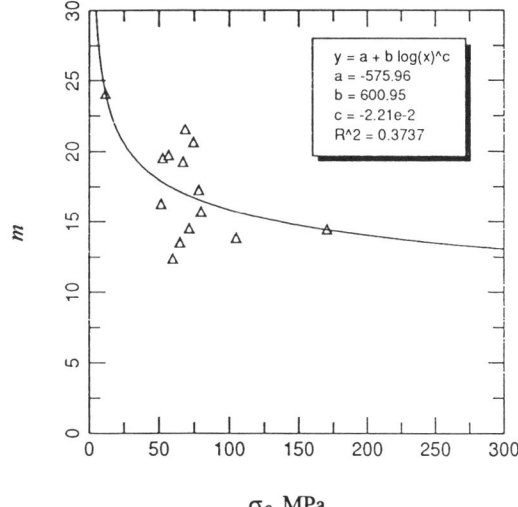

Figure 6. *m* in the Hoek and Brown criterion versus σ_c for sandstone.

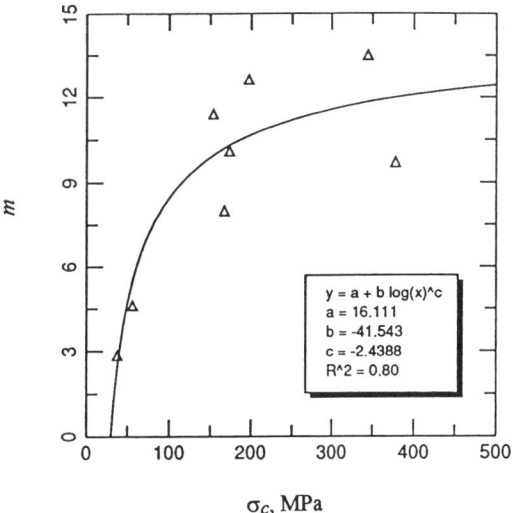

Figure 7. *m* in the Hoek and Brown criterion versus σ_c for shale.

Undisturbed rock masses:-

$$\frac{m_m}{m} = \exp\left(\frac{RMR - 100}{28}\right) \quad (7)$$

$$s = \exp\left(\frac{RMR - 100}{9}\right) \quad (8)$$

where m_m and s are the rock mass constants and m is the constant for the intact rock.

They recommended the application of the criterion only to isotropic rock mass i.e. when the volume of rock under consideration contains four or more closely spaced discontinuity sets and where none of these discontinuity sets is significantly weaker than any of the others. In this original criterion, the uniaxial compressive strength is used both in the *RMR* system and the criterion.

According to Hoek et al. (1992) the criterion predicts too high an axial strength for low values of σ_3 and also a finite tensile strength. They believe that the rock mass should have zero tensile strength. To overcome these deficiencies in the original Hoek and Brown criterion, they presented the following modified criterion:-

$$\frac{\sigma_1}{\sigma_3} = \frac{\sigma_3}{\sigma_c} + \left(m_m \frac{\sigma_3}{\sigma_c}\right)^a \quad (9)$$

where m_m and a are rock mass constants.

The values of $\frac{m_m}{m}$ and a depend upon the block size and surface condition. This modified criterion gives uniaxial compressive strength for any class of rock mass the value of zero.

3.1 Analysis of the original Hoek and Brown criterion

For rock mass, the Hoek and Brown criterion for intact rock can be modified by substituting uniaxial compressive strength of rock mass instead of uniaxial compressive strength of intact rock. The resulting equation is as follows:-

$$\frac{\sigma_1}{\sigma_{cm}} = \frac{\sigma_3}{\sigma_{cm}} + \left(1 + m_m \frac{\sigma_3}{\sigma_{cm}}\right)^{0.5} \quad (10)$$

To use this equation, σ_{cm} and m_m are required. From the original Hoek and Brown criterion for rock mass, the following equations have been derived:-

For disturbed rock mass:-

$$\frac{\sigma_{cm}}{\sigma_c} = [\exp(\frac{RMR - 100}{6})]^{0.5} \quad (11)$$

$$\frac{m_m}{m} = \frac{1}{(\frac{\sigma_{cm}}{\sigma_c})^{0.144}} \quad (12)$$

Figure 8 depicts the relationship given in Equation (12).

For undisturbed rock mass:-

$$\frac{\sigma_{cm}}{\sigma_c} = [\exp(\frac{RMR - 100}{9})]^{0.5} \quad (13)$$

$$\frac{m_m}{m} = \frac{1}{(\frac{\sigma_{cm}}{\sigma_c})^{0.3588}} \quad (14)$$

Figure 9 depicts the relationship given in Equation (14).

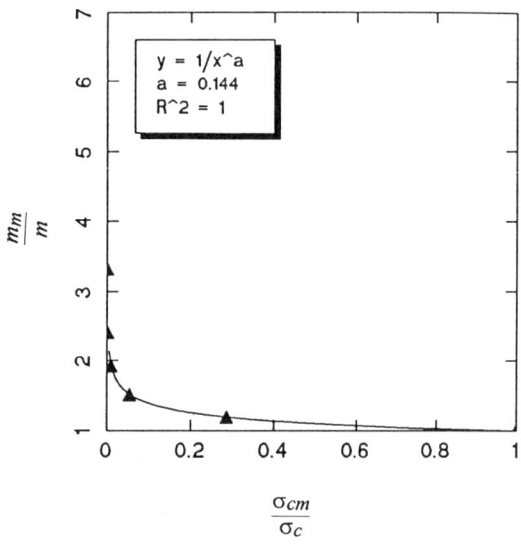

Figure 8. $\frac{\sigma_{cm}}{\sigma_c}$ versus $\frac{m_m}{m}$ for the disturbed rock mass for the modified Hoek and Brown criterion.

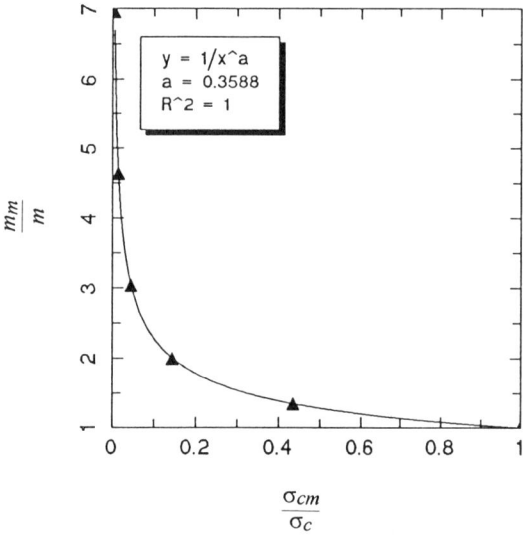

Figure 9. $\frac{\sigma_{cm}}{\sigma_c}$ versus $\frac{m_m}{m}$ for the undisturbed rock mass for the modified Hoek and Brown criterion.

The critical parameter in the modified criterion is the ratio between σ_{cm} and σ_c. According to Hoek and Brown (1988), the ratio depends upon RMR as well as condition of the rock mass i.e. disturbed or undisturbed. In 1992, Hoek et al. abandoned the use of RMR etc.

To properly assess the uniaxial compressive strength of rock mass, two factors have to be taken into account. The first factor is the effect of size on the uniaxial compressive strength of intact rock (in between the discontinuities) of the size under consideration and the second factor is the effect of discontinuities (number, orientation with respect to stress field, etc.) on the uniaxial compressive strength estimated taking into account the size effect (Vutukuri and Hossaini, 1991; Vutukuri, 1993). Once this is estimated, $\frac{m_m}{m}$ can be determined from Equations (12) and (14).

From the results obtained in laboratory experiments on discontinuous models of plaster and sandstone, the following relationships have been determined:-

For plaster:-

$$\frac{m_m}{m} = \frac{1}{(\frac{\sigma_{cm}}{\sigma_c})^{0.8942}} \quad (15)$$

For sandstone:-

$$\frac{m_m}{m} = \frac{1}{(\frac{\sigma_{cm}}{\sigma_c})^{0.4949}} \quad (16)$$

Figure 10 depicts the relationship given in Equation (16).

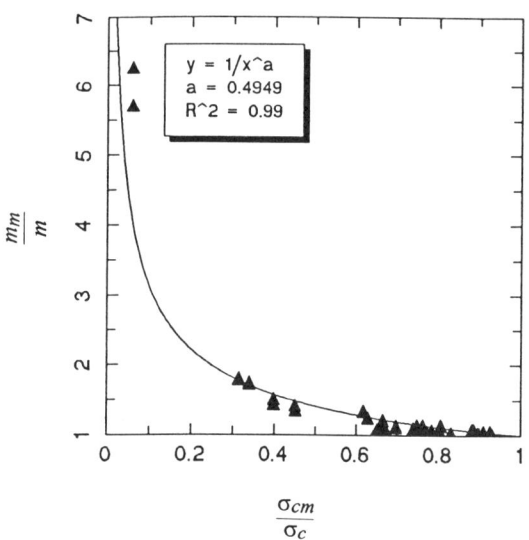

Figure 10. $\frac{\sigma_{cm}}{\sigma_c}$ versus $\frac{m_m}{m}$ for sandstone models tested in the laboratory for the modified Hoek and Brown criterion.

4 CONCLUSIONS

The parameter m in Hoek and Brown criterion for intact rock has been found to be dependent upon the uniaxial compressive strength, σ_c for a given rock type. Relationships between m and σ_c for various rock types have been given which can be used to estimate m. In the original Hoek and Brown criterion for rock mass, two parameters, namely, s and m_m are to be estimated.

However, Hoek et al. (1992) introduced a modified criterion for which two parameters m_m and a are to be estimated. This modified criterion gives uniaxial compressive strength for any class of rock mass the value of zero which is not true.

The modification suggested by the authors i.e. replacing m by m_m and σ_c by σ_{cm} in the original Hoek and Brown criterion for intact rock to apply to rock mass and analysis of the original Hoek and Brown criterion for rock mass gave relationships between $\frac{m_m}{m}$ and $\frac{\sigma_{cm}}{\sigma_c}$ for disturbed and undisturbed rock masses. It is not necessary to use these relationships and it is important to derive the relationships independently. In the original Hoek and Brown criterion for rock mass, $\frac{\sigma_{cm}}{\sigma_c}$ depends upon RMR. It is much better to estimate this ratio taking into account the effect of size and discontinuities. From laboratory model studies, relationships have been suggested for plaster and sandstone. More work is required on models of other rocks to determine the relationships for other rocks.

REFERENCES

Bieniawski, Z. T. 1974. Geomechanics classification of rock masses and its application in tunnelling. *Proc. 3rd Cong. Int. Soc. Rock Mech.*: 2A: 27-32.

Hoek, E. and Brown, E. T. 1980a. *Underground excavations in rock*. London: Instn. Min. Metall.

Hoek, E. and Brown, E. T. 1980b. Empirical strength criterion for rock masses. *J. Geotech. Engng. Div. ASCE* 106: 1013-1035.

Hoek, E. and Brown, E. T. 1988. The Hoek-Brown failure criterion - a 1988 update. *Proc. 15th Can. Rock Mech. Symp.*: 31-38.

Hoek, E., Woods, D. and Shah, S. 1992. A modified Hoek-Brown failure criterion for jointed rock masses. *Eurock 1992*: 209-214. London: Thomas Telford.

Johnston, I. W. 1985. Strength of intact geomechanical materials. *J. Geotech. Engng. Div. ASCE* 111: 730-749.

Shah, S. and Hoek, E. 1992. Simplex reflection analysis of laboratory strength data. *Can. Geotech. J.* 29: 278-287.

Vutukuri, V. S. 1993. Some aspects of the strength characteristics of intact and jointed rocks. *Proc. 14th West Japan Symp. Rock Engineering*: 105-114.

Vutukuri, V. S. and Hossaini, S. M. F. 1991. A critical review of predictive methods for estimation of compressive strength of pillars. *Proc. Int. Conf. Reliability, Production and Control in Coal Mines*: 262-271.

Vutukuri, V. S. and Hossaini, S. M. F. 1992. Assessment of applicability of four strength criteria for intact coal. *Proc. 6th Australia-New Zealand Conf. Geomechanics*: 280-285.

Failure criterion and Mohr's envelope for sedimentary soft rocks considering tensile strength

Kunio Hattori & Kenji Muranaka
Chubu Electric Power Co. Inc., Toshin-cho Higashiku, Nagoya, Japan

Shunsaku Nishie
Chuo Kaihatsu Co. Inc., Nishiwaseda Shinjukuku, Tokyo, Japan

ABSTRACT: This paper describes a method to determine both failure criterion and Mohr's envelope which best fit laboratory test results for several sedimentary soft rocks. The scope of soft rocks for this study is restricted to sandstones and mudstones. As concerns the characteristics of Mohr's envelopes, a difference was observed between sandstones and mudstones. Therefore, a attempt was made to express these two Mohr's envelopes by a single model which can best fit the laboratory test results. The proposed method consists of two characteristic equations for failure criterion and Mohr's envelope. These two equations took into account the tensile strength and adopted a power function. It can be concluded that this method to determine Mohr's envelope improves the Hoek&Brown's method, and also extended form of other conventional methods.

1 INTRODUCTION

In recent years, since increasing number of foundations have been placed on soft rock ground, it becomes of great importance to evaluate mechanical properties of soft rocks acculately.

General aspects and engineering characteristics of soft rocks in Japan were reported by Akai et al. (1978, 1979, 1993). Numerous failure criteria used for foundation stability analyses or elastic-plastic finite element analyses have been proposed.

As concerns relationships between the confining pressure and the strength of soft rocks, power function models were proposed by Hobbs(1966), Yoshinaka&Yamabe(1980), Adachi&Ogawa(1980), and Hoek&Brown(1980).

According to these discussions, it seems reasonably to conclude such linear equations as the Mohr-Coulomb's failure criterion are not suitable for soft rocks. But instead, non-linear equations as a power function may better describe the failure model for soft rocks.

Firstly, laboratory tests were conducted for sedimentary soft rocks including of sandstones and mudstones. Then, some of the conventional models proposed in the past were examined if they were applicable to the test results. The conventional models include those proposed by Griffith(1924), Fairhurst(1964) and Hoek&Brown(1980). None of them are found to be satisfactory.

In case of applying these methods, we found that it was very difficult to express both Mohr's envelopes of sandstones and mudstones by the identical model, it is because the strength characteristics are different between sandstones and mudstones.

So far as Mohr's envelope is concerned, sandstones are regarded as a granular material, while mudstones are regarded as a typical brittle material.

In practical, it would be convenient if Mohr's circles of these two kinds of soft rocks can be expressed by a single model.

Secondly, judging from this point of view, we studied the unique equation which best fit the Mohr's envelope of both sandstones and mudstones. In this stage, we focused on the strength characteristics in the range of low compressive stresses. Particular attention, moreover, was paid to the tensile strength so that the proposed model can be applicable at any stress levels.

Finally, we attempted to apply the proposed method to another soft rock. The result was satisfactory for us.

2 PROFILE OF SOFT ROCKS AND METHODS OF TESTS

Laboratory tests were conducted using specimen taken from Shizuoka prefecture in Japan. This soft rocks are sedimentary rock of Neogene deposits consisted of the alternative layers of sandstones and mudstones.

On the basis of the optical examination of rock cores obtained by a rotary core tube sampling technique, mudstones are genarally stiffer and stronger than sandstones. Undisturbed samples in drift excavated about 15 meters under the ground were used in laboratory tests.

In order to determine the material constants of shear strength, strength tests in laboratory were performed both so-called " Brazilian test " and consolidated-undrained triaxial compression test. All of the specimens used in triaxial test were saturated.

In this paper, stresses are described in terms of total stresses. The plot of the Mohr's circle for Brazilian tests represent the stress state at the center of the loaded thin disc. Utilizing the elastic solution under the plane stress condition analyzed by G.Hondros(1959), in the center point at fracture in Brazilian tests, the approximate values of principal stresses are given as follows :

$$\sigma_1 = 3\sigma_t \quad , \quad \sigma_2 = 0 \quad , \quad \sigma_3 = -\sigma_t$$

Where, $\sigma_1, \sigma_2, \sigma_3$ are the principal stresses and σ_t is the tensile strength ($\sigma_t > 0$).

3 SUMMARY OF BASIC PROPERTIES AND CHARACTERISTICS OF STRESS-STRAIN RELATIONS

3.1 Basic properties of soft rocks

Basic properties of soft rocks at this site are shown in Table 1, and comparison of basic properties between this site and the other sites are shown in table 2. As shown in table 1, ρ_t is the wet density, n_e is the effective porosity, and RB is the brittleness defined (q_u / σ_t). Where, qu is the uniaxial compression strength, and σ_t is the tensile strength. And also, w is the water content, and e is the void ratio, as shown in table 2.

As shown in table 1, the most noteworthy point among basic properties is the brittleness. That is, RB of mudstones is about 7.5, and it's value indicates general value of brittle materials. But, RB of sandstones is about 16.3, and its value is larger than mudstones'.

As mentioned latter, the difference of the brittleness between sandstones and mudstones caused difficulty to express Mohr's envelopes by the identical model for laboratory test results, in case of the conventional methods proposed in the past.

According to the comparison of basic properties as shown in table 2, uniaxial compression strength both of sandstones and mudstones indicates the middle value of the representative soft rocks in Japan.

Table 1. Basic properties of soft rocks at this site.

Kind of rocks	Basic properties of soft rocks		
	ρ_t (g/cm³)	n_e (%)	Brittleness RB(=σ_c/σ_t)
Sandstone	2.12	32.2	16.3
Mudstone	1.98	40.0	7.5

(note) Average value of test results

Table 2. Comparison basic properties of this site and the other sites for soft rocks.

Region	Kind of rocks	Basic properties		
		w (%)	e	qu (MPa)
This site	Sandstone	16.9	0.47	3.20*
This site	Mudstone	25.1	0.67	10.04*
Kobe-A	Mudstone	12.9	0.34	7.06
Kobe-B	Mudstone	18.4	0.49	2.63
Yokohama	Siltstone	32.6	0.94	2.06
Ohya-ishi	Tuffaceous	17.3	0.44	13.50

(note) *;Triaxial compressive strength at σ_3'=0.0

3.2 Characteristic of stress-strain curves

Among the mechanical properties in triaxial compression tests, stress-strain curves of the representative specimens are shown in fig.1 and fig.2.

As concerns axial strains at peak strength (ε_p), both ε_p of sandstones excepting for σ_3' = 3.43 (MPa) and mudstones, were obtained comparatively small strain value in the range of 1-4 (%).

Nevertheless, concerning stress-strain curve's patterns, a slight difference is observed between sandstones and mudstones. That is, while the former curve indicates from strain-hardening to strain-softening, the latter curve indicates perfect strain-softening.

So long as judging from characteristics of small ε_p or stress-strain curves, we evaluated that both of sandstones and mudstones could be classified brittle materials.

As concerns the dependence of the peak strength on the confining pressure, it was observed obviously a difference between sandstones and mudstones. That is, the dependence of mudstones was a little, but that of sandstones was obvious and similar to characteristics of granular materials.

According to stress-strain curve's patterns and the dependence of peak strength on the confining pressure, we understood that sandstone at this site had strength characteristics both of brittle materials and plastic materials such as granular materials.

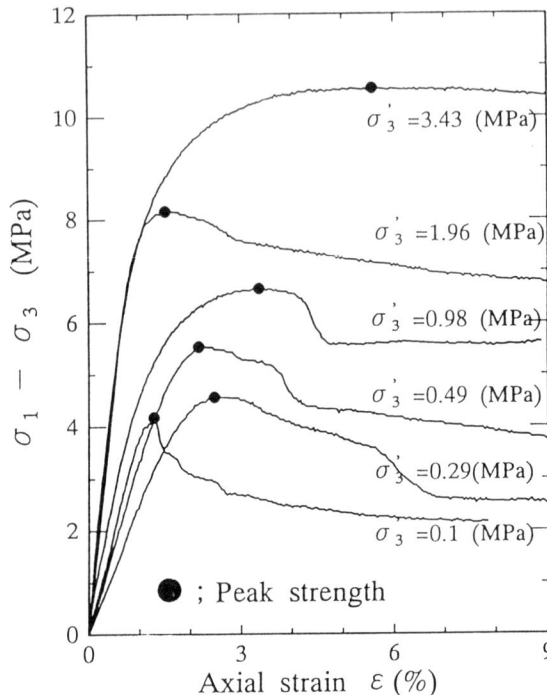

Figure 1. Stress-strain curve(sandstone)

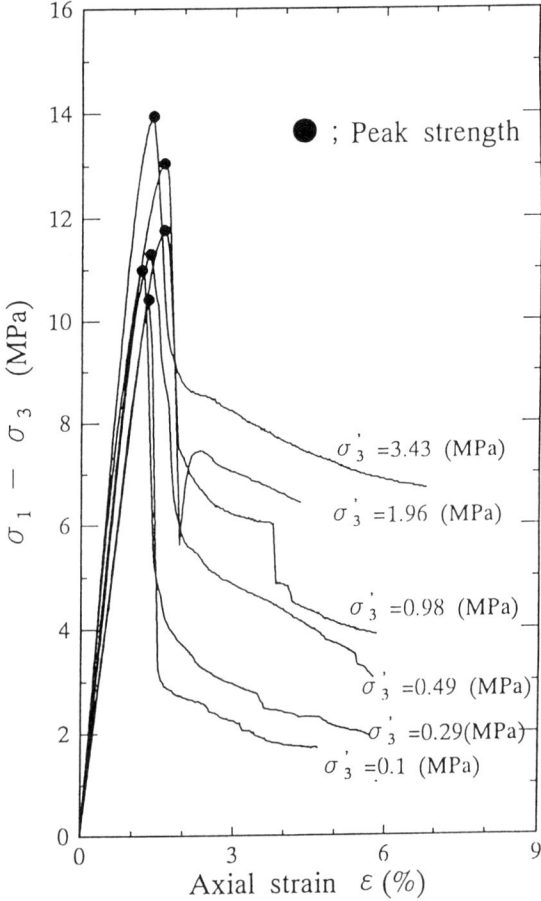

Figure 2. Stress-strain curve(mudstone)

4 APPLICATION TO TEST RESULTS BY THE CONVENTIONAL FAILURE CRITERIA AND MOHR'S ENVELOPES

4.1 Review of the conventional failure criteria and Mohr's envelopes.

A lot of failure criteria and Mohr's envelopes for laboratory test results have been proposed and published since many years ago.

Some of the conventional failure criteria and Mohr's envelopes are shown in table 3 and table 4.

Table 3. The conventional failure criteria

Type	Proposers	Failure criteria
Parabolic Function	Griffith model	$(\sigma_1 - \sigma_3)^2 = 8\sigma_t \times (\sigma_1 + \sigma_3)$
	Hoek&Brown	$(\sigma_1 - \sigma_3)^2 = m\sigma_c\sigma_3 + S\sigma_c^2$
Power Function	Hobbs	$(\sigma_1 - \sigma_3) = B \times \sigma_3^b$
	Yoshinaka & Yamabe	$\left(\dfrac{\tau_m}{\tau_{m0}}\right) = \alpha \times \left(\dfrac{\sigma_m}{\sigma_{m0}}\right)^\beta$
	Adachi & Ogawa	$\left(\dfrac{q}{p_0'}\right)_{peak} = \alpha_p \times \left(\dfrac{p'}{p_0'}\right)^{\beta_p}$ at $p' \le p_c$
	modified Hoek&Brown	$(\sigma_1' - \sigma_3') = \sigma_c \times \left(m_i \dfrac{\sigma_3'}{\sigma_c}\right)^a$

(note) σ_t (>0) is tensile strength

Table 4. The conventional Mohr's envelopes

Type	Proposers	Mohr's envelope
Parabolic Function	Griffith model	$\tau^2 = 4\sigma_t \times (\sigma + \sigma_t)$
	Fairhurst	$\tau^2 = \left\{\sqrt{\left(\dfrac{\sigma_c}{\sigma_t}\right)+1} - 1\right\}^2 \sigma_t(\sigma + \sigma_t)$
Power Function	Hoek&Brown	$\dfrac{\tau}{\sigma_c} = A\left(\dfrac{\sigma}{\sigma_c} - \dfrac{\sigma_t}{\sigma_c}\right)^B$ where, $\sigma_t = \sigma_c\left(m - \sqrt{m^2 + 4s}\right)/2$

(note) σ_t (>0) is tensile strength

According to both tables, typical features of these models can be described as follows.
(1) Relationships between peak strength and confining pressure, and relationships between shear stress and normal stress, are expressed by two models, parabolic type or power function type.
(2) The radixes used in these equations of failure criteria and Mohr's envelopes may include tensile strength.

As a consequence, it may be noted that the conventional modelings of failure criteria and Mohr's envelopes are basically classified with respect to two aspects; (1) mathematical expressions and (2) whether they take into account the tensile strength.

4.2 Application to test results, and problems for practical utilities.

In order to obtain such Mohr's envelopes as best fit the laboratory test results, applicability of the above conventional models has been examined.

The theory of Griffith(1924), the equations of Fairhurst(1964) and Hoek&Brown(1980) as shown in table 4 were used in this study. Here, a method of Hoek&Brown is expressed by the equation in the form of a power function, and the other two methods are expressed in the form of a parabolic function.

The obtained results of Mohr's envelopes are shown in fig.3. and fig.4. Each Mohr's circle in both figures is plotted representing the average of 10 specimens tested under the same confining pressure.

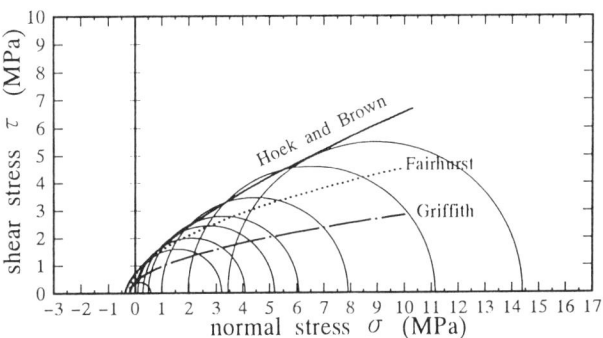

Figure 3. The conventional Mohr's envelopes of sandstones

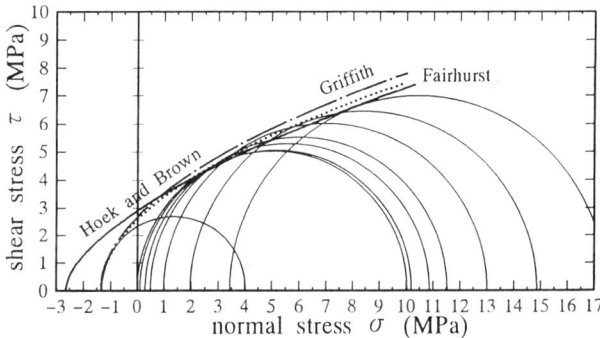

Figure 4. The conventional Mohr's envelopes of mudstones

According to these results, the following conclusions may be drawn concerning the conventional models for Mohr's envelopes.
(1) In the case of sandstones, Hoek&Brown's model expressed by a power function is better than the parabolic models proposed by Griffith and Fairfurst.

It seems that these different results are attributed to the strength characteristics of sandstone. Mohr's envelopes of sandstones are strongly governed by the Mohr's circles in the compression zone, whereas comparatively less influence is expected in the tension zone. This is a general feature of a brittle material.

We considered that methods of parabolic type might be not suitable for sandstone similarly indicated plastic characteristic such as granular materials.
(2) In the case of mudstones, more or less identical curves are obtained for three models in the zone of high compression stresses ($\sigma > 2.0$ MPa).

Shear stresses obtained by Hoek&Brown's method indicates comparatively greater values than the other methods. This reason is that tensile strength computed by the equation of the original Hoek&Brown are greater than the test results.

From the above results, it seems that the conventional models for Mohr's envelopes used in this study have the merits and demerits for the application to the test results.

It may be concluded that the conventional models were not satisfactory for us, because we required an identical method to express Mohr's envelopes for both sandstones and mudstones.

5 PROPOSAL FOR FAILURE CRITERION AND MOHR'S ENVELOPE CONSIDERING TENSILE STRENGTH AND ANALYSIS OF TEST RESULTS

5.1 Summary of the proposed method

According to the previous discussion, we found that the applicability of the conventional models for Mohr's envelopes used in this study to our laboratory test results was not satisfactory for us. Therefore, we attempted to express failure criterion and Mohr's envelope of sandstones and mudstones by the identical method.

In order to solve this problem, two characteristic equations are introduced based on the following assumptions.
(1) Power function is adopted in the unique equation.
(2) The formula of failure criterion is similar to that of Mohr's envelope.
(3) Mohr's envelope always passes $(-\sigma_t, 0.0)$ on the $\sigma-\tau$ plane.
Where σ_t is the tensile strength, σ is the normal stress, and τ is the shear stress.

5.1.1 Equation of failure criterion

Firstly, we assume that the general equation of failure criterion is expressed by equation (1):

$$(\sigma_1 - \sigma_3)_{peak} = A \times (\sigma_3 + \sigma_t)^n + B \quad (1)$$

As the state of stress is $\sigma_1 = 3\sigma_t$, $\sigma_2 = 0.0$, $\sigma_3 = -\sigma_t$ for the Brazilian test, equation (1) can be rewritten as equation (2):

$$(\sigma_1 - \sigma_3)_{peak} = A \times (\sigma_3 + \sigma_t)^n + 4\sigma_t \quad (2)$$

Where σ_1, σ_3 are the principal stresses at peak strength, and A, B, n are the material parameters, which can be obtained by the method of least squares from the triaxial compression test results. These material parameters are variables for the kind of soft rocks, and also depend on the triaxial compression strength or the instantaneous friction angle for a specified value of the normal stress.

5.1.2 Equation of Mohr's envelope

We assume the general equation of Mohr's envelope is given by the following equation:

$$(\tau/\alpha) = (\sigma + \sigma_t)^m \quad (3)$$

Where, both α and m are material parameters.

In order to determine eq. (3) from eq. (2), a numerical technique is required. Therefore, a general analytical solution for Mohr's envelope published by Balmer (1952) was used to determine the values of both α and m.

Normal stress (σ) and shear stress (τ) can be given by the following equations, which are obtained from a system of Mohr's circle by Balmer.

$$\sigma = \sigma_3 + \frac{(\sigma_1 - \sigma_3)}{1 + \frac{\partial \sigma_1}{\partial \sigma_3}} \quad (4)$$

$$\tau = (\sigma - \sigma_3) \times \sqrt{\frac{\partial \sigma_1}{\partial \sigma_3}} \quad (5)$$

Substituting equation (2) into equation (4) and (5), normal stress (σ) and shear stress (τ) can be rewritten in the following equations:

$$\sigma = \sigma_3 + \frac{(\sigma_1 - \sigma_3)}{2 + A \times n \times (\sigma_3 + \sigma_t)^{n-1}} \quad (6)$$

$$\tau = (\sigma - \sigma_3) \times \sqrt{1 + A \times n \times (\sigma_3 + \sigma_t)^{n-1}} \quad (7)$$

5.1.3 Determination of the Mohr's envelope

Utilizing equation (6) and (7), any the value of σ_i and τ_i corresponding to optional minor principal stress ($\sigma_{3,i}$) can be obtained easily.

Equation (3) can be rewritten in the following equation:

$$\log \frac{\tau}{\alpha} = m \times \log(\sigma + \sigma_t) \quad (8)$$

Let $X_i = \log(\sigma_i + \sigma_t)$ and $Y_i = \log(\tau_i)$, then solutions of least squares are given by the following equations.

$$m = \frac{\sum X_i Y_i - \frac{\sum X_i \sum Y_i}{k}}{\sum X_i^2 - \frac{(\sum X_i)^2}{k}} \quad (9)$$

$$\log \alpha = \frac{\sum Y_i}{k} - m \times \frac{\sum X_i}{k} \quad (10)$$

Where X_i and Y_i are the i-th values of X and Y, and k is the total number of each quantity.

From α and m given by equation (9) and (10), Mohr's envelope is determined. These series of procedure can be programmed easily on the personal computer.

5.1.4 Discussions about the proposed method

This proposed method must be one of variations as shown in table 3 or table 4. But the combination of equations used in this proposed method have not been published in the past.

It is obvious that this method is equivalent to one of the conventional methods to determine Mohr's envelope under the special conditions.

For example, this method is equivalent to methods of Faihurst in case of m=0.5, and also this method is similar to a method of Hobbs or modified Hoek&Brown in case of $\sigma_t = 0.0$.

Judging from this point of view, it may be said that this model is extended form of the conventional ones in respect of the determining of failure criterion or Mohr's envelope.

5.2 Analysis of laboratory test results utilizing the proposed method

In this section, we intend to the apply the proposed method to our laboratory test results, and to report the results of failure criteria and Mohr's envelopes obtained by the proposed model.

5.2.1 failure criteria for laboratory test results

Firstly, failure criteria for the average values of laboratory test results obtained by the proposed method were shown as the following equations:

sandstones: $(\sigma_1 - \sigma_3)_{peak} = 5.65 \times (\sigma_3 + 0.20)^{0.487} + 0.78$

mudstones: $(\sigma_1 - \sigma_3)_{peak} = 4.08 \times (\sigma_3 + 1.34)^{0.502} + 5.37$

Most characteristic point is that the n-value of power number as shown in eq.(2) was almost in the range of 0.40-0.60, and also the differences of the n-value between sandstones and mudstones were not so large. In this respect, the range of values obtained by this method was nearly same as results of Hobbs(1966).

5.2.2 Mohr's envelopes for laboratory test results

Secondly, the Mohr's envelopes obtained by the proposed method are shown in fig.5 and fig.6. Each Mohr's circle in both figures represents the average value of 10 specimens tested under the same confining pressure.

As shown in both figures, the Mohr's envelope fit Mohr's circle sufficiently.

In addition, so far as our laboratory test results are concerned, the proposed method is more suitable in modeling the test results compared to the conventional methods (see Figure 3 and 4).

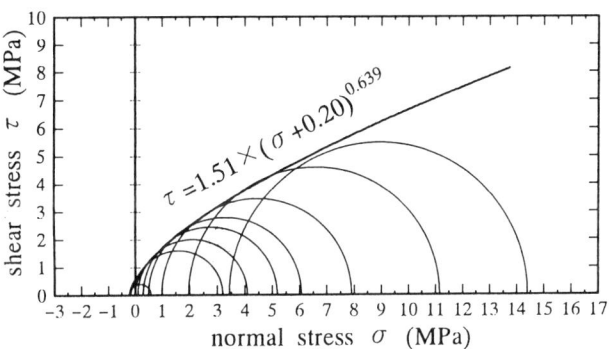

Figure 5. Mohr's envelope of sandstones obtained by the proposed method

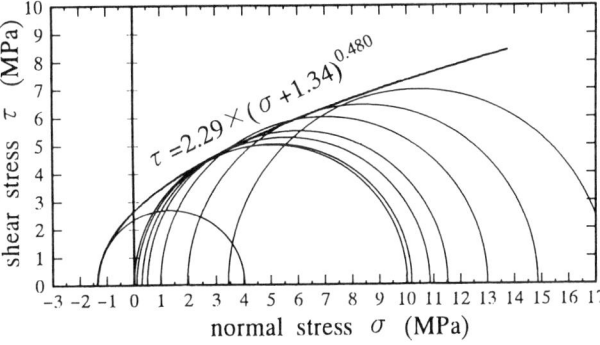

Figure 6. Mohr's envelope of mudstones obtained by the proposed method

6 ANALYSIS FOR OTHER SOFT ROCKS

In the preceding section of this chapter, we could understand that Mohr's envelopes of both sandstones and mudstones obtained by the proposed method were suitable for modeling Mohr's circles of test results.

So, we performed an example of analysis in order to evaluate the applicability of the proposed method for other soft rocks. The soft rocks used in this analysis was a mudstone of the Kobe-formations-A as shown in table 2, which has been reported by Yoshinaka and Yamabe(1980).

The result of this analysis is shown in fig.7. Here, special attention must be paid to the fact that the Mohr's envelope was obtained using the data excluding the Mohr's circle at $\sigma_3=6.42$(MPa) shown by a dotted circle in the figure.

It may be noted that the result is not so good as the ones shown in fig.5 and fig.6. It seems, however, that sufficient accuracy of modeling the Mohr's envelope is achieved for practical use.

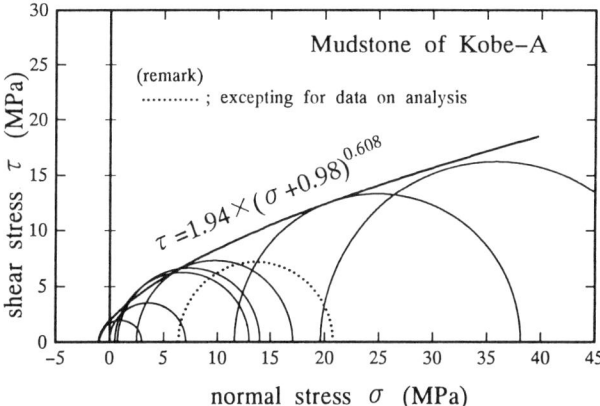

Figure 7. The example of Mohr's envelope obtained by the proposed method

7 CONCLUSION

The summary of results obtained by this study is described as follows.

(1) As concerns stress-strain curves, a slight difference is observed between sandstones and mudstones. But, judging from axial strain at peak strength or stress-strain curve's pattern, both rocks can be classified brittle materials.

(2) Although both sandstones and mudstones are brittle in the present, considering brittleness and the dependence of peak strength on the confining pressure, sandstones have strength characteristics both of brittleness and plasticity as granular materials.

(3) We applied the conventional models in order to determine Mohr's envelope to the laboratory test results. As a result, it seemed that they had both merits and demerits for the application. And we understood it was difficult to use them for our laboratory test results.

(4) We proposed a method consisting of two characteristic equations which could best model the laboratory test results of sandstones and mudstones. This two characteristic equations to determine failure criterion and Mohr's envelope are given by the following equations:

For failure criterion:

$$(\sigma_1 - \sigma_3)_{peak} = A \times (\sigma_3 + \sigma_t)^n + 4\sigma_t$$

For Mohr's envelope:

$$(\tau/\alpha) = (\sigma + \sigma_t)^m$$

(5) This proposed method improve the model by Hoek&Brown, and also a kind of the extended form of the conventional models used in this study.

(6) Applying the proposed method to both our laboratory test results and a result of other soft rocks reported by Yoshinaka and Yamabe(1980), we achieved the satisfactory modeling of Mohr's envelopes.

REFERENCES

K.Akai 1993. General aspects, Journal of Soils and Found., JSSMFE, No.429, pp.1-6, (in Japanese).

K.Akai,T.Adachi & K.Nishi 1978. Elasto-plastic behaviors of soft sedimentary rock (Porus tuff), Journal of Geotechnical Engineering, JSCE, No.271, pp.83-95, (in Japanese).

K.Akai,T.Adachi & K.Nishi 1979. Time dependent characteristics and constitutive equations of soft rock (Porus tuff), Journal of Geotechnical Engineering, JSCE, No.282, pp.75-96, (in Japanese).

Hobbs,D.W. 1966. A study of behavior of Broken Rock under Triaxial Compression and its Application to Mine Roadways, Int.J. Rock Mech.,Mining Sci., Vol.3, pp.11-14.

R.Yoshinaka & T.Yamabe 1980. Strength Criterion of Rocks,Soils and Found., JSSMFE, Vol.20, No.4, pp.113-126.

R.Yoshinaka & T.Yamabe 1981. A strength criterion rocks and rock masses, Proc. of the International Sympo. on Weak Rock,pp.613-618.

T.Adachi & T.Ogawa 1980. Mechanical Properties and Failure Criterion of soft sedimentary rock, Journal of Geotechnical Engineering, JSCE, No.295, pp.51-63, (in Japanese).

Griffith,A.A. 1924, The Theory of Rapture,Proc.,1st Int. Congr., Appl. Mech., pp.55-63.

Murrell,S.A.F. 1963. A Criterion for Brittle Fracture of Rocks and Concrete under Triaxial Stress and Effect of Pore Pressure on the Criterion, Proc., 5th Rock Mech.Sympo., pp.563-577.

Fairhurst,C. 1964. On the Validity of the 'Brazilian' Test for Brittle Materials,Int.J., Rock Mech. Mining Sci., Pergamon Press, Vol.1, pp.535-546.

Hoek,E. & Brown,E.T. 1980. Underground Excavations in Rock,Institution of Mining and Metallurgy, London.

Hoek,E.Wood,D. & Shah,S. 1992. A modified Hoek-Brown failure criterion for jointed rock masses, ISRM Sympo., Eurock'92, pp.209-214.

Hondros,G. 1959. Aust. J. appl. sci. 10, No.3, pp.243-268.

Obert,L. & Duvall,W. 1967, Rock Mechanics and The Design of Structures in Rock,John Wiley & Sons, Inc., pp.307-317.

Balmer,G. 1952. A General Analytical Solution for Mohr's Envelope, Amer. Soc. Testing Materials, Vol.52, pp.1260-1271.

Homogenization analysis of rock properties and rock bolts

Takashi Kyoya
Tohoku University, Japan

Naohiko Tokashiki
Ryukyu University, Japan

Toshikazu Kawamoto
Aichi Institute of Technology, Japan

ABSTRACT: The homogenization method is an effective mathematical tool to evaluate mechanical properties of materials which have complex geometry of the micro-structure. The essential viewpoint of the method is that such a material is regarded to have geometric periodicity of a characteristic micro-structure. In this paper, the homogenization method is briefly outlined, then two applications are presented: (1) quantitative evaluation of effect of pattern bolting installed into rock mass; and (2) analysis of mechanical properties of Ryukyu limestone which is a porous rock distributed in Okinawa, Japan.

1. INTRODUCTION

Rock bolts have been in use as effective method of reinforcement for a wide range of geotechnical construction works. However, it is hard to make quantitative evaluation of the effects of the rock bolts as to how much additional stiffness they introduce to rock mass.

Meanwhile, as for porous rocks and rock mass with distributed discontinuities it is difficult to evaluate their mechanical properties since their complex micro-structure.

The homogenization method is one of the mathematical methods which can evaluate the average property of a material with periodic micro-structure such as micro cracks or inclusions. The essential viewpoint required in such applications are that a porous rock and a rock mass reinforced by rock bolts are regarded as composite materials with geometric periodicity introduced by micro-cavities and rock bolts.

In this paper, after a brief introduction of the homogenization method is given, the change of elastic constants of the rock mass by installation of rock bolts is quantitatively evaluated and summarized in illustrative figures. And the next, mechanical properties of Ryukyu limestone which is a porous rock in Okinawa, Japan, is evaluated and the effect of porosity on the mechanical properties of porous rocks is discussed.

2. OUTLINE OF THE HOMOGENIZATION METHOD

Let us consider a body with periodic micro-structure as shown in Fig. 1, which is characterized by the domain Ω, fixed on the boundary $\partial\Omega_d$, subjected to body force f in Ω, traction t on the boundary $\partial\Omega_t$. The scale of an unit cell of the periodic structure is denoted by a small parameter ε.

The elasticity tensor and the body force are periodic functions which rapidly change in space with the small scale ε. Therefore,

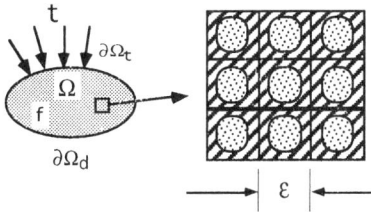

Fig. 1 Material body and its micro-structure

they are denoted by attaching superscript ε as $E^\varepsilon_{ijkl}(x)$ and by $f^\varepsilon_i(x)$. In this circumstances, displacement vector is also periodic function and is also denote by $u^\varepsilon_i(x)$.

Then, the variational form of the equilibrium equation is given by

$$\int_\Omega E^\varepsilon_{ijkl}(x)\frac{\partial u^\varepsilon_k(x)}{\partial x_l}\frac{\partial v^\varepsilon_i(x)}{\partial x_j}d\Omega \\ = \int_{\partial\Omega} t_i v^\varepsilon_i(x)dS + \int_\Omega f^\varepsilon_i(x)v^\varepsilon_i(x)d\Omega, \quad \forall v^\varepsilon_i(x) \quad (1)$$

Since the surface traction t_i is given on the boundary surface $\partial\Omega_t$ a priori, it is considered to be independent of small parameter ε. It is guaranteed that (1) has an unique solution if the elasticity tensor $E^\varepsilon_{ijkl}(x)$ is positive definite.

In order to handle such rapidly changing functions a new variables $y_i = x_i/\varepsilon$ which represents micro scale variable is introduced and the periodic functions are given in the forms:

$$E^\varepsilon_{ijkl}(x) = E_{ijkl}\left(x,\frac{x}{\varepsilon}\right) = E_{ijkl}(x,y)$$
$$f^\varepsilon_i(x) = f_i\left(x,\frac{x}{\varepsilon}\right) = f_i(x,y) \quad , \quad (2)$$
$$u^\varepsilon_i(x) = u_i\left(x,\frac{x}{\varepsilon}\right) = u_i(x,y)$$

The above functions are assumed to be Y-periodic (1-periodic) with respect to $y_i = x_i/\varepsilon$.

Let us search a displacement satisfying (1) in the following perturbed form:

$$u^\varepsilon_i(x) = u_i(x,y) = u^0_i(x) + \varepsilon u^1_i(x,y) + \varepsilon^2 u^2_i(x,y) + \cdots \quad (3)$$

This imply that an arbitrary admissible displacement for (1) is characterized by a small parameter ε as well as the material point x. The term $u^0_i(x)$ represent "averaged" displacement field since it depends only on the macro scale variable x and it remains in the limit $\varepsilon \to 0$. The terms $u^1_i(x,y)$, $u^2_i(x,y)$, $\cdots$ which vanish in the limiting process $\varepsilon \to 0$ represent perturbation of displacement field caused by micro structure.

Derivative of the above displacement is given by

$$\frac{\partial u_i^\varepsilon(x)}{\partial x_j} = \left(\frac{\partial u_i^0}{\partial x_j} + \frac{\partial u_i^1}{\partial y_j}\right) + \varepsilon\left(\frac{\partial u_i^1}{\partial x_j} + \frac{\partial u_i^2}{\partial y_j}\right) + O(\varepsilon^2) \qquad (4)$$

The averaged strain field is affected by the micro structure since the perturbation term $u_i^1(x,y)$ still remains in the limiting process $\varepsilon \to 0$.

Substituting the above expansions into (1) and taking the limit $\varepsilon \to 0$, we obtain the following two equations:

$$\int_\Omega \left[\frac{1}{|Y|}\int_Y E_{ijkl}(x,y)\frac{\partial u_k^1(x,y)}{\partial y_l}\frac{\partial v_i^1(x,y)}{\partial y_j}dY\right]d\Omega$$
$$= \int_\Omega \left[\frac{1}{|Y|}\int_Y \frac{\partial E_{ijkl}(x,y)}{\partial y_j}v_i^1(x,y)dY\right]\frac{\partial u_k^0(x)}{\partial x_l}d\Omega \qquad (5)$$
$$\text{for } \forall v_i^1(x,y)$$

$$\int_\Omega \left[\frac{1}{|Y|}\int_Y E_{ijkl}(x,y)\left(\frac{\partial u_k^0(x)}{\partial x_l} + \frac{\partial u_k^1(x,y)}{\partial y_l}\right)dY\right]\frac{\partial v_i^0(x)}{\partial x_j}d\Omega$$
$$= \int_\Omega \left[\frac{1}{|Y|}\int_Y f_i(x,y)dY\right]v_i^0(x)d\Omega + \int_{\partial\Omega}t_i v_i^0(x)dS \qquad (6)$$
$$\text{for } \forall v_i^0(x)$$

It is guaranteed that the following equation:

$$\int_Y E_{ijpq}(x,y)\frac{\partial \chi_p^{kl}(x,y)}{\partial y_q}\frac{\partial v_i(x,y)}{\partial y_j}dY$$
$$= \int_Y E_{ijkl}(x,y)\frac{\partial v_i(x,y)}{\partial y_j}dY \quad \text{for } \forall v_i(x,y) \qquad (7)$$

has an unique solution $\chi_p^{kl}(x,y)$ which is Y-periodic with respect to micro-scale variable y.. The vector function $\chi_p^{kl}(x,y)$ reflects the nature of the micro-structure of the unit cell, and is called "characteristic deformation".

Using the characteristic deformation, general form of the solution $u_i^1(x,y)$ for (5) is given by

$$u_i^1(x,y) = -\chi_i^{kl}(x,y)\frac{\partial u_k^0(x)}{\partial x_l} + \psi_i(x) \qquad (8)$$

in which $\psi_i(x)$ is an arbitrary vector depending only on the macro-scale variable x.

Substituting (8) into (6), we obtain the following macroscopic equilibrium equation on the domain Ω:

$$\int_\Omega E_{ijkl}^H \frac{\partial u_k^0(x)}{\partial x_l}\frac{\partial v_i^0(x)}{\partial x_j}d\Omega$$
$$= \int_\Omega f_i^H v_i^0(x)d\Omega + \int_{\partial\Omega}t_i v_i^0(x)dS \quad \text{for } \forall v_i^0(x) \qquad (9)$$

in which

$$E_{ijkl}^H = \frac{1}{|Y|}\int_Y\left(E_{ijkl}(x,y) - E_{ijpq}(x,y)\frac{\partial \chi_p^{kl}(x,y)}{\partial y_q}\right)dY \qquad (10)$$

and

$$f_i^H = \frac{1}{|Y|}\int_Y f_i(x,y)dY \qquad (11)$$

Comparing (9) with the original equation (1), it can be found that the elasticity tensor (10) and the body force (11) are employed instead of E_{ijkl}^ε and f_i^ε, respectively, the equation gives the averaged displacement field $u_k^0(x)$ of the material with periodic micro-structure. In this meaning, the tensor E_{ijkl}^H is called *the homogenized elasticity tensor* and the vector f_i^H is *the homogenized body force*.

Furthermore, we can approximate the displacements, the strains and the stresses by

$$u_i^\varepsilon(x) = u_i^0(x) - \varepsilon\chi_i^{kl}(x,y)\frac{\partial u_k^0(x)}{\partial x_l} + O(\varepsilon^2) \qquad (12)$$

$$e_{ij}^0(x,y) = \frac{1}{2}\left[\left(\delta_{ip}\delta_{jq} - \frac{\partial \chi_i^{pq}}{\partial y_j}\right) + \left(\delta_{jp}\delta_{iq} - \frac{\partial \chi_j^{pq}}{\partial y_i}\right)\right]\frac{\partial u_p^0}{\partial x_q} \qquad (13)$$

$$\sigma_{ij}^0(x,y) = E_{ijkl}^\varepsilon\left(\delta_{kp}\delta_{lq} - \frac{\partial \chi_k^{pq}}{\partial y_l}\right)\frac{\partial u_p^0}{\partial x_q} \qquad (14)$$

It should be noted here that the stress $\sigma_{ij}^0(x,y)$ given by (14) satisfies the equilibrium condition only in the averaged sense.

Concluding to say, the above homogenization method is summarized as follows: (i) solving (7) for a given unit cell of micro structure yields the characteristic deformation functions; (ii) the homogenized stiffness and the body force are obtained by (10) and (11), respectively; (iii) solving macroscopic boundary value problem (9) yields averaged displacement field; (iv) the displacements, the strains and the stresses of the material are approximated by (12), (13) and (14), respectively.

3. HOMOGENIZATION ANALYSIS OF ROCK BOLTS

Regarding the rock bolts regularly installed into rock mass as micro-structure of the mass, the stiffness can be calculated by the homogenization method. In this section increment of stiffness of rock mass added by rock bolts is quantitatively evaluated and summarized in illustrative figures.

The rock mass is assumed to be isotropic elastic material. In the analyses, Young's modulus of the rock mass is changed from 10,000 kgf/cm^2 to 100,000 kgf/cm^2 with appropriate intervals, while the Poisson's ratio is fixed at 0.25. As for a rock bolt, it is assumed that Young's modulus is 2,000,000 kgf/cm^2 and Poisson's ratio is 0.3, and that it have a diameter of 3cm. Spacing of installation is changed from 0.5m to 2.33m.

FE model of unit cell for calculation and the coordinate axes employed are presented in Fig. 2 and Fig. 3, respectively.

A rock mass becomes an anisotropic elastic medium by rock bolts, of which compliance matrix is given by

$$[C] = \begin{bmatrix} \frac{1}{E_x} & -\frac{\nu_{yx}}{E_x} & -\frac{\nu_{zx}}{E_x} & & & \\ -\frac{\nu_{xy}}{E_y} & \frac{1}{E_y} & -\frac{\nu_{zy}}{E_x} & & & \\ -\frac{\nu_{xz}}{E_x} & -\frac{\nu_{yz}}{E_x} & \frac{1}{E_z} & & & \\ & & & \frac{1}{G_{yz}} & & \\ & & & & \frac{1}{G_{zx}} & \\ & & & & & \frac{1}{G_{xy}} \end{bmatrix}$$

Calculated values of the coefficients in the above matrix are summarized in Table 1 and illustratively plotted in figures (Fig. 4 to Fig. 9). In the legends, E100 denotes that original Young's modulus of rock mass is to be 100,000 kgf/cm^2, and E50 to be E = 50,000 kgf/cm^2, and so on.

All coefficients rapidly changed when the spacing of bolts decreases from 1.0m to 0.5m. The stiffness in longitudinal direction E_x takes almost constant value for the same bolt spacing, while other coefficients changes their values depending on original stiffness of rock mass.

It is also found that the stiffness in longitudinal direction E_x can be well approximated by the one dimensional equation $E_x = (E_s A_s + E_r A_r)/(A_s + A_r)$ where subscripts s and r denote steel and rock, respectively, and A denotes the area occupied by the material in a cross section.

The change of Poisson's ratio is quite interesting. As the bolt spacing decreases ν_{yz} increases, while ν_{zx} and ν_{xy} decreases. It results from interaction between bolts($\nu = 0.3$) and rock mass ($\nu = 0.25$).

These result can be easily utilized in analyses of rock mass structures reinforced by rock bolts. For example, let us consider a circular tunnel reinforced by pattern bolts as shown in Fig. 10. In this case, the spacing of bolts gradually changes along radial direction. In the modeling, dividing the rock mass around the tunnel into some parts as shown in the figure, representative spacing of bolts can be chosen and the corresponding elastic constants can be determined from the curves. Therefore, the circular tunnel can be modeled by thick tube in which elastic constants gradually change.

Table 1 Change of mechanical parameters of rock mass by pattern rock bolting

Rock Mass E(kgf/cm²) & ν=0.25	Bolt Spacing (m)	Ex (kgf/cm²)	Ey, Ez (kgf/cm²)	Gyz (kgf/cm²)
E=100,000	0.5	103,640	100,490	40,120
G= 40,000	1.0	100,910	100,130	40,033
	2.1	100,210	100,030	40,008
E= 50,000	0.5	53,732	50,361	20,063
G= 20,000	1.0	50,933	50,096	20,017
	2.1	50,212	50,023	20,004
E= 20,000	0.5	23,789	20,261	8,026
G= 8,000	1.0	20,947	20,073	8,007
	2.1	20,215	20,017	8,002
E= 10,000	0.5	13,808	10,206	4,013
G= 4,000	1.0	10,952	10,063	4,004
	2.1	10,216	10,015	4,001

Rock Mass E(kgf/cm²) & ν=0.25	Bolt Spacing (m)	Gzx, Gxy (kgf/cm²)	νyz	νzx, νxy
E=100,000	0.5	40,147	0.252	0.243
G= 40,000	1.0	40,039	0.251	0.248
	2.1	40,010	0.250	0.250
E= 50,000	0.5	20,078	0.255	0.234
G= 20,000	1.0	20,021	0.251	0.246
	2.1	20,005	0.250	0.249
E= 20,000	0.5	8,032	0.262	0.213
G= 8,000	1.0	8,009	0.253	0.240
	2.1	8,002	0.251	0.248
E= 10,000	0.5	4,016	0.272	0.185
G= 4,000	1.0	4,006	0.257	0.230
	2.1	4,001	0.252	0.245

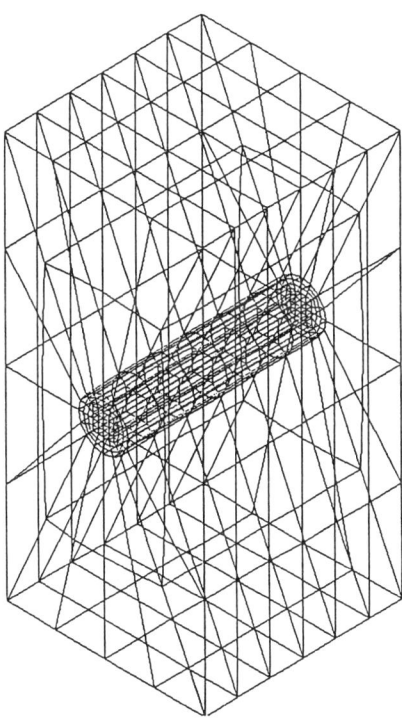

Fig. 2 FE model of an unit cell for homogenization analysis of rock bolts

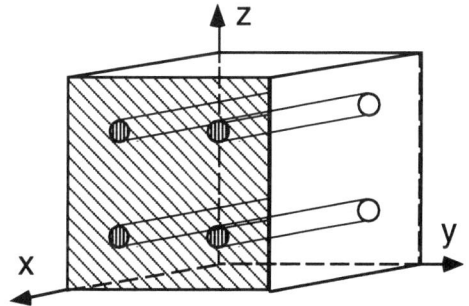

Fig. 3 Coordinate axes for rock mass with rock bolts

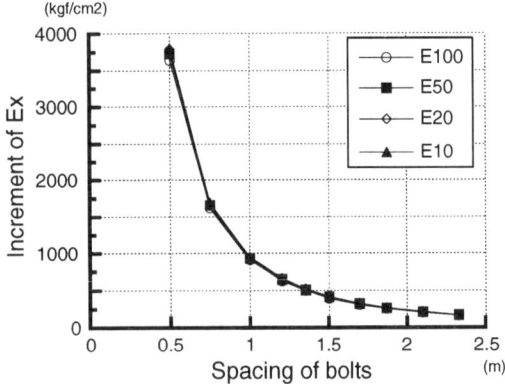

Fig. 4 Increment of the stiffness E_x

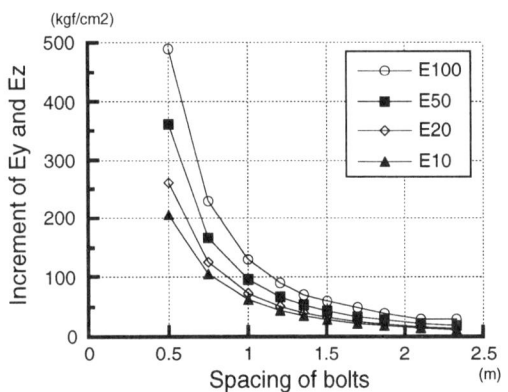

Fig. 5 Increment of the stiffness E_y and E_z

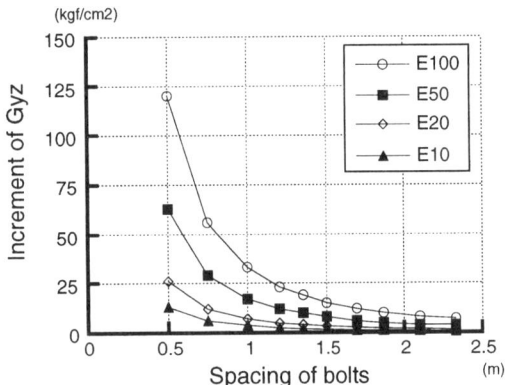

Fig. 6 Increment of the shearing stiffness G_{yz}

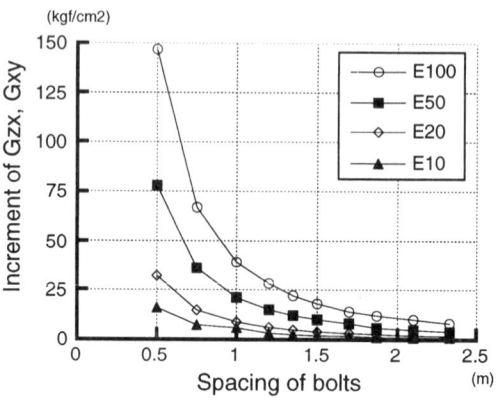

Fig. 7 Increment of the shearing stiffness G_{zx} and G_{xy}

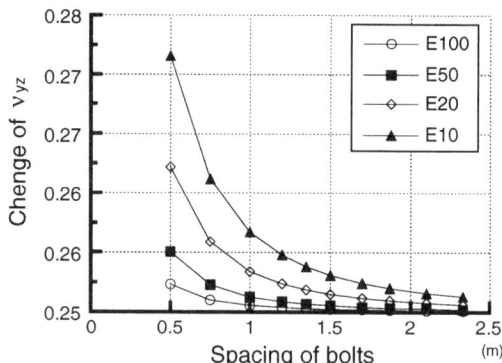

Fig. 8 Change of the Poisson's ratio ν_{yz}

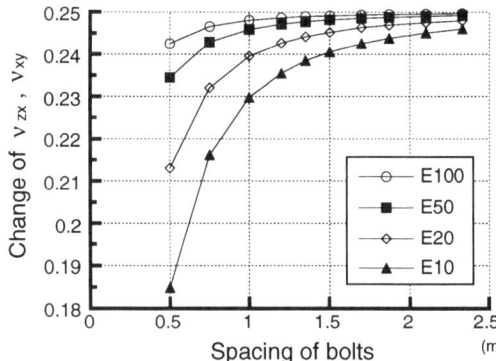

Fig. 9 Change of the Poisson's ratios ν_{zx} and ν_{xy}

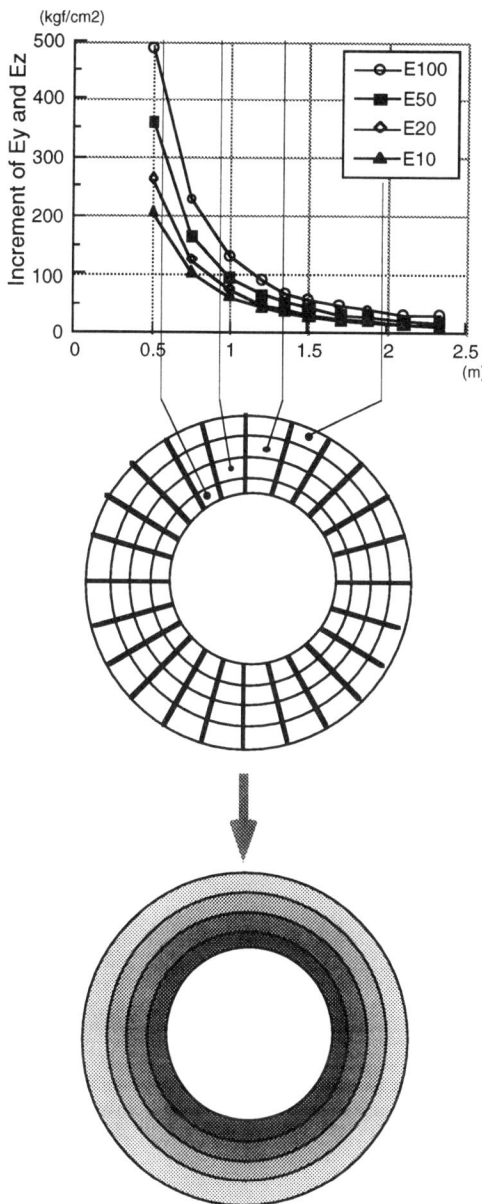

Fig. 10 Modeling of a circular tunnel reinforced by rock bolts

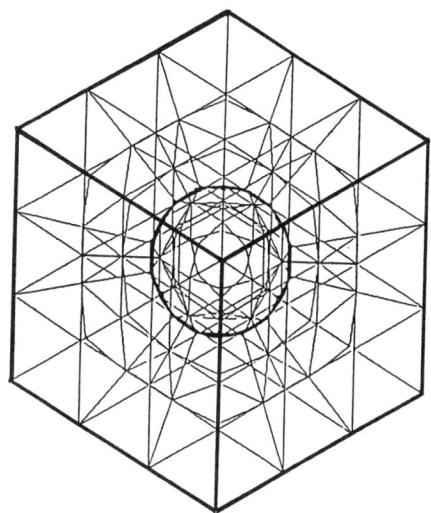

Fig. 11 FE model for a unit cell of Ryukyu limestone

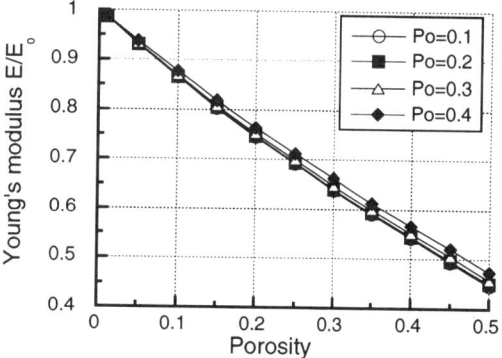

Fig. 12 Relationship between normalized Young's modulus and porosity

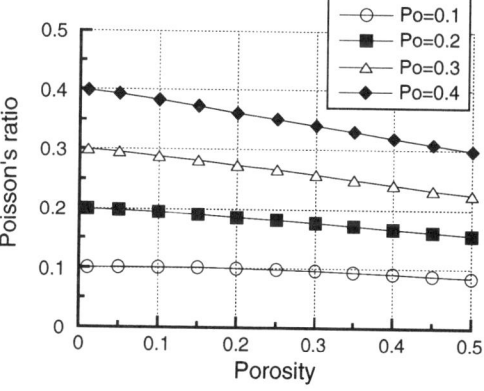

Fig. 13 Relationship between Poisson's ratio and porosity

4. HOMOGENIZATION ANALYSIS OF MECHANICAL PROPERTIES OF RYUKYU LIMESTONE

Ryukyu limestone is a porous soft rock in Japan which has a number of spherical or ellipsoidal pores. The diameter of the pores are distributed within a range from 1.8 mm to 2.5 mm, and the porosity (volumetric fraction of pores) varies in a range from 5% to 15% in the cylindrical specimens with a diameter of 5 cm and a height of 10 cm.

The Ryukyu limestone can be idealized as an assembly of a cubic unit cell which has a spherical pore in its center as shown in Fig. 11.

The matrix of the limestone is assumed to be an isotropic elastic material of which Young's modulus denoted by E_0 is 4.5×10^5 kgf/cm^2. Poisson's ratio of the matrix which is denoted by "Po" in the figure is changed from 0.1 to 0.4.

Calculated relationships between macroscopic elastic constants (Young's modulus E and Poison's ratio ν) and porosity are plotted in Fig. 12 and Fig. 13. In Fig. 12, the calculated macroscopic Young's modulus is normalized by stiffness of the matrix E_0.

The macroscopic Young's modulus and Poison's ratio decrease almost linearly as porosity increases. The rate of the change of Poisson's ratio depends on those of the matrix itself, while Young's modulus is not so affected by the matrix's Poisson's ratio. As for Ryukyu limestone, since its porosity distributes within 5% to 15% it is found that macroscopic Poisson's ratio is almost equal to that of the matrix, while macroscopic Young's modulus is strongly affected by porosity.

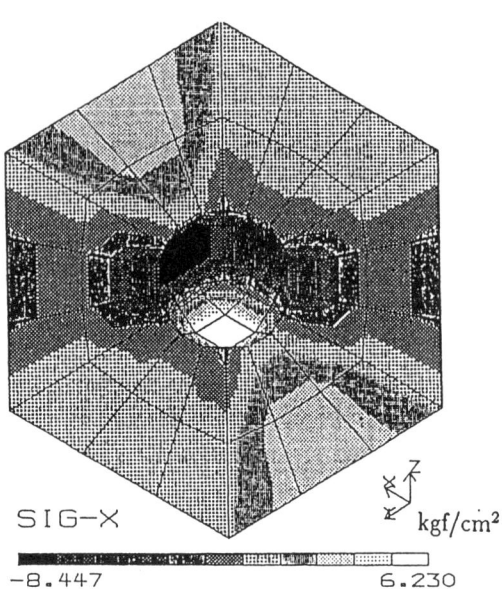

Fig. 14 Microscopic stress distribution in a unit cell

Table 2 Experimented strengths of specimens and calculated maximum microscopic tensile stress around a pore

Specimen No. (Ryukyu limestone)	Yield strength (kgf/cm^2)	Max. micro tensile stress (kgf/cm^2)		
		5%	10%	15%
B1	324	6.7	9.7	11.6
B2	409	8.5	12.2	14.7
B3	370	7.7	11.0	13.3
B4	300	6.2	9.0	10.8
B5	380	7.9	11.3	13.6
B6	280	5.8	8.4	10.0
B7	270	5.6	8.1	9.7
B8	460	9.6	13.7	16.5

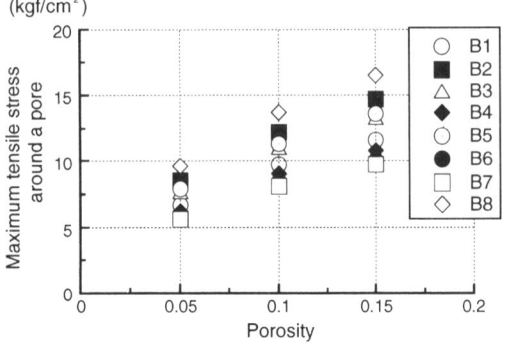

Fig. 15 Maximum microscopic tensile stress around a pore

A series of uniaxial compression tests were carried out on cylindrical specimens of Ryukyu limestone and various yield strengths of them were obtained. Meanwhile, FE analyses for each specimen at the yielding uniaxial load were carried out, then, microscopic stress in an unit cell were calculated by POSTMAT for three kinds of porosity of 5%, 1% and 15%. A calculated microscopic stress state in an unit cell which locates near the lateral surface at middle height of the specimen is presented in Fig. 14.

The yield strengths obtained by uniaxial compression tests and calculated maximum microscopic tensile stress around pores for various porosity is summarized in Table 2 and are plotted in Fig. 15. It is quite interesting that the strength of the Ryukyu limestone obtained by experiments scatter in a wide range from 270 to 460 kgf/cm^2, however, calculated maximum microscopic tensile stress varies in a small range from about 6 to 17 kgf/cm^2. This fact imply that the strengths of porous rocks are strongly affected by their micro-structure formed by pores.

CONCLUSION

In this paper, the homogenization method is applied to quantitative evaluation of the effect of pattern bolting. Amount of stiffness increment of rock mass by rock bolts are calculated and presented in illustrative figures.

The homogenization method is also applied to analysis of mechanical properties of a porous rock, Ryukyu limestone. Comparison of the experimental data and calculated results imply that the strengths of porous rocks are strongly affected by their micro-structure formed by pores.

These application imply that the homogenization method will be a powerful tool to analyze mechanical properties of complex rocks and rock mass.

REFERENCES

[1] E. Sanchez-Palencia, Nonhomogeneous media and vibration theory, Lecture Note in Physics, No.127, Springer, Berlin (1980)
[2] J. L. Lions, Some methods in the mathematical analyses of systems and their control, Science Press, Beijing (1981)
[3] N. Bakhvalov and G. Panasenko, Homogenization: Averaging processes in periodic media, Kluwer Academic Pub. (1984)
[4] J. M. Guedes and N. Kikuchi, Preprocessing and postprocessing for materials based on the homogenization method with adaptive finite element methods, Computer Methods in Applied Mechanics and Engineering, 83, pp. 143-198 (1990)
[5] T. Kyoya, An application of the homogenization method to beam-like structures, Proc. 5th Symp. Computational Mechanics, Japan Association of Science and Technology, pp. 249-256 (1991). (in Japanese)
[6] T. Kyoya and N. Tokashiki, Evaluation of mechanical properties of porous Ryukyu limestone by the homogenization method, Proc. 2nd Workshop on Characterization of Subsurface Cracks, Mining and Mineral Processing Institute of Japan, pp. 92-97, (1992). (in Japanese)
[7] T. Kyoya and T. Kawamoto, Quantitative evaluation of pattern bolting by the homogenization method, Proc. Int. Symp. Assessment and Prevention of Failure Phenomena in Rock Engineering, pp. 641 - 646, Balkema (1993)
[8] T. Kawamoto and T. Kyoya, Some applications of homogenization method in rock mechanics, Proc. Seminar on Impact of Computational Mechanics to Engineering Problems, Sydney (1993).

ns on pressuremeter tests conducted
in rock ground

Kazuo Tani, Toshiro Okamoto & Koichi Nishi
Central Research Institute of Electric Power Industry, Chiba, Japan

ABSTRACT: A pressuremeter test has become increasing popular as an effective technique for geotechnical site investigation in order to evaluate mechanical properties of ground at depths. For ordinary pressuremeter analyses, the tested ground is not only presumed to be uniform, but the mode of deformation is also assumed to be axisymmetric under a plane strain condition. Rock masses, however, are jointed in most cases not necessarily assuring a uniform deformation. Moreover, even in a least jointed sedimentary soft rock, large numbers of evidences have been reported of developing tensile cracks during pressuremeter expansion tests.

CRIEPI has recently developed a self-boring pressuremeter probe; a novel technique was introduced for measurement of borehole wall displacements, which makes it possible to monitor the deformation pattern of the expanding borehole. Typical results are presented herein demonstrating the influence of such mechanical discontinuities as pre-existent joints and induced tensile cracks. When tests are conducted in jointed rock ground, the expansion of the tested borehole may happen to become extremely non-uniform from the very beginning of the tests. On the other hand, when tests are conducted in fairly uniform sedimentary soft rock ground, a cylindrical expansion of the circular borehole in early stages of the test suddenly turns into an elliptic shape. This may possibly suggest that two tensile cracks have developed radially in opposite directions.

Since ordinary pressuremeter analyses assume an axisymmetric mode of deformation, it should be borne in mind that a uniform expansion must be guaranteed if these analytical modeling is to be justified.

1 INTRODUCTION

A pressuremeter test, a uniform pressure being applied to a surface of a drilled borehole of a finite length, has become increasingly popular these days. It is generally thought to be an effective technique for geotechnical site investigation especially to characterize the ground at depths. Because of presumably one-dimensional nature of the problem with a rather straightforward boundary condition, this test is considered as one of the simplest field tests for theoretical back-calculation of material properties of interest (Wroth, 1984). Recognized as a boundary value problem, the test analysis requires an appropriate modeling supposed to be consistent to the reality of the test situation. Almost all the conventional interpretations of the pressuremeter problem assume a cylindrical expansion of a circular borehole of an infinite length, i.e. an axisymmetric mode of deformation under a plane strain condition. This point might have been taken for granted as a paramount prerequisite, and have seldom, if ever, been examined carefully.

Rock masses, however, are jointed in most cases; hence they are regarded as discontinua typically behaving as a non-uniform anisotropic material. As a consequence, a uniform expansion of a tested borehole is not necessarily guaranteed even on application of a uniform pressure.

With present knowledge of technology, it appears to be unpracticable for various reasons to establish an appropriate analytical model for an individual test location taking into account the influence of all the relevant mechanical discontinuities. Firstly, there is no effective method, from technical and cost-effective points of view, to identify all the significant discontinuities in the ground in question. Secondly, modeling such mechanical discontinuities in rock masses is certainly a daunting task with little information available at an early stage of the site investigation where pressuremeter tests are normally carried out. It is therefore common practice to discard the idea of modeling each discontinuity in the rock, but instead the jointed rock mass is regard as a quasi-continuum material in order to evaluate the equivalent continuous properties. However, it should be borne in mind that, from the underlying principle as mentioned above, the deduced material properties from the tests are only representative of rock mass deformation of more or less continuous nature. This implies that the pressuremeter expansion should not involve any significant movements of the relevant rock along particular discontinuities. With due consideration, it is of great importance to ensure that the tests are conducted in such a way as not deviating excessively from the model presumed for the analyses. In practice, it appears that little attention has been paid in this respect.

Even in least jointed sedimentary soft rocks, a number of evidences have been reported of developing tensile cracks during pressuremeter expansion tests. Haberfield & Johnston (1989) conducted a series of model pressuremeter tests in a modified triaxial cell using cylindrical specimens of a cemented, consolidated mudstone powder. Tani et al. (1993) also carried out similar model tests using specimens of a cemented, consolidated mortar; an attempt was made to simulate a self-boring practice employing a real pressuremeter of a reduced size. Since these artificial materials are homogeneous and continuous in nature, the tests are ideal to validate the analytical model developed for a uniform rock mass; otherwise, using natural soft rocks for the specimens, the test result will be undoubtedly affected by possible inclusion of various mechanical inhomogeneity.

Both studies reported that, in most cases, the specimens' failures were observed with tensile cracks propagating in radial directions. In general, two or three major cracks were spaced evenly around the cavity, approximately 180° or 120° apart respectively. For the case of two tensile cracks, secondary major cracking was occasionally observed to take place in the orthogonal directions to the previously developed ones. This resulted in four major tensile cracks propagating crosswise from the borehole. Between these major tensile cracks, numerous numbers of minor cracks, also developed in radial directions, were sometimes detected close to the cavity wall.

Once such tensile cracks start to develop when the circumferential stress of the ground adjacent to the cavity wall reaches the tensile strength of the material, the mode of deformation cannot be axisymmetric any longer. As a consequence, it may be concluded that application of the ordinary pressuremeter analyses should be restricted to the results obtained not later than the moment for tensile failure to take place at the cavity wall. It becomes, therefore, of categorical importance to identify this limiting state in the test, otherwise misleading material properties will be deduced from the results.

There are several research works carried out in order to investigate the influence of the induced tensile cracks on pressuremeter test results. Typical examples of physical model studies are mentioned above and some attempts are also made of numerical modeling by finite element methods (e.g. Haberfield & Johnston, 1990). Few results, however, have been reported so far of the test data obtained in the field to verify their findings.

Central Research Institute of Electric Power Industry (CRIEPI) has recently developed a self-boring pressuremeter probe. Its remarkable feature is a new measuring system of borehole wall displacements, which makes it possible to monitor the deformation pattern of the expanding borehole. Typical test results are presented herein demonstrating the influence of such mechanical discontinuities as pre-existent joints and induced tensile cracks. In the end, the standard practice of pressuremeter tests at CRIEPI is described designed to cope with this problem.

2 PRESSUREMETER PROBE

Figure 1 illustrates a self-boring pressuremeter probe developed at CRIEPI. This is specifically designed for tests in rock ground. The expanding portion of the instrument is 140mm in diameter and 1120mm long. Through a 2mm thick rubber membrane, it can apply up to 40MPa of pressure to the borehole wall.

The strong point of this apparatus is a novel technique introduced for measurement of borehole wall displacements. A total of eight proximeter transducers, mounted 45° apart in the middle section of the pressurizing length, can monitor the movements of the outermost surface elements of the probe, i.e. metal sheaths which are pressed against the borehole wall (see Figure 2). The measured results are presumed to be most representative of the borehole wall response; thereby no correction is required for errors of various sources.

Besides improved accuracy of cavity strain evaluation, detailed examination of each individual transducer's record provides additional information concerning what is going on around the expanding borehole. This would allow still more sophisticated interpretation of the test. For instance, if a uniform deformation takes place around the cavity, the measured displacements must be identical in all directions. On the other hand, if the influence of pre-existent joints or induced tensile cracks becomes significant, the displacement pattern may be found to be very irregular or non-symmetric.

The detailed description about the apparatus can be found in Tani et al. (1995a & 1995b).

3 TEST IN SEDIMENTARY SOFT ROCK

3.1 Site specification

The test site is covered by about 10m thick Quaternary terrace deposit of silt and sand which is underlain by Pliocene deposits. This thick formation is mostly comprised of uniform soft rock with sporadically interbedded tuffaceous fine sand or mud layers which are lightly cemented.

The total unit weight $\gamma_t=15.5\sim17.0 kN/m^3$ and the water content $w=45\sim60\%$ are rather constant within the depths of 20~50m. The in-situ elastic wave velocities $V_s=450\sim600 m/sec$ and $V_p=1400\sim1800 m/sec$ were measured by the down-hole method (geophysical borehole logging). Moreover, the undrained shear strength $\tau_f=1.5\sim2.2 MPa$ was obtained by triaxial compression tests conducted under effective confining pressures equivalent to the in-situ effective overburden stresses.

3.2 Pressuremeter test

Pressuremeter tests were conducted in the depths of 20~50m at intervals of 5m. A typical result of the test carried out at the depth of 29.0m is shown in Figure 3(a) in the form of a pressuremeter curve, i.e. a pressure~cavity strain relation. The cavity strain ε_c denotes the average displacements measured by the eight proximately transducers divided by the initial borehole diameter. For this test, since the pressuremeter probe was inserted into the ground by a self-boring technique, the diameter of the probe itself is taken as the initial borehole diameter. Measured pressure is corrected for the effect of membrane rigidity to calculate the pressure p applied to the surface of the borehole.

Two small unload-reload cycles of different amplitudes $\Delta p=0.3$,

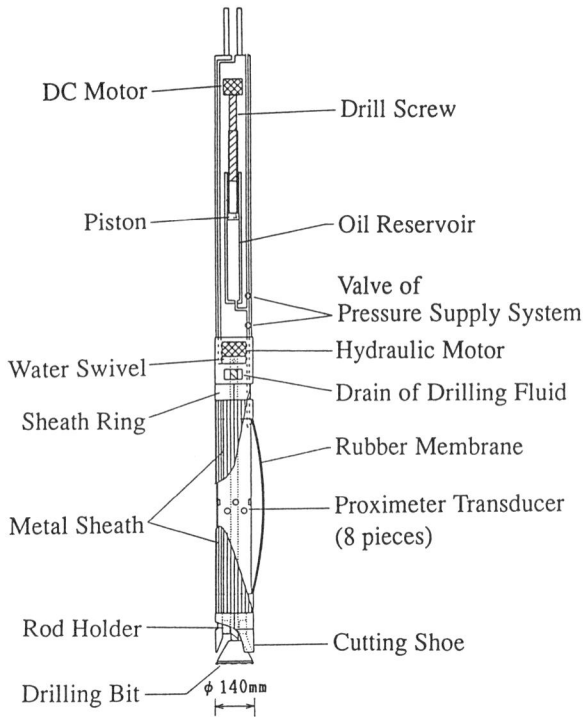

Figure 1. CRIEPI self-boring pressuremeter

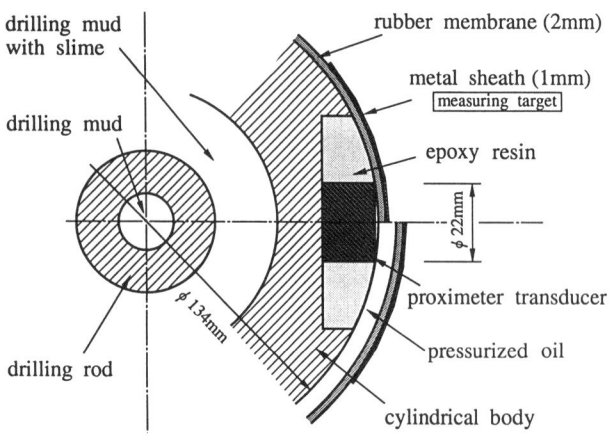

Figure 2. Measuring system using proximeter transducers

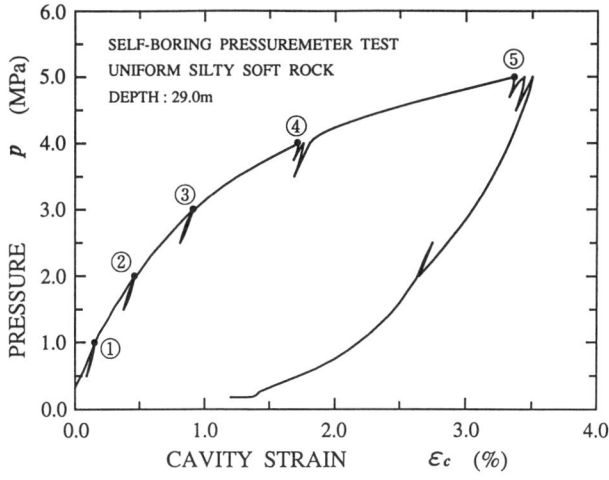
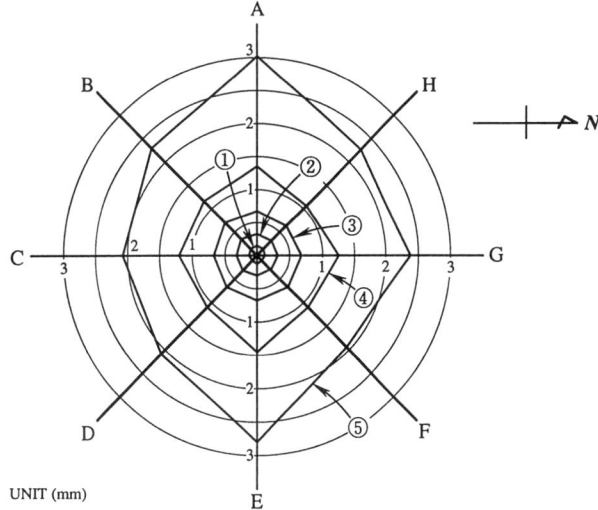

Figure 3. Typical example of a pressuremeter test conducted in uniform soft rock ground
(a) Pressuremeter curve   (b) Displacement patterns of the expanding borehole

0.5MPa were included at every 1MPa, shown by ①~⑤ in the figure. At the first three stages ①~③ ($p \leq 3$MPa), the unload-reload loops demonstrate rather recoverable elastic behavior. On the contrary, at the later stages ④&⑤ ($p \geq 4$MPa), the pressure~strain curve would not form a closed loop any longer leaving irrecoverable plastic deformation during the course of each cycle.

Figure 3(b) illustrates the displacement patterns of the borehole wall measured by the proximeter system immediately before each parcel of two unload-reload cycles. It is noted that, while the borehole wall displacements in all eight directions were fairly uniform at the first three stages ①~③ ($p \leq 3$MPa), the displacements in the east and the west directions (denoted as E and A respectively in the figure) were prominent at the later stages ④&⑤ ($p \geq 4$MPa). In other words, the shape of the borehole was kept circular in the beginning up to the threshold pressure somewhere between 3MPa and 4MPa, but later at the higher pressures, it turned into an elliptic one. This transfer of the deformation mode from the axis- to the plane-symmetry is probably attributed to the two radial tensile cracks developed from the borehole wall to opposite radial directions, which form the plane of symmetry (Tani et al., 1993).

The following is a possible interpretation of this particular test. As the pressure applied on the borehole wall increases, the circumferential stress at the cavity wall tends to decrease. When unload-reload cycles are included before this stress reaches the tensile strength of the ground material, the resultant pressure~strain relations at the cavity wall will become hysteretic curves (i.e. closed loops). If the ground is homogeneous and isotropic as expected for fresh sedimentary soft rocks, the borehole expansion should be axisymmetric.

Once tensile failure, however, takes place at the cavity wall, two or three tensile cracks start to develop in the radial directions at equal spacing of 180° or 120° respectively. It should be noted that, unlike a uniformly distributed plastic zone in all directions for the case of shear failure, propagation of tensile failure is directional. With this change in the mode of deformation, the borehole expansion is not axisymmetric any longer. If the number of the principal major cracks happens to be two, the borehole expansion should be planisymmetric; hence the displacement pattern can be elliptic. When unload-reload cycles are included at this stage, the resultant pressure~strain relations at the cavity wall may not form a closed loop but exhibit accumulating plastic deformation. The reason for this is because a significant area of the ground adjacent to the mouth of the tensile cracks may have undergone considerable deformation (Haberfield & Johnston,

1989). Some other interpretations include creep deformation as well as the effect of softening and/or weakening after failure being observed.

It must be clearly stated here that, the above-mentioned interpretation does not always hold and is true for this particular case shown in Figure 3. In many other tests, interpretations of the results can be rather complicated but not too confusing as compared to the tests in jointed rocks. When the pressure~strain relation obtained by each individual proximeter transducer is to be examined carefully, one may often encounter some peculiar records. For tests conducted in fairly uniform soft rocks, data obtained by the majority of the eight proximeter transducers usually demonstrate continuous expansion for monotonically increased pressure, i.e. smooth pressure~strain curves are obtained. A few of them, however, may indicate a discontinuous response, mostly a leap in expanding displacements at some applied pressures; as a result, pressure~strain curves become broken. And surprisingly, even a rebounding response may be recorded, although the average record of all transducers still indicates steady expansion.

On the basis of the above discussion, it may be justified to conclude that such a discontinuous response is a kind of tell-tale signs suggesting the very moment of tensile failure at the cavity wall. Since this peculiar event is usually observed to take place in a moment involving considerable deformation in the ground, the tensile failure at the cavity surface should bring about cracking of a considerable length rather than that of a minute length. It is because the cracks tend to develop instantly until the mechanical system would achieve a new equilibrium state. Then, these incipient cracks will propagate gradually into the ground with subsequent increase in applied pressure. This vital information about crack development might have been overlooked without direct point measurements of radial displacements in several directions.

As already described, the circular shape of the borehole while expanding uniformly at low pressure level may be transformed into an irregular one after discontinuous response being recorded by some transducers. This change, however, may not be always so appreciable as demonstrated in Figure 3(b). This is partly due to the fact that the principal major crack pattern is not necessarily bisectional but can be trisectional, and also due to the fact that, even for the case of bisectional crack patterns, subsequent cracking may develop a crosswise crack pattern. With a limited number of the transducers, such multi-directionally protrusive patters of borehole deformation cannot be perceived accurately. Furthermore, the measured record of each transducer is representative of the average movements of a few strips of metal

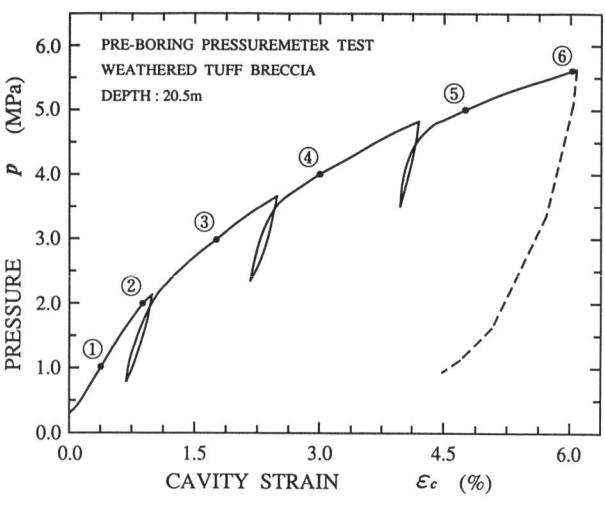 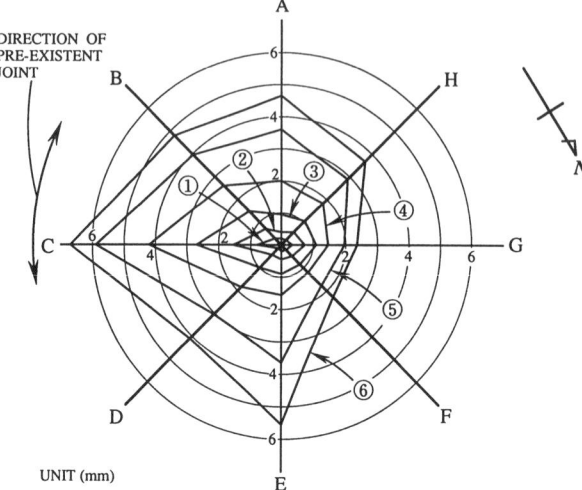

Figure 4. Typical example of a pressuremeter test conducted in weathered rock ground
(a) Pressuremeter curve   (b) Displacement patterns of the expanding borehole

sheathes immediately in front of the sensor's head. The size of the measuring target is of the order of tens of millimeters which are considerably larger than the most active area of the crack development. Hence, to some extent, very localized deformation could be masked or even missed with this system of the point measurements.

4 TEST IN WEATHERED TUFF BRECCIA

4.1 Site specification

The test site is covered by a few meters of peaty soil and a diluvial silt layer to the depth of 4.4m. The underlying layers are thick Miocene deposits of tuff breccia which are mostly comprised of subangular andesite fragments not larger than 20mm but occasionally include larger ones over 50mm. Within this rock formation below 4.4m, the upper layers, down to the depth of 18m, are found to be heavily weathered. Layers in the shallow depths, especially to 14m, are significantly decomposed and includes decayed fragments. Through the alteration zones around the depths of 18~22m, the weathering effect fades away to observe fresh host rock at deeper layers. Moreover, a thick fracture zone is encountered at the depth of 26~27m.

Within the depths of 13~35m, while the unit weight γ_t=19.4~22.3kN/m^3 increases, the water content w=16~29% decreases gradually with depth reflecting the weathering effect. The in-situ elastic wave velocities V_s=710~1130m/sec and V_p=1960~2600m/sec were measured by the suspension method (geophysical borehole logging). Undrained shear strength τ_f=1.3~3.0MPa was obtained by triaxial compression tests conducted under effective confining pressures equivalent to the in-situ effective overburden stresses.

4.2 Pressuremeter test

The pressuremeter tests were conducted in the depths of 13~35m at intervals of approximately 3m. A typical result of the test carried out at the depth of 20.5m is shows in Figure 4(a) in the form of a pressuremeter curve. For this test, the pressuremeter probe was inserted into the ground by a pre-boring technique. The initial borehole diameter for calculating the cavity strain ε_c, therefore, is assumed to be the diameter of the core barrel used to prepare the borehole. The drilling bit and cutting shoe are the largest parts whose nominal diameter is 0.5mm greater than that of the pressuremeter probe itself.

This particular zone of interest is in the layer of lightly altered tuff breccia which is abundant in mostly subangular andesite fragments of 10~60mm in diameter. Huge subrounded fragments whose diameters larger than some 100mm are occasionally encountered. Softening due to the influence of alteration and weathering is probably responsible for relatively low values of the in-situ elastic wave velocities V_s=710~730m/sec and V_p=1960~1990m/sec measured by the suspension method.

For this particular test, the pressuremeter probe was inserted twice into the ground; the first time for drilling the borehole and the second time for the pressuremeter expansion test, i.e. the pre-boring technique. This is to examine the surface of the drilled borehole by a borehole television. On completion of the drilling, the borehole was washed with clean water to improve the visibility. Then a TV camera was lowered in the borehole down to the test location.

Figure 5 schematically illustrates the borehole surface developed at the south direction. Within the tested length of 1.12m (between 19.94m and 21.06m in depth), five joints are identified to be spaced out 0.2~0.3m. Only one of them is an open joint located in the middle of the tested length. The visible aperture is 1mm wide, but no filling materials are observed in the gap. The rest are closed or with fine apertures.

There is another discontinuity in the top with a relatively wide aperture of 2mm. This encircles an extraordinarily large fragment greater than some 150mm.

The orientation of discontinuities relative to the pressuremeter axis largely is known to control the pressuremeter expansion tests. If the pressuremeter axis is perpendicular to a discontinuity, there will be little influence as all the deformation tends to take place in the plane parallel to the relevant discontinuity. However, such oblique discontinuities as intersecting the pressuremeter axis at acute angles will definitely alter the problem to a significant extent from the one assuming the ground as a uniform continuum. For the case of a vertical borehole as this study, discontinuities with greater dips are of serious matter with this respect.

As far as those outcropping discontinuities on the tested borehole surface are concerned, all the joints are observed to decline with moderate dips ranging from 25° to 50° as shown in Figure 5. The only exception is the steepest discontinuity in the top with a dip of 74°. It is, however, located at, or nearly off, the end of the zone in question with only a third of its intersection observed on the borehole surface below the top end of the tested length. Furthermore, this probably envelops a large fragment rather than extends laterally as a plane to a considerable distance.

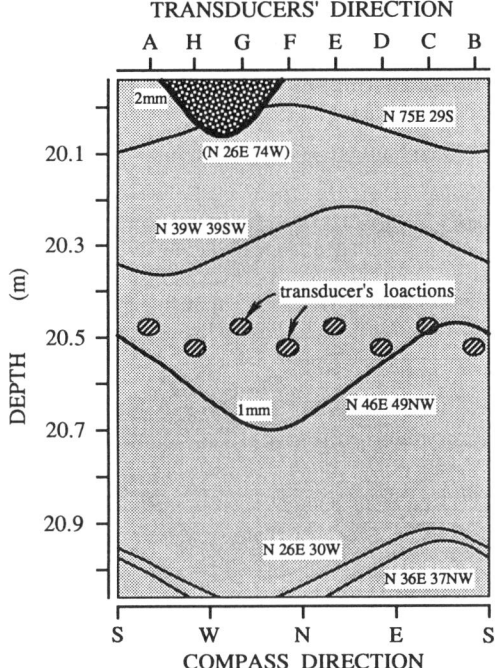

Figure 5. Borehole TV observation of the prepared borehole immediately before the test shown in Figure 4

Consequently, one might expect that the general feature of the problem may not deviate too much from a uniform expansion of a cylindrical cavity in a pseudo-continuum material.

As shown in Figure 4(a), a small unload-reload cycle of an amplitude $\Delta p=1.0$MPa was included at every 1.5MPa. Since the test was accidentally terminated with a membrane puncture at the maximum pressure of 5.63MPa, these pressure cycles were made at no more than three stages. The pressure~strain curves during these pressure cycles are observed to form closed loops at all three stages with a tendency of higher stiffness in the unloading phase and more marked non-linearity in the reloading phase for the cycles included at higher pressure levels.

Figure 4(b) illustrates the displacement patterns of the borehole wall measured by the proximeter system at different pressure levels indicated by the encircled numbers ①~⑥. Unlike the tests conducted in a uniform soft rock as shown in Figure 3(b), the expansion of the borehole is extremely non-uniform from the very beginning of the test. The displacement in the southeast direction (denoted as C in the figure) is found to outdistance those in the other directions to a considerable extent. Although this tendency may fades away with increasing pressure, the ratio of maximum to minimum displacements at the highest pressure level ⑥ is still around three.

It is of great interest to note that, with the information obtained by the borehole TV inspection (see Figure 5), the particular transducer which was oriented to the projecting direction of the displacement pattern was located just on the discontinuity with an aperture of 1mm. Local failure may have possibly taken place around the mouth of the open joint even at low pressures due to stress concentration. This probably led to highly localized deformation around the tested borehole, which is thought to be responsible for the greatly non-uniform displacement patterns measured even in early stages of the test. As higher pressures being applied on the borehole wall, more and more volumes of the rock masses tend to get involved in the test. As a result, this initially localized deformation may become less pronounced in due course; thereby the measured displacement pattern on the boundary may be rectified to a uniform one.

After the pressuremeter probe was retrieved, the tested borehole was washed again with clean water. Then, the TV camera was lowered down to the tested location in order to inspect the aftermath of the test brought to an end by a membrane puncture. A good imprint of the metal sheaths was observed as many straight lines parallel to the borehole axis at nearly equal spacing. The previously open joint in the middle (N 46E 49NW) was found to have closed or filled with pieces of broken fragments. This is a clear evidence of local failure around the mouth of the pre-existent open joint.

As far as the general features of the influence of pre-existent joints are concerned, the above-mentioned interpretation is true for most cases. This is because a strip of the metal sheaths can easily span over the mouths of open discontinuities, unless their outcrops are oriented more or less parallel to the pressuremeter axis and their apertures are sufficiently large for a strip of metal sheaths (12mm wide) to squeeze into the aperture. However, there is enough evidence to prove that, in some cases, the non-uniform deformation around the borehole can be further encouraged with increasing pressure applied to the borehole. If the gaps within the metal sheaths becomes so wide as allows the rubber membrane to protrude into the discontinuities, the wedging action is likely to take place. This would certainly promote the discontinuities to open leading to still more non-uniform deformation around the borehole. Such membrane protrusions through the gaps of the metal sheaths are clearly recorded either on the rubber surface as distinct impression or as permanent convexities of the membrane itself left by enormous deformation. It may be said that, a rather thin membrane of 2mm thick used in this study, to make sure a uniform pressure, has an additional advantage; the rubber works as an impression packer to record the vital information with respect to discontinuities in the ground. Membrane puncture or tear is an inevitable consequence for these tests to end.

On the basis of the above discussion, it may be concluded that pre-existent mechanical inhomogeneities as joints or fractures are responsible for the non-uniform expansion of the borehole during pressuremeter tests in jointed rocks. The degree of non-uniformity is typically dependent on the orientation of the joint systems relative to the axis of the pressuremeter. In most cases, it may be crucial at low pressure level, although uniform expansion can be encouraged as yielding of ground proceeds with increasing pressure. What matters most is the deviation from the analytical model assuming uniform expansion. The deduction of material properties from the average pressure~strain relation (i.e. the conventional type of ordinary pressuremeter curve) can be very misleading without any serious consideration on validity of the analytical model. This implies that, the most popular analysis to evaluate stiffness values from the initial slope of the pressuremeter curve should be applied with reserve if tests are conducted in jointed rock ground.

It should be added that, interpretation of pressuremeter tests in jointed rock masses is generally not so straightforward as explained above. When each transducer's record is examined carefully, one may always get embarrassed to encounter peculiar records which demonstrate very irregular responses. In fact, there are varieties of cases and some of them are quite inexplicable. Complete interpretation can only be made possible if all the discontinuities in the tested ground are to be characterized. As a consequence, it is strongly recommended that, if the test is thought to be affected by some particular mechanical inhomogeneities to a serious extent, the results should be discarded. It should be reminded again that the deduced material properties can only be justified if the test does not deviate excessively from the analytical model.

5 SUMMARY

For analyzing the pressuremeter test data, it is generally assumed that the rock mass is linearly elastic, isotropic, and homogeneous. Within these assumptions, this test method can provide useful data for rock masses. The deduced material properties are regarded as equivalent continuous properties (ISRM, 1987). Little attention, however, has been paid so far to what degree the test is actually carried out in such a way as conform to the assumptions introduced for the analytical models.

Typical results are presented in this paper demonstrating the influence of such mechanical discontinuities as pre-existent joints and induced tensile cracks. When tests are conducted in jointed rock ground, the expansion of the tested borehole may happen to become extremely non-uniform from the very beginning of the tests. On the other hand, when tests are conducted in fairly uniform sedimentary soft rock ground, a cylindrical expansion of the circular borehole in early stages of the test may suddenly turn into an elliptic one. This may possibly suggest that two tensile cracks have developed radially in opposite directions.

Such irregular patterns of the borehole wall displacements are, in fact, strongly dependent on the orientation and of the measuring system. An exact picture of non-uniformity in borehole wall displacements can be evaluated or possibly exaggerated if the projecting directions coincide with the transducer's orientation. The opposite is true. If discontinuities in the rock happen to be misaligned to the transducer's orientation, the measured results cannot justify the extent of non-uniform deformation around the borehole but only demonstrate its averaged and smoothed image of the borehole wall movement. Surely the directional resolution can be improved by increasing the number of transducers. Measured outcome of uniform expansion by the point measurements in several directions does not necessarily validate the analytical model, It should be noted, however, crucial information about discontinuities can be best obtained by this direct measuring system.

Since ordinary pressuremeter analyses assume an axisymmetric mode of deformation, it should be borne in mind that a uniform expansion must be guaranteed if these analytical modeling is to be justified. The following procedure is now the standard practice of a self-boring pressuremeter tests conducted at CRIEPI. Special attention is paid not to miss any kinds of tell-tale signs indicating inhomogeneities in the tested ground and anomalous response not conforming to the theoretical implications.

(1) A pilot borehole of a smaller diameter is drilled by diamond coring technique in order to obtain continuous cores from the depths of interest. Usually, a 66mm core barrel in outer diameter is used as a drilling bit. The recovered cores are completely logged with emphasis on joints, fractures and other mechanical inhomogeneities. Rotary sounding technique may be useful collecting various data during the course of drilling (Miyazaki et al.,1990). Additionally, the surface of this pilot borehole may be observed by a TV camera in order to identify any significant discontinuities pre-existent in the ground.

(2) The test location is selected based upon the fractured characteristics of the recovered cores and on inspection of the drilled wall by a TV camera. The acquired records of rotary sounding technique may also provide valuable information including the drilling rate, the thrust and torque exerted on the drilling bit, and the flow rate and pressure of the drilling mud.

(3) The pressuremeter probe is inserted in the borehole to the test location by a self-boring technique. Rotary sounding technique may be useful collecting various data during the course of drilling.

(4) The pressuremeter test is carried out. Multiple pressure cycling to progressively higher load may be an effective technique to obtain additional information as concerns if elastic responses are there.

(5) The pressuremeter probe is pulled out of the borehole. As soon as the probe is retrieved, the outer surface of the expanding section should be examined. The metal sheaths and the rubber membrane, in particular, must be carefully speculated to see if any valuable information might be imprinted on them.

(6) Finally, the tested borehole surface may be observed by a TV camera again. The purposes are to detect any evidences of localized failures or displacements along the pre-existent joints, to locate the directions of possible tensile cracks, and to discover whatever clue to help understand what happens in the tested ground during the test.

Sophisticated interpretation of pressuremeter tests in rock masses cannot dispense with the knowledge about discontinuities in the tested ground including pre-existent joints and induced tensile cracks. Carefully designed test procedure will definitely lead to still more accurate evaluation of the material properties from the tests in characterizing the ground in question.

REFERENCES

ISRM. 1987. Suggested methods for deformability determination using a flexible dilatometer. *J. Rock Mech. Sci. & Geomech. Abstr.*, Vol.24, No.2, pp.123~134.

Haberfield, C.M. & Johnston, I.W. 1989. Model studies of pressuremeter testing in soft rock. *Geotech. Testing J.*, GTJODJ, Vol.12, No.2, pp.150~156.

Haberfield, C.M. & Johnston, I.W. 1990. A numerical model for pressuremeter testing in soft rock. *Geotechnique* 40, , No.4, pp.569~580.

Miyazaki, T., Sakai, K., Kaneko, S. & Yukawa, H. 1990. In situ test by cyclic loading boring. *Proc. 25th Japan National Conf. on SMFE*, pp.153~154 (in Japanese).

Tani, K., Nishi, K., Okamoto, T. & Yoshida, Y. 1993. Deformation characteristics of soft rock obtained by pressuremeter test. *Proc. 25th Sym. on Rock Mech.*, Japan Soc. Civil Eng., pp.546~550 (in Japanese).

Tani, K., Nishi, K. & Okamoto, T. 1994. Deformation characteristics of soft rock by pressuremeter test. *Proc. Int. Sym. on Pre-failure Deformation of Geomaterials*, Vol.1, pp.203~206.

Tani, K., Nishi, K. & Okamoto, T. 1995a. A new measuring method of borehole wall displacements for pressuremeter tests. *Proc. 4th Int. Sym. on Pressuremeters*.

Tani, K., Nishi, K., Okamoto, T. & Yoshida, Y. 1995b. Development of a self-boring pressuremeter for rock ground. *CRIEPI Report*, to be published (in Japanese).

Wroth, C.P. 1984. The interpretation of in situ soil tests. *Geotechnique* 34, No.4, pp.449~489.

Non destructive detection of micro cracking in rock

M. P. Luong
CNRS Laboratoire de Mécanique des Solides, Ecole Polytechnique, Palaiseau, France

ABSTRACT: This paper describes a non destructive testing technique in use for detecting the occurrence of damaging non linearities, and a specific data reduction procedure to assess damage accumulation. As damage accumulates in rock, its strength diminishes. It is assumed that non linear energy is a measure of rock damage. The damaging process of the rock specimen has been detected by analysing the signal evolution of ultrasonic pulses propagating through the specimen that is subjected to increasing loads. A non linear analyser based on a multidimensional Fourier Transform permits to separate the linear and non linear parts of the mechanical response of rock material. Experimental results evidence a threshold of change of its dissipative behaviour, thus defining a limit of material stability.

1 INTRODUCTION

Interest in the non destructive testing for rock infrastructure has increased very much in recent years, because considerable discrepancy exists between the strength measured in the rock structure and the standard strength determined on drilled core samples from the same mass.

In seismic zones, earthquake damage to structures is often caused by large, permanent deformations of the ground. In all types of soils and rocks, these deformations are mostly due to shear failure.

In rock mechanics, various geological formations are being investigated as possible sites for underground storage of oil and gas, and underground repositories of radioactive wastes. Much attention has been given to the saline deposits due to their widespread distribution, relative accessibility and ease of excavation. The mechanisms of rock salt deformation and fracture are thus fundamental for processes of strata control, for the support of excavations and for the construction of salt cavities (Hardy and Langer 1981). Rock salt generally displays failure characteristics intermediate between ductility and brittleness. This soft rock deforms plastically, but it is extremely notch sensitive. In the presence of a flaw, brittle fracture often occurs with the onset of yielding. When subjected to dynamic or vibratory loadings, the failure mechanism of soft rock consists primarily in the formation and propagation of micro cracks.

Information about the location and significance of structural defects needed as a basis for maintenance decisions, including the extreme case of removal from service, can be obtained through inspection and non destructive evaluation. A non destructive testing technique is needed not only to detect damaging micro cracking, but also to characterise weakness locations for the assessment of their severity and prescription of repair procedures.

The paper describes a special testing device in use for detecting the occurrence of damaging non linearities and a specific data reduction technique to assess damage accumulation. As damage accumulates in a structural system, its strength diminishes. It is assumed that non linearity, that disturbs ultrasound propagation, is a measure of rock damage. As the energy is dissipated by the rock structure, the residual strength decreases. The damaging process of rock specimen has been detected by analysing the signal evolution of ultrasonic pulses propagating through the specimen that is subjected to increasing loads. A non linear analyser based on a multidimensional Fourier Transform permits to separate the linear and non linear parts of the rock mechanical response.

2 EXISTING TECHNIQUES

Common experimental methods have been traditionally used to obtain information concerning deformations, strains, structural integrity and failure mechanisms in the rock structure, using strain gauges, photo elasticity, moiré, ultrasound and radiography, as well as acoustic emission and thermographic methods. Unlike most metals that are mass homogeneously produced, the properties of a rock structure are not unique.

In most rocks, both the acoustic velocity and the attenuation vary greatly. According to Koltonski and Malecki (1958), granite, for instance, depending on its grain structure can reach values from 1.7 to 5.0 km/s. This invites the use of non destructive testing (NDT) which can be applied directly to the part being examined rather than a representative sample. Until today, a practical method, that can effectively and economically detect micro cracked zones in rock structures in a field environment, has yet to be developed.

The NDT method most extensively used at present is probably the ultrasonic technique. Several researchers (Jones 1952, L'Hermite 1954, and Sell 1959) have used sonic and ultrasonic method to study the fracture of concrete and reported that crackling noises occurred at about 25% to 75 % of ultimate load. It was recognised that these significant changes in the property of concrete are probably caused by micro cracking, resulting in the increase of volume under compression and the increase of Poisson's ratio.

The "discontinuity stress" (Newman 1968) has been used to describe the end of the quasi elastic behaviour of concrete, characterised by:

a) mortar cracks begin to extensively develop and to form a continuous network,

b) volumetric strain stops decreasing and Poisson's ratio starts to sharply increase, and

c) beyond this threshold, long-time load eventually leads to failure. As compared to ultimate strength, discontinuity stress is much less affected by external factors, such as the rate of loading, the details of loading arrangement and the stiffness of the testing machines. Consequently, discontinuity stress, rather than ultimate strength, should be used to establish the failure criterion of rock materials.

The first attempts to measure the discontinuity stress were made using ultrasonic pulse velocity (Robinson 1967 and Vile 1968), unfortunately without giving details on the method of such acoustic measurement. Since the change of pulse velocity is a continuous process, the determination of the discontinuity stress will have to be arbitrary and subjective.

The present work aims to describe a non destructive testing technique in use for detecting the occurrence of damaging non linearities and a specific data reduction procedure to assess damage accumulation.

3 THEORETICAL BACKGROUND

The salient feature of rock behaviour originates from their internal micro cracking announcing failure. Ultrasound propagation in a damage rock material, unlike in an ideally perfect solid, is generally accompanied with attenuation and dispersion. Attenuation refers to the diminishing of wave intensity or wave amplitude as a wave propagates through a damaged medium, while dispersion refers to the shape distortion of a wave due to the frequency dependence of the effective wave (phase) velocity. Both attenuation and phase velocity are measurable quantities, and the amount of change in the attenuation and phase velocity can be correlated to the level of damage states.

This feature is advantageously exploited in this ultrasonic non destructive evaluation, relying on the mechanism of ultrasound propagation through rock materials. This provides an ideal mean for detecting and characterising micro cracking in damaged rock material in the infrastructure. Techniques utilising ultrasonic waves are especially appealing because of the direct connection between the characteristics of the wave propagation and the damage states of a solid (Achenbach 1990). There are many factors affecting wave attenuation and dispersion, such spreading of a wave beam, scattering, absorption due to various mechanisms and mode conversion, resulting in partitioning of the energy among two or more wave modes, each travelling at its own velocity. The present analysis only considers, in the framework of small perturbation, the propagation mechanism of a small wave through an elastic plastic solid presenting locally a softening behaviour.

Within the theory of plasticity, several postulates that guarantee mechanical stability for an infinitesimal volume have been suggested, based either on energy (Drucker 1951, Hill 1958) or wave propagation considerations (Mandel 1964). Drucker's stability postulate based on thermodynamical considerations, and Hill's maximum work principle for a rigid plastic material, require the second increment of plastic work,

$$d^2W^p = d\sigma_{ij} d\varepsilon_{ij}^p$$

to be positive to guarantee stability and to ensure the plastic flow is associated. These postulates are mathematically powerful and their use is widespread for constitutive modelling of most of engineering materials.

The stability condition was analysed and discussed by Mandel (1964). He showed that Drucker's postulate was a sufficient, but not a necessary condition for a material to be stable. On the basis of the assumption that a stable material is able to propagate a small perturbation in the form of waves, Mandel proposed a necessary condition for stability. Mandel (1963) showed that a wave, moving with a velocity Ω, can propagate in a material of mass density ρ characterised by an elastic plastic matrix A, along the direction α, if and only if all the eigenvalues λ of the propagation matrix M are positive, where

$$d\varepsilon_{ij} = A_{ijkl} d\sigma_{kl}$$

$$M_{ik} = A_{ijkl} \alpha_j \alpha_l = \rho \Omega^2 [\gamma_i]$$

$$k = 1, 2, 3$$

$$\lambda_k > 0$$

$[\gamma_i]$ denotes the discontinuity of acceleration.

If one of the eigenvalues λ is negative or null, the corresponding components of the perturbation cannot propagate. This implies instability, and the possible appearance of strain localisation, sliding or micro cracking along a certain direction.

4 MICRO CRACKING IN ROCK MATERIALS

The strength of rock materials represents the maximum resistance that is capable of developing under the particular system of loading. The strength does not therefore represent the stress with which the structure begins to be damaged by loading.

Quasi brittle failure of rock material is characterised by the appearance of discrete discontinuities that can range from a single crack to any degree of fracturing. The fracture process that accompanies quasi brittle failure, partly or totally unloads the stress field. If the stress field is purely tensile, the opening of cracks is sufficient for total unloading. In the case where the stress field is predominantly compressive, unloading is much more complicated: crack formation is not sufficient for unloading, as the material is reduced to a number of continuous sub regions, still in partial contact with each other.

During failure, the rock material ceases to be a continuum and becomes a mechanical structure. Typically after reaching the failure condition, a quasi brittle rock material presents:

a) a discontinuous structure,
b) the ability nevertheless to sustain shear stresses,
c) a non elastic bulking wherein shear strain is accompanied by large volumetric strains,
d) the dissipation of energy through frictional work (Luong 1993), and
e) a failure which is strain history dependent.

It has been recognised that these particular properties have important consequences:

i. Unloading towards the free surfaces is a more complex process than for an elastic plastic continuum,
ii. The unloading process generally operates slower than for a comparable elastic plastic manifestation,
iii. Quasi brittle materials submitted to a rapid compressive loading can sustain higher loads than their static strength and for a time longer than if they were purely plastic, and
iv. This salient behaviour of quasi brittle failure in compression, can give rise to what appears to be a higher dynamic strength and apparent strain rate effects.

In the "Shear Friction" method of connection design in concrete construction (Millard and Johnson 1984), cracks are assumed to have occurred at disadvantageous locations within the region of connection. Reinforcement is needed to transfer shear, normal force and moment across the cracks when the connection is loaded.

Tests for grouted bolts and split-set rock bolts on shear transfer at cracks have been carried out by several workers using pre-cracked rock specimens restrained by the embedded reinforcement, employing a wide variety of test parameters such as initial crack widths, axial stresses normal to a crack surface and sectional areas of reinforcing bars.

In addition, the interest in the constitutive relations of cracks in rock materials is considerably enhanced during the development of analytical and numerical studies on cracked materials, introducing the dilatancy ratio and the frictional coefficient in order to reflect in a realistic manner the typical characteristics such as shear transfer due to aggregate interlock, cross effect such as crack dilatancy and frictional contact slip, and path dependence.

Thus the stress-strain behaviour and the ultimate strength of rock materials are governed by a process of continuous micro cracking, particularly associated with the interfacial region between cement and the aggregate particles. Schematically, micro cracking in rock materials under increasing uniaxial compressive stress can be roughly viewed in three stages evidenced by infrared thermography (Luong 1993):

i. No significant cracking activity is observed in this stage; a quasi elastic response to loading ends beyond a range from 20 % to 50 % of the nominal compressive strength for a rock salt.

ii. Non linearity of the stress-strain curve as well as an increase in apparent Poisson's ratio becomes obvious; under vibratory loading, subsequent energy dissipation occurs for values of stress ranging from 50 % to 75 %.

iii. Damage in rock salt significantly increases by coalescence of weak zones, crack patterns may propagate in an unstable manner; failure of concrete in compression occurs when continuous cracks develop into macroscopic failure surfaces.

There exist two main methods of micro crack detection:

1) Direct methods: visual sighting of cracks, microscopic observation of surface cracks, X-radiography, etc., and

2) Indirect methods: surface strain measurements, ultrasonic pulse methods, acoustic emission technique, holographic interferometry, cumulative energy dissipation and change in deformation modulus.

Among the indirect methods, the ultrasonic methods are mainly applicable to crack detection, while the acoustic emission technique is better used for the detection of crack initiation, though the overall extent of internal disruption may be indicated by the integration of the recorded acoustic signals.

Measurement of the elastic moduli using ultrasound has become fairly routine. The connection between the speed of sound c, density ρ and Young's modulus E is strictly valid for an ideally linear elastic and homogeneous material; it remains valid for heterogeneous materials so long as a key assumption is satisfied, namely, the wavelength is large compared to any characteristic length of the material.

Acoustic emission AE is applied in a broad sense to the sounds that are internally generated in a material subjected to loading. AE testing is relatively simple, needing only a sensor (usually piezoelectric), basic signal-analysis equipment (amplifiers, filter, data-processing equipment) and some means for loading the specimen. There is wide variety of mechanisms known to be responsible for AE generation, ranging from dislocation motion to crack propagation. Therefore, the basic AE monitoring techniques are applicable to a wide class of materials and structures (Ouyang et al 1991). Unfortunately the interpretation of results still remains a delicate affair.

5 NON LINEAR ANALYSIS

Linear analysers based on one-dimensional Fourier Transforms (FTs) and modal analysis programmes are popular tools used by experimentalists in linear structural dynamics.

In this experimental work, an input-output non parametric approach (Liu and Vinh 1991) has been chosen to portray the non linear behaviour of rock materials subjected to loadings.

A non linear functional Volterra series has been used to detect the occurrence of the non linear behaviour of rock subjected to increasing loadings. In this functional, the total response of the system y(t) is decomposed into components of various orders.

$$y(t) = y_1(t) + y_2(t) + \ldots + y_n(t) \tag{1}$$

Each component is defined by a functional

$$y_n(t) = \int_{-\infty}^{+\infty} \ldots \int_{-\infty}^{+\infty} h_n(\tau_1, \tau_2, \ldots, \tau_n) \prod_{k=1}^{n} x(t - \tau_k) \, d\tau_k \tag{2}$$

The first order component is described by a linear convolution

$$y_1(t) = \int_{-\infty}^{+\infty} h_1(\tau) \, x(t - \tau) \, d\tau \tag{3}$$

where x(t) denotes the input function and $h_1(\tau)$ the first order impulse response, describing the linear behaviour of the system.

The other components require more than one time variable and multidimensional signal processing is needed. As an example, let us describe the second order (non linear) component of the response. From Eq.(2), we have the following expression:

$$y_2(t_1, t_2) = \int_{-\infty}^{+\infty} \ldots \int_{-\infty}^{+\infty} h_2(\tau_1, \tau_2) \, x(t_1 - \tau_1) \, x(t_2 - \tau_2) \, d\tau_1 \, d\tau_2 \tag{4}$$

where τ_1, τ_2 are the two time variables and $h_2(\tau_1, \tau_2)$ denotes the second order impulse response.

Within this mathematical framework, two dimensional Fourier Transform is appropriate for the study of the rock material behaviour in the frequency domain.

$$\mathcal{F}[h_2(\tau_1, \tau_2)] = \int_{-\infty}^{+\infty} \ldots \int_{-\infty}^{+\infty} h_2(\tau_1, \tau_2) \exp(-j\omega_1\tau_1 - j\omega_2\tau_2) \, d\tau_1 \, d\tau_2 \tag{5}$$

This expression defines the second order transfer function $\mathcal{H}_2$ with two circular frequency variables ω_1 and ω_2. $\mathcal{F}$ is the Fourier Transform operator.

$$\mathcal{H}_2(\omega_1, \omega_2) = \mathcal{F}[h_2(\tau_1, \tau_2)] \tag{6}$$

In this application, a single impulse has been used as input

$$x(t) = a\,\delta(t - T)$$

Then from (1) and (2), we obtain:

$$y(t) = \sum_{i=1}^{n} y_i(t) =$$

$$\sum_{i=1}^{n} \int_{-\infty}^{+\infty} \ldots \int_{-\infty}^{+\infty} h_i(\tau_1, \tau_2, \ldots, \tau_i) \prod_{k=1}^{i} a\,\delta(t - T - \tau_k) \, d\tau_k$$

$$= \sum_{i=1}^{n} a^i h_i(t - T) \tag{7}$$

From the single impulse of m different test magnitudes, Eq.(7) gives:

$$y^m(t) = \sum_{i=1}^{n} a_m^i h_i(t - T) \tag{8}$$

where m denotes the mth test.

Under matrix form, Eq.(8) is as follows:

$$\{y\} = [a]\{h\} \tag{9}$$

where

$$\{y\} = \left(y^1(t), y^2(t), \ldots, y^m(t)\right)^T \tag{10}$$

$$\{h\} = \left(h_1(t - T), h_2(t - T), \ldots, h_n(t - T)\right)^T$$

$$[a] = \begin{bmatrix} a_1^1 & a_1^2 & \ldots & a_1^n \\ a_2^1 & a_2^2 & \ldots & a_2^n \\ \ldots & \ldots & \ldots & \ldots \\ a_m^1 & a_m^2 & \ldots & a_m^n \end{bmatrix} \tag{11}$$

or

$$\{h\} = [a]^{-1} \{y\} \tag{12}$$

This equation gives the impulse response of various orders with only one time variable.

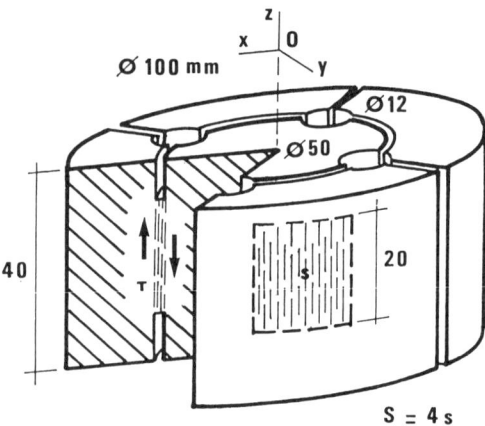

Figure 1. Mode II direct shear testing.

Figure 2. Ultrasound testing on a rock salt specimen.

Figure 3. Ultrasound testing on a sandstone specimen.

6 EXPERIMENTAL RESULTS

The proposed technique has been applied on two different types of rock: a soft rock (rock salt) and a hard rock (sandstone) subjected to a static mode II direct shearing (Figure 1). These geomaterials are composed of particles, mineral grains, that are bonded together by a combination of their intermolecular forces and mineral cements, and that are physically interlocked to varying degrees. These bonded masses may be isotropic, homogeneous or not. The heterogeneity may exhibit a definite pattern, or it may be random. Systems of cracks, such as joints and bedding planes, usually divide the material mass into discrete blocks that may be regular in shape and size, or extremely irregular with non planar separations and random sizes.

Cracks thus transmit significant amounts of shear by means of aggregate interlock also termed interface shear. A shear displacement or slip roughly parallel to the plane of the crack is essential to mobilise shear transfer by aggregate interlock. This displacement is accompanied by a displacement in the normal direction or dilation owing to rough asperities of aggregate and mortar tips projecting across the sliding path. Shear transfer by aggregate interlock is of frictional nature with the normal compressive forces being provided by the embedded reinforcement.

Thus the resistance against shear loading originates in three components: cohesion, shear friction and dilatancy. As the resistance given by friction and dilatancy increases with increasing normal stress σ, so also does the shear strength at failure τ_f. Therefore the strength can be represented in the Mohr diagram by an inclined line which rises with the increasing σ. The normal stresses across the interface in direct shear tests can be applied in three different modes, which results in three different boundary conditions for this problem:

1. It can be kept constant, which corresponds to zero normal stiffness across the interface,
2. It can be varied such that there is no crack dilation, which corresponds to infinite normal stiffness, or
3. It can be allowed to vary as the crack slides and dilates, which corresponds to variable finite normal stiffness. The latter type depicts to the most practical situation as there is always a finite normal stiffness provided by the embedded hard particles.

This also means that the shear and normal stresses (τ,σ) as well as the displacements (δ_t,δ_n) are generally coupled and vary according to some stress or displacement history.

A mode II shearing testing device (Luong 1989) has been used to detect the occurrence of micro cracking during shear fracture of rock salt (Figure 2) and sandstone (Figure 3). Typical characteristics such as shear transfer due to aggregate interlock (crack dilatancy), cross effect (frictional contact slip) can be determined in order to describe the failure mechanisms including strength and the evolution of failure. The described experimental procedure has been used to analyse the non linear behaviour of these rocks subjected to increasing static shear loadings.

The pulse transmission methods, using video scan piezoelectric transducer V150 (0.25 MHz), have been applied as a non destructive testing, in conjunction with a pulser-receiver Panametrics 5052PR that provides high energy broadband performance. The pulser section produces an electrical pulse to excite a piezoelectric transducer, which emits an ultrasonic pulse. The pulse travels through the specimen that is subjected to a given static shearing, to a second transducer acting as a receiver. The transducer converts the pulse into an electric signal that is then amplified and conditioned by the receiver section and made available for non linear analysis.

Signal records have indicated the evolution of the ultrasonic pulse travelling through the specimen that is loaded by increasing static loads (Figures 4 and 5).

They have evidenced (1) the occurrence of micro cracking that modifies the longitudinal part (the faster) of the pulse signal and (2) the effects of micro cracking that affect its transversal part (the slower).

The data processing technique in use provides the variation of non linear energy normalised by the linear energy as a function of stress levels (Figures 6 and 7).

$$NL = \frac{\int \left(|h_2(t)|^2 + |h_3(t)|^2 + ...\right) dt}{\int |h_1(t)|^2 dt}$$

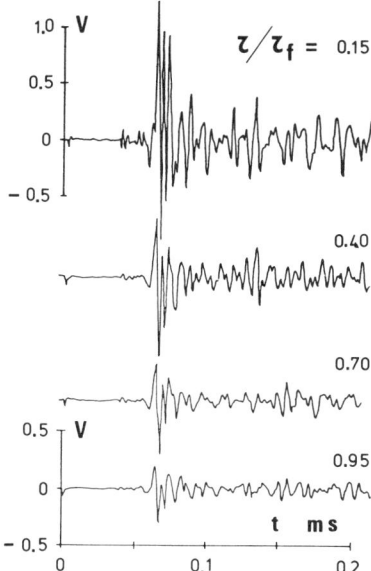

Figure 4. Output signals recorded on rock salt specimen.

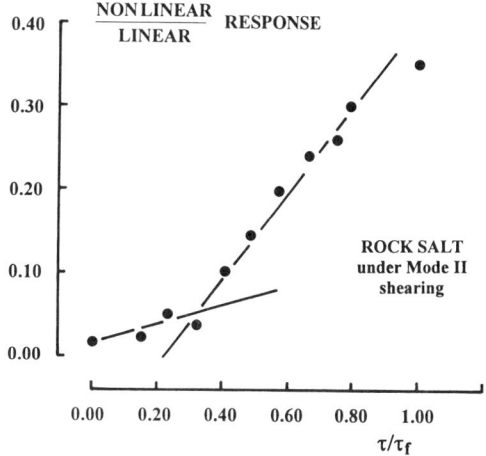

Figure 6. Evolution of the non linear response of rock salt.

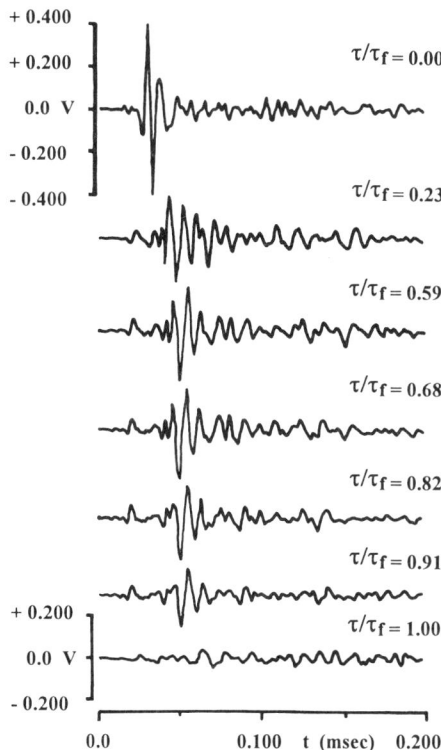

Figure 5. Output signals recorded on sandstone specimen.

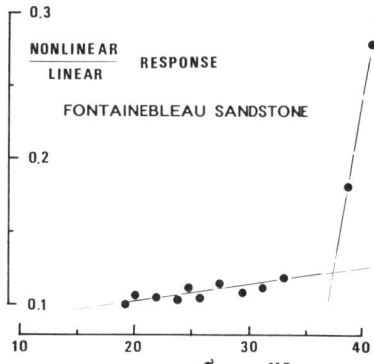

Figure 7. Evolution of the non linear response of sandstone.

The slope change of the experimental curve suggests that the sandstone specimen presents two different behaviours: the former is quasi elastic and stable, the latter may lead to a sudden failure caused by the extension of micro cracking that generates non linear effects affecting the ultrasonic wave propagation mechanisms.

Non linear dynamic analysis of rock structures designed to code static loading, responding to severe earthquakes, have given an indication of the order of the inelastic deformations required, and that the ductility demand concentrates in the weak part of frames. The capacity design procedure for earthquake resistant frames is to detail the beams and column bases so that a dissipating mechanism of ductile energy can develop by plastic hinging there during a severe earthquake, while sufficient reserve strength is provided so that the probability of column yielding, shear or bond failure, is low.

The traditional "ductility factor μ", commonly used for the design of earthquake resistant structures (Rosenblueth 1980), seems to be defined in reference to this threshold of non linearity appearance, similarly either in concrete or in rock materials. Ductility factors were used in a way that involves a general reduction in the design spectrum. Hence their definition is not very clear: some reasonable assessment of the allowable ductility factor is required. It must be aware of the differences between the various kinds of ductility factors involved in the response of structures to dynamic loading: the ductility factor of a member such as the rotational hinge capacity at a joint in a flexural member, the ductility factor of a floor or story in a building, and the overall ductility factor of the structure for use in the computation of base shear from the response spectral values.

Experimental results suggest the following definition for the ductility factor:

$$\mu = \frac{\sigma_k}{\sigma_f}$$

where σ_k denotes the loading threshold of change of non linear regime and σ_f the ultimate loading at failure.

7 CONCLUDING REMARKS

The proposed technique of damage evaluation using ultrasound propagation characteristics in non linear micro cracked media involves careful examination of areas where defects are most likely to occur. The critical areas can be identified by analysing the structure and the service history of similar environments. The application of ultrasound scanning to inspection of rock structures relies on the fact that during the process of micro cracking, the rock material is locally unstable and hence obstructs the propagation of ultrasound pulse.

An input-output non parametric procedure, based on ultrasonic pulse propagation and using a non linear analyser for data reduction, has been chosen to portray the non linear behaviour of rock materials subjected to static loadings.

The non linear analyser reveals to be very useful for the detection of the micro cracking process announcing the occurrence of propagating damage. It can be used to monitor non destructively and continuously the whole fatigue damage process of rock infrastructure so that the damage mechanisms can be quantitatively explained.

The experimental results provide a basis for selecting a ductility factor in earthquake, shock or blast resistant design, because ductile hysteretic response is a manner of transforming and dissipating the seismic energy or accumulating damage during successive events.

ACKNOWLEDGEMENT - The collaboration of Dr. H. Liu is gratefully appreciated.

REFERENCES

Achenbach, J.D. 1990. From ultrasonics to failure prevention. *Elastic Waves and Ultrasonic Nondestructive Evaluation*, ed. S.K. Datta et al, North-Holland, Amsterdam.

Drucker, D.C. 1951. A more fundamental approach to stress-strain relations. *Proc. 1st US Nat. Congress Appl. Mech.*, 487-491.

Hardy, R.H.Jr and Langer, M. 1981. The mechanical behaviour of salt. *Proc. 1st Conf. Pennsylvania*, Trans. Tech. Pub.

Hill, R. 1958. A general theory of uniqueness and stability in elasto-plastic solids. *J. Mech. and Physics Solids*, 6: 236-249.

Jones, R. 1952. A method of studying the formation of cracks in a material subjected to stresses. *British J. Applied Physics*, V 3(7), 229-232.

Koltonski, W. and Malecki, I. 1958. Ultrasonic method for the exploration of the properties and structure of mineral layers. *Acustica*, 8: 307-314.

L'Hermite, R. 1954. Present day ideas on concrete technology, 3rd Part, The failure of concrete. *Bull. 18, RILEM*, 27-39.

Liu, H. and Vinh, T. 1991. Multidimensional signal processing for nonlinear structural dynamics. *Mechanical Systems and Signals Processing*, 5(1): 61-80.

Luong, M.P. 1989. Fracture behaviour of concrete and rock under mode II and mode III shear loading. *Fracture of concrete and rock - Recent developments*, Elsevier Appl. Sc., 18-26.

Luong, M.P. 1993. Infrared thermographic observations of rock failure. *Comprehensive Rock Engineering*, ed. J.A. Hudson, Pergamon Press, 4(26): 715-730.

Luong, M.P. and Liu, H. 1992. Nonlinear dynamic analysis of damage in concrete. *Proc. 1st Int. Conf. Fracture Mechanics of Concrete Structures*, June 1-5, Colorado, USA.

Mandel, J. 1963. Propagation des surfaces de discontinuité dans un milieu élastoplastique. *Proc. Int. Symp. Stress Waves in Anelastic Solids*, Brown University.

Mandel, J. 1964. Conditions de stabilité et postulat de Drucker. *Proc. IUTAM Symp. Rheology and Soil Mechanics*, Grenoble, France, 58-68.

Newman, K. 1968. Criteria for the behavior of plain concrete under complex state of stress. *Proc. Int. Conf. Structure of Concrete*, Cement and Concrete Association, London.

Millard, S.G. and Johnson, R.P. 1984. Shear transfer across cracks in reinforced concrete due to aggregate interlock and dowel action. *Mag. Concrete Research*, 36(126): 9-21.

Ouyang, C., Landis, E. and Shah, S.P. 1991. Damage assessment in concrete using quantitative acoustic emission. *J. Engng. Mech.*, 117(11): 2681-2698.

Robinson, G.S. 1967. Behavior of concrete in biaxial compression. *J. Structural Div., ASCE*, 93(ST1), 71-86.

Rosenblueth, E. 1980. Design of earthquake resistant structures. *Pentech Press*, London, 142-194.

Sell, R. 1959. Investigation into the strength of concrete under sustained loads. *Bull. 5, RILEM*, 5-13.

Vile, G.W.D. 1968. The strength of concrete under short-term static biaxial stress. *Proc. Int. Conf. Structure of Concrete, Cement and Concrete Association*, London, 131-145.

Rock Foundation, Yoshinaka & Kikuchi (eds) © 1995 Balkema, Rotterdam. ISBN 90 5410 562 3

The examples of the new foundation evaluation system and permeability tests for weak rocks

Katuya Hirata, Takesi Sugiyama, Masaru Karasawa & Masato Ueda
Chuo Kaihatsu Corporation, Tokyo, Japan

ABSTRACT:We have had many problems in sampling cores or in performing Lugeon water tests on weak rock or sedimentary soft rock. For the purpose of solving these problems, we developed the new ground assessment system with the Foam Boring Method (F.B.M.) and the High Quality Permeability Testing System (H.Q.P.T.S.).

Moreover, we tried to enlarge the system over other measurements, and tests and geophysical loggings in borehole, for example as the purpose of getting accurate critical pressure values and distribution of Lugeon values as the ground assessment system.

On this paper, we comment the trial examples of the investigation in A, B dam site, in which we show permeability distribution map amended with the Lugeon water tests data by the ground assessment system. In relation to the comments, we indicate rational curtain grouting zone and ways, and riffer to basal geotechnique imformation of a dam basement cutting line on the basis of properties of foundation rocks.

1 INTRODUCTION

In the case of previous drilling methods for weak or soft rock mass foundations, we can't obtain reliable core sampling and smooth bore hole forming and perform enough complete drilling core judgments and in-situ permeability tests and several tests consist with other measurements and loggings in borehole.

Now we can realize that these strong points by the system combinated between the Foam Boring Method and H.Q.P.T.S.;① accurate drilling core sampling, ② physical and mechanical tests for the undisturbed drilling core ; ③ smooth bore hole forming , ④ defence of leakage from the packkered borehole wall; ⑤ accurate borehole load tests and several geophysical loggings.

We had been added the actual results , first in case of the Foam Boring Method we had been running experiments and operations already for about 10 years,and succeedingly in the case of H.Q.P.T.method equally about 3 years.

We will have the purpose of best effective results so we explained already as systemalizing these equipment and methods.

As under we illustrate in order;① these system's contents;② definite examplestudies;③ conclusion and next solutions question to problems.

2 THE ASSESSMENT SYSTEMS FOR WEAK OR SOFT ROCK MASS

2.1 The assessment systems contents

The new foundation assessment systems are formed by High Quality Permeability Testing System (H.Q.P.T.S) and several tests consist fo other measurements and geophysical loggings in borehole(Figure 1.).

Especially H.Q.P.T.S for weak or soft rock mass as the theme in this paper were developed by us as the purposes of next ; ① The new developed Foam Boring Method forms undisturbed and few uneven bore hole.② In the case of previous drilling methods, we can't control of sending water pressure to Lugeon test sections in borehole, and defence of leakage from the packkered borehole wall, but the new assessment system solves these problems. ③ Particularly under low pressure of Lugeon water test sections we can get accurate critical pressure values, Lugeon water values , reliable borehole load tests,geophysical loggings and physical and mechanical tests for undisturbed drilling cores.

2.2 The Foam Boring Method(F.B.M.)

Sure sampling of drilling cores and forming little disturbed borehole walls, which are strong points of F.B.M.,enables us to perform reliable borehole load tests,various kinds of geophysical well logging, and physical propreties and mechanical tests.

The inside of a dotted line shows the ground assessment system.

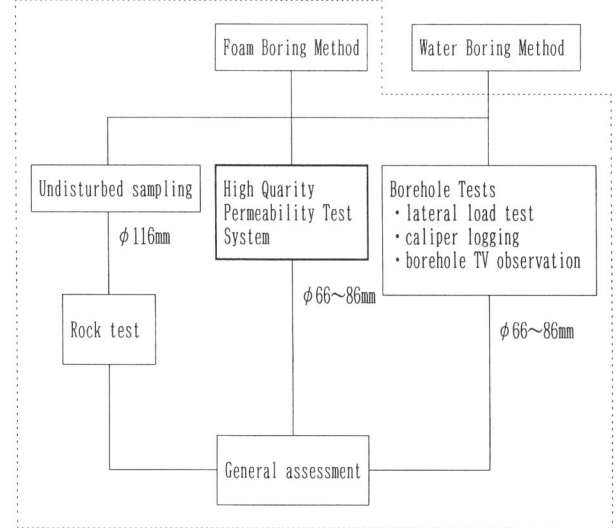

Figure 1. A flowchart of the ground assessment system.

Concretely, using "foam" as the drilling fuluid instead of "water" has effects on increasing the faculty of slime exclusion and efficiency in cooling bits. Through the whole method, We pursue high qualities from the drilling equipments of scaffolding, drilling machines, rods and drilling bits to preservation of the sampling cores.

Figure 2. shows the equipment of F.B.M. and Figure 3. is a flowchart of the working order of F.B.M.

2.3 High Quality Permeability Testing System(H.Q.P.T.S)

This system is consisted of the checking borehole and pure H.Q.P.T.method. Next we show the working arrangements. By the equip 2 compact water pressure sensors upper and lower part of packer for defense of leakage water, we can several tests consist of Lugeon water tests, in-situ permeability tests,spring water pressure tests , pore water pressure tests etc. in borehole,and we can get high reliable data. So H.Q.P.T.S. got great advancement of reliability of several tests comparing it with the previous methods.

Main characteristic of H.Q.P.T.S. are next descriptions.(cf.Figure 4.,5.)
 1. The direct measurements and controlling of water pressure particularly with high level quality under low pressure of Lugeon water test sections.
 2. In the case of H.Q.P.T.S. water pressure loss in water injection rods is no problems, hydrostatic pressure tests are possible at low water level places.
 3. By the check of leakage water of upper water pressure sensor of packer for defense, we can get accurate low critical pressure about $0.1 \sim 0.3$ MP.
 4. By using about double (this firm ratio) length packer, more accurate defense of leakage water through the contact face of rock mass arround packer is possible.
 5. So H.Q.P.T.S. gives great reliable data comparing with the case of previous methods, and easily reveals in-situ permeability mechanism.
 6. The accurate measurement of pore water pressure tests under test sections are possible.
 7. The accurate measurement of spring water pressure repeat tests under test sections are possible.

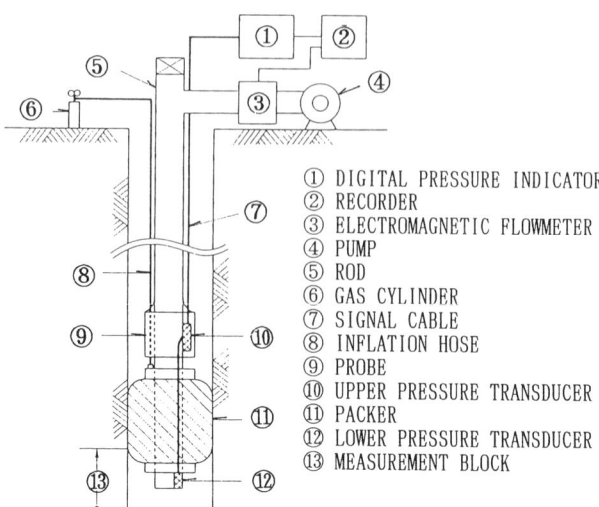

Figure 4. The organization of H.Q.P.T.S.

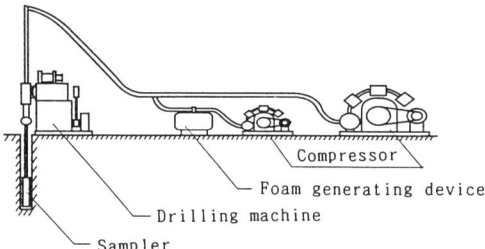

a. System structure of Foam Boring method

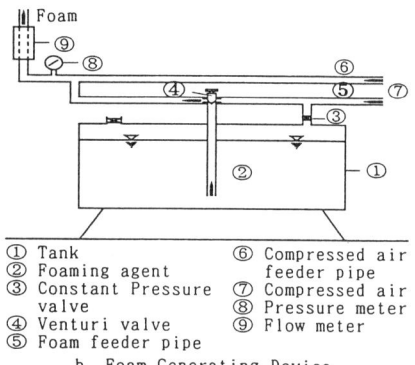

b. Foam Generating Device

Figure 2. The equipment of The Form Boring Method.

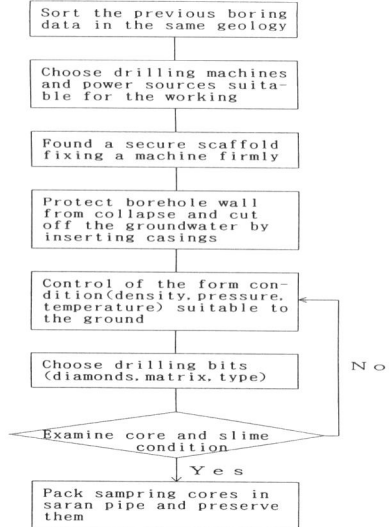

Figure 3. A flowchart of the working order of The Form Boring Method.

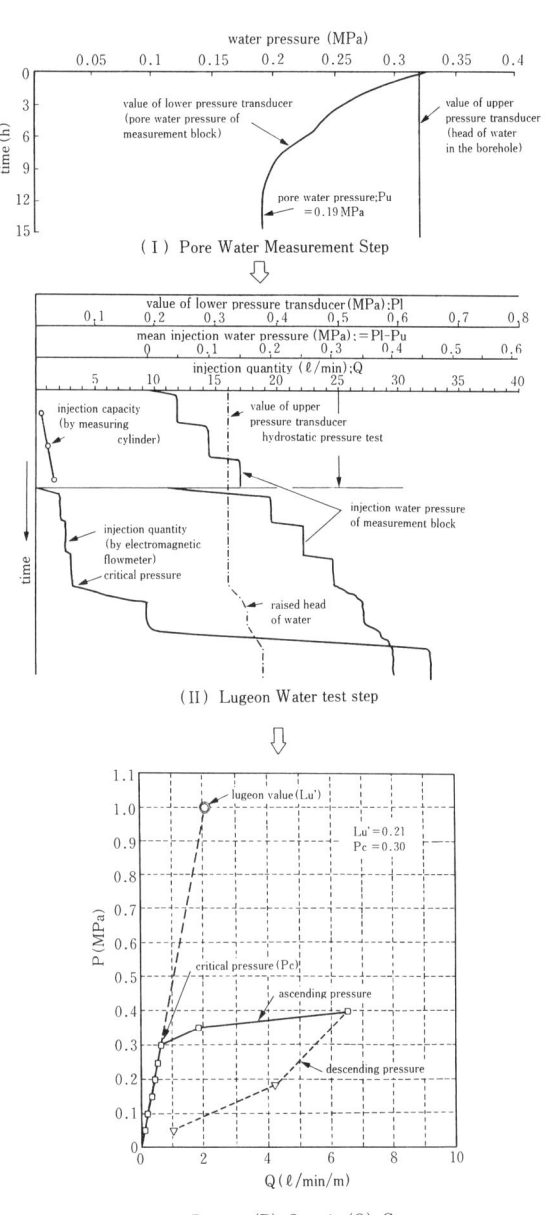

Figure 5. The measurement example of Lugeon water test

3 A CASE STUDY OF THE SYSTEM

3.1 A trial example of undisturbed sampling and High Quarity Permeability Test

1) Abstract
"A" Dam is located in the Green Tuff area in northeast Japan, and planed for a concrete gravity dam on Tertiary tuff and sandstone foundation. Landslide configuration on the left bank of the river required stability analyses for the Cut off slope and evaluation for permeability of foundation rocks.

2) Mechanical tests for drillng core specimen
In the previous boring investigation, core samplings of alterated clay or loose gravel were very difficult. Therefore, geological structure on the left bank was not clearly revealed, especially little geological information on the landslide was indicated. Now we obtained 100 percent cores by the form boring in the area to conjecture the depth of slide plane and age. Moreover, undisturbed sampling cores which can bear physical properties and mechanical tests enable us to perform triaxial compression tests with various stress conditions, turned to gain technological properties for design(Table 1.). With these data we could study stability analyses and countermeasure works for Cut off slope appearring at the dam abutment.

Table 1. Technological Properties

	kind of rocks	$C(kN/m^2)$	$\phi(°)$	$\gamma_w(g/cm^3)$
triaxial compression test (UU)	pumice tuff	130	6.2	1.644
	tuff breccia	119	2.3	1.851
triaxial compression test(CUB)	tuff breccia	100	14.1	1.851

3) High Quality Permeability Test (Lugeon water test)
On previous investigations, permeability tests (Lugeon water test) in soft tuff boreholes were not performed because of collapse along borehole wall, leakage from the packkered wall part, primary breaking due to maximum 0.5 MPa drilling water pressure. We performed Lugeon water tests in these soft tuff foundations using H.Q.P.T.S. As a result, except for strongly weatherd surface layer approximatery 5. meters deapth, distribution of acurate Lugeon values in soft pyroclastic rocks has been revealed. These results made it possible to design rational grouting from the dam abutment to river bed. Figure 6. shows comparison between Pressure-Quantity curves on two cases of Lugeon tests approximately in the same rock level boreholes(drilled using previous method and the F.B.M.). In the case of using the F.B.M. and H.Q.P.T.S., We acquired dense measurment points at low pressure zone and acurate critical pressures, which resulted in securing optimum low Lugeon values fitting the practical rock mass conditions because of no leakage from the packkered borehole walls.

Also Figure 7. shows comparison between Lugeon values in two boreholes.

Figure 8. is comparison distribution of Lugeon values in the left bank of "A" damsite, based on both previous method and new foundation evaluation system. The area was concerned to be high permeabity ground owing to difficulties of high quality permeability tests.

Tests using F.B.M. and H.Q.P.T.S. revealed low permeability, approximately 1 l/min/m/MPa in Lugeon value, of foundation except for weatherd surface layer,in spite of low critical pressure, 0.15 ～ 0.50MPa, be due to low concretion of rocks themselves.

These results contributed remarkably to select the grouting method and to limit Cut off inprovements by grouting, and are highly expected to a lot of savings of grouting costs.

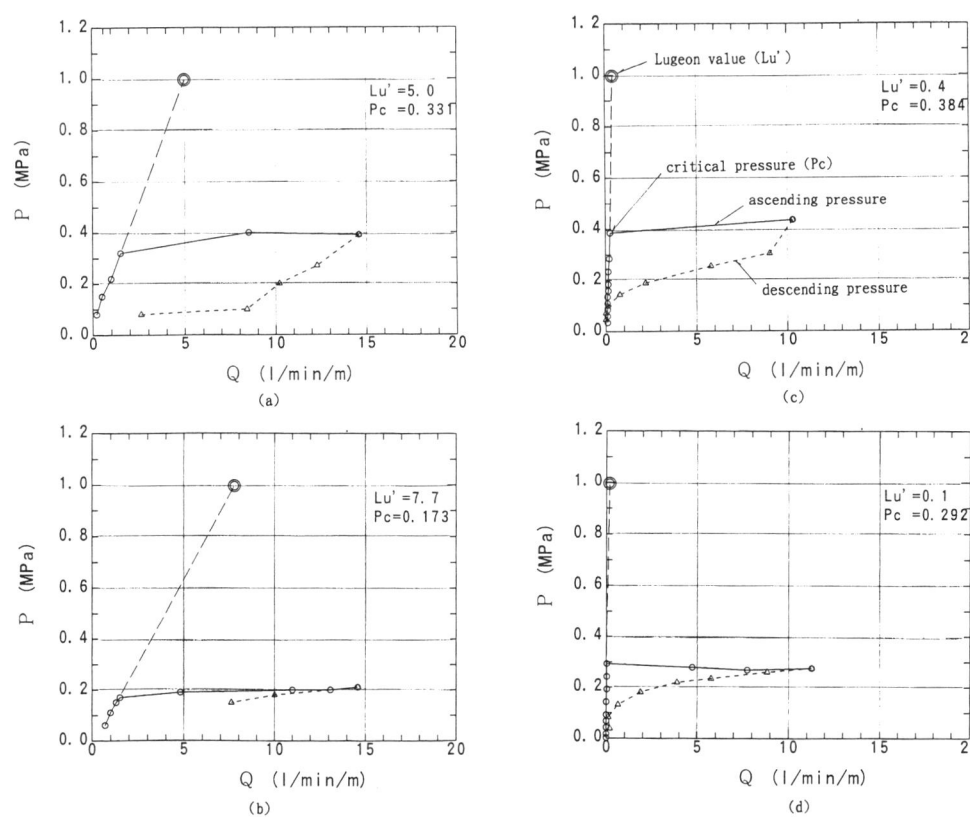

Figure 6. Comparison between Pressure-Quantity curves in two boreholes

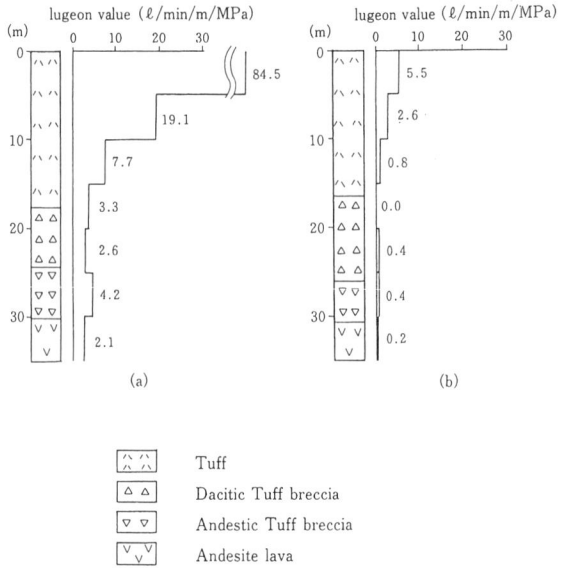
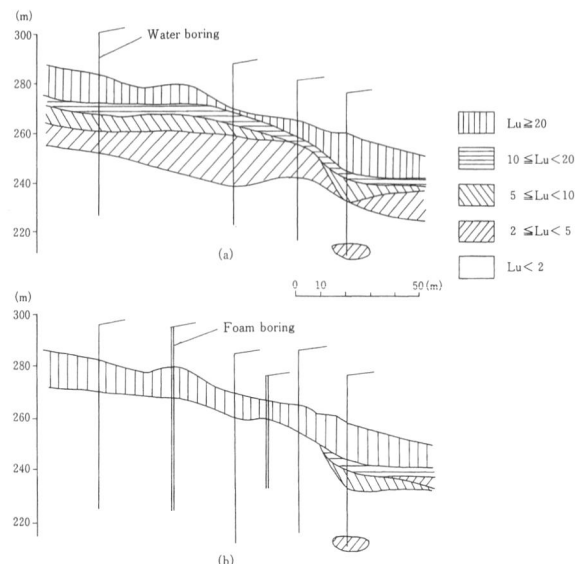

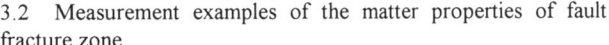

Figure 7. Comparison between Lugeon values in two boreholes

Figure 8. Comparison distribution maps of lugeon values

3.2 Measurement examples of the matter properties of fault fracture zone

1) Abstract

"B" Dam is located "Yamizo cordillera" southern area in the northeast Japan, and planed for a rockfill dam foundation on Mesozoic Accretionary Complexes formed of sandstone and conglomerate and slate and chert.

At the damsite leftside of the river, the fault fracture zone has a width of about10 meters and the direction of fault is the upper and lower sides of the river.

Here for the design of core zone and Inspection gallery in a rockfill dam, we put F.B.M. and H.Q.P.T.S. to gain the foundation cutting line examinations and deformation characters, the foundation grouting engineering data from permeability character.

As the result of comparison with the boring by previous methods, we could meet success with getting of accurate load tests and Lugeon water tests by forming smooth borehole, and we adviced important data of the foundation cutting line examination and the foundation grouting engineering zone data by the physical land mechanical tests for undisturbed drilling cores.

2) The results of analyses

The purposes of using F.B.M. in fault fracture zone are first complete samplings of fractured rock fragments and gouge because we could not get reliable sampling by previous methods. Secondly we can perform load tests in the smooth and undisturbed borehole which are checked by borehole scanner. Thirdly we obtain useful data by the physical and mechanical tests for undisturbed drilling cores in our laboratory.

In Figure 9. we show The Classification Map of Rock Mass and Distribution Map illustrated using previous Lugeon water tests in the fault fracture zone at the the axis of dam leftside of the river.

And in Photograph 1. we compared the cores of F.B.M. with the cores of previous drilling method in about same points .

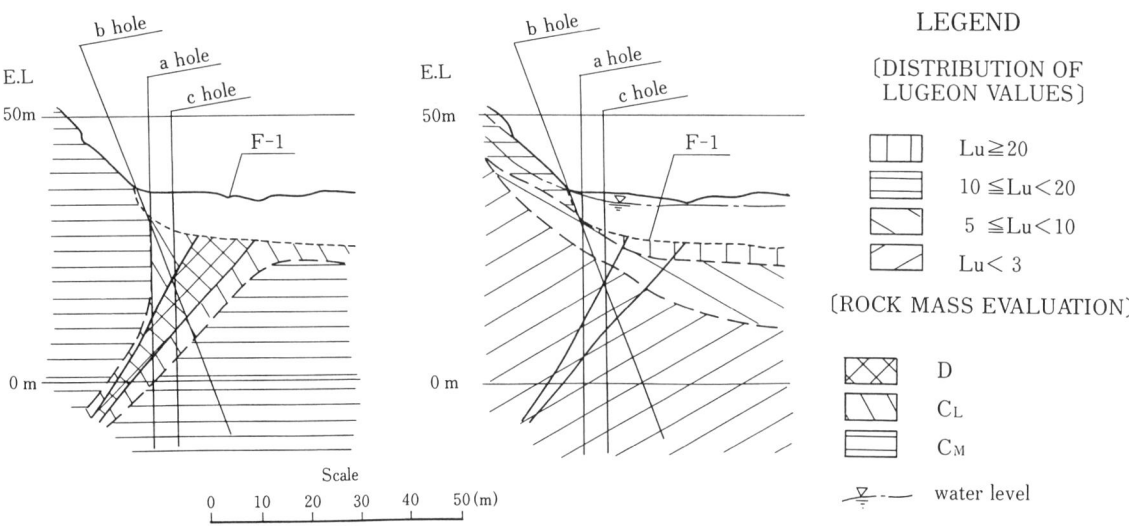

Figure 9. The classification map of rock mass and distribution map of the fault fracture zone at B damsite

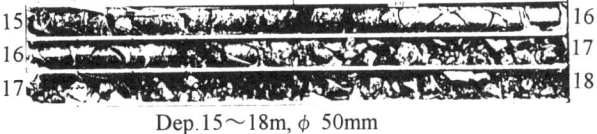

Dep.15〜18m, φ 50mm
The photograph of previous drilling method cores

Dep.15〜18m, φ 100mm
The photograph of Foam boring cores

Photogtaph 1. The compared photograph of 2 boring methods at B damsite

And the borehole load tests by previous drilling methods were very difficult in the fault fracture zone due to collapse of borehole wall.

In the case of using F.B.M. , we could gain particularly increasing number of reliable measurements and values caused by the stability of borehole wall , comparing with previous drilling methods .

In Table 2. we indicates the list of borehole load test results by using only F.B.M. in the fault fracture zone at B damsite.

Table 2. The list of load test results in borehole in the fault fracture zone

item	range(MP)	the class of rock mass	note
modulus of deformation D	150〜230	D	The member of rock in F-1 fault fracture zone form mainly slate, secondary sandstone・chert are inclued. This fracture zone fall under low velocity zone and the rock mass velocity shows 2.3Km/s.
	170〜440	CL	
modulus of elasticity E	460〜780	D	
	330〜2,620	CL	

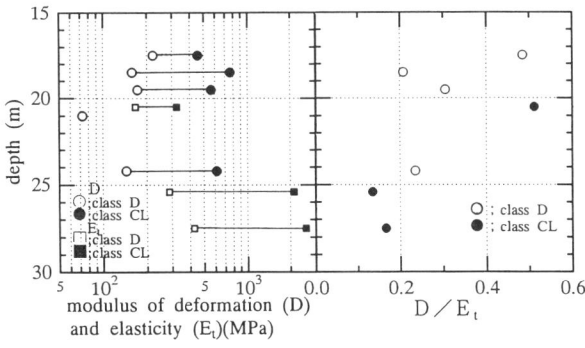

① The Correlation Illustlation of Depth and modulus of deformation D and modulus of elasticity E
② Correlation Illustlation of Depth and D/E

Figure 10. The correlation Illustlation of depth and modulus of deformation D by The New Foundation System in the fault fracture zone

The uneven measurement values indicates the degrees of fracturing and the changes of grain size that were reflected by these.

In Figure 10. we showed the correlation illustlation of depth and modulus of deformation D.

3.3 The tests in the laboratory

We perform the physical and mechanical tests for undisturbed drilling cores by F.B.M.. The accurate data of permeability character are obtained, by means of the contract macro in-situ message (for example Lugeon water tests) with micro message in limitted small section (for example permeability tests), particularly by the tests for undisturbed drilling cores in laboratory. Still more by the thinking together of other several data by the physical and mechanical tests, we believe the possibility of accurate analyses of permeability mechanism or leakage mechanism and the important judgment information to provide grouting materials, grououting hole intervals or grouting gel time.

By the results of physical tests, (the gravel fractions in the fault fracture zone are classifyed gravel size soil or sand size soil according to the ratio of gravel amount.) Further, we can know the small natural water contents brought from small void ratio , because of the good distribution of gradation size depend on dense hard rock fragments.

And as a matter of course the c , φ of gravel rich fraction sections in the fault fracture zone are bigger than those of gravel poor fraction sections.

The Coefficient of permeability K_{15} of undisturbed drilling core by F.B.M. in fault fracture zone are idicate $3.47 \times 10^{-6} \sim 5.22 \times 10^{-4}$cm/s in 14〜17m depth of gravel rich fraction sections , and $1.33 \times 10^{-6} \sim 2.17 \times 10^{-5}$cm/s in 24〜25m depth of gravel poor fraction sections in Figure 11. These sections come under the Lugeon water values of each 15〜20m depth and 20 〜25m depth sections of neighboring inclined boreholes, and the former is 4.2l/min/m/MP ≒ 4×10^{-5}cm/s ,the latter is 0.8l/min/m/MP ≒ 1×10^{-5}cm/s, the Lugeon water test sections have remarkable variety of rock facies and also fine fraction ratioes.

But the bottom limit of coefficient of permeability K_{15} of the former is samples of rich fraction , and also the upper limit is samples of fine poor fraction. Thus different permeability character consist in between the short sections of permeability tests in laboratory and the long sections of Lugeon water tests, properly show a little the phenomena of different permeability character. Therefore in the case of grouting , the grouting milk injections are inclined toward the high void ratio rock mass. Reversely the grouting milk injections are very difficult in the sections which are in contact with cracksand rich gouge, so injections of light density grouting and chemical grouting are very effffective.

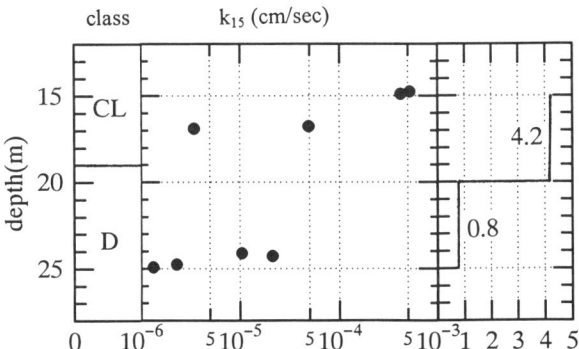

Figure 11. The contrast illustlation of permeability character compare laboratory tests with in-situ Lugeon water tests

Table 3. The list of laboratory test results of samples of fault fracture zone

Result of laboratory soil tests				Rich sections of gravel fraction	Poor sections of gravel fraction
Physical tests	General	Wet density ρ_t (g/cm³)		2.17~2.34	2.28~2.34
		Dry density ρ_d (g/cm³)		2.10~2.23	2.08~2.25
		Void ratio e		0.21~0.28	0.20~0.35
		Degree of saturation S_r (%)		35~68	53~72
	Texture	Gravel fraction 2~75mm (%)		59~69	33~44
		Sand fraction 75μm~2mm (%)		24~29	35~47
		Silt fraction 5~75μm (%)		3~8	8~18
		Clay fraction under 5μm (%)		1~5	4~14
		Maximum grain size (mm)		27~38	19~26.5
Dynamic Tests	Triaxial compression tests	Condition		$\overline{CU}$-test	
		Effective stress	c' (MP)	0.04	0.07
			ϕ' (°)	40.2	28.9
Coefficient of permeability K_{15} (cm/s)				3.47×10^{-6} ~ 5.22×10^{-4}	1.33×10^{-6} ~ 2.17×10^{-5}

cf. The each tests are 4 samples both rich gravel fraction sections and poor gravel fraction sections.

4. CONCLUSION AND SOLVABLE THEME OF FUTURE

As mentioned above by F.B.M. and H.Q.P.T.S. that are included our assessment system for rock mass, we riffer to get that high reliable Lugeon water tests and the exsamples of borehole load tests and laboratory test in the fault fracture zone. The facts indicate that these systems for weak or soft rock mass are effective by carrying of Lugeon water tests systems. This time we mentioned about the carryed exsamples for weak or soft rockmass, but The Assessment System for Rock Mass is also very effective against morewide geological field's objects and other several measurements. These many samples had been carryed already, but we don't mention them on this paper.

F.B.M. had been enough completed and have large advantage as the principle and engineering methods, on the contrary we have the next theme to be solved from now on.

1. The system units are not compact.
2. The drilling diameter is large for instance ϕ 116~120mm.
3. The drilling speeds are not fast to get the aim of complete core sampling, so the cost are not cheaper than that of comparison with the boring by previous methods because of most careful sampling.

Therefore we will make better these checked theme in the future.

And the H.Q.P.T.S. have the theme that (the improvement and miniaturizing on the ground measurement systems are the remained), while underground measurement systems are complete enough.

Accordingly we will make more effective combination of F.B.M. and H.Q.P.T.S. through many actual results. Further, we will aim to analyze the mechanism of permeability of rock mass by the increaseing reliability in many analysis examples, and build up the safer and more economical grouting methods.

At last we thank deeply the persons concerned who gived us pleasantly allowance of printing in this paper by ordered real survey examples, and the persons who are internal house and outside the company including advice to and inspection of this manuscript.

References

MASARU,K.,TETUYOSI,T.,MOTOHARU,T.(1987):JET FOAM BORING METHOD,The 22th Japan National conference on Soil Mechanics and Foundation Engineering (in Japanese),pp.107-108.

MASARU,K. and MOTOHARU,T. (1989):Sampling Technique Using the JFT FOAM BORING METHOD, Japan National Conference on Japan Society of Engineering Geology (in Japanese),pp113-116.

MASARU,K.,TETUYOSI,T. and MOTOHARU,T.(1990):An Example of Soil Inverstigation for a Dam Foundation Subsurface Using JFB Method,The 25th Japan National Conference on Soil Mechanics and Foundation Engineering (in japanese),pp155-156.

Rock mass classification of weathered granite and evaluation of geomechanical parameters for bridge foundation of large scale structure in the Honshu-Shikoku Bridge Authority

K. Ishikawa
Chuo Kaihatsu Corporation, Tokyo, Japan

H. Ochi
Bridge and Offshore Engineering Association, Tokyo, Japan

S. Takada
Honshu-Shikoku Bridge Engineering Co. Ltd, Tokyo, Japan

ABSTRACT: Rock mass classification for Honshu-Shikoku Bridge is performed to classify the engineering properties of foundation rock mass in designing the foundation and analyzing its stability. This classification of granite is to be determined quantitatively and objectively, not only by unaided-eye observation, but also by the use of a number of measuring indexes. These indexes consist of borehole logging, borehole loading test and rock test results. Principal among the parameters required to determine the bearing capacity of the foundation rock mass are deformation modulus, strength parameter and creep ratio. We chose to use one of the three methods for evaluation of bearing properties according to the importance of structures, time and stage of survey and design, etc.

1. INTRODUCTION

The geological survey by the authority for the Honshu-Shikoku Bridge, a strait-crossing structure to link the Japanese mainland with Shikoku over the Inland Sea (Setonaikai), consisted mainly of boring studies and borehole measurements as well as of rock mass classification for foundation design purposes.

Rock mass testing is the surest and most effective of means to direct judgment of the bearing capacity of foundation rock masses. Since frequent use of this method was difficult depending on site conditions, cost requirements, etc., however, a rock mass test was carried out on a land section where the same kind of rock mass was distributed as the sea bed to obtain data on the strength and deformation properties according to rock mass classification. The relations of these physical values from borehole measurements, rock mass tests, etc., versus the rock mass properties and rock mass classification were weighed up closely for evaluation of the mechanical parameters by rock mass classification.

On the basis of regarding the properties of a granite rock mass as "a mass of cracked rocks" for purposes of rock mass classification, and taking the work mass properties to depend on the combination or size of two scales -- 1) the degree of rock hardness (the scale of hardness), and 2) the state of rock cracking (the scale of cracking) -- we used a measurement index capable of indicating individual classification scales, in addition to unaided-eye observation work.

A method was proposed for determining the rock mass classification by measured values.

The present work, in connection with rock mass classification, considered the relation of rock mass properties to their engineering characteristics, then suggested the deformation properties obtained with measurement indexes as well as the classification of rock masses having their own strength properties, and studied how to infer the bearing capacity of rock masses that would correspond to the accuracy of geological surveys, the importance of structures, etc.

Figure 1 shows a block diagram of how the survey and analysis work was done.

The classification of rock masses, being stated here, is closely related to various engineering properties, is inseparably associated with the methods of foundation design and stability analysis, and is different in nature from any rock mass classifications in a general sense.

2. THE RELATION OF ROCK MASS CLASSIFICATION VERSUS ROCK MASS PROPERTIES AND MEASURING INDEXES

As rock mass classification generally means ranking the quality of rock masses, the present document notes that our rock mass classification was performed with the ultimate object of being able to classify the engineering properties of a foundation rock mass in designing the foundation and analyzing its stability.

That is proper evaluation of a foundation rock mass by this rock mass classification should lead to appropriate stability calculation, so rock mass classification becomes extremely important in making a reasonable foundation design.

2.1 Rock mass classification with unaided-eye observation

There are two elements of reference for the classification of boring cores; the degree of rock hardness (namely, the degree of weathering and deterioration), and the state of crack growth (namely, the form of core). The two are combined to classify and judge them.

There are three elements for classifying test pits; the degree of rock hardness, the degree of rock crack, and the properties of cracks. The three are combined to classify and evaluate them.

Table 3 is a comparison of the boring cores given in the preceding table with the classifications from test pit observation, with additional reference to their relation with the principal indexes.

2.2 Relation of rock mass properties versus measuring indexes

Granite, according to the processes of weathering progress from intact rocks, crack propagation, deterioration or softening, and growth into weathering soil, can be placed in one of the four different types; (1) a crack-family hard rock zone, (2) a crack-family soft rock zone, (3) a hard, granular rock zone, and (4) an earthy soft rock zone. Figure 2 shows a schematic diagram of them. Granite belongs in comparatively many cases to zone (1) and zone (4). As the measuring indexes vary intricately according to the degree or rock mass cracking and the degree of influence by hard-cracking and the degree of influence by hardness, their relation with the measuring parameters may be shown in Figure 2.

Table 1. Rockmass classification by boring core observation (Granite)

Class	Color Tone	① Degree of Hardness	② Degree of Weathering and Deterioration	③ Condition of Cracks	④ Shape of Core	Remarks
A	Bluish gray milky grey	Extremely hard. Metallic sound when struck by hammer. Below 2 cm/min with D.B..	Generally fresh crack surface. No weathering.	Few cracks,at a spacing of 20-50cm.	Rod-long columnar shape,sampling is done in sizes longer than about 30 cm.	
B	Milky grey brownish grey(light).	Hard. Light metallic sound when struck by hammer 2-4 cm/min with D.B..	Generally fresh Weathering along cracks. Deteriorated part shows brown.	Mainly 5-15cm of cracks spacing Partly opened.	Short columnarrod shape,sampling is done in sizes generally shorter than 20cm.	③④ are A, but ①② are B. ①② are A, but ③④ are B.
CH	Brownish grey-greyish brown(light).	Medium hard. Dull sound when struck by hammer. Hardness so as to be able to leave an incision with a knife Above 3cm/min with D.B..	Weathering developed along cracks. Feldspar,etc.,party discolored and deteriorated.	Cracking developed,clay is sandwiched in the openings. Haircracks developed easy to crack.	In the from of pieces of large sized rocks pieces,generally smaller than 10cm,many of them smaller than 5cm. Can be returned to original shape.	Short columnar shape but weathering developed and soft.
CM	Greyish brown light yellowish brown	Slightly soft-hard Easily broken when struck lightly by hammer. Able to be marked by finger nail. Suitable for D.B.. excavation.	Weathering developed except part of inside rock. Feldspar,mica,etc.,are generally weathered.	Cracks developed below 5cm of space. Clay is sandwiched in the openings.	In the form of rock pieces-small pieces(breccia),easy to break. Many of them are not round. Difficult to return to original shape.	
CL	Light yellowish brown yellowish brown	Soft. Very friable even with finger. Can be drilled by M.C..	Although weathering developed inside the rock,the rock structure remains. Quartz remains intact and unweathered.	Many cracks,but clay content developed, closely adhered.	In the from of small pieces,rock pieces remain. Easy to break even by fingers, forming powder. No circular core.	Samples were taken from the central portion of the crushed rock zone.
D	Yellowish brown.	Extremely soft. Very friable and tends to powderize. Can be drilled by M.C.without water.	Weathering developed uniformly. Decomposed granite Rock pieces slightly remain.	No crack because of developed clay content.	Residual soil form.	No samples can be taken in crushed zone nor in clay zone.

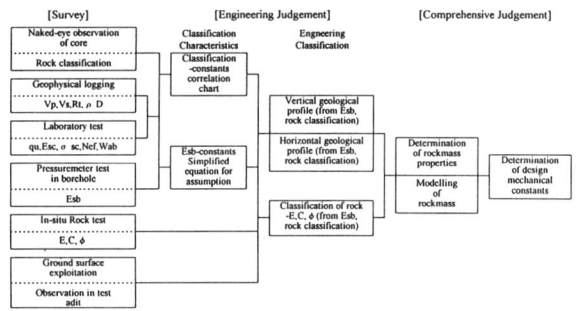

Figure 1. Block diagram of survey,engineering judgement and comprehensive judgement for granitic properties

Table 2(a). Rockmass classification by naked-eye observation in the test adit

Rock Class	Subdivision	Observation in the Test Adit / Condition of Rock
A	A, I, a	Fresh and hard,no deterioration in the rock-forming minerals. Crack spacing larger than 50cm. Cracks are closely adhered,no deterioration nor discoloration.
B	A, II-III,b	Hard:Rock color is light brown.Crack spacing about 15-50cm. Limonite adhered along cracks.
CH	B, III-IV,b-c	Relatively hard:Biotite and plagioclase are somewhat deteriorated.Crack spacing about 5-30cm. Very thin clay is sandwiched along the opening.
CM	C,IV-V,c	Breaks when struck by hammer.Deterioration of plagioclase developed.Crack spacing smaller than 15 cm.Clay is sandwiched along the opening face.
CL	C-D,III, a-b;C,IV-V,d	Biotite turns golden color,but quartz particles are hard. Plagioclase is deteriorated. When struck by hammer breaks into pieces.Crack spacing smaller than 5cm.
DH	D, II-III,b,D, III,a-b	Can be broken by hand.It is easy to break by hammer. Biotite turning to golden color,and brown in the periphery.Particles are hard, forming small,sand-like pieces. Apparent spacing of cracks becomes wider.
DM	E1, I-II, b-c;E1, II, b	Breaking by hand,it becomes sand-like remaining crystal of quartz and potassium feldspar. Mica loses its crystal form and plagioclase is mostly deteriorated. Apparent spacing of cracks becomes even wider.
DL	E2, I, c	Breaking by hand,mostly becomes powder,expect for party sand from.Most feldspar is deteriorated and becomes clayish soil. Original joint planes become indistinguishable.

Table 2(b). Criteria for subdivision of rockmass classification in the test adit

① Hardness classification

Class	Criteria for Judgement
A	When struck by hammer,rock piece cannot be broken easily, with metallic sound. Fresh, no deterioration of rock-forming minerals.
B	When struck by hammer, makes metallic sound-resonant sound.Joints are adhered, fresh.
C	Rock becomes broken when struck lightly by hammer,making resonant sound. (Smashing by finger-pressure for more than 20times,rock piece keeps almost intact)
D	Crushing by finger-pressure barely being possible,each piece is hard with feld-spar remained in the periphery of the quartz. (fragmental-sandy)(Rock pieces become broken by 7-10times finger crushing with more than 70% medium-small pieces)
E_1	Crushed when squeezed with finger,remaining particles of quartz and potassium feldspar. (Pieces become broken by 3-5 times finger crushing with 30-50% in powder form,50-90% in small pieces)
E_2	Generally in powder form when crushed by finger-pressure in the palm partly sand form.(Pieces become broken by 1-3times finger crushing with more than 50-70% in powder form)

② Classification by crack spacing

Class	Judgement Criteria
I	Larger than 50cm
II	50～30 cm
III	30～15 cm
IV	15～ 5 cm
V	Smaller than 5cm

③ Classification by cracking condition

Class	Judgement Criteria
a	Closely adhered,no deterioration or discolouring.
b	Adhesion of limonite along adhered cracks or very thin clay (brown in color) is sandwiched.
c	Deterioration along crack,about 1-2cm clay (white-greyish white) is sandwiched.
d	Opening.

Table 3. Relation between borehole and test adit measurements by rockmass classification

Rock class	Borehole Measurements and Observation					Test Adit Measurements and Observation								
	Esb (MPa)	Core hardness	Core shape	RQD (%)	Vpb (km/s)	Hardness	Crack spacing	Cracking condition	Sh (num.)	Yh (mm)	Dh (mm)	St (%)	N_R (%)	ρ t
D_L	5〜30	E_2	VI	0	<1.2	E_2	I	c	<4	<27	>100	—	35〜43	1.90
D_M	30〜80	E_1	VI	0	<1.5	E_1	I〜II	b〜c	5〜2	28〜32	70〜99	—	20〜35	2.10
D_H	80〜150	D	V〜VI	0〜10	1.5〜2.5	D	II〜III	a〜b	13〜25	33〜36	30〜69	<10	14〜20	2.20
C_L	150〜300	C〜D	IV〜V	0〜25	2.0〜3.3	C〜D	IV〜V	d a〜b	26〜39	>37	<29	11〜20	10±	2.40
C_M	300〜600	C	IV	0〜50	3.0〜4.2	C	IV〜V	c	40〜49	—	—	21〜30	5±	2.50
C_H	600〜1,200	B	III	25〜75	4.1〜5.0	B	III〜IV	b〜c	50〜60	—	—	>31	3±	2.60

Remarks: Sh: Shore hardness, Yh: Soil hardness by Yamanakas type penetrator, Dh: Penetration length by concrete drive bit, St: Rebound value by Schmidt Hummer test, N_R: Rockmass porosity, ρ t: Wet density

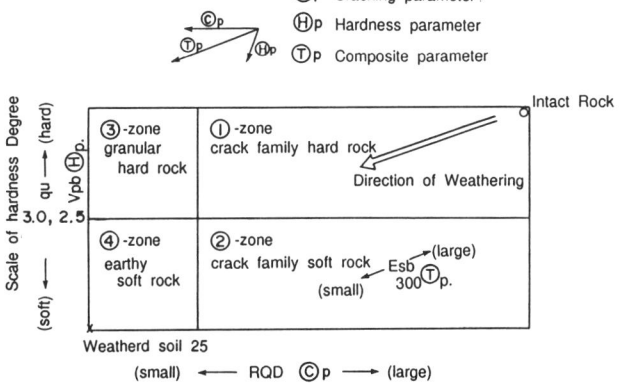

Figure 2. Weathering model by hardness and cracking parameter model

And the relation of the elements of effect on engineering properties by measuring indexes and rock mass properties bear to the degree of effect is shown in Table 4.

2.3 Relation of rock mass classification versus measurement values

Shown in Figure 3 is the composite relation of rock mass classification versus borehole physical (electricity, density, velocity) logging, borehole loading test, and rock test. Putting together the relation of the calculated values of rock mass porosity N_R and saturation S_r, as shown in the figures, we obtained the relation of measurement values according to rock mass classification as shown in Table 5. This table indicates that rock mass classification is to be determined quantitatively and objectively, not only by unaided-eye observation, but also by the use of a number of indexes.

2.4 Relation of deformation properties versus strength properties according to rock mass classification

The relation of deformation properties versus initial shear stress was studied in a shear test on the same material to obtain the following formula (see Figure 4).
Also, the following formula was obtained in a bearing capacity test (see Figure 5):

Table 4. Classification factor and degree on rockmass properties exert influence on measurement index parameters and engineering properties

Classification of rockmass Engineering properties parameters	Crack family type						Earthy granular type					
	③ - zone soft rocks			① - zone hard rocks			④ - zone soft rocks			② - zone hard rocks		
	Ⓗ	Ⓒ	Ⓣ	Ⓗ	Ⓒ	Ⓣ	Ⓗ	Ⓒ	Ⓣ	Ⓗ	Ⓒ	Ⓣ
Esb	◎	○	◎	◎	○	◎	◎	○	◎	◎	○	◎
RQD	○	○	○	△	◎	△	○	△	△	△	△	△
core shape	○	△	○	○	◎	○	◎	○	◎	△	◎	○
core hardness	◎	△	△	◎	△	○	◎	△	○	◎	×	×
core classification	○	○	◎	○	○	◎	○	○	◎	○	○	◎
Vpb	◎	○	◎	◎	○	○	◎	○	◎	○	○	○
Vsb	○	○	○	△	○	×	△	○	△	△	○	△
Rt	○	○	○	○	○	◎	○	○	◎	○	○	○
ρ t	○	○	×	○	△	×	○	○	○	○	△	△
Esb/qu	×	○	×	×	◎	×	—	—	—	×	○	×
Esb/Esc	×	○	×	×	◎	×	—	—	—	×	○	×
qu	○	×	△	◎	×	△	—	—	—	—	—	—
Sh	○	×	△	○	×	△	○	×	○	×	—	×
Nef	◎	△	○	◎	×	△	◎	○	◎	◎	○	◎
strength parameter	◎	△	△	△	◎	○	◎	△	○	◎	○	○
deformation parameter	○	△	○	○	○	◎	◎	△	○	○	△	○

Note: degree of effect (◎ remarkable, ○ moderate, △ little, × nothing, — unknown)

$$Es = 250 \, \tau_0^{1.160} \quad (1)$$
$$q_d = 0.032 \, Es^{1.199} \quad (2)$$
$$q_y = 0.016 \, Es^{1.199} \quad (3)$$

where q_d = ultimate bearing capacity
q_y = yield bearing capacity
Es = deformation modulus

Studying these results, we determined, as given in Table 6, the relation of mechanical parameters according to rock mass classification. In addition, the relation of actual shear stress values versus estimated values to be adopted in design.

3. JUDGING SUPPORT PROPERTIES

3.1 Estimating geomechanical properties by the subject matter of investigation

Principal among the parameters required to determine the bearing capacity of bridge foundation rock masses are three elements; deformation modulus, strength parameter, and creep ratio.
These can be estimated through direct mechanical tests (e.g. in-situ rock mass loading test, shear test, etc.) of the rock mass,

Table 5. Engineering properties and borehole measurement

Rock class	Core observation		Borehole measurement			Material values of core and rockmass			
	Core shape	Core hardness	Vpb (km/s)	Rt (Ω-m)	Esb (MPa)	ρ_C (t/m³)	ρ_R (t/m³)	n_C (%)	n_R (%)
D_L	VI	E2	1.5~1.8	1~ 4	5~ 30	1.7 ~2.1	1.55~1.75	33 ~ 58	50~64
D_M	VI	E1	1.8~2.2	4~ 7	30~ 80	2.1 ~2.3	1.75~2.0	19 ~ 33	37~50
D_H	V~VI	D	2.2~2.7	7~ 12	80~ 150	2.3 ~2.5	2.0 ~2.15	11 ~ 19	27~37
C_L	IV~V	C~D	2.7~3.3	12~ 20	150~ 300	2.5 ~2.55	2.15~2.3	6 ~ 11	20~27
C_M	IV	C	3.3~4.0	20~ 50	300~ 600	2.55~2.6	2.3 ~2.4	3.5 ~ 6	15~20
C_H	III	B	4.0~4.8	50~120	600~1200	2.6 ~2.65	2.4 ~2.5	2.0 ~ 3.5	11~15
B	II		>4.8	>120	>1200	2.65~2.66	2.5 ~2.6	1.1 ~ 2.0	8~11
A	I					2.66~2.67	2.6 ~2.65	0.65~ 1.1	6~ 8

or indirect physical surveys and tests (e.g. borehole measuring, geophysical logging, etc.) A variety of forms are available, however, for these surveys and tests, while estimated values and methods of estimation vary depending on design accuracy, etc.

Therefore, we chose to use one of the three methods for evaluation of bearing properties shown in Figure 6, according to the importance of structures, time and stage of survey and design, etc.

Method A calls for estimation by geological observation at the installation site of structure foundation or at the location of a similar rock mass and by borehole measuring and rock mass testing as well. This method was used mainly for foundation rock mass surveys of important and large-scale structures such as strait crossing bridges, at the stage of detail design.

Method B calls for geological observation and borehole measuring at the field site of structure foundations, as well as for use of separately specified mechanical parameters according to rock mass classification. This survey method, which employs boreholes, is an effective procedure for studying submarine rock masses, and was used most frequently in surveying the strait communication bridge.

Method C is one that estimates by using rock mass classification table (see Tables 1 and 2) based on boring cores at the field site of foundations and borehole evaluation with the unaided-eye.

This method, being a handy quick procedure, and left to engineering evaluation on the part of geological engineers, has been employed through its mastery, though open to question

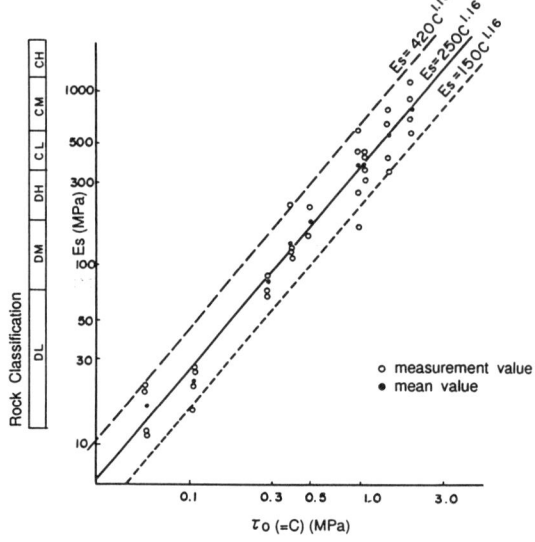

Figure 4. Relation between deformation modulus and initial shear strength by in-situ block shear test

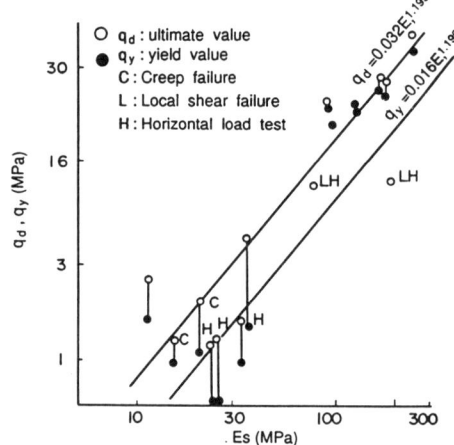

Figure 5. Relation between bearing capacity and deformation modulus by in-situ load test

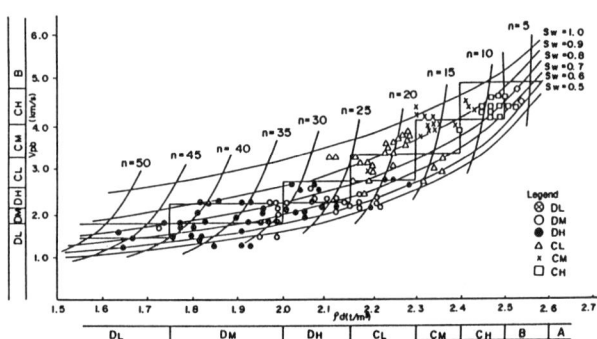

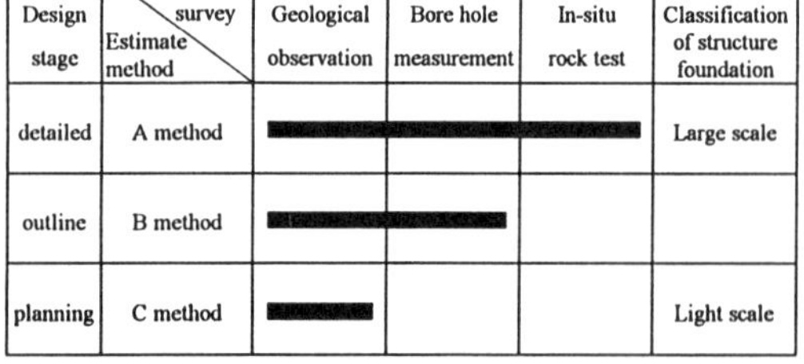

Figure 3. Relation between ρ_d and Vpb by rockmass classification (as parameters of n_R and S_r)

Table 6. Geomechanical parameters by rockmass classification

Rock Class	Deformation parameters			Strength parameters		Bearing capacity (φ 40cm)		
	Esb (MPa)	Es (MPa)	E_o (MPa)	C (KPa)	φ (deg.)	qd (MPa)	qy (MPa)	qa (MPa)
D_L	5~30	12.5~75	25~150	24~170	30	3.3~50	1.7~25	>0.73
D_M	30~80	75~180	150~360	170~360	35	50~140	25~70	>11
D_H	80~150	180~330	360~660	360~600	37.5	140~280	70~140	>31
C_L	150~300	330~600	660~1,200	600~1,000	40	280~570	140~285	>62

Design stage	Item of survey / Estimate method	Geological observation	Bore hole measurement	In-situ rock test	Classification of structure foundation
detailed	A method	■■■	■■■	■■■	Large scale
outline	B method	■■■	■■■		
planning	C method	■■■			Light scale

Figure 6. Relation between estimation method, item of survey and design stage

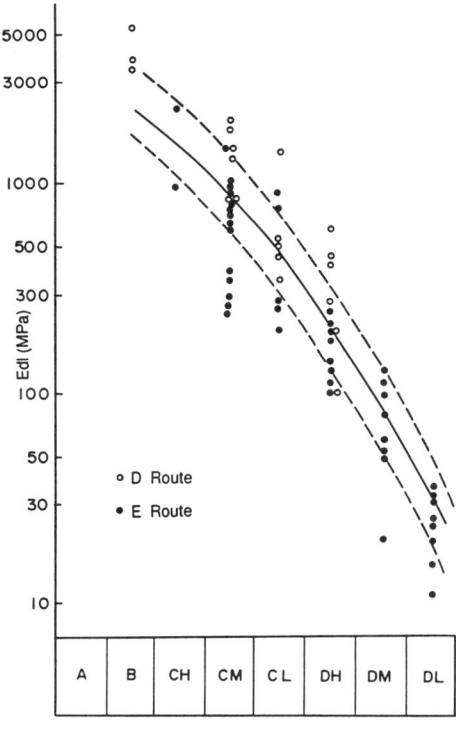

Figure 7. Relation between rockmass classification and deformation modulus by in-situ load test

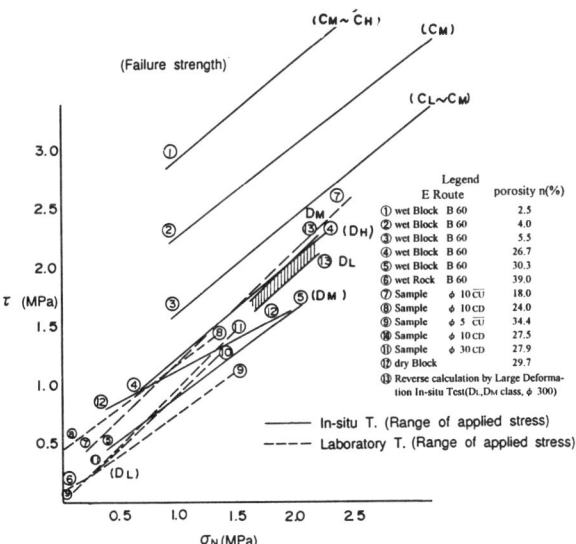

Figure 8. Strength parameters on laboratory and in-situ strength tests by rock mass classification

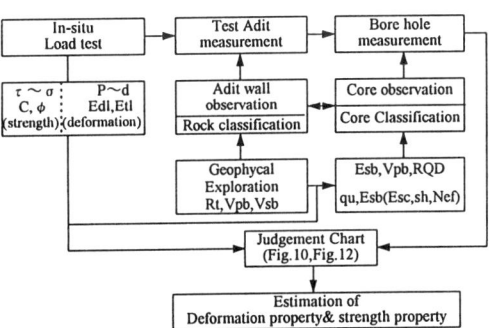

Figure 9. Flow chart of estimation for deformation property and strength property

accuracy-wise, mainly on land-section small-scale bridges and at the stage of general survey.

The relation of rock mass testing done by Method A versus rock mass classification is shown in Figures 7 and 8.

The test results, after being supplemented by technological evaluation, were either established as measurement values themselves or multiplied by particular transition factors for use in design work.

3.2 Estimating deformation modulus E

(1) Rock mass classification concerning deformation properties and estimation of deformation moduli were performed by one of the following methods, according to the importance and survey stage of structures.
1) Method A: Estimation that mainly uses measurement value derived from in-situ or similar-rock-mass plate loading test results.
2) Method B: Estimation that carries out borehole measuring and uses the deform classification of Table 7 (a).
3) Method C: Estimation that resorts to Table 7 (b) using rock mass classification based on rock mass and core observation with the unaided-eye.

Next, for translation of design value E from measurement value E, we decided to use the translation coefficient α of Table 8 in deriving ordinary and earthquake moduli of elasticity E_N and E_E in a standard loading range (B = 30 cm) from value E as estimated from various measurements and rock mass classification tables.

(2) Borehole measurement toward estimation of deformation properties

A procedure for estimating rock mass deformation properties mainly by borehole measuring is shown in Figure 9. The procedure is one in which the results of rock mass testing, borehole physical logging, and borehole measuring are compared to estimate deformation moduli according to rock mass classification by utilizing classification elements such as borehole measurement values E_{sb}, V_{pb}, RQD, q_u and coreshape, etc.

Figure 10 represent the relation of E (indicating a general scale), V_{pb} (indicating a hardness scale), and RQD (indicating a crack scale) used for that study. This was employed to estimate the deformation moduli following Figure 11 (a) (b).

The value E in the table is an index of direct and principal general scale, but in the case of hard rock masses rated at V_{pb} and $E_{sb} \geq 300$MPa were adopted as an index of hardness degree and an index of crack degree, respectively, for classification purpose (see Figure 12(a)).

3.3 Estimating strength moduli (C, φ)

(1) Similarly to the preceding section, we used one of the following three methods.
Method A: Estimation by rock mass shear testing carried out in survey pits.
Method B: Estimation by borehole measuring and using the strength classification readings of Table 9 (a).
Method C: Estimation by classifying rock masses with unaided-eye observation and using the strength classification readings of Table 9 (b).

(2) The procedure for estimating strength properties with borehole measuring by Method B is shown in Figure 9.

While the comparison of rock mass testing, borehole physical logging, and borehole measuring results is similar to that of deformation properties, the rock mass properties wee divided into two; an earth-to-granular rock group and a crack rock

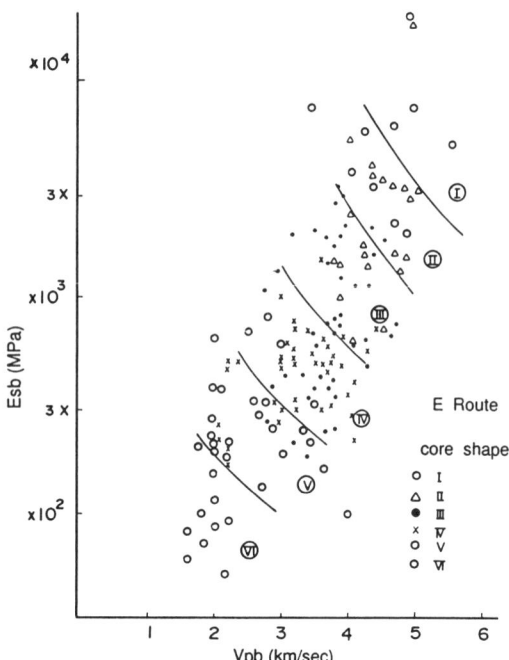

Figure 10. Relation between Vpb and Esb as parameters of core shape

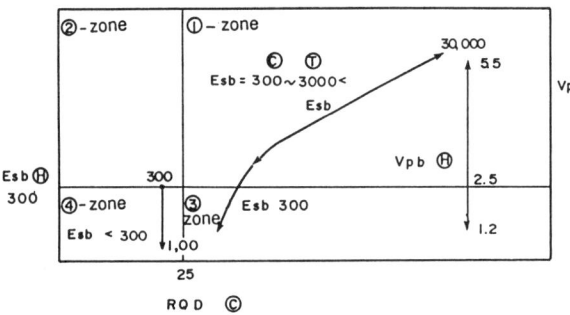

Figure 11(a). Indication of deformation property and effective measurement index by weathered type (B-method)

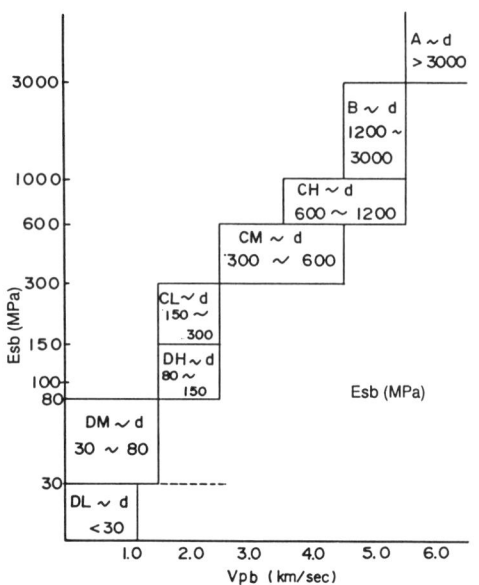

Figure 11(b). Judgement chart for estimation and classification of deformation property

Table 7(a). Deformation modulus on indication of deformation classification (B-method)

Indication of deformation classification	Range of deformation modulus Esb (MPa)	Representative value of Esb (MPa)
A~d	over 3,000	5,000
B~d	1,200~3,000	2,000
C_H~d	600~1,200	800
C_M~d	300~600	450
C_L~d	150~300	200
D_H~d	80~150	100
D_M~d	30~80	50
D_L~d	5~30	10

Table 7(b). Deformation modulus on rockmass classification by core observation (C-method)

Rockmass classification by observation	Representative value of Esb (MPa)	Range of estimated value of Esb (MPa)
A	2,000	over 2,000
B	800	800~3,000
C_H	450	450~1,200
C_M	200	200~600
C_L	100	100~300
D_H	50	50~150
D_M	10	10~80
D_L	5	5~30

Table 8. Conversion table of E-values used for design value

Deformation modulus by following test methods		Coefficient α	
		Ordinary condition E_S	Earthquake condition E_D
Deformation modulus by plate load test (ϕ 30cm) Edl		1.0	—
Tangent modulus by plate load test (ϕ 30cm) Etl		—	1.0
Deformation modulus by borehole test Esb(MPa)	Classification		
	D_L~D_M~d (Esb≤80)	2.5	5.0
	D_H~C_L~d (Esb≤300)	2.2	4.5
	C_M~C_H~d (Esb>300)	2.0	4.0

Table 9(a). Strength parameters on indication of strength classification (B-method)

Strength classification	Estimated value		Representative value	
	C (MPa)	φ (deg.)	C (MPa)	φ (deg.)
A~s	2~3	>50	2	50
B~s	1.5~2.5	45~50	1.5	45
C_H~s	1.5~2.0	40~45	1.5	45
C_M~s	1.0~1.5	>40	1.0	40
C'_M~s	0.5~1.5	40~45	0.5	40
C_L~s	0.5~1.0	35~40	0.5	40
C'_L~s	0.5~1.5	35~40	1.0	35
D_H~s	0.1~0.5	37~45	0.1	37
D_M~s	0.01~0.08	35~42	0	35
D_L~s	0~0.08	30~35	0	30
D'~s	0.5~1.5	30~35	0.5	30

Table 9(b). Strength parameters on rock mass classification by core observation (C-method)

Rockmass classification by observation	Representative value		Estimated value	
	C (MPa)	φ (deg.)	C (MPa)	φ (deg.)
A	1.5	45	1.5~3.0	>45
B	1.5	45	1.5~2.5	45~50
C_H	1.0	40	1.0~2.0	40~50
C_M	0.5	40	0.5~1.5	40~45
C_L	0.1	37	0.1~1.0	35~40
D_H	0	35	0~0.5	35~40
D_M	0	30	0~0.8	30~35
D_L	0	30	0~0.8	30~35

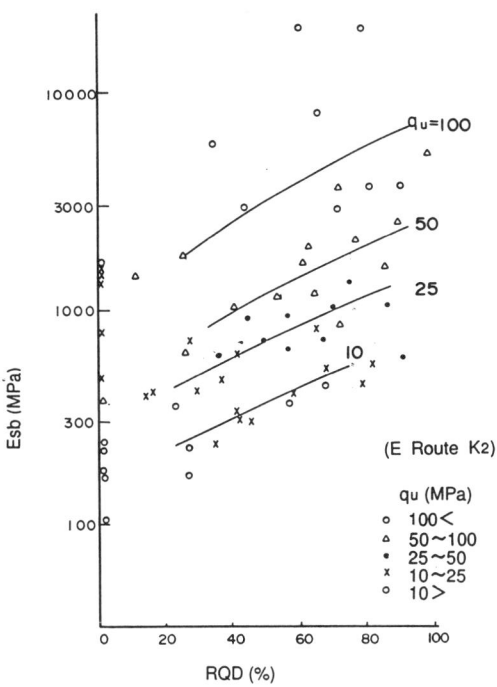

Figure 12(a). Relation between deformation modulus and RQD (as parameter of hardness degree by qu)

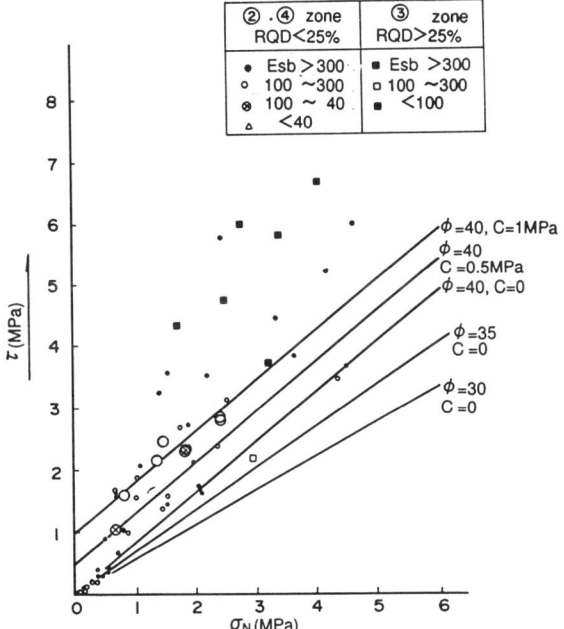

Figure 12(b). Strength parameter by in-situ and bore hole test (as measurement parameters of Esb, qu and RQD)

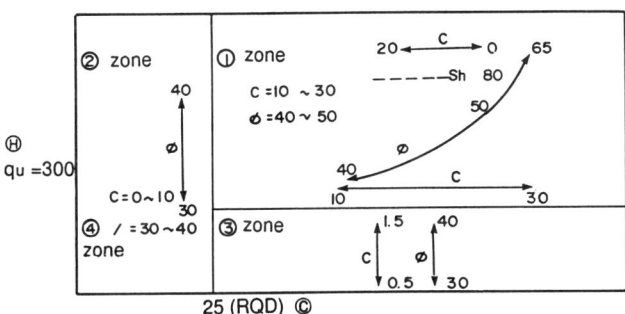

Figure 13. Indication of strength parameter and effective measurement index by weathered type

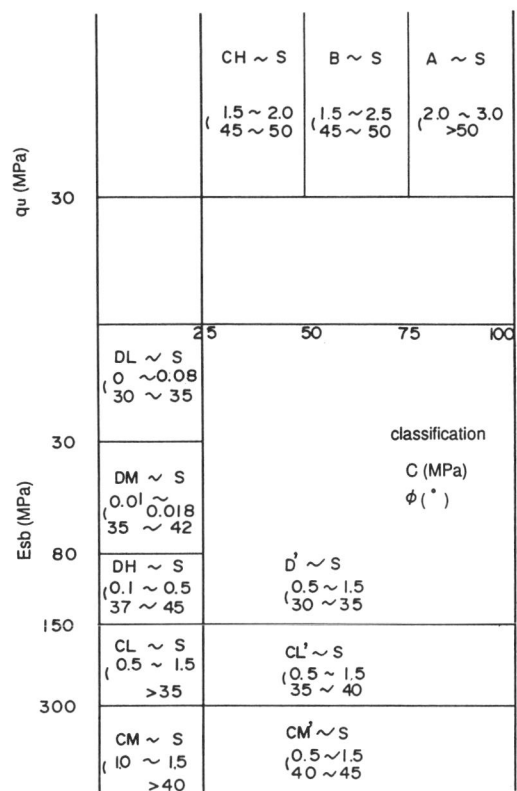

Figure 14. Judgement chart for estimation and classification of strength parameter

group, which were classified by respective measuring indexes (in the case of a crack rock group, qu, RQD, Esb). The study examples consulted for classification are shown in Figures 7 and 8 (see Figure 12 (b)).

The classification processes are detailed below.

1) First, since RQD < 25% represents the border between "very poor" and "poor" as distinguished by Deer (1967), this crack density value divides rock masses roughly into two groups, earth-to-gravel (soft, much weathered, brittle) ones and cracked ones (generally hard) ones.

2) Concerning RQD ≥ 25%, rock mass properties vary between masses which consist of hard rocks, being governed by crack quality (rock degree), and those which consist of soft rocks, being subject to the effect of rock hardness of softness (weathered degree).

3) This determines whether there exists an inter-relation between RQD and crack quality, qu=30MPa is used as a border for roughly dividing into soft rocks and hard rocks.

4) Hard rocks with cracks rated at RQD ≥ 25% and qu ≥ 30 MPa are classified into three by RQD that can become a principal index.

5) Hard rocks rated at RQD < 25% are classified into three by Esb since the loosening of rock masses and the hardness or softness of rocks that relate to strength properties are well reflected in E.

6) Gravel-to-earth rock masses rated at RQD and composed of much weathered soft rocks are classified into three Esb, holding that strength properties and deformation properties correspond well.

7) The above classification procedures are used to classify rock masses into eight A~s, B~s, CH~s, CM~s, CL~s, DM~s, DL~s.

3.4. Estimating creep ratios

One of the following methods was used.

Method A: Estimation that mainly uses actual measurement values from rock mass creep loading test.

Method B: Estimation that carries out borehole measuring and utilizes the deformation classification readings of Table 10 (a).

Method C: Estimation that makes unaided-eye observation and utilizes Table 10 (b).

To estimate creep deformation properties, long-time (180 days) borehole rock mass creep testing and triaxial compression testing of cored samples were carried out. We considered the validity of a visco-elastic model (a Voigt model) from the test results, and derived the visco-elastic moduli by rock mass class.

Table 10(a). Creep parameters on indication of deformation classification (B-method)

Deformation classification	Creep ratio (representative) G^*/G	Coefficient of viscosity (representative) η (MPa·min/cm^2)
A~d	—	—
B~d	—	—
C_H~d	0.5	4.9×10^{14}
C_M~d	0.5	1.5×10^{12}
C_L~d	0.5	1.5×10^{11}
D_H~d	0.83	2.6×10^9
D_M~d	0.83	8.7×10^8
D_L~d	1.0	9.2×10^7

Table 10(b). Creep parameters on rockmass classification by core observation (C-method)

Rockmass classification by observation	Creep ratio (representative) G^*/G	Coefficient of viscosity (representative) η (MPa·min/cm^2)
A	—	—
B	—	—
C_H	0.5	5×10^{14}
C_M	0.7	5×10^{12}
C_L	0.7	5×10^{11}
D_H	0.9	5×10^9
D_M	1.0	1×10^9
D_L	1.0	1×10^8

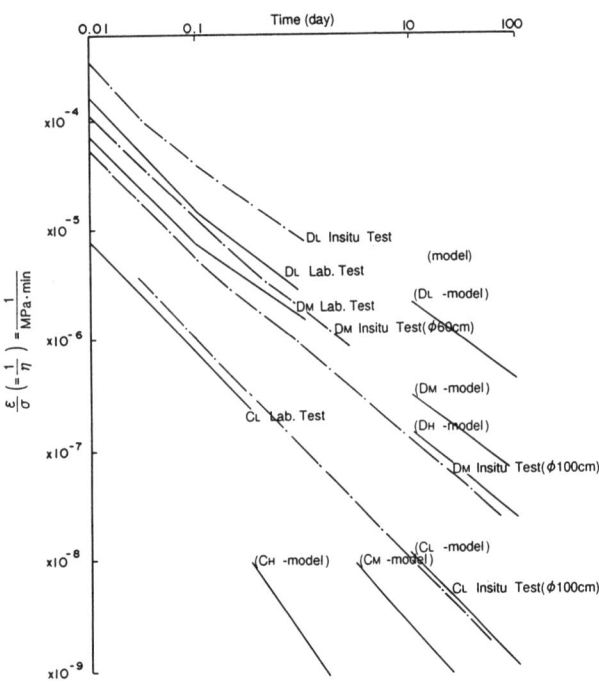

Figure 15. Relation between creep time and strain rate by in-situ and laboratory tests (as parameter of rockmass classification)

4. CONCLUSIONS

The conclusions from the present research are stated below.

1) Rock mass properties can be indicated by two classification elements, or the state or degree of rock mass cracks (hereinafter crack scale), and the degree of weathering and hardness/softness (hereinafter, hardness/softness scale), to form a model of the rock mass properties.

2) Next, measuring indexes that greatly affect the above classification elements can be adopted to make clearer the relation of rock mass classification versus engineering properties.

3) Various classified engineering properties, such as deformation properties and strength properties, can be estimated by a rock mass classification method using measurement values.

4) This method of estimation has made it possible to select the accuracy of survey according to the importance, survey stage, etc., of structures.

To sum up, the integration of survey and design work by the flow from survey accuracy through rock mass classification to the setting of design ground parameters concerned with deformation and strength properties is the first attempt and proposal of its kind as well. In addition, this research is currently adopted within the Honshu-Shikoku Bridge Authority as "a procedure for judging support properties."

5. ACKNOWLEDGMENT

The present research, one of the achievements in the geological survey on the Honshu-Shikoku Bridge that was started in 1971, has received numerous pieces of kind instruction and cooperation from many interested agencies and sources, including the Construction Ministry, universities, and geological consultants. We have also enjoyed kind information from the learned members of inquiry and study committees on geology and ground established at the Japan Society of Civil Engineers and the Japanese Society of Soil Mechanics and Foundation Engineering. This we profoundly appreciate.

REFERENCES

Ishikawa K. et al. 1981. Geological investigation utilizing borehole measurements system and judgment of engineering properties for weathered granite: International Symposium on Weak Rock. Tokyo. vol.2 p.387~392

Japan Society of Civil Engineers. 1980. Technical Report on Geological Investigation and Seismic Design for Honshu-Shikoku Junction Bridge

Geological investigation and evaluation of foundation bedrock for long-span bridges in Japan

S. Yamamoto
Honshu-Shikoku Bridge Authority, Tokyo, Japan

Y. Kimura
Public Works Research Institute, Ministry of Construction, Tsukuba, Japan

K. Takahashi
Kiso-Jiban Consultants Co., Osaka, Japan

K. Miyajima
Chuo Kaihatsu Corporation, Tokyo, Japan

ABSTRACT: Long-span bridges in Japan are constructed mainly on sedimentary or weathered soft rocks at straits or bay areas. The contact pressure of long-span bridge foundation is remarkably larger than those of ordinary bridge foundation. Consequently, several geological investigations are performed to obtain precise mechanical characteristics of these soft rocks. As many undersea foundations are constructed for these bridges, a geological investigation system consisting of drilling and borehole measurements is established. Further, in-situ rock tests at land site are conducted to determine geotechnical parameters. For design of substructure, design ground model and geotechnical parameters are set up based on comprehensive evaluation of these investigation results. As the case of geological investigation and evaluation for long-span bridge foundation, Honshu-Shikoku Bridges built on sedimentary soft rock, Kobe Foundation and weathered soft granite are summarized and pointed out future problems.

1. FOUNDATION BEDROCK FOR LONG-SPAN BRIDGES IN JAPAN

Most long-span bridges in Japan are suspension bridges and cable-stayed bridges. As shown in Figure 1, many such bridges have been constructed in the country's major bays Tokyo Bay, Osaka Bay, Ise Bay, etc. and across marine straits such as the Seto Inland Sea, Kanmon Strait, etc. In many of these cases, undersea foundations have had to be constructed. Most of these long-span bridges have been constructed on soft rock consisting of Tertiary sedimentary formations and weathered Tertiary granite. This has necessitated gaining a detailed understanding of the mechanical characteristics of these soft rocks through geological investigations and tests, since the contact pressure of long-span bridge foundations is considerably greater than that imposed by conventional bridges.

Long-span bridge foundations are of various types, including spread foundations, laid-down caisson foundations, caisson foundations, continuous wall foundations, multi-column foundations, and pile foundations. The following is a summary of typical undersea foundations for long-span bridges in Japan:
(1) All laid-down caisson foundations are constructed on soft or hard rock. The surface layer excavation is not very deep, except in the case of the Akashi Kaikyo Bridge and the Seto Bridges.
(2) Caisson foundations are constructed in less than 15 m of water, while pneumatic caisson foundations are used up to a maximum water depth of 52 m. Where deeper foundations are required, the open caisson method of construction is used.
(3) The bell-type foundation, one variation of the pile foundation, is used where the water depth is between 15 and 25 m.
(4) Multi-column foundations are often employed where the water depth is between 5 and 20 m. The piles are mostly less than 20 m long, except in the case of the Yokohama Bay Bridge, where they were 70 m.

A summarized description of the foundation structures for the Akashi Kaikyo Bridge, which will be the world's longest suspension bridge when completed, is given in Table 1.

2. OUTLINE AND DETAILS OF GEOLOGICAL INVESTIGATIONS AND TESTS

2.1 Outline of geological investigations

The long-span bridges in the Honshu-Shikoku Bridge system have foundation sizes and contact pressures of the order of 50 to 100 m and 1 and 2 MPa, respectively. This leads to the following problems when implementing geological investigations leading to the design of foundations:

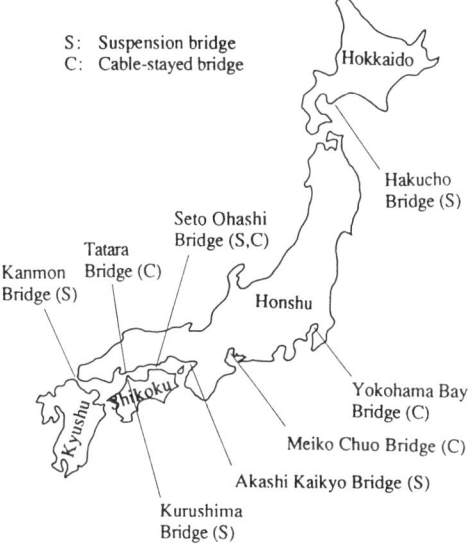

Figure 1. Location of main long-span bridges in Japan

Table 1. Outline of foundation structures

Foundation	Anchorage 1A	Tower pier 2P	Tower pier 3P	Anchorage 4A
Bottom level of foundation	T.P.-60 m	T.P.-60 m	T.P.-57 m	T.P.-20 m
Dimension	W59m, L85m	Dia. 80 m	Dia. 80 m	W59m, L85m
Foundation type	Direct	Direct	Direct	Direct
Bearing ground	Kobe F. Alt. of sandstone and mudstone (soft rock)	Akashi F. Pleistocene sand and gravel	Kobe F. Sandstone (soft rock)	Granite (hard rock)
Construction method	Continuous wall Dia. 90 m	Prefabricated caisson	Prefabricated caisson	

(1) The strength and deformation characteristics of the load-bearing layer sedimentary soft rock, weathered soft rock, diluvial gravel, etc. are not thoroughly understood.
(2) Although it is necessary in many cases to construct undersea foundation structures, there is great difficulty in verifying the strength and deformation characteristics of load- bearing rock below the surface.
(3) There are as yet no established analytical methods for determining the bearing strength and deformation of soft rock when loaded with large foundations. We have now developed a method of geological investigation for use in such circumstances, the procedure of which is outlined below.

1) The preliminary investigation consists of establishing a bridge route and construction plan. This is done by collecting existing topographical and geological data, carrying out geological surveys, etc. Further, an outline investigation is carried out to narrow down the route according to topographical and geological conditions. This consists of undersea geophysical prospecting (sonic prospecting) and the implementation of a small number of drillings, etc. A layout plan of the structures is then determined and a detailed investigation plan established.

2) The main method of investigation employed for the detailed investigation of load-bearing ground for undersea foundations is drilling. Drilling is carried out by means of all coring to determine geological formation. In addition, the tests and measurements described below are carried out to determine material properties and rock classification.
· Geophysical logging P and S wave logging, electrical logging, density logging, and caliper logging
· In-situ tests pressuremeter tests and standard penetration tests
· Laboratory tests soil tests and rock tests

3) The in-situ tests are carried out onshore using ground similar to that at the undersea location of the foundations. The correlation between the actual undersea bearing layer and the onshore test location can be ascertained after comparing and examining the results of core observations, geophysical logging, borehole loading tests, etc. carried out at both locations.

4) The geotechnical design parameters are established after a comprehensive examination of the results of the various tests. At the same time, the examinations described below are carried out which are needed in the analysis of ground behavior are carried out.
· Comparison of results of borehole loading tests and plate loading tests. (Determination of conversion factor)
· Examination of deformation and failure models of the ground by analyzing the behavior of the loading plate
· Examination of applicability of analysis techniques to the actual structure

5) To maintain control over construction and to verify the investigations, confirmatory inspections of the load-bearing ground and monitoring of the ground and structures are carried out. Monitoring consists mainly of measuring deformation, and forecasts of settlement and deformation are made by means of reverse analysis, making the results available for construction control and verification of the adequacy of the geotechnical design parameters.

Figure 2 illustrates the procedure from geological investigations to foundation design. As explained, the ground model and geotechnical design parameters are established according to the results of various geological investigations and tests, and the foundation layout and dimensions are established after carrying out an analysis of bearing capacity and deformation. The analytical model is set up by making appropriate modifications to the basic model to reflect local soil conditions.

2.2 Details of geological investigations and tests

a. Bridge scale and geological investigations
In Table 2, the geological investigations carried out for existing straits bridges in Japan are classified according to the maximum contact pressure applied to the load-bearing layer, which is a measure of the scale of the bridge.
(1) In the case of bridges where the maximum contact pressure was between 300 and 600 PA, physical and dynamic tests are used in combination with core drilling. Core tests are common in the case of soft sedimentary rock. In some cases, however, plate loading tests and bearing capacity tests were carried out in test adits.
(2) In the case of bridges where the maximum contact pressure was between 600 and 1000 PA, core drilling and borehole loading tests, physical logging for rock classification, and physical and mechanical tests of core samples were implemented. In addition, plate loading tests and block shear tests were performed in onshore test adits.
(3) In the case of large-scale bridges where the maximum contact pressure exceeded 1 MPa, systematic geological investigations were performed. Modeling of ground and estimation of geotechnical parameters are performed by means of comprehensive investigation including numerous core drillings physical logging, core tests, in-situ rock test, etc.

b. Soft rock type and geological investigations
The foundations for long-span bridges in Japan are built on either soft sedimentary rock or weathered soft rock. Since the origin and properties of these two main rock classifications differ, the respective investigation techniques given below are used.
(1) Sedimentary soft rock typically consists of Tertiary sandstone and mudstone and it has a well-developed stratified structure. Since material properties are relatively uniform and core samples can be taken easily, the mechanical properties obtained in laboratory tests can be readily applied to the actual rock. In addition, more accurate evaluations of material properties can be achieved by performing in-situ tests, such as borehole loading tests.
(2) On the other hand, weathered soft rock, as represented by weathered granite, exhibits heterogeneous weathering and the continuity of material properties is poor. Also, since core sampling is often difficult and suffers from the effects of discontinuity planes, numerous drillings and borehole tests have to be performed to grasp the degree of weathering and to evaluate material properties through in-situ tests.

c. Tests to obtain geotechnical parameters.
The geotechnical parameters needed to design bridge foundations vary greatly, but in the main they include deformation characteristics, strength parameters, and creep deformation characteristics, both in ordinary conditions and during seismic motion.

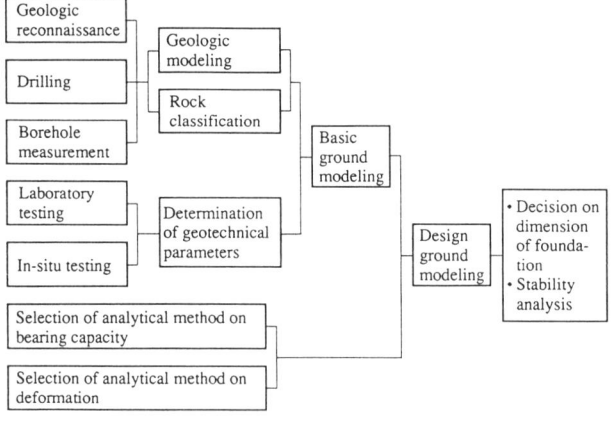

Figure 2 Flow from geotechnical investigation to design of substructure

Table 2. Present conditions of geological investigation for strait crossing bridges

Maximum contact pressure (10kN/m²)	Drilling		SPT	PMT	Well logging	Lab test	Rock test	
	Nos.	Total length (m)					Deformation	Strength
30-60	3-25	44-1420	O	⊙	O	⊙	O	△-O
60-100	18-73	350-3700	O	⊙	O	⊙	O	O
>100	26-73	1600-2900	O	⊙	O	⊙-O	O	O

SPT: Standard penetration test, PMT: Pressuremeter test
⊙: frequently adopted, O: adopted, △: rarely adopted

Table 3. Design geotechnical parameters, and their testing method

	Geotechnical parameter	Sedimentary soft rock		Weathered soft rock	
		Symbol	Testing method	Symbol	Testing method
Static condition	Strength parameter	C_{cd}, ϕ_{cd}	Triaxial test under CD condition	C_s, ϕ_s	In-situ rock shear test
	Deformation parameter	E_S	Pressuremeter test or triaxial compression test	E_S	Plate loading test or pressuremeter test
		ν_S	Triaxial or uniaxial compression test	ν_S	Triaxial or uniaxial compression test
	Creep deformation parameter	G^*	Triaxial drained (undrained) test	G^*	Plate loading test or triaxial compression test
		G_3		G_3	
		η_3		η_3	
Dynamic condition	Strength parameter	C_{cu}, ϕ_{cu}	Triaxial test under $\overline{CU}$ condition	C_d, ϕ_d	In-situ rock shear test
	Deformation parameter	E_D	Pressuremeter test	E_D	Plate loading test
		G_o	Shear wave logging	G_o	Shear wave logging
		G	Cyclic triaxial test	G	Cyclic triaxial test
	Damping parameter	D	Cyclic triaxial test	D	Cyclic triaxial test
Unit weight		γ', γ_t, γ_{sat}	Physical property test	γ', γ_t, γ_{sat}	Physical property test or density logging

Table 3 gives a summary of the test methods used to obtain the various geotechnical parameters in the case of the Kobe Formation (soft sedimentary rock) and weathered granite (weathered soft rock). These rocks form the load-bearing layer for the Honshu-Shikoku Bridges; test methods for these bridges varied according to the type of soft rock.

d. Drilling surveys

In the case of the Honshu-Shikoku Bridges, drilling was carried out in a grid pattern with spacing of between 15 and 30 m, allowing preparation of a geological profile in both the axial direction and perpendicular to the bridge axis. The surveys reached between 10 and 20 m around the foundation, so the results could be used as reference data in excavation of the foundation structures. Drilling depth was between one and two times the foundation width, at which point the stress transferred into the ground can be considered small compared with the contact pressure at the base of the foundation. In addition, to check for weak zones in the rockmass and geological structure, a small number of deep drillings were also carried out.

The unusual feature of these drilling surveys is that the engineering properties of the load-bearing ground were judged from numerous borehole loading tests and geophysical loggings in addition to core observations and core sample tests. The spacing of borehole loading tests was between 3 and 4 m on average. The modulus of deformation as obtained through borehole loading tests reflects the deformation characteristics of the ground on the data, but can also be used as an important index in comparing ground layers when judging the engineering characteristics as well.

Physical logging consisted of various measurements including speed, electrical, density, and reflection logging, yielding useful data for comparison of ground layers and engineering properties of the ground. In particular, shear wave logging was used as the basis for aseismic design.

Table 4. Quantities of geological investigation for main long-span bridges

Bridge name	Drilling		Pressuremeter test (point)	Geophysical logging (hole)	Laboratory test (nos.)
	Number	Total length (m)			
Akashi Kaikyo Bridge*	50	4,096	958	27	798
Bisan Seto Bridge	131	5,090	1,240	131	905

* Quantities of detailed investigation at Phase-1.

Table 4 summarizes the drilling surveys carried out for the Akashi Kaikyo Bridge and the Kita- and Minami-Bisan Seto Bridges, which are representative of the long suspension bridges in the Honshu-Shikoku Bridge system.

e. In-situ rock tests

Since the foundations of long-span bridges are typically constructed on soft rock, and the majority of them are undersea structures of great depth, it is difficult to carry out in-situ tests. Instead, mechanical properties are judged on the basis of laboratory tests on borehole core samples and borehole loading tests, etc. It is difficult, however, to take undisturbed samples from weathered soft rock, and it is also not possible for laboratory evaluations to take account of the effects of discontinuity planes within the rock. To obtain the required data, therefore, in-situ tests are generally carried out on similar onshore rock. The following in-situ tests are typical:

Table 5. In-situ rock tests performed at Honshu Shikoku Bridge site

Object	Kind of test	Test method	Test results
Static deformation characteristics	Plate loading test	A rigid circular plate with 30-300 cm in diameter, is used for the test. Multi-cycle method is adopted for the loading procedure	• Deformation modulus • Elastic modulus • Creep coefficient
Creep deformation characteristics	Creep test by plate loading	Long-term loading test is performed by plate loading test. A constant load is applied during 10-60 days.	• Rheological parameters • Creep coefficient
Bearing capacity	Bearing capacity test by plate loading	A rigid circular plate with 30 cm in diameter is used for the test. Monotonous increasing load is applied until the failure or yield of the ground is observed.	• Yield load • Failure load
Strength characteristics	Block (rock) shear test	Block/rock shear test is carried out applying vertical and shearing loads. Test block size is 60 x 60 cm in width and length.	• Shear strength parameter (c, ø)
Dynamic deformation characteristics	Dynamic plate loading test	A rigid circular plate of 60 cm in diameter combined together with a vibrator is utilized. Cyclic loads with different frequency are applied.	• Dynamic elastic modulus • Damping ratio • E-γ, h-γ relationships
	Cyclic shear loading test	Cyclic shear test is performed applying alternate cyclic shear loads combined together a constant vertical load. Dynamic deformation is obtained increasing dynamic loads of 25 cycle.	• Dynamic shear modulus • Damping ratio • G-γ, h-γ relationships

- Static and dynamic plate loading tests
- Static rock shear tests
- Bearing tests by plate loading
- Creep deformation tests by plate loading

The purpose and details of the in-situ rock tests performed in the case of the Honshu-Shikoku Bridges are summarized in Table 5. These tests have certain special features attributable to the fact that the bearing layer is mainly soft sedimentary rock and weathered soft rock, as follows.

(1) Since the bearing layer consists of soft rock, long-term creep deformation and stability in the case of earthquake become important design issues. For this reason, creep deformation tests and dynamic deformation tests were performed.

(2) Where large foundations are to be constructed in soft rock and diluvial gravel layers, there is no established stability analysis method. Consequently, the results of static and dynamic loading tests were analyzed through simulations and examined the evaluation method of rock material properties as input data and adequacy of stability analysis method, etc.

(3) Due to the presence of discontinuity planes and material property non-homogeneity in the load-bearing layers, strength and deformation characteristics are affected by the size of loading plate. Consequently, large-scale loading tests using various loading plates between 30 and 300 cm were performed to quantify the effect of size.

Since there are not that many large-scale bridge structures such as the Honshu-Shikoku Bridges, records of construction are scarce and the adequacy of bearing layer evaluation and design techniques has not been fully verified. For this reason, special large-scale loading tests were performed during the design of the Honshu-Shikoku Bridges.

1 In-situ tests using pneumatic caisson

In-situ shore-based tests on ground similar to the load-bearing layer yield the mechanical characteristics of unsaturated ground, so the question remains whether or not the characteristics of the saturated undersea foundation layer match them. In the case of the Akashi Kaikyo Bridge, the Tertiary Kobe Formation forms the foundation layer. A pneumatic caisson 8 m x 8 m in horizontal section was laid down about 80 m offshore, as shown in Figure 3, and static and dynamic deformation tests and shear test were performed on the rock exposed by the caisson. This yielded direct measurements of the mechanical properties of the undersea foundation layer.

2 Large-scale loading tests

In preparing an aseismic design method for the Akashi Kaikyo Bridge and the Kurushima Bridges, the latest understanding of seismic input and dynamic soil-structure interactions, etc., was adopted to prepare the design method. The same design methods were applied to the Tatara Bridge, which will be the world s largest cable-stayed bridge when completed. This calculation of aseismic design was theoretical, however, based on various assumptions and engineering judgments. To verify the aseismic design at the bridge location, loading tests were performed using large concrete blocks (as a foundation model) weighing about 3 MPa on heavily weathered granite, as shown in Figure 4.

The aim of these tests was to check the characteristics that proved problematic in preparing the aseismic design: the kinetic spring of the bedrock, the damping modulus, ground behavior, and residual deformation under extreme eccentric loading, etc. Cyclic loading, alternate loading, and shear loading, etc., were carried out on heavily weathered granite of classes C-D.

3. MODELING OF GROUND AND DESIGN GROUND PARAMETERS

3.1 Concept of stability analysis

No continuous and rational constitutive equation for the stress-strain relationship, as necessary in carrying out stability analysis, has yet been established. For this reason, bearing capacity and deformation are calculated separately in actual design practice. While bearing capacity is based on the equilibrium of forces in the limit equilibrium state, deformation is based on a calculation of elastic deformation and creep deformation under ordinary conditions (conditions of use).

Since the contact pressure on the foundation layer is considerable in the case of long-span bridges, a detailed examination of the stability analysis for the foundation is necessary. Table 6 shows the relationship between design level and analysis method for the Honshu-Shikoku Bridges. Further, Figure 5 illustrates the procedure for making a judgment of bearing capacity.

3.2 Rock classification

The purpose of rock classification in bridge foundation design is to grade the foundation layer from an engineering viewpoint and then to model the ground for use in the design of foundation structures and in establishing design ground parameters. Where long-span bridges are concerned, there are certain special features of the rock classification process: (1) the process centers on a classification by means of boring cores, making it applicable to investigations of undersea rock; and (2) classification is based not only on core observations, but also on various material properties and numeric measures of the scope of classification factor and score of classification factor.

(1) Soft sedimentary rock

There is often no need for a classification in the case of sedimentary rock, which is relatively young in geological terms, since weathering has progressed little and there are few fractures. Consequently, in the case of the Honshu-Shikoku Bridges, material properties were evaluated using geological rock facies

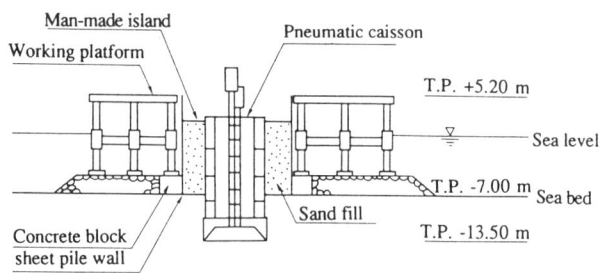

Figure 3. In-situ rock test for seabed utilizing a pneumatic caisson

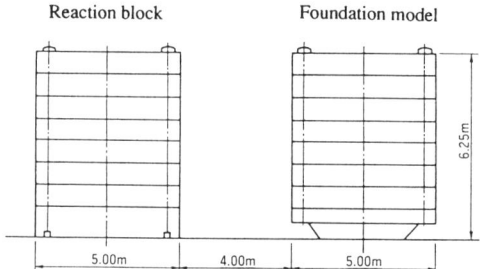

Figure 4. Arrangement of large scale loading test

Table 6. Relationship between analysis method and design level

	Ultimate vertical bearing capacity		Elastic deformation		Creep deformation	
Design level	Rough estimation	Detailed estimation	Rough estimation	Detailed estimation	Rough estimation	Detailed estimation
Analysis method	Slices method	Slices method or RBSM	Linear elastic FEM	Linear elastic FEM	Simplified creep calculation method	Linear elasto-plastic FEM

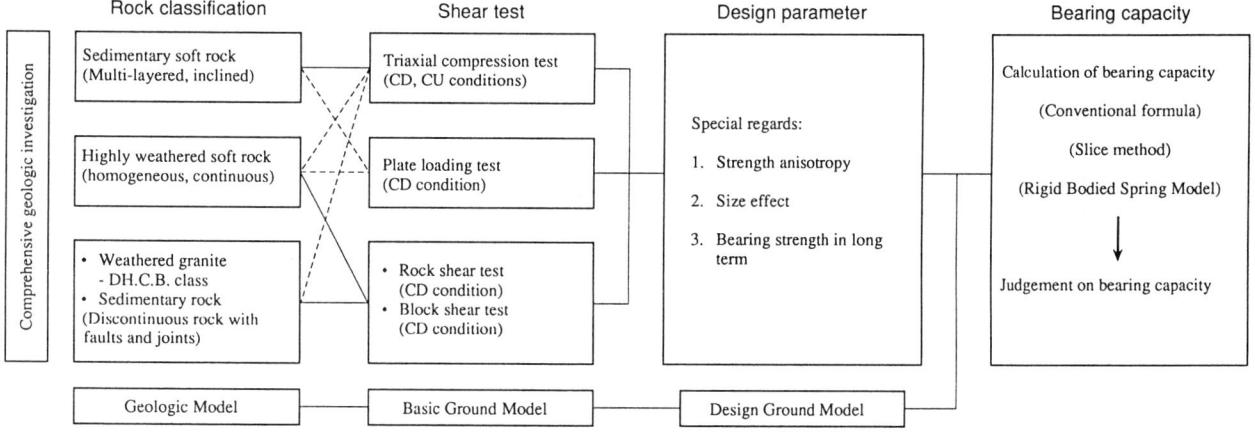

Figure 5 Flow of judgment on bearing capacity

classification only. In contrast, in the case of the Tertiary soft sedimentary rock under the Hakucho Bridge and the Ikutsuki Bridge, the core hardness of rock pieces, the core shape, the nature of fractures, and the RQD were taken as classification factors, and a rating method was devised in which a total score was obtained after scoring for each factor. On this basis, the rock was divided into six classes.

(2) Weathered soft rock

Where the foundation layer consists of weathered soft rock of hard rocks, classification is usually into six classes by means of Tanaka s classification method, which is often used at dam sites in Japan. The Japan Highway Public Corporation treats classifications A, B, CH, and CM in this method to be hard rock, while the CL and D grades are taken to be soft rock in the design of foundations.

On the other hand, the Honshu-Shikoku Bridge Authority classifies weathered granite on the basis of drilling surveys, since its many long-span bridges are on a weathered granite foundation. This classification method is characterized by the fact that, although reference is made to the Tanaka's classification, heavily weathered class D rock is subdivided into three DH, DM, and DL so as to make things easier when the bearing stratum is of class D. Additionally, judgment factors used for classification are not only unaided-eye core observations, but also various objective material properties. That is, the classification indexes include the P-wave velocity (Vpr) taken by velocity logging, the resistivity (Rt) taken by electrical logging, the density (rhoR) found by density logging, and the rockmass porosity (nr) and modulus deformation (Esb) found by means of borehole loading tests. Other classification factors are density (rhoc), porosity (nc), and ignition loss (IgL) as determined in laboratory tests of bored core samples. Table 7 shows the relationship between the subdivisions of class D rock and these indexes.

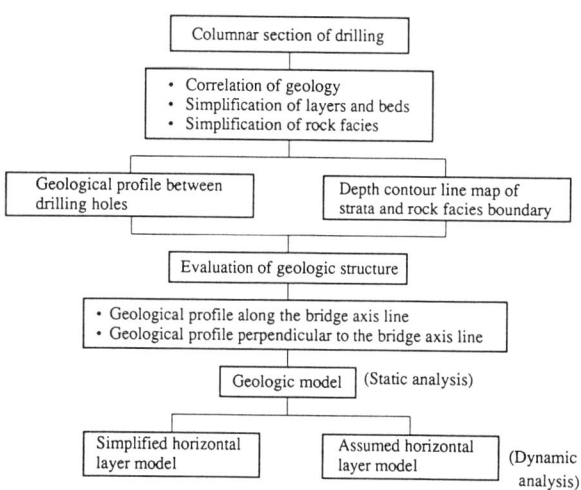

Figure 6 Procedure of geologic modeling

3.3 Ground model and geotechnical parameters

In carrying out a rational design of a bridge foundation, it is necessary to derive an adequate model of the foundation load-bearing stratum. Initially, the ground model which forms the basis of the design process is prepared in consideration of the foundation dimensions using the results of the geological investigations. This is called the basic ground model. This ground model is then corrected so as to conform to the chosen method of numerical analysis, taking into consideration the geological structure, the design level, and the particular features of the analysis method, etc. This corrected version is called the ground model for analysis.

The ground parameters associated with the ground model are

Table 7. Index properties for rock classification of weathered granite

Rock class	V_{pr} (km/s)	R_t ($\Omega \cdot m$)	E_{sb} (100 PA)	ρ_c (10 KN/m^3)	ρ_R (10 KN/m^3)	n_c (%)	n_R (%)	I_{gL} (%)
D_L	1.5 - 1.8	1 - 4	50 - 300	1.7 - 2.1	1.55 - 1.75	33 - 58	50 - 64	>3
D_M	1.8 - 2.2	4 - 7	300 - 800	2.1 - 2.3	1.75 - 2.0	19 - 33	37 - 50	0.8 - 1.2
D_H	2.2 - 2.7	7 - 12	800 - 1500	2.3 - 2.5	2.0 - 2.15	11 - 19	27 - 37	0.6 - 1.0
C_L	2.7 - 3.3	12 - 20	1500 - 3000	2.5 - 2.55	2.15 - 2.3	6 - 11	20 - 27	—
C_M	3.3 - 4.0	20 - 50	3000 - 6000	2.55 - 2.6	2.3 - 2.4	3.5 - 6	15 - 20	—
C_H	4.0 - 4.8	50 - 120	6000 - 12,000	2.6 - 2.65	2.4 - 2.5	2.0 - 3.5	11 - 15	—

Note: V_{pr}: P-wave velocity of rockmass, R_t: resistivity of rockmass, E_{sb}: deformation modulus of pressuremeter; ρ_c: density of core, ρ_R: density of rockmass, n_c: porosity of core, n_R: porosity of rockmass, I_{gL}: ignition loss of sample

established according to the results of mechanical tests on a rockmass or the intact rock. The method of mechanical testing is selected in consideration of the ground's failure mode, drainage conditions, etc. Further, in consideration of the size effect, the measured value is converted to a design value.

(1) Soft sedimentary rock

In this section, the modeling of the ground and the establishment of design ground parameters is described, taking the Akashi Kaikyo Bridge as an example. First, based on drilling data obtained from the foundation layer, the sedimentary rock along the bridge axis was zoned, and a ground model prepared by specifying ground parameters for each zone. Next, an analysis model was prepared by inspecting the geological profile and ground parameters to make the analysis easier from the viewpoint of foundation design. Figure 6 shows how the basic ground model was prepared.

The Kobe Formation, which constitutes the main foundation layer under the Akashi Kaikyo Bridge, is a multi-layered formation consisting of sandstone and mudstone. To simplify this geological structure for modeling purposes, the ground was divided into 2-5 zones with reference to similarities in the composition of rock facies, geophysical logging data, results of laboratory test, etc. Although certain parts of the Kobe Formation and the bedrock are inclined, all were treated as being horizontal in the ground model.

The design values applied to a ground model such as this one are of necessity the material properties of the geological zone rather than the individual material properties of the ground layers. In the case of the Akashi Kaikyo Bridge, therefore, the strength and deformation parameters used in design were the average values of material properties within each geological zone. It was decided that the strength characteristics of the Kobe Formation would be established according to the results of triaxial compression tests. These test values were obtained under consolidated drained conditions for the ordinary case, and under consolidated undrained conditions for the earthquake case. The actual strength characteristics used in design were obtained by reducing these triaxial compression test values based on a technical judgment of progressive failure, strength anisotropy, size effect, long-term strength, etc., as shown in Figure 7.

The deformation parameters under ordinary conditions were obtained from the results of borehole loading tests, calculating the average value for each geological zone and lithologically. The design deformation parameters are based on plate loading test values, as shown in Figure 8. These values were converted into the design deformation parameters by multiplying them by a conversion factor obtained from the borehole loading tests.

The dependence of dynamic deformation characteristics at the time of an earthquake on strain was obtained from the results of cyclic triaxial tests. The viscoelasticity modulus of creep deformation was established from the results of creep triaxial tests and borehole loading tests.

Table 8 shows the ground model for the Akashi Kaikyo Bridge 2P tower foundation.

Figure 7 Flow of determination of shear strength parametr for design in Kobe Formation

Table 8. Basic ground model for 2P site of Akashi Kaikyo Bridge

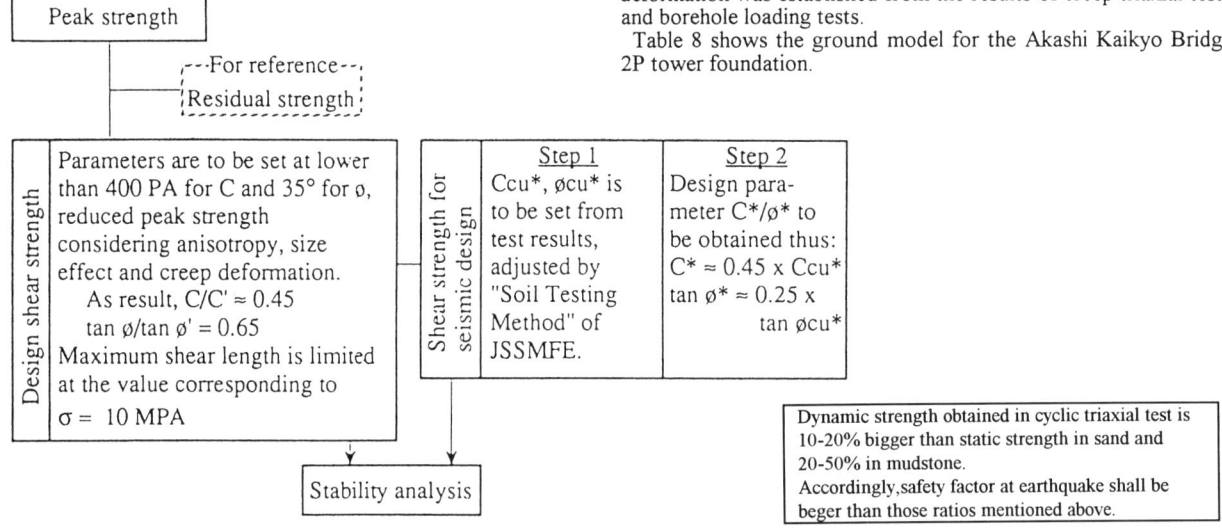

(2) Weathered soft rock

The method of modeling weathered soft rock is basically akin to that for soft sedimentary rock, except that the engineering features shown below have to be taken into consideration.

1) Since weathered soft rock such as granite has no layer structure, the boundaries between rock classifications have complex shapes.

2) Where weathering is particularly deep, the boundary between rock classifications is approximately parallel to the ground surface, and it is possible to model it as parallel multi-layered ground as in the case of soft sedimentary rock.

3) Since weathered rock has weak zones, such as faults induced by tectonic movements, such weak areas often cannot be disregarded.

The basic model for weathered soft rock is generally prepared by assuming that once the rock classification is performed, the strength and deformation characteristics can be unequivocally determined. Figure 9 shows the ground model for Kurushima Bridge anchorage 4A.

Mechanical characteristics related to strength and deformation differ in the case of weathered granite from those of the Kobe Formation, and they were established according to the results of in-situ block shear tests and plate loading tests. However, the strain dependence of dynamic material properties during seismic motion was based mainly on the results of cyclic triaxial tests.

In establishing strength parameters for use in design, the design value was arrived at by reducing actual measurements by a suitable factor analogous to that used in the case of the Kobe Formation. In particular, the design value of cohesion was made about 1/3 of the actually measured value under ordinary conditions and about 1/2 of the actually measured value under dynamic condition.

4. CONCLUSION

Geological investigations of the foundation load-bearing layer for long-span bridges in Japan has been systematized with the implementation of the Honshu-Shikoku Bridge Project. Undersea geological investigations are carried out, centering on drilling surveys. The effectiveness of this systematic investigation method has been demonstrated by a verification procedure adopted during the construction phase of the Honshu-Shikoku Bridge Project.

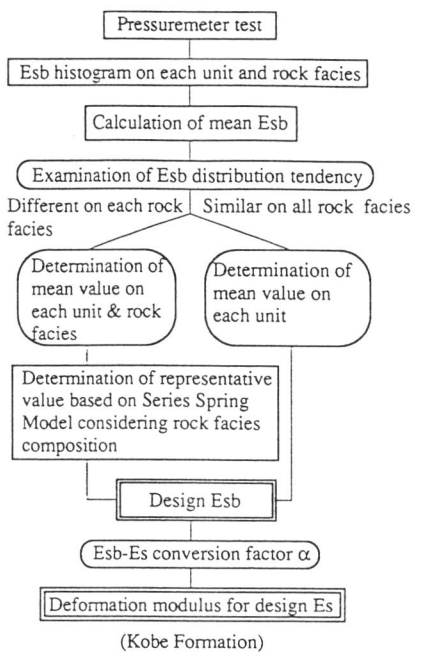

(Kobe Formation)

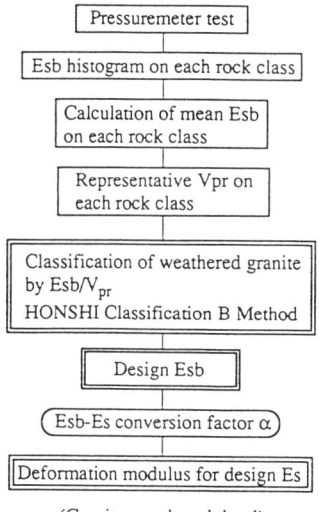

(Granite-weathered, hard)

Figure 8 Flow of determination on deformation modulus for design

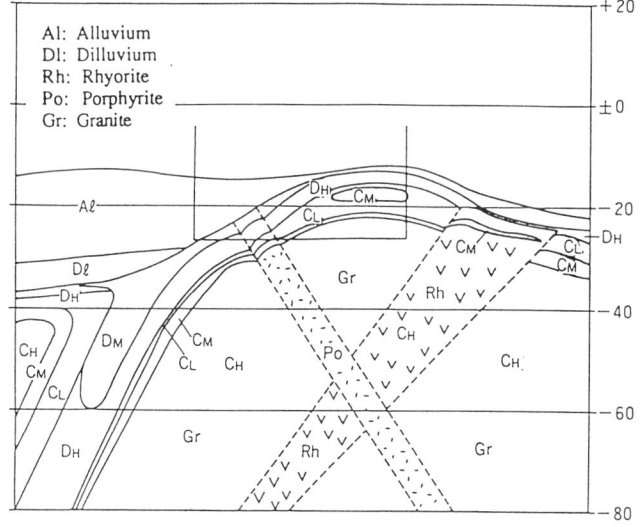

Figure 9. Basic ground model for 4A site of Kurushima Bridge foundation

Table 9. Geotechnical parameters for design of Kurushima Bridge foundation

Rock class	E_{sb} (100 PA)	Static condition								Dynamic condition						
		ρ (10 KN/m³)	E_s (100 PA)	C (100 PA)	ø (deg.)	v_s	G^* (100 PA)	G_3 (100 PA)	η_3	E_D (100 PA)	C (100 PA)	ø (deg.)	v_d	V_s (km/s)	G_0 (100 PA)	h (%)
D_L	50-300	2.0	150	0.2	30	0.42	75	75	9.2×10^3	300	0.3	30	0.46	0.5	5,100	—
D_M	300-800	2.1	750	1.0	35	0.40	375	450	8.7×10^3	1500	1.5	35	0.44	0.8	13,710	10-18
D_H	800-1500	2.2	1500	2.0	37.5	0.38	750	900	2.6×10^{10}	3000	3.0	37.5	0.42	1.0	22,450	10-18
C_L	1500-3000	2.3	3000	3.0	40	0.35	1500	3000	1.5×10^{12}	6000	5.0	40	0.40	1.2	33,800	10-18
C_M	3000-6000	2.35	6000	5.0	40	0.33	3000	6000	1.5×10^{13}	12000	8.0	40	0.38	1.5	53,950	—
C_H	6000-12000	2.4	12000	8.0	42	0.30	8000	16000	4.9×10^{13}	24000	12.0	42	0.38	1.7	70,780	—

Thanks to this investigation, it was shown that there is no significant deviation between predicted values and actual measurements related to the geology and engineering properties of the foundation rockmass. Further, measurements of the deformation behavior of completed foundation structures and the load-bearing layer are gradually demonstrating the adequacy of the investigation method.

In conclusion, the undersea geological investigation system established during the course of the Honshu-Shikoku Bridge Project is judged to be sufficiently accurate. In the future, it will be necessary to further develop the geophysical prospecting techniques used. Particular emphasis is necessary as regards geotomography, test methods, and stability analysis methods both for in-situ borehole tests and laboratory mechanical tests, etc., so as to achieve more accurate evaluations of the foundation layer.

REFERENCES

Honshu-Shikoku Bridge Authority 1988. Report on determination of geotechnical parameters for design of Akashi Kaikyo Bridge foundation, Tokyo

Japan Society of Civil Engineers 1984. Soft Rock Engineering --- Fundamentals and Case Studies for investigation, Design and Construction, Tokyo.

Japan Society of Civil Engineers 1992. Evaluation of Soft Rock --- Application to Investigation, Design and Construction ---, Tokyo

Nakano, M. et al. 1993. Present condition and feasibility of bridge foundations under deep water. Proc. 20th Japan Road Conference, 1014-1015.

Takahashi, K. et al. 1981, Investigation of Bearing Capacity of Foundation Ground of Honshu-Shikoku Bridges, Case History Volume, Proc. 9th ICSMFE, Tokyo; 1-50

Yamada, K. and S. Fukunaga 1991. Design properties of bedrock under Kurushima Bridge foundation (For Ryoke type weathering granite), Honshi Technical Report, 58: 34-42.

Yamada, K., M. Yamagata, and S. Yamamoto 1993. On-site Loading Test for Bedrock of Tatara Bridge, Honshi Technical Report, 68: 28-37.

Yamagata, M. and K. Miyajima 1985. Undersea ground investigation for Honshu-Shikoku Bridges, Preprint, Symp. for Undersea geological investigation, Japan Soc. Eng. Geology, 17-22.

Yoshida, I. et al. 1988. Foundations for Honshu-Shikoku Bridges, Proc. 2nd Int. Symp. Field Measurements in Geomechanics, Rotterdam; Balkema 451-461.

The relationship between ground characteristics and P- and S-wave velocities measured by the PS sonic logging system

T. Aizawa & K. Sasaki
Suncoh Consultants Company Limited, Tokyo, Japan

ABSTRACT: The PS sonic logging system is a digital logging device equipped with a sonde which can transmit and receive P-waves and S-waves separately. The sonde, which consists of two pairs of differential-type dipole receivers and one pair of dipole sources, can continuously measure, with a high degree of precision, the P- and S-wave velocities along the longitudinal axis of a borehole at intervals of 1 m.

This study established that the P- and S-wave velocities at intervals of 1m determined by this system represent the water content of ground formations. It was also found that the S-wave velocity and the modulus of deformation are highly and positively correlated.

It was considered that these P- and S-wave velocities could be used as indices in the modelling process that is required for a numerical analysis of ground.

1. INTRODUCTION

The PS-Sonic logging system (Suncoh Consultants Company Limited) provides a means of measurement for continuously recording the waveform of P- and S-waves propagated through the ground at specified intervals in the borhole. (Hayasi et al., 1989; zaitu et al., 1989) We have been examining the results of geophysical exploration, in-situ rock mass tests and laboratory rock tests, as part of their investigations into geotechnical properties which are to be used as indices in the modelling process that is required for the numerical analysis of ground.

The P- and S-wave velocity values recorded by the system satisfy the following requirements of the index:
1. the measurements should be made continuously along the borehole axis.
2. they should be accurate and should be taken at intervals of 1m.
3. they should be closely related to various elasticity constants, such as the modulous of deformation and of rigidity.

The results of the various examinations conducted, and based upon the data obtained by the system, are given below.

2. PS-SONIC LOGGING SYSTEM

2.1 Outline of the system

The system measures P- and S-waveforms which are transmitted along the bofehole. The P- and S-wave velocities are taken with a high degree of accuracy, at intervals of 1m along the borehole axis.

The system comprises a ground equipment and a sonde, the latter being lowered into the borehole, where it transmits or receives waves. The sonde consists of a pair of dipole transmitters and two pairs of dipole receivers, which are located 1m apart. The ground equipment comprises a digital waveform recorder, a source- and receiver-mode controller, and a personal computer which controls the action of these units and stores the data obtained. (see Figure 1.)

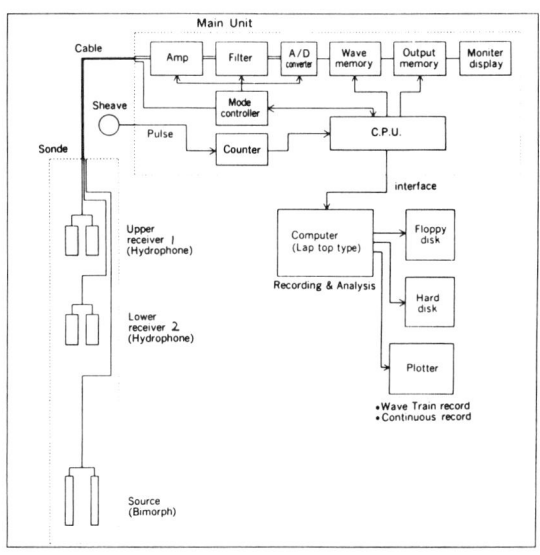

Figure 1. Block diagram of the PS Sonic Logginng System.

The elements of the system are described below.
〈Dipole transmitter〉
1) transmits pulses (at any frequency)
2) this single source can transmit P- and S-waves separately
3) high transmission power

〈Differential type dipole receiver〉
1) frequency band = 100 Hz to 20 KHz
2) receives P- and S-waves separately in modes suited to transmission mode
3) high receiving sensitivity

〈Ground instrument〉
1) transmitter and receiver modes can be set to either P- or to S-waves
2) gain, filter, and delay time can be set individually for P- and S-waves
3) capable of recording both wave trains, i.e. can

either measure the P- and S-waves when the sonde is stationary or can carry out continuous measurements when the sonde is in motion
4) full waveform can be recorded digitally
5) measured waveform can be displayed on a monitor at a time

2.2 Measurement(transmission and reception of P- and S-waves)

The measurements were made by transmitting and receiving the P- and S-waves separately. As can be seen in Figure 2, the P-waves are transmitted when two bimorph plates placed parallel to the axis of the borehole are moved simultaneously towards the borehole wall respectively, generating opposite polarity forces to the wall. To transmit S-waves, the axisymmetric two bimorph plates are moved in parallel directions, to generate shear deformation to the borehole wall. To receive waves, two cylindrical hydrophones are located along the borehole axis, on either side of a diaphragm, as is shown in Figure 3. When P-waves are transmitted, the parallel polarity waves are recorded by summing the waveforms received by the individual elements. The S-waves are transmitted, and the reversal polarity waves are recorded by establishing the difference between the waveforms received by the individual elements, and offset cancelling thereby the parallel polarity wave. These operations enable the P- and S-waves to be selected and extracted separately. The transmitter was located 1 m from the receiver 1, the lattr being 1 m away from the receiver 2 (see Figure 4.).

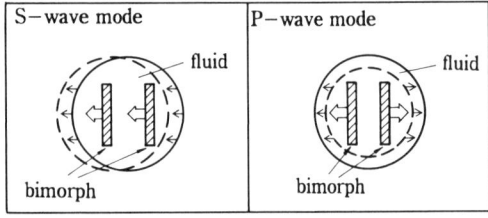

Figure 2. Dipole source elements motion of (a)S-wave mode and (b)P-wave mode. Arrows indicate exciting accoustic presser.

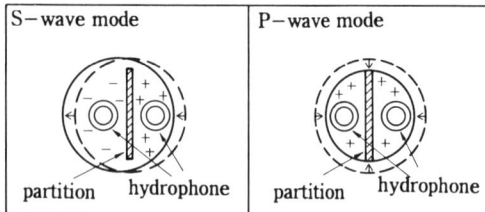

Figure 3. Dipole receiver observation patterns. (a) S-wave mode and (b)P-wave mode. Plus amd minus symbol is a distribution of the accoustic pressure inside borehole. The Arrous indicate mortion of the borehole, perpendicular to the borehole axis.

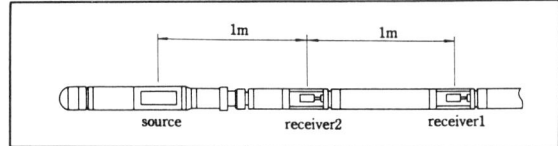

Figure 4. Location of transmitter and receivers.

2.3 Calculation of velocities

Velocity (V) is calculated by means of the following formula:

$$V = a / \Delta t$$

where Δt : time difference
a : distance of two receiver

where Δt is the initial or peak time differences taken by the upper and lower receivers. The outputs of P- andS-waves are obtained separately. The P-wave velocity can be calculated from the initial time difference, whereas the S-wave velocity is calculated from the peak time difference, because the initial time difference is not clear.(see Figure 5.)

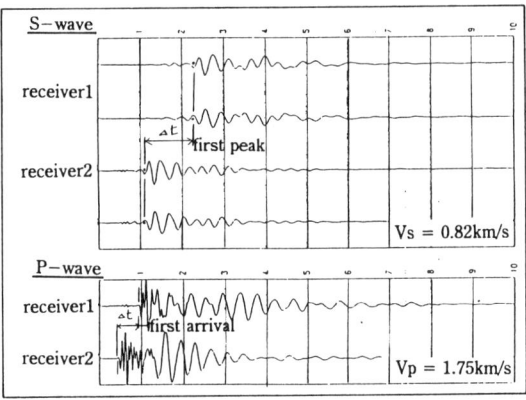

Figure 5. Observed waveform example.

3. MEASUREMENT WAVEFORM

The waveforms recorded by this system are shown in Figures 6 and 7. The records displayed in Figure 6. are those corresponding to the slow formation (borehole fluid velocity $Vs < 1.5$ km/s). The geological structure consists of Miocene alternating beds of sandstone and mudstone. When the S-wave velocity is slower than the acoustic velocity of the borehole fluid, the tube wave (a guide wave transmitted along the borehole wall) dominates after P-waves (Polet et al.,1991). However, simple S-waves are measured, because their amplitudes are minimized by the dipole transmitter and receivers of this system. The records illustrated in Figure 7. are those for the fast formation (borehole fluid velocity $Vs > 1.5$ km/s).

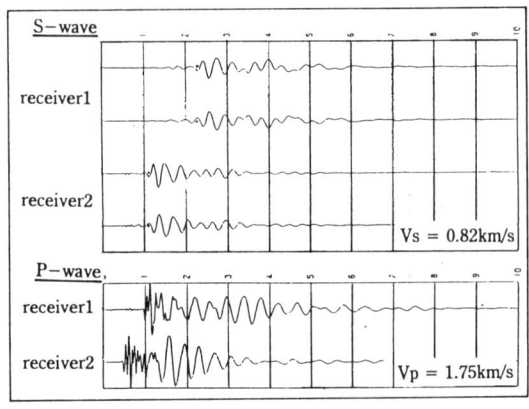

Figure 6. P- and S-wave observed in slow formatins.

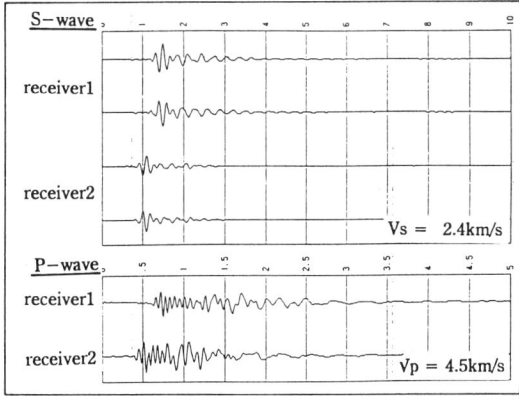

Figure 7. P- and S-wave observed in fast formatins.

The geological structure consists of granite. When the S-wave velocity is slower than the acoustic velocity of the borehole fluid, refracted S-waves which may be produced by critically refracted P-wave, generate noise which affects the measurement of the S-wave. However, the system, consisting of dipole transmitter and receivers produces no such noise, and are capable of recording simple S-waves. The recording of large-amplitude of P-waves may indicate that the dipole transmitter is sufficiently powerful.

4. EXAMINATION OF MEASUREMENTS

4.1 P-wave velocity (Vp) and S-wave velocity (Vs)

Figure 8. shows the distribution of Vp and Vs values obtained with this system. The measurements were taken for four different formations, in which granite, rhyolite, and andesitic tuff breccia were distributed. All measurements were made in boreholes drilled to a depth of 100m or less. In Figure 8, the circles symbol show measurement values taken in a borehole at depths greater than that of the unconfined aquifer, while the plus symbols show measurements taken above the level of the unconfined aquifer while infusing water. Although, in a borehole, the saturation and non-saturation of bedrock with groundwater does not take place entirely below or above the aquifer respectively, they can be used as criteria of saturation or of non-saturation. The distribution groups of Vp and Vs values can be distinguished by using the depth at which measurements are taken, namely, above or below the aquifer in a borehole, as a borderline. This indicates that the distribution of Vp and Vs is more significantly affected by the water content than by differences in rock.

It is generally known that, although the P-wave velocity is perturbed with perturbation of saturation, the S-wave velocity mainly depends on the framework of the rock structure, and is little influenced by the change of saturation (kitunezaki 1986). Figure 9 illustrates in a different way the relationship between Vp/Vs and S-slowness (inverse of velocity) shown in Figure 8. Although the plus symbols are dispersed, the line which joins the lower end figures of the circle symbols distribution is able to separate these two groups. It seems that this line is taken as the borderline between bedrock saturated with groundwater and unsaturated.

4.2 Relationship between Vp or Vs and the modulus of deformation

There was a positive correlation between P- and

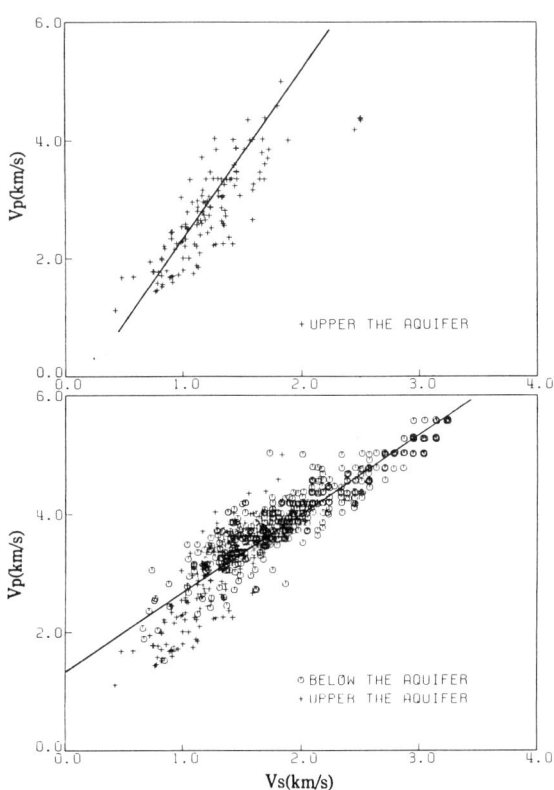

Figure 8. The crossplot of Vp and Vs.

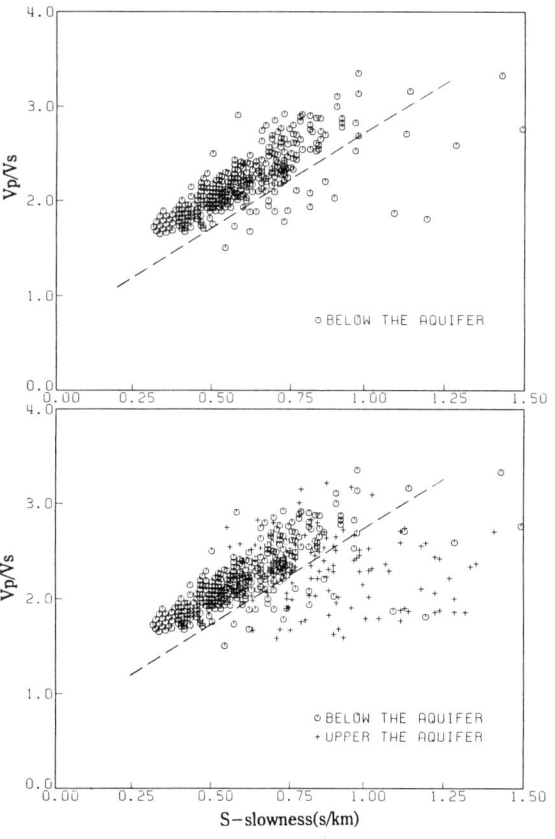

Figure 9. The crossplot of Vp/Vs and 1/Vs.

S-wave velocity measurements taken by the PS sonic logging system and the modulus of deformation measured with a pressure meter. Figure 10 is a correlation diagram, using P- and S-wave velocities measured at intervals of 1m at between approximately 50-100m below the sea bottom in eight boreholes drilled within a radius of 50 m, the modulus of deformation being measured at intervals of approximately 3 m. The ground in which the measurements were taken was a sandstone layer of Miocene. This layer was composed of fine- and medium-grained sandstone with interbeds of mudstone, muddy fine-grained sandstone and tuff.

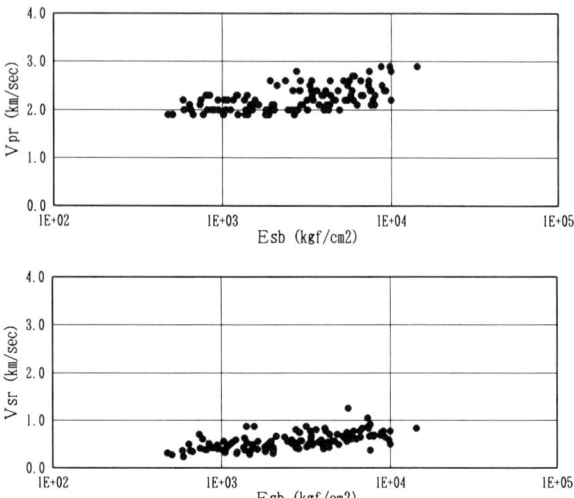

Figure 10. Correlation between the modulus of deformation and P- and S-wave velocities

It was observed that there was a positive correlation between the modulus of deformation and the P- or S-wave velocity, and that this correlation was higher with the S-wave than with the P-wave velocity. Data relating to the sandstone with interbeds of muddy sandstone are also plotted in this figure, but no difference in correlation was observed.

It was considered that this high correlation was due to the fact that the data were obtained in a limited area of ground composed mainly of Miocene sandstone. However, if an investigation were to be carried out in a limited area, and were to be subjected to limitations in funds and time, it would be possible to conduct the PS sonic logging of all boreholes, and thereafter to measure the modulus of deformation only in selected boreholes. It was felt that the modulus of deformation could be estimated from the S-wave velocity, by applying the relationship between the S-wave velocity and modulus of deformation values obtained at the selected boreholes to the S-wave velocities measured in all the boreholes.

5. APPLICATION TO GEOENGINEERING

5.1 Precautions to be observed when determining P- and S-wave velocities through the ground

The P- and S-wave velocities through the ground, as determined from the values measured with the PS sonic logging system, can be used in the process of aseismic design. These velocities are usually measured at intervals of 100 cm. When the velocity (interval velocity) over distances ranging from a few meters to several tens of meters is to be determined from measurements taken at intervals of 100 cm, care must be taken to avoid changing the travel time of P-and S-waves over the entire distance.

The interval velocity(V) can be expressed with the following formula:

$$V (m/s) = \sum_{i=1}^{n} h_i / \sum_{i=1}^{n} (h_i/v_i) \qquad (1)$$

where h_i (m) : section of measurement
v_i (m/s): the measurement value

Because the PS sonic logging system usually measures the velocity at intervals of 100cm, the above formula (1) can be expressed as follows:

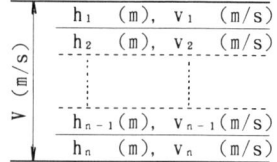

$$V = n / \sum_{i=1}^{n} (1/v_i)$$

$$n / V = \sum_{i=1}^{n} (1/v_i)$$

These formulas are similar to those applying to that of serial springs.

5.2 Indices for modelling the ground for nemerical analysis

Before a construction which includes underground carries out, it is necessary to carry out various exploratory tests in order to establish the geological and engineering characteristics of the ground (static, dynamic and hydraulic constants).

However, there are wide variations in the majority of the measurements thus obtained, because of the lack of quality in the geological and engineering characteristics of the ground, and this may lead to a number of personal errors and abnormal values. Moreover, because of the kind of the ground, of on-site measurement environments, and of limited time and funds, the required data may, in many cases, not be obtained.

Numerical analysis, on the other hand, requires that the scope of an analysis be divided into elements which represent engineering characteristics, by taking full account of ground classification, geological structure, hysteresis stress, the shape and scale of structures and analytical precision. At this time, care must be taken to avoid the alteration of geological characteristics. The physical property values of all these elements should then be established.

We attempted to model the ground by using measurements obtained with the PS sonic logging system. The conceptual flow sheet of the modelling is shown in Figure 11.

The S-wave velocity determined by the PS sonic logging method has been used as an index parameter of classification for the following reasons:
1) The system can complete the measurements in a

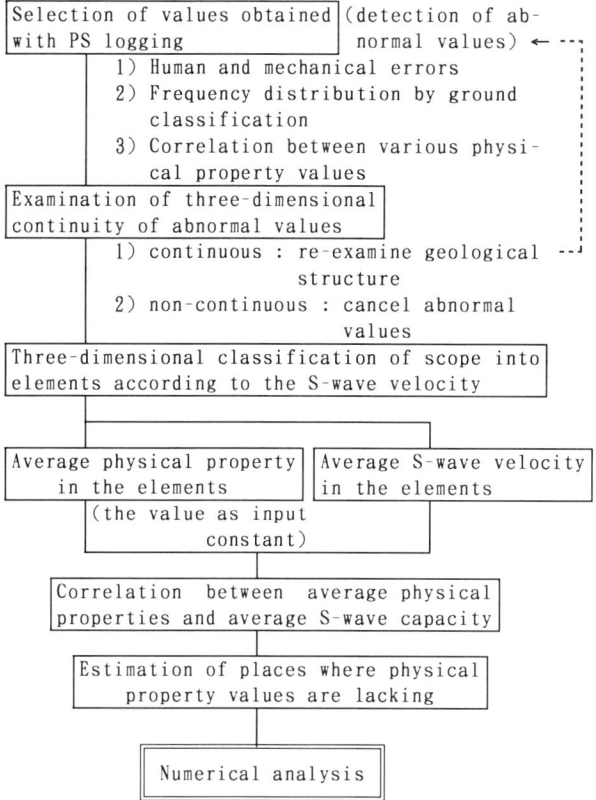

Figure 11. Flow of the modelling

relatively short time
2) The index parameters for individual elements need to be continuous figures, or figures measured at the smallest possible intervals. This system can measure S-wave velocities at intervals of 1m or less.
3) The S-wave velocity should be correlated with other engineering constants, particularly with the modulus of deformation, which is an important component of the FEM analysis.

However, PS sonic logging poses the following problems:
1) Boreholes must be filled with fluid when measurements are carried out
2) Precise measurement is essential, because the acceptable range of measurement values is very small.

PS sonic logging is considered to be capable of producing important index parameters of classification, although its use is subject to certain constraints.

6. CONCLUSION

The examination of the relationship between P- and S-wave velocity values measured by the PS sonic logging system, and of the relationship between them and other test results established that:
1. The S-wave velocity is highly correlated with the modulus of deformation, which is an important factor in a numerical analysis of the ground.
2. Information about the saturation or non-saturation with ground water can be obtained from the relationship between Vp and Vs.
3. The S-wave velocity obtained with the PS sonic logging system is considered to be suitable for use as an index parameter for the ground model.

REFERENCES

Hayasi, H. & T. Aizawa, & K. Morimoto, 1989. Development of P.S-wave logginng system(part1). Proceedings of the 80th SEGJ confference, 82-86

Zaitu, T. & H. Hayasi, & E. Isii, & T. Aizawa, 1989. Development of P.S-wave logginng system(part2). Proceedings of the 80th SEGJ confference, 87-92

Kitunezaki, T. 1986. Velocity and attenuation of elastic waves in water-filled layers with partial gass saturation and their practical significance. Geophysical Exploration 47, 5, 307-320. (in Japanese)

Polet, F.L. &C.H. Chen 1991. Accoustic waves in borehole. CRC Press

A quick and accurate strength profile along rock cores with the rock strength apparatus

Dominique M. Fourmaintraux & Cécile Lasserre
Elf-Aquitaine, Pau, France

Emmanuel Detournay & Andrew Drescher
University of Minnesota, Minneapolis, Minn., USA

Abstract

The Rock Strength Apparatus allows to determine mechanical strength parameters of rocks (cohesion and internal friction angle) on the basis of the forces measured on a cutting tool machining a small grove along the rock core surface and a phenomenological model of rock cutting developed early . The core isn't disturbed and can be reused . Accurate measurements are highly reproducible with a one centimeter lengthscale resolution .The RSA can provide a high resolution log of mechanical strength parameters directly measured on cores samples .

1 BACKGROUND

The concept of Rock Strength Apparatus is simple : it consist of maesuring the forces required to scratch a grove in the rock core surface along his axis at a constant low depth of cut (from 0.2 to 2 mm) using a sharp and/or blunt PDC cutter (PDC = Polysynthetic Diamond Crystal) . Interpretation is based on a phenomenological model for rock cutting developed by Detournay and Defourny (1992) and derived from cutting experiments performed by Glowka (1987) at Sandia N.L. and Almenara (1992) at Imperial College . The model hinges on the assumption that rock cutting is actually a combination of two processes allowing simple relationships : a "pure" stricto sensu cutting process allowing that the force on the cutting face is proportional to the cross sectional area of the grove ,and a frictional contact underneath the cutter implying a friction law for the forces across the wearflat .

1.1 .Specific energy , penetration strength

Forces measured on the cutter support are referenced to the rock surface : the tangential fore Fs is parallel to the cutting direction and the normal force Fn is orthogonal to the surface .

Considering the case of a sharp cutter edge , the contact area of the cutter is equal to A ,the cross sectional area of the grove. Its allows to define as Fairhurst and Lacabanne (1957) and Warren(1989) for the drilling performances :

-how the rock widerstand to the penetration (or the penetration strength ,S), $S = Fn / A$

-the specific energy (E) which is the ratio of the energy consumed to the volume of destroyed rock $E = Fs / A$

both having the dimension of stress (MPa or MJoule / m3) .

For perfectly sharp cutter , $E = e$ = "intrinsec cutting specific energy" (ICSE) .

For a blunt cutter , a wearflat occures underneath the cutter , and energy is dissipated in frictiona contact at the interface between the wearflat and the newly machined rock surface . Specific energy increases linearly with the force across this interface (so with the normal force S)

$E = Eo + \mu *S$ (figure 1)where μ is the friction coefficient of the interface .

We can express a cutting efficiency coefficient $h = e / E$. $h = 1$ for perfectly sharp cutter , while $h = 0$ would characterizes the limiting case where the cutter is merely sliding on the rock surface .

1.2.cutting face friction

The value , Eo, of the ordonate intercept in the (E , S) plane , is given by $Eo = e (1 - \mu*z)$,

where z is a parameter characterisitic of the sole cutter , depending of the rake angle value of the cutter,q, and of the frictional coefficient y of the interface between the cutting face of the cutter and the destroyed rock in front of this cutting face , $z = tang (q+ y)$.

This frictional coefficient was very similar for all the tests performed with the same cutter types and appears to depend only of the PDC types. From our tests on various types of rocks with the same sharp PDC cutter type at a rake angle q =15 degrees , the y value was 19 degrees , and the z value was 0.69(figure 1) . From the various tests performed for PDC cutter with

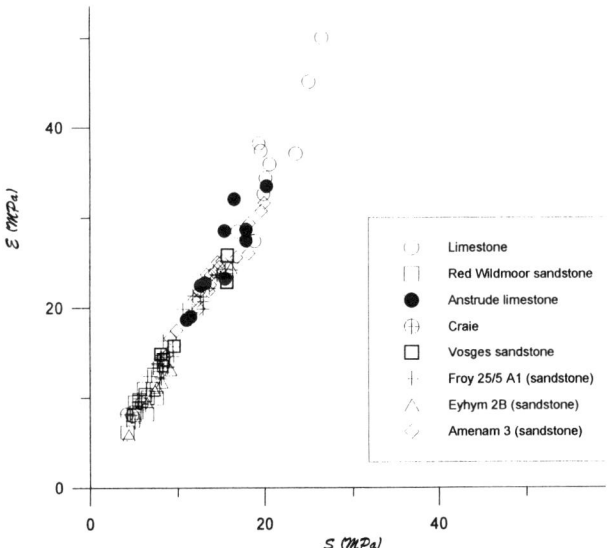

Figure 1 -Tests with sharp cutter on 8 differents types of rocks .

a rake angle usually equal to 15 to 20 degres , z varies between 0.7 to 0.9 .

1.3. cutting line , friction line , cutting point

Onto the plane (E , S) , this coefficient define (figure2)

the "CUTTING LINE " $E = (1/z)*S$

with the slope 1/z ; all the data from sharp cutter tests on all rock , will fall along this line.

The others parameters characteristic of the rock (E , and μ) are easily imaged onto the (E , S) plane too from the data from blunt cutters tests performed on the same rock at the same depth of cut : they will fall along

the "FRICTION LINE " $E = Eo + \mu*S$.

This friction line intersect the cutting line at

the "CUTTING POINT" whose ordonate is e, the intrinsec cutting energy value .

They more the wearflat is large , they more the test data fall far from the cutting point along the friction line for friction dissipate energy.

For cutter with constant wearflat area , they more the depth of cut increases (in the indicated range 0.2 to 2 mm) , they more the test data fall near from the cutting point along this friction line , for the cutter efficiency is better .

2. EXPERIMENTAL RESULTS

During the test instantaneous values fo forces Fs and Fn are measured at a sampling rate of 50 measurements for 1mm cutting length . The first data processing is to average the measurements recorded along various groves machined all around the same core sample fragment . For exemple its possible to perfome until eigth groves 1 cm width along a 4 cm diameter core sample ; for a 10 cm long sample that representes more than 40 000 individuals values . Usually 2 or 3 groves are sufficient for the results are very consistent from a grove to the other.

2.1. rock cohesion measurements

For the analysis of the mechanical strength of superposed

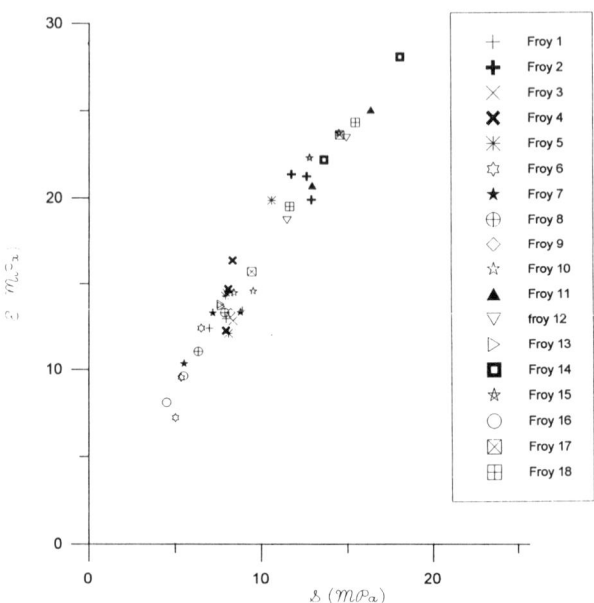

Figure 3 - Tests with sharp cutter on the 18 Froy sandstones samples .

ccessive oil producing rock layers , 18 samples were selected along the cores from the FROY sandstones reservoir in the North Sea between 3170m and 3240m depth . The data from sharp cutter tests on the 18 samples give the cutting line of figure 3 with the range 6 < e < 28 MPa for Intrinsec Cutting Specific Energy. The same analysis was performed for the AMENAM sandstones reservoir of the Nigerian offshore between 3690 to 3740 m depth . Data of 11 tests give the same slope (1 / z) = 0.67 for the cutting line (figure 4) and the range 14 < e < 32 MPa for ICSE .

After the cutting tests with the RSA have been performed , the samples were submitted to "classical"uniaxial compression test (UCS).

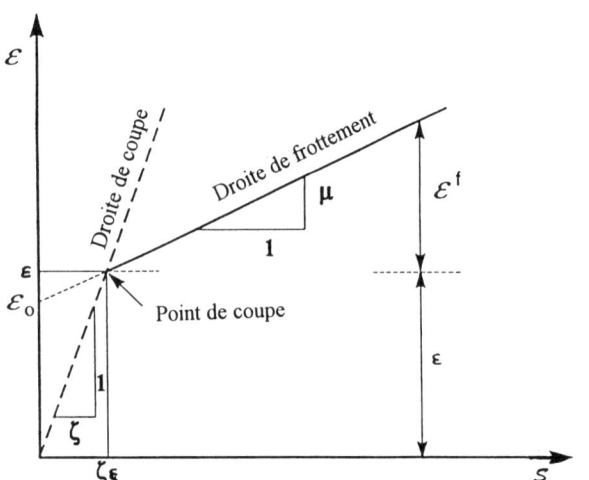

Figure 2 - Elements of interpretation in the (E , S) plane (see text) : Droite de coupe = cutting line ; droite de frottement = friction line ; point de coupe = cutting point .

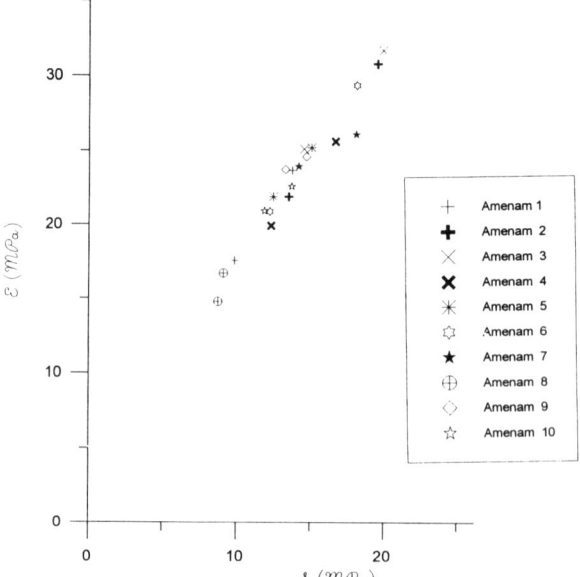

Figure 4 - Tests with sharp cutter on the 10 Ammenam sandstones samples .

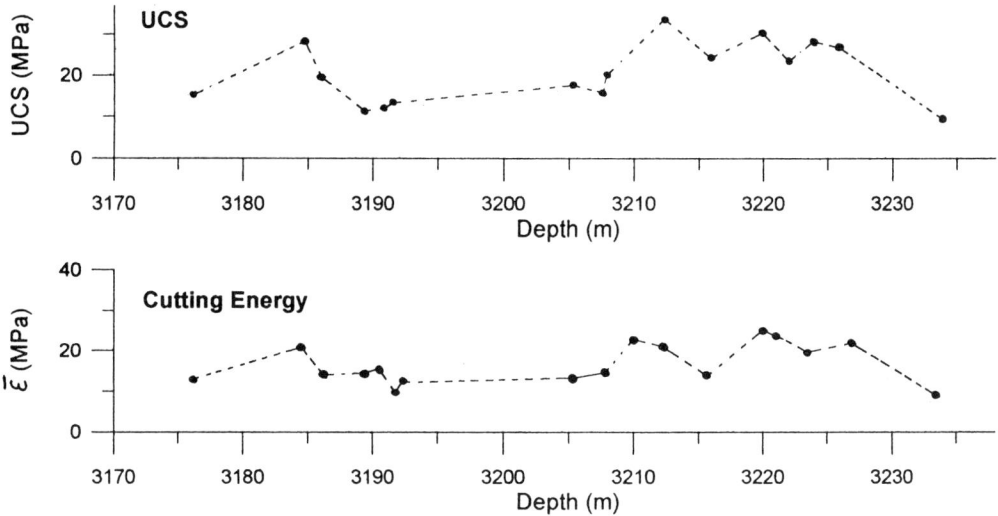

Figure 5 - Comparaison of mechanical strength profils along the Froy reservoir sandstones based on UCS tests values and Intrinsec Cutting Specific Energy values from RSA tests.

The figure 5 shows how the intrinsec cutting specific energy e values are able to describe the strength variations profil along the FROY reservoir thickness and the closed correlation with the UCS values.

Triaxial compression tests were performed too in this rocks. As we can expected, the internal friction angle ,f, for this quartz-rich sandstones have a low variability between 32 to 39 degres. Assuming a simple Mohr-Coulomb criterium for this materials, with

UCS = 2 C (sin f / 1 - cos f)

variations of UCS value reflecte mainly variations of the cohesion value C of the rock materials.

2.2. Interpretation

This results can be interpreted logically at the light of comparing rock destruction process during the test and observations of the petrographical microtexture of this rocks by mean of the SEM (scanning electronic microscop). This rocks are made of fine to medium coarse (0.1 to 1 mm) grains matrix of quartz partly bonded by a cement(giving the rock his cohesion) containing clays and intergranular silica, which is very much weak than the grains matrix. The depth of penetration of the PDC cutter edge (0.2 to 2 mm) is of the same order of magnitude as the grains diameters .The cutter broke the intergranular cement bonds without fracturing the too much strong quartz grains : the rock is just "decohesioned" along the exact gauge of the cutter width and the "cuttings" pushed in front of the cutter are made of isolated unbroken quartz grains. Using the specified range of depth of penetration (d < 2mm) any "chipping " process as described in the literature about rock drilling (Chaput, pers. com.) doesn't occure , just "grinding" process take place.

We have compared the specific energy values with the average grains size values obtained for the two reservoirs. A classic result is find : They more the grain diameter is low, they more the specific energy (= the cohesion) is high.

rock	grain size (µm)	specific energy e (MPa)
FROY SST	500 to 800	9 to 12
	100 to 200	13 to 25
AMENAM SST	600 to 700	16
	200 to 300	21 to 25

This preliminary results ought to be confirmed on others petrographical facies and could be improved by using more elaborated quantitativ parameters of rock texture.

However, we considere that the INTRINSEC CUTTING SPECIFIC ENERGY e is a valuable measure of the COHESION C of the rock material and the RSA provides a valuable mean for an accurate logging of the mechanical strength of the rocks facies along a coring.

2.3. Spatial resolution

An other processing is to consider the discrete values recorded during one test. It is then possible to display the "log" of the specific energy variation along the core sample at different lengthscale resolution by moving average procedure, using window of different sizes. The forces signals are digitally recorded at a sampling rate of 50 points / mm. On the figures 6 and 7, while logs at 1 mm resolution show the discrete failure events, logs at 1 cm lengthscale resolution appears to be highly reproducible and reflect an intrinsec variation of the rock cohesion : the "peak" value occuring at abcissa 30 corresponds to a small mineralized discontinuity (detected only by optical observation and punctual waves velocities increasing measurements) forming a stronger hetrogeneity ..

This last result is of great interest because it underlines the possibility for the RSA to allow to determine rock cohesion variations and / or strength heterogeneities in rock, with a 1 cm lengthscale resolution without the strength measurement being disturbed nor influenced by the proximity of the strength heterogeneity.

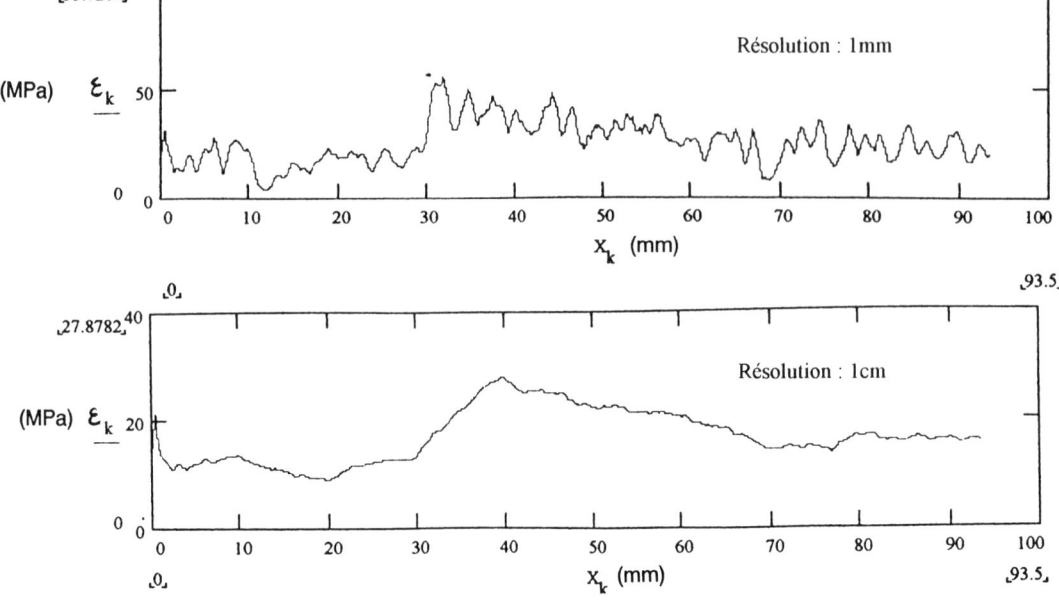

Figure 6 - Logs of the Intrinsec Specific Energy ε (MPa) along a core sample of Froy SST respectively at a 1mm (above) and 1cm (bottom) lengthscale resolution (above) ; cutting depth is 1mm .Note the "peak" at distance X= 30 mm from the origine corresponding to a stronger heterogeneity.

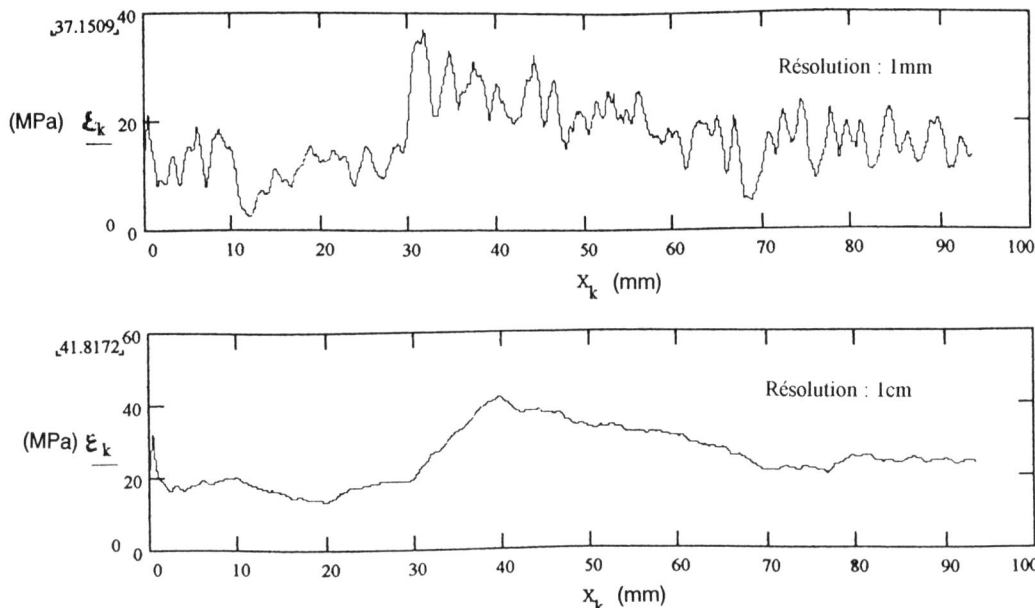

Figure 7 - Same as figure 6 ,but cutting depth is 1.5 mm.

2.4. Rock internal friction angle

Data from tests in the same rock with blunt cutter with various wearflat or cutting depth fall along the so-called FRICTION LINE (figures 8 and 9)

$$E = E_0 + \mu S$$

whose slope μ is defined as the frictional coefficient ot the interface between the cutter wearflat and the rock surface "just cut" underneath the cutter . It appears that the values of μ calculated from tests series on various sandstones are in the same range as the internal friction angle ,phi, values deduced from triaxial tests:

Rock	arc tan μ	f
Red Vosges SST	36	34
Redwilmoor SST	35	31<
Froy SST	32 to 39	30 to 35

How the internal friction angle is mobilized within the friction process of the cutter on the "just cut" rock surface ?

Explanation could be find in the microstructure of the boundary layer of this rock surface after the grinding process occured.

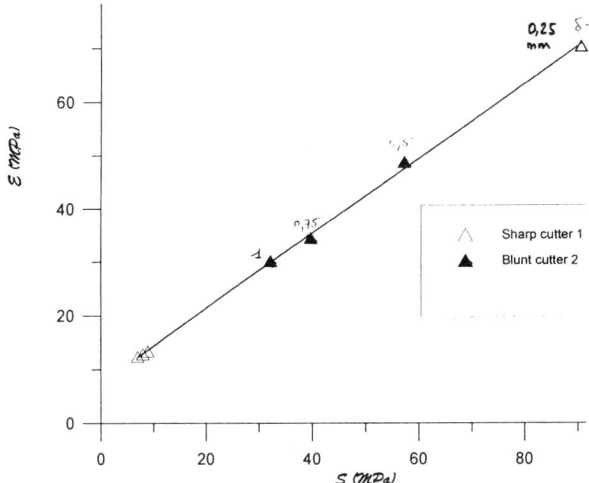

Figure 8 - Determination of the friction line by tests with sharp and blunt cutters on the same sample (Froy 1) at increasing cutting depth .: specific energy decreased .

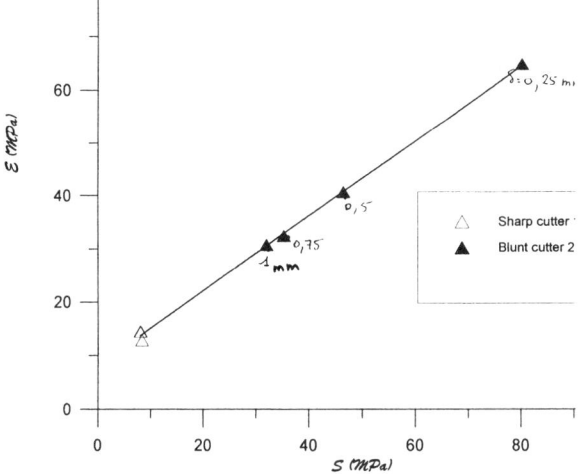

Figure 9- Same as figure 8 but sample Froy 3 : note the very similar slope of the friction line .

First observations show that the rock under the grove floor is damaged to a depth more or less equal to three times the average grains diameter: Grains often fractured and decohesionned are compacted .The response of such a damaged rock layer to the normal loading by the cutter could be imagined as based on lgrains small relative displacements and rotations , all micromecanisms corresponding to the macroscopic notion of internal friction .This analysis will be developed by accounting data from others rock types.

Hovever a log of the internal friction angle with the same resolution as the cohesion could also be performed by using a blunt cutter with a very small depth of cut. Forces acting across the wearflat would be very high compared to the forces transmitted by the cutting face , and the friction angle could simply be deduced from the orientation of the cutter force .

3.CONCLUSIONS

The experiments with the Rock Strength Apparatus RSA show that quasi continuous logs of" Cohesion" and of "internal friction angle" at a centimetric lengthscale resolution are possible without destroying nor disturbing the core . The apparatus would be build as a portable one : This logs could be easily obtained on the site using the cores directly out of the core barrel

Acknowledgements

The authors thank the Elf-Aquitaine Conpany for kind cooperation.

REFERENCES

ALMENARA J.R. and DETOURNAY E. ," Cutting experiments in sandstones with blunt PDC cutters ", Eurock'92 , pp. 215-220 , T. Telford pub. ,London, (1992).

CHAPUT E. , "Observations and analysis of hard rocks cutting failures mechanisms using PDC cutters " (1991) not puplished , pers. com.

DELIAC E. , "Optimisation des machines d'abattage à pics ", , Thèse de Doctorat ès-Sciences Physiques (Ph.D.Thesis) , Université P.& M. Curie , PARIS , (1986)

DETOURNAY E. and DEFOURNY P. , "A phenomenological model of the drilling action of drag bits ", Int. J. Rock Mech.Min. Sc.,29 (1) ,pp.13-23 , (1992).

FAIRHURST C. and LACABANNE W. D. , "Hard rock drilling technics ", Min. Quarry Engng , pp.157-16, 194-197 , (1957).

GLOWKA D. , "Development of a method for predicting the performance and wear of PDC drill bits", Tech. Report SAND 86-1745, Sandia Nat. Lab. , Albuquerque , New Mexico , (1987).

LASSERRE C. ," Rock friction apparatus : realisation de tests de coupe sur roches à l'aide d'un outil PDC" , Rapport de fin d'etudes, Institut des Sciences et Techniques Geophysiques et Geotechniques , Université P.&M. Curie , PARIS , (1994)

WARREN T. and SINOR A. , "Drag bit performance modeling", SPE Drilling Engng , June , pp128-136 ,(1989).

Representative elementary volume in natural discontinuous rock masses

Kenichirou Suzuki, Tohoru Kuwahara, Makoto Maruyama & Kunioki Hirama
Obayashi Corporation, Technical Research Institute, Japan

Abstract: There are two problems to be overcome in applying the theory of continuum mechanics and hydrology of porous media to natural discontinuous rock masses. The first is to determine whether or not the hypothesis of homogeneity of material properties over the whole rock mass may be assumed, since the discontinuous nature of rock reveals local heterogeneities. The second is to quantitatively evaluate errors in estimating material properties through in-situ investigations of joint geometry. Local heterogeneities of in-situ tests are attributed to the spatial distribution of discontinuities. The current approach is based on the classical continuum mechanics hypothesis. Solving these problems is important for the design of large and important structures founded on rock masses.
An approach to the study of REV through joint geometry is described. The natural joint geometry of five sites with different geological histories are surveyed. The applicability of error indices of joint geometry is confirmed for natural discontinuous rock masses.

1. INTRODUCTION

In recent decades, with the advent of large and increasingly important structures founded on rock masses, there has been a need for a more quantitative and accurate assessment of the deformability and permeability of rock masses. A serious problem to be overcome is to determine the effects of discontinuities such as faults, joints, and fissures. It is well known that the geometry formed by the spatial distribution of discontinuities dominates mechanical and hydraulic properties of a rock mass. Some literatures have reported the local heterogeneities of deformabilities(Charrua-Grace, 1979) and permeabilities(Shimo and Kamemura, 1986) of rock masses. The complexity of rock structures prevents separate treatment of all the heterogeneities.

Rapid advances in computer technology have made large-scale computations of geological problems feasible. However, development of numerical modeling techniques taking relevant field data into account has not paralleled this advance because of the geometrical complexity of discontinuities in rock masses that lead to the local heterogeneities of deformabilities and permeabilities. In the design stage, the current approach is to assume homogeneity and to model a discontinuous rock mass as a continuum or a porous medium with equivalent deformabilities and permeabilities. However, through considering the discontinuous nature of rock, it seems more sensible in estimate equivalent deformabilities and permeabilities over volumes as large as possible to eliminate the effects of local heterogeneities.

This has given rise to the concept of Representative Elementary Volume (REV)(Bear and Bachmat, 1984; Cuisiant and Haimson, 1992). When the sampling size is large enough to eliminate local heterogeneities of material properties, REV has been used. The REV is then defined as the smallest volume for which there is equivalence between the continuum material and the real rock. This concept has been recently applied in estimating hydraulic properties(Long and Witherspoon, 1985) and in the method of estimating errors of joint geometry in volumes smaller than the REV using the crack tensor concept(Oda, 1988; Oda et al. 1986). As these works are conducted on computer-generated homogeneous fracture patterns, it is necessary to validate the concept in actual rock masses.

This paper confirm the applicability of a method of evaluating the degree of error in estimating joint geometry through investigating actual rock masses. It confirms the applicability of this method, and then provides a better tool for judging whether or not a volume(area) investigated may be REV and the degree of error if the volume is smaller than the REV. The degree of error in evaluating rock mass deformability is also estimated. For these purposes, joint surveys were performed for five different rock types: Andesite, Schist, Limestone, Shale, and Gneiss rock masses, and for microcracks in a weathered granite. The followings are discussed:

(1) The REV may be determined for actual rock masses in terms of joint geometry.

(2) The errors in crack tensors estimated for volumes smaller than the REV may depend on the volumes surveyed.

2. CRACK TENSOR AND ERROR TENSOR

2.1 Crack tensor

Many experimental and theoretical studies have proved that the hydro-mechanical behavior of rock masses must depend on at least the following geometrical properties of discontinuities; (1) shape and dimension, (2) Position and density, (3) Orientation, and (4) Aperture. It is also important to note that their properties do not affect rock mass behavior independently, but interactively. Several assessment systems have been proposed(Bieniawski, 1974; Barton et al. 1974). However, these systems are generally deficient because they fail to account for the resolution at which numerical modeling is performed. Oda(1982) proposed a tensor quantity called a crack tensor which has been introduced as an index measure to explicitly express crack geometry, and can be determined from field data on the basis of the stereological method. Here, we use a two-dimensional crack tensor expressed by the following equation.

$$F_{ij} = \frac{1}{S}\sum_{k=1}^{m} r^{(k)2} n_i^{(k)} n_j^{(k)} \quad \ldots (1)$$

where S is the surveyed area, $n_i^{(k)}$ is a unit normal vector of the k-th joint on a fixed coordinate and $r^{(k)}$ is the trace length of the k-th joint. This includes the information for the joint density, which contains the number and length of joints, and the anisotropy of joint geometry. To describe the mechanical properties of discontinuous rock masses, it is important to be able to take both the average deformability and anisotropy into account. Thus, we use the crack tensor concept here.

2.2 Error tensor

The REV is defined as the smallest volume for which there is equivalence between the continuum material and the real rock.

The classical continuum theory may therefore not be valid in the volume smaller than REV. Hydraulic and mechanical properties of discontinuous rock masses should be defined in the volume (V) in which the joint distribution is statistically homogeneous when they are treated as "continua" or "porous media". The REV is also defined with regard to joint geometry in the same volume. Therefore, the volume in which a crack tensor has a stationary value can be regarded as the REV. If F_{ij} calculated in a restricted volume (V'<V) is rewritten as F'_{ij}, it includes a statistical bias as follows(Oda, 1988).

$$\delta F_{ij} = F'_{ij} - F_{ij} \qquad \ldots (2)$$

where δF_{ij} is called the error tensor. As the restricted volume V' increases to V, δF_{ij} decreases to 0. Thus, we statistically define that V' at $\delta F_{ij} = 0$ is the REV, from the standpoint that a discontinuous rock mass can be regarded as a continuum or a porous medium with equivalent property characterized by F_{ij}. The error tensor δF_{ij} involves the information for both joint density and anisotropy of joint geometry.

2.3 Relative scale and relative error

When dealing with discontinuities in rock masses and rocks, we are confronted with different structural levels ranging from microcracks to joints, fractures, and faults. Thus, a relative scale is defined as a scale of observation normalized by the mean length of joint trace to discuss any joint geometries which are formed by different geological structures.

e datacompThe error tensor δF_{ij} depends on the value of F_{ij}; for instance, ten percent error, $\delta F_{ii} = 1$, in a heavily jointed rock mass, $F_{ii} = 10$, corresponds to a hundred percent error in a lightly jointed one, $F_{ii} = 1$. Therefore we would use the following invariants normalized by each invariant of crack tensor at a homogeneous scale as indices of error(Oda, 1988).

$$Er^{(0)} = \left(\frac{\delta F_{ii}^2}{F_{ii}^2}\right)^{1/2} = \left(\frac{\delta F_0^2}{F_0^2}\right)^{1/2} \qquad \ldots (3)$$

$$Er^{(2)} = \left(\frac{\delta F_{ij} \delta F_{ji}}{F_{ij} F_{ji}}\right)^{1/2} \qquad \ldots (4)$$

where F_0 and δF_0 in eq.(3) are the first invariants of tensors corresponding to a porosity of a jointed rock mass (a joint density), which is the first approximation of joint geometry and the error in terms of porosity, respectively. F_{ij} and δF_{ij} in eq.(4) describe the anisotropy of joint geometry and the error taken into account because of an anisotropy. Thus, two indices evaluate the REV in terms of isotropic and anisotropic joint geometry respectively. These parameters are determined on each site and the relationship between the indices of errors in terms of joint geometries and the relative scales is discussed.

3. SITE GEOLOGIES AND DETERMINATION OF ERROR TENSOR

At site T, where the bedrock geology is mainly tuffaceous shale of acidic volcaniclastic rock of the Cretaceous period of the Mesozonicera, an exposed river bed is sketched as a two-dimensional joint map as shown in fig.1. The joints are affected by linearment of circumference. The two-dimensional crack tensor and mean trace length of joints are determined on this joint map, which assumes that joints are distributed homogeneously. The next step was to set sub-regions of arbitrary area and position, as shown in fig.1, and to determine the crack tensors in each sub-region. The error tensors and relative errors corresponding to their first and second invariant are calculated using eqs. (2), (3), and (4), respectively. Fig.2 shows examples of sub-regions and their crack tensors, which shows that the discontinuous nature of the rock mass is quite different both qualitatively and quantitavely

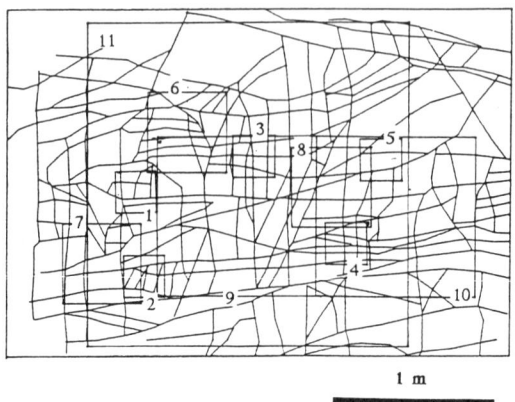

Figure-1 Joint Map and Sub-Regions of Site T
Area ; 5.52m², Mean Trace Length ; 0.78m

Area Root(Area) Relative Scale	F_{ij} and F_0	Joint Map and Scale
A = 5.52 m² L = 2.35 m L/r₀ = 3.01	$F_{ij} = \begin{pmatrix} 11.77 & 1.48 \\ 1.48 & 5.40 \end{pmatrix}$ $F_0 = F_{ii} = 17.16$	1m 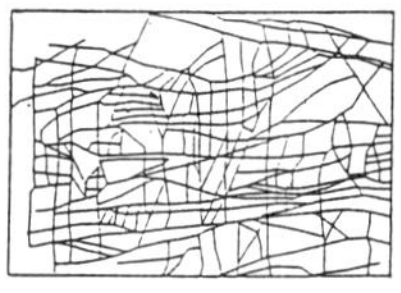 mean trace length r₀ = 0.78 m
A = 4 m² L = 2 m L/r₀ = 2.56	$F_{ij} = \begin{pmatrix} 11.00 & 1.46 \\ 1.46 & 5.10 \end{pmatrix}$ $F_0 = F_{ii} = 16.10$	1m 
A = 1 m² L = 1 m L/r₀ = 1.28	$F_{ij} = \begin{pmatrix} 7.79 & 1.04 \\ 1.04 & 3.01 \end{pmatrix}$ $F_0 = F_{ii} = 11.80$	0.5m 
A = 0.25 m² L = 0.5 m L/r₀ = 0.64	$F_{ij} = \begin{pmatrix} 4.46 & 1.42 \\ 1.42 & 3.32 \end{pmatrix}$ $F_0 = F_{ii} = 7.78$	0.2m 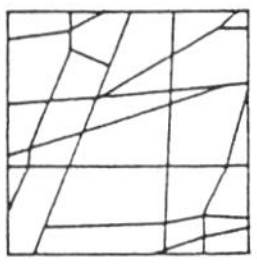
A = 0.125 m² L = 0.25 m L/r₀ = 0.32	$F_{ij} = \begin{pmatrix} 4.46 & -0.04 \\ -0.04 & 1.82 \end{pmatrix}$ $F_0 = F_{ii} = 6.45$	0.1m 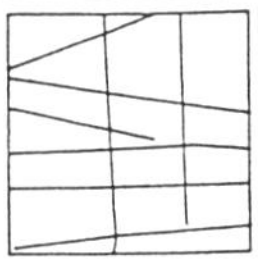

Figure-2 Examples of Joint Map of Sub-Regions of Site T

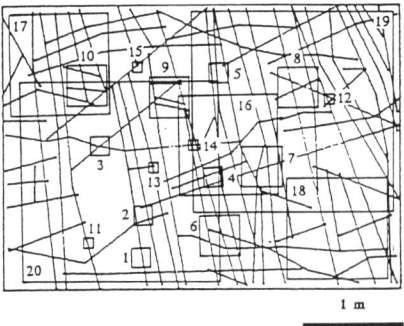

Figure-3 Joint Map and Sub-Regions of Site Na
Area ; 2.80m², Mean Trace Length ; 1.43m

when the surveyed areas are varied. As the sub-regions are smaller, the anisotropy of joint geometry is varied, as well as the joint density.

At site Na, where the bedrock geology is andesite with a large number of cooling joints as shown in fig.3, we adopted two methods to set sub-regions. One was the method used at site T. The other was to set sub-regions directly on an outcrop by taking photographs. The sub-regions set directly on the outcrop are extended from 10 cm to 200 cm in stages. These sub-regions look completely different discontinuous natures. These two methods of setting sub-regions are different in the cut-off length of joint traces. The former is a constant cut-off length depending on the joint map and the latter is a varied cut-off length. That is, the cut-off length of the joints which can be traced on each photograph become shorter as the sub-regions become smaller. Usually we can trace shorter joints as the surveyed regions become smaller. In estimating the joint geometry for the construction scale, we may neglect truncation errors since the effect of truncated joints is a little. Fig.4 shows that the joint densities described by the first invariant of crack tensors plotted against the scales of sub-regions. ● shows F_0 calculated in each photograph and ○ shows F_0 calculated in sub-regions arbitrarily arranged in the biggest photographed area. The difference between ○ and ● is attributed to truncated joint trace length. On the following discussion for the stationary characteristic of the joint geometry against the surveyed area, it seemed to be able to use the joint tracing method to set sub-region on the biggest joint maps as shown by the ○ symbols in fig. 4, since their difference is a little enough to be neglected.

Site Nb consists of blackschist and greenschist. It contains lateral faults and normal faults whose maximum width is about 15m, as well as subsequently developed small-scale faults and joint sets. Fig. 5 shows a joint map of the site and sub-regions. Fig.5(a) shows the left dam abutment, (b) shows the dam center,

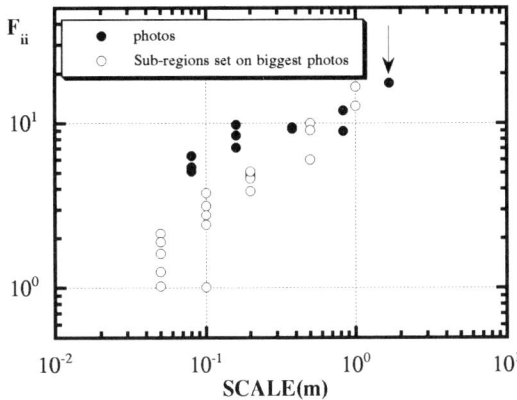

Figure-4 The relationship between scale and density of joints described by the first invariant of crack tensor

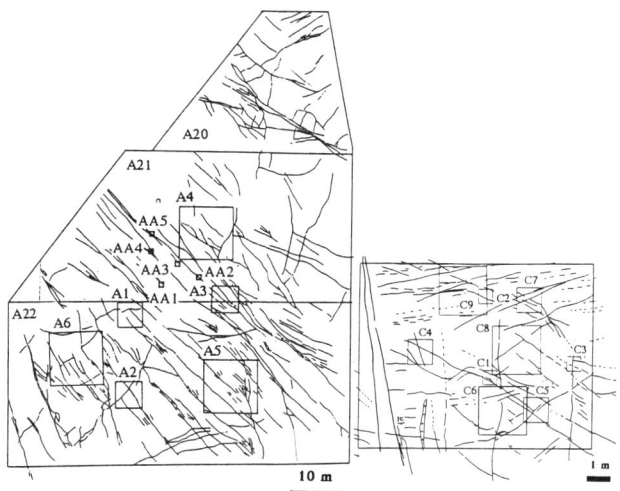

Figure-5(a) Joint Map and Sub-Regions of Site Nb
Area ; 3827 m², Mean Trace Length ; 6.00 m

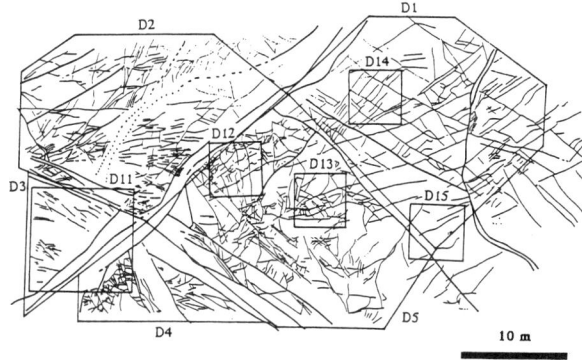

Figure-5(b) Joint Map and Sub-Regions of Site Nb
(dam center) Area ; 1482 m², Mean Trace Length ; 1.72 m

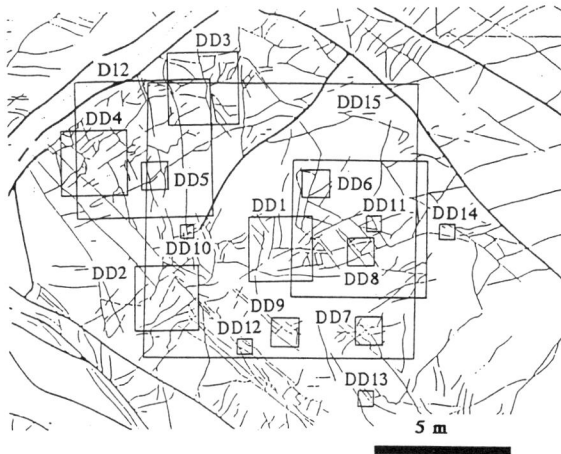

Figure-5(c) Joint Map and Sub-Regions of Site Nb
(Magnified view of the center)

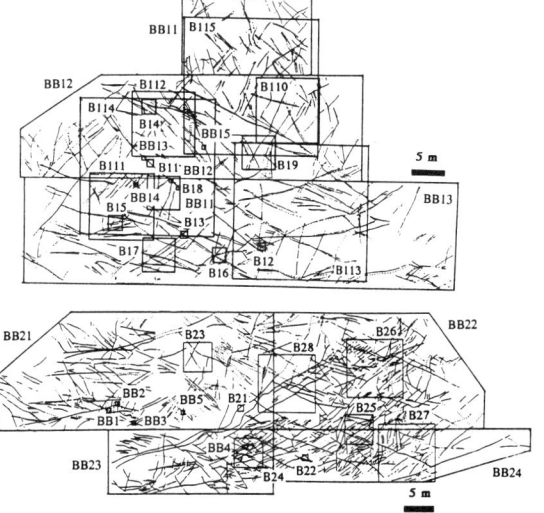

Figure-5(d) Joint Map and Sub-Regions of Site Nb (Right Bank)
Upper Part : Area ; 1823 m², Mean Trace Length ; 3.89 m
Lower Part : Area ; 1983 m², Mean Trace Length ; 2.34 m

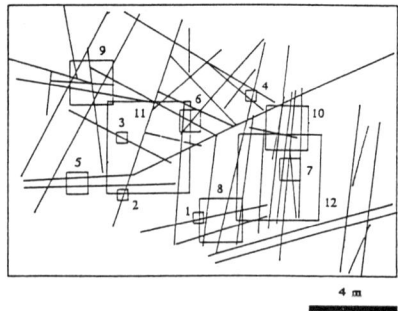

Figure-6(a) Joint Map and Sub-Regions of Site O(Vertical Cliff)
Area ; 240 m², Mean Trace Length ; 5.72 m

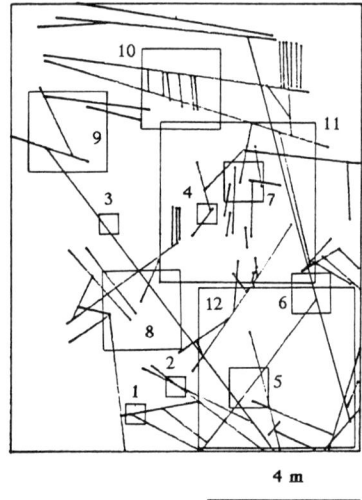

Figure-6(b) Joint Map and Sub-Regions of Site O(Horizontal Outcrop) Area ; 120 m², Mean Trace Length ; 2.02 m

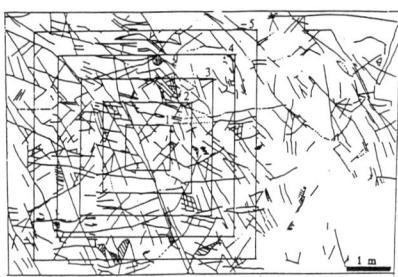

Figure-7 Joint Map and Sub-Regions of Site K
Area ; 54 m², Mean Trace Length ; 0.50 m

(c) shows a magnified view of part of the center, and (d) shows right dam abutment. An E-W strike sinistral fault and a NW-SE strike dextral fault cross a little to the left of the axis of the dam. Main faults decline about 80 degrees. These faults comprise displaced Neogene-period volcanic rock and, since tectonic relief and linearment have occured along them, volcanic activity seems to have ceased before the Quaternary period. Around the crossing point of the two main faults, a fracture zone has formed, bedrock has been altered, and many small-scale joints have formed.

At site O, where the bedrock is limestone, a 12 m x 10 m area on the horizontal exposure and a 12 m x 20m area on a vertical cliff are surveyed. The rock mass at this site has relatively few joints on the ground surface but many joints are spread on the cliff face. Twelve sub-regions are set in each joint map, as shown in figs. 6(a) and (b). There are no particular joint orientation and the mean trace lengths are 2.02 m of horizontal outcrop and 5.72 m of vertical cliff face.

At site K, bedrock consists of gneiss. A joint map is synthesized by sketches of 1 m square areas and square sub-regions are set on it, as shown in fig. 7. At this site, fifty-two 1 m diameter circular sub-regions are also set. The first invariants of two-dimensional crack tensors determined at each sub-region are distributed as shown in fig. 8, which indicates that there is local heterogeneity at this site.

Fig. 9 shows the traces of microcracks of weathered granite and eighteen sub-regions. The mean trace length of these microcracks is 2.41 mm.

4. RESULTS AND DISCUSSION

Two-dimensional crack tensors and error tensors described by eqs.(1) and (2) were determined in sub-regions of both arbitrary area and position in the joint maps shown in figs. 1 to 9. The calculated results are shown in table 1. These results include the relative errors calculated on sub-regions set on the joint maps with the mean trace lengths (r_0) ranging from a few millimeters to six meters and the scales of observation areas (L) ranging from a few centimeters to several ten meters. This means that the relative errors are calculated against the relative scale(L/r_0) ranging from 0.09 to 12.0. Figure 10 shows that an error occurring in estimating porosity (F_0) of the rock mass decreases with expansion of sub-regions. The ordinate is the error index described by eq.(3) and the abscissa is the relative scale of the sub-region. This figure includes all the results in table 1. Data described by ■ symbol at $L/r_0=2$ is shown as the range of fifty-two values. For example, consider a rock shear test whose standard size is 60 cm x 60 cm. Since the mean trace length of the joints is 1.72 m at the dam

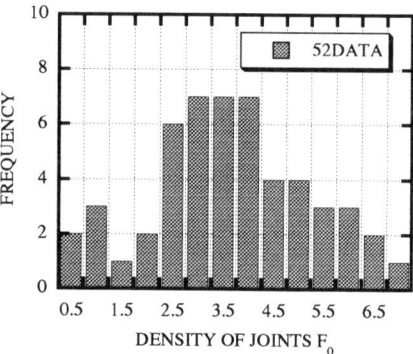

Figure-8 Distribution of Joint density determined in sub-regions of 1 m diameter circles

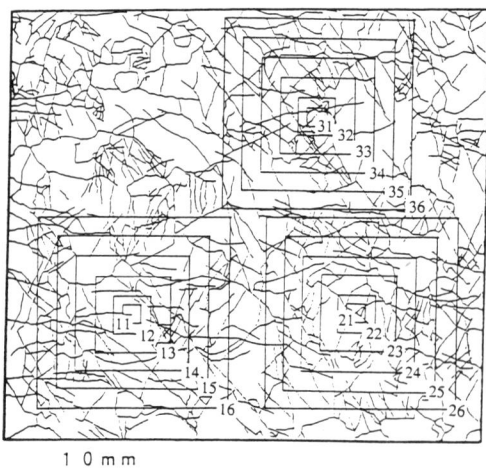

Figure-9 Joint Map and Sub-Regions of Microcrack of Granite
Area ; 6.89 cm², Mean Trace Length ; 2.41 mm

center of site Nb, the relative scale of the test area is 0.35, and then the estimation of the joint geometry include a 30 to 80 percent error with regard to the porosity of the rock mass, assuming that joints distribute isotropically. To estimate the joint geometry with about 20 percent error, it is necessary to survey an area which is four to five times as wide as the mean joint trace length. The numerical results of the investigation on joint traces distributed homogeneously and two-dimensionally by Oda et al.(1986) also showed the necessity of a survey area twice or three times as large as the mean joint length to obtain joint geometry with less than 10 percent error. The result from the investigation of actual rock masses is as a little larger than those from numerically simulated joint distributions. This figure includes the results obtained from different rock types and geological properties, which are (1) small-scale faults and joint sets directly concerned with large-scale faults, (2) joints affected by linearments of circumference, (3) cooling joints which formed just after the rock was formed, and (4) microcracks in a weathered granite. It is worth mentioning that the convergence of $Er^{(0)}$ with L/r_0 represented quite different geological conditions, and was in accord with the results obtained from the investigation of rock masses with complex geometries and several kinds of geological histories and with that from numerical experiments. In the same way, fig.11 indicates the relationship between the error index $Er^{(2)}$ taken into account by joint geometry anisotropy and the relative scale of the sub-region. Data described ● symbols in fig.11 indicate less convergence than in fig.10, which means the area surveyed was not large enough in terms of anisotropic joint geometry because small scale faults and joints were greatly affected by large-scale geological structures.

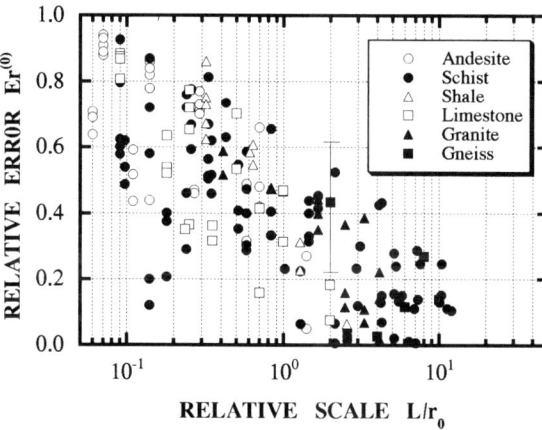

Figure-10 Relative Error Reduction with Relative Scale (the error associated with density of joints)

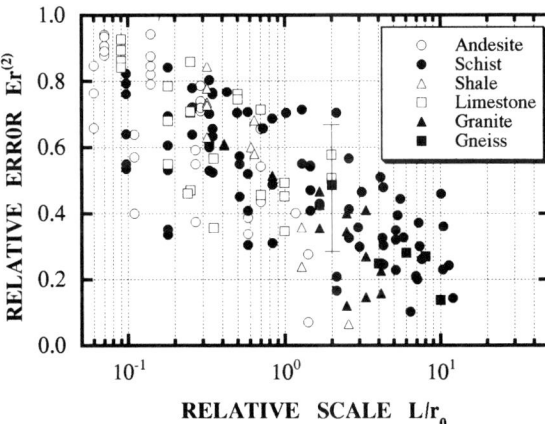

Figure-11 Relative Error Reduction with Relative Scale (the error associated with anisotropic geometry of joints)

Table-1 Calculated Results of Relative Scales and Relative Errors

Site & Rock	L/r_0	$Er^{(0)}$	$Er^{(2)}$	No.of Sub-Region
SITE T Shale	0.32	0.62 - 0.86	0.63 - 0.84	1 - 5
	0.64	0.55 - 0.61	0.58 - 0.68	6 - 8
	1.28	0.23 - 0.31	0.24 - 0.36	9 - 10
	2.56	0.06	0.07	11
SITE O (Horizontal) Limestone	0.25	0.37 - 0.77	0.47 - 0.86	1 - 4
	0.50	0.53 - 0.70	0.74 - 0.76	6 - 7
	0.99	0.31 - 0.47	0.35 - 0.49	8 - 10
	1.98	0.08 - 0.18	0.51 - 0.58	11 - 12
SITE O (Vertical) Limestone	0.09	0.81 - 0.88	0.84 - 0.93	1 - 4
	0.18	0.52 - 0.64	0.55 - 0.79	5 - 7
	0.35	0.32 - 0.36	0.36 - 0.57	8 - 10
	0.70	0.16 - 0.42	0.46 - 0.71	11 - 12
SITE Na (Map) Andesite	0.06	0.64 - 0.71	0.66 - 0.85	1, 7, 11
	0.11	0.44 - 0.59	0.40 - 0.64	2, 8, 12
	0.27	0.46 - 0.47	0.38 - 0.59	3, 9, 13
	0.58	0.32 - 0.49	0.34 - 0.39	4, 14
SITE Na (Photographs) Andesite	0.07	0.88 - 0.94	0.88 - 0.94	11 - 15
	0.14	0.44 - 0.86	0.79 - 0.94	1 - 5
	0.29	0.70 - 0.77	0.71 - 0.79	6 - 10
	0.70	0.42 - 0.66	0.44 - 0.66	16 - 18
	1.40	0.05 - 0.27	0.07 - 0.28	19 - 20
SITE Nb (Right bank) Schist	0.09	0.58 - 0.93	0.55 - 0.92	AA1 - AA5
	0.10	0.50 - 0.63	0.55 - 0.64	C1 - C3
	0.18	0.22 - 0.41	0.36 - 0.71	C4 - C5, C7
	0.35	0.47 - 0.63	0.64 - 0.77	C6, C8 - C9
	0.83	0.33 - 0.66	0.31 - 0.69	A1 - A3
	1.02	0.21	0.71	C
	1.67	0.42 - 0.45	0.42 - 0.43	A4 - A6
	6.37	0.01	0.10	A21
	6.45	0.13	0.32	A1
	6.91	0.11	0.21	A22
SITE Nb (Left bank) Schist	0.14	0.12 - 0.87	0.25 - 0.88	BB1 - BB5
	0.24	0.29 - 0.76	0.29 - 0.76	BB11 - BB15
	0.26	0.58 - 0.78	0.64 - 0.78	B11 - B13
	0.51	0.35 - 0.55	0.45 - 0.57	B14 - B16
	1.29	0.06 - 0.80	0.55 - 1.24	B17 - B19
	2.57	0.02 - 0.05	0.33 - 0.57	B110 - B112
	4.11	0.43	0.51	B11
	5.14	0.02 - 0.28	0.23 - 0.35	B113 - B115
	7.06	0.01	0.20	B12
	7.33	0.14	0.30	B13
	0.43	0.63 - 0.73	0.66 - 0.77	B21 - B22
	2.14	0.01 - 0.52	0.17 - 0.70	B23 - B25
	4.27	0.07 - 0.43	0.24 - 0.48	B26 - B28
	7.21	0.29	0.37	B23
	7.6	0.25	0.26	B24
	10.4	0.15	0.23	B22
	12.0	0.10	0.14	B21
SITE Nb (Center) Schist	0.33	0.50 - 0.81	0.53 - 0.80	DD10 - DD14
	0.58	0.29 - 0.59	0.30 - 0.71	DD5 - DD9
	1.45	0.31 - 0.44	0.41 - 0.55	DD1 - DD4
	2.95	0.23 - 0.67	0.36 - 0.70	D14 - D15
	3.00	0.12	0.30	D13
	3.11	0.30	0.46	D12
	4.90	0.27	0.54	D11
	5.30	0.24	0.39	D3
	5.50	0.13	0.44	D4
	5.78	0.15	0.33	DD15
	10.1	0.13	0.46	D2
	10.4	0.25	0.36	D15
	11.3	0.11	0.24	D1
SITE K Gneiss	2.00	0.13 - 0.96	0.41 - 0.96	11 - 62
	2.00	0.43	0.49	1
	4.00	0.03	0.25	2
	6.00	0.12	0.24	3
	8.00	0.27	0.27	4
	10.0	0.14	0.14	5
Microcrack Granite	0.41	0.52 - 0.96	0.61 - 0.96	11, 21, 31
	0.83	0.47 - 0.88	0.50 - 0.87	12, 22, 32
	1.66	0.35 - 0.44	0.36 - 0.47	13, 23, 33
	2.49	0.12 - 0.37	0.12 - 0.40	14, 24, 34
	3.31	0.07 - 0.39	0.15 - 0.41	15, 25, 35
	4.15	0.02 - 0.22	0.16 - 0.25	16, 26, 36

CONCLUSION

The applicability of a method of evaluating the degree of error in investigating actual rock masses is identified. The discontinuous natures of rocks investigated are formed by the quite different geological histories and range from micro-scale to rock mass scale. Therefore, the following are concluded generally.

(1) In representing in-situ tests and analytical models of discontinuous rock masses, the index measures calculated from error tensors which evaluate the representation can be applied to the actual rock mass.

(2) The REVs in natural discontinuous rock masses are dependent on the scale investigated and the density of discontinuites, such as heavily jointed or lightly jointed. By using the relative scale normalized by mean trace length of joints and the relative errors normalized by the invariants of a crack tensor, any joint geometries can be discussed at the same time.

(3) To evaluate the joint distribution with less than twenty percent error, joint surveys need to be carried out on a scale four to five times larger than the mean trace length of joints. If it is smaller, the degree of the error can be estimated from these results.

The rock mass deformability and permeability are affected by the joint geometry. Therefore, these relationships between relative scale and relative error are related to the interpretation of their scale effects.

REFERENCES

Barton, N., Lien, R. and Lunde, J., 1974. Engineerings classification of rock masses for the design of tunnel support rock mechanics 6, Springer-Verlag : 189-236.

Bear, J. and Bachmat, Y., 1984. Transport phenomena in porous media-Basic equations, Fundamentals of Transport Phenomena in Porous Media, Martinus Nijhoff Publishers : 3-61.

Bieniawski, Z.T., 1974. Geomechanics classification of rock masses and its application in tunneling, Proc. 3rd Int. Cong. Rock Mech., Vol. 2A : 27-32.

Charrua-Graca, J. G., 1979. Dilatometer Tests in the study of deformability of rock masses, Int. Proc. 4th Congr. ISRM, Vol.2, Montreux

Cuisiant, F. D. and Haimson, B. C., 1992. Scale effect in rock mass stress measurements, Int. J. Rock Mech. Min. Sci. & Geomech. Abstr., Vol. 29, No. 2 : 99-117.

Long, J. C. S. and Witherspoon, P. A., 1985. The relationship of the degree of interconnection to permeability in fracture network, Journal of Geophysical Research, Vol. 90, No. B4 : 3087-3098.

Oda, M., 1982. Fabric tensor for discontinuous geological materials, Soils and Foundation, Vol.22, No.4 : 96-108.

Oda, M., 1988. A method for evaluating the representative elementary volume based on joint survey of rock masses, Can. Geotech. J., 25 : 440-447.

Oda, M. , Hatsuyama, Y. and Takano, M., 1986. Minimum size of jointed rock masses to evaluate their hydro-mechanical properties, Proceedings of the 18th symposium on rock mechanics : 126-130. (in Japanese)

Shimo, M. and Kamemura, K., 1987. On the equivalent rock mass permeability considering spacial distribution in well test data, Proc. of 7th Japan Symposium on Rock Mechanics : 229-234. (in Japanese)

Scale effect of the rock

V.A. Mansurov
Institute of Physics and Mechanics of Rocks, Bishkek, Kyrghyzstan

ABSTRACT: Physical base of scale effect in the rocks is offered for discussion. It uses the kinetic ideas about of the failure process, developed by T. Ekobori and S.N.Zhurkov. The present results of the experiments on the rocks samples distinguished in volume in four decimal places. Obtained nonlinear dependencies changes of strength and deformation characteristics versus volume of samples in the process of their compressive loading. It was explained from the above mentioned points. Conformity to natural laws of radiation of accumulated elastic energy in the process of cracking on the various scale levels is found. Distinguished similarity of radiation elastic energy on the different scale levels. This statement is confirmed by observation in mines and also is used in processing of earthquakes catalog.

INTRODUCTION

It is necessary to know the laws of scale effects (SE-the effect, produced by the size of the body in conditions of its failure) when applying the laboratory results of rock testing to the field ones and when predicting strength characteristics and the value of energy, emitted by the object. Scale effect is the topic of numerous research works [1-7]. Our paper deals with the qualitative theoretical analysis of SE from the viewpoint of failure concept as a process, restricted by critical crack concentration. The results of such analysis are compared with experimental data in respect to rock failure in different sample volumes.

THEORETICAL CONCEPT

By now it has been discovered that in the loaded rocks thermoactivated generation of stable (initial) cracks is responsible for the increase of the failure scale and is characterized by two main stages. On the first stage, we have a delocalized accumulation of non-correlated initial cracks in the volume of the body, followed by spontaneous statistical clustering and generation of larger cracks which form failure nucleus. On the second stage, comparatively small, failure nucleus is developed on the account of joining up with correlated cracks, generated in the field of increased stresses, induced by clusters and make mechanically unstable magistral crack. Change of stages is marked by formation of a cluster with critical size i*, after which speed of a statistic growth of cluster becomes lower than that of correlated crack generation [8]. According to [8], number of n_i clusters, containing i of initial cracks with size r, at their concentration C in the body with volume V, is

$$n_i \approx \frac{VC}{i^{3/2}}\left(\frac{e}{k}\right)^i, \quad k = c^{-1/3}r^{-1} \quad (1)$$

where e - is the basis of the natural logarithms. As you see from (1), clustering is controlled by the value of dimensionless parameter k, having the meaning of an average distance between initial cracks $C^{-1/3}$, referred to their size r. The first cluster is rather large, it appears having size i* under

$$k^* \approx e\left[V_i^{-3/2}(er)^{-3}\right]^{1/(i^*+3)} \quad (2)$$

Expression (2) has the meaning of concentrational criterion for clustering. Simultaneously k* has the meaning of the volume cracking limit or initial cracks threshold limit, since as it was mentioned above, critical cluster formation stops the crack accumulation in the volume. It is noteworthy that in the expression (2) during i* growth $k^* \to e \approx 3$ independently of the size of initial cracks, stress state etc. This result is well-known on the example of different materials, including rocks [9] and allows to regard the concepts mentioned above as the basis for explanation for strength characteristics. Let us regard SE manifestations from the mentioned positions. Its dependence on the volume of the body V is shown in expression (1) and reflects additivity of the number of statistic independent events, leading to clustering. As a results of (1) SE of concentration threshold of failure appears (2) , with growth of V average relative space between initial cracks k* in the failed volume increases. This effect was observed in experiments for different rocks [10].

Dependence k* (V) may be displayed in different characteristics of failure, e.g. it may lead to SE of irreversable rupture deformation ε_p , restricted by cracking. Value of such deformation is proportional to crack concentration C so, that with taking into account (1)

$$\varepsilon_p \propto k^{*-3}$$

By force of SE of K* (2) value ε_p (3) also depends on V, namely,

$$\varepsilon_p(V) \propto V^{-3/(i^*+3)}$$

i.e. during volume extenuation rupture deformation decreases. This kind of SE was observed in [10] for different rocks testing (porphyryt, marble, coal, quartz, diabase, iron ore).

SE of k* also influences the value of rupture stress, i.e. strength σ_p. Really, with the stress increase with constant rate, $\dot\sigma$ strength, $\sigma = \dot\sigma\tau$ where τ time before failure. According to the stated above, τ is defined by the equation

$$C(\tau) \sim C^* = \left(k^*r\right)^{-3}$$

The dependence in the left part denoting concentration of cracks C versus time t under the given loading regime is

$$C(t) = C_0 \int_0^t \frac{dt}{\tau_0 \exp\left(\frac{u_0 - \gamma\dot\sigma t}{KT}\right)}$$

where $C_0 \sim r^{-3}$ - maximum possible concentration of places in the body, where crack generation is possible (size r), τ_0, u_0, γ -parameters of Zhurkov's formula [5,11], T - absolute temperature, K - Boltzman's constant. Summing up the expressions, we come to the following expression for the strength σ_p:

$$\int_0^{\sigma_p} \frac{d\delta}{\sigma\tau_0 \exp\left(\frac{u_0 - \gamma\sigma}{KT}\right)} = k^{*-3}$$

Hence by integrating we find

$$\sigma_p = \frac{1}{\gamma}\left(u_0 - KT \ln \frac{KTk^{*3}}{\gamma\sigma\tau_0}\right)$$

With taking into account SE of the value of k* (2) from the formula (4) it follows that during volume increase its strength σ_p decreases, namely when

$$\sigma_p(V) = const - \frac{3KT}{i^* + 3} \ln V \qquad (4a)$$

Elastic energy, emitted during failure also is characterized by SE:

$$E \approx \frac{\sigma_p^2}{2Y} V \qquad (5)$$

where Y - elasticity modulus. From (5) it is clear that during volume extension energy emission increases. This statement is a trivial one, since energy emission E is proportional to V (5). However, dependence E(V) is not a linear one by force of strength SE, leading in expression (5) to the dependence $\sigma_p(V)$. Strength SE effect in energy emission becomes more vivid during transition to the volume density of energy emission

$$\frac{E}{V} \approx \frac{\sigma_p^2}{2Y} \qquad (5a)$$

which as it is seen, qualitatively reproduces strength σ_p (4a), i.e. with volume extension energy emission density decreases. Concentration threshold of failure is displayed once more in the dependence of maximum local energy emission E_i on V during crack development. E_i is proportional to L - crack size. In conditions of cracking due to statistic mechanism L is proportional to the cluster size i, i.e. $E_i = f(i)$, where f is some increasing function its looks depends on cracking geometry, which we do not regard here. During clustering of initial cracks cluster distribution according to the sizes i according to (1) is

$n_i \sim VCi^{-2/3}$ Hence, maximum cluster size is $i_m \sim (VC)^{-2/3}$.

Correspondingly, maximum local energy emission SE

$$E^*_{i_m}(V) \propto f(V^{2/3}) \qquad (6)$$

so, the latter value grows with the volume increase.

EXPERIMENTAL DATA

Experiments were conducted on prismatic granite, marble, sandstone samples of different volume with geometric semblance in their linear dimensions - height a and basis rib b with relation a/b=2. The least volume was equal to 7×10^{-6} m^3, the largest - 0.3m^3. This allowed to investigate samples in the range of 3 - 4 orders change of the volume. Their loading was performed on the testing machine PR - 500t, and the largest samples with the volume 0.3 m^3 were loaded on testing machine PR - 1000t. Deformation rate was constant from experiment to experiment and was equal to 1×10^{-6} s^{-1}. In the process of loading three components of strain were registered, as well as the applied load and AE signals in accordance with the methods described in [12]. Experimental data - are represented in table 1, and in figures 1 - 5. From graphs in figure 1 and table 1, and figure 1 - 5. From graphs in figure 1 and table 1 it follows that strength and maximum longitudinal strain decrease with increasing volume, though this occurs in different ways for different materials. The smallest change of strength is observed for homogeneous and relatively plastic marble (curve 3, Figure 1a). For granite (curve 1), a sharper decrease in strength is observed at sample volume $V > 10^{-3}$ cm^3. The maximum strain for rocks decreases

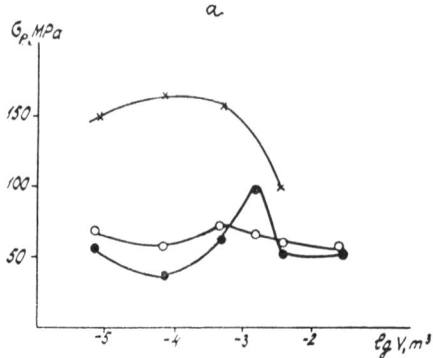

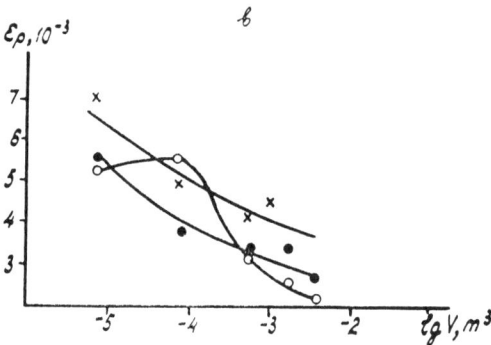

Fig. 1. The dependences of strength and maximum strain versus the volume of specimen V: 1-granite, 2-marble, 3-limestone.

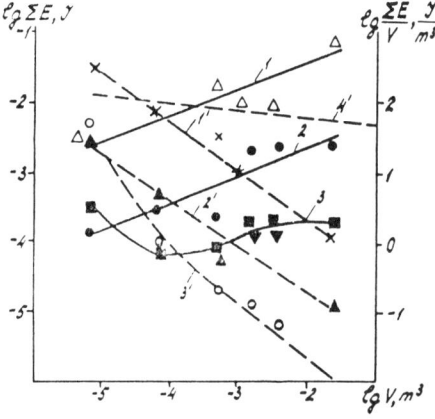

Fig. 2. The dependences of integral AE energy E and volumetric density of AE energy versus the volume of specimen V: 1,1'-granite, 2,2'-limestone, 3,3'-marble.

systematically too, and this is evidence of an increase in the localization of the failure process with an increase in volume.

During AE investigation first of all, it was of interest to make it clear how generation of elastic energy changes versus the sample volume as well as volumetric density of AE energy, $\Sigma E_i / V$. For this case, in laboratory experiments, energy of AE signals may be estimated with accuracy to the coefficient α as follows: $E = \alpha A^2 T$, where A. T. where correspondingly the amplitude and duration, measured by the envelope [12]. Values of all AE signals energies were summarized up to the sample macrofailure. These values are given in Figure 1 versus sample volume. Every point is the result of average value for 4 samples, therefore discrepancy of integral AE energy values is large enough. For one and the same sample size it difference during process of the deformation and failure reached 4.

We shall now consider data concerning the emission of elastic impulse energy. Table 1 contains the integral AE energy. Naturally, it increases with an increase in volume. However, volumetric density of AE energy is more informative. For all rocks it decreases as the volume increases (Figure 2), and this is evidence of failure localization and confirms the conclusions made during the analysis of maximum strains. It is interesting to note, that for a brittle rock (granite), this fall is sharper than for plastic one (marble).

Nonlinear changes in $\Sigma E_i / V$ were observed for limestone over the same range of sample sizes, in proportion to the number of faults. This results is interesting as it shows the ability of AE to reflect structural differences of the rocks, coupled with dimensions of the area being deformed. We also note that some more characteristic differences in AE can be observed in this area - in amplitude spectra, in collective events, and so on.

The main value of the results obtained is in their extrapolation to large volumes. It is interesting to evaluate the maximum possible elastic impulse which can appear in an observed volume when it falls. To obtain such results we made the following graphs (Figure 3a) [13]. The energy values of individual impulses, E_i, were graphed on the horizontal axis, and the total energy, ΣE_i, emitted by such quanta as $n_i E_i = \Sigma E_i$ on the vertical axis. The dotted line corresponds to the boundary case when The whole energy is emitted by a single impulse. The inter/section of the experimental lines with the dotted line gives the maximum possible value of the energy impulse of failing sample of a given volume. Maximum values of energy, E^*, obtained by such a method are shown on Figure 3b with the sample volume. As it is seen, in the similar volumes maximum possible energies E^* differ for the rocks more than by the order. The relationship for granite is linear in a double logarithmic plot, with slope 0.5 which enables us to write $E^* \infty V^{1/2}$. The exponent 1/2 also agrees with the localization of failure. It is worth nothing that obtaining such a relationship for various rocks allows one to make qualitative estimates of failure processes for various of their mechanical properties. However, for the other two investigated rocks, we failed to obtain results which would have enabled us to write an expression analogous to equation for E^*, because of nonlinearity of the graph of log E^*/logV.

These results indicate that it is possible to obtain values by extrapolation for larger volumes, though this probably demands a wider range of volumes so that specific structural peculiarities do not significantly distort the total law.

THE DISCUSSION OF THE RESULTS

Let us compare the experimental data with the theoretical concept, about volume of the body V versus characteristics of its failure, stated above. One can see, that in the sphere of larger volumes behaviour of the rocks investigated is totally concordant with the theory: with V increase maximum strain ε_p, strength σ_p, volumetric density of AE energy decrease, and AE energy itself increases. Note, that the point is in qualitative concordance only. Qualitative manifestations of SE depend on the critical cluster size i^*. So, strength SE as it is seen from (4) is expressed the more so, the less is i^*, and disappears under i^*. According to [9],

$$i^* \infty \frac{\xi}{\tau}\left(\frac{KT}{2\gamma\sigma} \ln \frac{V}{\gamma^3}\right)^2$$

where ξ - is the crack curvature radius, depending on plastically of the stuff, the rest parameters of formula (7) are named above, So, due to (7) manifestations of the regarded SE increase during stuff embrittlement, ξ decrease, T decrease, σ increase and body size decrease (value V/r^3). This conclusion is well concordant with the known knowledge about SE [1 - 7]. Experimental data shows that in the sphere of average volumes strength SE inversion is observed, i.e. its increase with the body size increase takes place. This effect is well-known for powder materials [14] and is connected with porosity, weakening the section. Such explanation is confirmed by relatively strong inversion of SE in ioos (friable) limestone compared to more compact granite and marble (Figure 1).

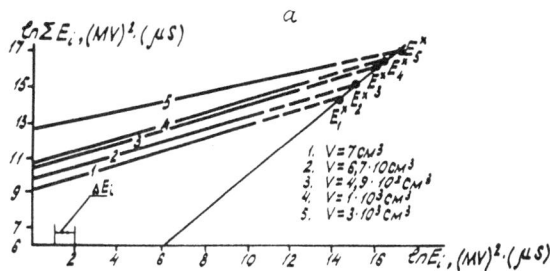

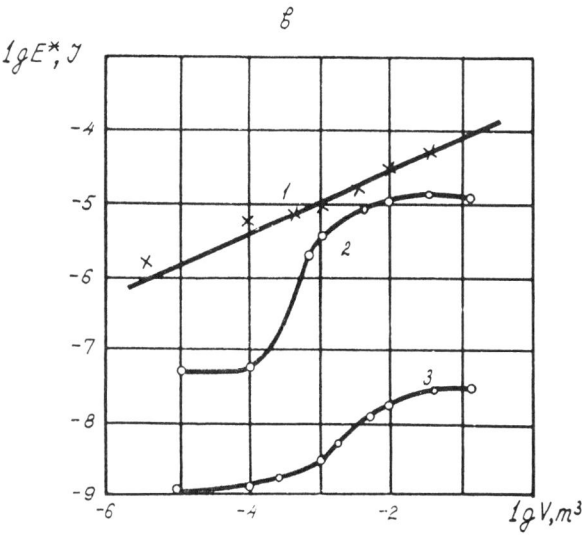

Fig. 3. (a) Change of integral AE energy with the energy of individual impulses E for granite samples of various volumes. (b) Dependence the elastic impulse energy E^* on volume V for: 1 - granite samples; 2 - limestone; 3 - marble.

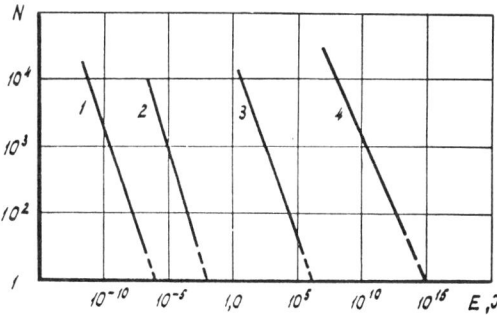

Fig. 4. Reccurent intervals for dynamic phenomena at various scales: 1- the failure of granite samples in a testing machine; 2- the failure of pillars in the Karnasurt mine (Kola peninsula); 3- the catalog of rock bursts for the North Urals bauxite mine (Ural mountains); 4- the catalog of earthquakes between 1955-1978 in the Nurek dam region (Tajikistan).

The theoretical and experimental results (table 1, (7), Figure 3) are especially important for rock burst prediction in rock massives. So, if granite during volume expansion did not have any structural peculiarities (inclusions, layers etc.), it would be possible to predict rock bursts for the mine conditions (10^5 - 10^6 J in volumes 10^6 m^3) these are usually blocks with the average linear size 100m; such bursts are registered in mines. But it is not yet the upper limit. In conditions of the higher pressures the effect may be stronger. For analysis relation between maximum energy (E*) which may be emitted in a single impulse of failure and total energy, emitted from the failed volume, it is convenient to make a plot in coordinates of single acoustic signal energy and summary energy (Figure 3a). Having in view strong change of these parameters we used binary (double) logarithmic coordinates. It is noteworthy that in (5, 5a, 6) total elastic energy of the sample is given, and in experiments we worked only with its share, spent for AE. Evidently, for all the sizes of the body there is a direct proportional relation between them, since AE energy during process of the deformation and failure is mainly determined by acoustic signals with high amplitudes and low frequencies, which practically do not attenuate for these samples.

TABLE 1

LOADING OF ROCK SAMPLES WITH DIFFERENT VOLUME

N	lg V, m	Strength, MPa	Maximum strain, x10	Total AE energy, lg E, J	Maximum AE energy, lgE*, J	Quantity of AE signals
			GRANITE			
1	-5.2	151.5	7.1	-2.7	-2.7	26088
2	-4.2	165.0	5.0	-2.2	-2.3	47394
3	-3.3	159.0	4.2	-1.8	-2.1	31240
4	-2.5	171.0	3.7	-0.4	-1.6	72569
5	-1.5	-	-	-2.0	-	4144
			LIMESTONE			
1	-5.2	55.0	5.6	-3.8	-5.0	6571
2	-4.2	36.0	3.8	-3.5	-4.9	2624
3	-3.3	64.0	3.4	-2.6	-4.2	4436
4	-2.8	104.0	3.4	-2.6	-2.6	2232
5	-2.4	54.0	2.7	-2.6	-1.9	8396
6	-1.5	54.0	-	-2.6	-1.5	11068
			MARBLE			
1	-5.2	69.0	5.2	-3.4	-5.4	15650
2	-4.2	58.0	5.5	-4.2	-5.4	12075
3	-3.6	75.0	3.2	-4.0	-5.3	4100
4	-2.8	69.0	2.6	-4.0	-4.6	2396
5	-2.4	62.0	2.2	-3.6	-4.4	3693
6	-1.5	55.0	-	-3.8	-	511

In order to widen the dimensional range of objects being investigated in laboratory, AE studies were conducted by complex investigations for samples of 1m on the big press(testing machine, enabling us to create an axial load up to 50000ton.) in the Institute of High Pressure Physics, Russian Academy of Sciences [15]. However, these samples only exceeded the volume of the largest of those studied in our experiments by 3-4 times, so we did not notice any significant qualitative differences.

The results of deformation and AE studies on larger natural samples, pillars in mines, seem to be more interesting. The loading was carried out through the explosive elimination of neighboring pillars, i.e. in conditions of sudden additional loading. Naturally, this does not allow comparison with the laws of deformation and energy emission under controlled loading conditions. Nevertheless, some comparisons can be made.

The best comparison for all objects investigated is carried out by the reccurent intervals. Figure 4 shows graphs of AE reccurent intervals for all the elastic impulses emitted up to the moment of failure of sample or an object.

First of all, let us note the qualitative similarity of the reccurent intervals for objects with differing volumes and of various rocks. Quantitative differences appear following comparison of the slopes of these graphs for different rocks. So, at a first approximation, we can assert the qualitative similarity in the process of fault accumulation and energy emission at all scale levels. Quantitative differences are caused by physical and mechanical properties of the rocks and massives and by stress conditions.

CONCLUSIONS

The result of obtained prove the necessity of studies in a wide range of samples and objects sizes. The investigations that have been carried out enable us to make some qualitative conclusions:
(i) Qualitative similarity is observed in the processes of deformation, failure and energy emission in a wide range of sizes of the objects being deformed.
(ii) An increase in the volume loaded leads to an increase in failure process localization.
(iii) An increase in the loaded object volume also increases the maximum energy possible in one impulse.

REFERENCES

1. Chechulin B.B. 1963. Scale factor and statistic nature of metals strength. Moscow, Mellurgy, p.189 (in Russian).
2. The discussion of scale effect nature. Factory laboratory, vol.27, pp.318-323 (in Russian),1961.
3. Ivanov A. G., Mineev V.N. 1975. About the scale criterion at brittle failure of constructions. The Far East Academy of Sci. of the USSR. vol.220, pp.575-581(in Russian).
4. Petrov V.A., Savitsky A.V. 1975. Thermofluctuational nature of size effect of strength. The Far East Academy of Sci. of the USSR, vol.224 , N4, pp. 806-809 (in Russian)
5. Koifman M. I. 1962. About the effect of sizes on the strength of rock specimens. In "The investigation of physic and mechanic characteristics of rock as applied to the tasks of the rock pressure management". Moscow, The USSR Academy of Sci. Publ., pp. 6-14 (in Russian).
6. Bieniawski Z., 1968. The effect of specimen size on compressive strength of coal. Int. J. Rock Mech. and Min. Sci., N4,pp325-336.
7. Hodgson K., Cook N. 1970. The effects of size and stress gradient on the strength of rock. Proc. 2nd Congr. Int. Soc. Rock Mech., Beograd, vol.259,pp31-34.
8. Petrov V. A. 1986. The Mechanism, kinetics and prediction of heterogeneous materials failure. In "Mechanics of composite materials", vol.5. pp.940-943 (in Russian).
9. Zhurkov S.N., Kuksenko V. S. and Petrov V. A. 1981. Physical principles of prediction of mechanical failure. The Far East Academy of Sci. of USSR, vol. 259, N 6,pp.1350-1353 (in Russian).
10. Petrov V.A. Gorobets L. J. 1987. The size effect of concentrational threshold of failure. News of the USSR Academy of Sci. (physics of the Earth), N 1, pp.95-98 (in Russian).
11. Zhurokov S.N., Kuksenko V.S., Petrov V.A. Saveljev V.N. and Sultonov U. 1977. About the prediction of rock failure. News of the USSR Academy of Sci. (Physics and the Earth), N 6, pp.11-18 (in Russian).
12. Mansurov V. A. 1994 Laboratory experiments: Their role in the problem of rock burst prediction. In "Comprehensive rock engineering", vol.3, pp.745-771.
13. Kuksenko V.S., Mansurov V. A. 1986. The localization of failure in rocks in various scale levels. Rock Sci. of the USSR, N2, pp.49-55.
14. Troschenko V. T., Rudenko V.M.1965. The strength of metal-ceramic materials and methods of its determination. Kiev, Technics, p.125 (in Russian).
15. Sobolev G. A., Koltsov A. V. 1988. Large scale simulation of preparation and forerunners of Earthquakes. Moscow, Science, p.205 (in Russian).

Scale effect on the deformability of layered rocks

Z.Y.Yang & C.S.Liu
Department of Civil Engineering, Tamkang University, Taiwan

ABSTRACT: The larger tested volume having more joints demonstrates lower deformation modulus. There exists a certain volume (representative elementary volume; REV) of rock masses beyond which the scale effect is negligible. A method, based on the laboratory tests to determine the maximum closure of a joint and the elastic deformation of the intact rock under uniaxial compression strength, will be proposed to predict the REV of scale effect on the deformability of layered rocks. Two main conclusions are drawn: (1) The secant modulus including more closure behavior reveals marked scale effect than the tangent modulus. (2) A steady deformation modulus (at REV level)reaches when the frequency of joints is more than 30 per meter in a jointed rock sample.

1 INTRODUCTION

The design of engineering structure requires a knowledge of the constitutive laws governing the layered rock masses. The rock mass consists of the intact rock and the weakness defects such as joints. The deformability of jointed rocks is mainly depended on the configurations and the mechanical behavior of joints for a rock foundation. Two approaches are usually used to analyze this complex composite mass. In one method, the whole rock mass is divided into two components of intact rock and discontinuities (joints). That is, the behavior of intact rock and joints within the mass is considered individual. Thus the yielding behavior is more closely to the real behavior of rock masses. However, this method is uneconomical due to the complicated geological structures. An alternative efficient method of analysis is to set to work the jointed mass as an equivalent homogeneous material.

A typical behavior of the whole rock mass, such as load-deformation behavior, is assumed based on the result of in-situ test or laboratory model test . The behavior resulting from any test scale is not yet match the behavior of prototype structure except for full scale tests. The test volume is different in the field tests depended upon the testing types and loading area, such as in the case of plate bearing test for a rock foundation. The larger tested volume including more joints always demonstrates lower deformation modulus, called scale effect. For one engineering structure size, there exists a certain volume (representative elementary volume; REV) of rock masses beyond which the scale effect is negligible.

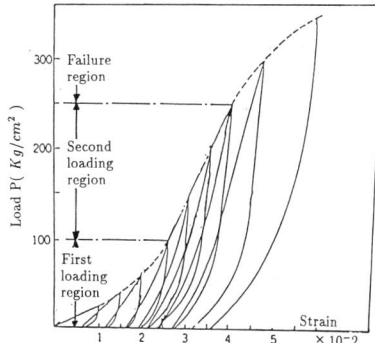

Figure 2: The stress- strain behavior of rock masses in field tests (Fukushima 1993).

The size of REV or considered range to consider rock masses as a continuum is not identical among many investigators. For example, a ten times volume larger than the area of loading plate is suggested for layer rocks (Chen 1987). The REV dimension is approximately an order of magnitude larger than the joint spacing so that enough joints exist in the rock mass to validate the averaging continuum description. A method, based on the laboratory tests to determine the maximum closure of a joint and the elastic deformation of the intact rock, will be proposed to predict the diminishing point (at REV condition) of the scale effect on deformability for layered rocks.

2 DEFORMATION BEHAVIOR OF ROCK MASSES

2.1 Deformation modes of rock masses

Franklin et al.(1992) have stated that the characteristic of stress-strain curve for jointed rocks, as shown in figure 1, can be divided into three parts before peak strength. (1) During the initial loading, the jointed rock progressively becomes denser as joints close, and becomes stiffer and less deformable. The load-displacement curve thus shows an upward concavity (region A) reflection this increasing stiffness at higher loads. (2) At intermediate stress levels, the behavior of most rocks becomes linear elastic (region B). All the joints have almost closed, so strain increments become proportional to the applied stresses. (3) At stresses increasing higher, the joints start to shear. The stress- strain curve becomes concave downward (region C). Fukushima(1993) also presents a result of plate bearing test as shown in figure 2. This result reveals that the low modulus of deformation increases from the beginning of the loading to close of openings in first loading region. A constant modulus appears in second loading region and finally reach the failure region.

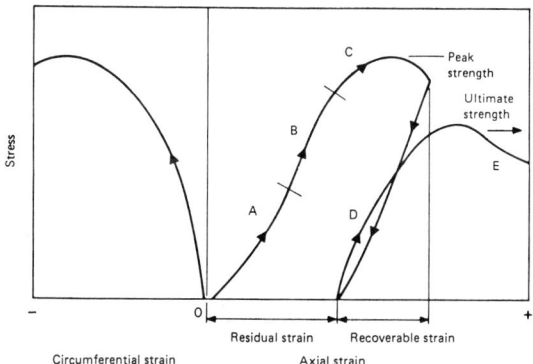

Figure 1: The stress- strain behavior of a rock mass illustrated by Franklin et al. (1992).

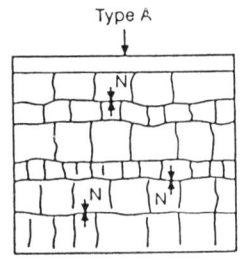

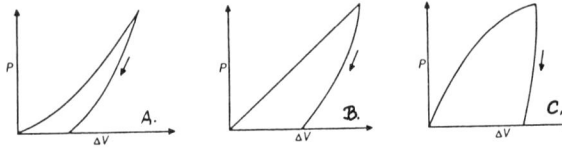

Figure 3: Deformation modes of rock masses (Barton 1986).

Three types of stress- strain behavior are noticed in field tests for rock masses with different joint configuration (Barton 1986). These deformation behavior of rock masses depends on the relative magnitudes of closure, shear and dilation of joints as shown in figure 3. For the horizontally bedded rock, the load-deformation curve obtained from large plate bearing tests resemble the normal closure behavior for rock joints. In this case, the closure behavior of joints is dominated the concave- shape curve of the rock mass at the low stress levels. While the applied load is further increased, the stress- strain behavior of rock masses reveals a linear elastic deformation or fracture followed the closure behavior.

2.2 Closure behavior of the single rock joint

The complex deformation behavior of rock mass can be more understood when the individual component of intact rock and joints is separated. In most cases, the component of intact rock material can be considered as a linear elastic body. On the other hand, Goodman(1977) has illustrated joint closure in compression taking account of the combination of the elastic behavior of intact rock and joint closure. The normal stress-closure behavior of joints is a highly nonlinear concave-shape curve, as shown in figure 4. The curve is characterized by a soft response at the beginning of loading and stiffening at higher levels of stress. The nonlinear behavior results from increasing contact areas and numbers of contacts as the joint is compressed. However, there is a limit called the maximum closure to the amount of compression. The amount of joint closure tends to the asymptotic maximum value when the closure behavior of joint is almost complete and acts as elastic material. These data of joints can be conveniently given by the closure test in the laboratory.

3 PROPOSED METHOD TO DETERMINE THE REV

3.1 Relationship of deformation modulus and joint number in a rock mass

The common cause of a scale effect is the existence of defects and joints in rock masses. As mentioned, the joint plays a major role in the deformation behavior of rock masses, especially for the layered rocks. In this paper, the REV is thus defined to be reached when the influence of joint (defect) closure on the deformability of rock masses is negligible as applied load increasing. The intact rock and rock masses show the same mean value of the deformability moduli, since the tight joints play no significant role in the rock mass deformation.

The superposition of intact rock and joint deformation leads to the deformability of rock masses. Thus the total deformation (ϵ_m) of a rock mass is assumed to equal the summation of intact rock and joint deformation for the horizontal layered rocks in this paper.

$$\epsilon_m = \epsilon_i + \epsilon_j \quad (1)$$

where ϵ_j is total strain of joints corresponding to the total normal closure δ_j and the properties are the same for each horizontal joint. ϵ_i is the elastic strain of intact rock material due to an amount deformation of δ_i. In additional, the height of the rock mass is L throughout this method.

Therefore, a secant modulus (E_m) from zero to the applied stress σ including most part of joint closure behavior can be defined by

$$E_m = \sigma/\epsilon_m = \sigma/(\epsilon_i + \epsilon_j) \quad (2)$$

To a certain stress level σ_n^{max} normal to joint plane, the maximum closure δ_{max} of the joint is completed. At this situation, the behavior of joints is transformed from nonlinear joint closure deformation to the elastic contact behavior and acting as the intact rock material. The effect of joint closure on the deformation behavior of the whole rock mass can be negligible. The total strain of joints ϵ_j is then equal to the summation of maximum closure ϵ_{max} (δ_{max} is the corresponding maximum closure) for each joint. That is,

$$\epsilon_j = N \times \epsilon_{max} \quad (3)$$

where N is the joint number within the considered volume. Combining equation(2) and equation(3), the equation(2) is reduced to the following form.

$$E_m = \frac{\sigma}{[\epsilon_i + N \times \epsilon_{max}]} \quad (4)$$

Alternatively, it can be expressed by

$$E_m = \frac{\sigma}{\epsilon_i [1 + N \times (\epsilon_{max}/\epsilon_i)]} \quad (5)$$

or, expressed by

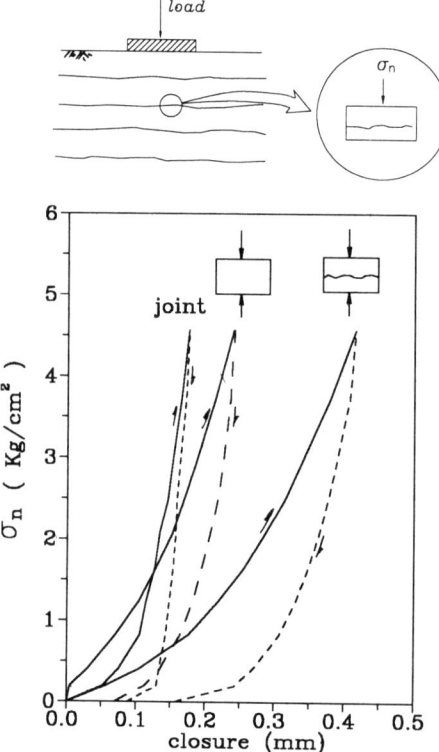

Figure 4: Closure behavior of a single joint.

$$\frac{E_m}{E} = \frac{1}{[1 + N \times (\epsilon_{max}/\epsilon_i)]} \quad (6)$$

In a closure deformation form, it can be given by

$$\frac{E_m}{E} = \frac{1}{[1 + N \times (\delta_{max}/\delta_i)]} \quad (7)$$

where E is the deformation modulus of intact rock material and equals to the value of σ/ϵ_i. From equation(7), it is obvious that the deformability of rock masses is depended on joint closure properties and joint number (joint frequency N/L). The deformation modulus of the layered rock is decreasing as the joint number increasing, as the parameters study shown in Figure 5.

3.2 Procedure for determining the REV in laboratory tests

To determine the relationship of deformation modulus and joint numbers for a rock mass, a closure test for the single joint with identical properties in the mass (as shown in figure 4) must be performed first. The closure behavior of a single joint can be obtained from the difference in the deformation between intact rock and jointed rock under uniaxial compression. From the test results, the maximum closure of a joint δ_{max} and the corresponding elastic deformation of intact rock δ_i are obtained. Through the equation(7), the asymptotic value of joint number at REV level is determined for the laboratory specimen scale.

4 TEST PROGRAM OF LAYERED ROCK MASSES UNDER COMPRESSION

4.1 Model material

A model material consists of mixtures of plaster, sand, and water is selected. The ratios of plaster : sand : water are 1 : 0.25 : 0.92 by weight. The mixing and curing of the model specimens are carefully controlled to obtain reproducible material properties. All the specimens are cured in a constant temperature and humidity chamber for 5 days at the temperature of $25°C$ and the relative humidity of 55%. The tangent modulus of deformation is 12000 Kg/cm^2, secant modulus is 8400 Kg/cm^2 and Poisson ratio is 0.2 for the intact model material.

4.2 Artificial joints and the closure behavior

A double-blade guillotine is designed (Huang and Yang 1991) to generate the artificial joints in this investigation. The cutting speed of the guillotine is controlled at a constant value of 1 mm/min for generating model joints with identical surface characteristics. Figures 6 illustrate the typical surfaces and JRC values of the artificial joint generated by the guillotine. Figure 4 depicts the closure behavior of the model joint and the maximum closure (δ_{max}) is 0.018 cm at normal stress σ_n^{max} 4.57 Kg/cm^2.

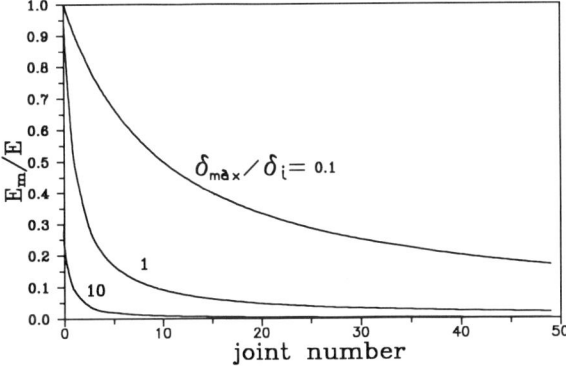

Figure 5: Deformation moduli vary with the joint number for a layer rock.

Also, the elastic deformation δ_i of intact material is 0.024 cm corresponding to this normal stres level obtained from these curves.

4.3 Uniaxial compression tests of layered rock masses

A series tests of rock masses with horizontal joints under uniaxial compression have been performed by Lin(1993). The dimension of cylindrical specimen is 5.4cm in diameter and 15cm in height. The number of joints ranges from 1 to 5 in the rock masses and the typical stress- strain curve is also plotted in figure 7. A tendency of softening in deformation modulus is observed and is more deformable for layered rock masses as joint number increasing. From these curves, the secant deformation moduli from zero to the maximum closure stress σ_n^{max} (4.57 Kg/cm^2) can be calculated. While the applied load is larger than σ_n^{max}, the slopes of stress- strain curves are almost linear and parallel to the intact rock (N=0).

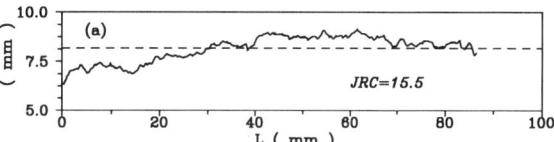

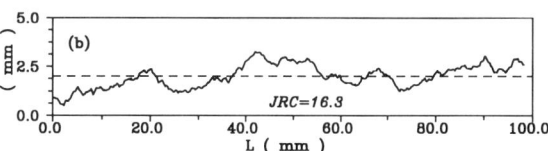

Figure 6: The profiles of an artificial extension joint along two orthogonal direction.

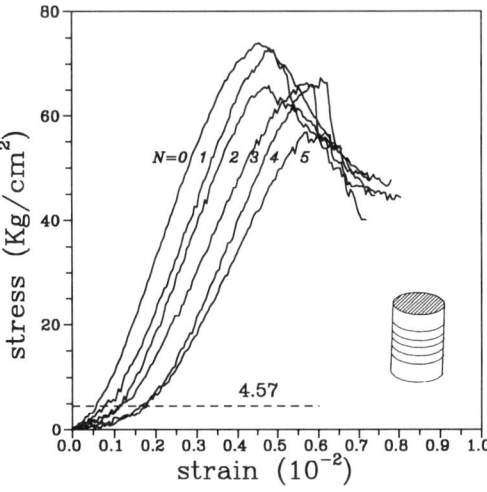

Figure 7: The stress- strain curve of layered rocks with different joint numbers.

5 VERIFICATION OF THIS PROPOSED METHOD BY TEST DATA

5.1 Verified by laboratory testing data

According to the equation(7), the tendency of secant modulus decreasing as joint number increasing can be calculated. The predicted result of modulus is plotted to be compared with the above testing data. As illustrate in figure 8, it can give a good approximation of the predicted and the testing results. The deformation modulus is reduced to 35% that of the intact mass. The diminished point of scale effect on deformation modulus for this masses at REV level is reached while the joint number is about 5. Equivalent, the joint intensity is about 30 per meter (5/0.15M) for this rock joints. Therefore, the REV assumption is reasonable at which the joint frequency is more than 30 per meter in a rock mass specimen.

5.2 Verified by field testing data

In a normal case, a test in the field cannot be a full scale prototype test. Instead of this, testing conditions must be only applied on a part of the whole volume which we call a sample. The testing results can only describe the behavior of the body, if the volume of the sample is equal or larger than the REV. Thus the REV is the smallest exchangeable part of a homogeneous area which is independent of test sites. The REV depends on the size of the considered homogenous area corresponding to the engineering structure size. Both the REV and the considered volume are characterized by the size and the distribution of basic elements of rock masses in the considered volume (Natau 1990).

The joint spacing is the basic element for rock foundation scale in a regularly jointed mass. For example, the considered volume or test volume which is influenced by the area and the shape of plate bearing tests. Straightforward, the test volume in the laboratory is the size of the sample. The stress in the field test is generally distributed unevenly from the maximum value at the loading points to zero far away. The test volume was defined by Heuze(1980) as that one included within a stress iso-line is 2% of the applied load magnitude. For a plate loading test with a circular plate, the test volume is a sphere with a radius four times that of the loading plate. The number and size of field test can be reduced and determined correctly, if the relation between field test and laboratory test is more understood. However, the method to extrapolate the result of field test form laboratory is not recognized to our experience.

Figure 9 shows the data of field test from several literature by Williams(1980) to detect the scale effect on rock masses deformation. It is revealed that the deformation modulus for stronger mudstone (A,B case) is decreased with joint number increasing. To reach the REV level, the joint frequency is also about 20-30 per meter in the field rock mass and is matching to the results of laboratory tests.

Therefore, the proposed method may be approximately used to predict the scale effect on deformation for layered rocks preliminary. In order to neglect the scale effect by only one test in the field test, the test volume of jointed rock masses is required to having the joint frequency 30 per meter at least. The size of loading plate is chosen according to the joint spacing to satisfy the mentioned test volume required. The plate size calculated by this concept may be too large. Thus a few field tests of different load areas are still necessary.

5.3 Applied to dipping rock formation

The deformability of rock masses is general depended on the dip direction of joints due to the anisotropic stress distribution. Is this proposed concept to determine the joint frequency at REV level available? In this investigation, the FLAC program is used to check the variation of the test volume (defined by 0.1 applied load) influenced by the joint direction.

As illustrate figures 10, it is concluded that the pressure bulb is strongly anisotropic according to the direction of joints. The test volume, arbitrarily choose the area within 0.1q and normalized by that of isotropic rocks, is about the same magnitude for different joint dips as shown in figure 11. The test volume is independent of joint dips, while the dip angle is less than 50

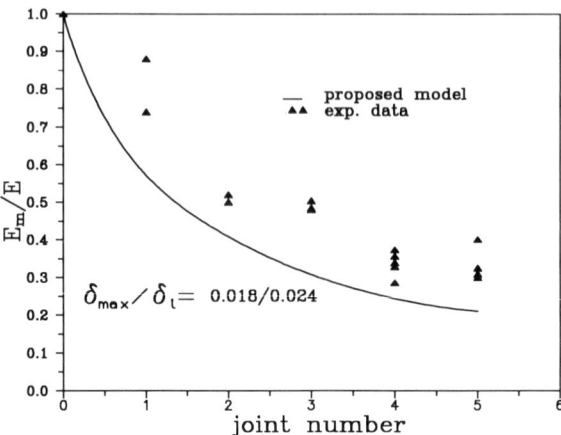

Figure 8: Comparison of secant modulus by proposed model and tests.

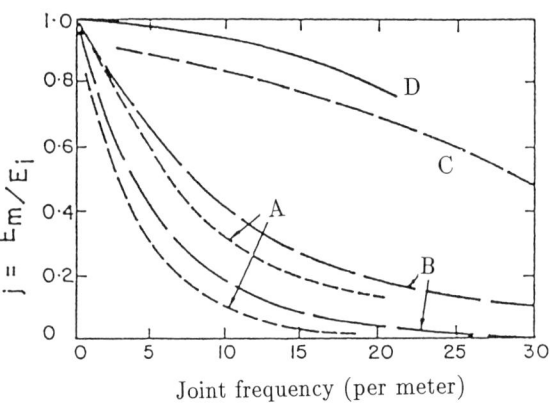

Figure 9: Scale effect on deformation modulus by field testing data (Willams 1980). A,B: mudstone; C,D: sandstone and shale

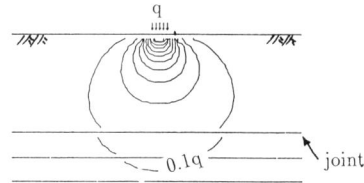

(a) dip angle=0°

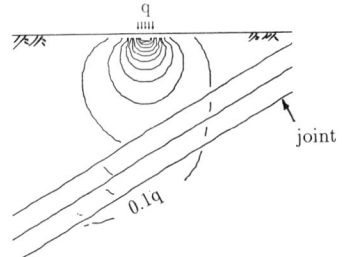

(b) dip angle=20°

Figure 10: The anisotropic stress distribution (10%q) varied with joint direction by FLAC.

degree. The larger test volume calculated may be affected by evident joint slip deformation, while dip angle is larger than 60 degree. A similar relation between test volume and joint dips for anisotropic rocks using ABAQUS pursued by Chen(1993) is also plotted in figure 11. If we can tentatively neglect or further clear the anisotropic character on scale effect, the above concept would be adopted to estimate the REV in dipping formation. However, more studies are required in this opinions.

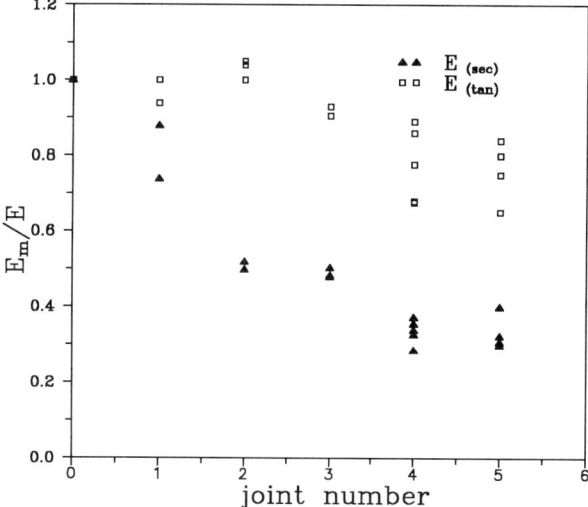

Figure 12: Scale effect on secant modulus and tangent modulus of mass models.

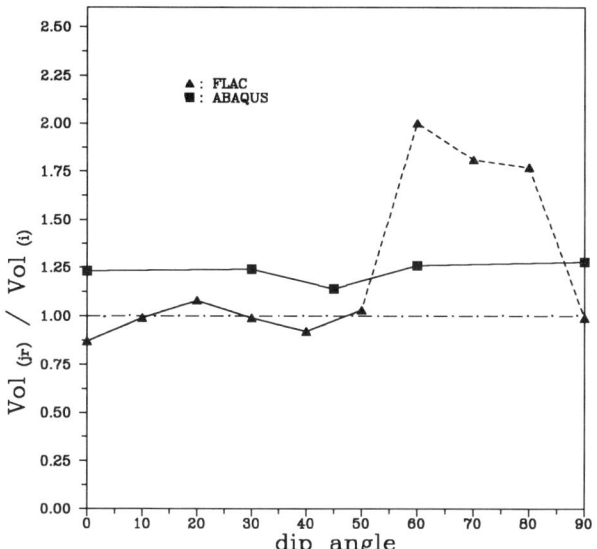

Figure 11: The test volume (10%q) varied with joint direction.

6 DISCUSSION AND CONCLUSION

At present, there are also two opposite opinions on the scale effect of plate loading areas. (a) Modulus increases with the plate loading area, since the improved characteristics of rocks in deeper pressure zone. (b) Modulus becomes decreasing with the loading area, since the more number of joints is included. Iofis et al.(1993) state, from field test data, that the modulus of mass is increasing as loading area increasing for the mass having small deformation modulus (approximate $E \geq 80\ Kg/cm^2$). On the contrary, the deformation modulus is decreasing as loading area increasing for high modulus of rock masses.

For the model masses in this investigation, an attempt comparison between the secant modulus at low stress level and the tangent modulus within the elastic deformation range is plotted as figure 12. From this figure, it reveals that the scale effect of secant modulus is more marked than tangent modulus. The tangent modulus is trivial reduced to 80% that of the intact mass. This distinction may depend on the contributed shares of shearing and closure deformation (depend on the dip of joint) in the masses. Thus to further clarify the scale effect, the method of calculating deformation modulus may be taken into account.

Some conclusions are drawn from this research program: (1) The deformation modulus of rock masses decreases as the joint frequency increases and the influenced degree is depends on calculating method. (2) The secant modulus including more closure behavior reveals marked scale effect than the tangent modulus. (3) The joint number at REV can be calculated based on the closure behaviors of the joint and the intact rock. (4) A steady deformation modulus (at REV level)reaches when the frequency of joints is more than 30 per meter in a jointed rock sample. (5) The scale effect on the deformability for dipping rock formation may be independent of the joint orientation tentatively, although the stress distribution is anisotropic.

REFERENCES

Barton, N.R. 1986. Deformation phenomena in jointed rock. Geotechnique 36, No.2: 147-167.

Chen, E.P. 1987. A computational model for jointed media with orthogonal sets joints, Sandia report86-1122, Albuquerque.

Chen, C.H. 1993. A study on the mathematical deformation model of jointed rocks. Ms Thesis, National Taiwan University, Taipei.

Fukushima, K. 1993. Scale effect on the underground openings according to their constructional sequences. Proc. the 2nd Int. Workshop on Scale Effects in Rock Masses, Lisbon: 93-99.

Franklin, J.A. & M.B. Dusseault 1992). Rock Engineering. McGraw- Hill Publ. Inc.

Goodman, R.E. 1976. Methods of Geological Engineering in Discontinuous Rock, ASTM Publ., St. Paul, West.

Huang, T.H. & Z.Y. Yang 1991. Generation of artificial joint and mechanical properties, Proc. of the 32nd U.S. Symp. on Rock Mech., Oklahoma: 1105-1114.

Heuze, F.E. 1980. Scale effect in the determination of rock mass strength and deformability. Rock Mech. and Rock Engr. Vol.12: 167-192.

Iofis, I.M. and V.I. Rechiski 1993. Studies of scale effect in plate loading tests. Proc. the 2nd Int. Workshop on Scale Effects in Rock Masses, Lisbon: 241- 249.

Lau, K.C., Crawford, A.M. & K.A. Smith 1984. The validity of equivalent rock mass models. Proc. of the 25th US Symp.on Rock Mech.: 261-268.

Lin, H.S. 1993. Influence on the strength of rock masses by horizontal joints. Ms Thesis, National Taiwan University, Taipei.

Natau, O. 1990. Scale effects in the determination of the deformability and strength of rock masses. Proc. the 1st Int. Workshop on Scale Effects in Rock Masses, Loen: 77-88.

Williams, A.F. 1980. Effect of jointing on rock mass modulus. Int. Conf. on Structural Foundation on Rock, Sydney: 29.

Scale dependency of rock mass properties

A.P.Cunha
Laboratório Nacional de Engenharia Civil (LNEC), Lisbon, Portugal

ABSTRACT: Rock masses being basically inhomogeneous and discontinuous media, the results of the laboratory and in situ tests for the determination of rock, joint and rock mass properties are affected by the volumes involved in the tests, a size dependency that is called Scale Effect.

The paper aims at getting a synoptic overview on the scale effects in the determination of deformability, strength, hydraulic conductuvity and internal stresses of rock masses. A precise definition of Scale Effect is presented, the concept of Representative Elementary Volume is discussed, and the variation of mean values and scattering of rock, joint and rock mass properties with the sample size are analysed. Attention is drawn on factors that influence the test results like sampling bias, sample disturbances, different techniques used, which may lead to incorrect conclusions concerning apparent scale effects.

1 INTRODUCTION

The determination of rock, joint and rock mass properties always faces the spatial variability of both the rock material and the discontinuity network, thus being affected by the dimensions of the volumes involved in the tests (Scale effects).

This applies for in situ and laboratory test results, and points out that rock mechanics should be regarded as a statistical mechanics, where the properties, evaluated over a sufficient number of tests, must be defined by a mean or most probable value, and by deviation measures, which account for the scattering of the results.

In fact, rocks, joints and rock masses tested at different scales often reveal trends for decreasing both the average values and the scattering of results with the increasing size of the sample. Opinions on the causes of the observed scale effects are sometimes divergent. For instance some argue that the decreasing strength-size relationships of intact rock are introduced by stress-gradients due to non-uniform loading and sample shape influences. Others interpret the strength reductions on the basis of the Weibull statistical volume effect, stored strain energy, etc. Concerning the shear strength of joints, the consensus is that a scale effect exists, in the broad sense that laboratory measurements cannot be directly extrapolated to the behaviour of in situ non-planar joints.

Investigations on potential scale effects in stress measurements appear to show a larger scattering of results at small scales, but no significant trends in terms of mean magnitudes. As regards the evaluation of hydraulic properties, it appears to show a real scale effect in hydraulic conductivity, the representative volume REV only being reached at large scales, which ensure representative interconnection of the conductive joints. The complicating influence of joint shearing may need special consideration, due to the scale dependence of the shear-dilation-conductivity coupling.

On the other hand, the statistical analysis often is disturbed in practice by sampling bias, sample disturbances, different experimental conditions or sample geometries which may lead to wrong conclusions regarding scale effects. There is a trend not only to sample the worst parts or the easiest ones, but also, because of the costs involved, the number of tests usually decrease notoriously as one moves from the laboratory tested intact rock or single joint to the rock mass scale. The reality of most projects is such that only a very few large scale tests can be conducted, and even with the greatest care, they may be unrepresentative of the variations in the rock mass. Furthermore, both deformation and failure modes are dependent not only on the natural block size, properties of boundary joints and rock flaws but also on the stress levels and on the relation between the prototype dimensions and the natural block sizes. This fact emphasizes the difficulties in selecting appropriate test sizes and extrapolating the measured property values to the scale of the rock engineering problems.

2 THE SCALE EFFECT CONCEPT

Rock masses are essentially discontinuous and heterogeneous media. Discontinuity begins at the microfissure scale and extends to the various types of joints and faults at the macroscopic level. Heterogeneity is a concept that summarizes the randomly changeable pattern of the number, size, individual properties and geometric distribution of the mineral components inside the intact rock. At the broad scale of rock masses, heterogeneity may be synonymous not only of lithological changes but also of different weathering and fracturing patterns.

As the sample dimension increases, tests may involve successively the intact rock, the jointed rock and the rock mass (Fig. 1). Even if theoretically homothetical, and concerning the same formation, those specimen will show, when subjected to similar tests, different results, as they involve different heterogeneity and jointing features. A variation in the mean values and in the scattering of the results with the specimen size should then be expected.

In fact, it has been experimentally established that geometrically homothetical samples of the same rock or rock mass, when submitted to loads following the same similarity conditions, show characteristics that are not constant but a function of the sample size. This variation is called Scale Effect. In other words, sets of samples of the same universe, with different sizes, show statistical distributions with different parameters, concerning the same property (definition of the ISRM Comission on Scale Effects in Rock Mechanics, Cunha 1990).

For practical purposes the fundamental problem arising is consequently the establishment of reliable links between tests at different scales, for a given rock mass and a given rock engineering property. The importance when one needs to extrapolate results obtained from inexpensive and quick tests on small volumes, very often only laboratory tests, to the scale of

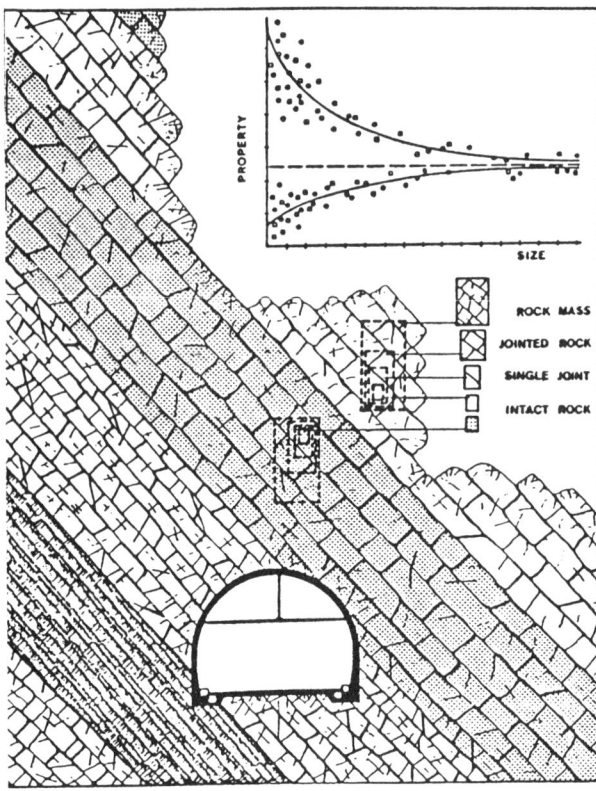

Figure 1. Generical representation of scale effects

the real heterogeneous and discontinuous rock mass and of the rock engineering projects, thus replacing the unknown safety factors wrapped up in the so called sound engineering judgement by a quantitative probabilistic approach of the representativeness of the results concerning different test volumes, does not need to be emphasized (Cunha, 1990).

It is worth noting that any analysis of potential scale effects exclusively on the basis of mean values, setting aside the evolution of the scattering of the results with the sample size, is insufficient and may lead to wrong conclusions. In fact, for several rock, joint or rock mass properties, there is experimental evidence of a decreasing trend of such scattering with the increasing of test size, even if the mean values do not appear to be sensitive to the test scale. The general conclusion that may be drawn is that as the size of the test specimens decreases, the number of tests to be performed must increase in order to ensure equivalent representativeness in the statistical analysis.

The experimental data allow also the conclusion that for a given rock mass, and at least for some relevant properties in rock engineering, a volume can be reached beyond which, in practical terms, tests will provide scale-free property values. Such volume, the smallest one that can be considered as representative, for that property, of the behaviour of the rock mass, or in other terms the smallest volume for which there is equivalence between the ideal continuum material and the real rock mass (Haimson, 1990), is commonly called Representative Elementary Volume (REV). It is expected that the REV will stabilize at different volumes, for different properties of a given rock mass, and may change significantly from a rock mass to another, concerning the same property.

Anticipating a legitimate comment, it may seem paradoxical to mention material properties mixed up with scale effects. As a matter of fact, in pure mechanics, a true material property is defined at a point and does not involve any scale dependency, and thus should not vary with testing procedures, whatever they are: sample geometry, loading conditions, specimen size. Thus, the assumption that properties will remain unchanged at any range of scales can be true if, and only if, the material itself remains also unchanged at those scales.

In fact there is no unbearable contradiction between pure mechanics and applied rock mechanics, where we deal with changeable geotechnical natural materials. The basic concepts of the theoretical mechanics have to be kept as guidelines and necessary frameworks for rock engineering, though some practical adjustments have to be adopted: a point property in theoretical mechanics has to be determined in rock engineering over a finite volume, the infinitely small to be converted into a macroscopic domain and the derivatives to be changed into finite differences. Whatever the scale of the test, in most cases the material may change; so, the test results for determining material properties, even keeping constant all the experimental factors, have to change accordingly. Scale effects do arise then and impact the engineering judgement.

3 SCALE EFFECTS IN THE DETERMINATION OF THE DEFORMABILITY AND STRENGTH OF ROCK, JOINT AND ROCK MASSES

Although in literature a number of contradictory opinions and observations about scale effects in the determination of rock strength can be found, an overwhelming majority of experimental data points out that a strength reduction occurs with the increase of specimen size for most rocks and the same applies to the scattering of the results. A collection of experimental data from several laboratory tests on unjointed rock is presented (Fig. 2) in a dimensionless form, by dividing the strength of the individual sample by the strength of a standard 50 mm in diameter specimen (thus eliminating differences due to variations in moisture content, shape and loading rates, which are usually the same for a given data set).

Scale effects in the determination of the deformability of intact rock also have an intimate connection with rock inhomogeneity and micro-fracturing. We believe that one of the most systematic and representative studies on the subject was

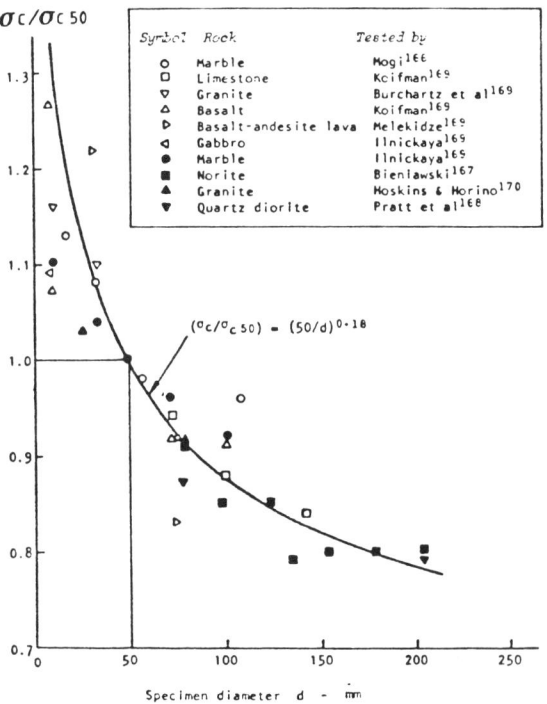

Figure 2. Scale effects in the determination of the intact rock strength (after Hoek and Brown 1980)

carried out by Peres-Rodrigues (1990) on cubes, 40 cm in edge, of three different rocks (limestone, gneiss and granite) from which it was experimentally drawn the important conclusion that the mean values of the moduli of elasticity of the rock are independent of the volume of the sample size. What depends on the specimen size are the deviations, which decrease as the sample dimensions increases.

Cunha (1991)followed this path for a rock mass and in situ tests: he analysed the results of using at the same site, with a given joint frequency, alternatively one, two or three large flat jacks, corresponding to test volumes of 10-30 m^3, and he found mean values of the same magnitude, with no increasing or decreasing trend with the volume. However, a closer look at the dispersion of the results showed that the scattering of the moduli decreased with increasing test volume, which is a demonstration of scale effects in the rock mass deformability and led to a general conclusion: as the size of the test specimen decreases, the number of tests to be performed should increase in order to ensure the same statistical representativeness of results (Fig. 3).

Concerning the mechanical behaviour of individual joints, the description of the shear strength has historically begun by the linear Coulomb strength envelope, which relies on a pair of constants C, ϕ.

It was Patton (1966) who proposed a bilinear law of friction, pointing out the existence of different mechanisms during the shear path, in the first part of which, for low stress levels, the strength is mainly governed by the overriding of asperities and induced dilation while for higher stresses, the shearing through the asperities prevails. A transient regime between both straigth lines led naturally to a curved strength envelope, for whose description several empirical laws have been proposed. A worldwide promotion has been granted to the empirical relantion defining the peak shear strength (Barton 1990, Bandis 1990) as a function of two parameters JRC and JCS plus the basic friction angle, which draw the attention on the importance of the sample size for the mechanical properties

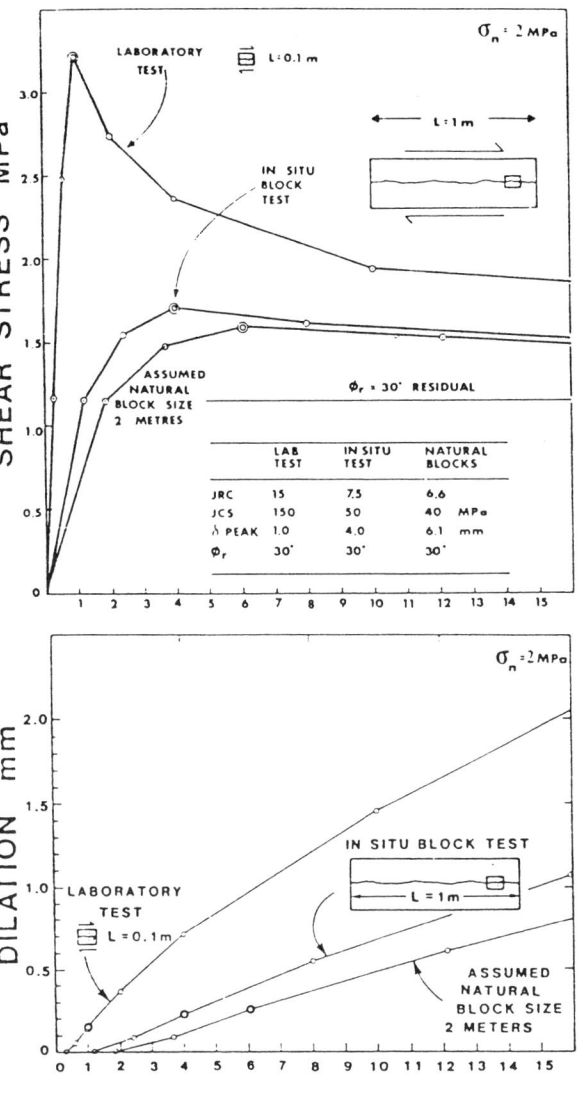

Figure 4. Scale dependent shear displacement-shear stress and dilation of joints (after Barton 1990)

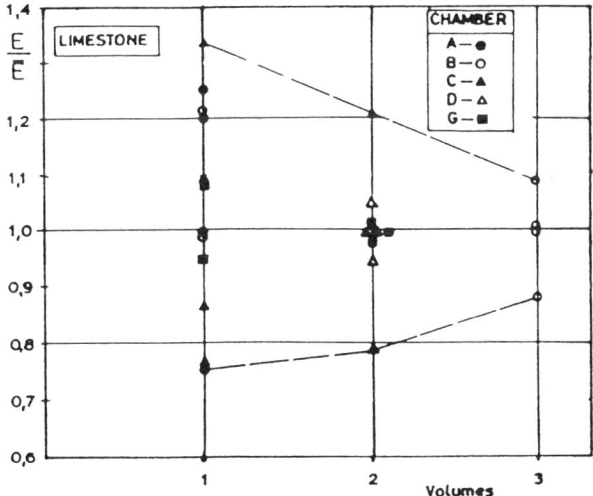

Figure 3. Mean value and scattering of deformability moduli vs. test volumes

of joints. In fact Barton and Bandis have produced empirical expressions scaling the JRC and JCS values. For describing the peak shear stiffness, Barton and Bandis have suggested an empirical relation where scale effects are reflected in a direct way through the explicit consideration of the specimen length.

In a similar way also the estimate of the displacement needed to reach peak shear strength is scaled and, as a consequence, also dilation is described as a scaled phenomenon (Fig. 4).

A fundamental experimental research on the mechanical behaviour of joints, important for the understanding of scale effects, was the one caried out by Bandis, pointing out clearly (Fig. 5) a decrease of peak strength as the sample size increases. At the same time, a change in the type of the shear stress-shear displacement diagrams appears, which moves from a brittle pattern to a more ductile one, thus increasing the peak displacement with increasing specimen size. Also a decrease of shear stifness with the increase of the sheared area is shown in the diagram.

It should be emphasized that the magnitude of the scale effects is directly related ro roughness pattern. The higher the roughness of the joints, the greater the scale effects, as a general trend. Whenever the joints are filled with low strength and highly deformable material and the width of the filling is sufficient to prevent wall contacts and thus wall roughness to play its role, no significant scale efects can be observed (Bandis 1990, Cunha 1990).

Series of joint shear tests on granite specimens with areas from 200 to 9 600 cm^2 were performed by Yoshinaka et al. (1993) and presented on a paper containing very relevant experimental data about scale effects on shear strength and deformation of rock joints. A decreasing of shear strength with

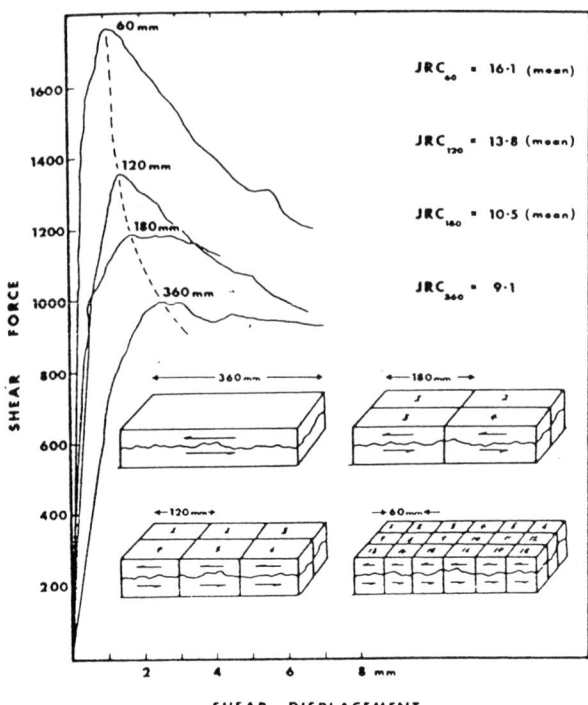

Figure 5. Influence of block size in the shear behaviour of joints (after Bandis 1990)

increasing joint areas, expressed in a bilogarithmic scale by a linear function is shown in Fig. 6 as long as asymptotic shear strength values with area are pointed out using bilinear scales. As regards the scale effect in shear stifness also a decreasing trend with increasing area was found, although less notorious than the previous one. Experimental evidence of an increase in maximum joint closure with area was also given, together with an opposite variation of the initial normal stifness. Finally the authors presented direct measurements by picture process analysis of the effective contact areas during the shear process, which changed with different normal stresses. Also a scale effect relating the effective decrease of sheared area and the increase of joint area was pointed out.

As concerns scale effects in the determination of the mechanical behaviour of joints, Cunha (1990) has reviewed several site testing programs and studied the variation and scattering of the shear strength and stifness of joint sets with the sheared areas, from the laboratory to the in situ scale. The author found for each joint set a decreasing trend of the mean values and scattering with increasing size. By defining for a given site an experimental law (straight line in a semilogarithmic scale) relating mean strength and joint areas, the author made an extrapolation of the mean joint strength at laboratory scale ($A=100$ cm^2, $\phi=40°$) and in situ test scale ($A=5\,000$ cm^2, $\phi=32°$) to a large joint with an area of 500 m^2 and found $\phi=12.8°$ which is a good demonstration of the importance of the scale effects in shear strength (Fig. 7). However, the author remarked that at the larger scale the waviness may play an important role, the asperities commanding shear being different from those at the test scales and thus the shear behaviour of large joints might be less dramatic than it looked.

Also Muralha and Cunha (1992) using empyrical exponential functions with positive asymptotes for the description of joint strength as a decreasing function of the area, performed numerical analysis of the stability of joint slopes with homothetical joint geometries, and checked the probability of the different failure scenarios which may increase significantly when scale effects are considered.

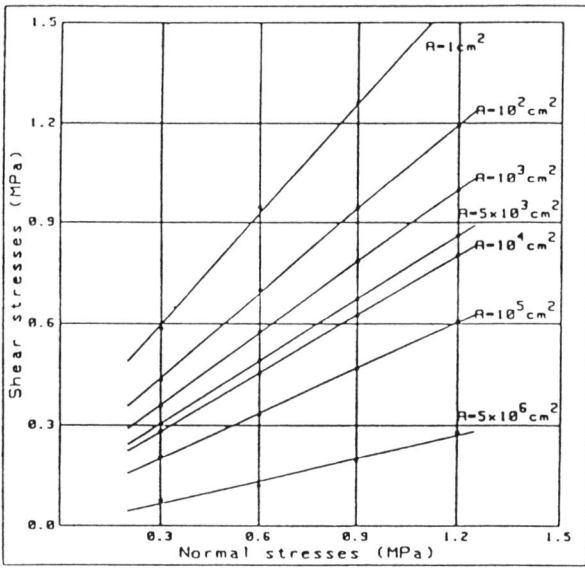

Figure 7. Decreasing strength envelopes for increasing joint areas

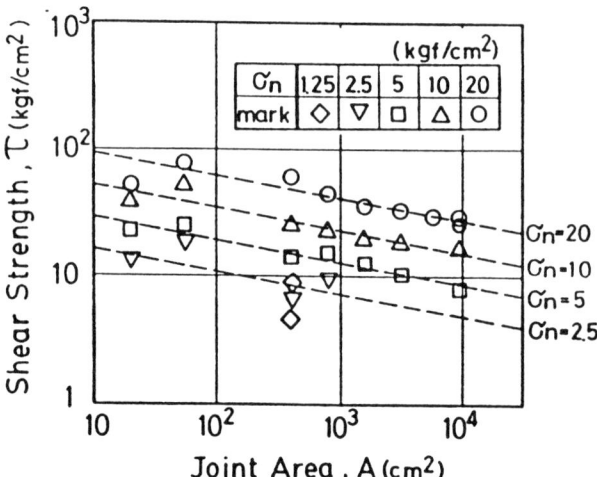

Figure 6. Relation between peak shear strength and joint area

4 SCALE EFFECTS IN ROCK STRESS MEASUREMENTS

The special lectures by Haimson (1990) and Haimson and Cuisiat (1993) on the scale dependency of in situ rock stress measurements distinguish the scale effect linked to the scale dependency of rock parameters needed for the interpretation of rock stress measurements like the tensile and compressive rock strength and the scale effect linked to the volume of rock mass involved in the stress determinations, according to the specific technique used. Besides reviewing the main techniques for rock stres measurements (overcoring techniques, flat jacks, hydraulic fracturing and borehole breakouts), the authors make comparisons between stress measurements with those methods, involving different scales, concluding that in most cases the stresses measured by separate techniques can be compared, despite the differences in scale of individual tests. Anyhow a warning is given by the author against the use of rough regional stress characteristics to substitute for local measurements in major projects.

In the opening lecture of the Loen workshop Cunha (1990) emphasized the great variability of the stress fields inside

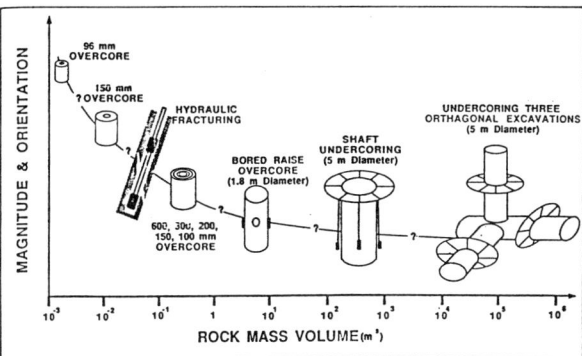

Figure 8. Investigation program on scale effects in stress measurements

the rock masses, as a consequence of their inhomogeneous--discontinuous nature. Stresses being point properties, stress concentrations and reliefs due to discontinuities or changes in rock stiffness are a reality and stress changes from block to block and in the same block can not be ignored. An average value being generally required as input for model analysis, a large number of small scale tests or a smaller number of larger volumes is thus required, for keeping the same reliability level. It is also suggested, given the influence of scale and of the testing techniques, that more than one technique shall be used, whenever possible, at each site, for corroboration of results.

Both Barton and Cunha (1990) quoted the important investigation program on scale effects on stress measurements of AECL (Martin et al. 1990) emphasizing, according to the authors, who have performed several overcoring stress measurements in boreholes with diameters in range 96 mm - 600 mm, that no significant scale effect trend was identified beyond the decreasing trend of the scattering of results with increasing diameter.

Therefore, it may be concluded that the curved line in Fig. 8, referring to potential stress measurements at different scales, does not stand as a variation with test size of the mean values (which tend to be constant) but as an upper boundary of experimental results at diferent scales, or a deviation parameter for a certain probability of occurrence, attesting the decreasing of scattering with test volume increasing.

5 SCALE EFFECTS IN THE DETERMINATION OF HYDRAULIC PROPERTIES IN ROCK MASSES

Hydraulic conductivity of the rock masses is strongly dependent on the frequency and interconnectivity of the joints. In fact, only fracture intersections allow water to move from one discontinuity to another and thus to establish the water flow within the rock mass. As a consequence, the rock mass hydraulic conductivity is critically dependent on the hydraulic properties of the single joints, and these, on the other hand, are a function of its geometrical properties, namely the roughness and aperture.

In fact, laminar flow through a joint is proportional to the cube of its hydraulic aperture and joint aperture changes with stress in such a way that a small change of aperture may cause significant changes of the magnitude and principal directions of permeability.

Conductivity estimates from packer tests are the normal basis for stochastic continuum models of sparsely fractured rock. Considering that the scale of such tests are often much smaller than the scale on which the rock acts as an equivalent porous medium, the extrapolation of those test results can be highly questionable.

Carlsson et al. (1990), after referring that predictions on water inflow based on borehole investigations frequently differ from the inflow encountered during excavations, one of the possible causes being scale effects, emphasized that the most common methods used in the hydraulic investigation being small scale tests, if the results of such tests are averaged and the resulting hydraulic conductivity is used for predicting the water inflow in the engineering projects, the predicted values may exceed the measured ones, occasionally by one or more orders of magnitude. Also one should expect a rapid decrease of standard deviations with increasing scale.

As regards Cunha's report (1990), it is emphasized the importance of joint conductivity, the interconnection of the joints being fundamental to ensure water flow inside the rock mass. Only after a certain volume is reached, involving a representative number of joints, can test results expect not to be affected by test volumes. The Representative Volume, however, depends on joint frequency and concerns a volume that can go far beyond the block size of the rock mass. Before the REV a potential decreasing trend of mean values and scattering with volume increasing should be expected.

In his special lecture, Gale (1993) calls the attention on the fact that comparisons of permeability data from tests that range from laboratory to regional scales need to consider the fundamental differences in the methods used in the tests. In the laboratory we have a complete sample of the core and the microcrack and fissure conduits that control the permeability. In the borehole, our test tool is placed within the sample resulting in poor control on the boundary conditions. In addition, the measured permeability is a reflection of the probability of the borehole encountering a certain permeability feature for its given length and orientation, the consequence of which is the wide range or variabilitiy in the measured in situ permeabilities.

6 CONCLUSIONS

Summarizing scale efects in the mechanical behaviour of intact rock, the available data point out a decrease of the mean compressive strength and scattering with the increase of the specimen size. As regards deformability, it has been ascertained that the mean deformability moduli are independent of the specimen size but the deviations decrease as the size increases.

At the scale of the rock masses, it appears that strength reduces as the specimen size increase, a trend similar to that of intact rock and individual joints, from which the strength of rock masses is dependent. However, potential changes in the failure mode can modify rock mass behaviour and do not allow an universal statement on the matter.

As regards the deformability of the rock masses, theoretical and experimental studies converge in the definition of an independency of the mean values of the deformability moduli from the test volumes, for a constant joint intensity. However, the scattering of the results decreases as the test volumes increase. Thus it is expected that the measured values will converge to the mean, for a finite volume, thus defining the Representative Elementary Volume of the rock mass.

Concerning the joint mechanical behaviour at laboratory scale only unevenness, the first component of roughness, is represented, or, at the most, the small scale waviness, while the large scale waviness not even at the in situ test scale can often be involved. This changing size of significant asperities along the joints, as sample size increases and different effective stress concentrations on the wall contacts during shear, have the major responsabilities on the scale effects observed.

Both the peak strength and shear stiffness of a joint decrease as the sample area increases, the greater the roughness of the joints the greater the scale effects. These laws appear to be valuable both for single joints and the mean properties of joint sets. Concerning the peak displacement it has been experimentally settled that it increases with the increase of the sample size. Scale effects have also been reported on normal closure and dilation. For joints with thick filling material, scale effects may be null.

As regards scale effects in the determination of internal stresses in rock masses, it seems that a consensual position can be found: the different techniques available (overcoring, flat jacks, hydraulic fracturing, borehole breakouts) if suitably applied, in spite of the notorious difference of scales, can lead to compatible results, that is, to similar average values, although an evidence of scale effects as regards the scattering of the results should be expected to occur. Thus, the smallest the test scale the greatest the need for a larger number fo tests, in order to ensure similar quality of the statistical inference. It is worth noting that while large scale tests give the average local stresses, required generally as an input for model analysis, smaller scale tests are measuring the pointwise stresses, with their real concentrations and reliefs, due to discontinuities or changes in rock stiffness, inside each block or from block to block. Given the scale effects and the difference between the methods for rock stress measurements, the use of more than one technique at each site is commendable, namely for important projects or for rock engineering problems being quite sensitive to the state of stress in the rock mass.

Finally, as regards scale effects in the determination of hydraulic properties of rock masses, it has been ascertained the major importance of joint conductivity as regards the rock matrix, as well as the extreme influence of the normal stress and the aperture on joint conductivity and the need for the interconnectivity of the joints in order to ensure water flow inside the rock mass. It means that only after a minimum volume is reached, involving a representative number of joints, test results can give up being affected by test volumes. The Representative Elementary Volume is thus a fundamental concept regarding the hydraulic behaviour of rock masses which depends also on joint intensity and concerns a volume that can go far beyond the block size of the rock mass.

Some doubts remain on several points, like the hydraulic conductivity tensor concept, applied at the different scales of the jointed rock masses, the representative elementary volume concept, namely the validity of the concept and the physical possibility of being determined. The scale variation of permeability, flow mechanisms, hydraulic-active joints, etc. are subjects for which a discussion is expected and necessary in the next future.

Gathering all the experience on the problems of scale effects, accumulated prior to and during the ISRM Commission activity, it may be concluded that in spite of some recent advances in our understanding, we still miss general solutions for important problems like:
- Definiton of laws for describing the variation of the mean values and the scattering of rock, joints and rock mass properties with the sample size.
- Determination of the test sizes beyond which scale-free values can be found for a given property and rock mass (representative volumes).
- Definition of relationships between laboratory and in situ tests for optimizing their use in estimating the properties of jointed rock masses, taking into account the scale effects.

For reaching such targets additional research is needed, conveyed not by a Commission but preferably through consistent lines of investigation leading to Ph. D. theses. Guidelines to be followed should include:
- Critical review of data on scale effects in Rock Mechanics.
- Experimental and theoretical studies on scale effects regarding rock, joint and rock mass properties.
- Analysis of case histories involving the consideration of scale effects, either in the design or emphasized by back-analysis.

A better understanding of the problem of scale effects will provide a better selection of the size, type, location and number of tests, the geotechnical parameters used in model analysis will be selected in a more scientific basis, namely following probabilistic approaches, and the solution of the rock engineering problems will become less expensive and more accurate.

7 ACKNOWLEDGMENTS

The ISRM Comission on Scale Effects in Rock Mechanics organized in Loen, Norway, June 1990, the 1st. International Workshop on Scale Effects in Rock Masses, the proceedings of which included 6 special lectures and 24 papers from 52 authors, spanning the five continents.

Within the Comission expectations the idea that by gathering the international expertise at the Workshop an updated state-of-art on scale efects in the determination of rock mass properties could be obtained, some improvement of the basic knowledge on the subject would be achieved and new research on such an important matter could be attracted. The live discussion engaging many of the about 150 attendants confirmed the scientific sucess of the Loen Workshop.

Three years later, the ISRM Commission organized in Lisbon a second International Workshop on Scale Effects in Rock Masses, in order to publish, present and discuss new contributions concerning the theme, and to evaluate the results of the research carried out or reactivated after Loen, about scale effects in the determination of rock mass properties.

Four special lectures and a number of interesting papers, most of them from authors that had not submitted comunications to the 1st. Workshop, were selected among a total of 43 papers and 88 authors, and were presented and discussed before almost 200 participants.

The ISRM Commission on Scale Effects in Rock Mechanics will end the activity at the term of the actual period of appointment, that is, with the Tokyo Congress, 1995. As President of the Commission, the author wishes to thank all the colleagues, that by the submission of papers, oral presentations, comments and discussions have encouraged the Commission in his work and definitely contributed to emphasize the importance of scale effects on engineering judgement whenever for rock engineering problems, the extrapolation of data concerning rock mass properties, determined under current experimental conditions and by current size tests, is required.

REFERENCES

Bandis, S.C. 1990. Scale effects in the strength and deformability of rock and rock joints. Special lecture. 1st. Int. Workshop on scale effects in rock masses, Loen.

Barton, N. 1990. Scale effects or sampling bias?. Closing lecture. 1st. Int. Workshop on scale effects in rock masses, Loen.

Cunha, A.P. 1990. Scale effects in Rock Mechanics. Opening lecture. 1st. Int. Workshop on scale effects in rock masses, Loen.

Cunha, A.P. 1993. Scale effects in rock engineering - An overview of the Loen Workshop and other recent papers concerning scale effects. Special lecture. 2nd. Int. Workshop on scale effects in rock masses, Lisbon.

Gale, J.E. 1993. Fracture properties from laboratory and large scale field tests: Evidence of scale effects. Special lecture. 2nd. Int. Workshop on scale effects in rock masses, Lisbon.

Haimson, B.C. 1990. Scale effects in rock stress measurements. Special lecture. 1st. Int. Workshop on scale effects in rock masses, Loen.

Peres-Rodrigues. 1990. About LNEC experience on scale effects in the deformability of rock masses. 1st. Int. Workshop on scale effects in rock masses, Loen.

Yoshinaka, R., Yoshida, J., Arai, H., Arisaka, S. 1993. Scale effects on shear strength and deformability of rock joints. 2nd. Int. Workshop on scale effects in rock masses, Lisbon

The scale and creep effects on strength of welded tuff

Noriyuki Yuki & Sumio Aoto
Kawasaki Geological Engineering Co., Ltd, Japan

Yoshihiro Ogata
National Institute for Resources and Environment, Japan

Ryunoshin Yoshinaka
Saitama University, Japan

Mitsuo Terada
Tochigi Prefecture, Japan

ABSTRACT: In the underground quarries excavated Ohya stone that is a typical soft rock, very large scale cave-ins occurred at the last few years. Therefore, it is important and urgent subject to evaluate the stabilities of underground quarries. The authors carried out the laboratory tests to the specimen size and the long term loading on these properties, in order to obtain the basic data in solving the failure mechanism of pillars and rockwall surrounding the underground quarries.
The following results were obtained by these laboratory tests.
1. The unconfined compressive strength and Young's modulus increase with the enlargement of specimen size. This relates to the content of pumice fragment which are included into the specimens of Ohya stone.
2. If the stress ratio is on and under 50% of the unconfined compressive strength of pillars, pillars are relatively stable.

1 INTRODUCTION

At Ohya town in Tochigi prefecture, Japan, the rhyolitic welded tuff called "Ohya stone" is excavated mainly by the room and pillar method. Ohya stone is a typical soft rock and includes many pumice fragments called "Miso" at Ohya district. And these pumice fragments are composed of clay minerals and chalcedony.

The underground quarries are from tens of meters to one hundred and tens of meters in depth. The ordinary sizes of rectangular pillars are 10 m square in width and 10 m to 20 m in height. But a few pillars reach 50 m in depth.

At the last few years, very large scale cave-ins occurred in these quarries. Therefore, it is important and urgent subject to solve the failure mechanism of pillars and rockwalls surrounding quarries, and to evaluate the stabilities of quarries. Especially, it is necessary to suppose the mechanical properties and the long term strengths of large rectangular pillars on the laboratory tests of small specimens.

The authors carried out the laboratory tests to search for the states of pumice fragments which are included into Ohya stone, the unconfined compressive strength, tensile strength and Young's modulus of several sizes of the square and circular columnar specimens, and the long term strength of a size of circular columnar specimen.

In this paper, we report the influence of the size and shape and the content of pumice fragments on the mechanical properties, and the long term strength of Ohya stone.

2 GEOLOGICAL CHARACTERISTICS OF OHYA STONE

2.1 Geology of Ohya stone

The Ohya formation which produces Ohya stone is mainly made of pyroclastic rocks caused by Green Tuff movement in the Neogene Miocene. And the Ohya formation can be divided stratigraphially into the lowest, lower, middle and upper member(Nakamura, etc., 1982). Ohya stone belongs the middle and upper members.

A few of fracture patterns like small discontinuity exist in the Ohya formation, but there are no large scale fractures. Therefore, the ohya formation can be said to be the massive and continuous(Yoshinaka, etc., 1984).

In these tests, we used Ohya stone with the highest quality called " fine grained stone " in belonging the upper member, because cave-ins occurred mainly in the fine grained stone.

2.2 Content of Alteration Mineral

Photo 1 shows a side surface of 30 cm cube of Ohya stone, and the black shadows on the surface are the pumice fragments called " Miso " at Ohya district. The size of pumice fragment has been

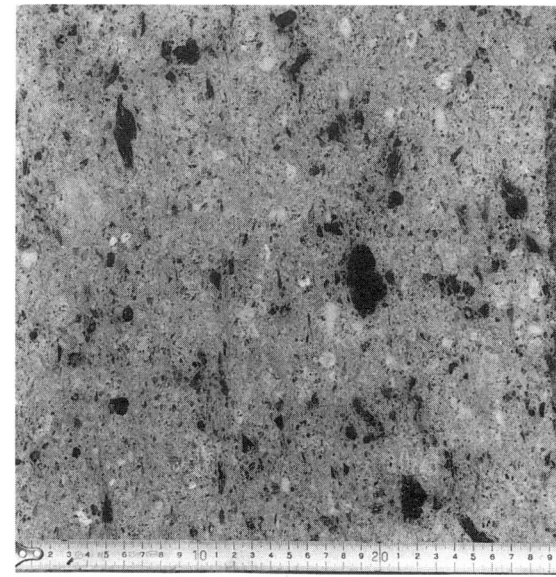

Photo 1. State of a surface Ohya stone of 30 cm cube.

determined on such photographs by the point count method. That is, the area of pumice fragment has been measured first. Then, the diameter of the circle whose area was equal to that measured has been determined. Here, this diameter is named the apparent diameter.

Figure 1 shows the frequency of the apparent diameter. In this figure, it is known that the most frequent diameter is about 8 mm and the maximum diameter is a little less than 30 mm.

On the other hand, it is said that the volumetric content of pumice fragment of Ohya stone is nearly equal to the content in the same block on a surface(Kobayashi, etc. 1994). Therefore, we measured the content of pumice fragment by the same method.

At the first, the square elements were divided on the surface of Ohya stone at intervals of 3 cm or 5 cm. Then, the content of pumice fragment was measured every elements of 3 cm, 5 cm, 10 cm, 15 cm or 30 cm square.

Figure 2 shows the relationship between the content of pumice fragment and the width of the enclosed block on a representative surface. It is known that the average content is approximately constant, but the maximum and the minimum

Table 1. The mechanical properties every shapes and sizes of specimens.

shape	size width or diameter mm	size height mm	number of specimen	unconfined compressive strength(qu) MPa	coefficient of variation of qu %	Young's modulus (E_{50}) GPa	coefficient of variation of E_{50} %
square column (1)	30	60	12	7.81	11.4	2.18	19.6
	50	100	12	6.33	16.0	1.61	25.6
	100	200	12	7.96	6.7	2.56	9.6
	150	300	12	8.45	7.2	2.87	11.9
	300	600	3	9.69	5.6	3.32	1.8
	400	800	3	9.87	2.7	3.75	2.1
circular column (1)	30	60	13	8.65	12.3	2.86	13.0
	50	100	15	8.01	14.4	2.78	23.6
	100	200	12	9.25	5.7	3.30	5.4
	150	300	12	9.65	9.2	3.52	8.8
circular column (2)	30	60	14	8.28	16.3	2.57	14.9
	50	100	15	8.61	10.1	2.90	8.6
	100	200	13	8.84	5.3	2.89	4.4
	150	300	12	9.42	2.6	2.99	8.0

(1):loaded horizontally to the depositional surface
(2):loaded vertically to the depositional surface

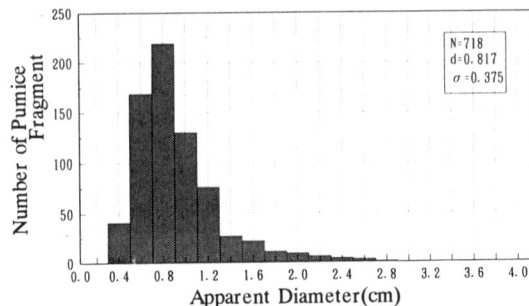

Figure 1. The frequency of the apparent diameter of pumice fragment. n represents whole number of pumice fragments on 12 surfaces of two blocks of Ohya stones.

content is noticeably different between the different areas of elements. That is to say that the smaller the area of element is, the more the maximum content of pumice fragment is. And the minimum content of pumice fragment gets smaller with small areas. This tendency is notable with the element of 3 cm and 5 cm square with which the maximum content increase rapidly.

3 SCALE EFFECT ON MECHANICAL PROPERTIES

3.1 The Method of Tests

Several blocks of Ohya stone were extracted in a underground quarry of about 60 m in depth. And the specimens were formed as specified shapes and sizes. The shapes of specimen are a square column and a circular column. The sizes of specimen are 3 cm, 5 cm, 10 cm, 15 cm, 30 cm and 40 cm in which for the square column and 3 cm, 5 cm, 10 cm and 15 cm in diameter for the circular column. The heights of specimens are twice the width or the diameter (See Table 1).

Every precaution was taken for the keeping the initial water content in the underground quarry at all stages from the extraction to the test of Ohya stone, because the mechanical properties of soft rock such as Ohya stone alter greatly with the water content in the specimen.

The loading on the specimen was performed by the stress control method, applying stress with a rate of 49 kPa/sec. The loading direction was horizontal to the depositional surface for square columns, and horizontal and vertical to the depositional surface for circular columns.

The deformation of the circular column was measured by the cross type wire strain gauge which was cemented on the side surface of specimen and the linear variable differential transformer(LVDT) which was fixed between two loading platens.

And the deformation of the square column was measured by LVDT only.

3.2 Results of the tests

Table 1 shows the results of these tests. And figure 3 shows an example of stress-strain curve of the square columns in loading on the horizontal direction to the depositional surface. In this figure, the solid lines represent the strain measured by the wire strain gauge which was cemented on the side surface of specimen, and the broken lines represent the strain which was converted from the displacement measured between two loading platens. Here, the former is named the local strain, and the latter is named the average strain.

It is known in figure 3 that the local strain is larger than the average strain up to the failure. But the tangential Young's moduluses which were determined at 50% of stress level (E_{50}) were scarcely different between two kinds of measuring methods. On the other hand, the local strain was measured in a narrow area of bigger specimen. Therefore, we opted for the tangential

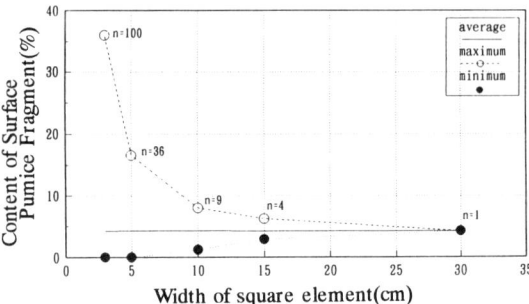

Figure 2. The distribution of the content of pumice fragment. n represents the number of element on a surface.

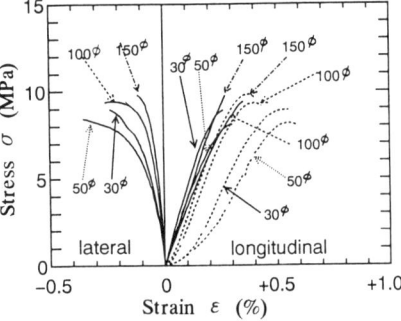

Figure 3. Stress-strain curve of circular column. Loading is the horizontal direction to the depositional surface.

Young's modulus at 50% of stress level on stress-average strain curve.

Figure 4 shows the relationship between the width and the unconfined compressive strength of the square column. The unconfined compressive strength tends to increase with the enlargement of the width of specimen. However, the unconfined compressive strength decrease with the enlargement of

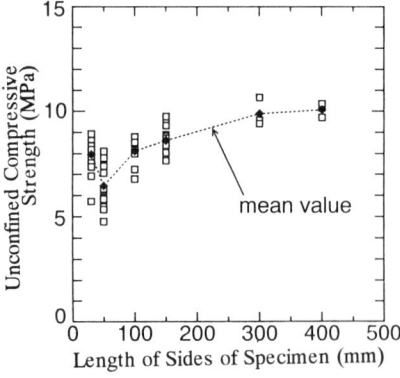

Figure 4. The relationship between the width of specimen and the unconfined compressive strength.

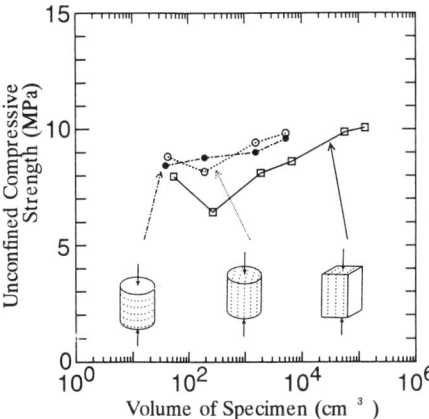

Figure 5. The relationship between the volume of specimen and the average of unconfined compressive strength.

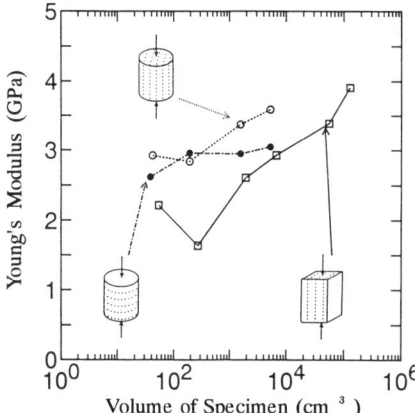

Figure 6. The relationship between the volume of specimen and the average of Young's modulus.

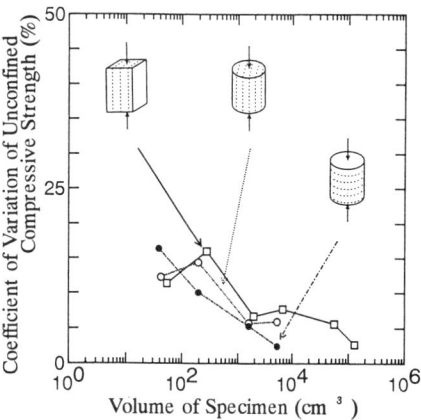

Figure 7. The relationship between the volume of specimen and the coefficient of variation of the unconfined compressive strength.

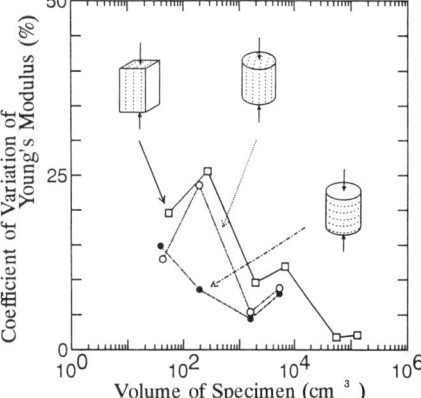

Figure 8. The relationship between the volume of specimen and the coefficient of variation of the Young's modulus.

the width to the specimens with 3 cm to 5 cm square.

From these results, the relationship between the unconfined compressive strength and the size of specimen is the same as the results of Kitaoka, etc. (1977) and Cunha(1993a, 1993b) to the sizes on and under 5 cm square, but is opposed to those to the sizes on and over 10 cm square.

Figure 5 and 6 show the relationship between the volume of specimen and the averages of unconfined compressive strength and Young's modulus every sizes of specimens. From these figures, it is known that, equal in the volume of specimen, the unconfined compressive strength and Young's modulus of the circular column are larger than those of the square column. Besides, when the load was acted on the horizontal direction to the depositional surface, these relationship to the circular column are the same as to the square column. But when the load was acted on the vertical direction to the depositional surface, the unconfined compressive strength and Young's modulus increase monotonously with the enlargement of the size of specimen. It is supposed that the mechanical properties of Ohya stone are some influenced by the depositional surface.

Figure 7 and 8 show that the relationship between the coefficient of variation of the unconfined compressive strength and YOung's modulus and the volume of specimen. These figures indicate that the coefficient of variation of the mechanical properties to the sizes on and under 5 cm in width or diameter are extremely larger than those to the sizes on and over 10 cm.

The followings are supposed in these tests of the content of pumice fragment and the scale effect on the mechanical properties of Ohya stone. The mechanical properties of Ohya stone increase with the enlargement of the size of specimen. And the coeffcient of variation of the mechanical properties increase with the reduction of the size of specimen. On the other hand, the difference between the maximum and the minimum content of pumice fragment grows big with the reduction of the size of specimen(See figure 2).

These results indicate that smaller the size of specimen is, the more the influence of pumice fragment on the mechanical properties is. Therefore, it was suggested to have to use the specimen of the large size, in order to evaluate the mechanical properties of the large scale pillars of Ohya stone.

4 LONG TERM STRENGTH

4.1 *The Method of Tests*

The creep tests were carried out in order to evaluate the long term strength of Ohya stone. That is to define the critical stress that the pillar will not failure for a infinite period, and the pillar will not eternally failure, and the relationship between the elapsed time to creep failure and the stress ratio.

The specimen was the circular column of 5 cm in diameter and 10 cm in height. The specimen was submerged in the water into thermostatic chamber which was maintained at $20 \pm 0.1°C$ for constantly keeping the temperature and the water content during the tests. In this case, the water was collected in the underground quarry in which extracted the blocks of Ohya stone.

The stress ratio was determined on the basis of the average of the unconfined compressive strength which was 8.82 MPa. Therefore, the stress ratios were from 20 to 90% to 8.82 MPa.

4.2 *Results of Tests*

Figure 9 shows that the relationship between the strain rate and the elapsed time at the stress ratio of 41% to 87%. It is known that the starin rate rapidly decreases up to the stationary level just after applying load. Thereafter, the strain rate inversely increases up to the failure. However, the creep failures have been observed within the time range of the last tests on and over the stress ratio of 60% or so. Figure 10 shows the relationship between the stress ratio and the minimum strain rate. It was

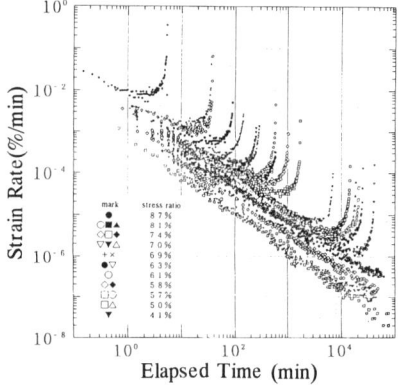

Figure 9. The relationship between the strain rate and the elapsed time.

Table 2. The evaluation of the stability of several underground quarries.

cavern No.	scale of cavern (m)	overburden (m)	height of pillar(m)	ratio of pillar(%)	stability of cavern
1	160×125	65	10	0.265	stable
2	156×163	65	15	0.268	cave-ins
3	93×154	35	20	0.150	cave-ins
4	93×175	25	36	0.211	stable
5	148×120	50	10	0.281	stable

tried to determine the critical stress by using the figure 10. That is, the minimum stress ratio can be approximately expressed as the equation (1).

$$\dot{\varepsilon} = 15.340\,\sigma_r - 15.738 \quad (1)$$

$\dot{\varepsilon}$: the minimum strain rate (%/min)
σ_r : stress ratio $\sigma/\sigma_c \times 100$ (%)
σ_c : unconfined compressive strength (MPa)
σ : stress acted on specimen (MPa)

Akai, etc. (1979) have defined this critical stress as static yield that corresponds to the stress for the minimum strain rate(%/min). That is to say the static yield stress of Ohya stone is 4.51 MPa(stress ratio of 51%)according to the figure 10.

Figure 11 shows the relationship between the elapsed time up to failure and the stress ratio. As results of the synthetic consideration of the dispersion of the unconfined compressive strength and so on, the equation (2) can been extrapolated from figure 11. But, the long term data over hours are shown because of the limited test period. Besides, if is assumed that

$$\sigma/\sigma_c = -0.066\,t_\sigma + 0.81 \quad (2)$$

t_σ : elapsed time to failure (hour)

this equation is able to been applied for a infinite period, the failure would have occurred on every specimen at any stress ratio.

Therefore, it has been assumed that a long term creep curve was extrapolated with a solid line on the basis on the static yield stress. According to this curve, the stress ratio that will failure after about thirty years is 50%. This roughly confirms the results presented by Yamashita, etc. (1994).

The results that have been surveyed in some of large underground quarries of Ohya stone are indication on figure 11. The filled triangles(▲)represent the pillars failed, and the circles (○)represent the stable pillars. Table 2 indicates the results of the evaluation of the stabilities of pillars in several underground quarries. Besides, the numbers which are written by the symbols in figure 11 correspond to the numbers quarries in table 2. The failure of pillar has not been observed at the stress ratio on and under 50%.

5 CONCLUSION

The conclusion of these tests can be summarized as follows:
1. Ohya stone includes the pumice fragment whose apparent diameter is 8mm average.
2. The average content of pumice fragment is about 4%. But

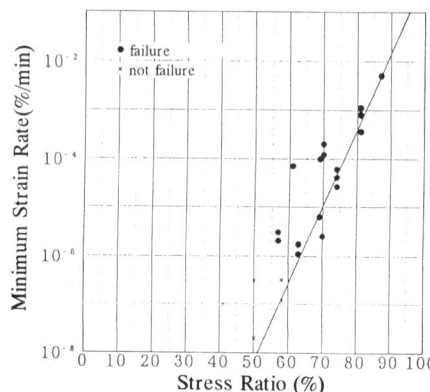

Figure 10. The relationship between the minimum strain rate and the stress ratio.

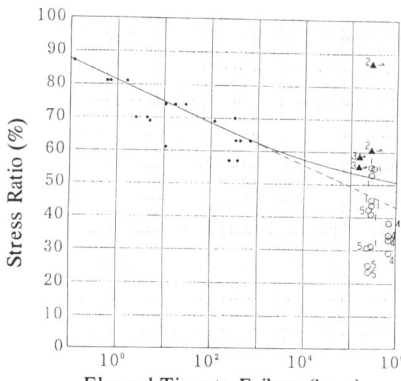

Figure 11. The relationship between the stress ratio and the elapsed time to failure.

the difference between the maximum and the minimum content of pumice fragment grows big with the reduction of the specimen size.
3. The unconfined compressive strength and Young's modulus increase with the enlargement of the specimen size. Except for specimen with 3cm to 5cm in width or diameter, such a tendency is opposed to the general theory.
4. The coefficient of variation of the mechanical properties increase with the reduction of the specimen size.
5. In case of Ohya stone, the mechanical properties and the content of pumice fragment is closely related to each other.
6. If it is assumed that the static yield stress corresponds to a stress ratio of 50% as the results of the creep tests, the pillars in the underground quarries of Ohya stone are stable at the stress ratio on and under 50%.

REFERENCES

Akai, K., Adachi, N., and Nishi, Y. 1979. Japan Society of Civil Engineers, No. 280, pp. 75-87.
Cunha, A. P. 1993a. Scale Effects in Rock Masses 93, ed. Cunha, Balkema, pp. 3-14.
Cunha, A. P. 1993b. A general report on the papers submitted to the Lisbon Workshop, Scale Effects in Rock Masses 93, ed. Cunha, Balkema, pp. 27-35.
Kitaoka, M., Tanaka, S., Kusunoki, K., and Endo, G. 1977. Japan Mining Industry's Journal, No. 1067, pp. 1-6.
Kobayashi, T., and Yoshinaka, R. 1994. Japan Sosiety of Civil Engineers, No. 487/III-26, pp. 31-40.
Nakamura, Y., Matsui, S. and Suzuki, A. 1982. Utsunomiya city, Bulletin of the factory of Education Utsunomiya University, No. 31, section 2, p. 105-116.
Yamashita, T., Sugimoto, F., Yamanouchi, M., and Kawabe, K. 1994. Shigen-to-Sozai, Vol. 110, No. 11, pp. 875-882.
Yosinaka, R., Kikuchi, K., Fujieda, M., Ono, U., Ohnishi, M., and Ohhashi, S. 1984. Proc. of ISRM syposium Cambridge, U. K., p. 97-104.

Mechanism of scale effect in rock joint

R. Yoshinaka & S. Arisaka
Saitama University, Japan

K. Sasaki & J. Yoshida
Suncoh Consultants Co., Ltd, Tokyo, Japan

ABSTRACT: Purpose of this investigation presented herein is to study the mechanism of scale effect in rock joint. We made direct shear test and measurment of joint roughness to find the characteristics of discontinuous plane. It was found that Z_2, the distribution of roughness angle and sheared area are major factor affecting the rock's shear characteristics.

1 INTRODUCTION

Characteristic form of a discontinuous plane is the contact between rock surfaces with roughness. Measurement and analysis of the roughness of rock surface have been considered as the most important prameter of the discontinuous plane and have been taken up in many studies concerning mechanical and hydraulic features of rock surfase. Therefore, methods of quantitation of surface roughness have been also proposed in in large numbers which can be divided into three main methods.

The first is a method to taken up as an index such statistical data as the mean value or standard deviation of roughness hight and its inclination.
The second is a method to make a waveform analysis which displays spectrally the roughness by Fourier transform used in frequency analysis. The third is a morphological method using Fractal geometry.

Since the substantial mechanism of the shering at discontinuous plane of rock is the destruction and slip of surface roughness, it is possible to calculate the apparent shear strength of the discontinuous plane if the actual shearing area is obtained. ladaniy and Archambault (1970) used a contact area ratio "A_s" as a parameter for the computation of shear strength. Barton and Bandis (1980) indicated that the scale effect of shear strength at discontinuous plane was due to the difference of true shear area. Though some trials were made in the past to directly measure the contact area of discontinuous plane, they were mainly conducted in the vertical deformtion test and there were no practical direct measurement of the contact area made at the time of shearing.

Basing on the result of shear test for granite test specimens conducted before now, we made measurement of surface roughness to find an index which has an influens on the strength/deformation characteristics of discontinuous plane and measured the true sheared area to get an amount which directly shows the magnitude of shear. In this report, we considered the presumed mechanism of scale effect obtained from above-mentioned measurement.

2 SCALE EFFECT OF SHEAR STRENGTH

Tests were made for the cracked plane of granite specimen. Fig.1 (a) and (b) shows the variation of peak shear strength for shear area of discontinuous plane. Broken line in the figure (a) shows these relations using following two equations representig pallalel regression curves.

$$\log_{10} \tau = a + b \log_{10} (A/A_0) \quad \cdots \cdots (1)$$

$$a = c + d \log_{10} \sigma_n \quad \cdots \cdots (2)$$

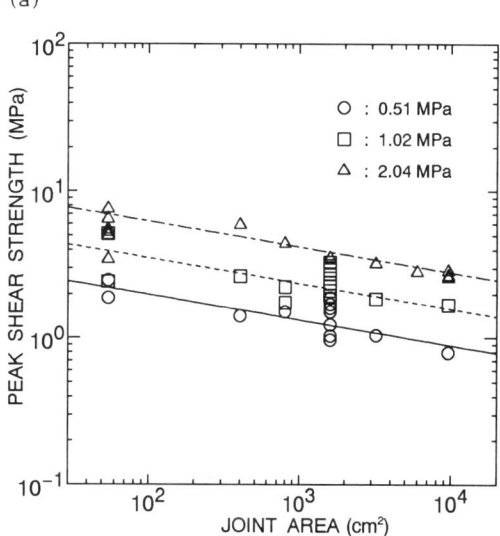

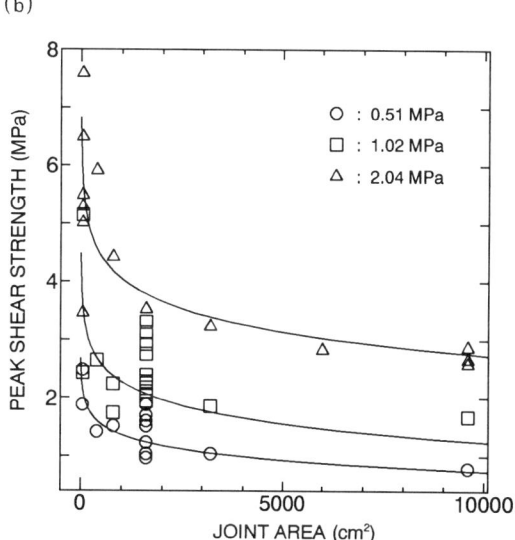

Fig.1 Relation between peak shear strength and jiont area

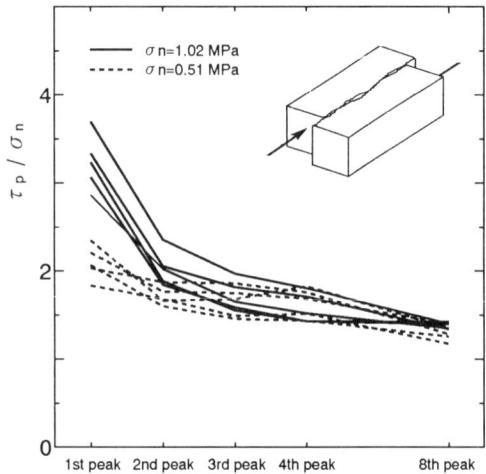

Fig.2 Change of peak shear strength during the repeated shear test

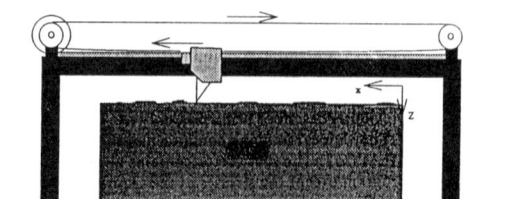

Fig.3 Concept figure of measuring system for roughness surface

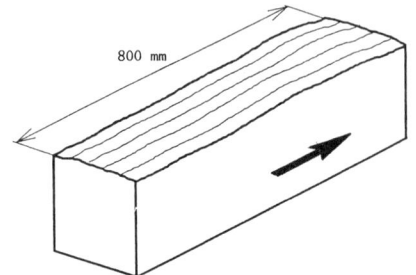

Fig.4 Measuret lines on a test specimen

Herewith, the regression curve shown by the above-mentioned equation (1) and (2) becomes an destruction envelopeof power function type given in the following equation (3).

$$\tau = k \cdot \sigma_n^{\alpha} \cdot (A/A_0)^{\beta} \qquad (3)$$

where k, α and β, are constants. It was determined that $k=1.14$, $\alpha=0.83$, and $\beta=-0.175$. τ and σ_n in the unit of MPa, A and $A_0=1$ in the unit cm².

Fig.2 shows the change of peak shear strength during the repeated shear test for each test specimens having a sectional area $A = 1600 cm^2$ respectively. This figure shows that the peak shear strength of granite test piece gets near a definite strength value, that is the residual strength of its cracked plane after repeated shearing. This curve has a similar tendency to that of the change of peak shear strength depending on the scale effect.

3 MEASUREMENT AND ANALYSIS OF SURFACE ROUGHNESS

As for the measurment of surface roughness in the vertical and horizontal directions were made with laser displacement meter and magne-scale respectively. Fig.3 shows an conceptional figure for them. Fig.4 shows measurment lines on a test specimen. One specimen has eight measurement lines.

Some methods have been proposed to quantitate surface roughness and using these methods correlation between surface roughness and physical properties of discontinuous plane has been investigated. Tse and Cruden (1976) proposed following indices and found a corellation between these indices and frictional coefficient of surface;

$$CLA = \frac{1}{M}\sum_{i=1}^{M}|z_i - \dot{z}| \qquad (4)$$

$$RMS = \left[\frac{1}{M}\sum_{i=1}^{M}(z_i - \dot{z})^2\right]^{\frac{1}{2}} \qquad (5)$$
$$= Z_1$$

$$Z_2 = \left[\frac{1}{M}\sum_{i=0}^{M-1}(\frac{z_{i+1}-z_i}{\Delta x})^2\right]^{\frac{1}{2}} \qquad (6)$$

In this study, Z_2 which gave good correlation with JRC, the index of shear strength used in the study conducted by Tse and Cruden, and the distribution of roughness angle proposed by Rengers (1970) were used.

(a)
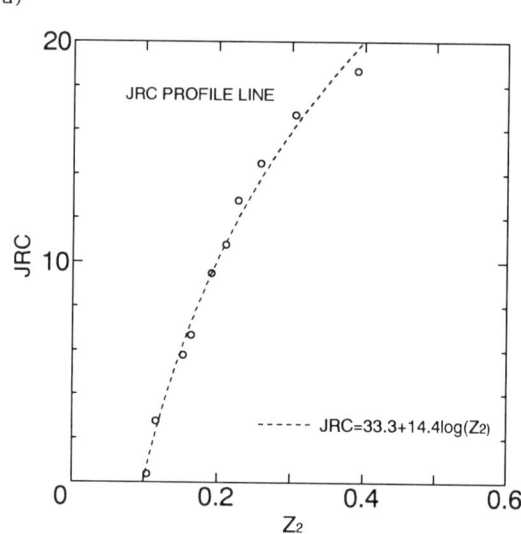

(b)
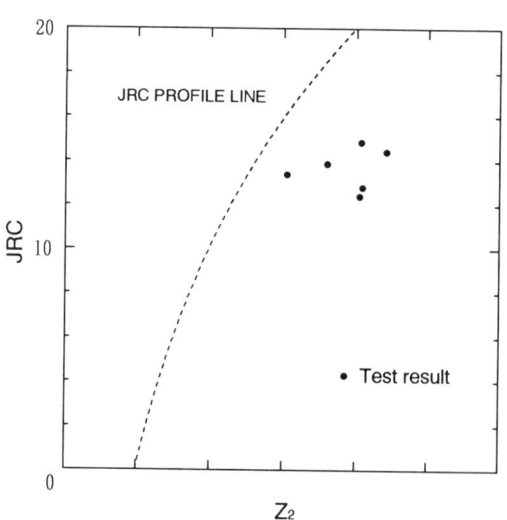

Fig.5 Analytical results of Z_2

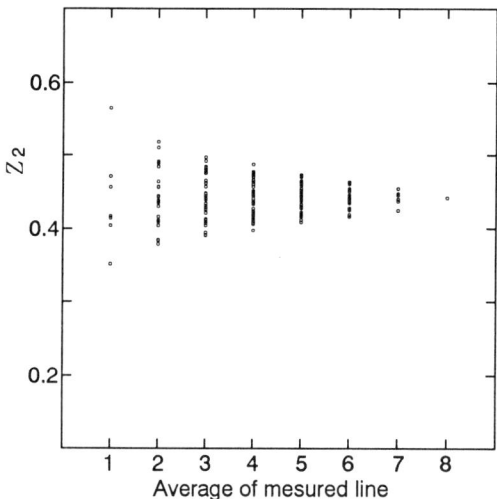

Fig.6 Average value of Z_2 on enght measurment lines

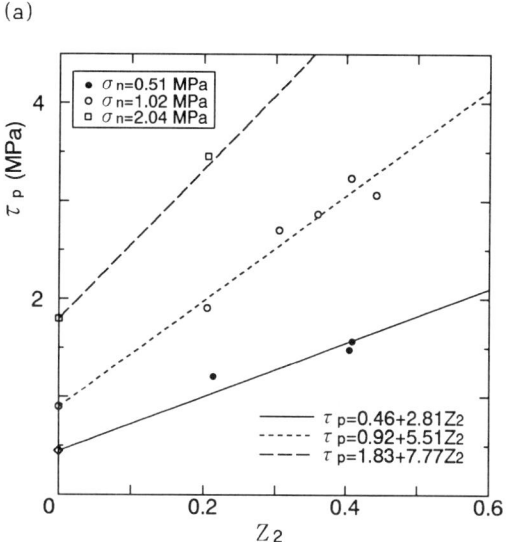

(a)

(b)

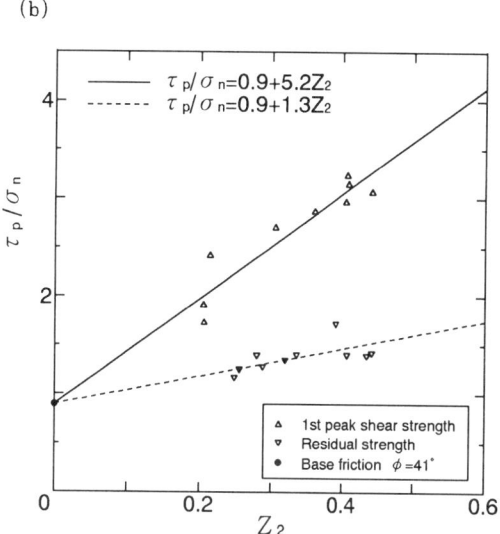

Fig.7 Relation between shear strength and Z_2

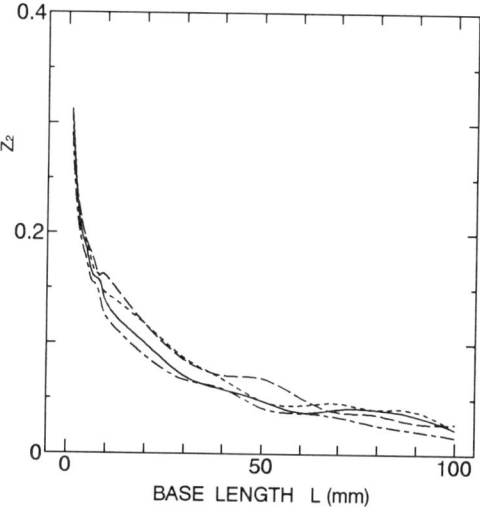

Fig.8 The valiation of Z_2 of measuring base length

Z_2 was analyzed with minimum measurment pitch of 1mm of make-up gage as a reference base length, because the value of Z_2 changes depending on the measurement base length. Fig.5 shows the analytical results of Z_2. (a) gives Barton's results concerning J R C profiles and the curve shows the approximation of respective correlation using power fonction. (b) gives the results for a shearing area $A = 1600 cm^2$. Approximate curve obtained from respective J R C profiles are plotted.

These results show that J R C profiles gives the same correlation with Tse and Cruden. However, the data obtained from granite test specimens tend to deviate from the J R C value presumed from abovementioned correlation. Values proposed by Barton and Cruden(1976) for J R C plofiles were the counted back values obtained from the direct shear test on specimens of small size. Therefor, if due consideration is paidto the scale effect, it shoud take larger J R C values. For it, we can see the following two reasons;

Firstly, a discontinuous plane in the natural conditions has no complete contact, but has interference by water and fillings. Barton's formula to calculate J R C values is on the supposition that the increment of shear strength obtained by removing the effects of fundermental frictional angle and vertical stress for the surface strength from the shear strength is wholly due to the surface roughness. Therefore, the conditions of contact at the discontinuous plane will be included in J R C values. Consequently, the counted back J R C values obtained from the shear strength should be recognized as an index of dynamic roughness and its degree of contact.

Secondly, the scatting of Z_2 on each measurement lines should be taken into consideration. Fig.6 shows values of Z_2 on eight measurement lines obtained from a shear plane and the mean values for two to eight measurement lines. On each measurement lines, Z_2 has a scatting of 0.46 to 0.56. It is thinkable that at the presumption of peak shear strength a scattering of about ± 0.5 for mean values of Z_2 obtained from eight measurement lines may slip out J R C values.

Fig.7 (a) and (b) shows a scattering of shear strength of a test specimen having a shearing area of $A = 1600$ cm² corrected with Z_2. In the figure (b), the solid line shows the peak shear strength at the first time and the broken line shows the residual strength after a repeated shear test. Since this figure clearly shows the existence of positive correlation in each case, the approximate expression shown in the figure may be utilized.

Fig.8 shows the valiation of Z_2 corresponding to the change of measuring base length. It is made known

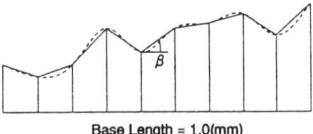
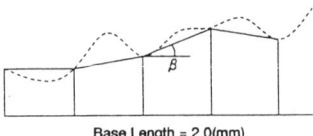
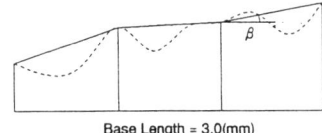

Fig.9 The roughness angle coming from base length

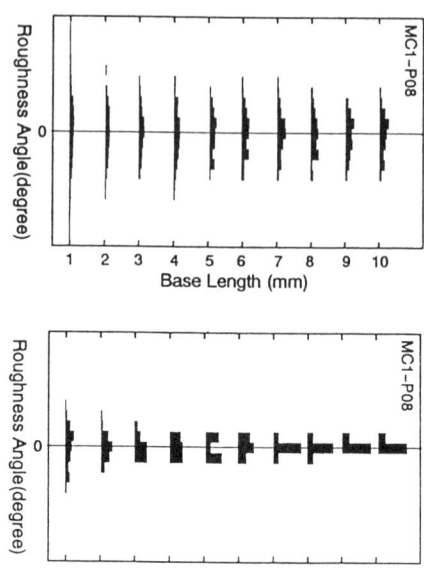

Fig.10 An example of the roughness angle distribution

from this figure that Z_2 also has a ceration scale effect. It is thinkable that shear strength of a discontinuous plane of a rock may be presumed by measuring a certain length of Z_2 on the said plane using above-mentioned two relations.

Further, using the distribution of roughness angle of Rengers (1970), we analyzed a method to calculate dilation caused by shearing on the surface roughness of discontinuous plane. Fig.9 shows a state of decrease of the roughness angle with the increment of the base length. Fig.10 shows an example of distribution of roughness angle. In this figure, distribution of roughness angle is shown with histogram in the case where base length is changed with 1mm- and 10mm-pitch for the ranges of base length of 1mm to 10mm to 100mm respectively. These distribution chart shows that the roughness angle for each base length gives a normal distribution and the concentration of distribution grows with the increment of base length.

Schneider (1976) expressed the relationship between the shear stress and the dilation angle with following formula.

$$\tau = \sigma_n \tan(\phi_b + \frac{dv}{dh}) \quad \cdots (7)$$

where, ϕ_b and dv/dh indicated the frictional angle of the flat plane and the tangential gradient concerning dilation.

$$\frac{dv}{dh} = k \cdot \sigma_{SD} \quad \cdots (8)$$

where, k is constant and σ_{SD} is the standard deviation of roughness angle.

Fig.11 shows the roughness angle counted back to coincide with the experimental values basing on the relation between shear stress and dilatancy. In this case, the counted back roughness angle is made dimensionless using each standard deviation of the distribution. This figure shows the result of $k = 1.0 \sim 1.5$ to obtain a good agreement with experiment. Before we have obtained $k = 2.2 \sim 3.0$ in the shearing area $A = 800 cm^2$.

In Fig.12, the standard deviation of roughness hight is plotted in the case where the measurement length of a specific section is changed and it is reconizable that the standard deviation of roughness tends to grow with the increment of the length of measurement section. Such change of the standard deviation of roughness distribution in relation to the length of measurement section proves that the swell

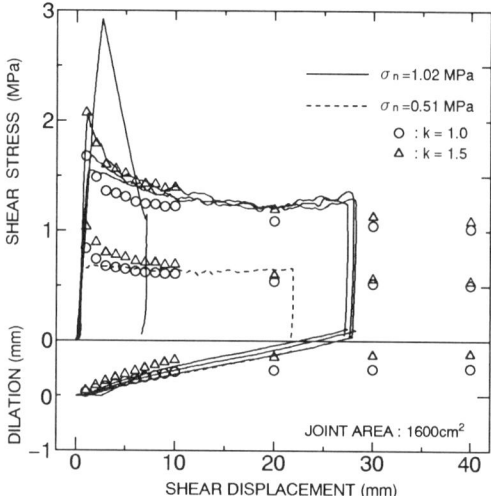

Fig.11 Estimated τ vs. U from ruoghness angle

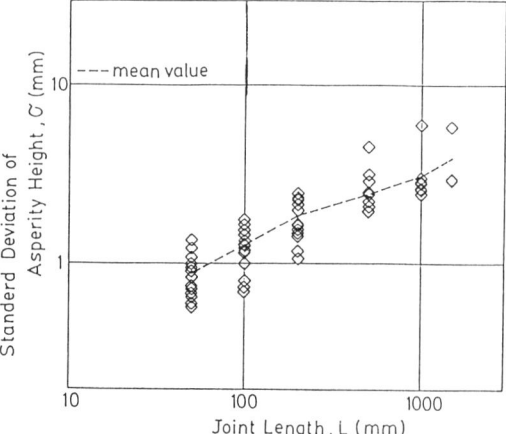

Fig.12 The standard deviation of roughness hight to change of measuring base length

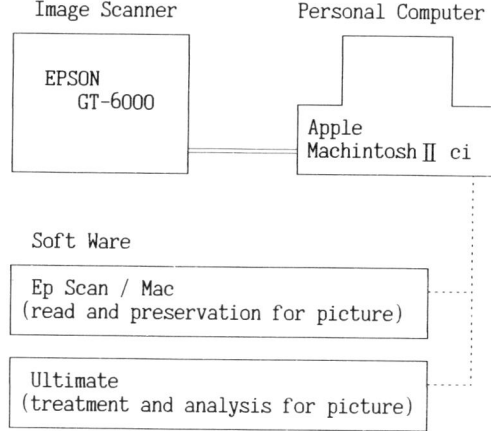

Fig.13 Concept figure of measuring system

Table 1 Results of picture prosesing analysis

A σ_n	MS (400)	MM (800)	ML (1600)	LL (9600)
0.51	0.87	0.62	0.49	0.31
1.02	2.50	1.48	1.32	0.93
2.04	5.84	2.50	1.76	1.37

(σ_n : MPa , A : cm^2)

becomes larger with increment of the section, and it may constitute one of the primary factor for the scale effect. On a discontinuous plane with several tens meter of length a large and gentle undulation with a wavelength of several meters is observed. However, on a discontinuous plane with several meters of length only unevenness of small scale is obtained. Therefore, a discontinuous plane of large scale has a greater individual contact section than that of the smaller scale, but it has a smaller contact area on the whole.

4 MEASUREMENT AND ANALYSIS OF SHEARED AREA

This study measured the real sheared area, which is the part of cutting during shear roading.

Before the commencement of the measuring work of the sheared area, we curved the discontinuous plane with China ink to distinguish the chipped portion on the said plane caused by the shear test. The granite test specimen was black on the whole and the sheared portion on the discontinuous plane looked white. Then, we took pictures of discontinuous planes to get image data. To calculate the area of sheared area, we utilized the image analysis system with aid of personal computer.

Fig.13 shows the outline of the hardware and software of the system. Photographs taken were read as the digital data by the help of image scanner and stored in the personal computer. Table 1 gives the result of culculation by which the ratio of the area of the sheared portion A_s to the whole shear area A was obtained. It is recognizable that the rate of A_s decreases with the lowering of normal stress and the increment of the shear area.

Fig.14 shows relationship between the peak shear strength and the sheared area at the time of first shear test. It is seen that good correlation exists between both. Basing on this relation it is imagined that the greater part of the shear strength at the time of the first shear test is governed by the breaking, with smaller part by the slip of the roughness on the discontinuous plane. It is recognizable in this figure that the sheared area A_s also showed similar scale effect and dependenth on the normal stress to the shear strength. Fig.15 shows the relation between the shear area and the sheared area. These show a tendency to decrease for each normal stress, with the result that their tendency well coincides with that of the scale effect for the shear strength as is shown in Fig.2.

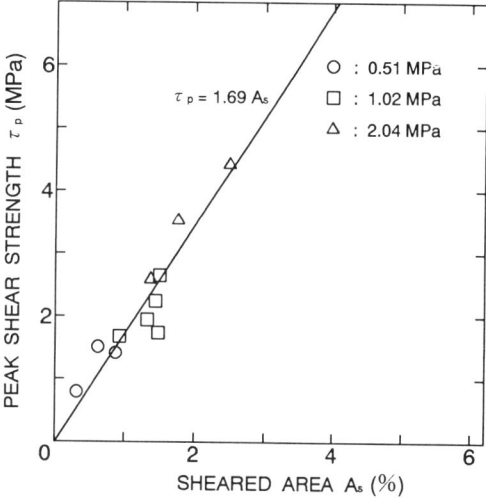

Fig.14 Relation between the peak shear strength and the sheared area

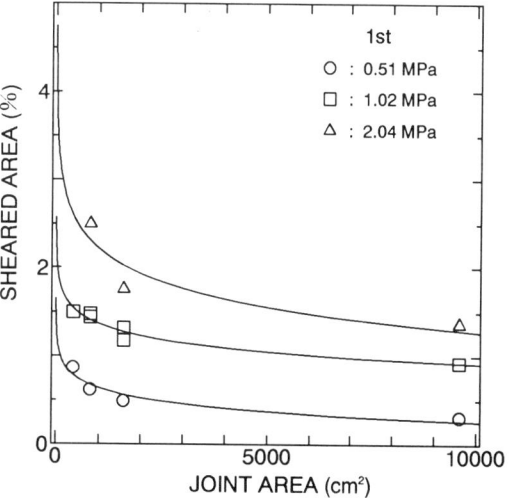

Fig.15 Relation between the shear area and sheared area

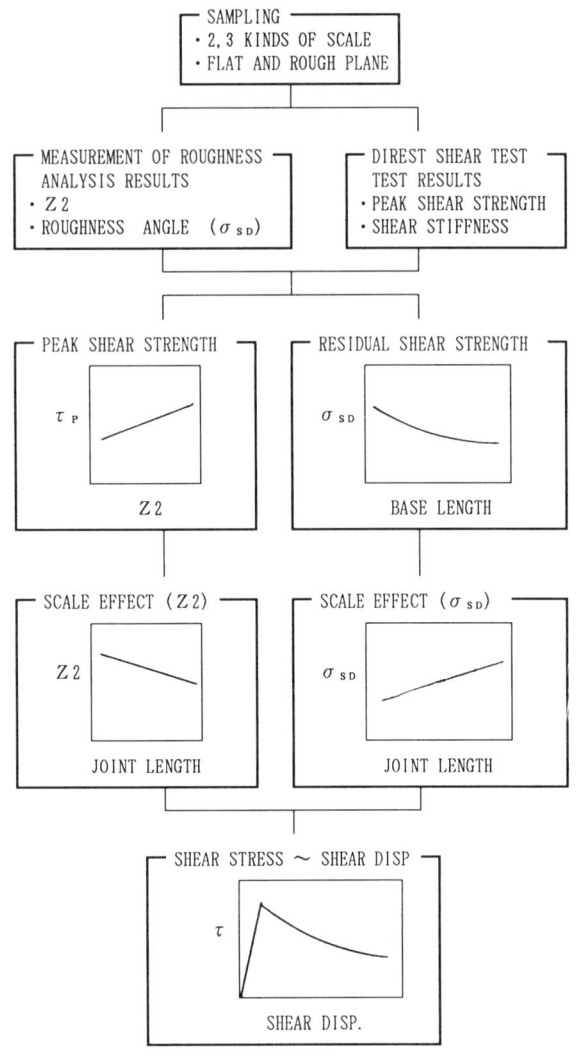

Fig.16 Estimating flow for shear characteristics

5 CONCLUSIONS

Laboratory test and measurement of roughness surface were excuted to clarify the mechanism of scale effect in rock joint.

Shear strength decreases in the form of power function with increment of the shearing area. We indicated the characteristics of the surface roughness of joint using the measured roughness indices (Z_2 and roughness angle) and showed the relation between joint roughness and the scale effect. Also, the contact area of the discontinuous plane is a major factor affecting the scale effect. Fig.16 shows the estimating flow for shear characteristics. We will provide the basic data to use for numerical analysis and practical application.

REFERENCES

Barton, N.R. & Bandis, S. 1982. Effect of Block Size on U.S. Sympo. on Rock Mech., pp.736-760.

Chunha, A.P.(ed.) 1990. Scale Effect in Rock Masses. Proc. 1st Int. Workshop, Loen.

Jaeger, J.C. & Cook, N.G.W. 1976. Foundamentals of Rock Mechanics.(3rd. Ed.)

Krsmonvic, D. & Popovic, M. 1966. Large Scale Field Tests of Shear Strength of Limestone. Proc. 1st Congr. ISRM, Vol.1, pp.733-799.

Pratt, H.R., Black, A.D. & Brece, W.F. 1974. Friction and Deformation of Jointed Quartz Diorite. Proc. 3rd Congr. IRSM, Denver, II-A, pp.306-310.

Rengers, N. 1970. Influence of Surface Roughness on the Friction Properties of Rock Planes. Proc. 2nd Cong. ISRM, Vol.1, pp.229-234.

Tse, R. and Cruden, D.M. 1979. Estimating Joint Roughness Coefficients Int. J. Roch. Sci. & Geomech. Abstr. Vol.16, pp.303-307

Yoshinaka, R., Yoshida, J., Shimizu, T., Arai, H., & Arisaka, S. 1990. Experimental Study on Scale Effects of Shear Strength. Proc. 22th Symp. on Rock Mechanics, JSCE, pp.211-215.

Yoshinaka, R., Yoshida, J., Shimizu, T., Arai, H., & Arisaka, S. 1990. The Influence of Roughness and Degree of Interlocking on Strength Characteristics. Proc. 22th Symp. on Rock Mechanics, JSCE, pp.206-210.

Yoshinaka, R., Yoshida, J., Shimizu, T., Arai, H., & Arisaka, S. 1992. Experimental Study on Scale Effect of Shear Strength. Proc. 24th Symp. on Rock Mechanics, JSCE, pp.216-220.

Yoshinaka, R., Yoshida, J., Arai, H., & Arisaka, S. 1993. Scale Effect on Shear Strength and deformability of rock joints. Proc. Scale Effect in Rock Masses 93, pp.143-149.

B: Design and analysis

Stability analysis of the LAXIWA arch dam abutment on complex rock formations

Zhou Wei Yuan, Yang Rouziong & Yan Guang Ri
Tsinghua University, Beijing, People's Republic of China

Zhang Ming Yao & Wu Xi
Northwestern China Design Institute, Xian, People's Republic of China

ABSTRACT: In the left bank of the Laxiwa Arch dam abutment, there found a possible rock slide with large deformations going on at present, which is designated by 2nd rock slope body. In this paper, its stability is examined, using TFINE program which is developed at Tsinghua University as a tool to analysis rock stability engineering problem. The interatction between dam and 2nd slope block is studied using TFINE- nonlinear elasto-plastic, fracture and damage F.E.M. program, and the stability evaluation of foundation bedrocks is also examined in the paper. TFINE program could be used to analyze the integral stability for large dams. A case study of LAXIWA arch dam is presented in the paper.

1 Introduction

Laxiwa arch dam with its maximum height more than 240m, is one of the largest dams in the world. In the dam site, most of the stones are sound granite, which offer a nice background for the high arch dam. However, some largely deformed rock blocks and precipitious rocks were found on the left abutments of the arch dam as shown in Fig.(1). Among them, the so-called 2nd rock slide bolck is located close to the left dam surface, and is crucial for arch dam stability. So two topics are put forward for engineering, they are given as below:

(1) The stability evaluation for 2nd rock slide body.
(2) Its effect to dam stress distribution and stability of left abutment.

The first one is studied in this paper using TFINE program. Two cases including pre and post the construction of the dam are given.

The second problem is studied by sensitivity analysis method, i.e. altering the elasticity modulus of 2nd rock block, while the reservoir is impounding. The influence of 2nd block upon dam could be made clearly by different stiffness. All the researchs were obtained using TFINE F.E.M. program.

The stability evaluation of rock abutments is examined and presented in the paper.

2 3-d Finite element stability analysis

The 2nd rock slide block sustain 1.35 million M^3, is located on the left abutment, inclining to bedrock and downstream. The block has a Hf_4 fault as bottom and L_{145} fault, F_{29} fault as lateral surface, as shown in Fig(1). The F_{29} with openning fissures strikes N.W direction. Two loading cases are considered for dam stability calculation: natural condition without dam, and dam reservoir is impounding. A comparison of dam stress is shown in table 1. The stress distribution and safety factor in the surface of faults F_{29} L_{145} and Hf_4 are given in Fig(2) and Fig(3). They indicate that faults F_{29} and Hf_4 are in shear-compression state with about 2.0MPa maximum nomal stress and 0.9MPa maximum shear stress. Tesile stresses exist in some points in F_{29} with maximum value of 0.4MPa.

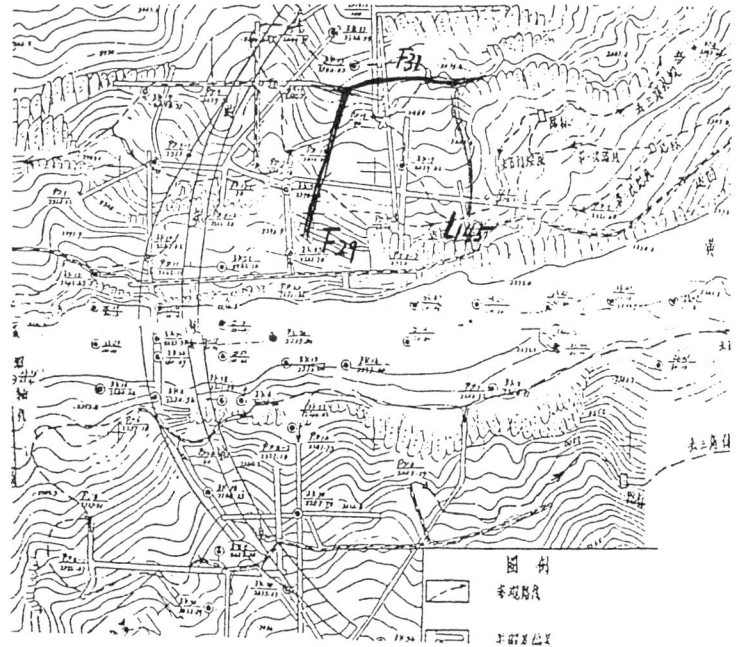

Fig.1 2nd slide block

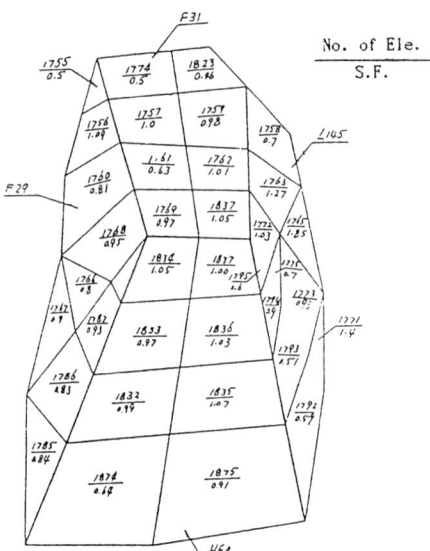

Fig. 2 Nature case

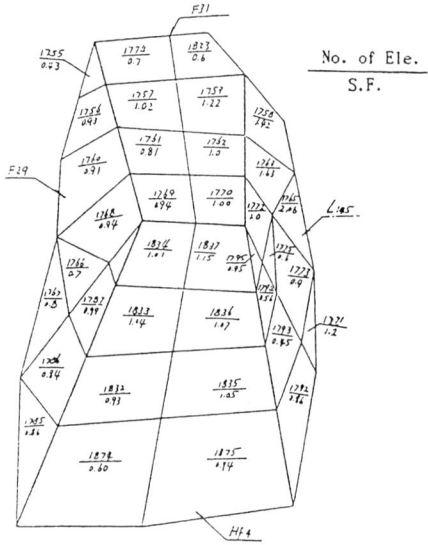

Fig. 3 Loading case

(1) Displacement Analysis

Tab1. A comparison of dam stress distribution (0.1MPa)

cases	upstream surface				Downstream Surface			
	σ_1	σ_3	Max arch stress	Max catl stress	σ_1	σ_3	Max arch stress	Max catl stress
E = 5000 MPa	22.5	-57.2	-56.7	-2.6	-5.2	-80.5	-55.2	-55.5
E = 1.7×10^4 MPa	26.8	-56.4	-57.6	+0.2	-5.1	-79.5	-55.16	-56.3

The displacements of Fault 29 are given in Tab.2. The relative displacements at the higher position of F_{29}, above the Ele. 2620, tend to be tensile under nomal water load with value of 0.05mm perpendicular to the fault surface.

However, at the Ele. 2540, it is compressed with 0.044mm, and is approaching zero below Ele. 2500m.

The displacements of F_{29} Hf_4, L_{145} are shown in Fig.(2).

Tab2. A comparison of deformations of left rock abutment on 2nd block (mm)

Elev.(m)	E = 5000MPa	E = 1.7×10^4MPa
2460	1.4	1.31
2430	2.6	2.6
2400	4.7	4.68
2360	7.5	7.48
2320	8.9	8.99

All the displacements are very small. They show that in the upper part, only tensile displacements are found. whereas, compressed displacements are existed in the intersection of F_{29} and Hf_4.

2nd block generally deform perpendicalar to the river, with a deformation of 2.5mm and along the river with 2.0mm.

Differences of stresses on the surface of 2nd rock block before and after the dam construction and reservoir full, is so small that 2nd rock block could be considered to be self-stabilized.

(2) Point safety factor and Yielding area of abutments.

The safety factors and yielding area of fault surface are shown in Fig (2,3,4). They are all approximate to 1.0. The element safety factors of the fault Hf_4 has a yeilding part near the dam, they are almost less than 1.0. The downstream parts are stable and their S.F.S are larger than 1.0. From the above analysis, it could be observed that, the dam exert very small load to 2nd Block, in other word, the dam will not cause damage or sliding state to the 2nd block.

To evaluate stability, the point safety factor is used conveniently and clear. Let F_A define as safety factor, it is an index for evaluating stability against sliding. In the limit equilibrium state, the following equations hold:

$$\sum_{i=1}^{n} \frac{A_i}{F_A} = 1 \quad (1)$$

where, n is the number of faults, A_i the cross section area, f the friction coefficient, and c. the cohesion. The point safety factor and yielding area expressed in faults like F_{29} t_{145},

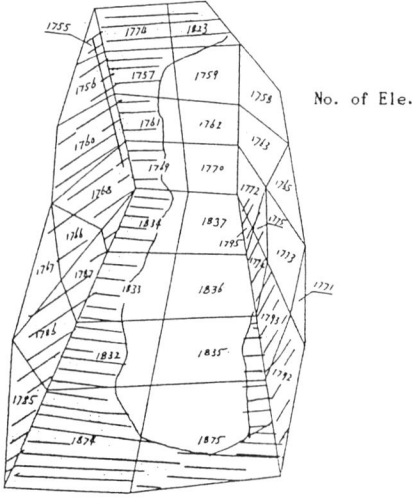

Fig. 4 Plastic Area under loading

Hf$_4$ etc, indicate that the main possible sliding direction is along the intersection line of Hf^4 and F$_{29}$. Their safety factors are:

F$_A$ = 1.09 and 1.04 respectively.

It is in a stable critical state, in other words, no crucial influence from dam construction have been found. No important actions from the dam have been exerted to the 2nd rock block as shown in Fig.4.

3 Sensitiveness Analysis

On altering f and c, the curves following formula (1), are shown in Fig.5, and Fig.6, the K-C curve is varied very slowly. That means, the K is not sensitive to the variation of c. However, the K-f line behaves an abrupt curve, which indicates that the S.F is subject to f.

4 The influence of 2nd rock block to dam stress distribution

Different values were given to the 2nd block, the stresses in dam express no variation due to the existance of 2nd block.

In Fig.7. are shown the comparison of stress distribution corresponding to different Elasticity Modulii of 2nd rock block.

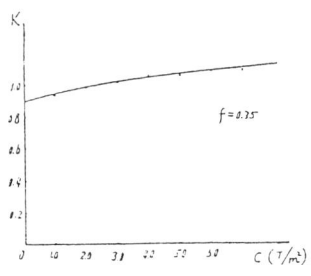

Fig. 5 K - C curve

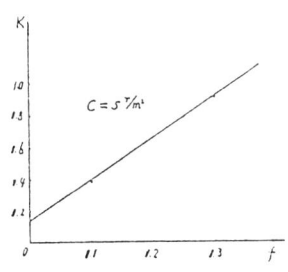

Fig. 6 K - f curve

5 Fracture mechanics model for rock-concrete like material which lie in upstream tensile region of Arch dam.

On assumming that rock material is of multi-porous one, its fracture intensity factor could be explained as in Fig(8) and shown as:

$$K_I = \frac{2}{\pi}\sqrt{\pi a} \cdot \frac{\sigma_1}{\sqrt{1-(\frac{a}{b})^2}}$$

where a   the radius of crack
       b   the radius space span
       σ_1 the farfield tension

From self-consistant principle,

$$U_d = \int_\Omega G d\Omega = \frac{1-\mu_0^2}{E_0}\int_{\Omega_0}(K_I^2 + K_{II}^2 + \frac{K_{III}^2}{1-\mu_0})d\Omega \quad (2)$$

where U$_d$ - strain energy from cracks
       G - Energy release rate.
       For tensile crack,

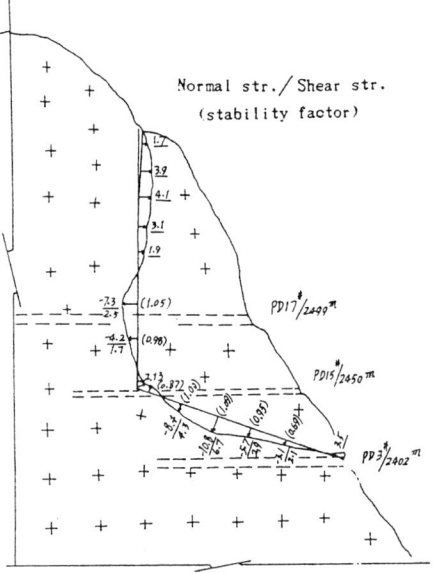

(a) Nature case

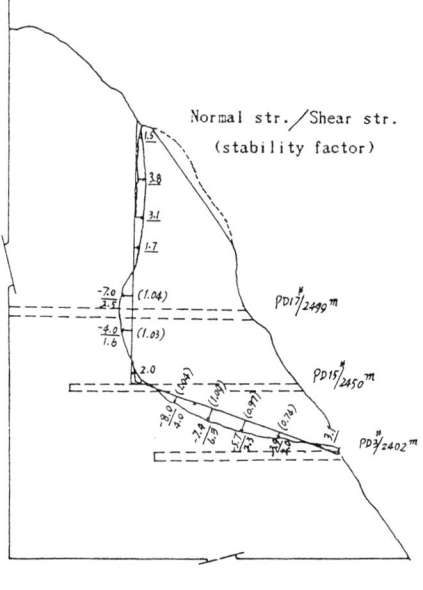

(b) Loading case

Fig.7 Stress distribution

$$U_d = \frac{1-\mu_0^2}{E_0}\int_\Omega K_I^2 d\Omega$$
$$= \frac{1-\mu_0^2}{E_0}\int_\Omega \frac{1}{1-(\frac{a}{b})^2}\frac{4}{\pi^2}\pi a\sigma_1^2 2\pi a da$$

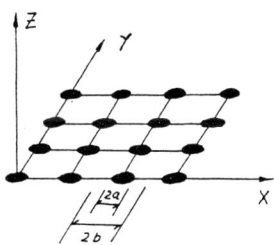

Fig.8. Multi crack system stress

$$= \frac{4(1-\mu_0^2)}{E_0} b^3 [ln\frac{1+\frac{a}{b}}{1-\frac{a}{b}} - 2(\frac{a}{b})]\sigma_1^2 \quad (3)$$

and then

$$\frac{E_0}{E_{o-d}} = 1 + 8(1-\mu_0^2)\rho^{(i)} b^3 [ln\frac{1+\frac{1}{b}}{1-\frac{a}{b}} - 2(\frac{a}{b})] \quad (4)$$

where E_{o-d} is equivalent elasticity along crack,
$\rho^{(i)}$ density of cracks along crack direction.
Assuming the main damage variable is $d^{(i)}$, then

$$d^{(i)} = 1 - \frac{1/E_0}{1/E_0 + \xi_y} = 1 - \frac{1}{1+2A\rho^{(i)}a^3 M(\frac{a}{b})}$$

where ξ_y is the porous flexibility.

$M(\frac{a}{b})$ is the coefficient for multi crack inter action.

The fracture propagation condition for multiporous material could be obtained and expressed as

$$K_I = \frac{2}{\pi}\sqrt{\pi a} E_0 \varepsilon_1 \frac{1-d^{(i)}}{\sqrt{1-Z^2}} \quad (5)$$

where $Z = \frac{a}{d^{(i)}}$ and the $d^{(i)}$ is the main damage variable.

In Fig(8), the relation between K_I and Z is shown.

when $K_I = K°_{Ic}$, the crack initiate propagation and Z_c could be defined. as shown in Fig.(8)

$$\frac{dk_I}{dZ} \begin{cases} < 0 & Z < Z_c & \text{stable propagation} \\ = 0 & Z = Z_c & \text{critical state} \\ > 0 & Z > Z_c & \text{unstable propagation} \end{cases} \quad (6)$$

6 A case study of Laxiwa arch dam crack propagation

From the above mentioned definition of damage, a flexibility of damage C^{o-d} is defined as follows:

$$C^{o-d}_{ijkl} = C°_{ijkl} + \frac{2}{E_0} A_{ijkl} + \frac{1}{2E_0}(B)$$

where A, B are of forth rank and second rank tensors respectively.

In Tab.3, the C^{o-d}_{ijkl} of bed rock of Arch dam Laxiwa are given. In Fig.(1), is shown the general gelogical map of Laxiwa arch dam. Using, F.E.M., the crack length in upstream of dam is calculated and described.

$$l = 6.75m$$

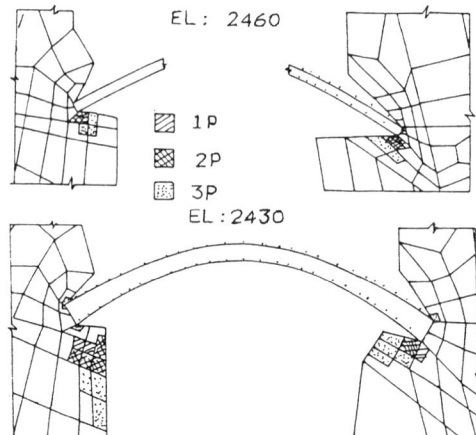

Fig.9. Evaluation of bedrock strength of Laxiwa Arch dam

This crack length is satisfied with dam safety. In Fig.(9), the crack progation area could be found using. F.E.M. According to the above described calculation, the determination and evlauation of bed rock quality could be obtained. The computer system TFINE has been the major tool to evaluate the dam stability, and bed rock design.

Tab 3. C^{o-d} Damage tensors of Laxiwa bed rock (granite)

average length of joints (m)	Density ρ No./m³	average width tc cm	C^{o-d} $10^{-3}(M^2/T)$					
2.85	0.129	0.206	33.19					
			-0.76	7.24				
			-0.4	-0.4	2.02			
			0.27	0.48	0.003	8.4		
			-0.01	-0.08	-0.07	0.18	5.42	
			0.34	0.003	0.45	-0.04	0.34	7.69
2.43	0.15	0.19	7.74					
			-0.474	3.22				
			-0.397	-0.39	1.902			
			-0.327	-0.36	-0.001	5.72		
			0.001	-0.55	-0.005	0.042	4.85	
			0.081	-0.53	0.097	-0.002	-0.202	5.4

7 Conclusion

The very crucial problem of abutment rock determination is examined in this paper. The authors proposes a fracture length evaluation for dam crack. A computer program system is used for evaluation of dam fracture. This system is used instead of the conventional decision by purely experience of engineers.

Acknowledgement– The presented paper was supported by the laboratory of Institute of Geology, Sinca Academy, and Scientific Foundation of China.

Reference

[1] O.C. Zienkiewicz, B.Best, C. Dullage and K.G.Stagg Analysis of Nonlinear problem in Rock Mechanics with particular Reference to Jointed Rock Systems. Proc. 2nd Int. Cong. of Rock Mech. Vol.3, 1972.

[2] H.Horii and S. Nemat-Nasser, Compression-Induced Microcrack Growth in Brittle solids: Axial splitting and shear failure, J. Geophysical Res., Vol.90. 1985.

[3] Zhou Weiyuang, Yang Yanyi, A Three Dimensional Nonlinear Finite Element Analysis of Arch dam with Fracture Damage Models. Int. Symp. on Analyical Evaluation of Dam Related Safety Problems. Copenhager. Proc. 1989.

Strength and deformability evaluation for jointed rock foundations of large dams

S. A. Yufin, V. N. Burlakov & M. G. Zertsalov
Moscow State University of Civil Engineering, Russia

ABSTRACT: Development of numerical modeling methods for rock structures demands improvement of techniques for evaluation of strength and deformability parameters of jointed rock masses. Rock foundations of several large dams on the territory of the former USSR have been investigated. For evaluation of stress/strain relationships for jointed rocks a new method based on the development of the energy theory of strength is presented. Numerical modelling techniques are tested and verified against in-situ experiments in jointed rock.

I regard rock mechanics without engineering geology as a wrong track. An immense number of examples could be cited to confirm the correctness of this assertion. After all, whether we achieve success in our design and our work relating to rock depends almost exclusively on how well we have recognized the geological situation and how well we have adapted our design and our work to that situation. And the geological situation decides which tools we can employ with advantage at the construction site and which mathematical tools ought to be employed during design.

Leopold Müller-Salzburg (1983).

1 INTRODUCTION

In many scientific fields, processes that are dealt with seem to follow well-defined natural laws, and reliable predictions can be made on the basis of theory or physical experiments alone. The same is true for many civil engineering applications involving man-made materials such as steel and concrete because their properties can be controlled or estimated reliably. The geotechnical engineer has an enormous disadvantage in comparison, since he must work with natural geologic materials. The properties of these materials are generally variable and often unpredictable or even unknown.

Getting a true picture of the actual conditions at a site is often impossible or impractical to achieve. Thus, it is frequently necessary to make assumptions regarding the real behaviour of the materials or of important details of rock mass structure, the degree of rock fissuring, the magnitude of in-situ stresses in the rock and so on.

Because of this, many geotechnical engineering problems do not lend themselves to solution strictly on the basis of numerical analyses or/and physical experiments alone. Other sources of information and other approaches to problem solving are required. This is where judgement and modelling come into the picture. These two concepts are very closely interrelated in rock mechanics and foundation engineering.

2 MODELING CONCEPTS

In the context of this report we will distinguish four types of models: engineering-geological, geomechanical, physical and numerical. It is generally adopted practice to define the engineering-geological model as a systematic description of the rock mass reflecting the nature of it's structure and state, and the types of rocks composing it. This description is based on analyses and generalisation of the in-situ data. Taking into consideration the genesis and geological age of the rock, a representation of known laws of rock mass formation and analogous rock masses investigated for similar projects; statistical, probabilistic or other approaches are used to fill in unknown details.

Creating the engineering-geological model is obviously the most important step in the design of any engineering structure, and this is especially true for the design of dams on rock foundations.

Based on the engineering-geological model, geomechanical

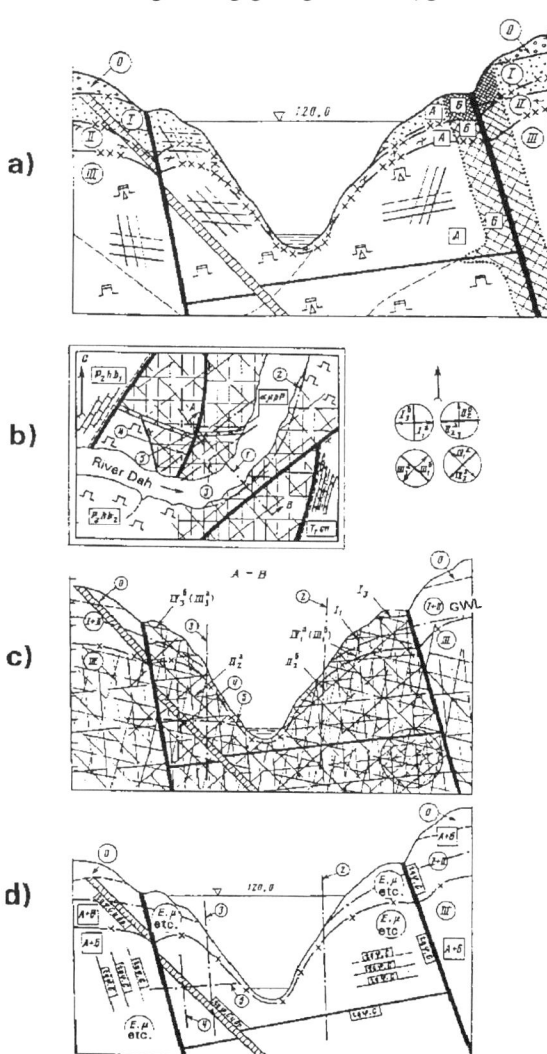

Figure 1. General engineering-geological model of the foundation of the Hoa-Binh Dam: a) cross-section, b) plan; and geomechanical models of the same site: c) model of jointing, d) model of rock properties. (After Fishman & Miroshnikova, 1984).

models can be created. These models are usually defined as "two or three dimensional representations which adequately portray the geological features and the physical and mechanical properties of the rock mass" (Ukhov 1975, 1979).

Several geomechanical models may be derived from one engineering-geological model depending on the goal and technique of further analyses, but there is principal difference between the source (engineering-geological model) and the outcome - problem-dependent geomechanical models. Where the first collects fundamental knowledge about the rock mass, the last ones serve special purposes for specific types of analyses.

The difference between these two types of models is clear from Figure 1 where the general engineering-geological model (a and b) has been used for creation of two geomechanical models: one for consequent discontinuum analysis (c - model of jointing); and another for continuum analysis (d - model of distribution of rock properties).

Based on the geomechanical model, physical and numerical models can be developed. Here it is very important to note that these models usually represent not only the foundation, but also the structure, and should be considered as models of rock-structure interaction, or, in the most complete sence for dams, models of dam-foundation-water flow interaction because of the obvious importance of taking into account ground and reservoir water in solving the problem of dam design.

Recent advances in numerical modeling made physical modeling almost absolete. However, in physical models one can clearly observe the failure of the structure or foundation, which is one of the reasons such models are still attractive and still in use.

The whole set of procedures involved in rock engineering may be summarised in the flow-chart in Figure 2. A suitable sequence of operations in the back analysis computerized system is presented by Barton et al (1994). Development of a knowledge-based system with comprehensive geomechanical data bases on the regional level is now in progress (Melnikov et al 1995).

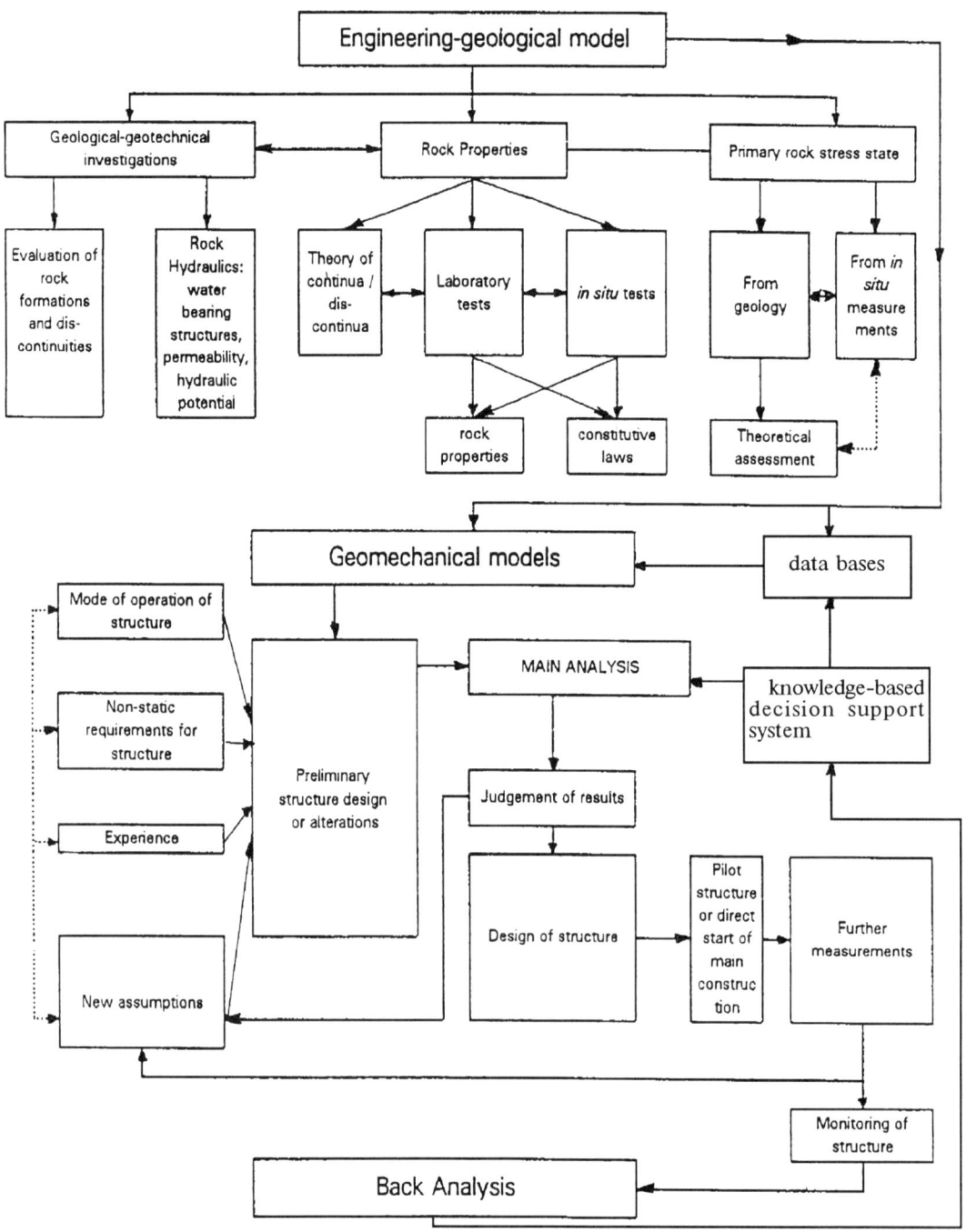

Figure 2. Flow-chart for design of structures on rock foundations or/and in rock. (modified after Gysel, 1987).

3 IN-SITU TESTS OF ROCK FOUNDATIONS OF DAMS

During the period of 1960-1980 considerable number of large dams had been constructed on rock foundations in Yakutia, Siberia, Central Asia, Caucasus and other regions of the former USSR. For Russian specialists this was principally new experience after decades of dam construction on soils.

Large volume of engineering-geological and geomechanical investigations involved in this construction noticeably put forward development of rock mechanics as a science and provided conditions for forming of several notable Russian schools of rock mechanics professionals. One of such schools has been established at the Moscow State University of Civil Engineering (MSUCE, former Moscow Institute of Civil Engineering - MICE). Recognition of practical importance of the four-models concept already described herein came there in the early sixties.

Even poor quality rock may sustain considerable loads, thus providing conditions for design and construction of important large-scale structures such as dams. Reliability of engineering decisions taken during the design and construction process largely depends on the concepts on which such decisions are based and quality of research involved. Reliability and safety of large dams remarkably depend both on qualitative and quantitative characteristics of investigations carried out to form engineering-geological models. Special attention should be given to site regioning, choosing structural blocks in rock foundations, ear-marking large joints and evaluating their descriptive parameters.

Geomechanical investigations include laboratory and in-situ tests for obtaining strength and deformability properties of foundation rock. Such investigations have been carried out for different types of rock ranging from semi-rock in foundations of dams on rivers Stryi (Ukraine) and Viliui (Yakutia) to hard sandstones in the foundation of the Kirovsk Dam on the river Talas in Kirgizia. In-situ plate-bearing tests with conrete pads and tests on rock pillars were used for deformabiliy and strength evaluation. These tests were accompanied by geophysical exploration, which permitted to localize large joints and structural blocks in foundations.

Back analyses of such tests may provide valuable data for evaluation of constitutive parameters of the rock foundation in question. Figure 3 shows typical load-unload curves for cylindrical plate 0.6m in diameter and axisymmetric FEM mesh for numerical modeling of tests.

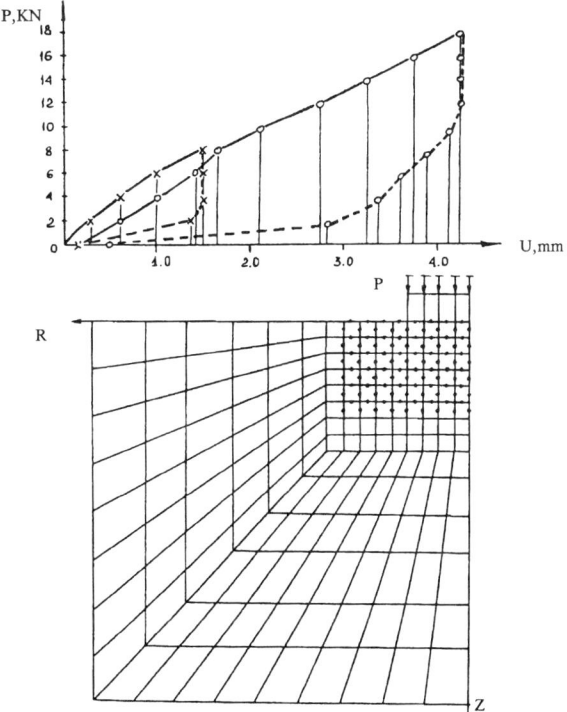

Figure 3. Typical load-unload curves for a plate-bearing (normal loading) test and FEM mesh used in back calculations.

For every plate-bearing test stress/strain relations may be interpreted as shown on Figure 4. Dependence of vertical displacements V on normal stress σ is shown on Figure 4c. After stabilization of deformations at the base of the plate, increasing shear load τ is applied as shown on Figure 4b. Gradual increase of τ causes increment of horizontal (shear) deformation U as shown on Figure 4a. Increase of τ and U is accompanied by contraction or dilation and leads to increments of vertical deformations V during shear. Dependence of vertical deformations on horizontal ones is shown on Figure 4d. Set of graphs on Figure 4 usually is termed as "rock passport"

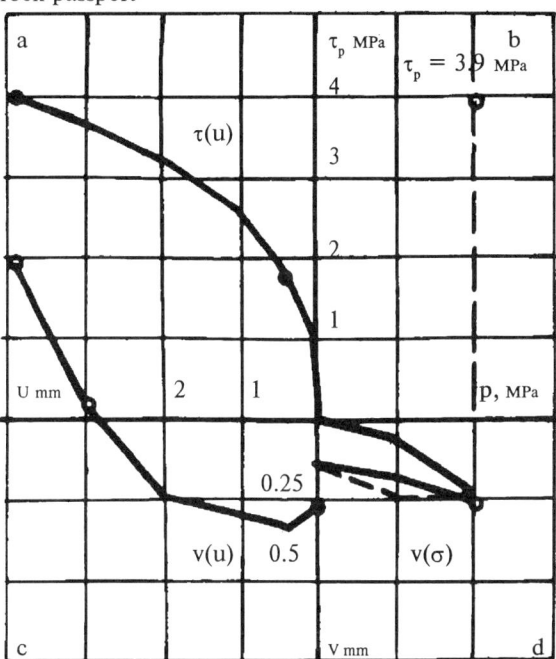

Figure 4. Graphs of stress/displacement relations for metamorphic shales.

These graphs can be also obtained as results of laboratory tests on rock samples. Relations between stresses and displacements at shear and stress and strain invariants can be written as follows:

$$\sigma = a_1 (\sigma_1 + \sigma_2 + \sigma_3)/3$$

$$\tau = a_2 \sqrt{(\sigma_1-\sigma_2)^2 + (\sigma_2-\sigma_3)^2 + (\sigma_3-\sigma_1)^2} \quad (1)$$

$$v = b_1 (\varepsilon_1 + \varepsilon_2 + \varepsilon_3)$$

$$u = b_2 \sqrt{2/3 (\varepsilon_1-\varepsilon_2)^2 + (\varepsilon_2-\varepsilon_3)^2 + (\varepsilon_3-\varepsilon_1)^2}$$

where a_1, a_2 depend on the lateral pressure coefficient, and b_1, b_2 on the thickness of the zones of compaction and shear.

Modulus of deformation can be evaluated as based on results of plate bearing tests using graph V=V(σ) on Figure 4c or as:,

$$D = \omega_p (1-v^2) b \, \Delta\sigma/\Delta v \quad (2)$$

where ω_p -dimensionless coefficient, v-Poisson ratio, b-size of the plate.

Shear resistance for every tested plane can be calculated as:,

$$\tau_p = \sigma (tg\phi + \eta_p)/(1 - tg\phi \cdot \eta_p) + C_p \quad (3)$$

where $tg\phi$, η_p can be taken from graphs $\tau=\tau(U)$ on Figure 4a and $V=V(U)$ on Figure 4c.

Results of shear tests, represented as a Coulomb-type dependence $\tau_p = \tau(\sigma)$ show large dispersedness of experimental points. As an example Figure 5 shows test results for metamorphic shales in the foundation of the Andizhan Dam.

Large amount of plate bearing tests along with geophysical investigations provide necessary data for formation of the geomechanical model of jointed rock foundation. However such investigations are expensive and time-consuming. Therefore there is a trend to partly replace in-situ geo-

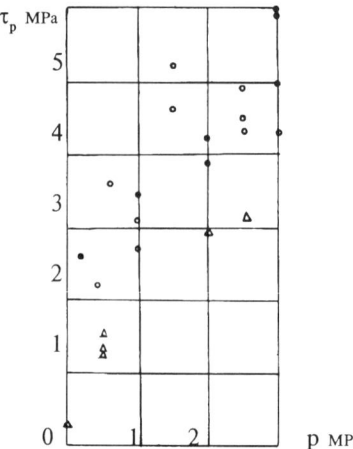

Figure 5. Dependence $\tau_p=\tau(\sigma)$ for metamorphic shales. o - averagely jointed; Δ - highly jointed.

mechanical tests with numerical modeling. Real jointed rock mass may be represented in each typical structural block by the equivalent block, which should have same displacements on the borders with other blocks, as the original.

In one of approaches used in MSUCE real jointed rock block structure is represented by the equivalent media with thin elliptical joints. Special software has been created for establishing parameters for such media, which for any chosen type of loading provides stress/strain nonlinear relations, representing real behaviour of the rock block.

Considerable difficulties arise with evaluation of mechanical properties of large joints at the borders of structural blocks. Shear strength and deformability of such joints largely depend on the compressive strength of rock and roughness of the joint walls as well as on filling and aperture of joints in natural conditions. During engineering-geological investigations profiles of joints are recorded. Special computer programs permit to represent real profiles of joints by schematic equivalent, for which set of graphs shown on Figure 4 can be established.

Another approach for evaluation of elements of the geomechanical model is perfection of classical constitutive equations based on the energy theorems.

4 STRENGTH CRITERION

The most acceptable criterion for description of soil and jointed rock properties is the energy one. The energy criteria of strength have been developed by von Mises, Hencky, Botkin, Rowe and many others. In contrast to these authors it is postulated, that strength is a function of a certain deformation potential, dependent on compressive stresses and density of rock (Burlakov 1994).

For estimation of rock density changes with the increment of the volumetric strain, density coefficient m, equal to relation of the volume of hard particles to the whole volume of the sample is proposed. Parameter m depends on

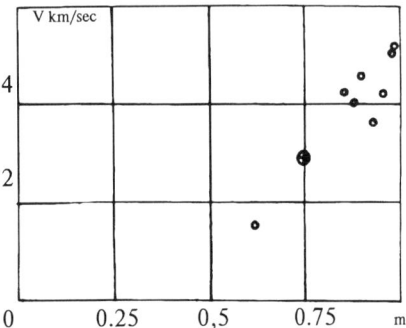

Figure 6. Dependence of wave velocities on coefficient m for metamorphic shales.

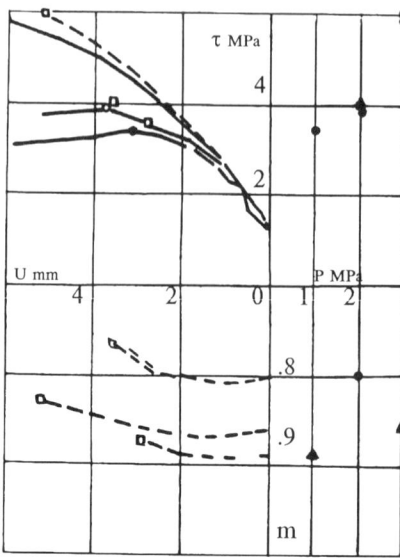

Figure 7. Graph $\tau=\tau(u)$, $m=m(u)$/ o - experimental, □ - calculated.

the porosity coefficient e:

$$m=1/(1+e) \quad (4)$$

or on skeleton density ρ_{sk} and density of hard particles ρ_s:

$$m=\rho_{sk}/\rho_s \quad (5)$$

Relation between density coefficient m and volumetric strain may be written as:

$$m=m_o/(1-\varepsilon_v) \quad (6)$$

where m_o - initial value of m. This initial value may be calculated using equations (4) or (5) from known initial values of porosity coefficient e_o or ρ_{sk}. The value of m can be estimated for any given volume of the rock mass by ultrasonic logging or by measuring velocities of seismic waves.

Relation between wave velocity and density can be evaluated in the laboratory. Example of such dependence for metamorphic shales is shown on Figure 6. Assuming, that full potential energy, accumulated in a sample or in a rock mass depends on density coefficient m and on the first invatriant of stress tensor $\sigma_1=1/3(\sigma_1+\sigma_2+\sigma_3)$ and considering (1) we can write $Э=Э(\sigma,m)$.

Plate bearing tests (or tests on rock pillars) imply raise of the external forses p,τ until the moment of collapse. The work of external forces can be represented as: $A=W+И$ where $W=\sigma V(u)$ -work of the constant vertical load, $И = \tau(u)du$- work, spent on shear deformations. When the limit value of $И_p$ at $\tau=\tau_p$, $И=И_p$ is achieved, the collapse is encountered. The limit value of $И_p$ will be proportional to the potential energy

$$И_p =\alpha\, Э(\sigma,m) \quad (7)$$

Development of shear deformations is accompanied by contraction or dilation, i.e. there is raise in Э, when vertical displacements develop downwards, or Э is diminishing if vertical displacements show deconsolidation in the shear zone. If function $Э(\sigma,m)$ is known, then one always can evaluate $И_p$ and respectively limit values of τ_p, U_p, i.e. known are graphs $\tau= \tau(u)$ and $v=v(u)$. Respectively for each value of τ it is possible to find displacement V, and for known displacement U we can find stresses τ and displacements V.

Figure 7 shows dependences between stresses and deformations analogous to Figure 4, but vertical displacements are changed for density coefficient and it's changes during shear. Experimental results and calculated values for metamorphic shales in the foundation of the Andizhan Dam are presented by Burlakov (1994).

5 NUMERICAL INTERPRETATION OF A LARGE-SCALE IN SITU TEST

To show that transition from geomechanical models to numerical ones is not a straightforward procedure let us analyse one of the well documented cases (Yufin et al 1992) where rock structure and properties were known quite well prior to numerical modeling and in addition results of in situ rock experiments were available.

During the design and some early stages of construction of the Zelenchuki hydraulic power station in Southern Russia (Northern Caucasus), several large scale tests were conducted in situ; loading rock mass with cylindrical hydraulic pressure devices 1.8m in diameter and to maximal pressures of up to 3.0 MPa, which was considered representative both in scale and in pressure range for real pressure tunnels and pressure shafts and surface structures as well.

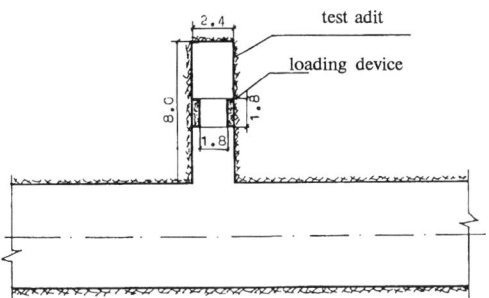

Figure 8. Set-up for a large scale pressure test.

Tests were performed in tunnel adits 8.0m long and 2.4m in diameter excavated in the walls of larger and longer tunnels. Test set-up is represented on Figure 8. The one described herein was carried out at approximately 50m distance from the surface, thus eliminating possible influence of weathering. In the middle of the adit, a concrete lining with an inner diameter of 1.8m and 1.8m long was erected and 8 axial slots ("6" on Figure 9) were left so that pressure from the loading device could be fully transmitted to the surrounding rock.

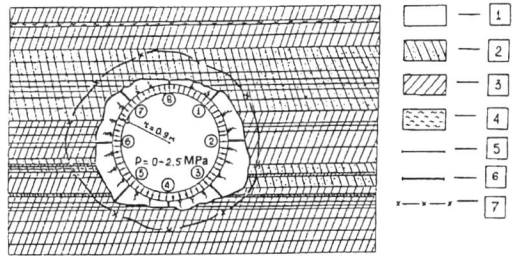

Figure 9. Rock structure and positioning of measurement points.

The pressure device itself consisted of a cylindrical rigid frame with 8 rubber cushions 1.5m long placed between the frame and the concrete lining in the test adit. Constant pressure could have been maintained in each segment of the device for quite a long period of time by the servo-controlled electric pump units.

The structure of the rock mass at the test site along with location of measuring points (inside the slots) is presented on Figure 9. Ultrasonic profiling was used to evaluate the border of the excavation-induced disturbed zone (EDZ) in the rock around the adit ("7" in Figure 9) and several measurements with Goodman jack-type devices were carried out prior to excavation of the adit.

The rock properties were evaluated with good reliability and three representative groups of rock properties were selected for further numerical analyses and design of structures:
- fine-grained sandstones with clayish basalt-type cement (E=7800 MPa, R_c=31.1 MPa, R_t=4.7 MPa, "2" in Figure 9);
- inter-layered sandstones and aleurolites with coal/mica intrusions (E=5600 MPa, R_c=19.0 MPa, R_t=2.9 MPa, "3" in Figure 9);
- thin coal layer (E=2400 MPa, R_c=9.5 Mpa, R_t=1.4 MPa, "4" in Figure 9);
Properties obtained for the concrete lining were E=10000 MPa, R_c=15.0 MPa, R_t=2.5 MPa, "1" in Figure 9.

Approximately 58% of the joints had an aperture of 0.1mm; for 20% aperture was less than 0.1mm; and for 12% aperture was more than 0.1mm. These are shown in Figure 9 with lines ("5"). The mechanical parameters of the joints were unknown and had to be evaluated during numerical modeling of experiments. These tests were conducted in the pressure range of up to 2.5 MPa in 0.5 MPa steps.

STATAS/DYNAS FEM computer code (Yufin 1979) was used for the numerical analyses. A major part of the FEM mesh is shown in Figure 10. For comparison and selection for further use, a continuous model and discrete model were analyzed. In the discrete model, "joint" or "contact" finite elements were placed between "solid" elements.

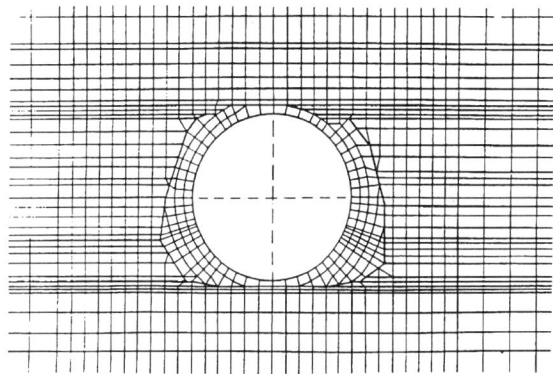

Figure 10. Central fragment of the finite element mesh.

Four sets of calculations will be discussed. In the first set the rock was considered to be linearly elastic with the above listed properties. The Mohr-Coulomb parameters for joints were gradually reduced until the best possible fit with in-situ measurements was reached at C=0.2 MPa and ϕ=14° (lines "1" on Figure 11).

In the second set the whole rock mass was considered to be continuous, transversely-isotropic, linear elastic with E_1=7000 MPa, E_2=3500 MPa, and ν_1=ν_2=0.22. The results are shown as "2" on Figure 11.

In the third set, continuum transversely-isotropic model of rock mass from the previous case was assigned elasto-plastic properties. A deformation model was selected in the form of three intersecting yield surfaces within an associated flow rule accounting for tension cut-off on horizontal planes. The results are presented as "3" on Figure 11. The last, fourth set of calculations was performed as a combination of the first and third model - with elasto-plastic rock layers and explicit modeling of joints. The results are shown as "4" on Figure 11.

The measured in situ displacements are presented in Figure 11 as "5". In horizontal directions all four numerical models produce acceptable results. This was not the case with vertical displacements, perpendicular to joints. Only the last, fourth result is close to measured values in the upper part of the adit, but none of the models give a good representation of the real rock mass behaviour in the lower part of the adit. The cause of this was present of water in the joints below the adit which could not be accounted for within a reasonable range of Mohr-Coulomb parameters in the joint models.

Of course, it was possible to achieve a perfect fit merely by back-calculating needed parameters so that the desired results were obtained, but judgement should prevail over "simplicity" of such a solution, and only physically reasonable parameters should be used in the analysis -hence the necessity of the knowledge-based system in the scope of general considerations represented on Figure 2. Apparently, for the case described, coupled rock-flow analysis (which is being recently developed inside the shell of the STATAS/DYNAS code) is needed and the whole example shows that numerical representation of a rock mass is not a direct procedure even in cases where the rock is investigated very carefully.

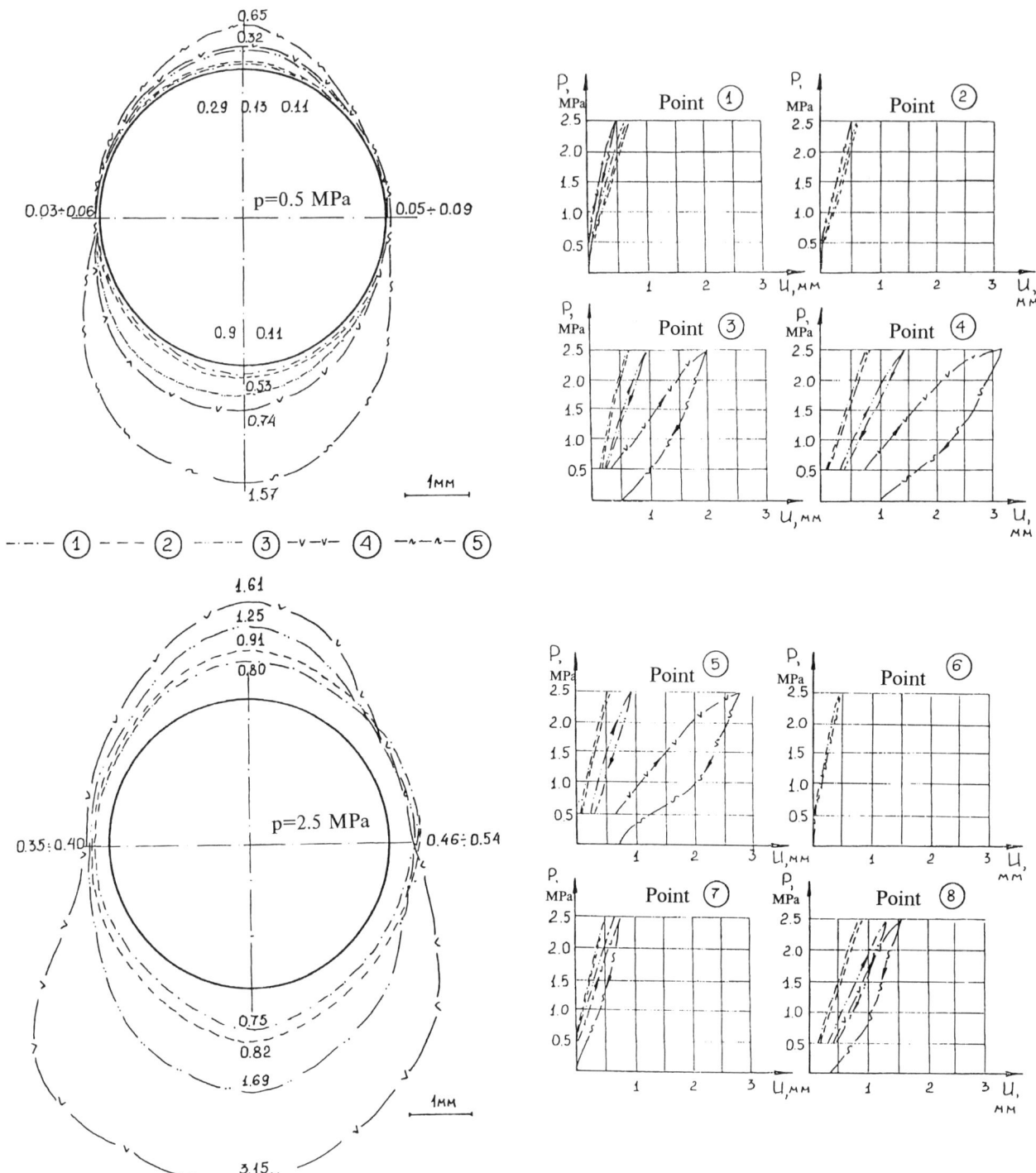

Figure 11. Comparative results of four sets of numerical modeling and in situ measurements. left - contour displacements for 0.5 and 2.5 MPa loads; right - load-unload curves for different measurement points.

6 REFERENCES

Barton, N., Yufin, S.A., Swoboda, G. 1994. Engineering decisions in rock mechanics based on numerical modeling. Computer Methods and Advances in Geomechanics. Siriwardane & Zaman (eds.). Rotterdam: Balkema, p.2221-2228.

Burlakov, V.N. 1994. Development of the energy theory of strength. Paper presented for the International Symposium on New Development in Rock Mechanics and Engineering. Shenyang, P.R.China: Northeastern University.

Fishman, Yu.A. & Miroshnikova, L.S. 1984. Experience in development and application of engineering-geological models in hydraulic engineering. Engineering Geology. Moscow: USSR Academy of Sciences, v.5, p.24-37 (in Russian).

Gysel, M. 1987. Design methods for structures in swelling rock. Proc. 6th ISRM Congress, Montreal. Rotterdam: Balkema, p.377-381.

Melnikov, N.N., Konukhin, V.P., Kozyrev, A.A., Yufin, S.A., Barton, N., By, T.L. 1995. Geomechanical aspects of radioactive waste disposal in deep formations of hard crystalline rocks. Proc. 8th ISRM Congress, Tokyo. Rotterdam: Balkema.

Tsytovich, N.A., Ukhov, S.B., Burlakov, V.N. 1970. Failure mechanism of fissured rock based upon displacement of loading plate. Proc. 2nd ISRM Congress, Beograd. Rotterdam: Balkema.

Ukhov, S.B. 1975. Rock foundations of hydraulic structures. Moscow: Energy publishers, 264p. (in Russian).
Ukhov, S.B. 1979. Principles used in developing geomechanical models of rock masses for solving engineering problems. Proc. 4th ISRM Congress, Montreux. Rotterdam: Balkema, p.691-696.
Yufin, S.A. 1979. Numerical analysis of rock structures considering material nonlinearities. Proc. 20th US Symposium on Rock Mechanics, Austin, TX., p.265-272.
Yufin, S.A., Alipova, G.S. & Zelensky, V.D. 1992. Engineering-geological substantiation of pressure tunnels design. Engineering Geology. Moscow: Russian Academy of Sciences. v.4, p.87-95 (in Russian).

Stability and deformation analysis of complex rock foundations of several large dams and hydropower stations in China

Ge Xiurun, Feng Dingxiang, Gu Xianrong & Feng Shuren
Institute of Rock and Soil Mechanics, Chinese Academy of Sciences, Wuhan, People's Republic of China

ABSTRACT: On the basis of several examples of engineering projects of large dams and hydropower stations constructed on complex rock foundations, problems of sliding stability and deformation are analysed. A new method for safety factor against sliding is proposed. At the same time a new technique for evaluating initial geostresses through the in situ measured data of displacement is presented as well.

1 INTRODUCTION

The research on geoengineering problems in the construction of dams and hydropower stations is closely related with their rational design and safe operation. Projects of rock foundations in hydraulic structures involve a wide range of problems such as seepage, rock mass quality appraisal, reinforcement and improvement of rock mass. However, among them, stability and deformation are of great importance. As a rule, for the structure built on the rock foundation with weak intercalations, the problem of stability against sliding is of predominant significance, and forms a great concern for the designers. The serious interlayer dislocations and deformation in rock mass are discovered in a number of cases of power station projects in China when deep foundation pits are excavated in layered rock mass with weak intercalations. How to analyse the cause of this deformation, how to evaluate and predict the possible order of magnitude is also an important subject to be studied.

On the basis of several large dam and hydropower station projects in China, which have been completed or are under construction, we emphatically analyse the problem of stability and deformation, that is to say, the stability against sliding of dams built on the rock mass having weak intercalations, and the deformation of foundation pits in power stations. And the examples of analysis and calculation are also given. In this paper a "vectorial sum" method for safety factor against sliding is presented, which forms a new method due to its remarked difference from conventionally used methods. In this paper is also given a mechanical model and a calculation method for interlayer dislocations in the excavation in layered rock mass with weak intercalations. This method was used for the first time in the Gezhouba power station project, Yangtze River, China. The initial geostresses derived by means of this method are in good agreement with in situ measured data subsequently obtained. In recent years it found another successful application in Tianshengqiao hydropower project in China.

2 ANALYSIS OF STABILITY AGAINST SLIDING IN DEEP ROCK FOUNDATION AND APPRAISAL OF SLIDING RESISTANCE OF ROCK MASS

Here the in situ tests, numerical simulation and appraisal methods are given for the stability against sliding along the deep mudded intercalations and the sliding resistance of rock mass behind the dam in the Second Channel Sluice, Gezhouba project, Yangtze River, China.

The site of the project is situated on the lower mouth of the Three Gorges. It has the largest installed capacity among the hydropower stations in China. The Second Channel Sluice is the main discharge structure, which contains a 27-bay spillway. The sluicegate chamber has a dimension of $65 \times 18.5 \times 40 m^3$. The largest hydrolic head is 27m. The water thrust acting on the gate is 1000 tones per meter. The rock foundation is formed by clayey siltstone. The rock layers lie nearly in horizontal direction with a very small dip angle of 4-8°. Among rock layers there are many mudded interculations. The depth of the main intercalation is about 10m below the chamber floor. The shear strength on the layer surfaces is low, therefore the stability against sliding along deep mudded intercalations becomes one of the main problems for this project. How to estimate the initial sliding resistance of rock mass behind the dam constitutes the key for solving this problem.

The physico-mechanical parameters for the bed rock and the mudded intercalations are obtained from in situ and laboratory tests. The deformation modulus for the clayey siltstone is 10-16 GPa, the wet compression strength is 8-10MPa; the shear strength is $\varphi = 26.5°, c = 0.05MPa$; the shear strength along the intercalations is $\varphi = 24.2°, c = 0.03MPa$.

2.1 Rigid body limit equilibrium analysis with consideration of resistance of rock mass behind the dam

According to the design standard, two sliding safety factors can be adopted:

$$K = \frac{f \Sigma W}{\Sigma p} \quad ; \quad K' = \frac{f' \Sigma W + c' A}{\Sigma P}$$

where ΣW is the total normal force; ΣP is the total tangential force, f shear friction coefficient; f' friction coefficient against shear failure, c' cohesion, A the section area of the dam foundation. In the above expressions no account is taken of the sliding resistance of the downstream rock foundation. Because of the considerable depth of the intercalations in the project, the rational results can be obtained only by modification of the above methods and consideration of this portion of sliding

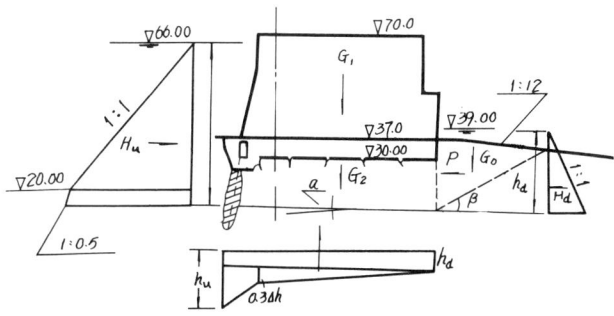

Figure 1. Loading.

resistance in the limit equilibrium analysis. Otherwise, the safety factor would become smaller at larger depth of intercalations, which is apparently irrational.

2.2 Formula for safety factor against sliding

Fig 1 gives a presentation for calculating loads. In consideration of sliding resistance of rock mass when the slide of rock mass downstream from the gate chamber takes place, the limit equilibrium formula is as follows:

$$K = \frac{f(\Sigma V\cos\alpha - \Sigma H\sin\alpha + P\sin\alpha) + P\cos\alpha}{\Sigma H\cos\alpha + \Sigma V\sin\alpha} \quad (1)$$

where f is friction coefficient, ΣV is the vertical load above the intercalation, $\Sigma V = G_1 + G_2 - W$; ΣH is the total horizontal load above the intercalation, P is the rock mass resistance; α is the dip angle of the intercalation. The sliding stability has a standard: $K \geq 1.30$.

2.3 Evaluation of rock mass resistance P downstream from the chamber

The rationality of calculation of P strongly affects the results of stability analysis. Here it is assumed that the resistance P is in the horizontal direction and can be calculated through wedge limit equilibrium. According to geological conditions the most unfavorable fracture plane is taken as the first damage plane, and the related c and φ is chosen. The downstream vertical plane in close vicinity of the gate is taken as the second damage plane. In order that a certain safety margin can be allowed in resistance calculation, the value of c for the first fracture plane is taken as zero, the resistance to shearing in the second fracture plane is negnected. Thus,

$$P = G_0 tg(\beta + \varphi) \quad (2)$$

where P is the rock mass resistance; G_0 the total vertical force on resistance block, β the angle of the first fracture plane to the horizontal direction, φ the friction angle of the first fracture plane ($\varphi = 19.3°$).

2.4 Large scale in situ tests for rock resistance

The resistant force of the rock mass downstream from the chamber is essential for its sliding stability. For rationally evaluating the bearing capacity of the resistant rock mass behind the sluicegate, Yantze Valley Planning Office has carried out large scale in situ tests at the sites. There are 2 specimens. The specimen I has a dimension of $11.5 \times 1.7 \times 2.35 m^3$, the specimen II has a size of $9.5 \times 1.7 \times 2.35 m^3$. The uniform load acting on the top of specimen I is 0.05MPa. No external load is applied on the specimen II.

Fig 2 indicates the specimen I. Fig 3 indicates the in situ stress-displacement curves of the two resistance blocks. The test damage strength of blocks I and II are 1.53 and 1.23MPa, respectively. The proportion limit is 0.4—0.5MPa, the yield strength 0.8—0.9MPa. In the limit equilibrium analysis the mean value of P is 0.3—0.36MPa, which is the third of the in situ yield value. Therefore the value of P used in the above analysis includes a margin and is on the safe side.

2.5 Numerical simulation and elastic-brittle-plastic non-linear analysis on the resistance block tests

The test process of resistance block I and II are simulated numerically. In Fig 4 is shown the meshing in the numerical simulation. The size is the same as that of in situ tests. The mechanical parameters are taken from the in situ tests. A main mudded surface called 202_4 ($c = 0.005MPa$, $tg\varphi = 0.2$) exists between the test block and the bed rock. There are also 3 horizontal layer surfaces with considerable shear strength: $c = 0.03MPa$, $tg\varphi = 0.88$. Joint elements are used for simulating these weak surfaces, with the elasto-plastic properties and non-tension considered. The elastic-brittle-plastic model is adopted for rock elements (Fig 5a). The behaviour is elastic before the peak strength is reached. The stress drop occurs when the peak value is passed. The characteristics of brittle damage are simulated by stress transmission. When stresses fall to the residual srrength, they are simulated by means of elasto-plastic properties. When shear damage, that is to say, the brittle damage takes place and stresses drop abruptly, the first invariant of stress tensor for the element concerned is assumed constant. This means that volumetric deformation keeps unchanged at the stress drop. The original stress circle is reduced to a concentric circle tangential to the residual strength envelope (Fig 5b). Through evaluating the difference of these two circles can be obtained the values of released stresses, which then are conversed into related nodal forces and readjustment is performed. The above-mentioned is the basic idea we use in treatment of the brittle properties.

Fig 6 gives the relation between the horizontal thrust and the area of damage zones of specimens I and II, which are obtained by the numerical simulation. It can be

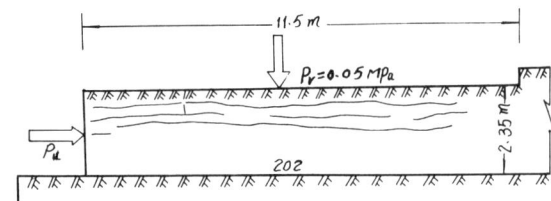

Figure 2. Resistant block test.

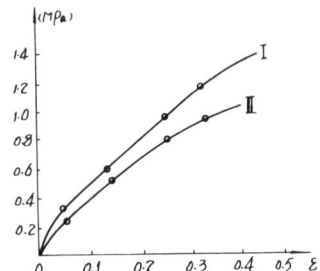

Figure 3. Stress-strain curves of resistant block.

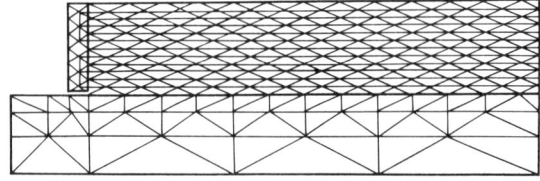

Figure 4. Meshing of resistant block.

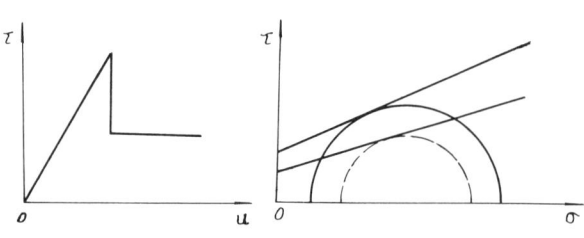

Figure 5. (a)τ~u curve.　　(b)τ~σ relation.

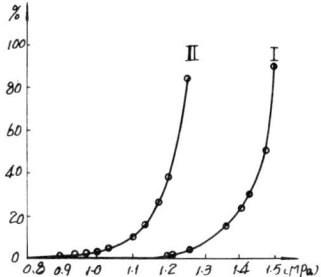

Figure 6. Curve of damage area VS. horizontal thrust.

Table 1. Comparison of test data to calculation results.

Resistance block		I	II
Vertical load (MPa)		0.05	0
P (MPa) at the begining of the damage of main mudded surface	in situ test	0.208	0.0532
	calculation	0.1	0.067
P (MPa) when 202_4 becomes a complete fracture surface	in situ test	0.406	0.399
	calculation	0.5	0.4
P (MPa) when damage of blocks takes place	in situ test	1.534	1.24
	calculation	1.48	1.26

seen that there is a good agreement. Test block II taken as an example, the damage develops accelerately (Fig 6) when the horizontal trust reaches 1.1 MPa. This basically agrees with the characteric thrust P=1.064MPa recorded in situ, when the serious damage occurs.

Table 1 gives the comparison of test data of the main mudded surface and the whole resistance block to the related calculation results. Obviously, they are in good agreement.

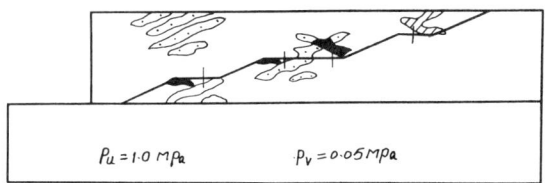

Figure 7. Damage of the resistant block with inclined fractures.

When the horizontal thrust on a unit area reached 0.467MPa a sliding surface of zigzag shape composed of inclined fracture surfaces and layer surfaces is formed. When the thrust rises to 0.567MPa, the opening of the sliding surface occurs. The damage of rock blocks begins at 0.7MPa of the thrust. The damage zones in the resistance block containing zigzag fracture planes at 1.1MPa of the horizontal thrust per a unit area is shown in Fig 7.

It is obvious that the value of P used in limit equilibrium is far below the capacity of the rock mass behind the gate even if declined joint surfaces exist.

2.6 FEM analysis on stability against sliding in the large depth in the Second Channel Sluice

The elastic-brittle-plastic non-linear analysis is the same as the above-mentioned. The plane problem is considered. Numerous schemes of reinforcement such as deep cutoff walls, seepage prevention slabs have been calculated. The length of calculation range in the upstream direction is twice the floor length, while the range arrived downstream at the anti-scour wall. The depth of the range equals the floor length. For the sake of space only the FEM results of the scheme of the cutoff wall is given (Fig 8). The so-called deep cutoff wall is made of concrete at the front of the chamber and goes past the mudded intercalations. From the stress distribution in the mudded intercalations it can be seen that within the section of 20m which is behind the chamber the resistance of the mudded intercalation is less than the sliding force, therefore it is a yield section, though the safety factor is about 2.0. According to the FEM results the resistance force of the rock mass behind the chamber is the third of the total resistance, therefore it is important in the analysis.

2.7 Foundation treatment

Because the resistant rock mass of the bays 1-6 in the Second Channel Sluice is weak and have inclined cracks, the safety degree for these bays is lower compared with others'. On the basis of the FEM results comprehensive measures of reinforcement and improvement including piles are taken, which are shown in Fig 9 (Xu et al., 1983). Observed data show that after the impoundment in the Gezhouba project the horizontal displacement of the gate chamber is 1-3mm, the vertical settlement 3-5mm, the relative dislocation of the mudded intercalation nearly zero. All of them is less than the related design values at the same level by 25-50%.

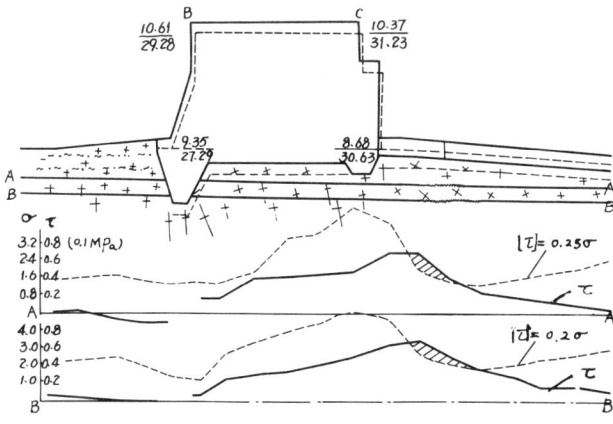

Figure 8. FEM results for the deep cutoff wall.

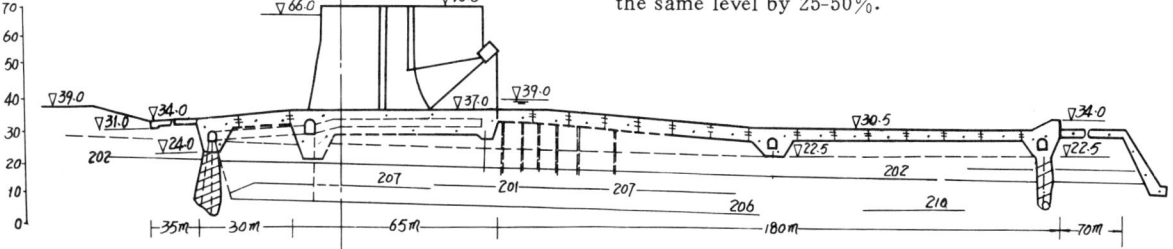

Figure 9. Treatment of sluicegate foundation.

3 MODEL TEST AND NUMERICAL SIMULATION FOR SLIDING STABILITY OF A GRAVITY DAM

Through loading tests and numerical simulation on the model dam the study is to be made on the process of the damage, which takes place at the sliding plane composed of two weak intercalations in the deep portion of the foundation. The verification of the numerical analysis is also to be peformed (X. R. Gu et. al, 1987).

3.1 Model test on sliding stability of the gravity dam

The body and foundation of the model dam are made of gupsum. The height of the model dam is 0.6m, and thickness 0.1m. In the rock foundation there are 2 weak intercalations, which exert a predominant effect on sliding stability. The physico-mechanical parameters are given in Table 2.

Table 2. The physico-mechanical parameters.

	E (MPa)	V	Compression strength (MPa)	tgφ	C (MPa)	γ (10^3N/m^3)
Rock	1650	0.2	1.9	0.87	0.4	6
weak inter.	10	0.2		0.317	0	

In the test, first of all, a vertical pressure of 4000N is applied to simulate self-weight. Then the horizontal loads are exerted in a step-wise way (Table 3) until failure takes place.

Table 3. Horizontal loading steps with standard thrust $P_0 = 1323N$.

step	1	2	3	4	5	6	7	8	9
p/p_0	0.25	0.5	0.75	1.0	1.25	1.5	1.75	2.0	2.25

The test indicates that slide begins at the overload of 1.25 and the final loss of stability takes place at the overload of 2.25.

3.2 Numerical simulation of the model dam test

The non-linear behaviours of joints which are used for simulation of weak intercalation of model dam foundation are shown in Fig 10.

The mesh of deformation corresponding to the 9th loading step ($P/P_0 = 2.25$) is shown in Fig 11.

Relative displacement between the corresponding points on the top and bottom of weak intercalations are related to the increasing horizontal loading steps as shown in Fig 12 and Fig 13. Fig 13 indicates that sliding plane A opens gradually with loading increasing. As for plane B, the situation is complicated: opeing of the edge section increases as load increases, while the other section remains in compaction.

3.3 Progressive failure of sliding surface in rock foundation

The progressive failure of sliding surface is shown in Fig

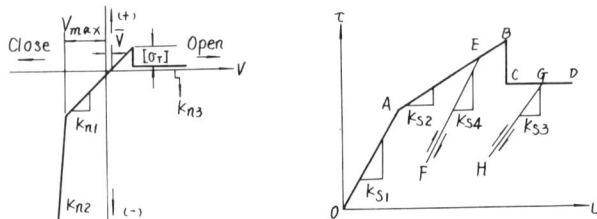

Figure 10. The mechanical non-linear properties of weakplanes (a) Normal Deformation model;
(b) Shear displacement model;

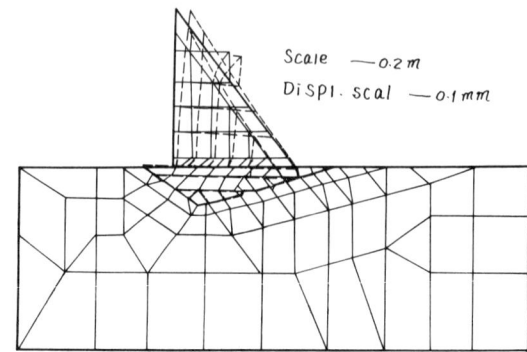

Figure 11. Mesh deformation of the model dam.

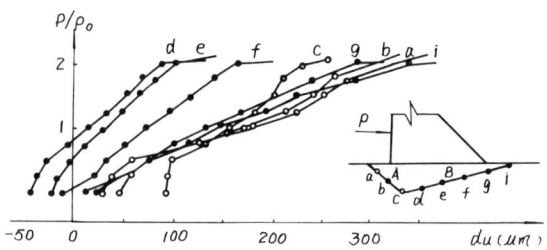

Figure 12. Tangential displacement of intercalation.

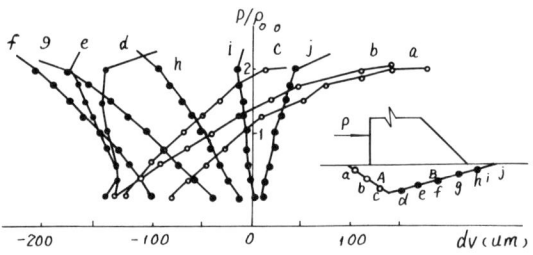

Figure 13. Normal displacement of intercalation.

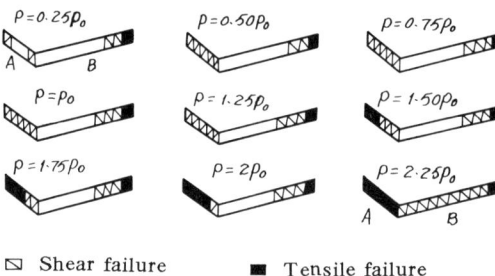

Figure 14. Progressive failure process.

14. It coincides very well with the model testing data, indicating the complete loss of sliding stability of model dam foundation at $P = 2.25P_0$.

4 A NEW METHOD OF CALCULATING THE SAFETY FACTOR FOR STABILITY AGAINST SLIDING

4.1 Conventional method based on the principle of "algebraic sum"

When the distribution of the normal and tangential stresses σ and τ on the potential sliding surfuce are obtained in some way, say, in FEM analysis, it is needed to evaluate the safety factor against sliding. Usually the following formula is adopted:

$$K = \frac{\sum_{i=1}^{n}(\sigma_i f_i \Delta L_i + C_i \Delta L_i)}{\sum_{i=1}^{n} \tau_i \Delta L_i} \quad (3)$$

where

σ_i, τ_i normal and tangential stress of element i on the sliding plane;
f_i, c_i friction coefficient and cohesion for material of element i;
ΔL_i length of element i along sliding plane;
n total number of elements on sliding plane.

However, the conception of the formula based on the principle of "algebraic sum" seems to be questionable.

As is well known, only if the potential sliding surface is a portion of circular arc, then the numerator of the expression can be regarded as the sum of moments of resistance forces with respect to the center of the arc, and the denominator as the sum of moments of sliding forces with respect to the same center. With the radius eliminated in the expression by reduction, the numerator and the denominator become the algebraic sums of resistance forces and sliding forces on the sections of the potential sliding surface. It is apparent that the expression (3) has clear-cut physical meaning only when the sliding surface is a circular arc, or nearly a circular arc, or a straight line segment.

4.2 A new conception of calculating the safety factor for stability against sliding

The new conception presented by the first author of this paper is based on the basic of the following considerations.

1. It is known from the practice that in many occasions the potential sliding surfaces in the rock foundations and slopes can be simplified, as a rule, into multiple line segments.
2. The different sections on the potential sliding surface differ in their role. Each surface usually are divided into two basic portions: sliding-resistent portion and sliding portion.
3. All the forces acting on the potential sliding surface are vectors. It is right to make a comparison after the resistance and sliding forces are superposed respectively, but this superposition should be vectorical, i. e., the vectorial sum rather than the algebraic one is to be used.
4. This superposition should be made with respect to a base plane. It is logical to select the main resistane portion on the sliding surface as the base plane.

4.3 New method for calculating K on the basis of vectorial sum

The vectorial sum can finally lead to the following formula the derivation of which is omitted for the sake of space:

$$K = \frac{\sum_{i=1}^{n}\sigma_i \Delta L_i \sin\alpha_i + \sum_{i=1}^{n}\Delta\sigma_i \Delta L_i f_i \cos\alpha_i + \sum_{i=1}^{n} c_i \Delta L_i \cos\alpha_i}{\sum_{i=1}^{n}\sigma_i \Delta L_i \sin\alpha_i + \sum_{i=1}^{n}\tau_i \Delta L_i \cos\alpha_i} \quad (4)$$

where α_i is the angle of the main sliding-resistent plane selected to the sliding section i. When the potential sliding direction is in quadrant I or IV (Fig 15a) the α_i in clockwise sense is negative. When this direction is in quadrant II or III (Fig 15b) the α_i in counterclockwise sense is negative. The compressive σ_i is positive, the sliding resistent τ_i is positive, as shown in Fig a, b.

4.4 Application I

The safety factor and the horizontal thrust are evaluated with the two methods, respectively, in the above-mentioned example of gravity dam. The relation curves are shown in Fig 16. Generally, the values of K obtained through the "vectorial sum" method are less than that through the "algebraic sum" method. But they tend to become identical at the limit state. Values of K from the "vectorial sum" have a nearly linear relation with the horizontal thrust. This may reflect the variation tendancy more realistically and makes it possible to obtain the maximal overload at K = 1 through calculation of several values of overload.

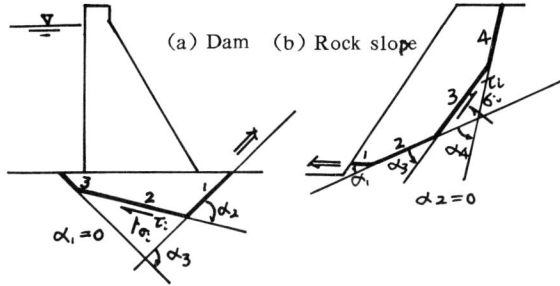

Figure 15. Potential sliding surface

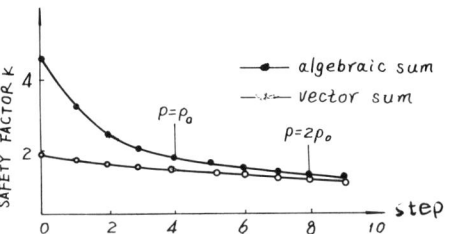

Figure 16. Safety factors against sliding.

4.5 Application II

Ankang dam is a gravity dam on the Han River, the largest tributary of Yangtze River. It has very complicated rock foundation which contain 5 fault belts and 6 weak intercalations. The distribution of principal stresses obtained from plane FEM non-linear analysis is shown in Fig 17.

The safety factors for selected types of possible glide tracks are listed in Table 4.

Table 4. Safety factors against sliding of Ankang Gravity Dam.

Method	Composition of sliding planes		
	$F_2-J_1-F_3$	$F_2-J_1-F_4$	$F_2-J_1-J_2$
Conventional method (Algebraic sum)	3.26	2.45	1.99
New Method (Vector sum)	2.00	1.85	1.70

It can be found that the dam is in a stable operation condition under normal loads. The discontinuity set of $F_2-J_1-J_2$ forms the dangerest potentical sliding planes for the dam rock foundation.

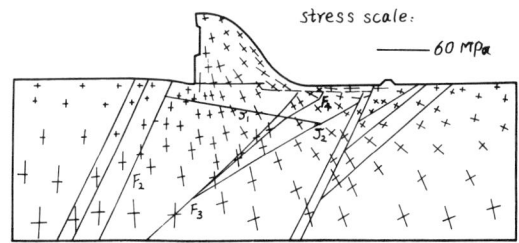

Figure 17. Principal stress distribution.

5 DISPLACEMENT ANALYSIS OF THE ROCK FOUNDATION PIT FOR THE GEZHOUBA HYDROPOWER PLANT №2

After the foundation pit of the Gezhouba hydropower plant №2 was excavated to its initial configuration remakable horizontal dislocations of the residual wall of presplit holes and the large bore holes were discovered along the weak intercalation and the maximum dislocation along the weak intercalation № 212 is about 80mm (Ge et. al. 1982). Fig 18 shows the sliding and displacement in a cross section of the foundation pit.

The region is composed of monocline in structure. The rock strata belong to the Lower Cretaceous series strata and have strikes of N30°E, dips of SE, and dip angles of 6°—8°. The rocks are chiefly argillaceous siltstones with little sandstones and ten more layers of weak intercalations. The relief of the initial geostresses due to excavation is the main factors which caused the large deformation of the foundation pit.

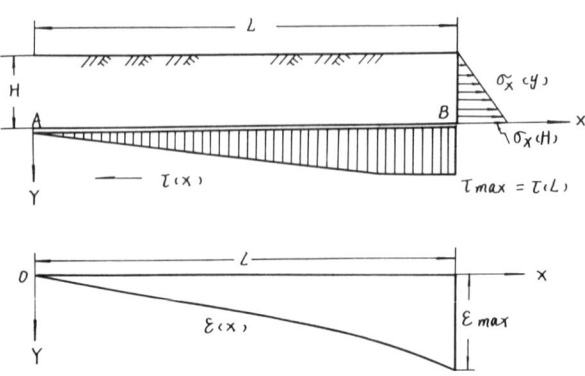

Figure 19. Distribution of $\varepsilon(x)$.

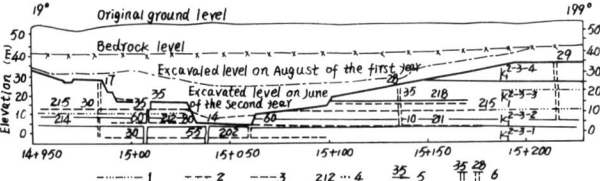

Figure 18. Constructed profile showing dislocations of the rock masses along the weak intercalations
1. Argillaccous siltstone; 2. Psephitic claystone; 3. Claystone; 4. Notation for weak intercalations; 5. Direction of rock-mass dislocation and displacements; in mm; 6. Notation for C 1000mm bore boles.

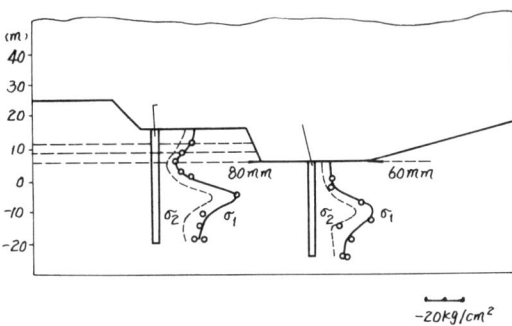

Figure 20. Distribution of geostresses σ_1 and σ_2.

5.1 The method for estimating the initial geostress based on the displacements of foundation pit observed in situ

First, on the basis of characteristics of horizontal layered rock mass a mechanical model is established for theoretical analysis, as shown in Fig 19a. The distribution of the longitudinal strain ε is given in Fig 19b.

According to the model we have

$$\int_0^H \sigma_x(y)dy = \int_0^L \tau(x)dx \quad (5)$$

where L and H are the length and depth of stress disturbance zone respectively.

Assume that

$$\varepsilon(x) = \varepsilon_{max}\left(\frac{x}{L}\right)^2 \quad (6)$$

$$\tau(x) = \tau_{max}\left[1 - \left(\frac{L-x}{L}\right)^2\right] \quad (7)$$

and ΔL indicates the sliding distance of rock mass along the weak intercalation. Then we have

$$\Delta L = \int_0^L \varepsilon(x)dx = \frac{2}{3}\varepsilon_{max} \cdot L \quad (8)$$

$$\varepsilon_{max} = \frac{\bar{\sigma}_x(y)}{E} = \frac{1}{E}\frac{\int_0^H \sigma_x(y)dy}{H} \quad (9)$$

If $\sigma_x(y)$ distributes linearly and the cohesion C is negnected, then

$$\sigma_x(H) = \sqrt{8\Delta L E \gamma tg\varphi} \quad (10)$$

$$L = \frac{6E\Delta L}{\sigma_x(H)} \quad (11)$$

According to in situ tests and measured data, E, ΔL tgφ and γ are 1660MPa, 0.08m, 0.3 and 0.24MN/m³. From (6) and (7) we have $\sigma_x(H) = 2.77$MPa, the effect range of dislocation L=288M.

5.2 The stress measurment in situ

In order to check and demostrate the direction and amount of the residual geostress of rock the stress measurement has been done in situ and three boreholes have been especially set up. By means of the overcoring technique for every borehole from the ground surface to the depth of 40m the stress measurement was carried out at 8—10 point. The data are shown in Fig 20.

The in situ measured data are $\sigma_1 = -3.1$MPa and $\sigma_2 = -2.3$MPa, which are in good agreement with the calculation value. In Tianshengqiao hydropower project the in situ measured data are $\sigma_1 = 13.9$MPa, $\sigma_2 = 9.8$MPa, $\sigma_3 = 1.6$MPa, while the calcuted $\sigma_1 = 14$MPa. These two examples indicate that the above analysis method has a universal significance.

REFERENCES

Xu, L. X. et. al. 1983. Sliding stability of foundation rock with shear zones. Proceedings of 5th Int. Congress on Rock Mechanics. Melbourne, Vol. 1. p205—208.

Gu, X. R., Li, J. & Ge, X. R. 1987. Analysis of stability against sliding of a gravity dam by means of microcomputer. Proceedings of 2nd EPMESC, Guangzhou, Vol 3, p474—477.

Ge, X. R., Feng, D. X. & Yang, J. L., 1982. The elasto-visco-plastic analysis for rock displacement of the foundation pit of a water power plant. Rock Mechanics, Vol 15, p145—161.

Interaction between dam structure and rock foundation: A case-history and analysis

Sijing Wang
Institute of Geology, Academia Sinica, Beijing, People's Republic of China

ABSTRACT: In this paper a theoretical topic of interaction between dam structure and rock foundation is discussed. A case-history is given on the rock deformation of rock abutment in Meishan reservoir. The analysis shows that an appropriate coordination of the structure-foundation system should be considered in the optimal design and construction.

1 INTRODUCTION

In an adequate design of large scale building on or/and in rock, not only the structure behavior but also the properties of rock foundation should be taken into consideration. However, some of the accidents occurred in many engineering projects are caused not fully due to misunderstanding the geologic structure but due to lack of knowledge on the interaction between the engineering structure and the rock mass structure, and due to mistakes in prediction of the realistic working state of rock mass after its coordination with the action of the engineering structure (Gu Dezhen and Wang Sijing, 1982 a,b). Moreover, the recent study has shown that the structure-foundation interaction may be considered as a part of human-nature relationship which is a major factor of rational planning of large scale engineering projects, and it dictates the process of environmental change, hazard occurrence and success of environmental protection, as well as of engineering planning (Wang Sijing, 1987; Wang Sijing,1990; Wang Sijing 1992). In this paper, a case-history of rock deformation and dam cracking in Meishan reservoir is introduced and analysed in detail, and some experiences and principles in rock envineering have been derived from it, taking the structure-foundastion interaction and the instability process of the system into account.

2. GEOLOGY OF THE DAM SITE

The Meishan reservoir is located in Anhui Province, China, on the Shihe river. It is a reservoir for irrigation and flood control. The main structure is a concrete multiple-arch dam of 88.2m high. The dam is composed of 15 buttresses and 16 arches with dam crest of 443 m long. Two gravity blocks connect the dam with rock abutments on both banks(Fig.1). The buttresses are hollow with a void space between two walls. Arches are constructed on the upper stream face of buttresses. The basements of arches are enlarged into platform. The rock deformation happened at the right abutment of the dam in the section from 14 th to 16 th arch. The construction of the dam was completed in 1956, and water impounding started in 1958. The deformation of rock abutment occurred in two months later after the highest level of water in reservoir was reached in 1962.

The bank slope of dam abutment is quite steep, reaching 45-60. On the front of upper stream of dam a lateral gully deeply

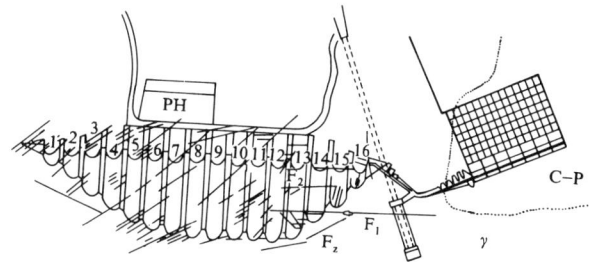

Figure1. Foundation Geolgy of Meishan Dam
C-P Schists;   Granite; F.Fault; PH. Power House

cut in the rock mass forming a cliff of 20 m high below the basement of dam.

The rock abutment is composed of granite which spreads in east-west orientation with length of 7 km and width of 1 km. The dam site is located near the eastern border of the granite massive, where granite is in contact with metamorphic serires of Carboniferous-Permian period. This is the reason of occurrence of extensive subhorizontal joints in the rock abutments. The main minerals composing the pink-reddish granite of dam foundation are fieldspar, quartz with mica and few magnitite. The fresh rock is hard and compact. The uniaxial strength of saturated granite reaches 100 Mpa and higher. The thickness of weathered rock is about 1-5 m with color changing into yellowish and increased jointing.

In the granite massive two sets of fractures striking in north-north-east and east-west orientation are dominant (Fig. 2). In the right abutment some faults and shear zones, such as F1 F2 F74, exist. The rock mass structure is characterized by a joint system comprising two sets of steep joints and one set of flat joints.

Until the rock deformation of the abutment happened the dam site was evaluated as being of good geological conditions, and safety of the dam was beyond any problem. The reasons considered are as follows: (1) bedrock was fresh and competent enough; (2) the faults and shear zone were treated by deep excavation and concrete replacement; (3) the drainage effect due to the void in the hollow buttresses between walls and the void in arches between buttress would be sufficient to reduce the uplift and seepage pressure in foundation; (4) the grouting screen was made along the axis of arch foundation.

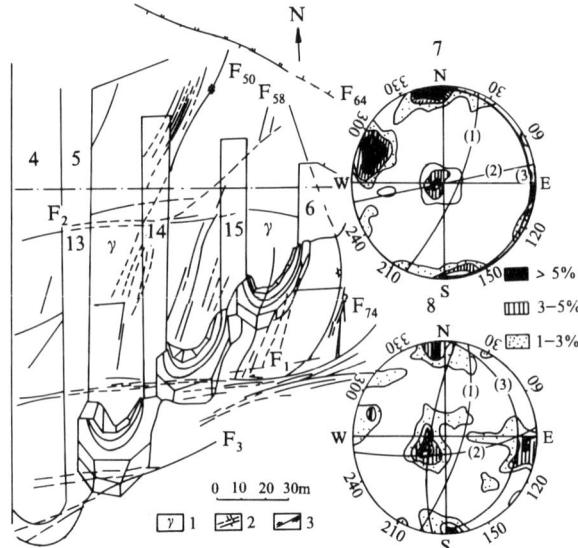
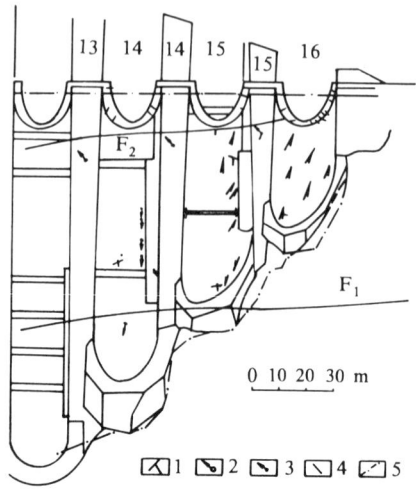

Figure2. Fracture Pattren of Granite Mass
1.Granite; 2.Joint; 3.Fault; 4.Arch; 5.Buttress; 6.Gravity Dam; 7.Joint System in Upper Stream (545 Joints); 8.Joint System in Downstream (511 Joints)

Figure3. Sliding Deformation of Right Abutment
1.Opening of Joints, Seepage; 2.Increase of Uplift; 3.Tilting of Lam Buttress; 4.Cracks; 5.Rock Cracks

3. ROCK DEFORMATION OF THE DAM ABUTMENT

The deformation of rock mass in the right abutment took place after sudden increase seepage flow on 6th November, 1962 and the development of rock deformation and damage of dam was fast within a few days. The progressive deformation was stopped by empting the reservoir urgently.

3.1 Seepage Flow in Dam Foundation

A large number of springs with seepage water flowing out from the dam foundation rock in the void space between buttresses were observed prior the extensive deformation. In general the seepage happened along joints and shear zones (Fig. 3). Locally it took place along the contact between rock and concrete. The total amount of seepage flow reached 70 l/s. The seepage pressure measured in a leaking borehole used for consolidation grouting reached 0.31 MPa.

3.2 Opening and Sliding along Rock Joints

The obvious joint opening and rock cracking were observed in the foundation rock of dam abutment (Fig. 4). rock sliding happened along subhorizontal joints with dipping angle less than 10. The opening of steep joints reached 17-20 mm. The fractures striking in east-west direction, such as F1, were open in the upper stream of dam axis and closed in the down stream, as this was shown by fault F2. The cracks and opening joints were of stepwise form in foundation rock. Locally the cracks propagated along rock-concrete contact and in concrete.

3.3 Deformation of Dam Structure

According the monitored results large displacements of buttresses 12-14 were observed. The maximum values for buttress 13 reached 42.04 mm down slope and 14.35 mm downstream. After water releasing from reservoir the buttresses moved reversely, and residual displacement

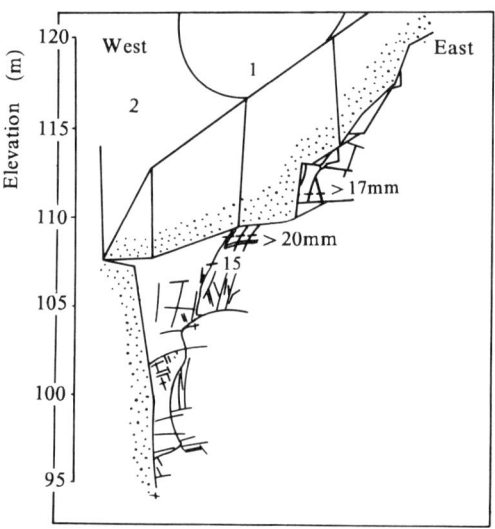

Figure4. Cracking in Rock Abutment
1.Arch; 2.Buttress

occupied 80% of the total displacement. The observation also showed an uplift of buttresses with tilting towards downstream and down slope. The displacement caused torsion deformation of the arches (Fig.5).

3.4 Cracking of Dam Structure

On 8th-13th, November 1962, follwing large amount of seepage some cracks appeared on the top of the dam with opening of working joints. 12 cracks were observed on the arches. The longest one observed on the arch No.15 reached 26 m in length. The basement platform of arches also suffered by cracking.

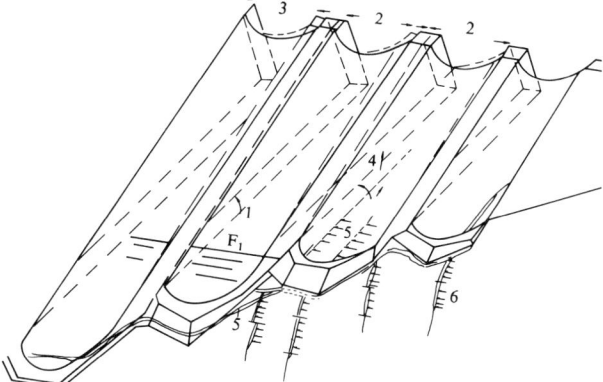

Figure 5. Defomation of Multiple Arch Dam
1.Torsion; 2.Tension; 3.Compression; 4.Uplift; 5.Cracks; 6.Seepage Pressure

3.5 Extent of Structure-founation Deformation

The deformation was limited by faults F1 on the upper stream and F2 on the downstream, and the arches No.12-No.16 were be involved in the deformation. The depth of deformation measured by water injection tests reached 15 m in rock mass (Fig. 6).

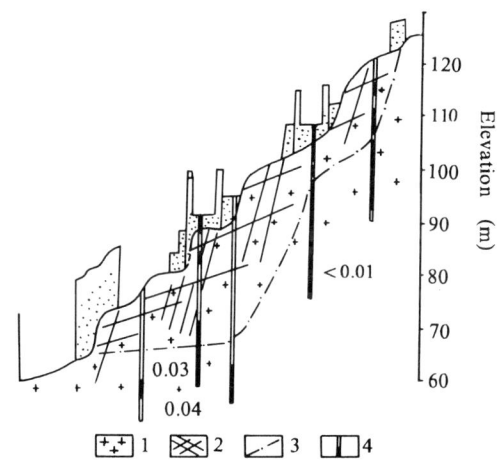

Figure 6. Depth of Sliding Deformation
1.Granite; 2.Jonits; 3.Failure Zone; 4.Borehole, unit water absorption (blank > 0.01/min, black < 0.01/min)

4. ANALYSIS OF DEFORMATION-MECHANISM AND PROCESS

4.1 Basic Conditions of the Deformation

The rock structure with 3 sets of joints comprised the basic condition of deformation. Meanwhile, the type of multiple arch dam, which is fragile in resistance against lateral loading and rock movement, was unfavorable condition for stability.

4.2 Formation of Seepage Pressure

The deformation and cracking of rock abutment were proceeded following the formation of high seepage pressure in rock mass. The mechnaism of forming seepage pressure can be explained by the theory of structure-foundation interaction.

In general the joints striking in north-north-east orientation could lead water from the reservoir into rock foundation. However, in an excavated to fresh rock foundation, which then isolated by grouting screen the joints are unlikely to play such role. The situation may change and becomes quite different, when the dam buttresses are constructed on the slope as an inclined foundation. In this case, two zones of different stress state may be created in the foundation. The zone of tensile stress is located inside the foundation, while the compression zone is located on the side near the surface of slope foundation (Fig.7). Under the action of tensile stress the steep joints striking in north-north-east direction tend to open. The joint opening can lead water from the reservoir into the foundation and may cause local failure of grouting screen. While the joints tend to be closed in the zone of compressive stress. The joints striking in east-west direction are also in compression. Therefore, drainage of inflowed water is blocked. Thus a high lateral seepage pressure and uplift are gradually built up and causes the breakout of the rock deformation and damage of dam structure.

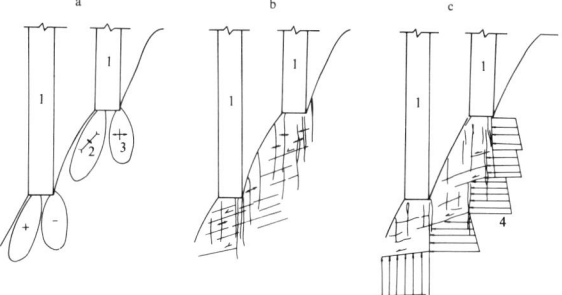

Figure 7. Mechanism of Foundation Deformation
a. Stress State of Rock Abutment; b. Deformation of Joints; c. Sliding Deformation
1.Buttress; 2.Compressive Stress; 3.Tensile Stress 4.Seepage Pressure

4.3 Developing Process of Rock Deformation

The formation of high seepage pressure and consequent deformation shows that the process is accompanied by the structure - foundation interaction. The whole process can be summarised by the following steps.
(1) Change of stress state in the foundation
The zone of tensile and compressive stresses are formed due to buttress loading on a slope foundation.
(2) Change of seepage regime in foundation
Due to change of stress state the permeability of rock mass becomes more heterogeneous. The joints striking in NNE orientation rock and demaging the grouting screen, while the drainage is blocked by the compression zone. Thus a high seepage pressure is formed.
(3) Sliding of rock mass of dam abutment
Under the high seepage pressure and uplift the rock foundation tend to slide along the subhorizontal joints towards the river slide and downstream.
(4) Deformation of dam structure
The dam structure is damaged by tilting, torsion causing

cracking of arches and buttresses.

(5) Stopping of deformation by dewatering reservoir

The emergency was taken by emptying reservoir through a bottom outlet. The deformation was stopped and a collapse of the Meishan dam was prevented.

The process of deformation mentioned above shows a typical strcuture-foundation interaction. The deformation of rock foundation was not caused directly by low bearing capacity, but indirectly by change of stress state and seepage regime during operation of the dam.

5. ROCK STRENGTHENING AND REMEDIAL MEASURES

The systematic remedial measures were taken for strengthening rock and protection against further deformation including additional grouting, drainage, consolidation, prestressing anchorage and retaining support (Fig. 8)

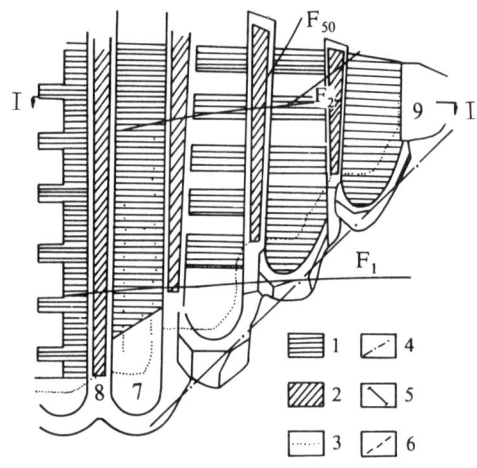

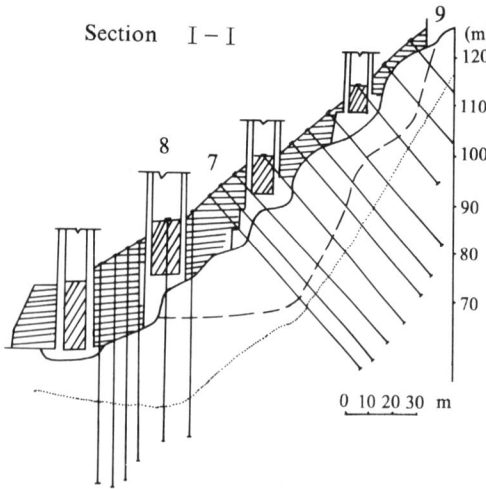

Figure8. Measures of Treatment of Rock Abutment
1.Support Wall; 2.Concrete; 3.Consolidation Grouting; 4.Curtain Grouting; 5.Prestressed Anchorage; 6.Depth of Failure; 7.Arch; 8.Buttress; 9.Gravity Dam

(1) New grouting screen of 2 rows with boreholes of 35-40 m deep in the section from arch No.12 to gravity block.
(2) New drainage system of boreholes of 20-25 m in depth behind arches and in buttress foundation.

(3) Consolidation grouting of the deformed area with boreholes of 10-23 m deep and 1.5 m in spacing.
(4) Retaining support walls in the void of arches and full concreting of void of buttress.
(5) Prestressing anchorage of 20-30 m deep with tension of 1.6-2.4 MN.

After the remedial treatment was taken, the impounding restarted in 1965, and the water in the reservoir reached its highest level. According to the monitoring report, the displacement of dam structure was limited in a safe range. The horizontal displacement of buttress towards down stream was 0.57-2.35 mm, and subsidence in 0.36-1.66 mm. No increase of displacement was observed, when the water level reached its maximum once again in 1984. The piezometric observation shows a low seepage pressure which was less than 10 m. And at present the reservoir is still normal in operation.

6. CONCLUDING REMARKS

(1) The prediction of the behavior of structure - foundation system should be based on the consideration of the interaction between structure and foundation, and not based only on the natural state before construction.
(2) A monitoring system working during construction and operation is very important to warn if any problems appear. The early measures of lead to prevent catastrophic accident.
(3) The grouting system may be damaged due to new stress condition and other process. In this case an additional grouting should be arranged after operation starts.
(4) For most hydro-engineering the seepage problem is an important factor in the structure - foundation interaction, therefore, the drainage of groundwater and water tighting measures should be carefully arranged.
(5) To ensure the whole safety some structural measures should be designed taking the case of emergency into consideration. For example, in case of Meishan dam a bottom outlet, which allows to dewater the reservoir to a low elevation, is essential to mitigate a potential disaster.

7. ACKNOWLEGEMENT

The author is grateful to the Natural Science Foundation of China for grant of the research project (49232050).

REFERENCES

Gu Dezhen and Wang Sijing, 1982, Fundamentals of Geomechanics for Rock Engineering in China, Rock Mechanics, Suppl. 12, 75-87

Gu Dezhen and Wang Sijing, 1982, On the Interaction between the Engineering Construction and Geologic Environment in the Man-made Lake Area, Proc. of 4th Congress of International Association of Engineering Geology, New Delhi, Vol. VII, 57-64

Wang Sijing, 1987, Fundamental Problems of Environmental Engineering Geology in Mountainous Areas-A General Review, Proc. of International Symposium on Engineering Geological Environment in Mountainous Areas, Beijing, Vol.2, 381-394

Wang Sijing, 1990, Engineering Geomechanical Analysis of Rock Foundation, Science Press, 1-371(in Chinese).

Wang Sijing, 1992, Human Activity and Natural Disasters, Development in Geoscience (2), Science Press, 407-416

Stability of foundations on jointed rock – Case studies

Duncan C. Wyllie
Golder Associates, Vancouver, B.C., Canada

ABSTRACT: Where structures are founded on sound but jointed rock in steep terrain, the stability of the foundation may be of much more concern than its bearing capacity. Instability can arise as the result of sliding or toppling of blocks of rock in the foundation formed by continuous intersecting joints, bedding planes or faults. Factors that influence stability include topography, shear strength of fracture surfaces, water pressures, non-vertical and vibrating structural loads, earthquake forces and long term weathering of the rock. The design should also take into account equipment access and productivity since this will influence the cost of construction in steep terrain.
The paper presents case studies of a transmission tower, a cableway tail tower, a railway bridge and a highway bridge which illustrate many of the conditions listed above.

1. INTRODUCTION

Where structures are founded on sound but jointed rock in steep terrain, the stability of the foundation may be of much more concern than the bearing capacity. This condition is particularly prevalent where there are significant non-vertical loads on the structure, such as dams and spillways. Where the fractures are oriented to form blocks that could slide or overturn, it may be necessary both to anchor the structure to the foundation and to reinforce the foundation to prevent movement of blocks of rock. This paper describes the foundations of structures where special attention was paid to foundation stability, as well as the particular requirements of access, constructability and cost in terrain where detailed investigations and the use of heavy construction equipment was not possible.

2. ROCK MECHANICS OF FOUNDATION STABILITY

Figure 1 shows two examples of foundations located on a bench cut in a steep slope where there is a potential for instability because the foundation contains continuous fractures that dip out of the slope. In the planar failure a) there is a single fracture striking parallel to the slope face, while in the wedge failure b) there are two fractures that intersect under the foundation with the line of intersection dipping out of the slope. The stability of the potentially sliding blocks in the foundation are determined by standard rock mechanics principles, in which all of the parameters described below have to be considered (Hoek and Bray 1981; Wyllie 1992).

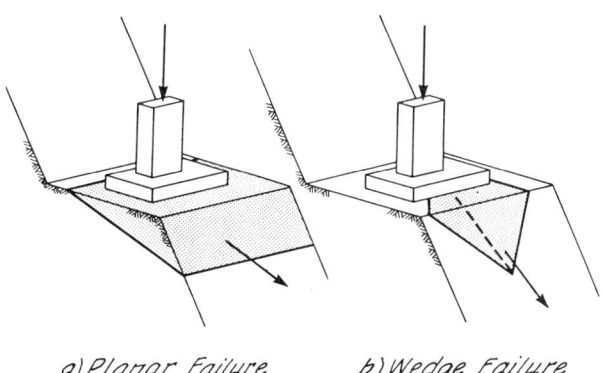

Figure 1. Potentially unstable foundations formed by continuous fractures dipping out of the face.

2.1 Fracture characteristics

The orientation, the position and continuous lengths of fractures will determine the shape, position and dimensions of the wedge, the kinematics of which can be examined. Of these three parameters, the orientation and position of the fractures can often be determined with reasonable reliability from mapping and/or drilling. However, the continuity of the fractures, which influences the size of the block, can only be estimated. For foundations the conservative assumption is usually made that critically oriented fractures are continuous.

2.2 Shear strength of rock fractures

The shear strength of any potential sliding fracture is expressed in terms of the cohesion and friction angle of the surface. Factors which influence shear strength are the compressive strength and roughness of the rock surfaces; it is usual to express the friction angle as the sum of the basic rock friction and the surface roughness. Another factor is the presence of infillings which may be strong and cohesive and will enhance stability, or have low strength and will be detrimental to stability. The possible reduction in strength with time due to weathering may also need to be taken into account.

2.3 Ground water

The primary effect of ground water is to introduce uplift and driving forces on the fracture surfaces of the foundation blocks that will be detrimental to stability. Ground water pressures may be produced by precipitation increasing the piezometric pressure in the foundation, or by flooding of the foundation and then rapid draw down. The foundation would be designed to resist ground water forces if drainage cannot be assured.

3. DESIGN

The projects described in this paper include a cableway tail tower which is a temporary structure to be used only during the construction of a dam, and a railway bridge that has already has an active service life of 100 years and is expected to continue in service for another 100 years. These two extremes illustrate the different approaches that need to be taken in the investigation and design of foundations.

3.1 Design Life

In the case of the tail tower with a design life of about two years, a thorough investigation of the complex geological conditions was justified because a high degree of reliability, under high but well defined non-vertical loads, was required. However, the design loads did not include earthquake motion because it could reasonably be assumed that there would be no earthquakes during the construction of the dam. Also, a simple stabilization method, comprising surcharge of the foundation with rock fill, was selected because this could be rapidly constructed with local material and would need little maintenance.

In contrast to the tail tower, the railway bridge had effectively been designed with very high factor of safety. It has successfully withstood a very significant increase in live load in terms of magnitude, frequency and impact loading between its original use by small steam trains to the present traffic of diesel electric locomotives hauling 100 tonne coal cars. Furthermore, the foundations have sustained the effects of severe winters with many freeze thaw cycles, as well as river scour. Because this bridge is of great importance to railway operations, regular inspections are carried out, and it has been necessary to reinforce the foundations at least three times. This reinforcement work has involved the use of reinforced concrete buttresses where possible, and corrosion protected anchors, so that the design life of the foundations will match that of the bridge.

3.2 Bearing Capacity

For the projects described in this paper, the strength of the rock in the foundation is greater than the concrete in the structure so bearing capacity of the material in the foundation is not an issue in design.

3.3 Stability Analysis

Figures 2 and 3 illustrate the principles of the design of stabilization measures for a foundation where sliding failure on pre-existing fractures is kinematically possible. In this example of a transmission tower foundation (Figure 2), there is a non-vertical load applied to the cables (s), which in combination with the vertical weight of the tower (W_T), produces a resultant force (T) which is a destabilizing force on the foundation. The tower is anchored to a slab of rock, the base of which is formed by a continuous joint dipping at an angle ψ_p out of the slope face. The forces acting on the base of the slab are the weight of the slab (W), and the shear resistance comprising the cohesion and friction angle ($\phi+i$). There is also water in the fracture at the base of the foundation slab that produces an uplift pressure (U). In this example the earthquake motion is simulated as a pseudostatic horizontal force with magnitude αW, where α is the earthquake acceleration relative to gravity acceleration.

The stability of the foundation slab under the applied external forces can be examined by resolving these forces normal (σ) and parallel (τ) to the sliding surface. For example, the resolution of the foundation weight and the structural load respectively are as follows:

$\sigma_W = W\sin(90-\psi_p)$, $\tau_W = W\cos(90-\psi_p)$, and
$\sigma_T = T\sin(\psi_T-\psi_p)$, $\tau_T = T\cos(\psi_T-\psi_p)$.

Figure 3 illustrates the resolved forces plotted on a Mohr diagram, together with the shear strength modeled as a straight line with cohesion intercept and slope ($\phi+i$). This diagram shows that the foundation block, with the water pressure acting on the base, is marginally stable, that is, the available shear strength (τ_a) is just greater than the total shear stress acting on the base of the slab (τ_t). However, the effect of the horizontal earthquake force and the load of the tower is to greatly increase the shear stress acting down the base of the foundation slab so that sliding would occur.

In this case stability of the foundation slab can be achieved by installing tensioned cables anchored below the base of the slab. The magnitude and direction of the anchor force (A) can be determined to produce a required factor of safety (FOS), where:

$$FOS = \tau_a/\tau_t$$

Figure 3 shows that there is an optimum inclination $[(\psi_A+\psi_p)=(\phi+i)]$ of the anchors that produces the required factor of safety with the minimum anchoring force (A_{opt}). However, in this case the optimum anchor inclination is above the horizontal which could cause difficulty in fully grouting the anchors. If the anchors are installed at an angle below the horizontal in order to facilitate grouting, then there is an increase in the anchoring force to produce the same factor of safety, i.e. $A > A_{opt}$.

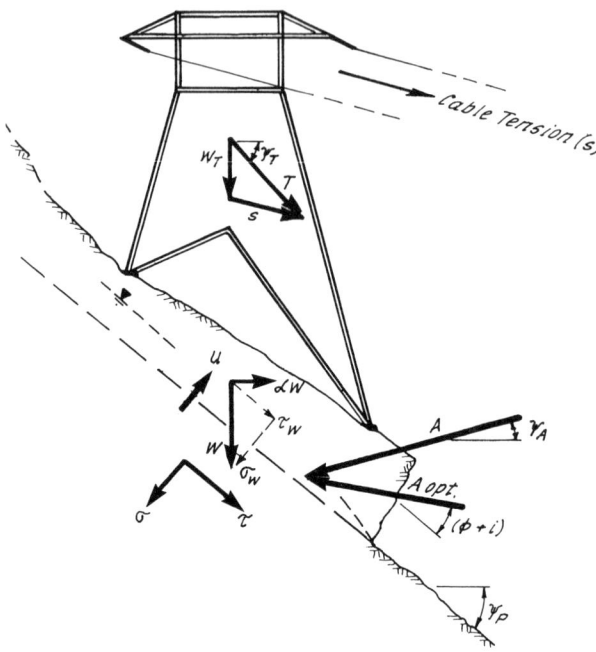

Figure 2. Forces acting on transmission tower foundation.

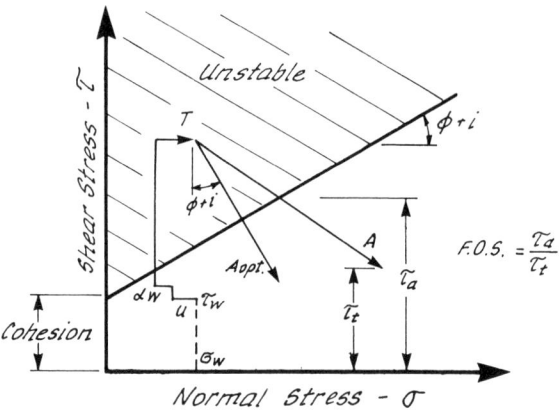

Figure 3. Resolution of foundation forces on Mohr diagram to find Factor of Safety.

The principles illustrated in Figures 2 and 3 can be applied to a wide range of geological and loading conditions in order to investigate sliding stability. However, it is also be necessary to check overturning stability for a loading condition such as that shown in Figure 2 where the structural load acts above the foundation.

3.4 Drainage

The vector of the water force U on Figure 3 shows that this force acts directly to decrease the normal force acting on the potential sliding plane. In these circumstances, drainage of the foundation using horizontal drain holes to reduce uplift pressures is worthwhile, although the effective operational life of such measures should be considered in design.

3.5 Earthquake forces

The effect of earthquake motion on foundations such as those shown in Figures 1 and 2 is to cause displacement and possibly failure of the sliding blocks, depending on the magnitude and duration of the shaking. In the example shown in Figures 2 and 3, the earthquake motion has been modeled by the pseudo-static method in which a constant force equivalent to that induced by the earthquake acceleration is applied to the foundation for the design life of the structure. This is a conservative design method that can result in high construction costs. A more realistic design procedure is to use the Newmark method which calculates the displacement produced by the transient earthquake motion (Newmark 1965, Jibson 1993). Using this method the foundation can be designed so that the displacement is approximately within the operational limits of the structure

4. CONSTRUCTION

In the design of the foundations described in this paper the particular site conditions of each had to be considered to ensure that an economical method of construction could be employed. In the rugged terrain where most of this work was carried out, the following were some of the factors that were taken into account.

4.1 Exploration

Where there is no access for exploration equipment, the site conditions have to be assessed from surface mapping only. Under these conditions, it is sometimes necessary to adopt conservative designs to allow for possible poor foundations conditions such as weathered or weak rock in the bearing area, or low shear strength of fractures on which sliding could take place. In some cases diamond drilling was possible, although it was necessary to bring the drill in with a helicopter which required the use of a lightweight drill that could be broken down into convenient payloads. In all the diamond drilling, NQTT triple tube coring (45 mm diameter) was used to minimize damage and disturbance to the core.

4.2 Blasting

On many sites it is necessary to carry out some blasting to prepare a bearing surface and/or remove areas of unsuitable rock. In all cases careful blasting techniques were used to minimize damage to the rock outside the limits of the excavation.

4.3 Equipment Access

For anchored foundations, the maximum size of anchor that is used is often dependent on the maximum size of drill that can be mobilized to the site. If helicopter access is possible only, then it is likely that the heaviest drill that can be used may limit the hole diameter to about 100 mm and the anchor diameter to about 50 mm. Under these conditions it may also be advisable to use strand anchors that are more flexible and easily handled than long, rigid bar anchors.

5. FOUNDATION STABILITY - CASE EXAMPLES

The following is a brief description of the design and construction of foundations in which the principal design feature was stability rather than bearing capacity.

5.1 Transmission tower

The construction of a transmission tower for a power line in rugged, mountainous terrain in northwestern British Columbia, Canada, required one tower anchored to a rock face about 200 m above the river valley (Figure 4). The rock in the foundation comprised very strong, fresh granite containing continuous exfoliation joints dipping at about 40 degrees out of the slope face. There were also two orthogonal joints sets that formed a series of slabs, about 2 m to 4 m thick, with overhanging faces. Because of the dip of the ex-foliation joints was close to the friction angle of the rock surfaces, there was a high risk of sliding failure if the slabs were to break loose. The existing slope was formed by a series of faces formed by previous sliding failures.

The tower was an "angle" tower between two tangent segments of the line so the tension in the cables produced a significant downslope load and overturning moment at the top of the tower. The resultant structural load on the foundation comprised the cable tension together with the wind load on the cables, and the weight of the tower. This is also a high seismic area so earthquake accelerations on both the tower and the foundation were considered in the design.

A further destabilizing force was the water pressures developed along the base of the foundation slab. This is a very high precipitation area so it could be assumed that significant heads could occur, and that drainage of the foundation would be beneficial.

The exfoliation joint forming the base of the foundation slab was tight but clean and contained no infilling. Therefore it was assumed that the shear strength of this surface would comprise only friction, with no cohesion. It was also assumed that the two conjugate joint sets, which would form release surfaces at the back and sides of the foundation slab, would not contribute to the shear resistance.

A detailed investigation of the site showed no evidence of movement or instability of the slab prior to construction. In order to assess the existing stability of the foundation slab, it was assumed that, in geological time, the slab had withstood simultaneously both the design earthquake acceleration and high water pressures. Figure 3 shows a factor of safety slightly greater

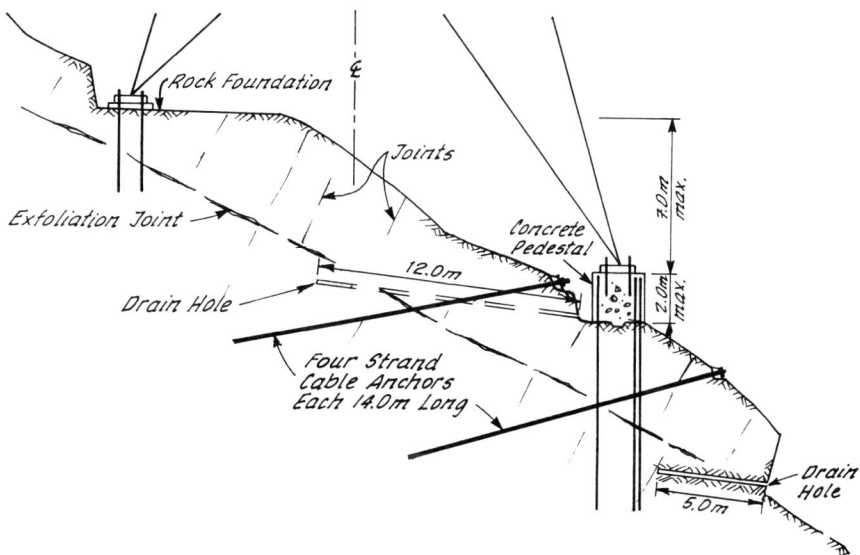

Figure 4. Transmission tower foundation secured with multi-strand cable anchors.

than 1 for these conditions. The addition of the structural load caused the factor of safety to drop below 1 and it was necessary to anchor the foundation slab in order to achieve a factor of safety of at least 1.5. The anchors would be installed through the foundation slab with the bond zone in sound rock below the foliation plane.

In determining the capacity and number of these anchors, it was necessary to keep in mind the difficult access to this site. All equipment would have to be lifted to the site by helicopter and this limited the maximum drill hole diameter to about 100 mm, and the diameter of the corrosion protected anchor assembly to about 50 mm. In order to meet both the design and construction requirements of the site it was found that a total of 14, four strand anchors were required, each with a design tensile capacity of 500 kN at 50 % of the ultimate strength. Strand anchors were preferred over bar anchors because of their lower weight for the same strength, and ease of handling a compact rolled bundle.

The design also called for the installation of drain holes through the foliation plane above and below the tiebacks. It was important that a row of drainholes be located above the anchors because the grout used to anchor the anchors would seal the lower half of the foliation plane and inhibit drainage.

5.2 Cableway tail tower

The construction of the Revelstoke gravity dam on the Columbia River in British Columbia, Canada (Foster 1986) required the use of a cableway spanning the river to place the concrete (Figure 5). The tail tower of the cableway was 22 m high and located on a curved track high on the left bank of the river. The weight of the cables, which were about 700 m long, and the loaded skip produced a substantial overturning moment which was counteracted by a counterweight on the back of the tower. The resultant load was 1.26 MN per meter length of track acting at an angle of 78 degrees below the horizontal.

The foundation for the rail track carrying the tower was on a 33 m wide bench cut into the steep rock face above the river. The rock in the foundation was strong schist containing foliation planes with continuous lengths of several tens of meters, that dipped out of the face and towards the river at an angles ranging from 10 to 20 degrees. Many of the foliation planes were filled with a firm to stiff sandy silt with a thickness up to about 200 mm. Direct shear tests of the silt infilling showed a friction angle of about 36 degrees and zero cohesion.

There was the possibility of ground water pressures developing along the foliation planes, particularly during rapid snow melt in the spring when drainage of the slope was inhibited by freezing of the face. However, these pressures were likely to be low because of the drainage ditch at the back of the bench prevented the development of significant heads.

Analyses of the shallow rock slabs in the foundation were carried out to examine the effect of the inclined tower load on the stability against sliding on the silt filled fractures. The analyses showed that the factor of safety was sensitive to variations in the dip of the foliation planes (ψ_p): the factor of safety dropped from 1.3 when ψ_p = 20 degrees, to 1.05 when ψ_p = 25 degrees. It was considered that a factor of safety of 1.3 to 1.5, with no earthquake loading, would be suitable for this temporary structure; this would require some improvement in sliding resistance.

The stabilization method selected was to place rock fill on the front on the bench. The weight of the fill increased the normal stress acting on the sliding planes and increased the factor of safety to an acceptable value of about 1.5. This stabilization method was inexpensive and quick to install because of the ready availability of the fill material. Furthermore, the fill would be well drained so additional water pressures would not develop in the foundation.

5.3 Railway bridge

Figures 6 and 7 shows the retaining walls supporting the approach fills of a 40 m long steel truss railway bridge spanning a narrow, steep sided canyon on the Fraser River in Canada. The walls and fills are on narrow benches cut into the bedrock and are located close to the crest of steep slopes. The weather at the site typically consists of wet summers and long, cold winters resulting in the development of water and ice jacking pressures in fractures in the rock that progressively loosen blocks of rock on the faces of the slopes.

The rock in the foundation is a very strong, fresh granite. The predominant fracture set is an exfoliation joint that dips downstream at an angle of between 40 and 50 degrees; the joints typically have continuous lengths of several tens of metres and spacing of about 2 to 4 metres. There are also two orthogonal joint sets having continuous lengths of about 1 to 2 m, that divide the rock mass into approximate cubic or slab shaped blocks.

The primary traffic on this line is heavily loaded unit trains carrying commodities such as coal, wheat and potash. The combined train dead and live impact loads is 220 kN/lineal metre of track; this load is equivalent to a mass of rock with a height of about 3 m. The traffic loading, combined with the weathering of the rock, had resulted in slight downslope movement of blocks of rock along the exfoliation joints, and in some cases failure of these blocks to create overhangs. A series of stabilization measures was undertaken to reinforce the foundations.

On the **north abutment** a concrete buttress, anchored to the rock, was poured to support the base of the foundation and prevent loosening of the rock in this area (Figure 6). It was not practical to support the upper part of the foundation with a similar

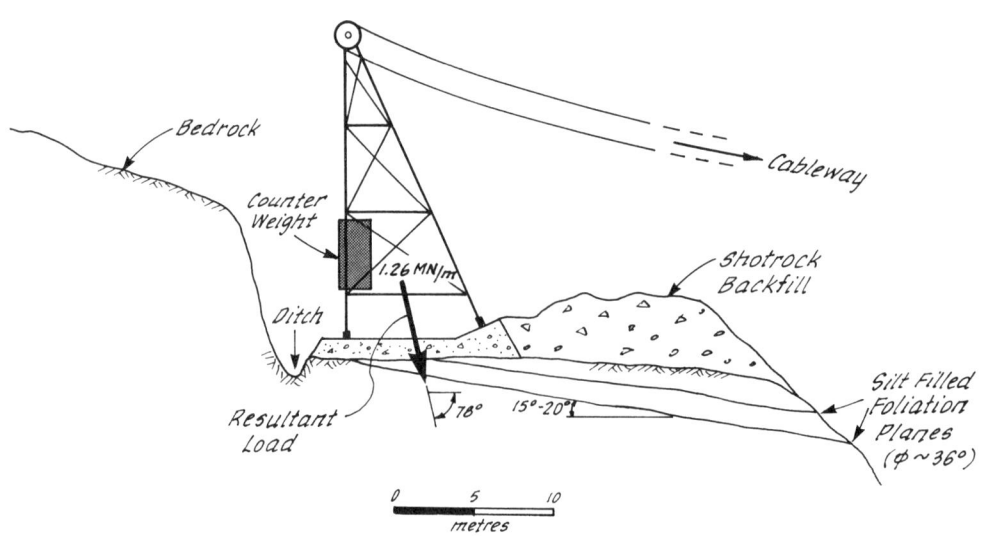

Figure 5. Stabilization of cableway tailtower foundation with shotrock surcharge.

concrete buttress because of the large quantity of concrete required, and the difficulty of drilling under the overhang to install anchors required to resist the overturning moments on the buttress. The alternative adopted was to install a total of five tensioned anchors through the overhang, with the bond zone located in sound rock already supported by the buttress. In order to accommodate the geometry of the site, the anchors were 16.8 m long and inclined at 40 degrees below the horizontal. Each anchor comprised a bundle of six strands, each strand 16 mm in diameter, with the total tensile load applied to each bundle was 1 MN. The strand bundle was anchored with non-shrink cement grout, and a 6.1 m long bond zone was used in a 75 mm diameter hole drilled with a percussion bit.

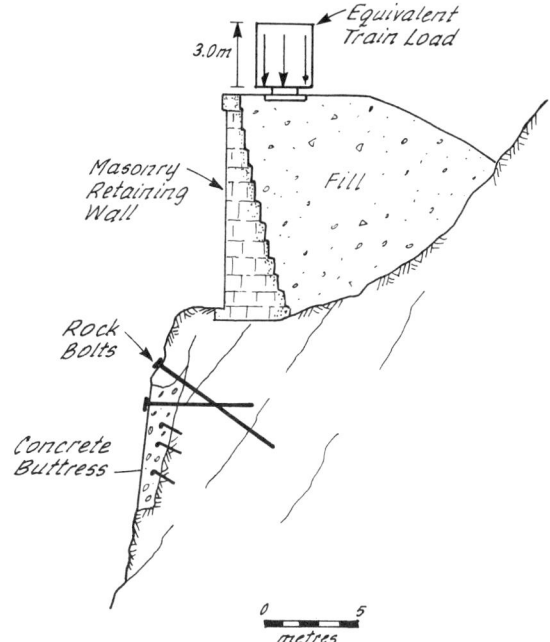

Figure 7. South approach of railway bridge - cavity in foundation supported with concrete buttress and rock bolts.

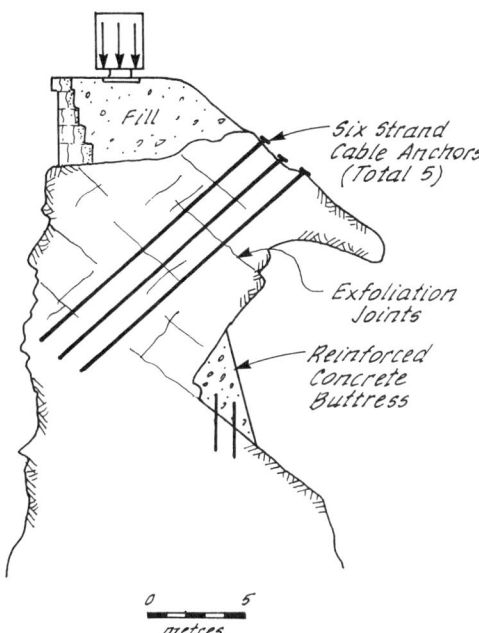

Figure 6. North approach of railway bridge - overhang supported with concrete buttress and multi-strand anchors.

On the **south abutment** the failure of a slab of rock about 7 m high had undermined the foundation of the retaining wall supporting the approach fill. This had caused the wall to rotate and the top had moved laterally by about 50 mm. Stabilization of the foundation comprised the construction of a reinforced concrete buttress in the cavity with the top surface of the concrete in close contact with the underside of the overhang. To prevent overturning of the buttress, fully grouted dowels connected the concrete to the sound rock in the back of the cavity. In addition, a series of 8 m long rock anchors, each tensioned to 360 kN, were installed through the overhang above the buttress.

In undertaking this work it was necessary that all the work could be conducted with no disruption to rail traffic, and that the stabilization work could be accomplished using light, mobile equipment that could operate within the limited work space available. Furthermore, there was no road access to the site and no space to operate a crane. These difficult construction conditions were met by having the work carried out with the men and equipment suspended from ropes.

5.4 Abutments of steel arch highway bridge

Figure 8 shows the configuration of a 220 m span steel arch, four lane highway bridge constructed across a canyon known as Dry Gulch in British Columbia, Canada (Rawlings and Wyllie 1986). The rock at the site was a very strong, coarse grained fresh to slightly weathered granodiorite; the structural geology comprised a number of major faults, dykes with sheared margins, and several sets of persistent joints, in places closely spaced with zones of weathered rock. Slope movement on these planes of weakness had produced areas of open jointing.

Construction of the bridge required substantial excavations to create a series of level benches for the bearing surfaces for both the vertical columns supporting the approach spans, and the abutments for the arch. The three design criteria for these excavations were as follows. First, that there be adequate bearing capacity for the applied loads; this required removal of all weathered, loose and broken rock in the bearing areas. Second, that the benches supporting the footings be stable; this required that the benches be large enough for a minimum 1 m set back distances between the front of the footing and the crest of the bench, as well as reinforcement of the crests with fully grouted dowels and shotcrete. Third, the slopes of the foundation excavation would be stable so that there would be no slope failures that would impact the foundations.

With respect to bearing capacity, the bearing pressures ranged from 550 kPa for the vertical footings, to 700 kPa for the arch footing. The rock was readily able to withstand these pressures, even where closely fractured, without any significant displacement.

With respect to foundation stability, Figure 8 shows the dimensions of the two excavations and the benches cut at each footing level. On the north abutment, a deep channel, which was a glacial meltwater feature infilled with sands and gravels, was discovered. The infilling material had inadequate bearing capacity and was removed which required that the footing at elevation 1151.5 m was lowered to elevation 1145 m. In addition, the crests of the benches were reinforced with 25 mm steel dowels fully grouted with resin, or cement grout for the longer dowels, and steel fibre shotcrete. On the south abutment, the rock in the foundation areas was more massive than that on the north abutment, and there was little need for reinforcement of the rock.

With respect to stability of the excavation slopes, a 1 m wide tension crack and substantial area of disturbed rock were discovered on the south abutment once all the colluvium had been stripped from the site. It was found that movement had occurred along a fault plane dipping towards the gulch, and that the area of instability did not intersect the foundation area. However, in order to form a stable excavation it was necessary to remove the unstable rock by cutting the slope at an angle of 50 degrees rather than the planned angle of 70 degrees.

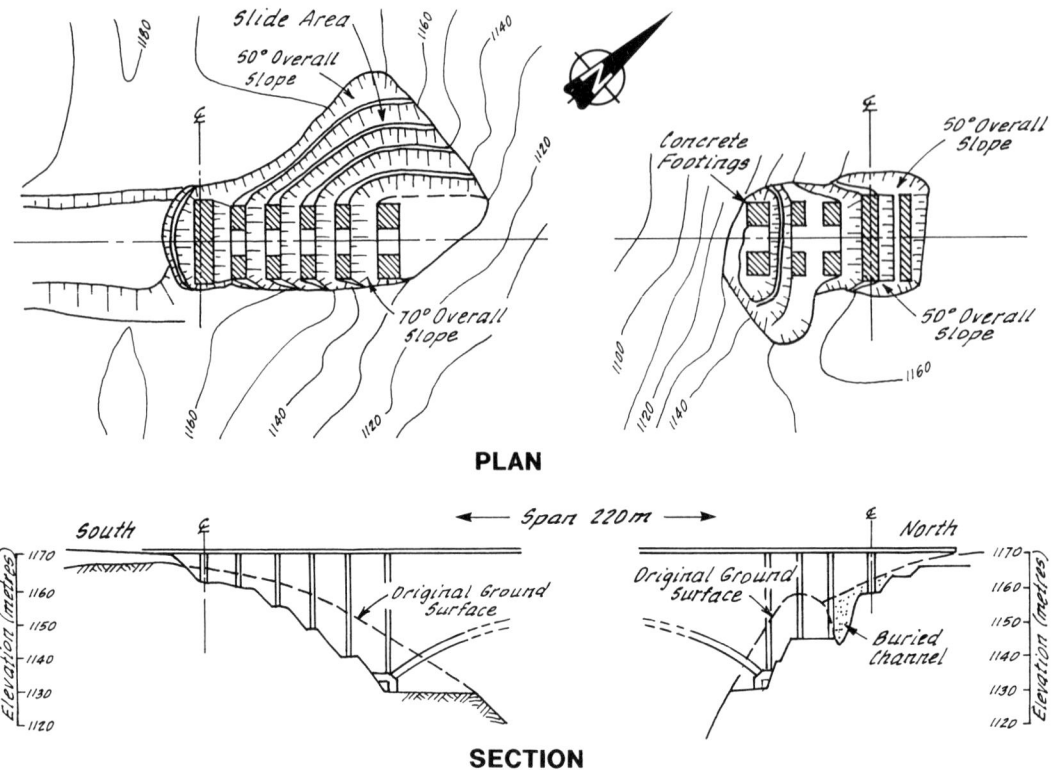

Figure 8. Plans and sections of foundation excavations showing areas of instability

References

Foster, J. W. 1986. Geological problems overcome at Revelstoke. *Water Power and Dam Construction*, July: 53-8, August: 42-5.

Hoek, E and Bray, J. 1981. *Rock Slope Engineering*. London: Institute of Mining and Metallurgy, 402 pages.

Jibson, R. W. 1993. Predicting earthquake-induced landslide displacements using Newmark's sliding block analysis. *Transportation Research Record*, Washington, DC, No. 1411, 9-17.

Newmark, N. M. 1965. Effect of earthquakes on dams and embankments. *Geotechnique*, 15(2): 139-60.

Rawlings, G. E. and Wyllie, D. C. 1986. Bridge abutments on rock. *Proc. Specialty Conf. on Transportation Geotechnique*. Vancouver Geotechnical Society, Vancouver, Canada, 16 pages.

Wyllie, D. C. 1979. Fractured bridge supports stabilized under traffic. *Railway Track and Structures*, July 1979, 29-32.

Wyllie, D. C. 1992. Transmission tower foundations for Kemano Completion Project, British Columbia. *Proc. 45th Canadian Geotechnical Conf.*. Toronto, pages 106-1 to 106-10.

Wyllie, D. C. 1992. *Foundations on Rock*. London, U. K.: Chapman and Hall, 381 pages.

ns **A procedure of aseismic design for Akashi Kaikyo bridge foundations**

A. Nitta & S. Yamamoto
Honshu-Shikoku Bridge Authority, Tokyo, Japan

S. Masuda & R. Isoyama
Japan Engineering Consultants Co., Ltd, Tokyo, Japan

ABSTRACT: The world's largest suspension bridge, the Akashi Kaikyo bridge with a center span of 1,990 meters, has been under construction and is scheduled for completion in the year 1998. Two gigantic anchorages and two tower piers were already built, and main cable of the bridge has been just carried across between Honshu main isle and Awajishima isle. This paper describes a procedure for estimating seismic response used in the design of foundations for the Akashi Kaikyo Bridge. Design seismic motion for the bridge foundations is determined at the bedrock based on the seismic activity around the bridge site and the seismic response of the foundations is computed by considering dynamic interaction effects between the foundations and their surrounding soil. Example calculations applied the proposed procedure to the foundation of a main tower pier are also presented in this paper.

1 INTRODUCTION

Since Honshu main isle and Shikoku isle are separated by the Seto Inland Sea, available means to connect traffic between two islands had been by either ship or aircraft, which is restricted by meteorological conditions. To remove this inconvenience in the daily life of people and to facilitate the balanced development of the country, the construction of three bridge routes, A, D and E (refer to Figure 1), connecting two isles was come to a decision.

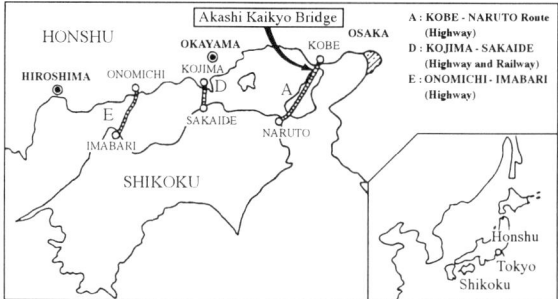

Figure 1. The Routes of the Honshu-Shikoku Bridges

Among the above routes, the Kojima-Sakaide route (D route) with a length of about 10 km was completed in March 1988. Three suspension bridges with around 1,000m center, two cable stayed bridges and other bridges constitute this route. Five of the nine major bridges planned on the Onomichi-Imabari route (E route) were completed by the year 1991 while the Ohnaruto bridge on the Kobe-Naruto route (A route) was completed in 1985. Construction work of the Akashi Kaikyo bridge, one of two suspension bridges on the Kobe-Naruto route, started in 1986 and is scheduled to be completed in 1998. The bridge will be the world's longest suspension bridge with a center span of 1,990 m and a total bridge length of 3,910 m. All of the bridges on the above three routes are planned to be completed by the end of this century.

The bridges in the Honshu-Shikoku routes have been generally designed and constructed according to the Aseismic Design Standard for Honshu-Shikoku bridges (The Honshu-Shikoku Bridge Authority, 1977). However, in the case of the Akashi Kaikyo Bridge, a new design procedure was required to develop, mainly because of its large scale, importance and construction site conditions such as ground conditions and seismic activity. This paper summarizes the seismic design procedure for rigid foundations specially developed for the Akashi Kaikyo Bridge.

2 OUTLINE OF AKASHI KAIKYO BRIDGE

The Akashi Kaikyo Bridge is a highway bridge currently being constructed across the Akashi Kaikyo (Akashi Strait). This Strait is about 4km in width and is located between Honshu main isle and Awajishima isle. An artist's conception as built drawing and side view of this bridge are shown in Figures 2 and 3, respectively (Kashima, Yasuda, Kanazawa and Kawaguchi 1987).

As shown in Figure 4, the geological composition comprises, in due order from the top, an alluvial stratum above an upper diluvial stratum, Akashi stratum, Kobe stratum and granite stratum. Whereas the other Honshu-Shikoku Bridges use the granite stratum for a supporting ground of the foundations, the Akshi Kaikyo Bridge is compelled to use the diluvial stratum, the Akashi stratum and the Kobe stratum since the granite stratum is extremely deep. These strata are relatively new formations and are relatively unconsolidated (Takahashi, Miyajima, Kashima, Yamamoto, Yamagata and Aizawa 1977).

Figure 5 shows the distribution of earthquake epicenters and their magnitudes around the bridge site in the past 100 years. It can be seen from the figure that several earthquakes of a magnitude 8.0 class occurred along the Pacific Coast (about 150 km south of the site) and that many earthquakes of a magnitude 6 to 7 class occurred elsewhere in the region. It seems that the seismic activity in this region is rather strong compared to other regions in Japan.

Because of the geological conditions, large scale of the bridge and seismic activity in the region, etc., the required foundations were, therefore, expected to be unprecedentedly large rigid foundations.

3 ASEISMIC DESIGN PROCEDURE FOR FOUNDATIONS

3.1 Conventional Procedure and its Objectives

In Japan, the design of a road bridge with a span not longer than 200m is required to comply with the Specifications for Highway Bridges (Japan Road Association 1980). In view of the fact that the span of each major Honshu-Shikoku bridge exceeds 200 m because of the located conditions, however, the bridge design was generally carried out in accordance with the design standards established independently by the Honshu-Shikoku Bridge Authority. Therefore, the aseismic design of other bridges has so far been carried out by utilizing one of the Authority's standards, Aseismic Design Standard Applied to Honshu-Shikoku Bridges (Honshu-Shikoku Bridge Authority 1977). According to the standards, a rigid body foundation such as a steel caisson, is modeled as a 2-degree of freedom system with rocking and swaying modes (see Figure 7) and seismic response of the foundation is computed by using the response spectrum based on

259

Figure 2　Akashi Kaikyo Bridge (as built drawing)

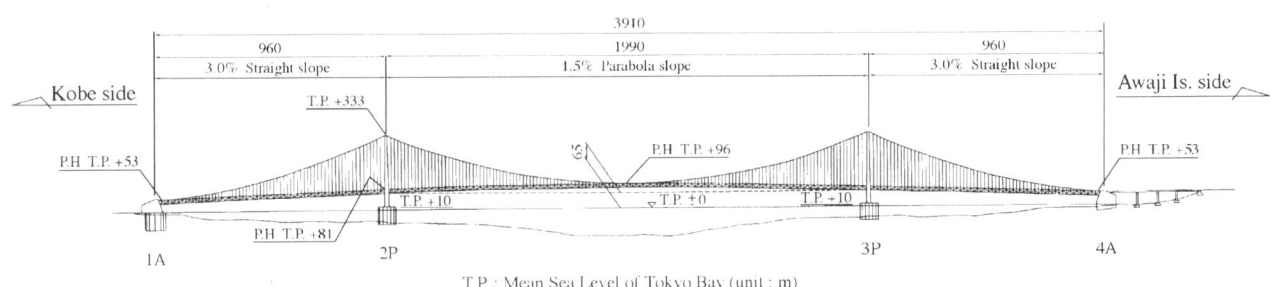

Figure 3　Side View of Akashi Kaikyo Bridge

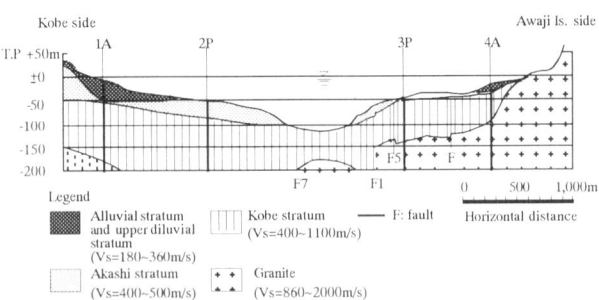

Figure 4　Geological Profile

the following conditions.

(1) Values of soil spring coefficient which expresses the restoring force between the foundation and the ground are determined based on results of a proper field test, such as a static plate loading test.

(2) A damping ratio of the foundation model is assumed to be 10% regardless of the size of the foundation and ground conditions.

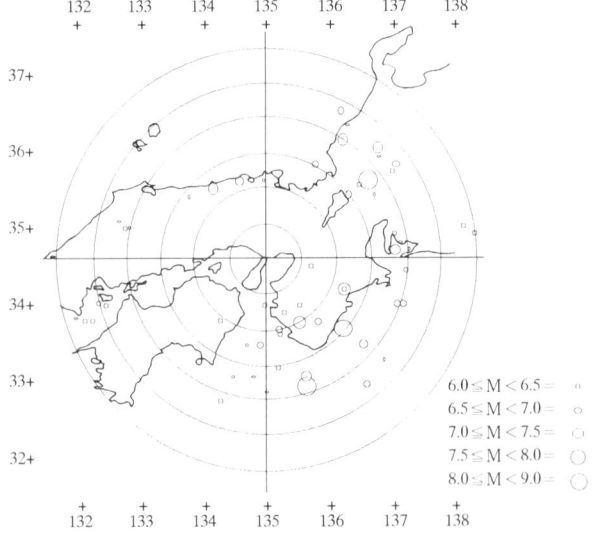

Figure 5　Epicenters and Magnitudes of the Past Earthquakes (1885 to 1979 year)

(3) The seismic motion to be applied is specified as an acceleration response spectrum of the free field motion on the rock site.

However, when the Akashi Kaikyo Bridge is compared with other bridges which have so far been constructed by the Honsyu-Shikoku Bridge Authority, the supporting ground is unconsolidated and relatively soft and the foundations form a huge structure. It was, therefore, considered inappropriate to apply the conventional aseismic design standard as described above and consequently a new aseismic design manual that takes into account the dynamic interaction between the ground and the foundation was provided for the design of Akashi Kaikyo Bridge (Japan Society of Civil Engineers 1983).

In the conventional aseismic design standard, the intensity of the input seismic motion is specified to be 180 cm/s^2 on the bearing ground of the foundation and the acceleration response spectrum used for dynamic analysis is also provided for. However, as these were laid down in 1974 and more information regarding the characteristics of seismic motion has since been clarified, they have had to be reviewed (Japan Society of Civil Engineers 1983).

3.2 Concept of New Aseismic Design Procedure

Figure 6 shows the flow chart of the aseismic design procedure for the foundations adopted in the manual of the aseismic design for the Akashi Kaikyo Bridge and is based on the following concept.

(1) The response of the foundations should be obtained as the response of corresponding 2-degree of freedom system with rocking and swaying modes using the response spectrum method.

(2) The model in (1) should be established taking into consideration the dynamic interaction effect between the foundations and the ground (see Figure 7).

(3) In order to evaluate the dynamic interaction effect between the foundations and the ground, the local soil conditions around the foundations should be taken into account.

3.3 Dynamic Interaction between Foundations and Ground

Recent progress in dynamic soil-structure interaction research shows that the interaction effects are composed of two effects, i.e. effective seismic motion and dynamic (complex) stiffness of the foundations (see Figure 6).

The effective seismic motion, sometimes called as the kinematic interaction effect, is defined as the response of massless foundations exposed by seismic excitations. The response of massless foundations or the effective seismic motion will differ from the seismic motion recorded at the free field surface which is used as an input motion of the conventional dynamic response analysis. These differences consist of a filtering effect of the horizontal motion, a decrease of the amplitude in the high frequency domain in general and the occurrence of rotational motion (see Figure 7). The effective seismic motion is a function of geometry of the foundations, especially the ground, which can be evaluated by utilizing the numerical analysis techniques, such as a finite element method.

In the conventional design standards, the spring coefficient which expresses the restoration force is evaluated as static force using results of a static loading test. For the dynamic condition, however, it is well recognized that the restoration force acting on the foundations is defined as a function of frequency and is called the dynamic or complex stiffness and is generally as described in the following equation.

$$K(\omega) = k(\omega) + i\omega c(\omega)$$

where $K(\omega)$ is dynamic stiffness, $i = \sqrt{-1}$, ω is frequency, $k(\omega)$

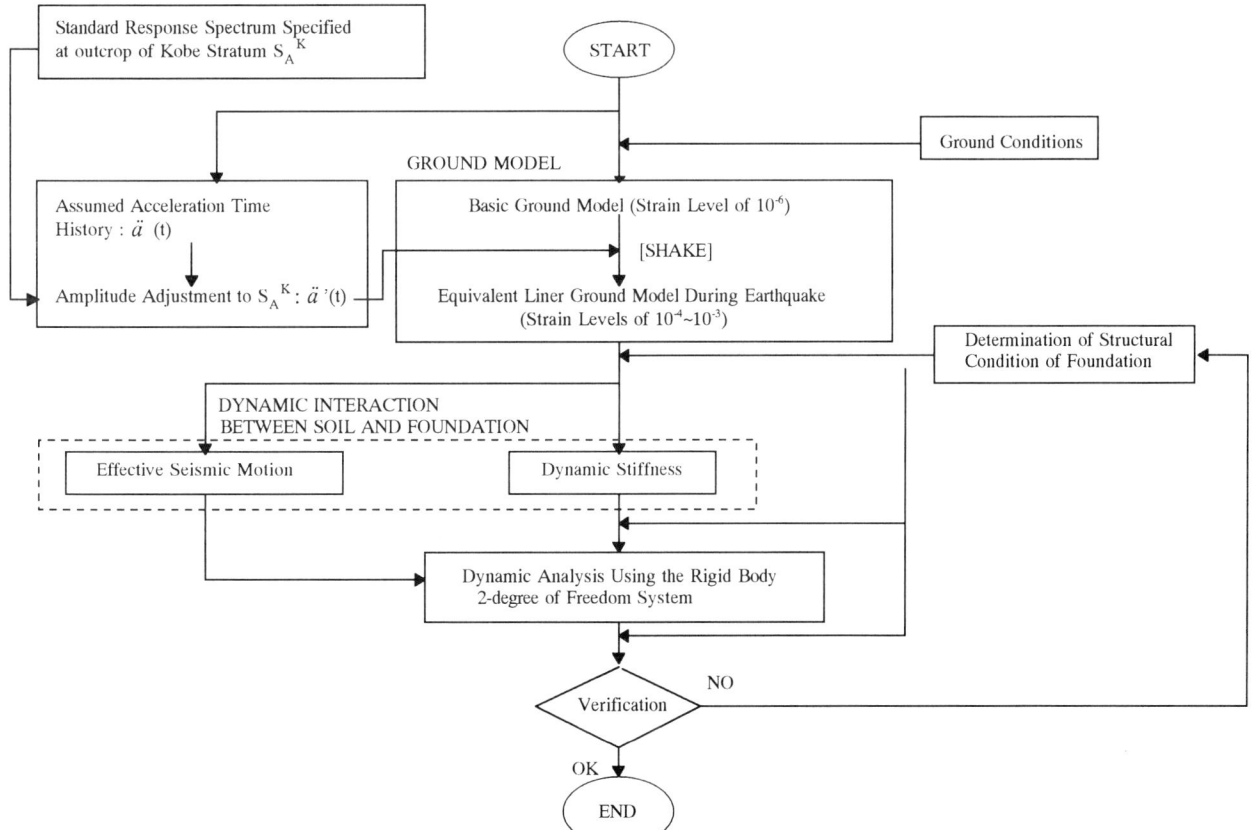

Figure 6 Procedure of Aseismic Design for Akashi Kaikyo Bridge Foundations

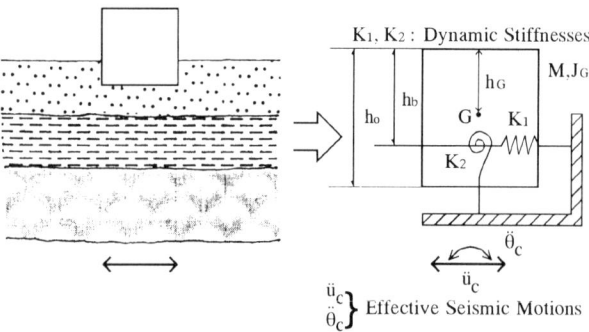

Figure 7 Dynamic Response Analysis Model

is equivalent spring coefficient and $c(\omega)$ is equivalent damping coefficient. The dynamic stiffness can be evaluated by either, a) finite element method, or b) continuum formulation method based on the elastic wave propagation theory (hereinafter referred to as FEM and CFM, respectively) (Harada, Kubo and Katayama 1981, Harada, Kubo and Katayama 1983).

3.4 Dynamic Analysis of Rigid Body Foundations

Assuming that the dynamic interaction between the foundations and superstructure is negligible because of differences in the vibration characteristics, motion for a rigid body is expressed by the following equation.

$$\begin{pmatrix} m & o \\ o & J \end{pmatrix} \begin{pmatrix} \ddot{u}_D \\ \ddot{\theta}_D \end{pmatrix} + \begin{pmatrix} K_{11} & K_{12} \\ K_{21} & K_{22} \end{pmatrix} \begin{pmatrix} u_D \\ \theta_D \end{pmatrix} = -\begin{pmatrix} m & o \\ o & J \end{pmatrix} \begin{pmatrix} \ddot{u}_C \\ \ddot{\theta}_C \end{pmatrix} \quad (1)$$

or in a matrix form:

$$[m]\{\ddot{u}_D\} + [K]\{u_D\} = -[M]\{\ddot{u}_C\} \quad (2)$$

where m and J are mass of rigid body and its moment of inertia of mass, K_{ij} is dynamic stiffness, u_D and θ_D are horizontal and rotational displacements of rigid body, and u_C and θ_C are effective seismic motions (displacements of horizontal and rotational components. In order to use equation (1) or (2) in the design practice, the response spectrum procedure that can be applied to equations (1) or (2) is required (Kashima, Harada and Isoyama 1986). If response spectra of effective seismic motions, $\ddot{u}_C$ and $\ddot{\theta}_C$, are prepared (see next chapter), the maximum response can be approximately obtained in the following manner. (In the following foundations, dynamic stiffness $K_{ij}(\omega)$ is assumed to be constant, that is, independent to ω for the procedure. A means to make $K_{ij}(\omega)$ constant is mentioned in chapter 4).

Assuming that the response $\{u_D\}$ in equation (2) is composed of two parts $\{u_D^{(1)}\}$ and $\{u_D^{(2)}\}$ corresponding to the seismic excitation $\ddot{u}_C$ and $\ddot{\theta}_C$ respectively, the following equations can be obtained.

$$[M]\{\ddot{u}_D^{(1)}\} + [K]\{u_D^{(1)}\} = -[M]\{r_1\}\ddot{u}_D \quad (3)$$

$$[M]\{\ddot{u}_D^{(2)}\} + [K]\{u_D^{(2)}\} = -[M]\{r_2\}\ddot{\theta}_D \quad (4)$$

where $\{r_1\}=\{1,0\}^T$ and $\{r_2\}=\{0,1\}^T$ are influence vectors. By applying the conventional response spectrum procedure for equations (3) and (4) respectively, the maximum response $\{u_D^{(1)}max\}$ and $\{u_D^{(2)}max\}$ can be evaluated. Using $\{u_D^{(1)}max\}$ and $\{u_D^{(2)}max\}$, the maximum response $\{u_D max\}$ can be estimated as follows;

$$\{u_D max\} = \sqrt{\{u_D^{(1)^2}max\} + \{u_D^{(2)^2}max\}} \quad (5)$$

3.5 Design Seismic Motion

The standard response spectrum commonly applied to the Akashi Kaikyo Bridge foundations is defined at the outcrop of the Kobe stratum (see Figure 4) as an envelope of the acceleration response spectra obtained under the following conditions (see Figure 5).

(1) An earthquake of a magnitude 8.5 is assumed to have occurred off Shikoku-Kii Peninsula 150 km to the south of the bridge site.

(2) Earthquake occurrences within a radius of 300km of the site are probalistically evaluated (Seismic Hazard Analysis), and the acceleration response spectrum corresponding to a return period of 150 years is selected.

The standard acceleration response spectrum corresponding to 5% damping ratio specified in the above-mentioned procedure is indicated in Figure 8. The response spectrum is applied to the upper boundary of the Kobe stratum (in the form of time history) at each foundation site and the effective seismic motion is determined using the finite element method which is described later.

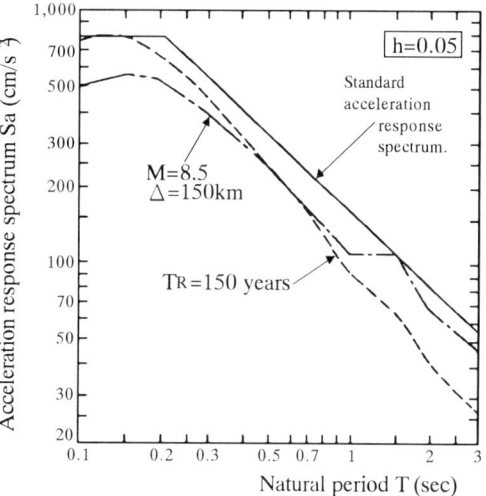

Figure 8 Standard Acceleration Response Spectrum

4 EXAMPLE CALCULATIONS

This chapter gives example calculations for foundation in accordance with the procedure described in the previous chapters. The caisson foundation of 2P is selected for this purpose (see Figure 3).

4.1 Analysis Conditions

The geometry of the 2P caisson is indicated in Figure 9. In order to investigate the effects of embedment to the response of the foundation, three embedment depths, a = 5, 10 and 15m, are examined. Figure 9 also shows a finite element model which is used in an evaluation of the effective seismic motion. The transmitting boundary is provided at the side of the stratum in order to prevent the wave reflection in the model. The shear moduli and the damping ratios of the soil elements are determined using the results of the equivalent linear one-dimensional wave propagation analysis (SHAKE).

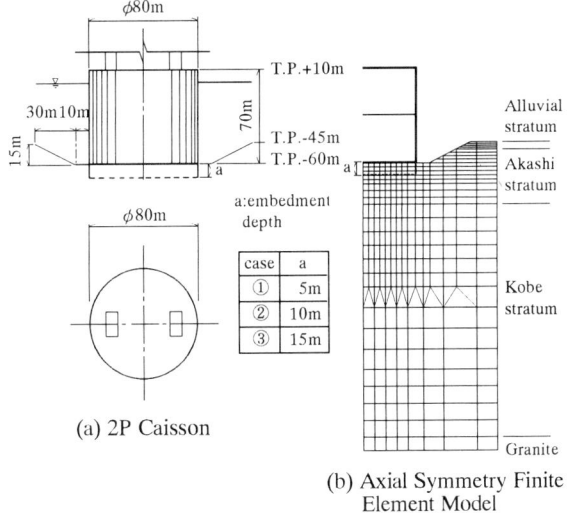

(a) 2P Caisson

(b) Axial Symmetry Finite Element Model

Figure 9   2P Caisson and its Finite Element Model

4.2 Effective Seismic Motion

Although the effective seismic motion can be obtained by applying the seismic excitation in a form of time history at the bottom of the FEM model shown in Figure 9 (b) as the response of a massless caisson, the following alternative procedure is used in order to investigate the effect of the kinematic interaction in terms of the embedment depth. If the response of the bottom of the rigid foundation to the bottom of the FEM model and the response of the free field surface to the bottom of the FEM model are obtained, a relationship of those two responses may be expressed as,

$$F_{B/A}^u(\omega) = F_B^u(\omega)/F_A(\omega) \quad \text{and} \quad F_{B/A}^\theta(\omega) = F_B^\theta(\omega)/F_A(\omega) \cdots (6)$$

where $F_A(\omega)$ is the frequency response of the free field surface to the bottom of the FEM model which is obtained by the one-dimensional propagation theory, and $F_B(\omega)$ is also the frequency response of the bottom of the caisson to the bottom of the FEM model. $F_B^u(\omega)$ and $F_B^\theta(\omega)$ are obtained in terms of the massless rigid foundation by using the FEM model shown in Figure 9 (b). u and θ indicate the corresponding components of the effective seismic motion. $F_{B/A}^u(\omega)$ is called as a filter for the effective seismic motion.

The effective seismic motion can be calculated in the following form.

$$u(\omega) = F_{B/A}^u(\omega)A(\omega) \quad , \quad \theta(\omega) = F_{B/A}^\theta(\omega)A(\omega) \cdots (7)$$

where $u(\omega)$ and $\theta(\omega)$ are the fourier spectra of the effective motion, and $A(\omega)$ is the fourier spectrum of the free field ground motion. $A(\omega)$ can be obtained using SHAKE. In the analysis of SHAKE, acceleration time history with the characteristics of which are adjusted to the standard acceleration response spectrum (Figure 8) is applied to the upper boundary of the Kobe stratum.

Figure 10 shows absolute values of the filter $F_{B/A}^u(\omega)$. As can be seen in the figure, the larger the frequency, the more substantially the amplitude if the filter in a horizontal direction is reduced, and the deeper the embedment, the more remarkable this trend becomes. Table 1 summarizes the maximum value of effective seismic motion at the bottom of the foundations obtained by the above-mentioned procedure. It is found that the maximum acceleration in horizontal direction decreases as the depth of the embedment increases. Meanwhile, although the rotational component increases, it is confirmed that effect of rotational component is negligible in the case of foundations with shallow embedment.

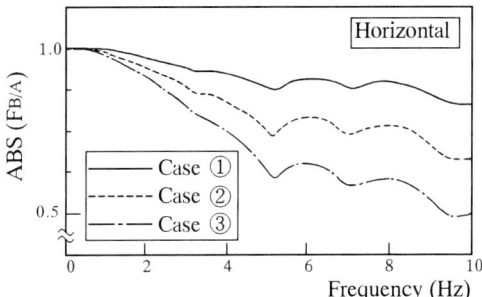

Figure 10   Filter $F_{B/A}^u(\omega)$ for Effective Seismic Motion

Table 1   Peak Values of Effective Seismic Motion

Case Component	F.F*	①	②	③
Horizontal Acc.(cm/s^2)	251	231	211	192
Rotational Acc.(rad/s^2×10^{-2})	—	0.09	0.34	0.75

* Free field motion

The acceleration response spectra of effective seismic motion for 5 % damping ratio are shown in Figure 11 with those of the free field motion and the case where no embedment is made. In addition, Figure 12 indicates the ratios of the acceleration response spectrum of effective seismic motion in the case of a response spectrum of case ① = 1.0. It can be seen from these figures that the acceleration responses decrease with the increase of the depth of embedment and, for example, in a period of 0.5 seconds the response in case ③ becomes 92% of that in case ① (Figure 12). It can also be seen from Figure 11 that the response spectrum of the free field motion is remarkably large compared to other cases.

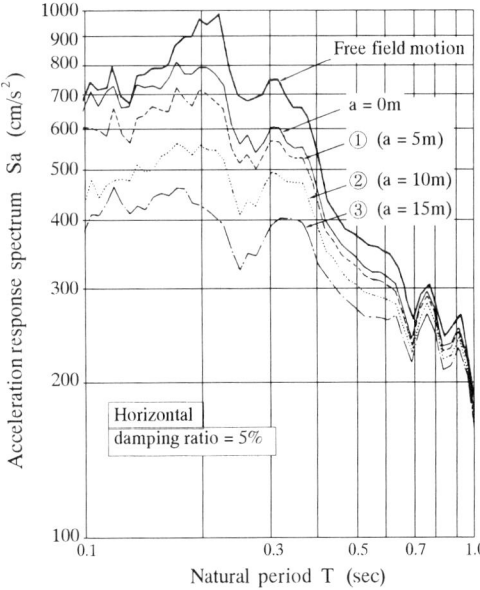

Figure 11   Acceleration Response Spectrum of Effective Seismic Motion

4.3 Dynamic Stiffness

Figure 13 gives an example of dynamic stiffness for rocking mode $K_{22}(\omega) = K_{22}(\omega) + i\omega c_{22}(\omega)$ when the depth of embedment is zero. This dynamic stiffness is calculated by FEM in this case. CFM proposed by Harada et al.(Harada, Kubo and Katayama

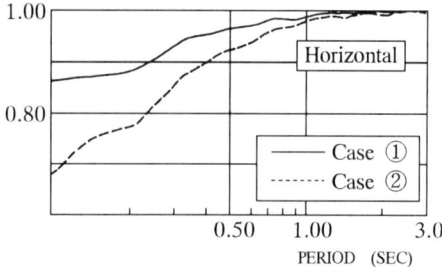

Figure 12　Response Spectrum Ratio when Case ①=1.0

Table 2　Results of Dynamic Response Analyses

	a= 0m	a = 5m	a = 15m
Natural period (sec)	0.469	0.475	0.504
Damping ratio (%)	7.4	7.7	8.6
Response Acc.* (g)	0.265	0.262	0.227

* at the center of gravity

5. CLOSING REMARKS

In order to execute a rational aseismic design for the foundations of the Akashi Kaikyo Bridge, a new procedure has been proposed taking into account the dynamic interaction effects between the ground and foundations. Although not shown in this paper, verification of the proposed procedure has been made using a more accurate numerical analysis procedure (FEM) and also experimental data, which indicates that the proposed procedure is able to estimate the response of the foundations reasonably well when used in design practice.

1981, Harada, Kubo and Katayama 1983) through the theoretical analysis of the elastic medium surrounding the structure, can also be applied. Although a more accurate solution is obtained by FEM rather than by CFM, CFM is adopted for the design of the Akashi Kaikyo Bridge foundations because of its simplicity and economy which is desired in the design practice. It is confirmed that the difference between the results obtained by both methods is minor in a practical sense.

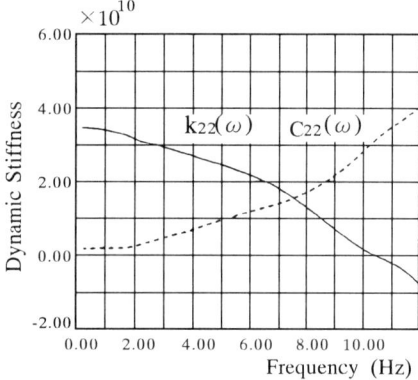

Figure 13　Example of Dynamic Stiffness (by FEM)

ACKNOWLEDGMENTS

The authors would like to express their thanks to the working group on Earthquake Resistant Design (Leader: Prof. E. Kuribayashi) of the Subcommittee on the Study and Research Pertaining to Earthquake Resistant Foundations for Honshu-Shikoku Bridges (Chairman: Prof. K. Kubo), the Japan Society of Civil Engineers, who were responsible for the examination of this new aseismic design procedure.

The authors would also like to convey their appreciation for the valuable guidance extended by Prof. T. Harada of Miyazaki Univ. and the personnel concerned of the Earthquake Disaster Prevention Department of the Public Works Research Institute, Ministry of Construction.

4.4 Response of Rigid Body Foundation

Using the effective seismic motion and dynamic stiffness discussed previously, response analyses were carried out based on equations (3) to (5) with respect to the three cases where depths of embedment are 0, 5 and 15 m.

Applying equations (3) to (5) in the response analysis, it is impossible to directly use the frequency-dependent stiffness, such as shown in Figure 13. Therefore, the predominant frequency ω_1 of the foundation is determined in advance using the model shown in Figure 7 with frequency-dependent dynamic stiffness. Frequency-independent dynamic stiffness is determined such manner as $K = K(\omega_1) = k(\omega_1) + i\omega c(\omega_1)$. Using these stiffnesses, the damping ratio and the natural period of the system can be calculated by the complex eigenvalue analysis.

The results of the response analysis are summarized in Table 2. From these results, the following comments can be made.

(1) The natural period lengthens in accordance with the depth of the foundation embedment. This is because, given the constant elevation of the foundation's crown, the center of gravity moves upwards and the mass increases in accordance with the increased embedment depth.

(2) The damping ratio increases in accordance with the increased embedment depth due to the effect of the underground dissipation of wave energy as a result of taking the dynamic stiffness into consideration.

(3) The horizontal acceleration at the center of gravity of the foundations decreases with the increase of the embedment depth. This tendency is considered to be combined effects of effective seismic motion (decrease with depth), damping ratio and the natural period.

REFERENCES

Kashima, S., Yasuda, M., Kanazawa, K. and Kawaguchi K. 1987. Earthquake Resistant Design of Akashi Kaikyo Bridge, Third Workshop on Performance and Strengthening of Bridge Structures, in Tsukuba, Japan.

Takahashi, K., Miyajima, K., Kashima, S., Yamamoto, N., Yamagata, M. and Aizawa, R. 1977. Investigation of Bearing Capacity of Foundation Ground of Honshu-Shikoku Bridge, Proc. of 9th ICSMFE.

Japan Road Association 1980. Specifications for Highway Bridges, Part V Aseismic Design. (in Japanese)

Honshu-Shikoku Bridge Authority 1977. Aseismic Design Standard Applied to Honshu-Shikoku Bridge. (in Japanese)

Japan Society of Civil Engineers 1983. Study and Research Report in Earthquake Resistant Design for Honshu-Shikoku Bridges : pp50 - 53. (in Japanese)

Harada, T., Kubo, K. and Katayama, T. 1981. Dynamic Soil-Structure Interaction Analysis by Continuum Formulation Method, Report of the Institute of Industrial Science, the University of Tokyo, Vol. 29, No. 5.

Harada, T., Kubo, K., Katayama, T. 1983. Dynamic Stiffness of Embedded Cylindrical Rigid Foundation, Proc. of JSCE, No. 339. (in Japanese)

Kashima, N., Harada, T. and Isoyama, R. 1986. Response Spectra Including Dynamic Soil-Structure Interaction Effects for Seismic Design, 7th Japan Earthquake Engineering Symposium. (in Japanese)

Investigation and seismic stability evaluation of rock foundation of nuclear power plants in Japan

Hiroshi Ito
Central Research Institute of Electric Power Industry, Chiba, Japan

Youichi Momose
Chubu Electric Power Co., Inc., Aichi, Japan

Yuichi Shibagaki
Kansai Electric Power Co., Inc., Osaka, Japan

ABSTRACT: In Japan, the reactor buildings of nuclear power plants are constructed on bedrock. This paper describes the results of a survey relating to current reactor building foundations as well as the process for determining the seismic stability, from the test stage through to the evaluation. In considering this process, two representative examples of recently completed plants are introduced, one that was constructed on a hard rock foundation and the other on a soft rock foundation.

1 INTRODUCTION

In Japan, where earthquakes often occur, not only are reactor buildings of nuclear power stations supported on rock foundations, but they are also built using exacting aseismic design and evaluation methods in order to give them a much higher tolerance to earthquake activity than general important structures. The basic philosophy and procedures are prescribed by the Committee for Nuclear Safety of the Science and Technology Agency in their "Regulatory Guide for Aseismic Design of Nuclear Power Reactor Facilities", and in the Special Safety Committee for Nuclear Reactors' publication "Guidelines for the Safety Evaluation of the Geology and Foundation Ground for Nuclear Power Plants". For actual purposes, the above principles are incorporated into such publications as the "Standardization of Seismic Resistance Evaluation Methods for the Foundation Ground of Nuclear Power Plants" produced by the Japan Society of Civil Engineers, and the Japanese Electrical Association's "Technical Guidelines for Aseismic Design of Nuclear Power Plants JEAG4601-1987" and these are generally applied. These guidelines and evaluation methods have developed into their present form as a result of the experienced gained in the design and construction of the many power plants constructed in Japan since the first Tokai Nuclear Power Station in 1959, and also as a result of vibration tests on actual plant and scale models, and the progress in seismology and methods of structural analysis.

More specifically, the various facilities of a nuclear power plant are divided into four classes, As, A, B, and C, with respect to their degree of importance for resisting earthquakes. The nuclear power plant buildings, the foundations for which the highest class of importance "As" applies, are the most strictly evaluated with regard to their seismic stability. For this reason, based on the most thorough geological and foundation ground surveys, with respect to a basic seismic motion S_2 for each site, which is the motion brought about by the design earthquake assumed from past earthquakes and active faults, an evaluation of the stability, considering such things as the bearing capacity of the foundation ground during an earthquake, ground settlement, and sliding, is made. For the evaluation of stability, the results of geological and foundation ground surveys and various tests are used, and the foundation ground model is made and the physical values are determined. As for the evaluation methods, the well established and easily applied conventional sliding surface method, or the static FEM analysis method (seismic intensity method), that can be applied where the ground conditions, loading conditions, and initial conditions are complex, and the dynamic FEM analysis method (response analysis) are commonly used.

In this paper, survey and test methods that are suited to the characteristics of the bedrock, and the aseismic stability evaluation process, are introduced with respect to hard rock and soft rock sites that exhibit the physical and mechanical properties that are typical of nuclear power station foundations to be found in Japan.

2 OUTLINE OF NUCLEAR REACTOR BUILDINGS IN JAPAN

In Japan at the current time, there are 45 commercial reactors (approx. 37,200 MW) in use in nuclear power generating stations, and a further seven (approx. 7,900 MW) are under construction. In Table 1 are shown the ground characteristics of the construction works started within the previous ten years. As shown in the table, foundation ground can basically divided into two major groups, one comprising sedimentary soft rock that exist in places such as Kashiwazaki and Hamaoka, while at the other hard rock is to be found, for example at Onagawa and Ooi. For the purposes of differentiating between the two, soft rock is taken to have an unconfined compressive strength of less than about 10 MPa-20 MPa, while hard rock is that having a strength higher than this. The characteristics of the rock mass made of hard rock varies according to the hardness, characteristics of the rock mass joints, and degree of weathering. For this reason, surveys and evaluations of the rock mass have to be performed with due consideration to these characteristics. On the other hand, soft rock tends to be of generally uniform geological characteristics, and the results of rock tests and rock mass tests are often the same. For this reason, data of the various surveys and tests based on the rock is obtained and the evaluation of the foundation ground is often carried out with reference to this data. In such cases however, it is necessary, either experimentally or by analysis, to confirm the fact that the rock mass characteristics can be adequately evaluated from the rock tests alone.

The stability evaluation is carried out with regard to bearing capacity, settlement, and sliding whether the rock is soft or hard, but for actual purposes the evaluation is based on the sliding stability. In other words, in order to select a foundation ground with the requisite bearing capacity at the initial planning phase, as the bearing capacity is rarely a problem, it is usual to confirm this by carrying out in-situ bearing capacity tests. Also, to evaluate the stability with regard to settlement, it is general to use the ground deformation properties obtained from in-situ testing in conjunction with elasticity theory.

Table 1. Outline of the mechanical properties of the recent NPP foundation ground

Power Station		Output (Mw)	Reactor Type	Rock Type	Strength of Rock Mass		Deformation of Rock Mass			Velocity of Elastic Wave		Unit Weight γ_t(kN/m³)	q_u (MPa)	S_2 (gal)	Note
					C(MPa)	φ(°)	D(MPa)	E_s(MPa)	E(MPa)	Vp(km/s)	Vs(km/s)				
Tomari	No.1	579	PWR	tuff breccia	1.17~2.41	45.9~46.3	1,800~7,100	2,100~9,200	2,400~8,500	3.14~3.41	1.45~1.87	21.8~23.1	8.1~24.2	370	in operation
				tuff	1.37	57.2	2,600~6,500	2,900~8,000	3,900~7,900			21.7	25.9~27.0		
	No.2	579	PWR	tuff breccia	2.07~2.83	45.5~50.3	1,800~4,200	1,600~4,900	2,600~4,800	2.82~3.41	1.53~1.88	20.6~22.3	13.0~24.1	370	in operation
				andesite lava	1.39	50.4	1,000~3,100	1,500~4,400	1,900~5,100			26.0	65.5		
				tuff	1.04	49.7	700~2,200	1,000~2,500	1,200~3,100			19.8~21.1	15.9~39.1		
				pumice-tuff	0.97	42.2	800~900	900~1,100	1,000~1,300			18.2	6.5		
Onagawa	No.2	825	BWR	sandstone	0.50~1.75	43~47	500~850	1,000~1,800	1,100~2,000	5.2	2.5	24.7~26.6	12.6~140.6	375	under construction
				shale	1.61	46	1,100	2,200	2,300			26.4~27.4	32.3~94.3		
	No.3	825	BWR	sandstone	0.80~1.32	50~54	520~800	1,200~1,900	1,200~1,900	5.1	2.5	25.5~26.5	48.3~153.2	375	planning phase
				shale	0.77	32	460	1,500	1,300			25.8~27.4	28.1~102.7		
Kashiwazaki	No.2	1,100	BWR	mudstone	0.36	43	656	819	772	1.71	0.52	16.8	3.19	450	in operation
	No.5	1,100	BWR	mudstone	0.30	41	597	753	714	1.64	0.53	17.4~20.2	2.42~3.36	450	in operation
	No.6	1,356	ABWR	mudstone	0.38	40	439	510	498	1.54	0.48	17.4~20.2	3.36~5.05	450	under construction
	No.7	1,356	ABWR	mudstone	0.38	38	341	392	415	1.59	0.49	17.5~20.2	3.04~4.96	450	under construction
Shika	No.1	540	BWR	andesite (homogeneous)	1.02	52.9	19,000	37,000	39,000	3.4~4.6	1.4~2.16	26.6	140~148	490	in operation
				andesite (breccia)	0.85	52.5	21,000	27,000	35,000			22.5	17.1		
				tuff breccia	0.86	58.6	19,000	24,000	31,000			23.0	19.3		
Hamaoka	No.4	1,137	BWR	altenation (sandstone and mudstone)	0.77	34	850~2,230	1,100~2,760	850~2,180	2.0~2.1	0.7~0.8	20.4~20.7 (mudstone) 21.2~21.3 (sandstone)	10.4 (mudstone) 2.61 (sandstone)	600	in operation
Ooi	No.3	1,180	PWR	fine quartz diorite	1.65~2.10	50.3~60.3	967~14,500	13,900~20,400	1,720~17,700	4.5	2.1	27.4	143.9	405	in operation
				diabase	1.41~2.19	38.9~56.1	1,370~3,830	1,540~4,470	1,960~5,240			28.8	103.4		
	No.4	1,180	PWR	fine quartz diorite	1.65~2.10	35.1~56.1	3,030~12,000	3,040~15,000	3,750~15,000	4.8	2.4	27.5	107.9	405	in operation
				diabase	1.41~2.19	38.9~56.1	1,370~3,830	1,540~4,470	1,960~5,240			28.0	124.5		
Shimane	No.2	820	BWR	black shale	0.95~1.16	53.9~57.3	930~3,170	1,020~4,480	860~4,720	3.63	1.64	25.6	80.8	380	in operation
				tuff breccia	1.59	62.4	2,510~8,510	10,880~12,800	9,600~10,240			24.5	58.5		
				tuff	0.53~1.79	37.9~55.9	370~8,850	350~15,150	500~16,460			24.8	84.2		
Monju		280	FBR	granite	1.8~4.1	39~49	240~6,310	560~8,830	580~8,830	4.3	1.9	25.0~25.9	18.1~78.8	466	under construction

3 HARD ROCK SITE (OOI NUCLEAR POWER PLANT)

3.1 Outline of the power plant and geology

The Ooi Power Plant is located at about the middle of Japan's central island of Honshu, on the Japan Sea side. An outline of the power plant can be seen in Table 2. The reactor building foundations are supported on hard bedrock. This bedrock is paleozoic igneous rock, comprising fine grain quartz diorite and diabase. The unconfined compressive strength of these rocks is approximately 100 MPa.

Table 2. Profile of nuclear power plant

No. of plant	No.3	No.4
Company	Kansai Electric Company	
Location	Ooi-cho, Ooi-district, Fukui-Prefecture	
Reactor type	PWR	PWR
Output of electricity	1,180 MW	1,180 MW
Starting construction	1987.5	1987.5
Starting operation	1991.12	1993.2

3.2 Surveys and tests in the region of the reactor location

To adequately determine the condition and structure of the foundation ground in the vicinity of the reactor location, 188 borehole surveys were made providing a total boring core length of 12,400m. Also, 1,350 m of exploratory adits were excavated to enable direct observation of the foundation ground and at the same time rock and rock mass tests were carried out to determine the physical properties of the NPP foundation ground. In addition, various other surveys, such as trench surveys were performed as required. The main surveys and tests are shown in Table 3.

Table 3. Contents of survey and test

Items of test	Contents of test
1) Laboratory test of rock a. physical test	· unit weight, rate of effective porosity, rate of water absorption, velocity of ultra sonic wave propergation, etc.
b. mechanical test	· uniaxial compression test · triaxial compression test · tension test, etc.
2) In-situ test of rock mass a. elastic wave velocity test in adit	· total observation length : about 1500m · observation span : 2.5m
b. deformation test	· plate loading test (16 points; φ 80cm, 30cm)
c. bearing capacity test	· plate loading test (4 points; φ 30cm)
d. shear test	· block shear (16 points, scale;60cm×60cm×60cm · rock shear (11 points, scale; 60cm×60cm×60cm
e. PS-logging	· total survey length ; 600m · observation span : 2m
f. boring pressure meter test	· operation in two bore hole with 300m depth
g. shumit rock hammer test	· observation span : 50cm in adit
h. in-situ permiability test	· operation in two bore hole with 300m depth · interval length of test ; 5m

3.3 Bedrock classification and evaluation of engineering characteristics

For the purposes of investigating the foundation ground of the reactor facility, it is necessary to categorize the various geological elements and obtain a firm grasp of the distribution of rock having the same engineering characteristics. Particularly at hard rock sites such as this one, it is essential that the rock classification adequately reflect the results of the rock mass tests.

From the fact that the major portion of the ground at the Ooi power plant site is igneous rock, the rock mass classification method (JSEG 1992) proposed by the Central Research Institute of the Electric Power Industry, which is often used in Japan with respect to hard rock, was adopted. In the exploratory adit, the rock mass was examined with respect to three basic parameters, i) degree of weathering, ii) characteristics of rock mass joints, iii) rock hardness (hammer tapping aural test) to provide the classification. This is shown in Table 4. Considering the above three elements of classification, finer classifications were prescribed for the boring core, as shown in Table 5. The classification of the rock mass for this site was carried out using the above two classification methods in accordance with the principles of Figure 1. Within this classification, the classification of engineering characteristics, necessary for the evaluation of the stability of the foundation ground, was determined based on rock and rock mass test results. The strength and deformation characteristics were decided as shown in the flow chart of Figure 2, from consideration of the distribution (distributed proportion) of each sub-division for each classification of rock mass. The physical values used for stability analysis were thus determined as shown in Table 6.

3.4 Evaluation of foundation ground stability

The safety of the foundation ground with respect to sliding was investigated by conventional methods and FEM (static and dynamic) analysis, using a model constructed on the basis of the engineering characteristic classifications described above. The model used for the analyses can be seen in Figure 3.

In the static analysis, a calculated seismic force based on the story shear coefficient $3C_I$ prescribed by the "Technical Guidelines for Aseismic Design of Nuclear Power Plants" was applied, and a seismic force equating to a seismic coefficient of 0.24 was applied in the vertical direction. A seismic coefficient of 0.2 in the horizontal direction, and 0.1 in the vertical direction

Table 4. Criteria of quality division of rock mass in adit

Class	Description
B'	·Rock is fresh and hard. ·Joints are closely adhered, no limonite. ·Joints spacing is more than 50cm. ·When struck by hammer, makes metallic-sound.
C'_H	·Rock is generally fresh and hard. ·Opening percentage of joints is about 50% in common and limonite adhered along joint surface. ·Joints spacing is from a few centimeter to 50cm in common. ·When struck by hammer, makes a dull metallic-sound.
C'_M	·Rock is slightly weathered and somewhat softened. ·Joints is almost opening and limonite adhered along joint surface. ·Joints spacing is from a few centimeter to 30cm in common. ·Sound by hammer blow is dim.
C'_L	·Rock is slightly weathered and softened. ·Joints is opening, rock blocks go to pieces by light hammer blow. ·Sound by hammer blow is dim.
D'	·Rock is strongly weathered. Rock texture can recognizes, but is soft. ·Scrubbing by hammer crumbles and become small pieces or sand form. ·Sound by hammer blow is remarkably dim.

Table 5(a). Classification of properties of boring core

hardness weathering	shape of core (length of core)	fracture condition	subdivision
① ·fresh color ·vary hard	[I] columnar (more than 50cm)	[a] adhered	b'
	[II] columnar (10~50cm) or short columnar (less than 10cm)	[a] adhered	C'h
	[III] block shape	[a] adhered	C'h
		[b] opening (adhesion of limonite in opening part and no pollution of limonite in the rock.)	C'm
② ·fresh color and weathering color along the fracture ·hard	[I] columnar (more than 50cm)	[a] adhered	b'
	[II] columnar (10~50cm) or short columnar (less than 10cm)	[b] opening (adhesion of limonite in opening part and no pollution of limonite in the rock.)	C'h
	[III] block shape	[b] opening (adhesion of limonite in opening part and no pollution of limonite in the rock.)	C'm

Table 5(b). Classification of properties of boring core

hardness weathering	shape of core (length of core)	fracture condition	subdivision
③ ·weathering color ·half hard	[II] columnar (10~50cm) or short columnar (less than 10cm)	[b] opening (adhesion of limonite in opening part and no pollution of limonite in the rock.)	C'h
		[c] opening (adhesion of limonite in opening part and pollution of limonite in the rock.)	C'm
	[III] block shape	[c] opening (adhesion of limonite in opening part and pollution of limonite in the rock.)	C'ℓ
④ ·slightly recognition or rock forming minerals and grains. ·soft-hard	[III] block shape	[d] sand or clay	d'
	[IV] sand or clay	[d] sand or clay	d'

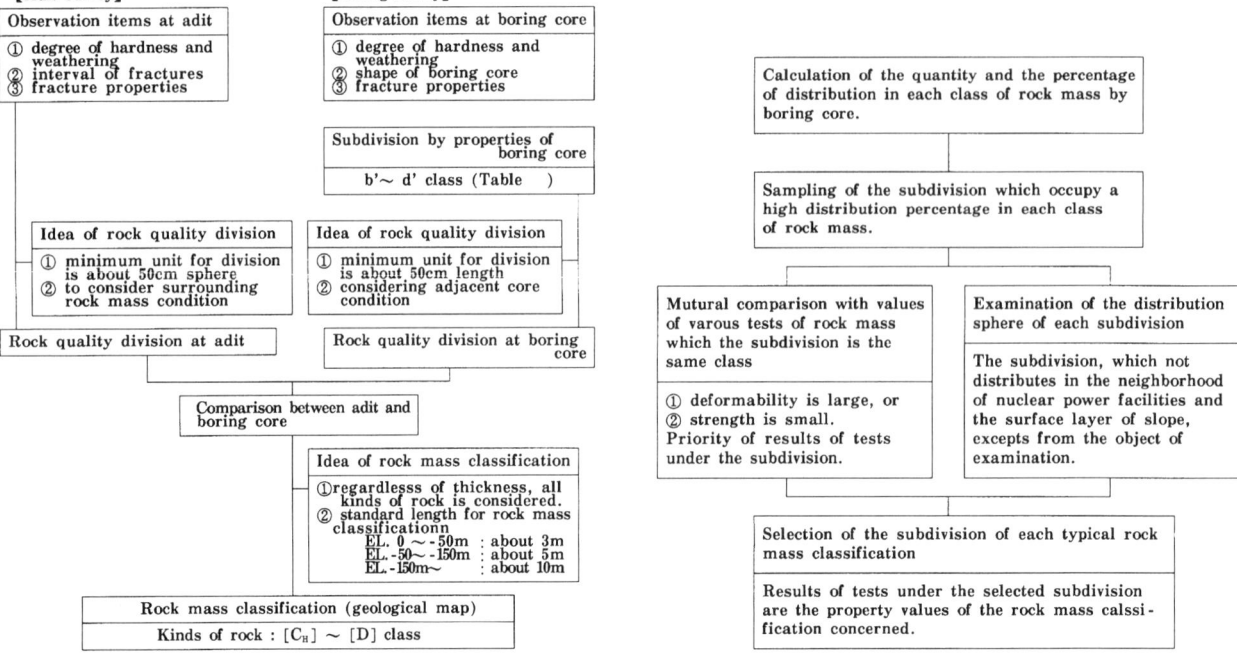

Figure 1. Flow chart of rock mass classification

Figure 2. Idea of strength and deformability of rock mass classification (engineering classification)

Table 6. Mechanical properties for stability analysis of NPP foundation

Kinds of rock mass		static					dynamic				
		γ_t (kN/m³)	E (MPa)	ν_s	τ_o (MPa)	ϕ (degree)	G (MPa)	τ_o (MPa)	ϕ (degree)	h (%)	ν_d
fine quartz diorite	[C_H]	27.2	14,400	0.23	2.1	60.3	13,000	2.1	60.3	3.0	0.34
	[C_M]	26.7	2,680	0.23	1.65	50.3	11,000	1.65	50.3	3.0	0.34
	[C_L]	26.6	990	0.23	0.2	35.1	2,900	0.2	35.1	3.0	0.34
	[D]	17.1	24	0.40	0.072	17.4	400	0.072	17.4	4.0	0.37
diabase	[C_H]	28.8	3,490	0.26	2.19	56.1	16,000	2.19	56.1	3.0	0.34
	[C_M]	28.7	1,810	0.26	1.41	38.9	11,000	1.41	38.9	3.0	0.34
	[C_L]	27.4	620	0.26	0.2	35.1	1,900	0.2	35.1	3.0	0.34
	[D]	17.1	24	0.40	0.072	17.4	400	0.072	17.4	4.0	0.37
slate	[D]	17.9	21	0.40	0.064	18.3	380	0.064	18.3	4.0	0.37
gabbro	[C_M]	29.6	5,900	0.26	2.19	56.1	13,000	2.19	56.1	3.0	0.33
Talus deposits		19.7	33	0.40	0.022	26.7	290	0.022	26.7	4.0	0.43
Terrace deposits		22.5	39	0.40	0.034	25.0	630	0.034	25.0	4.0	0.45
fracture zone		19.8	$E = 12.52\sigma_v^{0.89}$ $G = 4.47\sigma_v^{0.89}$	0.40	0.078	19.5	$G = 33.5\sigma_v^{0.75}$	0.078	19.5	3.0	0.37

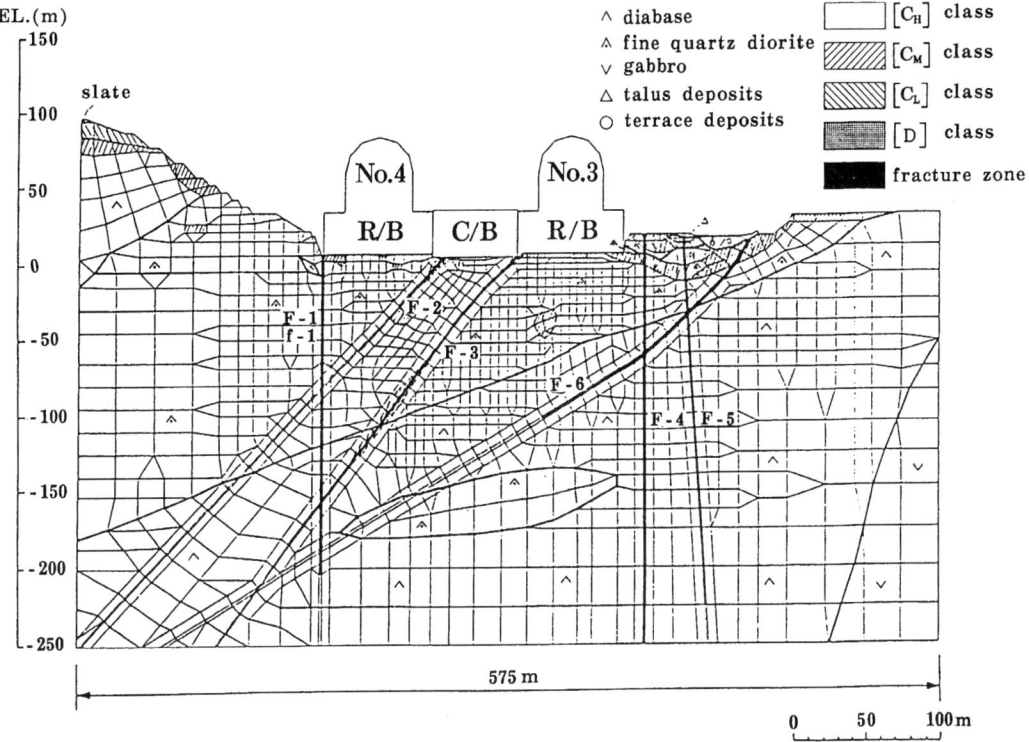

Figure 3. FEM model for stability analysis

were applied to the foundation ground. On consideration of the strength characteristics of the ground, a non-linear static analysis was carried out.

In the dynamic analysis, an input seismic motion based on the basic seismic motion S_2 (design limit earthquake) was applied to the model lower edge. The vertical seismic force was the same as used for the static analysis.

From the various analyses, the values for sliding safety factor were: from the conventional method (Bishop Method) above 5.7, from the static FEM analysis above 6.5, and from the dynamic FEM analysis above 6.9. From these results it was confirmed that an adequate safety factor with regard to sliding existed for the foundation ground for the seismic force considered. Also, from the results of bearing capacity tests and analyses, it was confirmed that adequate safety existed with regard to support force and ground settlement during an earthquake.

4 SOFT ROCK SITE (HAMAOKA NUCLEAR POWER PLANT)

4.1 Outline of the power plant and geology

The Hamaoka Power Plant is located at about the middle of Japan's central island of Honshu, on the Pacific Ocean side. An outline of the power plant can be seen in Table 7.

The important structures such as the reactor buildings are supported directly on the soft bedrock.

The rock mass at the site is a sedimentary soft rock of the Miocene series of the Neogene system, comprising alternating layers of mudstone and sandstone. It contains only a few joints together with bedding planes, which are more or less uniformly distributed.

The average thicknesses of the mudstone layers and sandstone layers are 13 cm and 6 cm respectively at the site for the reactor building base mat, with mudstone becoming gradually predominant with depth. The dip angle of the bedding plane is approximately 10°.

The unconfined compressive strengths of the mudstone and the sandstone are about 10 MPa and 2 MPa respectively.

Table 7. Profile of Hamaoka nuclear power plant

No. of plant	No.4
Company	Chubu Electric Company
Location	Hamaoka-cho, Ogasawara district, Shizuoka-prefecture
Reactor type	BWR
Output of electricity	1,137MW
Starting construction	1989.8
Starting operation	1993.3

4.2 Surveys and tests in the region of the reactor location

To adequately determine the geological conditions and structure of the foundation ground in the vicinity of the reactor location, 23 borehole surveys were made providing a total boring core length of 4,250 m. Also, 300 m of exploratory adits were excavated and rock tests were carried out to determine the physical properties of the rock mass. The surveys and tests were performed with particular emphasis on determining the characteristics of the soft rock. An outline of these tests is given below.

I) Laboratory tests

Tests were carried out to determine the physical and mechanical properties of the mudstone and sandstone.

Triaxial compression tests were performed on specimens in their natural state at seven different confining pressure levels

(σ_3 = 0, 0.1, 0.3, 0.5, 0.8, 1.3, and 2.0 Mpa).

For the mudstone and sandstone, the Mohr-Coulomb failure criteria for each region (Figure 4) are as follows:

Mudstone
Region ①
$\tau = 2.25 + \sigma \tan 42°$
Region ②
$\tau = 4.08 + \sigma \tan 19°$

Sandstone
Region ①
$\tau = 0.33 + \sigma \tan 62°$
Region ②
$\tau = 1.10 + \sigma \tan 33°$

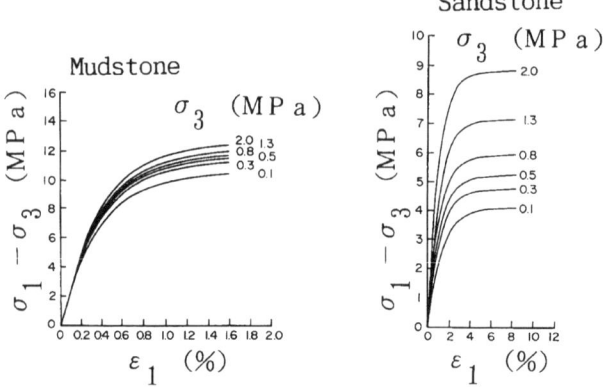

Figure 5. Mudstone and Sandstone Stress-Strain Curves

II) Large scale block shear tests

The large block shear tests were performed to determine the shear strength of the foundation rock directly beneath the reactor building. To determine the shear strength of the alternating layered rock mass, large scale blocks, 1.2 m long by 1.2 m wide and 0.6 m thick were tested as shown in Figure 6.

The rock mass was first subjected to an applied load of 0.5 MPa, equivalent to the dead weight of the reactor building under normal conditions. However, as earthquakes result in a variation of normal stresses, four normal stress levels, 0.1, 0.4, 0.7, and 1.0 MPa, were applied during the tests.

Following the increase in normal stress from the initial level of 0.5 MPa to each predetermined level, a shear load was then applied and increased at a rate of 0.15 MPa/min until failure occurred.

Figure 7 shows the relation between the normal stress and the shear stress for the sliding plane parallel to the bottom of the block. Generally, the shear strength tends to increase with an increase in the normal stress. From a regression analysis of the data with a linear function, the cohesion σ_0 was found to be 0.77 MPa with a friction angle of 34°.

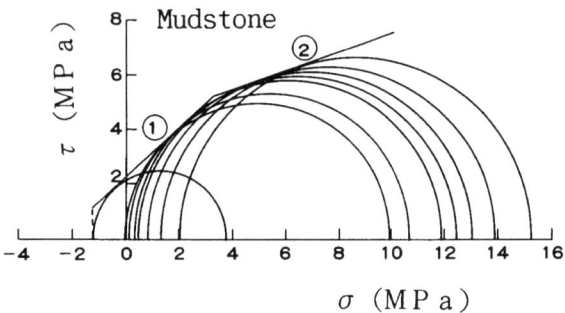

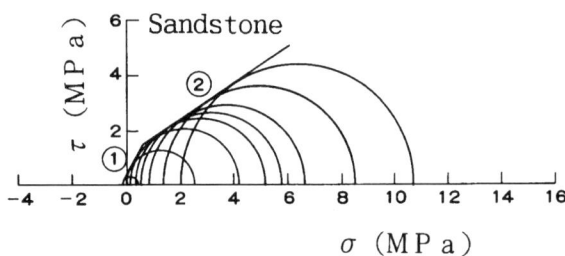

Figure 4. Mudstone and Sandstone Failure Envelopes

The deformation properties of the rock samples were approximated based on the stress-strain curves obtained from the triaxial compression tests, using the following exponential

$$\sigma_1 - \sigma_3 = (A\sigma_3 + B)\left\{1 - e^{-\varepsilon(C\sigma_3 + D)}\right\} \quad (1)$$

$$E = (A\sigma_3 + B)(C\sigma_3 + D)e^{-\varepsilon(C\sigma_3 + D)} \quad (2)$$

equations (Ito & Kitahara 1982).
where
A, B, C, and D are constants
σ_1 : axial stress
σ_3 : confining stress
ε_1 : axial strain
E : modulus of elasticity

Figure 5 shows the stress-strain curves for mudstone and sandstone approximated by equation (2).

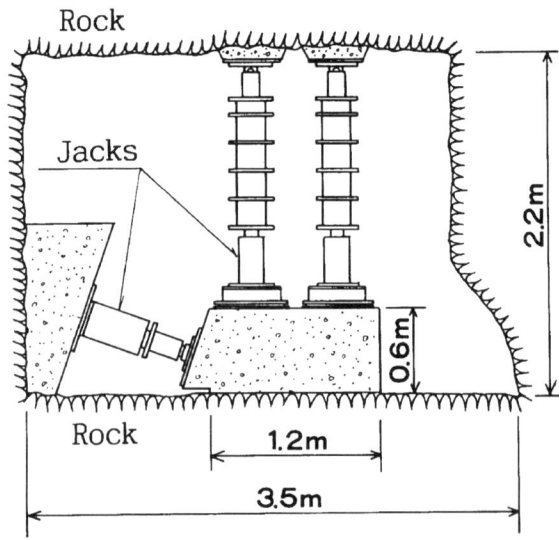

Figure 6. Block Shear Test Schematic

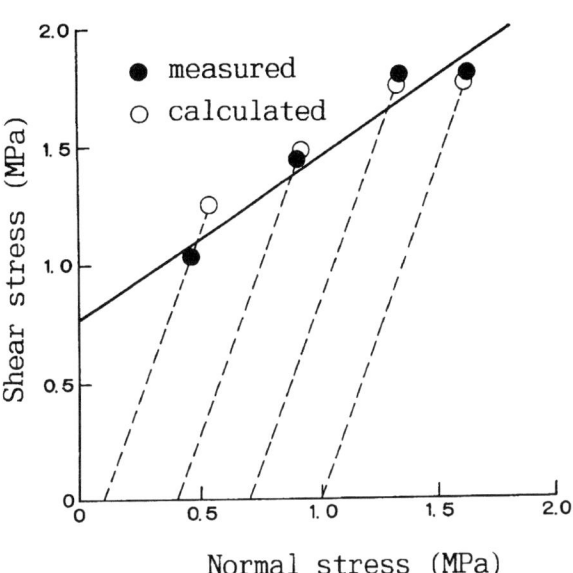

Figure 7. Comparison of Calculated and Measured Values of Shear Strength

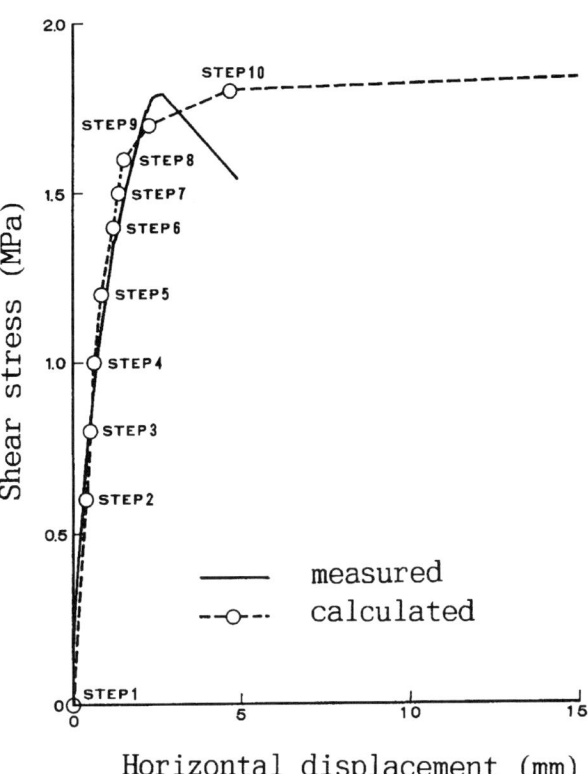

Figure 8. Comparison of Calculated and Measured Shear Stress - Horizontal Displacement Curves (σ N = 0.7 MPa).

III) Analytical simulation of the large scale block shear test

FEM analysis of the tests was carried out, taking into consideration the alternating mudstone sandstone structure of the rock mass, and a comparison made of the calculated results with the data obtained from in-situ deformation and strength tests.

In the FEM analysis, an in-situ alternating layered structure model was used to model the rock mass in the close proximity of the concrete block, and an orthotropic model was used to model the remainder of the analytical domain.

For the mechanical property values of materials in the analyses, the properties obtained from laboratory tests were used. For the deformation properties, the non-linearity due to the initial effective stress σ_0' and strain ε was taken into account. For the strength properties, the ratio ($\tau a/\tau$) of the shear strength τa under the initial effective stress σ_0' and the activated maximum shear stress τ were used to predict failure.

In the FEM simulation, an incremental non-linear analysis was performed. The shear load was incremented, and for each increment, every element was checked for yielding, and the stiffness of yielded elements was reduced in subsequent steps to simulate the propagation of the failure of the rock mass.

The calculated curves of shear stress and the block horizontal displacement by FEM are compared with those of the tests in Figure 8. As can be seen, the curves are in good agreement demonstrating the ability to predict the deformation properties of the rock mass by FEM analysis.

The point of abrupt and large deviation on the stress-displacement curve was taken to be the shear strength of the block in the FEM simulations (Figure 8) and it is compared with the in-situ test results in Figure 7.

Good agreement was achieved between the calculated and measured values and this verifies the possibility of estimating the shear strength of in-situ blocks using the deformation and strength properties obtained in laboratory tests and taking into account the structure of the rock mass.

4.3 Stability evaluation of the reactor building foundation ground

Using the geological profile based on the survey results and the physical values used in the analysis, the foundation ground sliding stability was investigated by carrying out a conventional analysis (Bishop Method) and FEM analysis (static and dynamic). The FEM model is shown in Figure 9 and the physical values used for the analysis are shown in Table 8.

For the static analysis, the ground was simulated by an orthotropic model and the physical values used were the mechanical characteristics obtained from tests on the rock. For the dynamic analysis, the elastic shear coefficient from elastic velocity of the rock mass was used taking into consideration strain dependence.

The static seismic force was taken to be the same as in 3.4 above. For the dynamic seismic force an input force based on the basic seismic motion S_2 was applied to the model bottom surface. For the vertical seismic force, a seismic coefficient of 0.31 was applied. Using the strength determined in the large scale block shear tests the safety factor during an earthquake was found to be, from the conventional method above 4.8, from static FEM above 5.8, and from dynamic FEM above 2.6, from which adequate safety with respect to sliding was confirmed for the seismic forces applied.

Also, from the results of bearing capacity tests and analyses, it was confirmed that adequate safety existed with regard to bearing capacity and ground settlement during an earthquake.

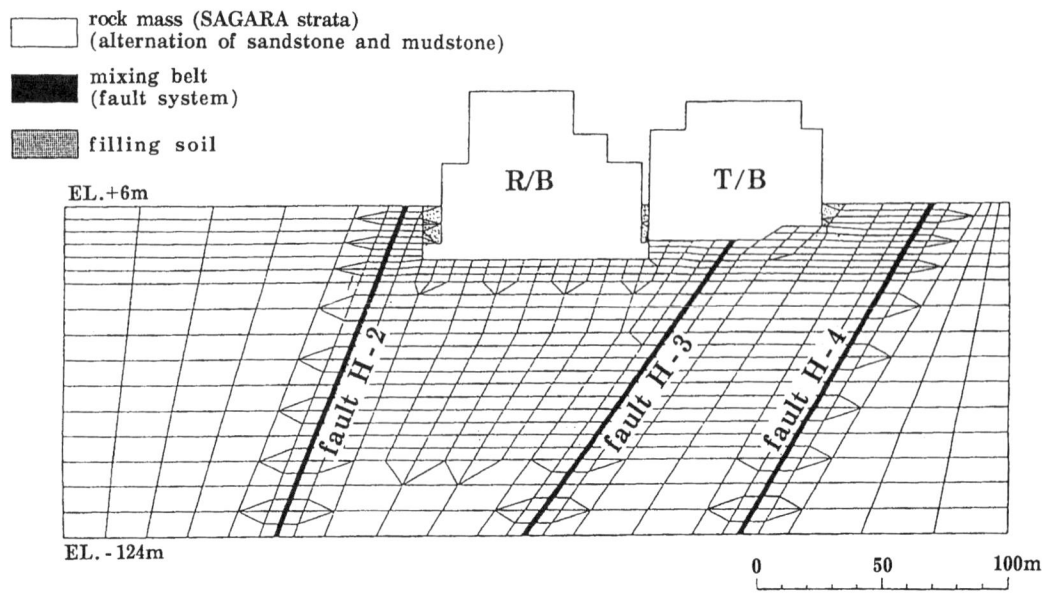

Figure 9. FEM model for stability analysis

Table 8. Mechanical properties for stability analysis of NPP foundation

division of rock mass	static analysis			dynamic analysis				strength	
	γ_t (kN/m³)	E (MPa)	ν_s	γ_t (kN/m³)	G_o (MPa)	G/G_o	ν_d	rock	rock mass
mudstone (Sagara strata)	20.7	(under low stress condition) $E = a_1 \cdot b_1 e^{-\varepsilon \cdot b_1}$ $a_1 = 0.411\sigma_o + 10.18$ $b_1 = 0.038\sigma_o + 26.64$ (under high stress condition) $E = a_2 \cdot b_2 e^{-\varepsilon \cdot b_2}$ $a_2 = 0.094\sigma_o + 11.39$ $b_2 = -0.108\sigma_o + 27.802$	0.23	20.9	(depth) 0~10m;770 10~20m;960 20~30m;1,110 30~60m;1,260	(shear strain) $\gamma=10^{-6}$;1.0 $\gamma=10^{-5}$;1.0 $\gamma=10^{-4}$;0.90 $\gamma=10^{-3}$;0.61	(depth) 0~10m;0.42 10~20m;0.42 20~30m;0.42 30~60m;0.41	(low stress) τ_o =2.25MPa ϕ=42° (high stress) τ_o =4.08MPa ϕ=19°	τ_o =0.77MPa ϕ=34°
sandstone (Sagara strata)	21.2	(under low stress condition) $E = a_3 \cdot b_3 e^{-\varepsilon \cdot b_3}$ $a_3 = 1.48\sigma_o + 2.60$ $b_3 = -1.689\sigma_o + 9.587$ (under high stress condition) $E = a_4 \cdot b_4 e^{-\varepsilon \cdot b_4}$ $a_4 = 0.239\sigma_o + 4.03$ $b_4 = -0.089\sigma_o + 8.988$	0.33	20.9	60~100m;1,430 100~200m;1,730 200~300m;2,050		60~100m;0.41 100~200m;0.39 200~300m;0.37	(low stress) τ_o =0.33MPa ϕ=62° (high stress) τ_o =1.1MPa ϕ=33°	τ_o =0.77MPa ϕ=34°
mixing belt of fault system	20.4	$E = a_5 \cdot b_5 e^{-\varepsilon \cdot b_5}$ $a_5 = 0.09\sigma_o + 0.83$ $b_5 = -0.244\sigma_o + 9.952$	0.48	20.4	(depth) 0~30m;300 30~60m;480 60~100m;660 100~200m;960 200~300m;1,250	(shear strain) $\gamma=10^{-6}$;1.0 $\gamma=10^{-5}$;1.0 $\gamma=10^{-4}$;0.91 $\gamma=10^{-3}$;0.50	0.48	τ_o =0.36MPa ϕ=21°	

(note) ε: Max. principal strain, σ_o: Mean principal stress

REFERENCES

Ito, H., & Y. Kitahara 1982. *Proceedings of a symposium on numerical models in geomechanics,* Zurich: 677-686. Rotterdam: Balkema.

Japan Society of Engineering Geology 1992. Rock mass classification in Japan. *Engineering geology*, special issue: 18-19.

Stability evaluation of large-scale bridge foundations built on sedimentary soft rock

K. Kanazawa & K. Kawaguchi
Honshu-Shikoku Bridge Authority, Tokyo, Japan

S. Matsumoto & T. Sanbyakuda
Oriental Consultants Co., Ltd, Japan

ABSTRACT: The Akashi Strait Bridge, which will connect Kobe City and Awaji Island, is a suspension bridge that will have the longest span in the world. The bridge foundations have been planned to be constructed on the sedimentary soft-rock and diluvial gravel strata, which will act as the bearing layers. These strata differ in their bearing and deformation characteristics from weathered granite and Izumi strata, which have been the bearing strata of bridges constructed in Japan so far. For these reasons, to evaluate foundation stability, methods for setting the ground coefficients by appropriately considering the soft rock in the sedimentary stratum (testing methods) and simple techniques of analyzing bearing capacity and displacement were studied. As a result, the slice method and RBSM method were adopted for bearing capacity analysis, and a simple calculation method for creep displacement was applied for displacement analysis. With its focus on the design of the Akashi Strait Bridge, this paper describes the viewpoints, procedures and techniques for evaluating the stability of the foundations of large-scale bridges.

1 GEOLOGY AKASHI STRAIT

The substructures that support the Akashi Strait Bridge consist of anchorages at both ends (1A and 4A in Figure 1) and main-tower foundations constructed under the sea (2P and 3P).

As shown in Figure 1, the geological makeup of Akashi Strait consists of a stratum of granite bedrock covered irregularly by other strata, i.e., Kobe Stratum, which belongs to the Miocene epoch of the Tertiary period, Akashi Stratum that extends from the Pliocene epoch of the Tertiary period through to the Pleistocene epoch of the Quaternary period, and diluvial and alluvial topsoil. The Kobe Stratum is a sedimentary stratum of soft rock composed mainly of sandstone and mudstone. However, on the Honshu side, while this stratum alternates between sandstone and mudstone, the stratum on the Awaji Island side is rich in coarse-grained arkose sandstone; thus there is a substantial difference in the lithofacies of both sides. The Akashi Stratum is a semi-solid stratum of sand and gravel containing boulders 200 mm or larger in diameter as well as fine to coarse gravels, and it is thickly deposited in the area near main tower foundation 2P.

Of the bridge foundations, Anchorage 1A and Main-tower Foundation 3P were planned to be constructed on the sedimentary soft-rock stratum (Kobe Stratum) as the bearing ground.

2 STRESS-STRAIN CHARACTERISTICS OF SEDIMENTARY SOFT-ROCK STRATUM (KOBE STRATUM)

The stress-strain diagram of the middle-grade sandstone of the Kobe Stratum, a result of triaxial compression tests under consolidated-undrained conditions (CU conditions), shows a noticeable tendency forwards so-called strain softening. That is, the shear strength reaches its maximum with relatively little strain; thereafter, the shearing stress gradually decreases with an increase in the strain, thus reaching a limit. On the other hand, the relationship between pore water pressure and strain is such that pore water pressure increases, as shear increases. However, this relationship becomes inverted at the stage before shear reaches its maximum strength, with pore water pressure gradually decreasing and becoming negative in regions of larger strain. The natural water content of the Kobe Stratum is 10 to 15%, with moist soil having a unit weight of 21.6 to 22.6 kN/m³.

Examination of the triaxial compression test results of mudstone under CU conditions revealed that the mudstone of the Kobe Stratum behaves differently from sandstone in terms of stress-strain relationship and pore water pressure-strain relationship depending on the existing conditions. On the other hand, this mudstone has the same tendencies as sandstone for pore water pressure and for lateral pressure to increase with strain

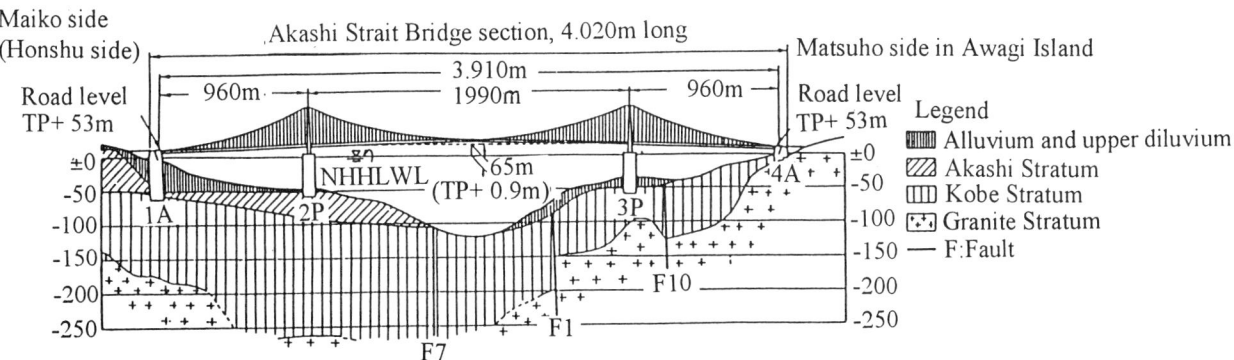

Figure 1. Geological composition and location of foundations of the Akashi Strait Bridge

softening. Furthermore, in spite of the mudstone, the Kobe Stratum has a low natural water content of 11 to 19% and a large wet unit weight ranging from 20.6 to 22.6 kN/m³.

In general, the strength characteristics of both the sandstone and mudstone of the Kobe Stratum can be said to be largely affected by pore water pressure.

3 GROUND COEFFICIENTS OF SEDIMENTARY SOFT ROCK USED FOR EVALUATING STABILITY

3.1 Ground Coefficients and Testing Methods

The ground coefficients used for evaluating stability include: (1) strength coefficients (C, φ), (2) deformation-related coefficients (E, ν), (3) creep coefficients (G, η), (4) unit weight (γ), and a (5) damping coefficient (D). Appropriate values for these coefficients are determined by the type of ground and loading conditions and must be used in practical analysis. Table 1 shows the ground coefficients to be used for the bearing capacity and foundation ground deformation analyses for the Akashi Strait Bridge, and the testing methods used to determine these coefficients. The following describes a practical approach for setting the strength coefficients and creep coefficients of the sedimentary soft-rock stratum (Kobe Stratum).

(1) Strength coefficients

The strength coefficients used to analyze the bearing capacity of the ground were determined by considering the characteristics of the ground and differences in the rate and duration of the variation of internal ground stress and the differences in behavior of pore water pressure when there is an earthquake. In the Kobe Stratum, which alternates between mudstone and sandstone, it can be assumed that there is no excessive pore water pressure under normal conditions. Therefore, it was decided to use strength coefficients obtained by tests under consolidated-drained conditions (CD tests) for analyses for normal conditions. For analyses for earthquake conditions, the strength coefficients were obtained by tests under consolidated-undrained conditions (CU tests), since it was assumed that the ground would become undrained because of a rapid loading rate. In this case, however, the consolidated-undrained shear strength that corresponds to the internal ground stress under normal conditions is determined at different points in the ground, on the assumption that the variation in the internal ground stress from an earthquake (stress variation when an earthquake occurs) does not affect ground strength.

The methods for determining shear strength under normal and earthquake conditions based on the above-mentioned viewpoints can be summarized as shown in Table 2.

Table 1. Relationship between Test Methods and Ground Coefficients Used for Analyses of Bearing Capacity and Deformation

		Ground Coefficient		Test Method
Under normal conditions	Strength coefficients	Akashi Stratum Kobe Stratum	C_{cd} ϕ_{cd}	C_{cd} and ϕ_{cd} obtained by triaxial compression test under consolidated-drained conditions, or effective stress indications, C and φ (equivalent to consolidated drained conditions)
		Granite	C_d ϕ_d	Shear strength coefficient obtained by bedrock test
	Deformation-related coefficients	E_s		By cavity loading (triaxial test)
		ν_s		Triaxial compression test
	Creep coefficients	G^*		Triaxial drained (undrained) creep test
		G_3		
		η_3		
	Unit weight	γ γ_t γ_{sat}		Geophysical test
Under earthquake conditions	Strength coefficients	Akashi Stratum Kobe Stratum	C_{cu} ϕ_{cu}	Shear strength coefficient obtained by triaxial compression test under consolidated drained conditions
		Granite	C_d ϕ_d	Shear strength coefficient obtained by bedrock test (same as that at normal time)
	Deformation-related coefficients	E_D		Cavity loading test (applicable to RBSM)
		G_0		Velocity logging
		G		Repeated triaxial test
	Damping coefficient	D		Repeated triaxial test
	Unit weight	γ γ_t γ_{sat}		Geophysical test
For a storm or ship collision	Strength coefficient	Akashi Stratum Kobe Stratum	C_{cu} ϕ_{cu}	Shear strength coefficient obtained by triaxial compression test under consolidated drained conditions
		Granite	C_d ϕ_d	Shear strength coefficient obtained by bedrock test
	Deformation-related coefficients	E_D		Twice that under normal conditions ($2E_s$)
		ν_D		Same as that under normal conditions
	Unit weight	γ γ_t γ_{sat}		Geophysical test

Table 2. Drain conditions and shear strength

	Normal Conditions	Earthquake Conditions
Drain condition	CD	UU
Strength for bearing capacity analysis	CD strength	CU strength Note
Shear strength	$\tau_f = C_{CD} + \sigma' \tan\phi_{CD}$	$\tau_f = C_{CU}^* + \sigma' \tan\phi_{CU}^*$

Note: Effective stress does not vary (consolidate) with load variation when an earthquake occurs. Therefore, σ' under earthquake conditions was set equal to consolidation strength under normal conditions.

When setting the coefficients for ground strength for earthquake conditions, the rate of variation in internal ground stress, earthquake duration, behavior of excess pore water pressure, and the effects of repeated loading and the loading rate must of course be considered. However, the accurate determination of ground coefficients that take these effects into account is practically impossible when the ground is heterogeneous in nature and composed of complex alternating layers (such as the Kobe Stratum). Consequently, it was decided that the simple methods described above would be applied.

(2) Creep coefficients
In the past, creep coefficients were briefly examined and it was presumed that the allowable vertical bearing capacity of the Akashi and Kobe strata would be around 1.77-1.96 MPa, based on the results of plate bearing tests for the land section. This approximate value for the vertical bearing capacity was determined on the supposition that no progressive creep would occur. This value was, however, derived from small-scale loading tests, and the bearing capacity of large 50- to 100-m long foundations was underestimated. In addition, although the load that will cause progressive creep is around 75% of the ultimate load, it is considered in the stability evaluation for vertical bearing capacity described hereafter that progressive creep will not occur, since a safety factor of 3 or more is intended for the ultimate bearing capacity. Therefore, it was decided to grasp creep displacement characteristics by focusing on the long-term displacement of foundations when the Kobe Stratum is the bearing layer.

Creep displacement characteristics can generally be described by the Voigt model. In this paper, the creep coefficients are determined by a 3-element Voigt model that expresses shearing strain as shown in Figure 2. The creep coefficients for ground with poor water permeability, such as the Kobe Stratum, were obtained by undrained creep tests. In this case, however, the undrained creep tests do not include any consolidation effect. Therefore, this effect should be evaluated by making comparisons with the results of drained creep tests. Then, on the basis of this evaluation, the creep coefficients should be calculated to correspond with the drained condition.

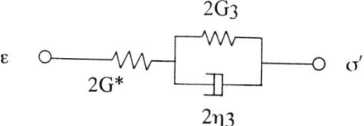

where:

ε = shearing strain = $\{\frac{1}{2G^*} + \frac{1}{2G_3}[1-\exp(-\frac{G_3}{\eta_3}t)]\} \cdot \sigma'$

σ' = deviator stress
G^* = coefficient of immediate shearing deformation
G_3 = coefficient of delayed shearing deformation
η_3 = viscosity coefficient
t = time

Figure 2. 3-Element Voigt model

3.2 Ground Model

In determining the ground coefficients based on the testing methods stated in 3.1 above, a simplified basic ground model of the Kobe Stratum, which is composed of intricate alternating layers of soil, is built by grouping the alternating soil layers and determining their respective ground coefficients. Table 3 shows the foundation of main tower 3P constructed in the Kobe Stratum, which serves as the bearing layer.

4 METHODS OF EVALUATING FOUNDATION STABILITY

The foundations of suspension bridges built in the past that connect Honshu (the largest island in Japan) with Shikoku (the fourth largest island in Japan) had their stability checked based on the premise that the bearing layer of the foundations was bedrock located deeper than Neogene layers, such as a granitic layer. However, the bearing layer of the foundations for the Akashi Strait Bridge is either Kobe or Akashi Stratum, which has bearing and deformation characteristics different from the traditional bearing layer for foundations. Furthermore, the foundations of the Akashi Strait Bridge are rigid and relatively wide compared with their embedded depth, and their bottom faces bear a large part of the load. Based on this, it was considered inappropriate to apply the traditional methods for evaluating stability to the present case, and new methods for evaluating stability were studied. The following description introduces the methods for evaluating stability against bearing, sliding and overturning that were used in designing the rigid foundations of the Akashi Strait Bridge.

4.1 Acting Forces from the Foundation to the Ground

The forces that act from the foundations to the ground are represented by a horizontal force, vertical force and overturning moment. As for the main loads to derive these acting forces,

Table 3. Basic Ground Model of Main Tower Foundation 3P

Stratum				Geophysical Properties					Strength Characteristics				Deformation Characteristics						
Name	Stratum Zoning	Depth (T.P.-m)	Thickness (m)	γsat (kN/m³)	γ' (kN/m³)	Vs (m/s)	Vp (m/s)	νD	Under Normal Conditions		Under Quake Conditions. For a Storm or Ship Collision		Under Normal Condition					For a Storm or Ship Collision	
									Ccd (MPa)	φcd (rad)	C*CU (MPa)	φ*CU (rad)	Deformation Coeff.		Creep Coeff.			Deformation Coeff.	
													Es (MPa)	νs	G* (MPa)	G*/G₃	G₃/η₃ (1/min)	ED (MPa)	νs
Alluvium	Al	10		19.2	9.4	180	1 300	0.49	-	-	-	-	82.4	0.38	-	-	-	164.8	0.38
Kobe Stratum	K3p-1	4		22.4	12.6	380	2 000	0.48	0.147	0.524	0.588	0.349	431.5	0.45	40.2	0.35		863.0	0.45
	K3p-2	14		22.3	12.5	470	2 100	0.47	0.196	0.611	1.257	0.436	578.6	0.36	96.1	0.06		1 157.2	0.36
	K3p-3	14		22.9	13.1	560	2 300	0.47	0.098	0.611	0.834	0.524	617.8	0.32	110.8	0.11	0.18 x10⁻⁴	1 235.6	0.30
	K3p-4	6		22.9	13.1	590	2 500	0.47	0.147	0.611	0.932	0.524	666.9	0.36	135.3	0.08		1 333.7	0.36
	K3p-5	12		21.6	11.8	450	2 100	0.48	0.049	0.436	0.343	0.262	431.5	0.45	45.1	0.35		863.0	0.45
	K3p-6	1		23.1	13.3	730	2 600	0.46	0.147	0.611	0.932	0.524	627.6	0.35	123.6	0.08		1 255.3	0.35
Weathered Granite Stratum	Gr'	36		23.0	13.2	860	2 900	0.45	0.098	0.646	Ccd=0.098	φcd=0.646	882.6	0.41	295.2	0.50	5.3 x10⁻¹⁰	1 765.2	0.41
Granite Stratum	Gr			25.5	15.7	2 000	4 500	0.38	0.490	0.698	Ccd=0.490	φcd=0.698	980.7	0.25	627.6	0.50	4.9 x10⁻¹²	1 961.3	0.25

dead load, earth pressure, hydrostatic pressure, buoyancy, tidal current force, and eccentric load due to construction errors were taken into consideration. Subsidiary loads such as wave force, effects of earthquakes, loads in construction work, wind load, and ship collision load were also taken into account.

To analyze the effects of earthquakes on the acting forces, dynamic stability was introduced to quantitatively evaluate as much as possible effective earthquake vibrations that take into account foundation shape and dimensions, the dynamic characteristics of peripheral ground when an earthquake occurs, the dynamic interaction between the foundations and ground as determined by the shape of their interacting parts, and the dynamic spring and underground-dissipative damping. Using a rigid-body vibration model with 2 degrees of freedom to reflect dynamic stability (Figure 3) as the analytical model, the acting forces from the foundations to the ground were calculated by a response spectrum method.

the view is taken that the internal stress variation in the ground when an earthquake occurs does not affect ground strength. Consequently, the internal ground stress for a stationary load was calculated, and the consolidated undrained shear strength corresponding to the calculated internal stress was cumulated over all the blocks to evaluate the bearing capacity. The effect of the eccentric inclination of foundation loads was taken into account as inclined loads within their effective width.

In the detailed examination stage the RBSM (Rigid-Body Sprig-Model) method, which can calculate ultimate vertical bearing capacity and fracture mode with higher accuracy than the slice method, was adopted to evaluate bearing capacity.

Table 4. Safety factors for ultimate bearing capacity

Design Load Condition		Safety Factor
On completion of construction	Under normal conditions	3.0
	Under earthquake conditions	2.0
	For a storm or ship collision	2.0
For construction work		2.0

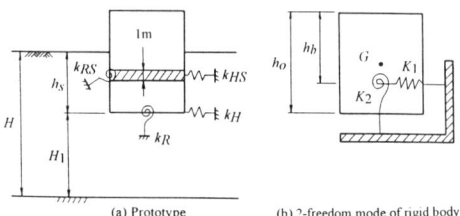

(a) Prototype (b) 2-freedom mode of rigid body

Notes: k_{HS}, k_{RS}: horizontal and rotational springs of every ground layer,
- k_H, k_R: horizontal and rotational springs on foundation bottom,
- K_1, K_2: horizontal and rotational springs in a rigid-body model with 2 degrees of freedom that reflect the characteristics of time-series response analyses for a prototype,
- G: Position of center of gravity.

Figure 3. Modeling of dynamic stability

4.2 Method of Evaluating Stability for Vertical Bearing

(1) Procedure of evaluating stability for vertical bearing
Foundation stability for vertical bearing was decided to be conducted basically according to the flow shown in Figure 4. The vertical ground reaction for a foundation bottom was set at less than the allowable vertical bearing capacity of the bearing ground, which was obtained by dividing the ultimate vertical bearing capacity of the ground by a safety factor shown in Table 4.

(2) Ultimate bearing capacity of ground
For both the simple and detailed examinations of the ultimate vertical bearing capacity, the slice method was adopted, as it is practical and can take the effects of multi-layered ground into consideration. The slice method enables the evaluation of bearing capacity under different drain conditions of the ground under normal and earthquake conditions as described above, and can consider the effects of a large eccentric inclined load.

The slice method calculates bearing force from a condition of ultimate equilibrium for sliced blocks formed on an arbitrary slip line. Figure 5 shows a model for bearing capacity calculation by applying the slice method. Particularly, for calculating the bearing capacity of the sedimentary soft rock by the slice method,

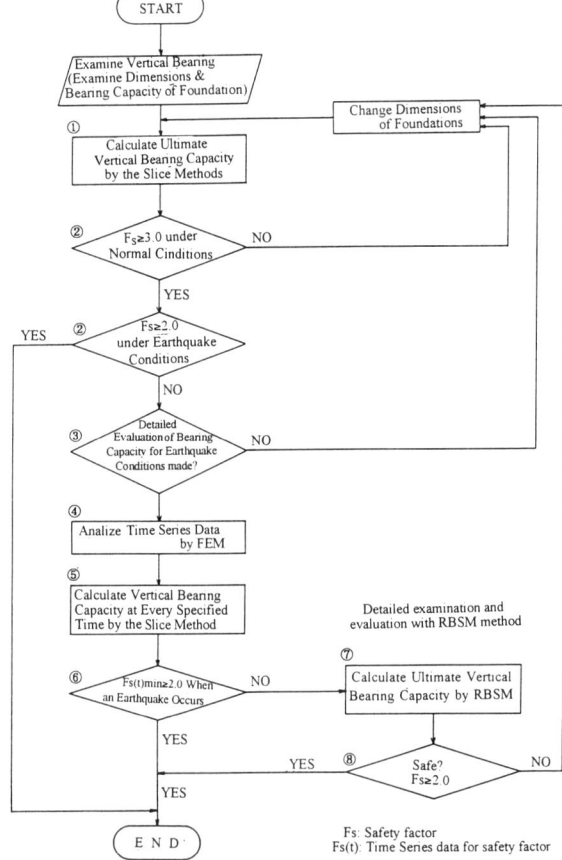

Figure 4. Flow of stability evaluation for vertical bearing

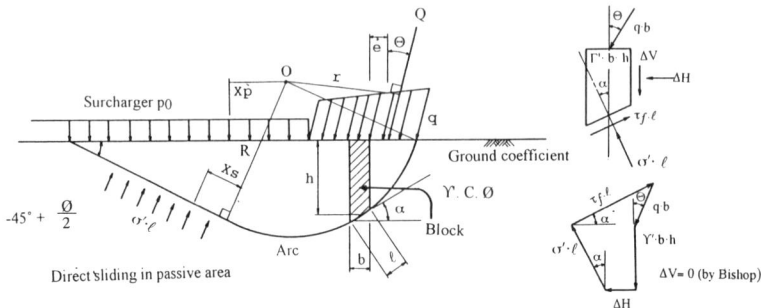

Figure 5. A model for bearing capacity calculation applying the slice method

The RBSM method is a limit analysis model designed for treating so-called discontinuities, such as sliding, fracture development and contact problems, when examining the bearing capacity of the ground.

4.3 Method of Evaluating Stability against Sliding

The stability of the ground against sliding was examined on the premise that the horizontal force acting on a foundation bottom is, in principle, less than the allowable shearing resistance of the ground on a foundation's bottom. The allowable shearing resistance on a foundation's bottom is obtained by dividing the shearing resistance calculated according to the following formula by the safety factor shown in Table 5.

$H_u = C_B A' + V \tan\phi_B$

where:
H_u = shearing resistance of foundation
C_B = cohesion of ground
ϕ_B = angle of internal friction
A' = contact area
V = vertical force acting on foundation bottom

Table 5. Safety factor against sliding

Design Load Condition		Safety Factor
On completion of construction	Under normal conditions	2.0
	Under earthquake conditions	1.2
	For a storm or ship collision	1.2
For construction work		1.2

The calculation of the shearing resistance for a foundation borne by the Kobe Stratum by applying the above formula was carried out using the friction angles ϕ_B and cohesion C_B shown in Table 6. To set the values of ϕ_B and C_B, particular attention should be paid to decreases in C_B owing to increases in sagging and cracking near the ground's surface associated with excavation for a structure's foundation. Generally, for the sliding resistance of a foundation bottom, $\tan\phi_B$ is 0.6 and C_B is zero in the case where the ground is rock and the foundation is concrete. However, this assumes that there is tertiary soft rock and rock of poor cohesion, and therefore it would be illogical to obtain the ground coefficients of the actual bearing stratum (Kobe Stratum) via testing. Furthermore, the results of block shear tests include many cases of shear failures occurring in the ground. For these reasons, ground strength at the bottom of a foundation was decided to be used in the calculations.

Table 6. Friction angle and cohesion of Kobe Stratum used for evaluating stability against sliding

Design Load Condition	Type of Ground	Friction Angle ϕ_B	Cohesion C_B
Under normal conditions (in construction)	Akashi and Kobe Strata	ϕ_{cd}	C_{cd}
	Granite	ϕ_d	C_d
Under earthquake conditions (in construction)	Akashi and Kobe Strata	0	$C_{cu}* + \sigma_v \tan\phi_{cu}*$
For a storm (in construction) For a ship collision	Granite	ϕ_d	C_d

4.4 Method of Evaluating Stability against Overturning

To maintain bridge stability during and after construction in order to prevent bridge topping, the resultant force should be located at the core of the bottom section of the foundation. In the case of an earthquake, storm, or ship collision, the resultant force should be positioned such that the center point of the bottom section of the foundation receives compression force. However, in the case of an earthquake, there is a fairly large margin of safety when only on working evaluating stability via the verification of the resultant force's location. If prescribed conditions are not satisfied, a more detailed evaluation of stability against toppling is carried out that considers such things as the toppling speed.

4.5 Results of Stability Evaluation

Based on the above-mentioned concepts for ground coefficients and stability evaluation, the stability of Anchorage 1A and Main-tower Foundation 3P (which is constructed on a bearing layer of sedimentary soft rock) was checked for the case of an earthquake. The results in Table 7 show that the bearing capacity of both 1A and 3P are in a severe situation. However, detailed evaluation using the RBSM method proved that sliding surfaces at the fringe of a foundation eventually stabilized (sufficient sliding safety factors were secured). These results verify that 1A and 3P will stabilize with cylindrical foundations 85-m and 78-m in diameter, respectively.

Table 7 Earthquake stability-evaluation results for Anchorage 1A and Main-tower Foundation 3P

		1A	3P	Allowable value
Maximum response acceleration, at center of gravity, A max (m/s^2)		0.95	2.17	
Forces on bottom	Vertical force, V (kN)	7,446,000	5,445,000	
	Horizontal force, H (kN)	2,105,000	1,880,000	
	Moment, M(J)	1.041×10^{11}	1.011×10^{11}	
Eccentricity, e(m)		14.0	18.6	
Angle of inclined loading, θ(rad)		0.276	0.333	
Reaction force of foundation bottom ground q max (MPa)		3.079	3.795	
Bearing	Ultimate vertical bearing capacity, Qv(kN)	17,647,000	13,231,000	
	Bearing safety factor, Fs	2.37	2.43	≥2.0
Sliding	Shear resistance, Hu(kN)	5,004,000	6,111,000	
	Sliding safety factor, Fs	2.38	3.25	≥1.2
Overturning	Eccentricity, d ($\phi/2-e)/\phi$	0.336	0.262	≥0.206

Note: ·The values in the above table were obtained by setting the dimensions of the substructure.
·The effects of an earthquake on 1A and 3P were analyzed by dynamic FEM analyses and with the CFM model, respectively.
·The ultimate bearing capacity was analyzed by the slice method.

5. METHOD OF ANALYZING FOUNDATION DISPLACEMENT

Foundation displacement must not unfavorably affect a superstructure. Particularly, in the case where a large-scale foundation is constructed on soft Kobe Stratum, creep displacement must be seriously considered since long-term displacement of the foundation will be a problem. The following below explains the analytical techniques for calculating the displacement of the foundations of the Akashi Strait Bridge, and a summary of a simple method for calculating creep displacement used in this study as a practical alternative technique is given.

5.1 Analysis of Elastic Displacement

For the analysis of elastic displacement under normal conditions, elastic FEM analysis was adopted for its applicability to the multilayered Kobe Stratum and for its analytical accuracy. Displacement due to earthquakes was analyzed via the response spectrum analysis using a model of dynamic rebound force described in Section 4.1.

5.2 Analysis of Creep Displacement

For rough estimates of creep displacement, the above-mentioned simple displacement calculation method (which is outlined later on) was applied. For detailed study, linear-viscoelastic FEM analysis was applied. The FEM analysis was applied for detailed examination because approximate creep displacement is predictable from elastic displacement calculations, and because analytical results of acceptable accuracy can also be obtained via the simple creep displacement calculation method.

5.3 Outline of Simple Creep Displacement Calculation Method

The principle of the simple creep displacement calculation method is to obtain creep strain for specific times by using a creep model that can represent the viscoelastic behavior of the ground, and to calculate the creep displacement by integrating the strain over the range of the depth affected by internal ground stress. The viscoelastic behavior of the ground was taken into account for the deviation component in the strain calculation using a three-element Voigt model as shown in 3.1 (2), and the volumetric component of the strain was treated as elastic.

The internal ground stress was calculated by combining an assumption on the angle of the load dispersion with a method for making an approximate calculation of stress that has the stress correspond to an uneven load distribution, as explained below:

(1) Assumption for load dispersion: It is assumed that the reaction effect the ground in terms of depth is three times the width of the foundation at an angle of $\theta=0.524$ rad.
(2) Method of for approximately calculating stress given an uneven load distribution: The load distribution is arbitrarily divided and stress dispersion considered by load in the respective divisions.

Furthermore, since the load that acts on the foundation varies with the lapse of time (from the start of the construction to its completion), the hysteresis of displacement with time was determined for every load division, and the hysteresis of divisions was accumulated to obtain the displacement for the whole period.

In this calculation method, the distribution of displacement determined along the bottom of a foundation is usually not linear. However, because the foundation can be considered to be a rigid body, the distribution of displacement along the bottom was replaced with a straight line on the basis of the displacement determined at the middle and both ends, and was compensated as a rigid foundation.

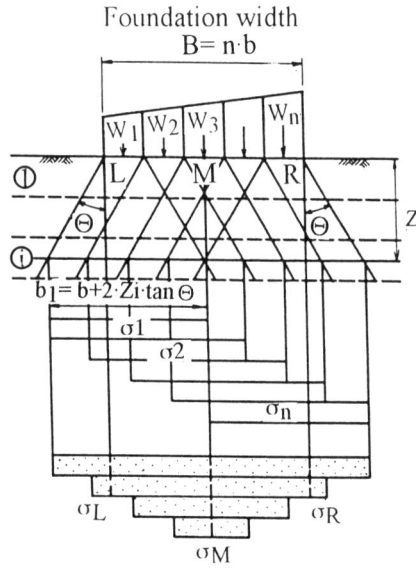

Figure 6 Method of calculating internal ground stress

Figure 7 shows an example of a trial calculation made by the above-mentioned method for the creep displacement of Anchorage 1A of the Akashi Strait Bridge. The estimate of creep displacement is after 100 years of public service. Given these same conditions, the anchorage was subjected to viscoelastic FEM analysis, and the displacements calculated by both methods were compared. The results of the comparison are shown in Table 8.

The results of the simple creep displacement calculation method are somewhat larger than those of the viscoelastic FEM analysis, owing to the inability of the simple method to take into account the effects of the ground's condition, for the simplicity of the ground model, and to the circular shift of the rigid foundation. Nevertheless, the simple method was adopted because it can be considered to be sufficiently applicable at the stage of rough estimation.

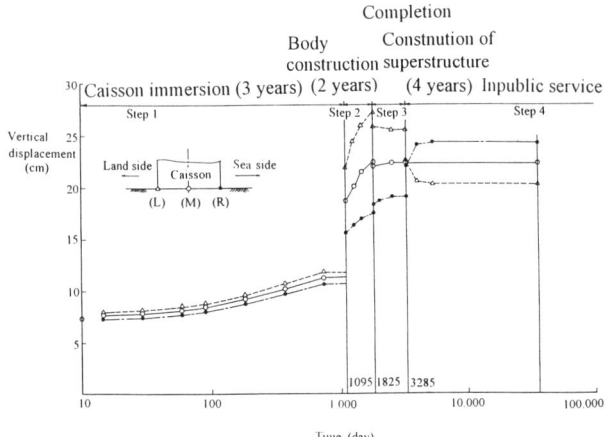

Figure 7 Change in vertical displacement at caisson bottom according to simple of creep displacement calculation method

Table 8 Comparison of vertical displacements of caisson bottom calculated by the two analytical methods

(unit: cm)

	Analytical Technique	Vertical-Displacement		
		Land Side	Middle	Seaside
Elastic displacement	Viscoelastic FEM	7.6	11.4	15.0
	Simple creep	13.1	14.3	15.4
Creep displacement (100 years)	Viscoelastic FEM	2.8	5.5	7.9
	Simple creep	7.3	8.1	9.0
Total displacement	Viscoelastic FEM	10.4	16.9	22.9
	Simple creep	20.4	22.4	24.4

REFERENCES

Manual for Stability Calculation of Substructures of Akashi Strait Bridge (Draft), and Its Commentary, Honshu-Shikoku Bridge Authority, March, 1988.

Manual for Earthquake-Resistant Design of Akashi Strait Bridge (Draft), and Its Commentary, Honshu-Shikoku Bridge Authority, March, 1988.

Report of Setting of Ground Coefficients for Designing of Akashi Strait Bridge, Honshu-Shikoku Bridge Authority, March, 1988.

A unified design method for rock anchor foundations of super-high pylons

Ö. Aydan
Tokai University, Department of Marine Civil Engineering, Shimizu, Japan

S. Komura
Chubu Electric Power Co., Research Institute, Nagoya, Japan

S. Ebisu
Okumura Corporation, Research Institute, Tsukuba, Japan

T. Kawamoto
Aichi Institute of Technology, Toyota, Japan

ABSTRACT: The demand for the use of rock anchors as foundations in many geotechnical engineering structures such as pylons, dams, suspension bridges, has been increasing in recent years. Particularly, the design of rock anchors as foundations for pylons is one of the intensive fields of research since the present design concept is based on the design of anchors in soils. The authors have initiated a research program concerning laboratory, in-situ experimental studies and, theoretical and numerical studies on the mechanics and load bearing capacity of rock anchors. From this research program, some guidelines are drawn and a unified design method for rock anchor foundations is proposed. In the present paper, the authors describe this unified design method and give some examples of its application.

1 INTRODUCTION

There is an increasing demand for the use of rock anchors as foundations in many geotechnical engineering structures such as pylons, dams, suspension bridges, etc. Particularly, the design of rock anchors as foundations for pylons is one of the intensive fields of research since the present design concept is based on the design of anchors in soils, and it is overconservative and costly. Fig. 1 shows a typical design of pylons utilising the rock anchors as foundations against uplift due to wind and seismic loadings. The use of rock anchors as foundations is very cost effective since the volume of excavation is small as compared with the pile-type foundations. This is an important element in foundation design since many pylons are constructed in mountainous areas where the construction is difficult in view of material transport, excavation etc. Nevertheless, the experience on the use of rock anchors as foundations is few and their long-term performance is in doubt as little experimental data available. Therefore, a fundamental research program is necessary on the short and long-term performance of rock anchors as foundations.

The authors have initiated a research program concerning laboratory, in-situ experimental tests and, theoretical and numerical studies on the mechanics and load bearing capacity of rock anchors. On the basis of the outcomes of this research program, some guidelines are drawn and a unified design methodology for rock anchor foundations is proposed. In the present paper, this unified design method is described and some examples of applications of the method are given.

2 FAILURE FORMS OF ROCK ANCHORS

In this section, we briefly describe the failure modes observed in-situ pull-out tests and their possible mechanisms and the associated conditions with these failure modes. Failure modes of anchors are closely inter-dependent with the anchor type, kind of medium and loading conditions and introduced surfaces of weakness due to constructional and structural reasons. In general, the failure of anchors may involve one or a combined form of the following failure modes (Fig. 2):

1- *Failure of intact medium* along
    i-) a shear failure plane (band), or
    ii- Tensile failure plane (band),

2- *Failure along an surface of weakness*, and

3- *Failure of anchor material in tension*.

2.1 *Failure of intact medium* (truncated conical form)

Although this is a commonly quoted form of failure of anchors installed soil-like medium, it is an unlikely form of failure for cylindrical anchors (see Aydan et al. (1993) for reasoning).

2.2 *Failure along a surface of weakness*

As anchors is a man-made structure, some surfaces of weakness are indispensibaly introduced. As a result, these surfaces have lower adhesive strengths as compared with that of the medium. Since applied loads transferred into the medium by shearing through these interfaces, failure takes by shearing at one of interfaces (i.e. *Grout-tendon* interface (GT), *Grout-rock* interface (GR), and *Grout-sheat-grout* interface (GSG) (if a corrosion protection sheat is used).

On the mechanism and fracturing state along these interfaces were previously investigated and discussed in detail by the authors and classified in their earlier publications (Aydan 1989, Aydan et al. 1990, Ebisu et al. 1993).

2.3 Failure of steel bar

The main element transferring the load applied by the super-structure to the ground is steel bar. The common failure of steel bar is a tensile failure. It should be noted that the partial failure of interfaces or medium may take place before the failure of the steel bar.

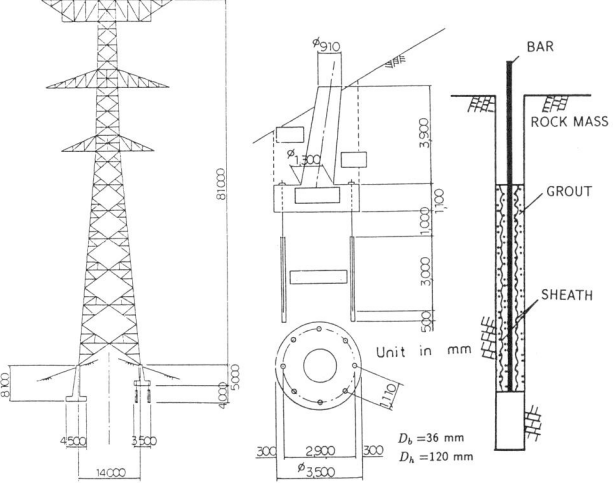

Fig. 1 An example of pylon foundation design

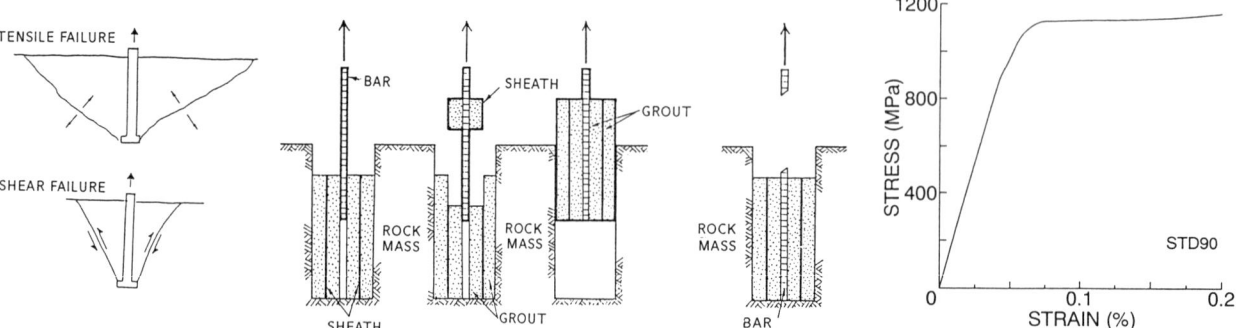

Fig. 2 Failure modes of rock anchors

Fig. 3 Uniaxial stress-strain behaviour of steel (STD90)

3 MECHANICAL BEHAVIOUR OF ROCK ANCHOR SYSTEMS

3.1 Mechanical behaviour of steel

The mechanical behaviour of steel is well known for many centuries. Its mechanical behaviour is generally elasto-plastic. Fig. 3 shows a stress-strain curve of steel obtained in a uniaxial tensile test which has become commonly used as tendons of rock anchors in recent years.

3.2 Mechanical behaviour of grouting material and interfaces

Since the interfaces are found as cylindrical surfaces in rock anchor systems, it is difficult to study their mechanical behaviour. Therefore, we have rolled these surfaces onto two dimensional planes and carried out both short term and long term tests. The configurations of samples tested by the authors are shown in Fig. 4 and typical short-term shear responses for each interface are shown in Fig. 5.

Considering the extreme conditions, winds such as typhoons have a period of 10 seconds with a velocity of 30-50 m/s. As the shear strength of interfaces decreases with cyclic loads of longer period, we selected a sinusoidal cyclic loading with a period of 20 s. Fig. 6 shows a typical cyclic response of the GSG interface under a normal stress of 4 MPa at a selected stress level. Fig. 7 shows the S-N responses for the GSG interface under a normal stress of 4 MPa.

3.3 *Mechanical behaviour of rock mass*

Although the mechanical behaviour of intact rocks is well known for about a century, the mechanical behaviour of rock masses are less understood since it is very expensive to carry tests on them *in-situ*. However, their mechanical behaviour would be similar to intact rocks. The rock mass around boreholes in which rock anchors installed behaves in an elasto-plastic manner. Fig. 8 shows a response of rock mass obtained in a borehole-jacking test in fractured sandstone.

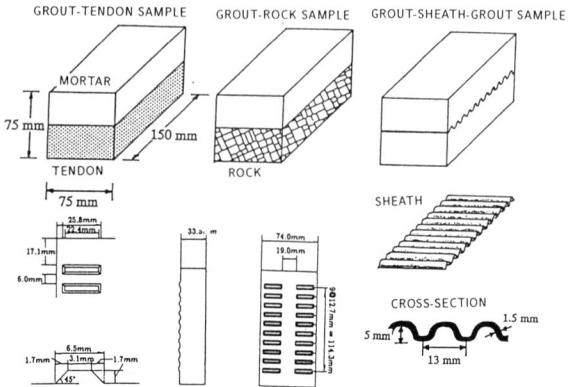

Fig. 4 Configuration of samples of interfaces used in direct shear tests

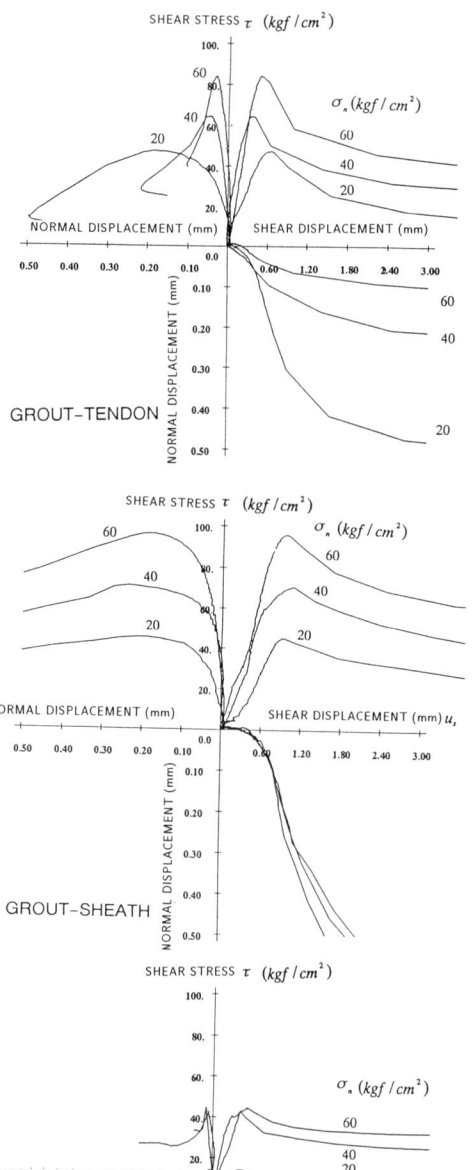

Fig. 5 Responses of interfaces in direct shear tests

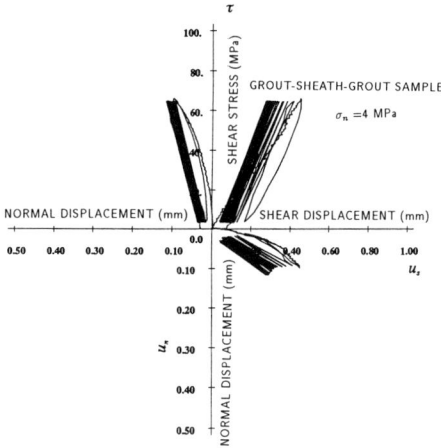

Fig. 6 Cyclic response of the GSG interface

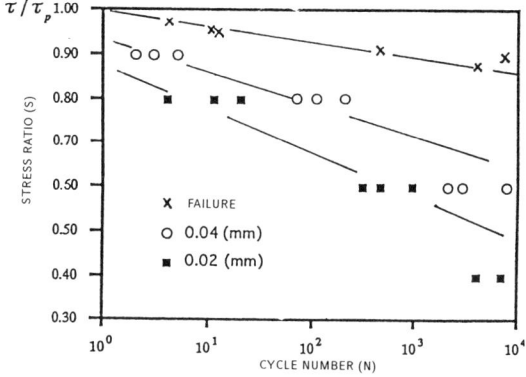

Fig. 7 The S-N relation for the GSG interface

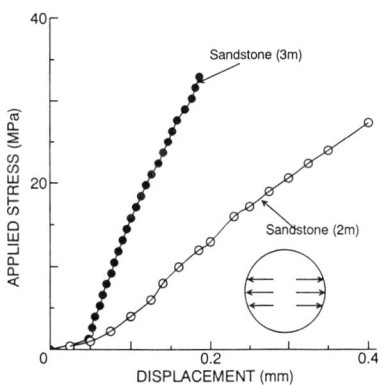

Fig. 8 The response of rock mass around a borehole jacking test

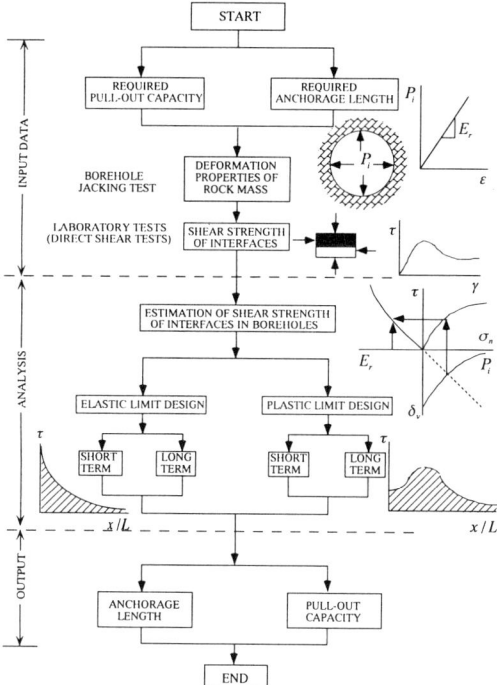

Fig. 9 Flow chart of the design method for rock anchors

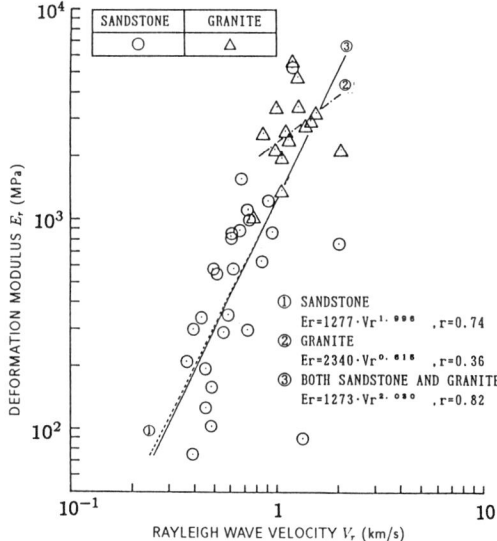

Fig. 10 Relation between deformation modulus and Rayleigh wave velocity

4 A DESIGN METHOD FOR ROCK ANCHORS

The design procedure is illustrated in Fig. 9 and is described in the following sub-sections.

4.1 Characterization of rock mass

The outcrop surveying is generaly recommended but the detailed information is generally difficult to obtain since the rock mass is usualy covered with top soil. Therefore, the borehole drilling is necessary to get more information. We have used RQD and other parameters to assess the rock mass. Our investigations show that most of the available techniques are not satisfactory for assessing rock masses as anchor foundations.

We have used various geophysical exploration tecniques to assess the rock mass (Ebisu et al. 1992). Among these techniques, we have found out that the exploration technique based on the Rayleigh waves is the most sensitive to the structure of rock masses and it suits for the site-investigation of near-surface masses as it yields the most detailed information on the physical state of rock mass. Besides these, the technique is the most cost-effective one and easy to operate on the site since it does not require any borehole. We have also obtained a good correlation between the deformability of rock masses by the borehole jack test and the Rayleigh wave velocity V_r as shown in Fig. 10. For measuring the deformability of rock mass, it is presently recommended to use the borehole jack test or the pressuremeter test in principle at each pylon site. However, in the future, we have been considering to use the correlation between the Rayleigh wave velocity V_r and deformability of rock masses, as an indirect assessment of the deformability of rock mass, which is expected to reduce the cost of site-investigation.

4.2 Characterization of steel

Mechanical properties of steel can be obtained from standard tests. The data on mechanical properties of steel will be generally the same unless the type of steel is varied.

4.3 Shear strength of interfaces in boreholes

The normal (radial) stress on interfaces during pull-out

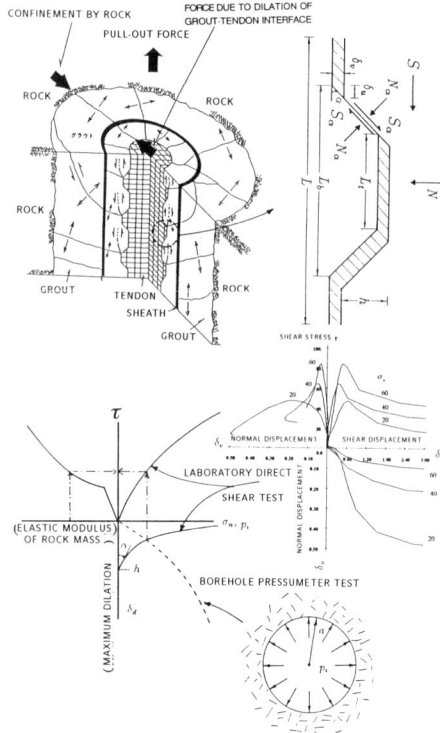

Fig. 11 Mechanical model for estimating the shear strength of interfaces in boreholes

tests of rock anchors or rockbolts is an important element for estimating the shear strength of interfaces in boreholes. The normal stress is likely to be a function of the stiffness of the surrounding medium since experimental results indicates that the effect of the stiffness on in-situ tests plays a big role on the interface strength. However, the assessment of the normal stress is still an unsolved problem. For the specification of the shear strength of interfaces in boreholes, we suggest to utilize the direct shear tests results and the radial response of the surrounding medium around the borehole. The fundamental concept of the procedure is illustrated in Fig. 11. When both curves shown in the figure intersect, that intersection point is assumed to correspond to the effective normal stress to be observed in rock anchor systems. The normal stress (radial) - displacement response is a function of the deformability of surrounding medium. If the interface yield criterion is a function of normal stress (i.e. Mohr-Coulomb criterion), the interface strength will be a function of the deformability of the surrounding medium. As noted from the figure, the shear strength of interface will increase as the rock deformation modulus of rock increases. In the followings, we demonstrate this conceptional model in terms of mathematical expressions.

As it is well known from the theory of elasticity, the following relation holds between the internally applied radial pressure and radial deformation of the hole in an infinite medium:

$$u = A\sigma_n, \quad A = \frac{1+\nu}{E}a \quad (1)$$

where ν: Poisson's ratio; E: elastic modulus; a: radius of hole. Let us assume also that the functional form of peak dilatancy of interface is of the following form

$$u_d = \frac{B}{1+C\sigma_n^{av}} \quad (2)$$

where B and C are experimental constants and σ_n^{av} is averaged normal stress over a typical wave length of the interface. Using following conditions:

$$u_d = u_o, \quad \frac{du_d}{d\sigma_n^{av}} = -\tan i_a \alpha \quad \text{at} \quad \sigma_n^{av} = 0$$

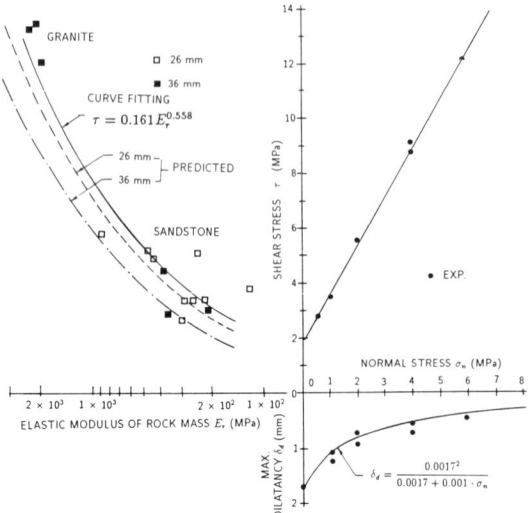

Fig. 12 Comparison of predicted shear strength in boreholes with in-situ measurements

where i_a is asperity inclination; u_o is height of asperity; α is a parameter to adjust the units, the coefficients B and C are obtained as

$$B = u_o, \quad C = \frac{\tan i_a \alpha}{u_o}$$

With the above relations, Eq. (2) becomes

$$u_d = \frac{u_o^2}{u_o + \alpha \tan i_a \sigma_n^{av}} \quad (3)$$

The contact state of an interface with a periodic asperity pattern can be illustrated as shown in Fig. 11. The contact area decreases as the interface dilates and the contact area can be written as

$$L_i^c = L_i - \frac{u_d}{\tan i_a} \quad (4)$$

Average normal stress on the interface in terms of actual normal stress is expressed as

$$\sigma_n^{av} = \frac{L_i^c}{L_b}\sigma_n \quad (5)$$

Inserting relations given by Eqs. (1,2,5) in Eq. (3), we finally obtain the following identity

$$\alpha^2 \tan^2 i_a \sigma_n^{av3} + 2u_o\alpha \tan i_a \sigma_n^{av2} + \\ (1 - A^*L_i\alpha)u_o^2 \tan i_a \sigma_n^{av} - A^*(L_i \tan i_a u_o^3 - u_o^4) = 0 \quad (6)$$

where $A^* = 1/(A \cdot L_b)$. The solution of the above identity yields the desired values of σ_n^{av}.

Application to tendon-grout interface

The asperity inclination of the tendon is 45°. Consequently, L_i is equal to asperity height u_o from the geometry. Inserting these values in Eq. (6) yields:

$$\left(\alpha^2 \sigma_n^{av2} + 2u_o\alpha\sigma_n^{av} + (1 - A^*L_i\alpha)u_o^2\right)\sigma_n = 0 \quad (7)$$

Then the solutions are

$$(\sigma_n^{av})_1 = 0, \quad (\sigma_n^{av})_{2,3} = \frac{u_o}{\alpha}\left[-1 \pm \sqrt{A^*\alpha u_o}\right]$$

Since we are interested in the positive values of the normal stress, the selected solution for the normal stress is

$$(\sigma_n^{av})_3 = \frac{u_o}{\alpha}\left[-1 + \sqrt{\frac{E}{1+\nu} \cdot \frac{\alpha u_o}{aL_b}}\right]$$

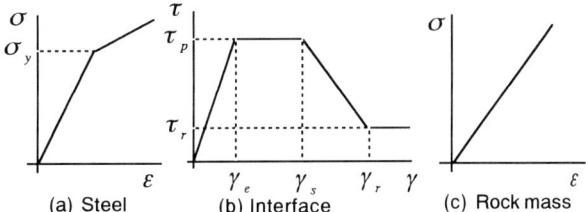

Fig. 13 Assumed mechanical behaviour of materials in rock anchor systems

The shear strength of tendon-grout interface in terms of average stress is obtained as

$$\tau_{tg} = 1.85 + \sigma_n^{av} \tan(58.78)$$

Parameters in Eq. (3) have the following specific values

$$\alpha = 0.001 \quad u_o = 0.0017, \quad L_b = 0.0128, \quad \nu = 0.25$$

Using this concept, the measured peak shear strength in pull-out tests performed in-situ as a function of the deformability of rock mass and predicted strength curves are plotted in Fig. 12 and compared with each other.

4.4 Estimation of uplift capacity of anchor systems

There are, in general, four fundamental approaches for the design of anchors, namely, 1- standards and emprical approaches, 2- limiting equilibrium approaches, 3- closed form solutions, and 4- numerical methods.

The state of art on methods for estimating uplift capacity of rock anchors have been reviewed by the authors. From this review, we found that the standards and limiting equilibrium approaches are unsatisfactory. For the design of rock anchors and rockbolts, Aydan et al. (1985a, 1985b, 1989) have developed several theoretical solutions for various kind of failure forms interfaces and yielding of steel bar. The final solutions for the pull-out capacity of rock anchors by considering the elastic and elasto-plastic behaviours of grout-tendon interface are only given herein (Fig. 13). For more complex conditions, it is suggested to use the numerical techniques such as the one developed by Aydan (1989).

Elastic limit (Fig. 14(a))

$$P_0 = \frac{\tau_p^{gt} \alpha E_a}{K_g} \tanh \alpha L, \quad \alpha = \sqrt{\frac{2K_g}{E_a r_a}} \quad (8)$$

where τ_p^{gt} is the peak shear strength of the grout-tendon interface, r_a radius of anchor, r_h radius of borehole, r_0 radius of the fictious rigid boundary ($r_0 = 2L$), E_a elastic modulus of tendon, G_g shear modulus of grout, G_m shear modulus of medium,

$$K_g = \frac{G_g}{r_a \ln(\frac{r_h}{r_a})} \frac{K-1}{K}, \quad K_m = \frac{G_m}{r_h \ln(\frac{r_0}{r_h})} \quad K = \frac{G_m}{G_g} \frac{\ln(\frac{r_h}{r_a})}{\ln(\frac{r_0}{r_h})} + 1.$$

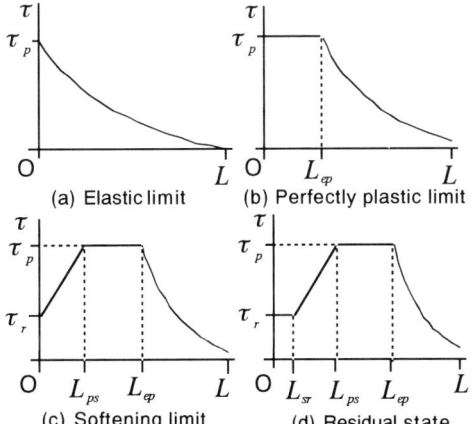

Fig. 14 States along the grout-tendon interfaces

Perfectly plastic limit (Fig. 14(b))

$$P_0 = \tau_p^{gt} \left[(\xi_r \frac{E_a L_{ep}}{K_g} \frac{L_{ep}}{r_a} \right] \quad (9)$$

where L_{ep} is the length of plastic region measured from the anchor head.

Softening limit (Fig. 14(c))

$$P_0 = \tau_p^{gt} \left[\frac{p E_a \beta \eta}{K_g} \frac{1 - \eta \cos p L_{ps}}{\sin p L_{ps}} \right] \quad (10)$$

where

$$p = \sqrt{\frac{2K_g}{E_a r_a \beta}} \quad \beta = \frac{\xi_r - \xi_s}{1 - \eta}$$

and L_{ps} is the distance of the point of transition from the perfectly plastic behaviour to the softening behaviour.

Residual state (Fig. 14(d))

$$P_0 = \tau_p^{gt} \left[\frac{p E_a \beta}{K_g} \frac{1 - \eta \cos p(L_{ps} - L_{sr})}{\sin p(L_{ps} - L_{sr})} + \frac{2\eta}{r_a} L_{sr} \right] \quad (11)$$

where L_{sr} is the distance of the point of transition from the softening behaviour to the residual plastic behaviour.

For estimating the uplift capacity of anchors, one of the formula presented above can be used depending upon the designer's choice. The design of rock anchors involve not only the mechanical considerations but also economical, constructional, environmental considerations and the safety concerns upon the failure in each country. From the mechanical point of view, the design of anchors involves the followings:

1-) Determination of diameters of bars and boreholes,
2-) Determination of spacing of anchors, and
3-) Determination of anchorage length.

Borehole diameter is governed by the constructional conditions and the existence of corrosion protection sheath. Diameters of the bars and spacing are governed by the sizes of the bars and the allowable intensity of pull-out load per anchor. From our own experimental program, we found that, if the anchor spacing is 30 times the anchor diameter, the interaction between anchors almost disappears. Once the diameters of the borehole and bar and the spacing of anchors are specified, then the problem is to determine the anchorage length for a given load or the pull-out capacity for a given anchorage length.

The long term behaviour of rock anchors under cyclic loading can be evaluated by taking into account the degradation of the shear strength of interfaces as a function of cycle numbers (Aydan et al. 1994). This approach brings a new concept, based on the principles of mechanics, to the rock anchor design methodology.

The yield strength of interfaces under cyclic loading conditions is taken into account as a convolution of the short term strength yield criterion τ_s and a S-N function $S(N)$ for cyclic loading of interfaces:

$$\tau(N) = \tau_s * S(N) \quad (12)$$

The function τ_s can be one of yield criteria appropriate for interfaces of rock anchors. The function $S(N)$ is suggested to be of the following form:

$$S(N) = A + Be^{-\frac{N-1}{N^*}} \quad (13)$$

The specific values of the constants in the above formula determined from tests for the *grout-tendon* interface are:

$$A = 0.75 \quad B = 0.25 \quad N^* = 22500$$

With the use of above approach, Fig. 15 shows the degradation of the shear strength of the *grout-tendon* interface as a function of cycle number.

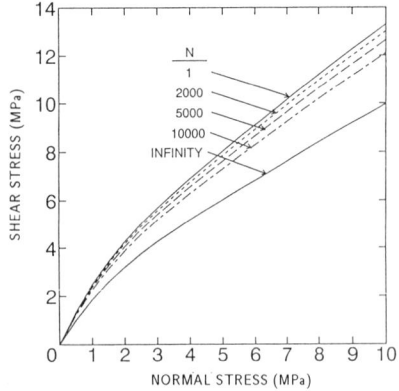

Fig. 15 The degradation of the shear strength of grout-tendon interface

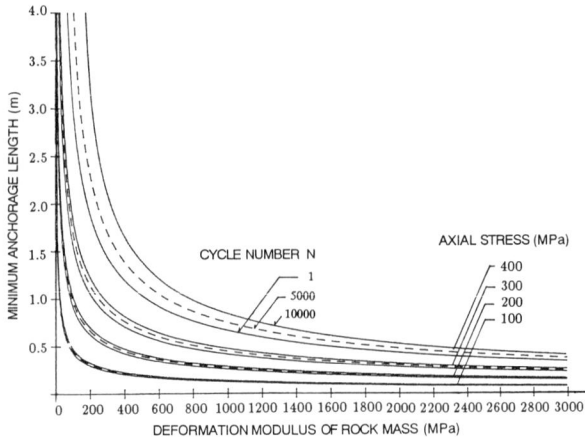

Fig. 17 A design nomogram for rock anchors (elastic limit state)

5 APPLICATIONS AND DISCUSSIONS

To check the validity of the present design method for rock anchor foundations of pylons, we have followed the steps described in the previous sub-sections and applied it to in-situ pull-out tests at two locations, which are typical rock masses found in the Chubu district of Japan. The rock masses were jointed sandstone and granite. In-situ explorations have shown that the sandstone was classified as weak rock, and granite as a medium hard rock. The deformation modulus of the sandstone ranged between 100 to 1000 MPa with an average of 400-500 MPa. The deformation modulus of granitic site was between 1000 and 6000 MPa with an average of 2000-3000 MPa. The anchorage length of anchors were 30, 100 and 300 cm in sandstone site and 300 cm in granitic site. 30 cm anchors had a load bearing capacity of 100-250 MPa while 100 cm long anchors failed at a load level between 250-550 MPa. 300 cm long anchors were all failed by the failure of steel bars in both sites. Fig. 16 shows the load bearing capacity of rock anchors as a function of the ratio of anchorage length to bar diameter.

Design nomograms were prepared for selected limit states. Fig. 17 shows an example of the design nomogram in which the necessary anchorage length of rock anchors were calculated as a function of the elastic modulus of rock masses for various axial force levels, by using the formula for the elastic limit state together with the experimentally obtained shear strength formula shown in Fig. 11. In the same figure, the long term relations as a function of cycle number are also shown. As an actual application of the present method, the design load per anchor has been currently selected as 32 tonf and the anchorage length as 3 m. As seen from the figure, the anchors should remain elastic under the current design load for s anchorage length of 3 m in a great range of rock mass conditions.

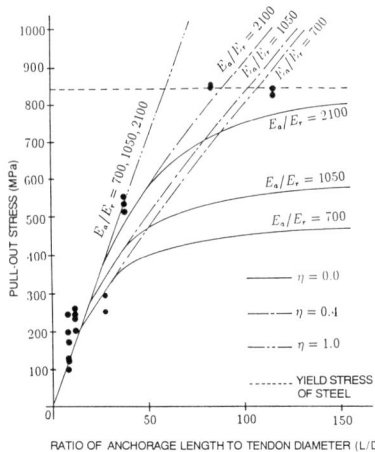

Fig. 16 Comparison of pull-out capacity predictions with measurements

6 CONCLUSIONS

From our research program concerning laboratory, in-situ experimental studies and, theoretical and numerical studies on the mechanics and load bearing capacity of rock anchors, some guidelines are drawn and a unified design method for rock anchor foundations is proposed. The design method proposed in this paper is expected to result in an increase on the use of rock anchors as pylon foundations and bring a considerable reduction in the construction cost as compared with that of the conventional pylon foundation design.

ACKNOWLEDGEMENTS

The authors acknowledge the Chubu Electric Power Co. for the permission to publish some of the outcomes of the research project on rock anchors and point out that the opinions expressed in the paper are not necessarily those of the company.

REFERENCES

Aydan, Ö., Ersen, A., Ichikawa, Y. & Kawamoto, T. 1985a. A new analytic solution for stresses, strains and displacements in / along rockbolts (in Turkish). *Madencilik*, Ankara, 24(3):27–36.

Aydan, Ö., Ichikawa, Y. & Kawamoto, T. 1985b. Load bearing capacity and stress distributions in/along rockbolts with inelastic behaviour of interfaces. *The 5th Int. Conf. on Num. Meths. in Geomechanics*, 2:1281–1292.

Aydan, Ö. 1989. The stabilisation of rock engineering structures by rockbolts. *Doctorate Thesis*, Nagoya University.

Aydan, Ö., Ichikawa, Y., Ebisu, S., Komura, S. & Watanabe, A. 1990. Studies on interfaces and discontinuities and an incremental elasto-plastic constitutive law. *Int. Conf. on Rock Joints*, Loen, 595–602.

Aydan, Ö., Ebisu, S. & Komura, S. 1993. Pull-out tests of rock anchors and their failure modes. *Int. Symp. on Assessment and Prevention of Failure Phenomena in Rock Engineering*, Istanbul, 285-294.

Ebisu, S., Komura, S., Aydan, Ö. & Kawamoto, T. 1992. Characterization of rock masses for rock anchor foundations. *Int. Conf. Fractured and Jointed Rock Masses*, ISRM, Lake Tahoe.

Ebisu, S., Aydan, Ö. & Komura, S. 1993. Mechanism of interface failure of rock anchors. *Int. Symp. on Assessment and Prevention of Failure Phenomena in Rock Engineering*, Istanbul, 677–686.

Sloping rock layer foundation of bridge structure

N. Ogata
Japan Highway Public Corporation, Japan

S. Gose
CTI Engineering, Tokyo, Japan

ABSTRACT : As the construction of expressway comes to a period to construct cross expressways, the oppotunity to construct bridges in mountain area becomes more and more in Japan Highway Public Coporation. For the geographical, geological and constructional limitation, deep foundation piles constructed in soft rock layer are used in most of cases. In this paper, the considerations of the design method of deep foundation pile were introduced as well as the results of in-situ loading test that was done to verify the appropriateness of the design method.

1. Introduction

Up to the end of 994, there have about 5600 *km* expressways that belong to *Japan Highway Public Corporation(JH)* being in operation. And, it is planned to the beginning of 21 century, to construct 11520 *km* expressways. The ratio of total length of bridges in mountain area to the total length of all the expressways, including those under construction or planning, is increasing as the increase of expressway construction. For the geographical, geological, and constructional limitation, large foundations are used in most of the cases. As the feature of mountain area bridges, the foundations are mostly constructed on/in inclined ground. The ground layers are usually sandy soil, soft rock or hard rock. Foundations are spread foundations or large deep foundations.

In this paper, it is intending to introduce the considerations about the design method of large deep foundation. In-situ loading test was done to check the validity of design method. Comparison was made with test results.

2. Basic Considerations of the Design Method [1]

The deep foundations in this paper means those founded on inclined grounds which have inclining angle larger than 10 degrees. The flowchart of design is given in Fig. 1. The design method is changed for the design of sectional forces of members and of displacement, and for the check of horizontal stability, that is, the first is based on the calculation of beams on elastic bearing, in other words, elastic design method, and the second is based on plastic calculation that considers the plastication zones, in other words, plastic design method. The design should satisfy the following items:

1). the vertical subgrade reaction occurred in the bottom of deep foundation should not exceed the allowable bearing capacity.
2). the up-limit value of shear resistance occurred between the bottom surface of deep foundation and ground layer is taken as the maximum value of the shear resistance occurred in the bottom of deep foundation.
3). the displacement of deep foundation should not exceed the allowable displacement.

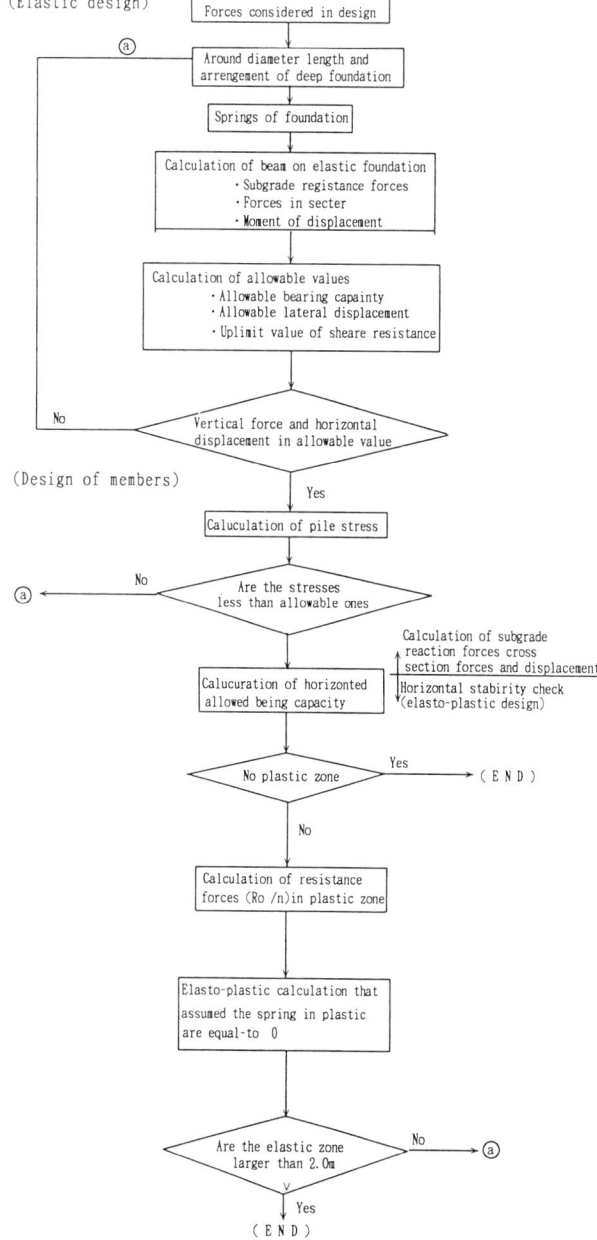

Fig. 1 Design Flowchart

4). the total horizontal subgrade reaction in front of deep foundation to each depth should not exceed the horizontal allowable bearing capacity in the depth.

The allowable displacement is given in Table 1 and the example types are illustrated in Fig. 2.

Table 1 The allowable displacement of deep foundation

	δ (cm)
Normal time During Earthquake	D*0.01 D: the diameter of pile in deep foundation (cm) But, the maximum is taken as 5 cm both in normal and earthquake condition. For abutment, it is taken as 1.5 cm in normal condition.

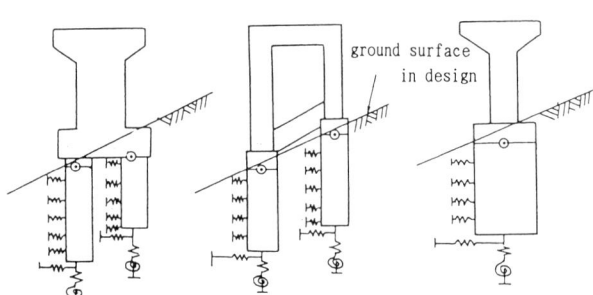

(a). Foundation with footing (b). Frame type foundation (c). Pier type foundation

Fig. 2 The check position of allowable horizontal displacement

3. Calculation of subgrade reaction, sectional forces and displacement

3.1. Subgrade reaction coefficient

(1) Horizontal subgrade reaction coefficient (K_H)

$$K_H = K_{H0} \cdot \left[\frac{B_H}{30}\right]^{-3/4} \quad (1)$$

where

K_H: Horizontal subgrade reaction coefficient (kgf/cm^3)

K_{H0}: Horizontal subgrade reaction coefficient equivalent to a value of plate bearing test using a rigid disk of 30 cm in diameter, and it is obtained by using the following formula, if estimating it from a modulus of deformation to be found as a results of soil quality tests and surveys (kgf/cm^3)

$$K_{H0} = \frac{1}{30}\alpha E_0$$

E_0: modulus of deformation of a soil layer at a point in issue, measured or estimated as shown in Table 2.

α: a coefficient to be used for estimating a coefficient of subgrade reaction, as given in Table 2.

B_H: equivalent loading width of foundation (cm) to be obtained by means of the following formula.

$\sqrt{D/B}$ $(1 > \beta l)$, $\sqrt{A_H}$ $(1 \leq \beta l)$

D: loading width (cm) of a foundation intersecting orthogonally a load-working direction

β: characteristic value of deep foundation pile (cm^{-1})

A_H: loading area (cm^2) of a foundation intersecting orthogonally a load-working direction.

(2). Vertical subgrade reaction coefficient (K_V)

$$K_V = K_{V0} \cdot \left[\frac{B_V}{30}\right]^{-3/4} \quad (2)$$

where

K_{V0}: taken the same value as K_{H0} (kgf/cm^2)

B_V: equivalent loading width (cm) of a foundation to be obtained by means of following formula.

$$B_V = \sqrt{A_V}$$

Table 2 E_0 and α

Modulus of deformation E_0 (kgf/cm^2) to be obtained by means of the following methods	α	
	Normal time	During Earthquake
A value equal to 1/2 of a modulus of deformation to be obtained from a repetitive curve of a plate bearing test using a rigid rigid disk of 30 cm in diameter	1	2
Modulus of deformation to be measured in bore hole.	4	8
Modulus of deformation to be obtained by means of unconfined or triaxial compression test of samples	4	8
Modulus of deformation to be estimated from $E_0=28N$ using the N value of a standard penetration test	1	2

3.2. Calculation method

On premise that a computer is used, the computation of a beam on elastic bearing layer is made in the method.

In case of inclined ground, a modification of the horizontal subgrade reaction coefficient to be obtained from horizontal ground layers should be made based on following formula.

$$\begin{vmatrix} K_h = 0 & (0 \leq \alpha \leq 0.5) \\ K_h = (0.3\log_{10}\alpha + 0.7)K & (0.5 \leq \alpha \leq 10) \\ K_h = K & (\alpha > 10) \end{vmatrix} \quad (3)$$

where,

K_h: coefficient of horizontal subgrade reaction that considered the inclination of ground (kgf/cm^2)

α: ratio of horizontal thickness of soil to the surface of inclined ground to the diameter of deep foundation. The maximum value is taken as

$$\alpha_{max} = 10$$

K: coefficient of horizontal subgrade reaction of a horizontal layered ground to be obtained based on Eq. 1 (kgf/cm^3).

The above equation is based on FEM computations that considered the effect of the thickness of foundation on inclined ground (horizontal distance from the body of foundation to inclined ground surface). It is only used in case that the coefficient of horizontal reaction is known only for horizontal layered ground.

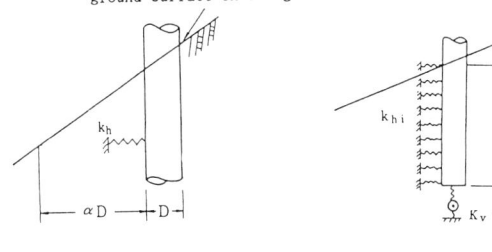

Fig. 3 Fig. 4

Springs needed in the calculation of a beam on elastic bearing can be computed based on the following Equations.

(1) Horizontal spring (K_h)
$$K_h = k_h \, D \, \triangle L \qquad (4)$$
where,

k_h: coefficient of horizontal subgrade reaction (kgf/cm^3) that considered the effect of inclination of ground and of neighboring foundation and is obtained based on Eq. 3.

D: diameter of deep foundation (cm)

$\triangle L$: the length of interval of springs (cm)

(2) Vertical spring at bottom of foundation (K_V), in case that whole section are effective.
$$K_V = k_V \, A \qquad (5)$$
where

k_V: coefficient of vertical subgrade reaction (kgf/cm^3)

A: bottom area of foundation (cm^2)
$$(= \pi D^2/4)$$

(3) Rotational spring at bottom of foundation (K_r), in case that whole section are effective.
$$K_r = k_V \, I \qquad (6)$$
where,

I: moment of inertia of the bottom section of foundation (cm^4)
$$(= \pi D^4/64)$$

The standard interval of spring ($\triangle L$) is taken as 50 cm.

(4) Shear spring at bottom of foundation (K_S), in case that whole section are effective.
$$K_S = k_S \, A \qquad (7)$$
where

k_S: coefficient of shearing spring (kgf/cm^3)

A: bottom area of foundation (cm^2)

The horizontal, vertical and shear subgrade reactions occurred in a position of horizontal spring are computed based on the Eqs. 8, 9 and 10, respectively.

(a) Horizontal subgrade reaction
$$R_i = K_{hi} \, \delta_i \qquad (8)$$
where,

R_i: horizontal subgrade reaction occurred in a position of horizontal spring (tf).

K_{hi}: horizontal spring (tf/m)

δ_i: horizontal displacement (m)

(b) Vertical subgrade reaction
$$q = \frac{N}{A} \pm \frac{M}{I} \cdot \frac{D}{2} \qquad (9)$$
where

q: vertical subgrade reaction stress (tf/m^2)

N: vertical reaction acted on bottom (tf)

M: moment acted on bottom ($tf \cdot m$)

A: sectional area of bottom (m^2)

I: moment of inertia (m^4)

D: diameter of bottom section (m)

If negative stress is occurred in calculated subgrade reaction stress, modification should be made based on effective area.

(c) Shear resisting force
$$S_b = K_S \, \delta_b \qquad (10)$$
where,

S_b: shear resisting force acted on bottom (tf)

K_S: shear spring at bottom (tf/m)

δ_b: horizontal displacement at bottom (m)

4. Vertical allowable bearing capacity of ground

4.1 Vertical allowable bearing capacity

Vertical allowable bearing capacity at bottom of deep foundation is calculated based on following Equations.
$$q_a = \alpha \cdot q_{a0} \qquad (11)$$
$$q_{a0} = \frac{1}{2}(q_{d0} - \gamma_2 D_f) + \gamma_2 D_f \qquad (12)$$
where,

q_a: vertical bearing capacity at bottom of deep foundation (tf/m^2)

q_{a0}: vertical bearing capacity at bottom of deep foundation (tf/m^2) based on assumed horizontal layered ground.

α: deduction coefficient caused by inclination of ground. The value can be obtained from Fig. 5.

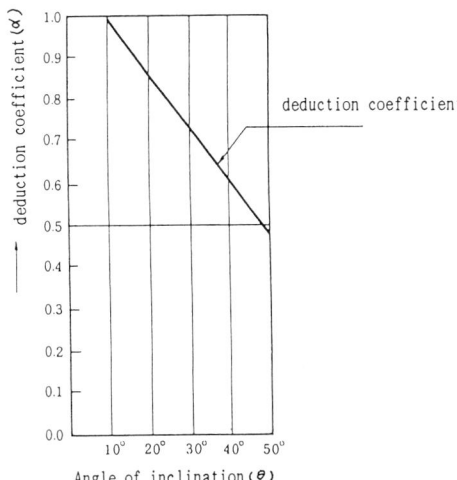

Fig. 5 Deduction coefficient

q_{d0}: ultimate bearing capacity to be determined from the bearing layer in the assumed horizontal ground (tf/m^2) and being calculated by means of Eq. 13.

n: safety ratio (Table 3).

D_f: effective embedment depth in bearing layer.

γ_2: unit weight of ground soil above the bottom of foundation (tf/m^3)

Table 3 Safety factor

	Normal time	During earthquake
Safety factor	3	2

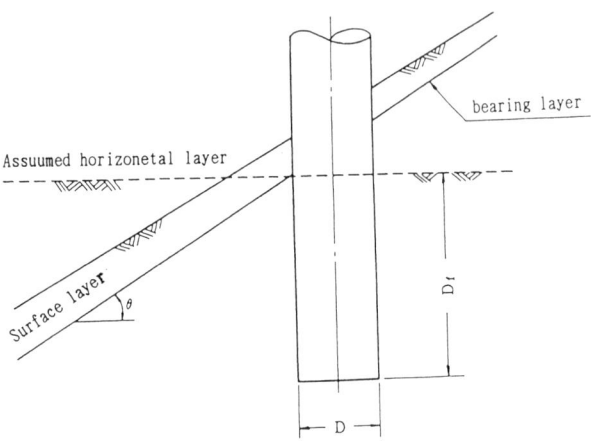

Fig. 6 Effective embedment depth

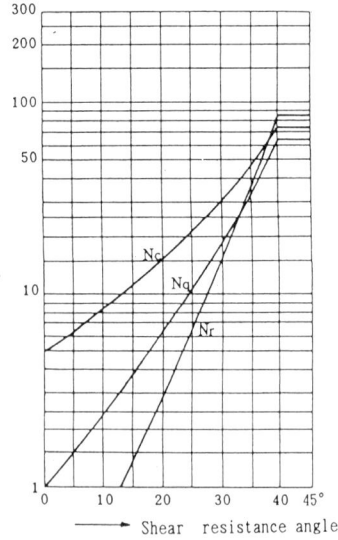

Fig. 7 Bearing capacity coefficient

4.2 Limited bearing capacity determined from ground

Limited bearing capacity of the layer at the bottom of foundation in a assumed horizontal ground is determined based on the synthetic estimation of those obtained from static mechanics formula and plate bearing test. The static mechanics formula is given in following.

$$q_{d0} = 1.3 C \cdot N_c + 0.3 \gamma_1 D \cdot N_r + \gamma_2 D_f N_q \quad (13)$$

where

q_{d0}: Limited bearing capacity of the layer at the bottom of foundation in a assumed horizontal ground (tf/m^2)

c: cohesive strength of the layer under the bottom of foundation.

γ_1: unit weight of the layer under the bottom of foundation. (tf/m^3)

γ_2: unit weight of the layer above the bottom of foundation. (tf/m^3).

D: diameter of the bottom section of foundation.

D_f: effective embedment depth in bearing layer (m).

N_c, N_r, N_q: bearing capacity coefficient (Fig. 7).

5. Horizontal allowable bearing capacity of ground

5.1 Horizontal allowable bearing capacity

Horizontal allowable bearing capacity for ground in front of a foundation is taken as the result of ultimate bearing capacity in the same position divided by a safety factor as given in Table 3.

$$R_{qa} = \frac{R_q}{n} \quad (14)$$

where,

R_{qa}: horizontal allowable bearing capacity (tf)

R_q: ultimate bearing capacity from Eq. 15 (tf)

n: safety factor (Table 3).

5.2 Limited horizontal bearing capacity

The ultimate horizontal bearing capacity is taken as the minimum value of shear resistance on the linear sliding surface as shown in Fig. 8. The equation is as follows.

$$R_q = \frac{W(\cos\alpha + \sin\alpha \cdot \tan\phi) + C \cdot A}{\sin\alpha - \cos\alpha \cdot \tan\phi} \quad (15)$$

where,

R_q: ultimate horizontal bearing capacity (tf)

W: the weight of soil blocks above sliding surface (tf).

A: area of sliding surface (m^2)

ϕ: internal friction angle of ground soil (°).

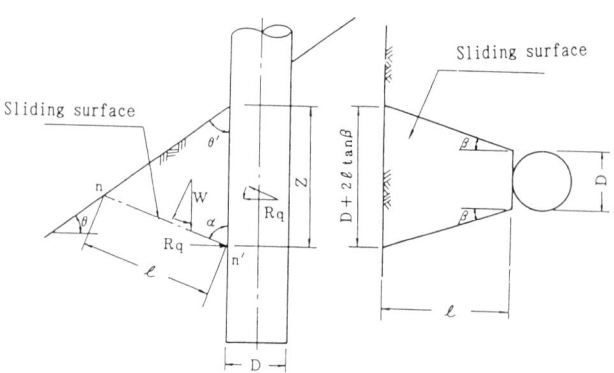

Fig. 8

C: cohesive strength of ground soil (tf/m^2).

α: angle of sliding surface. Generally, it is taken as
$$\alpha = 45 + \phi/2 + \theta/2$$

β: extending angle of sliding surface, usually, it is taken as $\beta = 30 + \phi/3$ for soft rock, and $\beta = \phi/3$ for hard rock.

5.3 Resistance in plastic zone

The total ultimate resistance from plastic zone is taken as the shear stress on sliding surface caused by weight of plastic zone and calculated by means of following equation.

$$R_0 = \frac{(\cos\alpha_0 + \sin\alpha_0 \cdot \tan\phi_s) \cdot W_0 + C_0 \cdot A}{\sin\alpha_0 - \cos\alpha_0 \cdot \tan\phi_s} \quad (16)$$

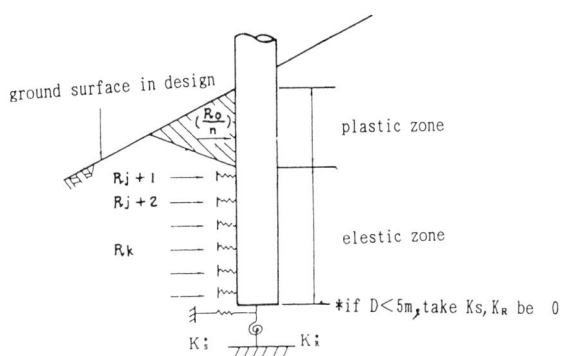

Fig. 10 Stability in horizontal direction (elasto-plastic solution)

Table 4 Shearing constants after plasticization

	soil ~ soft rock (CL)	middle hard rock (CM or above)
Cohesive force C_0	$C_0 = C$	$C_0 = 0$
Frictional angle ϕ_s	$\phi_s = \phi'$ ($\phi' = \phi \leqq 30°$)	$\phi_s = 2/3\,\phi'$ ($\phi' = \phi \leqq 30°$)

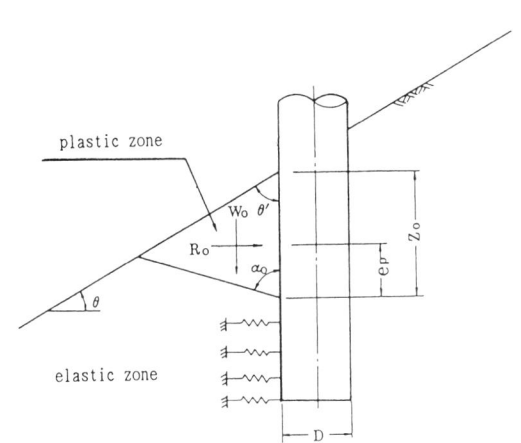

Fig. 9

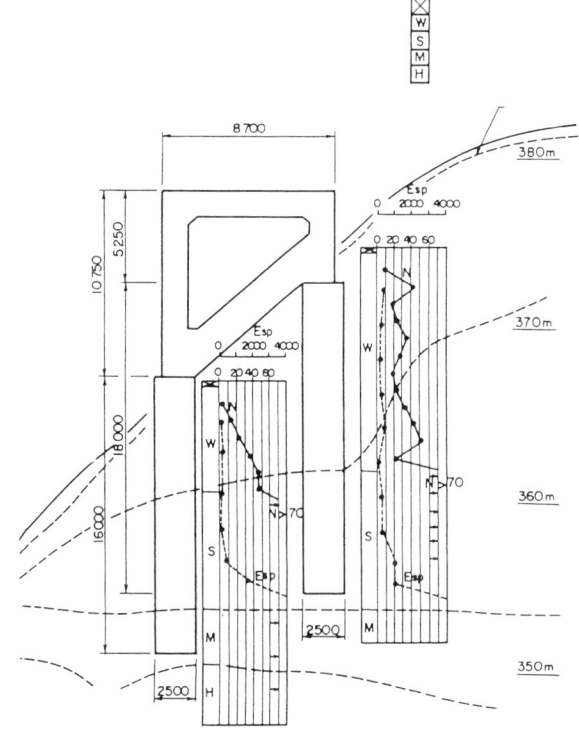

Fig. 11 Layout of ground layers and structure

Table 5 Test results of rock, predicted C and ϕ (mountain side, in deep foundation at depth 16.5m)

Specimen No.	Velocity of ultrasonic		Dynamic Poisson ratio	Unit weight (tf/m^3)			Water absorbing ratio (%)	Effective pore ratio (%)	Unconfined compression strength (kgf/cm^2)	Dynamic elastic modulus (kgf/cm^2)	Static elastic modulus (kgf/cm^2)	Color of specimen
	P wave	S wave		Natural	Wet	Dry						
1	4.39	2.39	0.29	2.55	2.58	2.49	3.68	9.14	126.4	3.83×10^5	1.55×10^5	Black
2	4.11	2.22	0.29	2.35	2.43	2.27	7.08	16.08	190.6	3.06×10^5	8.36×10^5	white
Avg.	4.25	2.31	0.29	2.45	2.51	2.38	-	-	158.5	3.45×10^5	1.19×10^5	-

Cracking coefficient C_r	C/C_0	ϕ/ϕ_0	ϕ based on Specification for substructures of road bridges (°)				$C_0 = \frac{q_u}{2}\sqrt{\frac{1-\sin\phi}{1+\sin\phi}}$ (tf/m^2)	C considering cracking C (tf/m^2) Depth 16.5m
			Depth					
			2.5m	6.5m	12.5m	16.5m		
0.77	0.07	0.6	32	34	52 ↓ 45	71 ↓ 45	262~395	18.3~27.7

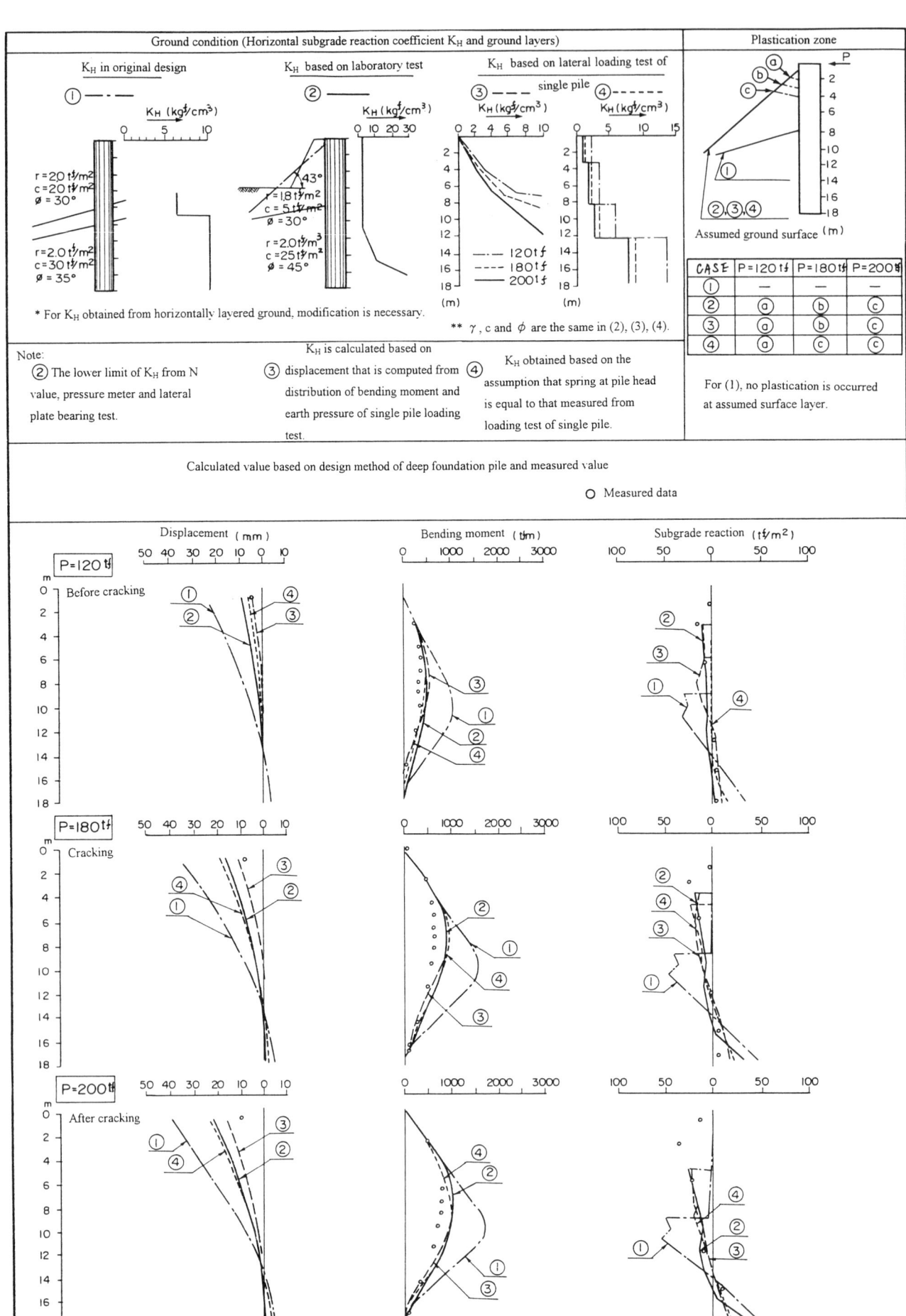

Fig. 12 Comparison of designed and measured data

The position e_p of R_o is computed by use of following equation.

$$e_p = \left[\frac{1}{4} + \frac{1}{12\left\{1 + \frac{2}{3D} \cdot \frac{\sin\theta' \cdot \tan\beta}{\sin(\theta' + \alpha_0)} \cdot Z_0\right\}}\right] \cdot Z_0$$

where,

R_o: ultimate resistance in plastic zone (tf);

W_o: the weight of rock in plastic zone (tf);

α_0: sliding angle that causes ultimate horizontal resistance;

ϕ_s and C_o: frictional angle and cohesive strength between plastic and elastic zones, respectively, and is given in Table 4.

5.4 Check of horizontal stability

Eq. 17 is followed in the check of horizontal stability. Also, even Eq. 17 is satisfied, the embedment depth in elastic bearing zone should be longer than 2 m on the consideration of safety (Fig. 10).

$$R_{qak} \geq \left(\frac{R_0}{n}\right) + \sum_{i=j+1}^{k} R_i \quad (17)$$

where,

R_{qak}: horizontal allowable bearing capacity at a position the number k spring (tf);

(R_o/n): resistance from plastication zone that considered the safety, in that R_o is calculated from Eq. 16 and n is safety factor that is equal to 3 at normal time and 3 during earthquake;

$\sum_{i=j+1}^{k} R_i$: the total reaction from the position of spring $j+1$ to k,

i.e., the total reaction in elastic zone (tf).

6. Comparison with in-situ loading test results [2]

In-situ bearing test has been done on following purposes:

(1). to understand the behavior of lateral loaded deep pile foundations constructed on inclined slopes;

(2). to understand the applicability of cohesive strength, internal friction angle, later subgrade reaction coefficient and resistance of ground to be obtained by means of ground survey and lateral plate bearing test, respectively, to the analysis of real behavior of foundation; and

(3). to verify the appropriateness of the design method of deep pile foundation.

In this paper, the loading test results of mountain side pile (Fig. 11, single pile, pile head free) will be compared (the loading test was done on whole structure). The test results at a depth of 16.5 m (soft rock) are illustrated in Table 5.

The comparison of displacement, bending moment and subgrade reaction coefficient based on design method has been done with those of measured ones. The results are illustrated in Fig. 12. The subgrade reaction coefficient to be obtained from different experimental methods is compared with the one measured from in-situ loading test of foundation. The following conclusions are obtained from the results:

(1). The calculated values of displacement, bending moment and subgrade reaction in Cases 2, 3 and 4 that considered the effects of surface layer are a little bit large. However, good agreement is obtained with measured data. Especially, the results of Case 3 that uses the back calculated k value present the agreement being the best.

As the plastic zone developing (based on calculation), at $P=200t$, good agreement is shown between calculated and measured subgrade reactions. However, the calculated value of the position of maximum earth pressure (-3.0 m below pile head), after plastication, is deeper than that of measured one.

The above results show that if a suitable k value of surface

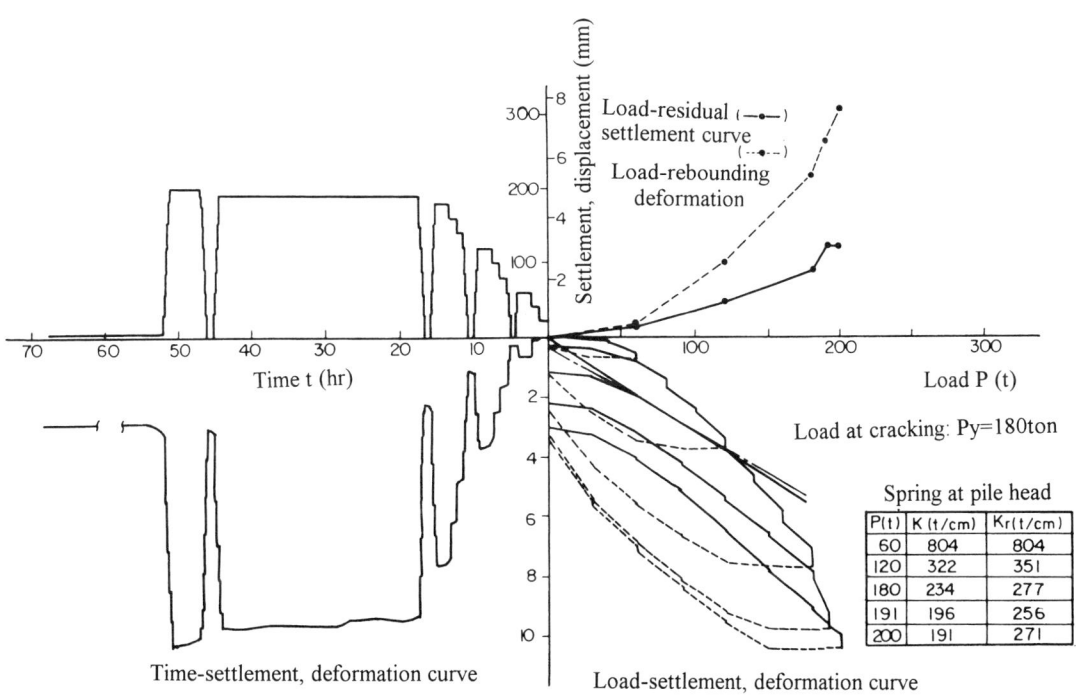

Fig. 13 Load-displacement-time curve

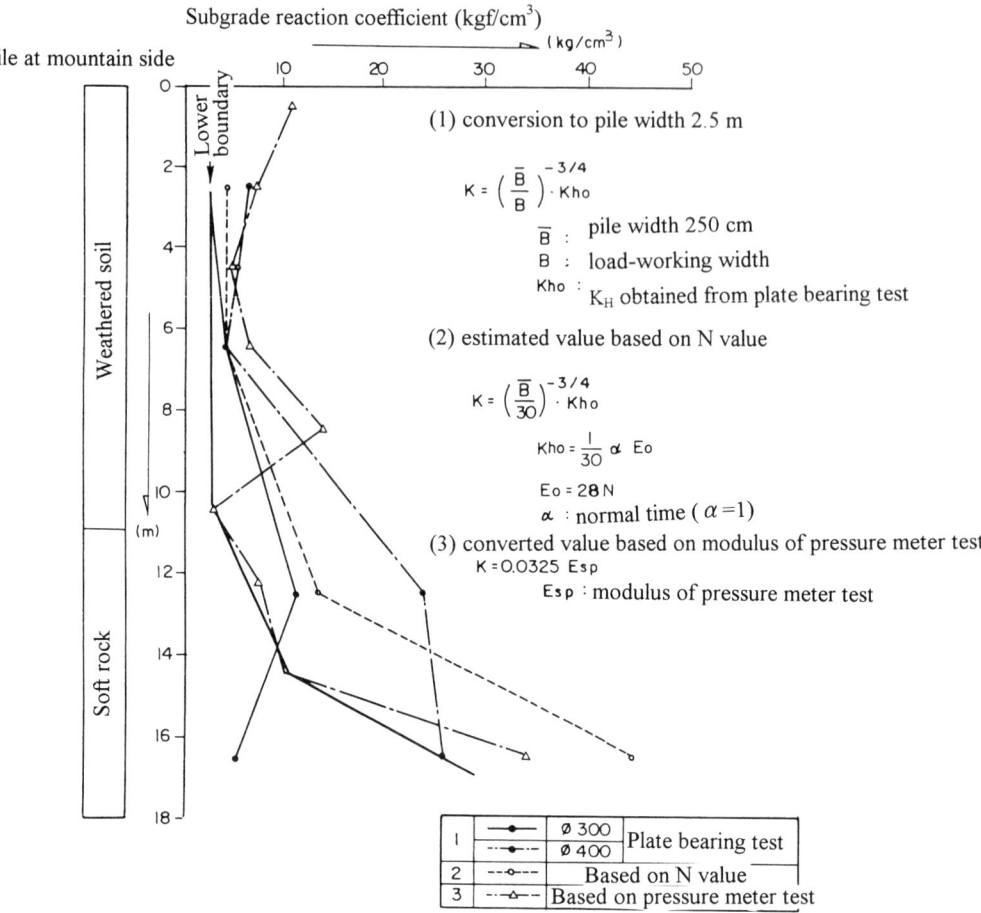

Fig. 14 Subgrade reaction coefficient obtained from different methods

layer is considered in a design, the design will obtain a result its behavior is quit near the real one, also, the design is at safety side.

The shearing constants of surface layer in this calculation are assumed to be, $c=5$ tf/m^2 and $\phi =30°$. From the results of measured earth pressure, it is considered that the resistance of this layer is larger than the assumed one.

(2). Although clipping is considered neither for ground nor for pile body and the measured data are from short term loading test in this design, it gives a displacement larger than the measured one. For the deep foundation constructed on inclined slope will bear long term lateral load, it is necessary to consider the clipping behavior in the selecting of k value.

(3). The design that neglects the effect of surface layer causes the results quit larger than the measured ones. Therefore, it is necessary to consider its effect, especially, in the case that the thickness of surface layer is relatively large.

For reference, Figs. 13 and 14 illustrate the load-displacement-time curves and subgrade reaction coefficient obtained from different methods, respectively.

Reference:
1). Japan Highway Public Corporation, (1990,7), *Design Method*, vol. 2.
2). Japan Highway Public Corporation, (1978,2), *Research Report on Design of Deep Foundation Pile*.

Deep excavation of soft rock by vertical NATM

Ryoji Ito & Kohei Watanabe
Shimizu Corporation, Japan

Atushi Takagi, Mamoru Ueno & Kenji Nakasita
Tokyo Gas Co., Ltd, Japan

ABSTRACT : Vertical NATM, which can secure the stability of the mudstone layer with shotcrere and rockbolts, was apllied to the excavation works of LNG inground storage tanksat Negishi Terminal of Tokyo Gas. 200,000kl LNG inground storage tanks are now under construction. The scale of excavation work was 76m in diameter and 57m in depth, which is the world's largest among of vertical excavation of soft rock using NATM.
The stability of the mudstone layer was confirmed by the non-linear rock analysis and the loosened zone was estimated to determine the arrangement of rockbolts. Based on the monitoring results, the behavior of the mudstone layer can be analytically estimated with accuracy.

1 INTRODUCTION

In 1970, the first LNG inground storage tank with a capacity of 10,000kl was completed at Negishi Terminal of Tokyo Gas Co. in Yokohama. Two LNG inground storage tanks with a capacity of 200,000kl are now under construction and these tanks are to be the world's largest.

At the stage of excavation work, the cut-off slurry wall method is generally adopted as a retaining wall against the ground in which a soft reclaimed layer or alluvium layer exists. However at the Negishi site, soft rock, so-called mudstone layer exists close to the ground surface. Therefore Vertical NATM (New Australian Tunneling Method) is applied as retaining structure, taking advantage of the self-supporting characteristics of mudstone. Vertical NATM has been already applied to five tank projects and the excavation work of the 200,000kl inground storage tanks with 76m in diameter and 57m in depth was recently completed.

Based on the observed results of the previous projects, the behavior of mudstone layer can be analytically estimated with accuracy.

This paper presents the design and the actual excavation work carried out by Vertical NATM. The comparison of analysis and observation is also described.

2 THE OUTLINE OF LNG INGROUND STORAGE TANK

2.1 Soil Properties

At Negishi Terminal, a very soft silt layer (Fc) about 9m in thickness exists under the ground surface, and the flat mudstone layer (Kac) exists below this silt layer. the ground water level is relatively high of 1.5m below the ground surface. The mudstone layer at this site is fissure-free, consistent and impermeable. The mudstone exhibits about 25 kgf/cm² of unconfined compressive strength and about 1.0% of strain at the point of failure. Generally, the slaking phenomenon affects the overall stability of the ground during vertical excavations. However, the mudstone at the Negishi site has very high slaking durability. The distribution of the unconfined compressive strength of mudstone is shown in Fig.1.

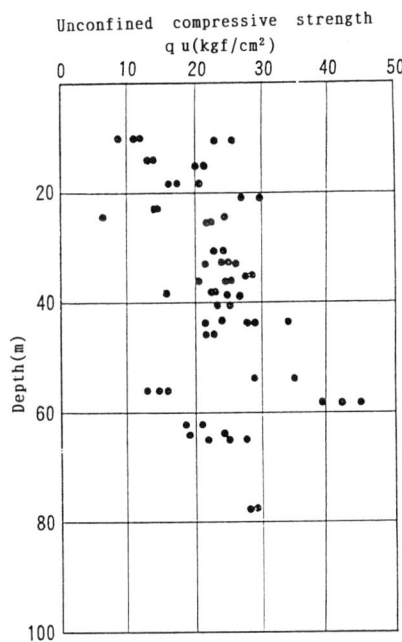

Fig.1 Unconfined Compressive Strength of Mudstone

2.2 200,000kl LNG Inground Storage Tank

The general structure of 200,000kl LNG inground storage tank is shown in Fig.2.

The configuration of the tank is cylindrical and the dimensions are 68m in diameter and 55.1m in liquid depth.

After the completion of excavation work, the bottom slab and the side wall of reinforced concrete were constructed. High strength concrete (the design strength of 600kgf/cm² is about twice that of the conventional concrete) was adopted to the side wall of this tank. The Membrane and Insulation were attached to the inner surface of the tank body. The membrane maintains the liquid and gas tightness and the insulation preserves the cryogenic condition of the tank interior. A steel concrete composite doomed roof was placed on the top of the side wall to keep the inner gas tightness. In the surrounding area of the tank, heating systems are provided to control ground freezing.

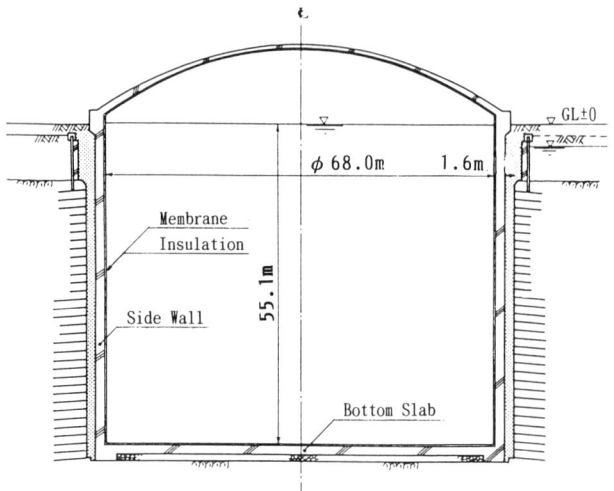

Fig.2 General Structure of 200,000KL LNG Inground Storage Tank

2.3 Development of Vertical NATM

NATM with shotcrete and rockbolts is generally applied to rock tunneling and is a simple and economical method. In the late 1970's, the excavation work with a ground anchor retaining system was carried out for the inground tank project at Negishi Terminal. In the middle of the 1980's, Vertical NATM was first applied to 60,000kl inground tank and the excavation work was successfully completed. As a result, Vertical NATM is now a standard excavation method at Negishi Terminal. The excavation work of the 200,000kl inground tank, 76m in diameter and 57m in depth, is the world's largest scale among of soft rock excavation by Vertical NATM. The application of Vertical NATM at the Negishi site is shown in Fig.3.

3 DESIGN OF VERTICAL NATM

Vertical NATM is a rational method which utilizes the most advantage of the self-supporting characteristics of mudstone. It is important that the stability of mudstone layer is analytically confirmed and the loosened zone is estimated to determine the arrangement of rockbolts.

3.1 Rock Stability Analysis

Non-linear rock analysis was conducted to evaluate the stability of mudstone layer during excavation. The finite element analytical model for excavation is shown in Fig.4.

As for the analytical model, an axial symmetrical model of 120m in side boundary and of 165.5m in depth boundary was adopted. The mudstone was modeled into two layers (Kac1 and Kac2), taking the difference in strength into consideration. The rockbolts were not modeled in this analysis. The analytical constants are shown in Table 1. The constants of strength were determined from the results of tri-axial compression test and tensile test.

The modulas of deformation was determined from in-situ tests and monitoring results of the previous project and is mentioned in chapter 5.

Table 1 Mechanical Properties of Mudstone

	Kac1	Kac2	Unit
Unit weight	1,900	1,900	kgf/cm^3
Shear strength	3.4	4.5	kgf/cm^2
Tensile strength	1.5	2.3	kgf/cm^2
Modulas of deformation	6,000	12,000	kgf/cm^2
Poisson's ratio	0.3	0.3	—

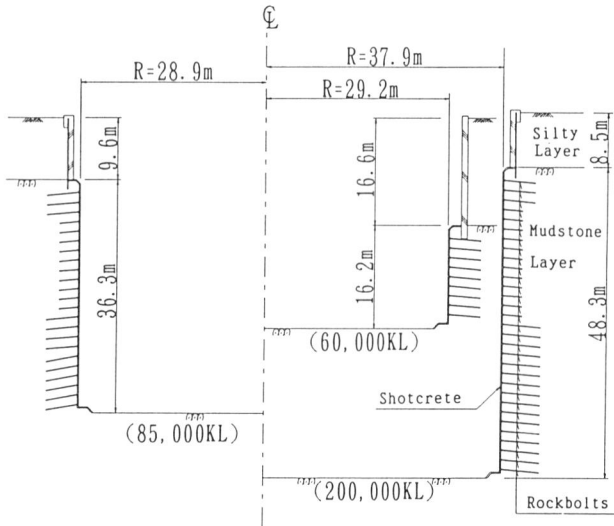

Fig.3 Vertical NATM at the Negishi Site

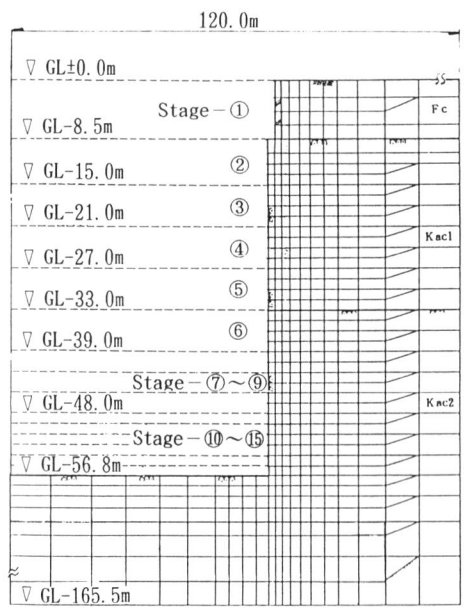

Fig.4 FEM Analytical Model for Excavation

As for the mechanical properties of mudstone, the mudstone is assumed to be non-linear elastic according to stress level. The non-linear behavior of mudstone is modeled by using the non-dimensional parameter R which is called fracture severity. (The fracture severity method was proposed by Central Research Insitute of Electric Power.) The parameter R shows the stress margin ratio against failure at the generated stress condition. R is defined as equation (1) and is also shown in Fig.5. The point R = 1 indicates the limit of linear elasticity. The range $0 < R < 1$ indicates non-linear elasticity and the modulas of deformation E decreases according to the parameter R. The parameter Fs is related to R as the inverse of stress level and represents local stress safety factor. Both R and Fs are regarded as state parameters which indicate the stability during excavation.

$$R = k \cdot \min(d_1/D_1, d_2/D_2) \quad (1)$$
$$Fs = \min\{D_1/(D_1-d_1), D_2/(D_2-d_2)\} \quad (2)$$
$$R \geq 1 : E = E_0 \quad (3)$$
$$0 < R < 1 : (E - E_f)/(E_0 - E_f) = R^{1/a} \quad (4)$$
$$R = 0 : E = E_f \quad (5)$$

E_0 : initial modulas of deformation
E_f : final modulas of deformation
a : coefficient of envelope line

where k is constant that indicates the limit of linear elasticity (the yield strength = the peak strength × $(k-1)/k$) and k is 2~3 on the mudstone.

Based on analytical results, two methods are used for judging the stability. The zone in which both judgements are not satisfied, is defined as "loosened zone". The judgements for stability are as follows :

(1) Judging from how far the stress circle is from the rupture envelope line ;
    $Fs \geq 1.5$ and $R \leq k \cdot 1/3$
(2) Judging from the tensile stress ;
    $\sigma_3 \leq 1.0 kgf/cm^2$

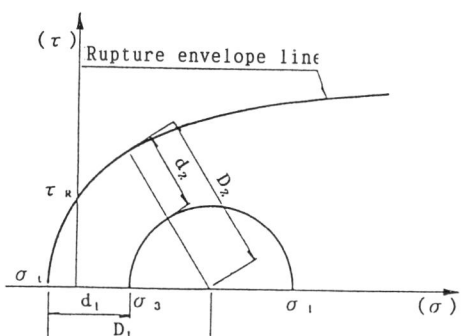

Fig.5 Failure Criteria of Mudstone

3.2 Arrangement of Rockbolts

The analytical results predicted the loosened zone about 3.5m deep from the excavated surface at the final excavation stage. The predicted loosened zone is shown in Fig.6.
The rockbolt length in the predicted loosened zone was determined to be 6m, which includes the anchorage length of 2.5m to safer side. The anchorage length was determined to be 2.5m from the rockbolt pull-out test at the previous project.

Taking the reinforcing effect of rockbolts into consideration, the stability of the vertical excavated surface was calculated by means of the Slip Plane Method to determine the arrangement of rockbolts. The loosened zone was assumed to decrease into the residual strength ($c=0, \phi=35°$) which is 1/3 ~ 1/4 of the peak strength. The rockbolts were assumed to increase the shear resistance on the sliding plane and the reinforcing effect of rockbolt τ_B was as follows.

$$\tau_B = P_y\{\mu \cos(\theta - \alpha) + \sin(\theta - \alpha)\}/A$$

where
P_y : yield strength of rockbolt
α : installed angle of rockbolt
μ : coefficient of friction
θ : angle of sliding plane against the vertical surface
A : area of slip plane.

All sliding planes in the loosened zone were assumed by changing the angle of the sliding plane and the stability of the vertical excavated surface was confirmed as shown in Fig.6. As a result, $\theta = 30°$ was the most critical sliding plane and was nearly equal to $(45° - \phi/2)$.

The top part of the rockbolts was determined to reinforce the embedded layer of the retaining wall where steel sheet piles were driven 2.0m deep into the mudstone layer before the excavation work. The detail of the rockbolt is shown in Fig.7.

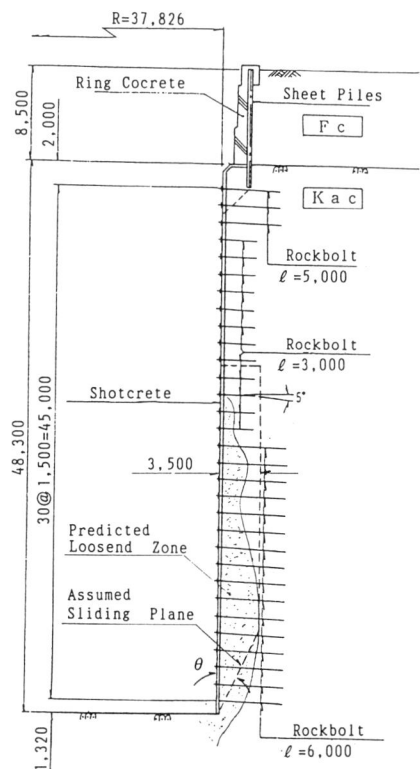

Fig.6 Predicted Loosend Zone at Final Excavation Stage

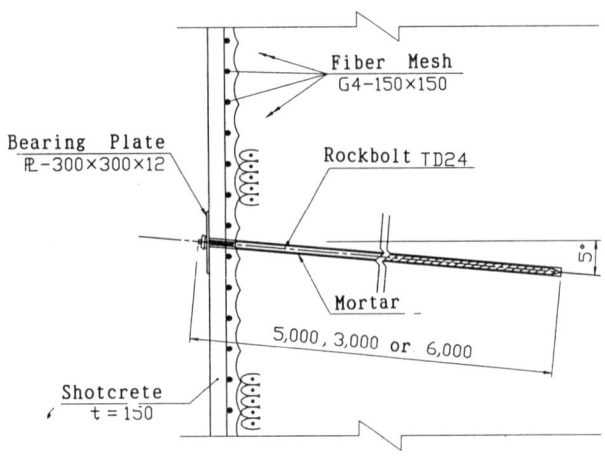

Fig.7 Detail of Rockbolt

4 Excavation Method by Vertical NATM

After the excavation of the soft silt layer using the retaining wall of steel sheet piles and ring concrete, the excavation of the mudstone layer was carried out with protecting the excavated surface. The protection was performed on the excavated surface in the manner of shotcrete with reinforcing mesh and rockbolts. Automated equipment for protecting the excavated surface was introduced in the actual work of Vertical NATM during excavation.

The system of this equipment is composed of three independent robots : "mesh setter", "mortar shooter" and "rockbolt setter". the automatic system is shown in Fig.8 and Photo. 1. These robots play the roles of putting the reinforcing mesh to the vertical excavated face, spraying the mortar on it, and inserting the rockbolts. These robots are hung from the rails set on top of the retaining wall and moved along the inner excavated surface. When they move around, one cycle of the work is completed. Such cyclic work procedures were repeated until 57m in depth. With this automatic system, the works were so efficient and the safety is highly improved.

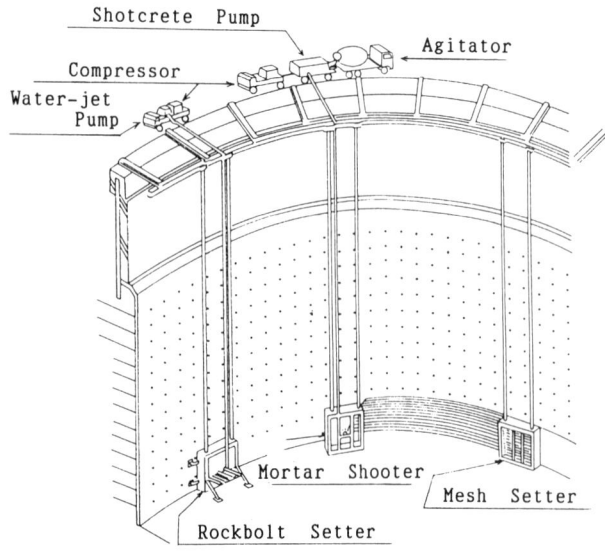

Fig.8 Automatic Systems for Vertical NATM

The monitoring items are shown in Table 2. Among these items the horizontal displacement was monitored with inclinometers which were installed in 8 directions. The axial forces in rockbolts were monitored in 3 directions and at 6 levels. The excavation control was conducted by using these monitoring items. The control values was determined by the result of the predicted analysis.

The excavation work for the 200,000kℓ LNG inground tank was successfully completed in December 1993 after 12 months of excavation. The excavation of the tank is shown in Photo. 2.

Table 2 Monitoring Items

Monitoring Items	Monitoring Points	Interval
Stress of concrete	Ring concrete	once a day
Horizontal displacement	Excavated surface	twice a week
	Top of ring concrete	twice a week
Axial Force	Rockbolt	once a day

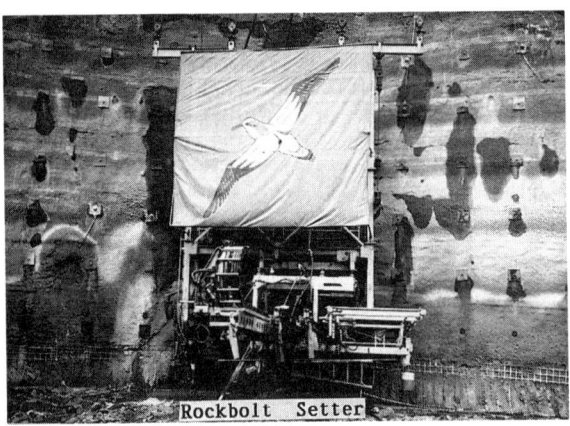

Photo.1 Automatic Systems for Vertical NATM

Photo.2 Excavation of 200,000kℓ LNG Inground storage tank

5 Comparison of Analysis and Observation

At the previous project of 85,000kl LNG inground storage tank, the observed horizontal displacement of the excavated surface was smaller than 1cm for the excavation dimension of 58m in diameter and 45m in depth. As the observed result showed the different pattern and value from the initial analysis result, the FEM feed-back analysis was conducted. The revalued constants were the modulas of deformation E and the initial coefficient of lateral pressure K_0.

E values obtained from various tests are shown Table 3. E values should be based on the smaller strain level generated in the actual field. Therefore E value at unloading obtained from bore-hole lateral load test was adopted to the feed-back analysis and is also nearly equal to one half of the value of the dynamic modulus of elasticity obtained from elastic wave exploration.

As K_0 value was uncertain, the parameter study was conducted in the feed-back analysis to simulate the horizontal displacement. As a result, $K_0=0.75$ at the deep part and $K_0=1.0$ at the shallow part were obtained from simulating that horizontal displacement at the shallow part was larger than the deep part.

It was assumed that such deformation pattern was caused by the reduction of ground arch action. At the cylindrical excavation, the ground arch action is decreased at the shallow part in proportion as the inner diameter increases.

Based on the feed-back analysis result of 85,000kl LNG inground storage tank, the prediction analysis of 200,000kl LNG inground storage tank was conducted. The comparison of analysis and observation is shown in Fig.9. As for the horizontal displacement, the analytical value matched very well with the observed value.

Table 3 Modulas of Deformation of Mudstone

	Kac1	Kac2
Design	6,000	12,000
Elastic wave test	9,900	20,000~23,000
Borehole load test	—	9,600~12,300
uniaxial compressive test	3,000	5,000~6,000

Unit:kgf/cm²

In the excavation work of 200,000kl inground storage tank, an elastic wave test was carried out to examine the range of loosened zone, which was predicted 3.5m in depth. Consequently it was confirmed that the wave velocity decreased in the range of about 2m and the physical change of mudstone has occurred due to the excavation. The predicted loosened zone was proved to be on the safer side as shown in Fig. 9. The actual strength of mudstone was higher near the bottom level of excavation than the design strength. Using the core sampled at the bottom level of excavation, the revalued strength of mudstone was applied to the feed-back analysis. The zone of $R<1$ in the feed-back analysis matched with the range in which wave velocity decreased.

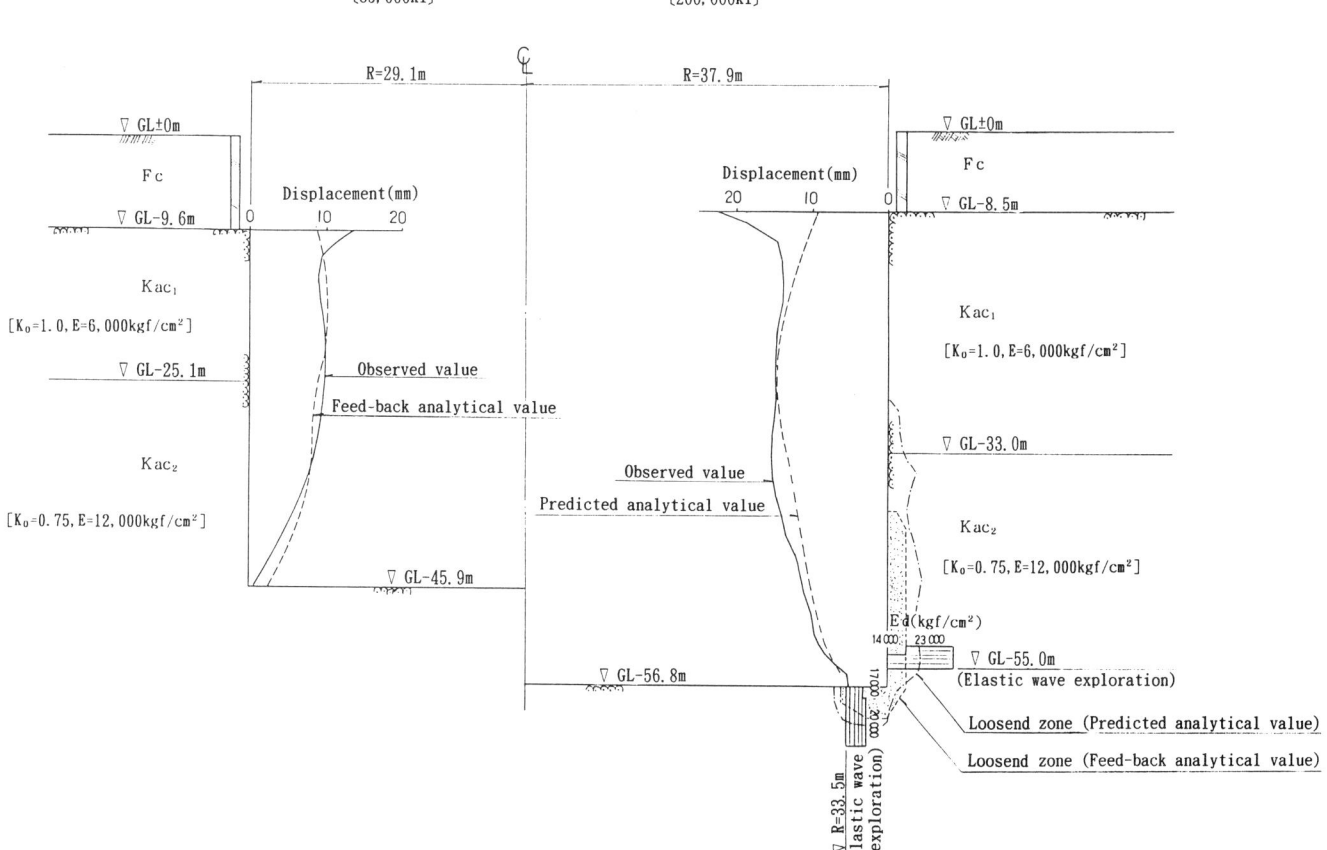

Fig.9 Comparison of Analysis and Observation

6 CONCLUSION

The mudstone at the Negishi site is regarded as continuous material and minor cracks and joints have been determined to be negligible. The overall deformation can be predicted as elastic deformation in the smaller strain level. On the other hand, the excavated surface shows non-elastic behavior and the prediction of loosened zone is important for the design of rockbolts. The behavior of the mudstone layer at the excavation work can be predicted with accuracy by the analytical method described in this paper.

REFERENCE

Hayashi M. & Hibino S. 1968, Progressive analysis of relaxation around under ground excavating space, Central Research Institute of Electric Power Industry Report.

Nakano M. & Nishikawa Y. 1993, The study on the observed results during the large scale cylindrical excavation with shotcrete and rockbolts, Proc. Int. Sympo. on Hard Soils - Soft Rocks, Athens.

Ito R., Watanabe K., Nakano M., Ueno M. and Nakashita K. 1994, Comparion of Observation and Analaysis in the cylindrical excavation of mudstone layer, The 9th Domestic Sympo. on Rock Mechanics, Tokyo.

State of the arts on the super-high-rise building foundation in Japan

Yoshiaki Nagataki, Shigeru Iiboshi & Kazuki Aoshima
Technology Research Center, Taisei Corporation, Japan

ABSTRACT: This report briefly describes the state of the arts of superstructure and foundation structures of super-high-rise buildings in Japan. Specially description is provided of seismic design methods which have been considered to be an important issue for many years.
The subject of wind force presenting a greater problem than seismic force, as is witnessed in recent super-high-rise buildings, is also discussed.

Super-high-rise buildings have traditionally been supported by mat foundations in the case of hard ground. Since super-high-rise buildings employing a pile foundation built on reclaimed land have increased, an additional description is provided concerning these foundation types and design methods. Finally, the outlook of the foundation for future super-high-rise buildings along with the direction in which they are likely to progress is discussed.

1 PRESENT STATE OF SUPER HIGH RISE BUILDING FOUNDATION

1.1 Characteristics of Super-highrise builings

Ikeda,T et al(1988) made following several important statements on super-high-rise buildings. Although there are no strictly defined standards for the term super-high-rise building, buildings having a height of 100 meters or more and roughly 25 stories or more are generally referred to as super-high-rise buildings. The first super-high-rise building in Japan is the Kasumigaseki Mitsui Building completed in 1968. This building stands 147 meters high and has 36 stories.

Until 1963, the maximum allowed height of buildings according to the Building Standards Act was, as a general rule, 20 meters in residential districts and 31 meters in non-residential districts. Thus, when attempting to construct a building having a large total floor area, the building had to be built so as to cover as much as of the building site as possible since it was not allowed to exceed 31 meters in height according to the above law. This resulted in the occurrence of various obstacles. However, the curtain was finally lifted on the age of super-high-rise buildings following abolition of height restrictions in 1963 accompanying the progress of architectural technology.

Among buildings exceeding 60 meters in height, with respect to those of which structural calculations was made using dynamic analysis and other methods for confirming the safety of the building in terms of structural yield strength, these buildings must be approved by first being evaluated by the Steel Frame Structure and High-Rise building Structural Evaluation Committee of the Foundation Japan Architectural Center, and then reviewed by the Building Technology Screening Committee of the Ministry of Construction.

In contrast to lower structures employing a rigid structure, the upper structures of super-high-rise buildings incorporate a flexible structure. Two of the main factors behind the coming of the current age of super-high-rise buildings are the accumulation of seismic recordings using strong motion seismographs, and the development of computerized dynamic design techniques.

According to dynamic analysis using the above-mentioned seismic recording, it was found that during an earthquake, although considerable seismic force is generated in low-story, rigid buildings, seismic force is not generated to a significant extent in high-rise buildings employing a more flexible structure.

The upper structures of super-high-rise buildings typically consist of a steel rigid frame structure having a central core portion. Some reasons for employing this type of structure are indicated below.

(1) A flexible structure can be realized by using a steel frame offering high strength and ample tenacity.
(2) In the case of super-high-rise buildings, the weight of the building can be reduced to minimize the weight load applied to the foundation and lower stories as much as possible.
(3) Excellent productivity is achieved since the steel frame can be produced in the factory and simply assembled at the construction site.

More recently, however, super-high-rise buildings have appeared that employ a reinforced concrete pre-cast structure in which reinforced concrete members are prefabricated in the factory or fabricated at the construction site followed by assembly.

1.2 Foundations of Super-High-Rise buildings

The foundations conventionally employed in super-high-rise buildings have consisted nearly exclusively of direct foundation types as seen in the super-high-rise buildings located in the Shinjuku district of Tokyo, the secondary center of the city. More recently, however, since other large-scale buildings including super-high-rise buildings are being built at waterfront areas, pile foundations and foundations combining piles and diaphragm walls are being used in place of the direct foundations used conventionally.

Moreover, with respect to pile foundations, since top driven piles are hardly used in consideration of environmental problems, cast-in-place piles are used in the majority of cases in their place.

Various types of these cast-in-place piles are being used. Of the cast-in place piles, recently enlarged bottom cast-in-place piles developed for the purpose of providing greater end bearing capacity, and enlarged top cast-in-place piles developed and wrapped steel pipe cast-in-place piles to oppose the pile head moment generated during an earthquake having frequently been employed.

In addition, there have also recently been examples of buildings employing permanent anchors in the vertical direction for the purpose of preventing collapse of high-rise buildings during an earthquake and preventing floating caused by buoyancy. Moreover, permanent anchors in the direction of inclination are being used for preventing collapse of buildings constructed on steep inclines.

At this point, the bearing ground and foundation types of super-high-rise buildings will be examined by region. In looking first at the bearing ground of super-high-rise buildings, Sasao describes that the bearing ground in the Tokyo and Osaka regions consists almost entirely of diluvial gravel strata referred to as the Tokyo gravel stratum and Tenman gravel stratum, respectively. In addition, Sasao,H(1977) describes that the bearing ground in the Yokohama region consists of consolidated silt of the Tertiary Period Kazusa strata group (usually referred to as the "mud stone stratum").

In addition, with respect to foundation types, perhaps due to the increasing number of cases recently in which super-high-rise buildings are being built on the waterfront, the number of super-high-rise buildings using pile foundations or a combination of diaphragm walls and pile foundations is on the increase.

2 GROUND SURVEYS, TESTING AND MEASUREMENT FOR SUPER-HIGHRISE BUILDINGS

2.1 Present State of Surveys and Testing Methods

Survey methods naturally differ between the case of supporting with piles directly on diluvium and supporting with piles by driving through soft ground as in the case of waterfront areas.

1) Case of Direct Foundations on Diluvium
    Sakaguchi,O(1979) discribes Case of Direct Foundations on Diluvium as :
  (1) Survey of Static Properties
Examples of current location tests include the standard penetration test (N value), acquisition of undisturbed samples and the plate bearing test. Examples of indoor tests include physical tests (density, grain size, moisture content) and dynamics tests (uniaxial and triaxial consolidation).
  2) Survey of Dynamic Properties
Examples of current location tests performed include PS logging, density logging and microtremor measurement. These are performed to determine dynamic properties and predominant period of the ground. Examples of indoor tests include the dynamic triaxial test and hollow torsional shearing test. These are used to determine the dynamic deformation properties and damping properties of the soil.

  (2) Case of Supporting with Piles on Soft Ground
    Kawasaki,T(1991) discribes case of Supporting with Piles on Soft Ground as :
  1) Survey of Static Properties
Examples of current location tests include the standard penetration test (N value), acquisition of undisturbed samples, plate bearing test, underground horizontal bearing test, ground water level and local water permeability test. In addition, load tests may also be performed on the piles when necessary.

Examples of indoor tests include physical tests (density, grain size, moisture content) and dynamics tests (uniaxial and triaxial consolidation).
  2) Survey of Dynamic Properties
Examples of current location tests performed include PS logging, density logging and microtremor measurement. These are performed to determine dynamic properties and predominant period of the ground. Examples of indoor tests

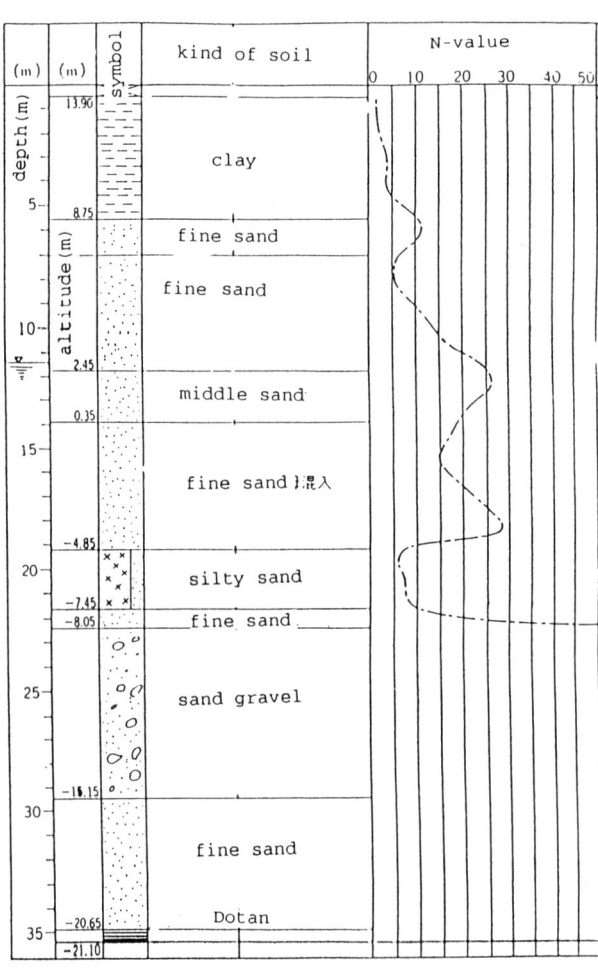

Figure 1 Results of soil investigation

include the dynamic triaxial test and hollow torsional shearing test. These are used to determine liquefaction strength as well as the dynamic deformation properties and damping properties of the soil.

2.2 Survey and Measurement Examples

1) Case of a Direct Foundation having the Tokyo Gravel Stratum for the Bearing Stratum
Mutoh,K(1966) discribes case of a Direct Foundation having the Tokyo Gravel Stratum for the Bearing Stratum as :

(1) Ground Summary
Figure 1 shows results of soil investigation. Earth filling and a volcanic clay stratum (Kanto loam formation) are accumulated to a depth of roughly 5 meters from the surface, and the Tokyo gravel stratum, the bearing ground, is distributed at a thickness of roughly 6 meters from A.P. to 10 meters.

(2) Results of Plate Bearing Test
Figure 2 shows results of plate loading test on TOKYO gravel layer.
In looking at the load-settling curve obtained from the results of the plate bearing test on the Tokyo gravel stratum, the amount of settling is roughly 5 mm when a load of 300 t/m^2 (2940kPa) is applied, and has not reached the breaking load.

(3) Results of Standard Penetration Test
Figure 1 shows stndard penetration test.
The N values of the Tokyo gravel stratum and lower Tokyo layer are 100 or more according to the standard penetration test.

(4) Results of Microtremor Measurement
Figure 3 shows power spectrum of micro-viblation measurement.
The predominant period is 0.4 second according to the results of microtremor measurement.

2) Case of a Direct Foundation having the Consolidated Silt of the mud stone stratum for the Bearing Stratum
Yamazaki,K(1991) discribes case of a Direct Foundation having the Consolidated Silt of the mud stone stratum for the Bearing Stratum as :

(1) Ground Summary
Figure 4 shows section of soil.
Ground composition consists of a horizontal distribution of an earth filling stratum, alluvium stratum and the Kazusa strata group in order from the surface. Earth filling stratum (F) is a viscous soil consisting of a mixture of loam, silt, rock and brick. The alluvium stratum (As) consists of silty fine sand, while the diluvium stratum (Tm, s) consists of alternating strata of consolidated silt and sand in the Kazusa strata group.

(2) Results of Standard Penetration Test
In figure 4, the N value of the earth filling stratum is 1 to 50, and that of the Kazusa strata group, the bearing stratum, is 50 or more.

(3) Results of Uniaxial Consolidation Test
There is no increasing trend in the direction of depth down to roughly GL-90 of the Kazusa strata group, and the mean is 33.8 kgf/cm^2.

(4) Results of Triaxial Consolidation Test
The values for Ccu and fcu of the Kazusa strata

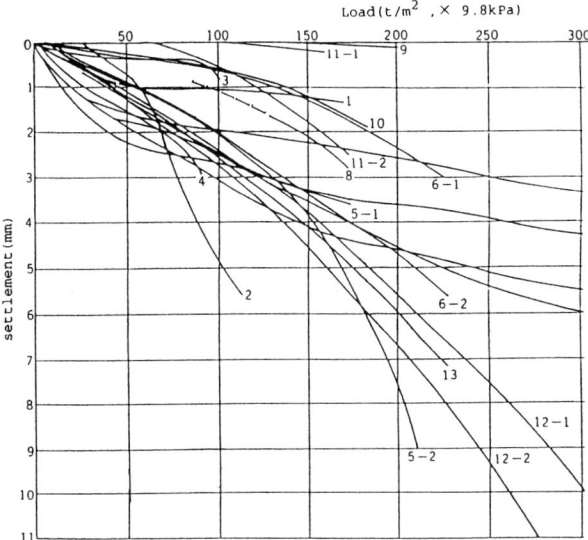

Figure 2 Results of plate loading test (TOKYO gravel layer)

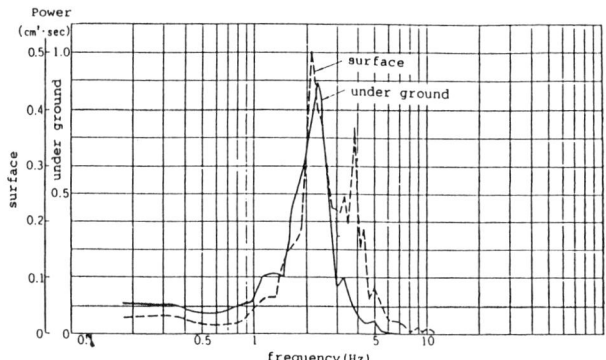

Figure 3 Power spectrum of microviblation

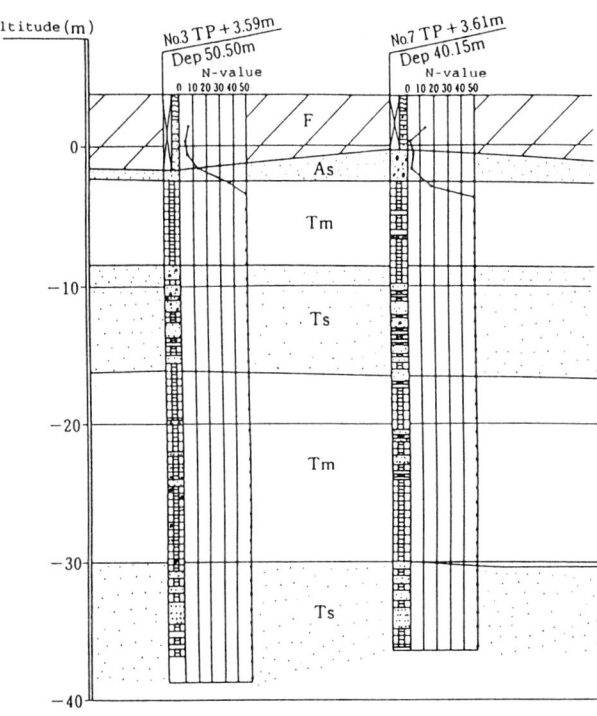

Figure 4 Soil column

group are 1.5 kgf/cm² and 38 degrees, while those values for the siltstone stratum are 7.5 kgf/cm² and 36 degrees according to the results of the triaxial consolidation test under consolidated, non-drainage conditions.

(5) Results of PS Logging
The mean velocity of the P and S waves of the diluvium stratum (Tm, s) are 1.8 km/s and 0.6 km/s, respectively.

3 DESIGN METHODS FOR SUPER-HIGHRISE BUILDING FOUNDATION

3.1 Ground Survey

(1) On-the-spot Survey
In general, exploratory boring, current position tests, underground logging and microtremor measurement are performed. Exploratory boring consists of the plate bearing test and acquisition of undisturbed samples. Examples of current position tests include the plate bearing test, measurement of underground horizontal load, and measurement of ground water level. In addition, PS logging and density logging are performed as a part of underground logging. Moreover, in the case of pile foundations, vertical and horizontal load tests are performed on the piles as necessary.

(2) Indoor Soil Tests
Indoor soil tests are broadly divided into physical tests, mechanical tests and dynamic tests.

Physical tests include tests for density, grain size, moisture content and liquid plasticity limit. Mechanical tests consist of uniaxial, triaxial and consolidation tests. In addition, dynamic tests include the performing of the dynamic deformation test according to the dynamic triaxial test. Liquefaction tests are also performed when a sand stratum is present.

3.2 Foundation Selection

Fukui discribes Foundation Selection as :

Examples of foundation types for super-highrise buildings that can be considered include (1) direct foundations (including floating foundations), (2) friction pile foundations, (3) bearing pile foundations, and (4) combination of continuous underground wall and bearing pile foundations.

In normal cases, the type of foundation is selected based on the following ways of thinking while taking into consideration the particular ground conditions.

(1) Direct foundations are used when the bearing power of the ground is large in comparison with the weight of the building. Thus, it is necessary that the stratum to serve as the bearing stratum be relatively shallow.

(2) Friction pile foundations are used when attempting to bear the weight of the building by the frictional force of the piles. These foundations are usually used in cases in which the bearing stratum is usually extremely deep and the use of bearing piles is uneconomical. However, this type of foundation is not used to support currently existing super-high-rise buildings.

(3) Bearing pile foundations support the building load in the lower diluvium and so forth after piles are driven through the earth filling stratum and alluvium. This type of foundation is used when the bearing stratum is deeper than the flooring level. This type of foundation has recently been widely used during construction of super-high-rise buildings on coastal reclaimed land.

3.3 Load Conditions

(1) Long Time Loading
The entire weight of the building, including the movable load, is used.

(2) Short Time Loading
The following two types of loads are used with respect to earthquakes and external wind force.

Level 1: Load of a magnitude that is predicted to be applied once during the service life of the building.

Level 2: Load of the maximum magnitude that has the possibility of occurring.

For the design load, the response to a level 1 external force is calculated followed by determination of the shearing force and overturning moment of each stratum.

3.4 Design Methods

Figure 5 shows design flow of foundation of Super highrise building.
Yamazaki,S(1991) discribes Design Methods as :

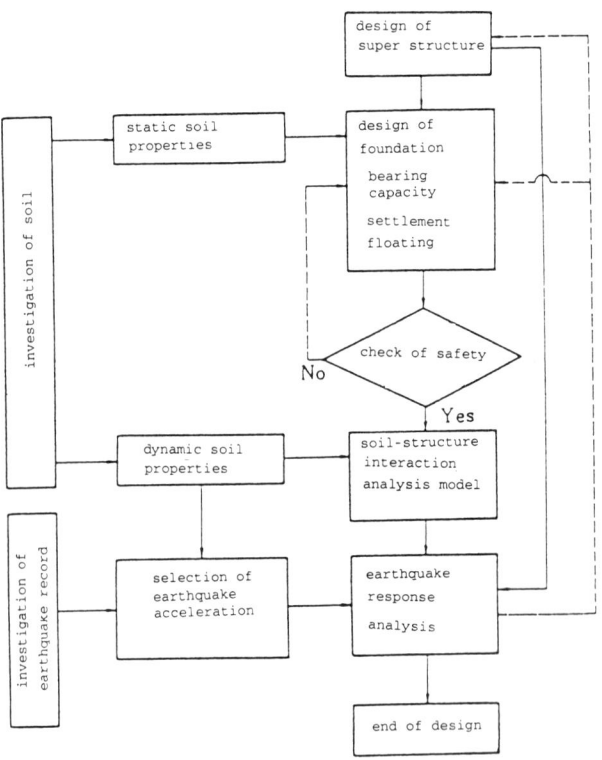

Figure 5 Design flow of foundation of Super highrise building

[1] Direct Foundations
(1) Setting of Soil Unit Bearing Caparsity
Although the setting of said unit bearing capacity varies somewhat according to the designer, a recent example of a method employed for the setting is described below.

The allowable unit bearing capacity of the ground is determined according to Building Foundation Structural Design Guidelines based on the results of soil tests, and 1/3 of this value is taken to be the long time allowable soil unit bearing capacity. Moreover, 1/2 of the above value is taken to be the short time allowable soil unit bearing capacity. Finally, yield strength is confirmed by the plate bearing test.

(2) Settling Analysis
The mat slab and footing beams of a direct foundation are replaced with a frame stress analysis model. The ground resiliency of the governing area of each joint is then determined, and these values are formed into an elastic support model that is applied to the joints.

The amount of settling is determined by adding the design load to this model, after which this amount is confirmed to be within the allowable amount of settling and deformation angle.

Although various methods are used for determining ground resiliency depending on the designer, examples of such methods include those using Steinbrenner's approximate solution and Boussinesq's solution.

Finally, the mat slab is designed by determining the amount of deformation and stress using FEM analysis integrating the mat slab and the ground.

(3) Examination of Overturning
External force obtained from the results of analyzing the response of earthquakes and external wind force is added to static factors to examine the safety of the building with respect to overturning.

[2] Pile Foundations
Kawabata,K(1989) discribes Design Methods as :
1) Allowable Unit Stress Design
Allowable unit stress design is performed corresponding to the stress resulting from the vertical load and the stress resulting from design shearing force.

(1) Examination of Vertical Allowable Bearing Capacity
The vertical allowable bearing capacity of the pile foundation is determined according to Building Foundation Structural Design Guidelines to examine the safety of the building with respect to long time loading.

(2) Examination of Extraction Resistance
Extraction resistance is determined according to Building Foundation Structural Design Guidelines to confirm that the building is free of floating with respect to short time loading.

(3) Examination of Pile Horizontal Bearing Capacity
The horizontal allowable bearing capacity of the pile foundation is determined according to Building Foundation Structural Design Guidelines to examine the safety of the building with respect to short time loading.

2) Ultimate Design
Ultimate design is performed for stress equal to 1.5 times the design shearing unit stress with respect to the bearing capacity and extraction force of the piles.

The following parameters are examined in accordance with retained yield strength and deformation performance in building aseismic design.

(1) Ultimate yield strength in the vertical direction
(2) Ultimate yield strength in the horizontal direction
(3) Ultimate extraction yield strength

3.5 Aseismic Design of Foundations

(1) Dynamic Design Methods
According to Design of Aaseismatic Structures, a model is made of the super-high-rise building for a vibration system of the super structure, foundation structure and ground. Response is then calculated when a seismic wave is input

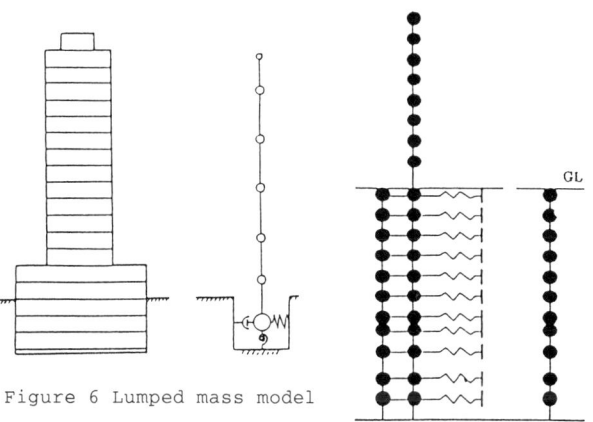

Figure 6 Lumped mass model

Figure 7 Penzien model

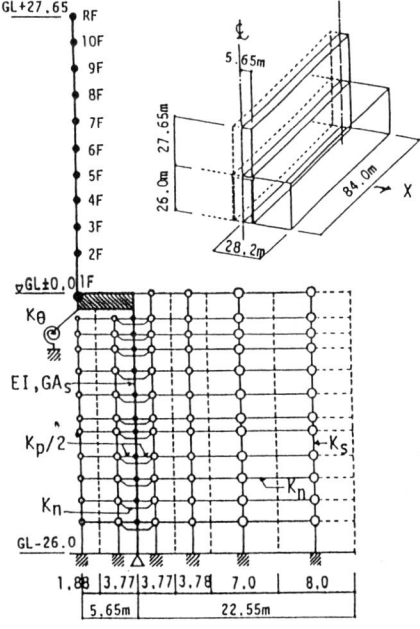

Figure 8 Lattice model

with an acceleration waveform to the lowest portion of the building. Maximum acceleration, shearing force and displacement response value are determined to confirm the safety of the building.

(2) Calculation Models
Examples of calculation models that are used include (1) lumped mass model, sway locking model, (2) Penzien model, and (3) Lattice model. Figure 6,7,8 show these models.

(3) Input Seismic Waves
Ideally, although it is best to use strong motion recordings monitored at the building site, in actuality, the El Centro wave and Taft wave in the U.S., and the Hachinohe wave in Japan, are used setting the maximum velocity to 25 kine in the case of level 1 and 50 kine in the case of level 2. In addition, there are also cases in which a base wave is produced by analytically removing ground properties from the monitored waves, after which this base wave is then input to the base and foundation bottom of expected construction values. More recently, the use of coastal waves, the Yokohama wave and so forth has been proposed which are determined based on the mechanism by which earthquakes occur.

(4) Response Displacement Method
The response displacement method is a technique involving static calculation of the stress of a pile foundation during an earthquake, and takes into consideration the inertial force from the building and the forced displacement from the ground. This method has begun to be used in static studies of pile foundations of super-high-rise buildings that are supported by piles on soft ground.

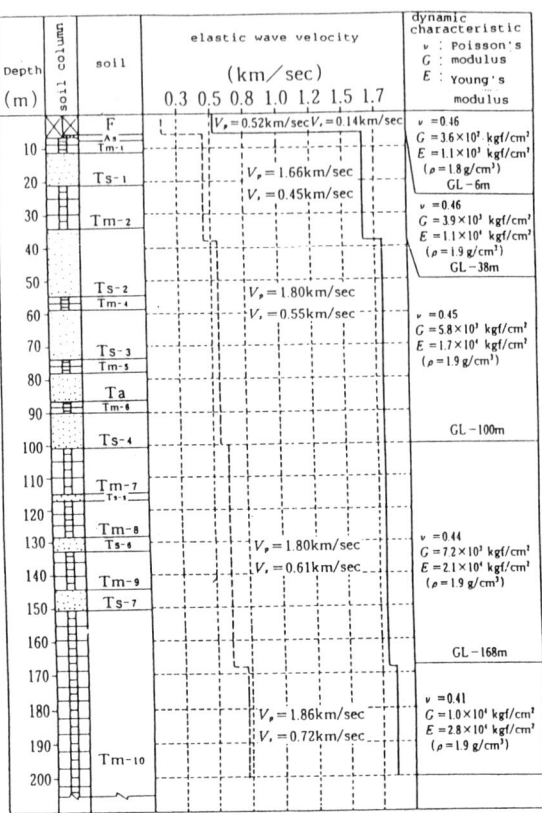

Figure 10 Results of soil investigation

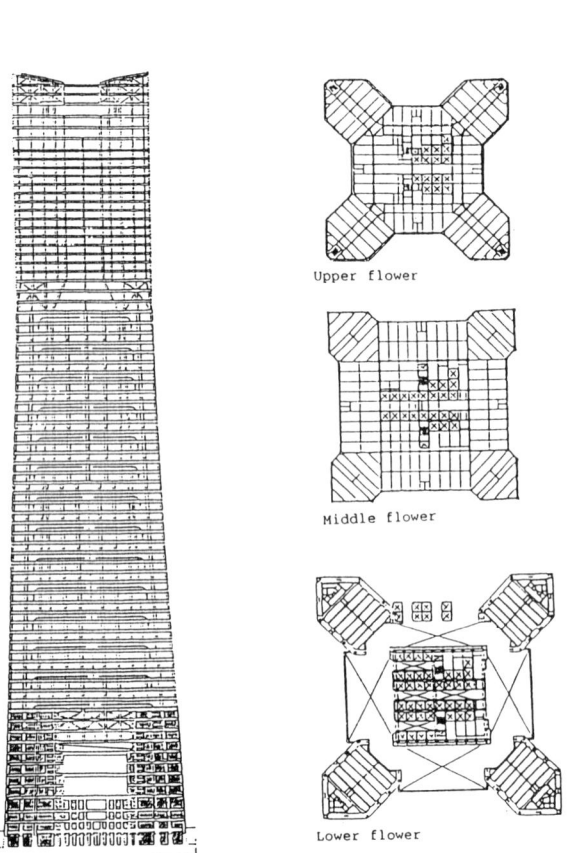

Figure 9 Plan and Section of building

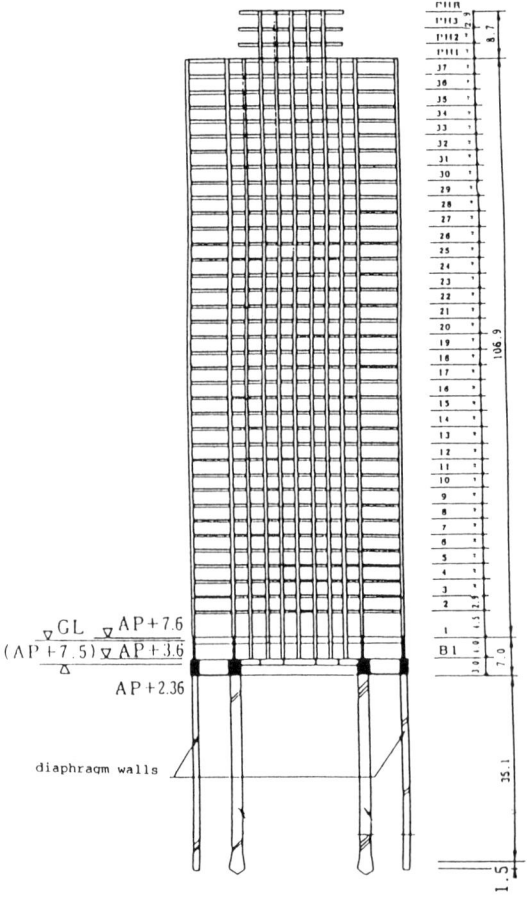

Figure 11 Section of building

4 DESIGN EXAMPLES

4.1 Direct Foundation Type

Figure 9 shows plan and section of direct foundation type building. Figure 10 shows results of soil investigations.

- Bearing Ground: Kazusa strata group

- Building Overview: 70 stories above ground, 3 stories below ground, maximum height: 296 meters, weight: 330,000 tons, foundation flooring location: G.L. -24 meters

- Structural Plan: Steel structure with double tube construction

- Foundation Plan: Direct foundation
  Allowable ground unit bearing capacity: 300 to 500 tf/m^2
  Design long time allowable unit bearing capacity: 100 tf/m^2
  Design short time allowable unit bearing capacity: 200 tf/m^2

- Analytical Model: Plane frame structural analysis by ANSYS
  Ground vertical elasticity: 1.3 kgf/cm

- Maximum Ground Pressure: Long time 89.3 tf/m^2
  Short time 154.0 tf/m^2

4.2 Pile Bearing Type

Figure 11,12 show section and foundation plan of pile bearing type bulding. Figure 13 shows soil column of this ground.

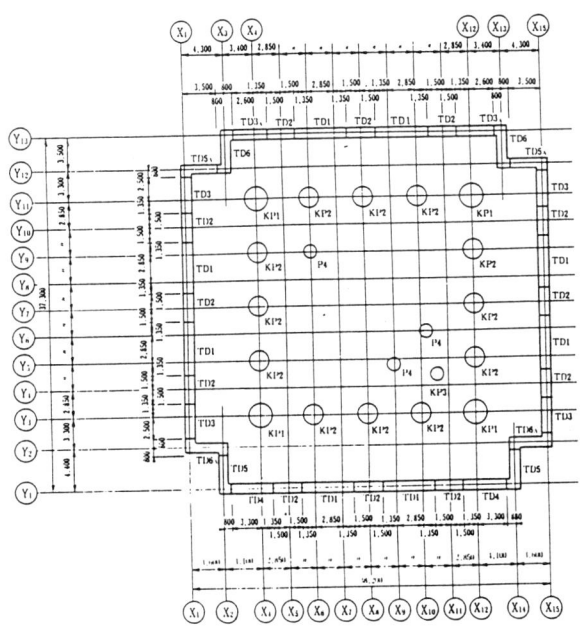

Figure 12 Plan of foundation

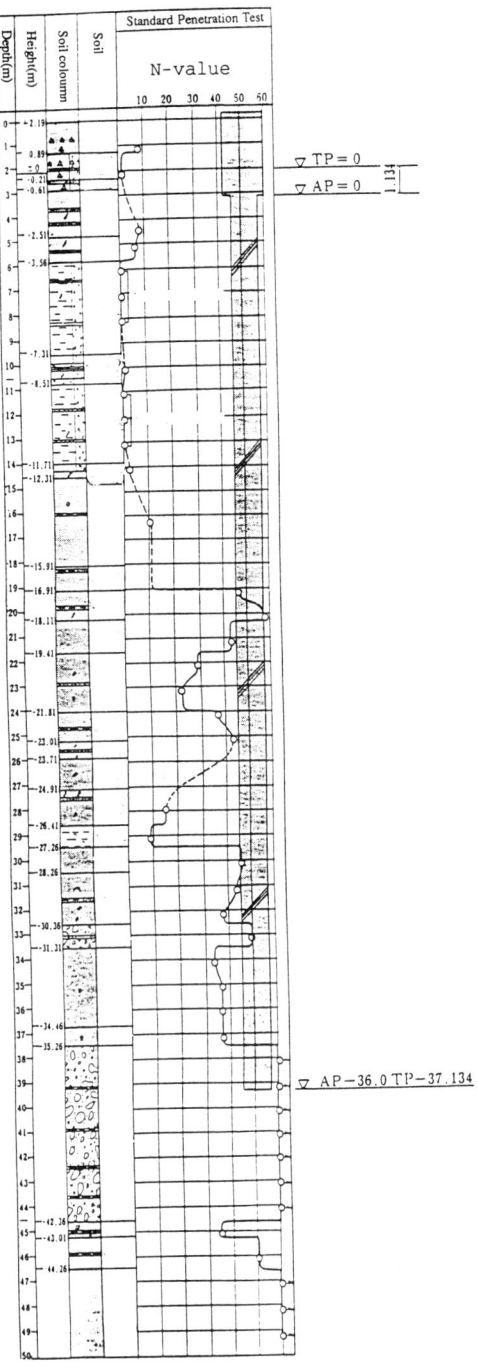

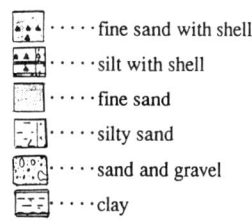

Figure 13 Soil column

Bearing Ground: Tokyo gravel

Building Overview: 37 stories above ground, 1 story below ground, maximum height: 119 meters, foundation flooring location: G.L. -7 meters

Structural Plan: RC pre-cast construction combining the use of expanded bottom piles and diaphragm wall

Foundation Design: Vertical long time allowable unit bearing capacity: 193-223 tf/m^2
Vertical ultimate unit yield stress: 497 to 574 tf/m^2

5 FUTURE OUTLOOK FOR SUPER-HIGHRISE BUILDING FOUNDATIONS

Since urban development along coastlines is likely to progress even more in the future, there is the possibility that super-high-rise buildings will appear on reclaimed land sites that are larger in scale than those currently in existence.

Since building weight will naturally increase as the size of the building increases, it is likely that foundations supported by piles or diaphragm walls driven into hard bearing strata in the manner of many existing buildings will naturally become quite common.

As a potential problem in this case, if current bearing capacity calculation methods are employed, there is the possibility of severe circumstances being encountered in the case of bearing by piles of diaphragm walls.

For example, when negative friction occurs in piles resulting from settling due to consolidation in the case of soft ground along coastlines, this action ends up reducing bearing capacity. Thus, there will be a need for a construction technique that both economically and efficiently reduces this negative friction.

In addition, various other construction techniques that enable bearing force to be increased will also have to be devised.

REFERENCES

Fukui, K. et al. 1991.1. Foundation Concepts in Super-High-Rise buildings, *Kisokou (Foundation Construction)*, pp. 23-33.(in Japanese)

Japan Architectural Society. 1988.1. Building Foundation Structural Design Guidelines, pp. 119-122.(in Japanese)

Japan Architectural Society. 1990.10. Retained Yield Strength and Deformation Performance in Building Anti-Earthquake Design, pp. 204-205.(in Japanese)

Japan Architectural Society. 1993.2. Kanto Branch: Design of Anti-Earthquake Structures, pp. 450-453.(in Japanese)

Ikeda,T.and Yasutomi,K. et al. 1988.12. The Super-High-Rise building Mini-Encylopedia, Kodansha Publishing.(in Japanese)

Kawabata, K. 1989.3. Design and Construction Examples of the Large Cast-in-place concrete pile expanded Bottom Construction Method, *Kisokou (Foundation Construction)*, pp. 62-69.(in Japanese)

Kawasaki, T. 1991.1. Ground Surveys for Super-High-Rise buildings on Waterfronts, *Kisokou (Foundation Construction)*, pp. 11-16.(in Japanese)

Mutoh, K. 1966.3. Approach to Super-High-Rise Architecture, *Kajima Research Publishing*, pp. 47-50.(in Japanese)

Sasao, H. 1977.2. Foundation Planning for Super-High-Rise buildings, *Kisokou (Foundation Construction)*, pp. 27-35.(in Japanese)

Sakaguchi, O. 1979.5.. Ground Surveys for Super-High-Rise buildings, *Kisokou (Foundation Construction)*, pp. 12-21.(in Japanese)

Yamazaki, S. et al. 1991.1. Foundation Design of Landmark Tower Commercial Districts 21 and 25 of Minato Mirai, *Kisokou (Foundation Construction)*, pp. 34-41.(in Japanese)

Non-linear elastic stress and coupled fluid flow analysis of anisotropic rock masses

Tadashi Yamabe & Masanobu Oda
Saitama University, Japan

Keisuke Maekawa
PNC, Japan

ABSTRACT : Rock masses, which commonly contain a large number of discontinuities like joints, are treated as homogeneous, anisotropic porous media. The corresponding permeability tensor k_{ij} is formulated in terms of a symmetric, second-rank tensor N_{ij}, which is dependent on the geometry of related joints, and a current stress state σ_{ij}. And an elastic compliance tensor for jointed rock masses is formulated by treating each crack as a set of parallel planar plates connected by two non-linear springs. The stiffness values of the springs are chosen such that the stress- and size-dependence can be expressed in accordance with the experimental evidence. Two-dimensional non-linear elastic stress and coupled fluid flow analysis using an elastic compliance tensor and a permeability tensor estimated from the joint survey was conducted. On the basis of this study, non-linear constitutive equation and coupled fluid flow relation provide a powerful tool to understand the deformation and seepage phenomena under rock foundation.

1 INTRODUCTION

The change of elastic and fluid flow properties induced by stress change is an important factor in the design of rock foundations, especially in the case of discontinuous rock masses. Ground water flow through geological discontinuities is believed to be the most significant problems. Field tests can be a practical solution to overcome the present difficulty. Large scale hydraulic conductivity tests were carried out to investigate the hydraulic properties of a low permeable, jointed granite (cf. Wilson et al. 1983). Pratt et al.(1977) measured an in-situ elastic and transport properties as a function of compressive stress for a variety of load paths on a jointed granite near Laramie, Wyoming. The test block in-situ was 2.8 m square and 2.6 m deep. And the specimen contained 3 vertical joints which are parallel to a set of well-developed microstructures. The loads were applied to the specimen by flatjacks on each of four sides. Figure 1 shows an example of stress-induced change of flow rate observed for a major joint. Flow rate decreased by a factor of 4 as the load was applied normal to joint, but then increased under uniaxial stress parallel to the same joint. This is the case of stress induced anisotropic permeability.

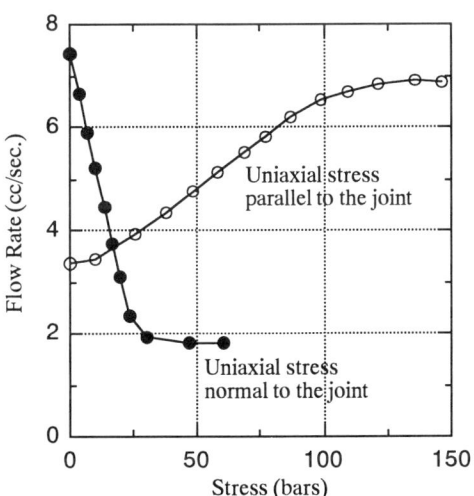

Fig.1 Stress-induced anisotropy of permeability along a joint (after Pratt et al., 1977).

In addition to such field tests, many theoretical (or numerical) studies were also done to give a sound basis for predicting the overall permeability of rock masses; e.g., Snow (1969), Long et al.(1982).

Among others, Oda, Hatsuyama and Ohnishi (1987) have also proposed a theory in which discontinuous rock masses are treated as homogeneous, anisotropic porous media. The corresponding permeability tensor was formulated in terms of in-situ measurable quantities such as orientation data of cracks projected on a Schmidt's equal area net and maps of crack traces visible on excavated walls. By extending the previous study, we will propose a set of equations which make it possible to analyze the deformation and coupled flow behavior of jointed rock masses under rock foundations.

2 ELASTIC COMPLIANCE TENSOR FOR ROCK MASSES WITH RANDOM CRACKS

In order to represent the effects of discontinuity in rock masses, the crack can be replaced by an elastically equivalent set of parallel planar plates connected by two elastic non-linear normal and shear springs. Using the spring model of cracks, an elastic compliance tensor was formulated by Oda(1986). Here, the brief outline is repeated.

2.1 Shear and normal stiffness

Many experimental studies have been reported to investigate the effects of crack geometry on the elasticity and failure criterion (e.g. Bandis et al. 1983, Yoshinaka and Yamabe 1986). And there are some empirical equation to represent the shear stiffness G of joints. These equations have a characteristics that G depends not only on the size but also on the stress in a very complicated manner.

Oda (1986) linearized these equations with respect to the normal stress. Yamabe et al. (1987) slightly changed these equation and obtained the following equation for the averaged shear stiffness.

$$\bar{G} = \frac{G_0 + G_1 \sigma_{ij} N_{ij}}{r} = \frac{G(\bar{\sigma}, N)}{r} \quad (1)$$

The summation convention is adopted through the paper, if any subscript appears twice. This is not the case, however, when superscripts are in parentheses.

In equation (1), it should be noted that the stiffness G must vary according to the orientation of $\boldsymbol{n}$ relative to the applied stress. And the stiffness is averaged with respect to the orientation. A probability density function $E(\boldsymbol{n})$ is introduced such that $E(\boldsymbol{n})d\Omega$ gives the probability of the unit normal $\boldsymbol{n}$ oriented inside a small solid angle $d\Omega$. The stiffness G is then averaged over the entire solid angle. Here, N is a symmetric tensor depending only on the distribution of $\boldsymbol{n}$, and represented by the following equation.

$$N_{ij} = \int_\Omega n_i n_j E(\boldsymbol{n}) \, d\Omega \tag{2}$$

The second rank tensor N_{ij} can be calculated if the density of poles normal to cracks, which is commonly given as poles on Schmidt's equal area net, is available. G_0 and G_1 in equation (1) are the material constants for the shear deformation behavior of joint set.

Next consider the normal stiffness of the joint. Increase of the normal stress produces the corresponding displacement which causes closure of crack aperture. Bandis et al.(1983) have concluded, on the basis of the very careful examination of their experiment on the relation between normal stress and the corresponding displacement of jointed sample, that the curve is best fitted to a hyperbolic function. Rearranging the equation proposed by Bandis et al.(1983) on a single joints, Oda (1986) obtained the averaged normal stiffness of jointed rock masses as follows:

$$\bar{K} = \frac{K_0 + C\sigma_{ij}N_{ij}}{r} = \frac{K(\bar{\sigma},N)}{r} \tag{3}$$

where, an aspect ratio C is introduced as a measure of crack shape; i.e. $C = r/t$, t = thickness of joints. This parameter has a physical meanings that larger crack tends to have a wider initial aperture. Accordingly, it can be assumed that the ratio is constant through cracks being in the given rock mass. Here, N is also used to average the stress on an entire region of jointed rock masses. And K_0 in equation (3) is the material constant for the normal deformation behavior of joint set.

2.2 Elastic compliance tensor

A straight scanline is set in a representative elementary volume in order to derive the stress strain relation for jointed rock mass. When it is stressed, then the scanline is displaced to a new position due to the elastic deformation. The displacement is continuous in the solid matrix, while discontinuous displacements occur at the positions where the scanline intersects cracks. Accordingly, two sources of elastic strain can be distinguished. The first is the elastic strain, $\bar{\varepsilon}_{ij}^{(s)}$, of the solid matrix. And the second is the additional term due to the discontinuous displacements at cracks, $\bar{\varepsilon}_{ij}^{(c)}$. The corresponding total strain is assumed to be additive. Using the stiffness values of equations (1) and (3), the discontinuous displacements are summed along the scanline to give

$$\bar{C}_{ijkl} = \left\{ \frac{1}{K(\bar{\sigma},N)} - \frac{1}{G(\bar{\sigma},N)} \right\} F_{ijkl} + \frac{1}{4G(\bar{\sigma},N)} \left(\delta_{ik} F_{jl} + \delta_{jk} F_{il} + \delta_{jl} F_{ik} + \delta_{jk} F_{il} \right) \tag{4}$$

where F are the even-rank (second- and forth-rank) tensors depending on number of cracks per a unit volume, $E(\boldsymbol{n},r)$ = density function to show the statistical distribution of $\boldsymbol{n}$ and r. F is represented by the following equation:

$$F_{ijkl} = \frac{\pi\rho}{4} \iint_\Omega r^3 n_i n_j n_k n_l E(\boldsymbol{n},r) \, d\Omega \, dr \tag{5}$$

And C satisfies the following symmetry conditions:

$$\bar{C}_{ijkl} = \bar{C}_{jikl} = \bar{C}_{ijlk} = \bar{C}_{klij} \tag{6}$$

The solid matrix is assumed to be isotropically elastic. Then, the corresponding elastic compliance tensor M for solid matrix becomes

$$\bar{M}_{ijkl} = \frac{1+\nu}{E} \delta_{ik} \delta_{jl} - \frac{\nu}{E} \delta_{ij} \delta_{kl} \tag{7}$$

where E and ν are Young's modulus and Poisson's ratio of the solid matrix respectively.

Using equations (4) and (7), a non-linear elastic stress-strain relation is finally obtained by taking into account the overall effect of cracks on the elasticity as follows:

$$\bar{\varepsilon}_{ij} = \bar{T}_{ijkl} \bar{\sigma}_{kl} = \left(\bar{M}_{ijkl} + \bar{C}_{ijkl} \right) \bar{\sigma}_{kl} \tag{8}$$

This compliance matrix $\bar{T}_{ijkl}$ has a following symmetric characteristics.

$$\bar{T}_{ijkl} = \bar{T}_{jikl} = \bar{T}_{ijlk} = \bar{T}_{klij} \tag{9}$$

The inverse form of $\bar{T}_{ijkl}$ is used in the finite element analysis.

3 HYDRAULIC PROPERTIES OF DISCONTINUOUS ROCK MASSES AND PROCEDURE FOR COUPLED SOLUTIONS

3.1 Permeability tensor

In this paper, changes in stress and corresponding changes in hydraulic conductivity tensor are formulated for a rock foundation problem as follows;

$$k_{ij} = \frac{k_{ij}^{(0)}}{1 + k \, \sigma_{ij}' N_{ij}} \tag{10}$$

where k is a non-dimensional scalar, $k_{ij}^{(0)}$ is the initial permeability tensor and k_{ij} is the permeability tensor corresponding to the current stress states, σ_{ij}' is the effective stress tensor, respectively. Note that also in equation (10) $\sigma_{ij}' N_{ij}$ represents the effect of induced stress and crack orientation observed in rock masses. This term means the average normal stress component for jointed rock mass.

3.2 Formulation of finite element procedures for non-linear stress and coupled fluid flow analysis

The constitutive relation of coupled phenomena for discontinuous rock mass is represented by equations (8) and (10). The results of discretized relation for finite element analysis is formulated as follows;

$$\begin{bmatrix} K & -C \\ -C^T & -\Delta t \theta \bar{K} \end{bmatrix} \begin{Bmatrix} U^n \\ P^n \end{Bmatrix} = (1-\theta) \Delta t \begin{Bmatrix} \dot{F}^{n-1} \\ Q^{n-1} \end{Bmatrix} + \Delta t \theta \begin{Bmatrix} \dot{F}^n \\ Q^n \end{Bmatrix} + \begin{bmatrix} K & -C \\ -C^T & -\Delta t(1-\theta) \bar{K} \end{bmatrix} \begin{Bmatrix} U^{n-1} \\ P^{n-1} \end{Bmatrix} \tag{11}$$

where,

$$K = \int_\Omega B^T D B \, dV, \quad \bar{K} = \int_\Omega G^T \frac{k}{\gamma_w} G \, dV, \quad C = \int_\Omega B_v^T \bar{N} \, dV$$
$$\dot{F} = \int_{\partial\Omega_t} N^T \dot{t} \, dS, \quad Q = \int_{\partial\Omega_q} \bar{N}^T \frac{\dot{q}}{\rho_w} dS + \int_\Omega G^T k w \, dV \tag{12}$$

K is the non-linear elastic stiffness matrix, $\bar{K}$; the permeability matrix, C; the coupled matrix, F; the load vector, Q; the flux vector, Δt; the time increment for finite difference scheme in time domain. In the following numerical analysis, $D = \bar{T}_{ijkl}^{-1}$ is used as stress-strain relations, where $\bar{T}_{ijkl}$ is the compliance tensor given by equation (8). Then the system of equation (11) represents the non-linear elastic stress and coupled fluid flow media equivalent to the jointed rock masses.

4 NUMERICAL RESULTS OF STRESS - INDUCED CHANGE OF PERMEABILITY UNDER ROCK FOUNDATIONS

In order to see if equation (11) is applicable to interpret the flow and deformation behavior of jointed rock mass, two idealized numerical cases were analyzed. The first case contains the horizontal joints in the rock mass, and the second case is the real rock mass model that the data was taken from the in-situ outcrops.

Figure 2 shows the mesh division of finite element analysis and its boundary conditions. In this figure the hydrostatic pressures are shown by arrows, and the uniform settlements are loaded from the center of top surface. The bottom plane is impermeable. The nodes at bottom plane are fixed to the vertical displacements.

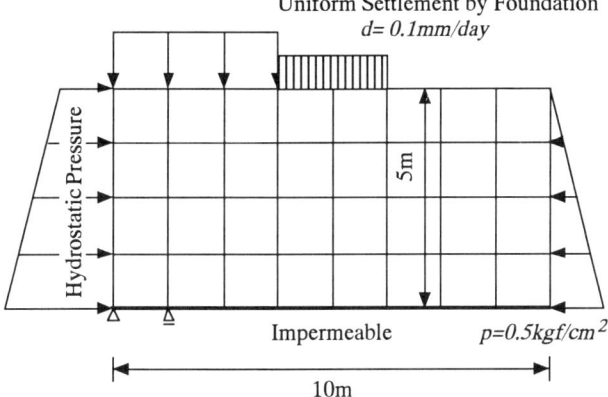

Fig. 2 Mesh division of finite element analysis and its boundary conditions.

4.1 Numerical results for idealized case.

The input parameters for idealized rock mass model is shown in table 1. Also the corresponding geometrical features of jointed rock mass are represented by the matrix form through the concept of crack tensor. These tensors are equivalent to the horizontal joint sets. The components of hydraulic conductivity tensor $k_{ij}^{(0)}$ in equation (10) for horizontally jointed rock mass are determined as shown in table 1. In this case, the load-displacement relation calculated at the given settlement position in figure 2 is shown in figure 3. In this figure, the non-linear (concave downward) curve is obtained. Figure 3 also contains the other load-displacement relations that are for the intrinsic rock (without joint) and the case of vertically jointed rock mass. It is reasonable that the displacement for horizontally jointed rock mass gives the larger displacement compared with the vertically jointed rock mass. Although these cases are extreme of course, the existence and geometry of joints seriously affects the behavior of fluid flow and deformation.

Flow vectors for the horizontally jointed rock mass is given in figure 4. It is evident that almost all of vectors have the tendency of horizontal flow.

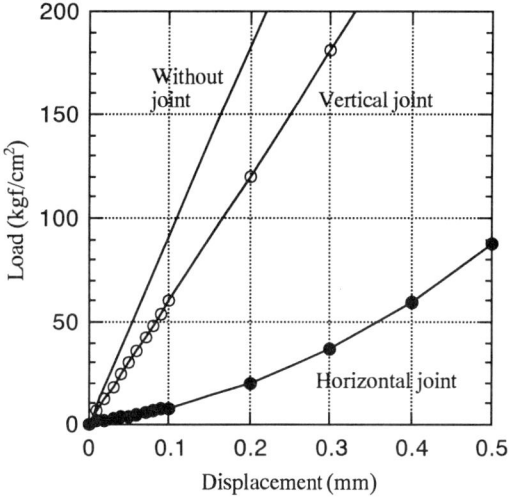

Fig.3 Load-displacement relation at rock foundation of Fig.2

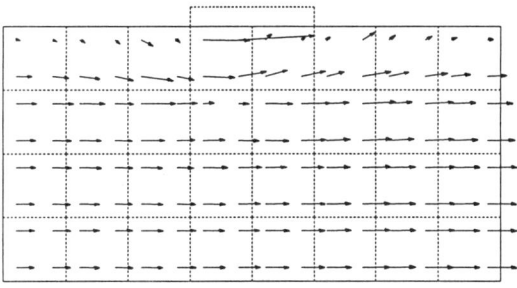

Fig.4 Flow vectors for the horizontal joints at 10 days after loading.

4.2 Numerical results for real rock mass model.

A complete set of geological data was provided by Saitoh et al. (1989). Using these published data and the developed numerical method, the non-linear elastic deformation and coupled fluid flow behaviors under rock foundation were simulated. The input data is given in table 2. It can be seen that the components of the hydraulic permeability tensor have an anisotropic characteristics. The ratio of principal permeability is about 1.5. The calculated results of flow vectors is given in figure 5.

Table 1. Input parameters for idealized rock mass model

Young's Modulus of Rock Matrix	E	9.80×10^5 kgf/cm²
Poisson's Ratio of Rock Matrix	ν	0.30
G_0 in Equation (1)	G_0	6.24×10^5 kgf/cm²
G_1 in Equation (1)	G_1	3.00×10^5
K_0 in Equation (3)	K_0	6.24×10^5 kgf/cm²
C in Equation (3)	C	1.00×10^4

$$N_{ij} = \begin{bmatrix} 1.0 & 0.0 \\ Sym. & 0.0 \end{bmatrix}, \; F_{ij} = \begin{bmatrix} 10.0 & 0.0 \\ Sym. & 0.0 \end{bmatrix}, \; F_{ijkl} = \begin{bmatrix} 10.0 & 0.0 & 0.0 \\ & 0.0 & 0.0 \\ Sym. & & 0.0 \end{bmatrix}$$

$$k_{ij}^{(0)} = \begin{bmatrix} 1.04 & 0.0 \\ Sym. & 10.42 \end{bmatrix} (\times 10^{-6} cm/sec)$$

Table 2. Input parameters for real rock mass model

Young's Modulus of Rock Matrix	E	2.57×10^5 kgf/cm²
Poisson's Ratio of Rock Matrix	ν	0.28
G_0 in Equation (1)	G_0	3.03×10^5 kgf/cm²
G_1 in Equation (1)	G_1	3.00×10^5
K_0 in Equation (3)	K_0	3.03×10^5 kgf/cm²
C in Equation (3)	C	1.00×10^4

$$N_{ij} = \begin{bmatrix} 0.72 & 0.0 \\ Sym. & 0.28 \end{bmatrix}, \; F_{ij} = \begin{bmatrix} 5.12 & 0.05 \\ Sym. & 1.95 \end{bmatrix}, \; F_{ijkl} = \begin{bmatrix} 3.26 & 1.42 & 0.44 \\ & 2.82 & 0.49 \\ Sym. & & 1.01 \end{bmatrix}$$

$$k_{ij}^{(0)} = \begin{bmatrix} 4.94 & 0.0 \\ Sym. & 7.20 \end{bmatrix} (\times 10^{-6} cm/sec)$$

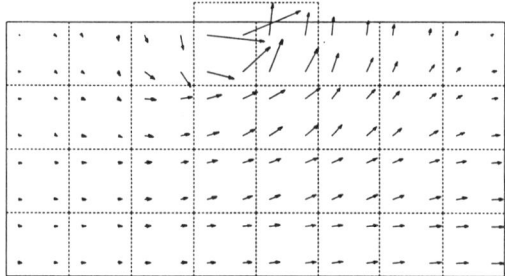

Fig.5 Flow vectors for the real rock mass model at 10 days after loading.

5 CONCLUSIONS

An non-linear and anisotropic poro-elasticity for discontinuous rock mass was formulated by treating each crack as an elastically equivalent parallel planar plates connected by two springs. The complex geometry of cracks, which is rather in actual rock mass, was explicitly taken into account by means of the crack tensor.

REFERENCES

Bandis, S.C., A.C.Lumsden and N.R.Barton 1983. Fundamentals of rock joint deformation., *Int. J. Rock Mech. Min. Sci. & Geomech. Abstr.* 20 : 249 - 268 .

Long, J.C.S., J.S. Remer, C.R. Wilson and P.A. Witherspoon 1982. Porous media equivalents for networks of discontinuous fractures. *Water Resour. Res.* 22 : 645-658.

Oda, M. 1986. An equivalent continuum model for coupled stress and fluid flow analysis in jointed rock masses. *Water Resources Res.* 22 : 1845-1860.

Oda, M., Y. Hatsuyama and Y. Ohnishi 1987. Numerical experiments on permeability tensor and its application to jointed granite at Stripa mine, Sweden. *J. Geophys. Res.* B92 : 8037-8048.

Pratt, H.R., H.S. Swolfs, W.F. Brace, A.D. Black and J.W. Handin 1977. Elastic and transport properties of an in-situ jointed granite. *Int. J. Rock Mech. Min. Sci. & Geomech. Abstr.* 14 : 35-45.

Saitoh, T., M. Oda, T. Yamabe and K. Kamemura 1989. Estimation of permeability tensor based on field investigation. *Proc. 21th Symp. on Rock Mech. JSCE* : 386-390.

Snow, D.T. 1969. Anisotropic permeability of fractured media, *Water Resour. Res.* 5 : 1273-1289.

Wilson, C.R., P.A. Witherspoon, J.C.S. Long, R.M. Galbraith, A.O. Dubois and M.J. Mcpherson 1983. Large-scale hydraulic conductivity measurements in fractured granite, *Int. J. Rock Mech. Min. Sci. & Geomech. Abstr.* 20 : 269-276.

Yamabe, T. and M. Oda 1987. Parameter study for elastic analysis of jointed rock masses by crack tensor. *Proc. of Japanese Soc. Civil Engr.* 3-382 : 121-130.

Yoshinaka, R. and T. Yamabe 1986. Joint stiffness and the deformation behavior of discontinuous rock. *Int. J. Rock Mech. Min. Sci. & Geomech. Abst.* 23 : 19-28.

Bearing capacity of rock foundations

T. Ramamurthy
Indian Institute of Technology, New Delhi, India

ABSTRACT: Rock being a jointed mass, in the estimation of its ultimate bearing capacity often greater conservatism is exercised. To arrive at a more realistic value of bearing capacity, laboratory tests on jointed foundation blocks were conducted for the case of surface strip loading. The inclination, location and strength of joint were varied. Based on the findings of tests on jointed specimens, the bearing capacity is linked to the unconfined compressive strength of jointed rock. Expressions have been presented to estimate the failure deformation, strength and modulus values of jointed rocks in confined state as well for the analysis of foundations resting at deeper depths.

1 INTRODUCTION

Foundations are often constructed on rocks which are rarely homogeneous and intact; rock masses are mostly hetrogeneous and discontinuous, having cracks, fissures, joints, faults and/or bedding planes with varying degree of strength along these planes of weakness. All the load from a structure has to be transmitted to the rock through individual spread footing, strip, mat or pile tip. When the spacing of joints is very wide, the bearing capacity of individual footing or pile tip may be estimated as that of an intact rock using the classical theories of Prandtl, Terzaghi, Hansen, Bell, etc. as in the case of soils. Heavily fractured rock with closed spacing of joints may also be treated as a dense granular mass. The only problem is the testing of such a mass in the laboratory or in the field to evaluate shear strength parameter (c,ϕ). When a foundation mass is likely to fail in shear along two or three planes of weakness, the stability is assessed by considering the summation of shear strength to the summation of shear stresses developed along these planes due to the imposed load on the rock. Such cases wherein well defined foundation block boundaries are established, some solutions exist in literature (Ramamurthy & Bhatia, 1978).

By considering rock mass as a cohesive material, as per Terzaghi theory one would get safe bearing capacity with a factor of safety of 3 for surface loading same as the unconfined compressive strength of the rock mass. It will be independent of the width of the loaded area. If the rock is considered only as a frictional material, then the bearing capacity depends on the density, width of the loaded area and the bearing capacity factor and cannot be related to unconfined compressive strength of rock mass directly. When rock mass fails in shear often the rupture surface is controlled partly by the intact mass and partly by the joints. Model studies were carried out in order to ascertain the mode of failure of rock in the presence of a joint located either at the middle or at one end dipping under-side of the footing (i.e. footing resting on the hanging wall); the influence of the inclination of the joint with the vertical and the reduction in the shear strength along the joint were also experimentally investigated. The ultimate bearing capacity of the jointed rock has been linked to the unconfined compressive strength of the rock. Based on vast experimental data, the relationship of unconfined compressive strength of jointed rock has been related to that of the intact rock through a joint factor. Similar relationship was extended to deformation modulus. These relations have been extended to predict the bearing capacity of jointed rock and also its deformations for the shallow as well as for deeper foundations.

2 EXPERIMENTAL STUDY

Block of plaster of Paris with borex and water were cast and cured for seven days. Joint was created by making a notch and breaking the block along the desired joint inclination. By this process a weak tight joint retaining the natural broken surface was created. By locating the joint at the mid line of the footing or at one edge of the footing load tests were conducted. In all 32 tests on intact, tight joint with varying inclination and joint with thin mica powder layer were conducted to study the influence of inclination of weak plane and change of shear strength along the joint plane.

The patterns of ruptures under the strip loading with a weak inclined plane located at the middle or at the edge of the footing are shown in Fig. 1 for various inclination of joint plane. With increase of inclination of joint, the extent of the fractured mass extends along the joint. This extent of rupture surface varies from 1.5 to 2.5 times the width of the footing, Fig. 1a. For the centrally located joint, the rupture occurred on the foot wall as well as on the hanging wall. When the joint plane was located at the edge of the footing and dips under the footing, the rupture of mass is essentially in the hanging wall.

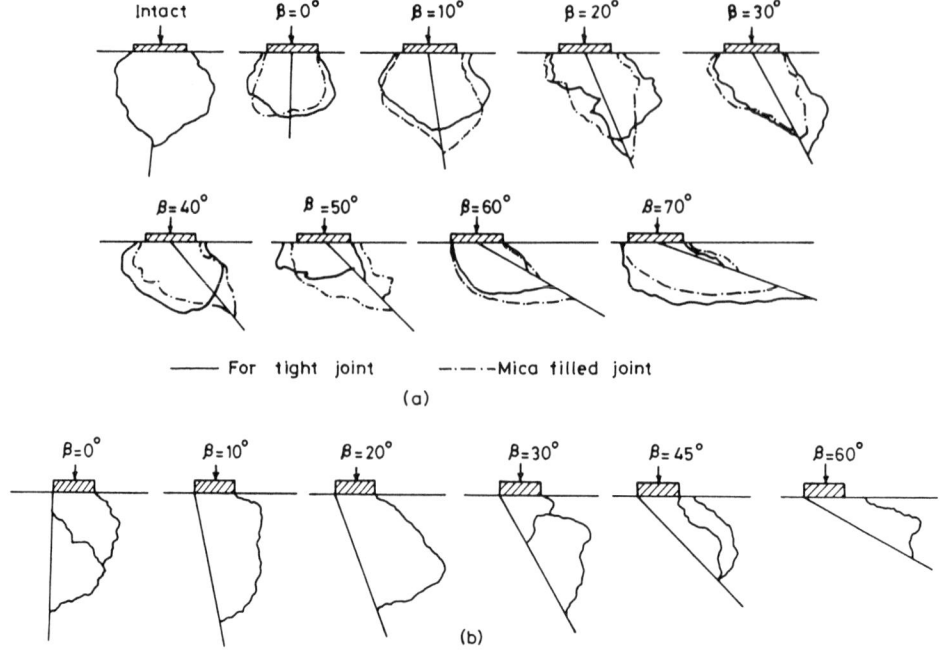

Figure 1. Modes of foundation failure with position and inclination of joint.

The rupture surface extends to about 3 times the width of the footing Fig. 1b. The spread of the rupture surface is less in this case compared to that in the case of centrally located joint. With the increase of the joint inclination with the vertical (ß) located at the centre, the ultimate bearing capacity (B_{cj}) decreases; it is about 1.4 times the unconfined compressive strength of intact rock (σ_{ci}) (UCS) for a vertical joint and decreases to about 0.90 times the UCS when ß = 70°, Fig. 2. For the joint located at one edge of the footing and dipping under the footing, these ratios reduced to 1.1 at ß = 0° to 0.75 for ß = 70°. For the joint located at the edge of the footing a reduction of about 25% is expected in the bearing capacity of the foundation compared to the location of joint at the middle of footing. The ultimate bearing capacity in the case of intact rock is about 1.4 times its unconfined compressive strength.

Figure 3 shows the variation of the bearing capacity of the jointed case in comparison to the unjointed one for varying joint inclination. The reduction was about 30% for ß = 0° to ß = 70°. It is of interest to note that the variation of bearing capacity with the inclination of joint did not exhibit any U-shaped or shoulder shaped variation as is generally observed in the case of UCS variation with ß. Continuous decrease in bearing capacity with ß suggests the influence of some confinement on bearing capacity even for surface footing.

When thin mica powder layer was introduced along the tight joint for its central location, the ratio of bearing capacity with UCS of intact mass at ß = 0° was about 0.64 and it continuously reduced to about 0.27 at ß = 70° (Fig. 4). The ratio of bearing capacity of mica filled joint to that of the intact mass varies from 0.58 at ß = 0° to about 0.25 at ß = 70°, Fig. 5.

Figure 6 shows the variation of failure strain of the jointed mass in comparison to

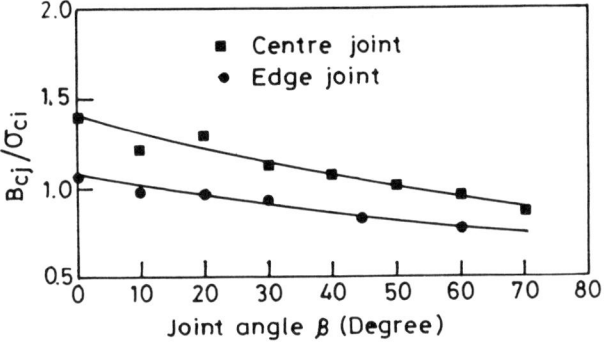

Figure 2. Variation of bearing capacity with inclination and location of joint.

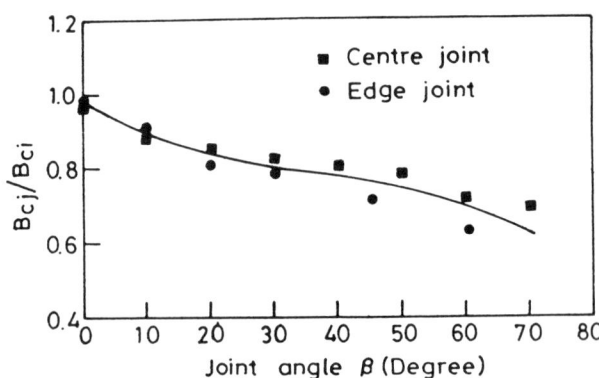

Figure 3. Comparison of bearing capacity of jointed to the intact mass due to inclination and location of joint.

that of the intact mass ($\epsilon_{1j}/\epsilon_{1i}$) with the variation of inclination of joint. The strain was calculated by assuming the depth of zone involved in rupture as 2 times the width of the footing. This strain ratio

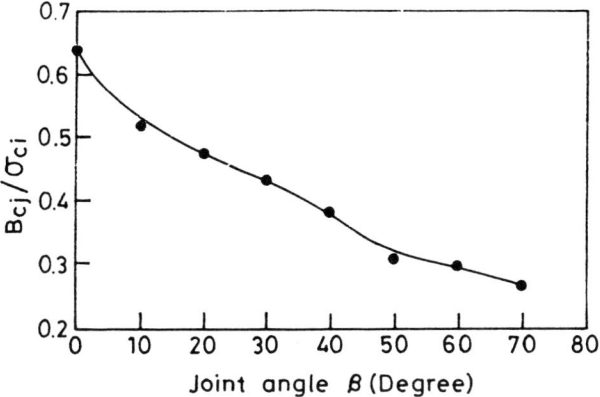

Figure 4. Variation of bearing capacity of mica filled joint for its central location.

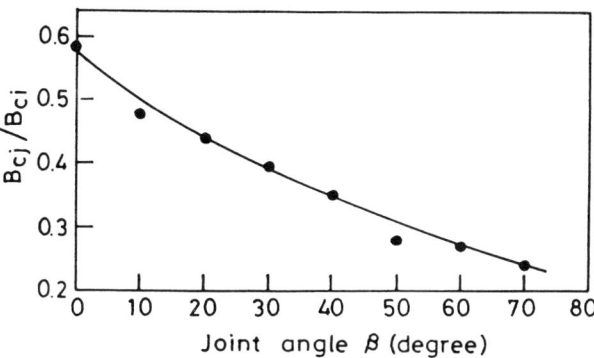

Figure 5. Comparison of bearing capacity of mica filled joint with intact strength.

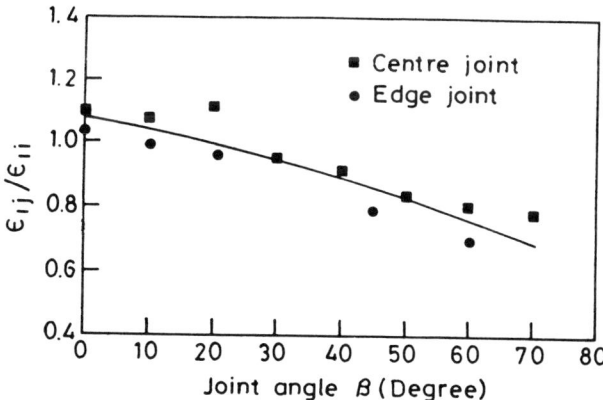

Figure 6. Variation of failure strain with inclination and location of joint.

varies from 1.1 at ß = 0° to 0.70 at ß = 70°. In the case of joint with mica, these ratios increase to 1.4 for ß = 0° and 1.0 for ß = 70°. When the dip of the joint plane is low one may expect lower failure strains.

From the foregoing it could be inferred that the bearing capacity and failures strains of the jointed mass should in fact be linked to the corresponding values of the jointed rock mass at least corresponding to those of vertical joint because of their gradual decrease with the increase of the inclination of the joint. The following discussion will enable to arrive at the suitable values of jointed rocks in general.

3 DISCUSSION

3.1 Unconfined compressive strength of jointed rock

Till the end of 1980 only two approaches were available to estimate the compressive strength of jointed rock based on the rock mass classification. These were by Bieniawski (1973) and Hoek and Brown (1980). Bieniawski in his Rock Mass Rating (RMR) suggested cohesion (c) and friction angle (ϕ) for various ratings of rock mass. By adopting these values of c and ϕ the unconfined compressive strengths (σ_{cj}) of the rock mass for varying values of RMR were calculated using the relationship

$$\sigma_{cj} = \frac{2 c \cos\phi}{1 - \sin\phi} \quad \text{as per Mohr-Coulomb}$$

criterion and are presented in the Table 1.

This table suggests very low values of compressive strength varying from about 2 MPa for RMR more than 81 to less than 0.26 MP for RMR less than 20.

The strength criterion and the relationships suggested by Hoek and Brown (1980) between m,s, and RMR based upon unconfined compressive strength of Punguna Andesite field test results yield:

$$\sigma_{cj}/\sigma_{ci} = \sqrt{s}; \quad s = \exp[\frac{RMR-100}{6.0}] \quad (1)$$

where i and j refer to intact and jointed rocks respectively. This Eqn. 1 suggests that the unconfined compressive strength of the rock mass depends not only on RMR but also on the UCS of the intact rock. In other words the strength of the rock mass is linked to that of the intact rock through RMR. That is to say, $\sqrt{s}$ is a reduction factor or a weakness coefficient when applied to the compressive strength of the intact rock, UCS of the rock mass is obtained. In these relations (Eqn. 1), the compressive strength of the intact rock has been considered twice-firstly in the rating of RMR and secondly in obtaining the rock mass strength. Table 2 gives the reduction factors for various values of RMR.

Tables 2 gives unconfined strength of rock mass several times that given by Table 1. Figure 7 compares the compressive strengths as per Hoek and Brown with those by Bieniawski for intact strengths of 200 and 100 MPa. Large differences are very obvious.

A similar observation was also made by Krauland et al. (1989) with regards to Laisvall and Langsele mines, Swedan. The strength of the pillers were markedly different, from the values predicted by Bieniawski's rock mass rating and by Hoek and Brown theory. Such glaring differences in the actual to the predicted unconfined compressive strengths of jointed rocks leave no other option but to carryout extensive testing of rock specimens varying the inclination of joint set, varying strength along the joint sets and joint frequency.

To achieves this objective extensive laboratory study was carried out on

Table 1. RMR and strength values (Bieniawski 1973)

RMR	100-81	80-61	60-41	40-21	< 20
c kPa	> 400	300-400	200-300	100-200	< 100
ϕ (deg)	> 45	35-45	25-35	15-25	< 15
σ_{cj} MPa	> 1.97	1.97-1.18	1.18-0.64	0.64-0.26	<0.26

Table 2. RMR and reduction factor (from Hoek and Brown 1980)

RMR	100	80	60	40	20	10	0
$\sqrt{s}$	1	1.9×10^{-1}	3.6×10^{-2}	6.7×10^{-3}	1.3×10^{-3}	5.5×10^{-4}	2.4×10^{-4}

Table 3. Values of n for different joint inclinations, $\beta°$

β	n	β	n
0	0.82	50	0.30
10	0.46	60	0.46
20	0.11	70	0.64
30	0.05	80	0.82
40	0.09	90	0.95

different grades of plaster of Paris, sandstones and granite by varying joint frequency and joint inclination under unconfined and confined conditions by various researchers (Yaji 1984, Arora 1987 and Roy 1993); the results of Brown 1970, Brown and Trollope 1970 and Einstein and Hirschfeld 1973 have been analysed. It has been found that the compressive strength of a jointed rock could be linked to that of the intact rock through a factor called-Joint factor. This factor combinedly accounts for the influence of joint frequency, joint inclination and strength along the joint. Tests conducted with joints with different inclinations of joints (with and without gauge) in test specimens (Roy 1993) have also been considered in evaluating the compressive strength of jointed rock and the parameters involved in the strength criterion of such rocks. The data of a few thousand test specimens confirm the findings reported by Ramamurthy 1993, Ramamurthy and Arora 1994.

The unconfined compressive strength (σ_{cj}) of jointed rock is expressed as (Fig. 8):

$$\sigma_{cj}/\sigma_{ci} = \exp[-0.008\ J_f] \quad (2)$$

where σ_{ci} is the compressive strength of intact rock specimen.

$$J_f \text{ is Joint Factor} = J_n/r.n. \quad (3)$$

J_n is joint frequency i.e. number of joints per meter depth of rock.

n is a coefficient for the effect of joint inclination. The value of n are given in Table 3.

r is the coefficient for joint strength and is the friction coefficient along the sliding joint set/joint.

The hatched lines in the Fig.8 indicate the range of experimental data.

The joint which has the maximum value of J_f is the critical one to experience sliding. In fact J_f could be estimated per meter depth of rock mass. Wherever the J_f values are higher, this region is weaker. It is infact a coefficient of weakness to be applied to the intact rock to obtain the strength of jointed rock.

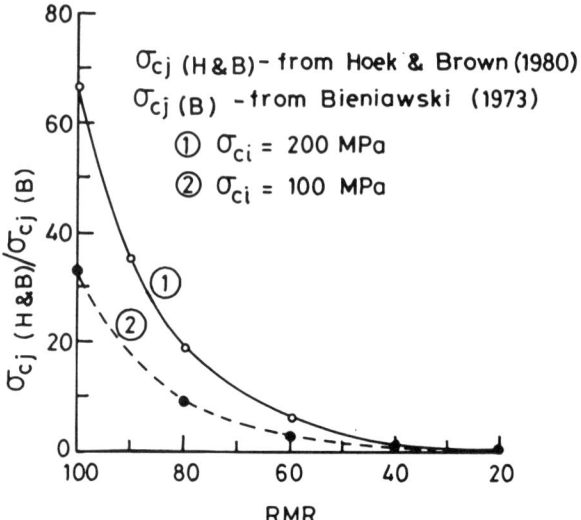

Figure 7. Comparison of uniaxial compressive strengths from classification.

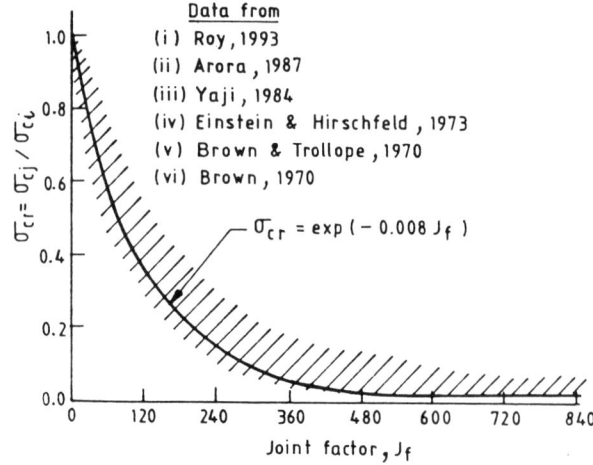

Figure 8. Variation of compressive strength of jointed mass with joint factor.

The strength criterion for jointed rocks is

$$\frac{\sigma_1 - \sigma_3}{\sigma_3} = B_j \left(\frac{\sigma_{cj}}{\sigma_3}\right)^{\alpha_j} \quad (4)$$

σ_1 and σ_3 are major and minor principal stresses respectively.

σ_{cj} is the unconfined compressive strength of jointed specimen

B_j and α_j are constants.

Further α_j and β_j are obtained as follows (Figs. 9 and 10):

$$\alpha_j / \alpha_i = (\sigma_{cj}/\sigma_{ci})^{0.5} \quad \text{and} \quad (5)$$

$$\beta_i / \beta_j = 0.13 \exp[2.037 \, \alpha_j/\alpha_i] \quad (6)$$

α_i and β_i are the values of similar constants obtained by testing intact specimens of the rock under different confining pressures (Ramamurthy 1986, 1993).

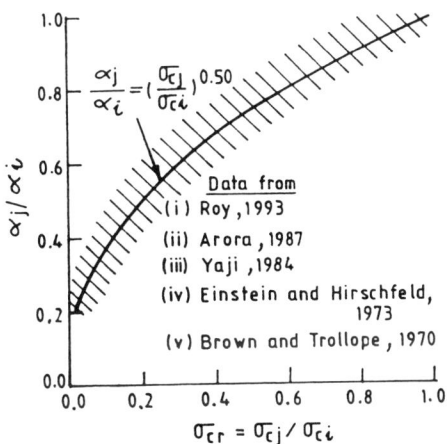

Figure 9. Variation of α, constant of strength criterion with jointed mass strength.

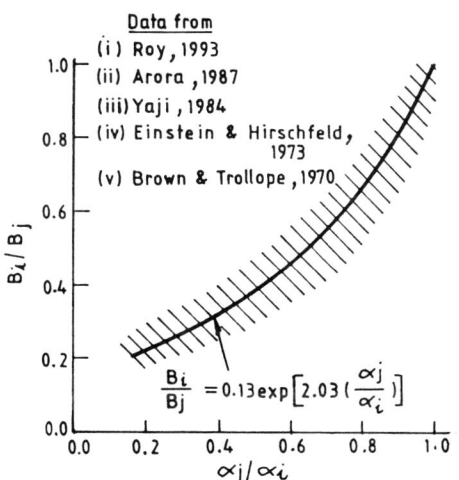

Figure 10. Variation of β, constant of strength criterion with constant α.

Now reverting back to the question of bearing capacity of jointed rock, from Eqn. 2, the unconfined compressive strength for the laboratory blocks with tight joint and mica filled joint was obtained for vertical joint. The Bearing capacities of the jointed block for various values of ß by were normalized by unconfined compressive strength of the block for the vertical joint. This ratio varied from 2 at ß= 0° to 1.7 at ß = 70°. Similar ratios for the mica filled joint varied from 2.4 at ß = 0° to 1.0 at = 70°. In both the cases the average values are about 1.70.

Therefore on the basis of the foregoing analysis one may assume that the ultimate bearing capacities of jointed rocks to be about 1.7 times the unconfined compressive strength of the rock having vertical joint. For shallow foundations of spread footing or tip of pile directly resting on rock surface, a factor of safety of 2 to 3 may be adopted depending upon the design requirement or variation of rock properties to obtain the safe bearing capacity of the jointed rock. In extreme case one may treat unconfined compressive strength of rock estimating for the case of vertical joint system as the ultimate bearing capacity and adopted lower factor of safety to obtain safe value of bearing capacity.

3.2 Deformation modulus

The deformation modulus of the jointed rock was linked to that of the intact rock by the following expression (Fig. 11),

$$E_{tj}/E_{ti} = \exp[-0.0115 \, J_f] \quad (7)$$

Since one often observes near linear stress-strain response of rocks in the unconfined state, by dividing Eqn. 2 by Eqn. 7 one obtains the ratio of failure strains jointed to intact rock as

$$\epsilon_{1j}/\epsilon_{1i} = \frac{\sigma_{cj}}{\sigma_{ci}} \cdot \frac{E_{ti}}{E_{tj}} = \exp[0.0035 \, J_f] \quad (8)$$

The values of J_f for the tight jointed blocks varied from 42 for ß = 0° to 57 for

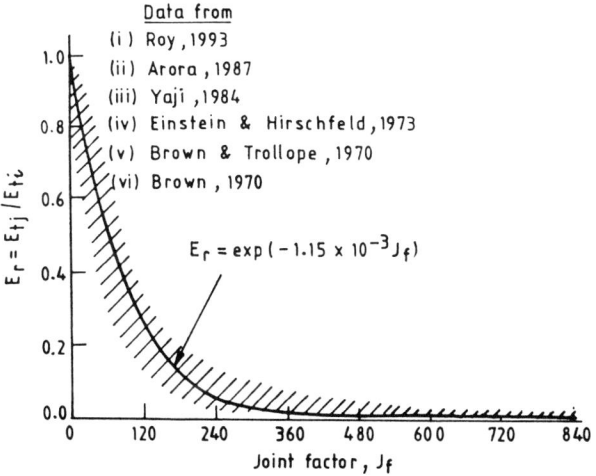

Figure 11. Variation modulus of jointed mass with joint factor.

$\beta = 70°$ and for the case with mica filled joint the corresponding values were 155 and 189 respectively. For these values, using Eqn. 8, $\epsilon_{1j}/\epsilon_{1i}$ varies from 1.16 to 1.95 in the case of unconfined compression test. Whereas from the bearing capacity test the maximum value of this ratio was 1.4. Certainly in the bearing capacity tests, some degree of confinement influenced the failure strain ratios. Therfore one may consider Eqn. 8 to give a fair indication of the failure strain of the foundation jointed rock. Further that for any deformation analysis one could adopt Eqn. 7 in estimating the modulus of jointed rock by evaluating the joint factor of the rock mass. If modulus values are required under any confining pressure the following expression may be adopted (Ramamurthy 1993, Ramamurthy and Arora (1994) for the cases of intact and jointed rocks.

$$E_{tio,jo}/E_{ti3,j3} = 1-\exp[-0.1 \frac{\sigma_{ci,j}}{\sigma_3}] \quad (9)$$

where 0 and 3 refer to $\sigma_3 = 0$ and $\sigma_3 > 0$ cases.

4 CONCLUSIONS

Based on the tests carried out on foundation blocks with and without joint located at the middle or at the edge of surface strip footing and the data of large number of tests on specimens of rocks and model materials, the following inferences are drawn:

1. The unconfined compressive strengths of rock masses obtained from classification system namely from RMR as per Bieniawski and as per Hoek and Brown seems to leave considerable margin of doubt. For the same RMR value several times variation could be expected in the values of compressive strength of rock mass.

2. The ultimate bearing capacity of surface footings on rock mass may be considered as 1.7 times the unconfined compressive strength of jointed rock having vertical jointing. The compressive strength of jointed rock could now be estimate with the help of joint factor and unconfined compressive strength of intact rock.

3. The deformation of the rock mass under a shallow/surface footings may be estimated from the expression linking the failure strains of the intact and jointed rocks through joint factor - a coefficient of weakness due to joint system.

4. The modulus value of jointed rock can be estimated for use in the deformation analysis of foudnation rock based on the modulus of intact rock and the joint factor.

5. Expressions are also available based on large number of tests on different rocks and model materials to estimate the effect of confinement on compressive strength of rock mass and also on the modulus values for use in the analysis of rock mass under deeper foudnations.

6. The relationships presented will enable to analyse the foundation rock mass for its stability.

REFERENCES

Arora, V.K. 1987. Strength and deformation behaviour of jointed rocks, Ph.D. Thesis, Ind. Inst. of Tech., Delhi.

Bieniawski, Z.T. 1973. Engineering classification of jointed rock masses. Trans. S. Africa Instn. Civ. Engrs., 15:12: 335-344.

Brown, E.T. 1970. Strength of models of rock with intermittent joints. J. ASCE, 96:SM6: 1935-1949.

Brown, E.T. and Trollope, D.H. 1970. Strength of model of jointed rock, J.ASCE, 96:SM2 : 685-704.

Einstein, H.H. and Hirshfeld, R.C. 1973. Model studies in mechanics of jointed rocks, J. ASCE, 99: SM3: 229-248.

Hoek, E. and Brown, E.T. 1980. Empirical strength criterion for rock masses. J. ASCE, 106:GT9: 1013-1035.

Krauland, N. Soder, P. and Agmalm, G. 1989. Determination of rock mass strength by rock mass classification - some experiences and questions from Boliden mines, Int. J. Rock Mech. & Min. Sci. & Geomech. Abstr., 26:1: 115-123.

Ramamurthy, T. 1986. Stability of rock mass. Eight Indn. Geotech. Soc.Annual Lect., Ind. Geotech. J., 16:1: 1-73.

Ramamurthy, T. 1993. Strength and Modulus responses of anisotropic rocks. Comprehensive Rock Engg., ed. by JA Hudson, Pergamon Press, U.K.; 1: 315-329.

Ramamurthy, T. and Arora, V.K. 1994. Strength prediction of jointed rocks in confined and unconfined states. Int. J. Rock Mech. Min. Sci. & Geomech. Abstr. 31:1: 9-23.

Roy, N. 1994. Engineering behaviour of rock masses through study of jointed models. Ph.D. Thesis, Ind. Inst. of Tech., Delhi.

Yaji, R.K. 1984. Shear strength and deformation response of jointed rocks. Ph.D. Thesis, Ind. Inst. Tech., Delhi.

The influence of compression and tensile loads on water permeability on rock foundations

E.S. Kalustian
Institute 'Hydroproject', Moscow, Russia

E.G. Gaziev
'Vodstroi' Company, Moscow, Russia

ABSTRACT: The results of field investigations of tensile strength of fractured rock masses in the foundation of a number of concrete dams and the influence of compression and tensile loads on water permeability of rock foundations are considered.

1 INTRODUCTION

Constantly increasing requirements for safety of dams (Safety assessment... 1994) may be satisfied at improvement of our notions of dam behaviour together with the foundation including properties of rock masses. These problems involve tensile strength of the rock mass under conditions of a pressure flow.

At the designing of concrete dams on rock foundations it is customary to consider the design schemes (models) with independent acting force factors: hydrostatic loads and associated uplift pressure, the load of the structure, inertial forces at seismic effects of a given intensity, ice, wind, temperature and some other loads.

However the analysis of failure at the Malpasset arch dam in France in 1959 once more shows that not all force parameters should be considered as independent and, in particular, the uplift water pressure in fractures of the rock mass depends to a considerable degree either on compression or tensile stresses acting in the rock mass (Londe 1966). This phenomenon was confirmed repeatedly by many scientists (Jimenez 1963; Bernaix 1967; Serafim 1967; Matsumoto 1987). At the present time topicality of studies of fracturing regularities has been enhanced because fracturing presented problems of great concern in operation of a number of dams (Kalustian 1994) particularly at the Kolnbrein arch dam in Austria (Lombardi 1991).

It is quite obvious that a decrease of water permeability of the rock mass results from the closing of fractures at compression of the rock mass and an increase in water permeability results from the opening of fractures under the action of tensile stresses. In this case coefficient of permeability varies with the square of the opening width of fractures (Gaziev, 1973) which explains a sharp increase of water permeability under the action of tensile stresses on the rock mass.

Until the present time the problem of the role of tensile stresses in the rock foundation is limited by the assumption of the opening of the tension fracture under the pressure face.

For substantiation of the answer to the problem of an actual role of tensile stresses in the rock foundation of concrete dams it is essential to carry out in situ large-scale tests of water permeability of fractured rock foundations at alternating loads and to clear up a question of the capacity of the rock mass to take tensile stresses.

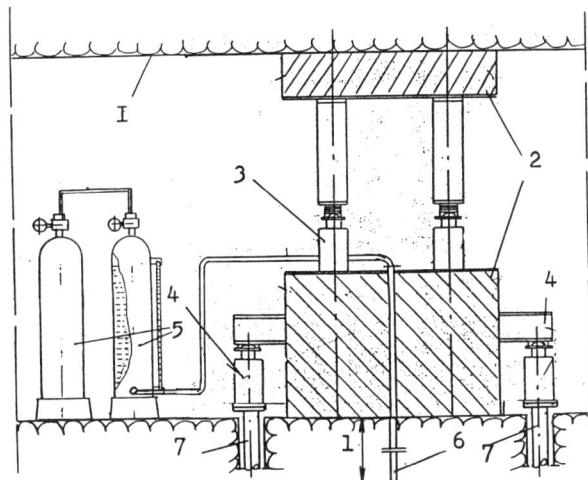

Figure 1. Field test apparatus for strength and permeability properties of rock foundations: 1 - cavern; 2 - concrete test and support block; 3 - system for compression load; 4 - system for tension load; 5 - permeability apparatus; 6 - bore hole diam. 42 mm; 7 - support for jack

The "Hydroproject" Institute performed a series of field studies of strength of rock masses in foundations of a number of concrete dams in Georgia (Khudoni arch dam), in Iraq (Al-Baghdadi dam), in Kirghizia (Tashkumyrskaya dam) and in Viet Nam (Hoa Binh dam) by the procedure developed by Kalustian E.S. (Kalustian 1994a) which made it possible to determine the tensile strength of rock masses and to reveal the effect of compression and tensile stresses on water permeability of rock foundations of concrete dams.

2 THE PROCEDURE OF FIELD INVESTIGATIONS

Normally for determination of deformation properties of rock foundations cocnrete test blocks are used which makes it possible to define the relationship between deformation of the rock mass and the applied compression stress.

It was decided to use these traditional test procedures for determination of the rock mass strength in tension and for revealing of the effect of comrpession and tensile stresses on water permeability of the rock mass.

A concrete test block was equipped with jacks to create both tensile and compression stresses. Additionally a special seepage apparatus was worked out to deliver water under the test block (into a bore hole) under pressure (Fig. 1) reaching 2.5 MPa with measurement of water flow with the help of a volumetric procedure (Kalustian 1994a).

The sequence of operations was as follows:
- the block was slightly loaded by compression (of about 1.5-2.0 MPa);
- water pressure was supplied under the block and a seepage flow and vertical deformations of the block were recorded;
- compression stresses on the rock mass were decreased;
- gradually jacks lifting the block were put into operation which made it possible to bring the compression stress to zero (to compensate for the weight of the concrete block) and then to create a tensile stress on the rock mass.

A seepage flow and vertical displacements of the test block were recorded at each step of loading.

3 THE RESULTS OF FIELD INVESTIGATIONS

The first and a very important inference of the performed tests and investigations is the fact that even at small compression stresses applied under the test block the water pressure being several times exceeding the compression stress does not give rise to the uplift pressure under the test block which would be registered by an increase in pressure in jacks or by indicators measuring vertical displacements of the plate.

The second important conclusion is the tensile strength in fractured rock masses. In this case as it is demonstrated in Table 1 the scatter of the obtained tensile strength values (in spite of their small absolute values) is less considerably than the values of modulus of deformation.

Table 1. The obtained experimental values of modulus of deformation and values of tensile strength of a number of rock masses.

Types of rock	Tuff-siltstone, tuff-sandstones		Tuff-breccia*)		Limestones	
Dam	Khudoni		Hoa-Binh		Al-Baghdadi	
Parameters	E_d GPa	R_t MPa	E_d GPa	R_t MPa	E_d GPa	R_t MPa
	12.67	0.028	5.8	0.166	4.02	0.14
	6.26	0.030	22.2	0.30	1.54	0.11
	2.11	0.034	14.4	0.59	3.38	-
	2.60	0.018	4.4	0.13	3.38	0.16
	3.60	0.012	2.4	0.135	3.75	0.12
	2.32	0.007	26.6	0.35		
		0.043	12.6	0.055**)		
		0.052	7.3	0.035**)		
		0.098				
Average	4.93	0.036	12.6	0.278	3.2	0.13

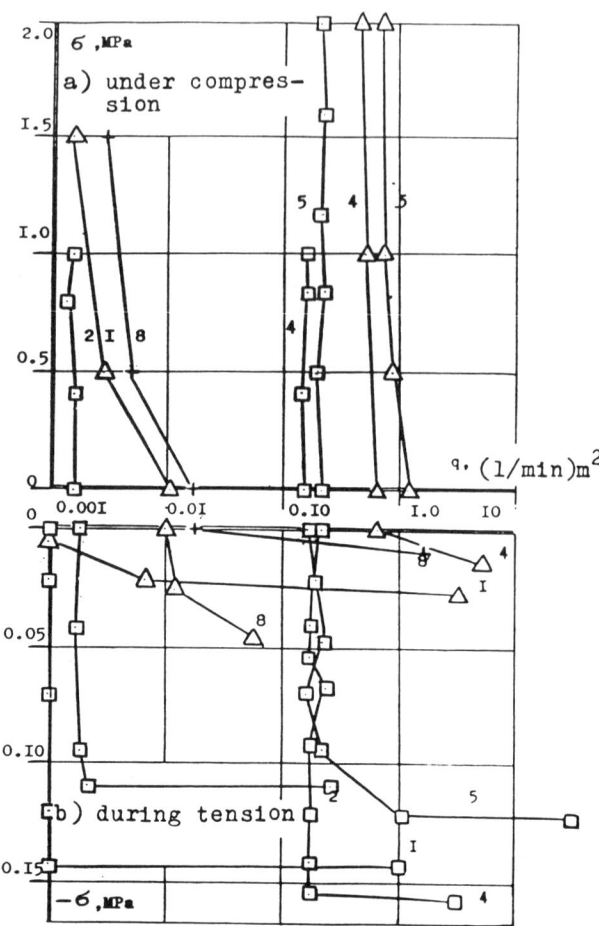

Figure 2. Rock foundation permeability vs compression and tension loads: △ - tuff-sandstones and lava breccias of Khudoni arch dam foundation; □ - limestones of Al-Baghdadi concrete dam foundation; + siltstone of Tashkumyr dam. Figures show numbers of test blocks

*) The tests were carried out at the section of grouted rocks;
**) The contact of the concrete block and the foundation was disturbed at goruting.

The tests demonstrated that the grouting of the rock mass may substantially increase its tensile strength.

The performed investigations and tests showed that the values of tensile strength obtained in the laboratory on core samples taken from the rock mass are about two orders of magnitude higher than the tensile strength of the fractured rock mass obtained in situ conditions on cocnrete blocks of 1 m² area. For instance, for the effusive rocks of the foundation of the Khudoni arch dam the tensile strength of the core samples is 4.90 MPa which is by 136 times higher than the average strength of the rock mass - 0.036 MPa.

The results of studies of the effect of compression and tensile stresses on water permeability of rock foundations are shown in Fig. 2. The demonstrated relationships point to a sharp increase of water permeability of the rock mass at tensile stresses of the specified tensile values being normally 60-70% of the rock mass tensile strength.

4 CONCLUSIONS

1. The fractured rock masses are characterized by resistance to tensile stresses, i.e. embody tensile strength.
The performed in-situ determinations of the tensile strength of rock masses demonstrated rather stable results.

2. Water permeability of rock foundations depends greatly on values of applied stresses. At compression of the foundation by rather slight loads permeability remains at a rather low level or practically it is negligible. A sharp increase in water permeability of the foundation and seepage are observed at a change of a sign of stresses and attainment of a tensile stress of a certain limit comprising normally 60-70% of the tensile strength of the rock mass.

3. The abovementioned allows one to give up the conventional model of computation of concrete dams on rock foundations with unchangeable curves of uplift pressure on the foundation and the dam abutment and to transfer to the new model in which the uplift pressure is to be taken into account where tensile stresses exceed the tensile strength of the rock mass for which the tests are required by the considered procedures. This will give a more justified determination of external forces acting on the structure.

REFERENCES

Bernaux, J. 1967. Etude géotechnique de la roche de Malpasset. Dunod, Paris.

Gaziev, E.G. 1973. Rock mechanics in civil engineering. Stroyizdat, Moscow.

Jimenez Salas, Uriel, S. 1964. Some recent rock mechanics testings in Spain. VIII Congress on Large Dams, Edinburgh. Q.28. R.53.

Kalustian E.S. 1994. Requirements for concrete dams safety on the basis of their rock foundations diagnostics. Gidrotekhnicheskoe stroitel'stvo, No 5, Moscow.

Kalustian, E.S. 1994a. Method of definition of permeability characteritics for hydraulic structures rock foundations. Patent 2021588 (Russia). Bulletin izobreteniy, No 19, Moscow.

Lombardi, G. 1991. Kolnbrein dam: an unusual solution for an unusual problem. Water Power and Dam Cosntruction, June.

Londe, P. & F. Sabarly 1966. La distribution des permeabilites dans la fondation des barrages voûtes en fonction du champ de contrainte. Primer Congrès de la SIMR, 8.6, Lisboa.

Matsumoto, N. & Y. Yamaguchi 1987. Deformation of foudnation and change of permeability due to full placement in embankment dams. Proced. VI ISRM Congress, vol. 1, Balkema, Rotterdam.

Safety assessment and improvement of existing dams. 1994. Trans. XVIII ICOLD. Vol.1, Q. 68. Durban.

Serafim, J.L. 1967. Influence of interstitial water on the behaviour of rock masses, University of Wales Swansea, USA.

Allowable bearing capacity of typical rock classes depending on RMR and compression strength of intact rock

A. Serrano
Universidad Politécnica, Madrid, Spain

C. Ollala
Laboratorio Geotecnia, Cedex, Madrid, Spain

ABSTRACT: Using the Hoek and Brown non linear failure criterion and the characteristics method (Sokolovskii, 1960 and 1965), the ultimate and allowable load of shallow foundations are determined by Serrano and Olalla (1994, 1995). A simplified chart is proposed under the hypotheses of strip load, horizontal surface, no overburden pressure and no inclined loads, depending on RMR and m_o values. Typical values of the ranges of safety factors to be used with these rock media are proposed for a failure probability less than 10^{-4} and 10^{-3}. Anisotropy arguments are pointed out and outlined. A review is made of several real examples of fixing allowable bearing capacity for shallow foundations in Spanish civil works.

1 INTRODUCTION

The determining of the allowable capacity of spread foundation on rock masses has traditionally been considered taking into account previous experience, using empirical criteria, or by applying local or national codes (Goodman, 1989; Wyllie, 1992).

This practice gives rise to certain simplifications. For example, except when fixed values are taken from official codes, it is generally accepted that the allowable bearing pressure is a certain percentage of the uniaxial compressive strength (σ_c). This percentage usually ranges from 20% to 60%, depending on the rock type and the local conditions (Jiménez Salas et al., 1975, Canadian Geotechnical Manual, 1985, ROM 05.94, 1994).

In an earlier paper (1994a), the authors defined a methodology for bearing capacity quantification, based on the Hoek and Brown (1980) rock failure equations and by applying the characteristics method for solving the differential equation systems which govern the stress field. On the basis of works developed by Sokolovskii (1960, 1965) and Serrano (1976) for solving the plasticity problem, three different regions (Karman and Prandtl types) are considered and the boundary conditions defined by the external load system and the geometric configuration are included. The mathematical expressions are integrated for a close form solution, using the "instantaneous friction angle" concept as was originally defined by Serrano (1976).

2 SYNTHESIS OF THE METHOD

A) General case.

In the most general case, the validity of the method for diverse rock types is accepted under different arguments for a variety of weathering situations (Groups I, IV and V) and rock types, as can be seen in Fig. 1 (Serrano and Olalla, 1995).

Two new parameters are used to define the rock strength behaviour, β and ζ, where

$$\beta = \frac{m\sigma_c}{8} \quad (2.1)$$

$$\zeta = \frac{8s}{m^2} \quad (2.2)$$

The ultimate load (P_h) is directly obtained from the following expression, by means of charts similar to that shown in Fig. 2.

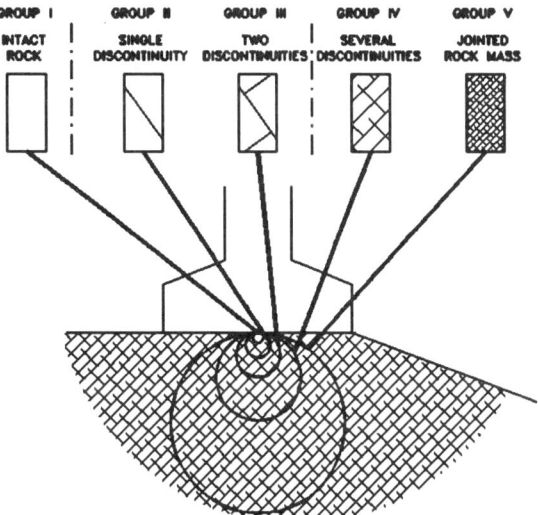

Figure 1. SIMPLIFIED REPRESENTATION OF THE INFLUENCE OF SCALE ON THE TYPE OF ROCK MASS BEHAVIOUR MODEL WHICH SHOULD BE USED IN DESIGNING SHALLOW FOUNDATIONS ON ROCK SLOPE.

It gives the value of N_β as a function of the slope inclination, the load inclination and the equivalent to the overburden pressure, (σ_{01}^*), as follows:

$$P_h = \beta (N_\beta - \zeta) \quad (2.3)$$

where

$$\sigma_{01}^* = \sigma_1/\beta + \zeta \quad (2.4)$$

The three parameters involved in the Hoek and Brown failure model (m; s; and σ_c) are transformed and related to the RMR index, through expressions obtained by the above-mentioned authors (1988). The parameter m_o, is used (m_o identifies the m value, for the intact rock matrix) as follows:

$$m = m_o \exp \frac{RMR - 100}{a} \quad (2.5)$$

$$s = \exp\frac{RMR - 100}{b} \quad (2.6)$$

where

$a = 28$ and $b = 9$; for an undisturbed rock mass (e.g. carefully blasted or machine excavated rock), and
$a = 14$ and $b = 6$; for a disturbed rock mass, (as in slopes or blast-damaged rock).

A simple computer program can also be used (Serrano and Olalla, 1994b).

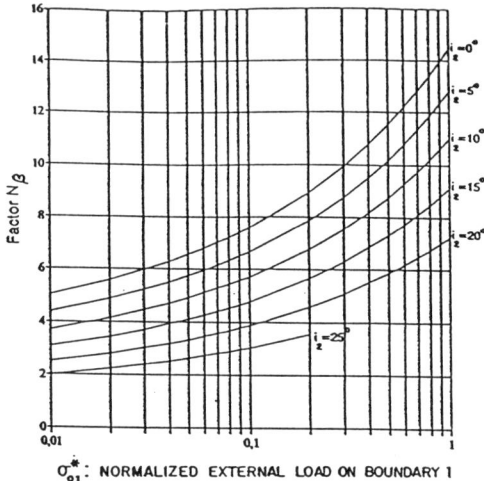

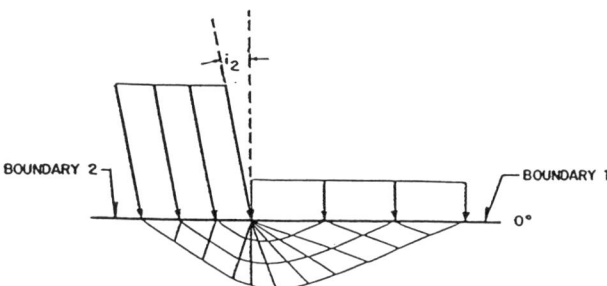

Figure 2. Values of the load inclination factor (N_β) dependant on the normalized external load on Boundary 1 and the inclination of the load on Boundary 2 (horizontal surface; $\alpha = 0°$; $a = 28$ and $b = 9$)

B) Simplified case

The problem is in its simplest form when the foundation surface is horizontal, there is no overburden pressure ($\sigma_{ol}^* = 0$), and the external load is vertical. All the equations become simpler and a new relationship can be established, which is similar to the most widely used Codes and geotechnical engineering practice:

$$P_h = N_\sigma \cdot \sigma_c \quad (2.7)$$

where N_σ is a dimensionless factor.
On the basis of the authors' theory, values of N_σ are proposed in Fig. 3. depending on RMR magnitudes, for values of m_o ranging from 7 to 25, assuming that $a = 28$ and $b = 9$.

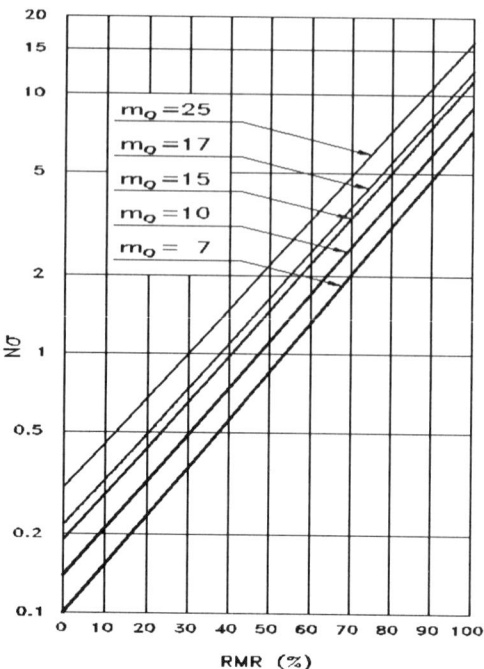

Figure 3. Relationship between N_σ and RMR

3 SAFETY FACTORS AND ALLOWABLE LOADS

Under the homogeneity-heterogeneity conditions defined in Fig. 1 as being valid for use in the proposed method, it is necessary to take into consideration a safety coefficient (F) to determine the allowable loads. This safety coefficient is linked to two partial and supplementary aspects ($F = F_m \cdot F_p$);

- (F_m), which takes into account the uncertainties intrinsic to the plastic model used here. This partial safety coefficient of the model (F_m) will be linked to the brittleness of the rock. If the behaviour of the rock can be described as plastic, the value of this partial safety coefficient will be equivalent to one. This is the case for Groups IV and V with all rock types, and for Group I with soft rocks.

- (F_p), which takes into account the statistical variations of the parameters involved. The greater the scatterings, the greater the values corresponding to this partial safety coefficient.

General statistical studies carried out by the authors, on the basis of hypotheses with reference to the distributions of the parameters m_o; RMR and σ_c of the type to the Beta and Gamma functions, indicate a strong dependence of the value of (F_p) on the respective values of the COV, as can be seen in Fig. 4, for a failure probability of below 10^{-4} (Serrano and Olalla, 1995), and in Fig. 5 for a probability of below 10^{-3}.

4 ANISOTROPY

There are many situations in nature, where it is not possible to adopt the homogeneity hypothesis for rock masses. A typical case occurs when there is a family of surfaces, generally assumed to be plane and parallel surfaces, in which the shear strength may clearly be below that of the rock mass. Such a phenomenon makes it necessary to include this type of plans, where an active zone, a transition zone and a passive zone, are jointly produced in the failure mechanism definition. This situation is graphically depicted in Fig. 6, together with its implications in the definition of the characteristic lines.

Until the anisotropic model for the determination of the ultimate load is available, it is recommended that use be made of the

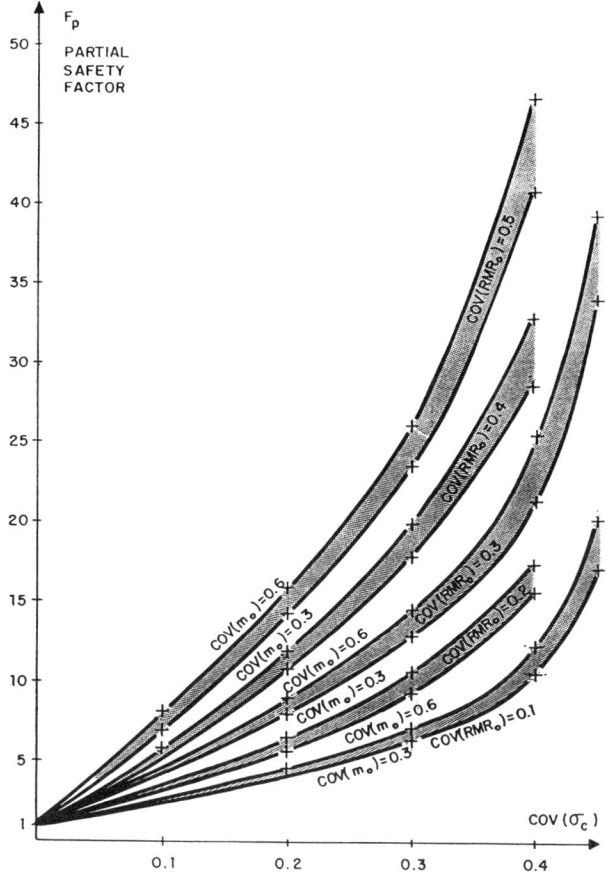

Figure 4. Partial safety factors (F_p) for a probability of failure less than 10^{-4}

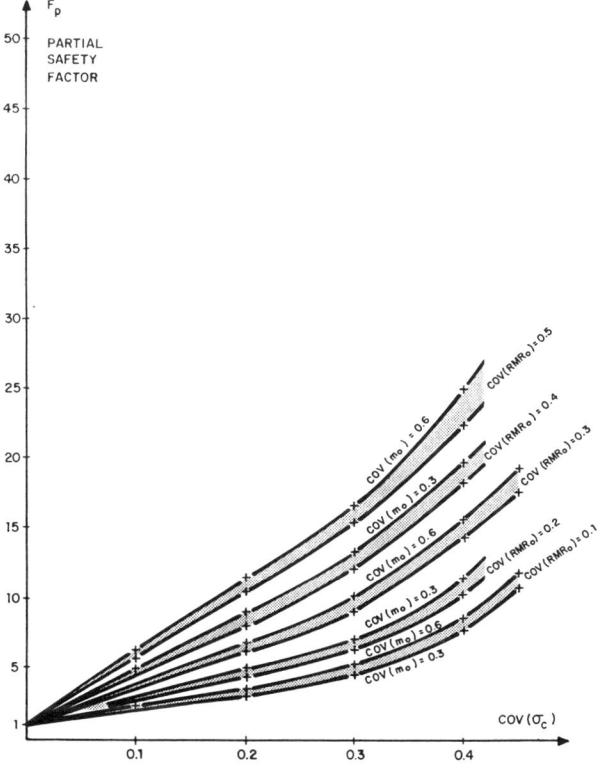

Figure 5. Partial safety factors (F_p) for a probability of failure less than 10^{-3}

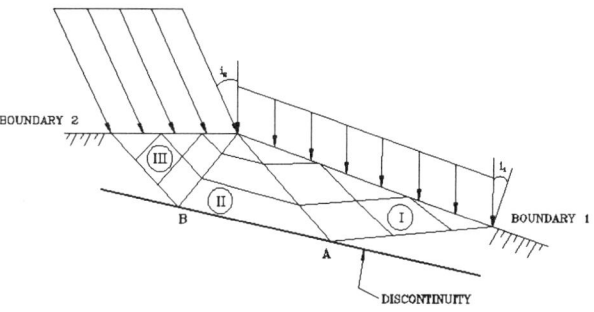

Figure 6. A simplified anisotropy failure mechanism

unconfined compressive strength, as measured in laboratory tests, on samples with the least favourable schistosity orientation.

5 EXAMPLES

Some specific applications of the method to Spanish civil works are shown below. The beneficial effects of the weight of the foundation ground itself have not been included in any of these applications, so additional safety coefficients must be added, if necessary.

In all cases, partial safety coefficients are used that have been deduced for a failure probability below 10^{-4}.

5.1 Fourth bridge spanning the River Guadiana (Badajoz)

A bridge spanning the River Guadiana has been constructed in the city of Badajoz, close to the frontier between Spain and Portugal. It has a large central pile, 92 m. high (see photograph). The foundations were laid by sinking two 18 m. diam. circular footings into an extremely weathered and fractured metamorphic rock mass, basically consisting of gneiss with a degree of schist.

The foundations were laid at least 1.5 m. into the rock, below the alluvium, and were designed to withstand a load of 11,000 Tons per footing, thus producing an average work load of 430 kPa. It acts on a horizontal surface with an inclination angle of 10°.

The average calculation values assumed were as follows:

RMR = 30; COV (RMR) = 0.15
σ_c = 1,800 kN/m²; COV (σ_c) = 0.25
m_o = 25; COV (m_o) = 0.6
σ_1 = 60 kN/m² (12 kN/m³ x 5 m)

The ultimate bearing capacity value deduced, (for intermediate parameters of β = 462; ζ = 0.0008), is 2,800 kN/m² which, for a safety coefficient of 6.5, yields the 430 kN/m² required for the design.

5.2 Viaduct spanning the River Arnoia (Orense)

In the province of Orense, Galicia, a 1,100 m. viaduct is being constructed on 23 piles to take a highway over the River Arnoia. The surface of the valley is covered with varying thicknesses of alluvium and "jabre", a residual granite soil, overlying granite which shows different degrees of weathering. The average values that are most significant and representative of the conditions which have enabled the ultimate bearing capacities (P_h) and allowable bearing capacities (P_{adm}) of the rock to be determined, are shown in summarized form in the following table.

PILES NUMBER	WEATHERING DEGREE	RMR	σ_c (MPa)	DEPTH (m)	P_h (MPa)	F_p	P_{adm} (MPa)
6	II-III	45	50	2,5	31.14	22	1.415
7	II-III	40	15	3	8.61	13	0.662
15	III	35	15	3	7.28	12	0.607
16 & 17	III-IV	30	5	2	2.32	7.5	0.309
18I	IV-V	25	5	2	2.00	7	0.285
18D	IV	25	7.5	4	3.18	7.5	0.424
20 a 22	III	30	15	2	5.85	11	0.532
23	III-IV	30	10	2	4.13	9	0.458

WEATHERING DEGREE	RMR	P_h (MPa)	F_p	P_{adm} (MPa)
II-III	45	14,2	18	0.789
III	40	11,9	17	0.700
III-IV	35	10,1	15	0.673

In all cases, the following data have been taken: $m_o = 9.5$ (Betournay et al. 1991), a slope inclination of 10° and a work load inclination of 5°.

Piling has been used to found the remaining piles.

Bridge spanning the River Guadiana (Badajoz)

5.3 Viaduct spanning the River Lagos (Málaga)

This is a hyperstatic bridge over the River Lagos, and is part of the Mediterranean Motorway. Shales (quartzitic, greywackes and quartz-shale), all of which are fractured in the first few metres, lie under the different Quaternary soils of varying thicknesses upon which the 295 m. bridge is constructed. With a degree of weathering of III-IV and II-III, their average unconfined compressive strength value is 26 MPa. Assuming a value of $m_o = 10$ and a range on the RMR Index of between 35 and 45, the accompanying table shows the values adopted for designing the direct foundations. The slope inclination is 22° and the overburden pressure value is 43 kN/m².

5.4 Viaducts on the Atlantic Motorway

The stretch of the Atlantic Motorway presently under construction in the province of Lugo, Galicia, includes numerous viaducts that are needed to span the different river beds, where rock outcrops close to the surface have made it possible to use direct foundation methods. The table below shows a synthesis of some of the most significant data for two of these viaducts.

The following table has been prepared to show the most relevant geotechnical calculation data and the results of the calculations in terms of ultimate and allowable bearing capacity under verticality hypotheses.

5.5 Summary

By applying the method to different cases, the following aspects can be highlighted;
- the versatility of the procedure
- the possibility of including the basic parameters involved (rock type, RMR Index, unconfined compressive strength, overburden pressure and load inclination)
- the ultimate bearing capacity and the allowable load values are consistent with Geotechnical Engineering practice under normal circumstances.

VIADUCT	LENGTH (m)	PILES NUMBER	ROCK TYPE	m_o	FOUNDATION DEPTH (m)	WEATHERING DEGREE
SAN MIGUEL	160	4	Shales	10	2.5	V-(IV)
SAN PEDRO	284	14	Shales & schists	10	2.0	II-III

VIADUCT	RMR	σ_c (MPa)	SLOPE INCLINATICON	p_h (MPa)	F_p	(MPa) P_{adm}
SAN MIGUEL	15	2.5	10°	1.44	5	0.288
SAN MIGUEL	25	2.5	10°	1.85	6	0.308
SAN PEDRO	50	5.0	15°	5.50	12	0.458
SAN PEDRO	50	15.0	15°	14.03	15	0.935
SAN PEDRO	70	5.0	15°	10.71	13.5	0.793
SAN PEDRO	70	15.0	15°	29.58	17.5	1.690

6 FURTHER RESEARCH NECESSARY

It is considered that there are at least six lines of basic research that should be explored if greater insight is to be obtained into the procedure described here:

1. Validation of the method for different rock masses, with a view to comparing the theory used, with real bearing capacity values, or in the absence thereof, comparison with scale models which recreate the most important factors involved in the failure.
2. Experimental definition with different rock materials, on the basis of a correct interpretation of laboratory test results, which makes it possible to determine the ultimate bearing capacity of a rock medium, and especially the range of variation of the parameter m_o of Hoek and Brown, depending on the type of material.
3. Identification of the statistical functions of density which are most suitable for representing the variations of the different parameters involved, (m_o, RMR and σ_c). Especially the need to define the coefficients of variation (COV) which enable one to quantify with one single parameter, the scattering of the real values in each situation. Consequently it will be possible to allocate the partial safety coefficients (F_p), that must be included if a specific failure risk is to be reached.
4. Develop the specific theory referred to here, which takes into account the heterogeneous nature of the rock medium and that incorporates the presence of parallel plans of weakness, significantly less resistant, which applies the homogeneous hypotheses herein assumed, to anisotropy situations.
5. Compares the methodology proposed here, with the most usual work load valuations for different types of rock masses in shallow foundations.
6. Incidence of the calculations on the inclusion of the ground weight, when the foundation size is sufficiently great to be of significance, at least in relative terms.

ACKNOWLEDGEMENTS

The authors would like to express their thanks to the "Dirección General de Carreteras" of the MOPTMA, the "Consejería de Obras Públicas" of the C.A. Extremadura, and the consulting companies EPTISA, IDEAM, INYPSA and TAGSA for providing the data presented in this paper, and for giving their permission to publish it.

REFERENCES

Betournay, M.C., Gorski, B., Jackson, R. and Gyenge, M. (1991). "*New Considerations in the Determination of Hoek and Brown Material Constants*", Proceedings of the VII International Conference Rock Mechanics, Aquisgran, Germany, Vol. I, pp. 195-200.

Canadian Geotechnical Society (1985). "*Canadian Foundation Engineering Manual*". Bitech Publishers Ltd. Vancouver, B.C. Canada.

Goodman, R.E. (1989). "*Introduction to Rock Mechanics*". John Wiley and Sons. pp. 1.562. U.S.

Hoek, E. and Brown, E.T. (1980). "*Empirical Strength Criterion for Rock Masses*". Journal of Geotechnical Engineering Division, American Society of Civil Engineers, Vol. 106, GT9, pp. 1013-1035.

Hoek, E. and Brown, E.T. (1988). "*The Hoek-Brown Failure Criterion, a 1988 Update*". Proceedings 15th Canadian Rock Mechanics Symposium. Univ. of Toronto. October.

Jimenez Salas, J.A., Justo, J.L. and Serrano, A.A. (1975). "*Geotecnia y Cimientos I. Mecánica del Suelo y de Rocas*", 1st edn,. Rueda, Madrid, España.

Ministerio de Obras Públicas (1994). "*ROM 0.5-94. Recomendaciones Geotécnicas para el Proyecto de Obras Marítimas y Portuarias*". Ente Puertos del Estado. Madrid, España.

Serrano, A.A. (1976) "*El Sólido Plástico. Separata Cap. IV*". Geotecnia y Cimientos II. Mecánica del Suelo y de Rocas, 1st edn, pp. 105-413. Rueda, Madrid, España.

Serrano, A. and Olalla, C. (1994a). "*Ultimate Bearing Capacity of Rock Masses*". International Journal of Rock Mechanics and Mining Science. Vol. 31, n° 2, pp. 93-106.

Serrano, A. and Olalla, C. (1994b). "*Carga de Hundimiento en Macizos Rocosos*". Monografía. CEDEX-MOPTMA. pp. 1-82.

Serrano, A. and Olalla, C. (1995). "*Allowable Bearing Capacity in Rock Foundations, based on a Non Linear Criterion*". Int. Journal of Rock Mechanics and Mining Science. Submitted to be published.

Sokolovskii, V.V. (1960). "*Statics of Soil Media*". Butterworth, London, U.K.

Sokolovskii, V.V. (1965). "*Statics of Granular Media*". Pergamon Press, London, U.K.

Wyllie, D.C. (1992). "*Foundations on Rock*". Chapman & Hall. London, U.K.

Bearing capacity of soft rock foundation on in-situ bearing capacity tests under inclined load

Atusi Nitta & Shigeki Yamamoto
Honsyu-Shikoku Bridge Authority, Tokyo, Japan

Tamaki Sonoda & Takayasu Hosono
Dia Consultants Co. Ltd, Tokyo, Japan

ABSTRACT One of the important factors in carrying out aseismic design of long-span bridge foundations is proper evaluation of the strength, that is the bearing capacity characteristics, of foundation ground under extreme eccentric inclined loads acting during an earthquake. This evaluation currently consists of determining material values to be used for design by comparatively studying and technically interpreting ground strength coefficients obtained from in-situ shearing tests and bearing capacity tests and from laboratory tests.

This article summarizes the results of (1) in-situ shearing tests of blocks (base 300 mm x 300 mm) subjected to vertical loads simulating inclined loads, (2) bearing capacity texts using a φ 300 mm loading plate and (3) laboratory cyclic triaxial compression tests on boring cores carried out on weathered granite in order to investigate the ultimate bearing capacity of ground subjected to inclined loads. Also, these test values were compared with bearing capacity values calculated using the bearing capacity formulate given in 'Specifications for Highway Bridges'. As a result it was shown that for the weathered granite tested this time, the bearing capacity formula of plastic theory as used for rigid foundations with no penetration can be applied in the evaluation of the bearing characteristics of ground under extreme inclined loads.

1. INTRODUCTION

One of the important factors to carry out seismic design for a long-span bridge foundation is suitable evaluation for ground strength, bearing capacity, under inclined load with eccentricities. Design values are mostly evaluated in consideration of engineering judgments from the result of in-situ shear tests and bearing tests, comparing with that of laboratory tests for the time being.

Although Komada (1967) researched on ultimate bearing capacity of the ground and Shioi et al.(1975) conducted in-house model experiments in Japan, no in-situ verification test has been done yet. Those researches are studied for a rigid foundation without embedment, and bearing capacity is derived from the solution of the Kötter's equation (based on a theory of plasticity) taking account of self-weight of soil and an angle of internal friction.

Referring to those researches, 1) in-situ shear tests of a block under the vertical load with an inclination, 2) bearing capacity tests with a φ 300mm loading plate, 3) triaxial tests of boring core samples were conducted to Hiroshima type weathered granite in Ikuchi-Island as one of the work for the Honsyu-Shikoku Bridge project. In the test site, weathered granite which is classified C_L, D_M according to the standard of Honsyu-Shikoku Bridge Authority is distributed, however D_H classified rock as same as the ground of Tatara-Oohashi was tested.

In this study, the results of those tests were summarized and compared with the bearing capacity calculated according to " *Specification for Highway Bridges* ", then the shear strength of the weathered granite and the ultimate bearing capacity were discussed.

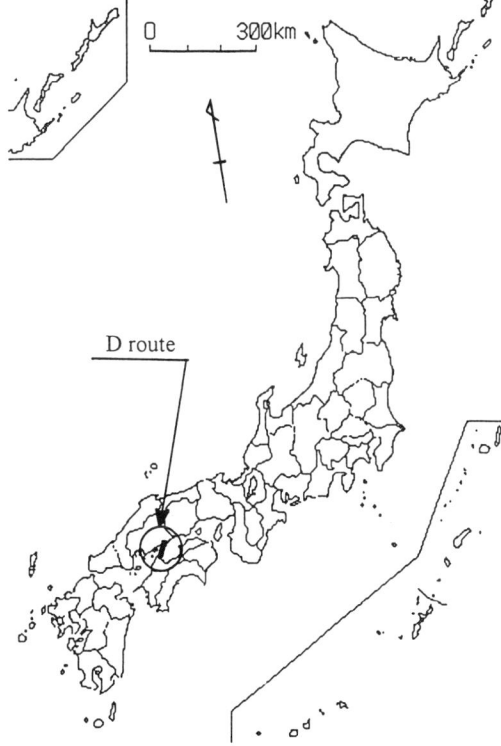

Fig.1 D route (Nishi-Seto highway)

as shown in Fig. 2 and has the approximate properties in Table 1. Among the several tests, the in-situ shear tests and the bearing tests by plate loading were conducted for D_H class near-by the reinforced concrete block T2 of the large-scale rock tests done by the author et al.(1993).

2. TEST SITE AND METHOD OF TESTING

2.1 Test site

The test site is located in the site of the access road to Ikuchi bridge which was constructed in Innoshima, Hiroshima-ken on the route of the Nishi-Seto Highway connecting between Omichi-shi, Hiroshima-ken and Imabari-shi, Ehime-ken(Fig.1). The rock ground excavated in the depth of 5m of the natural slope was tested. The rock is weathered granite C_{M-H} to D_L class

2.2 Method of testing

(1) Block shear tests

As block shear tests, six shear block were tested under the

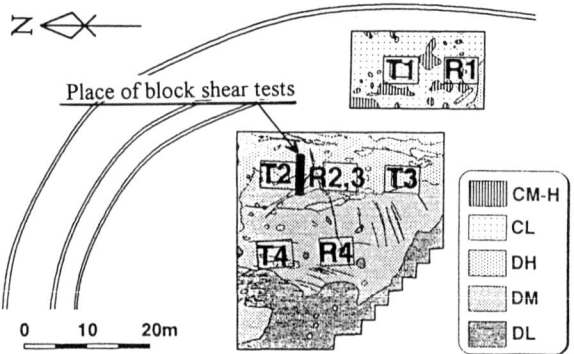

Fig.2 Rock classification map of test site

initial vertical stresses as shown in Table 2. The vertical stresses were determined through the wide range from low stresses in which have linear relationship between a vertical stress and a shear stress to high stresses in which have non-linear

Table.1 Rock class, S-wave velocity (Vsf), densities (ρd) and modulus of deformation (Esb) on test site

Test site	Rock class	Vsf (m/s)	ρd (g/cm³)	Esb (Mpa)
T1	CL	700	2.25-2.50	192-555
T2	DH-DM	370	2.00-2.35	62-200
T3	DH-DM	400	2.05-2.30	58-150
T4	DM	300	2.05-2.30	37-124

Table.2 Initial vertical stresses of block shear tests (Mpa)

Block	Primary test	Secondary test					
No.1	0.4	0.2	0.4	1.0	2.0	----	----
No.2	5.0	0.2	0.6	1.0	2.0	3.0	9.5
No.3	7.0	0.2	0.6	1.0	2.0	3.0	----
No.4	7.0	0.2	0.6	1.0	2.0	3.0	----
No.5	9.0	0.2	0.6	1.0	2.0	----	----
No.6	11.25	0.2	0.6	1.0	2.0	3.0	----

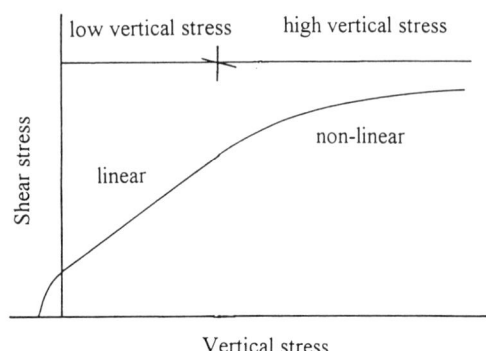

Fig.3 Representation of the relation between a vertical stress and a shear stress

relationship between them (Fig.3), assuming various load angles ($\tan\theta = \tau/\sigma v$). The tests included both ordinary shear tests and bearing capacity tests under inclined load. Explaining in other words, if the vertical load is small, it causes shear failure parallel to the block base. But if vertical load becomes large, failure progresses to the inner rock. Eventually ultimate bearing capacity under inclined load can be obtained.

At first, deformation tests under vertically cyclic load were conducted by means of the equipment shown in Fig. 4. As for shear, after a certain vertical load was applied, a shear load was applied with the loading speed 50pa/min by the jack with 15 degree angle to the horizontal until the ground failed. Then, the peak strength was obtained. (Hereunder, the first shear test is named a primary test, and the peak strength is named primary shear strength.) After this, the magnitude of vertical load was changed variously, a shear load was applied under each load case in the same way as the primary test in which the peak strength was obtained, and the friction resistance of each case was obtained (Hereunder, this test is named a secondary test, and the friction resistance is named secondary shear strength.)

(2) Bearing capacity tests
Bearing capacity tests was performed to examine a loading pattern as shown in Fig. 5.

(3) Triaxial tests (consolidated and undrained condition)
Triaxial tests were conducted using specimens with 65 mm diameter and 130mm height under consolidated and undrained condition. The effective constrained pressure P' was changed to 0.1, 0.3, 0.6, 1.2 and 2.4 Mpa for each test and with loading speed 0.05%/min.

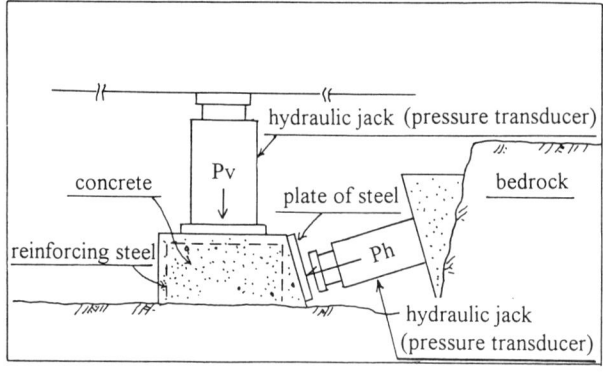

Fig.4 Epuipment of in-situ shear test

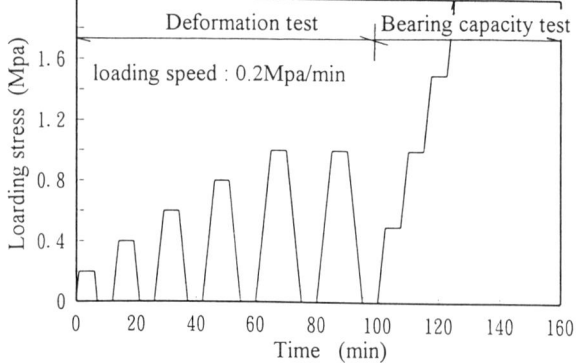

Fig.5 Load pattern of plate loading tests

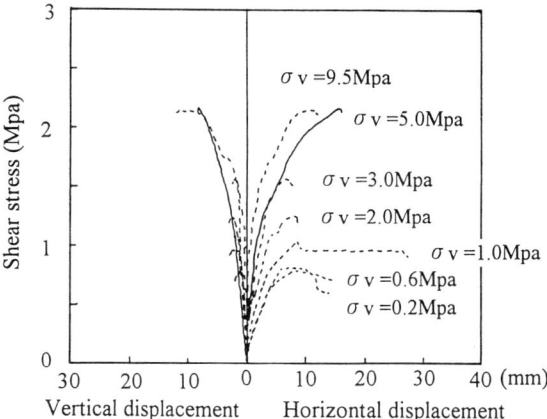

Fig.6 Shear stress - Displacement diagram

3. TEST RESULTS

(1) Block shear tests

An example of the shear stress and the displacement curve obtained from the shear tests is shown in Fig.6. The continuous line shows the result of the shear tests performed to obtain the first primary shear strength. The dotted line shows the shear stress and the displacement curve from which the secondary shear strength was obtained. It was found that the secondary shear strength became large as the vertical load increased.

Fig.7 shows the relationship between the primary shear strength and the secondary shear strength in the case of low vertical stress. The failare envelop line from the relationship gave cohesion C=0.45Mpa and internal friction angle $\phi=38°$.

The relationship in all the test data shows in Fig.8. The relationship between both strength is seen as the fact that the secondary shear was approximately smaller than the primary strength in the range of low vertical stress. However no significant deference was found in the range of high vertical stress. It is considered that the failure progressed in the primary test in the range of large vertical load before the primary shear strength began to be effective, and that plastic behavior dominated near the peak value as shown in the secondary test.

Additionally, the relationship between the vertical stress and the shear stress in the case of low vertical stress (No.1 block) was plotted near the failure envelope line in the figure as the result of the triaxial tests and shown that the shear stress tended to increase as the vertical stress became large. In the large range of vertical stress, however, it was found that the shear stress did not increase even if vertical stress became large. It is considered that the failure was a mainly shear failure parallel to the base of the block when vertical stress was low, and that the failure was a failure due to the lack of bearing capacity when vertical stress was high

(2) Bearing capacity tests

The relationship between the stresses and the displacements in the plate loading tests was plotted in Fig. 9, and the ultimate bearing capacity resulted in 18Mpa.

(3) Triaxial tests

Mohr circles obtained from the triaxial tests and a failure envelope line were shown in Fig.10. The cohesion and the angle of internal friction of the peak strength were respectively Ccu=0.28Mpa and ϕcu=41.3° those of the residual strength were Ccu=0.1Mpa and ϕcu=38.1° in the same way.

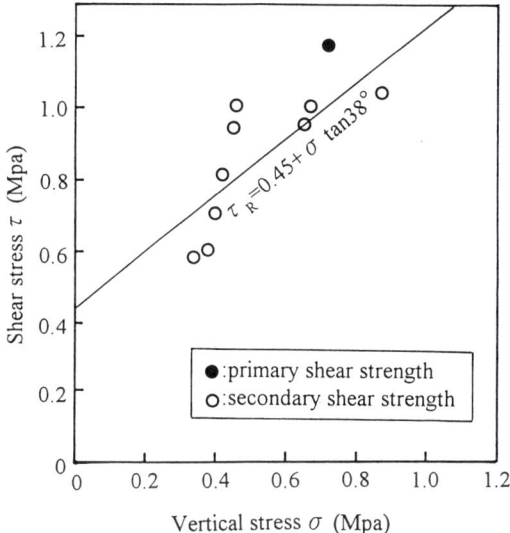

Fig.7 Shear stress - Vertical stress diagram in case of low vertical stress in block shear tests

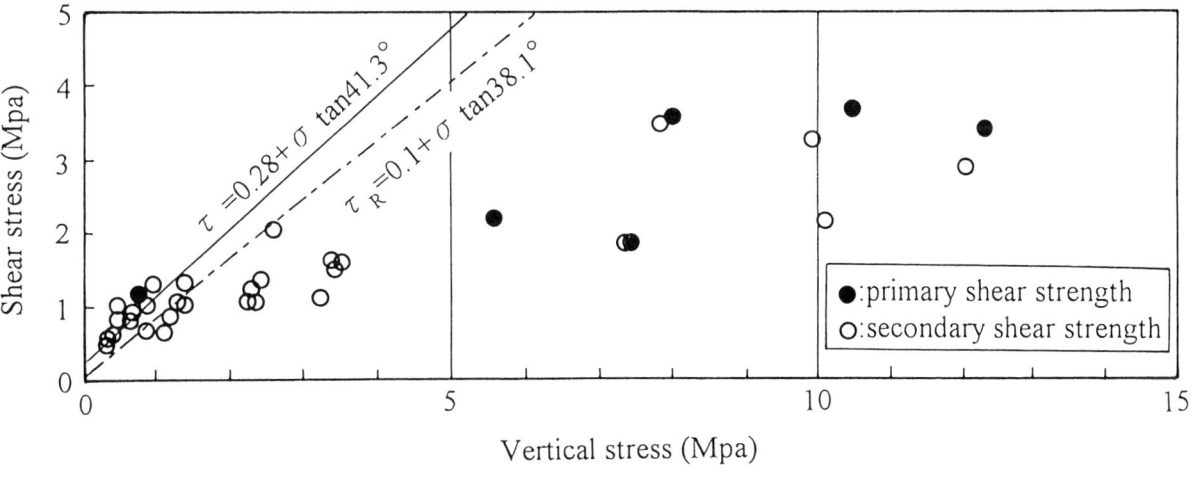

Fig.8 Shear stress - Vertical stress diagram in block shear tests

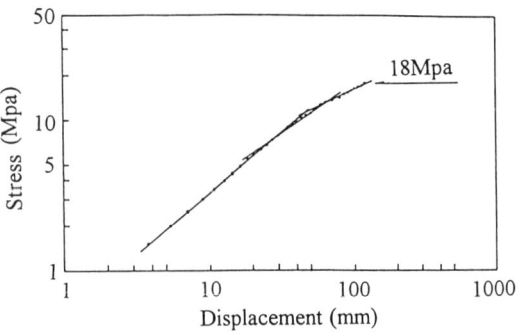

Fig.9 Stress-Displacement diagram in bearing capacity test

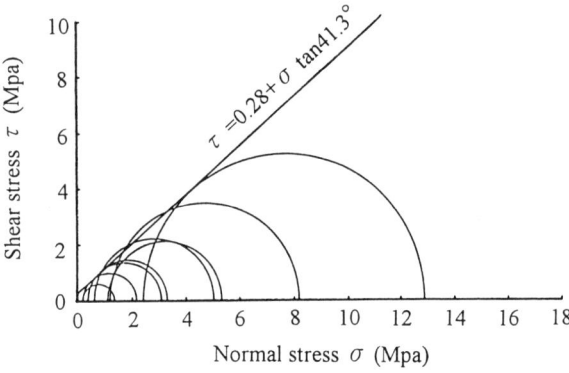

(a) Peak strength

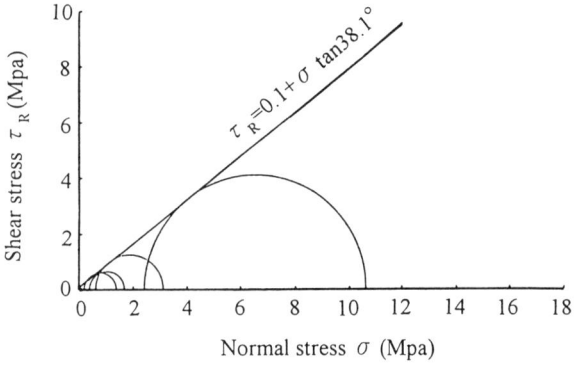

(b) Residual strength

Fig.10 Mohr-Coulomb's failure criterion

4. BEARING CAPACITY UNDER INCLINED LOAD

Bearing capacity under inclined load can be gained from integrating the Kötter's equation, the basic equation of a theory of plasticity. The following equation is an ultimate bearing capacity equation of the ground in "*Specification for the Highway Bridges*" and also of the integral solution with self-weight and an angle of internal friction.

$$Qu = A' \times \{\alpha \times k \times c \times Nc + k \times q \times Nq + 1/2 \times \gamma_1 \times \beta \times B' \times N_\gamma \}$$

where

Qu : ultimate bearing capacity in consideration of inclined loading (MN)
c : cohesion of base soil (MN/m^2)
q : surcharge load (MN/m^2), $q = \gamma_2 \times D_f$
A' : effective contact area (m^2)
γ_1, γ_2 : unit weights of the bearing stratum and of surrounding soil (MN/m^3)
B' : effective contact area in account of load with an eccentricity (m), $B' = B - 2e_B$
B : width of base (m)
e_B : eccentricity of load (m)
D_f : effective depth of foundation (m)
α, β : shape factor
k : coefficient of depth effect
Nc, Nq, N_γ : bearing capacity factors in account of inclined load

The dotted lines in Fig. 11 shows the relationship between the shear stress and the vertical stress at failure calculated from the above equation in accordance with the relationship between the shear stress and the vertical stress in the block shear tests. The angle of internal friction used in the calculation was that of the residual strength of the triaxial tests, $\phi = 38°$ and the cohesion was changed in the range from 0.2 to 0.6 Mpa since no significant deference was observed in the results of the primary test and that of the secondary test. In this case, bearing capacity factors were those of Sokolovski (1965) and Komada. Also, in the figure, the ultimate bearing capacity was plotted in terms of vertical stress (shear stress τ =0Mpa) with the results of other site C_L, D_M.

The results of the block shear tests and the bearing tests in this site agreed with that of calculation by the equation of bearing capacity in the case of the angle of internal friction ϕ =38° and cohesion C=0.2 to 0.3Mpa. Therefore, it is

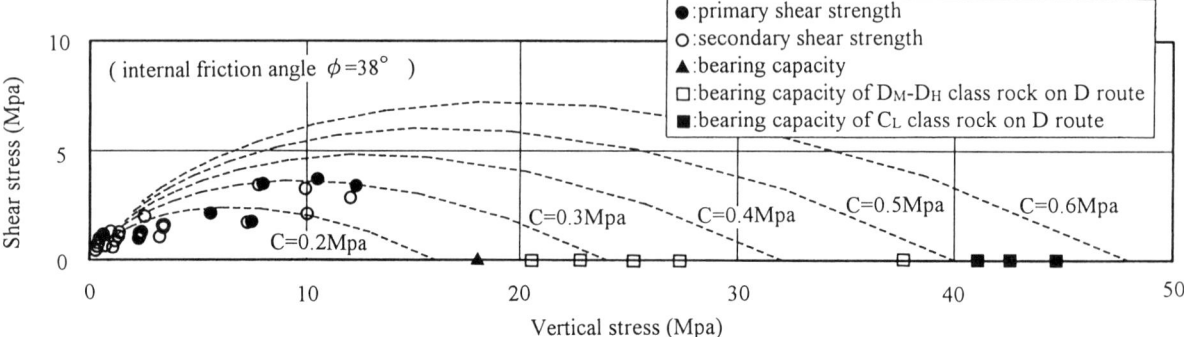

Fig.11 Shear stress-Vertical stress diagram in ultimate bearing capacity by bearing capacity formula

understandable that the equation of bearing capacity can be used to evaluate comprehensively both test results for weathered granite in this site.

5. CONCLUSION

Subjecting to weathered granite, in-situ block shear tests, bearing capacity tests and laboratory triaxial tests were conducted and the following points were concluded for ultimate bearing capacity of the ground under inclined load.

(1) The relationship between the primary shear strength and the secondary shear strength, defined in this paper, in the block shear tests was seen as the fact that the secondary shear strength was smaller than the primary shear strength in the range of low vertical stress, on the other hand, both were nearly equal in the range of high verticalstress.

(2) The relationships among the primary shear strength, the secondary shear strength and the vertical stress in the block shear tests showed that shear stress tended to increase as vertical stress increased in the low range of vertical stress but shear stress did not increase even if vertical stress increased in the high range of vertical stress.

(3) Mentioning the results of the block shear tests and the bearing capacity tests, the equation of bearing capacity taking account of an angle of internal friction and soil properties obtained from triaxial compression tests can be used to evaluate comprehensively both test results.

REFERENCES

Japan Road Association. (1990) " Specification for bridge design, I Common Specification and IV Substructure". pp. 211 to 228

Komada, K. (1967). "Stability of Direct Foundation, Foundation of Structure", J.S.C.E.

Shioi, Y., Asanuma, H., and Sugizaki, M.(1975). Study on critical bearing capacity for shallow foundation of rigid body. Tech. Note of Public Works Research Institute. Division of Foundation Engineering, Public Works Research Institute, Ministry of Construction, Japan.

Yamada, K.,b Yamagata, M., and Yamamoto, S. (1993). On-site loading test for bedrock of Tatara Bridge. Honshi Tech. Report Vol. 17, No.68, pp28-37

Sokokovski, V.V. (1695). "Statics of Granular Media". Pergamon Press, Oxford.

Bearing capacity of foundation on rock mass with weak layer

P. Miščević
Faculty of Civil Engineering, University of Split, Croatia

I. Jašarević
Faculty of Civil Engineering, University of Zagreb, Croatia

ABSTRACT: This paper deals with a strip surface footing on a rock mass with a weak layer. The calculations of the bearing capacity were performed simultaneously by two computer programs using the finite element method and the finite difference method with independent development of the models. The calculation was performed for different widths of the footing according to the width of the layer. The axis of the layer was placed on the edge of the footing and in the center of the footing, with different inclinations from a vertical. In the iteration procedure used in the calculation, special attention was paid to the stepping up (load increase) in the determination of "local" and "global" failures in the subgrade. The results show a significant decrease in the value of the bearing capacity for some widths and positions of a weak layer, as well as an important influence of a differential settlement (between the edges of the footing) on the determination of the collapse loads. The analyzed problem is illustrated by an example of a bridge footing (a bridge over Rijeka Dubrovacka, Croatia).

1 INTRODUCTION

Foundation engineering in Dalmatia (Croatia) deals with two dominant problems, karst areas and flysch layers. In both cases there are many planes and zones which can be determined as "weak layers" considering the foundation problems. A displacement of the footing, at a small distance from the location of a "weak" layer may cause different reactions in the subgrade and may also lead to changes in the value of the bearing capacity.

In the karst areas there are many weaknesses which cannot be determined as joints, because of their size compared to the dimensions of the footing. If they are long enough and wide enough, filled with clay, then the characteristics of clay are dominant for the behavior of the zone. All those kinds of wide joints, thin caverns, etc., can be modeled as "weak" layers, when compared to limestone, which form the karst structure.

A dominant characteristic of flysch is that its structure is made of relatively thin layers (from a few centimeters to a few meters), with significant changes in the characteristics between layers. These changes can range from the rocklike clay (soft rock) or even clay, to a layer of limestone. In most cases, under such conditions, there is a layer which is weaker than the main mass.

The ultimate bearing capacity of the footing in these conditions cannot be calculated with classic theories; hence, numerical modeling was used in the paper. For every problem two independent models were developed for two computer programs used in the analysis. The calculations were performed by computer programs : Z_SOIL - Zace Services Ltd., Switzerland (the finite element method); and FLAC - Itasca Consulting Group, Inc., USA (the finite difference method).

Although all parameters used in the analysis represent the characteristics of the previously described rock masses, the conclusions are valid for other similar problems.

2 PARAMETERS USED IN THE ANALYSIS

2.1 Parameters of shear strength and deformation

The properties of the rock masses described in the introduction were used in the analysis. The values of Young's modulus (E) and Poisson's ratio (ν) were used as average values from the published results of the measurements.

For the homogeneous and isotropic rock mass the ultimate bearing capacity can be expressed by the function:

$$q = f(B, L, D, c_R, \varphi_R, E_R, \nu_R, \gamma_R, c_c, \varphi_c, E_c, \nu_c, \gamma_c)$$

where are:
B, L - dimensions of footing
D   - depth of foundation
c_R, c_c - cohesion of rock and concrete
φ_R, φ_c - angle of internal friction
E_R, E_c - Young's modulus
ν_R, ν_c - Poisson's ratio
γ_R, γ_c - unit weight of rock and concrete

For a nonhomogeneous rock mass described in the introduction, the bearing capacity is also a function of the position, inclination (α) and width (b) of the layer. Even in this case the shearing strength (of the main rock mass and a layer), described in this paper by cohesion (c) and internal friction (φ), has a dominant influence on the results. Accordingly, the range of values characteristic for the described rock masses was analyzed on one preliminary model (with characteristics: $\alpha=30°$; b/B=0.5; a layer axis on the edge of the footing), in order to choose the parameters for the analysis. The applied model and the changes of the value of the bearing capacity (q) with the change in the shear strength are presented in Figure 1. and Figure 2.

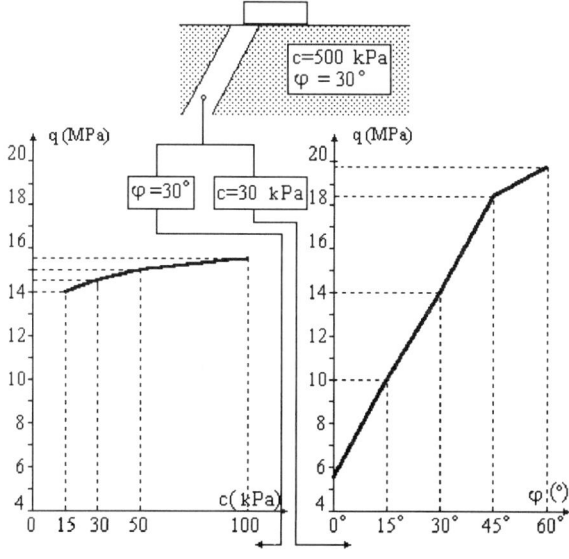

Figure 1. Change of the bearing capacity dependent on shearing strength of a "weak" layer in the preliminary model

If the shearing strength of the "weak" layer is analyzed (in the used model), as shown in Figure 1, the angle of internal friction has a dominant influence on the bearing capacity. However globally a change in the value is within the 20% range. For the main "strong" mass high values of cohesion do not have an extreme influence, while with an increase in the angle of internal friction, the bearing capacity rises rapidly (Figure 2).

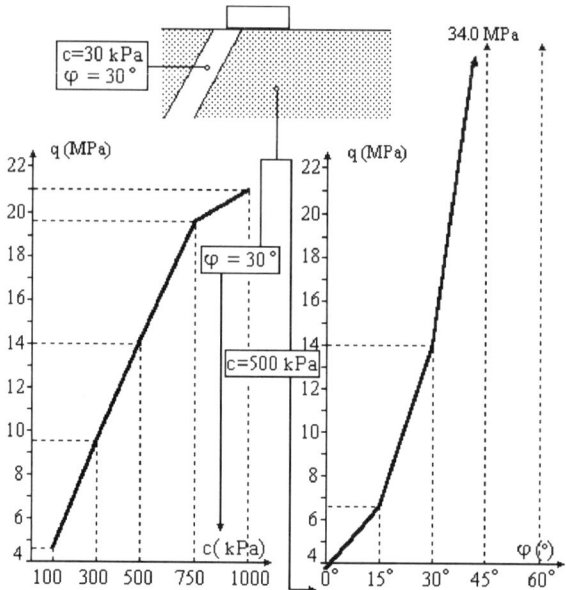

Figure 2. Change of the bearing capacity dependent on the shearing strength of the main rock mass in the preliminary model

After comparing the average properties of the rock mass described in the introduction with the results of the presented analysis, (in order to emphasize the influence of the "weak" layer for the purpose of the main analysis) the following properties were chosen:
- limestone from the flysch layer or karst area, defined as the main "strong" material in the paper:

E_1= 2.0 GPa
v_1= 0.3
c_1= 500 kPa
φ_1= 30°
γ_1= 22 kN/m³
K_1= 1.66 * 10⁶ kPa
G_1= 7.77 * 10⁵ kPa

- clay from the flysch layer, defined as a "weak" material in the paper:

E_2= 0.5 GPa
v_2= 0.3
c_2= 30 kPa
φ_2= 30°
γ_2= 20 kN/m³
K_2= 4.16 * 10⁵ kPa
G_2= 1.9 * 10⁵ kPa

According to the requirements of the FLAC computer program, the elastic bulk modulus (K) and shear modulus (G) were calculated from Young's modulus and Poisson's ratio with a correlation from the elastic theory.

The footing was assumed to be of concrete with reinforcement, modeled as an elastic material without weight with the following properties:

E_c= 30.0 GPa
v_c= 0.25
K_c= 2 * 10⁷ kPa
G_c= 1.2 * 10⁷ kPa

The rock mass was modeled as elastic-plastic with Mohr-Coulomb yield criteria for plane strain conditions. The initial stress field was calculated from the weight of the mass due to gravity without the influence of the footing weight.

2.2. *Analyzed parameters*

The paper is mainly concerned with the influence of the following parameters on the ultimate bearing capacity:
- position of the layer (axis) (Figure 3):
    a) on the edge of the footing
    b) in the middle of the footing
- inclination of the layer (α), (Figure 4): α = -60°; -30°; 0°; 30°; 60°;
- width of the layer (b) related to the width of the footing (B), (Fig. 4): b/B= 1/4; 2/4; 3/4;

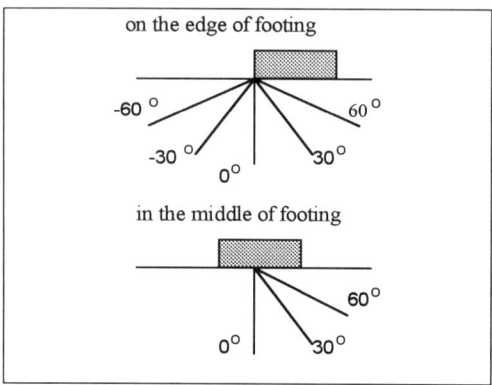

Figure 3. Positions of the layer axis (α)

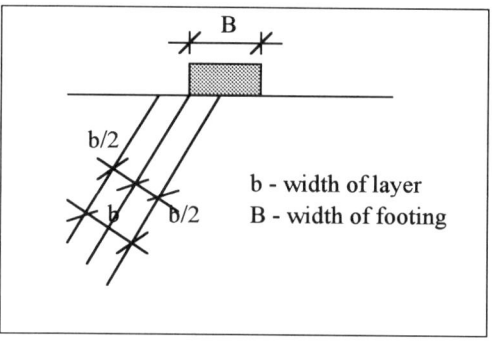

Figure 4. Width of the layer (b) / width of the footing (B)

2.3 *Determining collapse loads*

In both computer programs the load step increment was defined, while unbalanced forces and displacements of the footing edges were observed. An example of the analyzed models with the position of the applied load is shown in Figure 5.

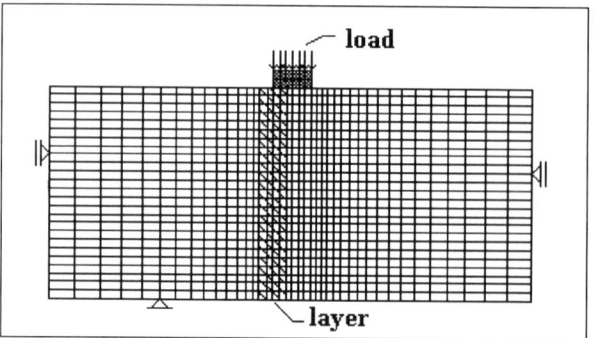

Figure 5. Example of the model

The load for which unbalanced forces started to increase rapidly and for which the increase in displacement was three times greater than in the previous step, was determined as collapse load. The precision used in the stepping was 100 kPa. In the case of the homogeneous and isotropic rock mass with the characteristics of the "main strong" material (in the analysis), the calculated ultimate bearing capacity was q_{ult}=18.5 MPa. This value was used for the deduction of a change of the bearing capacity according to the analyzed characteristics of a "weak" layer.

3 RESULTS

The loads which determine the ultimate bearing capacity are presented in the following figures:
- Figure 6 - layer axis on the edge of the footing
- Figure 7 - layer axis in the middle of the footing

All figures also show where the differential settlement, between edges of the footing, is dominant for estimating the ultimate bearing capacity. The points are marked with:
- DS - differential settlement is dominant
- W - differential settlement occurs, but is not dominant
- unmarked - no significant differential settlement

The area of the dominant differential settlement, formed with the points "DS", is shadowed in the figures.

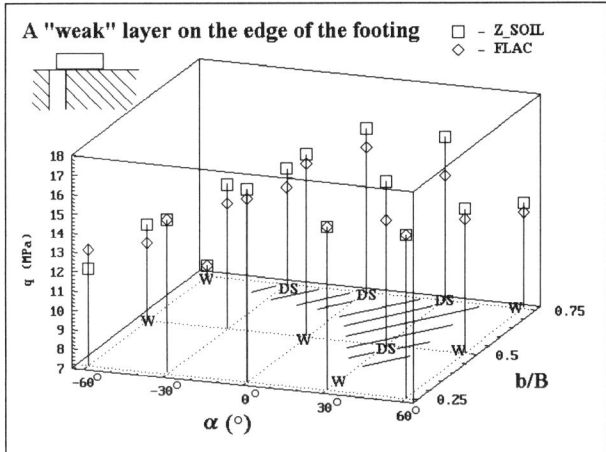

Figure 6. The changes of the ultimate bearing capacity (q) for a "weak" layer on the edge of the footing

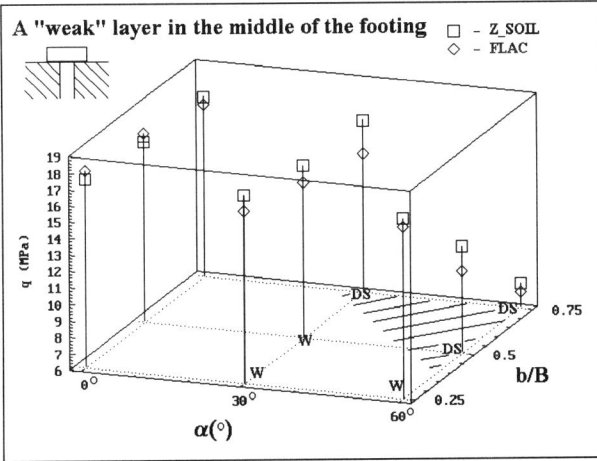

Figure 7. The changes of the ultimate bearing capacity (q) for a "weak" layer in the middle of the footing

If the results, presented in Figure 6 and Figure 7, are compared with the ultimate bearing capacity estimated for the main "mass", it can be concluded:

- for b/B>0.75; -30°>α>30°; the layer axis in the middle of the footing, there is a significant decrease in the bearing capacity value and dominant influence of settlements;

- for b/B>0.5; -30°>α>30°; the layer axis on the edge of the footing, there is a significant decrease in the bearing capacity value or the influence of differential settlements become dominant;

- for other investigated combinations the value of the bearing capacity is slightly changed and differential settlement is not dominant.

4 EXAMPLE

The analyzed problem is illustrated by an example of the bridge over the Rijeka Dubrovacka in Croatia, near Dubrovnik. Figure 8 presents the shape of the footing and the numerical model used in the calculations. The width of the footing is B=11.0 m, and the width of a "weak" layer is b=8.5 m. The foundation problem can be approximately presented with a correlation b/B=0.75 and an inclination of the layer α ≈ 0°.

The material properties of the rock mass are defined by (material 1 - "weak" layer; material 2 - main rock mass):

K_1= 0.94*10^6 kPa K_2=12.5 *10^6 kPa
G_1= 0.43*10^6 kPa G_2= 5.77*10^6 kPa
c_1= 75 kPa c_2= 180 kPa
φ_1= 32° φ_2= 40°
γ_1= 22.0 kN/m^3 γ_2= 19.0 kN/m^3

Static loads (from the construction above) used in estimating the footing dimensions :
M = 22678.0 kNm/m'
V = 5045.0 kN/m'
H = 457.0 kN/m'

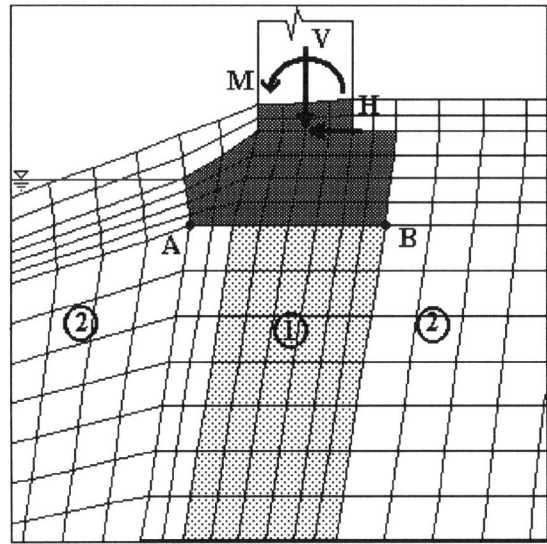

Figure 8. Numerical model of the presented example; dimensions on the figure are distorted.

The ultimate bearing capacity for a homogeneous mass (material 2) calculated with the numerical model (only for compressive vertical stress) was q = 18.5 MPa. For a vertical loading and the nonhomogeneous mass (from Figure 8), the bearing capacity was assumed to be q = 12.75 MPa. The decrease in the value is in a good agreement with the results of the presented analysis (Figure 6 & 7).

According to the ultimate bearing capacity, the safety factor for the used static load (including the moment and horizontal force) for a homogeneous mass was assumed to be F_S=12. For a nonhomogeneous mass the safety factor is estimated to be F_S=10. These values were obtained without taking into account the differential settlement. As Figure 9 shows (settlement of points A and B on the edges of footing), for estimating a reliable safety factor in the presented example, differential settlement is dominant. When the differential settlement was analyzed, the real safety factor used in estimating dimensions of the footing was $F_S=4$.

The displacement vectors resulting from the static load shown in Figure 10 can be used in calculating the footing rotation angle, which is recommended in the Eurocode 7 as a measure of differential settlement.

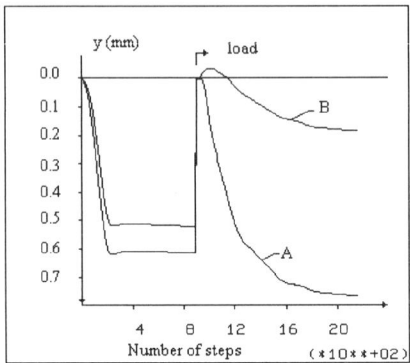

Figure 9. Vertical displacement of points A and B according to static load (see Figure 8)

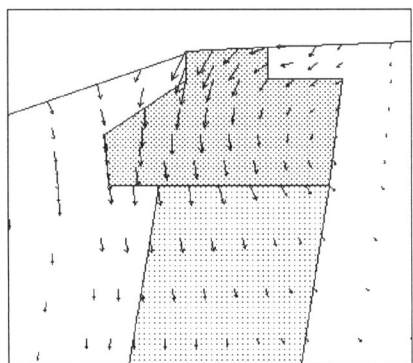

Figure 10. Displacement vectors

For a nonhomogeneous rock mass with a "weak" layer (described in the example) the principal stress tensors are shown in Figure 11. Compared with the solution for the homogeneous isotropic mass, there is a redistribution of stresses into a "stronger" part of the rock mass (see the rounded mark in Figure 11). The value of the ultimate bearing capacity depends on the "shape" of this redistribution.

5 CONCLUSION

All presented results of the ultimate bearing capacity are highly dependent on parameters of deformation and shear strength used in the calculations. Nevertheless, in practice they represent a useful base for judgment in cases where field conditions are not equal to those predicted from previous investigations. For example, if the footing is strictly dimensioned with properties of a "strong" mass and there is an unexpected highly different "weak" layer, there is always a question as to whether or not recalculation is necessary (according to the determined characteristics of a layer).

For the ultimate bearing capacity estimated by numerical modeling the results are dependent on the range of load increment in the stepping.

For example, in case of a "weak" layer in a "strong" mass on the edge of a footing, with $b/B>1/4$, depending on the value of the load increment, the result can be either a "local" or a "global" collapse load, for the same rules.

The term "local" means reaching the bearing capacity of a "weak" material, recorded as numerical instability, before the redistribution of stresses into a "strong" mass. Both terms are graphically presented in Figure 12. With a big load increment there is a possibility of noting only a "global" collapse, without any information about a "local" collapse.

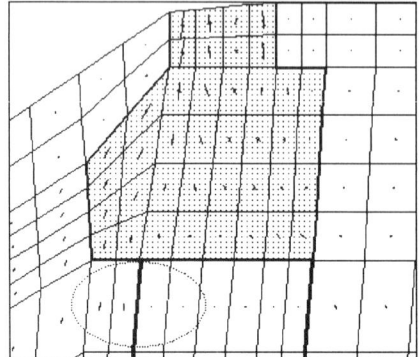

Figure 11. Principal stress tensors for static load (example of the redistribution of stresses is emphasized with the rounded mark)

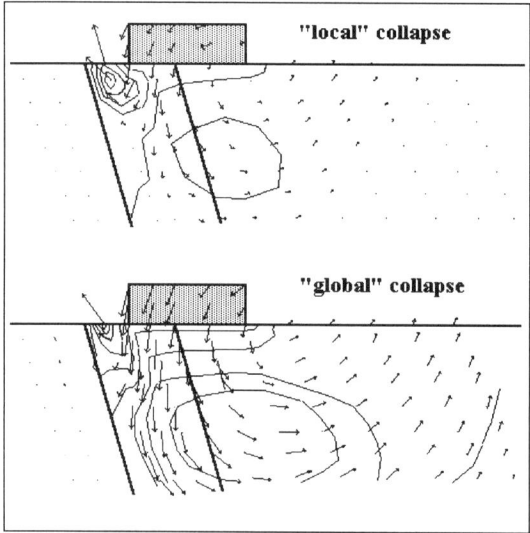

Figure 12. A "local" and a "global" collapse

The analysis presented in the paper can be improved by modeling the joints between a layer and a rock mass, as well as between the footing and a rock mass. Although in this case the characteristics of the joints are required (additional amount of parameters in calculation), for some dimensions and positions of the "weak" layer calculated the displacements become more reliable.

REFERENCES

FLAC (Fast Lagrangian Analysis of Continua) 1991. *User's manual, Volume I & II*. Itasca Consulting Group, Inc., Minneapolis, Minnesota, USA.
Goodman,R.E. 1980. *Introduction to Rock Mechanics*. University of California at Berkeley: John Wiley & Sons.
Liao,J.J. & Amadei,B. 1991. Surface loading of anisotropic rock masses. *J. of Geotech. Engrg.* vol. 117(11):1779-1802.
Miščević,P. & Jašarević,I. 1995. Ultimate bearing capacity of strip footing on layered rock mass by numerical modeling, *Proceedings, Second International Conference on Mechanics of Jointed and Faulted Rock, April 10-14, 1995., Vienna, Austria*. Rotterdam :A.A. Balkema: 619-624
Wittke, W. 1990. *Rock mechanics (theory and applications with case histories)*.Berlin:Springer-Verlag.

Stability analysis of jointed rock foundations by discontinuous deformation analysis

T. Sasaki & D. Ishii
Kajima Corporation, Tokyo, Japan

Y. Ohnishi
Kyoto University, Japan

R. Yoshinaka
Saitama University, Japan

ABSTRACT : The Discontinuous Deformation Analysis (DDA) was developed by G.H.Shi and R.E.Goodman. It solves a finite element type of mesh where all the elements are isolated blocks and bounded by pre-existing discontinuities under kinematics conditions of dynamic and quasi-static motion. The authors applied the elasto-plastic constitutive laws based on Drucker-Prager's yield criteria in the DDA theory. The stability of jointed foundation models with different joint angles is analyzed and compared with the experiment of (Maury, 1970), Gaziev et al.(1971) and analytical theory. The numerical results obtained by DDA agreed with those expected.

1 INTRODUCTION

The Discontinuous Deformation Analysis (DDA) was developed by G.H.Shi and R.E.Goodman(1985). It solves a finite element type of mesh where all the elements are isolated blocks and bounded by pre-existing discontinuities under kinematics conditions of dynamic and quasi-static motion. The theory classified in the category of a kind of hybrid displacement type model of the finite element method (K.Washizu, 1972). The authors developed the elasto-plastic constitutive laws based on Drucker-Prager's yield criteria, rock-bolt and connected elements and improved the penalty method for penetrations of each block. The theory is based on the updated Lagrangian formulations, hence, Jaumann derivative is introduced for the superposing of stress of each increment reason to adjust the block rotations. The developed method is applied for the stability analysis in the jointed rock foundations.

2 BASIC THEORY

Since DDA is a kind of FEM as described chapter 1, the same procedures as FEM are employed. But, unknown parameters are defined at the center of a block. These are rigid body displacements, rigid body rotation and strains. The penalty function method is introduced for penetrations of each block by using least square energy evaluations, and it solves under the condition of no penetrations. The characteristics of DDA theory are as follows.
(1) The equilibrium equations are established by minimizing the total potential energy.
(2) The forward and backward analysis are available.
(3) The dynamic and quasi-static motions are available.
(4) An arbitrary constitutive laws can be introduced as FEM.
(5) An arbitrary contact conditions, boundary conditions, load conditions, initial strains, inertia forces, and rock bolting can be introduced as FEM.

2.1 Block deformation and displacement variables

Assuming each block has constant stress and constant strains with in small time steps, the displacement (u,v) of any point (x,y) of a block can be represented by six displacement variables.

$$\{D_i\} = \{ u_0\ v_0\ r_0\ \varepsilon_x\ \varepsilon_y\ \gamma_{xy} \} \quad \cdots \cdots \cdots (2.1)$$

where,
$\{D_i\}$: the unknown vector,
u_0, v_0, r_0 : the rigid body displacement and rotation,
$\varepsilon_x, \varepsilon_y, \gamma_{xy}$: the strains of block i as shown Fig. 1.
The displacement functions of arbitrary positions of block i, u and v, are expressed by parameters of the coordinate x, y of a block which is the same as the triangular constant strain element of FEM.

$$u = a_1 + a_2 x + a_3 y \quad , \quad v = b_1 + b_2 x + b_3 y \quad \cdots \cdots (2.2)$$

The unknown parameters of Eq.(2.1) are expressed by using the displacement functions of Eq.(2.2).

$$u_0 = a_1 + a_2 x_0 + a_3 y_0 \quad , \quad v_0 = b_1 + b_2 x_0 + b_3 y_0 \quad \cdots (2.3)$$

$$\gamma_0 = \frac{1}{2}(\frac{\partial v}{\partial x} - \frac{\partial u}{\partial y}) = \frac{1}{2}(b_2 - a_3) \quad \cdots \cdots \cdots (2.4)$$

where,
x_0, y_0 : the coordinate of a point at rigid body displacement in which is the center of gravity of a block in general.
The stretch and shear strains are expressed by Eq.(2.5).

$$\varepsilon_x = \frac{\partial u}{\partial x} = a_2\ ,\ \varepsilon_y = \frac{\partial v}{\partial y} = b_3\ ,\ \gamma_{xy} = (\frac{\partial v}{\partial x} + \frac{\partial u}{\partial y}) = (b_2 - a_3) \quad \cdot (2.5)$$

Substituting Eq.(2.3) and Eq.(2.4) into Eq.(2.2), the unknown parameters of Eq.(2.1) are expressed Eq.(2.6).

$$\begin{Bmatrix} u \\ v \end{Bmatrix} = \begin{bmatrix} 1 & 0 & -(y-y_0) & (x-x_0) & 0 & (y-y_0) \\ 0 & 1 & (x-x_0) & 0 & (y-y_0) & (x-x_0) \end{bmatrix} \{D\} = [T_i]\{D_i\} \quad \cdot\cdot (2.6)$$

where,
$[T_i]$: the block deformation matrix.
Eq.(2.6) expresses the relationship between deformation and block shapes and block shape is updated by using Eq.(2.6) on each time step.

2.2 Strain energy and stiffness matrix of blocks

In the two-dimensional case, the strain energy of block e is given by Eq.(2.7).

$$\Pi_e = \iint \frac{1}{2}(\varepsilon_x\sigma_x + \varepsilon_y\sigma_y + \gamma_{xy}\tau_{xy})dxdy$$
$$= \iint \frac{1}{2}\{\varepsilon_i\}^T\{\sigma_i\}dxdy = \frac{1}{2}\{\varepsilon_i\}^T[E]\{\varepsilon_i\}dxdy \quad \cdots (2.7)$$
$$= \frac{S}{2}\{\varepsilon_0\}^T[E_0]\{D_i\}$$

where,
$[E]$: the stress-strain constitutive law,
$\{\varepsilon_i\}$: the strain vector,
$\{\sigma_i\}$: the stress vector.

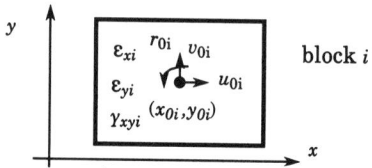

Figure 1. Coordinate system and unknown parameters.

$$\{\sigma_i\} = \{\sigma_x \sigma_y \tau_{xy}\}^T \; , \; \{\varepsilon_0\} = \{0\;0\;0\;\varepsilon_x\varepsilon_y\gamma_{xy}\}^T \quad \cdots (2.8)$$
$$\{\sigma_i\} = [E]\{\varepsilon_i\} \; , \; \{\varepsilon_i\} = \{\varepsilon_x\varepsilon_y\gamma_{xy}\}^T \quad \cdots\cdots (2.9)$$
$$[E] = \frac{E(1-\nu)}{(1+\nu)(1-2\nu)}\begin{bmatrix} 1 & \nu/(1-\nu) & 0 \\ \nu/(1-\nu) & 1 & 0 \\ 0 & 0 & (1-2\nu)/(2(1-\nu)) \end{bmatrix}$$
$$[E_0] = \begin{bmatrix} [0] & [0] \\ [0] & [E] \end{bmatrix} \quad \cdots\cdots (2.10)$$

E : Young's modulus,
S : the area of a block,
$[0]$: the 3×3 zero matrix,
$[E_0]$: the 6×6 expanded stress-strain matrix.

Minimizing the potential energy of a block as Eq. (2.7), the block stiffness matrix is given by Eq. (2.11).

$$[k_{rs}]_{ii} = \frac{\partial^2\Pi_e}{\partial d_{ri}\partial d_{si}} = \frac{S}{2}\frac{\partial^2}{\partial d_{ri}\partial d_{si}}\{D_i\}^T[E_0]\{D_i\} \quad \cdots (2.11)$$
$$= S[E_0] = [K_{ii}] = [K_e]$$

$[K_e]$ is the 6×6 matrix and consists of only the two parameters which are S and the stress-strain constitutive law of $[E]$.

2.3 *Stiffness matrices of rockbolts*

Fig. 2 shows that arbitrary two points (x_i, y_i) and (x_j, y_j) in blocks i and j are connected by a rockbolt. Let u_i, v_i, u_j, v_j, each be displacements of points i and j in x and y directions respectively, the deformation of the rockbolt is given Eq.(2.12).

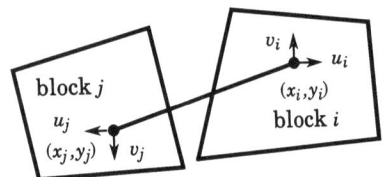

Figure 2. Modeling of a rockbolt.

$$dl = (1/l)[(u_i, v_i)\{l_x, l_y\}^T - (u_j, v_j)\{l_x, l_y\}^T] \quad \cdots (2.12)$$
$$l = \sqrt{(x_i - x_j)^2 + (y_i - y_j)^2} \quad \cdots (2.13)$$
$$l_x = (1/l)(x_i - x_j) \; , \; l_y = (1/l)(y_i - y_j) \quad \cdots (2.14)$$

where,
l : the length of the rockbolt,
l_x, l_y : the direction cosines.

Let S_b the stiffness of the rockbolt, the stress of rockbolt is defined Eq.(2.15).
$$f = -S_b(dl/l) \quad \cdots (2.15)$$
The strain energy of the rockbolt is given by Eq.(2.16).
$$\Pi_R = -(1/2)fdl = (S_b/2l)dl^2 = (S_b/2l)\{D_i\}^T\{E_i\}\{E_i\}^T\{D_i\}$$
$$-(S_b/l)\{D_i\}^T\{E_i\}\{G_j\}^T\{D_j\}$$
$$+(S_b/2l)\{D_j\}^T\{G_j\}\{G_j\}^T\{D_j\} \quad \cdots (2.16)$$
where,
$\{E_i\}^T = [T_i]^T\{l_x, l_y\}^T \; , \; \{G_j\}^T = [T_j]^T\{l_x, l_y\}^T$

Minimizing the potential energy of Eq.(2.16), the stiffness matrix of rockbolt is given by Eq.(2.17)~(2.20).

$$[k_{rs}]_{ii} = \frac{\partial^2\Pi_R}{\partial d_{ri}\partial d_{si}} = \frac{S_b}{2l}\frac{\partial^2}{\partial d_{ri}\partial d_{si}}\{D_i\}^T\{E_i\}\{E_i\}\{D_i\} \quad \cdots (2.17)$$
$$= (S_b/l)\{E_i\}^T\{E_i\} = [K_{\lambda R}]_{ii}$$

$$[k_{rs}]_{ij} = \frac{\partial^2\Pi_R}{\partial d_{ri}\partial d_{sj}} = \frac{S_b}{2l}\frac{\partial^2}{\partial d_{ri}\partial d_{sj}}\{D_i\}^T\{E_i\}^T\{G_j\}\{D_j\} \quad \cdots (2.18)$$
$$= (S_b/l)\{E_i\}^T\{G_j\} = [K_{\lambda R}]_{ij}$$

$$[k_{rs}]_{ji} = \frac{\partial^2\Pi_R}{\partial d_{rj}\partial d_{si}} = \frac{S_b}{2l}\frac{\partial^2}{\partial d_{rj}\partial d_{si}}\{D_j\}^T\{G_j\}^T\{E_i\}\{D_i\} \quad \cdots (2.19)$$
$$= (S_b/l)\{G_j\}^T\{E_i\} = [K_{\lambda R}]_{ji}$$

$$[k_{rs}]_{jj} = \frac{\partial^2\Pi_R}{\partial d_{rj}\partial d_{sj}} = \frac{S_b}{2l}\frac{\partial^2}{\partial d_{rj}\partial d_{sj}}\{D_j\}^T\{G_j\}^T\{G_j\}\{D_j\} \quad \cdots (2.20)$$
$$= (S_b/l)\{G_j\}^T\{G_j\} = [K_{\lambda R}]_{jj}$$

2.4 *Stiffness matrices and external force vectors of bonding elements*

The authors developed a bonding element by the penalty method for the purpose of analyzing local stress and the strain of blocks or the bond of blocks like shotcrete as the lining of tunnels (Sasaki et al., 1994).

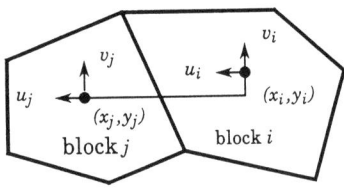

Figure 3. Bonded blocks.

Fig. 3 shows the relationship between blocks i and j which are bonded to each other. Relative displacement of arbitrary points (x_i, y_i) and (x_j, y_j) in bonded blocks i and j is defined as Eq. (2.21).
$$\{d_x, d_y\}^T = \{x_i + u_i - x_j - u_j, y_i + v_i - y_j - v_j\}^T \quad \cdots (2.21)$$
The relative displacement of Eq.(2.21) must be zero at the bonding face, hence, these conditions have to be given in Eq.(2.22).
$$\{d_x, d_y\}^T = 0 \quad \cdots (2.22)$$
The spring force by the relative displacement is given by Eq.(2.23).
$$\{F_x, F_y\}^T = -P\{x_i + u_i - x_j - u_j, y_i + v_i - y_j - v_j\}^T \quad \cdots (2.23)$$
Strain energy by the relative displacement is given by Eq. (2.24).
$$\Pi_c = -(1/2)(d_xF_x, d_yF_y) \quad \cdots (2.24)$$
Express the displacements of the points (x_i, y_i) and (x_j, y_j) in bounded blocks with the unknowns of blocks i and j.
$$\{u_i, v_i\}^T = [T(x_i, y_i)]\{D_i\} \; , \; \{u_j, v_j\}^T = [T(x_j, y_j)]\{D_j\} \quad \cdots (2.25)$$
Substituting Eq. (2.25) into Eq.(2.24), we get Eq.(2.26).

$$\Pi_c = (P/2)(\{D_i\}^T[T_i]^T[T_i]\{D_i\} - 2\{D_i\}^T[T_i]^T[T_j]\{D_j\}$$
$$+ \{D_j\}^T[T_j]^T[T_j]\{D_j\} + 2\{D_i\}^T[T_i]^T\{x_i - x_j, y_i - y_j\}^T$$
$$- 2\{D_j\}^T[T_j]^T\{x_i - x_j, y_i - y_j\}^T$$
$$+ \{x_i - x_j, y_i - y_j\}\{x_i - x_j, y_i - y_j\}^T) \quad \cdots \cdots (2.26)$$

Minimizing the potential energy of Eq.(2.26), the stiffness matrix of bonding elements is given as follows.

$$[k_{rs}]_{ii} = \frac{\partial^2 \Pi_c}{\partial d_{ri} \partial d_{si}} = \frac{P}{2} \frac{\partial^2}{\partial d_{ri} \partial d_{si}} \{D_i\}^T [T_i]^T [T_i] \{D_i\} \quad \cdots (2.27)$$
$$= P\{T_i\}^T\{T_i\} = [k_{\lambda c}]_{ii}$$

As same manner,

$$[k_{rs}]_{ij} = -P[T_i]^T[T_j]^T = [k_{\lambda c}]_{ij} \quad \cdots \cdots (2.28)$$
$$[k_{rs}]_{ji} = -P[T_j]^T[T_i]^T = [k_{\lambda c}]_{ji} \quad \cdots \cdots (2.29)$$
$$[k_{rs}]_{jj} = P[T_j]^T[T_j]^T = [k_{\lambda c}]_{jj} \quad \cdots \cdots (2.30)$$

External force vectors are given by Eq. (2.31) and (2.32).

$$\{f_r\}_i = \frac{\partial^2 \Pi_f}{\partial d_{ri}} = -P \frac{\partial}{\partial d_{ri}} \{D_i\}^T[T_i]^T\{T_i\}\{x_i - x_j, y_i - y_j\}^T \quad (2.31)$$
$$= -P[T_i]^T\{x_i - x_j, y_i - y_j\}^T = \{F_{\lambda c}\}_i$$
$$\{f_r\}_j = P[T_j]^T\{x_i - x_j, y_i - y_j\}^T = \{F_{\lambda c}\}_j \quad \cdots \cdots (2.32)$$

2.5 Movements and equilibrium equations of blocks

The strain energy of the contact spring by penetration between blocks i and j in Fig. 4 is evaluated by Eq.(2.33), where d is expressed by Eq.(2.34) with functions of coordinates of elements $\{e\}$ and $\{g\}$, and unknowns of blocks $\{D_i\}$ and $\{D_j\}$.

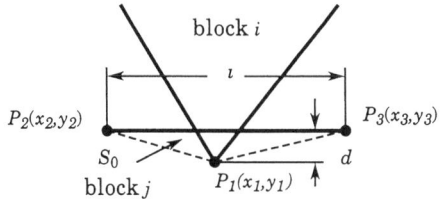

Figure 4. Block penetrations.

$$\Pi_P = \frac{P}{2} d^2 \quad \cdots \cdots (2.33)$$
$$d = \frac{S_0}{l} + \{e\}^T\{D_i\} + \{g\}^T\{D_j\} \quad \cdots \cdots (2.34)$$

where,

$$\{e_r\} = [(y_2 - y_3)[T_{ir}(x_1, y_1)] + (x_3 - x_2)[T_{jr}(x_2, y_1)]]/l \quad \cdots (2.35)$$
$$\{g_r\} = [(y_3 - y_1)[T_{ir}(x_2, y_2)] + (x_1 - x_3)[T_{jr}(x_2, y_2)]]/l$$
$$+ [(y_1 - y_2)[T_{ir}(x_3, y_3)] + (x_2 - x_1)[T_{jr}(x_3, y_3)]]/l \quad \cdots (2.36)$$

P : a penalty coefficient which has enough large positive value,
S_0 : the area of penetration shown as Fig. 4.

Substitute Eq.(2.34) into Eq.(2.33) and minimize Π_P under condition that the penetration d is not beyond a certain value.

Hence, the stiffness matrix of penetration is given by Eq.(2.37) ~(2.40).

$$[k_{rs}]_{ii} = \frac{\partial^2 \Pi_P}{\partial d_{ri} \partial d_{si}} = \frac{P}{2} \frac{\partial^2}{\partial d_{ri} \partial d_{si}} \{D_i\}^T\{e\}^T\{e\}\{D_i\} \quad \cdots (2.37)$$
$$= P\{e\}^T\{e\} = [k_{\lambda P}]_{ii}$$
$$[k_{rs}]_{ij} = -P\{e\}^T\{g\}^T = [k_{\lambda P}]_{ij} \quad \cdots \cdots (2.38)$$
$$[k_{rs}]_{ji} = -P\{g\}^T\{e\}^T = [k_{\lambda P}]_{ji} \quad \cdots \cdots (2.39)$$
$$[k_{rs}]_{jj} = P\{g\}^T\{g\}^T = [k_{\lambda P}]_{jj} \quad \cdots \cdots (2.40)$$

The external terms are given by Eq.(2.41) ~(2.42).

$$\{f_r\}_i = -\frac{\partial \Pi_P}{\partial d_{ri}} = -\frac{P S_0}{2} \frac{\partial}{\partial d_{ri}} \{e\}^T\{D_i\}$$
$$\quad \cdots \cdots (2.41)$$
$$= -\frac{P S_0}{2} \{e\} = \{F_{\lambda P}\}_i$$

$$\{f_r\}_j = -\frac{\partial \Pi_K}{\partial d_{rj}} = \frac{P S_0}{2} \frac{\partial}{\partial d_{rj}} \{e\}^T\{D_i\}$$
$$\quad \cdots \cdots (2.42)$$
$$= \frac{P S_0}{2} \{e\} = \{F_{\lambda P}\}_j$$

2.6 Friction condition of block surfaces and penalty method

Fig. 5 shows the concept of the criteria of penalty method for block penetration by Shi (1985) and Sasaki et al. (1994). Shi (1989) applied the penalty method to both normal penetration and the shear component. In this criteria, if shear stress is smaller than the fiction strength, no block stress conditions are caused due to the non-slip condition (locking) being applied to the block.

Hence, the authors introduced the shear stiffness K_s just for the shear component instead of the penalty term and the penalty method is applied only to the normal component.

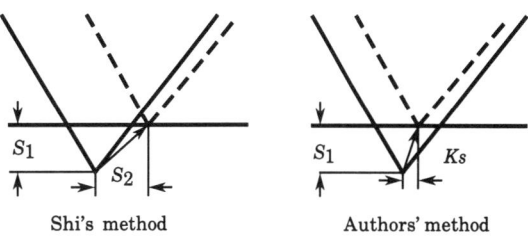

Figure 5. Criteria of the penalty method.

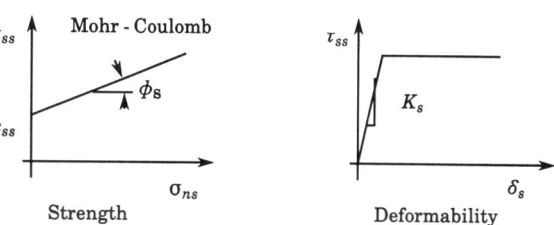

Figure 6. Friction condition of a block face.

Fig. 6 shows Mohr-Coulomb's friction conditions of a block face. The criteria of friction conditions of a block face are given by Eq. (2.43) and (2.44). In the case of

$\tau_{ss} < \sigma_{ns}\tan\phi_s + c_{ss}$,shear spring constant $= K_s$ $\cdot\cdot$ (2.43)
$\tau_{ss} \geq \sigma_{ns}\tan\phi_s + c_{ss}$,shear spring constant $=0$ $\cdot\cdot$ (2.44)

are applied.
where,
τ_{ss} : the shear stress of a block face,
σ_{ns} : the normal stress of a block face,
c_{ss} : the cohesion of a block face,
ϕ_s : the friction angle of a block face,
K_s : the shear stiffness of a block face.

2.7 Global governing equations and time integration

The equation of motion for dynamic problems is given by Eq.(2.45).

$$[M]\{\ddot{D}\} + [C]\{\dot{D}\} + [K]\{\Delta D\} = \{\Delta F\} \quad \cdots \cdots (2.45)$$

Rearranging Eq.(2.45) by displacement increment $\{\Delta D\}$,

$$[\tilde{K}]\{\Delta D\} = \{\tilde{F}\} \quad \cdots \cdots (2.46)$$

where,

$$[\tilde{K}] = \frac{2}{\Delta t^2}[M] + \frac{2}{\Delta t}[C] + [K] \quad \cdots \cdots (2.47)$$

$$\{\tilde{F}\} = \frac{2}{\Delta t}[M]\{\ddot{D}\} + \{\Delta F\} \qquad \cdots \cdots (2.48)$$

Δt : time step,

$[K]=[K_e]+[K_{\lambda R}]+[K_{\lambda C}]+[K_{\lambda P}]+[K_{PP}]$
: global stiffness matrix.

$\{\Delta F\}=\{F_{\sigma 0}\}+\{F_P\}+\{F_b\}+\{F_{\lambda C}\}+\{F_{\lambda P}\}$
: global external force vector,

$[K_e]$: the stiffness matrix of blocks,

$[K_{\lambda R}]$: the stiffness matrix of rockbolt,

$[K_{\lambda C}]$: the stiffness matrix of the bonding element,

$[K_{\lambda P}]$: the penetration matrix,

$[K_{PP}]=P[T_i]^T[T_i]$: the fixed point matrix,

$[M]= MS[T_i]$: the inertia force matrix,

M : the mass per unit area,

S: the area of a block,

$[C]= \eta\, S[T_i]$: the viscosity matrix,

η : the coefficient of viscosity,

$\{F_{\sigma 0}\}=S\{\sigma_0\}=S\{0\ 0\ 0\ \sigma_x\ \sigma_y\ \tau_{xy}\}$

$\{F_P\}=\{T_i\}^T\{f_x f_y\}^T$: the point loading vector,

f_x, f_y : the point load in x and y direction,

$\{F_b\}=\{f_x S\, f_y S\, 0\, 0\, 0\, 0\}^T$: the body force vector,

f_x, f_y : body force in x and y direction,

$\{F_{\lambda C}\}$:the external force vector of bonding,

$\{F_{\lambda P}\}$: the external force vector of penetrations.

$\frac{2}{\Delta t}[M]\{\ddot{D}\} = (\frac{2M}{\Delta t}(\iint[T_i]^T[T_i]dxdy)\{V_0\}$
: the inertia force vector,

$\{V_0\}$: initial velocity vector (only in case of dynamic analysis).

The velocity in the each time step is given by Eq. (2.49).

$\{V_i\} = \Delta t(\partial^2[D(t)]/\partial t^2)+\{V_{i-1}\} = (2/\Delta t)\{D_i\}-\{V_{i-1}\} \cdot \cdot (2.49)$

The global governing equation (2.45) is solved with small time steps of incremental forward differential equations. The total displacement vector $\{D\}$ is given by Eq.(2.50).

$$\{D\}= \Sigma\{\Delta D_i\} \qquad \cdots \cdots \cdots (2.50)$$

The iteration method is necessary when solving the penetration in case that d is not smaller than a certain value.

3 ELASTO-PLASTIC CONSTITUTIVE LAW OF BLOCKS

The authors introduced the elasto-plastic constitutive laws based on Drucker-Prager's yield criteria for a block shown in Fig. 7.

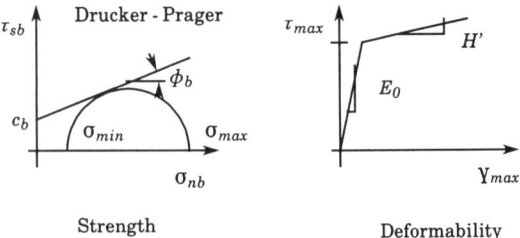

Figure 7. Strength and deformability of a block.

$$F = \alpha J_1 + \sqrt{J_2} - \bar{\sigma}/\sqrt{3} = 0 \qquad \cdots \cdots (3.1)$$

$$J_1 = \sigma_{ii}, \quad J_2 = S_{ij}S_{ij} \qquad \cdots \cdots \cdots (3.2)$$

$$\alpha = \sqrt{\frac{\sin^2\phi_b}{3(3+\sin^2\phi_b)}}, \quad \bar{\sigma} = c_b\{3(1-2\alpha^2)\} \quad \cdots (3.3)$$

where,

σ_{ii}: $\sigma_1+\sigma_2+\sigma_3$: the principal stresses,

S_{ij} : the deviatoric stresses,

J_1, J_2 : the first and second invariants of deviatoric stresses,

c_b: the cohesion of a body,

ϕ_b : the friction angle of a body.

The coordinates of blocks are updated at each time step. The stress increment with the rotation of the coordinate with block rotation Δr_0 at each time step is represented by Jaumann's co-rotational differential, hence, the objectivity of the stress is fulfilled. Then the stresses before and after the rotation are superposed by Eq.(3.4) and (3.5).

$$[\dot{\sigma}_{ij}] = [\Delta\sigma_{ij}]+[\sigma_{ik}][\Delta W_{kj}]+[\sigma_{jk}][\Delta W_{ki}] \quad \cdots (3.4)$$

$$[\sigma_{ij}]^{t+\Delta t} = [\sigma_{ij}]^t + [\dot{\sigma}_{ij}]\Delta t \qquad \cdots \cdots \cdots (3.5)$$

where,

$[\Delta W_{ij}]$: the rigid rotation increment tensor,

$[\dot{\sigma}]$: the stress rate tensor,

$[\Delta\sigma]$: the stress increment tensor.

This manipulation is important for the case of large spin motion on toppling analysis of slope for instance and the elasto-plastic analysis in which a block has the different principal axis of strain between each steps in particular.

4 APPLICATIONS AND DISCUSSIONS

In large scale rock foundations within the discontinuous planes existing in a rock mass, the behavior of deformation and stress distribution along structure foundations are strongly influenced by the pattern of the discontinuous planes. The presented methods are examined by using Boussinesq's analytical theory, and the experiment models of Maury (1970) and Gaziev(1971).

Table 1 given the material properties of the experiments. The applied external load is 4.17kgf/cm² for the 12cm width at the center of the upper part of the rigid block made by irons as shown in Fig. 8. Fig. 9～Fig. 12 show vertically and horizontally staggered models presented Maury(1970) through the use of 4cm ×4cm duralumin square rods. Here, in order to express the roller conditions along boundaries, the dummy slender blocks are attached along the boundaries in DDA models shown in Fig. 8. This is to cut the friction between inner blocks and fixed blocks of the outer boundaries.

Mohr-Coulomb's friction criterion is employed for the block surface. A large number penalty coefficient is applied to non-penetration conditions between blocks in the normal direction of the block surface. If shear stress of the block surface is greater than shear strength, the zero spring coefficient is employed for shear slip conditions.

Fig. 9 and Fig. 11 show the principal stress distribution of the continuum models in which the tensile strength is given for the friction conditions of a block surface. Fig. 10 and Fig. 12 show the principal stress distribution of discrete model of Model-A and Model-B.

Fig. 13 shows the vertical stress distribution along bottom boundary blocks of the vertically staggered model (Model-A, Discrete) and the result of Maury's experiment. Fig. 14 also shows the vertical stress distribution along bottom boundary blocks of the horizontally staggered model (Model-B, Discrete)

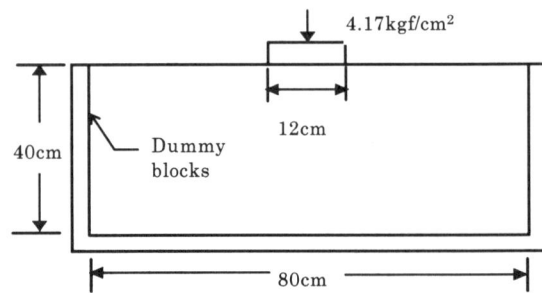

Figure 8. The boundary and loading conditions of DDA model.

Table 1. Material properties of the block model test(Maury,1970).

Components	Properties
Young's modulus	$7 \times 10^5 \text{kgf/cm}^2$
Poisson's ratio	0.34
cohesion of surface	0 kgf/cm^2
friction angle	40°
cohesion of body	elastic
internal friction angle	elastic

and the result of Maury's experiment. The vertical stress distribution of Model-B is concentrated greater than Model-A near the center of analytical region. The numerical results agree with the experiments approximately.

Fig. 15 shows comparison of the vertical stress distribution at the center of loading point for Model-A, Model-B and Boussinesq's analytical theory. The stress concentration of the numerical solutions (DDA) is smaller than analytical theory at the center of loading point. Thus, the reason of different stress distribution seems to be the fundamental difference between the discontinuum and continuum models, and the infinite region is assumed in the analytical theory.

Fig. 16 and Fig. 17 show the principal stress distribution of the 45° inclined model (Model-C) of the continuum model and the

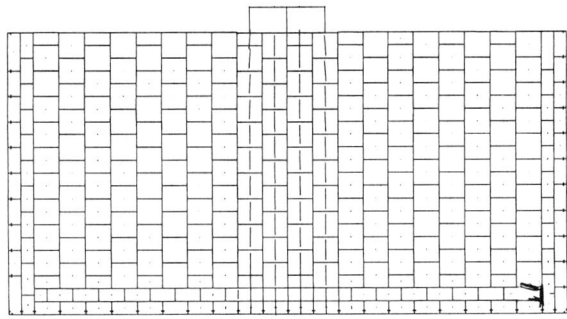

Figure 12. The horizontally staggered model(Model-B, Discrete).

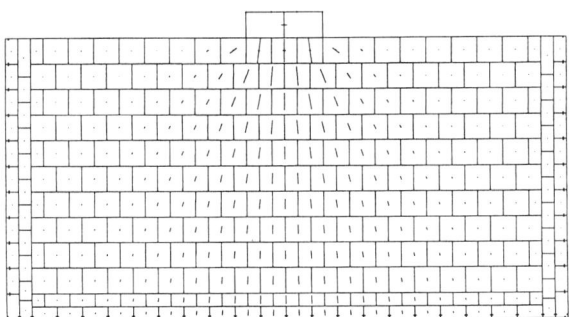

Figure 9. The vertically staggered model(Model-A, Continuum).

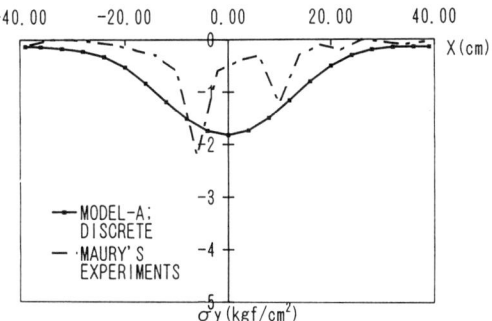

Figure 13. The vertical stress distribution along bottom boundary blocks (Model-A vs. Maury's experiments).

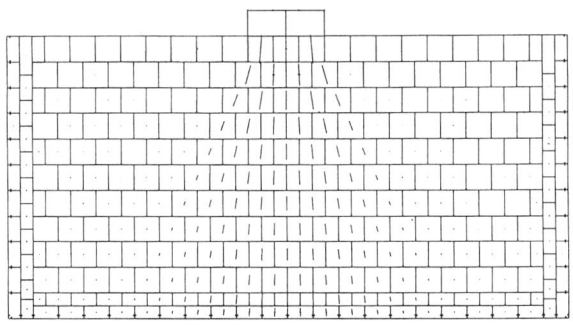

Figure 10. The horizontally staggered model(Model-A, Discrete).

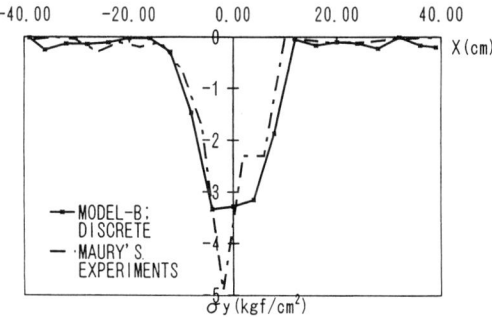

Figure 14. The vertical stress distribution along bottom boundary blocks (Model-B vs. Maury's experiments).

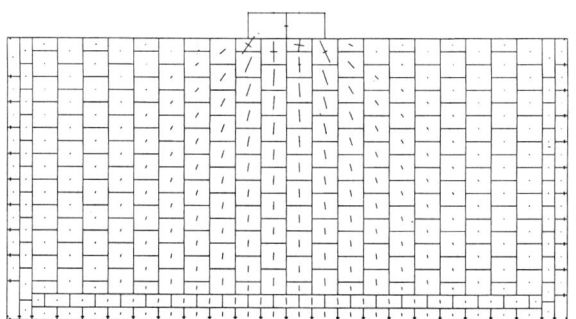

Figure 11. The vertically staggered model(Model-B, Continuum).

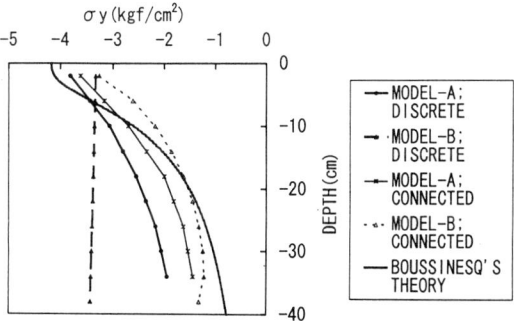

Figure 15. The vertical stress distribution along center of region.

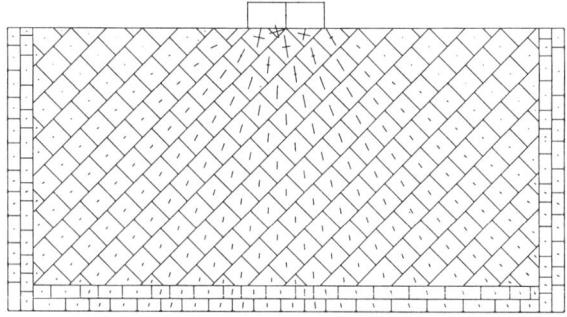

Figure 16. The 45° inclined model (Model-C, Continuum).

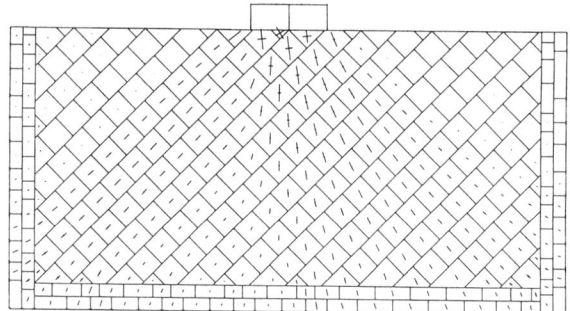

Figure 17. The 45° inclined model (Model-C, Discrete).

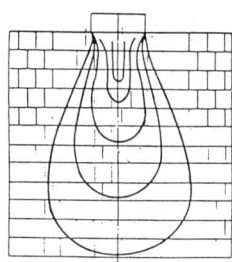 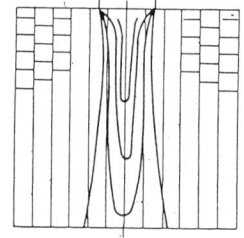

The vertically staggered model   The horizontally staggered model

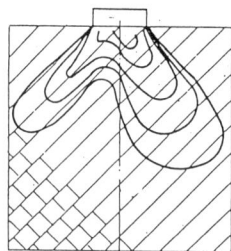 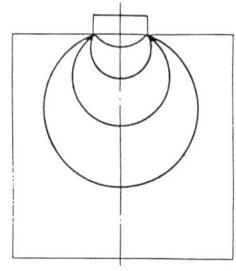

The 45° inclined model   The elastic model

Figure 18. Maximum principal stress contours of Gaziev's experiment results(1971).

discrete model presented by Gaziev(1971) shown in Fig. 18.

Fig. 18 shows the maximum principal stress contours of Gaziev's experimental results(1971) and the numerical results obtained by DDA agreed with those experiments.

5 CONCLUSIONS

The authors improved penalty method of DDA (Shi,1984) and developed both the elasto-plastic constitutive law and connected element. The applicability of the presented method was examined for the image of large scale rock foundations with jointed rock model. The results were also compared with the results of analytical theory, Maury's and Gaziev's experiments.

DDA models are extremely modeled as the discontinuum in jointed rock mass. The deformation of DDA depends strongly on the friction criteria of its block surfaces. The presented method is applicable for the discontinuum model through its controlling of friction criteria of block surfaces. Therefore, the investigations of friction mechanisms of block surface is important for practical use.

ACKNOWLEDGMENTS

The authors would like to acknowledge the assistance of Dr. G. H. Shi for the area of theory and his many informative discussions.

REFERENCES

Gaziev, E. G.& S. A. Erlikhman 1971. Stresses and strains in anisotropic rock foundation (model studies), Symposium Soc. Int. Mec. des Roches, Nancy, II-1.

Maury, V. 1970. Mécanique des milieux stratifiés, experiences et calculs, DUNDO, Paris.

Sasaki, T., Ohnishi, Y. & R. Yoshinaka 1994. Discontinuous deformation analysis and its application to rock mechanics problems. Prc. JSCE. 493/III-27 : 11-20 (in Japanese).

Shi, G. H. & R. E. Goodman 1984. Discontinuous deformation analysis. In C. H. Dowding & M. M. Singh (eds), Proc. 25th U. S. Symp. on Rock Mech. Evanstone, AIME : 269-277.

Shi, G. H. & R. E. Goodman 1985. Two dimensional discontinuous deformation analysis. Int. Journal Anal. Methods Geomech. 9 : 541-556.

Shi, G. H. & R. E. Goodman 1985. Discontinuous deformation analysis - A new method for computing stress, strain and sliding of block systems. In P. A. Cundall, R. L. Sterling & A. M. Starfield (eds), Proc. 29th U. S. Symp. on Rock Mech. 381-393. Minneapolis : Balkema.

Washizu, K. 1972. Variational principle of elasticity, JSSC, Baihuukan (in Japanese).

Stability analysis of jointed rock foundations by FEM using the multiple yield models

T. Sasaki, S. Morikawa & T. Matsukawa
Kajima Corporation, Tokyo, Japan

R. Yoshinaka
Saitama University, Japan

ABSTRACT : This paper describes the equivalent continuum modeling of deformabilities and the strength of jointed rock mass by finite element method. The authors classified the discontinuous planes of the cracks and joints in rock mass into the two types of non-oriented small cracks and oriented joint sets as presented by Yoshinaka et al. (1990). They proposed the elasto-plastic deformabilities and yield criteria, which is termed the multiple yield models, of the rock mass with non-oriented micro cracks and combined arbitrary oriented joint sets based on the compliance matrix method (Singh, 1973). The applicabilites of the presented method is examined using the stability analysis of the jointed rock foundation models presented by Maury(1970) and Gaziev(1971).

1 INTRODUCTION

The discontinuous planes of cracks and joints in rock mass are given important influences in the evaluation of local stabilities of rock foundations. They are also given important influence in the evaluation of the local stability of rock foundations, slopes and caverns during excavations. The factor of the proportions for the amounted deformation around excavated surface can be estimated by results of the in-situ test and measurements. The factors for the total quantity of deformation by the opening of cracks and the slip of joints in rock mass are large compared with the elastic deformation of intact rock.

The authors classified the discontinuous planes of the cracks and joints in rock mass into the two types of non-oriented small cracks and oriented joint sets (Yoshinaka et al. 1990) and assumed that the total deformation of rock mass is expressed by compliance (Hill,1963, Singh, 1973) as the superposing of the elastic deformation of intact rock, the opening of non-oriented small cracks, and the slip and opening of oriented joint sets.

The yield conditions of the rock mass are introduced from one of the three types of yield surfaces as the intact rock, the non-oriented cracks and the oriented joint sets in which the yield conditions employed satisfy the fastest one in those three modes, hence, this yield criteria is called the multiple yield models (Sasaki et al. 1994). The applicability of the presented method is examined using rock foundation models.

2 DEFORMABILITIES OF JOINTED ROCK MASS

Fig. 1 shows the concept of the stress change for near the excavated rock surface. The unloading condition is introduced for the perpendicular direction of the excavated surface and the loading condition is introduced for the parallel direction at the excavated surface. Fig. 2 shows the concept of the in-situ plate loading test and the modified simple relation of load-displacement. The reason that the slope at the start of loading and the final part of un-loading is smaller than the central part at the middle load points is that the slip, opening and closing of cracks and joints in the rock mass are induced by the stress redistribution with the actual tunnel scale excavation.

The reason that the slope of the first loading and the final part

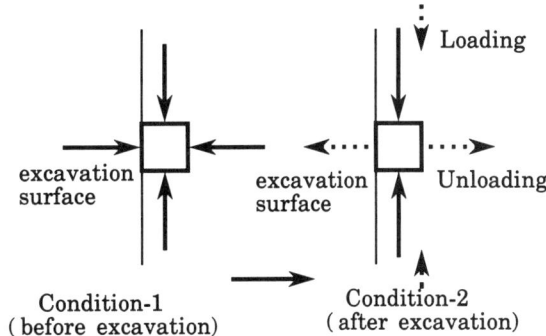

Figure 1. The stress change before and after excavated surface.

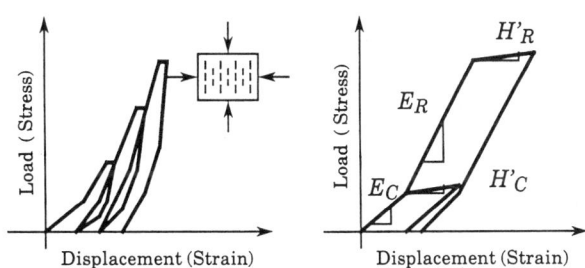

Figure 2. The load-displacement relation of plate loading test.

of un-loading in multiple loading tests is smaller than the slope of central part of loading in particular is that the closing of opening cracks and joints up to a certain stress level is occurred. Thus, cracks and joints in the rock mass are opening on the un-loading process, hence, the slip of the joints and yield of the intact rock can be considered by increasing shear stresses. The authors assumed that the discontinuous planes in rock mass can be classified into the two types of non-oriented micro cracks and oriented joint sets. And, the total deformation of rock mass can be expressed as the summation of the following component as shown in Fig. 3.

① the elastic deformation of the intact rock,
② the opening of non-oriented small cracks,
③ the slip and opening of oriented joint sets.

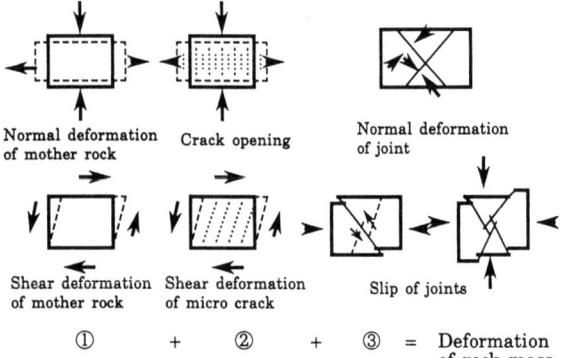

Figure 3. The components of deformation of a rock mass.

Fig. 4 shows a concept of stress-strain relations of the triaxial compression test. In the case of rock core or intact rock, there are no actualized cracks and discontinuous planes which are the governing deformation characteristics of rock mass, hence, the isotropic deformability is assumed by the experimental results under certain or higher confining pressures.

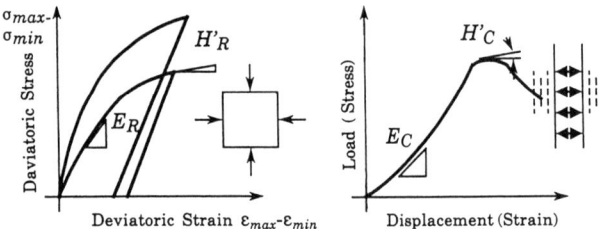

Figure 4. The deformation of the intact rock.

Figure 5. The deformation characteristics of the borehole loading test.

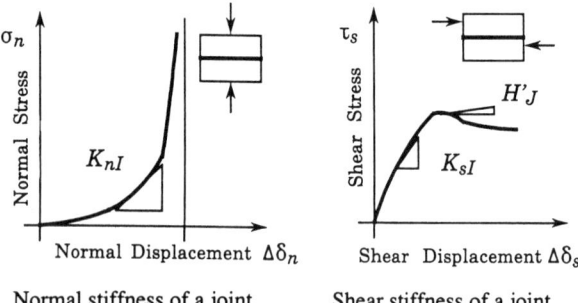

Normal stiffness of a joint Shear stiffness of a joint

Figure 6. The deformation characteristics of a joint.

Fig. 5 shows the conceptual relation of the load-deformation of the borehole loading test. It can be estimated that the non oriented micro cracks are caused by stress release around the borehole face of the circular directions. Fig. 6 shows the conceptual relation of load-normal and shear deformation of a joint. In this case, the shear stiffness is smaller than the normal stiffness of a joint, hence, the influence of shear deformation of a joint on the total rock deformation can be estimated very large. Fig. 7 shows the serial models of the deformation factors in elasto-plastic as the multiple yield models. In the case of jointed rock mass, it can be expressed by the combination of (1)～(5) in Fig. 7. The yield criteria of the jointed rock mass is assumed to be the first mode of

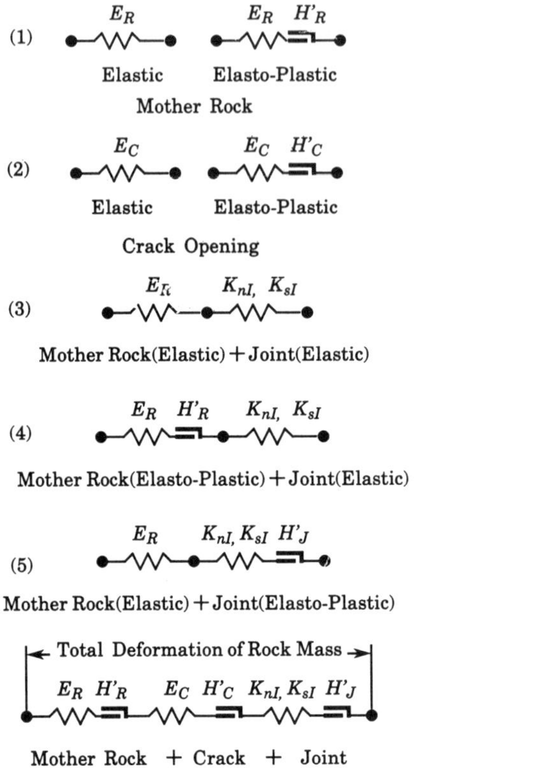

(1) E_R Elastic E_R H'_R Elasto-Plastic
Mother Rock

(2) E_C Elastic E_C H'_C Elasto-Plastic
Crack Opening

(3) E_R K_{nI}, K_{sI}
Mother Rock(Elastic) + Joint(Elastic)

(4) E_R H'_R K_{nI}, K_{sI}
Mother Rock(Elasto-Plastic) + Joint(Elastic)

(5) E_R K_{nI}, K_{sI} H'_J
Mother Rock(Elastic) + Joint(Elasto-Plastic)

Total Deformation of Rock Mass
E_R H'_R E_C H'_C K_{nI}, K_{sI} H'_J
Mother Rock + Crack + Joint

Figure 7. The multiple yield models (Sasaki et al., 1994).

the yield conditions of (①～③) as described before is introduced.

3 MULTIPLE YIELD MODELS

As shown in Fig. 7, in the multiple yield models (Sasaki et al. 1994), the strength and the deformability are defined for the three types of intact rock, non-oriented micro cracks and oriented joint sets. The above assumptions are introduced in elasto-plastic two dimensional plane strain constitutive law as follows.

3.1 Strength and deformability of intact rock

Fig. 8 shows the model of deformation characteristics for a rock mass containing non-oriented micro cracks. The deformability of rock mass is assumed to be isotropic, and the rock mass has initial deformability E_R under the stress conditions which are satisfied between a certain seating stress (S_S) and the yield stress (condition-1). The yield condition adopted the Drucker-Prager's yield criterion expressed by Eq.(1) is shown in Fig. 9.

$$F = \alpha J_1 + J_2^{1/2} - \overline{\sigma}/\sqrt{3} = 0 \quad \cdots \quad (1)$$

$$J_1 = \sigma_{ii}, \quad J_2 = S_{ij}S_{ij}, \quad \overline{\sigma} = C_R(3(1-2\alpha^2)) \quad \cdots (2)$$

$$\alpha = \sqrt{\frac{\sin^2\phi_R}{3(3+\sin^2\phi_R)}} \quad \cdots \cdots (3)$$

where,

σ_{ii} : $\sigma_1 + \sigma_2 + \sigma_3$: the principal stresses,
S_{ij} : the deviatoric stresses,
J_1, J_2: the first and second invariants of deviatoric stress,
C_R : a cohesion of the body,
ϕ_R : a friction angle of the body.

The hardening coefficient H'_R as the gradient of the stress-plastic strains curve is adopted as the final gradient of the stress-strain relation of the triaxial compression tests approximately.

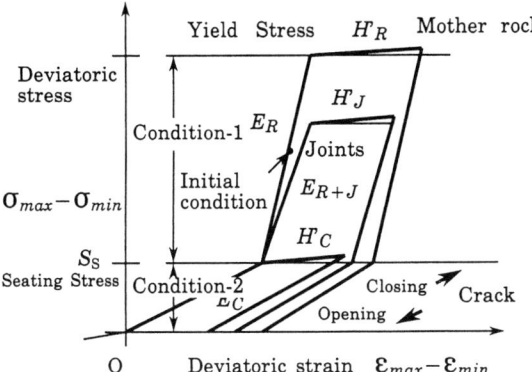

Figure 8. The deformation characteristics of a intact rock mass with non-oriented micro cracks.

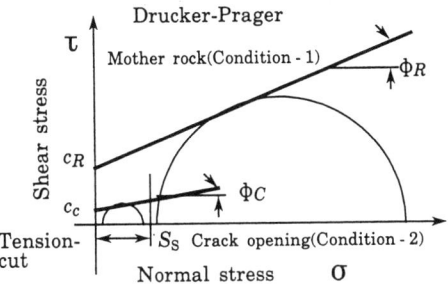

Figure 9. The strength of a intact rock mass with non-oriented micro cracks.

3.2 Yield condition of tension stress

A tension-cut yield condition is employed for the tension stress in the direction of the opening of cracks and joints as Eq. (4).

$$F_t = S_1 - S_t = 0 \qquad (4)$$
$$S_1 = (\sigma_x + \sigma_y)/2 + \sqrt{(\sigma_x - \sigma_y)^2/4 + \tau_{xy}^2} = \sigma_n \qquad (5)$$

where,
σ_n: the normal stress of cracks or joints,
S_t: the tension strength of cracks or joints.
The associated flow rule in which the plastic potential assumed to the same function as the yield surface is employed.

3.3 Strength and deformability of micro cracks

The opening of the micro cracks in rock mass is assumed under the condition of the minimum principal stress which is smaller than a certain stress S_S (seating stress) as shown in condition-2 of Fig. 8. The modulus of deformation E_2 is employed as the gradient of the borehole loading test at the first loading deformation on the monotony loading and is smaller than the modulus of deformation of the mother rock.

The strength is assumed to be isotropic and based on Drucker-Prager's yield condition similar to the intact rock of condition -2 as shown in Fig. 9. The hardening coefficient H'_C as the final gradient of the stress-plastic strain relation is employed approximately from the borehole loading test same as for the intact rock. In this case, when the micro cracks are open and the stress satisfies the yield condition, the quantity of crack opening is defined approximately by Eq. (6) which is presented by Oñate et al. (1988) as shown in Fig. 10. In this case, the direction of the opening of micro cracks coincides with the direction of the minimum principal stress. The apparent quantity of the micro crack opening is assumed that the plastic strain of the direction of opening is expressed by multiplying the plastic strain by the representative width B of the element.

$$\{\Delta\varepsilon^{cr}\} = B[T]\{\Delta\varepsilon_P\} \qquad (6)$$
$$\{\Delta\varepsilon^{cr}\} = \{\Delta\varepsilon_n^{cr}, \Delta\varepsilon_t^{cr}, \Delta\varepsilon_{nt}^{cr}\}^T \qquad (7)$$
$$\{\Delta\varepsilon_P\} = \{\Delta\varepsilon_x^P, \Delta\varepsilon_y^P, \Delta\varepsilon_{xy}^P\}^T \qquad (8)$$

where,
$\{\Delta\varepsilon^{cr}\}$: the mean quantity of opening of cracks of an element,
$[T]$: the direction cosine matrix of cracks,

$$[T] = \begin{bmatrix} \cos^2\theta & \sin^2\theta & \sin 2\theta/2 \\ \sin^2\theta & \cos^2\theta & -\sin 2\theta/2 \\ -\sin 2\theta & \sin 2\theta & \cos 2\theta \end{bmatrix} \qquad (9)$$

$\{\Delta\varepsilon_P\} = \{\Delta\varepsilon_T\} - \{\Delta\varepsilon_e\}$: the plastic strain increment vector of an element,
$\{\Delta\varepsilon_e\} = [D_e]^{-1}\{\Delta\sigma\}$: the elastic strain increment vector of an element,
$\{\Delta\varepsilon_T\}$: the total increment strain vector of an element,
B: the representative width in an element,
θ: the angle of the minimum principal stress from the x axis,
n: the perpendicular direction of cracks,
t: the parallel direction of cracks,
nt: the shear direction of cracks.

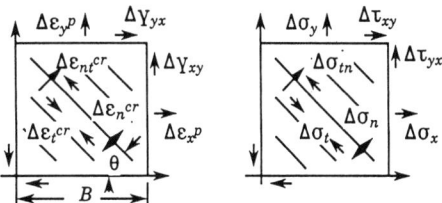

Figure 10. The apparent quantity of a micro crack opening.

3.4 Deformability of joint sets

The characteristics of rock mass deformation with joint sets are expressed by introducing the compliance matrix presented by Hill (1963) and Singh (1973). The total strain of the rock mass contained in combined arbitrary oriented joint sets is expressed by the summation of strains of the mother rock and each joint set as shown in Fig. 11.

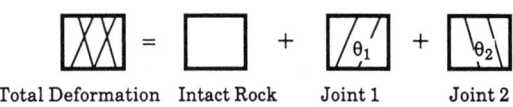

Figure 11. The deformation of a rock mass with joint sets.

Thus, it is assumed that the joints are distributed periodically and the volume of each joint set is ignored in comparison with the volume of mother rock. And it is assumed that the stresses of the mother rock and joints coincide. The stiffness matrix of joint set No. I in the local coordinate system is transformed for the global coordinate system by using the coordinates transformation matrix of Eq.(10).

$$[F_I] = [T_I]^T [C_I][T_I] \qquad (10)$$

The stiffness matrix of joint set I in the local coordinate system is expressed by Eq. (11).

$$[C_I] = [K_I]^{-1}/S_I \qquad (11)$$

The total strain vector with joint sets on the global coordinate system is expressed by Eq. (12).

$$\{\varepsilon_T\} = \sum\{\varepsilon_I\} + \{\varepsilon_R\} = [\sum[F_I] + [E]]\{\sigma\} \qquad (12)$$

The total compliance matrix of the rock mass with joint sets in the global coordinate system is expressed by Eq. (13).

$$[C]=[D]^{-1} \quad\quad\quad\quad (13)$$

where,
$\{\varepsilon_I\}$: the strain vector of the joint set No. I,
$\{\varepsilon_T\}$: the total strain vector of the rock mass,
$\{\varepsilon_T\} = \{\varepsilon_{xT}, \varepsilon_{yT}, \varepsilon_{xyT}\}^T$, $\{\sigma\} = \{\sigma_x, \sigma_y, \sigma_{xy}\}^T$
$[E]$: the compliance matrix of the mother rock,
$[C]$: the stress-strain constitutive law of the rock mass with joint sets,
$[T_I]$: the coordinates transformation matrix of the joint set No. I,

$$[T] = \begin{bmatrix} \sin^2\theta & \cos^2\theta & -2\sin\theta\cos\theta \\ -\sin\theta\cos\theta & \sin\theta\cos\theta & \cos^2\theta - \sin^2\theta \end{bmatrix} \quad\quad (14)$$

$[K_I]$: the stiffness matrix of the joint set No. I,

$$[K_I] = \begin{bmatrix} K_{nI} & 0 \\ 0 & K_{sI} \end{bmatrix} \quad\quad\quad\quad (15)$$

S_I : the spacing of the joint set No. I,
θ_I : the angle of the joint set No. I in the counter-clockwise direction from the x-axis.

The normal stiffness K_{nI} and shear stiffness K_{sI} of the joint set No. I is provided by assuming the imaginary joint width, hence, those strains correspond one to one in the experiments as shown in Fig. 6.

The stress vector of the joint set No.I in the local coordinates system is expressed by Eq.(16).

$$\{\sigma_I\} = [T_I]\{\sigma\} \quad\quad\quad\quad (16)$$

The strain vector of the joint set No. I in the local coordinates system is expressed by Eq.(17).

$$\{\varepsilon_I\} = [C_I][T_I]\{\sigma\} = [C_I]\{\sigma_I\} \quad\quad (17)$$

These stress components are expressed by Eq.(18).

$$\{\sigma_I\} = \{\sigma_{nI}, \tau_{sI}\}^T \quad\quad\quad\quad (18)$$

The strain components are expressed by Eq.(19)

$$\{\varepsilon_I\} = \{\varepsilon_{nI}, \gamma_{sI}\}^T \quad\quad\quad\quad (19)$$

where,
σ_{nI} : the normal stress of the joint set No. I,
τ_{sI} : the shear stress of the joint set No. I,
ε_{nI} : the normal strain of the joint set No. I,
γ_{sI} : the shear strain of the joint set No. I.

The stress and strain of the each joint set are determined directly from the above definitions.

3.5 Yield condition of joint

The Mohr-Coulomb's yield condition defined by Eq.(20) is employed for the arbitrary oriented joint sets and shown in Fig. 12.

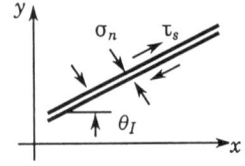
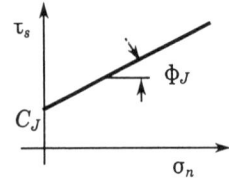

Figure 12. Strength of a joint.

$$F_s = |\tau_s| - (C - \sigma_n \tan\phi_J) \quad\quad\quad\quad (20)$$

$$\tau_s = \tau_{xy}\cos 2\theta_I - ((\sigma_x - \sigma_y)\sin 2\theta_I)/2 \quad\quad (21)$$

$$\sigma_n = (\sigma_x + \sigma_y)/2 - ((\sigma_x - \sigma_y)\cos 2\theta_I)/2 - \tau_{xy}\sin 2\theta_I \quad (22)$$

where,
τ_s : the shear stress of the joint,
σ_n : the normal stress of the joint,
ϕ_J : the friction angle of the joint.

When the stress conditions of a joint is in plastic, the hardening coefficient H'_J as the final gradient of the stress-shear deformation relation is employed as shown in Fig. 6 (condition-3).

If plural joints exist in the rock mass, one of the yield mode of them is employed in which the stress conditions of yield is satisfied faster than the others in the element.

4 APPLICATIONS AND DISCUSSIONS

In order to examine the applicability of the presented method, the authors carried out the stability analysis of the jointed rock foundation models presented by Maury(1970) and Gaziev(1971). Maury used a 4cm×4cm duralumin blocks and Gaziev used plaster blocks for the test materials. However, the geometrical shapes of the Gaziev's test is unclear in his paper. Hence, in this research, the geometrical dimensions in the numerical models is employed from Maury's test and the block patterns are employed from Gaziev's test. The analytical conditions and finite element mesh are shown in Fig. 13 presented by Maury (1970).

The materiel properties of the models are given in Table 1. The normal and the shear spring constants of the staggered block patterns was determined by Singh's (1973) investigations.

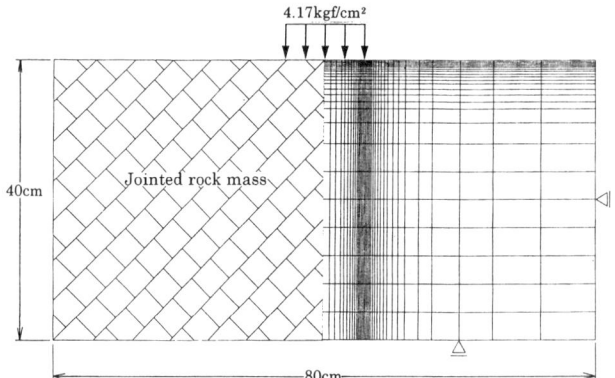

Figure 13. Analytical conditions and FEM mesh.

Table 1. Material properties of blocks (Maury, 1970).

Components	Properties
Young's modulus of blocks	7×10^5 kgf/cm^2
Poisson's ratio of blocks	0.34
Cohesion of block surfaces	0 kgf/cm^2
Friction angle of block surfaces	40°
Normal stiffness of joints	$K_n = 7 \times 10^5$ kgf/cm^3
Shear stiffness of joints	$K_s = 7 \times 10^3$ kgf/cm^3
Stress concentration factor for normal stress	$B_n = 1.0$
Stress concentration factor for shear stress	$B_s = 13.5$

Fig. 14 shows the maximum principal stress contours of Gaziev's experiment results. The numerical models of the isotropic, $\alpha =0°$ (vertically staggered), $\alpha =30°$, $\alpha =45°$, $\alpha =60°$ and $\alpha =90°$ (horizontally staggered) models are also analyzed as Gaziev's experiments, here, α is the angles of the block patterns of counter-clock wise from x axis shown in Fig. 14. Fig. 15~Fig. 20 show the maximum principal stress contours of the numerical results of the analysis in which the material properties of the mother rock and the joints are assumed to be elastic, respectively. By comparing these with the Gaziev's test results shown in Fig. 14, it can be seen that the overall match is good except $\alpha =45°$ model.

The reason that the numerical results of the $\alpha=45°$ model are different from the test results that the characteristics of the anisotropy of the element stiffness are symmetric for the analytical region and the loading conditions in the case of $\alpha=45°$ model, hence, the contours of the numerical result is symmetric as shown in Fig. 18.

Fig. 21~Fig. 25 show the plastic zone of the each models using the Mohr-Coulomb's yield criteria of the joint. In this case, the mother rock is elastic and the yield criteria applied only for the joint stresses in the element. From these results, it can be seen that the plastic zone expansion differs depending on the direction of the joint set angles and configuration. Fig. 26 show the reaction forces along the downward boundary of the numerical results. Looking at these, it can be seen that they are qualitatively well matched. The greatest reaction force at the center of the load points is the $\alpha=90°$ model, the smallest one is the isotropic model and the $\alpha=0°$ model is intermediate. The $\alpha=30°$ model and the $\alpha=60°$ model are symmetric to each other with respect to the center line.

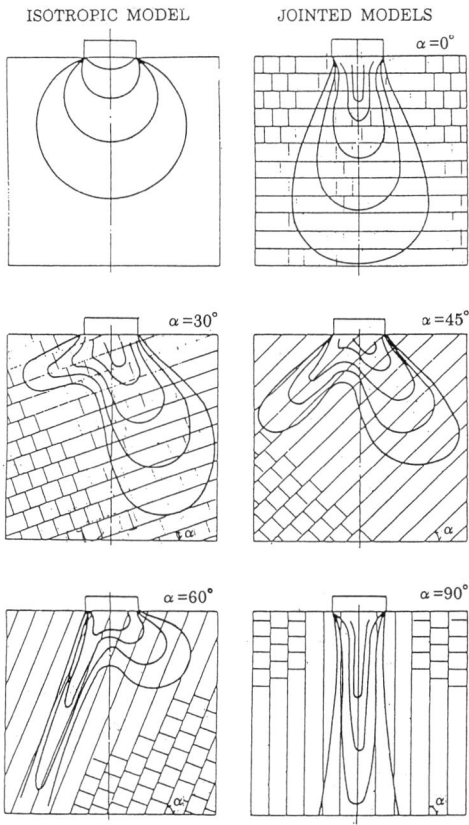

Figure 14. Maximum principal stress contour of Gaziev's experiment results(1971).

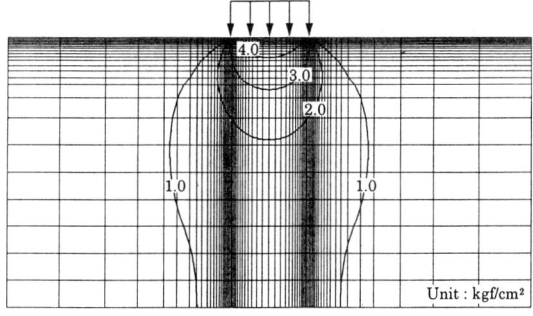

Figure 15. Maximum principal stress contour of isotropic model.

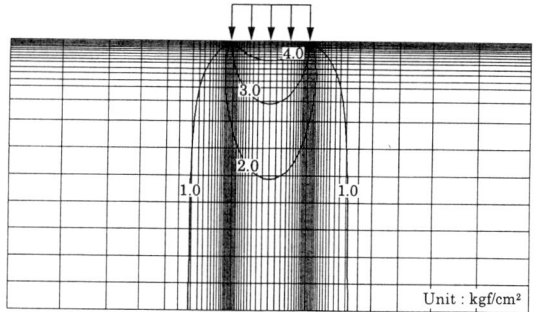

Figure 16. Maximum principal stress contour of $\alpha=0°$ model.

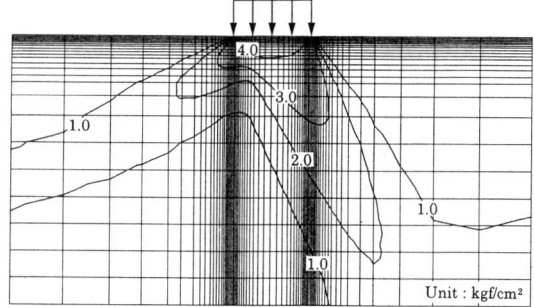

Figure 17. Maximum principal stress contour of $\alpha=30°$ model.

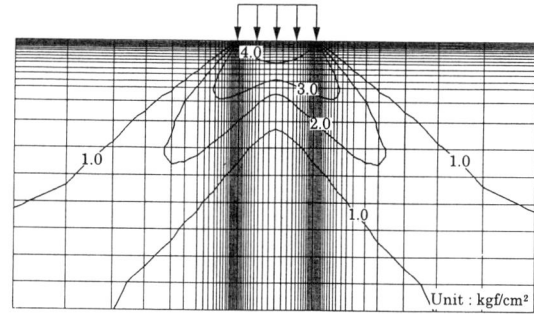

Figure 18. Maximum principal stress contour of $\alpha=45°$ model.

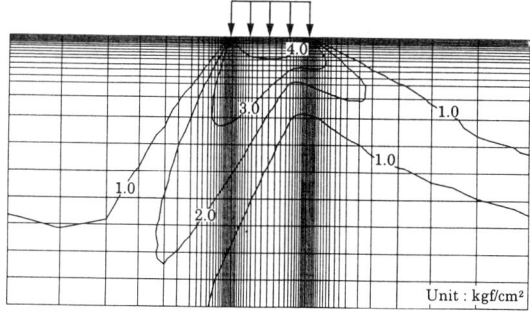

Figure 19. Maximum principal stress contour of $\alpha=60°$ model.

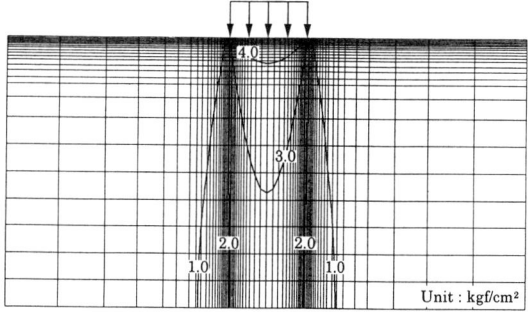

Figure 20. Maximum principal stress contour of $\alpha=90°$ model.

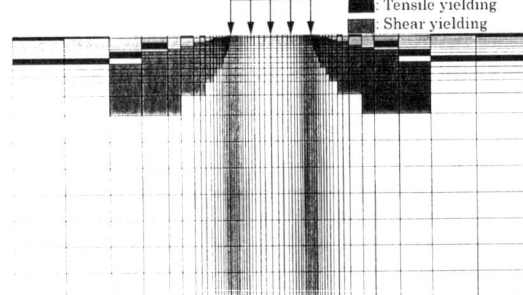

Figure 21. Plastic zone of $\alpha = 0°$ model.

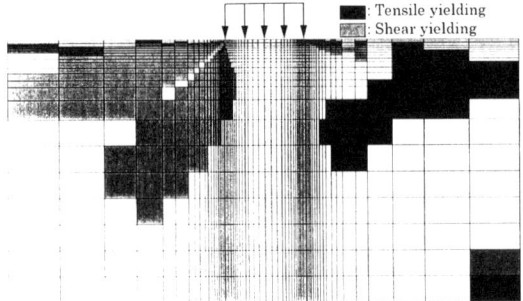

Figure 22. Plastic zone of $\alpha = 30°$ model.

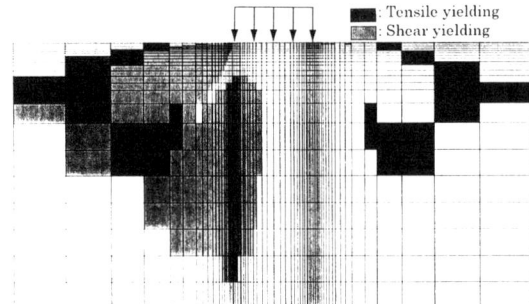

Figure 23. Plastic zone of $\alpha = 45°$ model.

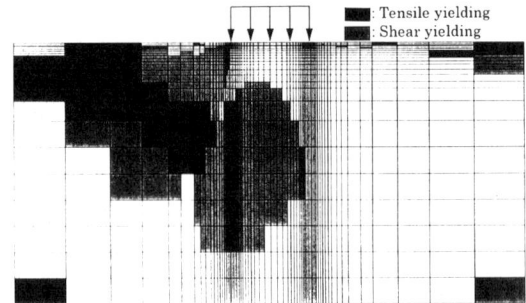

Figure 24. Plastic zone of $\alpha = 60°$ model.

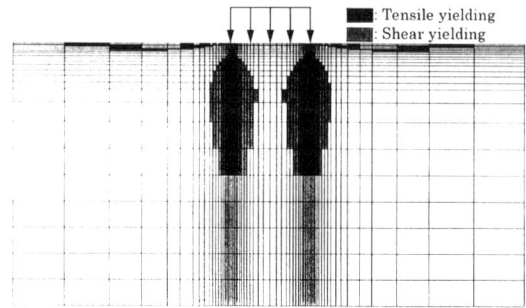

Figure 25. Plastic zone of $\alpha = 90°$ model.

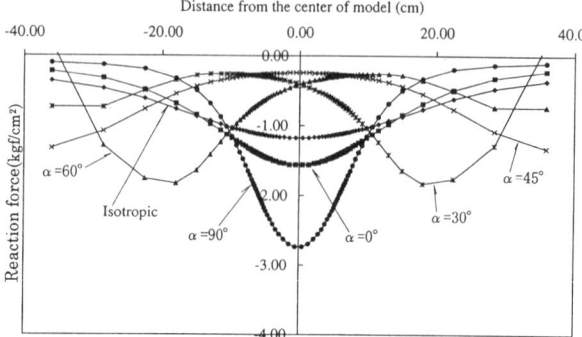

Figure 26. Reaction stresses along the bottom boundary.

5 CONCLUSIONS

The authors presented the equivalent continuum model of FEM by using compliance matrix method for the jointed rock mass. And they also presented the multiple yield models in which the three yield modes as the intact rock mass, the non-oriented micro cracks and the oriented joint sets are employed.

The presented methods are examined by using the Maury's and Gaziev's experiment results which simulate jointed rock mass. The results showed that this method and the tests matched well and that this method is useful for analyzing these types of jointed rock mass. In actuality, however, the jointed rock mass in general is assumed to be more partially and doubtfully connected than completely discontinuous. In such a case, it is advantageous to use the equivalent continuum model to be analyzed with just one mesh by changing the input properties.

Therefore, the spring constants used in the discontinuous model must be skillfully set. Whether used for continuous or discontinuous models, these values are major factors in the largest fluctuation in results. Hence, a method must be developed that finds the discontinuous model spring constants for actual rock mass. To accomplish this, many tests and measurements must be performed and these must be analyzed to accumulate experience. The research in this paper is but one proposal for an analysis model, and its applicability to and usefulness for actual rock mass must be confirmed in future.

REFERENCES

Gaziev, E. G. & S. A. Erlikhman 1971. Stresses and strains in anisotropic rock foundation (model studies), Symposium Soc. Int. Mec. des Roches, Nancy, II-1.

Hill, R. 1963. Elastic properties of reinforced solids: Some theoretical principles, J. Mech., Phys. Solids : 357-372.

Maury, V. 1970. Mécanique des milieux stratifiés, experiences et calculs, DUNDO, Paris.

Oñate, E. ,Oller, S., Oliver, J. & Lubliner, J. 1988. A constitutive model for cracking of concrete based on the incremental theory of plasticity, Eng. Comput., Vol. 5 : 309-319.

Sasaki, T., Yoshinaka, R. & F. Nagai, 1994. A study of the multiple yield models on jointed rock mass by finite element method, Proc. of JSCE, No. 505/III-29 : 59-68.

Singh, B. 1973. Continuum characterization of jointed rock mass, Part I - The constitutive equations, Int. J. Rock Mech. Min. Sci., 10 : 311-335.

Yoshinaka, R., Yoshida, J. & T. Yamabe, 1990. Studies of the geometricity of discontinuous planes of rock and its modeling, Proc. JSSMFE, Vol. 30, No. 3 : 161-173.

In-situ seepage flow tests on jointed rock mass and its analysis

Kohkiohi Kikuchi & Yoshitada Mito
Kyoto University, Japan

Masao Nakada
Mitsui Construction, Co., Ltd, Tokyo, Japan

ABSTRUCT: The author had carried out 3 series of in situ permeability tests and examined the characteristic of seepage flow in jointed rock masses in the past study. In this study, in situ experiment has also been carried out and the test results are analyzed using the deterministic model and stochastic model of joint network. As the results, the suggested analyzing method is applicable to simulate the actual seepage flow in joints.

1. FOREWORD

Since permeability of hard rock masses is strongly affected by rock joints, it is essential to clarify the water flow in joints in order to clarify the seepage flow in rock masses.

In order to analyze seepage flow of rock masses, it is necessary to grasp ① joint distribution and ② characteristics of flow through joints. For the point ① some authors have been examining the modelling methods. Concerning the point ②, the characteristics of flow through joints are assumed that of parallel plate flow without the influential knowledgeas analyzing groundwater flow.

The authors had carried out the following 3 serieses of in situ seepage flow tests in order to know the characteristics of flow through actual joints[1].

a) Experiment on the flow through "super conductor" joint system
b) Experiment on the flow through single joint
c) Experiment on the flow through joint system

Based on the results in such 3 experiments, the authors suggested the 3-D analytical method for seepage flow using the 3-D probabilistic joint distribution model (disc shaped geometric model), by taking "channeling" phenomenon into account. The analytical results fitted the experimental result very well.

In this study, the applicability of the seepage flow analysis mentioned above is examined using the data obtained from the in situ experiment which is carried out in granitic rock.

2. THE PAST EXPERIMENTS

For engineering purpose, it is essential to clarify the water flow in joints in order to clarify the seepage flow in rock masses. Therefore, it is necessary to carry out a seepage flow analysis by means of the analytical model sufficiently considering a) Geometry of joint distribution and b)Characteristics of flow through joints.

The authors had carrried out 3 serieses of in situ seepage flow tests in order to know characteristics of actual flow through joints, and simulated the tests using the results and joint geometrical distribution.

2.1 Experiment on super-conductor (TEST 1)

Figure 1 shows the schematics of the TEST 1. The super-conductor joint system which strongly controls groundwater flow of the site was existing as shown in Figure 1.

At first, constant head permeability test was carried out using the borehole. The flux intothe super-conductor was measured and the flux out of the super-conductor was measured at 8 sectional part along the joint trace. Darcy's low is effective for the flux into and out of the joints and partial flux out of the joints. The flux into joint is nearly equal to the total flux out of joints.

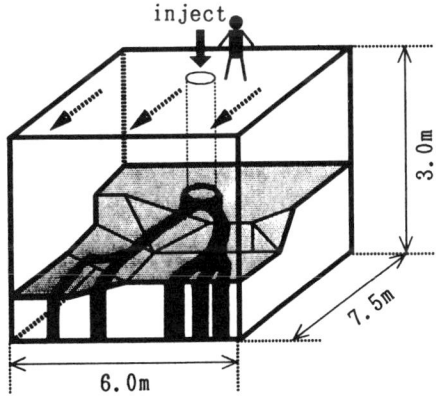

Constant head permeability test

Paths survey

Fig.1 Schematics of TEST 1
Experiment on "super-conductor"

Next, After injecting the grout milk from the borehole, the upper part of the rock mass on the superconductor was removed and the distribution and thickness of grout lamps was surveyed. Figure 2 shows the estimated paths through the joints. The ratio of area of paths to area of joint plane is distributed as shown in Figure 3.

Finally, the simulation of the permeability test was carried out. Figure 4 shows the segment element model of paths through joints. As the simple model expressed the flow of each element, the assembled pipe flow model is asuumed as shown in Figure 5. In the case that Poiseuille flow is assumed, the mean velocity in a pipe, v_p is given by;

$$v_p = \frac{g\, t_p^2}{32\nu} i \qquad (1)$$

where t_p ; joint aperture
g ; accelerated gravity
ν ; coefficient of kinematic viscosity
i ; hydroaulic gradient

Using the number of the pipes N_p, we obtained

$$Q = N_p \pi \left(\frac{t_p}{2}\right)^2 v_p = \frac{N_p \pi g t_p^4}{128\nu} i \qquad (2)$$

Using the area of the paths, A_p and the length of the paths, L_p, We asuume

$$N_p = \frac{A_p}{L_p t_p} \qquad (3)$$

Thus,

$$Q = \frac{A_p \pi g t_p^3}{128 \nu L_p} i \qquad (4)$$

1-D FEM analysis was carried out by establishing the head of borehole point and the flux out of joints using the field data. The head distribution of the node on joint trace obtained from the analysis is nearly equal to zero.

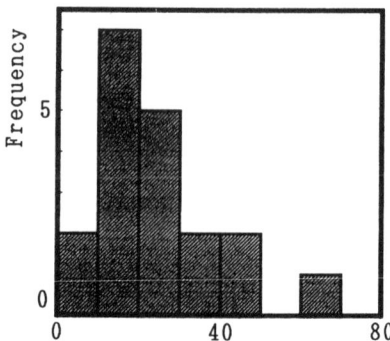

Fig. 3 Frequency distribution of the porosity on the joint plane (TEST 1)

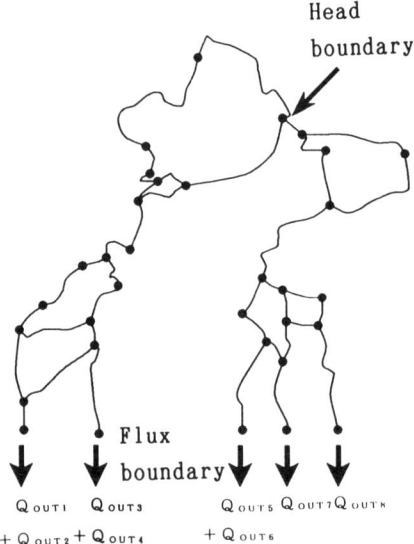

Fig. 4 Analysed segment model for the simulation on TEST 1

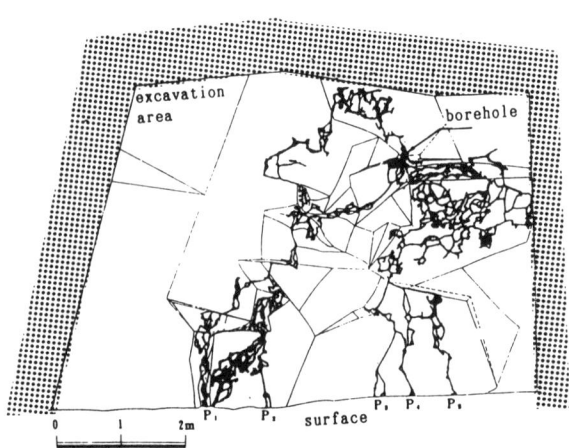

Fig. 2 Estimated flow channel (TEST 1)

*The area surrounded the lines is the joint surface

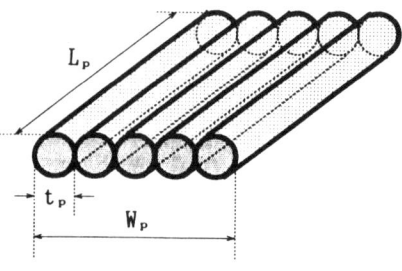

Fig. 5 Analysed model in the element (the assembled pipe flow model)

*The parameters of this model is obtained from aperture of joint, length of path area of path. The flow through these pipes is assumed Poiseuille flow.

2.2 Experiment on a single joint (TEST 2)

At fast the rock block was arranged as shown in Figure 6. A single joint is located in the mid of the block. Two sides of the block were covered with concrete in order to establish the boundary condition.

Constant head permeability test on the objected joint was carried out. Darcy's low is effective as well as TEST 1. As the result of flow route survey by using grout as the tracer, it could be recognized that flow structure showed complex channel (see Figure 7).

The test result was examined using the assembled pipe flow model. From equ.(4) the following is obtained.

$$t_p = \sqrt[3]{\frac{128 \nu L_p Q}{A_p \pi g i}} \quad (5)$$

Substituting the test results, $t_p = 0.63$ (mm) is obtained. This corresponds to the mean thickness of the grout, $t_p = 0.52$ (mm).

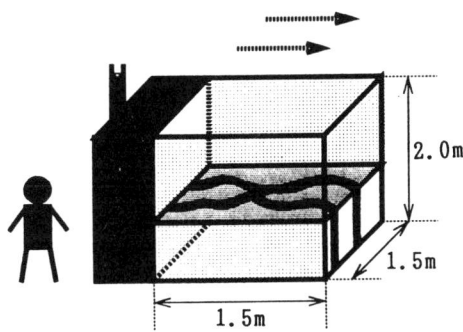

Constant head permeability test

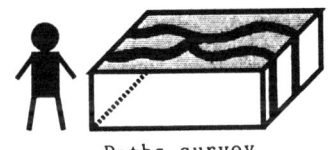

Paths survey

Fig.6 Schematics of TEST 2
Experiment on "Sigle joint"

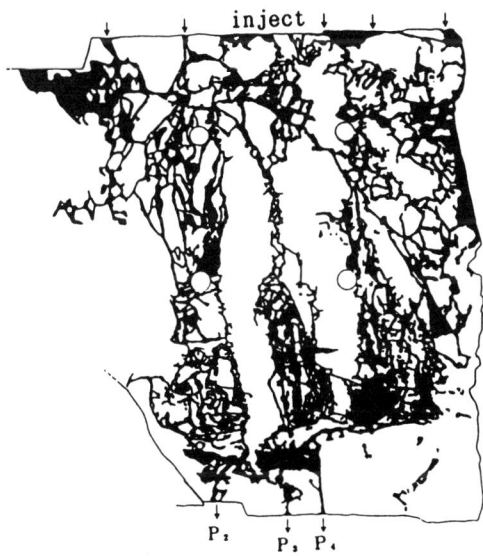

Fig.7 Estimated flow channel (TEST 2)

2.3 Experiment on joint system (TEST 3)

It is clear that seepage flow route in jointed rock masses becomes more complicated because of joints connection. The block specimen in which the joint system is distributed was made as well as TEST 2. After making a specimen (3m×2m×2m, see Figure 8), joint survey were carried out on the surface of the block.

Constant head permeability tests were carried out. As the result, Darcy's law is effective as well as TEST 1 and TEST 2. The seepage flow analysis (as mentioned in 4.) using stochastic model was carried out and the analytical results is well corresponded to the test result.

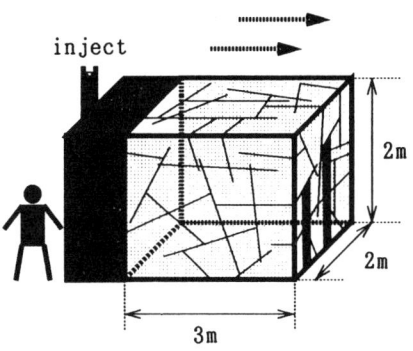

Fig.8 Schematics of TEST 3
Experiment on "Joint System"
(Constant head permeability test)

2.4 Knowledge obtained from 3 series of the TESTS

From the 3 series of the TESTS, the following Knowledges are obtained;

a) Seepage flow in jointed rock mass is mainly the flow through the joints.
b) The shape of the flow is so-called "channeling".
c) Darcy's law is effective under low hydraulic gradient condition.
d) Actual flow velocity is very high (e.g. approximately 1000 times of permeability in TEST 1)
e) The assemble pipe flow model is effective for expressing channeling in a joint.

3. EXPERIMENT ON JOINT SYSTEM (TEST 4)

The block specimen in which the joint system is distributed was made as well as TEST 3, and it is composed of massive hard granite (Cretaceous). Figure 9 shows the schematics of TEST 4. The size of the specimen is 1.3m×1.8m×1.5m. The chamber for the injection of the water was made by concrete so that one side of the block can be covered completely. Two sides and upper plane of the block were also coverd with concrete in order to establish the boundary condition.

3.1 Joint mapping

Joint trace maps of four sides and upper plane of the specimen were drawn and investigation of the joint properties (orientation, trace length, aperture and filling) was carried out.

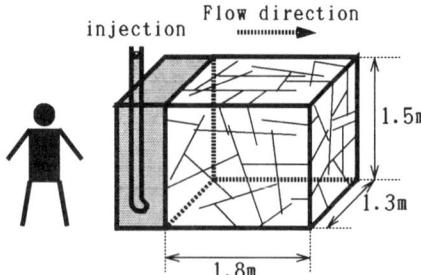

Constant head permeability test

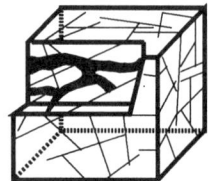

Paths survey

Fig. 9 Schematics of TEST 4
Experiment on "Joint System"

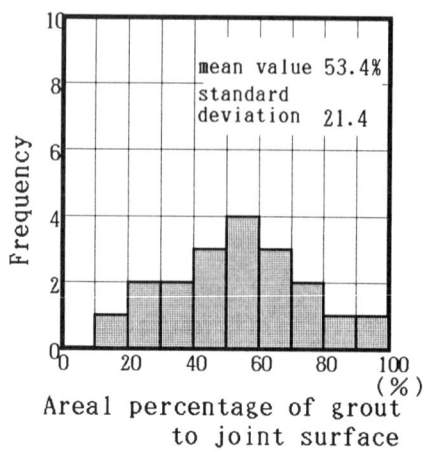

Fig. 10 Frequency distribution of the porosity on the joint plane (TEST 4)

3.2 Measurement of flux

The flux of the specimen is mesured as the function of head difference by measuring flux into and out of the specimen. The both fluxes show almost same value. As the results, the permeability of 1.54×10^{-4} cm/sec is obtained.

3.3 Measurement of flow velocity

The travelling time of water throughout the specimen is mesured by changing the water into tracer rapidly under the constant head condition (Hydraulic gradients is 3.61). As the results, tracer out of the joint has been observed when 248 seconds had passed after the injection.

3.4 Flow pathways survey

The grout milk (super fine cement) had been injected from the water chamber after the permeability test in order to carry out the mapping of the flow pathways in joints. After the sufficient time had passed, the excavation of the specimen and the mapping have been carried out. As the results, the followings are clarified.

a) So-called channeling pathways have been observed on joint planes.
b) Pathways had been occured between the intersected joints in a joint plane.
c) There is no pathways in the portion of filling materials
d) Frequency distribution of the porosity on the joint plane (the percentage of the area of pathways to the area of a joint plane) is close to nomal distribution as shown in Figure 10. Mean value is 53.4%.

4. THE METHOD OF ANALYSIS

According to the results of 3 series of the insitu tests, we assumed Darcy's law is still effective on seepage flow in jointed rock mass. Therefore, basic hydraulic model can be expressed as follows;

$$Q = KAi \quad (6)$$

where Q ; flux per unit time when flowing at section A
i ; hydraulic gradient
K ; permeable coefficient

Further, when analyzing seepage flow, of which flow paths are mainly in a joint system, it is considered the joint system model would be useful.

Figure 11 shows a flow chart of the system of analysis suggested in this study. The procedure is described as follows;

4.1 Modelling of joint system

In order to establish a model by statistical handling of joint distribution, it is necessary to estimate unknown parameters, which are able to provide distributions after clarifying characteristics of distributions, based on the result of in situ joint survey. The parameters for modelling are joint orientation, radius, 3-D density, aperture, and filling. The authors suggested 3-D joint distribution model based on these estimations[2].

4.2 Analytical method

For the 3-D network analysis, 1-D FEM analysis using segment elements can be used. In short, the authors assume the flow occurs in only connected joint traces on the joint plane, and after clarifying boundary condition of permeable/non-permeable, remove edge part having no flow, then establish a linearized analyzing model, of which nodal point is each crossing point.

When considering hydrauric model, it is clarifed that the assembled pipe flow explains the present condition better than parallel flat flow from the tests results.

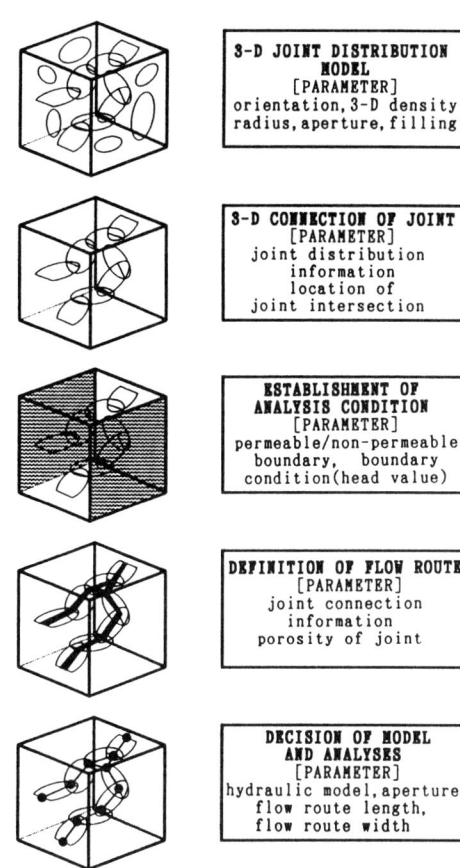

Fig.11 Flow chart of 3-D channeling network seepage flow analysis

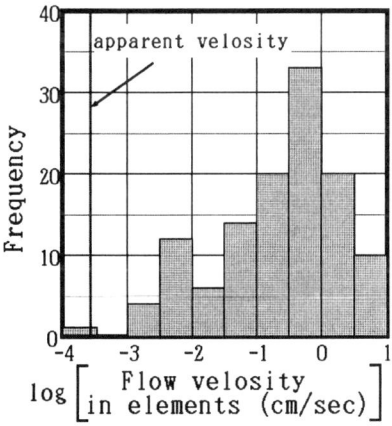

Fig.12 Frequency distribution of the calculated flow velocity in the element (TEST 4: deterministic model)

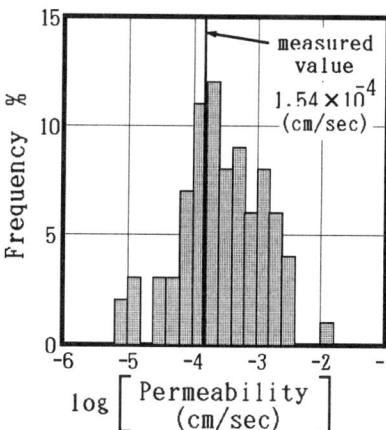

Fig.13 Frequency distribution of the calculated permeability (TEST 4: 100 stochastic models)

calculated and the permeability of 0.70×10^{-4} cm/sec is obtained. This value is almost close to the actual permeability, 1.54×10^{-4} cm/sec.

Figure 12 shows the frequency distribution of velocity in the element. The distribution form is almost lognormally. The fastest travelling time of water throughout the specimen, 81 seconds (34% of the actual value) is also calculated by Dijkstra method.

From the above examination, it is clarifed that the suggested analyzing system can give the almost appropriate estimation.

5. SIMULATION USING DETERMINISTIC MODEL

The authors applied the above-mentioned analyzing system to TEST 4 using the deterministic model which is obtained from the pathways survey in order to examine the applicability of the analyzing system.

As the boundary condition, the head values of inlet and outlet sides are given by the result in the case that the hydraulic gradients is 3.61. As the results of the simulation, total flux through the specimen is

6. SIMULATION USING STOCHASTIC MODEL

100 Stochastic models are generated by taking the joint distribution function which was obtaied from the preliminary joint mapping of the sides of the specimen into account in order to examine the modelling technique.

The simulation of TEST 4 is carried out as well as the simulation using the deterministic model. Figure 13 shows the frequency distribution of the 100 calculated permeabilities. The actual field data is almost close to the mode of the distribution.

Figure 14 shows the frequency distribution of the calculated velocity in the element of 100 models. The distribution form is almost similar to that of the diterministic model as shown in Figure 12.

The frequency distribution of the fastest travelling time of water throughout the specimen is as shown in Figure 15. The actual field data is almost close to the mode of the distribution as well as the permeability.

From the above examination, it is clarifed that the simulation using the stochastic model can give the almost appropriate estimation.

7. CONCLUSION

In this paper the authors have carried out in situ test for the purpose of examining the suggested seepage flow analysing system. The conclusion of this study is as follows;

(1) In situ experiment on seepage flow in jointed rock masses was crried out using the block specimen. As the results of joint mapping, measurement of flux and flow velocity, and flow pathways survey, it is clarified that Darcy's law is effective for the experiment in this study and channeling pathways

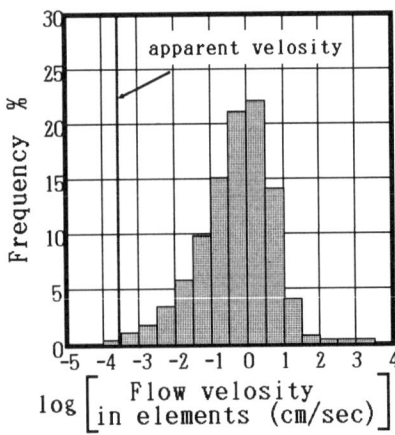

Fig.14 Frequency distribution of the calculated flow velocity in the element (TEST 4: 100 stochastic models)

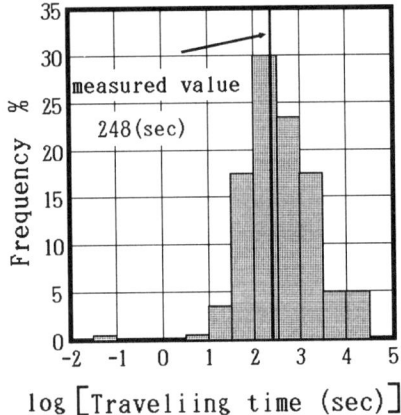

Fig.15 Frequency distribution of the calculated travelling time throughout the specimen (TEST 4: 100 stochastic models)

have been observed on joint planes.

(2) As the result of the simulation of the experiment using the deterministic model, it is clarifed that the suggested analyzing system can give the almost appropreate estimation.

(3) As the result of the simulation of the experiment using the 100 stochastic models, it is clarifed that the simulation using the stochastic model can give the almost appropreate estimation.

REFERENCES

1) Kikuchi.K and Mito,Y: Characteristic of seepage flow through the actual rock joints; Scale Effects in Rock Masses 93, Lisbon, Portugal, 1993.
2) Kikuchi.K, Mito.Y and Honda.M: Geotechnical modelling system of rock joint distribution; Internationanal Symposium on Rock Mechanics at Great Depth, Pau, France, 1989.

Seepage under concrete dam founded on rock formation using artificial neural networks

Yuzo Ohnishi & Mohamed Soliman
School of Civil Engineering, Kyoto University, Japan

ABSTRACT: The flow of water under a concrete dam, founded on rock formation, is one of the important aspects of the design and operation of the dam. This paper focused on the application of artificial neural network to characterize hydraulic properties of the porous media from a limited field data, in which these data were used to train the artificial neural network. The rock formation was assumed to be transformed into a continuum porous media. Seepage flow analysis were performed via finite element method (FEM). The same field data which used to train the neural network were used to perform the stochastic analysis of seepage under the concrete dam. The mean value of the data was used to conduct the deterministic analysis. The pressure distribution under the dam and the downstream discharge were calculated for each analysis and comparison between the reference, deterministic, stochastic and artificial neural network results indicated that neural network can be used as a promising tool for characterizing the porous media for seepage analysis under concrete dam. Therefore the proposed ANN maybe also a useful practical tool to characterize the geological media of large scaled foundation where available field data are very limited compared with the foundation structure scale.

1 INTRODUCTION

In recent years, there has been a growing interest in a class of computing devices that operate in a manner analogous to that of biological nervous systems. These devices, known as Artificial Neural Networks (ANN), or connections systems, are finding applications in almost all branches of science and engineering. ANN are massively parallel computational models for knowledge representation and information processing. As their names imply, ANN are inspired by the neuronal architecture and operation of the human brain. Because of their fundamental hardware similarity to that of the human brain, ANN have some unique, human-like capabilities in information processing. Probably, learning from examples is the most important capability of ANN, in which ANN are capable of learning complex, highly nonlinear relationships and associations from limited field data. The information and knowledge learned by the ANN is encoded and stored in the connection strengths of the network. The retrieval of the stored information is done routinely, by providing the network with an input pattern which acts as a key. This newly developed computing device is used to characterize the porous media properties from a limited field data. In which the limited field data can be used to teach or code the neural network and after teaching or coding, mapping or encoding is performed in order to map the whole area of interest.

Rock formation consists of rock mass, joints, fractures etc. and it is not easy to simulate such complex formation. Instead, continuum models can be used for joint-based analysis, after some transformation. In this study, we assumed that the rock formation was transformed into a continuum porous media, see for example the work of Cho (1988) among others.

In the present study, back-propagation ANN, Rumelhart et al. (1986) was utilized to characterize the permeability coefficient from limited field data. In order to test the developed technique, we conduct the deterministic and stochastic analysis of the same problem using the same data set. Figure 1 shows the schematic of the concrete dam founded on rock formation indicating the boundary conditions of the problem as well as the field data locations. For deterministic analysis, we use the mean value of the field data to conduct the FEM calculation. For stochastic analysis we use the data set to find the mean value and standard deviation of the data which were used to generate the permeability distribution for FEM calculation.

Section 2 of the paper deals with the development of the ANN, backpropagation algorithm. Section 3 deals with ANN application to civil and geotechnical engineering with special emphasize on seepage flow. Section 4 deal with the traditional stochastic approach to geotechnical engineering with specific emphasize to seepage problems. Discussion of ANN, deterministic and stochastic analysis for seepage under concrete dam, founded on rock formation, is dealt with in section 5.

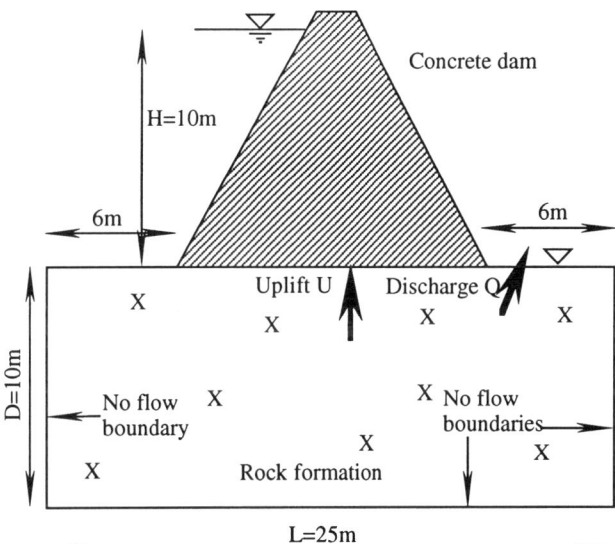

Figure 1 Schematic of concrete dam on rock formation and flow boundary conditions. X indicate the field data location.

2 ARTIFICIAL NEURAL NETWORKS

The ANN shown in Figure 2, consists of three layers, input layer of two units, hidden layer of 10 units and output layer of one unit. The three layers are connected with connections represented by the weight matrices and its biases. Training, teaching or coding of the neural networks is a major concern for its development, in which the networks determine the appropriate set of weights that makes it perform the desired function. There are many ways that this can be done; the most popular class of these algorithms are based on supervised training. Typically, supervised training starts with a networks comprising an arbitrary number of hidden neurons, a fixed topology of connections, and randomly selected values for the weights. The networks is then presented with a set of training patterns, each comprising an example of the problem to be solved (the inputs) and its corresponding solution (the targeted outputs). An algorithm, known as backpropagation ANN, to perform the above procedure can be stated as follow:

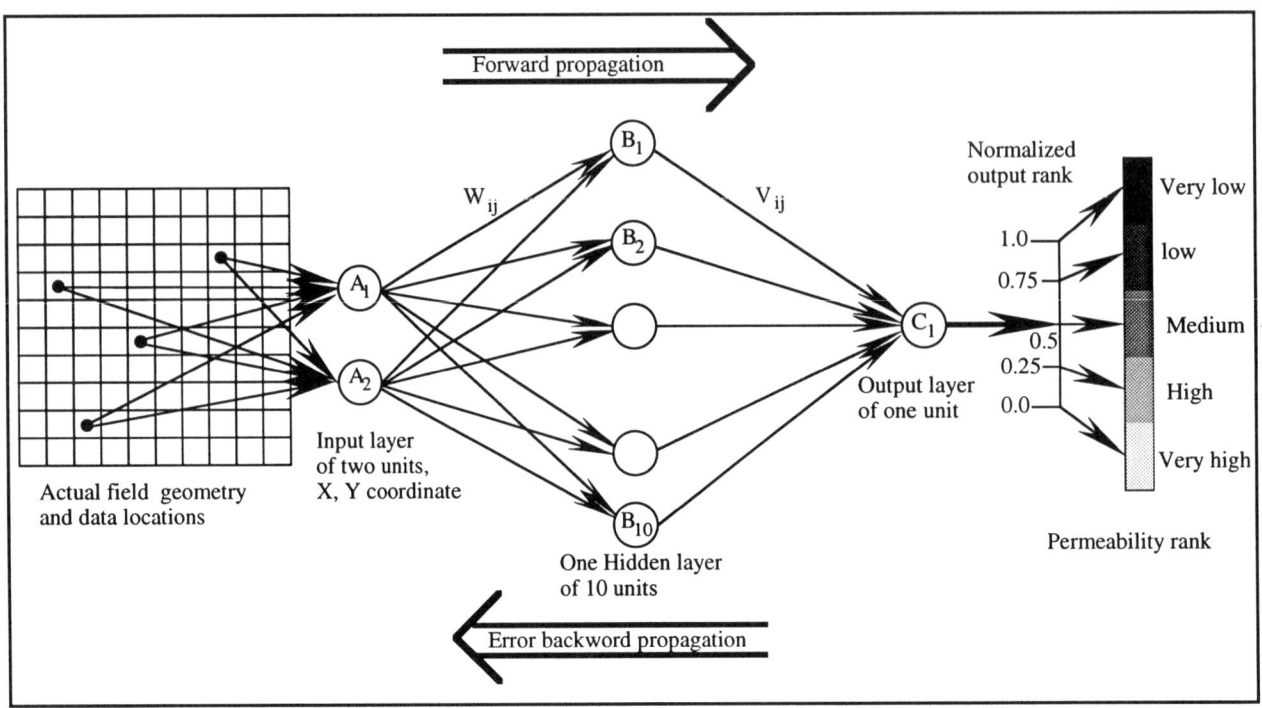

Figure 2 Schematic representation of ANN topology used for permeability characterization using backward propagation algorithm.

1- Initialize weights and biases;
Assign random values in the range [+1, -1] for all weights W_{ij} and V_{ij} and their biases, where W_{ij} is the weight matrix between input layer and hidden layer and V_{ij} is the weight matrix between hidden layer and output layer.

2- Present input and output target;
Present an input vector representing each pattern to be learned and its output target.

3- Propagate the simulation of the inputs;
Compute net input to hidden nodes by weighting and integrating the outputs (o_i) impinging on the nodes

$$net_i = \sum_j w_{ij} o_j$$

Compute the output of the nodes using a differentiable sigmoid nonlinear activation function

$$o_i = \frac{1}{1+e^{-(net_i + b_i)}}$$

For output nodes, follow the same procedure as for hidden nodes.

4- Backpropagate errors;
Start at output nodes and work back to hidden nodes. During this backward pass, compute an error signal δ_i for each of the output nodes. δ_i is determined by

$$\delta_i = (t_i - o_i) o_i (1 - o_i)$$

where o_i is the activation of the output node i; t_i is the desired output of node i.

The error signal is then backpropagated to each hidden node connecting with the output node. The error signal for each hidden node is determined by

$$\delta_i = o_i (1 - o_i) \sum_k \delta_k w_{ki}$$

where W_{ki} is the weight from hidden node i to output node k. The summation term means that the influence of error signals from all output nodes accumulates.

5- Adjust weight;
The weight on each connection is adjusted by the following learning rule

$$w_{ij} = \eta \delta_i o_j$$

in which η is the learning rate smaller than 1.0 that ensures the learning process is incremental. Internal node biases are adjusted in a similar way by assuming they are weights to a unit from a unit whose activation is always 1.0.

6- Repeat step 2 to step 5 whenever a new pattern is presented.

3 APPLICATION OF ANN TO SEEPAGE FLOW

3.1 ANN in civil and geotechnical engineering

The unique capabilities of the ANN are proving very useful in a wide variety of engineering applications. ANN were successfully applied to some of civil engineering problems, see for example ASCE Journal of Computing in Civil Engineering, *special issue* on Neural Networks (1994), where structure, transportation, river, and construction engineering problems were handled by the use of neural networks. Application to geotechnical engineering (soil and rock mechanics) only goes to early 1990s, Lee & Sterling (1992), and cover a limited number of applications, that is identifying probable failure modes for underground openings for tunnel design. Ghaboussi (1992) discussed ANN potential applications to some geotechnical engineering problems, such as soil classification, liquefaction potential of sand deposits, constitutive modeling and in situ material modeling. Zhou & Wu (1994) use ANN to analyze and interpret site investigation data.

3.2 ANN in seepage flow

In groundwater flow, Aziz & Wong (1992) applied the ANN to analyze well and aquifer hydraulics from pumping tests, Ranjithan et al. (1993) applied the ANN for screening groundwater reclamation and contaminants control. The developed ANN in section 2 was applied to characterize the geological media under a concrete dam, in which 9 field data of permeability coefficient were obtained randomly. The locations of these data are shown in Figure 3. The ANN was trained with the 9 data, in which the coordinate x and y were the input for each

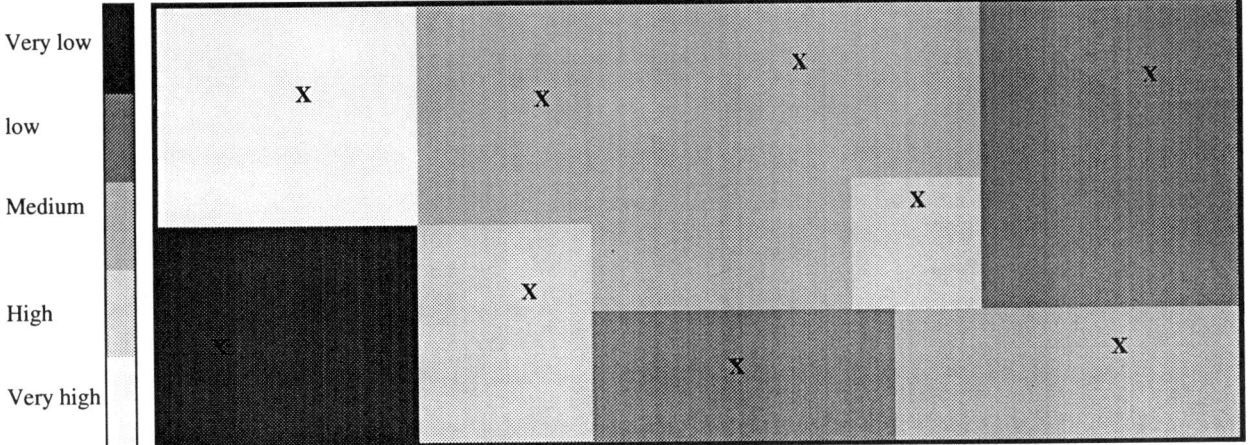

Figure 3 The original distribution of permeability where **X** indicate the location of the field data.

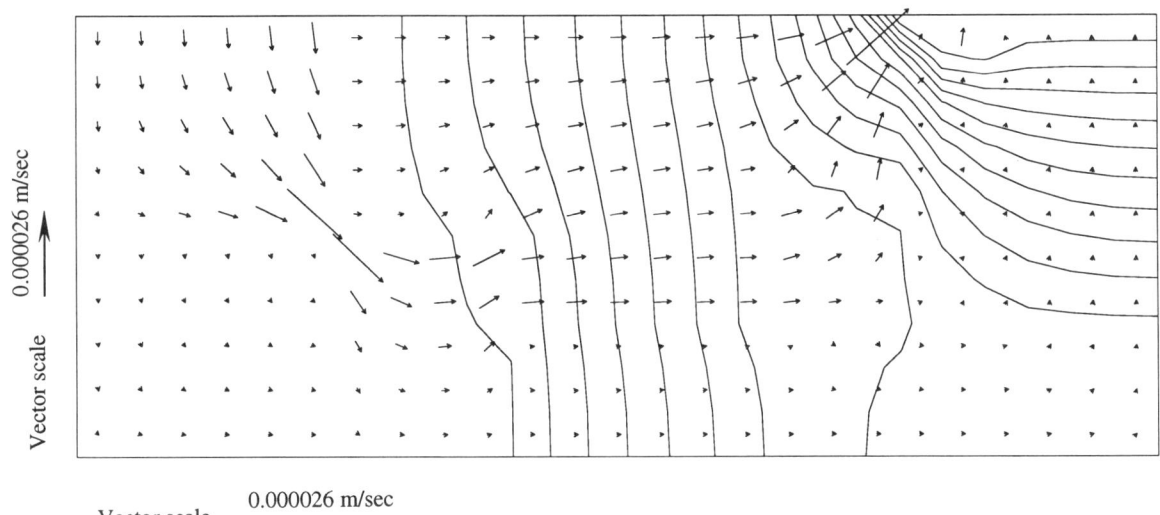

Figure 4 Water pressure distribution and velocity vectors of the reference model.
(Contour interval is 0.5m)

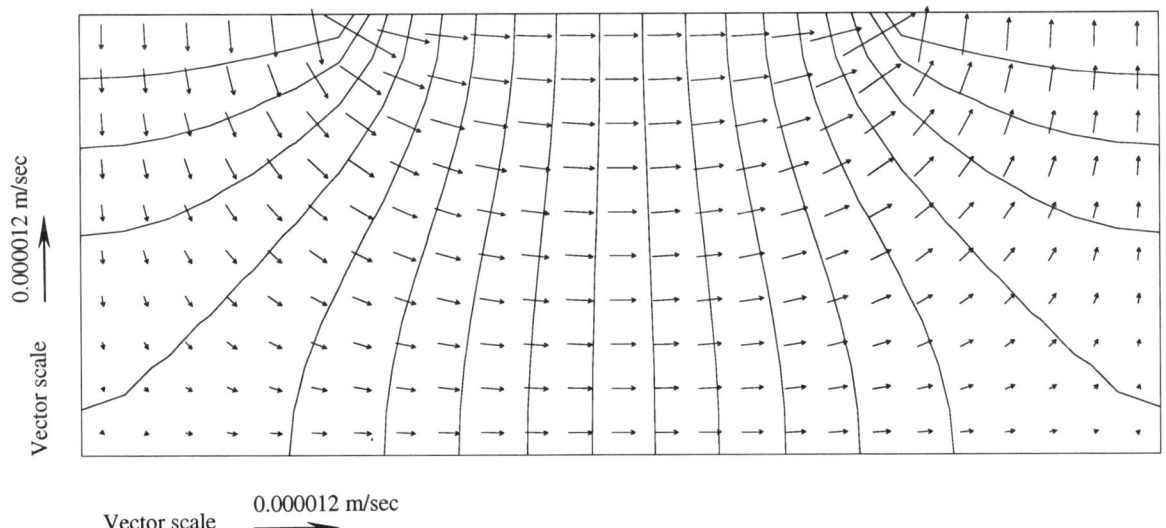

Figure 5 Water pressure distribution and velocity vectors of the average permeability.
(Contour interval is 0.5m)

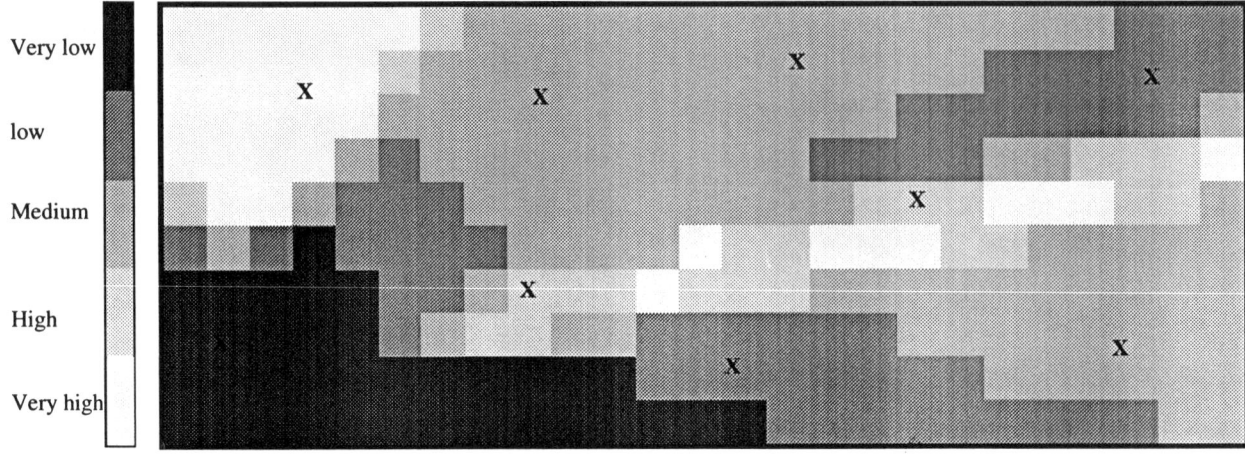

Figure 6 Permeability distribution of neural network characterization using 9 data. **X** indicate the location of the field data.

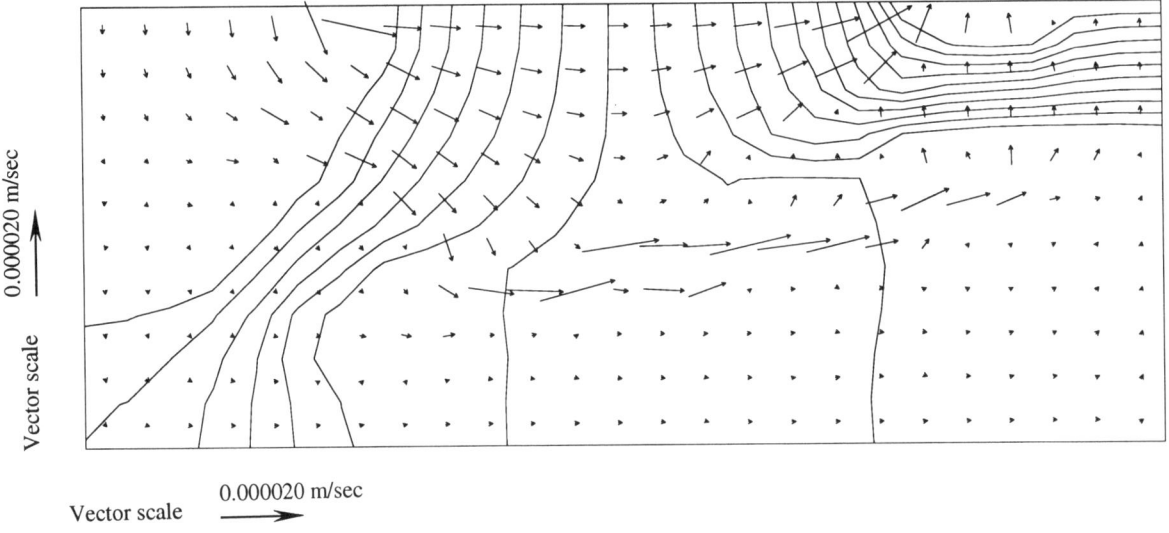

Figure 7 Water pressure distribution and velocity vectors of the ANN, permeability characterized using 9 field data. (Contour interval is 0.5m)

single data and the rank of the permeability coefficient was the output. The rank of the permeability in this case was set as very high, high, medium, low and very low, Figure 3. The input layer consists of two units, A_1 and A_2, Figure 2, corresponding to the x and y coordinate of each block center. The hidden layer consisted of 10 units, B_1 up to B_{10}, Figure 2. Output layer has only one unit, C_1, Figure 2. The number of unit in the hidden layer was determined after trial and error procedure by changing the number of units in the layer and observing the convergence of the net to specific error and/or reach a number of iteration steps. It was found that 10 units in the hidden layer, for this specific problem, meet the required criteria. That is to satisfy either a maximum error of 0.0001 or less than 100,000 iteration steps. After training the networks, encoding operation was conducted for all blocks over the area of interest. That was done by fixing the weight matrices W_{ij} and V_{ij} and their biases, and feeding the networks with input data and obtaining the normalized permeability rank, Figure 2. Which in turn transformed to the true permeability rank. Finally, the domain was characterized into 5 different materials of different permeability coefficients, Figure 6, which were used as input parameters to conduct the FEM calculation.

FEM calculations were performed using VPFLOW code, Soliman (1993) and Ohnishi et al. (1993 & 1994) and UNSAF code, Akai et. al. (1977). However, for the sake of simply presenting the results UNSAF code was used for velocity vectors plotting, due to the fact that VPFLOW code give the velocity vectors at 8 node element while UNSAF code provide an average velocity within each element. The water pressure distribution of the two codes as well as the exit discharge were same.

FEM mesh consists of 250 elements and 286 nodes. The dimension of the domain was set as 25 m length and 10 m depth. The head difference between upstream and downstream of the dam is 10 m.

Figure 3 shows the reference permeability distribution of 5 different materials. Figure 4 shows water pressure distribution and velocity vectors of reference analysis. Figure 5 shows the water pressure distribution and velocity vectors of deterministic analysis, which uses the mean value of the permeability coefficient in FEM calculation. Figure 6 shows the permeability distribution of ANN characterization of permeability coefficient into 5 materials. Figure 7 shows the water pressure distribution and velocity vectors of flow under the concrete dam, using the characterized permeability of ANN.

4 ANN VIA STOCHASTIC APPROACH TO SEEPAGE ANALYSIS

4.1 Stochastic approach to geotechnical engineering

Most geotechnical engineering, including seepage flow, analysis are deterministic in that the geologic properties used are assumed to be averaged values. Variations in the porous media properties

Figure 8 Permeability distribution of stochastic model using 9 data.

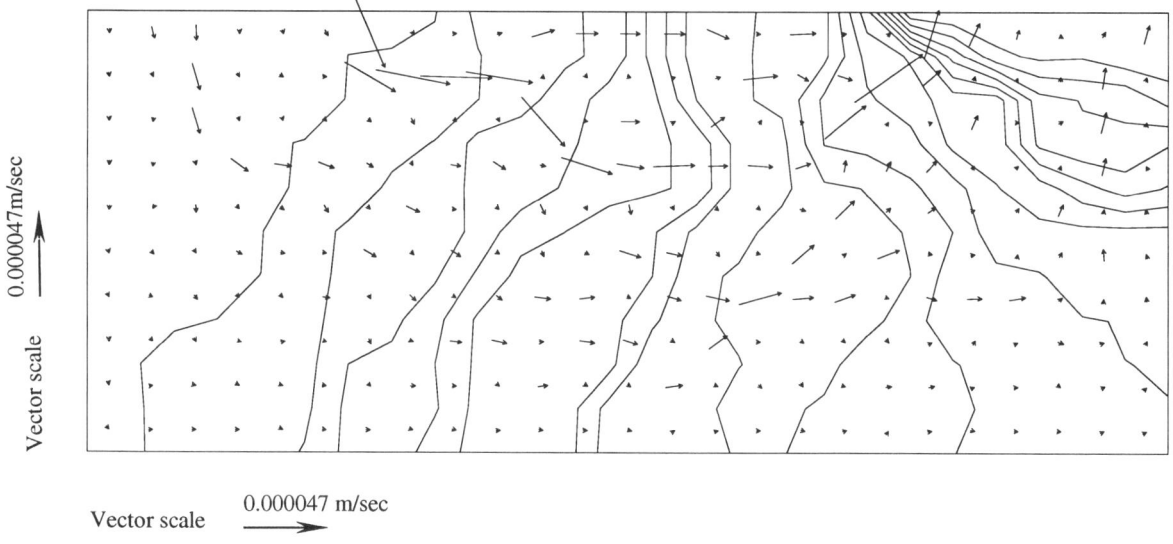

Figure 9 Water pressure distribution and velocity vectors of one of the stochastic realizations, permeability generated using 9 field data. (Contour interval is 0.5m)

are accounted for by the use of safety factors which are often applied arbitrarily to the computed results. This average approach to the definition of soil properties has tended to be applied not only to classical soil and rock mechanics calculations, but also to numerical computations using sophisticated numerical techniques such as the finite element method. Properties are usually assigned on the basis of a limited number of laboratory and/or in situ tests. In reality, these properties vary from point to point and can be determined deterministically only through numerous field tests. Since this is expensive and impractical, random field models can be used to represent the geomaterial. The parameters of these models can be estimated from a limited number of test results.

Mean porous media, soil and rock, are fairly well established. The data gathering has been motivated largely by the availability of random field simulation algorithms and their potential for producing useful results. The increased performance of computers has also enabled more detailed discretization of boundary value problems, and better modeling of stochastical properties of the input parameters. The finite element method is an ideal tool for modeling materials with a spatial materials with spatial variation in properties. Stochastic finite element analysis has been implemented in a number of areas of geotechnical interest, see e.g., Beacher & Ingra (1981) for stress analysis, Righetti & Harrop-Williams (1988) for settlement of foundation, Ishii & Suzuki (1987) for slope stability, Smith & Freeze (1979a, 1979b) for confined seepage and Grifiths & Fenton (1993) for seepage beneath water retaining structures.

Here we utilize the permeability field data to conduct the stochastic analysis in order to compare its results with ANN's one. The permeability distribution was log-normally distributed as also found in the literature.

4.2 Stochastic seepage analysis

Field measurements of permeability coefficient indicated an approximately log-normal distribution, see e.g. Hoeksema & Kitanidis (1985) and Sudicky (1986), which was applied to seepage analysis, Grifiths & Fenton (1993). Assume that Y is normally-distributed parameter defined as $Y = \log K$, where K is the permeability coefficient. Y is generated with a fixed mean value and standard deviation. In the present analysis we kept the mean value of permeability coefficient, μ_k, equal to 10^{-5} m/sec and its standard deviation, σ_k, equal to 10^{-2} m/sec. One of the realizations generated by this technique is shown in Figure 8. Figure 9 shows the water pressure distribution and velocity vectors of one of the same realization.

5 DISCUSSION

Figure 10 shows the water pressure distribution under the concrete dam for all conducted analysis. It is clear that ANN result is closer to the result of the reference case than the deterministic case and lay in the range of stochastic results. Table

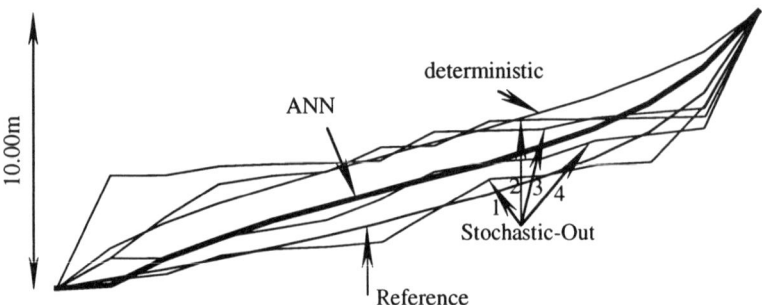

Figure 10 Water pressure distribution on concrete dam base for reference, deterministic, ANN and stochastic results.

Table 1 Discharge of the downstream of the dam for reference, deterministic, ANN and stochastic results.

	Q in m^3/s/m
Reference	0.00003897
deterministic	0.00004360
ANN	0.00006353
Stochastic -1	0.00003955
Stochastic -2	0.00003210
Stochastic -3	0.00005749
Stochastic -4	0.00007055

1 summarize the exit discharge downstream of the dam for all analysis, which indicated that the exit discharge for ANN case lay in the range of stochastic results and more than both deterministic and reference results.

Comparison between ANN results and both reference and stochastic analysis, Figure 10 and Table 1, we can conclude that ANN is a promising tool for characterizing hydraulic properties of geological media from limited field data, which maybe used as a practical tools in real geotechnical applications.

6 CONCLUSION

Artificial neural networks was applied to characterize the permeability coefficient of rock formation using a limited field data. Seepage flow under a concrete dam was analyzed with the developed technique via FEM. Mean value of he data set was used to conduct the deterministic analysis of the problem. The same data set, which were used to teach the ANN, was used to conduct the stochastic analysis of same problem. VPFLOW and UNSAF codes were utilized to conduct the FEM calculations. Comparison between the reference, deterministic, ANN and stochastic analysis, was made for two parameters, the water pressure distribution under the dam (uplift force) and the exit discharge downstream of the dam.

The comparison between a number of stochastic realizations, reference and deterministic results with that of ANN indicated that ANN is a promising device to characterize the geological formation from limited field data. Therefore the proposed ANN maybe also a useful practical tool to characterize the geological media of large scaled foundation where available field data are very limited compared to the scale of foundation structure.

REFERENCES

Akai, K., Y. Ohnishi & M. Nishigaki 1977. Finite element analysis of saturated-unsaturated seepage in soils. *Proc. Japan Soc. of Civ. Engng*, No. 264: p. 87-96.(*In Japanese*)

ASCE 1994. Neural Networks *Special issue J. Comput. in Civ Engng Am. Soc. Civ. Engrs* **8**, No. 2.

Aziz, A. R. A. & K. F. V. Wong 1992. A neural-network approach to the determination of aquifer parameters, *Ground Water* **30**: 164-166.

Beacher, G. B. & T. S. Ingra 1981. Stochastic FEM in settlement prediction. *J. Geotech Engng Am. Soc. Civ. Engrs* **107**, GT4: p. 449-463.

Cho, T. F. 1988. Continuum and discrete modelings of porous and jointed rock: Application to the design of near surface annular excavations. *PhD dissertation, university of Wisconsin-Madison*, p. 30.

Hoeksema, R. J. & P. K. Kitanidis 1985. Analysis of spatial structure of properties of selected aquifers. *Wat. Resour. Res.* **21**, No. 4: p. 563-572.

Ishii, K. & M. Suzuki 1987. Stochastic finite element method for slope stability analysis. *Struct. Safety* **4**: p. 111-129.

Ghabousi, J. 1992. Potential applications of neuro-biological computational models in geotechnical engineering. *Num Mod in Geomech*. Pande & Peitruszczak Eds, Vol. 2: p. 543-555. Rotterdam, Balkema.

Griffiths, D. V. & G. A. Fenton 1993. Seepage beneath water retaining structures founded on spatially random soil. *Géotechnique* **43**, No.4: p. 577-587.

Lee, C. & R. Sterling 1992. Identifying probable failure modes for underground openings using a neural network. *Int. J. Rock Mech. Min. Sci. & Geomech. Abstr.* **29**, No. 1: p. 49-67.

Ohnishi, Y. & M. Soliman 1993. Finite element analysis of velocity field for groundwater and its application. *J. Dam Engng. Japan Soc. Dam Engrs*, No. 11: p. 45-52 (*in Japanese*).

Ohnishi, Y., M. Soliman & A. Kobayashi 1994. Finite element analysis of groundwater velocity field. *Proc. Second Geotch. Engng. Conf. Cairo University*. p. 388-399.

Ranjithan, S., J. W. Eheart & J. H. Garrett, Jr. 1993. Neural network-based screening for groundwater reclamation under uncertainty, *Water Resources. Res.* **29**, No. 3: p. 563-574.

Righetti, G. & K. Harrop-Williams 1988. Finite element analysis of random soil media. *J. Geotech Engng Am. Soc. Civ. Engrs* **114**, GT1: p. 59-75.

Rumelhart, D. E., G. E. Hinton & R. J. Williams 1986. Learning internal representations by error propagation. Parallel Distributed Processing, Vol. 2, D. E. Rumelhart and J. McClelland, Eds. MIT press, Cambridge, Mass .

Smith, L. & R. A. Freeze 1979a. Stochastic analysis of steady state groundwater flow in a bounded domain. 1. One-dimensional simulations. *Wat. Resour. Res.* **15**, No. 3: p. 521-528.

Smith, L. & R. A. Freeze 1979b. Stochastic analysis of steady state groundwater flow in a bounded domain. 2. Two-dimensional simulations. *Wat. Resour. Res.* **15**, No. 6: p. 1543-1559.

Soliman, M. 1993. Finite element analysis of velocity field for groundwater. *Master of Eng. Dissertation, School of Civil Engng. Kyoto University*.

Sudicky, E. A. 1986. A natural gradient experiment on solute transport in a sand aquifer: spatial variability on hydraulic conductivity and its role in the dispersion process. *Wat. Resour. Res.* **22**, No. 13: p. 2069-2083.

Zhou, Y. & X. Wu 1994. Use of neural networks in the analysis and interpretation of site investigation data. *Jour. Computers in Geotechnics* **16**, No. 2: 105-122.

C: Monitoring and reinforcement

Strength and deformation behaviour of grout jointed sandstone – A laboratory study

R. K. Srivastava, M. Singh & A. K. Tripathi
MNR Engineering College, Allahabad, India

ABSTRACT : One of the important features of river valley projects in India is rock grouting. Most of the studies that have been carried out are related to strength behaviour and prediction in case of intact or jointed rocks. Understanding response of grout jointed rock and prediction of its strength behaviour is of great relevance. The present study is an endeavour in this direction. A laboratory study has been carried out on Indian sandstone. Planar joints at an angle of 30° and 45° from vertical have been created and filled with cement-fine sand grout (w/c = 0.7) of thickness 2, 3 and 4 mm. Triaxial tests have been carried out at confining pressures 0, 50, 75 and 100 kg/cm². Ramamurthy et. al strength criterion proposed for intact rocks have been modified and extended for strength behaviour prediction of grout jointed rocks.

1 INTRODUCTION

In most of the river valley projects, under investigation or construction stage in India, it has been observed that rock mass that is encountered is invariably discontinuous and contains planes of weakness like bedding planes, joints, fissures, faults etc. Several sites have to be rejected because of the importance and massive nature of the projects due to adverse geological conditions. But in several regions alternatives have been reduced due to the need of developmental projects at a particular site. Under the circumstances ground improvement techniques have to be resorted to. One of the most used ground improvement technique is grouting. Though the technique has been in use for long but still it is practiced more as an art than science. Techno-economic viability of grouting work is yet to be established. The data available on grouting performance is very meagre. In fact, large scale field tests before and after grouting are required to evaluate the actual performance of the grouting work carried out. But this is very expensive and time consuming. Further, as a first step, rational understanding is required to be developed under controlled conditions, such as in a laboratory, to formulate suitable grouting starategy for an actual project.

Keeping the above facts in view an investigation programme has been carried out on grout jointed sandstone from Vindhyan region of Uttar Pradesh, India. The strength behaviour of planar jointed rocks with variable grout thickness have been studied under triaxial conditions. Cement-fine sand grout (w/c 0.7) of thickness 2, 3 and 4 mm has been used with joint plane oriented at β = 30° and 45°. Further the failure criterion proposed by Ramamurthy et. al. (1985) for intact rocks have been modified (by including a factor which depends on the grout thickness) to predict the strength behaviour of grout jointed rocks. The grout thickness factor has been computed by using a curve enveloping σ_1 values so that major principal stress values are not under predicted. The sensitivity of results with respect to change in values of grout thickness parameters has also been discussed.

2 EXPERIMENTAL PROGRAMME

The sandstone rock specimen used are from Vidhyachal region of Uttar Pradesh, India and belong to Bhander Series of Upper Vindhyans. The rock is isotropic, light yellowish in colour due to presence of more silica and feldsper. Physical properties (e.g. density, sp. gr., water absroption and porosity) and strength indices (e.g. Brazilian and Point load strength, UCS of intact rock) have been determined first to characterise the rock.

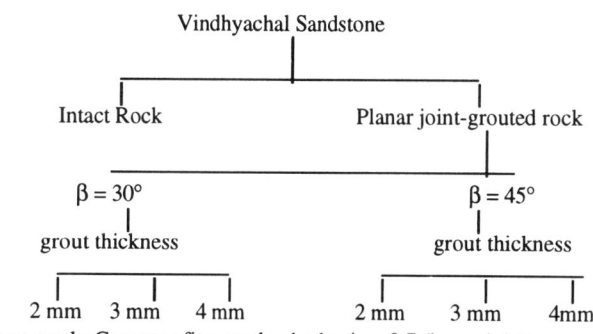

grout used : Cement : fine sand = 1 : 1 w/c = 0.7 (by weight)
Confining pressures used = 0, 50, 75 and 100 kg/cm²

Fig. 1 Schematic diagram-specimen tested.

Subsequently a series of tests have been carried out under triaxial stress conditions on various types of specimen as shown schematically in figure 1.

3 STRENGTH CRITERION

A number of strength criteria have been proposed and are in use for intact and anisotropic rocks. Some of the earlier widely used criteria for anisotropic rocks are by Walsh and Brace (1964), Jaeger (1960) (single plane of weakness theory -assuming that the rock behaves linearly with the applied load). Jaeger (1960) (variable cohesive strength theory) and Mclamore and Gray (1967) criterion (predicting non linear behaviour of anisotropic rocks).

Mclamore and Gray (1967) assume that the material fails in shear and has a variable cohesive strength τ_0, but constant values of internal friction tan φ. But Walsh and Brace (1964) assume that the failure is tensile in nature and that the body is composed of long, non-randomly oriented cracks that are superposed on an isotropic array of randomly distributed smaller cracks or Griffith's cracks. They further assume that fracture may occur through the growth of either the long or small cracks depending upon the orientation of the long crack system to the applied stress, σ_1. The above strength criteria can not be used for all types of rock because of their limitations. The practical utility is very less as a large number of tests are to be performed at different confining pressures and orientations, to evaluate the anisotropic rock strength.

Hoek and Brown (1980) developed empirical failure criterion for intact and anisotropic rocks using the non-linear failure enevelope predicted by Griffith's theory for plane compression and through process of trial of and error. It can be written as

$$\sigma_1 = \sigma_3 + (m_a \sigma_{ca} \sigma_3 + s_a \sigma^2_{ca})^{1/2} \quad (1)$$

Where, σ_1 and σ_3 are major and minor principal stresses respectively.

σ_{ca} = uniaxial compressive strength of rock with weak plane.

m_a and s_a are dimensionless constants which characterise the degree of interlocking between particles in rock mass containing a weak plane. For intact rock s = 1 and for completely broken rock mass s = 0. The range of variation of m is very wide and is believed to be a function of rock type and rock quality.

Yudhbir et. al. (1983) have also proposed an empirical criterion by modifying Bieniawaski's (1974) criterion to take into consideration the anisotropic rock behaviour and can be expressed as :

$$\sigma_1/\sigma_{ca} = A_a + B_a (\sigma_3/\sigma_{ca})^{\alpha_a} \quad (2)$$

where

A_a = constant depending on rock mass quality

B_a = rock material constant.

α_a = slope of the plot between $(\sigma_1/\sigma_{ca} - A_a)$ vs. σ_3/σ_{ca} on log-log scale.

σ_{ca} = uniaxial compressive strength of rock with a weak plane.

Ramamurthy et al. (1985) suggested a criterion for isotropic rocks as

$$(\sigma_1 - \sigma_3)/\sigma_3 = B(\sigma_c/\sigma_3)^\alpha \quad (3)$$

where

σ_c is the uniaxial compressive strength of rock

B = material constant.

α = slope of the plot between $(\sigma_1 - \sigma_3)/\sigma_3$ and σ_c/σ_3 on log-log scale.

Rao (1984) has critically reviewed and evaluated the applicability of the various strength criteria proposed by using data generated from carrying out tests on four Indian rocks and analysis of published data for more than 100 rocks. He has concluded that a better prediction of strength is possible by use of Ramamurthy et. al. (1985) proposed strength criterion. In the present study Ramamurthy et. al. (1985) criterion for intact rock has been modified and a criterion is proposed to predict strength behaviour of planar grout jointed rock with various grout thicknesses. The parameters B and α of Ramamurthy et. al. criterion have been replaced as follows.

$$B_j = B (\phi_g \cdot t + \phi_{gj}) \quad (4)$$

$$\alpha_j = \alpha(\theta_{gj} \cdot t^{\theta_g}) \quad (5)$$

where ϕ_g, θ_g are constants ; ϕ_{gj} and θ_{gj} depend on angle of joint orientation, t is grout thickness in mm.

B, α are constants as proposed in Ramanurthy et. al. criterion.

To compute the parameters θ_g, θ_{gj}, ϕ_g and ϕ_{gj}, the criterion is fitted intoexperimentally obtained data base by transforming it into linear form by taking log. Least square method is used and straight line is fitted into the transformed equation to evaluate parameters of the proposed criterion. The procedure is as follows :

Let the equation of the straight line be

$$y = a + bx \quad (6)$$

(x_i, y_i), i = 1 to n are n data points in which the line is to be fitted. Using least square method values of a and b are obtained as follows

$$an + b\Sigma x_i = \Sigma y_i \quad (7)$$

$$a\Sigma x_i + b\Sigma x_i^2 = \Sigma x_i y_i \quad (8)$$

Constants a and b are obtained from solving the above equations as

$$a = \bar{y} - b \cdot \bar{x} \quad (9)$$

$$b = (\Sigma x_i y_i - n\bar{x}\bar{y})/(\Sigma x_i^2 - (\bar{x})^2) \quad (10)$$

where, $\bar{x} = (\Sigma x_i)/n$

$\bar{y} = (\Sigma y_i)/n$

Using the values of a and b the values of B_j and α_j are calculated. The suitability of these values have been checked by using coefficient of correlation and coefficient of accordance. The coefficient of correlation is defined as.

$$\gamma^2 = \frac{(\Sigma x_i y_i - \frac{\Sigma x_i \Sigma y_i}{n})^2}{(\Sigma x^2_i - \frac{(\Sigma x_i)}{n})(\Sigma y_i^2 - \frac{(\Sigma y_i)^2}{n})} \quad (11)$$

In case of better correlation, γ^2 approaches towards unity. To evaluate the accuracy of prediction, coefficient of accordance can be used. It is defined as

$$\psi^2 = \frac{\Sigma(\sigma_{1exp.} - \sigma_{1cal.})^2}{\Sigma(\sigma_{1exp.} - \bar{\sigma}_{1exp.})^2} \quad (12)$$

Where, σ_{1exp} is the observed values and σ_{1cal} is the predicted value of σ_1. $\bar{\sigma}_{1exp.}$ is the average of σ_{1exp} values

Approach of coefficient of accordance towards zero is indicative of better prediction. To determine the values of the constants (θ_g, θ_{gj}, ϕ_g and ϕ_{gj}) which depend on B_j and α_j, procedure of trial and error and enveloping curve fitting has been adopted.

4 RESULTS AND DISCUSSIONS

4.1 Physical properties

The mean value of the physical properties of Vindhyachal sandstone are presented in table 1.

Table 1 : Physical properties of sandstone

Propetry		Value
Water absorption		2.81%
Sp. gravity		2.65
Density	(a) dry	2.41 g/cm^3
	(b) saturated	2.52 g/cm^3
	(c) bulk	2.47 g/cm^3
Porosity	(a) apprent	7.22%
	(b) Total	7.35%

4.2 Strength indices

The mean value of UCS, Brazilian and axial and diametrial point load strength tests results are presented in table 2. The classificaiton of this sandstone on Deere and Miller's (1966) chart is 'CM'. The ratio of $\sigma_c/\sigma_{tb} = 14$, $\sigma_c/\sigma_{(tp)d} = 23.5$ and $\sigma_c/\sigma_{(tp)a} = 13.5$ which are in the range of values published in the literature (e.g. Broch and Frankin 1972).

Table 2 : Strength indices Vindhyachal sandstone.

Strength Index		Value(Kg/cm^2)
σ_c		709.5
σ_{tb}	air dry	50.4
	saturated	44.9
$\sigma_{(tp)d}$	air dry	30.2
	saturated	18.1
$\sigma_{(tp)a}$	air dry	53.5
	saturated	36.5

4.3. Intact rock behaviour

The table 3 presents the values of peak stress (σ_1), and elasticity modulus E_t (obtained at 50% peak stress value) for various confining pressures. From the analysis of triaxial test data, the value of $c = 142.88$ kg/cm^2 and $\phi = 44.9°$ is obtained.

Table 3: Variation of σ_1, and Et with confining pressure-Intact rock.

σ_3 Kg/cm^2	σ_1 Kg/cm^2	E_t x 10^{5}Kg/cm^2
0	709.5	1.96
50	1015.1	2.57
75	1157.0	3.72
100	1255.6	5.80

4.4 Shear strength parameters

The values of c and ϕ for grouted rocks are presented in Table 4. It is observed that the variation in angle of internal friction for joint orientation is small but the value of cohesiom is quite sensitive to grout thickness and it increases with increasing grout thickness.

Table 4 : Shear strength parameters of grouted rocks.

grout thickness (mm)	$\beta = 30°$ c Kg/cm^2	ϕ	$\beta = 45°$ c Kg/cm^2	ϕ
2	33.7	26°58'	89.2	17° 48'
3	58.8	20° 37'	125.5	16°39'
4	80.0	16°27'	173.0	13° 20'

4.5 Strength criterion for grouted rock

The proposed strength criterion based on Ramamurthy criterion for intact rock is as follows

$$(\sigma_1 - \sigma_3)/\sigma_3 = B_j (\sigma_{cj}/\sigma_3)^{\alpha_j}$$

where

$$B_j = B (\phi_g \cdot t + \phi_{gj})$$

$$\alpha_j = \alpha(\theta_{gj} \cdot t^{\theta_g})$$

α and B are constants proposed by Ramammurthy et. al. (1985) for intact rock (for sandstone the values of $\alpha = 0.8$ and $B = 2.5$ have been taken).

From the analysis of data, following values of ϕ_g, ϕ_{gj}, θ_g and θ_{gj} have been obtained for planar grout jointed sandstone.

Table 5 : Values of constants.

Joint orientation	Constants			
	ϕ_g	θ_g	ϕ_{gj}	θ_{gj}
$\beta = 30°$	–0.168	0.554	1.326	0.432
$\beta = 45°$	–0.168	0.554	1.105	0.575

Table 6 shows the experimentally obtained and predicted values using proposed strength criterion and values of the constants.

Table 6 : Experimental and predicted values of σ_1.

β	Grout thickness t (mm)	σ_{cj} Kg/cm^2	σ_3 Kg/cm^2	σ_{1exp} Kg/cm^2	σ_{1cal} Kg/cm^2	Variation %
30°	2	109.1	50	240.1	243.40	1.38
			75	305.6	312.53	2.77
			100	372.1	374.81	0.97
	3	174.6	50	283.8	289.45	1.99
			75	327.5	354.62	8.18
			100	393.0	412.14	4.87
	4	218.3	50	305.6	310.62	1.64
			75	349.3	366.45	4.91
			100	414.8	415.51	0.17
45°	2	341.0	50	393.0	401.5	2.16
			75	480.0	484.03	0.84
			100	524.0	540.21	3.09
	3	349.0	50	437.0	455.27	4.18
			75	480.0	506.74	5.57
			100	540.0	551.56	2.14
	4	480.0	50	524.0	568.27	8.45
			75	544.0	595.42	9.45
			100	568.0	629.95	9.50

From a sensitivity analysis, the range of the constants in the proposed strength criterion is given in table 7.

Table 7 : Range of constants.

Constant	Range
ϕ_g	–0.158 to –0.181
θ_g	0.449 to 0.585
$\theta_{g30°}$	1.167 to 1.361
$\phi_{g45°}$	0.946 to 1.140
$\theta_{g30°}$	0.362 to 0.453
$\theta_{g45°}$	0.505 to 0.596

5 CONCLUSIONS

An attempt has been made in the present study to develop an understanding of strength behaviour and strength prediction in case of planar jointed grouted rocks (with joint orientations at 30° and 45° from vertical and grout thicknesses of 2, 3 and 4mm), following observations have been made from the present study.

1. The physical and strength properties indicate that the sandstone used in the present study is a medium strength rock that may be classified as CM from Deere and Miller (1966) chart. The ratio of $\sigma_c/\sigma_{tb} = 14$, $\sigma_c/\sigma_{(tp)d} = 23.5$ and $\sigma_c/\sigma_{(tp)a} = 13.5$ are obtained which are in the reported range of values. In case of intact rock, variation of σ_1 with σ_3 is non-linear. The value of Et also increases with increasing confining pressure as expected.

2. The value of peak strength σ_1 increases non linearly with σ_3. It is higher for higher grout thickness but with increasing confining pressures this difference decreases.

3. The specimen fail in general by sliding for $\beta = 30°$ and in case of $\beta = 45°$, by a combination of tensile splitting and shear across the joint for all confining pressures.

4. The variation in the value of angle of internal friction is small whereas, cohesion is very sensitive to the change in grout thickness.

5. Empirical strength criterion proposed by Ramamurthy et. al (1985) has been modfied and the proposed criterion, it is observed, pridicts reasonably good values. The variation being always on the safer side. There is a need to generate more data for different rock types and grout thicknesses to provide a range of values of constants in the proposed criterion for its wider applicability.

REFERENCES

Bieniawaski, Z. T. 1974. Estimating the strength of rock materials. J. S. Afr. Inst. Min. Metall. Vol. 74, No. 8 : 312-320.

Broch, E. and Franklin, J. A. 1972. The point load strength test. Int. J. Rock Mech. Min. Sci. Vol. 9 : 669-697.

Deere, D.U. and Miller, R.P. 1966. Engineering classification and index properties for intact rock. Teach. report, A.F.W. lab. New Mexico.

Hoek, E. and Brown, E.T. 1980. Underground excavations in rock. Institution of Mining and Metallurgy. London.

Jaeger, J.C. 1960. Shear failure of anisotropic rocks. Geol. Mag., Vol. 97 : 65-72.

Mclamore, R. and Gray, K.E. 1967. The mechnical behaviour of anisotropic sedimetary rocks. Trains. Am. Soc. Mech. Engr. Series B, Vol. 89 : 62-76.

Ramamurthy, T., Rao, G.V. and Rao, K.S. (1985). A strength criterion for rocks. Proc. Indian Geotechnical Conference, Roorkee, India : 59-69.

Rao, K.S. 1984. Strength and deformation behaviour of sandstones. Ph.D. Thesis, submitted to I.I.T. Delhi, India.

Walsh, J.B. and Braces, W.F. 1964. A fracture criterion for brittle anisotropic rocks. Geophysics Reasearch J., Vol. 69 : 34-49.

Yudhbir, Lemanza, W. and Prinal, F. 1983. An empirical failure criterion for rock masses, 5th Int. Cong. Rock Mech., Vol. 1 : B1-8, Melbourne.

Behaviors of Kaore Arch Dam and its foundation rock during test filling

H. Suzuki
Okumino Pumped-storage Plant Construction Office, Chubu Electric Power Co., Inc., Japan

ABSTRACT: In this paper, the behaviors of Kaore arch dam and its foundation rock at test filling, comparing measured results with numerical analysis are presented. Since this dam is constructed for upper reservoir of pumped-storage power plant, water level can be controlled artificially by pumping and power generation at a rate of 1m/day or less, and further more, hydrostatic load can be changed step by step between normal water level and low water level in a few days. First, this makes it possible that the behaviors of dam and foundation rock can be analyzed while temmperature influence on behaviors of dam can be almost neglected. Secondary, useful information of behaviors about dam and foundation rock are provided, such as elasto-plastic deformation of rock, that is, the effect of crack or loosened zone on the behaviors of foundation, just like in-situ deformation test.

1 INTRODUCTION

Kaore Dam, a concrete arch dam 107.5 m high, is an upper reservoir dam of Okumino pumped storage Power Plant (capacity: 1,500 MW).

The Okumino Power Plant is located about 40 km north of Gifu City in the western part of Gifu Prefecture, bordering on Fukui Prefecture. Its upper reservoir, the Kaore Dam, located in the Nishigahoradani River at the uppermost reach of the Itadori River of the Kiso River system, has a gross storage capacity of 17,200,000 m^3 and an active storage capacity of 9,000,000 m^3. The lower reservoir is Kamiosu Dam, a rockfill dam 98 m high, located at the uppermost reach of the Neo Higashidani River, with a gross storage capacity of 14,500,000 m^3 and an active storage capacity of 9,000,000 m^3.

The Okumino Power Plant generates electricity by taking advantage of a difference in height of nearly 500 m in a system that links the upper and lower reservoirs via a 2.5-km waterway tunnel (consisting of headrace, penstock and tailrace), involving an underground power station situated in the middle of the waterway tunnel. Since both Kaore and Kamiosu Dams have relatively small catchment areas (2.5km² for the former and 12.0 km² for the latter), these reservoirs do not have enough inflow after test filling.

In both upper and lower reservoirs, test filling commenced in October, 1993, and was successfully completed in December, 1994. A large number of measuring were carried out to monitor the safety of the dam and foundation rock, to analyze their behaviors during test filling, and to evaluate design techniques. This papermainly reports on the structural behaviors of Kaore Dam and its foundation rock, based on the results of measurements taken during test filling.

2 OUTLINE OF KAORE DAM AND MEASUREMENTS

2.1 Outline of Kaore Dam

Kaore Dam is a concrete arch dam, 341.2 m in crest length and about 400,000m³ in volume. Its reservoir has an effective depth of 26.5 m, a normal water level of EL.1,015.0 m, and a design flood level of EL. 1,016.5 m. With a crest length about three times the dam height, Kaore Dam has a relatively longer crest length than ordinary arch dams. The arch center line is basically three-centered in configuration, and the arch is asymmetric in shape due to the valley configuration. Figure 1 shows a ground plan and a typical cross section of the Dam.

The foundation rock of the Dam comprises Omodani rhyolite of the Cretaceus period of the Mesozoic era, with small partial intrusions of porphyrite. It is relatively strong and hard overall. In construction, the abutment was built on grade C $_\mathrm{H}$ (rock classifi-

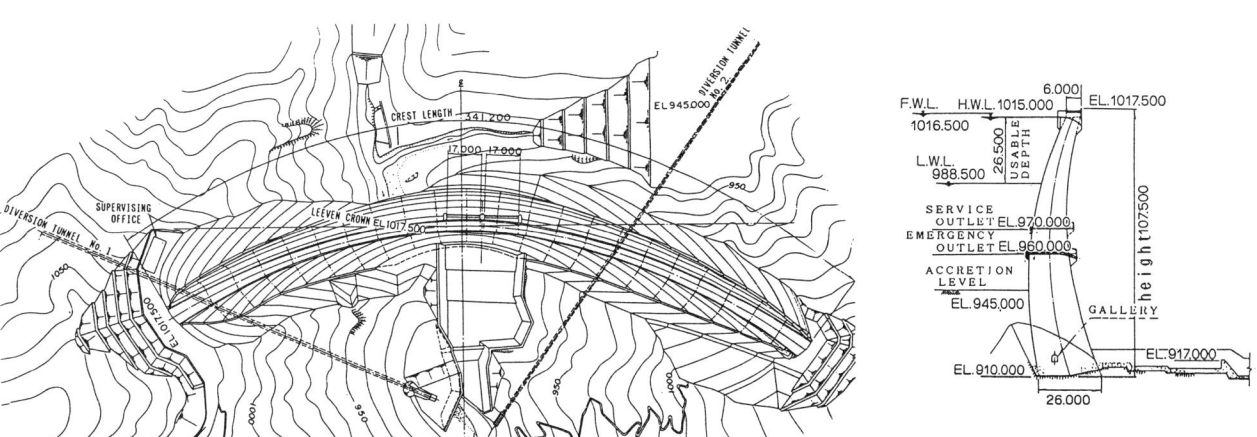

Figure 1. Plan and longitudinal section of Kaore dam

cation by the Central Research Institute of Electric Power Industry) bedrock. However, since grade C_M bedrock was found partially in the upper right bank on the downstream abutment side, dam thickness was increased in these places to reduce dam stress on the abutment.

Among a small number of fault zones found in the foundation rock, larger ones include fault zone A, which extends about 2 km almost parallel with the arch 15 m from the base toward the upstream (strike and dip: N64 - 68E/55 -60 SE).

There is no distinctive direction in which other fault zones of varying sizes run. Yet, on both the right and left banks, a weak concentration of fault zones is found running in an E - W direction toward a northern inclination of 60 - 80 degrees. On the right bank, this crosses the dam thrust at right angles; on the left bank, it runs parallel with the dam thrust toward the center of the foundation rock.

In addition to fault zone A, three fault zones were monitored carefully as to stability; they extend to the ground surface on the downstream dam abutment side on the right bank, although on a small scale. Figure 2 shows a representative geological slice at EL.945 m, a height about one-third the dam height.

2.2 Measurement

Figures 3, 4 indicate the types and locations of instrumentation installed in Kaore Dam and its foundation rock. The behaviors of the dam and its foundation rock, including contact area, were examined at three representative horizontal elevations, EL.990 m, EL.960 m and EL.930 m. All measurement results given below were recorded automatically by remote control. At present, measurement is taken on the hour as a rule, to allow necessary responses to water level changes due to pumping and power generation, in a daily cycle.

Measurement and measuring instruments used are summarized below.

(1) Dam temperatures
Dam temperatures were measured directly with embedded thermometers, strain meters or non-stress strain meters equipped with a thermometer, depending on location. In the horizontal direction, temperatures were measured at nine locations, including points 20 cm, 1 m, 2 m and 4 m from the dam surface and in the middle of the dam.
(2) Dam deflection
Dam deflection was automatically measured using plumblines at four locations, including two levels on the dam crown and one each on the right and left banks. As well, external deflection was manually measured at several locations downstream of the dam, with respect to off-dam references. In this report, external deflection is not discussed, due to limited space.
(3) Foundation rock deflection (reverse plumbline)
Reverse plumbline measurement was conducted at three locations, including the dam crown section and the right and left banks.
(4) Foundation rock deformation (horizontal rock deformeter, joint meter)
Measurement by horizontal rock deformeters was conducted in the direction of dam thrust at the normal water level on the downstream side and in the orthogonal direction on the upstream side. On the right bank, rock deformeters were installed at different depth, sandwiching the fault zone mentioned above, to monitor its behavior. A fixation point in the foundation rock was set at about 20 m from the rock-concrete interfaces on the abutments.

Rock joint meters were installed at three locations (upstream, downstream and in the middle) to measure deformation within a 5 m radius of the rock-concrete interfaces on the abutments. All rock joint meters measured deformation in the direction of dam thrust at the normal water level.
(5) Foundation rock deformation at riverbed (vertical rock deformeter in bedrock)
Vertical rock deformeters were installed at three riverbed locations, in inclined and vertical directions at each location, to estimate the effect of fault zone A on the behavior of the surrounding foundation rock. Deformeters were installed at four levels, with fixation points sandwiching fault zone A.
(6) Other
In addition to the above measuring instruments, dam stress and strain, leakage amount, uplift and behavior of the dam construction joint were measured. This report does not provide details, due to limited space.

3 ANALYSIS METHODS

The behavior of dam was analyzed using the trial load method, which had been used in the basic design stage, as well as the three-dimensional FEM, which was newly considered. The latter method was applied to a relatively detailed model to evaluate foundation rock measurement results.

Studies using models are outlined below.

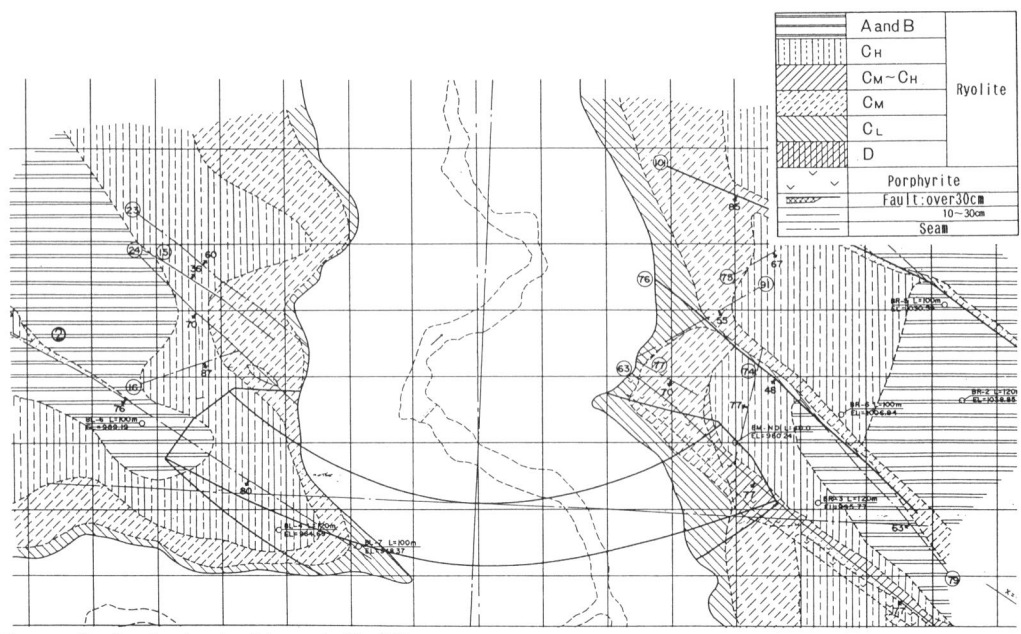

Figure 2. Geological slice at EL.945 m

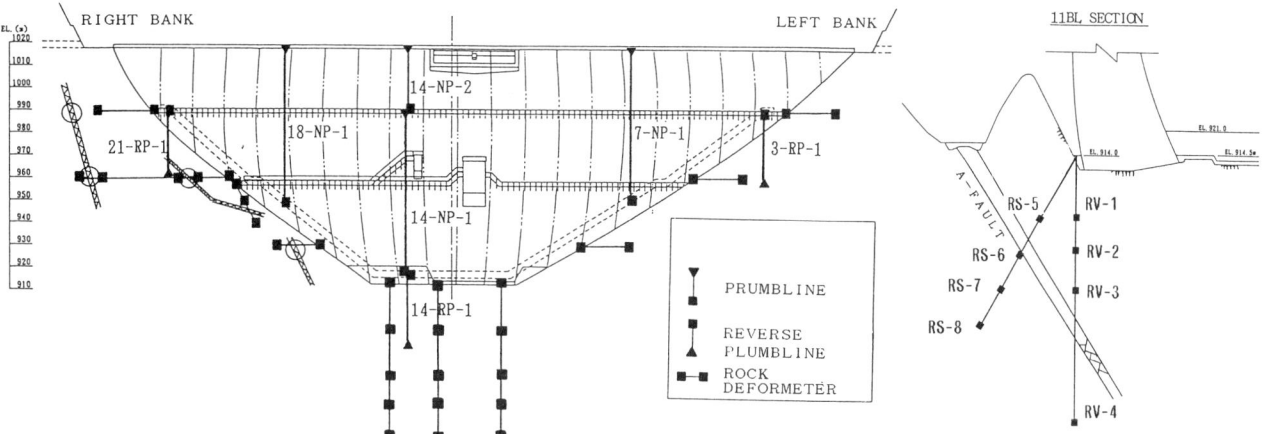

Figure 3. Locations of measuring instruments (development of downstream & longitudinal section)

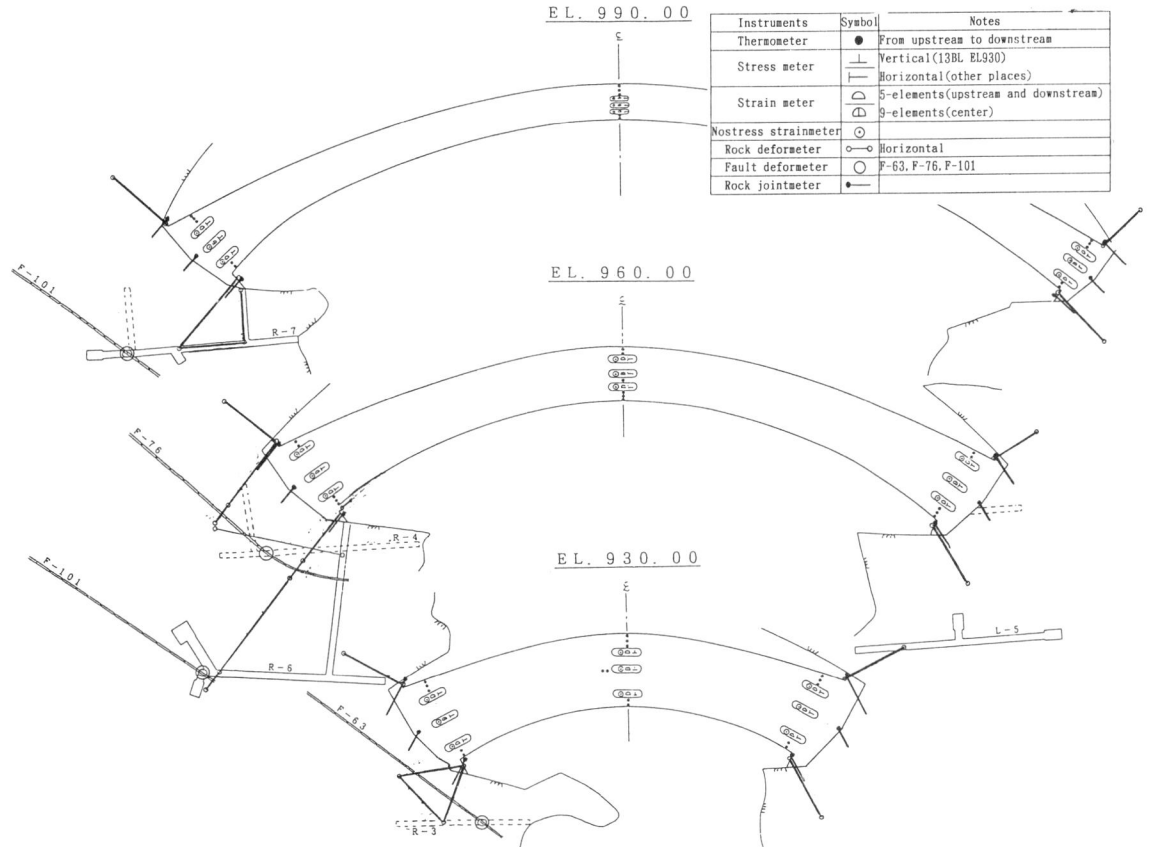

Figure 4. Locations of measureing instruments(plan)

3.1 Analysis by trial load method

Figure 5 shows a model of the trial load method. The principle of this method is omitted in this report. Table 1 shows material characteristics used in calculation, and analytic constants, such as temperature conditions. Loads taken into account in calculation

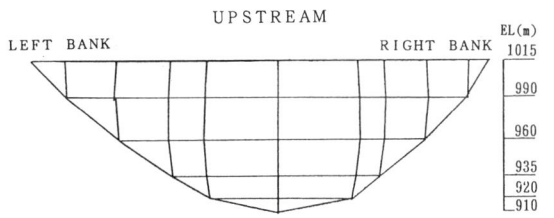

Figure 5. Model of trial load method

Table 1. Parameters in T.L.M and FEM

Concrete	Unit weight	2.35 (t/m³)
	Elastic moduls	2.1×10⁵ (kgf/cm²)
	Poissons ratio	0.2
	Coefficient of linear expantion	1×10⁻⁵ (1/℃)
Rocks	Elastic modulus	Right bank EL990m : 38,000 (kgf/cm²) Right bank EL1015m : 50,000 (kgf/cm²) Others 60,000 (kgf/cm²)
	Poissons ratio	0.25
Atmospheric temperature	Average in a year	9 ℃
	in summer	23 ℃
	in winter	−5 ℃
Water Temperature	Surface in summer	16 ℃
	Surface in winter	0 ℃
	Bottom	4 ℃

included hydrostatic pressure, thermal load and uplift. The material values used in the basic design were adopted.

3.2 Three-dimensional FEM

Figure 6 shows a model of the three-dimensional FEM, using the same material characteristics as in the trial load method. In this model, however, thermal load was not included; it is expected to be taken into account in the future. With regard to foundation rock, a decline in its rock elastic modulus in the right bank at a high elevation area and fault zone A were taken into consideration.

As for analysis programs, general-purpose ABAQUES and FENIX were selected and compared in terms of dam body models. With FENIX, into which high-precision 8-node elements were incorporated, analysis precision was equivalent to that by ordinary isoparametric 20-node elements, despite a smaller number of nodes. Since comparison confirmed these attributes, FENIX was finally adopted and applied to a model of 8-node elements.

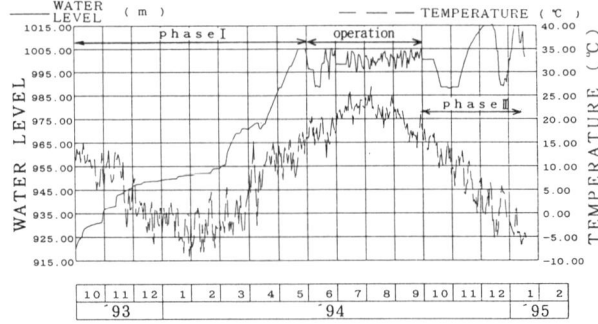

Figure 7. Reservoir water level and temperature

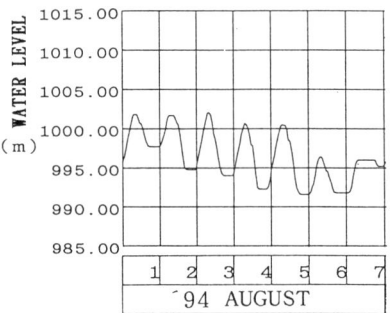

Figure 8. Reservoir watre level change in summer

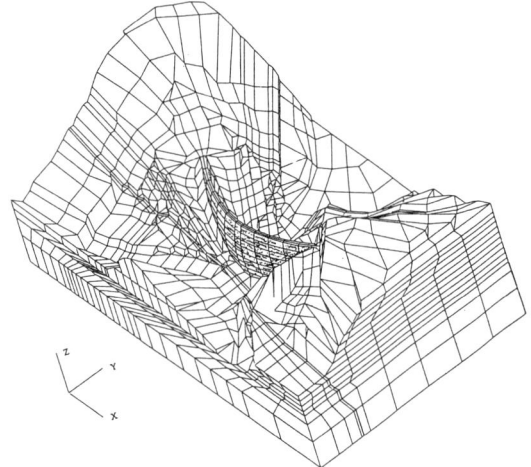

Figure 6. Model of three dimensional finite element method (FEM)

4. BACKGROUND OF TEST FILLING

Since the catchment area of Kaore Dam is very small in filling, the lower reservoir is filled first and pumped up to fill the upper reservoir. For this reason filling can be conducted while regulating daily water level rise according to a plan. As well, since the Okumino Power Plant is a pumped-storage power plant, the regulation of water level rise or drop in a daily to weekly cycle facilitates the observation of dam behavior.

Test filling commenced on October 1, 1993 and continued until the Dam was filled up to EL.1,005.2 m by May, 1994, the beginning of the summer flooding season (filling phase I). From June to September, 1994, following the commencement of partial operation by Units 1 and 2, operation tests using electric devices and water level changes were carried out within the range from EL.1,005.2 to the minimum operating level (summer operation period). Test filling resumed in November, 1994, and was completed on December 5, when the water level reached 1,015.0 m, the normal water level (filling phase II). Figure 7 shows daily changes in water level, air temperature and rainfall. Water level and air temperature data were taken at 9:00 a.m.

In the summer operation period from June to September, the water level changed in a daily cycle, although this is not indicated in the figure. Figure 8 shows a typical pattern of daily water level change.

5 MESUUREMENT RESULTS AND OBSERVATIONS

5.1 Dam temperatures

Among data from thermometers embedded in the dam at the EL.960 m crown section, Figure 9 shows, as representative examples, daily changes in dam temperature measured at locations 20 cm, 1 m, 2 m and 4 m from the downstream face and in the middle of the dam. From the surface to the inside, phase delay became larger, and amplitude smaller.

Figure 10 indicates temperature distribution inside the dam when filling began; in summer and winter; and when water reached the normal level. And further more, they show changes in temperature distribution when the water level rose by pumping during the first filling in phase I (April 18 - May 24), the operation tests in phase I (June 16 - June 20), and the final filling in phase II (November 8 - December 6), respectively. Although the temperature rose during filling phase I and dropped during phase II, the temperature was lower during phase I than during phase II, in absolute values. During the operation tests, the effects of only water level rise had to be examined, although it was not virgin load; since pumping was completed in only a few days, while the temperature showed almost no change.

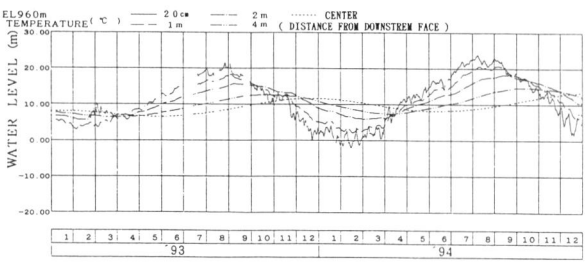

Figure 9. Comparsion of internal temperature of concrete

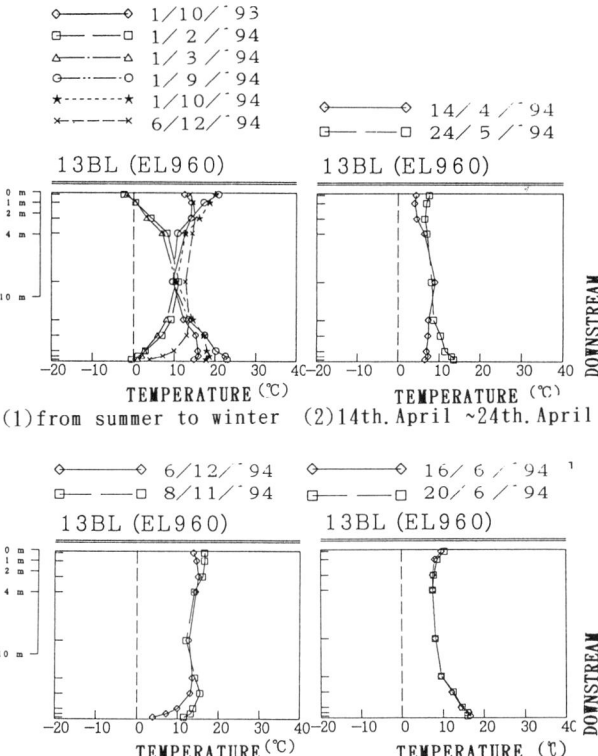

Figure 10. Distribution of internal temperature of concrete

5.2 Dam deflection

Figure 11 shows the relation between plumbline deformation and storage level, measured at the crown area (14 BL.), as a representative section of the dam. To compare measurement results with calculations, results by the trial load method and the three-dimensional FEM are shown together.

Until mid March, while the water level changed very little, the dam deflected severely toward downstream by about 25 mm at most. It is almost equal to the difference in amount of deflection between the temperature rising and falling periods as calculated by the trial load method. Considering that the mean temperature, measured at various locations ranging from the dam surface to the dam interior, peaked in October, when filling commenced, measured and calculated values were very close. To evaluate this deflection, actual differences in dam temperature distribution were linearly approximated and analyzed; point A in the figure indicates the calculated results of deflection. From this, it can be said that the measured deflection was caused by dam temperature change.

The bold line in the figure indicates deflection when the water level was on the rise due to pumping. The maximum deflection was about 50 mm in phase I and about 70 mm in phase II. Comparison between the two indicates that deflection gradients differed (that of phase II was larger than that of phase I) even at almost the same water level. This is believed to have been caused by the fact that dam temperature rising and falling tendencies, illustrated in Figure 10 were completely reverse in the two phases. The deflection gradient during the operation tests in phase I, shown with bold lines in the figure, indicates an intermediate tendency between the gradients of the two phases. The measured deflection gradient was lower than the calculated value, probably because the elastic modulus of the concrete (Ec) might be higher than the design value. The results of analysis by the trial load method, in which the elastic modulus was modified to 300,000 kg/cm^2, closely agreed with measured values.

5.3 Foundation rock deformation (reverse plumbline)

Figures 12, 13 show relations between storage level and deformation on the riverbed, and right and left banks. Taking as representative examples deflection in radial directions on the riverbed and in tangential directions on the right and left banks, basic deformation tendencies are summarized as follows:

(1) Filling phases I and II

The effect of the temperature drop until mid March was seen in deformation registering about 0.4 mm on the riverbed and left bank, and about 1 mm on the right bank. Deformation, by temperature change, was small at El.1,005 m and near the minimum operating level during about ten-day water level retaining period. This contrasts with the large deflection of the dam body itself during the same period. From this, it is considered that foundation rock deformation is more easily affected by long-term temperature change than by short-term temperature change.

Deformation during water level rise by pumping showed similar gradients in phases I and II; this also corresponded to FEM analysis. During a short-term water level drop (about 1 week; effect of temperature is small), a residual strain of 0.6 - 1 mm was found. During the water level rise in phase II, a sharp gradient similar to that of the water level drop was found below the level filled in phase I. While in the

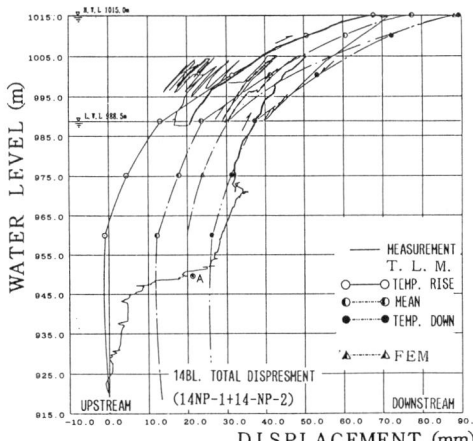

Figure 11. Relation between water level and radial diflection of crown cantilever, compared with analyses

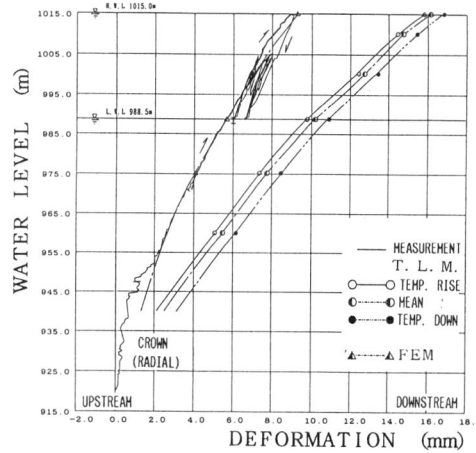

Figure 12. Relation between water level and rock deformation of at base, compared with analyses (reverse plumbline)

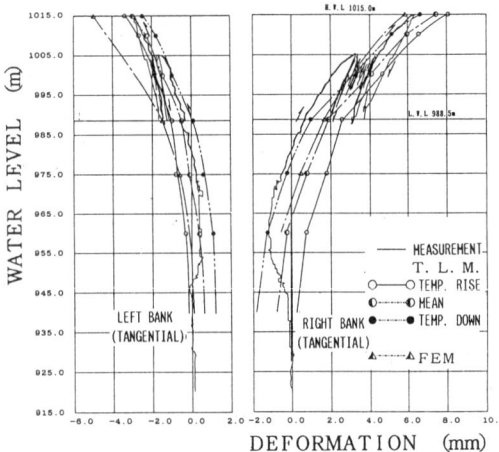

Figure 13. Relation between water level and rock deformation on the right and left banks, compared with analyses (reverse plumbline)

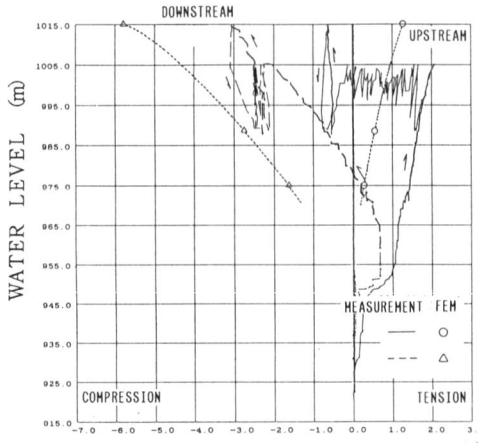

Figure 15. Relation between water level and rock deformation, compared with analyses (right bank at EL. 960m)

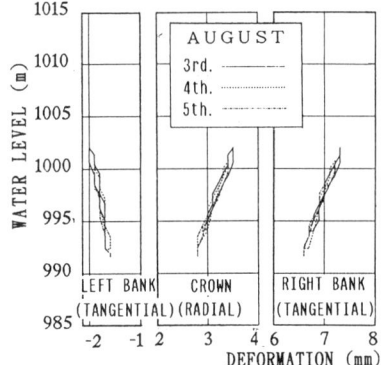

Figure 14. Relation between water level and deformation at base for three days in summer

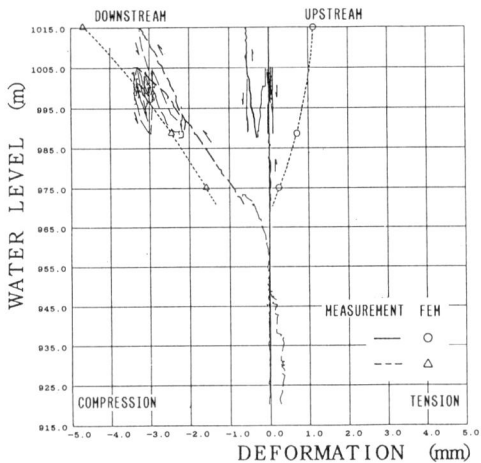

Figure 16. Relation between water level and rock deformation, compared with analyses (left bank at EL. 960m)

unfilled range, deformation occurred with a gradient that was an extension of deformation under virgin load in phase I. These results clearly reflect the elasto-plastic behavior of the foundation rock.

(2) Summer operation period
Figure 14 shows the foundation rock deformation measured on the hour for three consecutive days. Foundation rock deformation was slightly elastic and judged to be sound, from its hysteretic tendency.

5.4 Foundation rock deformation (horizontal rock deformeter, rock joint Meter)

Figures 15 - 18 show relations between water level and deformation, with measurement results at EL. 960 m as representative examples. Tendencies found are summarized below.
(1) Filling phases I and II
Residual strain occurred during the initial water level rise and the subsequent water level drop in phase I, showing elasto-plastic behavior. In phase II, deformation increased in the unfilled range, with a gradient similar to that under virgin load. On the whole, these tendencies agreed well with results by reverse plumblines. In comparison with analysis results obtained by three-dimensional FEM, foundation rock deformation by hydraulic pressure increase showed a very similar increasing tendency, although a little differences were found in absolute values due to seasonal temperature changes.

In terms of the effect of temperature drop until mid March, however, there was almost no deformation on the left bank, as compared to the right bank. It is not clear whether this was due to some problem with fixation points in the foundation rock or to dam be-

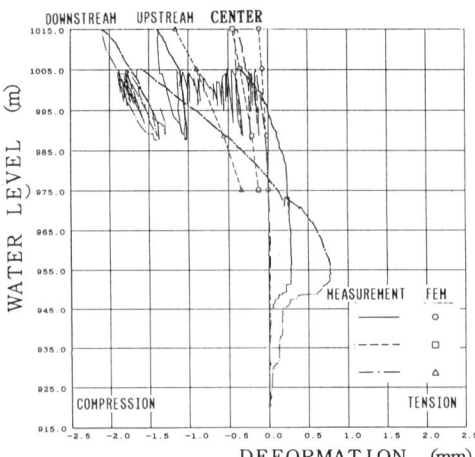

Figure 17. Relation between water level and rock deformation by jointmeter, compared with analyses (right bank at EL. 960m)

havior characteristics. Further observation of dam behavior during winter is seemed necessary.

The elastic modulus by in-situ tests was used in analysis. Since such values reflect deformation characteristics of foundation rock, analyzed results closely agreed with the measured values in the virgin

372

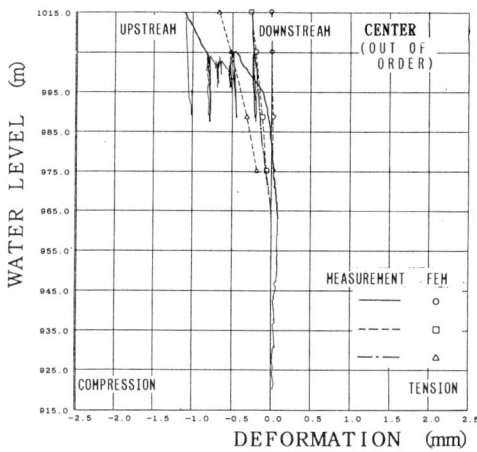

Figure 18. Relation between water level and rock deformation by jointmeter, compared with analyses (left bank at EL.960m)

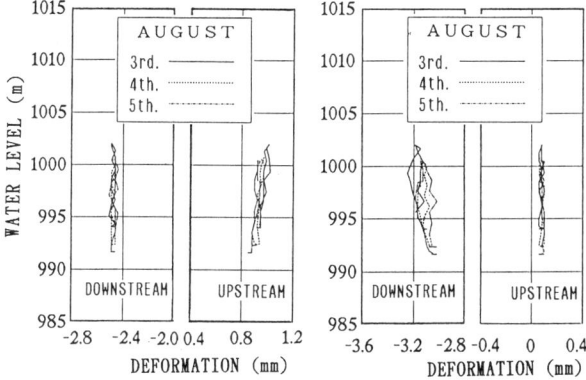

Figure 19. Relation between water level and rock deformation for three days in summer (EL.960m)

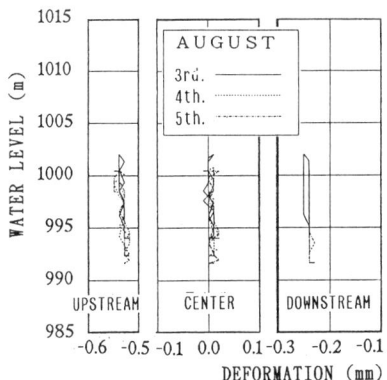

Figure 20. Relation between water level and rock deformation by jointmeter for three days in summer (right bank at EL.960m)

load range, which must be represented by a deformation modulus. Possible causes for this agreement include, as suggested by in-situ test resultsin Table 2, the determination of design values at the lower end of test results, or foundation rock treatment by sufficient consolidation grouting to a depth of 15 m.
(2) Summer operation period
Figures 19 - 21 show the results of deformation measurement on the hour for three consecutive days. According to these results , foundation rock deformation did not immediately follow rapid water level changes that occurred daily for 6 to 7 hours;rather, the fou-

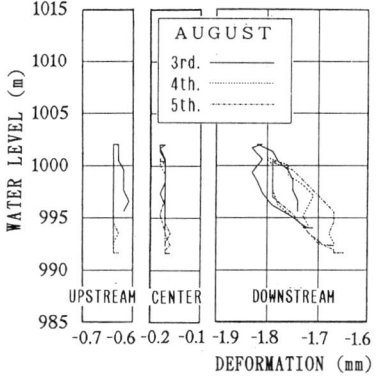

Figure 21. Relation between water level and rock deformation by jointmeter for three days in summer (left bank at EL.960m)

Table 2. Deformation modulus obtaiod by in-situ test

($\times 10^3$ kgf/cm²)

Classification	B	C_{II}	C_{M}
Elastic modulus	285.0~53.0	134.0 ~44.2	65.7~25.5
Average	129.9	63.7	39.3
Modules of deformation	231.0~26.5	88.3 ~18.2	38.4~11.0
Average	87.7	32.6	20.2
Number of samples	8	16	26

ndation rock showed behavior suggesting plastic deformation. It was not plastic deformation; however, the foundation rock is believed to have not been immediately affected by short-term load changes, since an elasto-plastic tendency was dete- cted in evaluation conducted on a yearly basis, as with seasonal changes.

5.5 Foundation rock deformation at riverbed (vertical rock deformeter in bedrock)

Since vertical rock deformeters had to be embedded in the foundation rock after foundation grouting, measurement commenced three months before filling. For this reason, measured values indicate the extension of ground, which had been compressed by the dead weight of the dam, triggered by hydraulic pressure.
With 11 BL. as example, Figures 22 and 23 show relations between deformation and storage level, and Figures 24 and 25 deformation distribution. Tendencies in these figures are summarized below.
(1) Filling phases I and II
As in the case of the foundation rock mentioned above, relations between water level and deformation suggest elasto-plastic behavior. In comparison with analysis results obtained by three-dimensional FEM, foundation rock deformation (extension side) caused by hydraulic pressure increase showed a very similar increasing tendency, in terms of relative deformation from the fixation point in the lower part in fault zone A. In terms of relative deformation at the fixation point in the upper part, measured values were larger than analysis values.
In deformation distribution diagrams, analysis results by three-dimensional FEM indicate that sectional deformation concentrated immediately above and below fault zone A. On the other hand, measured values indicate that deformation mostly occurred within a depth of about 20 m, irrespective of fault zone A. This condition closely agrees with the results of analysis, conducted on the assumption that the entire foundation rock was homogeneous, ignoring fault zone A, although absolute values differ. Almost identical tendencies were found at other locations where deformeters were installed, although those tendencies are not discussed here, due to limited space.
Such behaviors suggest that after the commencement

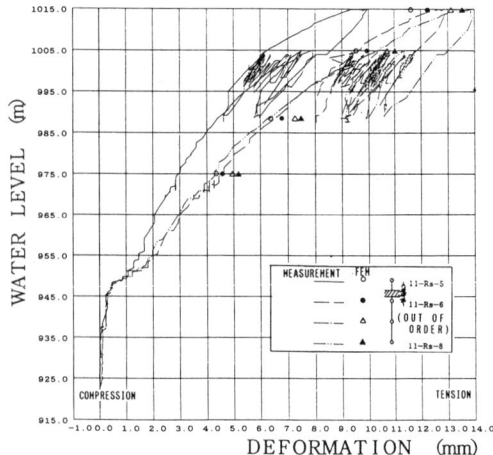

Figure 22. Relation between water level and vertical rock deformation, compared with analyses (11BL-RS)

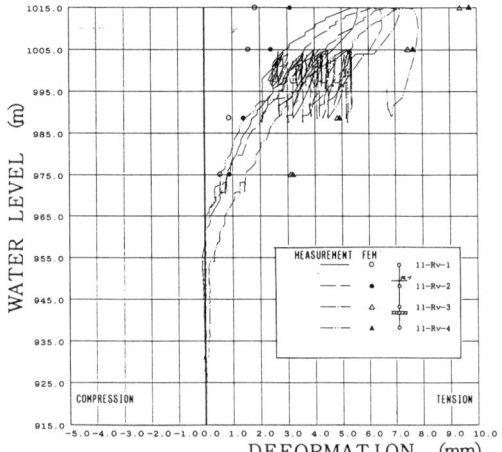

Figure 23. Relation between water level and vertical rock deformation, compared with analyses (11BL-RV)

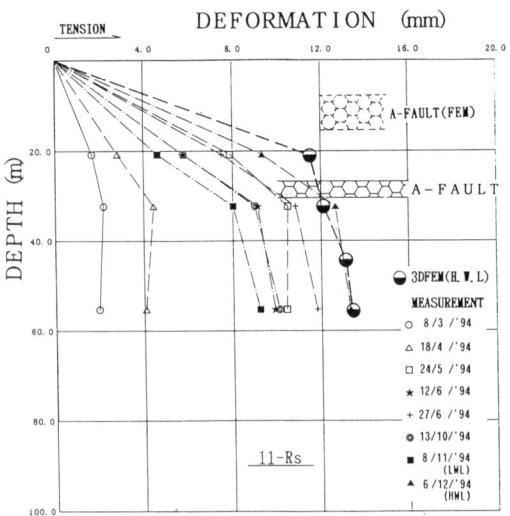

Figure 24. Distribution of vertical rock deformation, compared with analyses (11BL-RS)

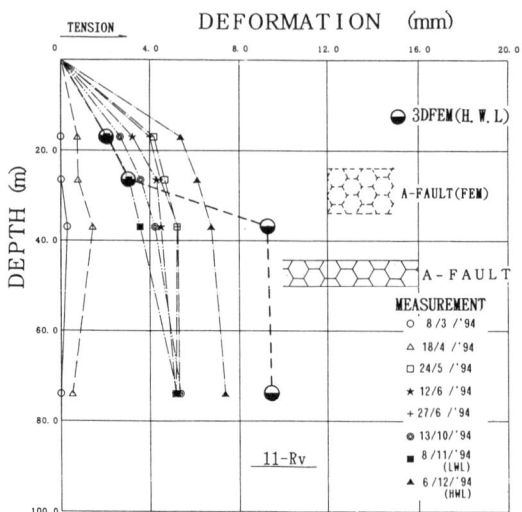

Figure 25. Distribution of vertical rock deformation, compared with analyses (11BL-RV)

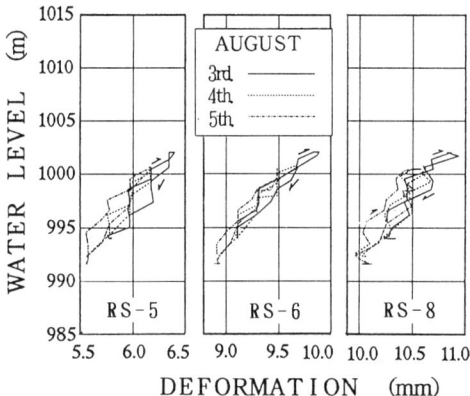

Figure 26. Relation between water level and vertical rock deformation for three days in summer (11BL-RS)

of filling, fault zone A did not experience excessive deformation and remained sound, like the surrounding parts of the foundation rock. Nevertheless, it is not clear whether this was because the design elastic modulus of fault zone A was underestimated, or whether fault zone rigidity increased due to plastic deformation caused by the dead weight of the dam before filling, since measurement was not conducted during construction.

(2) Summer operation period

Figure 26 indicates the results of deformation measurements measured on the hour for three consecutive days, with two points sandwiching fault zone A of 11 BL, as representative. The hysteresis suggests that the upper and lower parts delineated by fault zone A showed an elasto-plastic tendency while deformation in response to water level rise and drop, i.e. sound behavior for an arch dam foundation.

6 CONCLUSION

This report summarizes measurement results at Kaore Dam, whose test filling was completed in December, 1994. The measurement results revealed that the behaviors of the dam and foundation rock showed greater than expected proximity to analysis results based on design values. In the future, detailed analysis must be conducted concerning values indicating the structural properties of the dam and foundation rock, to evaluate measured values in a more theoretical manner.

Since filling was completed only recently, monitoring will be continued in order to monitor any changes due to aging, such as creep, in the dam and foundation rock.

In closing, the author wishes to thank Hazama Corporation's Technical Research Institute for their kind cooperation in FEM calculation.

On a quantitative management of dam grouting by real time analysis

Tamiharu Tashiro & Kenji Hayashi
Kajima Corporation, Tokyo, Japan

Kiyomi Mihashi & Kiminori Takahashi
Chemical Grout Co., Ltd, Tokyo, Japan

ABSTRACT: Although dams have usually been constructed on sounder rock foundation so far, the availability of potential sites decreases on account of frequent worse conditions in the recent geology.
The tendency brings up various problems on foundation treatment and increasing grouting amount as well. Especially grouting should play a vital role for reduction of grout amount under efficient and optimal specification and for prediction of the grouting effect.
The authors discuss a computer aided statistic analysis of the data of dam grouting on an assumption that the comparison of the data of the dam on the way with records of other dams can definitely predict the grouting effect quantitatively.
The paper reports the achievement especially in curtain grouting from an abundant data in the respective dam types, focusing on zoning the behavior, checking the specification, and evaluating the effect.

1 INTRODUCTION

Grouting is the basic, but major, work entailed in constructing a dam in order to stabilize the rock foundation and ensure proper water cut-off. However, the reality is that water permeability and rock improvement resulting from grouting has not been sufficiently clarified, at this point. Thus, much of management of the planning and construction works still remains to be largely dependent on experience.

The authors discuss how to analyze the data on foundation grouting through a statistical approach from a contractor's standpoint. The overall process will then be statistically analyzed to propose an ideal quantitative process control, by reviewing curtain grouting data of 10 dams.

2 OUTLINE OF EACH DAM

Table 1 summarizes the dam shape, foundation geology, and curtain grouting details. The breakdown consists of 1 arch dam, 6 gravity dams and 3 rock-filled dams, respectively. The following three characteristics are outlined in the table.

2.1 Dam (A) is an arch type which requires high pressure grouting to the very depths with large volumes of injections.

2.2 The 6 gravity dams stand on either the tertiary or Mesozoic and Paleozoic eras foundations, and the respective layers require fairly different grout take.

2.3 Rock-filled dams stand on the foundations in tertiary, Mesozoic and Paleozoic eras, however, the average cement injection for each is around 30 kg/m.

3 QUANTITATIVE ANALYSIS

The authors point out the following items for a quantitative analysis:

3.1 Zoning unusual conditions on the grouting map

Foundations of dams are not always brittle, other than in special cases when the overall foundation geology of the dam is composed of soft rock. Therefore, there is a need to combine data analysis with known geology information to point out unusual grouting conditions for review.
A correlation of Lugeon Number (hereafter called Lu) and cement injection volume(per meter)(hereafter called C) plays a vital role to extract zones into high Lu with low C zones and low Lu with high C zones.

Table 1. Outline of Each Dam

Dam	Type	Height (m)	Volume (10,000m³)	Geology	Grouting (m)	C (kg/m)	Pattern	P (kgf/cm²)
A	Arch	140	66	Neocene period.tuffbreccia,diorite.	108,343	55.7	3m/C*	50
B	Gravity	100	116	Green tuff from the tertiary period. welted tuff,tuffbreccia,basalt.	30,290	36.9	3m/C	25
C	Gravity	156	200	Green tuff from the teriary period. volcanic breccia,volcanic conglmeeate tuff.	117,436	33.5	3m/C	25
D	Gravity	81	29	Green tuff from Neocene period.spotted breccia,andsite,porphyrite.	19,204	19.8	1.5m/O*	25
E	Gravity	117	109	Mesozoic and paleozoic eras.amphibolite.	26,378	9.4	1m/O	30
F	Gravity	38	9	shale of the Cretaceous period of the early Mesozoic era.	10,446	20.5	1.5m/O	25
G	Gravity	42	9	complete crystal of the Cretaceous period of the early Mesozoic era.	11,952	7.6	3m/C	20
H	Rock-f	63	193	exposed to Tertiary green tuff alter- rations volcanic andsite.	69,915	32.5	1.5m/C	20
I	Rock-f	62	162	The left shore and riverbedd is composed of Tertiary,new Mesozoic breccia pyroclastic rocks.	72,735	33.9	3m/C	20
J	Rock-f	89.5	270	muddy fragmeted hornfels of the Mesozoic and paleozoic eras.	29,798	35.9	1.5m/C	30

* C:checkerd O:one row

3.2 Ensuring Proper Injection Specifications

We have tried to check whether the conventional specifications of dam grouting prove most proper, by comparing all possible data from each dam, and drawing figures, accordingly.

3.3 Evaluation of the injection effect

As a rule, the following guidelines have to be satisfied to determine the injection effects:
a. That 85-90% of the overall improvement goal has been attained for the target area.
b. That there are no areas that does not reach (less than half) the improvement goal.
c. There are no corrective portions that cannot attain the improvement goal in both horizontal and vertical directions.

a should be evaluated macroscopically from probability of excedence, while b and c microscopically from the Lugeon Map.

However, as the Lu indicates the value prior to grouting, we are unable to determine the Lu number after injection. Thus, we need to focus on such items in addition to the extra-probability figure and Lugeon map, as Lu per every split spacing, the mean value of C, and standard deviation in addition to the probability of excedence and the Lugeon map. Figure 1 clarifies the effect of improving and the correlation between the Lu before/after injection.

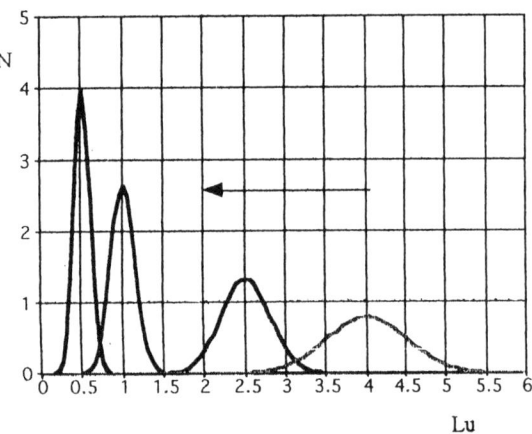

Figure 1. the Effect of Improving and the Correlation of Lu

4 ZONING OF UNUSUAL GROUTING CONDITIONS

4.1 The procedures

a. Studying the correlation of Lu and C, selecting the following two as Lugeon numbers.

L(10): Lugeon number under the pressure of 10 kgf/cm^2 (980kPa)
L(P): Lugeon number under maximum pressure.

Selecting the following two for the cement weight:
C: Cement weight per meter (kg/m).
C/P: Injection ratio (p indicates the maximum injection pressure).

b. Figure 2 indicates the correlation coefficient of each dam. The correlation between L(P) and C/P has proved most critical.
c. Seeking a regressive equation based on the above mentioned correlation between L(P) and C/P.
d. Picking up a high Lu and low C area and low Lu and high C and a certain deviation from the regressive equation. Plotting the former and latter on the map as the overstandard and the substandard, respectively.
e. Comparing the standard portion to the unusual conditions portion.

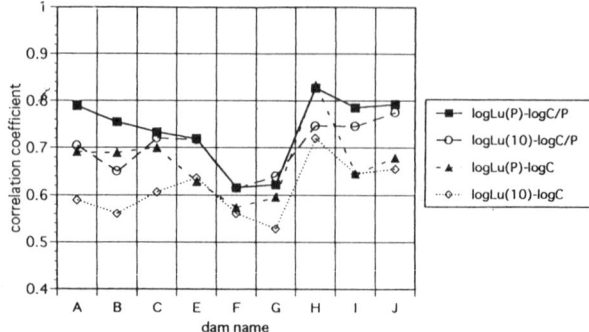

Figure 2. Comparison of Correlational Coefficient of Each Dam

4.2 The results

a. The total degree of concrete and rock-filled dams computed with the L(P):C/P regressive equation are indicated in Figure 3 and Figure 4, respectively.

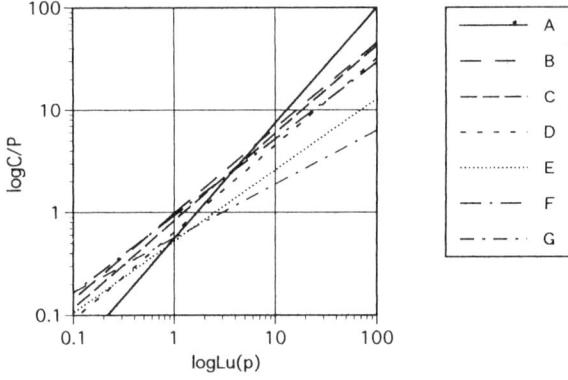

Figure 3. Comparison of Recovery Formula of Concrete Dams

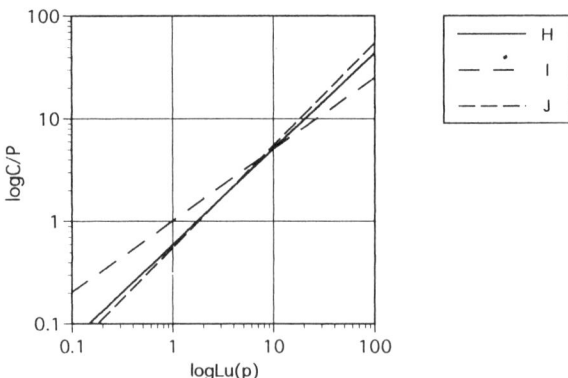

Figure 4. Comparison of Recovery Formula of Rock-Filled Dams

b. The regressive equation for the pilot holes and final designing of holes are indicated in Figure 5 and Figure 6, respectively.

c. The final designing of holes and sites of additional hole maps upon curtain grouting Dam C are indicated in Figure 7 and Figure 8, respectively, as an example of mapping unusual processes.

d. The sub/overstandard sites and faults mapped for consolidation grouting of the same Dam C are indicated in Figure 9 and Figure 10, respectively.
e. Table 2 indicates the general and unusual injection details of each dam.

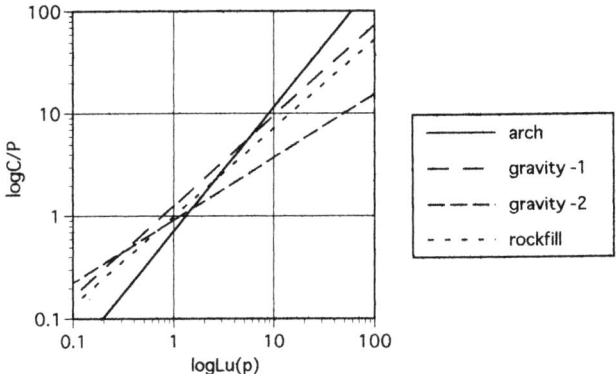

Figure 5. Comparison of Recovery Formula of Pilot Holes

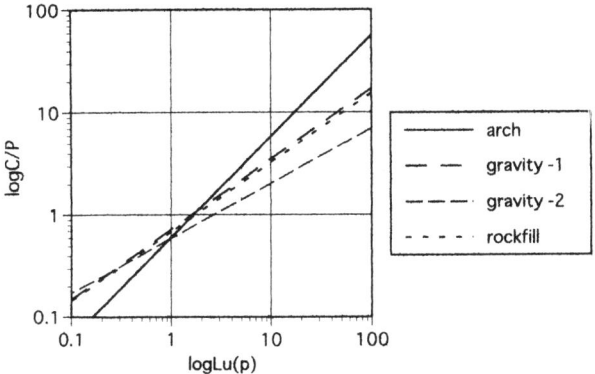

Figure 6. Comparison of Recovery Formula of Final Hole Design

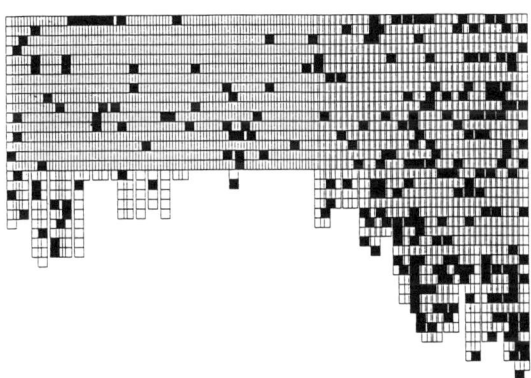

Figure 7. Map of Unusual Curtain Area in Dam C

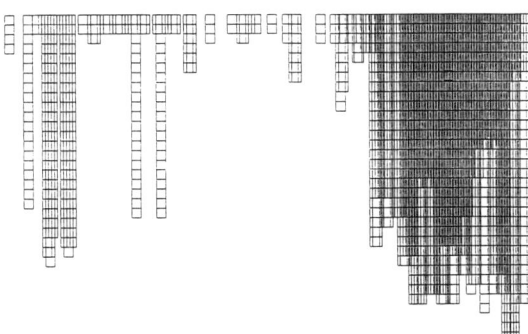

Figure 8. Map of Additional Curtain Holes in Dam C

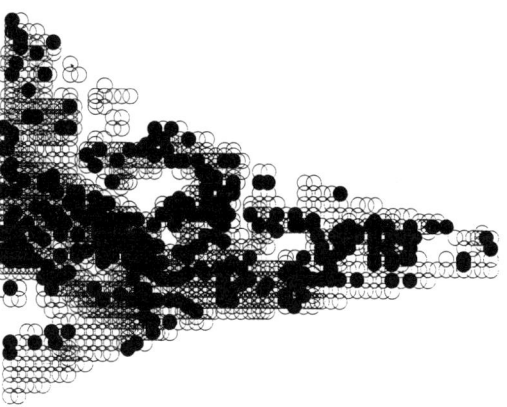

Figure 9. Map of Unusual Consolidation in Dam C

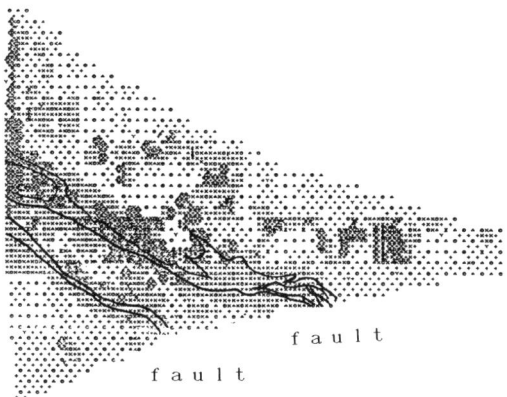

Figure 10. Map of Fault in Dam C

4.3 The details of the study.

a. According to Figure 2 we find that Dams E, F, G and J have fairly small correlational coefficient differences between L(10) and L(p). The geological foundation of these dam sites are the Mesozoic and Paleozoic eras, and are fairly hard rock foundations, other than Dam J, in which cement injection volume is also the same on the average.

b. Figures 3 and 4 shows the respective regressive equation and as for the arch dam, gravity dam (tertiary period), gravity dam (Mesozoic and Paleozoic eras) and rock-fill dam through grouting.

c. The injection volume for the same Lu was in the order of an arch dam, gravity dam (tertiary period), rock-filled dam and gravity dam (Mesozoic and Paleozoic eras).

d. According to Figures 5 and 6, the C is lowest in all groups with the progress of split spacing. This seems grouts are injected through large fractures.

e. It would be possible to forecast the improvement tendency even at the early stage of injection, up to the final design stage by comparing a dam with similar geological foundation and type.

f. According to Figures 7 and 8 we find that the substandard portions corresponds at the final hole design stage and the areas where extra holes are required.

g. According to Figure 9 and 10 we find that the substandard holes correspond to faults.

h. Table 2 is a comparison of the over/substandard of injection. this enables us to define the width/direction of seams and cracks. Dam J, for example, has a smaller upper-limit injection efficiency(C/Lu) compared to other dams, however, the lower-limit is a mean value. This seems to indicate that the width of the crack is smaller, but more numerous compared to other dams.

5 PROPER INJECTION SPECIFICATIONS

5.1 Injection Pressure

The following three tendencies are generally found when summarizing the mean per degree of Lu and C for the injection pressure of each dam, on a graph.

a. Type A (Figure 11. Refer to Dam B; Dams C, D, H & I are similar)
C increases as the injection pressure rises, while Lu decreases while the degree progresses.
Therefore, it is thought that the comparative injection pressure is appropriate.

b. Type B (Figure 12. Refer to Dam A; no similar dams)
C increases as the injection pressure rises, however, there is practically no drop in Lu.
The injection pressure is high compared to other dams.

c. Type C (Figure 13. Refer to Dam J; Dams E, F and G are similar)
C does not increase much with the rise in injection pressure, and there is practically no drop in Lu.
There is a possibility that the Lu will drop with higher injection pressure.

As indicated above, excessively high injection pressure results in wasting injection materials, and excessively low pressure results in wasteful, additional boring. Based on this fact, it is important to compare the injection pressure of other dams to decide upon the injection pressure that best suits the rock foundation of the dam.

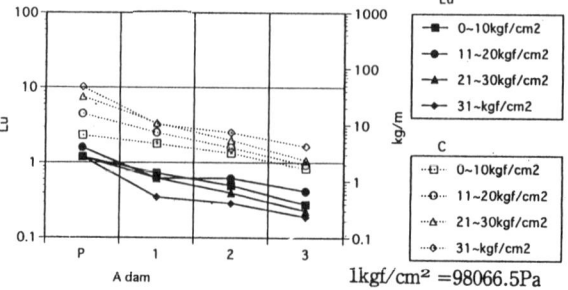

Figure 12. Lu Degree and Mean C Volume per Injection Pressure in Dam A

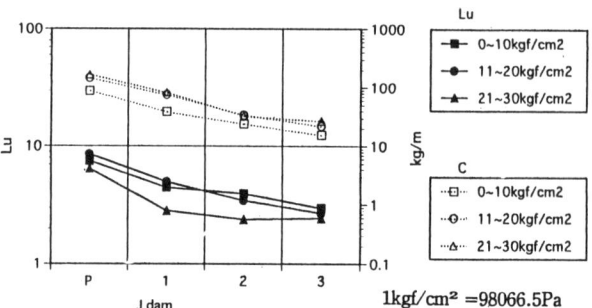

Figure 13. Lu Degree and Mean C Volume per Injection Pressure in Dam J

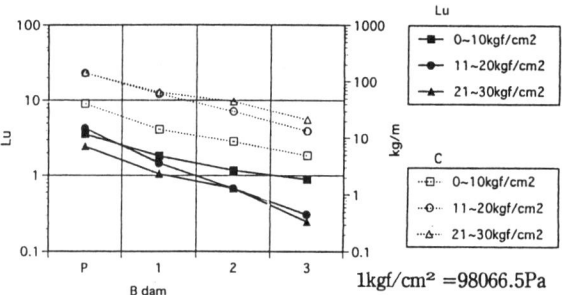

Figure 11. Lu Degree and Mean C Volume per Injection Pressure in Dam B

5.2 Injection Mix

Table 3 indicates an example of the final mixture ratio for dam A dam according to Lu. Table 4 indicates the 95% and 5% non-exceeding probability mixture of each Lu. Figure 14 indicates the common final mixture and starting mixture standard of each dam by Lu.
Figure 15 summarizes the 95% reliability of the final mixture by common Lu and pressure based on the injection pressures of the data.

The following was found as a result of this information.

a. According to Figure 14 the ideal starting mixture of C:W by Lu would be: 1:10 for Lu 0-10, 1:8-1:6 for Lu 10-20 and around 4 for 20 Lu and over.

b. It seems that the initial mixture standard for 9 dams is often excessively dilute mixtures, though this situation suited some dams.

c. According to Figure 15 we find that the final mixture becomes concentrated as the injection pressure rises, if the Lu is high. In other words, it is advisable to consider injection pressure and Lu upon determining the starting mixture.

d. Table 5 is an example of a starting mixture standard by injection pressure, summarized from the statistical data of the 9 dams studied.

e. Table 3 indicates the mixture switching standard. This makes it possible to add intermediate mixtures when the degree is extremely low upon switching the mixture, based on the final mixture degree by Lu and pressure information in the statistical tables.

Table 2. The General and Unusual Injection Details

Injection of Overstandard Sites (mean)

Dam	A	B	C	D	E	F	G	H	I	J
Lu	6.46	0.70	1.12	1.56	0.73	0.69	1.51	1.42	0.98	2.28
C	504.89	79.86	72.52	79.90	28.97	86.56	8.52	46.70	37.18	35.24
C/Lu	74.68	121.22	67.38	51.35	39.79	124.43	5.64	33.20	38.26	15.45

Injection of General Sites (mean)

Dam	A	B	C	D	E	F	G	H	I	J
Lu	4.41	0.54	0.64	0.78	0.46	0.85	1.05	1.04	0.81	2.06
C	84.87	11.32	12.31	5.66	6.05	10.68	5.30	10.58	10.42	14.28
C/Lu	20.42	22.03	20.15	7.29	13.09	12.59	5.09	10.23	12.94	6.95

Injection of Substandard Sites (mean)

Dam	A	B	C	D	E	F	G	H	I	J
Lu	4.57	0.66	0.83	1.54	0.61	0.47	1.10	1.21	0.97	2.45
C	26.11	4.64	6.50	3.61	3.39	2.02	3.50	5.69	6.64	11.46
C/Lu	5.73	7.20	8.01	2.35	5.55	4.31	3.22	4.70	6.82	4.69

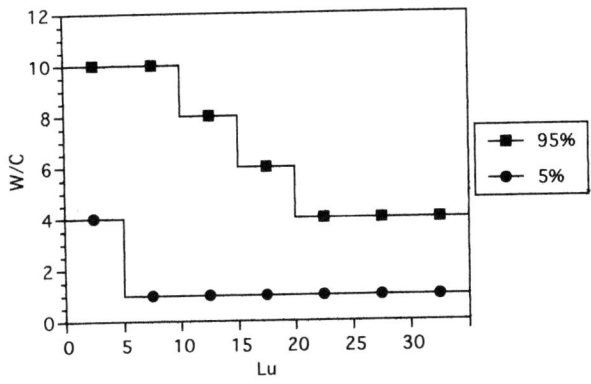

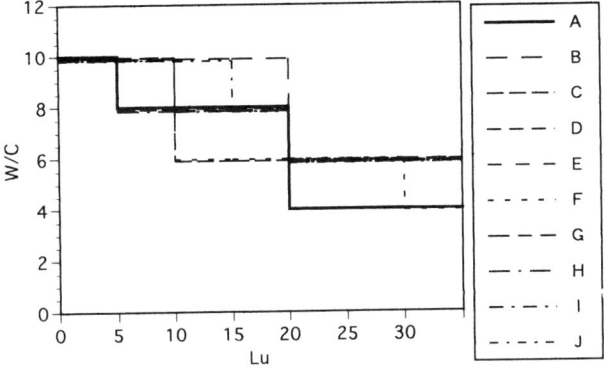

Figure 14. The Common Final Mixture and Starting Mixture Standard in Each Dam

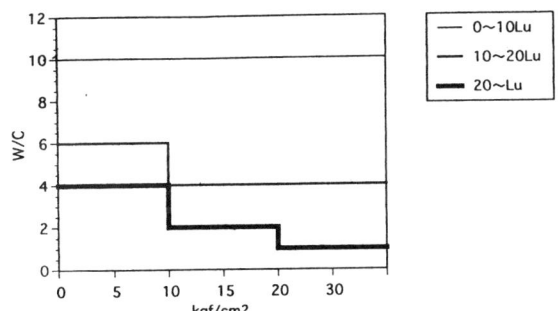

Figure 15. 95% Probability Final Mixture by Lu Degree and Injection Pressure

Table 3. Final Mixture Degree of Dam A (%)

Mixture (c:w) \ Lu	0~1	1~2	2~5	5~10	10~15
1:10	85	61	28	9	1
1:8	7	16	17	7	–
1:6	3	9	15	14	9
1:4	2	8	16	19	13
1:2	1	3	11	15	12
1:1.5	1	1	7	13	21
1:1	1	2	5	17	31
1:0.8	–	–	2	6	13
Total(%)	100	100	100	100	100
Data	962	855	136	661	179

Mixture (c:w) \ Lu	15~20	20~25	25~30	30~
1:10	–	–	–	–
1:8	1	–	–	–
1:6	4	–	–	2
1:4	11	6	8	–
1:2	20	15	8	–
1:1.5	16	15	13	8
1:1	30	29	38	47
1:0.8	18	35	33	43
Total(%)	100	100	100	100
Data	71	34	24	53

Table 4. 95% Non-Exceeding Probability of Finaly Mixture of each Dam

Dam \ Lu	0~1	1~2	2~5	5~10	10~15
A	10	10	10	10	6
B	10	10	10	10	10
C	10	10	10	8	6
D	10	10	10	10	8
E	10	10	10	10	8
F	10	10	10	10	8
G	10	10	10	10	8
H	10	10	10	8	8
I	10	10	10	8	4
J	10	10	10	8	6

Dam \ Lu	15~20	20~25	25~30	30~
A	6	4	4	1.5
B	6	4	1	1
C	4	4	1	2
D	8	8	6	–
E	6	4	4	2
F	6	6	4	4
G	6	4	4	4
H	6	4	4	4
I	3	4	4	3
J	4	2	2	2

5% Non-Exceeding

Dam \ Lu	0~1	1~2	2~5	5~10	10~15
A	4	2	1	0.8	0.8
B	2	1	1	0.8	0.8
C	2	1	1	1	1
D	10	1	1	1	1
E	10	6	2	1	1
F	2	1	1	1	1
G	6	2	1.5	1	1
H	4	1	1	1	1
I	6	6	4	1	1

Dam \ Lu	15~20	20~25	25~30	30~
A	0.8	0.8	0.8	0.8
B	0.8	0.8	0.8	0.8
C	1	1	1	1
D	1	1	1	–
E	1	1	1	1
F	1	1	1	1
G	1	1	1	1
H	1	1	1	1
I	1	1	1	1
J	1	1	1	1

Table 5. Example of Standard Starting Mixture

Lugeon Unit	Injection Pressure	Starting Mixture
0 ~ 10	0 ~ 10 (kgf/cm²)	1 : 10
	10 ~ 20	1 : 10
	20 ~	1 : 10
10 ~ 20	0 ~ 10	1 : 6
	10 ~ 20	1 : 4
	20 ~	1 : 4
20 ~	0 ~ 10	1 : 4
	10 ~ 20	1 : 2
	20 ~	1 : 1

1kgf/cm² = 98066.5Pa

6 EVALUATING THE INJECTION EFFECTS

6.1 Figure 16 indicates the mean and standard deviation of the Lu and C degree of each dam.
This figure gives us the following information.

a. The Lu tends to be the same for all dams other than A and J. The regressive equation is as follows:

$\log \sigma = 0.372 + 0.663 \log Lu$, at a 0.95 correlation, which is fairly high.

b. The drop in mean value was limited in Dam J. The mean value was fairly small, with a limited drop in standard deviation in Dam A, which requires a comparative study with other dams.

c. The C value was similar in all dams.

As a result, it would be possible to review design degree during construction by studying the progress of Lu and C, found to be different from the recovery formula as indicated in a. above.

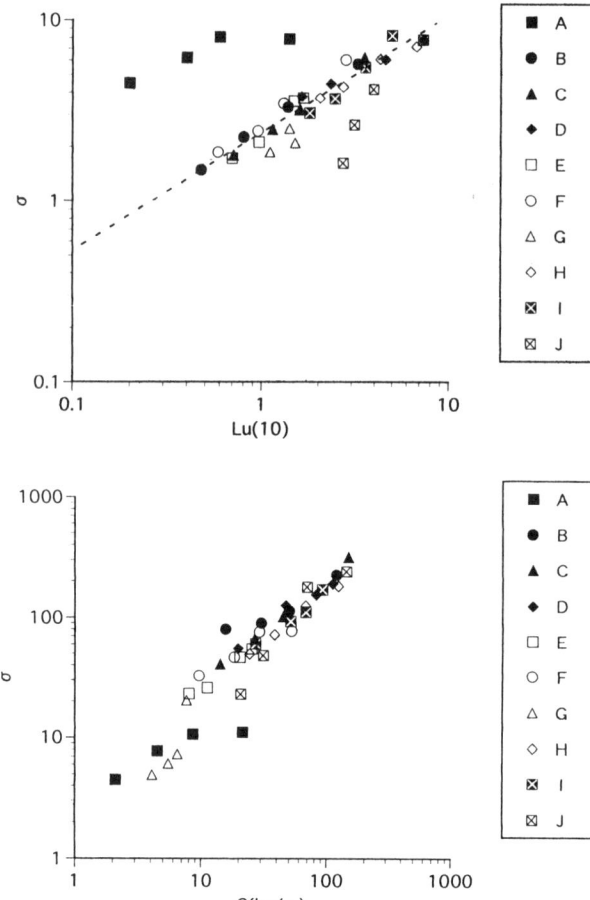

Figure 16. Mean and Standard Deviation of the Lu and C Degree

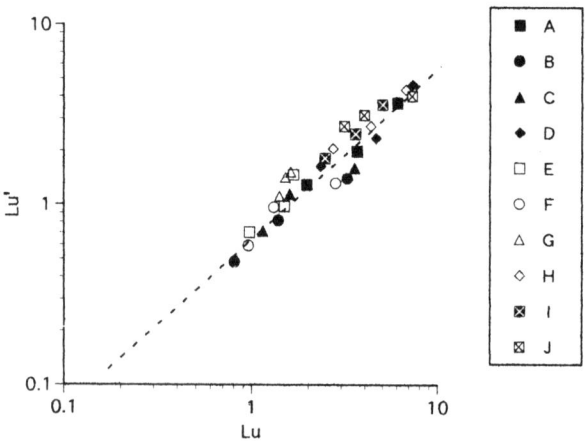

Figure 17. Pre-and-Post Injection Lu Degree of Each Dam

6.2 Figure 17 indicates the mean Lu of the subject Lu degree after injecting the mean Lu degree.

The following information was found based on this figure.
a. The differences in each dam are minimal and resemble each other to a degree.
b. Based on a linear recovery trend the correlation would be 0.95 with a recovery formula of:
log Lu ′ = 0.945 log Lu - 0.167

Therefore, it would be possible to forecast the Lu after injection at a high probability if there is a high correlation of yielding such reliability. Measures of some sort would be required if it does not follow this pattern.

7 CONCLUSION

The following dam grouting procedures are advised based on the quantitative analysis of grouting of rock foundation of dams.

7.1 Categorize the subject dam based on previously constructed dams with similar features (currently 10 dams).

7.2 Compare the improvement tendencies of similar dams to conduct the following progress control.

a. Specify unusual conditions for grouting an early stage for review.
b. Determine the proper injection specifications in terms of injection pressure and injection mixture based on the data of previously constructed dams with similar geological characteristics. Change the specifications if there are any unusual sequences.
c. The injection effect is determined based on the regressive equation computed from the graph indicating the mean and standard deviation. Make the decision for final improvements by forecasting the Lu after injection, for portional injections.

7.3 Accumulate data on the dams

A practical quantitative analysis would be feasible during execution if the above are observed. Future applications will focus on quantitative analysis by collecting and reviewing information on dams currently under construction and data on dams that have not been reviewed.

REFERENCES

Nagayama, I., Yoshinaga, T., Tsugaki, A. & Tashiro, T. 1984. Consideration on analysis and control of foundation of the dam: the 6th Japan Symposium on Rock Mechanics. Tokyo, Japan.

Nagayama, I. & Yoshinaga, T. 1984. Considerations on characteristics of grouting for foundation rock of the dam: the 16th Symposium on Rock Mechanics. Tokyo, Japan.

Tashiro, T., Hayashi, K., Mihashi, K. & Takahashi, K. 1994. On a quantitative management of dam grouting: the 9th Japan Symposium on Rock Mechanics. Tokyo, Japan.

Management of the excavation of foundation of the Ohta Dams constructed on soft rock

S. Yasufuku & M. Kageyama
The Kansai Electric Power Co., Inc., Japan

H. Kakuhara
NEWJEC Inc., Japan

1. FORWARD

The five dams given a general name of "Ohta dams" were constructed around the upper regulating reservoir of the Ohkawachi pumped-storage water power station (operation in 1992, Completion in 1994, 1,280 MW) by the Kansai Electric Power Company.

The location of the Ohta dams is the east-end of the Chugoku Mountains near the port cities of Kobe and Osaka in Japan. (Ref. Fig. 1)

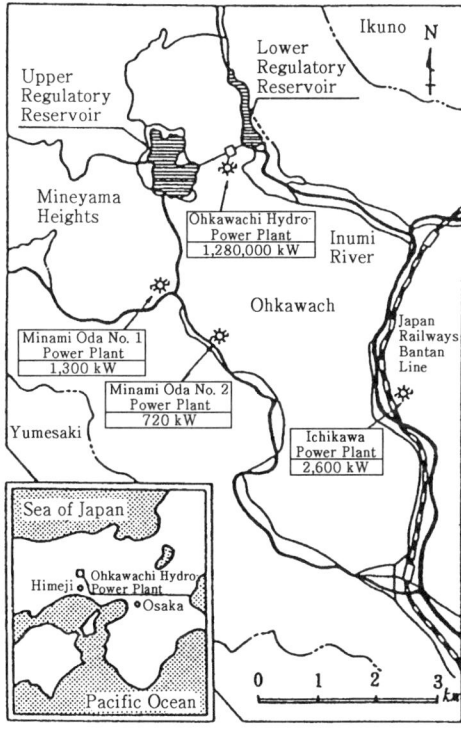

Fig. 1 Map of Power Plant

The heights of the five dams (Rock-fill with central core type) vary from 23.5 m to 55.5 m.

Ohkawachi power plant is the fourth large-scale pumped storage plant in the Kansai Electric Power Company following the plants of Kisenyama (1970, 466,000 kW), Okutataragi (1974, 1,212,000 kW) and Okuyoshino (1978, 1,206,000 kW). The dams of the upper regulating reservoir of the three previous power plants were all constructed by the rock-fill method. Compared to concrete dams the topographical and geological conditions of rock-fill dams are more advantageous in terms of strength of foundation and ability to deform. In addition, levee material is more easily obtainable in the vicinity of the dam site. Moreover, the core portion of the foundation excavation is CM class or higher (with portions of CL class at higher elevations), but for the rock portion, large rock masses which are stronger than banking material can be secured, provided that such organic materials as surface soil, talus cones and tree roots are eliminated. Thus, the foundation's excavation and management did not require a great deal of labor power.

The conventional method of foundation excavation was insufficient for the construction of the Ohta Dams, due to the presence of a thick layer of weathered rocks (D class rock mass). As countermeasures, a reinvestigation of levee gradient, counter-weight fills at the toe of slopes, and alternative excavation of soils of differing strengths were examined. Taking into account the volume balance of the banking soil and economic considerations, it was decided to excavate the rock portion of the foundation as deep as possible while still attaining the necessary shearing strength, and retaining the weathered rocks.

Excavation of Ohta Dam No. 3 began in May, 1989, with embankment of all dams completed in December, 1991. In June, 1992, water impounding was initiated, and commercial operation began in October of that year. Since then, all dams have been operating smoothly.

This paper will introduce the excavation management methods of the D class rock masses which are the foundation of the rock portion of the Ohta Dams.

2. GENERAL DESCRIPTION OF THE OHTA DAMS

The five Ohta Dams were constructed in tandem. The No. 1 and No. 2 dams control water flow on the Ohta River, which is a tributary of the Odawara River of the Ichikawa river system, and is located in the town of Ohkawachi in the Kanzaki district of Hyogo Prefecture. The No. 3, 4 and 5 Dams were built on the eastern saddle and right saddle of Ohta Dam No. 1, where the elevation is lower than the topography at the planned high water level (see Figure 2). Table 1 shows the planning factors of the Ohta Dams.

Figures 3 and 4 show standard sectional drawings and vertical sections of Ohta Dam No. 1, which has the highest dam height of the five dams, and Ohta Dam No. 3, which was reported herein as a management example.

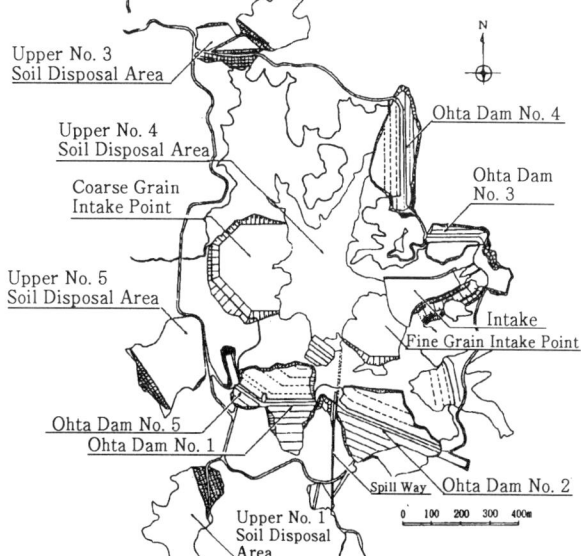

Fig. 2 General Site Plan of the Ohta Dams

Table 1 Design Data for the Ohta Dams

Dam	Name	Ohta No. 1 Dam	Ohta No. 2 Dam	Ohta No. 3 Dam	Ohta No. 4 Dam	Ohta No. 5 Dam
	Type	Center Impervious Wall Type, Rock-Fill Dam				
	Height of Dam	55.5m	44.5m	23.5m	26.0m	26.5m
	Length of Dam	175.3m	397.1m	208.0m	406.0m	108.5m
	Volume of Dam	693,000m³	957,000m³	172,000m³	491,000m³	119,000m³
Regulatory reservoir	River name	Ohta River, a tributary of the Odawara River of the Ichikawa River System.				
	Contributory area	1.64km				
	High water level	EL 798.00m				
	Maximum capacity of water kept	9,313 × 10³ m³				
	Efficient capacity of water kept	8,660 × 10³ m³				
	Available Depth	19.00 m				

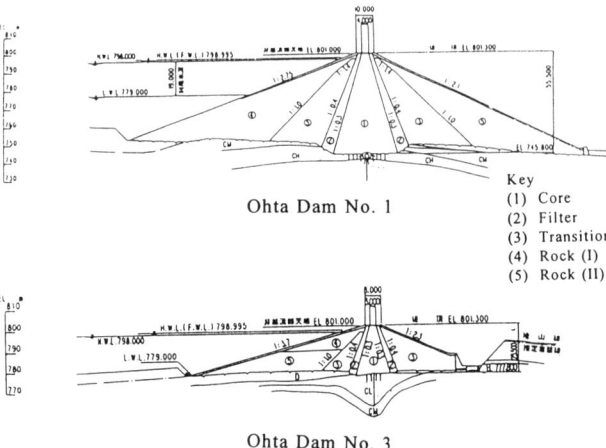

Fig. 3 Standard Sectional Drawings of the Ohta Dams (No. 1 and No. 3)

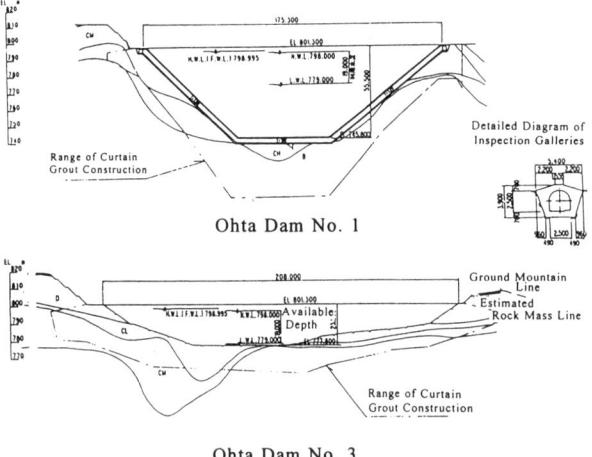

Fig. 4 Vertical Sections of Ohta Dams (No. 1 and No. 3)

2.1 Topography and Geology

In the neighborhood of the dam site, the ground shape appears to be a typical upheaval peneplain dotted with small hills of about 800 m in height. In general the inclines of the hillsides are slight. On the east side of the dam site is the dividing mountains of the Ohta and Inumi rivers. The eastern slope of the divide has a steep decline of 30 to 40 degrees, which continues down to the valley where the lower dam was constructed.

According to a geological survey, the area of the site is made up mainly of considerably thick distributions of porphyrite left by Ikuno layer of the Mesozoic Era, which is a characteristic of upheaval peneplains. This weathered layer is a formative substance from the Peneplain Era. Layers of as much as 10 m or more thick have been observed. Unlike the pure sand of granite, these weathered layers contain fine grains which resemble silt.

2.2 Physical constants of foundation

As stated above, there is a thick distribution of weathered rock (D class rock mass) around the dam site. In order to determine the physical constants of the D class rock mass, a variety of laboratory and on-site tests were carried out.

a. Density
Undisturbed samples (ϕ10 cm × 25 cm, ϕ5 cm × 21 cm) were obtained by vertical and horizontal test-pits method, and the densities were measured by using calipers. On-site densities were measured by sand replacement and water replacement methods using a ϕ25 cm × 25 cm pit. In the case of unobtainable sample with a steel sampler, lump samples were taken and density was measured by the paraffin paint method.

Figure 5 illustrates the relationship between density and underground depth. It can be seen in the cases of Ohta No. 1 - No. 4 Dams that the foundation soil density increases with depth, but this relationship is not so conspicuous in the case of the No. 5 Dam. In addition, the soil density of the No. 5 Dam is low compared to that of the No. 1 - No. 4 Dams.

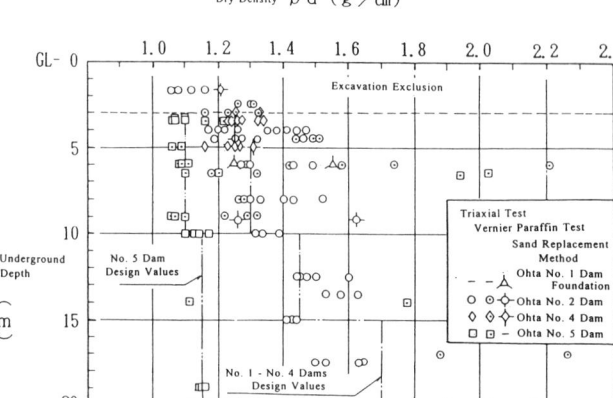

Fig. 5 Relationship between Dry Density and Depth

b. Shearing strength
Shearing strength was investigated using undisturbed samples in laboratory triaxial compression tests, on-site direct shearing tests, and on-site summary shearing tests.

The undisturbed samples also used for the density measurements. A testing machine owned by the Technological Research Institute of the Kansai Electric Power Company was used to conduct shearing tests under strain controlled (0.1%/min.), consolidated undrained conditions.

On-site direct shearing tests were conducted using a large machine typed of the city university of Osaka. A special feature of this shearing test machine is that it is equipped with a torque jack which keeps test specimens from tilting during shearing by producing moments in the opposite directions of the shearing moments (Figure 6).

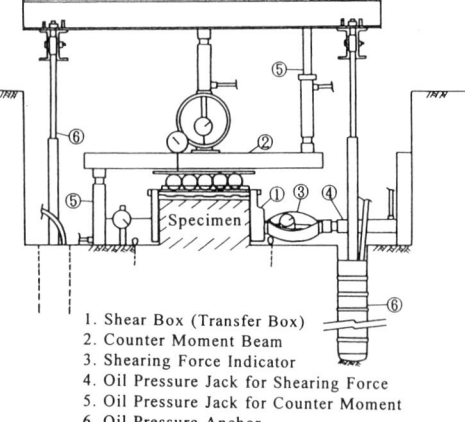

1. Shear Box (Transfer Box)
2. Counter Moment Beam
3. Shearing Force Indicator
4. Oil Pressure Jack for Shearing Force
5. Oil Pressure Jack for Counter Moment
6. Oil Pressure Anchor

Fig. 6 On-Site Direct Shearing Test Apparatus

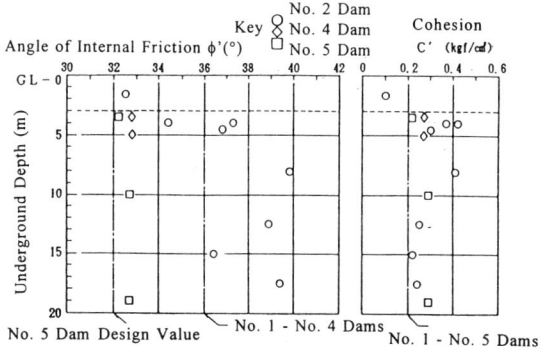

Fig. 7 Relationships of Angle of Internal Friction and Cohesion to Underground Depth

Figure 7 shows the relationships among the depth from the ground surface, and the angle of internal friction and the cohesion of soil. The Angle of internal friction tends to increase with the depth from the ground surface at the No. 2 Dam site. At the depths of less than 5 m, the lower limit value of the internal friction angle is 32°, while at depths of greater than 5 m, the lower limit value of the internal friction angle is 32°, while at depths of greater than 5 m, the lower limit of that is 36°. In contrast, the values of the angle of the internal friction at the No. 5 dam site keeps fairly the same 32° even as depth increases.

On the other hand, cohesion is unrelated to depth from the ground surface, and has a value over 0.2 kgf/cm² at depths of over 3 m.

Based on the above tests, the physical constants of the foundation rock of the Ohta Dams were determined as represented in Table 2.

Table 2 Design Values of Foundation D Class Rock Mass of the Ohta Dams

Foundation		No. 1 - No. 4 Dam Foundations				No. 5 Dam Foundation	
Physical Property		3<H≤5m	5<H≤10m	10<H≤15m	15m<H	5<H≤10m	10m<H
Specific Gravity		2.75	2.75	2.75	2.75	2.75	2.75
Void Ratio		1.20	1.12	0.90	0.62	1.50	1.39
Unit Volume (tf/m³)	Dry	1.25	1.30	1.45	1.70	1.10	1.15
	Wet	1.76	1.78	1.89	2.06	1.67	1.70
	Saturated	1.80	1.83	1.92	2.08	1.70	1.73
Shearing Strength	Angle of Internal Friction	32	36	36	36	32	32
	Internal Friction Coefficient	0.625	0.727	0.727	0.727	0.625	0.625
	Viscosity (kgf/cm²)	0.2	0.2	0.2	0.2	0.2	0.2
Permeability Coefficient (cm/sec)		1.0×10^{-4}	1.0×10^{-4}	1.0×10^{-4}	1.0×10^{-4}	1.0×10^{-4}	1.0×10^{-4}

3. THE MANAGEMENT PLAN OF THE FOUNDATION EXCAVATION IN OHTA DAMS

As related earlier, each of the Ohta Dam sites has areas of distribution of thick weathered rock layers. Ohta Dams No. 1 and No. 2 have respective heights of 55.5 m and 44.5 m, and the establishment of culvert type grout galleries is planned. Since these grout galleries will be made of rigid concrete, they will be less adaptable to transformations when set in soft rock. Concern over the possibility of cracks being formed in such a situation led to the decision to use in principle rock mass of CM class or above. The core portion is both hydraulically and dynamically the most important zone in a rock-fill dam, and in order to best control uneven settlement, settling of the foundation must be avoided as much as possible. Therefore, it was decided in principle to set into CL class rock or above. As for the rock portion itself, since it is more adaptable to transformations than other zones, D class rock mass was used for the foundation of the rock zone only.

3.1 Foundation excavation procedure of the Ohta Dams

As stated above, the design values of the foundation D class rock mass appear in Table 2. According to the table, the necessary average shearing strength can be achieved by excavation of 3 m for No. 1 - No. 4 Dams, and 5 m for No. 5 Dam. However, considering that these values were set based on tests on samples taken from limited areas such as the test pit above the dam shaft, excavation depths of 3 m or 5 m would not necessarily produce satisfactory results, in terms of both quality and cost-efficiency. Therefore, it was decided that the excavation depth be minutely worked out before commencing the excavation. The procedure for the foundation excavation of the Ohta Dams was planned as illustrated in Figure 8.

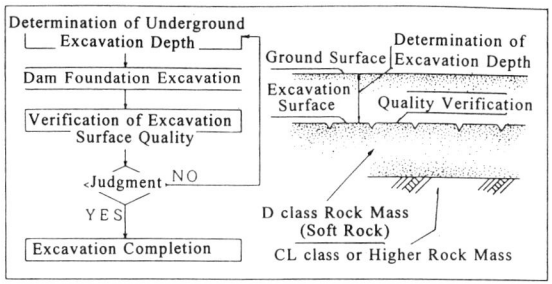

Fig. 8 Foundation Excavation Procedure of Ohta Dams

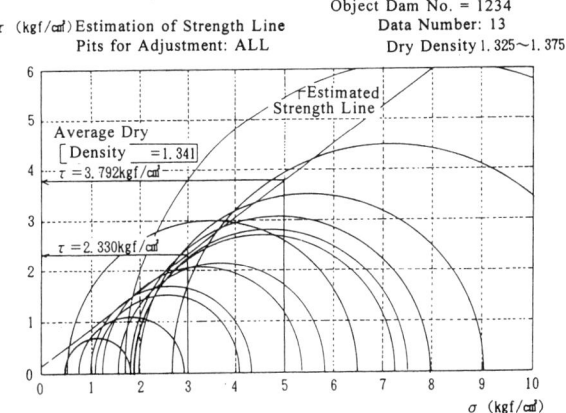

Fig. 9 Example of Adjustment of Results of Triaxial Compression Test Conducted on D Class Rock mass of Ohta Dams

3.2 Ascertaining the quality of excavation

Ascertaining the quality of the dam foundation excavation means confirming that the design is satisfactory in terms of fixed quantities. However, executing a design value such as shearing strength directly many times can cause problems with operations, cost-efficiency and in the work stage. Therefore, alternate values for shearing strength were introduced. In fact, it became necessary to look for other simple measurable material values.

The search began by arranging the data collected from the triaxial compression tests. The data arranging proceeded as follows:

(1) All data from the triaxial compression test results from 135 test specimens which tested in the range of 1.0 g/cm³ < dry density (ρ_{dt}) ≤ 1.7 g/cm³ (ρ_{dt}: dry density measured by the tube method) was extracted.
(2) Dry density (ρ_{dt}) was divided into 24 intervals of 0.025 g/cm³, and test data for separate layers was sampled according to ρ_{dt} values.
(3) C' and ϕ' were obtained from Mohr stress circles of the ρ_{dt} of each separate layer. Figure 9 shows one example of strength estimated from Mohr stress circles in a ρ_{dt} range of from 1.325 to 1.375 g/cm³.
(4) The relationship between shearing strength (τ) and dry density (ρ_{dt}) was discussed.

Figure 10 was attained from the above procedures. A strong correlation detected between shearing strength and dry density led to the discovery that dry density can be used as an alternate value for shearing strength of the foundation D class rock mass at the spot.

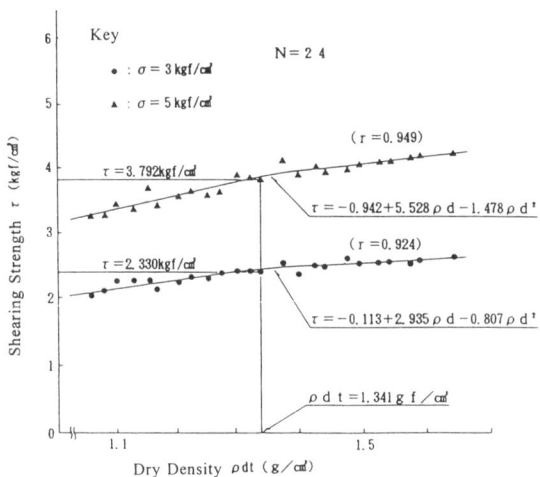

Fig. 10 Relationship Between Dry Density and Shearing Strength

Table 3 Ohta Dam Foundation Excavation Management Test Method (Proposed)

Classification	Test Method		Test Value	Material Values Estimable from Test Values
A	Simple Elastic Wave Exploration		Elastic Wave Speed	Rock Mass, Geological Structure, Dry Density
	Standard Penetration Tests		Number of Hits N Value	Nd Value, Dry Density
	Dynamic Penetration Tests		Number of Hits Nd Value	N Value, Dry Density
B	On-site Density Test	RI Logging	Dry Density ρdr	
		Replacement Method	Dry Density ρds	

Table 4 Correlation Between Test Values

Classification	Test Method	Test Value	Correlation				
			Vs	N value	Nd value	ρdr	ρds
A	Simple Elastic Wave Exploration	Vs		r=0.70	r=0.89	r=0.85	-
	Standard Penetration Tests	N Value	r=0.70		r=0.79	r=0.66	-
	Dynamic Penetration Tests	Nd Value	r=0.89	r=0.79		r=0.73	-
B	On-site Density Test	RI Logging	ρdr	r=0.85	r=0.66	r=0.73		r=0.98
		Replacement Method	ρds	-	-	-	r=0.98	

Next, it was decided to check the quality of the dam foundation excavation by studying records of in-situ tests and search methods to directly and indirectly find out dry density (ρ_{dt}). The results of these studies were evaluated comprehensively in terms of cost-efficiency, speed, and degree of reliability, and then narrowed down. In adherence to the procedure for the foundation excavation of Ohta Dams as illustrated in Figure 8, the depth of the Ohta Dam foundation excavation (h) was obtained from the tests in section A of Table 3, and the quality of the dam foundation excavation was ascertained by the test methods listed in section B. However, in order to find out whether reliable dry density (ρ_{dt}) values could be determined from the various values obtained from the several tests and searches, on-site confirmation tests were conducted in the neighborhood of the dam sites (Table 4). According to Table 4, a large correlation coefficient of r = 0.89 was found for simple elastic wave velocity (Vs) and dynamic penetration test (Nd) values. In addition, a high correlation of r = 0.85 is shown for Vs and (ρ_{dt}). On-site density tests were conducted using surface type RI and sand replacement methods which also resulted in a strong correlation of r = 0.98.

From the above, the following was concluded:
(1) There is a strong correlation (r = 0.92 to 0.95) between dry density and shearing strength. Dry density can be used to estimate shearing strength.
(2) Dry density cannot be accurately estimated from independently executed simple elastic wave explorations and dynamic penetration tests (the latter is more efficient than standard penetration tests). However, running the two tests together can produce a rough measure of dry density.
(3) For ascertainment of excavation surface quality, a precise understanding of dry density is necessary. However, it is also possible to apply RI logging and replacement methods as conducted in this case.

Following from (1), (2) and (3), the excavation depth (h) is determined by the simple elastic wave explorations, dynamic penetration tests and existing soil quality test results. RI logging will be the main method for ascertaining excavation quality, with the sand replacement method being used to check the RI logging test results.

3.3 Management flow of the foundation excavation of the Ohta Dams

For the ascertainment of the quality of the foundation excavation, it was decided that it was important for statistical evaluations of measured dry density values and soil quality evaluations to be run in parallel and that the quality ascertainment would be based on the overall evaluations. In addition, simple elastic wave explorations (pick up intervals of 2.5 m) were conducted on a line above a 30 m grid network built on the dam plot, while dynamic penetration tests were run at grid intersections. RI logging of the excavation was conducted at even more precise grid intersections of 15 m.

From the above results, the management flow of the foundation excavation of Ohta dams was drawn up (Figure 11).

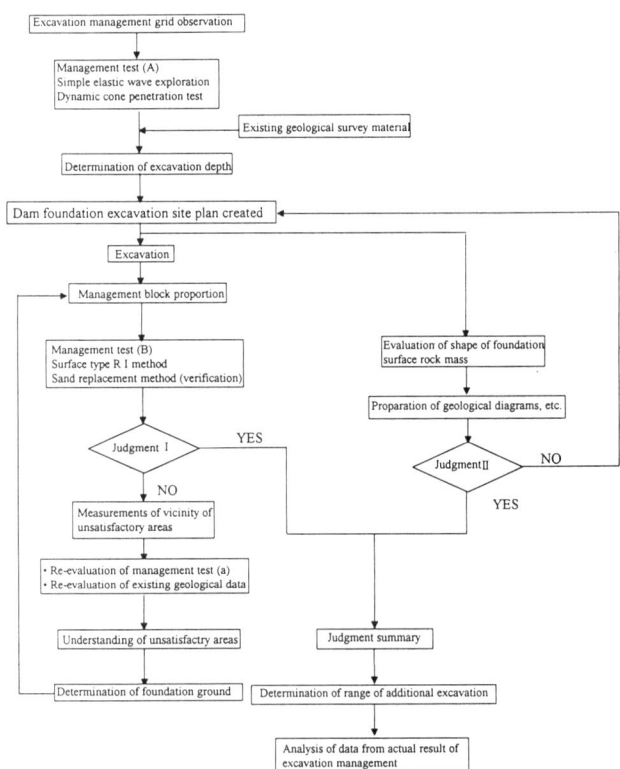

Fig. 11 Ohta Dam Foundation Excavation Management Flow

a. The establishment of standard management values

Based on the excavation depth (h) that was determined through the above stated methods, dry density of the excavated surface was measured using surface type RI method, and quality was ascertained. The standard values of each test value and measured value were determined by the following methods:

(a) Excavation depth (h)

Desired values for dynamic penetration tests and simple elastic wave explorations for determining excavation depth (h) were set in the following way:

Figure 12 illustrates frequency distribution of dry density (ρ_{dt}) obtained using the tube method near the proposed excavation depth in the neighborhood of the site (3 to 5 meters from the surface) (number of samples N = 80).

Standard values of dry density (tube method) at the foundation excavation were set at $-1/2\sigma$, $\rho_{pt} = 1.162$ g/cm^3. This is equivalent to a dry density of $\rho_{dt} = 1.122$ g/cm^3 under surface type RI testing (according to the correlation between the tube and the RI methods). This dry density is equivalent to a dynamic penetration test value of Nd = 13 times, or a simple elastic wave exploration of Vs = 160 m/sec.

(b) Ascertainment of excavation quality
Since the functioning stress level of the soil foundation of the Ohta Dams is σ = from 3 to 7 kgf/cm^2 and the design strength of D class rock mass is C' = 0.2 kgf/cm^2, $\phi = 32°$ (see Table 2), from the diagram (see Figure 10) of the relationship between shearing strength ($r = C + \sigma\tan\phi$) and dry density (ρ_{dt}), ρ_{dt} is found to be 1.09 g/cm^3. Therefore the standard management value at the excavation surface was set at $\rho_{dt} = 1.05$ g/cm^3.

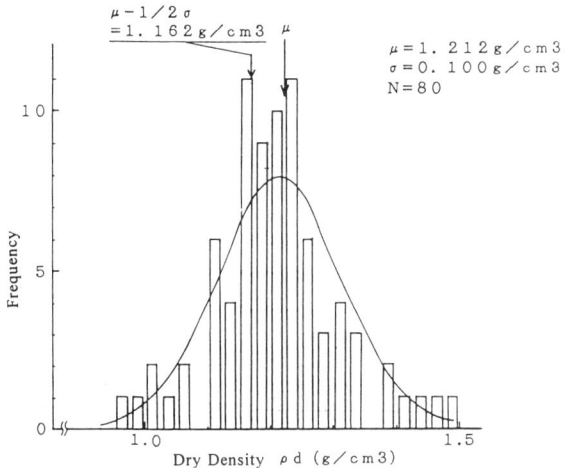

Fig. 12 Measured Dry Density Values and Distribution Management

b. Judgment I
(a) The individual measured dry density values obtained from surface type RI tests conducted at the excavation surface must not be lower than the standard value.
(b) The shape of the distribution of measured dry density values for each management block will be determined by the following formula:
$\overline{\rho d} - \sigma(\rho d) \geq$ (standard value)
$\sigma(\rho d)$: standard deviation of measured values
$\overline{\rho d}$: average value of measured values

c. Judgment II
Soil evaluations of points such as the shape of the rock mass of the dam foundation will be run.

(a) The structural shape of the rock mass will be evaluated (verification of rock structure).

(b) Joint condition of rocks will be evaluated (adhesiveness, seam condition, etc.).

4. FOUNDATION EXCAVATION MANAGEMENT OPERATION STATUS

In the case of Ohta Dam No. 3 (Fig. 3 and 4), embankment had first been completed (in October, 1990), according to the management flow of the foundation excavation. Operating results from the decision of excavation depth to the ascertainment of the quality of the foundation excavation surface are shown in Section 4.1 ~ 4.4.

4.1 Foundation excavation management test operating specification
Testing methods which accompanied foundation excavation management are as described previously. Specification for testing operating methods, appear in Table 5. Since dynamic penetration tests tax the operating capacity heavily, a mobile machine equipped with dynamics was used to improve efficiency. In the case of simple elastic wave explorations, portable personal computers were used to eliminate excess work in ascertaining waveforms.

4.2 Results of foundation excavation depth measurements
Figure 13 shows one example of measurement results at Ohta Dam No. 3. This is one 30 m traverse line, which is a unit of measurement, with both ends as lattice points of the grid network. In other words, it is a dynamic penetration test site. This was a case of design excavation lines which were a uniform 3 m deep and became shallower, but there were also cases of traverse lines which became deeper.

Table 5 Foundation Excavation Management Test Execution Methods

Classification	Management Test A		Management Test B
Test	Simple Elastic Wave Exploration (S Wave)	Dynamic Cone Penetration Test (Steel Study Type)	Dry Density Measurement (RI Method)
Specifications	·Traverse Line Length: 30m ·Pick-up Interval: 2.5m ·Earthquake generation method: plank hammering ·Earthquake generation positions (3): both ends and middle portion	·End Shape: Cone ·Hammer Weight: 63.5kgf ·Drop Height: 75cm ·Penetration Amount: 30cm	Surface Permeable Type Moisture Density Measure Made by Soil and Rock Company

4.3 Results of ascertainment of foundation excavation surface quality (Dry density measurement)
Excavation was conducted in accordance with the relevant ground plan based on the foundation excavation depth measurements. For the excavation surface, surface penetration type RI was used to measure dry density. Figure 14 is a distribution diagram showing those results. The minimum value was $\rho_{dr} = 1.115$ gf/cm^3, which satisfied all of the conditions of Judgment I.

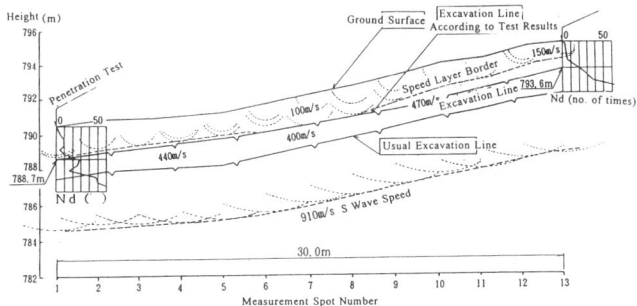

Fig. 13 An Example of Foundation Excavation Depth Measurement Results (Cross Section)

4.4 Verifying the relationship between dry density and shearing strength
Since Ohta Dam No. 3 was the first of the five dams to be constructed, the relationship between dry density and shearing strength (Figure 10) upon which the management method was based was verified by using the actual excavation surface. On-site direct shearing tests were conducted at points of lowest dry density. The using test machine the Osaka City University type is shown Figure 6. The results of these tests are shown in Figures 15 and 16. Satisfactory design values at points of minimum dry density were obtained, and Figure 16 proves the relationship between dry density and shearing strength.

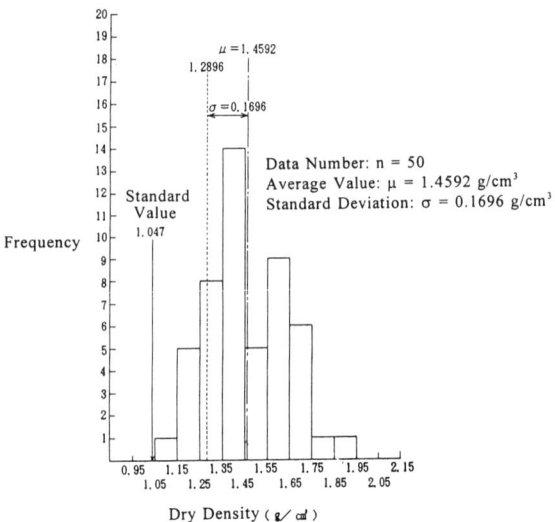

Fig. 14 Distribution Diagram of Measured Dry Density Values of Ohta Dam No. 3

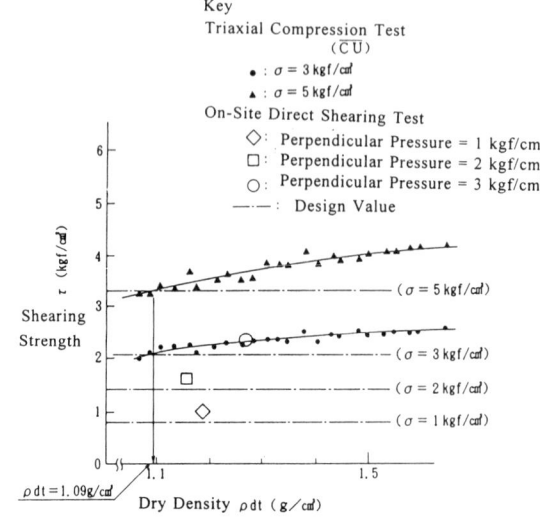

Fig. 16 Result of Inspection of the Relationship Between Dry Density and Shearing Strength

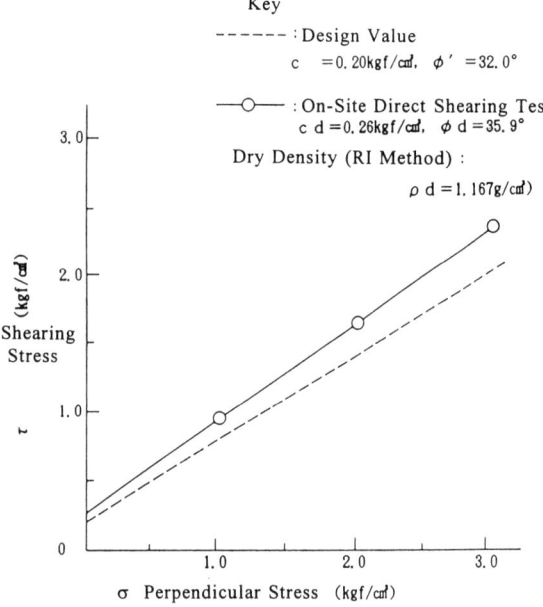

Fig. 15 Relationship Between On-Site Direct Shearing Test Results and Design Values

5. AFTERWORD

This concludes the introduction of the methods for determining excavation depth and ascertainment of excavation surface quality adopted for the D class rock mass (weathered porphyrite) which was the foundation of the rock zone of the rock-fill Ohta Dams. There were spots which became shallower or deeper than the originally decided uniform excavation depth, and it was possible to determine depth very precisely. Prevention of unnecessary excavation at points that had become shallower, and efforts to eliminate excess work at points which had become deeper made for an efficient operation. It is hoped that the methods herein discussed will be helpful for the future construction of rock-fill dams in areas with geological restrictions.

In closing, we express our deepest gratitude to the many people who extended to me their abundant guidance and cooperation in the development of these methods.

Treatment of soft rock masses used for the soil cement mixing wall at the Ohta Dams

S. Yasufuku & M. Kageyama
The Kansai Electric Power Co., Inc., Japan

K. Sakashita
Aoki Corporation, Japan

1. FORWARD

The Ohta Dams are a group of five rock fill dams with center impervious wall and levee heights varying from 23.5 to 55.5 m. They function as upstream dams which regulate reservoirs with respect to the Ohkawachi Pumped-storage Water Power Station, which was constructed by Kansai Electric Power Company (1,280,000 kW. Operation began in 1992, and construction was completed in 1994), and is located near the center of Hyogo prefecture.

The construction site of the power plant is located in the eastern end of the Kibi highland region, which is known for having typical topography of upheaval peneplain. The Ohta Dams are to be constructed on five sites, including the saddle of this upheaval peneplain region. In general, such peneplain regions are made up of distributions of weathered rock (soft rock) layers. According to geological surveys, these layers are thickly distributed about the Ohta Dam site.

The decision to build five separate center impervious wall-type dams was based on ground shape and geological conditions, and the comparative ease of gathering levee material in the vicinity of the dam site. In addition, for the dam foundation, aside from part of the installation at the left and right banks of the dam, CL class rock was used in principle for the core portion, and considering cost-efficiency, D class rock mass was used for the rock portion.

2. SUMMARY OF THE PROCESSING OF THE OHTA DAMS FOUNDATION

As stated above, because of the thick distributions of weathered rock layers at each dam site, various ideas for dam foundation and shaft portion water cut-off methods were discussed from the beginning of the design stage. First, based on the foundation excavation plan, and considering the existence of D class rock mass at the dam installation portions and the shaft portions, the grout tests were also run on the D class rock mass. According to the test results, the stage grouting was adopted for class C rock mass and above. After discussing various methods, it was confirmed that double tube double packer grouting would be sufficient for improving desired values. However, since the improvement of D class rock mass via double tube double packer grouting requires a great deal of labor power, the possibility of employing an underground continuous water cut-off wall was also discussed as a more reliable foundation water cut-off processing method. The following methods were discussed:

1. In principle, to set the core into rock mass of CL class or higher, and to process water cut-off at the dam foundation and the shaft portion via grouting.
2. To exclude excavation of D class rock at which fixed quality can be attained, and to set in the core portions together with using grouting for water cut-off at the dam foundation and the shaft portion.
3. Essentially the same idea as 1, except that water cut-off walls would be constructed in a portion of D class rock mass in the vicinity of the shaft portions, and water cut-off would be processed via grouting at areas of CL class rock mass or higher.

It was obviously necessary to fully discuss architectural feasibility and hydraulic stability regarding the use of water cut-off walls discussed in 3 above. Laboratory proportion tests and on-site structural and seepage flow analysis tests were conducted to verify that the desired quality could be satisfied before deciding to adopt the idea.

After discussion of the above three ideas in terms of construction feasibility, quality and cost-efficiency, the method of 1 was adopted for Ohta Dams No. 1, No. 3 and No. 4, and the 3 was adopted for the No. 2 and No. 5 Dams. After discussion of various types of water cut-off walls, it was decided to adopt column strip type underground continuous walls that could be adapted to deformations in D class rock mass for the pertinent dams. Figure 1 is a vertical section of Ohta Dam No. 2, and Figure 2 is a standard section of the points at which column strip walls were placed.

The following is a summary of the design of the dam foundation processing and construction of the column strip walls.

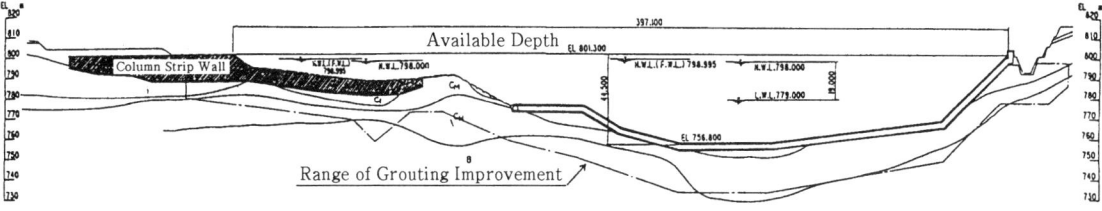

Fig. 1 Vertical Section of Ohta Dam No. 2

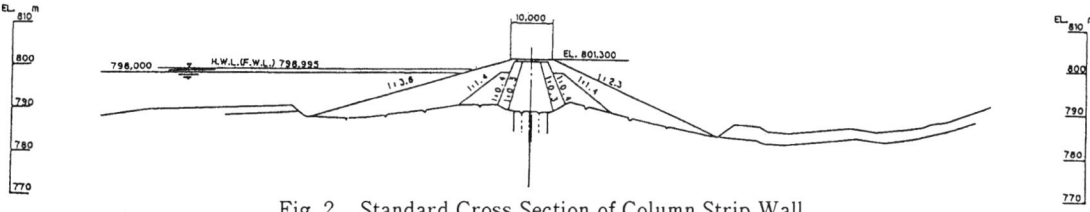

Fig. 2 Standard Cross Section of Column Strip Wall Employment Points of Ohta Dam No. 2

Table 1 Column Strip Wall Laboratory Test Combination Range and Results

Injected Milk Mixture	Cement Volume (kg/m³)	100	150	200	210	225	240	250
	Bentonite Volume(kg/m³)	5~30	5~30	5~30	10~30	10~50	10~30	10~30
	Water Volume (kg/m³)	160~300	270~300	300~631	321~628	316~623	311~618	308~615
	Injection Volume (ℓ/m³)	198~343	320~350	367~700	400~700	400~700	400~700	400~700
D Class Rock Mass	Dry Density (g/m²)	1.31						
	Moisture Content (kg/m³)	30~35						
Mixed Soil	Cement Volume (kg/m³)	82~94	113~125	122~162	128~157	137~168	147~180	153~187
	Bentonite Volume (kg/m³)	4~26	4~25	4~22	6~22	6~34	6~22	6~22
	Water Volume (kg/m³)	517~574	552~601	560~666	583~663	580~661	576~658	574~656
	Grain Weight (kg/m³)	1079~1226	987~1091	801~1058	800~980	900~980	800~980	800~980
	W/C (%)	553~693	442~532	347~545	371~5170	344~480	320~449	307~429
	W/S (%)	42~53	51~61	53~83	60~83	59~83	59~82	59~82
Specimens 4 Weeks	Triaxial Compression Test							
	Deformation Modulus E_{50} (kgf/m²)	370~850	373~1403	1007~2833	983~4167	1157~5533	1500~5533	1667~6450
	Viscosity C_u (kgf/m²)	1.53~3.00	1.47~3.30	3.31~4.70	1.95~4.20	2.32~5.49	2.37~8.05	3.04~8.62
	Angle of Internal Friction $\phi u(°)$	0.9~18.3	14.5~25.8	23.6~33.6	4.90~33.6	5.6~31.9	3.4~36.6	12.8~31.5
	Permeability Test							
	Permeability Coefficient K (cm/sec)	0.3~3.7×10⁻⁶	1.2~3.3×10⁻⁶	—	0.9×10⁻⁶	3.6~5.2×10⁻⁶	4.2×10⁻⁶	0.1×10⁻⁶

3. COLUMN STRIP WALLS (SMW METHOD)

3.1 Design discussions and various tests

a. Laboratory mix proportion tests

In order to fully understand the basic material values in connection with hydraulic and structural analysis discussions, laboratory tests (permeability, triaxial compression, etc.) were conducted. Portions loosened from the D class rock mass taken from the dam site test pits were altered into a naturally hydrated state. Tests were conducted using these portions and specimens which were prepared by using the mixed dirt resulting from evenly mixing and stirring cement milk (type B blast furnace cement). Table 1 shows the proportion range (weight of each material for 1 m³ of D class rock mass) and test results.

According to the above results, regarding the triaxial compression test, water-cement ratio as well as soil grain volume have an effect on the deformation modulus, viscosity and angle of internal friction. However, it was found that if the cement milk can be mixed evenly with the D class rock mass from the site, a flexible wall of deformation modulus between several hundred and several thousand kgf/cm² can be built. In addition, the coefficient of permeability for all combinations satisfied the order of 10⁻⁶cm/s, thereby fulfilling the quality requirements of water cut-off walls.

b. Seepage flow analysis

Based on the results of laboratory permeability tests, two-dimensional saturation and partial saturation seepage analyses were run. The cross section analyzed was the portion of the column strip wall with the highest water head, which was the tightest cross section of the left bank levee of Ohta Dam No. 2. According to the results of the analysis, the maximum current speed at time of high water level inside the core near the top portion of the wall is Vmax = 2.6 × 10⁻⁵cm/s. Similarly, at the bottom portion of the wall, which is made up of CL class rock mass, the maximam current speed is Vmax = 1.8 × 10⁻⁴cm/s. These speeds are much lower than the current speed at which the grain portion of the core material can be dragged loose, and do not present a problem.

c. Stress and transformation analyses

Laboratory combination tests conducted on column strip wall specimens were based on the results of the triaxial compression tests. Since the deformation modulus of the D class rock mass in the vicinity of the wall (deformation modulus E50 = 300kgf/cm³) and that of the wall itself were made to differ as little as possible, in order to preserve the required strength, the deformation modulus of the column strip wall was estimated between 1,000~2,000 kgf/cm³, and structural analysis was conducted. Furthermore, dynamic material values were established for three cases of column strip walls, as shown in Table 2. In addition, since the development of concentrated stress in the core near the top of the wall was anticipated, two types of combined shapes of the wall and the levee core, protruding and corresponding, were hypothesized as shown in Figure 3, and discussed in comparison. The results of the discussions are as follows:

Table 2 Dynamic Material Values of Column Strip Walls

Deformation Modulus E_{50} (kgf/cm²)	Poisson Ratio	Unit Weight (gf/cm³)			Viscosity C_u (kgf/cm²)	Angle of Internal Friction $\phi u(°)$
		Dry	Wet	Saturated		
1000	0.3	1.10	1.67	1.68	3.0	20.8
1500	0.3	1.17	1.72	1.73	3.9	23.7
2000	0.3	1.18	1.73	1.74	4.0	27.4

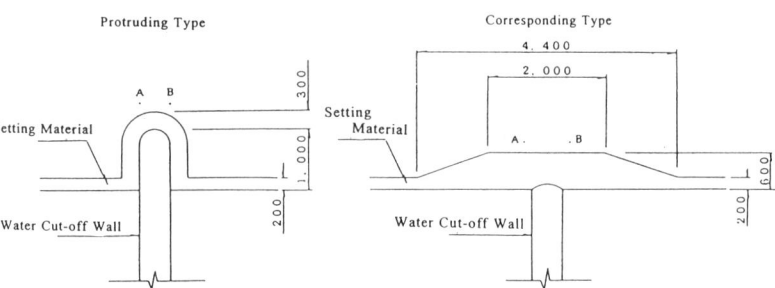

Fig. 3 Combined Shape of Column Strip Wall and Levee Core Portion

(a) Displacement

Table 3 shows the vertical displacement of the head portion of the wall. In the case of the protruding type wall, since stress concentration on the wall is high, the vertical displacement becomes slightly larger. In addition, the difference between the vertical displacement of the D class rock mass surface of the core site and the wall was 1 to 2 cm.

(b) Stress
Maximum principal stress distribution inside the column strip wall, in the case of a wall modulus of deformation having 1,500 kgf/cm², is as shown in Figure 4. Stress concentration at the wall head is 5 kgf/cm² for protruding type and 3 kgf/cm² for corresponding type, a difference of 2 kgf/cm². In addition, maximum principal stress distribution generation points near the root insertion portion, during earthquake at time of high water level, was in both cases found to be 9 kgf/cm². Furthermore, the maximum principal stress of the D class rock mass portion was about 2 kgf/cm², increasing the stress concentration level of the column strip watl by 3 to 4 times.

(c) Coefficient of local safety factor
The stress conditions of the column strip wall at the maximum points of the principal stress distribution and the local safety factor in the case of a corresponding type wall with deformation modulus of 1,000 and 2,000 kgf/cm² are depicted in Figure 5. Although the local safety factor decreases as the modulus of deformation of the wall increases, this does not present a problem. In order to also confirm the effect of a difference between the deformations of the wall and the D class foundation on the core portion of the levee, the local safety factor of the core zone directly above the wall head (the position is shown in Figure 3) was checked. The results are shown in Figure 6. In the case of the protruding type wall, the local safety factor decreased, and also fell below 1.0 at one portion.

From the above results the target range of the modulus of deformation from 1,000 to 2,000 kgf/cm² was set and the corresponding type wall was chosen as the combined head portion shape.

Table 3 Column Strip Wall Head Portion Vertical Displacement
Unit: cm

Shape of Head Portion	Stage of Analysis	E_{SMW} (kgf/cm²)		
		1000	1500	2000
Corresponding Type	Time of Construction Completion	2.79	2.46	2.22
	Time of High Water Level	2.81	2.48	2.24
	Time of Earthquake	2.64	2.32	2.09
Protruding Type	Time of Construction Completion	3.35	2.91	2.61
	Time of High Water Level	3.37	2.93	2.63
	Time of Earthquake	3.18	2.76	2.46

d. On-site validation tests
Prior to actual construction, in order to check construction feasibility, the construction quality of the column strip walls, and in order to determine the cement milk mixture, test construction using actual machines was conducted at the left bank of Ohta Dam No. 2, which was to be the site of construction of the main body, from October, 1988 to March, 1989.

The test were divided roughly into three types (No. 1 - No. 3). The construction element pattern of each test is shown in Figure 7, and the injected milk mixture is shown in Table 4.

When advance punching was conducted in the No. 1 test, superior construction feasibility was discovered when the triaxial mixing auger machine's drilling speed stabilized and vibrations were few. Good uniformity of the mixed soil was also found. In the No. 2 test, after hardening of the column strip wall (in principal, 4 weeks after construction), the column strip wall, including the construction joints, was diagonally bored through and the Lugeon test (pressure-type hydrostatic pressure test) was conducted. The average value of all five construction joints was 1.28 Lu. However, it was found that after a construction stoppage of about 10 days, the value was lower than 1.0 Lu. Furthermore, the No. 3 tests, as shown in Table 4, for wall bodies that were prepared every four combinations, the core was sampled with a Denison type sampler (bore of f98 mm), and triaxial compression tests were run. The results of these tests are as shown in Table 5; based on stress and transformation analyses, the calculated local safety probability was above Fs = 1.5, and the deformation modulus was also within the required quality range.

3.2 *Construction methods and machines*

Since N values over 50 were obtained from standard penetration tests on the deepest portion of the D class rock mass which makes up the column strip wall, and since in consideration of the grouting wrap of the ends of the wall, a CL class rock mass was inserted at one portion, and also in accordance with the on-site validation tests mentioned

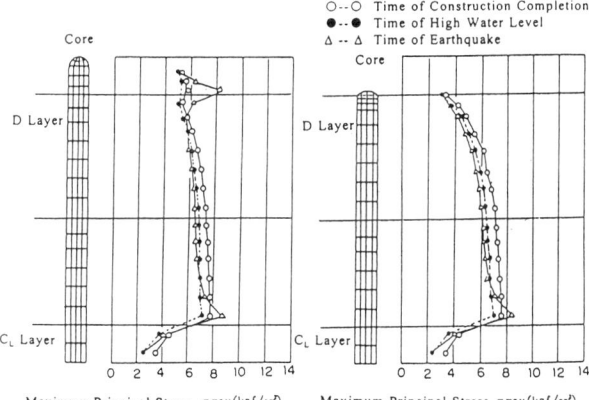

Fig. 4 Maximum Principal Stress Distribution of Column Strip Walls

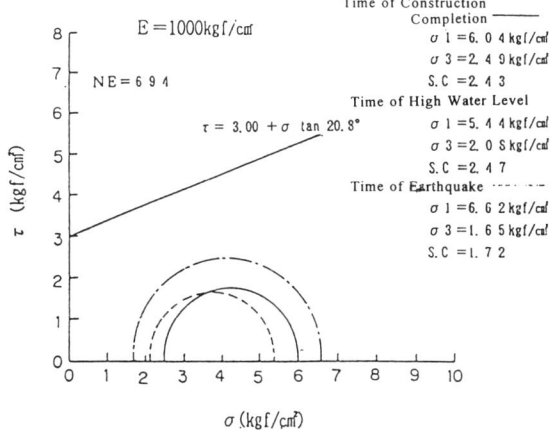
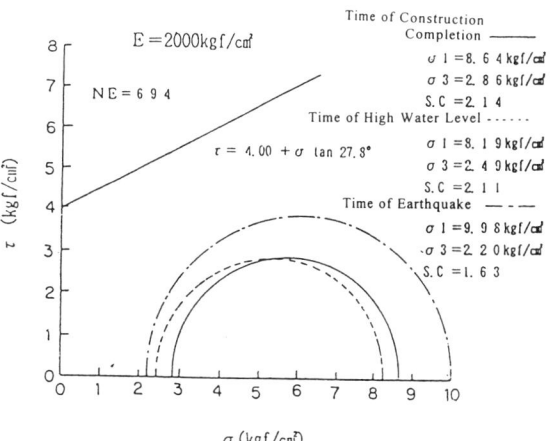

Fig. 5 Local Safety Factor of Column Strip Walls (At Points of Greatest Stress)

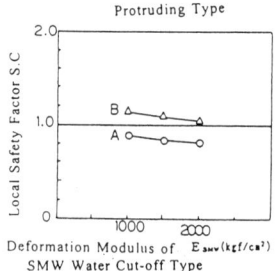
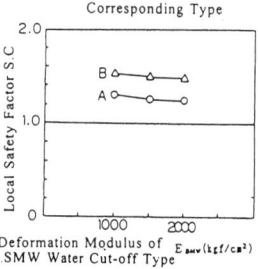

Fig. 6 Local Safety Factor inside Core Directly Above the Head Portion of Column Strip Walls

Table 4 On-Site Validation Test Injected Milk Mixture

Lead Punching	Injection Volume Cement Volume Bentonite Volume Water	350 ℓ/m³ 35 kg/m³ 14 kg/m³ 333 ℓ/m³	
		Case 1	Case 2
SMW	Injection Volume Cement Volume Bentonite Volume Water	700 ℓ/m³ 250 kg/m³ 30 kg/m³ 606 ℓ/m³	700 ℓ/m³ 200 kg/m³ 30 kg/m³ 622 ℓ/m³
		Case 3	Case 4
SMW	Injection Volume Cement Volume Bentonite Volume Water	400 ℓ/m³ 250 kg/m³ 30 kg/m³ 306 ℓ/m³	400 ℓ/m³ 200 kg/m³ 30 kg/m³ 322 ℓ/m³

No	Test Construction Element Pattern	Test Goal
1	With Lead Punching / Without Lead Punching	Depending on whether or not lead punching was conducted, the quality of construction feasibility at the time of wall preparation is ascertained.
2	A C / D B Construction procedure A → B → C → D Construction Procedure B: Constructed 3 Days After A C: Constructed 10 Days After B D: Constructed 20 Days After A	The estimated number of work stoppage days and its effect on the water cut-off ability of the construction joints is ascertained.
3	Case1, Case2, Case3, Case4	Based on the laboratory combination test, wall quality is ascertained after the four selected types of additive agent combinations are used for test construction.

Fig. 7 On-Site Validation Test Element Pattern

Table 5 Results of Triaxial Compression Test Conducted on Extracted Core of Column Strip Wall (No. 3 Test)

Case No.	Deformation Modulus E_{50} (kgf/cm²)	Viscosity Cu (kgf/cm²)	Angle of Internal Friction ϕu (°)	Safety Factor Fs
1	1500 (6 Weeks)	6.99	6.8	2.84
	1400 (6 Weeks)	3.27	23.9	1.87
2	1500 (7 Weeks)	3.04	13.6	1.50
	996 (6 Weeks)	3.73	8.2	2.10
3	2000 (6 Weeks)	3.10	33.9	1.88
	2033 (4 Weeks)	4.11	20.2	1.79
4	1023 (7 Weeks)	2.35	26.5	1.94
	1100 (6 Weeks)	0.84	41.0	1.73

discharged from the tip of the auger and stirred in situ. Then, the ends of this column strip were wrapped in order and continuous walls were prepared. Figure 8 shows the construction procedure.

3.3 *Construction management system*

SMW construction methods usually have many temporary uses, such as end dam wall filling, and there have been many cases of lax management of wall quality. However, it was necessary to manage the quality of the wall strictly, since aspects of the site, such as the foundation portion of the body of the dam, had to be prepared, and items such as adaptability to deformation had to be confirmed. Therefore, the construction management system described below was newly introduced:

a. Filling volume control system
Since the volume of injection milk per unit volume of in situ soil will greatly affect the quality of column strip walls, injection volume per unit drilling length must be fixed, and efforts to stabilize quality must be made. Therefore, this system was designed so that pump discharge volume could be controlled by responding to the auger's drilling speed. Under this system, information obtained during construction from the drilling machine and a flowmeter (drilling position, depth, drilling speed, injection volume, etc.) is processed by a personal computer and expressed on the monitor screen so that the construction status can be confirmed (see Figure 9). At the end of the works, an injection management sheet was prepared for each element based on this data.

b. Auger grade measuring system
The goal of this system was to confirm and manage perpendicularity, which is what controls the quality and shape of the column strip wall. According to clinometers installed in various places inside the auger, the angle of inclination (X, Y at right angles, 2 directions) are obtained, and from these the shape of the holes punched by the auger are verified. The construction status was confirmed via computer monitor and management sheets were produced in the same way as under the injection capacity control system (see Figure 10).

3.4 *Construction results*

Based on the results of the on-site validation tests, column strip walls were created on the left bank of Ohta Dam No. 2 from August to September, 1989, and on the extension of the right bank of Ohta Dam No. 5. Photograph 1 shows the construction status at Ohta Dam No. 2. Actual construction numbers appear in Table 6. Construction speeds of 1.0 m/min. for advance punching and 0.5 m/min. for transaxial preparation were standardized. The maximum daily volume of advance punching was 17 holes, with an average of 9 holes. For transaxial preparation the maximum was 10 elements (a construction lengthening of 9m),

above, it was decided to conduct advance punching using a uniaxial auger machine.

For this, the soil was dug up and loosened in advance, so that the stirring conditions were good at the time of triaxial mixing (wall preparation), a condition which improved the uniformity of quality and the vertical accuracy of the wall. After punching, while drilling the soil with the transaxial mixing auger machine, the injection milk was

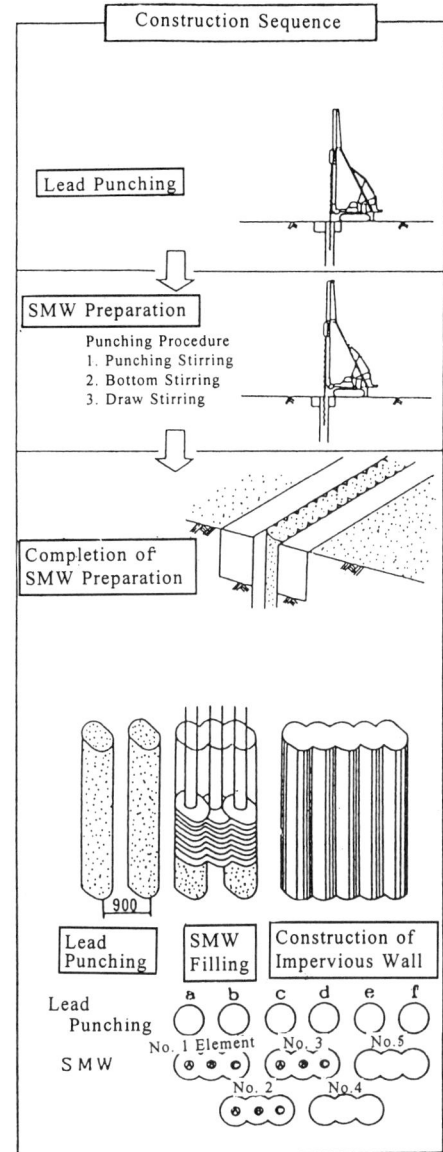

Fig. 8 Column Strip Wall Construction Procedure

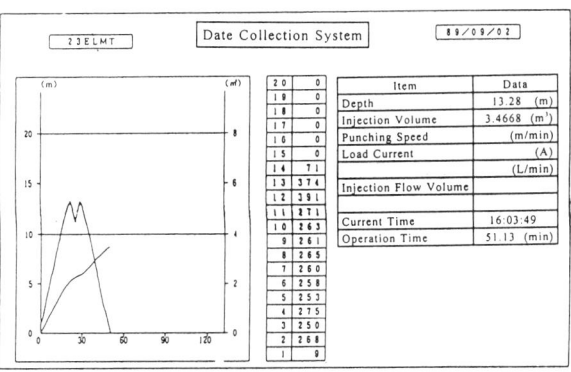

Fig. 9 Screen Expression of Status of Column Strip Wall Construction

executing an average construction of 6 elements.

In order to verify the quality of the wall in connection with construction, freshly mixed samples were extracted from every ten elements from an area of constant height in the prepared wall, specimens were produced, and triaxial compression tests and permeability tests were conducted.

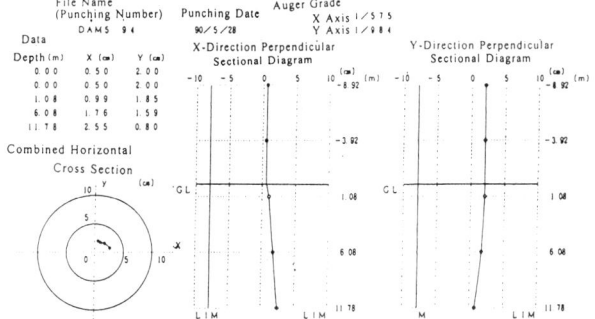

Fig. 10 Auger Grade Management Sheet

Photo 1 Column Strip Wall Construction

In addition, the above mentioned Denison type sampler was used to extract the core of the prepared wall, and the same tests were run. The results of the tests run on the core are shown in Table 7. Satisfactory results for adaptability to deformations, strength, and impermeability were obtained. In particular, the deformation modulus was near the lower limit of the desired range, and from the standpoint of its familiarity with D class rock mass, it is thought that a quality wall was created.

Table 6 Column Strip Wall Construction Actual Figures

Dam Name	Length	Punching Depth	Wall Fill Height	Wall Area
No. 2 Dam	177.3m	4.1~15.2m (average 11.3m)	1.4~13.2m (average 8.8m)	1,565m^2
No. 5 Dam	83.7m	10.0~17.0m (average 14.1m)	10.0~15.7m (average 13.1m)	1,098m^2

Table 7 Column Strip Wall Quality Verification Results

Element No. 2	Permeability Test	Triaxial Compression Test			
		Angle of Internal Friction (°)	Cohesion (kgf/cm^2)	Deformation Modulus (kgf/cm^2)	Local Safety Factor
145	1.08×10^{-5}	15.4	2.75	1075	1.51
165	8.44×10^{-6}	25.5	2.48	978	1.62
185	9.29×10^{-6}	14.9	3.45	1083	1.77
		16.0	3.37	1143	1.76

4. AFTERWORD

The Ohta Dams began with the excavation of Ohta Dam No. 3 in May, 1989, and the embankment of all dams completed in December, 1991. In June, 1992, water impounding to the reservoir was initiated, and commercial operation of the plant began in October of that year. Since then, all dams have been operating smoothly. Embedding gauges have been installed inside the column strip walls and the inside the D class rock mass in the vicinity and an evaluation of behavior that includes the vicinity of the column strip walls is in process, although the current status is in line with the original design idea. In the future, adjustments will be made on the basis of the results of discussions to be held. It is hoped that information helpful for future applications in similar geological areas will be obtained.

We express herein our deep gratitude for advices given by many organizations and persons concerned including Central Research Institute of Electric Power Industry (a corporation) in the design and execution of base treatment of Ohta Dams in which we had many things experienced for the first time.

Case study on the mechanical improvement of rock masses by grouting

Kohkichi Kikuchi & Yoshitada Mito
Kyoto University, Japan

Toshiyuki Adachi
Ministry of Construction, Tokyo, Japan

ABSTRACT: The aim of consolidation grouting in dam foundation is to improve permeability and mechanical properties of the foundation near the surface. Generally, the check for the improvement on permeability is usually carried out by using the check hole, however, the check for the improvement on mechanical propeties is not carried out. Therefore, the improvement on deformability could have not been applied to the mechaical design for dam foundation. The authors have made an in situ experiment in order to examine the grouting effects on deformability of rock masses. As the results, the followings were obtained;
1. The deformability of rock masses can be improved by grouting.
2. More deformable before grouting, higher improvement can be expected.
3. The rheological model is suggested in order to examine grouting effects.
4. The shear strength of rock masses can also be improved by grouting.

1. FOREWORD

Recently, with the lack of foundations that have high quality for constructions, more or less unstable rock masses would be selected as dam foundations.
Therefore, the improvement of rock masses (foundation treatment) is indispensable in order to secure the stability of foundations.

Consolidation grouting is mainly carried out at the foundation of concrete dam such as arch dam and gravitation dam. The aim of the consolidation grouting is to improve the deformation and intensity of rock masses by filling up crack with grout milk. It is also effective for controlling seepage flow at the part where the hydroulic gradient is the highest.

The effect of consolidation grouting is usually comfirmed by refering the results of permeability tests. In the case that the discontinuities in rock masses are filled with grout milk, the permeability would be low. Such a comfirmation method is not enough to grasp the mechanical properties of rock masses quantitatively. Therefore, the mechanical effects of grouting could have not been applied to the design of the foundation. If the degree of mechanical improvement of rock masses can be obtained quantitatively, the effective design could be carried out.

In this study, the authors has made an in situ experiment in order to examine the grouting effects on the mechanical properties (deformability and shear strength) of rock masses in Tertiary tuff breccia.

2. METHOD OF IN SITU TEST

In order to examine the grouting effect on the mechanical properties (deformability and strength) of rock masses, we have carried out an in-situ experiment as follows.

2.1 Outline of testing yard

The testing yard is close to Miyagase dam site. Miyagase dam is the one of the largest gravitation dams in Japan (under construction). A square field which length of one side is 10 meters is secured. The rock mass of the testing yard is composed of Tertiary tuff breccia. The uniaxial compressive strength of non-weathered intact rock is about 80 MPa. Joints which surface are weathered are distributed in dense. The layout of grouting holes is shown in Figure 1.

2.2 Surface survey

In order to grasp the condition of the rock mass, rock mass classification and joint survey have been carried out on the surface of the testing yard.

2.3 In-situ tests using the borehole

At first the borehole Primary holes (◎) have been dug with the length of 5 meters, borehole expansion test (Lateral Loading Testing), Lugeon water pressure test and grouting have been carried out. After grouting of the rest primary holes (○), the secondary holes (■) have been dug and the same tests and grouting have been carried out. In the same way, the tertiary (▲) and quaternary holes (●) have been tested.

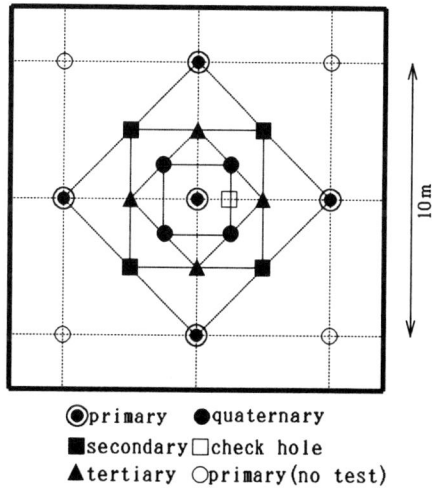

◎primary ●quaternary
■secondary □check hole
▲tertiary ○primary(no test)

Figure 1 Layout of grouting holes

After having finished these tests and grouting, we re-dug all the boreholes and made the borehole expansion test at the same position of the boreholes in order to compare the grouting effect on the mechanical properties of rock masses.

Borehole expansion tests have been carried out at 3 portions per borehole. The length of expansion part is 60 centimeters. The depths of the centre of expansion part are established 1.5 meters, 2.5 meters and 3.5 meters. The expansion apparatus is a lateral loading testing machine. The velocity of expansion is 0.5 MPa/min.

3. GROUTING EFFECT ON DEFORMABILITY

3.1 Modulous change

Figure 2 and Figure 3 show the frequency distributions of modulous of deformation and modulous of elasticity of before and after grouting, respectively. Concerning modulous of deformation, the range of values before grouting is under 1000 MPa and that after grouting, 500-1500 MPa. The same tendency is percieved in the distributions of modulous of elasticity.

Figure 4 and Figure 5 show the relationships between

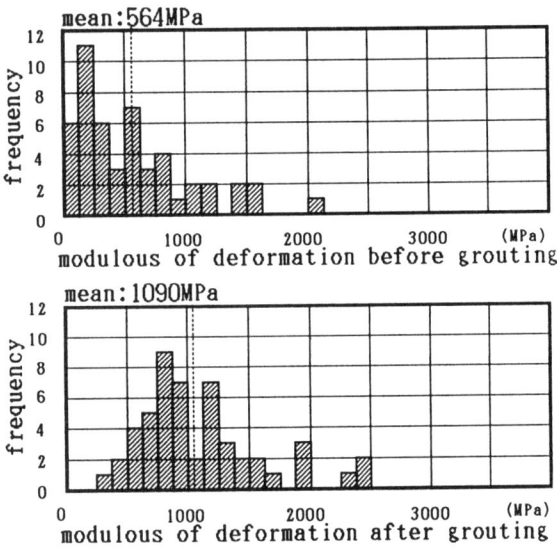

Figure 2 Frequency distribution of modulous of deformation

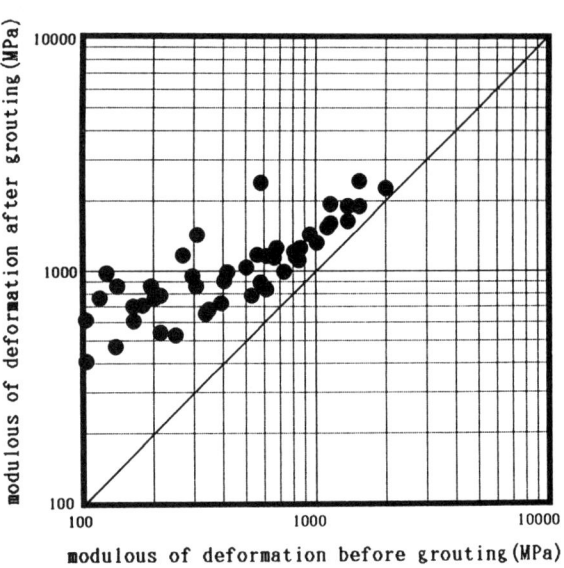

Figure 4 Relationship between moduli of deformation of before and after grouting

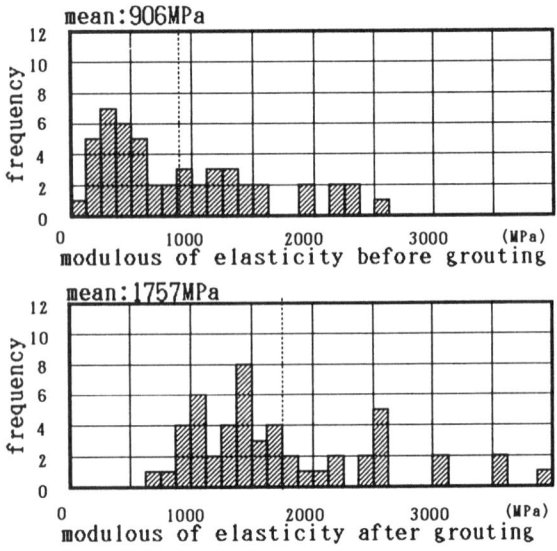

Figure 3 Frequency distribution of modulous of elasticity

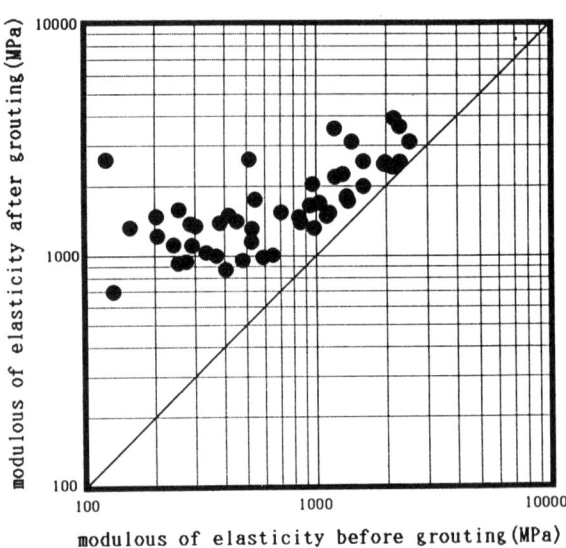

Figure 5 Relationship between moduli of elasticity of before and after grouting

moduli of deformation and moduli of elasticity of before and after grouting, respectively. Figure 6 and Figure 7 show the relationships between modulous of deformation and modulous of elasticity of before and after grouting, and improvement ratio (after/before).

From these figures, we can find a tendensy that in the case of poor rocks which have the lower mechanical propeties before grouting, the higher mechanical improvement would be expected after grouting, and in the case of good rocks which have higher mechanical propeties before grouting, the lower improvement would be after grouting.

Table 1 shows the mean value of improvement ratio of each grouting series. Mean improvement ratio of primary holes is higer than those of secondary, tertiary and quaternary holes. Table 2 shows the mean value of moduli of each grouting series. The values increaces as the grouting.

Table 1 Mean value of improvement ratio of each grouting series

modulous of deformation		modulous of elasticity	
holes	ratio	holes	ratio
primary	2.323	primary	2.241
secondary	1.765	secondary	1.833
tertiary	1.791	tertiary	1.999
quaternary	1.981	quaternary	1.775

Table 2 Mean value of deformability of each grouting series

modulous of deformation		modulous of elasticity	
holes	value(MPa)	holes	value(MPa)
primary	397	primary	682
secondary	599	secondary	934
tertiary	678	tertiary	963
quaternary	610	quaternary	1082

3.2 Mechanical model

Deformation of roka mass is the sum of deformations of intact rock and joints. In this paper the simple mechanical model is suggested as shown in Figure 8 in order to express grouting effects.

Elastic modulus of rock mass before grouting (E_B) is expressed by elastic modulus of intact rock (E_R), stiffness of joint (k_n) and joint intensity (I) as follows;

$$\frac{1}{E_B} = \frac{1}{E_R} + \frac{I}{k_n} \qquad (1)$$

Similarly, elastic modulus of rock mass after grouting (E_A) is expressed by E_R, k_n, I and elastic modulous of grouting effects (E_G) as follows;

$$\frac{1}{E_A} = \frac{1}{E_R} + \frac{1}{k_n / I + E_G} \qquad (2)$$

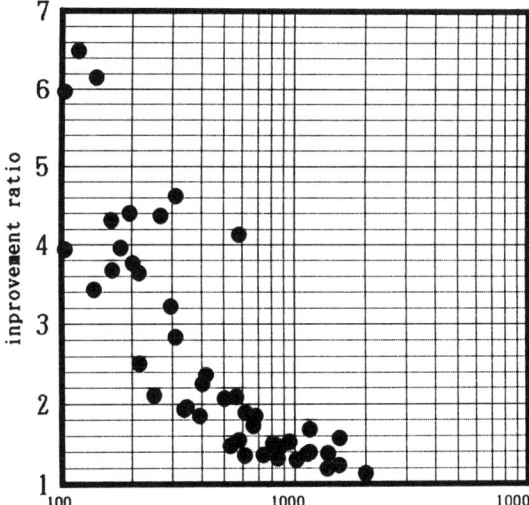

Figure 6 Relationship between modulous of deformation before grouting and improvement ratio

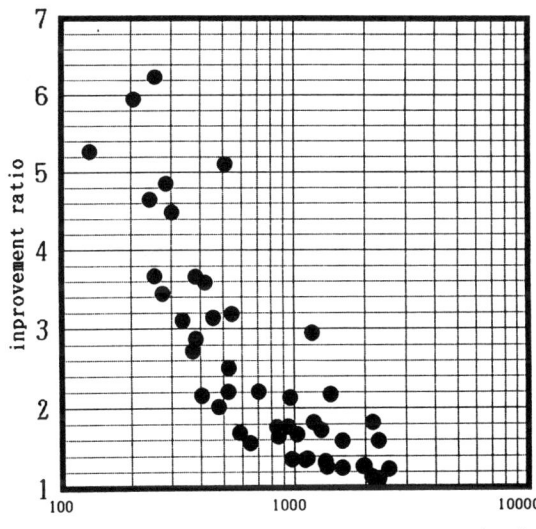

Figure 7 Relationship between modulous of elasticity before grouting and improvement ratio

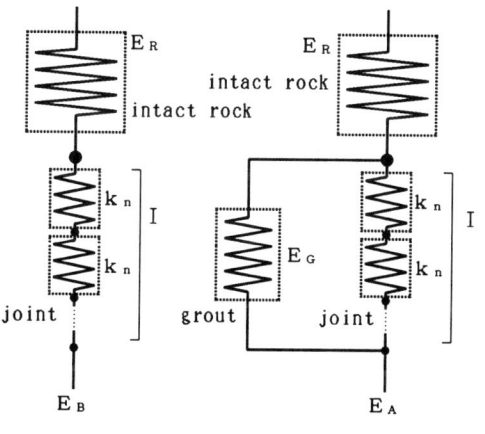

(a) before grouting   (b) after grouting

Figure 8 Mechanical model of grouting effect

Using equations (1) and (2), the relationship between E_B and E_A is expressed as shown in Figure 8(a) by treating k_n/I value as a parameter. Similarly, the relationship between E_B and E_A/E_B (improvement ratio) is expressed as shown in Figure 8(b). Hence E_G (is equal to 500 MPa approximately) is estimated by least square method.

The curevs in Figure 9 and Figure 10 fit the in situ experimental data very well. From this fact, we can find that mechanical property of joints would be improved by grouting.

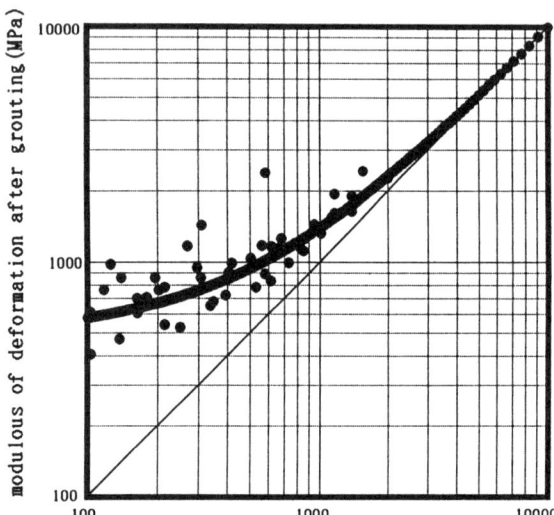

Figure 9 Relationship between moduli of deformation of before and after grouting obtained from the mechanical model of grouting effect

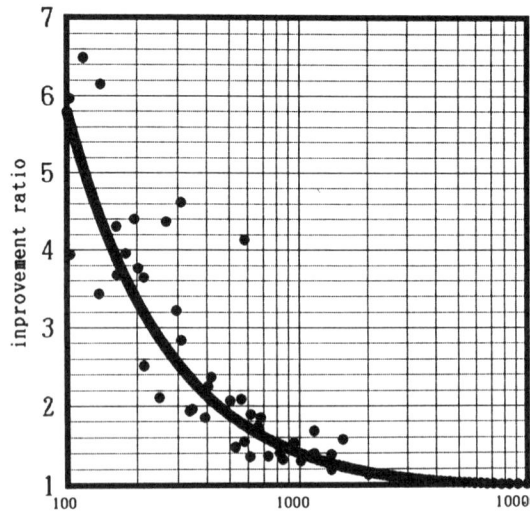

Figure 10 Relationship between moduli of elasticity of before and after grouting obtained from the mechanical model of grouting effect

4. GROUTING EFFECT ON SHEAR STRENGTH

In general, the foundation of concrete dam is designed taking shear strength of rock mass into account. if the grouting effects on shear strength could be clarified quantitatively, More economic design would be carried out because we can retrench excavation volume and concrete volume by not excavating the rock mass that can be improved by grouting.

It is difficalt to know the grouting effects on shear strength because we would not tested the rock mass which shear strength had be obtained. The rock mass specimen is to be crushed by shearing test. Therefore, the authors examine the grouting effects on shear strength using the yield stress data obtained from borehole expansion tests reffering Takeuchi and Ohhashi (1984)[1]. (Some authors[2] reported that the maximam stress is proportional to yield stress.)

Takeuchi and Ohhashi(1984) suggested the estimation method for shear strength from yield stress data. In the case of the rock masses which have the same mechanical properties, yield stress (P_y) is proportional to depth (z) as follows;

$$P_y = a z + b \qquad (3)$$

where a, b are constants. Friction angle (ϕ) and cohesion (c) are given by

$$\phi = \sin^{-1}\left(\frac{a-A}{a+A}\right) \qquad (4)$$

$$c = \frac{A b}{a-A} \tan\phi \qquad (5)$$

where

$$A = (\gamma_R - \gamma_w) \frac{\nu}{1-\nu} \qquad (6)$$

with γ_R denoting the unit volumetric weight of rock, γ_w the unit volumetric weight of water and ν Poisson's ratio.

Constants a, b are given by the estimation by least square method using the data obtained from the rock masses whose properties are almost same. In this study, two grades of rock masses is defined as follows;

Grade 1: Modulous of deformation is under 400 MPa.
Grade 2: Modulous of deformation is 400 to 1000 MPa.

Figure 11 and Figure 12 show the relationships between depth and yield stress of Grade 1 and Grade 2, respectively. We can find that yield stress has been increased by grouting. Table 3 and Table 4 show the shear strength before and after grouting of Grade 1 and Grade 2 which are estimated by the above-mentioned method. The significant effect on cohesion (0.4 MPa) can be percieved and the effect on friction angle can not be.

Although it is difficult to estimate friction angle using the Takeuchi and Ohhashi method because of insufficient (narrow) range of depth data, it can be considered that less change of friction angle is appropriate. Mito et'al(1994)[3] carried out the laboratory tests concerning the grouting effects on shear strength of the single joints and the similar results was obtained.

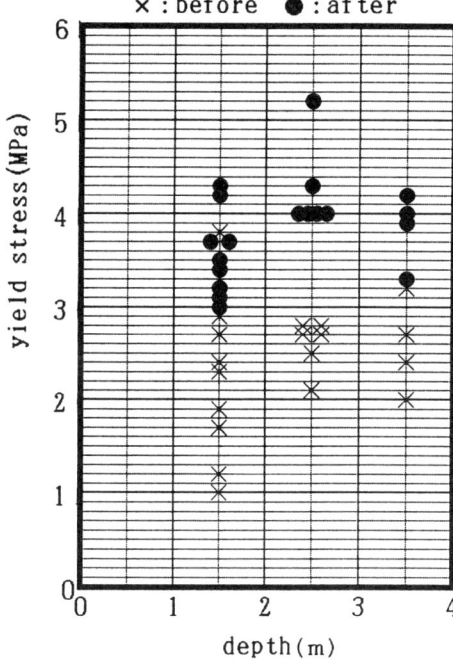

Figure 11 Relationship between depth and yield stress (Grade 1)

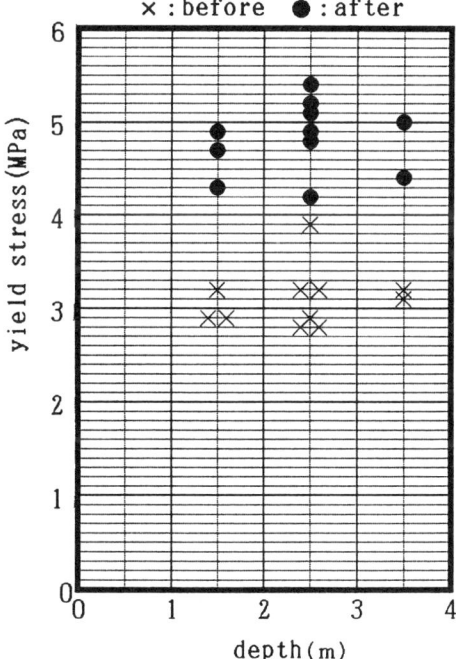

Figure 12 Relationship between depth and yield stress (Grade 2)

Table 3 Shear strength before and after grouting (Grade 1)

	cohesion(MPa)	friction angle(deg)
before grouting	0.679	30.2
after grouting	1.091	30.4

Table 4 Shear strength before and after grouting (Grade 2)

	cohesion(MPa)	friction angle(deg)
before grouting	0.840	33.3
after grouting	1.304	33.7

5. CONCLUSION

The authors have made an in situ experiment in order to examine the grouting effects on the mechanical properties (deformability and shear strength) of rock masses. As the results, the following conclusions were obtained;

1. Deformability of rock masses can be improved by grouting as well as permeability.
2. More deformable before grouting, higher improvement can be expected.
3. The rheological model is suggested in order to examine grouting effects. This model can be well expressed the test results.
4. Shear strength (especially, cohesion) of rock masses can be improved by grouting as well as deformability.

REFERENCES

1) Takeuchi.T and Ohhashi.T: Determination of the C and ϕ in-situ by the result of borehole load test; 6th Japan Symposium on Rock Mechanics, 1984.
2) Sakamoto.R and Inuzuka.H: Geological survey for Hirato-ohhashi bridge; Tsuchi-to-kiso JSSMFE Vol.22 No.6, 1974.
3) Mito.Y et al.: An experimental study on improvement by grouting for the strength of the joint surfaces; 9th Japan Symposium on Rock Mechanics, 1994.

Behavior of soft rock foundation of the Ohta Dams during construction

Hisashi Nakajima & Masahiro Idogaki
The Kansai Electric Power Co., Inc., Amagasaki, Japan

Tuyoshi Torii
Construction Project Consultant Inc., Japan

ABSTRACT: The Ohta Dams are located upstream of the Ohkawachi Pumped-Storage Water Power Station, which was constructed as the fourth pure pumped-storage water power station of Kansai Electric Power Co. The Ohta Dams consist of five rockfill dams of heights ranging from 23.5 m to 55.5m. Since thick weathered-rock layers are distributed at the dam site, column-type continuous cutoff walls were constructed in the bedrock underlying the core foundations under the right and left abutments for two dams out of the five. This paper reports on a study of the behavior of a cutoff wall and a dam during embankment and impoundment. Measurements confirmed that (1) the cutoff wall and the surrounding rock were deformed coherently; (2) displacements of the wall and the surrounding rock were mostly smaller than had been calculated at the design stage; (3) strains in the wall were sufficiently smaller than the critical strain; and (4) differences between the pore water pressure upstream of the wall and that downstream were smaller than had been calculated at the design stage. These results confirmed that the cutoff wall has sufficient stability and water-stopping performance.

1 INTRODUCTION

As one of the technologies it possesses in the field of dam construction on soft rock foundations, Kansai Electric Power Co. recently applied column-type continuous cutoff walls constructed by the soil mixing wall method (hereafter referred to simply as "cutoff walls") to dam foundations and confirmed sufficient levels of water-stopping performance, safety and economy of the cutoff walls.

The cutoff walls were used as the foundations for the main body of a group of dams upstream of the Ohkawachi Pumped-Storage Water Power Station.

Ohkawachi Station is a pure pumped-storage power plant located roughly at the center of Hyogo Prefecture. Construction of the power station began in December 1988, and then the first generator started up in October 1992. As shown in Figure 1, the power station has an upper regulating reservoir with five rockfill dams (Ohta Dams No. 1 to No. 5) and a lower regulating reservoir with a concrete gravity dam (102 m high). The maximum output of the power station is 1,280,000 kW.

The above-mentioned cutoff walls were constructed for Ohta Dams No. 2 and No. 5. This paper reports on a study of the behavior of Ohta Dam No. 2, based on the measurements and analyses carried out on the dam's No. 3 cross section, for the purpose of following up the new construction technology.

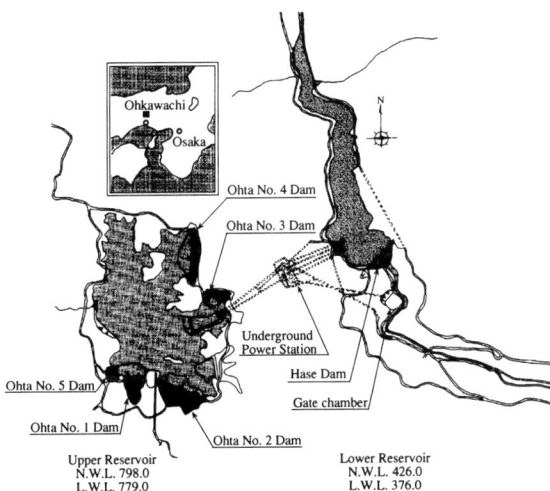

Figure 1. Plan of Ohkawachi Pumped-Storage Water Power Station

2 OUTLINE OF OHTA DAM NO. 2

The Ohta Dams were constructed on an uplifted peneplain. Geologically, Mesozoic porphyry belonging to the Ikuno group is dominant at the dam site. Thick layers of weathered rock (soft rock) characteristic of uplifted peneplains are also distributed in the area. All of the five dams have rockfill structures with core walls. As the criteria for foundation rock, CL and higher classes (under the rock classification by Central Research Institute of Electric Power Industry) were applied to the core sections except at some parts of the right and left abutments. Class D grades were applied to the rockfill areas (Yasufuku et al. 1995a).

For Ohta Dam No. 2, however, a cutoff wall was constructed as a part of the core foundation, leaving Class D rock on the left bank to reduce construction cost. Class CL or stronger foundation rocks were improved by grouting.

Basic dimensions of Ohta Dam No. 2 are shown in Table 1. Longitudinal profile and typical cross section of the dam near the left bank, whose core was founded on the Class D rock, are shown in Figures 2 and 3.

3 DESIGN AND CONSTRUCTION OF CUTOFF WALL

Before the construction of the cutoff wall, laboratory mix design tests on cutoff wall materials, structural analysis, seepage analysis and experimental construction were carried out to investigate the structural and hydraulic stability of the planned wall structure. As a result, it was made clear that a cutoff wall whose top does not project into the core is advantageous, because of minimum stress concentration, over a wall projecting into the core (by 1 m); and that the local safety factor for the core will not be lower than 1.0 if the modulus of deformation of the cutoff wall ranges between 1

Table 1. Data on Ohta No. 2 Dam

Dam	Type: Height: Crest length: Volume: Upstream slope: Downstream slope:	Central core type rockfill dam 44.5 m 397.1 m 957,000 m^3 1:3.6 (15.5°) 1:2.3 (23.5°)
Reservoir	Name of river: Catchment area: High Water Level: Normal Water Level: Reservoir Capacity: Active storage: Effective depth	Ohta River (a tributary to Odawara River, Ichi River System) 1.64 km^2 EL. 798.995 m EL. 779.000 m 9,319 × 10^3 m^3 8,660 × 10^3 m^3 19.00 m

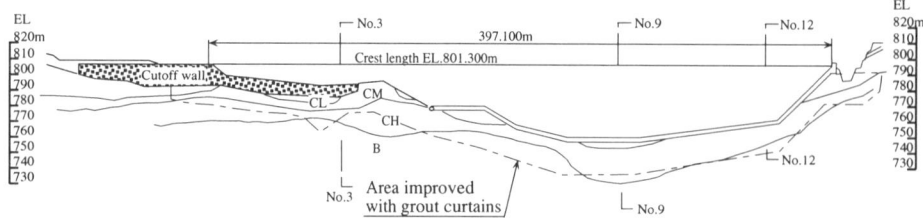

Figure 2. Longitudinal Profile of Ohta No. 2 Dam

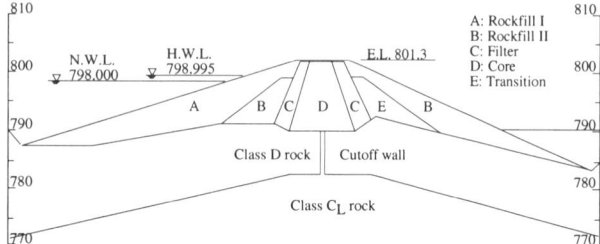

Figure 3. Typical Cross Section of Ohta No. 2 Dam with Cutoff Wall

Table 2. Measuring Instruments (No. 3 Section, Ohta No. 2 Dam)

Measurement Item	Instrument	Symbol	Number of Instruments
Pore water pressure	Electrical piezometer	P	4
Earth pressure	1-component earth pressure cell	E	3
Deformation	Rock displacement gauge	V	2
	Rock inclinometer	KU,KD	2
Strain in cutoff wall	Rubber-type strain gauge	S	2
Deformation of cutoff wall	Inclinometer	KS	1

$\times 10^5$ kN/m² and 2×10^5 kN/m². Thus, it was decided that the top of the cutoff wall should be kept at the same level as the bottom of the core, and that the wall's modulus of deformation should be kept between 1×10^5 kN/m² and 2×10^5 kN/m². It was also confirmed, from the results of the laboratory mix design tests, that the coefficient of permeability of the cutoff wall is not greater than 5×10^{-8} m/s (Yasufuku et al. 1995b).

To construct the cutoff wall, pilot holes were bored with a single-shaft auger. Then, a three-shaft mixing auger was used to drill the succeeding holes. At the same time, a mixture of cement, bentonite and water was discharged from the lower end of the auger and mixed in situ. The wall structure was formed by lapping the ends of the column rows thus constructed.

4 INSTRUMENTATION

Various measuring instruments were installed at the dam site to control the embankment work, monitor the behavior of the completed dam and collect technical information that could be used for future design.

At Ohta Dam No. 2, the measuring instruments installed in Sections No. 3, No. 9 and No. 12. In Section No. 3 of a height of 12.5 m, which is a representative cross section of the cutoff wall, the behavior of the foundation and the wall were measured (see Table 2). In Section No. 9 of a height of 44 m, which is a representative cross section of Dam No. 2, the behavior of the dam body was measured. In Section No. 12 of a height of 34.4 m, which is a representative cross section of the rockfill foundation resting on a thick layer of Class D rock, the behavior of the rock was measured.

Figure 4 shows the arrangements of the measuring instruments in Section No. 3 which is reported on in this paper. The depth of the cutoff wall in this cross section is 7.5 m.

For the measurements of the cutoff wall, three holes (ϕ127 mm) were bored at the center of the wall. A rubber-type strain gauge each was buried in the right and the left hole to measure vertical strain. Four inclinometers were buried in the middle hole at 1.5 m vertical intervals to measure horizontal displacement. Each of the rubber-type strain gauges buried in the wall consisted of six 1-meter-long hard urethane rubber cylinders with a diameter of 10 cm; two strain gauges (for steel) attached longitudinally to the cylinder so that one gauge faced upstream and the other downstream; stainless steel connectors to interconnect the cylinders; and a stainless steel weight to be attached to the lower end of the six-cylinder strain-measuring device. The modulus of elasticity of the urethane rubber used was about 1×10^5 kN/m²; this particular type of rubber was selected because its rigidity was close to that of the cutoff wall. Before the rubber-type strain gauges were used, laboratory tests were conducted to confirm their performance, including measuring accuracy, long-term stability, and ductility to follow the cutoff wall's deformations.

For the measurements of the Class D rock, three holes (ϕ66 mm) each were bored 1 m upstream and downstream from the center of the cutoff wall. Rock displacement gauges, inclinometers and electrical piezometers were buried in these holes. A rock displacement gauge each was installed in two holes, one upstream and the other downstream. The fixed point at the end of the displacement gauge was set at the depth of 11.5 m below the bottom of the core to measure relative vertical displacement of the Class D rock in the close vicinity of the cutoff wall above this level. A total of four inclinometers each were buried in two holes, one upstream and the other downstream, at vertical intervals of 2 m to measure horizontal displacement of the Class D rock in the close vicinity of the cutoff wall. Two electrical piezometers each were buried in two holes, one upstream and the other downstream, spaced vertically to measure pore water pressure upstream and downstream of the cutoff wall.

To measure stress concentration in the cutoff wall, earth pressure cells were buried in the core 1.2 m directly above the top of the wall, and also on the foundation rock surface 1 m upstream and downstream from the center of the wall.

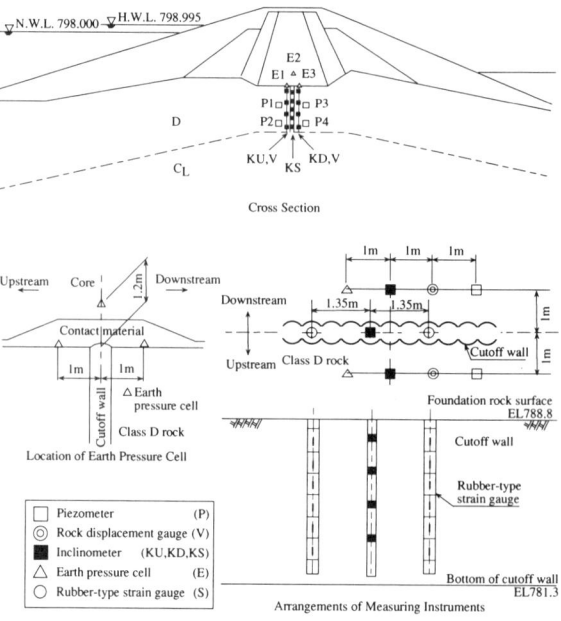

Figure 4. Instrumentation (Section No. 3)

5 MEASURED BEHAVIOR

5.1 Vertical displacement

Figure 5 shows vertical displacements of the cutoff wall and the Class D rock during the embankment work and also after its completion. Displacements of rock (V1, V2) indicate the relative displacements between the bottom of the core and the fixed point (at the depth of 11.5 m) on the displacement gauge. Displacements of the cutoff wall (S1, S2) indicate the relative displacements obtained by multiplying the average of six measured strains by the height of the cutoff wall (7.5 m) from six vertically-spaced points. Strains were measured with rubber-type strain gauges. Although the measuring span of the displacement gauge was 4 m longer than the height of the cutoff wall, no corrections of measured displacements were made for differences in measuring span. The reason is that errors were considered to be negligibly small because of a sufficient rigidity of the bedrock below the cutoff wall.

As shown in Figure 5, the vertical displacements (V1, V2) of the Class D rock correlate well with the progress of the embankment work, and the displacements converged almost completely on completion of the embankment. Vertical displacements on completion of the embankment were 13-15 mm. Vertical displacement of the cutoff wall increased with the progress of the embankment work, continued to increase gradually after the completion of the embankment, and then converged after initial impoundment. The cumulative vertical displacement of the cutoff wall was about 13 mm. Since the displacement of the wall on completion of the embankment was about 8 mm, roughly one-third of the total displacement occurred after the completion of the embankment. Gradual increases in displacement of the cutoff wall after the completion of the embankment are considered to have been caused by creep.

Figure 6 shows vertical displacements observed during the four months from July 1992 when the impoundment was started. During the initial impoundment, the vertical displacement (V2) of the Class D rock downstream of the wall increased, in the direction of elongation, by about 0.6 mm. During the same period, however, there occurred little vertical displacement of the cutoff wall (S1, S2). As far as these results are concerned, therefore, the influence of the initial impoundment on vertical displacement is not clear. During the period of water level variations, however, both the displacement of the Class D rock (V2) and that of the cutoff wall (S1, S2) occurred, that is, in the direction of elongation when the water level was rising and in the direction of compression when the water level was declining. Water level during this period lowered close to the bottom of the cutoff wall. The likely explanation is that decreases in buoyancy of the upstream rockfill and the foundation rock caused compression displacements. Vertical displacements due to water level changes of the rock downstream from the wall were about ±0.5 mm, and vertical displacements of the cutoff wall were about ±0.3 mm, which are less than one-tenth of the vertical displacements observed during the construction of the embankment. From these results, it can be said that during the period of water level variations the cutoff wall and the surrounding rock were elongated or compressed coherently.

5.2 Horizontal displacement

Figure 6 shows changes over time in horizontal displacement of the foundation rock surface and the top of the cutoff wall. The bottom elevation of the cutoff wall was defined as the fixed point (point at which horizontal displacement is zero) for the calculation, by use of the inclinometers, of the horizontal displacement of the top of the cutoff wall and the foundation rock surface. Horizontal displacements of the foundation rock surface downstream of the cutoff wall are not shown because of malfunctioning of some of the instruments used.

During the initial impoundment, the foundation rock surface and the top of the cutoff wall moved downstream by about 0.6 mm and 0.4 mm, respectively, because of the water pressure

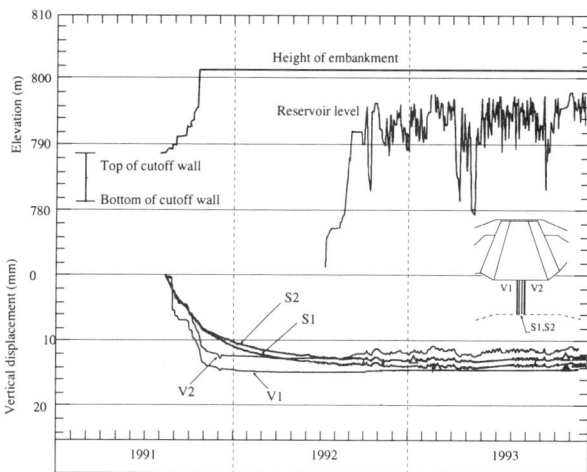

Figure 5. Changes over Time in Vertical Displacement of Cutoff Wall and Class D Rock

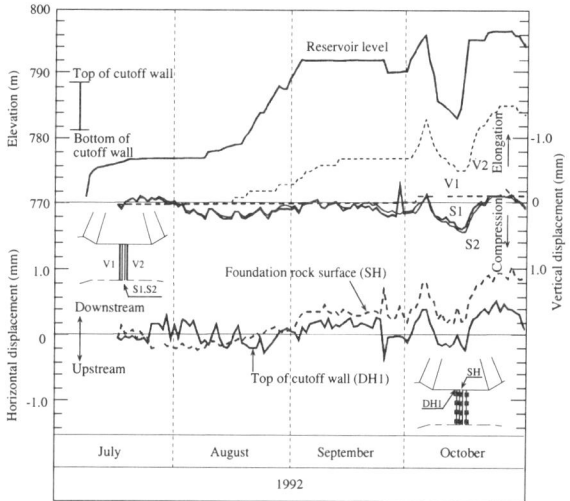

Figure 6. Changes over Time in Vertical and Horizontal Displacement of Cutoff Wall and Class D Rock

acting on the upstream face of the core, as shown in Figure 6. During the period of water level variations, the foundation rock surface and the top of the cutoff wall moved upstream when water level lowered, and downstream when it rose. Displacements of the foundation rock surface and the top of the cutoff wall were both around 0.8 mm. This indicates that during the period of water level variations, the cutoff wall and the surrounding rock were deformed upstream or downstream just like a single structure as the water level lowered or rose.

5.3 Pore water pressure

Figure 7 shows changes over time in pore water pressure in the Class D rock. As shown in Figure 7, during the embankment work there was no noticeable increase in pore water pressure in the Class D rock. During the period after completion of the embankment to the commencement of impoundment, the upper piezometers did not record any changes in pore water pressure because they were installed above groundwater level, while the lower piezometers recorded slight changes. This was probably caused by seasonal changes in groundwater level.

As shown in Figure 7, pore water pressure in the Class D rock increased as the reservoir was filled, but the amount of increase in pore water pressure downstream of the cutoff wall was smaller than that upstream. The pore water pressures were recorded on February 22, 1993, when the reservoir was almost full. There was

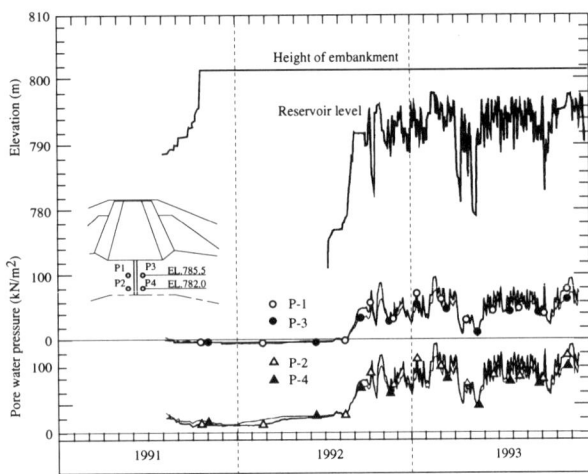

Figure 7. Changes over Time in Pore Water Pressure in In-Situ Rock

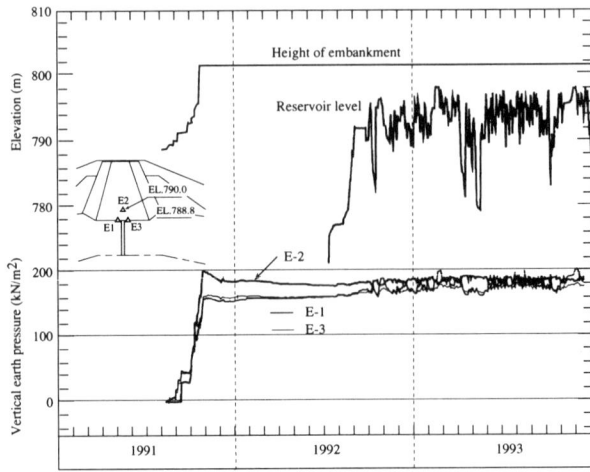

Figure 8. Changes over Time in Earth Pressure

a difference as large as 20 kPa (equivalent to 2 m of head) between the pore water pressure upstream of the wall and that downstream (see Table 6). Pore water pressure in the Class D rock during the water level variations followed the pattern of water level changes, but the amount of change downstream was smaller than that upstream.

5.4 *Earth pressure*

Figure 8 shows changes over time in vertical earth pressure in the core at a point 1.2 m directly above the cutoff wall and at the foundation rock surface at points 1 m upstream and downstream of the wall. Vertical earth pressure increased with the progress of the embankment work and converged on a certain value after completion of the embankment. However, the earth pressure in the core at the top of the cutoff wall decreased by about 10%. This can be explained as follows: Since the earth pressure cell was installed above the contact materials (highly moist and fine-grained core material), creep-like deformations caused stress redistribution to decrease the earth pressure. Earth pressure in the core at the top of the cutoff wall on completion of the embankment was about 1.2 times as high as the earth pressure at the foundation rock surface. From this, it was concluded that major stress concentration had not occurred in the cutoff wall.

6 COMPARISON WITH CALCULATED VALUES

6.1 *Analysis of stresses and deformations during embankment*

A two-dimensional finite element analysis of Section No. 3 was conducted. Table 3 shows material properties considered. A linear elasticity model was employed because the rigidity of the Class D rock and the cutoff wall was relatively high.

The strength parameters and the modulus of deformation of the embankment materials were determined through verification tests on actual materials used for banking, and density values were determined from an execution management test. The strength parameters and density of the Class D rock and the cutoff wall were determined through verification tests on undisturbed specimens taken from the site after the excavation of the foundation rock, and the modulus of deformation was determined on the basis of measured values of earth pressure and displacement. For the material properties that could not be verified through testing, the values used in the predictive analysis performed at the design stage were used.

Table 4 shows displacements of the Class D rock and the cutoff wall on completion of the embankment. The calculated values show relative displacements between the foundation rock surface and a point 7.5 m below the foundation rock surface (i.e., the bottom elevation of the cutoff wall).

Table 3. Physical Properties of Dam and Foundation Rock

Zone		Unit Weight		Water Content	Void Ratio	Strength Parameters				Modulus of Deformation	Poisson's Ratio	Coefficient of permeability
						During Embankment		During Impoundment				
		γ_t	γ_{sub}	W_n	e	c	ϕ	c', c_D	ϕ', ϕ_D	E	ν	k
		tf/m³	tf/m³	%		kN/m²	deg	kN/m²	deg	10²kN/m²		m/s
Class B Rock		2.60	1.60	5.0	0.19	3000	50	3000	50	100,000	0.25	1×10^{-9}
Class C_H Rock		2.60	1.60	5.0	0.19	1500	45	1500	45	60,000	0.25	2×10^{-7}
Class C_M Rock		2.60	1.60	5.0	0.19	1000	40	1000	40	25,000	0.25	5×10^{-7}
Class C_L Rock		2.24	1.24	10.0	0.39	300	36	300	36	5,000	0.30	5×10^{-6}
Class D Rock	< 5m	1.75	0.75	44.0	1.29	125	11	30	32	310	0.30	1×10^{-6}
	5~10m	1.83	0.83	39.0	1.12	190	15	20	36	680	0.30	1×10^{-6}
	10~15m	1.92	0.92	29.0	0.90	190	15	20	36	680	0.30	1×10^{-6}
	> 15m	2.08	1.08	24.0	0.62	190	15	20	36	680	0.30	1×10^{-6}
Cutoff Wall		1.64	0.64	63.0	1.87	430	26	430	26	1,300	0.30	5×10^{-8}
Contact Material		1.80	0.80	38.0	1.09	120	18	30	33	30	0.45	1×10^{-7}
Core		2.09	1.11	19.0	0.57	200	18	35	33	160	0.30	1×10^{-7}
Filter		2.17	1.29	4.3	0.26	30	35	0	35	200	0.30	1×10^{-5}
Transition		2.15	1.31	1.9	0.26	20	37	0	37	300	0.30	1×10^{-4}
Rockfill I		2.24	1.39	1.4	0.22	0	41	0	41	380	0.30	1×10^{-3}
Rockfill II		2.18	1.31	1.9	0.26	0	40	0	40	650	0.30	1×10^{-3}

Table 4. Comparison of Measured and Calculated Values for Cutoff Wall and Class D Rock During Embankment

Measurement Item	Measuring Point		Unit	Measured Value	Confirmatory Analysis	Predictive Analysis (Design Stage)		
						Modulus of Deformation kN/m^2		
						1×10^5	1.5×10^5	2×10^5
Vertical Displacement	Cutoff wall	Top of cutoff wall S	mm	13.2	17.1	25.0	21.6	19.2
	Class D rock	1 m upstream of cutoff wall V1		14.8	18.6	28.4	26.4	24.8
		1 m downstream of cutoff wall V2		12.8	18.5	28.2	26.2	24.6
Earth Pressure	Foundation rock surface	1 m upstream of cutoff wall E1	kPa	163	224	224	227	229
		1 m above cutoff wall E2		180	217	218	224	229
		1 m downstream of cutoff wall E3		164	224	223	226	228

Table 5. Comparison of Measured and Calculated Values for Deformations of Cutoff Wall and Class D Rock During Impoundment

Measurement Item	Measuring Point		Unit	Measured Value	Confirmatory Analysis	Predictive Analysis (Design Stage)		
						Modulus of Deformation kN/m^2		
						1×10^5	1.5×10^5	2×10^5
Vertical Displacement*1	Cutoff wall	Top of cutoff wall S	mm	-0.5	0.0	0.9	0.8	0.8
	Class D rock	1 m upstream of cutoff wall V1		0.0	0.2	0.4	0.3	0.3
		1 m downstream of cutoff wall V2		-1.0	-0.2	-0.1	-0.1	-0.1
Horizontal Displacement*2	Foundation rock surface	1 m upstream of cutoff wall KS	mm	0.6	0.7	3.2	3.2	3.2
		1 m above cutoff wall KU		0.7	0.8	3.2	3.2	3.2
		1 m downstream of cutoff wall KD		0.5	0.7	3.0	3.0	3.0

*1. Minus (−) signs for vertical displacements represent elongation.
*2. Displacement in the downstream direction due to water level rise is regarded as positive (+) displacement.

As shown in Table 4, the calculated vertical displacements of the Class D rock and the cutoff wall are about 30% greater than the measured values. This is thought to be because the moduli of deformation for these zones used in the analysis were too small. Table 4 also shows the results of the predictive analysis carried out at the design stage. As shown, the calculated values were greater than the measured values, and the modulus of deformation for the Class D rock and the cutoff wall had been underestimated.

The values of earth pressure obtained through confirmatory analysis are roughly 30% greater than the measured values. The reason is probably that the underestimation of the modulus of deformation of the rockfill and filter zones, and the resulting underestimation of stresses transferred to this zone, led to overestimation of earth pressure at the base of the core.

Figure 9 shows the distribution of strains in the cutoff wall. Values of strain obtained from the confirmatory analysis are greater than the measured strains. They do have a similarity, however, in terms of the pattern of strain distribution, suggesting that the ratio between the modulus of deformation of the cutoff wall and that of the surrounding ground was appropriate. The values of strain obtained through measurement and confirmatory analysis were sufficiently small as compared with the critical strains determined through unconfined and triaxial compression tests on specimens taken at the time of installation of the rubber-type strain gauges. It was thus confirmed that the cutoff wall was in a stable condition.

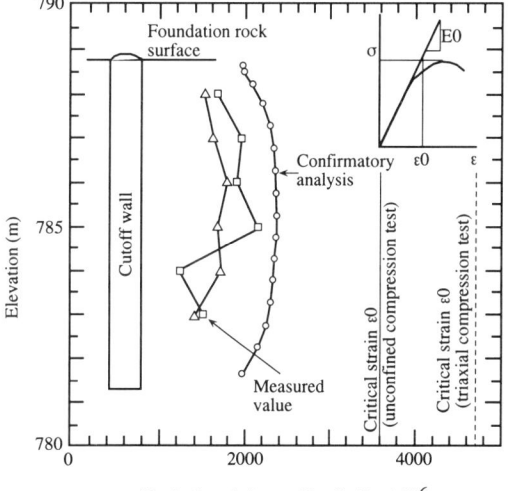

Figure 9. Vertical Strain Distribution of Cutoff Wall

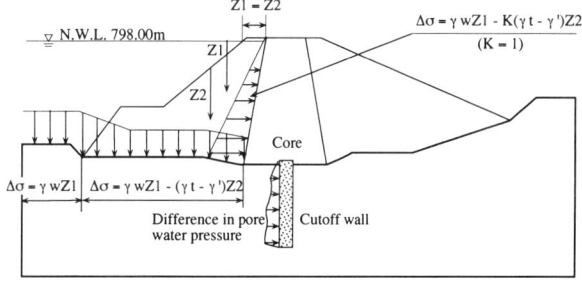

Figure 10. External Forces Considered in Analysis of Stresses and Deformations During Impoundment

6.2 *Analysis of stresses and deformations during impoundment*

In the analysis of deformation during impoundment, the stress on completion of the embankment was assumed to be the initial stress. Then, the external forces shown in Figure 10 were applied.

Table 5 shows vertical and horizontal displacements of the Class D rock and the cutoff wall during impoundment. Since vertical displacement could not be measured accurately during the initial impoundment, the measured values shown here indicate displacement increases which were observed while the water level was rising during the period of reservoir level variations. As shown in Table 5, the calculated horizontal displacements of the Class D rock and the cutoff wall agree well with the measured values. This result agrees with the actual behavior of the rock and cutoff wall—that is, the behavior of being slightly and repetitively deformed upstream and downstream like a single structure rather than separate ones.

Table 6. Comparison of Measured and Calculated Values of Pore Water Pressure in Class D Rock During Impoundment

Measurement Item	Measuring Point			Unit	Measured Value	Confirmatory Analysis (Steady Seepage Analysis)	Predictive Analysis (Design Stage; Steady Seepage Analysis)
Reservoir Level				m	EL.797.87 (1993.2.22)	EL.798.0 (N.W.L.)	EL.798.0 (N.W.L.)
Pore Water Pressure	EL.785.5	Upstream of cutoff wall	P1	kpa	97	91	81
		Downstream of cutoff wall	P3		77	66	54
	EL.782.0	Upstream of cutoff wall	P2		136	118	110
		Downstream of cutoff wall	P4		116	105	94

Table 7. Differences Between Pore Water Pressures Upstream and Downstream of Cutoff Wall Obtained from Steady Seepage Analysis Parameterizing the Coefficient of Permeability of Cutoff Wall

Measurement / Analysis	Coefficient of Permeability of Cutoff Wall (m/s)	Unit	EL.785.5			EL.782.0		
			Upstream	Downstream	Difference	Upstream	Downstream	Difference
Measurement		kpa	97	77	20	136	116	20
Analysis	Without cutoff wall		83	75	8	115	109	6
	5×10^{-8} (design value)		91	66	25	118	105	14
	5×10^{-9}		98	60	38	121	101	20

With respect to vertical displacement, the observation showed that both the cutoff wall and the Class D rock were elongated. On the other hand, the confirmatory analysis indicated compression displacements (although they were small), which contradict the actual behavior. The probable reason is as follows: to simplify the calculations, it was assumed that infiltration into the foundation rock had not begun. Actually, however, infiltration into the rock did occur as the reservoir level rose, thus reducing the effective stress in the foundation rock.

Table 5 also shows the results of the predictive analysis conducted at the design stage. As shown, the values of horizontal displacement obtained from the predictive analysis are greater than those obtained from the confirmatory analysis. This is because in the confirmatory analysis horizontal constraints at the right and left boundaries of the region considered were assumed to be constant, while in the predictive analysis it was assumed that there was no constraint.

6.3 *Seepage analysis*

Two-dimensional saturated and unsaturated seepage analyses of Section No. 3 were carried out. The coefficient of permeability for each zone is shown in Table 3. The coefficient of permeability of the cutoff wall was assumed to be equal to the design value (5×10^{-8} m/s). Before the analyses, it was confirmed that the coefficient of permeability of the specimens taken during the installation of the rubber-type strain gauges falls within the range between 3×10^{-8} m/s and 3×10^{-9} m/s.

Table 6 shows the calculated values of pore water pressure in the Class D rock when the reservoir water was at Normal Water Level (798.0 m), along with the measured values. The pore water pressures observed on May 22, 1993, when the reservoir level became close to N.W.L., show fair agreement with the values obtained through the confirmatory analysis. For reference, the results of the predictive analysis performed at the design stage are also shown in Table 6.

Table 7 shows differences between the pore water pressure upstream of the cutoff wall and that downstream. The analysis dealt with three different cases: the case where there is no cutoff wall, the case where the coefficient of permeability of the cutoff wall is equal to the design value (k = 5×10^{-8} m/s), and the case where the coefficient of permeability is smaller by one order (i.e., k = 5×10^{-9} m/s). According to this table, the difference between pore water pressure in the rock upstream of the cutoff wall and that downstream increases as the coefficient of permeability of the cutoff wall becomes smaller. In the case where the coefficient of permeability is equal to the design value, the difference in pore water pressure is 2.5 times greater than in the case where there is no cutoff wall. If the coefficient of permeability is smaller by one order, the difference is about 4 times greater than in the case where there is no cutoff wall. The observed differences between the pore water pressure upstream of the wall and that downstream fall within the middle range of the values calculated for the above-mentioned cases, namely, the case where the coefficient of permeability is equal to the design value and the case where the coefficient is smaller by one order. Thus, the water-stopping performance of the cutoff wall is well within the design limits, confirming a sufficiently low permeability of the wall.

7 CONCLUSIONS

This paper has reported on the study based on the observation and confirmatory analysis of the behavior of Ohta Dam No. 2 constructed on a soft rock foundation, during the period from embankment work to impoundment. As a result of a series of measurements and analyses, it was confirmed that (1) the cutoff wall and the rock were deformed coherently like a single structure; (2) displacements of the cutoff wall and the surrounding rock tended to be smaller than had been estimated at the design stage, and strains in the cutoff wall were sufficiently smaller than the critical strain; and (3) the difference between the pore water pressure upstream of the cutoff wall and that downstream was actually greater than had been calculated at the design stage. From these results, it was concluded that the cutoff wall had sufficient degrees of stability and water-stopping performance.

In closing, the authors would like to thank all the people concerned of Newjec, Inc. who conducted the measurements for this study.

REFERENCES

Yasufuku, S., M. Kageyama & H. Kakuhara. 1995a. Management of the excavation of foundation of the Ohta Dams constructed on soft rock. *Proc. International Workshop on Rock Foundations.*

Yasufuku, S., M. Kageyama & K. Sakashita. 1995b. Treatment of soft rock masses used for the soil cement mixing wall at the Ohta Dams. *Proc. Internatinal Workshop on Rock Foundations.*

Treatment methods for highly permeable and soft foundations of rockfill dam construction

Kiichi Kanazawa
Civil Design Office, The Electric Power Development Co., Ltd (EPDC), Japan

Junya Takimoto
Okinawa Seawater Pumped-Storage Power Plant Construction Office, EPDC, Kunigami, Japan

ABSTRACT: After decades of development, Japan is running out of dam sites for hydroelectric power generation with favorable topographic and geological conditions. In order to most economically build safe dams, it is essential to provide efficient and secure foundation treatments. In this paper, foundation treatments for rockfill dams built on soft foundations are discussed, by following the examples of Tedorigawa Dam and Ouchi Dam, constructed by the EPDC (Electric Power Development Co., Ltd.). At the same time, problems with the conventional method of cement grouting are discussed. Development of new foundation treatments for rockfill dams on soft foundations with high water permeability are then discussed, with particular reference to their application in the construction of Tadami Dam. Finally, methods for deciding properties for analyses are discussed, while comparing the analytical and instrumental results for each dam.

1. FOUNDATION TREATMENTS FOR LARGE ROCKFILL DAMS

Recently, construction of large rockfill dams on soft rock foundations have increasingly been seen in Japan. However, in such construction, there is always a possibility of deformation of the foundation, caused by the weight of the dam body and/or the impounding water. Therefore, careful investigation is required whenever adverse effects are anticipated at the points where the dam body and foundation make contact. Even when sufficient grouting work has been conducted prior to construction of a dam, considerable seepage flow from the abutment could occur, thus requiring additional grouting work.

Insufficient grouting work prior to construction of a dam and separation of cement milk were considered to be the main causes of the seepage flow, and they were regarded as somewhat unavoidable.

Other examples were encountered wherein seepage flow occurred at the initial impoundment stage because of cracks created by deformation of the abutment during construction of dams with soft-foundation.

In this chapter, the existing methods for foundation treatments for large rockfill dams constructed on soft rock beds are discussed.

1.1 Example of Tedorigawa Dam

1.1.1 Geology of the foundation

Tedorigawa Dam (height: 153 m, crest length: 420 m, volume of embankment: 10,100,000 m³) is a rockfill dam with a center imperious core having a gross capacity of 231,000,000 m3 of impounded water. See Figure 1 for a typical cross section of the dam.

The middle and upper part of the abutment at the right-bank side of the dam are composed of conglomerate. The riverbed and the abutment at the left-bank side of the dam are composed of gneiss. There are two kinds of gneiss divided by a large-scale fault which runs through the middle of the left-bank side abutment. This was the largest fault found in the foundation of the dam. The thickness of the fractured zone reached approximately 25 m. The fractured zone was composed of fault clay and sheared rock. See Figure 2 for a geological section along the dam axis.

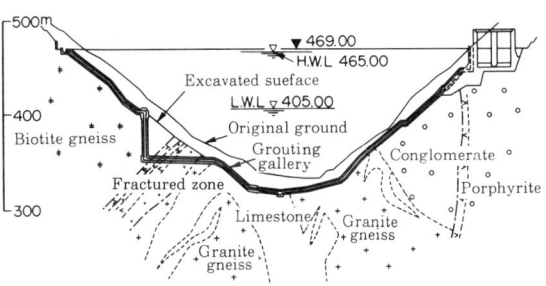

Figure 2. Geological section along the dam axis

1.1.2 Outline of the foundation treatments

The following were main issues during the foundation treatments of Tedorigawa Dam.
(1) Treating the large-scale fault at the left-bank side
(2) Treating the lime caverns at the riverbed
(3) Treating the open cracks at the upper part of the conglomerate at the right-bank side
(4) Installing the grouting gallery

The lime caverns at the riverbed (2) were successfully treated by conducting careful grouting work. The open crack at the upper part of the conglomerate at the right-bank side (3) was also successfully treated by performing careful grouting work and partially replacing it with a concrete diaphragm cutoff wall.

Treatment of the large-scale fault at the left-bank side (1) and installation of the grouting gallery (4) are discussed below.

The permeability coefficient of the ground within the fault was measured as between 2×10^{-8} and 3×10^{-7} cm/s according to the tests conducted on samples, and between 5×10^{-7} and 5×10^{-5} cm/s according to the tests conducted at the site.

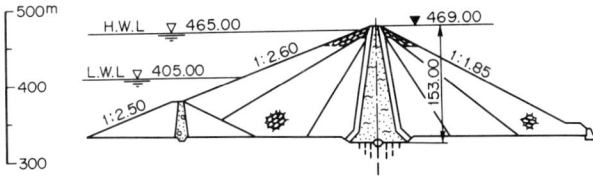

Figure 1. Typical cross section of Tedorigawa Dam

The deformation moduli were measured as between 12 and 41 MPa according to the plate-bearing tests. The deformation moduli obtained were more scattered: They were approximately 10 MPa at or near the surface and approximately 200 MPa in the deeper areas, according to the bore hole bearing test.

In order to determine the method for foundation treatment at the fractured zone of the fault, a linear deformation analysis based on the finite element method was conducted on the two-dimensional cross sections by taking up the above-mentioned properties for analysis. Figure 3 shows the properties for analysis.

Figure 4 shows the results of the analysis. According to these results, the maximum settlement was calculated as 41 cm. Table 1 shows some of the stress values within the foundation and their corresponding weights for the dam.

According to Table 1, the stress values were about the same as the weights for the dam for the rock bed of the right-bank side, whereas in the rock bed of the left-bank side, the stress values were smaller than the weights for the dam when measured within the fractured zone, and larger when measured at both sides outside the fractured zone.

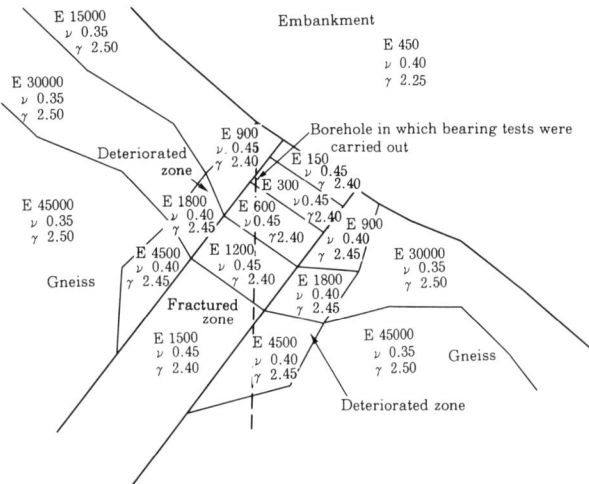

Figure 3. Mechanical properties of materials

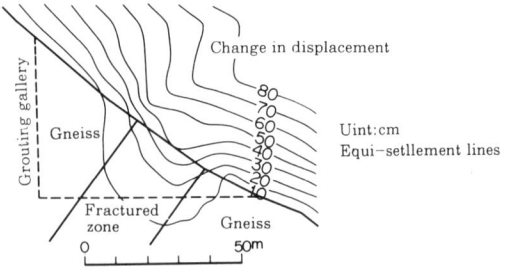

Figure 4. Results of the deformation analysis

Table 1. Some of the stress values within the foundation and their corresponding weights for the dam (Ouchi Dam)

	Left bank			Right bank
	1	2	3	4
Max. principal stress	4.66	1.12	5.2	2.97
Min. principal stress	0.83	0.78	1.77	1.61
Stress by dam weight	2.41	2.64	3.22	2.99

1. Above the fractured zone
2. Within the fractured zone
3. Beneath the fractured zone

Unit: MPa

Generally, a method which required excavation of a part of the fractured zone and replacing it with concrete has often been used when there was a fault in a hard rock foundation of a dam. In the case of Tedorigawa Dam, since the scale of the fractured zone was massive and the rock beds divided by the fractured zone were both weathered, such a method was considered to be unsuitable. As a result, the following method of foundation treatment was planned:

* Remove loose materials near the surface of the fractured zone.
* Conduct grouting work only within the fractured zone.
* At the fractured zone, use a method of construction that allows for the deformation which occurs during the embankment and/or impounding water.
* Conduct grouting work throughout the surface of the core at the top and bottom boundaries of the fractured zone.
* Increase the width of the core at the point where the fractured zone makes contact.
* Using filter materials, cover the surface of the fractured zone, exposing the foundation surface of the rock bed upstream and downstream.

According to the initial plan for Tedorigawa Dam, a culvert-type gallery would be installed. However, it was changed to a tunnel at the fractured zone. The reason for the change was that the stability of the grouting gallery would be impaired at the contact place of the excavation surface of the foundation and the core, thus causing other adverse effects. The settlement at the passing point of the culvert-gallery after the change was less than 15 cm, according to the analysis mentioned earlier.

1.1.3 Results of the foundation treatments

A total of 148 kg/m of cement was injected over a length of approximately 190,000 m (originally planned to be 43,000 m) for the foundation treatment for Tedorigawa Dam.

The changes in the length and width of the blanket grouting holes, the limestone treatment, and the length and series of holes in the upper middle part of the left-bank side were the main reasons for the increased volume of injected cement. In the case of Tedorigawa Dam, the curtain grouting work was conducted in the curtain row with the improvement value (set prior to the embankment) targeted at 3 Lugeon. However, the median of the permeability coefficient increased from a value of 2 to 3, and as much as 16% of the test results registered over 5. Therefore, in order to achieve the original target, regrouting work was performed for an injection length of 7,200 m. The average volume of cement for reinjection was 64 kg/m.

Extra attention should be paid to the phenomenon of increased permeability caused by the deformation of the foundation during grouting work when constructing a high rockfill dam on a soft rock bed. This could increase the uncertainty of the injection volume required when conducting grouting work on the foundation of a rockfill dam on a soft rock bed.

1.2 Example of Ouchi Dam

1.2.1 Geology of the foundation

Ouchi Dam (height: 102 m, crest length: 340 m, volume of embankment: 4,400,000 m³) is a rockfill dam with a center core. See Figure 5 for a typical cross section of the dam.

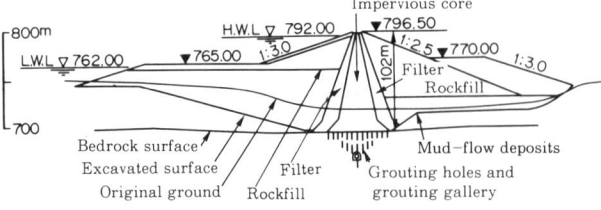

Figure 5. Typical cross section of Ouchi Dam

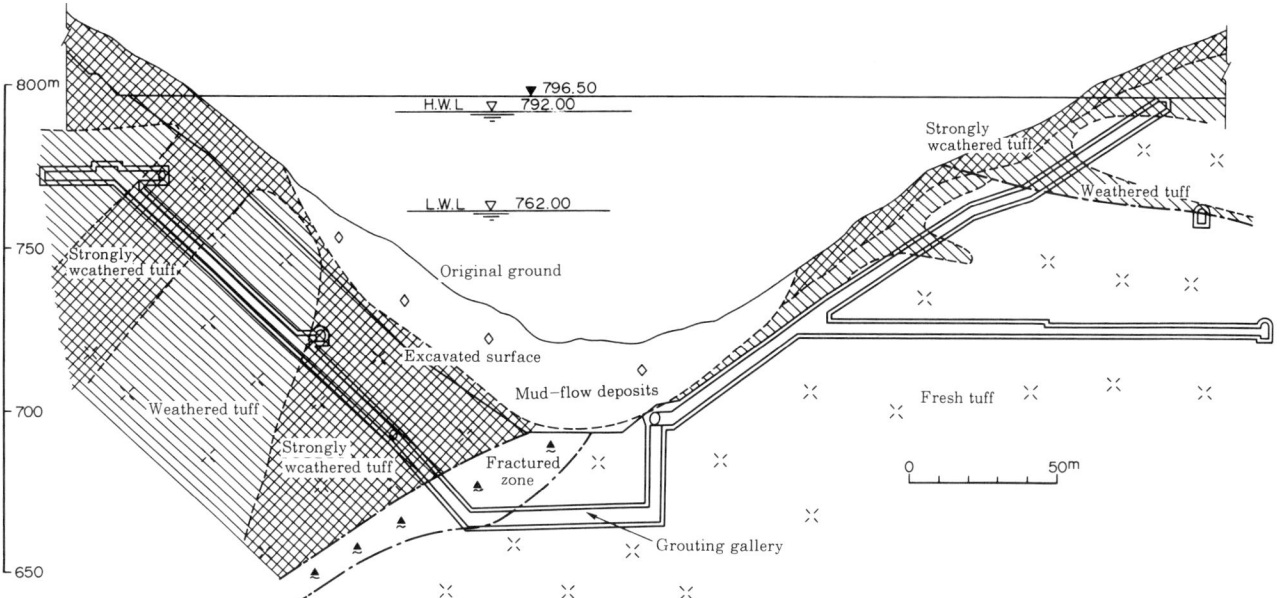

Figure 6. Geological section along the dam axis

The dam was constructed in a so-called "green tuff" area. The foundation of the dam consisted of tuff from the Tertiary period. Unconsolidated deposits of Quaternary period called mud-flow deposits covered the tuff to a maximum thickness of 30 m.

There was a fault running at an angle of approximately 25 degrees through the riverbed along the valley at the left-bank side. The thickness of the accompanying fractured zone, consisting of the mixture of clay and debris of tuff, was approximately 10 m. Owing to the effect of the fault in the riverbed, the tuff on the left-bank side had deteriorated. Tuff on the right-bank side had not been affected by the fault as much. See Figure 6 for a geological section along the dam axis.

1.2.2 Outline of the foundation treatments

The following were the main issues during foundation treatments at Ouchi Dam.

(1) Treating the large-scale fault in the riverbed and tuff, which had deteriorated because of the effects of the fault, on the left-bank side
(2) Treating tuff with many cracks on the right-bank side which was formed in a comparatively thin spine
(3) Installing the grouting gallery
(4) Treating the mud-flow deposits

Tuff with many cracks on the right-bank side (2) was treated successfully with a conventional technology of careful grouting work. The mud-flow deposits (4) were excavated, found unsuitable to remain, and removed from the foundation at the part of the core zone. In accordance with the detailed in-situ tests, the mud-flow deposits were permitted to remain in the foundation of the rock part as long as possible. At the same time, the design of the embankment was adjusted with the toe weight.

Special attention was required for treatment of the fractured zone in the riverbed fault and the rock bed, which had deteriorated because of the fault on the left-bank side for their soft and deformable features and high permeability. The treatment of this area (1) and installation of the grouting gallery (3) are discussed below.

The permeability coefficient of the rock was uniformly measured as 10^{-4} cm/s, according to the in-situ tests conducted within the rock foundation on the left-bank side. However, in the tests conducted on the undisturbed samples, they were measured as smaller figures by more than one decimal place. The discrepancy was presumably caused by the fact that the undisturbed samples gathered from the rock foundation did not necessarily represent the whole rock foundation in regard to permeability. The permeability coefficient within the fractured zone was measured as approximately 1×10^{-6} cm/s, according to the tests conducted on the undisturbed samples.

The deformation modulus of the rock foundation was obtained by the plate-bearing tests and the bore hole bearing tests. It was measured as approximately 800 MPa at the rock foundation of the right-bank side.

The deformation moduli were measured as approximately 40 MPa at the surface of the fractured zone of the fault, and between 100 and 200 MPa in the deeper area.

The deformation moduli on the left-bank side were measured as between 50 and 400 MPa. The dispersion was caused by random distribution of the fault's effect.

A forecast analysis of deformation on the foundation during construction of the dam was conducted. Linear-deformation analysis was conducted along a two-dimensional cross section of the dam axis.

Figure 7 shows the zone divisions for the calculations. A curve of the deformation moduli increasing as depth increased was used to estimate the deformation moduli. Table 2 shows the figures used for the calculations. The maximum values of displacement were 27-36 cm at the excavation surface of the foundation, and 21-27 cm at the center line of the gallery. The maximum values of settlement (vertical displacement) were calculated to be between 22 and 29 cm at the excavation surface of the foundation, and between 15 and 18 cm at the center line of the gallery.

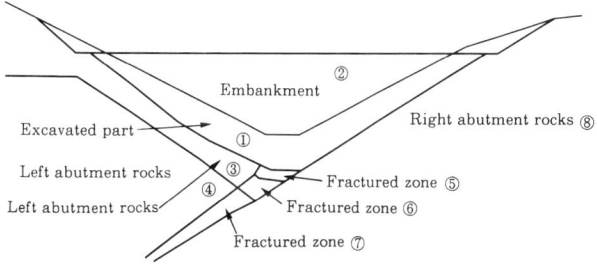

Figure 7. Zone divisions for the calculations (Ouchi Dam)

Table 2. The figures used for the caluculation (Ouchi Dam)

Zone	Specific gravity (t/m³)	Poisson ratio	Deformation Modulus MPa	
Excavated part	2.26	0.35	①	50
Embankment	2.22	0.45	②	30
Left abutment rocks	2.42	0.35	③ ④	200～350 300～380
Fractured zone	2.22	0.45	⑤ ⑥ ⑦	50～100 100～150 200～250
Right abutment rocks	2.42	0.30	⑧	1000

The characteristics of the dam foundation are summarized below according to the above-mentioned examinations:
(1) The rock bed on the right-bank side could be treated with conventional technology.
(2) Though the fractured zone in the fault showed low permeability in general, certain parts could have higher permeability.
(3) There were many places on the left-bank side rock bed showing over 10 Lugeon (permeability coefficient: approximately 1×10^{-4} cm/s). It is not easy to perform grouting work on such an area, since high-pressure injection is not feasible for a soft rock foundation. However, by conducting thorough grouting work, the area was demonstrably improved, and the values were brought down to under 10 Lugeon, even with low-pressure injection.
(4) Significant deformation of the foundation rock caused by the embankment load was expected.

According to the characteristics of the foundation rock discussed above, the following method was planned in order to treat the foundation:
* Conduct thorough grouting work mainly targeting improvement of ability to detain water, since the dam body could sustain deformation, according to the above-mentioned analysis. As much as possible, conduct grouting work from the surface of the ground prior to the embankment.
However, there was a possibility that part of the inside of the foundation rock would become loose because of the deformation caused by the weight of the embankment and/or the impounding water.
Therefore, during construction of the embankment, conduct grouting work at parts which were inaccessible from the surface. At the same time, inspect the area where grouting work had already been done, and perform regrouting as required. It should be noted that a restriction owing to the large load of the dam body was utilized inversely in this method of foundation treatment. Observe the foundation behavior during impoundment and regrouting perform as required.
* Install the grouting gallery along the dam axis. Install tunnels with a depth of between 20 and 25 m at the riverbed and the left-bank side. The depth had to be determined so the excavation of the tunnels would not affect the excavation surface of the dam. On the right-bank side, install the grouting gallery on the excavation surface of the dam. Install the short galleries upstream and downstream to facilitate the regrouting work.

1.2.3 Results of the foundation treatments

An average of 242 kg/m of cement was injected over a length of approximately 140,000 m for grouting work of Ouchi Dam.
Improvement has been made evenly on the right-bank side, the left-bank side and the fractured zone of the fault to approximately 3 Lugeon. The injection length on the left-bank side was 94,100 m (more than double the 40,400 m on the right-bank side). The discrepancy occurred from the difficulty in trying to improve the left-bank side, which was composed of soft rock foundation.
After the dam had been constructed, water permeability was improved at parts of the left-bank side by additional grouting work. This shows that regrouting is essential when the foundation is soft. At the same time, it is hard to anticipate the injection volume for difficult areas.

2. DEVELOPING NEW METHODS FOR FOUNDATION TREATMENTS

2.1 Necessity of new methods

Regarding foundation treatment on rockfill dams, grout injection with ordinary cement will become harder when the conditions of the foundation ground deteriorate. Conducting careful grouting and regrouting would not affect so much in regard to the construction costs in large-scale dams such as Tedorigawa Dam and Ouchi Dam. However, it would affect cost considerably in construction of small-scale dams. Therefore, the EPDC has been researching and developing new methods for foundation treatment, including construction of the bentonite-concrete diaphragm wall (which promotes complete water cutoff) and WMC (wet milled cement) grouting work (which is suitable for soft foundations and foundations with alluvial deposits).

2.2 Bentonite concrete diaphragm wall

Generally, it is very effective to add bentonite to concrete in order to increase the plasticity of the concrete. The EPDC had developed a diaphragm wall of large capacity for deformation and high water-cutoff properties by taking advantage of the ability on bentonite to preserve water. Using bentonite enables construction of an appropriate impervious structure for a dam on a foundation with the possibility of large deformations (such as on a foundation with alluvial deposits), because of its capacity for deformation.
Tests were conducted in order to determine the basic properties of analysis of bentonite concrete. Between 10 and 30 kg/m³ of bentonite was mixed in with a unit weight of cement of between 75 and 200 kg/m³ at a ratio of water to cement (W/C) of between 120 and 350 %.
Figure 8 shows the relationships between the content of bentonite and unconfined compressive strength according to the content of cement. The capacity for deformation is increased by adding bentonite, since it also increases the amount of water in the ratio of water to cement.

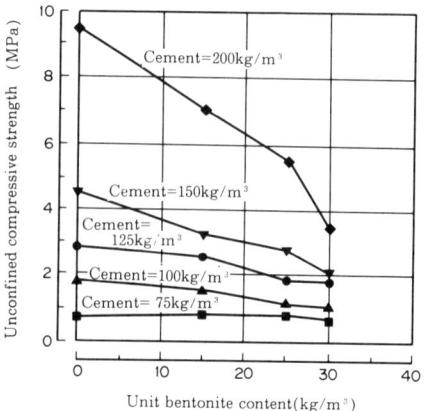

Figure 8. Relationships between the content of bentonite and unconfined compressive strength according to the content of cement

Figure 9 shows the relationship between unconfined compressive strength and the deformation modulus. Sufficient correlation can be observed between the two. The deformation moduli were obtained by triaxial compression tests.

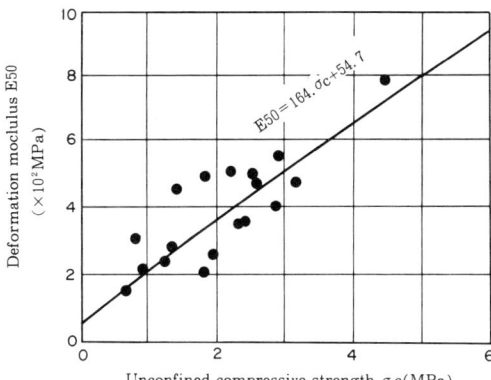

Figure 9. Relationship between unconfined compressive strength and the deformation modulus

Figure 10 shows the relationship between unconfined compressive strength and the permeability coefficient. The permeability coefficient decreases as unconfined compressive strength increases. The permeability coefficient is smaller, and water cutoff properties are higher, when greater amounts of bentonite are added to the cement, even though unconfined compressive strength is the same. In addition, it was confirmed that the permeability coefficient will decrease with time after starting the test. The durability of bentonite concrete was also confirmed.

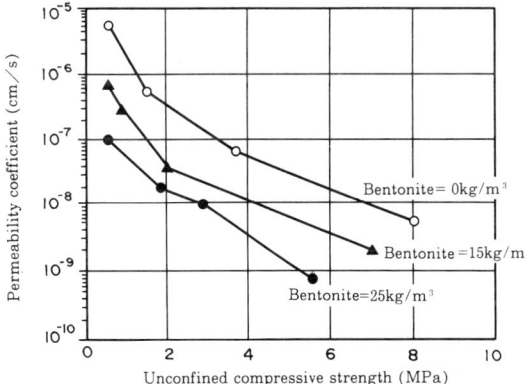

Figure 10. Relationships between unconfined compressive strength and the permeability coefficient

An amount of bentonite in the concrete, while aiming for the same deformation modulus as with all the alluvial deposits in the surrounding riverbed as much as possible, was chosen that would obtain high water-cutoff properties, stability of quality and workability. See Table 3 for the standard mix of bentonite concrete. Based on the figures in this table, in-situ tests at the dam were conducted in order to check the quality, workability, water cutoff properties and deformation capacity of a bentonite concrete diaphragm wall. The scale of the test was about half that of the original construction.

According to the results of the in-situ tests, it was confirmed that the bentonite concrete diaphragm wall had characteristics which were observed in lab tests and would not present any problems in regard to construction.

Table 3. The standard mix of bentonite concrete

Water-Cement ratio W/C (%)			223
Sand percentage s/a (%)			47
Unit weight (kg/m³)	bentonite	B	25
	Cement	C	125
	Water	W	279
	Fine aggregate	S	792
	Coarse aggregate	G	928

2.3 WMC grouting

Ordinary cement used for grouting a rock foundation is superior in many ways, in handling, strength, durability and cost. However, it is usually not effective for grouting of foundations with alluvial deposits, since the particle size is too large. We have developed a WMC grouting method which takes advantage of a slurry state of cement grout, converting the micromilling process of cement from the conventional dry type to a wet type, and conducted the operation at the construction site for greater efficiency.

This method is achieved by simply adding a wet micromilling process to the conventional process of injecting cement grout.

The wet-type micromill (hereinafter referred as the "micromill") pulverizes cement particles, which are passing through gaps of steel beads filled in the micromill rotating at high speed by adding the power of friction, compression, and shear, etc. Figure 11 shows the micromill. Particle size can be adjusted between the sizes found in ordinary cement (an average of 20 microns) down to as small as 2 to 3 microns by changing various conditions when pulverizing. Therefore, the WMC grouting method enables expansion of the injection to soft rock foundations and alluvial deposits with minute gaps, which are difficult to treat with cement and/or clay grout. At the same time, it is possible to reduce water permeability by using the WMC grouting method. WMC grout has almost the same elution, erosion, water pressure resistance and durability as ordinary cement.

Generally, the sleeve injection method is employed for injection of WMC grout.

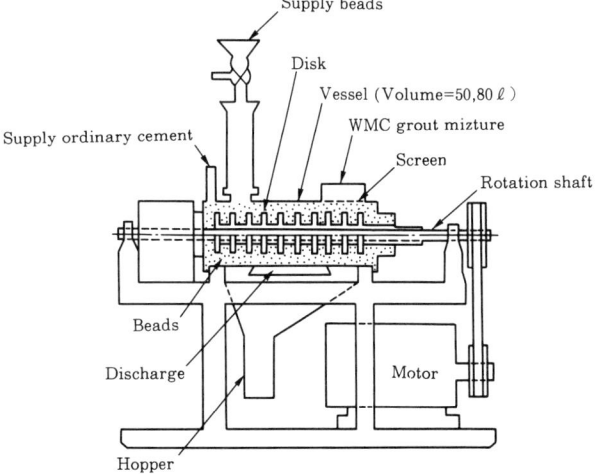

Figure 11. Micromill

3. APPLICATION TO TADAMI DAM

3.1 Design of Tadami Dam

Excavating and removing river alluvial deposits (which measured approximately 20 m at the site of Tadami Dam) was determined not to be economically feasible.

Therefore, it was decided to construct the dam directly on alluvial deposits by using a bentonite concrete diaphragm wall in order to intercept water at the foundation with a WMC blanket. WMC grouting was chosen as the method by which optimum control of the permeating current at the bottom of the core could be obtained.

See Figure 12 and 13 for a typical cross section of Tadami Dam and details of the top part of the diaphragm wall.

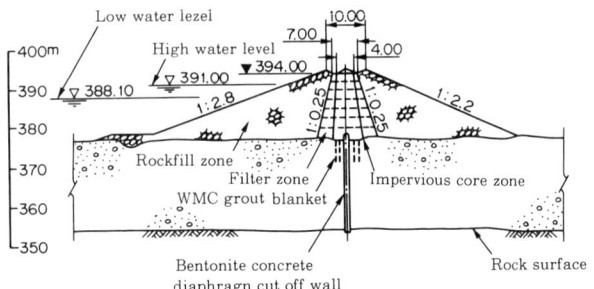

Figure 12. Typical cross section of Tadami Dam

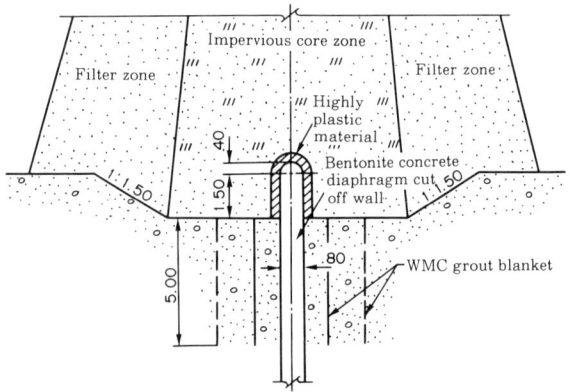

Figure 13. Details of the top part of the diaphragm wall

3.2 Construction of the bentonite concrete diaphragm wall

At this site, a panel-type diaphragm wall, which would have fewer joints and excel in plumbness and continuity of the wall body, was adopted. For the excavation system, the bucket type, which is suited for excavation of river deposits containing cobblestones, and which enabled cobblestones comparatively fast excavation, was adopted. Prior to bucket excavation, preboring was conducted using a donut-type rock auger to facilitate excavation.

Bentonite concrete was produced after adding facilities for manufacturing, storing, measuring and supplying bentonite slurry to a ordinary concrete batching plant. Addition of bentonite slurry was conducted using an automatic dispensing apparatus.

The bentonite concrete was transported in ordinary concrete mixers, and placement of the concrete was done by means of tremie pipes.

During the production of the bentonite concrete, in order to obtain concrete of uniform quality, detailed measurements were taken of the specific gravity and cohesion of the bentonite slurry and of the surface water of fine aggregate.

When placing the bentonite concrete, tests related to slump, compressive strength, and water permeability were performed to ascertain quality. The results of the quality-control tests are shown in Table 4.

Table 4. The results of the quality-control tests

Items			Results			
Name of tests	Units	Frequencies	Reference values	Min.	Max.	Ave.
Slump	cm	Once per 60m³ or element	18±2.5	16.0	18.9	18.1
Strength	MPa	3 samples per 60m³ or element	2.0±0.4	1.76	2.34	2.10
Permeability	×10⁻⁶ cm/s	Once per element	1.0	0.12	0.77	0.44
Unit weight	t/m³	3 samples per element	–	2.18	2.23	2.20

As a result of the tests, the permeability coefficient was found to average 4.4×10^{-7} cm/s. The control criterion for the permeability coefficient of the wall was 1.0×10^{-6} cm/s or less. Therefore, the result of the test showed that ample watertightness had been secured.

3.3 WMC grouting

The grout holes were made using a hydraulic rotary percussion drill. The grout manufacturing plant consisted of a mixer, a stocking pump, a micromill and an agitator. The cement used as the raw material was ordinary Portland cement.

Slurry concentration at the time of micromilling was set at cement to water = 1:2.8, the most efficient ratio, according to the results of studies made in the past. The retention time of slurry inside the vessel was 2 to 3 minutes. Steel wire with a diameter of 1.2 mm was cut into length of 1.2 mm and used as steel beads. The vessel was filled 95% or more with the cut wire.

The total amount of WMC grout produced for this construction was approximately 1,300 kilo-liters, whereas the capacity of the micromill was approximately 7.9 liters/min.

The quality of WMC grout produced was checked by collecting samples from the agitation tank at least twice daily, and measuring particle size and the specific gravity of the slurry and by bleeding, etc. The particle size of the WMC grout averaged 6.7 microns for D95, 4.7 microns for D85, and an average of 2.4 microns for D50. All were well within the control values. The particle-size distribution curve of the WMC grout is shown in Figure 14.

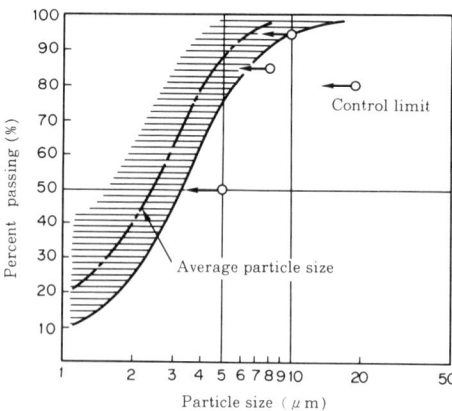

Figure 14. Particle-side distribution curve of WMC grout

The injection plant was composed of a slurry pump, an automatic water gauge, a grout mixer for adjusting concentration, a grout pump, and a pressure and flowmeter with an automatic recorder.

Water permeability tests were performed in order to

confirm the improvement effect after primary injection. When permeability tests indicated that adequate improvement effects were not obtained with primary injection, secondary injection was conducted by recracking the valve used in the primary injection. When secondary injection did not improve the situation, injection was conducted a third time after adding extra holes.

The grout injection recorded was injection length: 1,725 m, total number of grout holes: 335, total number of valves: 5,176, total injected volume: 2.75 million liters, total amount of cement injected: 450 tons, and average volume injected per valve: 530 liters (average cement volume: 260 kg/m).

At the difficult-to-improve right-bank side (permeability coefficient prior to injection: 2.3×10^{-3} cm/s), since adequate improvement was not obtained with primary injection, secondary injection and then injection with additional grout holes were performed. As a result, the probability of Lugeon values of 5 not being exceeded, became 88 percent, meeting the target value. The amount injected per valve was 456 liters (average volume of cement: 204 kg/m).

At the left-bank side, with a comparatively high water-permeability (1.7×10^{-2} cm/s), the improvement target was more or less reached with the primary injections. As a result of permeability tests after a slight amount of additional grouting, the probability of Lugeon values of 5 not being exceeded, became 93 percent. The amount injected per valve was 632 liters (average volume of cement: 346 kg/m).

As a result of the above-mentioned injections, it was confirmed that adequate improvement was attainable with the WMC grout when working not only on a foundation with permeability of 10^{-2} cm/s, but also on a foundation with permeability of 10^{-3} cm/s.

4. ISSUES ON FORECASTING DEFORMATION OF FOUNDATIONS

When deformation of the foundation is expected in construction of a rockfill dam on a soft rock foundation or foundation with alluvial deposits, it is essential to obtain an accurate displacement in order to obtain structural stability and water-cutoff ability.

Displacement is often over-estimated when designing a rockfill dam, since calculations tend to be done on the safe side.

4.1 Examples of Tedorigawa Dam and Ouchi Dam

In the case of Tedorigawa Dam, the settlement at the surface of the fractured zone was actually measured by using measuring instruments for rock-foundation displacement, installed at the fractured zone of the fault at the left-bank side. As a result, the actual settlement was found to be approximately 10 cm, which is considerably smaller than the figure estimated from the analysis (41 cm). In order to define the reasons for this deviation, the contrast between two-dimensional and three-dimensional analysis, effects caused by structural zoning of the embankment, and other geological effects were examined. As a result, it was found that geological conditions had more significant effects than differences in analysis methods.

The following is a comparison between estimated analysis values and actual measurements in regard to deformation of Ouchi Dam:

* Actual measurements were found to be smaller than estimated values in general. For example, the maximum settlement at the foundation of the dam had been estimated to be 22 cm, whereas it was found to be approximately 10 cm when it was actually measured.
* In particular, almost no settlement was measured at the foundation at the lower part of the fault.
* Actual measurement of the settlement at the center of the gallery was found to be considerably smaller than estimated. For example, the maximum settlement

volume had been estimated to be 15 cm, whereas it was found to be 2.5 cm when it was actually measured.

As mentioned above, actual measurements showed a tendency to be smaller than the estimated values. In the case of Ouchi Dam, grouting and embankment had evidently had a great effect on deformability of the rock foundation at the soft rock foundation of the left-bank side. Figure 15 shows the variation of cumulative relative frequency of deformation modulus at the left-bank rock foundation and the fractured zone at each stage of the construction. In conclusion, according to the transition in median figures, the deformation modulus was found to be 10 times greater.

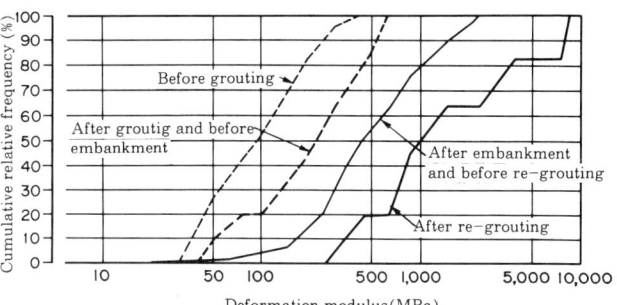

Figure 15. Variation of cumulative relative frequency of deformation modulus

4.2 Example of Tadami Dam

In the case of Tadami Dam, the actual settlement with the bentonite concrete diaphragm wall was about half of the estimation at the design stage. The results of deformation analysis is shown in table 5.

Table 5. The results of deformation analysis of Tadami Dam

	Deformation Modulus		Vertical Displacement	
	Diaphragm wall (MPa)	Alluvial deposits (MPa)	Diaphragm wall (mm)	Alluvial deposits (mm)
Design	500	170	21	25
Analysis	500	220~400	13.2	9.3
Measurement	−	−	13.7	9.6

Structural analysis was performed in order to clarify the reasons for the difference. During the analysis, the deformation moduli of the alluvial deposits were regarded as nonlinear, beam elements were adopted for the diaphragm wall, and joint elements by Goodman were introduced between the beams and triangular elements. The figure obtained by this method was approximately the same as the actual measurements of the settlement.

In the case of this dam, the design work was done according to the results of the plate-bearing test conducted at the surface of the alluvial deposits, and the deformation moduli were estimated too high.

Since a large dispersion of data was found in measurement of a strain of bentonite concrete, a test was conducted at the site, and the results of the test were used to estimate displacement, which were then applied in the design. Recently, it has been understood that the dispersion of data mentioned above could cause an estimate to be too high when the figures were used to obtain an axial strain based on the vertical displacement of cap and piston during unconfined and triaxial compression tests conducted at a soft rock foundation with a modulus of deformation similar to that of bentonite concrete. In addition, a new measurement method has been proposed to solve the problem. In the future, more accurate estimates of displacement will be available by using this new method, which will lead to rational design for the diaphragm wall.

5. CONCLUSION

When conducting foundation treatments of soft foundations with high permeability, it is often difficult to attain an accurate design using conventional cement injection methods, since there are many uncertain factors owing to the deformability of the foundation and the difficulty of improvement.

Establishing more reliable foundation treatment methods is greatly needed. Thus, the EPDC has been developing new methods, including the bentonite concrete diaphragm wall and the WMC grouting system.

The bentonite concrete diaphragm wall and the WMC grouting method were applied to the construction of Tadami Dam, and high accuracy was obtained. In short, they were proven to be effective.

The bentonite concrete diaphragm wall and the WMC grouting method can contribute to achieving a plan with less uncertainty for foundation treatments based on the effectiveness of these treatments, which were actually applied to Tadami Dam.

REFERENCES

Nishigori T. and Takimoto J. :"Foundation Treatment for Alluvial Deposits at Tadami DAM", 17th International Congress on Large Dams, Vienna Q. 66, R. 24, 1991

Tatsuoka, F. et al., :"Field and Laboratory Test Methods to Obtain Non-linear Deformation Characteristics of Sedimentary Soft Rocks," Tsuchi-To-Kiso Vol. 40 No. 11 Nov. 1992

Settlement of a pier foundation for Akashi-Kaikyo Bridge and its numerical analysis

M.S.A. Siddiquee
Bangladesh University of Engineering and Technology, Bangladesh (Formerly: University of Tokyo, Japan)

F. Tatsuoka, A. Inoue & Y. Kohata
Institute of Industrial Science, University of Tokyo, Japan

O. Yoshida & Y. Yamamoto
Honshu-Shikoku Bridge Authority, Tokyo, Japan

T. Tanaka
Meiji University, Japan

ABSTRACT: Settlement recorded during construction of a large pier foundation (3P), constructed on sedimentary softrock, for the world-longest suspension bridge, Akashi Kaikyo (Strait) Oh-hashi Bridge, was simulated by non-linear elasto-plastic FEM analysis. The material properties were evaluated based on elastic stiffness obtained by field seismic survey while taking into account the dependency of stiffness on shear stress and pressure levels obtained from triaxial compression tests on undisturbed samples.

1. INTRODUCTION

Akashi Kaikyo Bridge (**Fig. 1a**) is now under construction over Akashi Kaikyo (Akashi Strait), located 15 km west Kobe City, which will become the world longest suspension bridge when completed. The foundations for the bridge have been completed. They were not damaged by the Great Hanshin·Awaji Earthquake of January 17th 1995 with an earthquake magnitude of 7.2, the epicenter of which was located only a very short distance from the bridge (**Fig. 1b**).

Pier 2P was constructed on a lightly cemented stiff gravel (Kobe group), which is of Late Neogene to Early Quaternary underlain by a sedimentary soft rock of silt-to-sandstone (Kobe group), which is of Tertiary with a geological age of about four million years or more. Anchor 1A and Pier 3P were directly placed on Kobe group silt-to-sandstone, which was exposed by deep ground excavation (see Fig. 1a), underlain by a thick granite formation. Anchor 4A was constructed directly on a granite layer also after ground excavation.

The seismic stability of 2P was one of the major concerns in the design stage, the analysis for which was and will be reported by Tatsuoka et al. (1991 and 1995d).

The displacement (i.e., settlement and tilting) of the four foundations under static working loads during and after construction has been another major concern. Excessive displacements of the foundations during construction could be a sign of unacceptable instability of the foundation and possible continuing large displacements after construction which may damage the superstructure. Besides, too large displacements of the foundations during construction may lead to a serious modification of the structure details and the construction procedure of the superstructure.

The settlement of the foundations and that in the supporting ground along the central line of each foundation during construction have been recorded. The records for Piers 2P and 3P were numerically analyzed successfully by the authors as reported partly by Tatsuoka and Kohata (1995). This analysis is characterized by the use of the stiffness values of the geomaterials concerned that were evaluated based on the elastic stiffness obtained from field seismic surveys. In so doing, the dependency of stiffness on shear strain (or shear stress level) and pressure level were taken into account, which was obtained from laboratory triaxial compression (TC) tests on undisturbed samples. In the TC tests on silt-to-sandstone samples, axial strains were measured locally along the lateral surface of specimen for a range of strain from about 0.0001 % to about 1 % (Kohata et al., 1994, Tatsuoka and Kohata, 1995). The laboratory testing method is described in detail also by Tatsuoka et al. (1993, 1994, 1995b). Only the result of the analysis for Pier 3P (**Fig.3**), located on sedimentary soft silt-to-sandstone, will be reported herein.

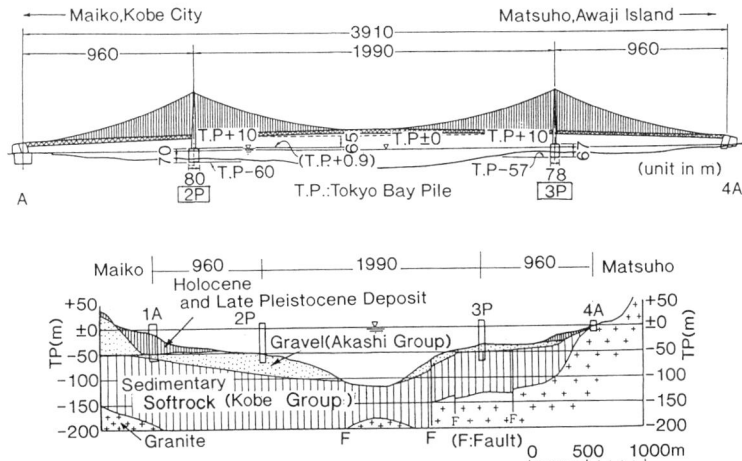

Fig. 1a) General view of Akashi Kaikyo (strait) Bridge

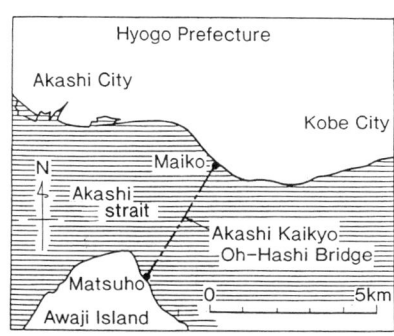

Fig. 1b) Location of Akashi kaikyo Bridge

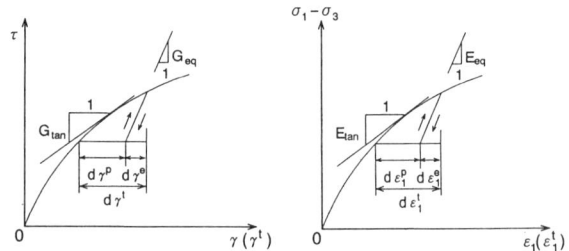

Fig. 2 Illustration of stress-strain relationship for a geomaterial

2. MODELING OF SEDIMENTARY SOFT ROCK

When following the conventional geotechnical investigation procedure, the stiffness of sedimentary softrock for the analysis of footing settlement is usually evaluated by field static loading tests, among which the plate loading test is most popular. In the design stage of this bridge, the Young's moduli E_{PLT} obtained based on the linear theory from unload/reload curves in plate loading tests were used. The use of unload/reload moduli, despite that the net change in the vertical stress in the ground upon the construction of the foundations is dominantly an increase, is based on past experiences. That is, PLT stiffness values obtained from primary loading are generally too small to explain footing settlements that were actually observed. Plate loading tests could not be performed on the ground surface on which these foundations were to be constructed in advance of deep ground excavation, and furthermore it was difficult to perform them at the sites under deep sea for Piers 2P and 3P. Therefore, the stiffness values used in the design were actually estimated by substituting the Young's moduli E_{BHLT} obtained based on the linear theory from primary loading curves in bore hole lateral loading tests (i.e., pressuremeter tests) performed at the sites of the foundations into an empirical relationship between E_{PLT} and E_{BHLT} values established for this silt-to-sandstone obtained by investigations performed at on-shore sites.

This method described above is considered practical due to its simplicity. However, the possible dependency of stiffness on strain level (or shear stress level) and pressure level is not considered, but a geomaterial is usually assumed to be a linear elastic material as it was in this case. It has been required, therefore, to develop a design procedure which can be applied to general cases and can hopefully simulate more accurately the behaviour of the foundations.

For the elasto-plastic FEM used in this study, the tangent Young's modulus E_{tan} and Poisson's ratio ν_{tan} for total strain increments and corresponding elastic Young's modulus E_{eq} and elastic Poisson's ratio ν_{eq} at a given stress state are required (**Fig. 2**). In this study, these quantities were assumed to be a function of the present stress state in terms of shear stress level and pressure level as described below, while the effect of stress histories were ignored.

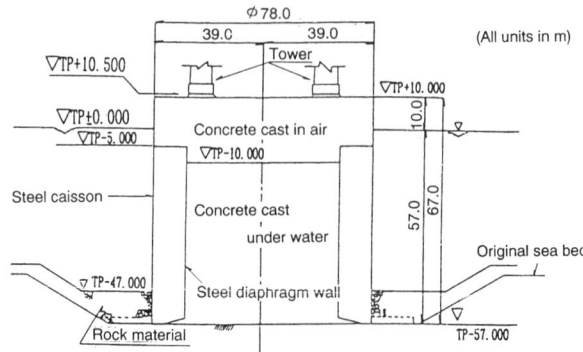

Fig. 3 Pier 3P for Akashi Kaikyo Bridge

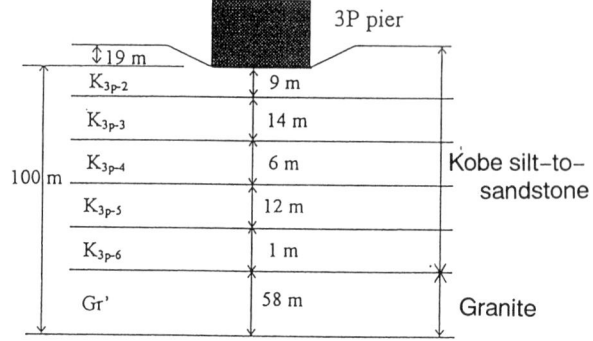

Fig. 4 Axi-symmetric model of Pier 3P and its supporting ground

2.1 YOUNG'S MODULUS AND POISSON'S RATIO

Young's modulus E_{tan} was obtained as:

$$E_{tan} = 2(1 + \nu_{tan}) \cdot G_{tan} \qquad (1)$$

The value of ν_{tan} used was different for drained and undrained conditions, which was assumed to be 0.2 and 0.46, respectively, under isotropic stress conditions. Furthermore, the Poisson's ratio for elastic strain increments ν_{eq} was assumed to be equal to their respective value of ν_{tan}. On the other hand, the shear modulus G_{tan}, which is independent of drainage conditions (Tatsuoka and Kohata, 1995), was given as:

$$G_{tan} = G_{eq} \cdot g(SL) \qquad (2)$$

The current elastic shear modulus G_{eq} was also assumed to be independent of drainage conditions. $g(SL)$ is the plasticity function, which will be explained later. G_{eq} is given as:

$$G_{eq} = E_{eq}/\{2(1 + \nu_{eq})\} \qquad (3)$$

E_{eq} and ν_{eq} are the elastic Young's modulus and Poisson's ratio at a given stress state in the field, which is obtained as:

$$E_{eq} = E^e \cdot f(SL) \qquad (4)$$

Table 1 Layering at the site of Pier 3P

Layer name	Layer thickness(m)	γ_{sat} (tf/m^3)	V_s (m/s)	Poisson's ratio, ν_u	Calculated σ_{v0} (kgf/cm^2)	Calculated E_f (kgf/cm^2)
K_{3P-2}	9	2.27	470	0.47	1.736	5116.76
K_{3P-3}	14	2.34	560	0.47	4.585	7488.0
K_{3P-4}	6	2.34	590	0.47	5.895	8311.0
K_{3P-5}	12	2.20	450	0.48	6.48	4545.91
K_{3P-6}	1	2.36	730	0.46	8.046	12833.1
Gr'	29	2.35	860	0.45	10.626	17735.30

f(SL) is the damage function representing the deterioration of the elastic Young's modulus due to shear deformation, which will be explained later. E^e is the undamaged Young's modulus, which was obtained from the undrained Young's modulus for elastic deformation obtained from the field seismic survey performed before ground excavation to a depth of 19 m at the site of the construction of 3P (**Fig. 4, Table 1**). This procedure is the heart of this analysis.

In the analysis, it was considered that the undamaged elastic Young's moduli in the vertical and horizontal directions became different, due to stress system-induced anisotropy as explained later, as:

$(E^e)_v = E_f \cdot x \cdot j(\sigma_v) \cdot \alpha \cdot \beta$ (5a)
$(E^e)_h = E_f \cdot x \cdot j(\sigma_h) \cdot \alpha \cdot \beta$ (5b)

$E_f = 2(1+\nu^e) \cdot (\gamma_{sat}/g) \cdot V_s^2$ (5c)

$x = 1.0$ for undrained E^e, and
$= (1+\nu_d)/(1+\nu_u)$ for drained E^e (5d)

$(E^e)_v$ and $(E^e)_h$ are the undamaged elastic Young's moduli for the vertical and horizontal directions, and ν_d and ν_u are drained and undrained elastic Poisson's ratios (0.2 and 0.46 were assumed).

Eq. 5 is based on the following findings. First, at strains of less than about 0.001 %, the deformation of sedimentary softrocks (and other geomaterials) is virtually elastic, and therefore, the stiffness determined dynamically and statically under otherwise the same conditions are essentially the same (Tatsuoka and Shibuya 1992, Tatsuoka and Kohata 1995, Tatsuoka et al., 1993, 1994, 1995a, b, e). This point is demonstrated also in **Fig. 5**. It may be seen from this figure that the initial Young's moduli E_{max} obtained from TC tests (**Fig. 6c**) is close to the corresponding E_f values. These TC tests were performed on undisturbed samples, which were obtained from the bottom of the excavation (- 61 m) for Anchor 1A. The fact that the average of E_{max} values is slightly lower than average of the E_f values at the concerned depth may be due to the effects of sample disturbance. The samples were consolidated isotropically to the original field effective vertical stress σ_{vo}.

A result typical of these TC tests is shown in Fig. 6. Other samples were consolidated isotropically to pressure other than σ_{vo}. In Fig. 6, "LDT" means the locally measured correct axial strains, while "external" means the externally measured axial strains including the effects of bedding error (see a figure inset in Fig. 6b). It may also be seen from Fig. 5 that the other Young's moduli are much smaller than the E_f and E_{max} values, due to : a) strains at which the stiffness was defined were much larger than the elastic limit strain (about 0.001 %) (for E_{50}, E_{BHLT} and E_{PLT}), and b) bedding errors between softrock surface and a loading platen (for E_{50}, E_{BHLT} and E_{PLT}). In particular, the E_{50} values scatter very large, while the average is considerably lower than the value of E_f and E_{max}. This point is described in detail by Tatsuoka and Kohata (1995). Besides, pressure levels in these tests may have been different from σ_{vo}, either smaller or larger than σ_{vo}, which obscures the comparison of these E values with the values of E_f and E_{max}.

The second rationale for Eq. 5, is that it can be reasonably assumed that the elastic Young's modulus in a certain direction is a unique function of the normal stress in that direction (Kohata et al., 1994, Tatsuoka and Kohata, 1995). According to this assumption, $(E^e)_v$ and $(E^e)_h$ values become different to each other at anisotropic stress states. The pressure level function $j(\sigma_v)$ was obtained based on the results of TC tests on undisturbed samples

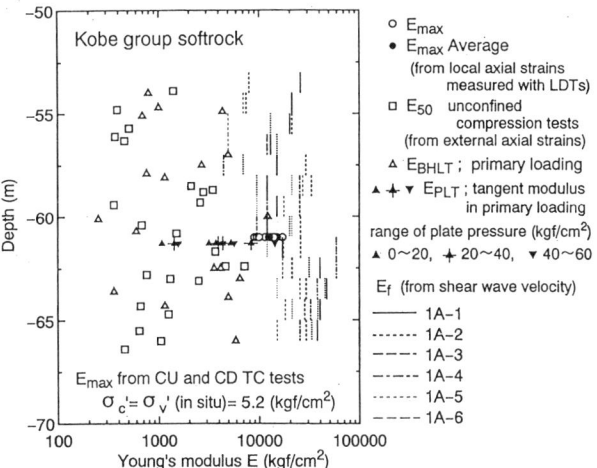

Fig. 5 Distribution of Young's moduli obtained by various field and laboratory tests at Anchor 1A site (Tatsuoka and Kohata, 1995)

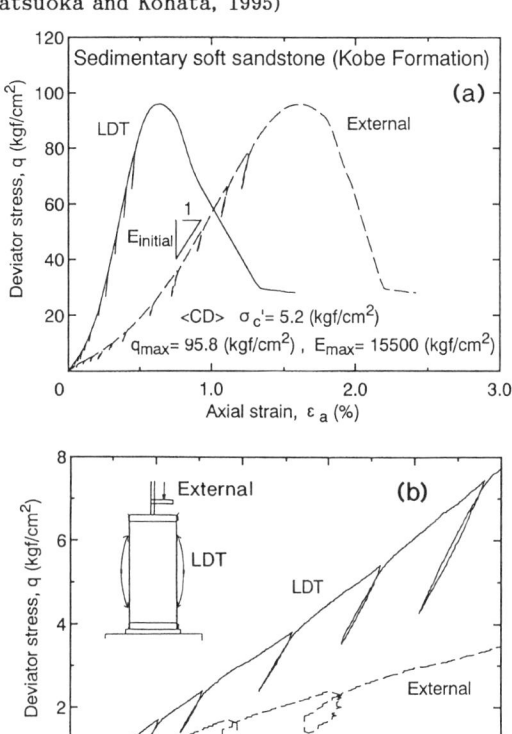

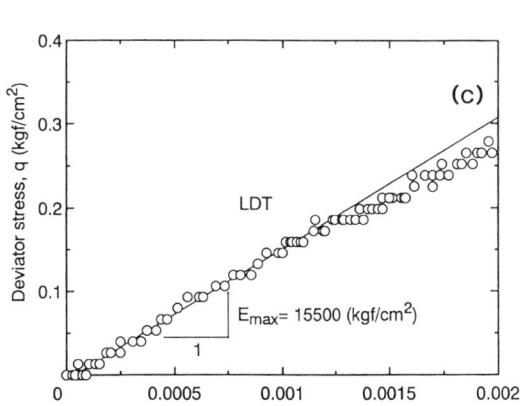

Fig. 6 Typical CD triaxial compression test for silt-to-sandstone (Kohata et al., 1994)

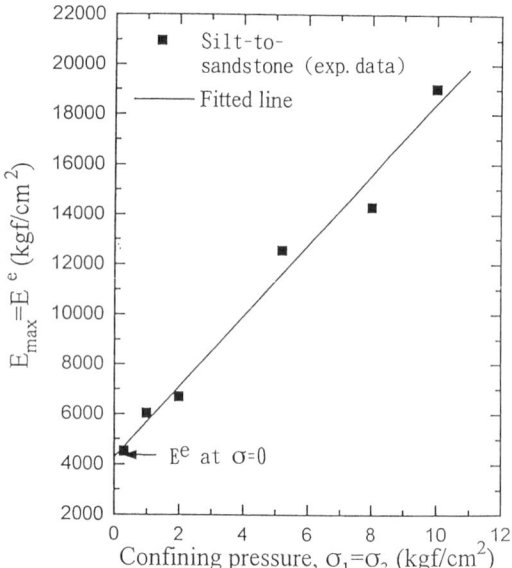

Fig. 7 Relationship between E_{max} (averaged) and σ_c from CD TC tests (Siddiquee et al., 1994, Kohata et al., 1994)

obtained from Anchor 1A site as shown in **Fig. 7**. The E_{max} value for each data point in this figure is an average of several data. Based on the pressure dependency of the average relationship between E_{max} and confining pressure shown in Fig. 7, with the field value E_f measured when $\sigma_v = \sigma_{vo}$, we defined:

$$j(\sigma_v) = 1 + 1,414 \cdot (\sigma_v - \sigma_{vo})/E_f \quad (6)$$

σ_v and σ_{vo} are the current and initial effective vertical pressures (kgf/cm^2), respectively.

In the triaxial compression tests performed in this study, many small unload-reload cycles were applied during otherwise monotonic loading to examine the change in elastic stiffness during triaxial compression. It may be seen from Fig. 6 that the peak-to-peak secant Young's modulus E_{eq} for a small unload/reload cycle, which represents the current elastic Young's modulus, increases with the increase in deviator stress q $(= \sigma_1 - \sigma_3)$. This increase is due to the increase in the pressure level (i.e., axial stress= $\sigma_1 = \sigma_v$) as modeled by Eq. 6. The effect of this pressure dependency is essential for realistic analyses of plate loading tests and full-scale behaviour of a proto-type foundation on this type of soft rock (Siddiquee et al., 1994).

2.2 α AND β

These are parameters, which are less than or equal to unity, represent respectively, the possible effects of the discontinuities (i.e., joints, cracks and faults) in a field rock mass and the effects of disturbance by excavation work. In this study, both values were assumed to be equal to unity. The assumption $\alpha = 1.0$ is one of the unique features of this analysis, and this assumption has been found valid for several construction cases for sedimentary soft mudstone (Tatsuoka and Kohata, 1995).

2.3 f(SL)

This is the damage function which takes into account the effect on the elastic deformation properties of the damage to the structure of softrock due to shear deformation. It was assumed that the damage can be measured as the decrease in the elastic shear modulus during triaxial compression (TC) corrected for the effects of pressure level. Ofoegbu and

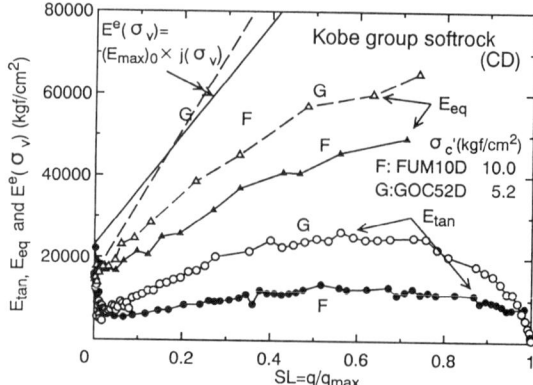

Fig. 8 Comparison among undamaged and measured elastic Young's moduli, E^e and E_{eq}, and tangent Young's modulus E_{tan} during two CD TC tests on Kobe sedimentary soft silt-to-sandstone (Kohata et al., 1994).

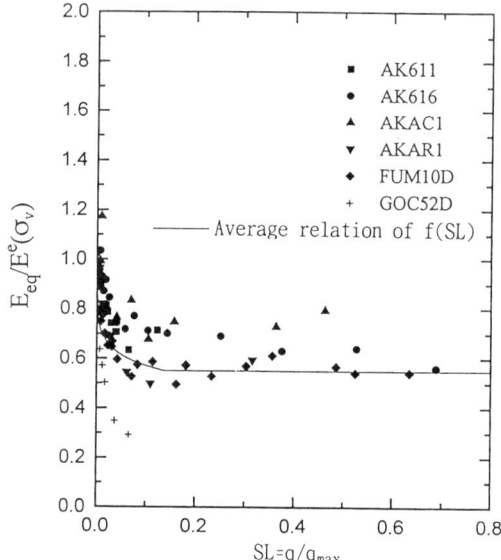

Fig. 9 Damage function $f(SL) = E_{eq}/E^e$ from TC tests of silt-to-sandstone (Siddiquee et al., 1994)

Curran (1991) adopted a similar method for intact hard rock core for which a unique pressure-independent undamaged elastic Young's modulus can be defined. The damage function for this softrock is given later. The shear stress level SL was defined as:

$$SL = q/q_{max} = \tau/\tau_{max} \quad (7)$$

where $\tau = q/2$, and q_{max} and τ_{max} are the compressive strength and the shear strength. The failure criteria obtained from a series of CD TC tests is;

$$\tau_{max} \text{ (kgf/cm}^2) = q_{max}/2 = 39.03 + 1.33 \cdot \sigma_3 \quad (8)$$

The initial stress state in the field before excavation was assumed to be isotropic. In this case, f(SL) decreases from f(0.0)=1.0 to f(1.0)=0.552 with the increase in SL. Very likely, the damage may not be isotropic; during triaxial compression, the elastic Young's modulus in the lateral direction decreases more rapidly than the elastic Young's modulus in the axial direction (Sato, 1994). However, for simplicity, it was assumed that f(SL) is isotropic with respect to the direction of σ_1 in the field.

The E_{eq} value in the vertical direction measured at a certain σ_v becomes smaller than the undamaged elastic Young's modulus at this σ_v value, which is obtained from Eq. 5 as:

$$(E^e)_v = (E_{max})_0 \cdot j(\sigma_v) \quad (9a)$$
$$j(\sigma_v) = 1 + 1,414 \cdot (\sigma_v - \sigma_{v0})/(E_{max})_0 \quad (9b)$$

where $(E_{max})_0$ is the maximum Young's modulus at the start of TC, where $\sigma_v = \sigma_{v0}$. This feature is demonstrated in **Fig. 8**, where the results of two CD TC tests on core specimens obtained from 1A site at a depth of 61 m that were isotropically consolidated to $\sigma_{v0} = 5.2$ kgf/cm² and 10.0 kgf/cm² are presented. It may be seen that with the increase in SL (i.e., with the increase in σ_v), the value of E_{eq} increases, but the undamaged Young's modulus E^e estimated based on Eq. 6 increases more rapidly. The ratio E_{eq}/E^e is defined as the damage function f(SL). The result of a series of TC tests and its average relation are shown in **Fig. 9**, which is:

$$f(SL) = (1.0 - 0.15 \cdot SL^{0.15})/(1.0 + SL^{0.25}) \quad (10)$$
$$(0.552 \leq f(SL) \leq 1.0)$$

2.4 g(SL)

This is the plasticity function, equal to the ratio of the elastic to total shear strain increments as a function of SL:

$$g(SL) = G_{tan}/G_{eq} = d\gamma^e/d\gamma \quad (11)$$

As it was assumed that the elastic and plastic Poisson's ratios at a given condition are the same with each other, we obtain:

$$g(SL) = G_{tan}/G_{eq} = E_{tan}/E_{eq} \quad (12)$$

In summary, we obtain:

$$G_{tan} = G^e \cdot f(SL) \cdot g(SL), \quad G^e = E^e/\{2(1+\nu^e)\} \quad (13)$$

The empirical equation of the combined function $h(SL) = f(SL) \cdot g(SL)$, called the non-linear function, was obtained based on the TC test results shown in **Fig. 10**, which is:

$$h(SL) = G_{tan}/G^e$$
$$= \{1 - SL + C \cdot (SL^2 - SL) + D \cdot (SL^3 - SL)\}/(1 + B \cdot SL)$$
$$(B = 9,674, C = 778 \text{ and } D = -2,740) \quad (14)$$

This model fits well the measured stress-strain relationships from TC tests (**Fig. 11**) (Siddiquee et al., 1994).

2.5 CROSS ANISOTROPY

The modeling based on Eq. 5 implicitly implies anisotropic elastic deformation properties under anisotropic stress condition. In this study, the simplest anisotropic model was employed, which is cross-anisotropy expressed by the compliance matrix shown in **Fig. 12**. Here, E and ν mean the elastic Young's modulus E_{eq} and Poisson's ratio ν_{eq}; e.g., E_v and ν_{vh} stand for $(E_{eq})_v$ and $(\nu_{eq})_{vh}$.

Although this matrix has six stiffness coefficients, they are fully obtained for a given stress state as follows. First, the values of E_v and E_h are obtained from Eq. 5. Second, a compliance matrix for elastic deformation should be symmetry from the requirements of energy conservation, which leads to:

$$\nu_{vh}/E_v = \nu_{hv}/E_h \quad (15)$$

Note that the compliance matrix for total strain increments used in this study is also symmetry, since an isotropic functions h(SL) is assumed. The simplest functions for the Poisson's ratios which satisfy Eq. 15 are:

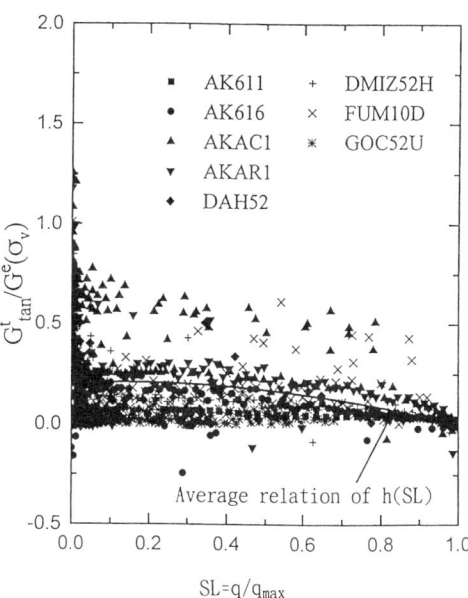

Fig. 10 Combined function $h(SL) = E_{tan}/E^e$ from TC tests of silt-to-sandstone (Siddiquee et al., 1994)

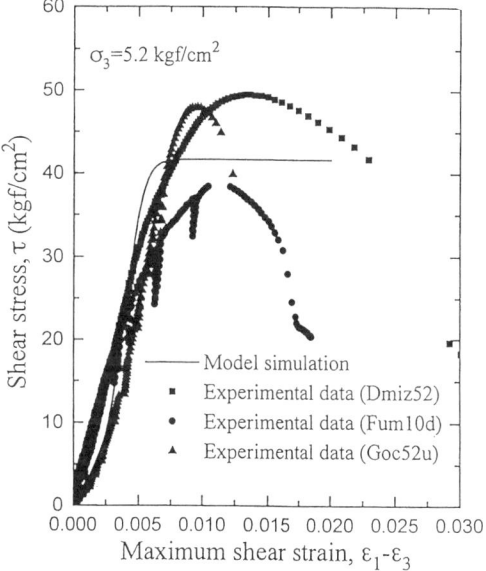

Fig. 11 Comparison of measured and modeled stress-strain relations from CD TC tests at σ_{v0}.

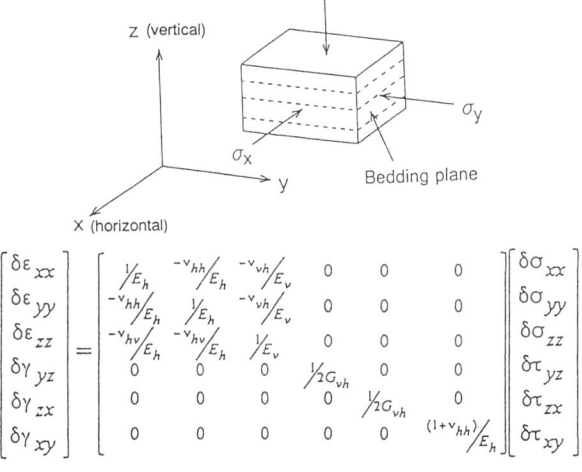

Fig. 12 Cross anisotropic compliance matrix

$$\nu_{vh} = \nu_0 / (E_h/E_v)^{0.5} \quad (16a)$$
$$\nu_{hv} = \nu_0 / (E_v/E_h)^{0.5} \quad (16b)$$

where ν_0 is the drained elastic Poisson's ratio under isotropic stress conditions, equal to 0.2. Then, ν_{hh} is naturally assumed to be equal to ν_0. Finally, G_{hv} is obtained as

$$G_{hv} = E_{45} / \{2(1 + \nu_d)\} \quad (17a)$$
$$E_{45} = (E_{tan})_v \cdot \{j(\sigma_{45})/j(\sigma_v)\} \quad (17b)$$

E_{45} is the elastic Young's modulus for major principal strain increments occurring in the direction at an angle of 45 degrees relative to the vertical. In this direction, the normal stress σ_{45} is equal to $(\sigma_h + \sigma_v)/2$.

The modeling method for the sedimentary soft silt-to-sandstone was applied also to the underlying granite layer. The errors due to this assumption may be small.

3. FEM

The characteristic features of this FEM analysis by means of the code developed by Tanaka and Kawamoto (1988), which is based on the dynamic relaxation technique, are described in detail in Siddiquee (1994). The summary is as follows:

1) An elasto-plastic isotropic hardening model is employed. This simplification may be valid for this case, where the footing settlement problem during monotonic loading is dealt with.

2) The plastic part is modeled by a non-associated flow rule based on a generalised Mohr-Coulomb yield and failure function Φ using a Drucker-Prager formula:

$$\Phi = \alpha \cdot I_1 + \{1/g(\theta)\} \cdot \sqrt{J_2} - K_{mob} = 0 \quad (18)$$

where:

$$\alpha = (2 \cdot \sin\phi_{mob}) / \{\sqrt{3}(3 - \sin\phi_{mob})\} \quad (19)$$

$$g(\theta) = (3 - \sin\phi_{mob}) / (2\sqrt{3} \cdot \cos\theta - 2 \cdot \cos\theta \cdot \sin\phi_{mob}) \quad (19)$$

in which θ is the Lode angle, I_1 and J_2 are the first and second stress invariants, and ϕ_{mob} is the mobilized internal angle of friction given by;

$$\phi_{mob} = \arcsin\{(\sigma_1^* - \sigma_3^*)/(\sigma_1^* + \sigma_3^*)\} \quad (20)$$

in which σ_1^* and σ_3^* are the equivalent major and minor principal stresses, given as:

$$\sigma^* = c \cdot \cot\phi + \sigma \quad (21)$$

ϕ is the angle of internal friction (i.e., the peak strength = 53.1 degrees). K_{mob} is the mobilized cohesion parameter given as;

$$K_{mob} = 6 \cdot c \cdot \cot\phi \cdot \sin\phi_{mob} / \{\sqrt{3}(3 - \sin\phi_{mob})\} \quad (22)$$

c is the cohesion intersect ($= 13.02$ kgf/cm^2). The following tension cut off was introduced;

$$\sigma^* = c/5 \quad (23)$$

One of the features of this model is that strain-hardening is controlled by shear strain increments, not by volumetric strain increments as Cam Clay model. The specific form of shear stress - shear strain relation is obtained by integrating Eq. 13.

The FEM analysis has been validated by simulating successfully the results of plate loading tests performed at the bottom of excavation (-61 m) at Anchor 1A site (Siddiquee et al., 1994).

Pier 3P is the steel caisson with a diameter of 78 m and a height of 67 m (Fig. 3), the inside of which was filled with concrete under water for about 11 months and in air for about 14 months. The level of the concrete crest during construction was assumed to be always level in the analysis, although it was not the case. The Young's modulus of 3.0×10^5 kgf/cm^2 for concrete was used. An axi-symmetric 3-D analysis was performed.

4. COMPARISON OF RECORDED AND SIMULATED FOOTING SETTLEMENT

Fig. 13 shows the time histories of the average pressure $(p)_{ave}$ at the pier base, the settlement s of Pier 3P and the associated in-ground settlements along the center line of the pier. **Fig. 14** shows the meshing for the FEM analysis of the settlement of Pier 3P. Around the interface between the sedimentary soft rock and granite layers, between which stiffness values are largely different, finer meshing was introduced so as to achieve smooth stress wave propagation. **Fig. 15** compares the recorded and simulated relationships. In the analysis, the undrained and drained conditions were simulated by using Poisson's ratios equal to 0.46 and 0.2, respectively. It may be seen that the simulated relation for the drained conditions is remarkably similar to the recorded one, which may suggest that the drained condition was prevailing in the ground.

As seen from Fig. 13 that after the end of casting concrete, the settlement s increased at a constant $(p)_{ave}$ as seen from Figs. 13 and 15. A similar, but less clear trend can be seen when $(p)_{ave}$ is about 5.0 kgf/cm^2. In the conventional design procedure, long-term time-dependent behaviour of footing settlement is predicted totally based on the creep theory using the result of drained triaxial creep tests. However, the comparison between the recorded and simulated behaviour suggests that this recorded creep-like behaviour is not totally due to the drained creep, but largely due to "consolidation" of soft rock. In fact, it is reported that the ratio of the "creep" compression to the instantaneous one for a given ground layer which occurred by a given increment $(p)_{ave}$ increases with depth (Nasu et al., 1993). This behaviour may be better explained by taking into account the consolidation of softrock.

It should be noted that a very good agreement between the $(p)_{ave}$ and s relations (Fig. 15) is somehow fortuitous, since the agreement between the

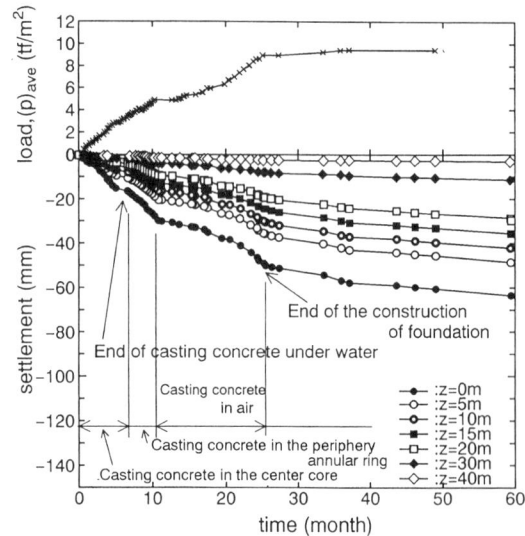

Fig. 13 Time histories of average pressure, settlement for Pier 3P and associated in-ground settlements along the center line of the pier

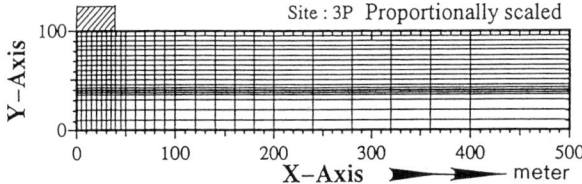

Fig. 14 Meshing for FEM analysis of Pier 3P

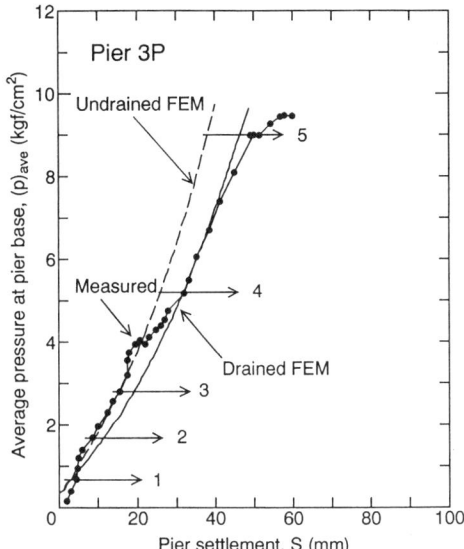

Fig. 15 Measured and simulated relationships between the average pressure at the foundation base and the settlement for Pier 3P.

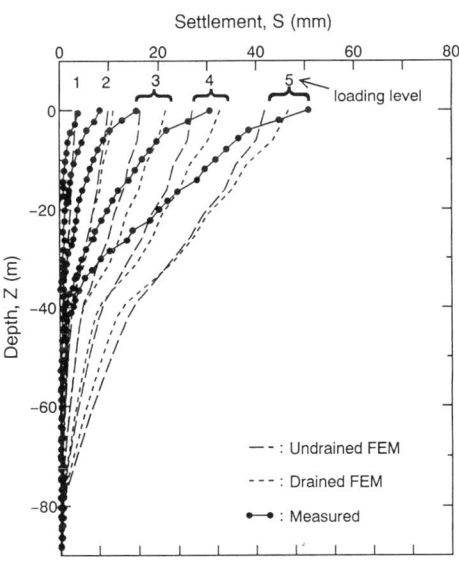

Fig. 16 Measured and simulated relationships between the in-ground settlement and the depth along the centerline of Pier 3P.

measured and simulated relationships between the in-ground settlements and the depth along the centerline (**Fig. 16**) is not as good as the relationship between $(p)_{ave}$ and s. The comparison shows that the stiffness has been over-estimated and under-estimated at shallower and deeper places, respectively. In the analysis, the non-linear function f(SL) was assumed to be independent of the pressure level, but it seems that f(SL) is pressure-dependent in such a way that f(SL) for a given SL is larger and smaller at deeper and shallower places than that given by Eq. 14. Further study will be needed to clarify this point.

The trend of the increase in the rate $d(p)_{ave}/ds$ with the increase in s seen in the simulated behaviour, which is also seen in the recorded behaviour, is due to that the effect of the pressure level dependency of stiffness prevail over that of the shear stress level dependency of stiffness. This behaviour results from that in the ground subjected to increasing footing load, the rate of the increase in pressure (i.e., σ_v) is much larger than the rate of the increase in stress ratio (i.e., σ_v/σ_h or SL). The importance to take into account the dependency of stiffness on shear stress level and pressure level when analysing different patterns of recorded footing settlement-footing pressure relations together with associated in-ground settlements for foundations constructed on different geomaterials is discussed by Tatsuoka et al. (1995, this symposium).

5. CONCLUSIONS

The sedimentary soft silt-to-sandstone, which is supporting directly Pier 3P for Akashi Kaikyo Bridge, was modeled based on the elastic Young's modulus obtained from the field seismic survey, while taking into account the dependency of stiffness on shear stress level and pressure level evaluated by triaxial compression tests performed on core samples.

The FEM analysis successfully simulated the footing load-settlement relation, but not very well the vertical distribution of settlement in the ground.

It is considered that the simulation method presented in this paper is a useful and consistent one which can be applied to general cases without assuming geomaterials having linear and pressure-independent stress-strain relations.

Based on the comparison of the simulated behaviour under fully drained and undrained conditions with the measured relationship, it is suggested that large part of the delayed settlement of Pier 3P may be due to the consolidation of soft rock at deeper places.

References

Kohata, Y., Tatsuoka, F., Dong, J., Teachavorasinskun, S. & Mizumoto, K. 1994a. Stress-state affecting elastic deformation moduli of geomaterials. Proc. Int. Sympo. on Prefailure Deformation Characteristics of Geomaterials. Sapporo. September. 1994 (Shibuya et al., eds.), IS-Hokkaido. '94, Balkema. Vol.I: 3-9.

Nasu, S., Yoshida, O., & Yamagishi, K. 1993. Layer Deformation of Akashi-Kaikyo-Bridge tower foundations; observation and back-analysis. Proc. Sympo. on Utilization of Deep Underground in Urban Area. JSSMFE. Tokyo. (in Japanese).

Ofoegbu, G., I. & Curran, J., H. 1991. Yielding and damage of intact rock. Canadian Geotechnical Journal. Vol. 28: 503-516.

Sato, Y. 1994. Fundamental study on dilatancy of rocks, Dr. Engnrg Thesis, Hokkaido Univ. (in Japanese).

Siddiquee, M. S. A., Tatsuoka, F., Hoque, E., Tsubouchi, T., Yoshida, O., Yamamoto, S. & Tanaka, T. 1994. FEM simulation of footing settlement for stiff materials. Proc. IS-Hokkaido '94, Balkema. Vol. I: 531-537.

Tanaka, T. & Kawamoto, O. 1988. Three-dimensional finite element collapse analysis for foundations and slopes using dynamic relaxation. Proc. of Num. methods in geomechanics. Insbruch: 1213-1218.

Tatsuoka, F., Kohata, Y., Mizumoto, K., Kim, Y.-S., Ochi, K. & Shin, D. 1993. Measured small strain stiffness of soft rocks. Geotechnical Engineering of Hard Soils - Soft rocks. (eds. Anagnostopoulos et al.). Balkema. Vol. 1: 809-816.

Tatsuoka, F., Yamada, K., Yasuda, M., Yamada, S. & Manabe, S. 1991. Cyclic undrained behaviour of an undisturbed gravel for a seismic design of a bridge foundation. Proc. 2nd Int. Conf. on Recent Advances in Geotech. Earthquake Engnrg and Soil Dynamics. St Louis. Vol. I : 141-148.

Tatsuoka, F., Sato, F., Park, C. S., Kim, Y.-S., Mukabi, J., N., & Kohata, Y. 1994. Measurements of elastic properties of geomaterials in laboratory compression tests. Geotechnical Testing Journal. ASTM. Vol. 17.No.1: 80-94.

Tatsuoka, F. & Kohata, Y. 1995. Stiffness of hard soils and soft rocks in engineering applications. Keynote Lecture. Proc. IS-Hokkaido '94. Balkema. Vol.II.

Tatsuoka, F., Lo Presti, D. & Kohata, Y. 1995a, Deformation characteristic of soils and soft rocks under monotonic and cyclic loads and their relationships. Keynote Lecture. Proc. Int. Conf. on recent Advances in Geotech. Earthquake Engnrg and Soil Dynamics. St Louis. (to appear)

Tatsuoka, F., Kohata, Y., Tsubouchi, T. & Ochi, K. 1995b. Stiffness of sedimentary softrocks evaluated by triaxial compression tests. Proc. 8th Int. Congress on Rock Mechanics. Tokyo. (to appear)

Tatsuoka, F., Kohata, Y., Tsubouchi, T., Ochi, K., Wang, L. & Inoue, A. 1995c. Stiffness of soft rocks tests in Tokyo Metropolitan Area - from laboratory to full scale behaviour -. Keynote Lecture Proc. Int. Workshop On Rock Foundation. Sept. Tokyo. Balkema. (this workshop)

Tatsuoka, F., Yoshida, O., Manabe, S., & Yamada, S. 1995d. Evaluation of displacement during earthquake of a bridge foundation on ground. Proc. First Int. Conf. on Earthquake Geotech. Engnrg. Tokyo. Balkema. (to appear)

Estimation of ground properties and behavior under construction of huge suspension bridge

K. Izumi & M. Ogiwara
Metropolitan Expressway Public Corporation, Japan

K. Nishida & H. Kameya
Oyo Corporation, Saitama, Japan

ABSTRACT : During the construction of the Rainbow bridge, displacement of two anchorages and their foudation ground was measured. To explain the observed bahavior, deformation characteristics of the foundation ground at small strain levels and time-dependent properties were studied experimentally. The deformation moduli mesured by triaxial compression tests with LDT and those back-calculated from the observed behavior were comparable when the strain levels were considerd. The displacements calculated by the 3D-FEM analysis, with stress-strain relations from triaxial compression tests and time-dependent properties from triaxial creep tests, were almost identical to observed values.

1. INTRODUCTION

There are increasing cases where observation of foundation structures resting on soft rock, as well as the foundation ground itself, is carried out as part of construction management (Yamagata 1993). The observational results are important not only as data for construction management but also as basic data for the rationalization of investigation and design methods.

Rainbow Bridge (Figure 1) is a trans-Tokyo Port suspension bridge with a total length of 798m. The geological profile of its construction site is presented in Figure 2. At 30m below the ground level lies a stratum mainly composed of a soft alluvial clay on the thick deposit of the stiff silt of Kazusa Group in Pleistocene and the Tertiary period. The bridge's foundation structures, such as anchorages and main towers, were constructed by means of the caisson method using this stiff silt as a supporting layer. The basic physical properties of the stiff silt are presented in Table 1. Its unconfined compressive strength is 2.5-4MPa. As heavy structures, such as the anchorages and the main towers, had to be constructed on soft rock, a thorough review was performed on the bearing capacity and displacement of the stiff silt (Tomizawa 1987). The contact pressure and lateral force on the foundation ground vary as the work progresses, namely, anchorage construction, cable laying, stiffening truss erection and completion of the bridge construction. In this light, during the construction of the pair of anchorages, long-term observation was made on the displacements in the foundation ground, pore pressure distribution and settlement of the foundation structures.

This paper reports on the results of this long-term observation as well as displacement prediction attempted in both the design and execution phases. To explain the observed behavior, strain level- and time-dependent deformation characteristics, which are inherent in soft rock, were studied. This paper also discusses numerical simulation attempted based on the study results.

Table 1 Properties of the stiff silt.

Texture	Sand/gravel content 6.1 %	Silt content 44.7 %	Clay content 49.2 %
Consistency	WL=60.1 %,	WP=26.9 %,	IP=33.2 %
Moisture content		Wn=30.7 %	
Wet density		$\rho t= 1.92$ g/cm^3	
Void ratio		e=0.86	
Consolidation yield stress		Pc=from 5 to 7 MPa.	
Unconfined compressive strength		qu=from 2.5 to 4 MPa.	

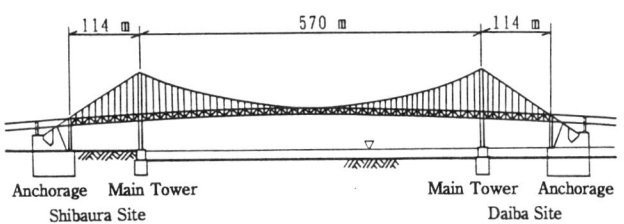

Figure 1 Side view of the Rainbow Bridge.

Figure 2 Geological profile of the Rainbow Bridge.

2. OUTLINE OF THE LONG-TERM OBSERVATION AND ITS RESULTS

This Section introduces the anchorage constructed on the Daiba side.

2-1 Outline of the long-term observation

The long-term observation was made on the following four items:
- Pore pressure of the foundation ground (by means of the MP system),
- Horizontal displacements (by means of the inclinometer),
- Vertical displacements (by means of the slidingmicrometer) and
- Settlements of the anchorages

Observation of the foundation ground carried out by inserting a probe into casings installed inside boreholes. This approach is more advantageous compared to those using buried systems for the following reasons.
(1) Continuous and multiple-point measurement is possible.
(2) This system is suitable for a long-term observation because the electric circuit is free of corrosion.

As is shown in Figure 3, the observation holes (boreholes) were excavated at 3.5m from the caisson's front and rear sides. The anchorages are composed of two sections ; a box-shaped lower section, caisson, of 70m wide and 45m deep, and an upper section which supports cables. The deep well method was adopted for the construction of the caissons along with the pneumatic caisson method to reduce air pressure. The stiff silt was excavated to the depth of about 8m to construct the foundation. The anchorage weighed approximately 280,000t at the time when concrete was filled in the caisson section, which increased by nearly 140,000t when the upper anchorage section was completed. Further, tensile force was induced by cable laying and stiffening truss erection. Figure 3 illustrates the approximate weight carried by the anchorage when it was completed.

2-2 Result of the long-term observation

Figure 4 shows variations with time in the vertical displacement and pore pressure of the foundation ground. The load at the foundation bottom level load varies not only in magnitude but in distribution as well with the progress of the work. The subsequent paragraphs discuss variations in displacements with the progress of the work.

(1) During the process of Caisson immersion

Vertical displacements of the foundation ground and pore water distribution during the process of caisson immersion are shown in Figure 5. At point ①, the tip of the caisson is located at T.P.-30m (T.P. is the depth with respect to Tokyo peil). As can be seen from the Figure, the over burden load on the foundation ground decreases as a result of excavations during the immersion process, causing a rebound to occur. Point ② indicates the state of the caisson immediately after resting on the foundation bottom. The use of the deep well method gave rise to decreases in the pore pressure and resultant compression of the upper portion of the foundation ground (TP-40-60m). During the process of filling concrete into the caisson, the foundation ground exhibited a complicated deformation behavior due to increases in the caisson weight combined with the variations in the pore pressure.

(2) During the process of the anchorage construction

Vertical and lateral displacements in the foundation ground during the process of the anchorage section construction are illustrated in Figure 6. The figure shows the increments in the displacements measured based on the initial displacements observed at the commencement of the anchorage construction.

Both the vertical and horizontal displacements increase with increases in the anchorage weight. The displacement measurement took place away from the anchorage, so the maximum value of vertical displacement is found at levels slightly deeper than the foundation bottom. Meanwhile in the case of the lateral displacement, the maximum value is found in the proximity of the caisson. Figure 7 illustrates the amount of

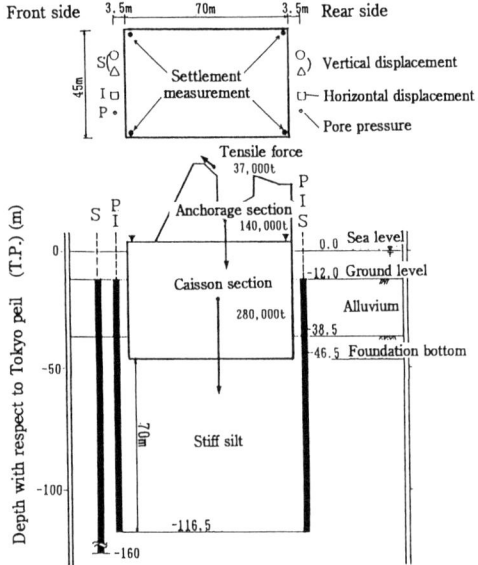

Figure 3 Anchorage and location of measurement systems.

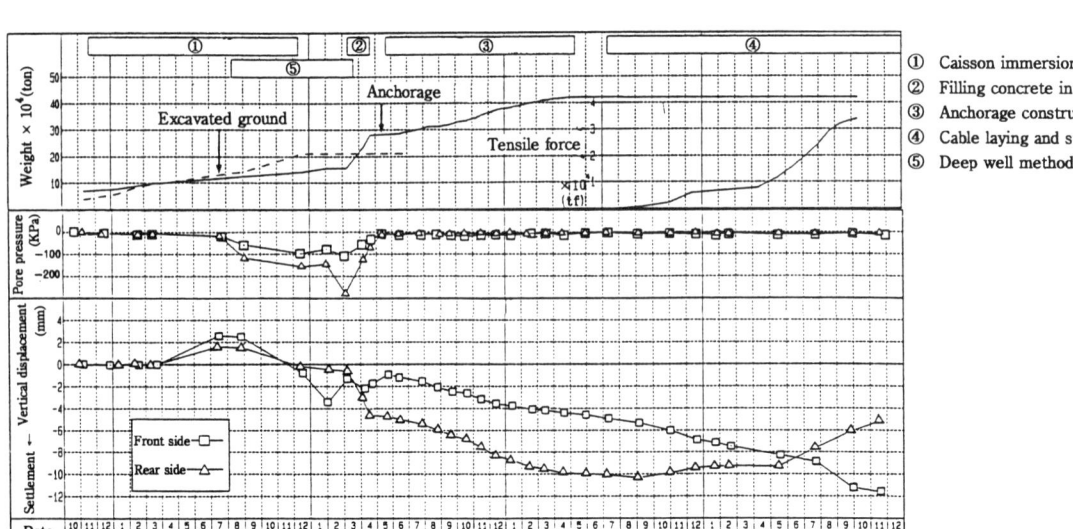

Figure 4 Vertical displacement and pore pressure at the level of the foundation bottom.

the settlement of the pair anchorages in relation to load. Reflecting the unevenly distributed weight, the magnitude of settlement becomes larger in the rear side of the anchorage. Both anchorages are almost identical in the load-settlement relationship.

(3) During the process of cable laying and stiffening truss erection

During the process of cable laying and stiffening truss erection, increases in the tensile force of the main cable cause rotational moment to occur. In Figure 8, the vertical displacements of the foundation ground during this process are presented. As can be seen from the Figure, increases in the tensile force of the cable cause settlements and upheavals in the front side and rear side of the anchorages, respectively.

The pore pressure remained stable throughout the process, except for the period during the pumping, which is an essential part of the deep well method, was carried out. The permeability coefficient of the stiff silt itself is $1-2 \times 10^{-8}$ cm/sec. However, a sand layer ranging in thickness from several centimeters to 2mm is distributed across the ground, and the continuity of this sand layer is, on the whole, confirmed favorable. Consequently, the sand layer functions as a drainage layer, and thereby excess pore water pressure produced during the work dispersed as the work progressed.

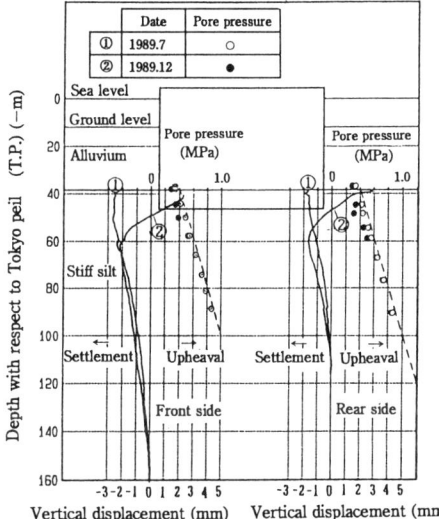

Figure 5 Distribution of vertical displacement and pore pressure during the process of caisson immersion.

3. EXAMINATION OF DISPLACEMENTS DURING DESIGN AND CONSTRUCTION PHASES

Major movements of the anchorage points in the anchorages cause the following problems:
(1) Negative effects on the navigational limit
(2) The irregular vertical alignment
(3) Decreases in the bridge rigidity due to deeper sags
(4) Positive bending in the stiffening truss
(5) Negative aesthetic effects due to angled sags

During the design phase, the amount of the displacement in the foundation structures was estimated, and the amount of the support point movement was determined, taking into consideration the upper structure design. These attempts, however, were based on roughly established construction conditions, so, in the construction phase, the amount of displacement of the foundation structures was reestimated taking into account the observational results. Based on the reestimation, the amount of displacement of the upper structures was calculated, and stresses on principal members were examined. The results proved that the completed anchorages had none of the anticipated problems.

According to the results of the reestimation, adjustments were made during the erection of the main towers, side towers, cable anchor frames and spray saddles so that the completed bridge would reflect its design requirements (more specifically, (a) the normal position and height of the spray points; (b) the normal position and height of the main towers and side towers).

In view of the fact that both vertical and horizontal displacements should be assessed displacements were examined using the FEM analysis. In addition, because it is necessary to assess long-term displacements, the amount of creep displacement in the foundation ground becomes unneglectable. From laboratory test results it was estimated that creep displacement equivalent to half the amount of displacement derived from the FEM analysis would occur within 100years from completion.

It is considered that the reestimation of the displacements not only during the design phase but during the construction phase enabled the rational construction of the bridge. However it is also true that the observed displacements were as small as one thirds of the estimated values. To enhance estimation accuracy, it is essential to establish a method to assess the deformation characteristics of the foundation ground that can explain the deformation behavior grasped through the long-term observation. The following examinations were conducted based on this idea.

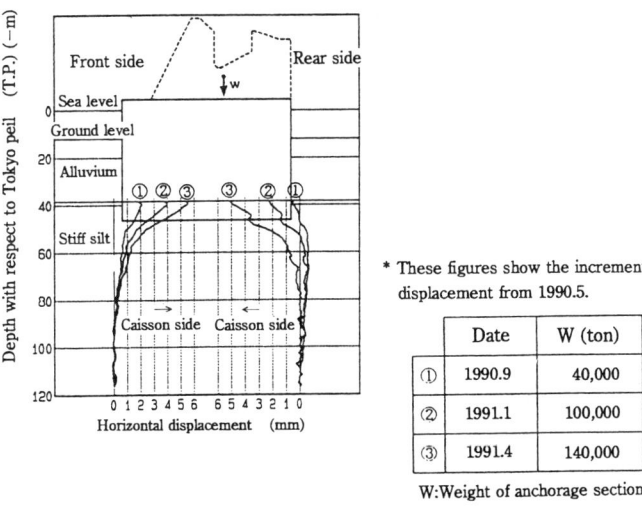
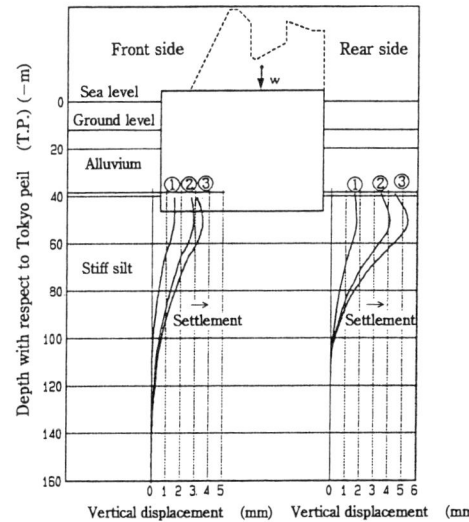

Figure 6 Distribution of vertical and horizontal displacement during the construction of anchorage.

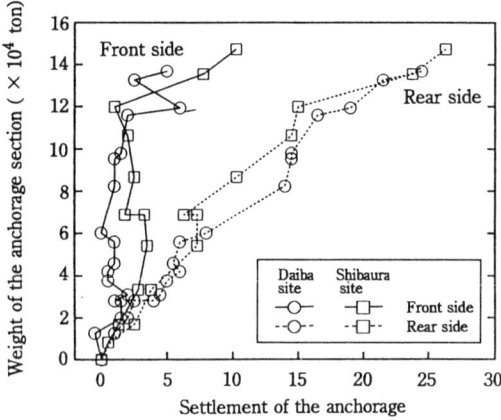

Figure 7 Settlement of the anchorage during the process of the anchorage construction.

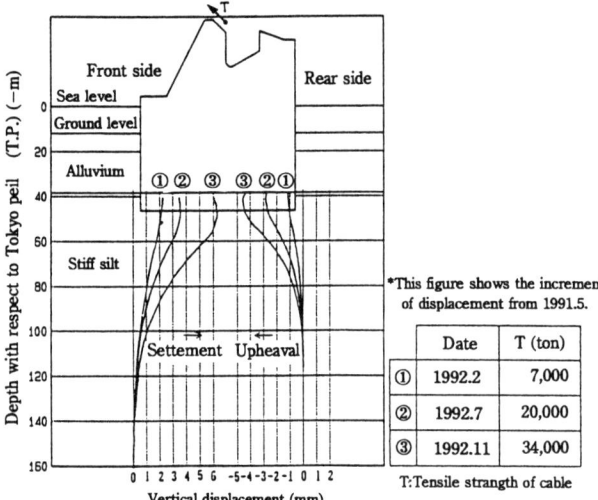

Figure 8 Distribution of vertical displacement during the process of cable laying and stiffening truss erection.

4. DEFORMATION CHARACTERISTICS OF THE FOUNDATION GROUND

The results of the vertical displacement measurement showed that the magnitude of the strain induced in the foundation ground by the anchorage construction was 10^{-5}–10^{-4}. In order to correctly assess the deformation characteristics in the small strain level, triaxial compression tests were conducted using LDT (Local Deformation Transducer (Goto 1991)), for the measurement of axial strain. The tests were conducted under consolidated-drained condition (CD condition), because throughout the construction period, the pore pressure in the foundation ground maintained a uniform level as stable as the hydrostatic pressure state. Furthermore, to consider creep deformation, triaxial creep tests were conducted under consolidated-drained condition (CD condition).

4-1 Results of the triaxial compression tests

Figure 9 illustrates the results of the triaxial compression tests as the secant modulus of deformation – axial strain relationship. For comparison purposes, the moduli of deformation obtained from suspension PS logging, pressuremeter tests and triaxial compression tests under consolidated-undrained condition (CU condition), as well as the modulus of deformation back-calculated from the long-term observational data were also shown in the figure.

The modulus of deformation derived from the triaxial compression tests (CD condition) coincides for the most part with that back-calculated from the long-term observational data in the case of strains at levels between 10^{-5} and 10^{-4}.
Meanwhile in the case of strains at a level of around 10^{-2}, the modulus of deformation coincides with that obtained from the pressuremeter tests. Thus, the modulus of deformation derived from the triaxial compression tests (CD condition) coincides for the most part with that obtained from in-situ tests and measurements when the strain levels are considered. The moduli of deformation derived from the triaxial compression tests under CU condition and suspension PS logging are larger than that obtained from the triaxial compression tests under CD condition. The triaxial compression tests (CD condition) were conducted under varying confining pressure. The test results showed that the effect of the confining pressures on the modulus of deformation was neglectable.

In Figure 10, the moduli of deformation derived from suspension PS logging and the moduli of initial deformation obtained from the triaxial compression tests (CD condition) were compared. The modulus of initial deformation derived

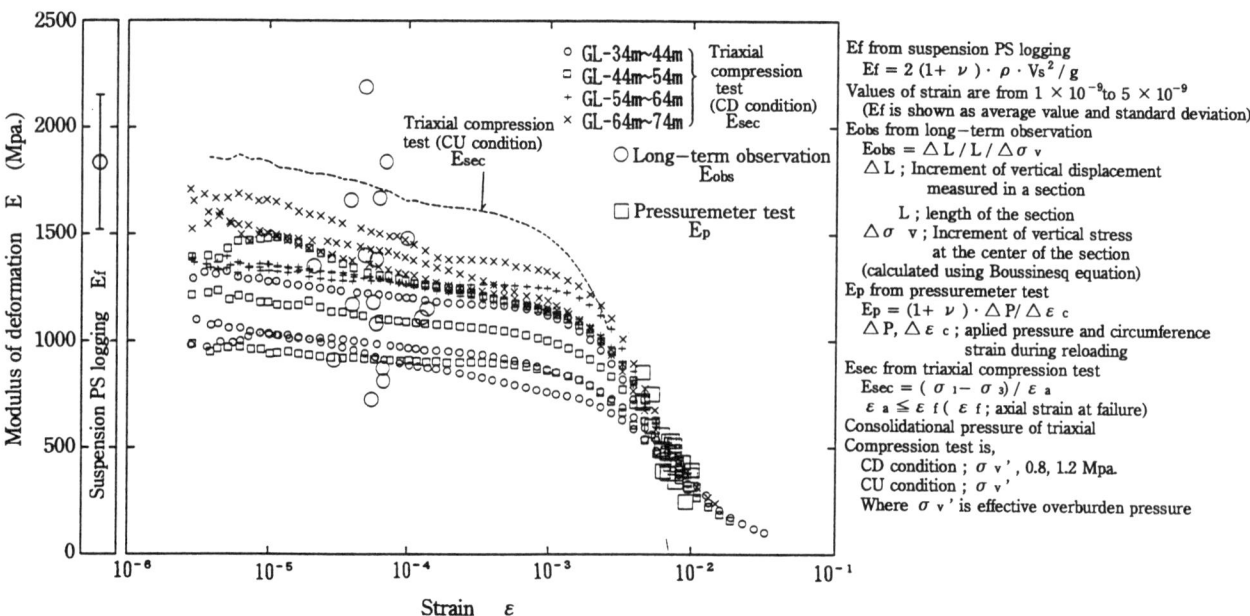

Figure 9 Relation between the modulus of deformation and strain.

Table 2 Results of the triaxial creep tests.

Depth G.L. −m	Elastic strain ε_e %	Creep strain ε_c %	Total strain $\varepsilon_t = \varepsilon_e + \varepsilon_c$	$\dfrac{\varepsilon_e}{\varepsilon_t}$
37.34	0.213	0.039	0.252	0.845
48.63	0.192	0.043	0.235	0.817
59.68	0.179	0.031	0.210	0.852
62.29	0.189	0.042	0.231	0.818
			Average	0.833

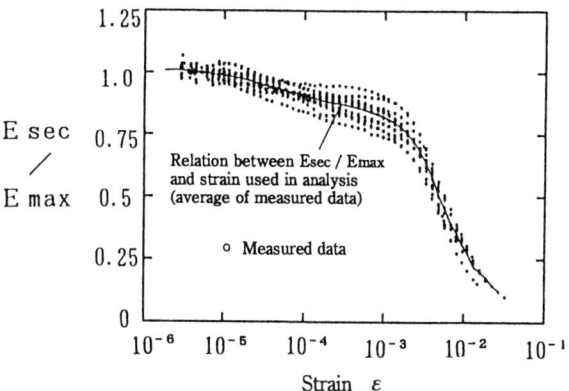

Figure 12 Normalized modulus according to strain level.

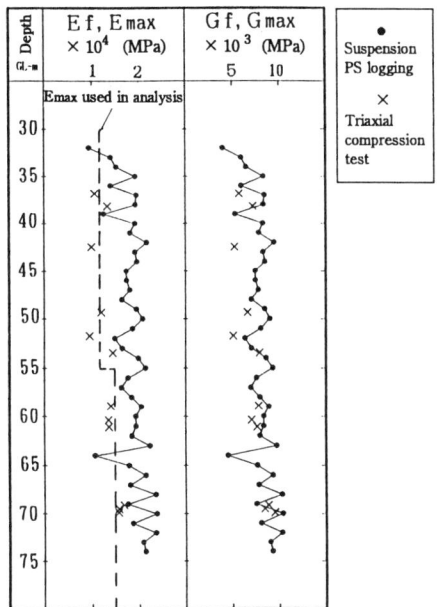

Figure 10 Distribution of the modulus of deformation.

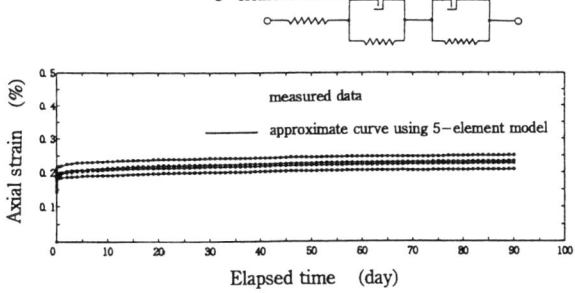

Figure 11 Relation between elapsed time and axial strain from triaxial creep test.

from the triaxial compression tests, E max, was determined as the secant modulus, where axial strain is at a level of 5×10^{-6} G max was calculated by means of the following equation.
   Gmax=Emax / (s • (1+ ν '))   −1)
Poisson's ratio, ν ',was found using the axial strain and volumetric strain obtained from using triaxal compression tests (CD condition).
Where, Ef and Gf were obtained from suspension PS logging. As can be seen from Figure 10, shear rigidity, G max, coincides with Gf. From this it can be said that the discrepancy between the moduli of deformation, E max and Ef, is due to drainage conditions.

4-2 Results of the triaxial creep tests

Samples were taken from four different depths in the same manner as the case of the triaxial compression tests. The samples were consolidated with an effective overburden pressure and then subjected to loading of 90days' duration at a fixed axial pressure level of 1.2MPa. The results were simulated using the 5-element visco-elasticity model, as shown in Figure 11 and Table 2. Although the ratio of elastic strain to creep strain is not uniform, it is still within the range between 0.17 and 0.22. The mean value of total strain (the sum of the elastic and creep strains) - elastic strain ratio was found to be α = 0.833.

5. DEFORMATION CHARACTERISTICS OF THE FOUNDATION GROUND GRASPED THROUGH NUMERICAL SIMULATION

To relate the deformation behavior obtained through the long-term observation to the deformation characteristics derived from the laboratory tests more practically, numerical simulation of the deformation behavior during the construction phase (2-2 (2)) was attempted.
Non-linear elastic analysis that considers strain level- and time-dependent deformations caused by creep was adopted. Strain level-dependency was grasped by normalizing the secant modulus of deformation of each specimen used in the triaxial compression tests (CD condition) using the modulus of initial deformation, and then by averaging the obtained values according to the strain level, as is shown in Figure 12. That is,
   Esec (ε) / Emax = f (ε)   −2)
There are some cases where the stress-strain relationship of soft rock is expressed by an established formula (Tatsuoka 1992). In this study, however, f (ε) is not treated as a function but as a numerical value of f (ε) corresponding to ε.
In simulation of the foundation ground, the foundation ground was treated as a uniform stiff silt, judging from the fact that the sand layer, being distributed in the stiff silt in a law ratio, is considered to have virtually no effect on the deformation behavior. As is shown in Figure 10, the modulus of initial deformation, E max, fluctuates in the depth direction. To make it easier, the modulus of initial deformation, E max, was distinctively determined for each layer, and further to consider the time-dependency of creep deformation, E max' was derived from the following equation.
   Emax' = Emax $\times \alpha$ −3)
Poisson's ratio was determined, from the triaxial compression test results, to be ν ' =0.12 (fixed). As mentioned in 4-1, in this study the effect of confining pressures on the modulus of deformation, for being not pronounced, was not taken into consideration. The structure was assumed to be a rigid body. The FEM analysis was attempted using a three-dimensional model shown in Figure 13. In the initial phase of the calculation, Emax' was given as the rigidity of the foundation ground. In the phase that follows, the rigidity of the foundation

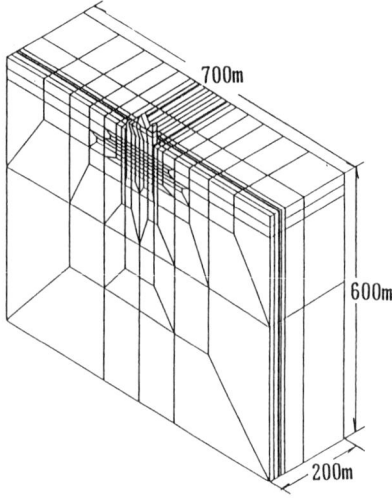

Figure 13 The finite element model.

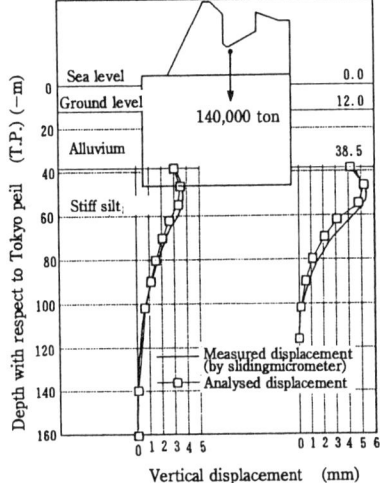

Figure 14 Comparsion between measured and analysed vertical displacement.

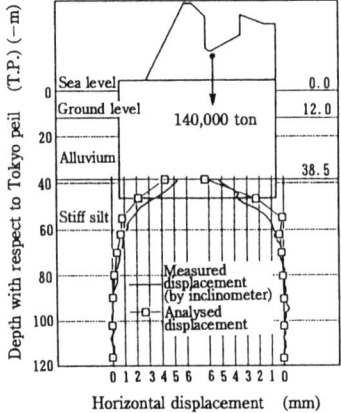

Figure 15 Comparsion between measured and analysed horizontal displacement.

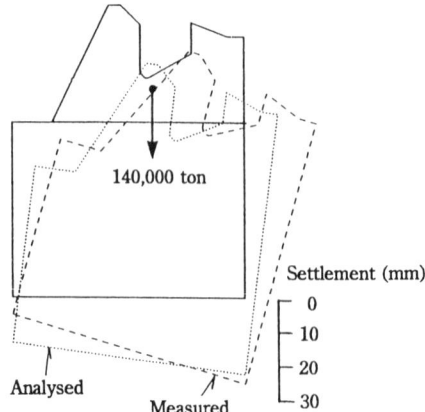

Figure 16 Comparsion between measured and analysed settlement of the anchorage.

In Figures 14 to 16, the analytical results are compared with measured values. As can be seen from the Figure 14 and 15, both the horizontal and vertical displacements obtained from the analytical results coincide favorably with the long-term observational results. These figures also well illustrate their disposition (horizontal displacements occur in the proximity of the caisson and vertical displacements occur, taking the form of upheaval, above the anchorage foundation level). As regard with the settlements, it can be said that the analytical results coincide, on the whole, with the measured values, considering the fact that the observed settlements contain such errors as the deformations of the anchorage.

6. CONCLUSION

This paper described the long-term observation and deformation estimation carried out during the construction of the Rainbow bridge. In order to accurately assess the deformation characteristics of the foundation ground in small strain level, the triaxial compression tests were conducted using LDT. The test results favorably coincided with in-situ tests and measurements. Numerical simulation attempted based on the test results also proved that the test results coincided, on the whole, with the long-term observational results.

REFERENCES

Yamagata, M.1993. Monitoring and Analysis of Seabed-layer Deformation at the Akashi Kaikyo Bridge Tower Foundation, Soils and Foundation, 41-2.

Tomizawa, S. 1987. The Properties of Stiff Clayey Silt (Dotan) as the Foundation of "the Tokyo Port Bridge (Tentative Name)", Soils and Foundation, 35-3.

Goto, S. et al. 1991 A simple gauge for local small strain measurement in the laboratory, Soils and Foundation, 31-1.

Tatsuoka, F. et al. 1992. Non linear stress-strain behavior of sedimentary soft rocks and its modelling, The 28TH Japan national conference on soil mechanics and foundation engineering.

ground was determined based on the strain distribution obtained from the results of the calculation performed in the preceding phase. Iterative calculation was continued until the difference in the strain distributions between the above two steps settles below the given value.

A consideration on the selection of excavation shape for large underground opening

S. Murakami
Chiba Institute of Technology, Narashino, Japan

ABSTRACT: The need for large underground openings increases year by year, as their application for various purposes are growing. Sections of caverns are expanding notwithstanding the complicated states of rock mass, especially in Japan. Determination of cavern section at the initial design stage is important affecting the whole work, as well as construction control now rapidly improving. This paper discusses methods of initial study for selection of shape and reports on some plans for relatively inexpensive and effective analysis in case where the quantity and preciseness of data on site does not fit for the sophisticated analysis. A pending subject of the studies is the evaluation of rock properties that vary according to excavation sequence. At this point, continued studies in various fields are to be expected.

1. Introduction

During the last forty years, many undreground openings, for hydropower stations, fuel storages, disposal yards were constructed, and also, possibly nuclear power plants will be built undreground in the near future.

In many cases, such caverns are accompanied by other caverns, such as machine-halls with surge chambers, tramsformer-rooms, or gate chambers.

Very often, such caverns are installed in parallel rows so as to be unrelieved from the interference with each other, because of limited layout space from geological and topographical conditions or, sometimes, from social conditions.

The sectional area of main cavern often exceeds 1500m²; its width exceeds 30m, caused by increase of unit capacity of turbines or pump turbines, and its height may exceeds 100m in the case of surge chambers.

These conditions of underground openings differ significantly from tunnels from the standpoint of rock behavior that arises from excavation works.

Concerning the locality of rock environment, contrary to tunnels which cover long distances, caverns are installed in limited areas.

In contrast to tunneling that needs to anticipate the rock properties and behaviors with the progress of excavation, underground openings need to know the variations of rock properties and behaviors following the enlargement of excavated sections.

2. Concept of shaping of underground openings

The shape of underground openings for hydropower stations is determined primarily from the original function required by the operation and maintenance, namely setting and overhauling of electrical and mechanical equipments for machinehall, transformer room, etc. and, the hydraulic stability and reduction of hammering shock of water for surge chamber etc.

Secondarily. The behaviors of surrounding rock mass are studied mechanically to get more stable and more suitable shape of caverns on condition that it satisfy the purposes mentioned above, because the condition of rock at sites selected for underground powerstations of large scale is generally adaptable to any shaping with some extent of strenghtening or protectiong that's practically applicable.

Apart from the above, the following matters must be examined.

1. Safety of work and sense of secnrity
The fear factor is not comparable with that of open air works and it often causes delays of construction schedule and expansion of construction costs when some troubles occur on the way of construction.

2. Construction economy.
Not only the amout of excavation, quantities of reinforcement and protection, but also all appurtenant works, difficulty and danger of works must be reckoned in.

On the whole, a minimum opening space is preferable as far as it conforms with the above-mentioned concept, disregarding, to some extent, mechanical disadvantages, be cause rock mass is not perfectly obedient to the artificial evaluation of properties based on in-situ testings conducted on some spots of the whole area.

In the past, there have been a few cases where inside strensthening by concrete lining or steel bracing support was able to prevent the deformation of wall occurring from fairly deep portions. which could not be properly handled with prestressed wires or rods and rock boltings. (hereafter "PS" is an abridgement of prestressed)

3. Recent trends in shaping of underground openings in Japan

In Japan, more than thirty underground openings of large scale have been constructed, mainly as machine halls and surge chambers of hydropower stations.

The shapes are originally mushroom type with reinforced concrete arch and side walls, being caused mainly by unfavorable condition of rock mass in the country.

Some caverns were excavated before reinforced shotcrete and *PC* rods and wires were applicable. At that time, there often occurred falling of boulders and unexpected displacement of side walls during the excavation.

As time passed, development of construction methods and protection measures based on mechanical analysis of rock behaviors made the construction of large caverns, even with strong in-situ stress succeeded, although temporary or partial extraordinary deformation takes place occasionally.

Recently so-called "observational construction methods" are often applied to large scale openings to assure that the control with mechanical analysis keeps pace in the field based on observational data that have been collected from time to time.

The original mushroom type that has been adopted in many cases but the egg type and bullet type are now being used.

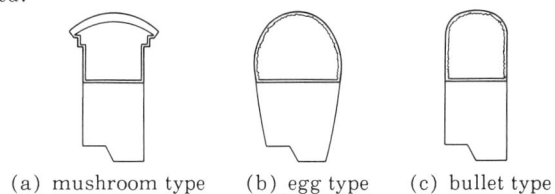

(a) mushroom type　(b) egg type　(c) bullet type

Figure 1. Typical shapes of cavern

The mushroom type has been pointed out the disadvantage in stability of surrounding rock and expansion of relaxed zones as concluded by mechanical analysis, especially loosening of arch abutment portions accompanied by stress increase in the arch concrete, though, on the other hand, it prevents further loosening of vault rock mass and convergence of side walls.

The egg type is evaluated the best from the standpoint of mechanical stability of surrounding rock mass, but on the other hand, it leads to increase of work amounts.

The bullet type is not familiar in Japan, though it is adopted in many conntries, usually in the case of good rock conditions, without any concrete lining.

Figure 1 shows the typical outline of these three types that are applied to similar unit capacity sizes of reversible pump-turbines and motor-generators.

The egg type has a larger cross sectional area exceeding the others by about 10%, resulting in potentially higher construction costs accompanied by reinforced shotcrete and appurtenant works. Farther, the excavation work must be elaborated to avoid leaving rugged edges in the side walls and vault surface.

The bullet type seems to be the most preferable when the rock condition is fairy good and in-situ stress does not deviate from the ordinary range. But if the rock is blocky and difficult to cut with smooth surfuce, increasing the shotcrete work may affect the overall construction economy or irregularity of the surface may cause unfavorable problems, similarly to the egg type, according to rock properties.

Further, if the horizontal component of in-situ stress is predominant, the increment of horizontal thrust may possibly induce cracks in the arch portion of shotcrete.

Needless to say, though arch concrete of the mushroom type should bear the thrust from deformation of side wall rock, the extent of stress increment in concrete and steels are predicted and measured in many cases, and reasonable designs of arch concrete are almost established in ordinary states of rock mass.

As stated above, these three types of shaping have their fortes and foibles, especially when the rock condition is complex as in Japan.

It is very important to select the shape of cavern in the initial design stage according with the geological condition to secure the easy, rapid and economical construction in the safety of workmen.

4. General concept of design of section of caverns

From many past experiences in construction mainly of underground power stations, some of the failures and accidents, helped the construction methods, and the control systems of the caverns to advance remarkably followed by development of mechanical, analitical techniques and measuring instruments.

Nowadays, almost all constructions of large caverns hire control systems, d, i, on the way of construction at each stage of excavation work, reexamine the stability of surrounding rock and redesign the reinforcement using data obtained by measurement up to that stage.

Many cases in various projects were granted the benefits of forseeing unexpected rock behaviors and taking countermeasures in advance.

Many engineers are engaged in such works and striving to improve the methods to realize safe and smooth construction in very cautious ways.

On the other hand, the cross sections or main configurations of caverns are discussed and determined mainly relating to the layout and dimensions of main equipments and adaptability against operation and maintenance.

Selection of cavern shape is often decided in accordance with the decisive judgement of experienced engineers by taking geological conditions into account.

In some cases, comparisons of the mechanical behaviors of surrounding rock mass concerning the stability of caverns are carried out, assuming various features, but finally resulting in comparing of the three types of cross sections.

The past comparative studies were conducted in such a way that a sort of visco-plasto-elastic (nonlinear) analysis is done regerding the following.

— progess of excavation is taken into account.
— modulus of deformation, poisson's ratio, shear strength, etc. are varied according to the state of stress in rock mass.
— reinforcement, like rock bolts, rock anchors etc, are designed to secure stability against sliding, assuming slide lines from the results of the above analysis or slip circles.
— analysis based on NATM theory assuming circular section is done to evaluate needed support force of rock anchors and to calculate displacement.

Comparison of types is made mainly accounting the range of loosening of rock surrounding the cavern, which is defined by the value of varied poisson's ratio. Figure 2 shows examples of such studies. (Mimaki, 1981, Harada, et al. 1991)

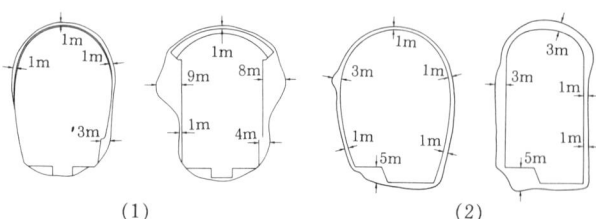

Figure 2. Comparison of relaxed zone of each type
(Mimaki.,1981)
(Harada.,1991)

These procedures have been successful as the initial designs for predicting the displacement and strain, stress of the surrounding rock mass, and no notable hindrancences are found for the execution being followed by aforesaid control techniques.

From the viewpoint that the determination of main features of caverns, like cross sections which affect the whole construction in economy and security, more practical studies are to be expected, because rock behaviors peculiar to the large caverns have become clear to some extent on the basis of past construction experiences.

The main features of the behavior of surrounding rock of large caverns that have been clarified expressly apart from the case of tunneling are as follows.

As the shapes are not close to circular, as its height/width rutio, exceeds or nears to 2.0 and its height, nearly vertical, the width are larger beyond comparison, as mentioned previously.

1. Distribution of in-situ stress to be released by excavation is not assumed to be uniformly distributed.

2. Condition of stress distribution during or after the excavation differ between the vault and side wall portions, depending upon the direction of principal in-situ stress.

3. In general, loosening zone is greater in the side wall portion than in the vault and the amount of opening of discontinuous planes trends toward a kind of "size effect". In the case of long span of releasing the stress, the degree of opening of such will be much larger than that of circular walls like tunnels.

Figure 3. shows the distribution of principal stress
i) when the ratio of horizontal stress to vertical stress, $\sigma_{ho}/\sigma_{vo}=2.5$ ii) when $\sigma_{ho}/\sigma_{vo}=1.0$

In the vault, stress concentrates close to crest, but along the side wall the flow of max. principal stress is far from the wall. There, the stress is mitigated as approaches the surface though stress condition $\{\sigma\}$ accords to the yield criterion.

According to many measurements of strains in rock mass, strains increase closer to the surface. But if the shear resistance has the edge over the sliding force on sliding plane or discontinuous plane, it does not show any destruction and the wall will be stable from collapse.

Though the rock mass where large caverns are built is generally mid-class or high-class with regard to mechanical properties, lots of observations in the surrounding rock such as borehole TV or seismic wave measurement, show that the enlarging of existing joints and opening of submerged cracks possibly play major role in the displacement as compared with the deformation of rock mass itself.

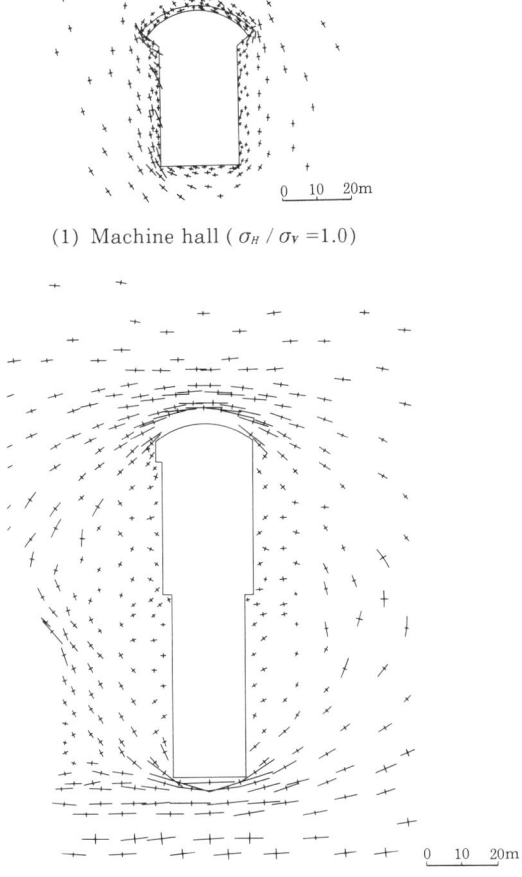

(1) Machine hall ($\sigma_H / \sigma_V = 1.0$)

(2) surge chamber ($\sigma_H / \sigma_V = 2.5$)

Figure 3. Principal stress distributim

The size of space that is released from constraining stress affects the degrees or rates of above-mentioned openings and displacement of surrounding rock. This was verified by observations in various caverns of different sizes to exceed the limit of three dimensional effects (Hibino et al., 1993)

The above facts about existing caverns, suggest that the minimizing of the space of excavation work and the prevention of movement along joints or discontinuous planes will be effective for securing the stability of the surrounding, especially high side walls.

On the other hand, vault portions exhibit different behaviors.

In many cases, the vertical displacement occurs within 10 m from the surface and during the excavation of vaults; it does not apparently increase further along the below excavation. On occasion, it even shows some decrease of subsidence caused by deformotion of side walls in case where lining of concrete is installed.

At present in Japan, there are no large caverns that have not been reinforced with rock bolts, rock anchors and shotcrete.

From the past, almost all engineers recognize that both rock bolts and anchors act very effectively to secure the stability of surrounding rock, especially to protect the surface or close to surface, from sliding or collapse.

But the quantitative evaluation of their effectiveness is not clear, but the case of unbond type *P S* anchors.

Generally speaking, it is said that the effect is not so notable for the hard rock because the strain of steel, enbedded in the rock, that produce the constraining force does not develop enough.

Many studies have been made to clarify the mechnical effects of rock bolts and bonded rock anchors, but those studies that resorting to elastic or elsto-platic analysis generally demonstrale effectiveness for weak rock but less effectiveness for hard rock.

For instance, figure 4 shows the stress distribution and axial displacement of rock mass in elastic zone, produced by constraining of 12 rock bolts of 5m length surrounding a 10m dia. circular tunnel when released from initial stress.

Even if max. axial force of rock bolt is 20 tonf, max. radial stress induced in rock is 0.3 kg/cm^2 and displacement 0.3 mm. Such amount is too small compared with in-situ stress in the fields. (Yamachi. 1994)

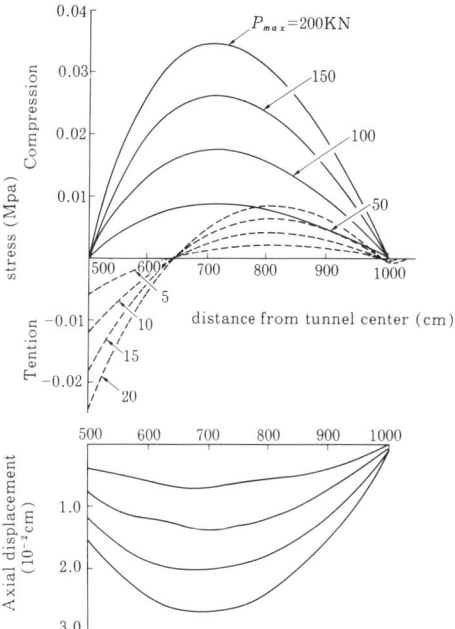

Figure 4. Axial force of rockbolt and stress in rock (Yamachi. 1994)

Non-linear elasto-plastic finite element analysis for a circular tunnel indicates that, when ground strength ratio (uni-axial strengh/in-situ stress) is 0.5, radial and tangential stress clearly increase in rock as the confining effect of rock bolts, but in the case of the ratio is 3.0, increase of stress by rock bolts can not be identified. (Hisatake., 1990)

As regards sizable vertical or that like side wall, measuremeuts of stress or deformation of rock bolts are not so many, though rock anchors for supporting the side walls used to be measured the process of increasing from initial to final axial force.

But rock bolts that are extracted from the rock mass are usualy deformed by elongation and bending. From these matter, functions of bolts other than that caused by elastic, plastic deformation of supporting rock, are assumed to be raised up,maybe, caused by displacement of discontinuous planes in rock mass.

Some experimental studies using models of various properties simulating jointed rock mass were executed and the influence of bolt setting angle to the discontinuities were examined. The results showed that the reinforcing effect of rock bolts consists of Tension effect and Dowel effect, both depend on bolt setting angles.

Further, the reiforcing effects by rock bolts were shown according to shear displacement and inducing measure of such effects into numerical analysis was indicated. (Yoshinaka et al. 1987, Hisatake. 1990)

Other laboratory tests were performed. simulating rock mass by a few kinds of material with different Young's modulus and at random joints.

The effects of rock bolts were clarified by the stress-strain relation undre uni-axial compression tests comparing with non-jointed model. The results show that reinforced model show apparent strengthening of deformation modulus, and the peak shear strength increases more remarkably in hard material than in soft material. From these results, it is suggested that numerical analysis for jointed rock should be done assuming the continuum reinforced by rock bolts taking into accout the effects of interaction between rockbolts and discontinuous planes. (Sakurai, et al.,1992)

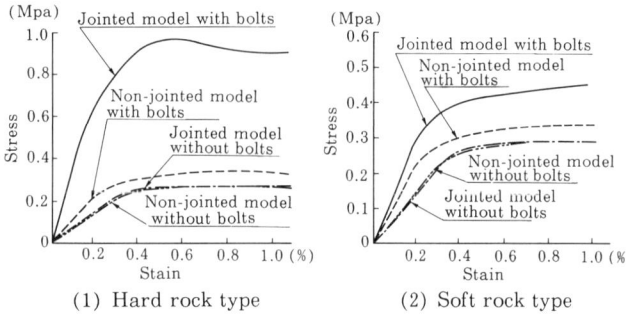

(1) Hard rock type (2) Soft rock type

Figure 5. Stress-strain curves obtained by model tests

In actual states, inasmuch as stress and strain measurements of the rockbolt itself are rather rare, many measurements by extensometre set in rock mass show the strain up to about 10^{-3} in the side wall close to the surface (Nozaki, et al. 1987)

Farther, this is less than peak of strain in the same hole.

These facts show that strain produced in rock along the rock bolt and strain of bolts are far more than those evaluated by continuum analysis and imply the effective action of rock bolts by their unexpectedly big axial force.

5. Consideration for initial design of underground openings

Excavation of underground openings of large section is similar to tunneling from the view point that it is the process of unloading in-situ stress while retaining the stability of the surrounding ground.

But unlike tunneling, underground openings are executed within a limited, but vast area and excavation is done step by step, necessitating to maintain the stability of stress released portion and anticipate forward behavior of rock caused by the subsequent excavation.

So evaluation of rock properties of preceding portions and its effect on the behaviors by following steps becomes necessary.

As said beforehand, the present design in initial stage for sizable underground caverns in rock mass as continium generally has recourse to visco-plasto-elastic analysis, taking account of the variation of rock properties, mainly deterioration following the developments of stress, strain according to the advance of excavation.

In the analysis, rock properties are assumed not to vary by reinforcements or such reinforcements are assumed to exist in the rock mass as elements inserted from the start of excavation.

Both are far from the fact, as bolts and anchors begin to act according to the deformation of surrounding rock, except the effect of pretension of them.

Practical excavation of caverns is performed step by step

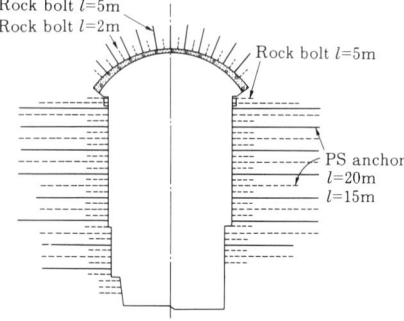

Figure 6. Reinforcement of cavern

from the vault to bottom divided into more than ten portions, with each step consisting of drilling, explosion, mucking, boring, setting of rock bolts, shotcrete and setting of PS anchors, pretension and grouting. At this stage bolts and anchors of the said portion have no effect.

Finally, in the usual case in Japan, rock bolts and PS anchors are installed very densely on the surface of vault and side wall, as shown in figure 6.

It resembles some features of group piles applied for foundation works.

Considering the above situation, the following are to be reckoned into the study of initial design.

1. In-situ stress should be taken as it is in the site, but if that is difficult, it is preferable to assume that vertical stress varies proportionally to overburdens. If in-situ stress is taken constant in the whole area body force by gravity must be reckoned in.

2. Resorting to elastic finite element or boundary element method in advance, grasp anticipated distribution of principal stress roughly along the side wall, and study the density and size of rockbolts and PS anchors to be counted in the following analysis.

3. Unloading is to take place for each step according to the sequence expected in the actual construction works.

4. Evaluate the effects of reinforcement expected to play an actual role and quantify the varied rock properties such as perfect plastic, softening or hardening, taking into accout the reinforcing effects.

5. Analysis should be made in accordance with the shape which is close to that of each step of excavation.

Except the analysis as continuum, several methods that take into account discontinuity and isotropy, may be preferable to realize the effects of reinforcements.

But for the stage of initial design, it seems to be difficult to grasp the whole facets of irregularity as to simulate the actual rock behaviors.

Owing to this concern, the analysis for discontinuous rock mass is not taken into account hereafter.

6. Process of the analysis

Ideally, the analysis at each atep, is to be done by the plasto-elastic analysis with viscosity including elements of rock bolts and anchors, preferably in 3-D models.

But for the purpose of initial design aiming to clarify the superiority between some types of cavern shape, it will be too costly in time and effort, and not validate, while informations for rock properties are not so abundant or accurate.

Further, it is difficult to reflect the effects of bolts to constrain the deformation in discontinuous rock mass, as verified in many laboratory tests, provided that the prevailing orientation of discontinuity is not specified.

The proposed analytical method for the initial design to determine the basic section, is a sort of progressive procedure following the expected orders of excavation, step by step, applying primarily the following measures.

6.1 Plasto-elastic analysis by the coupled boundary element-characteristics method (Sugawara 1988)

This is used to analyze the overall area behavior's that accrue due to excavating of the portion of a certain step.

The characteristics method is easy to execute if rock bolts and anchors, apart from pre-tension effect, are evaluated seperately in approximate method as, referred lately.

In this analysis, as Mohr-Coulomb's criterion is applied to discern between plastic and elastic zones, tension zones where the tensile stress exceed uni-axial strength must be treated seperately in the process.

As this analysis should be done by incremental unloading, firstly the elastic boundary element method is to be carried out and the elastic stress on the boundary, that is cavern surface, should be added to the stress calculated formerly on the boundary already released from in-situ stress.

Next, all surfaces are examined for accumulated stress to determine the assumed plastic zones where characteristic curves are to be sought inwards from the surface.

Here, be careful to evaluate the varied modulus of

deformation for elastic solution and varied yield criterion for elasto-plastic analysis for the reinforced zones by rockbolts and PS anchors already working, by means referred afterwards.

The solution derived by the process above is the boundary deformation of released portion up to this step, but it excludes the constraining effects of rock bolts and PS anchors to be induced during the unloading of the said step. Furthermore, necessary displacement of inner points should be calculated.

6.2 Calculation of final deformation of the step.

After the plasto-elastic analysis, it is necessary to count the mitigation of displacements caused by rock bolts and anchors. If unbond type PS anchors are effective in this step their effect must be evaluated seperately and treated as inner pressure to act on the surface of the excavated portion.

Here we assume, the partial linearity of rock behavior in regard to the interaction between rock bolts or anchors and in-situ stress release acting on the boundary, as tractions, and no slippery or yield occur between bolts or anchors and rock mass. Stress in bolts and anchors is supposed to be proportional to relative displacement as follows (Saitoh 1981)

$$\tau = c(\omega_b - \omega_r) \quad (1)$$

$$\sigma_b = -E_b \frac{dw_b}{dx} \quad (2)$$

$$\frac{d\sigma_b}{dx} = -\frac{2\tau}{r} \quad (3)$$

then

$$\frac{d^2 w_b}{dx^2} = \frac{2c}{E_b r}(\omega_b - \omega_r) \quad (4)$$

where τ : shear stress between bolt or anchor and rock
c : coef to be determined from rock test or existing data
ω_b : displacement of bolt or anchor
ω_r : displacement of rock around bolt or anchor
σ_b : axial stress of bolt or anchor
E_b : Young's modulus of bolt or anchor
r : radius of bolt or anchor
x : cordinate along the axis of bolt or anchor

Equation (4) can be solved numerically if the distribution of ω_r is known.
Then strain energy stored in bolt or anchor, in the step of excavatin is

$$E_R = \frac{\pi r^2}{2E_b} \int_0^L \sigma_b^2 \, dx \quad (5)$$

where L is length of bolt or anchor.
Next E_R is summed up for all bolts and anchors which act in the said step

$$\overline{E}_R = \sum_{i=1}^{n} E_{R_i} \quad (6)$$

where n : numbers of bolts and anchors

From the reciprocal theorem, we can install the following, as the relation between strain energy of bolts and the reduction of displacement on newly excavated surface of the said step.

$$\sum_{i=1}^{n} \int_0^l \tau \omega_r \, dx = -P_o \int_{S_i} \Delta u \, ds \quad (7)$$

where τ : shear stress around bolt and anchor in state of balance of the step
P_o : in-situ stress on Si
Δu : displacement on Si forced by bolts and anchors

The left side of the equation is twice the strain energy of bolts ($\overline{E}_R{'}$), but differs from the former calculated strain energy ($\overline{E}_R$) which accords with plasto-elastic analysis without bolts effect.

Figure 7 shows, the relation of $\overline{E}_R$, $\overline{E}_R{'}$ and Δu and ground characteristic curve for Si, boundary of step i.

Though the characteristic curve can not be settled strictly in this case, it is drawn approximately in consequence of a few times of plasto-elastic BEM-characteristics analysis. It is utilized to figure $\overline{E}_R$ and determine $\overline{E}_R{'}$ and Δu.

Also equivalent inner pressure P_R is determined seperately with which elastic BEM or FEM analysis is to be executed to compute final displacement and stress on the whole surface for step i.

Inasmuch as procedures that seek the final deformation of the concerned step aim to evaluate the reiforced properties in the following process, it will be unnecessary when the effect of reinforcement is too small to numerically evaluate, such as when all zones are within elasticity.

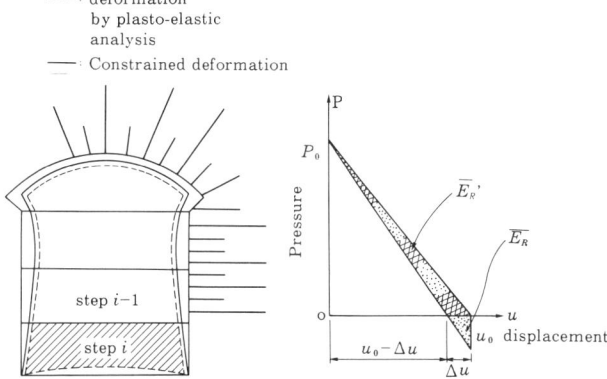

Figure 7. Displacement on Si step i.

6.3 Evaluation of properties reinforced by bolts and anchors.

In advance to progress to the next step of excavation, there is a need to evaluate the modulus of deformation and yield criterion of the portion reinforced precedently.

As was stated before, many numerical analyses of rock behavior with rockbolts, upto the present, have shown that reinforcing effects of rockbolts are not so remarkable.

But some investigation data of existing tunnels show that axial forces of bolts installed have a tendency to increase with tougher rock and confining pressure by rockbolts lift up the critical strain which seems to be the limit of stability of tunnels.

Unfortunately, from the shortage of observation and measurement concerning rockbolts the effects are not clear in the case of large caverns, apart from that data of PS anchors in some cases.

Accordingly two evaluation approaches are taken here, and engineers should ultimately judge the degree of reinforcement by examination of those results and by referring to other examples or the alike.
1) Numerical evaluation of modulus by back analysis
Being based on the results of former analysis, that is final displacements accumulated, back analysis by FEM is to be carried out using the displacements of nodal points, to estimate increased Deformation modulus as D_b for rock bolts zone together with PS anchors and D_p for PS anchors zone surrounding the former zone.
2) Evaluation by existing data or experiments
From existing 24 caverns of hydropower stations in Japan data of measurements were gathered and put in order.

Among them, strain of side wall and induced tension force of rock anchors per unit area converted to dimensionless value through dividing by critical strain and uni-axial strength of the rock respectively, have strong correlation (Nagai et al. 1993)

Figure 8 cited from the said report indicates that both cases of PS anchor only and anchor plus rockbolt differ to get the same strain ratio.

Then calculated axial force of a rockbolt to compensate for the difference must be 29 tonf/unit, far exceeding the ultimate strength of a bolt, which is about 18 tonf/unit.

This is thought to verify that rock mass with rockbolt varied its strength. From this data, the rate of increase of σ_c is estimated at about 1.6 times of its original.

Much more information is needed to evaluate the

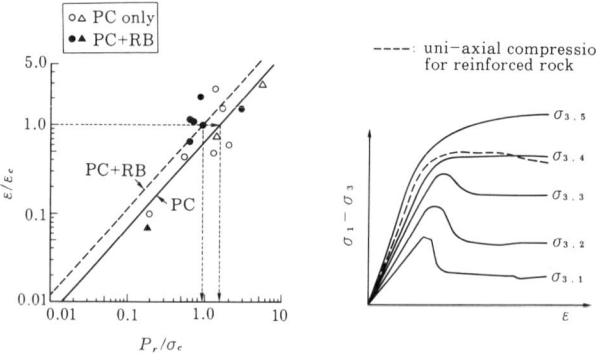

Figure 8. Strain-pressure relation (Nagai et al.,1993)

Figure 9. Compressive tests of rock with reinforcement and constraining force

reinforcement of bolting. Some experiments of models with bolts show a stress-strain relation as in Figure 9.

If the uniaxial compression of reinforced rock shows the dotted line equivalent to certain σ_3 of unreinforced rock, the resultant modulus of deformation and yield criterion would be clearly estimated.

6.4 Flow of procedures

The main outline of procedures based on the priciples mentioned herefore is shown in Figure 10.

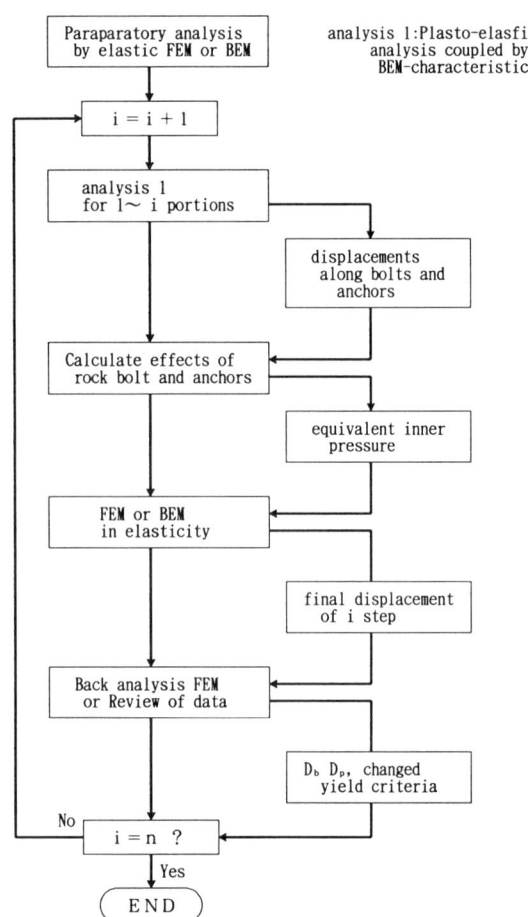

Figure.10 Flow diagram of analysis

7. Conclusion

Planned underground openings are increasing their sizes, and the layouts in which they are included are getting more various.

More over, in some case, unfavorable geometry can not be avoided.

There are many subjects to be studied and solved. This paper limited the subject to a single cavern concerning to the initial design concept.

The author think that more effort should be poured in the initial design to determine the shape of the cavern, in the manner to reflect the actual sequence of unloading and reinforcement induced step by step, because large caverns are generally impossible to realize without such procedures and treatments.

But in view of the real situation that data of rock properties at hand in initial stage are not enough or accurate to reflect the exact behaviors of rock mass and unexpected matters would often occur on the way of construction, the said analysis is to be comformable to the degree of accuracy but not derail from the main course of development. The proposed procedures were thougt out to attain this aim.

Though the importance was put on the evaluation of rock mass properties which vary caused by deterioration and strengthening in the course of excavation. some results of experimental studies and statistic studies were only cited in this paper.

But, at present, the situation is not enough to get firm ground of the said evaluation applicable to put into the analysis such as to be requested in this case.

Now, lots of data of existing underground power plants are being collected and put in order to grasp real rock behaviors and actual conditions of reinforcement, in large caverns.

Such results are expected to be used in the analysis proposed to realize proper anticipation of behaviors of the surrounding rock to attain the appropriate determination of the shape of caverns.

Reference

Harada,M., Katayama,T. and Yada,A. 1991. Design and construction of underground opening of Ōkouchi power station, Electric Power Civil Eyneering No.230, pp46～57

Hibino,S., Motojima,M. 1993. Rock behaviour during excavation of large scale cavern and proposal of lining design concept, Journal, geomechanical engineering, J.S.C.E.

Hisatake,M. 1990. Consideration on rock bolt effects in soft rock tunnel.

Mimaki,Y. 1981. Design of Imaichi underground power plant, Electric Power Civil Engineering No.173, pp24～38

Nagai,T., Kunimura,S. and Ikejiri,K. 1993. The evaluation of rock bolts in discontinuous rock masses, 25th Symposium on rock mechanics J.S.C.E. pp356～360

Nozaki,T., Ito,H. and Hibino,S. 1987. Consideration for stability measures for large scale caverns. Eletric Power Civil Engineering No.209 pp40～48

Saitoh,T. and Amano.S. 1981. Fundamental consideration on design of rock bolts reinforcement 14th Symposium on rock mechanics, J.S.C.E, pp76～79

Sakurai,S. and Kawashima,I. 1992. Modeling of jointed rock masses reinforced by rockbolts. Journal, geomechanical engineering, J.S.C.E No457 Ⅲ-21 pp147-150

Sugawara,K., Aoki,T. and Suzuki,Y. 1988. Elasto-plastic analysis of rock caverns by a coupled boundary element-characteritics method. Mininig and Metallurgical Institute. Japan

Yamachi,H. 1994. Study on the effect of rock bolt in tunneling and the optimal dimensioning of it. Kōbe University

Yoshinaka,R., Shimizu,T., Arai,H., Kato,E. and Arisaka, S. 1987. Reinforcing effect of rock bolt in rock joint. 7th Symposium on rock mechanics, J.S., Soil mechanics & Found.Eng. pp431～436

Behaviour of arch abutments anchored in rock under thermal loading

G. Ballivy & Moha Melouki
Civil Engineering Department, Université de Sherbrooke, Que., Canada

F. B. Slimane
École des Mines, Nancy, France

J. Tardif
Les Consultants SM, Longueil, Que., Canada

ABSTRACT: The performance of the four arch abutments of the Vélodrôme-Biodôme structure in Montreal is presented. It is documented by the data collected from more than 150 instruments installed during three different phases since 1975.

At two of the abutments, W and Z, the rock is composed predominantly of a highly fractured limestone with silty clay filling the vertical and horizontal joint planes in the first 15 meters. The tendon anchors were grouted in the sound rock below this zone and tensioned (average load of 6 MN) before decentering of the arches. The tensile loads are monitored by load cells, the settlements in fractured rock are monitored by extensometers, and the deformation of the concrete abutments is recorded by strainmeters, extensometers and tiltmeters (rotation movements).

It is shown in this paper that deformations induced in the arches by thermal expansion of the concrete structure are very significant. The abutments rotate along a transversal axis due to the increase of load. A direct effect of the variation of load in the tensioned grouted anchors and an increase in tension are observed.

1. INTRODUCTION

The Vélodrôme in Montreal's Olympic Park, has been built in 1974 - 1975 for bicycle racing (Fig. 1). Since 1992, this structure is housing the City's Biodôme. This large concrete structure has undergone several stages of geotechnical and structural instrumentation specifically concerned by the behaviour of the foundations anchored in the fractured bedrock. The main reason for the first round, which took place before and during construction, was to monitor the behaviour of the works and ensure conformity to the original design conception. The subsequent instrumentation was to identify factors influencing the behaviour of works. If anomalies were discovered, decisions could then be made regarding necessary interventions. The last instrumentation phase took place during 1991-1992 in connection with the changeover of the Vélodrôme to its present function as a Biodôme, where several different natural ecosystems have been recreated. The last phase was made necessary after the results of earlier sudies revealed the effects of various climatic and environmental conditions on the behaviour of the structure, and more specifically the foundations.

The roof is designed in the shape of a shield, or a maple leaf according to the architect, Roger Taillibert, and it is practically the entire structure which is founded on the bedrock (Fig. 1): four buttresses or abutments, W, X, and Y on one side and Z at the other end (Fig. 2-3) serve as support points and also tranfer the anchorage loads in depth at the sound rock level (Fig. 4).

2. GEOTECHNICAL CHARACTERISTICS

The rock base, composed of fine-grained shaly limestone, forms sub-horizontal beds of 10 to 20 cm in thickness. These beds alternate with thinner shale layers, whose thickness varies from 1 to 5 cm, in general. The upper zone is fractured (Fig. 1), the degree of which decreases with depth, and from East to West. Conditions near buttresses Z and W of the Vélodrôme (Fig. 2) are particularly unfavourable.

The geologic mapping of the excavation walls of one area of the Olympic site and a part of the Viau-Pie IX subway (Fig. 1) indicate the presence of quaternary glacial neotectonic [Durand and Ballivy, 1974]. The upper part of the rocky substratum has been displaced by glaciers along a bed of very soft shale (highly plastic clay material) whose thickness varies between 5 and 20 cm. This slippage could obscure the presence of open joints, such as those observed in the rock at the Vélodrôme site after the excavation of the post glacial soil-deposits, as described elsewhere [Ballivy et al., 1977].

The deepest slippage occurs between depths of 3 and 21 m. This poses difficulties in the design of the foundations, as efforts are almost parallel to the slippage surface and the bedding. This also partly influences the dimensionning of the active anchors which attach the buttresses to the deeper solid rock (Fig. 4).

Due to jointing and fracturing present within the rock mass (Fig. 1), intense cement grout injection of the zones below the buttresses was done, up to the anchoring level. The grout intake was around 5% of the volume of the treated rock mass.

3. STRUCTURAL DESIGN

The superstructure takes the form of a convex shell (Fig. 4), composed of three pairs of principal acrches made of prestressed concrete. Each pair connects buttress Z to one of the other three buttresses (W, X or Y). The entire superstructure was built using

Figure 1. Vélodrôme under construction and bedrock exposure

a system of falsework. The components of the arches were prefabricated on site and fitted together and post-tensionned. Areas between the arches were made from thin prestressed concrete slabs, and a system of narrow concrete members spanning between the arches to support the translucent panels. To bring the superstructure to the rest on the four buttresses, which provide the only support, the weight of the entire structure was transferred from the falsework by axially compressing each arch to lift it upward. This decentering operation involved many flat-jacks.

4. INSTRUMENTATION

- Initial phases

The analytical studies [D'Appolonia et al., 1976] showed that the performance of the buttresses would be acceptable for the predicted loads. However, it was considered necessary to proceed with the instrumentation of the foundation to better examine their behaviour and evaluate their performance during the tensionning of the anchors and the decentering. Figure 4 is a schematics of a typical buttress instrument setup.

In general, observations deriving from the surveillance system on buttress behaviour conform to predictions made by the numerical model employed during the design and planning of the different construction phases [D'Appolonia et al., 1976] and the decentering operation proceeded as expected.

In each buttress, comparison of the different extensometer readings verifies that only buttresses Z and W underwent a slight tilting towards the rear.

- 1984 instrumentation phase

Because buttress movements were revealed by analysis of the instrument readings (1975-1984), further documentation for better analyses was required. Four uniaxial tiltmeters (Fig. 4) were installed in 1984 in buttresses W and Z (two in each), to measure the transverse and longitudinal rotation of the two buttresses. From the data gathered, the general tendencies regarding the behaviour of the buttresses were identified:

- transverse rotation was very slight;
- longitudinal rotations of the buttresses fluctuated, translating to alternate rocking from front to back. These motions were linked to temperature, which influence the intensity of the thermal load transfered to the buttresses by the arches. In each case, the amplitude of the longitudinal rotations remains small (less than 2 minutes).

- 1991 instrumentation phase

In 1989, a comprehensive validation study of the behaviour of the structure was undertaken. Roof survey analysis of the Velodrome showed, unequivocally, the close relation between movement of the arches and climatic conditions, particularly temperature changes. In view of this observation and the later decision to modify the Velodrome into a Biodome, where climatic conditions (temperature and humidity) would vary greatly, it was decided to add complementary instrumentation to monitor deformation of the arches and buttresses. It was also decided to proceed with the establishment of a programmable data acquisition system connected by a modem to a microcomputer dedicated to this monitoring and to an alarm system.

This instrumentation, installed on the arches (Fig. 4) included the following:

- two series of three thermocouples, to determine the temperature profile inside and in the concrete and one outside (TC-7) on the roof;

- two probes inside, at the same level as the thermocouples and one outside to measure relative humidity;

- six tiltmeters, distributed in groups of three, on arches 1 and 2 to measure angular variations related to movement of these arches near buttress Z and one tiltmeter at the top of arch 2 (tilt no. 7);

- two convergencemeters, to monitor vertical movements between the floor and the arch 2 and correlate angular movements.

5. GLOBAL BEHAVIOUR ANALYSIS

- Roof

The recordered readings obtained by survey since 1975, and completed and extended by readings obtained every six hours from the tiltmeters (Fig. 5) and convergencemeters (Fig. 6), clearly indicate that the roof undergoes seasonal movement, particularly due to temperature variations, as illustrated on Figure 6. Considering the reactions of the relatively more rigid buttresses, the roof has a tendency to rise in summer and fall (contract) in winter compared to its initial position. The maximum amplitude of the measured displacement is on the order of 12 cm (survey point number 4) on arch 2 and 8 cm on arch 1 (Fig. 3).

- Buttresses

The tiltmeter readings installed on the buttresses (Fig. 4) indicate that they experienced slight longitudinal swaying which is apparently reciprocal, the maximum recorded amplitude is about

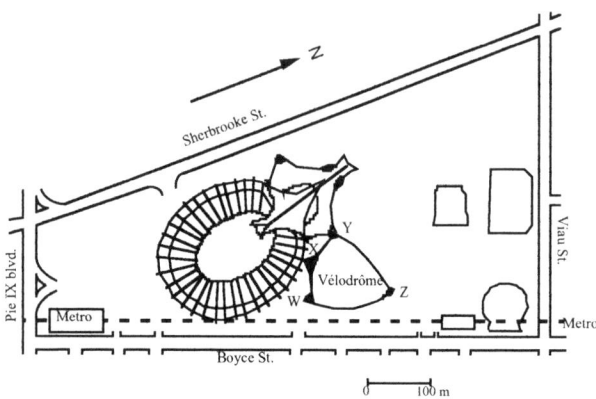

Figure 2. Site of the olympic games (Metro = subway alignment)

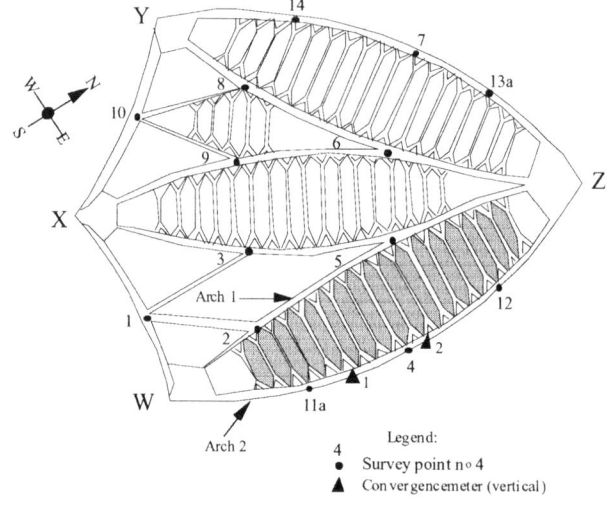

Figure 3. General arrangement

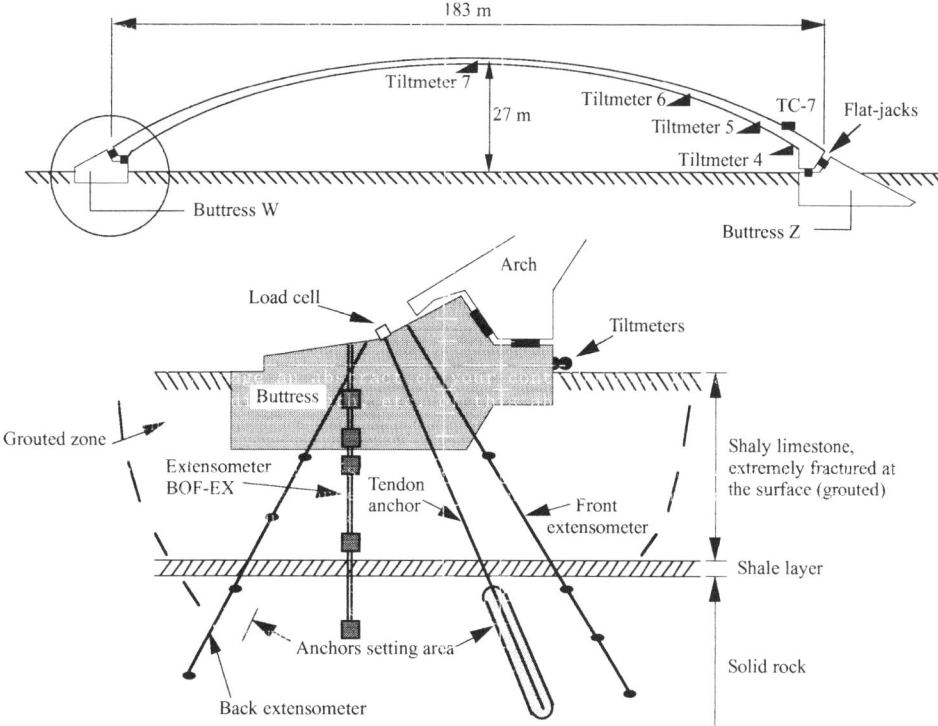

Figure 4. Instrumentation of arch 2

2 minutes. Changes in direction of the swinging (towards the back or front) would be related to variations in the load transferred to the buttresses by the arches. The origin of this variation is largely thermal [Ben Slimane, 1993].

The extensometers in place since 1975 have permitted monitoring of changes in the bedrock. All extensometers show a general increase in recorded displacements. The accumulated subsidence recorded by all extensometers amounts to less than 3 mm.

• Anchors loads

It appeared that the tension in the instrumented multi-tendon anchors was increasing or decreasing depending on their location in the buttresses and with time. A detailed description of the characteristics of these tensioned grouted anchors is presented elsewhere [D'Appolonia et al., 1976]. It has been decided to investigate as and validate the origin of tension increase and their incidence on the safety of the structure. Therefor a numerical modelisation was conducted to this end.

6. NUMERICAL MODELISATION OF BUTTRESS Z

A numerical modelisation by the finite elements method has been applied to the buttress Z which plays a major role in the structure. The rock is also severely fractured and due to a change in the design, a part of the initial excavation has been filled with a low strength concrete to level on the bottom of the excavation (material II on Figure 7).

The geometry of the model is illustrated on the Figure 7. It consists of 7 groups of materials and the elastic parameters of the bedrock determined by insitu dilatometer test which are kept constant in the simulations. However, it is considered the deterioration of the fill concrete (zone II) by reducing the modulus of elasticity from E = 20 GPa (zone I) to 15, 7.5 and 1 GPa, as it has been observed in 1991 in a drilling investigation that the zone II is deteriorating (Table 1).

The numerical modelisation (Fig. 8) was done using a special software CESAR [Laboratoire Central des Ponts et Chaussées, 1991]. The actual applied loads on the buttress are illustrated on the Figure 9, the components F_i are induced by the arch reaction and the C_i by the anchors' tension [D'Appolonia et al., 1976]; it is also illustrated qualitatively the displacements under these different loadings conditions. This model gives the same displacements as observed in the field in 1975 during the decentering process of the arches [D'Appolonia et al., 1976].

To simulate the contribution of the thermal loading, the values F_i were increased up to 30%. The rotating displacements were then increased (Fig. 9) so that the anchors which are located in the front of the buttress are over tensionned, and the others, in the back, are undertensionned. It appears that the load increase in the anchors is larger than anticipated when taking only into account the displacements measured in the rock foundation by the extensometers. So it is considered that the large mass of concrete of the buttress itself is also subjected to thermal volumetric changes: two extensometers have been installed in the concrete mass in 1992 (Fig. 4) specifically to monitor this eventuality.

7. CONCLUSION

The following major conclusions may be drawn from this long term monitoring (20 years period):

- Arch deformation is essentially cyclic and of thermal in origin, linked to seasonal temperature fluctuations. The influence of these deformations on the buttresses is obvious, but remains limited.

- Deformations in the foundations are restricted to the upper part of the fractured rock, the concrete fill (zone II) and the concrete buttresses (zone I). The buttresses undergo slight, apparently reversible, longitudinal swaying.

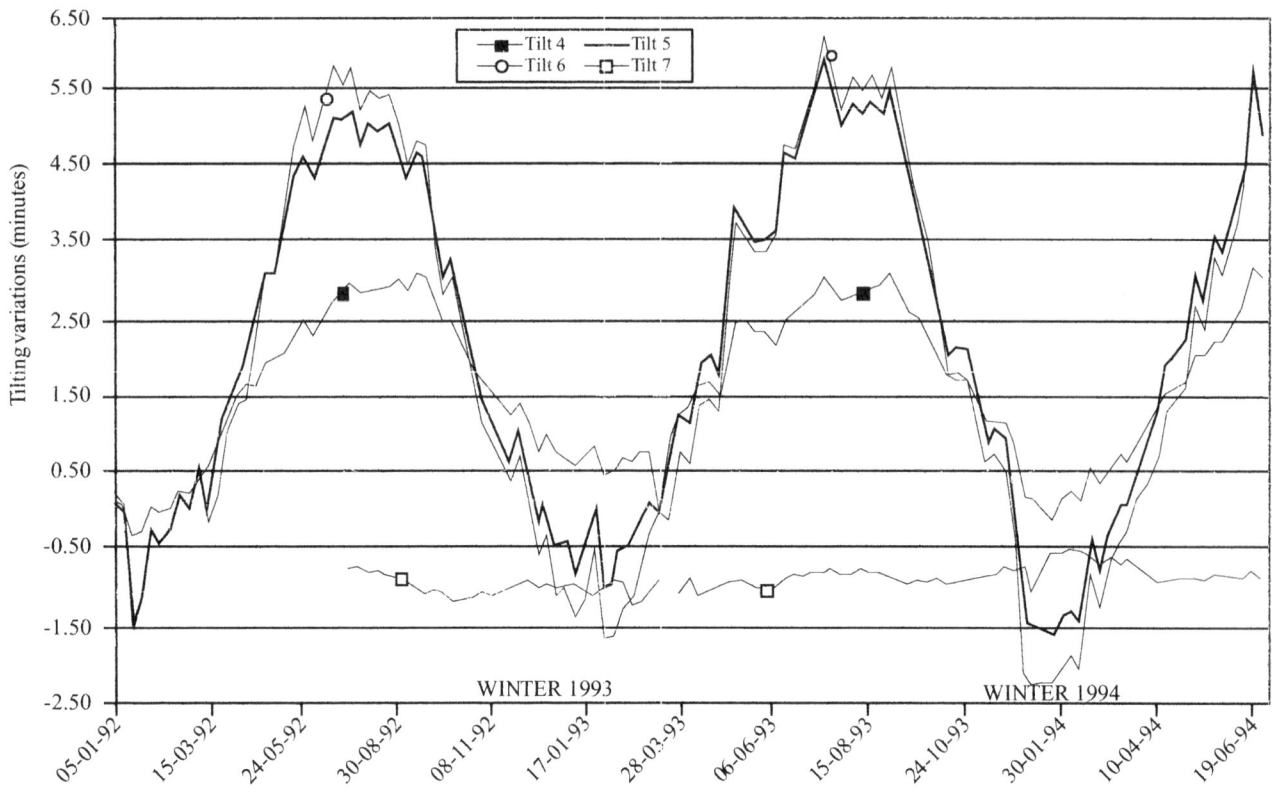

Figure 5. Variations of tilting on arch 2

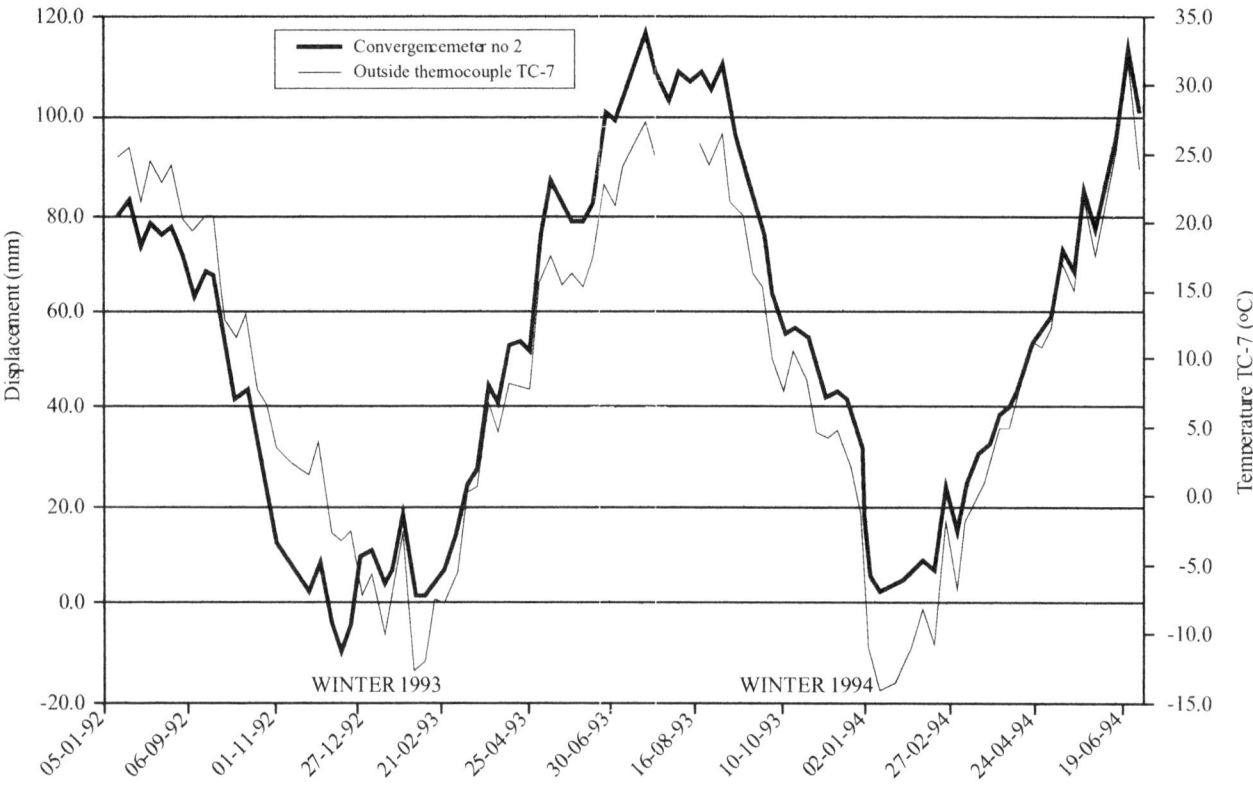

Figure 6. Variations of elevation of arch 2 (convergencemeter)

TABLE 1. Geomechanical parameters of the materials

Material no.	Type	E (GPa)	ν	γ (MPa/m)	Thickness (m)
I	Structural concrete	20.70	0.17	0.025	Variable
II	Fill concrete	20.7 / 15.0 / 7.5 / 1	0.20	0.025	Variable
III	Foundation rock	6.90	0.25	0.025	12.5
IV	Foundation rock	1.40	0.25	0.025	12.5
V	Shale	0.07	0.25	0.025	0.5
VI	Foundation rock	6.90	0.25	0.025	15.0
VII	Foundation rock	13.80	0.25	0.025	32.0

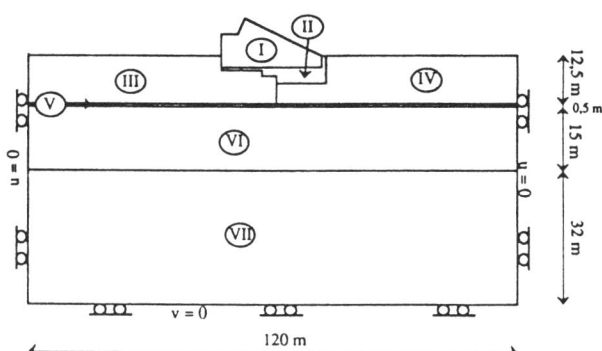

Figure 7. Geometry of the model (buttress Z)

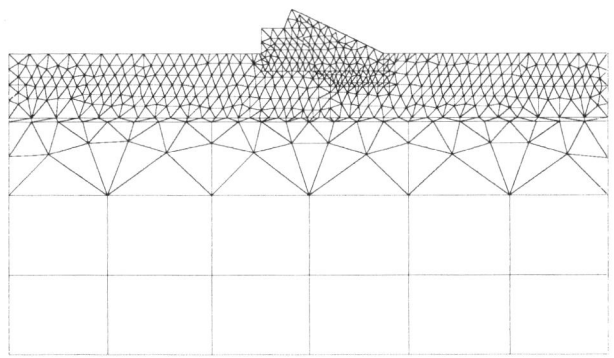

Figure 8. Finite element mesh (buttress Z)

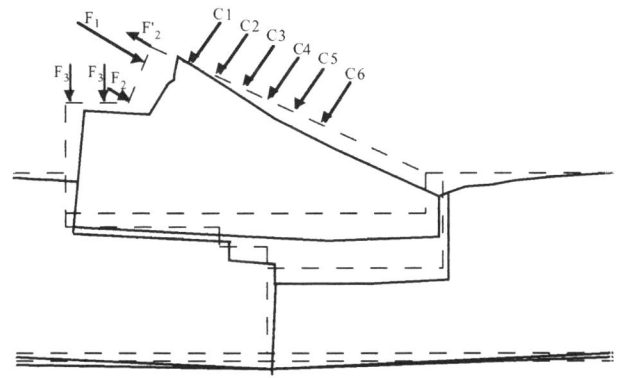

Figure 9. Applied loads and displacements (buttress Z)

- Different instruments installed in the concrete of the buttresses seem to indicate that decompression occurs. The cause and significance of this decompression requires further investigation, this is the object of the addition of recent complementary instrumentation (extensometer BOF-EX, Fig. 4).

- Fluctuations in the anchors loads, installed to assure buttress stability, result from a combination of several complex factors related to anchor/structure interactions. So far, increases in load recorded in some anchors remain well within the safe limits but it requires a close attention.

8. ACKNOLEDGEMENTS

The autors would like to thank kindly all those who have provided access to the entire data set gathered in the past twenty years, particularly the city of Montreal and the "Régie des installations olympiques".

REFERENCES

(1) Ballivy, G., Loiselle, A., Durand, M., Poitrier, M. (1977) *Caractéristiques géotechniques du secteur du parc olympique, Montréal*, Revue canadienne de géotechnique, vol. 14, p. 193-205.

(2) Ben Slimane, F. (1983) *Comportement à long terme des ancrages des butées du vélodrome de Montréal (Période 1975-1992)*, Rapport de stage, Laboratoire de mécanique des roches et de géologie appliquée, Université de Sherbrooke, 66 p.

(3) D'Appolonia, E., Shaw, D.E., Richard, J., Raynaud, D.A. (1976) *Finite Element Analysis of Arch Abutments*, Proceedings of a speciality conference, published by the American Society of Civil Engineers, University of Colorado, Boulder, Colorado, August 15-18, vol. 1, p. 55-81.

(4) Durand, M., Ballivy, G. (1974) *Particularités rencontrées dans la région de Montréal résultant de l'arrachement d'écailles de roc par la glaciation*, Revue canadienne de géotechnique, vol. 11, n° 2, p.302-306.

(5) Laboratoire Central des Ponts et Chaussées (1991) *Code de calcul CESAR Version 3.0*, Paris, France

Inspection and confirmation for undersea foundation rock of Kurushima Bridges

Y. Yanaka & Y. Hasegawa
Honshu-Shikoku Bridge Authority, Tokyo, Japan

N. Masui
Obayashi Corporation, Japan

ABSTRACT: The existing undersea foundation rock in Honshu-Shikoku Bridge Project has been confirmed by the diving operation after the excavation. However, at the Kurushima Bridges in this Project a rock grade of the undersea foundation was predetermined by the pregeological survey before the excavation. With the progress of excavation the rock grade was reviewed by the data supplied from the execution. When any difference might exist in a rock grade between before and during excavation, a detailed survey was conducted to evaluate the rock grade. As the result the rock excavation could be carried out without any impact on cost and schedule.

1 INTRODUCTION

The Kurushima Bridges is the main part of the Onomichi-Imabari Route of the Honshu-Shikoku Bridge Project as shown in Figure 1, which consists of nine long span bridges connecting ten islands. The Bridges, the first three continuous suspension bridges in the world as shown in Figure 2, has the total length of 4km and it is intended to promote the local development. It crosses three straits including an international navigation route and is located in one of the National Parks with numerous beautiful small islands.

The bridge scheme has paid much attention to harmony with a landscape by minimising modification of the existing topography and considerably small influence on a ship navigation. As the result the Bridges includes five undersea foundations in which the maximum water depth is 40m and the tidal current reaches 7 knot in an extreme condition.

The geology at the Bridges site is composed of an igneous rock such as granite, instrusive rock and so on overlaid by a quartenary deposit. Many joints as well as weathering along them exist in the igneous rock.

This paper presents the method of inspection and confirmation of a bearing rock for the anchorage 4A, which is the largest undersea foundation and serves as the centre anchorage for two suspension bridges.

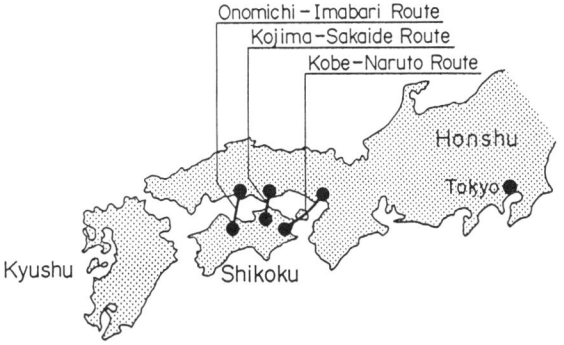

Figure 1. Routes of Honshu-Shikoku Bridge Project

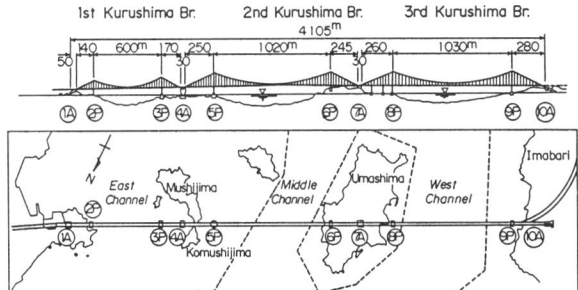

Figure 2. Plan of Kurushima Bridges

2 GENERAL VIEW ON PREGEOLOGICAL SURVEY

The foundation rock at the Kurushima Bridges has been classified into six grades depending upon the design parameters, as shown in Table 1. The classification of rock is based on the detailed survey with borings, laboratory tests on samples, in-situ tests on land and so on.

Table 1. Design parameters of rock grade

Rock grade	E_{sb} (MPa)	Ordinary condition				Earthquake condition		
		ρ_D (KN/m³)	E_s (MPa)	C_s (KPa)	ϕ (Degree)	E_D (MPa)	C_D (KPa)	ϕ (Degree)
D_L	5~30	20	15	20	30	30	30	30
D_M	30~80	21	75	100	35	150	150	35
D_H	80~150	22	150	200	37.5	300	300	37.5
C_L	150~300	23	300	300	40	600	500	40
C_M	300~800	23.5	600	500	40	1,200	800	40
C_H	800~1,300	24	1,200	800	42	2,400	1,200	42

The pregeological survey includes offshore borings, lateral loading tests and loggings in bore holes and laboratory tests. In-situ survey was conducted with a small self-elevating platform.

The anchorage 4A is located in the water between Mushijima and Komushijima where the water depth is 10 to 20m and an irregular seabed surface exists there.

Table 2 shows the details of the pregeological survey for the anchorage 4A, and the bedrock is classified into six grades as shown in Figure 3, which includes the assumed cross section of bedrock.

Table 2. Pregeological survey for anchorage 4A

Survey			Quantity
Boring	No. of holes		26
	Total depth		892m
In-situ test	Lateral loading test in bore hole	Monotonic load	223
		Alternating load	7
	Geophysical logging	Penetration	512
		Electrical	506
		Density	505
		S-wave velocity	508
		Borehole deformation	509
Laboratory test	Apparent density & Water content		83 series
	Uniaxial test		13 series
	CD triaxial test		3 series

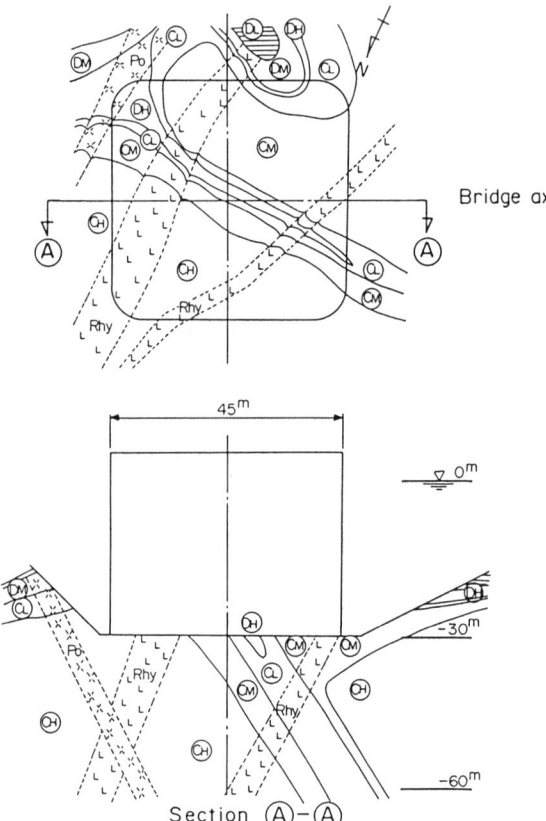

Figure 3. Rock grade at anchorage 4A based on pregeological survey

The bearing bedrock consists of a granite and instrusive rock, which are higher than grade C_L. Although grade C_L first appears at the water depth of 20m, the foundation bottom was set at -30m because of the deposit overlaid on the bearing bedrock.

The granite mainly consists of adamellite with medium and coarse grain size, whereas the instrusive rock has 4~10m thickness of a porphyrite and rhyolite.

In the pregeological survey several borings on the southside found a deteriorated layer($D_L \sim D_H$) continuously running downward from the foundation bottom surface. Based on a pendage and inclination of joints and seams observed in sampled cores as well as the survey on land it was assumed that the deteriorated layer might exist below the foundation bottom. Therefore much effort was made to confirm the degree of deterioration and area of this layer.

3 SURVEY AND CONFIRMATION OF FOUNDATION ROCK

The survey and confirmation of foundation rock has been carried out by classifying a rock based on the data obtained at each excavation stage and by checking it with the grade previously determined from the pregeological survey.

When any inconsistency exists in the grade of a rock, a detailed survey by inspecting data supplied from divers has been conducted to make the final decision on it. Figure 4 indicates the procedure of the survey and confirmation method employed in this project.

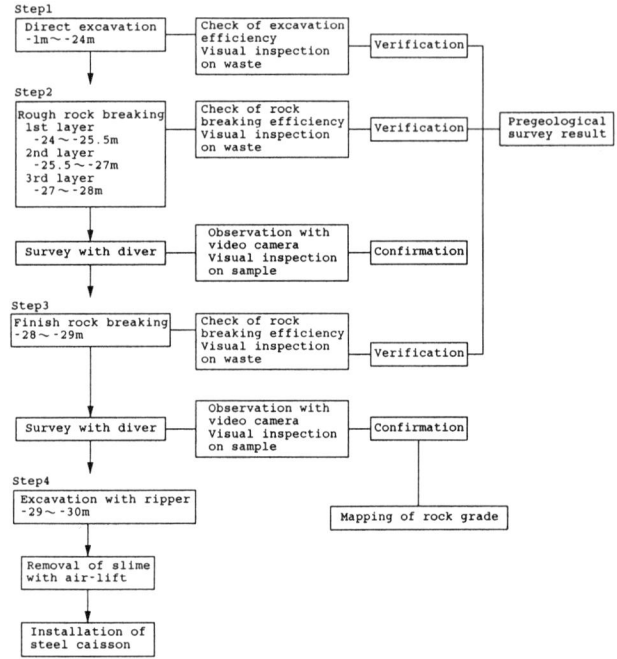

Figure 4. Steps in excavation and associated survey

3.1 Method of excavation

The undersea foundation of the Kurushima Bridges employs the laying-down caisson method, in which the rock is excavated by a grab bucket together with a rock breaking rod and the bottom surface is levelled by ripping with an ultra heavy grab bucket.

The method and procedure of undersea excavation is described as follows.

Step 1 (Direct excavation):
Excavation of deposit and grade C_L with grab bucket in 2m thickness.

Step 2 (Rough rock breaking):
Ripping grade C_M with free falling of rock breaking rod at each 1.5m depth and removal of waste with grab bucket.

Step 3 (Finish rock breaking):
Same as Step 2, except reduction in impact energy of breaking rod by limiting excavation depth and falling height because of minimisation of effect on the foundation rock.

Step 4 (Excavation with ripper):
Surface levelling with claws of the ultra heavy grab bucket from 1m above the bottom to the final depth, removal of waste with flat bucket and check of the elevation of excavated surface. Finish of bedrock surface by repeating the above operation.

Information based construction has been introduced to controlling surface roughness, location of excavation with grab bucket, falling height of rock breaking rod and depth of excavation.

Slime is removed by an air-lift after the excavation and the installation of steel caisson is followed.

440

3.2 Classification of rock based on construction data

In the direct excavation of Step 1 the excavation efficiency was introduced. This efficiency is the ratio of excavated soil volume to the nominal volume of bucket.

$$\text{Excavation efficiency} = \frac{\text{Excavated soil volume}(m^3)}{\text{Nominal volume of bucket}(m^3)}$$

In Steps 2 and 3 a rock breaking efficiency was mobilised. This efficiency means the falling height of rock breaking rod for each 10cm penetration depth based on the sum of falling height and penetration depth for each stratum.

Rock breaking efficiency

$$= \frac{\text{Sum of falling height}(m)}{\text{Sum of penetration depth}(m)/\text{penetration depth } 0.1m}$$

The correlation between these efficiencies and the rock classification is verified by a visual inspection of waste and laboratory tests.
  The visual inspection was conducted by following the rock classification based on the visual observation on cores obtained from the pregeological survey.
  The inspection classifies a waste into an appropriate grade depending upon the type of rock, hardness and degree of joint and weathering.
  The laboratory tests include apparent density (ρc) and effective porosity (Nf) which were conducted in numerous times at the pregeological survey and are closely related to the degree of weathering. The same samples are used in both visual inspection and laboratory tests.

3.3 Verification of rock classification

When any inconsistence in rock classification might be found between the construction data and pregeological survey, the detailed survey was mobilised by underwater observation with video camera and core sampling. The procedure of this survey is as follows. The rock surface was exposed by removal of waste at the bottom and a certain area (approx. 10m²) was enclosed with a rope. In each area a visual observation was carried out to check rock type, degree of joint and weathering and hardness by picking with a hammer. A high quality and watertight colour TV set with a wire line was available in this observation. In addition core sampling was also conducted for laboratory tests.
  Laboratory tests as well as visual inspection in air were employed for core samples. Rock classification was verified through the above survey.

3.4 Result of rock survey

Figure 5 indicates the distribution of rock breaking efficiency. Figures 6 and 7 show rock classification based on the visual inspection on waste during construction and the observation in the detailed survey respectively. Both of them also include the previous rock classification derived from the pregeological survey.
  When any difference in the rock grade might exist in between the visual observation and laboratory tests, priority is given to the visual observation. Grade C_H is included in grade C_M, because both of them are applicable to a bearing stratum.
  The rock breaking efficiency and visual observation demonstrated that the area with rock breaking efficiency of 2m or more was met with that of grade C_L and higher, which was considered as a bearing stratum.

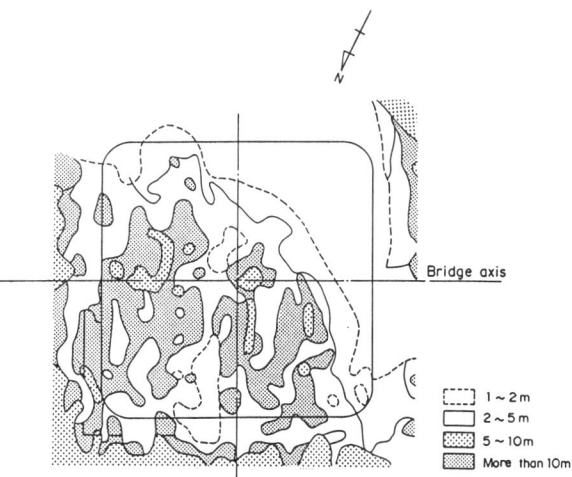

Figure 5. Rock breaking efficiency at-29m

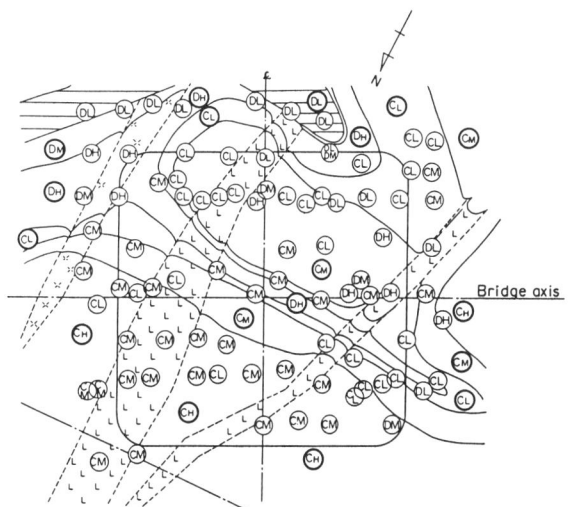

[Remarks]
1. Rock grade with double circles is based on the pregeological survey.
2. Grade C_H is included in grade C_M.

Figure 6. Rock grade based on visual inspection on waste at-29m during construction

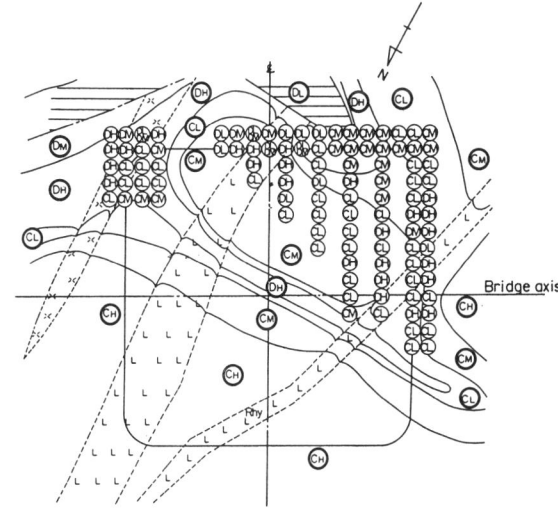

[Remarks]
1. Rock grade with double circles is based on the pregeological survey.
2. Grade C_H is included in grade C_M.

Figure 7. Rock grade based on visual observation on waste at-29m in detailed survey

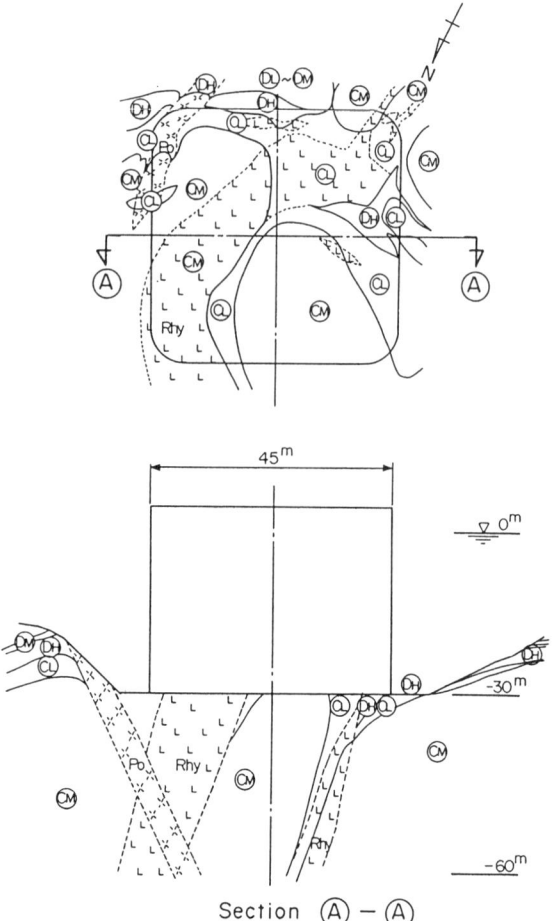

Figure 8. Rock grade after detailed survey

In accordance with the distribution of rock breaking efficiency and rock classification based on the observation on waste, it was found that the deteriorated area running in the east and west direction in the centre of the foundation was no-existent or rather small scale, and a lower grade rock than that predicted at the pregeological survey appeared at the southside of the foundation.

These findings were verified by the fact that the detailed survey through the video camera and visual observation on core samples found a seam clay in the southside.

Figure 8 shows the grade of rock after the detailed survey. In comparison with the rock grade at the pregeological survey the final one does not include a deteriorated continuous band of grade D_H but a lower grade in the southside.

The above modification in the rock grade seems reasonable provided the deteriorated band has the inclination in the opposite direction.

Table 3 shows the area ratio of each rock grade.

Although the pregeological survey overestimated rock grade because of reduction and increase in area of grade C_M and C_L respectively in the final survey, the total area ratio of grade C_L and higher leading to bearing strata was approximately same in the both surveys.

Table 3. Area ratio of each rock grade at foundation

Rock grade	D_M	D_H	C_L	$C_M \sim C_H$
E_s (MPa)	75	150	300	600~1,200
Area ratio based on pregeological survey (%)	1.8	8.0	11.9	78.3
Area ratio after excavation (%)	1.1	6.8	35.0	57.1

4 CONCLUSIONS

As the Kurushima Bridges employed the undersea rock excavation without blasting and the information based construction, reliable information and data were available with the progress of excavation and they led to easy confirmation on the execution from the early stage of construction.

By means of the above excavation method it was possible to minimise areas confirmed by a diving operation and to conduct surveys without any effect on the excavation schedule. As the efficiency of data obtained from the construction was confirmed in the early stage, the survey thereafter was more effective.

REFERENCES

Yamada, K. & Fukunaga, S. 1991. Design properties of bedrock under Kurushima Bridges foundation. Honshi Technical Report Vol.15 No.58

Yamagata, M., Hirayama, J., Hasegawa, Y. & Harada, M. 1992. The underwater investigation of foundation bedrock for Kurushima Bridges of Honshu-Shikoku Bridge Works. Tsuchi-to-Kiso Vol.40 No.11

Fukunaga, S., Yamada, K. & Miyajima, K. 1992. Bridge foundation rock evaluation at design stage and excavation results. The 27th Japan National Conference on Soil Mechanics and Foundation Engineering.

Application of SMA splitter to the breaking of cast-in-place concrete pile

Tsutomu Inaba
Nishimatsu Construction Co., Ltd, Tokyo, Japan

Minoru Niishida & Katsuhiko Kaneko
Kumamoto University, Japan

Kiyoshi Yamanouchi
Tokin Co., Sendai, Japan

ABSTRACT: A new kind of rock splitter has been developed and applied to the breaking work of construcion engineering. The SMA splitter consists of rods of TiNi shape memory alloy, a heating apparatus and a pair of two layered platens made of steel. The breaking process is as follows. The splitter is inserted into a borehole and clearance between the borehole wall and the splitter is adjusted. After this operation, by heating the TiNi alloys, they begin to lengthen until their original length. During this process, a recovery force is generated and the concrete or rock around the borehole is broken. The performance of the developed splitter has been examined by laboratory test and numerical stress analysis.

A field experiment for the application of SMA splitter to the breaking of cast-in-place concrete pile has been performed. The extra placing concrete of the pile has been easily splitted without loud noise. From this result, it has been confirmed that the present splitter is not lacking in terms of performance and is able to control the direction of crack propagation. In addition, the breaking work of the pile is completed safely and silently by using the splitter. We thus believe that the developed splitter has great promices as means for breaking of concrete and rock in the urban district.

1 INTRODUCTION

Various static breaking systems of rock and rock-like materials have been presented from the prevention of noise and vibration pollutions in the urban district in a past decade. Those are the utilization of fluid pressure cell, expansive agents, wire saw with diamond abrasive, water jet etc. (FUKUDA et al.,1987; IIHOSHI et al.,1987; NOMA et al., 1990), which good neither one thing nor the other. For example, the expansive agents is time-consuming and difficult to apply to hard rock because its expansion is attributable to the chemical reactions. The use of fluid pressure cell is required the complex procedure and limited in narrow working space because of lots of accessory equipments.

It has been well-known that the maximum tensile strength of rock and rocklike materials in the engineering fields is about 20MPa. On the other hand, the maximum recovery stress of deformed TiNi shape memory alloys has been recognized about 700MPa(Jackson et al.,1972).

In the present study, the application of TiNi alloys to the pressure source of the static rock breaker is proposed. First of all, the principle for designation of the rock breaker is described based on the material characteristics of TiNi alloy and the numerical analysis using Displacement Discontinuity Method (DDM). Subsequently, the results of laboratory experiments using granite and concrete specimens and the results of a field test and a test in a room of cast-in-place reinforced concrete pile in order to verify the breaking capacity are described.

2 CONSTRUCTION AND PERFORMANCE OF BREAKER

An example of trial products of the breaker is shown in Figure 1, which consists of 6 pieces of TiNi rod and a pair of two-layered wedge type plantens of steel (KANEKO and NISHIDA, 1989). The other products consists of 3 and 9 pieces of TiNi rod. For convenience, each breaker represents hereinafter B3, B6 and B9 corresponding to the number of TiNi elements. All breakers are designed for borehole of 45 mm in diameter and the length of B3, B6 and B9 is 10,15 and 20cm, respectively.

In addition to these, heating apparatus is required in practical operation as described below. The breaker is inserted into borehole of rock mass and then distance between the breaker and borehole wall is adjusted by sliding the inner platen. Subsequently, compressively prestrained TiNi elements are heated and they begin to lengthen until their original length associated with the reverse transformation of martensitic phase. During this, the TiNi element generates the recovery force and rock around the borehole is broken.

It is apparent that the breaking capacity of the present breaker dpends on the recovery force of TiNi element, which intimately relates to alloy composition and heat treatment. To determine the suitable TiNi element for the breaker, the recovery force and displacement of various TiNi alloys were investigated (NISHIDA et al., 1989). This experiment is important for practical use to determine recovery temperature range of TiNi element.

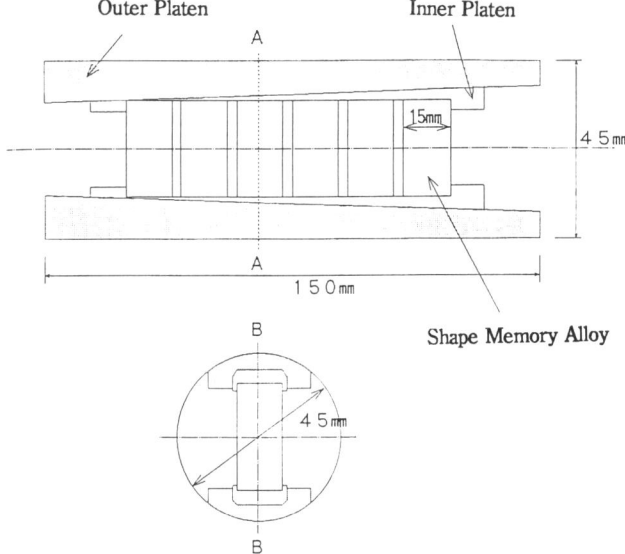

Figure 1(a). Section of B6-Type Breaker.

Figure 1(b). Overall view of B6-Type Breaker.

However, there has been no experiment of the shape memory behaviors under compressive stress so far as we know. All the specimens were 15mm in diameter and 29mm in length and the experiment was carried out in a MTS materials testing machine and a handmade thermostat chamber. Figure 2(a) shows an example load-displacement curve for TiNi element. The specimen is compressively loaded and unloaded as indicated by single-arrowed line in Figure 2(a) in the chamber at room temperature. After the recovery force reaches maximum by gradual heating up to 80°C, the amount of shape recovery

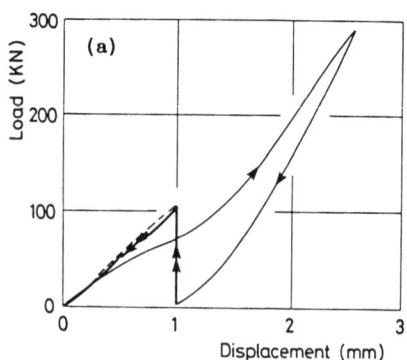
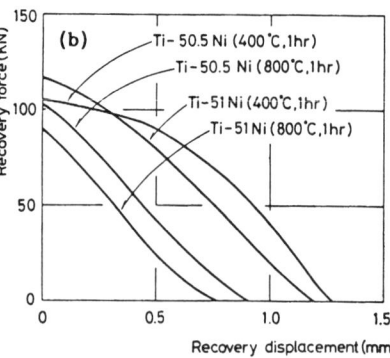
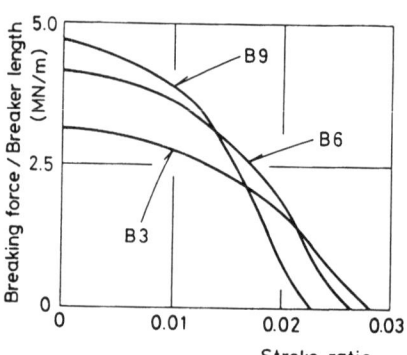

Figure 2(a). Typical example of load − displacement and recovery force − recovery displacement of TiNi alloy. (b)Effects of alloy compositions and heat treatment conditions on the recovery behavior of TiNi alloys.

Figure 3. Relation between breaking force and stroke ratio of ratio of trial products.

increase with decreasing the forces as indicated by double arrowed line in Figure 2(a). Figure 2(b) shows recovery force vs. recovery displacement of TiNi elements with different compositions and heat treatments. The element with high recovery force, large recovery displacement and long duration of recovery force is favorable for pressure source of the breaker.

From this view points, Ti−50.5 at%Ni aged at 400°C for 1 hour is the most suitable for our purpose at present. Recovery temperature range of this element is about 45 to 70°C. Figure 3 shows the breaking force and stroke ratio μ_0/R of each breaker using the element as mentioned above, where μ_0 is recovery displacement of the element and R is radius of borehole. The maximum breaking force per unit length of each breaker is 3 to 5 MN/m. The breaking force of B3, B6 and B9 is about 300kN, 500kN and 900kN, is maximum, respectively.

It is realized that the breakers have high loading capacity. The stroke ratio of each breaker attains 2.3 to 2.8%. This value is considered to be somewhat smaller but comparable with cubical expansion ratio of expansive agents. Lack of the stroke ratio can be easily overcame, even if the borehole wall surface is rough or the axis of the borehole deviates slightly in practical site, simply by sliding the inner platen.

3 NUMERICAL EVALUATION OF BREAKING FORCE

The breaking force of the present breaker depends on not only the strength but also the stiffness of rock mass since shape memory alloy is utilized as the solid pressure source of the breaker. To evaluate and develop the breaking force, it is important to understand the crack propagation process and the deformation state of the borehole wall (INABA et al., 1990). The numerical anlysis has been carried out based on Displacement Discontinuity Method (DDM)(CROUCH and STARFIELD, 1983). Analytical models are shown in Figure 4(a) and 4(b). The former is for the loading of single borehole and the latter is for the simultaneous loading of two boreholes. In both models the borehole wall and the crack are expressed by DD−element and the number of partition was 100 to 300 for (a) and 200 to 600 for (b). As the boundary condition, the traction is distributed in the area corresponding to the platen of the breaker. That is $\beta = \pi/4$. The crack length in the former model postulated an identical in both sides of the borehole and then an analysis performed on the barious crack length. Successive calculations performed on the propagation of the crack with larger stress intensity factor for the latter model in Figure 4(b). the stress intensity factor is estimated from opening displacement of the end element of crack (SHIBA et al., 1989). The change of borehole size is expressed with the displacement μ_0 of the top of platen.

In case of the single borehole loading (Figure4(a)), relation between the stress intensity factor K_I of the end of crack, the changing ratio of borehole size μ_0/R and the load per unit length is expressed by using the nondimensional crack length c/R as follows.

$$K_I = P/\sqrt{R} \cdot F^P(c/R), \quad \mu_0 = P/ER \cdot M^P(c/R) \qquad (1)$$

Where E is Young's modulus of rock, and $F^P(c/R)$ and $M^P(c/R)$ are illustrated in Figure 5(a) and 5(b), respectively, from the results of DDM analysis. Consequently, P^P and μ_0^P/R required for the crack length c^P are given by substitution of $K_I = K_{IC}$ to eq.(1).

$$P^P = K_{IC}\sqrt{R} \cdot F^P(c^P/R), \quad \mu_0^P/R = P^P/ER \cdot M^P(c^P/R) \qquad (2)$$

Where K_{IC} is the fracture toughness value of rock. From Figure 5(a)., $F^P(c/R)$ decreases as increasing c/R in the region of c/R<0.4. This is the characteristic tendency of the problem on pressurized hole with crack.

It indicates that the initiation of crack from the borehole wall relates the presence of latent preexisting crack.

When crack initiates at the point A corresponding to the length of latent crack, the crack propagates to the point A' at a stretch. These phenomena were confirmed in the laboratory experiments as described later. Subsequently, in case of the static propagation of crack after crack initiation in the region c/R>0.4, eq.(2) can be applied.

Figure 6 shows the relation between the stress intensity factor and the length of inner crack for various d/R values in Figure 4(b). The stress intensity factor shows minimum and then abruptly increases with increasing c/R. This increment results of the drastic propagation of crack due to the mutual interaction of inner cracks. Therefore, if crack propagates the point at which the stress intensity factor in minimum,

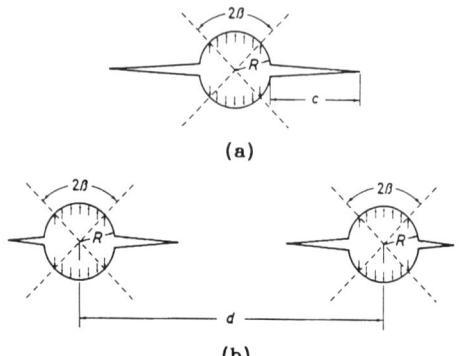

Figure 4. Analytical model for DDM.

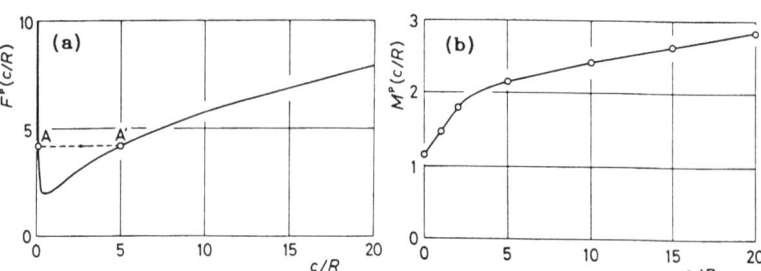

Figure 5. Nondimensional functions $F^P(C/R)$ and $M^P(C/R)$ for crack propagation of the model in Figure 4(a). See text in detail.

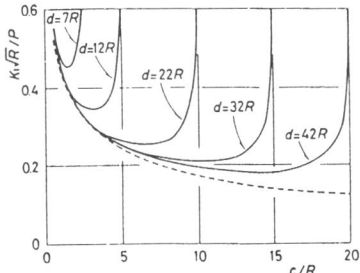

Figure 6. Stress intensity factor of inner cracks for the model in Figure 4(b).

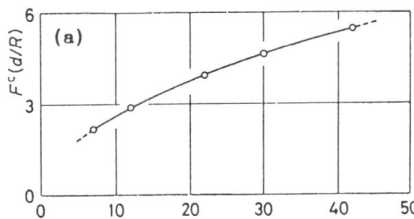
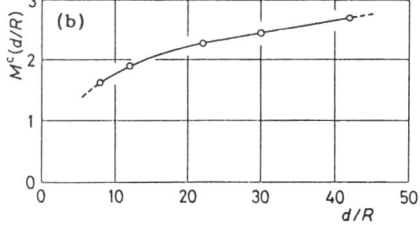

Figure 7. Nondimensional functions F^c (C/R) and M^c (C/R) for crack conection of the model in Figure 4(b). see text in detail.

cracks between two holes are connected as an inevitable consequence. The criteria of crack connection between two boreholes are given as folows

$$P^c = K_I \sqrt{R} \cdot F^c(d/R), \quad \mu_0/R = P^c/ER \cdot M^c(d/R) \quad (3)$$

where $F^c(d/R)$, and $M^c(d/R)$ are shown in Figure 7(a) and (b). For prctical evaluation of the breaking force of the present breaker, materials constants of medium hard, hard and ultra hard rock are given as follows.

Medium hard rock:
    $S_t = 2.5$ MPa, $E = 15$ GPa, $K_{Ic} = 0.5$ MPa$\sqrt{m}$
Hard rock :
    $S_t = 8.0$ MPa, $E = 35$ GPa, $K_{Ic} = 2.0$ MPa$\sqrt{m}$
Ultra hard rock :
    $S_t = 20$ MPa, $E = 60$ GPa, $K_{Ic} = 3.0$ MPa$\sqrt{m}$

where S_t, E and K_{Ic} are tensile strength, Young's modulus and fracture toughness value of the rock, respectively. Substituting these values to eq. (2) for the loading of single borehole, the breaking criteria with the present breaker are shown in Figure 8 by overlapping the breaking capacity in Figure 3. It can be concluded that the present breaker has sufficient breaking force even for the ultra hard rock.

4 LABORATORY EXPERIMENTS

To confirm the breaking capacity of the present breaker and the applicability of breaking criteria in the previous section, laboratory experiments were carried out by using concrete and granite specimens. Experimental conditions, materials constants, specimen size etc., are listed in Table 1. Heating apparatus used consisted of steel cover block and cartridge type electric heater of 110VA in maximum capacity. It has been confirmed that this apparatus can supply nearly constant heating rate to the TiNi elements in the preliminary experiment. The system composition of the breaker used to a series of the experiments is compact as shown in Table 2.

The experiment A is concerned with the breaking of thin wall. The specimen thickness is equal or less than the breaker length. Since the breaker is set the whole length of borehole, two dimensional condition is satisfied. In this case, the loading capacity of the breaker largely overcomes the estimated breaking load of the specimen as listed in Table 1. All the specimens were broken within 1 minute. The breaking completed before the reversion to the original length of the TiNi elements.

This indicates the stroke ratio of the breaker is sufficient as same as the loading capacity due to the wedge effect of two layered platens as explained above. In all the experiments performed in this study, the crack initiates at the edge of platen and propagates along the vertical direction of loading axis as shown in Figure 9.

In the experiment B the borehole length i.e., the specimen thickness, is larger than that of the breaker. The purpose of this experiment is to examine the three dimensional effect in thicker direction of the specimen.

The loading capacity of the breaker through the experiments B to D listed in Table 1 is divided the total loading capacity of the breaker by the borehole length. The specimens B−1 and B−3 were completely broken within 120 seconds. In case of the specimen B−2, the crack partially reached the bottom of the specimen but the breaking was not made completely. It is likely that the breaker was placed at the upper part of 10cm depth for the borehole length of 50cm. Therefore, it is necessary to put the breaker at deeper position of borehole for thicker specimen.

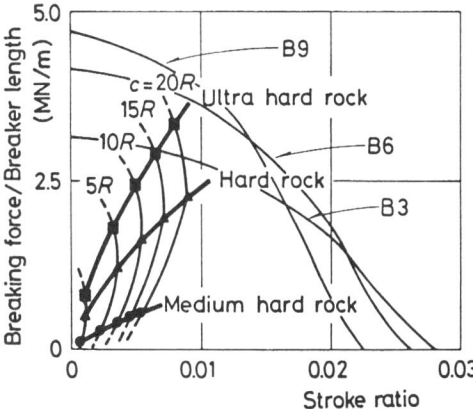

Figure 8. Comparison between breaking criteria and breaking capacity of trial products.

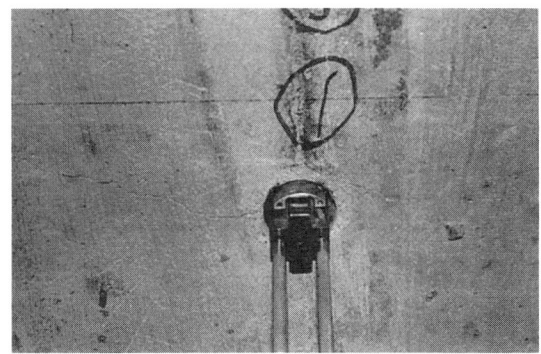

Figure 9. Feature of crack propagation.

The specimen used in the experiment C is granite. The loading capacity of the breaker used and the postulated breaking load of the specimen is almost the same but the specimen was broken completely about 160 seconds. The crack propagated at a moment. That is, the dynamical crack propagation effect is remarkable in the rock with high strength and brittleness.

The experiment D is an example of the partition of a mass of concrete to small pieces. The specimen was partitioned in the order of D−1 to D−3. Plural boreholes were drilled into the specimen D−1 and D−2 based on the breaking criteria in the previous section. For example, two boreholes with same horizontal level were drilled with the distance of 35cm is 4 vertical planes. The ratio of the borehole diameter and the distance between two boreholes was 7.8, which is the same as the criteria of expansive agents in practical use. The crack propagated through the postulated path and the specimen was completely divided. Figure 10 shows a feature of the specimen D−1 after breaking. The necessary time were 180 seconds for D−1 and within 60 seconds for D−2 and D−3. It is suggested that the necessary time for the partition in the experiment D is 10 minute at most, exclusive of drilling time. It is concluded that the rapid breaking can be made by using the

Table 1 Experimental conditions of laboratory tests for concrete and granite specimens.

Experiment	Material	Specimen	Size ,cm	Number of hole	Type of Breaker (Number of TiNi elements)	Loading capacity of used breaker	Posturated breaking load of specimen
A	Concrete I	A-1 A-2 A-3	30x30x10 50x50x10 50x50x10	1 1 1	B3(3) B6(6) B6(6)	3.1 6.3 3.1	0.25 0.35 0.35
B	Concrete I	B-1 B-2 B-3	30x30x30 50x50x50 50x30x30	1 1 1	B3(3) B6(6) B6(3)x2	1.0 1.3 2.1	0.25 0.35 0.25
C	Granite	C-1	30x30x30	1	B6(3)	1.0	0.99
D	Concrete II	D-1 D-2 D-3	95x95x74 95x95x37 95x37x37	2x4 2 1	B6(3)x8 B6(3)x2 B6(3)	0.66 0.85 0.85	0.50 0.50 0.25

Concrete I: S_t=2.5MPa, E=15GPa, K_{1c}=0.5MPa$\sqrt{m}$; Grnaite; S_t=10MPa, K_{1c}=2.0MPa$\sqrt{m}$
Concrete II: S_t=4.0MPa, E=25GPa, K_{1c}=1.0MPa$\sqrt{m}$

Table 2 System composition of breaker

Shape memory alloy	ϕ15 × L29mm Recovery power 10tf/piece Compression deformation: less than 20°C Recovery: More than 50°C
Loading plate	L=150mm, clearance adjustment by spacer
Heater block	40 × 20 × 135mm Use 2 cartridge heaters
Heater	ϕ6.3 × L250mm (heater part L48mm)
Power source unit	200V - 400W AC200V allowable current 40A Heater terminal 10 sets, with an energizing time setting timer.
Press	1500 × 800 × 500mm Max. 50t, common use 30t, constant compression auto releasing.

Figure 11. Insitu breaking of andesite.

Figure 12. Insitu breaking of sandstone.

Figure 10. Partition of a mass of concrete specimen D-1.

present breaker. In practical partition of a mass of rock and rocklike materials, the stepwide work described above can be shortened at a time by operation difference in time of heating apparatus.

5 FIELD EXPERIMENT

5.1 Application to Breaking of rock

For rock, a cobbing break experiment of the boulder of granite, sand stone or andesite was conducted also in the field of each area. Its result could break as the basic experiment for the energizing time of 2 or 3 minutes as shown in Figure 11,12. The Figure 11,12 is a case of andesite and sanstone, respectively. And a jack hammer was used for the breaking.

5.2 Application to breaking of cast-in-place concrete pile

In the work of cast-in-place concrete pile, the upper part of the pile must be removed to arrange in a certain form. Usually, the extra placing concrete is broken by hydropressure breaker. However, the breaking work becomes greater problem in urban district, because it generates loud noise. Therefore SMA splitter has been applied to the breaking of the pile.

As the object of an experiment, a case-in-place reinforced concrete diaphram wall was used as a pile (hereafter called a diaphram pile) and the wall thick was 1040mm, the main reinforcing bar(D25) pitch was 150 mm, the design strength of concrete was 24MPa, and a method to break and remove the extra placing part in a block shape was adopted.

To remove in a block shape, a preparatory work to separate adhesion and break in a vertical and horizontal direction is required. A feature of the diaphram pile and the preparatory work are shown in Figure 14.

The breaker for separate adhesion in a horizontal direction was used to cause some crack of 1m in the depth (wall thick). But because a prototype breaker is impossible to 1m of the depth, a new breaker was manufactured with a tier of three alloys and about three mm extension. The alloy and the new breaker are shown in Figure 13. A bore hole for insertion of breaker was required with about 100mm in diameter and rigid polyvinyl chloride pipe (107.8mm in diameter, hereafter called PVC pipe) was previously installed to the reinforcing cage before a concrete

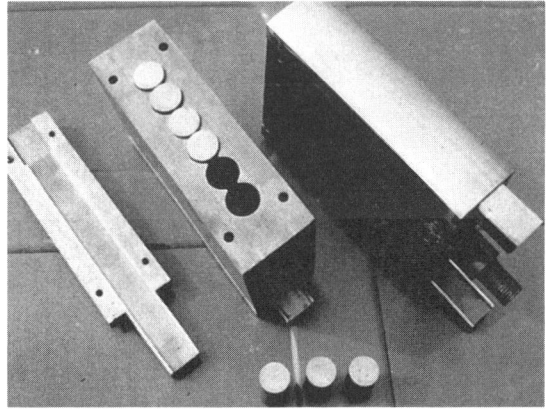

Figure 13. Shape memory alloys and new breaker.

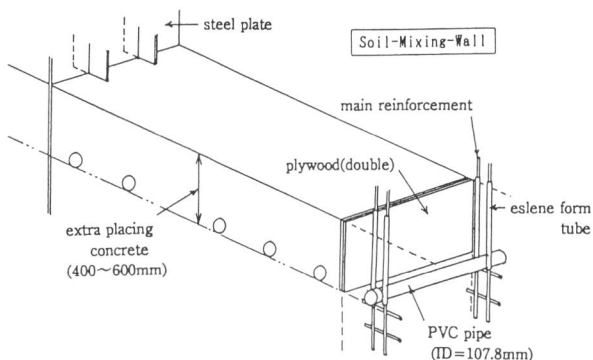

Figure 14. Feature of the diaphragm wall and the preparatory work.

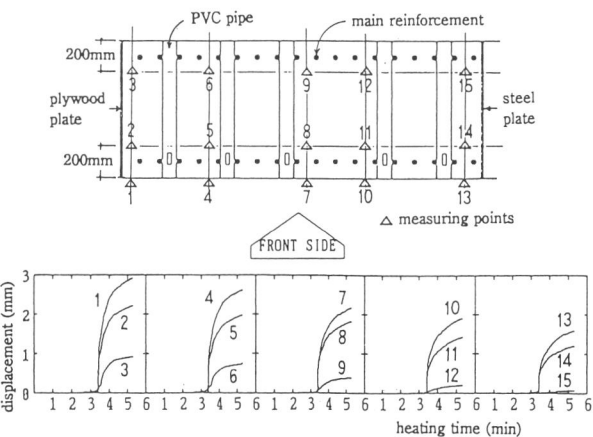

Figure 16. Measuring points and the results.

Figure 17. Situation of splitting and measuring.

placing in order to omit a boring work. The setting pitch of PVC pipe was taken as 750mm at the inserting position of a tremie pipe for concrete placing and 450mm at a ordinary part and 300mm at the corner parts. Figure 15 shows the state of setting a breaker in a PVC pipe.

The separation of adhesion in a vertical direction was provided with a partition steel plate used in a diaphram pile execution and two plywood plates in an intermediate position of an element(1m length) that is excuted at once. An eslene form tube was installed to the main reinforcing bar to eliminate the adhesion between the concrete and the main reinforcing bar.

As a result of the experiment with actual size specimens prior to a real execution, it was confirmed that crack progressed to 1 meter of the depth though the PVC pipe was removed, and so also in a real execution, a breaking work was performed by insertion of a breaker without drawing out of the PVC pipe.

A development of crack in a depth direction was verified by measuring the rising amount(displacement) of the crest of the pile for a few blocks. Example of the measured position and measured results are shown in Figure 16. Figure 17 shows the situation of splitting and measuring.

Although some difference is found in the rising amount by an influence of the adhesion of the pile back surface with the edge planing material in a

Figure 15. State of setting a breaker in a PVC pipe.

Figure 18. Situation after breaking.

Figure 19. Situation under lifting.

vertical direction, it is found out that crack generates in three minutes after the heating of alloy and develops to 1 meter in the depth in one breath.

After the breaking in a horizontal direction, the concrete pile was jacked up by a small hydraulic jack and then withdrawn through lifting by a heavy machine or broken by a concrete breaker. Figure 18, 19 show the situation after the jacking up of 10mm and under lifting, respectively.

6 Conclusion

The principle and structure of the static rock breaker using TiNi shape memory alloy were presented, and numerical and experimental investigations were performed on the breaking capacity of the breaker.

It has been confirmed that the trial products had extremely high loading capacity as static rock breaker. The proposed breaker was able to break a mass of concrete and granite in a short period.

In addition, the size of the present breaker was very compact and the accessory equipment was not required in comparison with the conventinal static rock breaker, except for the heating apparatus. It can be concluded that proposed breaker is a promising one for static breaker process of rocks and rocklike materials.

And so a wide range of its application can be expected, for example, rock foundation etc. This breaker is scheduled to add more improvement for practical use.

REFERENCES

CHIBA, T., K. KANEKO, Y. OBARA and K. SUGAWARA, 1989.
 Applicability of displacement Discontinuity Method to crack problems, J. of Mining and Materials Processing Institute of Japan, 105, 13, 981 -986.
CROUCH, S.L., and A.M. STARFIELD 1983.
 Boundary element method in solid mechanics, George Allen & Unwin Ltd., (London).
FUKUDA, K., H. KUMASAKA, Y. OHARA and ISHIJIMA, 1987.
 Proc. 6th Cong. ISRM, 1, 625-628.
IIHOSHI, S., K. NAKAO, K. TORII and T. ISHII, 1987.
 Proc. 6th Cong. ISRM, 1, 659-662.
INABA, T., K. KANEKO, M. NISHIDA, A. HIRATA, K. ISHIYAMA and K. YAMAUCHI, 1990. Study on static rock breaker using shape memory alloy, Proc. 8th Japan Symp. on Rock Mechanics, 175-180.
JACKSON, C.M., C.J. WAGNER and R.J. WASILEEWSKI, 1972.
 55-NITINOL-The alloy with a memory: its phgysical metallurgy, property, and applications, NASA-SP5110.
KANEKO, K., and M. NISHIDA, 1989.
 Static rock breaker using shape memory alloy, J Science & technology in Japan, 8, 31, 52-54.
NISHIDA, M., K. KANEKO, T. INABA, A. HIRATA and K. YAMAUCHI, 1988.
 Static rock breaker using TiNi shape memory alloy, Proc. 6the Int. Conf. on Martensitic Transformation, Sydney, 2, 711-716.
NOMA. T., S. KADOTA, H. MURAYAMA, S. UEDA, M. KONDO and K. TATSUNAMI, 1990.
 Development of static fracturing method of rock mass by using hydraulic pressure, Proc. 8th Japan Sym. on Rock Mechanics, 181-186.
T. INABA, K. KANEKO, M. NISHIDA and K. YAMAUCHI, 1991
 Static Rock Breaker using Shape Memory Alloy, Proc. of Int. Congress on Rock Mechanics in Aachen

Assessment of structural stability through measurements

H. Maleki
Spokane Research Center, US Bureau of Mines, Wash., USA

K. Hollberg
Tg Soda Ash, Granger, Wyo., USA

ABSTRACT: As part of its mission to improve mine safety, the U.S. Bureau of Mines implemented an extensive research program to characterize the sequence of damage to underground mine structures. Both static and geophysical measurements were obtained in the laboratory and in a western U.S. mine to study the in situ load-deformation behavior of floor material and to identify the location, timing, and growth of damage near mine excavations. This work was supplemented with three-dimensional numerical modeling to help explain the cause of damage.

It was shown that the stress field influenced the degree of damage to roof rock during excavation. Roof deformation and the extent of the damage zone grew rapidly as a result of failure in the mine floor. This failure was associated with pillar unloading and pillar shifting toward the abutment block, which increased compressive stresses in the mine roof near the pillar line. Laboratory and field measurements complemented each other, providing critical input for stability assessments and numerical models.

1 INTRODUCTION

As part of its mission to improve safety in U.S. mines, the U.S. Bureau of Mines (USBM) implemented an extensive measurement program to develop techniques for assessing changes in roof stability. Initial activities in a western U.S. trona mine utilized both static and tomographic techniques to develop a field method for assessing roof stability. These measurements (Maleki et al., 1993) also identified the interaction among roof, pillar, and floor structural elements and related the deterioration in roof conditions to premature failure of the mine foundation (floor).

To improve understanding of floor behavior, a laboratory and field measurement program was recently implemented at the same mine in an attempt to characterize in situ material behavior of the mine floor. These measurements were supplemented by both static and tomographic measurements in the mine roof and pillar and numerical modeling to better define the interaction among floor, pillar, and roof movements and the resultant stability problems.

In this paper, following a review of the mine's geological setting, foundation measurements (direct-shear tests, uniaxial compressive strength tests, in situ borehole shear tests, in situ plate-bearing tests, and tomographic imaging) and other measurements in the mine roof and pillar will be presented. Numerical modeling results will be used to verify the sequence and growth of damage in mine structures.

2 GEOLOGICAL SETTING AND MEASUREMENT PROGRAM

The field measurements were obtained in a western U.S. trona mine located in the Green River basin, Wyoming. The mine uses continuous miners and shuttle cars to develop a four-entry panel access with additional extraction on one side of the panel during retreat mining. Entries were 5 m wide, leaving 5-m-wide, 30-m-long pillars in place to support overburden.

The mineable seam is 3.3 m high and is located 426 m below the surface. The immediate roof consists of 1 m of trona and is overlain by a sequence of marlstone, trona, and shortite inclusions (fig. 1). Thin clay layers are present in both the seam and the roof. The immediate floor consists of marlstone with variable amounts of shortite crystalline inclusions.

Trona is the stiffest member of the strata. Young's modulus for trona is 17.24 GPa, six times higher than that of the marlstones in the mine floor. Specific gravity ranges from 2 to 2.4, and moisture content is 28 pct on the average.

The in situ stress field in the basin was measured using overcoring techniques and USBM deformation gauges in three boreholes. Measured stresses were as follows: east = 9.4 MPa, north = 15 MPa, vertical = 19 MPa, east-north shear = 2.8 MPa, north-vertical shear = 0.2 MPa, and east-vertical shear = 0.9 MPa. Both vertical and horizontal stresses were greater than expected from calculations of overburden weight. Similar measurements at the mine have confirmed the orientation of stresses, but magnitudes were 50 pct lower in the mine roof.

Earlier instrumentation projects in the mine (Maleki et al., 1993) identified large amounts of deformation that occurred first in the mine floor, possibly as a result of pillar penetration. However, there were not sufficient data to confirm the rock mass strength and the failure mechanism. In this investigation, an instrumentation program was designed to improve understanding of floor-rock behavior.

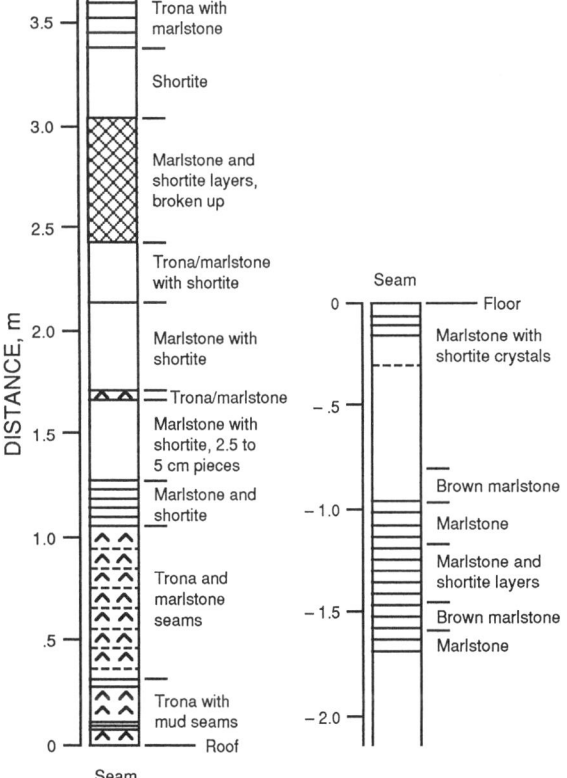

Figure 1. Corehole lithology.

Mining geometry and instrument layout is shown in figure 2. Most of the instruments were installed in room 1 and the adjacent pillar. However, the whole access panel was periodically mapped to record any obvious changes in strata fracturing, floor heave, rib falls, etc. At installation time, the face was 30 m from the cluster of instruments at mid-pillar. The development continued for 500 m before a 100-m-wide retreat face was set up and retreated toward the instruments. The following tasks were completed.
o Tomographic surveys in the mine roof and floor at 45- to 60-day intervals.
o Biweekly deformation measurements in the mine roof, pillar, and floor.
o Mapping and borehole observations and core testing for the mine roof and floor at the beginning of the monitoring program and at other selected times.
o Overcoring stress measurements within the pillar near completion of the monitoring program.
o Borehole shear tests in the mine roof and floor at the beginning of the program.
o Plate-bearing tests at the beginning of the program.

The tomographic layout is shown in figure 2. Accelerometers were attached to the mine roof and floor at 0.6-m spacings along the entry span. These accelerometers had flat frequency responses in the range of 1 to 6,000 Hz. Both an impact unit (Maleki et al., 1993) and blasting caps were used in boreholes as the energy source. They were energized in two source holes at 0.6 to 3.6 m into the roof and floor at 0.6-m spacings. The data acquisition system is described in a paper by Maleki et al. (1992).

Using seismic tomographic techniques (Schneider, 1990), one may construct a velocity or attenuation map of the zone of interest. In this study, a seismic travel-time tomographic survey of the mine roof and floor was implemented along the width of the entry to monitor changes in strata fracturing. Tomography is a mathematical inversion technique for imaging the areal distribution of a physical property of a solid object. Measurements of physical properties on the plane of interest can be solved for measurements obtained at the boundaries. A series expansion technique using an iterative, reweighted, least-squares method was used. Pixel dimension was approximately 50 cm in both the horizontal and vertical directions, which was compatible with the resolution for this study. For ray tracing, minimum time wave-front modeling was utilized (Schneider, 1990). A constant average velocity of the medium was used initially in the iterative process. First P-wave arrivals were picked for the tomographic calculations.

The resolution of the seismic method depends on the predominant wavelength in the signal. A frequency analysis of the recorded signal showed that the predominant frequency of the signal was 2 Hz. Since the seismic velocity of trona is about 4,200 m/s, the predominant wavelength was at best 2.1 m. A practical rule of thumb to determine resolution is that the seismic method should afford the resolution of features with a 1/4-wavelength dimension. The resolution of the system used in this study was at least 50 cm. The sampling frequency was 25 kHz, and the trace length was 40 milliseconds. The accuracy of triggering for this system was 40 microseconds. This time gave a 17-cm error in a medium of 4,200 m/s.

An array of extensometers was installed in the test area to measure relative roof movements. The extensometers were anchored from 1 to 5 m above the roof on 1-m spacings. Two cores were obtained from the roof and floor to measure physical properties. Roof appearance also was mapped to monitor development of fracture patterns.

Seam instruments consisted of vibrating wire stress meters placed in the pillars and abutment block and rib movement pins anchored 0.6, 1.5, 2.1, 2.7, and 3.3 m into the pillar. Relative lateral movements were measured using a tape extensometer attached to a 5-m-deep reference point in the abutment block. These stressmeters and floor movement pins were installed to monitor pillar response and roof-pillar-floor interactions. The pins were anchored diagonally below the pillars at depths of 0.9, 1.5, 2.1, and 3 m into the floor; tape extensometers and the same 5-m-deep reference point were used for these measurements.

USBM-developed deformation gauges and overcoring techniques were used in a horizontal hole to measure vertical stresses in the pillar at 0.5-m spacings; these measurements were critical for determination of residual strength of pillar-floor composite materials. Borehole observations and tomographic pillar surveys were carried out, but are not described in this paper.

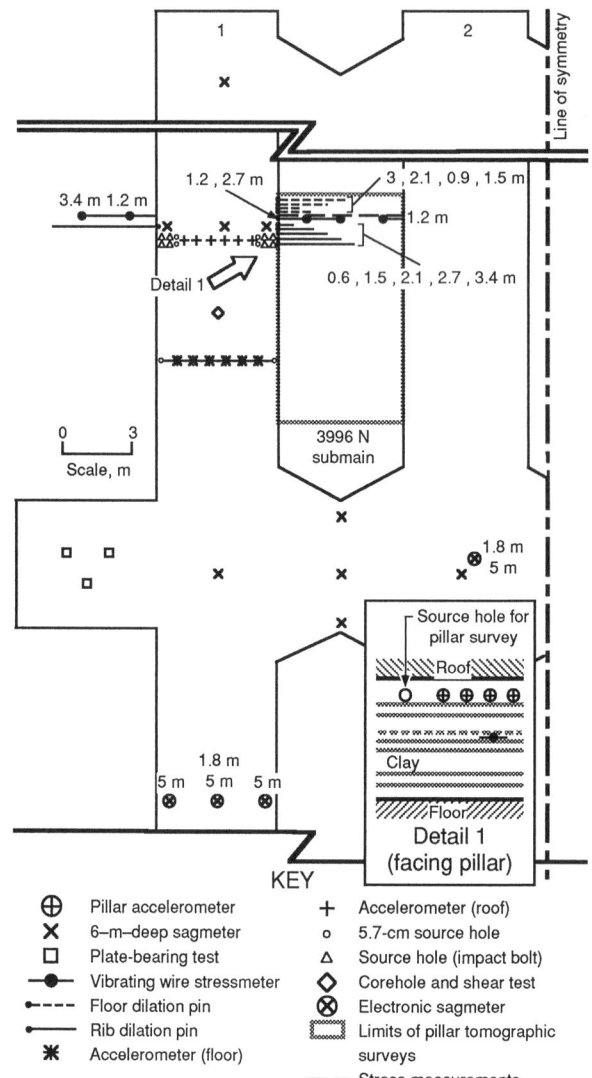

Figure 2. Instrument layout and mining geometry at time of installation.

3 FLOOR MEASUREMENTS

A number of laboratory and in situ measurements were obtained to characterize the load-deformation behavior of the floor strata. Laboratory tests consisted of direct-shear tests, uniaxial compressive strength tests, and sonic tests; these tests were completed on 50 mm in diameter samples obtained from the mine floor at the beginning of the field program. In situ measurements consisted of borehole shear tests, plate-bearing tests, tomographic measurements of velocity patterns, and lateral deformation measurements.

o Laboratory direct-shear tests

These tests were performed on marlstones of the immediate floor, which contained various amounts of shortite crystals. To obtain shear strength parameters (cohesion, angle of friction, and dilation angle), normal load was changed from 0 to 400 kPa within the limitations of the laboratory testing machine. Samples were tested with a controlled shear velocity of 0.5 mm/min.

Most samples deformed 1 mm in shear (0.2 strain) before they reached their peak shear strength. A few samples that contained clay veins, however, yielded to a residual strength

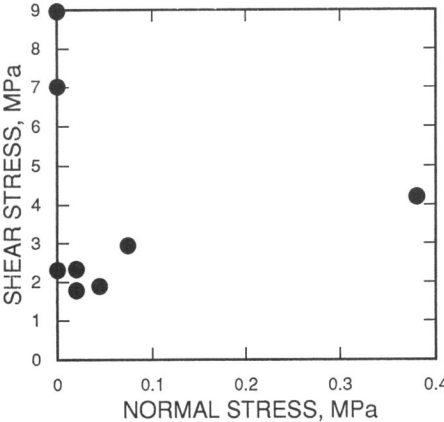

Figure 3. Direct-shear test results.

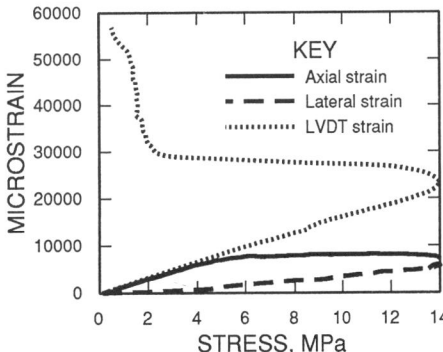

Figure 5. Typical stress-strain relationship for samples from mine floor.

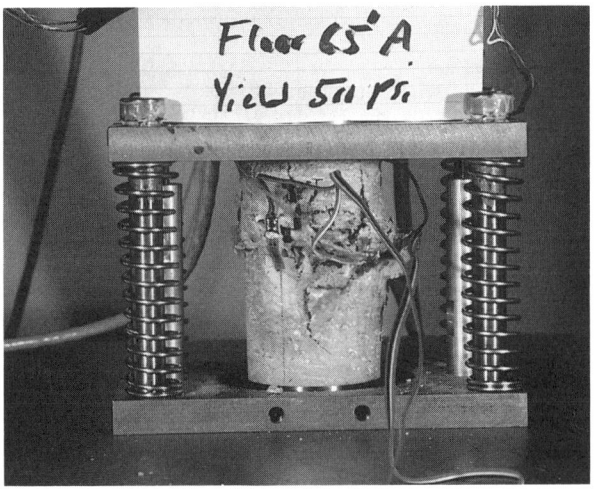

Figure 4. Typical condition of samples during unloading.

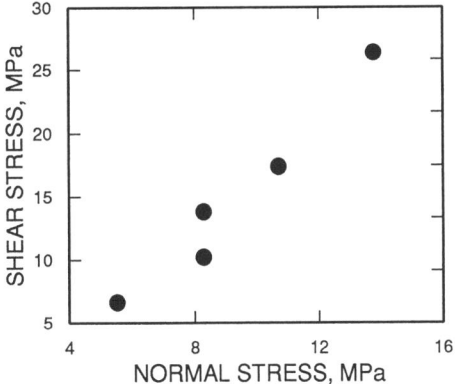

Figure 6. Normal shear stress relationship, borehole shear test.

equal to 84 pct of the peak strength. Most vertical expansion (dilatancy) was reached just prior to peak strength, but dilatancy was reduced significantly during sliding. The dilation angle was influenced by the location of shortite inclusions, causing a significant uplift for some samples prior to failure. Dilation angles of 44° to 0° were calculated during loading and unloading of samples.

Figure 3 presents normal shear-stress relationships for these tests; shear strength is not strongly influenced by normal stress and thus the strength appears to be primarily influenced by cohesion. The cohesion was between 2 to 3 MPa for most tests at low normal stresses.

o Uniaxial compressive strength
Six intact 50- by 100-mm samples were tested in a stiff testing machine to determine full load-deformation behavior of the floor samples. Sample deformation was monitored by axial and lateral strain gauges and two axial linear transducers, which were used to monitor average strain in the post-failure regime when the strain gauges were not functional. Figure 4 shows the testing equipment and sample conditions during unloading.

Typical results (fig. 5) indicated a peak compressive strength of 13.8 MPa. Other tests show peak compressive strengths reaching 18 MPa. Samples rapidly unloaded to a residual strength of 1.5 MPa at 2 to 3 pct strain. Unloading was associated with near-vertical cracks, confirming that strength was controlled primarily by sample cohesion. Young's modulus and Poisson's ratio were ranged between 2.7 to 8 GPa and 0.1, respectively.

o Sonic measurements
The P-wave velocities of core samples were measured in the laboratory prior to destructive tests and results were compared to in situ velocities obtained through tomographic calculations. There was a very good agreement between trona velocity (4,600 m/sec) calculated from the core and in situ. Floor rocks, however, exhibited 24 pct higher velocities in situ than in the laboratory; this result was attributed to one or any combination of the following: (1) stress-induced structural damage during excavation, (2) rock damage during sampling and preparation, and (3) higher confining stresses in situ.

o Borehole shear tests
The Iowa borehole shear test device was used to determine the shear strength parameters of floor rocks. The device consists of two grooved plates that are pushed into the borehole wall under a uniform normal load. Shear failure near the wall is induced by pulling the device out and recording the amount of deformation. The device was used in both intact and fractured zones that corresponded to core logs.

Typical results (fig. 6) for an intact zone provided an upper limit for the shear strength parameters that were in general agreement with laboratory results, i.e., a cohesion of 3.3 MPa and an angle of friction of 21°. Cohesion was significantly lower in the fractured zone.

o Plate-bearing tests
The plate-bearing device consisted of a 100-ton hydraulic jack installed between the roof and the floor, a 152-mm in diameter circular steel bearing plate, and a level for measuring the displacement of the hydraulic piston into the floor. A large plate was used at the roof contact to avoid any penetration into the roof. The tests were conducted in a fresh cut (fig. 2) after all debris was removed.

Two typical load-deformation behaviors were identified (fig. 7). The first was nonlinear behavior that changed into perfectly plastic behavior, and the second was linear behavior that reached peak bearing capacity and then dropped to a residual value at 16 mm of deformation. The peak stress range, for all tests, was from 7.5 to 20 MPa and the residual stress was 3 MPa.

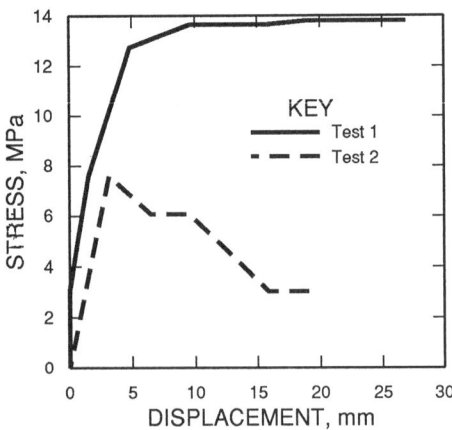

Figure 7. Typical load-deformation behavior, plate-bearing tests.

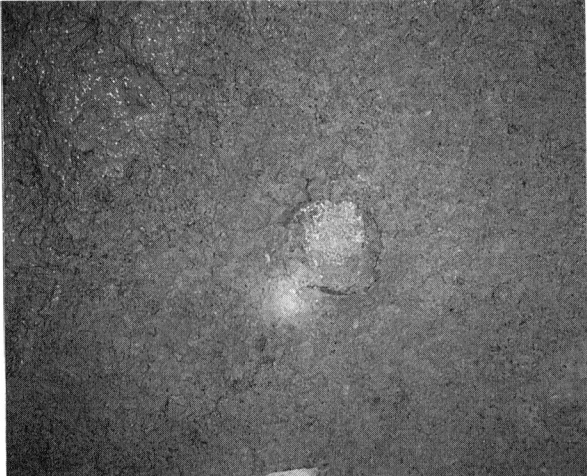

Figure 8. Failure pattern, plate-bearing tests.

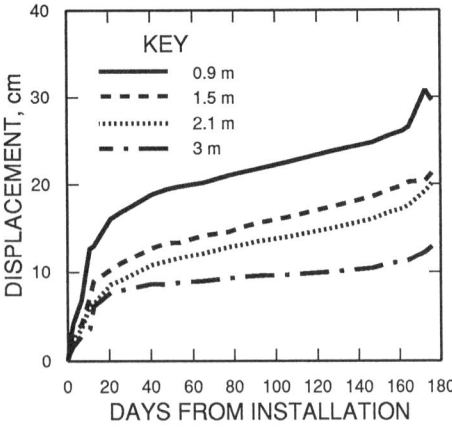

Figure 9. Lateral deformation history of mine floor.

The steel plate initially penetrated into the floor without any observed radial cracks (fig. 8); such behavior is common to porous, brittle material (Ladanyi, 1968) for which the mode of failure is structural collapse and little lateral strain at higher deformations exceeding 15 mm, however, in this test radial cracks appeared around the plate.

o Lateral floor deformation
Four deformation pins were anchored diagonally below the pillar at depths of 0.9, 1.5., 2.1, and 3 m into the floor to measure the amount of lateral movement of the floor rocks toward the excavation (fig. 2). The deepest anchor was approximately 2 m below the excavation horizon.

Three movement cycles were identified (fig. 9). (1) An accelerated cycle that occurred during the first 10 days after mining when pillar stress and penetration were maximum, (2) a steady deformation during which movement continued at a slower rate, and (3) another accelerated cycle beyond day 150 when secondary retreat mining resulted in load transfer to the pillar, which renewed penetration.

Lateral floor deformation exceeded elastic strains (1 pct) significantly, indicating that the failure process was initiated below the entire pillar width shortly after mining. The depth of the failure zone was at least 2 m because the amount of measured deformation significantly exceeded the amount of elastic deformation (1 pct strain) measured in the laboratory.

4 ROOF AND PILLAR MEASUREMENTS

An important part of this study was to assess the impact of floor failure on roof stability. This required obtaining detailed measurements of roof deformation history, changes in wave velocity in the roof, and periodic mapping. In addition, pillar stress and deformation profiles were measured to confirm inelastic deformation of the composite pillar and floor material.

o Roof deformation
Maximum roof relative deformation was measured between the collar of a hole and the deepest anchor (5 m into the roof). Roof movements were greatest in the No. 1 entry and least in the crosscuts.

Relative roof deformation is presented in figure 10 along the span of the No. 1 entry. The movements were always higher near the pillar line than near the center of the entry, indicating rock damage at this location.

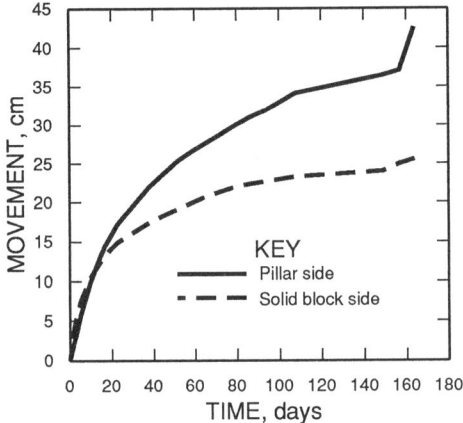

Figure 10. Relative deformation history of mine roof.

o Wave velocity pattern
Wave velocity images of the mine roof at the beginning and end of the monitoring program are presented to show the presence and growth of fracture zones (figs. 11 and 12). Although there was no visual sign of fracturing at the development stage (fig. 11B), wave velocities were lower near the pillar; this finding was in agreement with roof deformation measurements, indicating fracturing and structural damage near the pillar.

Wave velocity in the mine roof changed significantly (25 pct) during this 6-month monitoring period as damage to the rock mass increased because of the growth of fractures and increases in bed separations. Figure 12 illustrates the velocity pattern and ground conditions when the retreating face approached within 15 m (50 ft) of the instruments. As rock fracturing developed toward the pillar side, the roof behaved as a cantilever beam. In time, other fractures formed in the upper portion of the roof (near the solid block), forming a block of rock that was suspended from the upper strata by roof bolts. The operator installed sets of secondary support (wire mesh and 2.4-m-long bolts) and tertiary support (3.6-m-long bolts) to control block movement.

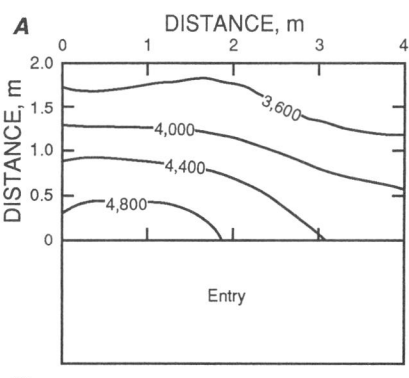

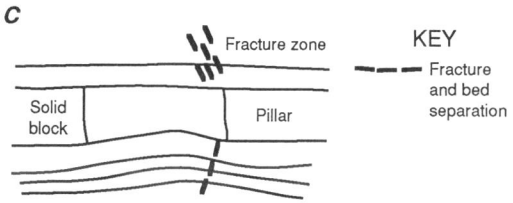

Figure 11. Velocity contours and ground conditions at development face positions. (A) Velocity image (M/A); (B) actual roof condition; (C) schematic of strata movements.

o Lateral pillar movement
The movement of the pillar with respect to the abutment block was measured with a tape extensometer attached between a reference point and anchors located 0.6, 1.5, 2.1, 2.7, and 3.4 m deep into the pillar. Note that the deepest anchor was set 0.5 m beyond the center (mid-width) of the pillar; this made it possible to determine whether the entire pillar was moving or just the ribs.
Although there was some differential movement between the anchors, the deformation pattern (fig. 13) indicated that the whole pillar shifted toward the abutment block by at least 50 mm. This behavior is not typical for an elastic pillar-floor system but can be represented with a Mohr-Columb strain-softening constitutive model for a mine floor, as described later.

o Pillar stress profile
A pillar stress profile was obtained using the overcoring stress measurement method when the retreating face was 50 m from the instrumented pillar. Ten overcores were obtained and tested in the biaxial chamber to obtain Young's modulus, which was used for calculation of the vertical and horizontal stresses.
Biaxial tests revealed a significantly lower Young's modulus for the first 1 m of the pillar than for the core of the pillar, indicating structural damage; this finding was in agree-

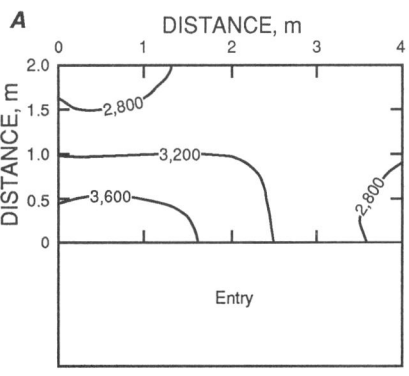

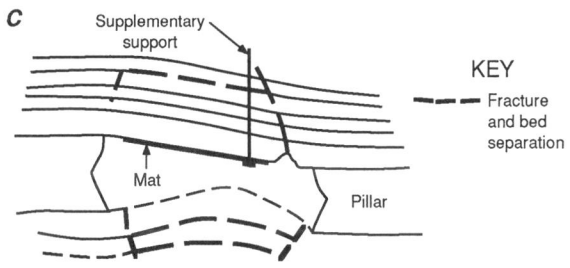

Figure 12. Velocity contours and ground conditions at retreating face position. (A) Velocity image (M/A); (B) actual roof condition; (C) schematic of strata movements.

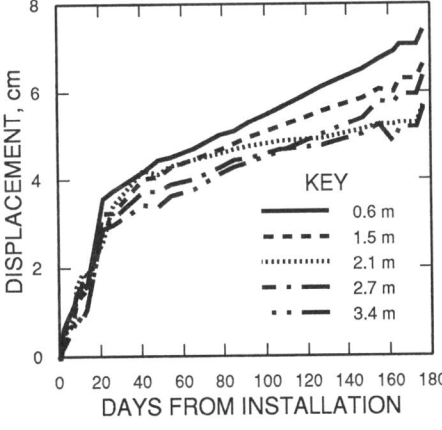

Figure 13. Lateral deformation history of mine pillar.

453

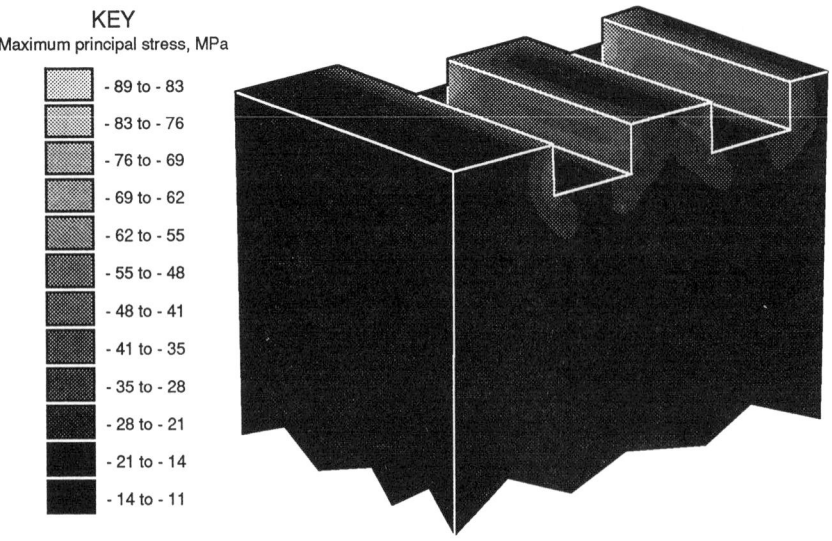

Figure 14. Model geometry and maximum principal stress contours, elastic analyses.

Table 1. Mechanical properties.

	Main roof	Seam and immediate roof	Floor
Shear modulus, GPa	2.1	6.4	0.8
Bulk modulus, GPa	3.2	19	1.7
Angle of friction	NA	NA	10°
Dilation angle	NA	NA	21° to 21.5°
Cohesion	NA	NA	7 MPa at $P_e = 0$
			1 MPa at $P_e = 0.008$

NA Not applicable.
P_e = plastic strain.

ment with field observations of rib movements and slabbing.

The vertical stress for an average pillar was 7.3 MPa, significantly below the elastic pillar stresses of 24 MPa. This result confirmed that the composite pillar and the floor had yielded, reaching a residual strength of 7.3 MPa. Such a mechanism was further explored using three-dimensional numerical models.

5 NUMERICAL MODELING

Laboratory and field measurements provided valuable information regarding the load-deformation characteristics of floor rocks and the extent of failure zones in the mine floor and fracture zones in the mine roof. In particular, it was shown that the marlstones of the mine floor were soft and weak, reaching peak strength and then exhibiting strain-softening characteristics. Shear strength was controlled by cohesion while the angle of friction and dilatancy were generally small near the peak strength.

Although roof rocks were initially damaged under the influence of a rather high stress field, the damage and amount of deformation grew in time as a result of large amounts of inelastic deformation in the mine floor. Floor failure significantly influenced load transfer through the pillar and contributed to pillar shifting toward the solid block.

To help explain observed ground conditions and the failure mechanism, a series of two- and three-dimensional numerical models were set up. The modeling methodology (Maleki et al., 1993) consisted of a simple elastic analysis followed by inelastic modeling of the mine floor, pillar, and roof. An analysis using Mohr-Columb plasticity with strain softening for the floor rocks best agreed with observed deformation but required input regarding changes in cohesion, angle of friction, and dilation as a function of plastic strain (Itasca, 1994). These data were not available until recently. The modeling was completed for development mining (day 150), but excluded any load transfer resulting from retreat mining.

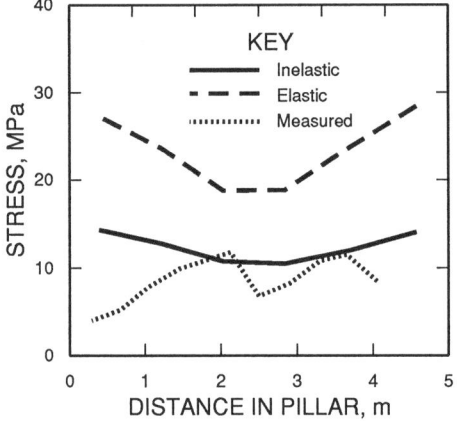

Figure 15. Comparison of measured and calculated vertical stresses on pillar.

Material property inputs are presented in table 1. Elastic properties were obtained from laboratory measurements while variations in cohesion as a function of plastic strain were obtained from combining uniaxial and plate-bearing test results. Direct-shear tests were used for determination of dilation angles. A relatively low angle of friction of 10° was used based on laboratory and borehole shear tests (0° to 21° range).

The variations in cohesion with plastic strain were important in influencing calculated stresses and deformation. An upper and lower bound for cohesion was estimated using (1) one-half of uniaxial compressive strength (7 to 9.6 MPa range) and (2) bearing-capacity tests and Terzaghi's equation for general shear failure (1 to 2 MPa) (Das, 1994).

Initially, a three-dimensional elastic analysis was completed to study areas of high stress concentrations and failure zones; the latter was obtained through post-processing of results using a Mohr-Columb failure criterion. Figure 14 presents the analyzed geometry and the maximum principal stress contours. The front, right, and back sides of the model are lines of symmetry, effectively modeling an area eight times larger than shown. The top (roof) of the model is removed so stress distributions in the pillar floor, entries 1 and 2, the abutment block, and the crosscuts can be viewed.

Average elastic pillar stresses were 24 MPa, significantly higher than measured stresses. This result confirmed that the pillar-floor material yielded, transferring loads to the solid block. In addition, the maximum principal stress concentration was highest near the pillar rib and intersections (fig. 14), contributing to rock damage at an early stage of excavation at this location.

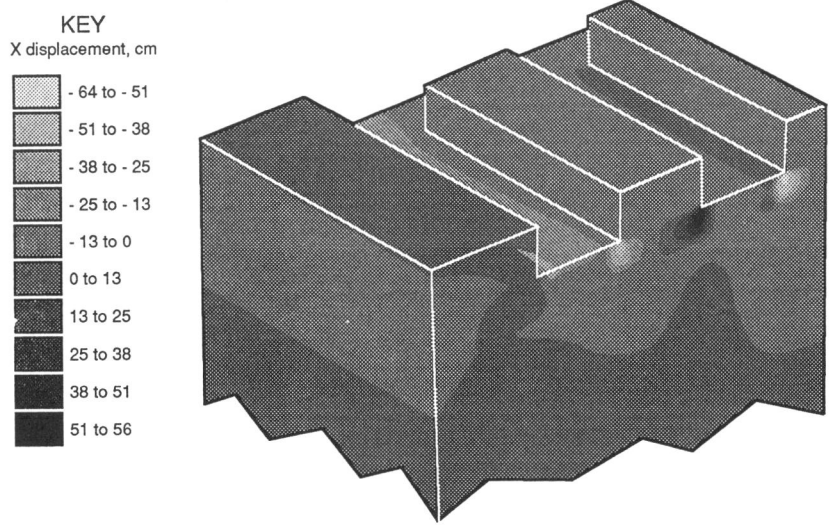

Figure 16. Lateral deformation contour, strain-softening model.

General calculated failure patterns and pillar stresses for the inelastic analysis were in good agreement with the measurements. In particular, the failure zone extended 2 to 3 m below the pillar, the pillar (fig. 15) transferred loads to the solid block, and lateral (x) the deformation of the whole pillar was toward the solid block (fig. 16).

Measured and calculated pillar vertical stresses are compared in figure 15 using the results from both elastic and inelastic modeling. Results confirm significant pillar unloading as a result of failure in the mine floor. Modeling predictions can be further improved by incorporating rib softening and including additional loads transferred because of retreat mining.

6 CONCLUSIONS

An assessment of the sources and the sequence of events leading to structural damage was identified using laboratory tests, field measurements, and numerical modeling techniques. Laboratory and field data were used to estimate in situ shear strength parameters for input to numerical models. Field measurements were also critical in identifying the initiation and growth of rock damage in the mine roof and floor.

It was shown that the roof was initially damaged under the influence of the stress field. The amount of roof deformation and the extent of the damage zone increased rapidly as a result of failure in the mine floor. This was associated with pillar unloading and shifting toward the solid block. This further compressed the roof along the pillar line (entry 1), increasing the extent of the damage zone. A three-dimensional Mohr-Columb plasticity model with strain softening agreed with the overall deformation and pillar stress patterns, providing a mechanism for explaining the cause of structural damage.

REFERENCES

Das, M.B. 1994. Principles of geotechnical engineering. 3rd ed., Boston: PWS Publishing, 672 pp.

Itasca 1994. Fast Lagrangian analysis of continua in three dimensions. Version 1.0, Minneapolis.

Ladanyi, B. 1968. Rock failure under concentrated loading. *10th Symp. on Rock Mech.*, SME.

Maleki, H., W. Ibrahim, Y. Jung & P.A. Edminster 1992. Development of in integrated monitoring system for evaluating roof stability. *Proc. 4th Conf. on Ground Control for Midwestern U.S. Coal Mines*, S. Illinois Univ., pp. 255-271.

Maleki, H., Y. Jung & K. Hollberg 1993. Case study of monitoring changes in roof stability. *Int. J. Rock Mech. Min. Sci. & Geomech. Abstr.*, Vol. 30, No. 7, pp. 1385-1401.

Schneider, W.A., Jr. 1990. A three-dimensional physical model applying tomographic inversion and seismic migration to the tunnel detection problem. Ph. D thesis, Colorado School of Mines.

Author index

Adachi, T. 393
Aizawa, T. 185
Aoshima, K. 299
Aoto, S. 219
Arisaka, S. 223
Aydan, Ö. 279

Ballivy, G. 433
Barton, N. 57
Burlakov, V.N. 235

Cunha, A.P. 213

Detournay, E. 191
Drescher, A. 191

Ebisu, S. 279

Faquan Wu 115
Feng Dingxiang 243
Feng Shuren 243
Fourmaintraux, D.M. 191
Fukazawa, E. 25

Gaziev, E.G. 317
Ge Xiurun 243
Gose, S. 285
Grossmann, N.F. 103
Gu Xianrong 243
Gutierrez, M. 57

Hasegawa, Y. 439
Hattori, K. 139
Hayakawa, T. 19
Hayashi, K. 375
Hirama, K. 197
Hirata, K. 163
Hollberg, K. 449
Hosono, T. 327
Hossaini, S.M.F. 133

Idogaki, M. 399
Iiboshi, S. 299
Inaba, T. 443
Inoue, A. 413
Ishii, D. 337
Ishikawa, K. 169
Isoyama, R. 259
Ito, H. 265
Ito, R. 293
Izumi, K. 421

Jašarević, I. 333

Kageyama, M. 381, 387
Kaji, Y. 75
Kakuhara, H. 381
Kalustian, E.S. 317
Kameya, H. 421
Kanazawa, K. 273, 405
Kaneko, K. 443
Karasawa, M. 163
Kashima, S. 119
Kawaguchi, K. 273
Kawamoto, T. 145, 279
Kikuchi, K. 349, 393

Kimura, Y. 177
Kishi, K. 25
Kobayashi, T. 91
Kohata, Y. 3, 413
Komura, S. 279
Kuwahara, T. 197
Kyoya, T. 145

Lasserre, C. 191
Li Xibing 99
Liu Deshun 99
Liu, C.S. 207
Luong, M.P. 157

Maekawa, K. 307
Makurat, A. 57
Maleki, H. 449
Mansurov, V.A. 81, 203
Maruyama, M. 197
Masuda, S. 259
Masui, N. 439
Matsukawa, T. 343
Matsumoto, S. 273
Matsumura, S. 127
Melouki, M. 433
Mihashi, K. 375
Mimuro, T. 91
Miščević, P. 333
Mito, Y. 349, 393
Miyajima, K. 75, 177
Momose, K. 25
Momose, Y. 265
Morikawa, S. 343
Murakami, S. 427
Muranaka, K. 139

Nagataki, Y. 299
Nakada, M. 349
Nakajima, H. 399
Nakasita, K. 293
Niishida, M. 443
Nishi, K. 151
Nishida, K. 421
Nishie, S. 139
Nishigaki, Y. 127
Nitta, A. 35, 75, 259, 327

Ochi, H. 169
Ochi, K. 3
Oda, M. 307
Ogata, N. 285
Ogata, Y. 219
Ogiwara, M. 421
Ohnishi, Y. 337, 355
Okamoto, T. 151
Ollala, C. 321
Osada, M. 109

Ramamurthy, T. 311

Sakai, T. 25
Sakashita, K. 387
Sanbyakuda, T. 273
Sasaki, K. 185, 223
Sasaki, T. 337, 343
Sasao, M. 119

Serrano, A. 321
Shibagaki, Y. 265
Siddiquee, M.S.A. 413
Sijing Wang 249
Singh, M. 363
Slimane, F.B. 433
Soliman, M. 355
Sonoda, T. 327
Srivastava, R.K. 363
Sugiyama, T. 163
Suzuki, H. 367
Suzuki, K. 197

Tada, K. 67
Takada, S. 169
Takagi, A. 293
Takahashi, K. 119, 177, 375
Takimoto, J. 405
Tanaka, T. 413
Tani, K. 151
Tardif, J. 433
Tashiro, T. 375
Tatsuoka, F. 3, 413
Terada, K. 25
Terada, M. 219
Tokashiki, N. 145
Torii, T. 399
Tripathi, A.K. 363
Tsubouchi, T. 3

Ueda, M. 163
Ueno, M. 293

Vutukuri, V.S. 133

Watanabe, K. 293
Watanabe, T. 19
Wu Xi 231
Wyllie, D.C. 253

Xu Gancheng 85

Yamabe, T. 307
Yamada, K. 127
Yamada, S. 119
Yamagata, M. 35, 127
Yamamoto, S. 35, 75, 119, 177, 259, 327
Yamamoto, Y. 413
Yamanouchi, K. 443
Yan Guang Ri 231
Yanaka, Y. 439
Yang Rouziong 231
Yang Xiangbi 99
Yang, Z.Y. 207
Yasufuku, S. 381, 387
Yoshida, I. 67
Yoshida, J. 223
Yoshida, O. 413
Yoshii, Y. 47
Yoshinaka, R. 91, 109, 219, 223, 337, 343
Yufin, S.A. 235
Yuki, N. 219

Zertsalov, M.G. 235
Zhang Ming Yao 231
Zhou Wei Yuan 231